CRC Handboo
of
Sample Size Guidelines
for
Clinical Trials

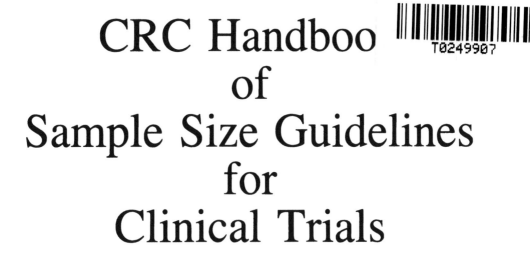

Author

Jonathan J. Shuster, Ph.D.
Professor of Statistics, University of Florida
Gainesville, Florida

and

Co-Principal Investigator
Pediatric Oncology Group Statistical Office
St. Louis, Missouri

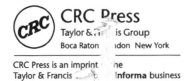

CRC Press
Taylor & ... is Group
Boca Raton   ...don New York

CRC Press is an imprint   he
Taylor & Francis ...   **Informa** business

First published 1990 by CRC Press
Taylor & Francis Group
6000 Broken Sound Parkway NW, Suite
300 Boca Raton, FL 33487-2742

Reissued 2018 by CRC Press

A Library of Congress record exists under LC control number: 92049333

Publisher's Note
The publisher has gone to great lengths to ensure the quality of this reprint but points out that some imperfections in the original copies may be apparent.

Disclaimer
The publisher has made every effort to trace copyright holders and welcomes correspondence from those they have been unable to contact.

ISBN 13: 978-1-138-10539-3 (hbk)
ISBN 13: 978-1-138-55842-7 (pbk)
ISBN 13: 978-1-315-15086-4 (ebk)

Visit the Taylor & Francis Web site at http://www.taylorandfrancis.com and the CRC Press Web site at http://www.crcpress.com

PREFACE

This book is primarily designed as a guide for physicians and biostatisticians working in clinical trials, where the primary objective is survival (or disease-free survival or progression-free survival etc). It will also be useful as a supplement to text material in courses dealing with biostatistics.

To project patient needs of a trial, the physician needs to provide "planning parameters". These include endpoints (eg 3 year survival); minimum difference the trial is to be sensitive to (eg 50% control vs 65% experimental); patient accrual rate (eg 55 patients per year are expected to be accrued); precision measures ("ALPHA" = probability of falsely concluding experimental therapy has efficacy given that it has none: typically .05) ("POWER" = probability of correctly concluding the experimental therapy has efficacy when the minimum difference is as above: typically .8 or .9 depending on the field); and anticipated rate of patient losses for non-study related reasons called "LOSSES TO FOLLOW-UP".

This Handbook is organized into three Chapters, an Appendix, and Tables. CHAPTER 1, *"How to Use the Sample Size Tables"*, is designed for users who simply want to know how many patients they need. This section is therefore a "Cookbook" approach, including a worksheet for obtaining the appropriate sample size for the trial. CHAPTER 2, *"Design and Analysis of Randomized Clinical Trials"*, is a non-technical discussion of the major elements of such trials. This section may be helpful for the user wishing to write a clinical protocol or a clinical research grant. A blueprint for protocol construction is included. CHAPTER 3, *"Derivation of the Statistical Results"*, contains a heuristic derivation of the statistical methods presented in CHAPTER 2. One full year of Calculus-based Mathematical Statistics is sufficient background to understand the material in CHAPTER 3. In addition, this Chapter contains some simple results on multiple comparisons and stratified analysis (in non-technical language). APPENDIX I contains a fairly detailed review of Mathematical Statistics, thereby making the text self-contained with respect to CHAPTER 3.

Summary of What you need to read:

1.   Cookbook: Give me the sample size needed. (CHAPTER 1)

2.   Clinicians: Help me write my Research Protocol. (CHAPTERS 1 and 2. For a discussion of "losses to

follow-up'', the reader might also consult Section 3.10 of CHAPTER 3. If the study has more than two treatments, see Section 3.12 of CHAPTER 3. Although that Section is far more technical than CHAPTER 2, most clinical investigators should be able to follow the main ideas. If after reading CHAPTER 2, "stratification" is contemplated, see Section 3.13 of CHAPTER 3.)

3.　　Biostatisticians: Entire text material.

Extensions and Limitations: While the major scope of this Handbook is for two treatment randomized trials, CHAPTER 3 (Section 3.12) does contain methods for the design and analysis of trials involving three or more treatments. This Handbook is limited in scope to therapeutic comparisons involving survival (or disease-free survival or progression-free survival, etc) as the primary endpoint. Analysis of prognostic factors are not discussed.

NOTE: In any Clinical Trial, it is in everyone's best interest that a team of researchers be assembled at the planning stage. This team should consist of clinical researchers, a biostatistician, and computer scientists. Laboratory researchers are often important members of the team.

THE AUTHOR

Jonathan J. Shuster, Ph.D., is currently a Professor in the Department of Statistics, University of Florida, Gainesville, Florida. He serves as Co-Principal Investigator of the Statistical Coordinating Center of the Pediatric Oncology Group, which at any time has at least 75 active multi-center clinical trials in childhood cancer. Apart from cancer, he has participated in major collaborative projects in Ophthalmology and Urology. He has published over 80 articles in methodologic research and in medical applications. His major statistical research interests include the design and analysis of clinical trials and of large epidemiologic studies.

Dr. Shuster has served on numerous NIH Study Sections including the Epidemiology and Disease Control Study Section #1 (1986—90). He served as Chair of that Study Section for the 1989—1990 fiscal year. It is largely through service on these study sections that Dr. Shuster recognized the need for this book.

In addition to his research efforts, Dr. Shuster has been active in teaching. He has taught service statistics courses to students in the health sciences, as well as elementary, intermediate, and advanced courses in statistics.

Dr. Shuster received his Ph.D. degree from McGill University, Montreal, Canada in 1969. Since then, he has been at the University of Florida. As of the spring of 1990, he had received nearly $10 million in grants, mostly from the National Institutes of Health.

The author wishes to express his appreciation to many kind friends who directly contributed to this project, most notably to Dr. Jim Boyett, to Dr. Dick Scheaffer, and especially to Dr. Jeff Krischer.

This book is dedicated to Isobel and Samuel Shuster, parents of the author, on the occasion of their golden wedding anniversary.

TABLE OF CONTENTS

Chapter 1

HOW TO USE THE SAMPLE SIZE TABLES

Introduction

It is vital that a clinical trial has a reasonable chance of achieving its objectives. Small

negative studies could be harmful, as they might not have had adequate power to find important

differences. Researchers might falsely conclude from the completed trial that a new therapy lacks

efficacy. Freiman et al (1978) reviewed a large number of published negative studies and concluded

that the vast majority were not adequately sensitive to clinically meaningful differences. To avoid this

problem, clinical trials must be well designed, and must have sufficient numbers of patients to answer

the study question.

In order to use the sample size tables, the user must supply several values called parameters.

The worksheet, Figure 2, will be helpful. Before getting too formal, let us have a glance at a

"Statistical Consideration" of a research protocol. At first, it probably will look quite technical, but we

shall unravel the details.

STATISTICAL CONSIDERATION

"Historically, the Institute has been able to accrue 125 patients per year with newly diagnosed

X-disease. Based on our prior studies, the three year survival rate is expected to be about 60% on the

standard therapy (or control therapy). In order to detect a 15% improvement under the experimental

therapy (three year survival of 75%), at "alpha"=.05 (one-sided) and 80% "power", a sample size of

204, half randomly assigned to each treatment, will be needed. This calculation assumes exponential

survival with all patients followed to death or termination point of the study. Allowing for a 10% loss

to follow-up, a figure consistent with past history, a revised sample size of 227 (1.8 years of accrual) is

needed. Patients will be followed until death, loss to follow-up, or the completion of the study (up to

4.8 years). The logrank test will be used to make the treatment comparison."

1.1   Identification of Parameters

(1)   Minimum Planned Follow-up: At how many years will the key survival comparison be

made?

3 years

(2)   Annual Accrual

125 Patients per year

(3)   Expected Accrual thru Minimum Follow-up

125 X 3=375 (This parameter is used in tables)

(4)   PCONT=Planning value for survival thru the minimum follow-up period (In this case 3

years) under the CONTROL treatment.

.6 (ie 60%) (This parameter is used in tables)

(5)   DEL=Planning value of improvement under the experimental treatment at the endpoint

(ie at three years in this case.)

.15 (ie 15%) (This parameter is used in tables)

(6)   ALPHA=Chance of falsely concluding experimental therapy is superior to the control,

when in fact it is equivalent to the control. In most studies, a value of .05 (ie 5%) is used.

There is no special reason to pick this value other than standardization. For further

discussion, see Chapter 2, especially as to one-sided versus two-sided tests. See also the

footnote at the conclusion of the chapter.

.05 (ie 5%) (This value is used in the tables)

(7) POWER=Chance of correctly concluding that the experimental therapy is superior to the control, when in fact the planning values (4) and (5) are correct. The most common values for power are 80% and 90% depending on the field. This is discussed further in Chapter 2 and in the footnote at the end of this chapter.

.80 (ie 80%) (This value is used in the tables)

(8) FACT (for factor). This asks the following question. Consider a typical day before and after the endpoint. A value of FACT is the ratio of the instantaneous death rate after the endpoint to that before the endpoint. The exponential distribution implies a value of 1.0. If you consider patients cured after three years, you would use a value of 0. However, the tables can be used for intermediate situations. For example, you might consider that the typical day more than three years out carries half the risk of the typical day less than three years out. In that case, you would use a value of FACT=.5 (or 50%). A value of

FACT=BIN is for comparison only. This gives the sample size required if you only looked the "Binary" event of surviving three years or not surviving three years. The comparison would ignore time of death and all data collected after the patient had made it to the three year point. This comparison is distinct from the logrank test and carries a different sample size. This alternate and less powerful test is discussed in Chapter 2 under the name "Kaplan-Meier." If you browse through the tables, you will note that for many situations, the FACT=.00 and FACT=BIN sample size requirements are very close. When FACT=.00, no deaths are expected after the endpoint.

FACT=1.0 (Exponential) (This value is used in the tables)

SUMMARY

(3) EXPECTED ACCRUAL THRU MINIMUM FOLLOW-UP=375

(4) PCONT=.60

(5)   DEL=.15

(6)   ALPHA=.05 (One-sided)

(7)   POWER=.80

(8)   FACT=1.0

To look up the sample size, we start from ALPHA and POWER. These are coded as follows:

To look up the sample size, on the basis of (6) and (7), we shall employ the TABLE 7. This is arranged in order of increasing expected accrual (3). Once the chart for ALPHA=.05, POWER=.80, and EXPECTED ACCRUAL=375 is located, locate the PCONT=.60 row and the DEL=.15 column.

The following six sample sizes:204,211,220,231,245, and 235 correspond respectively to FACT=1.0,.75,.50,.25,.00, and BIN. Since our FACT=1.0, we use

Sample Size= 204

The statistical section inflated the sample size to adjust for 10% losses to follow-up, for a total of 227 patients. By dividing by the accrual rate of 125 patients per year, this means that the accrual is expected to take 1.8 years, (22 months) and the study will require 4.8 years to complete.

*** Footnote on one-sided vs two-sided tests. In situations where you wish to compare two treatments, neither of which can be viewed as a control, you should run a two-sided test. Basically, you proceed exactly as above but replace the value of ALPHA by ALPHA/2. If the above study consisted of two experimental therapies, we would use ALPHA=.025. (TABLE 5) For the 45% vs 60% comparison, we would use PCONT=.45 and DEL=.15, with FACT=1.0. This would require 282 patients. For the 60% vs 75% comparison, we would use PCONT=.60 and DEL=.15 with FACT=1.0. This would require 250 patients. To be safe, we take the larger of these two numbers, 282. Allowing for 10% loss to follow-up, the final sample size would be 313 (2.5 years of accrual).

*** Further Reading. The sample size results of this handbook are similar to those of Schoenfeld (1981), Rubinstein et al. (1981), and especially Lachin (1981, where his equations (25) and (26) are employed). Another interesting method is due to Halpern and Brown (1987), who offer a computer program which can simulate the properties of the logrank statistic under any survival distributions provided by the user.

Finally, a method of Sposto and Sather(1985) is useful in some applications. That method yields similar results to those of this handbook.

REFERENCES

1. **Freiman, J. A., Chalmers, T. C., Smith H., and Keubler, R.,** The importance of beta, the type II error and sample size in the design and interpretation of the randomized controlled trial, *N. Engl. J. Med.*, 299, 690, 1987.

2. **Halpern, J. and Brown, B. W.,** Designing clinical trials with arbitrary specification of survival functions for the log rank or generalized Wilcoxon test, *Control. Clin. Trials*, 8, 177, 1987.

3. **Lachin, J.,** Introduction to sample size determination and power analysis for clincial trials. *Control. Clin. Trials*, 2, 93, 1981.

4. **Rubinstein, L. V., Gail, M. H., and Santner, T. J.,** Planning the duration of a comparitive clinical trial with loss to follow-up and a period of continued observation, *J. Chron. Dis.*, 34, 469, 1981.

5. **Schoenfeld, D.,** The asymptotic properties of nonparametric tests for comparing survival distributions, *Biometrika*, 68, 316, 1981.

6. **Sposto, R. and Sather, H. N.,** Determining the duration of comparitive clinical trials while allowing for cure, *J. Chron. Dis.*, 38, 683, 1985.

FIGURE 1. Codes For Tables

ALPHA=.005 and POWER=.80 TABLE 1

ALPHA=.005 and POWER=.90 TABLE 2

ALPHA=.01   and POWER=.80 TABLE 3

ALPHA=.01   and POWER=.90 TABLE 4

ALPHA=.025 and POWER=.80 TABLE 5

ALPHA=.025 and POWER=.90 TABLE 6

ALPHA=.05   and POWER=.80 TABLE 7

ALPHA=.05   and POWER=.90 TABLE 8

FIGURE 2. Worksheet For Sample Size Determination

One-Sided Studies

50% assigned to each treatment.

(1)	Expected Annual Accrual	_____	(Patients)
* (2)	Endpoint (? Year Survival) Minimum Planned Follow-up	_____	(Years)
(3)	Expected Accrual thru Minimum Follow-up	_____	(1)X(2)
(4)	PCONT (Control Planning Survival Probabilty to Endpoint)	_____	(Decimal)
(5)	DEL (Planned improvement in Survival to Endpoint under Experimental Treatment)	_____	(Decimal)
(6)	ALPHA=Probability of concluding experimental therapy superior when it is equivalent to control	_____	(Decimal)
(7)	Power=Probability of correctly concluding experimental therapy superior when DEL is as in (5)	_____	(Decimal)
(8)	FACT=Ratio of instantaneous death rate after the endpoint to that before the endpoint	_____	(Decimal)

From (6) and (7), use FIGURE 1 to identify table code. Next, use (3) to identify chart. CONT and DEL identify row and column. This identifies six sample sizes, one for each of five values of FACT and one for binary comparison.

(9)	Identify Initial Sample Size	_____	(Patients)
(10)	Estimated Fraction to be Lost to Follow-up	_____	(Decimal)
(11)	Subtract number in (10) from 1.0	_____	(Decimal)
(12)	Final Sample Size (9) divided by (11)	_____	(Patients)
(13)	Accrual Period (12) divided by (1)	_____	(Years)
(14)	Total Study Duration (2) + (13)	_____	(Years)

FIGURE 3. Modifications of Figure 2

1:   FOR TWO-SIDED TESTS

   (A)   Replace ALPHA by ALPHA/2 in (6).

   (B)   Replace PCONT by the lower of the two planning survival probabilities in (4)

   (C)   Replace DEL by the difference between the two planning survival probabilities in
         (5).

   (D)   Proceed as in Figure 2. (You may have to do this twice if the planning survival
         probability is known for one of the experimental therapies while the other is planned
         at from DEL worse to DEL better. If this is the case, do both calculations and use
         the larger of the two sample sizes.)

2:   FOR UNEQUAL PATIENT ALLOCATION (ie not a 50-50 randomization) (Valid for
     allocations from 33% to 67%)

   (A)   Find the sample size as if the allocation was 50-50 using FIGURE 2 (and FIGURE
         3-1 if needed)

   (B)   Adjust as Follows
         Allocation 2:1(67%) or 1:2(33%),Multiply result in (A) by 9/8
         Allocation 3:2(60%) or 2:3(40%),Multiply result in (A) by 25/24
         Allocation X:Y,Multiply result in (A) by $(X+Y)(X+Y)/(4XY)$
         Note that this sample size is always higher than a 50-50 allocation.

Chapter 2

DESIGN AND ANALYSIS OF RANDOMIZED CLINICAL TRIALS

This Chapter is a concise, non-technical introduction to the design and analysis of randomized clinical trials. For more detail, at a non-technical level, the reader is referred to the following excellent references: Pocock (1983), Buyse et al. (1984), and Meinert (1986). While the Tables of this volume apply to very specific types of trials, they are in fact applicable to the majority of major definitive trials run today. We shall discuss clinical trials in a more general framework, but restrict our analytic considerations to randomized trials. We shall discuss the trials under each of their four basic steps:

1. Formulate a therapeutic research question

2. Design a trial to answer this question

3. Conduct the trial as planned

4. Analyze the data and report the results

2.1 Formulation of the Therapeutic Question

This part of the trial is more medical than statistical. The question might be, for example, "Can Beta-Blocker X improve survival of newly diagnosed myocardial infarction patients over that of the standard Beta-Blocker Y?" This is the type of question which can form the starting point of a clinical trial. Note that there are two competing therapies proposed for this trial. While this book is devoted to two treatment trials, Section 3.12 provides extended use for more complicated questions involving three or more competing therapies.

Note that the research question is very loosely worded at this point. There will be considerable work to convert this question into a truly testable hypothesis.

2.1.1. One-Sided vs. Two-Sided Question

An important component of your research question is whether your question is "one-sided" or "two-sided". The following discussion will help make this determination.

When you conduct a clinical trial, you will usually put on trial one of the following statements

1. One-Sided Statement: The new (test) therapy is superior to the standard; or

2. One of the two therapies is superior to the other, but you have no vested interest as to which.

Sample One-Sided Questions

Drug companies, developing a new drug, are generally interested in one-sided questions. Their product may only be marketable if they can prove beyond a reasonable doubt that their product is superior to the standard. The competing therapies have asymmetrical interest.

Again, if any scientific researcher is considering the efficacy of adding a toxic agent to a standard regimen, and comparing this to the standard regimen alone, a one-sided question results. After the study is over, one would use the new therapy only if it proved to be significantly better than the standard. Whether the efficacies were not significantly different or the toxic agent had significantly reduced the efficacy, one would make the management decison not to use this toxic agent for future patients.

Sample Two-Sided Questions

There may be two standard therapies that are in widespread use, but no one really knows which is superior. The scientific issue here is simply, which is the better therapy? Interest is symmetric.

In some diseases, ther may be no known effective therapy. One might wish to test two new experimental therapies. Here too, a two-sided question is clear.

One-sided questions need fewer subjects than two-sided questions of equal precision. The "Zone of proof beyond a reasonable doubt" is concentrated in the zone where the test outperforms the control treatment. The two-side question must create two such zones, one corresponding to superiority of each treatment. Consequently, the region of superiority of each given treatment must be smaller in the two-sided question corresponding to the one-side question o the same precision.

The one-sided vs. two-sided question must be spelled out in advance as a vital part of the research question. Clearcut rationale for asymmetric interest must be given. In point of fact, most trials conducted in the medical school environment should be two-sided, unless there is a compelling reason for asymmetry.

2.2 Design of the Clinical Trial

In order to conduct a valid clinical trial, it is vital to develop a complete plan for its conduct. A manual of operations, called a "Protocol", should be written with the following elements:

A. Objectives of the Study (What you hope to accomplish.)

B. Background Information (Literature review of prior therapy and outcome. Include your own experience as well.)

C. Rationale (Why you feel you have good prospects to achieve your objectives.)

D. Patient Eligibility (A Precise definition as to who can participate in the trial.)

E. Treatment Plans (Full description of how patients will be treated under each competing therapy. This must include contingency plans for complications.)

F. Off-Study Requirements (Full description of events that place the patient off study. For example, patient may go off study at time of recurrence of disease, allergic reaction to the drug, or death.)

G. Data Submission Requirements (This describes the schedule of follow-up, the medical tests required and their timing, and the description and timing of the data form submission.)

H. Statistical Consideration. (This describes the actual design, including the endpoint of the study, whether stratification and randomization will be used, plans for interim and final analysis, and sample size projections. If the study is a "masked design", this is also discussed in this Section.)

I. Study Monitoring (A committee called the Data Safety and Monitoring Committee should be listed along with their meeting schedules and responsibilities.)

J. Appendix (This should include blank copies of all data forms and the informed consent form. Instructions on forms completion should also be included.)

K. Database Documentation (While this may not be a formal part of the protocol, it is a critical element of the design. A protocol cannot be compromised by turnover in computer staff.)

This author was once sent a computer tape of a trial completed some six years earlier, in order to re-analyze the data. Unfortunately, the senders lacked documentation of the data. Despite our best efforts, we could not crack the code with enough confidence to complete any analysis.

2.2.1 Statistical Considerations

In planning a trial, there are a number of issues that need to be considered in (H) above. First, the eligibility conditions define the "Target Population". This is the hypothetical pool of past, present, and future patients with well defined characteristics. Note that the actual patients who participate in the study are considered to be a representative sample from this target population. Our ultimate conclusions will be directed to the target population not to the sample. Of course, we use the sample data to make an inference about this target population. Such inferences are subject to sampling error, since only a portion of the target population has been sampled. This is much like the use of voter surveys to project election results.

Answers to the following question are needed to formulate the statistical consideration.

What is the single primary variable upon which the major comparison will be based?

It is essential to pin down a primary variable up front. If this is not specified, the data may be analyzed in a wide variety of ways, leading to investigator selection bias. Each analysis carries a sampling error, and hence the more analyses one performs, the greater the cumulative error effect. To control this, a primary single analysis should be specified. Others can be done, but as strictly secondary analysis. Example primary variables might be three year remission, five year survival, two year peak intra-ocular pressure below 23 mm hg, etc.

As mentioned in the Introduction, there are two types of primary variables considered in this Handbook. The first type is "Censored Survival Data". Time is measured from entry onto the study until failure, termination of the study, or loss to follow-up. The last occurs if the patient exercises his or her option to quit the study. The only survival information available are (1) the time from entry to last follow-up and (2) the patient status at last follow-up. The data are censored (or more precisely, right censored), since for survivors, we do not know their true survival times, although we do know that they survived at least a given number of days (eg 655+ days). The second type of data in which these sample size tables can be used is binary data (yes/no). For example, a study on operative deaths (whether or not the patient died during surgery), is not concerned with time of death, but only whether or not the patient survived the surgical procedure. Unlike the censored survival situation, follow-up is short and equivalent for all patients. In a censored survival study, involving a three year patient accrual phase plus a two year minimum follow-up (a five year total study), patients have a potential follow-up of from two to five years. The risk of a study death among earlier entries is higher than that among later entries, since the early entries are at risk longer. This needs to be taken into account in the analysis. The operative death problem is quite different. The only data used in that analysis is whether or not the patient survived surgery.

Will the study be randomized?

This is a controversial topic. Let us suppose that you plan to compare two therapies. You may have considerable experience with one of the therapies (the "control" or "standard") and little experience on the other (the "experimental" or "test"). You may ask, "Why not simply recruit patients to the experimental regimen and compare these to the existing control patients?" This is called an "Historical Control Study". There are both advantages and disadvantages of such a strategy. Historical control studies often require fewer patients than randomized studies, and hence can often be completed in less time and at less cost than concurrently controlled studies. Unfortunately, however, historical control studies cannot be viewed as definitive scientific evidence about the relative merits of the competing therapies. You cannot adequately attribute differences (or lack of differences) in outcome with differences in efficacy. Subtle changes in any of the following can hopelessly confound the results of an historical control trial: Changes in referral pattern, changes in personnel, changes in diagnostic techniques, changes in concomitant care, changes in public awareness programs, etc. In addition, the trial may have been motivated by disappointing results under the standard therapy. Disappointment is more likely when the sampling error under the historical control is negative than it would be if the sampling error is positive. In other words, suppose the true success rate is 60% under the standard. A 50% observed success rate in your sample of 50 historical patients could provide greater motivation to conduct a study than a 70% observed success rate.

One advantage of historical control studies, not often mentioned in publications, is the issue of gaining informed consent. It is very difficult to approach a patient in order to gain consent for randomization. Uncertainty in a physician can undermine the confidence of the patient. However, in truth, the same uncertainty is generally present in the historical control study, although it is often downplayed.

Finally, historical control studies are fully justified if the historical control success rate is virtually zero. In such cases, you are unwilling to subject patients to a hopeless therapy. Even here, however, there is no unanimous agreement among scientists. For example, in 1986, randomized studies were conducted in AIDS patients, using a control known to offer no hope for survival. Surprisingly, there was a strong moral argument for the randomization. Since only a limited number of patients were approved for initial testing of new agents, a randomized design could treat as many patients as a non-randomized design. Further, by careful interim comparison of treatment vs. control patients, the randomized study offers the best way to study non-fatal events. For additional information on randomized vs. historical controls, see Zelen (1983), Shuster et al. (1985), Pocock (1983), Gehan (1978 and 1982), Peto et al. (1976), Diehl and Perry(1986), and Dupont (1985). These articles present all sides of the issues.

In summary, a randomized study is recommended very strongly, except perhaps when the historical success rate is near zero. However, if you consider a randomized study and determine that the required sample size is too large, an historical control study might be considered. Prior to doing so, however, the possibility of running a randomized study with additional institutions should be seriously considered.

Note: Randomization is a more complex topic than one might at first believe. It is important that the investigator cannot predict what therapy the next patient will be assigned. The purpose of randomization is threefold. First, it prevents unconscious bias in treatment assignment as compared to non- randomized assignment. Second, the framework of statistical analysis relies on randomization. Finally, a properly set up randomization procedure offers an audit trail of treatment assignments. This protects the investigator if the results of a study are questioned by a review group or by journal editors.

Randomization Methods

Some randomization techniques are discussed in Pocock (1983, Chapter 5). Figure 4 presents a table of randomized treatment assignments, based on "simple randomization". Each successive entry is an independent flip of an unbiased coin (50% probability of heads). Heads represents treatment 1, while tails represents treatment 2. Figure 5 represents an alternate design, with "random length permuted blocks". The following is the mechanism for Figure 5. First, a balanced coin is flipped. If the coin is heads, then we generate a random "word" consisting of three 1's and three 2's. Each of the possible 20 words consisting of three 1's and three 2's has an equal chance of occurring. If the coin lands tails, then we generate a random word consisting of two 1's and two 2's. Each of the following six equally likely words is drawn: 1122, 1212, 1221, 2112, 2121, 2211 This procedure is repeated, until the figure is completed. These figures provide the serial assignment of treatments for successive patients.

Simple randomization (Figure 4) is very simple to employ, but may lead to rather substantial differences in the number of patients assigned to each therapy. To illustrate, after 813 randomizations in Figure 4, there are 443 patients assigned to Treatment #1 and 370 assigned to treatment #2, for a net difference of 73 patients. This is the peak difference over the 1200 patient assignments. The random length permuted blocks procedure keeps the numbers well balanced throughout the study, while it is quite unpredictable to the investigator, especially if the exact mechanism of randomization is masked. Randomization is discussed further under stratification and institutional balancing below.

Will the study be stratified?

Are there key "risk factors" that need to be balanced "prospectively" or "retrospectively"? There is always concern that a randomized design will place a disproportionate number of poor risk patients on one of the treatments. If one has historical information about such risk factors, they might be planned for in two ways. First, we may wish to stratify the trial prospectively. That is, we shall define patient subgroups according to the risk factors, and essentially run independent trials within each subgroup. Using random length permuted blocks within each subgroup, we can maintain good balance of the treatments across each stratum. This is a good design feature, but it comes at a price. Over stratification can reduce the precision. Peto et al. (1976) estimate that the effective sample size of the study is reduced by one for each stratum you create, as compared to the luckily perfectly balanced randomized trial. For example, a trial stratified on age group (4 levels), sex (2 levels), smoking history (3 levels), and serum cholesterol (3 levels) would create $4 \times 2 \times 3 \times 3 = 72$ strata. Clearly, a poorly designed study could be so heavily stratified that there would be very few patients in any one stratum. The purpose of stratification is to compare like patients with like patients. However, stratification limits the possibility of comparison, as the following example illustrates. Suppose that there are 100 patients, 50 in stratum 1 and 50 in stratum 2. Suppose that 50 are assigned to each treatment. The un-stratified design allows 50 treatment #1 patients each to be compared to 50 treatment #2 patients (2500 comparisons). The stratified design eliminates half of these, namely patients in different strata and different treatments. Clearly, there is a trade-off. We recommend only very limited stratification, based on important proven risk factors, be made.

The second method to deal with stratification is to adjust for the risk factors retrospectively in the analysis. This method is discussed in CHAPTER 3.13. The reader is further referred to Peto et al. (1977). No within-stratum balancing is done, but imbalances are corrected for in the analysis.

If the study is a multi-center study, will institutional balancing be used?

When conducting a multi-institutional study, especially if it is conducted by a large number of institutions, each with a relatively small number of patients, one should consider balancing the randomization within institutions. While it seems that this might create too many strata, especially if the study is also stratified according to risk factors, a method called "Marginal Balancing" can be used to avert this problem. Basically, marginal balancing, as discussed in Shuster et al. (1985), works as follows. Assign the patient according to prior assignments for the institution and stratum as follows:

Priority #1: Check how many patients have been assigned to each treatment at this institution. If the difference between the number of assignments exceeds the maximum allowable margin, (say 2), assign the treatment that has been assigned less often at that institution. For example, if at X-Hospital, 7 patients are on Treatment #1 and 4 patients are on Treatment #2, and the maximum allowable margin for randomization is 2, the next patient at this institution must receive Treatment #2.

Priority #2: Applies only if Priority #1 has not forced a treatment assignment. For the patient's stratum, check the number of patients assigned to each treatment. If the difference exceeds the maximum allowable margin for randomization, assign the less used treatment. (Note that the stratum maximum margin need not match that of the institution.)

Priority #3: Applies only if Priorities #1 and #2 have not forced a treatment assignment. Assign treatment according to the next unused entry on the random length permuted block assignment sheet. (For example, Figure 5.)

Bookkeeping: After each patient assignment, it is necessary to update the number of patients assigned to each therapy by institution and stratum. In addition, the assignment should be recorded on the random length permuted block assignment sheet by crossing off the first available entry matching the assigned treatment. This will be the next entry if Priority #3 applies.

Marginal balancing keeps institutions and strata balanced. It also keeps the overall study balanced because of the permuted blocks. One of the advantages of marginal balancing is that the study can be analyzed as if it was a single institutional study. In other words, one can ignore the stratification effect of institution in the analysis. This is especially important in studies conducted by a large number of institutions with a large number of strata. In practice, the "error" one makes in treating the study as a single institutional study is in the conservative direction. That is, there will be a slight tendency to underestimate the significance of the differences between the treatments (overestimate the "P-values"). However, this error is rarely of practical significance.

Marginal balancing is also important for non-statistical reasons in studies where some institutions contribute small numbers of patients. Each institution should gain experience on both therapies. Without institutional balancing, there is more than a 12% chance that an institution contributing four patients will place all four on the same treatment.

If the study is multi-institutional, will centralized randomization be employed?

Centralized randomization is strongly recommended for all multi-center trials. Clearly, if marginal balancing is used, a central office must have up-to-the-minute information on all prior treatment assignments. There are more important considerations, however. The integrity of the study is best insured through centralized randomization. No patient can get onto the study without contact with the central office. Eligibility can be verified by standardized criteria prior to acceptance of the patient. Without centralized randomization, the study is open to the possibility that early failures may never be placed on the study. Individual investigators may overlook the important task of forwarding the information on such patients to the central office. With centralized registration and randomization, patients must be put on study before starting treatment. Registration and randomization need not be done at the same time. Some studies will have both treatments identical for the first 6-12 weeks, followed by a divergence. All participants should be registered at the time of treatment start, but a callback for randomization should occur at the time when the treatments diverge. This callback will in fact improve the precision of the comparison, since the major comparison will be based on outcomes after the divergence point. Failures before the divergence point are not part of the study question, although they are vital to the estimation of the efficacy of each therapy. The treatment differences must occur after the divergence point, but the overall success rate of each therapy is measured from the start of treatment. You have the luxury of using all patients (regardless of assigned therapy) to assess the efficacy of the therapy from its starting point to the divergence point.

Will the study be masked (blinded)?

It is a good experimental technique to mask the identity of the treatment from both the patient and the investigator. (The more common term is blinded, but as one with experience in ophthalmic trials, your author considers that designation is a bit short sighted.) Masking is an especially important consideration if (1) patient psychological factors may be involved, (2) the response measurements involve subjective judgement on the part of the physician or the patient, or (3) the treatment can be put in a masked form, ethically and logistically. Contra-indications for masking include possible discomfort to the control patient; the need for the physician to perform expensive and/or painful medical tests to monitor one of the treatments and its side-effects; and the ability to mask the therapy. It would be unethical to perform sham surgery on a patient in order to mask the true procedure.

Some masking is better than no masking. For example, if it is unethical to mask the treating physician, it might be possible to have the critical observations made by a colleague who is indeed masked as to the treatment identity. A vital consideration in masked studies is that the master code, indicating who got each treatment, be securely maintained. More than one masked study has failed because the master code was lost! If such a loss occurs, nothing can be done to recover.

What methods of analysis will be used?

This will be discussed further in Section 2.4. However, the statistical section must spell out the exact statistical details as to what methodology will be used.

This needs to be specified in the protocol, as a protection to the investigators. Journal reviewers can sometimes criticize a study on the basis that it was not analyzed according to the design of the study. The best protection against this contingency is to submit a copy of the protocol with the manuscript. Often, the protocol itself undergoes peer review, and hence, there is strong support for your analysis.

What patients will be included in the analysis?

In most studies, there will be some severe violations of the protocol, often for reasons beyond the control of the investigators. The question one might pose is, "Will the major comparison be based on all compliant patients, or on all eligible patients?" There is now fairly unanimous agreement that the most convincing scientific analysis is based on all eligible patients. Once a patient is deemed to meet the entry requirements, that patient is included in the major analysis, whether or not the patient complies with the assigned protocol. This is often called "The intent to treat principal". Many journals have published policy that the major analysis must consist of all patients who meet the eligibility conditions, whether or not they comply with the protocol. Two important editorials supporting this point are

Zelen (1983) and Simon and Wittes (1985). As soon as subjective judgement is allowed to enter the picture, the potential for bias arises. Other analysis can be done as secondary evidence. However, severe bias can be induced by eliminating non-compliant patients. Compliance can have a strange relationship to treatment, apart from merely a disproportionate number of non- compliers within the two treatment groups. Non-compliers on one treatment may be at increased risk, while non-compliers on the other treatment may be at decreased risk. The "intent to treat principal" also implies that patients are analyzed on the treatment to which they were assigned, even if they actually received the other therapy! This may seem to be invalid at first glance, but it is the only way to prevent bias. The argument is that the Intent to Treat Principal closely resembles the real world application of the treatment. Poor compliance to a protocol will adversely affect its success. It is vital to keep track of compliance, but the major analysis should not utilize compliance. For additional detail on this topic, see Shuster et al (1985), Simon and Wittes (1985), and Haynes and Dante (1987).

What are the plans for interim analysis?

When conducting a long-term trial, there is considerable motivation to look at the results to date to see if the question can be answered early. Such a plan for example led to the early closure of the Multi-Institutional Osteosarcoma Trial of Link et al. (1986). That protocol specified times for interim analysis as well as conditions which would lead to early termination of the study. It needs to be pointed out that the decision to close a study early should be based on very compelling evidence. Generally speaking, the more interim analyses that are conducted, the greater the risk of false-positive results. If interim analyses are to be conducted, they should be conducted at a very small P-value as compared to a one-time final analysis. This can be put into a formal statistical methodology called "Group Sequential Methods" or in a more informal way, where the critical P-value is a small fraction of the final planning P-value. (eg .005 for interim analysis vs. .05 for the final planned analysis.) For more information on this topic see Armitage (1975), O'Brien and Fleming (1979) , and Tsiatis and Rosner (1985).

In any case, the protocol should spell out exactly when interim analyses will be conducted, and how decisions based on interim analyses will be made. If an interim analysis does in fact suggest that the study be terminated, the decision should not be made until the data have been reviewed carefully by the "Data Safety and Monitoring Committee". See Section 2.2.(I) discussed earlier.

2.3 Conduct of the Trial

Once the trial is opened to patient accrual, there are many elements which are required to ensure that the trial meets its objectives. Each institution must obtain annual approval from its Institutional Review Board (IRB) in order to register patients on the study. In order to place any patient on the trial, informed written consent must be obtained. The risks and benefits of the competing therapies must be explained as well as the objectives of the study. If the trial is randomized, the patient should know this before the treatment is assigned. There are some pre-consent randomization designs. See Zelen (1979 and 1982), Smith (1983), and Fletcher (1985) for more details. Such designs are highly controversial as these references point out. The vast majority of studies assume randomization after consent, unless the patient is incapable of giving informed consent, Abramson et al. (1985).

The patient becomes part of the trial as soon as consent is obtained. Further, once treatment is assigned, the patient is a part of that treatment group, regardless of whether or not the patient complies with the therapy. The investigator is ethically bound to follow the therapy as prescribed in the protocol, unless there is overwhelming evidence that violating the protocol is in the patient's best health interest. The patient assumes no legal commitment to follow the protocol, but does so in the spirit of physician-patient relationship.

Data Submission

It is the responsibility of each investigator to fully comply with all data submission requirements of the trial. Even the world's most outstanding statistician is limited by the quality and quantity of data submitted by the investigators. Even small amounts of missing data can bias a study.

Interim Analysis

A well -written protocol has provisions for interim analysis. However, it is important that while the trial is accruing patients, this information should be restricted to the Data Safety and Monitoring Committee. Investigators should not see interim treatment comparisons while patient accrual is ongoing. The committee is charged with the responsibility of determining if randomization remains ethical. With open reporting, selection bias could result from individualized analysis on the part of the investigators.

In some situations, it may be important to mask the treatment comparisons until all patients have completed therapy. For example, suppose that Treatment A consists of two years of Beta-Blocker X, while Treatment B consists of two years of Beta-Blocker Y, from the time of myocardial infarction. If an unblinded interim analysis shows a trend in favor of B, some investigators may want to switch their A patients to B. This could compromise the entire trial. The decision to switch or not switch should be in the hands of the Data Safety and Monitoring Committee, not the individual investigator.

Protocol Amendments

Once a protocol is actively accruing patients, amendments should be resisted. However, even the most carefully written protocols are not immune to unexpected results. For example, one might encounter toxicities, that had not occurred in pilot studies. Counter measures would have to be written into the protocol. If the counter measure has efficacy implications on one or both treatments, then the amendment may have the effect of forcing the investigators to start the accrual process over. The patients already accrued would not be used in the comparison of the treatments. Each amendment has to be sent back to the IRBs for approval. In addition, the Statistical Consideration and Consent Form will need modifications to incorporate these amendments. Amendments must be handled with as much precision as the original protocol.

2.4 Analysis and Reporting of the Trial

Upon completion of the accrual phase and minimum projected follow-up on all patients, the final analysis phase begins. The first step is to acquire all possible delinquent and missing information. Next, the analysis is completed as prescribed in the Statistical Consideration. Finally, the results of the trial are submitted for publication in a refereed medical journal.

2.4.1. Crucial Elements

The analysis described in the statistical section must be employed in the analysis. The sample size calculation only applies to the statistical methods detailed in the protocol.

Definition of P-value: The methodology of the statistical consideration provides an ordering of the possible outcomes in terms of how well they support a difference between the treatments (one-sided or two sided as specified in the design.) The P-value is interpreted as follows: Suppose that you repeated this study in another universe where the treatments were in fact equivalent. What is the maximum probability of observing evidence of a difference in this repetition that is as least as convincing as that observed in the actual trial? (The answer is the P-Value.) For example, suppose that the observed success rate of the test treatment was 46/84 while that in the control treatment was 34/81. The statement, which we shall discuss in more detail below is, "P = .05 one-sided." That is, if we repeated our protocol in a population where the success rates are equal, then there is at most a 5% chance of observing a difference of 12.8% or more in favor of the test treatment. (46/84-34/81 = .128=12.8%).

Methods of Analysis of Clinical Trial Data

We shall discuss three methods of analysis of clinical trial data. These include Sections 2.4.2, Binomial Comparison; 2.4.3, Kaplan-Meier Comparison, and 2.4.4, Logrank Comparison.

2.4.2 Binomial Comparison

The Binomial Comparison is applied to trials where the outcome is binary: Success or Failure. The following properties must be satisfied, at least conceptually, in order to apply this methodology:

P1: Trials are "INDEPENDENT". That is, the outcome of a given patient is not influenced by the outcome of any other patient. This methodology should not be used for matched studies, since the outcome of one patient will have more predictive value on the matched patient than it would on any other patient.

P2: The "Success Probability" within a given treatment is the same from patient to patient. That is, without knowing the patient characteristics in advance, the probabilities of success in the first, second, third, ... patients assigned to the given treatment are equal. For censored survival studies where patients enter serially and where the study is analyzed at a fixed point in time, this assumption is violated. Early entries are at risk for a longer period than later entries, and hence, have lower success rates than later entries. Of course, the success rate of each treatment need not be the same.

2.4.2.1   Large Sample Methodology

Let N1=Sample Size for the Standard or Control Therapy

S1=Number of Successes in the Control Sample

F1=Number of Failures in the Control Sample

P1=S1/N1=Fraction of Successes in the Control Sample.

Let N2=Sample Size for the Test Therapy

S2=Number of Successes in the Test Sample

F2=Number of Failures in the Test Sample

P2=S2/N2=Fraction of Successes in the Test Sample.

The difference in Success rates is P2-P1. We also need a quantity called the "Estimated Variance of P2-P1", denoted by:

$$V=\{P1(1-P1)/N1\}+\{P2(1-P2)/N2\}$$

The test statistic for which a decision about the therapies is based is:

$$Z=(P2-P1)/SE$$

where

$$SE=SQRT(V) \qquad\qquad (SQRT=Square\ Root)$$

The approximate "P-value" associated with Z is obtained from FIGURE 6 for one-sided comparisons and from FIGURE 7 for two- sided comparisons. The intuitive interpretation of the P-value is given in Section 2.4.1.

2.4.2.2 Confidence Intervals

Another more descriptive way to report the above information is via a CONFIDENCE INTERVAL. A "One-Sided Confidence Interval" is reported for DEL, the true target population difference in success rates, as:

$$DEL > P2-P1-W\ SE$$

where W is selected from Figure 8 according to the desired confidence. The intuitive interpretation of the Confidence Interval is given below Figure 8.

A "Two-Sided Confidence Interval" is reported as:

$$P2-P1+W\ SE > DEL > (P2-P1)-W\ SE$$

where W is selected from Figure 9 according to the desired confidence. The intuitive interpretation of the Confidence Interval is given below Figure 9.

2.4.2.3 Numerical Example (Test)

Suppose that the Control therapy had N1=81 patients of whom S1=34 were successes and F1=47 were failures. We have P1=34/81=0.420. Suppose further that the Test therapy had N2=84 patients of whom S2=46 were successes and F2=38 were failures. Hence, P2=46/84=0.548.

To calculate V, note that 1-P1=1-.420=.580 and 1-P2=1-.548=.452.

$$V=\{(.420)(.580)/81\}+\{(.548)(.452)/84\}=\{.2463/81\}+\{.2477/84\}=0.00304+0.00295=.00599$$

$$SE=SQRT(V)=.0774$$

and

$$P2-P1=.548-.420=.128$$

Hence,

$$Z=.128/.0774=1.653$$

From Figure 6, this translates to a P-Value of slightly less than 0.05, for a one-sided test. Had this been a two-sided study, we see from Figure 7 that the two-sided P-value is slightly less than 0.10. Finally, if the Experimental and Test labels had been reversed so that the Experimental group had 34 successes in 81 patients while the Control group had 46 successes in 84 patients, the One-Sided P-Value would have been approximately 0.95.

2.4.2.4 Numerical Example (One-Sided Confidence Interval)

If one was to construct a 95% One-Sided Confidence Interval for DEL, the target difference in success rates, one would employ FIGURE 8:

$$DEL > .128 - 1.645\{.0774\} = .128 - .127 = .01$$

That is, DEL> .01 with 95% confidence.

Notice that the value of Zero, the point of equal efficacy, lies just outside the confidence interval. (Had the P-value been greater than 0.05, Zero would have appeared within the confidence interval. The two-sided confidence interval will illustrate this.)

2.4.2.5 Numerical Example (Two-Sided Confidence Interval)

Had this been a two-sided question, a 95% two-sided confidence interval would be obtained from Figure 9 as follows:

$$.128 + 1.96\{.0774\} > DEL > .128 - 1.96\{.0774\}$$

That is .28 >DEL > -.02 with 95% confidence.

Had this been a two-sided experiment, the outcome would not have achieved statistical significance at the conventional .05 level. However, we are highly confident that the Test treatment ranges from trivially inferior to vastly superior. The value in the confidence region most favorable to the control treatment represents only a 2% advantage to the control. The confidence interval allows a management decision to prefer the test treatment for future patients, despite the fact that conventional significance of the two-sided trial was not achieved. The P-Value alone does not allow this type of management logic.

2.4.2.6 Small Sample Methodology

In order to apply the Large Sample Methodology, the following rules of thumb are often used:

Each of the following four quantities are greater than 5.0:

$$N1(S1 + S2)/(N1 + N2) = \text{"Expected Successes for Control"}$$

$$N1(F1 + F2)/(N1 + N2) = \text{"Expected Failures for Control"}$$

$$N2(S1 + S2)/(N1 + N2) = \text{"Expected Successes for Test "}$$

$$N2(F1 + F2)/(N1 + N2) = \text{"Expected Failures for Test "}$$

The term "Expected" refers to the idealized distribution of the actual patient results to the two groups, where the success rate would be the same in both treatment groups. There is nothing magic about the number 5.0. However, the approximations seem to be "adequate" at this number of patients.

Generally speaking, if a randomized clinical trial is designed to be sensitive to clinically relevant differences, the approximations made for large samples will be valid. However, in very small trials with low power, exact methods are needed. The two most widely used methods are Fisher's Exact Test and the Exact Unconditional Z Test. We shall discuss these briefly. For more detail, the reader should consult Armitage (1971), Suissa and Shuster (1985), and Shuster (1988).

Fisher's Exact Test

In the numerical example above, imagine that we have a deck of cards, distributed as follows: 80 Cards are labeled SUCCESS and 85 Cards are labeled FAILURE, for a total of 165 cards. Now suppose that we randomly distribute the cards to two players called Control and Test (81 to Control and 84 to Test). What is the probability of distributing 46 or more successes to the Test player? In other words, if we repeated the experiment on a different target population, where in fact the control and test therapies were equivalent, and if we happened to again get a total of 80 Successes and 85 Failures, what is the likelihood of obtaining evidence at least as convincing as that observed in favor of the test therapy? The answer is P=.068. (The approximation had P=.05 one-sided)

Note that this is not a true P-value, since the additional proviso is made that in the replication of the experiment, you must match the total number of successes with that observed (80). We call such a P-Value "an Exact Conditional P-Value" to remind us that it was calculated under an added artificial condition.

A two-sided analogue of Fisher's Exact Test is also possible. The observed difference in success rates was P2-P1=.128.

If we had seen 44 successes in 81 controls and 36 successes in 84 test (keeping the total successes at 80), then the difference in success rates would have been .115; while the difference between 45/81 vs 35/84 is .139. On a two-sided basis, if the test group is dealt either (A) 46 or more successes or (B) 35 or fewer successes, the evidence in favor of a difference would be at least as strong as that observed (.128).The probability that either (A) or (B) occurs in redeals is:

$$P=.120 \text{ (Conditional) vs } P=.10 \text{ (2-sided approx.)}$$

The P-value requires the same condition in the replication of matching total successes to that observed. Thus, this is not a true P-value, but rather a conditional one.

Exact Unconditional P-Value

The Exact Unconditional Z-test works as follows. The observed Z value was 1.653. What is the probability that if we had 81 independent success/failure observations for control and 84 success/failure observations for test, with the success rate identical for both control and test, that we would observe a Z-value of at least 1.653? The answer may depend on the value of that common success rate. The exact unconditional Z-test maximizes this probability over all possible identical values for the success rates (from 0 to 1). According to the definition, this yields the P-value. A microcomputer program (including hardware requirements), to perform the calculations is described in Shuster (1988). The one-sided version considers values above the observed Z of 1.653 in favor of the test treatment, while the two-sided version considers values above the observed value of 1.653 in favor of either treatment.

The exact unconditional P-values are .057 (vs approximation of .05 one-sided) and .108 (vs approximation of .10 two-sided). These agree remarkably well.

As Suissa and Shuster (1985) point out, the Exact Unconditional Z-Test generally outperforms Fisher's Exact Test. (Requires fewer patients for trials.)

2.4.3 Kaplan-Meier Comparison (Large Sample)

The Kaplan-Meier (1958) methodology is designed to estimate the probability of surviving for a given number of years, utilizing "Censored Survival Data". Unlike the Binomial methods above, the Kaplan-Meier technique recognizes the possibility that patients' times at risk will be unequal. However, the Kaplan-Meier technique makes the very important assumption that the prognosis of a patient who withdraws from the study is the same as that of a patient of the same "age" who remains on the study. This means for example, that the prognosis of a patient who was lost to follow-up after 6 months on the test therapy is the same after the six month point as that of a patient still on the test therapy at six months. It is therefore clear that in many studies, the validity of the Kaplan-Meier method rests heavily on obtaining virtually complete follow-up on all patients. Note that the above assumption is reasonable for patients whose follow-up is short, but current. For example, it is reasonable to assume that if a patient entered the study six months before it was to be analyzed he/she has the same outlook post six months as that of a patient who entered the study on the first day the study opened, four years prior to the analysis.

For each treatment, the Kaplan-Meier curve provides an estimate of the true probability of surviving at least T-days. The true probability represents the hypothetical target population of all past, present, and future patients who meet the eligibility conditions of the study. The estimate is based on the sample of patients assigned to the given treatment. It is subject to sampling error, since the actual sample assigned to the treatment is assumed to be a conceptual random sample from the target population. It also could be subject to selection biases, if the sample of patients cannot be viewed as a conceptual random sample from the target population. A good example where selection bias can occur, is where there is selective participation by some institutions in a multi-center trial. For example, institution X agrees only to place their advanced patients on a study.

2.4.3.1 Assumptions Underlying the Kaplan-Meier Curve

P1. All observations are "INDEPENDENT". That is, the outcome of a given patient is not influenced by that of another.

P2. Given the treatment assignment, the survival probability distribution of each patient over time is the same from trial to trial. In other words, patient prognosis on the test treatment is the same for early entries in the trial as later entries. The same holds true within the control treatment.

P3. The loss to follow-up rate is very small. Otherwise, we assume that patient losses are not in any way influenced by the patient's prognosis. Failure of this assumption is especially serious if one of the therapies is more toxic, leading to vastly differing withdrawal rates between the treatments.

Comment: Withdrawals for toxicity probably violate this assumption. It is therefore vital to keep track of study losses to toxicity.

2.4.3.2 Calculation of the Kaplan-Meier Curve

The Kaplan-Meier curve starts at DAY 0 at a value of 1.00(100%). On each succeeding day, we compute the value from the preceding day as follows:

Let   R=Number of patients left at risk on that day(At Risk)

F=Number of patient failures on that day (Deaths)

$$KM=KM \text{ (on previous day) } X \ (R-F) / R \tag{1}$$

The estimate is easily justified intuitively as follows. The estimate is built on a day-to-day basis, using the data for each day. To survive 100 days, a patient has to survive day 1, day 2,...., day 100. The logical estimate of the probability of surviving a given day, assuming that the patient was alive at the start of that day, is (R-F)/R, the ratio of the number of patients who survived that day to the number at risk on that day. The Kaplan-Meier estimate for day 100 is simply the product of the (R-F)/R values for days 1,2,3,...,100. An analogy to successive taxes will further motivate the multiplication. Imagine that you wish to send $1000 to a relative in Angola. The money is routed as follows. First it goes to Canada, which extracts 1%. Next, it goes to Cuba, which takes 13% of the balance. Finally, it goes to Angola, which charges 16% of what is left. We see that 99% survives the first taxation; of that, 87% survives the second taxation; and of that, 84% survives the final levy.

$$\text{Surviving Gift}=\$1000X.99X.87X.84=\$723.49$$

1. Canada: 1% of $1000.00 tax ; 99% of $1000=$990 survives

2. Cuba : 13% of $990.00 tax ; 87% of $ 990=$861.30 survives

3. Angola: 16% of $861.30 tax:84% of $861.30=$723.49 survives.

Note that the KM value in (1) stays the same as the previous day, unless at least one failure occurs. We shall demonstrate the curve concept for the Control Treatment (See Figures 10 and 11).

2.4.3.3 Numerical Example: See Figures 10 and 11.

Control Kaplan-Meier Curve

On Day 1, F=2 and R=186 and hence (R-F)/R=184/186.

On Day 2, only 184 patients are at risk, two having been lost on day 1. For Day 2, F=1 and R=184 and hence (R-F)/R=183/184.

On Days 3-6, no failures occurred and hence (R-F)/R=1.000.

On Day 7, F=1 and R=183, and hence (R-F)/R=182/183.

On Days 8-11, no failures occurred and hence (R-F)/R=1.000.

On Day 12, no failure occurred, but one survivor had only 12 days of follow-up. Here too, F=0 and hence (R-F)/R=1.000.

On Days 13-15, no failures occurred and hence (R-F)/R=1.000.

On Day 16, F=1 and R=181, and hence (R-F)/R=180/181.

This process continues to day 3454.

On Day 1, KM=184/186=.98925.

On Day 2, KM=.98925X183/184=.98387.

On Days 3-6, KM stays at .98387, but on Day 7,

KM=.98387X182/183=.97850

On Days 8-15, KM stays at .97850, but on Day 16

KM=.97850X181/182=.97309

This process continues to day 3454.

Wouldn't a computer be nice? There are many excellent programs for both mainframe computers and micro-computers.

2.4.3.4 Standard Error

A second important quantity relative to the Kaplan-Meier Curve is the "Standard Error". This latter quantity is used to obtain confidence intervals for fixed term survival as well as for comparison of fixed term survival for two competing treatments. There are several forms of the standard error, but the most convenient is due to Peto et al (1977, Statistical Note 6).

For a given day,

$$SE=KM \ SQRT\{(1-KM)/RX\} \qquad (2)$$

With KM as in (1) above. RX is the number at risk at the start of the following day. [Some users replace RX by the value of R from (1). However, in nearly all cases, the alternate SE is virtually identical to that of (2).]

One-sided and two-sided approximate confidence intervals can be obtained (using W from Figures 8 and 9 respectively) for a single fixed term survival as: True Survival >KM-(W SE) One-Sided Upper (Figure 8)

True Survival <KM+(W SE)	One-Sided Upper (Figure 8)
True survival <KM+(W SE)	One-Sided (Lower Figure 8)
KM-(W SE) <True Survival <KM+(W SE)	Two-Sided (Figure 9)

From Figure 11, at three years, we see that KM=51.6%(.516) for the Control Group. RX=91, the number at risk at the start of Year 4, ie the next day. Note that there was no failure in the control group (Figure 10) at day 365*3=1095. Hence, R may be taken as the number at risk at the start of year 4 (Interval 3-4 Years), that is, R=91. In this case, the alternate SE is the same as that computed in (2).

$$SE= .516 \ SQRT \ \{(1-.516)/91$$
$$= .516 \ SQRT \ (.484/91)$$

$$= .516 \ SQRT \ (.005319)$$

$$= .516 \ X \ .0729$$

$$= .038 \ (or \ 3.8\% \ as \ per \ Figure \ 11)$$

A One-sided 95% confidence interval for three-year survival is:

True control 3-year survival>.516-1.645X.038=.453

A Two-sided 95% confidence interval for three-year survival is:

.516-1.96X.038<True Control 3-year survival<.516+1.96X.038

that is between .442 and .590.

2.4.3.5 The Kaplan-Meier Comparison of Two Treatments

For the desired fixed term (eg 3 years), the following quantities are obtained: Control Group: KM1 and SE1, the Kaplan-Meier estimate and Standard Error at the desired time. Test Group: KM2 and SE2, the Kaplan-Meier estimate and Standard Error at the desired time.

Calculation of approximate P-value:

Let

$$Z = (KM2\text{-}KM1)/SQRT(SE1^2 + SE2^2) \tag{3}$$

Utilize Figure 6 (One-sided) or Figure 7 (Two-sided) to obtain the P-value.

In the numerical example of Figures 10 and 11, taken at three years,

KM1=.516 SE1=.038

KM2=.589 SE2=.036

$Z=(.589\text{-}.516)/SQRT(.038^2 + .036^2)$

$=.073/SQRT(.001444+.001296)$

=.073/SQRT(.002740)

=.073/.0523

=1.39

From Figure 6, the approximate P-value is .085, one-sided. (Had this been a two-sided comparison, the approximate P-value would have been .18. See Figure 7.)

2.4.3.6 Confidence Intervals for Difference in Fixed-Term Survival

Confidence Intervals for the difference in Fixed Term Survival can also be constructed. As in the Binomial example, they can be important in management decisions.

KM2-KM1 = Estimated difference in True survival at prescribed times (eg 3 years)

SED=SQRT(SE1^2+ SE2^2) is the Standard Error of this difference

A one-sided confidence interval is obtained using W from Figure 8 as:

True difference in fixed survival>(KM2-KM1) - W SED

A two-sided confidence interval is obtained using W from Figure 9 as:

(KM2-KM1) - W SED <True Difference <(KM2-KM1) + W SED

A one-sided 95% confidence interval in the numerical example of Figures 10 and 11 at three years is: (see calculation of Z above)

True difference in three year survival>(.073)-1.645 X.0523

> - .013

The corresponding two-sided 95% confidence interval at three years is:

$$-.030=.073-1.96\ X.0523 <\text{True Difference} <.073 + 1.96\ X.0523 =.176$$

As in the binomial case, the practical implication of the confidence interval over that of the P-value alone is clear. With high confidence, we can state that the test treatment has from a 3.0% disadvantage to a 17.6% advantage in three year survival, when compared to that of the control. If disadvantages of the order of 3% are not considered to be clinically significant, then the test therapy is preferred, even though conventional significance at $P<.05$ was not achieved.

2.4.4 Logrank Test

By far the most common method of comparing treatments in censored survival experiments is via the logrank test. Justification for the name is somewhat technical. (Interested readers might refer to Section 5 of Peto and Peto (1972) for the rationale behind the name). The major difference between the Kaplan-Meier and Logrank tests concern the question which each asks. The Kaplan-Meier comparison asks if the survival probabilities differ at one and only one pre-specified point in time (eg 5 Years). The logrank test asks if the survival curves differ. In other words, the logrank comparison looks at the entire available spectrum of time and attempts to distinguish between the treatments on the basis of overall performance in terms of survival.

For each day, starting the clock at randomization, the data used is:

N1=Number of patients on the control therapy at risk at the start of that day.

F1=Number of control therapy patients who fail on that day.

N2=Number of patients on the test therapy at risk at the start of that day.

F2=Number of test therapy patients who fail on that day.

If we are asking the question: Are the therapies equivalent in terms of survival, we would expect that given the total number of failures on that day, that they would tend to be distributed to treatment groups in proportion to the number at risk in that group. We calculate

$$E1=N1(F1+F2)/(N1+N2) \text{ and } E2=N2(F1+F2)/(N1+N2)$$

The Logrank test looks at F1 vs. E1 and F2 vs. E2 across days. Suppose that when one adds across days, the F1 total vastly exceeds the E1 total. That is, the control therapy has many more failures than one would expect for equivalent therapy. One would infer that the hypothesis of equivalence was in doubt, and accept the hypothesis that the test therapy was superior. Conversely, fewer than expected failures on the control therapy would lead to inferred superiority of the control therapy.

In order to formalize this into a statistical test, a quantity, V is calculated for each day as follows. Suppose a deck of (N1+N2) cards has (F1+F2) failures. If we shuffle this deck and assign N1 to player "Control" and N2 to player "Test", we would expect on average E1 failures to be assigned to "Control" and E2 to "Test". The quantity V is the "Variance" in the number assigned, that is the AVERAGE SQUARED DIFFERENCE between the randomly dealt assignment and the EXPECTED ASSIGNMENT.

If N1+N2>1 then

$$V=N1N2(F1+F2)(N1+N2-F1-F2)/\{(N1+N2)(N1+N2)(N1+N2-1)\}$$

If N1+N2 = 1 then V=0.

The logrank statistic is defined as:

$$Z=\{Sum(F1)-Sum(E1)\}/SQRT\{Sum(V)\} \ ,$$

where Sum is taken over each day for which patients are on study.

An approximate P-value is obtained from Figure 6 for one-sided tests and from Figure 7 for two-sided tests.

2.4.4.1 Important Observations with Respect to the Logrank Test

Prior to studying a particular example, we note the following:

1. Days on which no failure occurs make no contribution to Z. On such days, F1=0, E1=0 (since F1+F2=0), and V=0. Hence, the sum can be restricted to days on which failures occur.

2. If either N1=0 (only test patients left on that day) or N2=0 (only control patients left on that day), no contribution to Z is made. Clearly, if N1=0, then F1=0, E1=0, and V=0. If N2=0 then clearly the number of test therapy failures on the day (F2) is zero. This further implies from the above formulas that E1=F1 and V=0. The contribution to the Sum(F1) is counterbalanced by that to Sum(E1). Hence, such days have no impact on either the numerator or denominator of Z.

3. On many studies, only one failure will occur on any given day. If this occurs,

$$V=N1N2/\{(N1+N2)(N1+N2)\}=E1E2. \text{ (One Failure)}$$

4. E1+E2=F1+F2. (See formulas).

5. Provided that relatively few failures occur on any given day, an excellent approximation for Sum(V) is

$$Sum(V) \doteq Sum(E1)Sum(E2)/\{Sum(E1)+Sum(E2)\}$$

$$\doteq Sum(E1)Sum(E2)/(\text{Total Number of Deaths})$$

2.4.4.2 Numerical Example

Suppose that the control group survival (in days) is:

20,49+,122,245,301,332+,355+,378,398,402+,455,467+

while the experimental group survival is:

28+,122,255+,301+,344+,366,388+,409+,444+,477+,500,553+,

the survival in a randomized study of 24 patients. Only values followed by + indicate surviving patients.

The following table is a worksheet for the logrank test. Only days on which failures occur are included.

Day	N1	F1	N2	F2	E1	E2	V
20	12	1	12	0	.500	.500	.250
122	10	1	11	1	.952	1.048	.474
245	9	1	10	0	.474	.526	.249
301	8	1	9	0	.471	.529	.249
366	5	0	7	1	.417	.583	.243
378	5	1	6	0	.455	.545	.248
398	4	1	5	0	.444	.556	.247
455	2	1	3	0	.400	.600	.240
500	0	0	2	1	.000	1.000	.000
Sum		7		3	4.113	5.887	2.200

$$Z=(7-4.113)/SQRT(2.200)=2.887/1.483=1.95$$

From Figure 6, this translates to a one-sided P=0.026.

Sample calculation for day 122 — As can be seen from the data, there were 10 Control and 11 Test Patients alive at the start of Day 122. (Of the 12 controls, two did not reach day 122. One died at day 20 and another is alive but has only reached day 49.) (Of the 12 test patients, one is alive but has only reached day 28.) As can be seen from the data, one patient in each group failed on Day 122, and hence. F1=1 and F2=1. From the formulas for E1 and E2:

$$E1=N1(F1+F2)/(N1+N2)= (10X2)/21=20/21=.952$$

$$E2=N2(F1+F2)/(N1+N2)= (11X2)/21=22/21=1.048$$

$$V=N1N2(F1+F2)(N1+N2-F1-F2)/\{(N1+N2)(N1+N2)(N1+N2-1)\}$$

$$=(10X11X2X19)/(21X21X20)=4180/8820=.474$$

All other entries in the above table had only one failure on the day and hence for THOSE DAYS ONLY,E1=Proportion of patients on Control, E2=Proportion of patients on test, and V=E1E2.

2.4.4.3 Footnotes to Numerical Example and the Logrank Test

1. The approximation of Sum(V) in the above example is

$$Sum(V) \doteq Sum(E1)Sum(E2)/\{Sum(E1)+Sum(E2)\}$$
$$\doteq 4.113 \times 5.887/(4.113+5.887) = 2.421$$

Had we used this approximation, we would have obtained:

$$Z \doteq 2.887/SQRT(2.421)=2.887/1.556=1.86$$

instead of Z=1.95.

2. For large trials, despite the simplicity of the ideas, a computer would be most helpful in completing the analysis.

3. The validity of the approximate P-values depends on the accuracy of the "Normal Approximations". Generally speaking, for most applications, the approximations are acceptable if Sum(E1) and Sum(E2) both are five or more. The above example would be a borderline case.

As a check on the validity of the above one-sided P-value, we ran a small simulation study as follows. For each of 250 allocations, 12 patients were randomly reallocated to either the "Control" and 12 to the "Test" group. Using the same survival data, 4 of these reallocations had Z at least 1.95 in favor of the "Test" group. If there really was no target difference between the therapies, then any hypothetical allocation of the 24 patients into groups of 12 would be equally likely. Thus, of the 251 allocations (including the actual one), 5 (including the actual one) were 1.95 or higher. This yields a P-Value of about .020, one-sided. While as pointed out in Shuster and Boyett (1979) that this is a legitimate Exact "Conditional" P-value, it should be viewed as an approximation to complete enumeration of the 2.7 million possible reallocations of the 24 patients into two groups of 12.

A review of the definition of P-Values will be needed to grasp the next paragraph. See Section 2.4.1 (Crucial Elements).

Conditional P-values, as mentioned in the Section on Fisher's Exact Test, are similar to other P-values, except that they impose artificial restrictions on the replication process used in our definition of P-Value. For this simulation study, the restrictions are that the same 24 outcomes are obtained, and the actual replicated treatment assignment is randomly picked from the 251 assignments (250 from simulation study plus the actual observed assignment.).

4. For trials involving more than one treatment, see Section 3.12 of CHAPTER 3. Note that "Z sub ALPHA" and "Z sub BETA" are the upper 100ALPHA% and upper 100BETA% point of the Standard Normal Distribution (Figure 6).

5. If the trial is stratified, the analysis can be conducted as per CHAPTER 3, Section 3.13. This is a non-technical Section, but the notation is slightly different.

Corresponding Values for Each Stratum

	CHAPTER 2	CHAPTER 3 (Section 3.13)
Failures on treatment #1	Sum(F1)	O
Expected failures on treatment #1	Sum(E1)	X
Variance	Sum(V)	V

2.4.4.4 Proportional Hazards and the Logrank Test

While the logrank test is valid for testing the equality of two survival curves, it has large sample optimality properties, see Peto and Peto (1972), for the "proportional hazard" situation defined below. The sample size tables of this Handbook are computed under this model.

Proportional Hazards Model Assumption

Under a proportional hazards model, we assume that hazard ratio, the ratio of instantaneous death rates (Control:Test), does not depend on the time on test. To make this assumption concrete, suppose that the probability that an individual alive 6 months post randomization has a 1/1000 chance of dying on the next day under the test therapy and a 2/1000 chance of dying on the next day under the control therapy. Considering a day as an instant, the ratio of instantaneous death rates is:

$$(2/1000) / (1/1000)=2.0 \text{ at 6 months}$$

While the treatment specific instantaneous death rates are allowed to change over time in the proportional hazards model, the ratio is assumed to be constant over time. If the instantaneous death rate for the test group is 1/3000 at 2 years , we assume it is 2/3000 (double) at 3 years in the control group, since otherwise the ratio at 6 months would not be the same as that at 2 years.

This assumption is only used in obtaining the sample size tables. Failure of this assumption does not render the logrank test invalid as a device to compare the equality of two survival distributions.

REFERENCES

1. **Abramson, N. S., Meisel, J. D., and Safer, P.,** Deferred consent: a new aproach for resuscitation on comatose patients, *JAMA,* 255, 2466, 1986.

2. **Armitage, P.,** *Statistical Methods in Medical Research,* Blackwell Scientific, Oxford, England, 1971.

3. **Armitage, P.,** *Sequential Medical Trials,* Blackwell Scientific, Oxford, England, 1975.

4. **Buyse, M. E. , Staquel, M., and Sylvester, R. J.,** *Cancer Clinical Trials: Methods and Practise,* Oxford University Press, England, 1984.

5. **Diehl, L. F. and Perry, D. J.,** A comparison of randomized concurrent controls with matched historical controls: are historical controls valid?, *J. Clin. Oncol.,* 4, 114, 1986.

6. **Dupont, W.,** Randomized vs. historical clinical trials: are the benefits worth the cost?, *Am. J. Epidemiol.*, 122, 940, 1985.

7. **Fletcher R. H.,** Clinical trials, randomized consent and estimation (letter and response), *J. Chron. Dis.*, 37, 953, 1984.

8. **Gehan, E. A.,** Comparative clinical trials with historical controls, *Biomedicine*, 28 (Special issue), 13, 1978.

9. **Gehan, E. A.,** Randomized or historical control groups in cancer clinical trials: are historical controls valid?, *J. Clin. Oncol.*, 4, 1024, 1986.

10. **Haynes, R. B. and Dantes, R.,** Patient compliance and the conduct and interpretation of therapeutic trials, *Controlled Clin. Trials*, 8, 12, 1987.

11. **Kaplan E. L. and Meier, P.,** Nonparametric estimation from incomplete observatons, *J. Am. Stat Assoc.*, 53, 447, 1958.

12. **Link, M. P., Goorin, A. M., Miser, A. W. Green, A., Pratt, C., Belasco, J. B., Pritchard, J., Malpas, J. S., Baker, A. R., Kirkpatrick, J. A., Ayala, A. G., Shuster, J. J., Abelson, H. T., Simone, J. V., and Vietti, T. J.,** The effect of adjuvant chemotherapy on relapse-free survival in patients with osteosarcoma of the extremity, *N. Engl. J. Med.*, 314, 1600, 1986.

13. **Meinert, C. L.,** *Clinical Trials: Design, Conduct, and Analysis*, Oxford University Press, New York, 1986.

14. **O'Brien, P. C. and Fleming, T. R.,** A multiple testing procedure for clinical trials, *Biometrics*, 35, 549, 1979.

15. **Peto, R. and Peto, J.,** Asymptotically efficient rank invariant test procedures, *J. R. Stat. Soc. Ser. A*, 135, 185, 1972. (With discussion.)

16. **Peto, R., Pike, M. C., Armitage, P., Breslow, N. E., Cox, D. R., Howard, S. V., Mantel, N., McPherson, K., Peto, J., and Smith, P. G.,** Design and analysis of randomized clinical trials requiring prolonged observation of each patient. I. Introduction and design, *Br. J. Cancer*, 34, 585, 1976.

17. **Peto, R., Pike, M. C., Armitage, P., K., Peto, J., Breslow, N. E., Cox, D. R., Howard, S. V., Mantel, N., McPherson, K., Peto, J., and Smith, P. G.,** Design and analysis of randomized clinical trials requiring prolonged observation of each patient. II. Analysis and examples, *Br. J. Cancer*, 35, 1, 1977.

18. **Pocock, S. J.,** *Clinical Trials: A Practical Approach*, John Wiley & Sons, New York, 1983.

19. **Shuster, J. J.,** EXACTB and CONF : Exact unconditional procedures for binomial data, *American Statistician,* 42, 234, 1988.

20. **Shuster, J. J. and Boyett, J. M.,** Nonparametric multiple comparison procedures, *J. Am. Stat. Assoc.,* 74, 379, 1979.

21. **Shuster, J. J., Krischer, J. P. , and Boyett, J. M.,** Ethical issues in cooperative cancer therapy trials from a statistical viewpoint. I. A general overview, *Am. J. Pediatr. Hematol./Oncol.,* 7, 57, 1985.

22. **Shuster, J. J., Krischer, J. P., and Boyett, J. M.,** Ethical issues in cooperative cancer therapy trials from a statistical viewpoint. II. Specific issues, *Am. J. Pediatr. Hematol./Oncol.,*7, 64, 1985.

23. **Simon, R. and Wittes, R. E.,** Methodologic guidelines for reports of clinical trials, *Cancer Treatment Rep.,* 69, 1, 1985.

24. **Smith, W.,** Randomization and optimal design (with discussion by M. Zelen), *J. Chron. Dis.* 36, 609 1983. (Discussion 613—614; Rejoinder 615.)

25. **Suissa, S. and Shuster, J. J.,** Exact unconditional sample sizes for the 2 ×.2 binomial trial, *J. R. Stat. Assoc.,* 53, 447,1971.

26. **Tsiatis, A. A. and Rosner, G. L.,** Group sequential tests with censored survival data adjusting for covariates, *Biometrika,* 72, 365, 1985.

27. **Zelen, M.,** New design for randomized clinical trials, *N. Engl. J. Med.,* 300,1242, 1978.

28. **Zelen, M.,** Strategy and alternate randomized design in cancer clinical trials, *Cancer Treatment Rep.,* 66, 1095, 1982.

29. **Zelen, M.,** Guidelines for publishing papers on cancer clinical trials: responsibilities of Editors and Authors, *J. Clin. Oncol.,* 1. 154, 1983.

FIGURE 4. Simple randomization for 1200 patients.

```
122112222211111121112111221212112222111212211212
>>>>>>>>>>>>>>>>>>>>>>>>>>>>>>>>>>>>>>>>>>>>>>>>>>
211121122111222122211221112111221212111212111221
>>>>>>>>>>>>>>>>>>>>>>>>>>>>>>>>>>>>>>>>>>>>>>>>>>
122121121212221112211221221222122211222121222112
>>>>>>>>>>>>>>>>>>>>>>>>>>>>>>>>>>>>>>>>>>>>>>>>>>
211111212121111211112222121212222221111122112111
>>>>>>>>>>>>>>>>>>>>>>>>>>>>>>>>>>>>>>>>>>>>>>>>>>
111212222121211122211211122122122112122112112122
>>>>>>>>>>>>>>>>>>>>>>>>>>>>>>>>>>>>>>>>>>>>>>>>>>
111112111112221221112212221121221122221111112221 22
>>>>>>>>>>>>>>>>>>>>>>>>>>>>>>>>>>>>>>>>>>>>>>>>>>
222211111221222222111112111211112122122121212111
>>>>>>>>>>>>>>>>>>>>>>>>>>>>>>>>>>>>>>>>>>>>>>>>>>
222122221121212111221121221121112112222222111121
>>>>>>>>>>>>>>>>>>>>>>>>>>>>>>>>>>>>>>>>>>>>>>>>>>
112112121111111121221121212211212222121121212221
>>>>>>>>>>>>>>>>>>>>>>>>>>>>>>>>>>>>>>>>>>>>>>>>>>
211212121111211121221121211111122121211122111122
>>>>>>>>>>>>>>>>>>>>>>>>>>>>>>>>>>>>>>>>>>>>>>>>>>
212211211221212212212111122211111112111221211222
>>>>>>>>>>>>>>>>>>>>>>>>>>>>>>>>>>>>>>>>>>>>>>>>>>
211222212121222121111121211111212222111221211122
>>>>>>>>>>>>>>>>>>>>>>>>>>>>>>>>>>>>>>>>>>>>>>>>>>
112121112112211111212112122112122211122111211121 1
>>>>>>>>>>>>>>>>>>>>>>>>>>>>>>>>>>>>>>>>>>>>>>>>>>
122112222112222221111222111111212112112221211211
>>>>>>>>>>>>>>>>>>>>>>>>>>>>>>>>>>>>>>>>>>>>>>>>>>
111121112111221222121111122221211112111212111111
>>>>>>>>>>>>>>>>>>>>>>>>>>>>>>>>>>>>>>>>>>>>>>>>>>
111222221112122222121111212111121222211111111221
>>>>>>>>>>>>>>>>>>>>>>>>>>>>>>>>>>>>>>>>>>>>>>>>>>
222212211112121112212222112222222221221122112112 2
>>>>>>>>>>>>>>>>>>>>>>>>>>>>>>>>>>>>>>>>>>>>>>>>>>
212112221221121122112222111111221112121212222222
>>>>>>>>>>>>>>>>>>>>>>>>>>>>>>>>>>>>>>>>>>>>>>>>>>
111212211121222211222221121222221211221112121212
>>>>>>>>>>>>>>>>>>>>>>>>>>>>>>>>>>>>>>>>>>>>>>>>>>
222212222122112112222222222211211222111222211221
>>>>>>>>>>>>>>>>>>>>>>>>>>>>>>>>>>>>>>>>>>>>>>>>>>
122112221112221112212211111121211121221122221111
>>>>>>>>>>>>>>>>>>>>>>>>>>>>>>>>>>>>>>>>>>>>>>>>>>
111211221211221222122222122122112121221111112111
>>>>>>>>>>>>>>>>>>>>>>>>>>>>>>>>>>>>>>>>>>>>>>>>>>
211121211121112212221111111212221112122111122112
>>>>>>>>>>>>>>>>>>>>>>>>>>>>>>>>>>>>>>>>>>>>>>>>>>
211211211112221111222221121112112222221112122222
>>>>>>>>>>>>>>>>>>>>>>>>>>>>>>>>>>>>>>>>>>>>>>>>>>
```

FIGURE 5. Random length permuted block lengths of four or six patients.

```
221112212121221211212112121122122121121122122112
>>>>>>>>>>>>>>>>>>>>>>>>>>>>>>>>>>>>>>>>>>>>>>>>>>
211112221221121212112212122112211212211122121211
>>>>>>>>>>>>>>>>>>>>>>>>>>>>>>>>>>>>>>>>>>>>>>>>>>
221221221112122111221122211121222211211122122211
>>>>>>>>>>>>>>>>>>>>>>>>>>>>>>>>>>>>>>>>>>>>>>>>>>
212121121222121121112221212211211221112212221211
>>>>>>>>>>>>>>>>>>>>>>>>>>>>>>>>>>>>>>>>>>>>>>>>>>
112122122122111222112212111212212122112121212211
>>>>>>>>>>>>>>>>>>>>>>>>>>>>>>>>>>>>>>>>>>>>>>>>>>
121212211221112212122121211221121212212121111222 1
>>>>>>>>>>>>>>>>>>>>>>>>>>>>>>>>>>>>>>>>>>>>>>>>>>
211222112212112211111221121222211121212111221221
>>>>>>>>>>>>>>>>>>>>>>>>>>>>>>>>>>>>>>>>>>>>>>>>>>
221122211122121122111221212122112221112211121122
>>>>>>>>>>>>>>>>>>>>>>>>>>>>>>>>>>>>>>>>>>>>>>>>>>
122211211212221122122121112221221112122221111112
>>>>>>>>>>>>>>>>>>>>>>>>>>>>>>>>>>>>>>>>>>>>>>>>>>
222211212211112122211222111221121121221121221221
>>>>>>>>>>>>>>>>>>>>>>>>>>>>>>>>>>>>>>>>>>>>>>>>>>
121122222111112222211111221121222121122111222211
>>>>>>>>>>>>>>>>>>>>>>>>>>>>>>>>>>>>>>>>>>>>>>>>>>
121122121212221112121112221212211122112222111122
>>>>>>>>>>>>>>>>>>>>>>>>>>>>>>>>>>>>>>>>>>>>>>>>>>
212122211111122211212212122121212121212121122112
>>>>>>>>>>>>>>>>>>>>>>>>>>>>>>>>>>>>>>>>>>>>>>>>>>
122121211122122112221121122211122122211122121111
>>>>>>>>>>>>>>>>>>>>>>>>>>>>>>>>>>>>>>>>>>>>>>>>>>
122221122121212112121221112122112122121122111222
>>>>>>>>>>>>>>>>>>>>>>>>>>>>>>>>>>>>>>>>>>>>>>>>>>
222111212211212121121121222212112111222111222112
>>>>>>>>>>>>>>>>>>>>>>>>>>>>>>>>>>>>>>>>>>>>>>>>>>
211122212122211111222111221122122211222111221211
>>>>>>>>>>>>>>>>>>>>>>>>>>>>>>>>>>>>>>>>>>>>>>>>>>
221211212112211212112122112211222111221211112122
>>>>>>>>>>>>>>>>>>>>>>>>>>>>>>>>>>>>>>>>>>>>>>>>>>
111222221121222111112212121211212211221221112222
>>>>>>>>>>>>>>>>>>>>>>>>>>>>>>>>>>>>>>>>>>>>>>>>>>
112121121112222111212121212121122211121122211212
>>>>>>>>>>>>>>>>>>>>>>>>>>>>>>>>>>>>>>>>>>>>>>>>>>
211122221121212121121212122112222111111222112212
>>>>>>>>>>>>>>>>>>>>>>>>>>>>>>>>>>>>>>>>>>>>>>>>>>
221121122112121211212212211122112221211212112122
>>>>>>>>>>>>>>>>>>>>>>>>>>>>>>>>>>>>>>>>>>>>>>>>>>
211212211211222121211212122121211122121222111122
>>>>>>>>>>>>>>>>>>>>>>>>>>>>>>>>>>>>>>>>>>>>>>>>>>
211222112111122212212212112121111221222111212121 2
>>>>>>>>>>>>>>>>>>>>>>>>>>>>>>>>>>>>>>>>>>>>>>>>>>
222111212111212211222121122121211112221211112211 22
>>>>>>>>>>>>>>>>>>>>>>>>>>>>>>>>>>>>>>>>>>>>>>>>>>
```

FIGURE 6. One-sided P-values for normal curves.

IF Z WITHIN : P=	****	IF Z WITHIN : P=	****	IF Z WITHIN : P=	****	IF Z WITHIN : P=	****	IF Z WITHIN : P=
> 3.090: .001	****	3.090 & 2.878: .002	****	2.878 & 2.748: .003	****	2.748 & 2.652: .004	****	2.652 & 2.576: .005
2.576 & 2.512: .006	****	2.512 & 2.457: .007	****	2.457 & 2.409: .008	****	2.409 & 2.366: .009	****	2.366 & 2.326: .010
2.326 & 2.257: .012	****	2.257 & 2.197: .014	****	2.197 & 2.144: .016	****	2.144 & 2.097: .018	****	2.097 & 2.054: .020
2.054 & 2.014: .022	****	2.014 & 1.977: .024	****	1.977 & 1.943: .026	****	1.943 & 1.911: .028	****	1.911 & 1.881: .030
1.881 & 1.852: .032	****	1.852 & 1.825: .034	****	1.825 & 1.799: .036	****	1.799 & 1.774: .038	****	1.774 & 1.751: .040
1.751 & 1.728: .042	****	1.728 & 1.706: .044	****	1.706 & 1.685: .046	****	1.685 & 1.665: .048	****	1.665 & 1.645: .050
1.645 & 1.598: .055	****	1.598 & 1.555: .06	****	1.555 & 1.514: .065	****	1.514 & 1.476: .07	****	1.476 & 1.440: .075
1.440 & 1.405: .08	****	1.405 & 1.372: .085	****	1.372 & 1.341: .09	****	1.341 & 1.311: .095	****	1.311 & 1.282: .100
1.282 & 1.227: .11	****	1.227 & 1.175: .12	****	1.175 & 1.126: .13	****	1.126 & 1.080: .14	****	1.080 & 1.036: .15
1.036 & 0.994: .16	****	0.994 & 0.954: .17	****	0.954 & 0.915: .18	****	0.915 & 0.878: .19	****	0.878 & 0.842: .20
0.842 & 0.806: .21	****	0.806 & 0.772: .22	****	0.772 & 0.739: .23	****	0.739 & 0.706: .24	****	0.706 & 0.674: .25
0.674 & 0.643: .26	****	0.643 & 0.613: .27	****	0.613 & 0.583: .28	****	0.583 & 0.553: .29	****	0.553 & 0.524: .30
0.524 & 0.496: .31	****	0.496 & 0.468: .32	****	0.468 & 0.440: .33	****	0.440 & 0.412: .34	****	0.412 & 0.385: .35
0.385 & 0.358: .36	****	0.358 & 0.332: .37	****	0.332 & 0.305: .38	****	0.305 & 0.279: .39	****	0.279 & 0.253: .40
0.253 & 0.228: .41	****	0.228 & 0.202: .42	****	0.202 & 0.176: .43	****	0.176 & 0.151: .44	****	0.151 & 0.126: .45
0.126 & 0.100: .46	****	0.100 & 0.075: .47	****	0.075 & 0.050: .48	****	0.050 & 0.025: .49	****	0.025 & 0.000: .50
0.000 & -.025: .51	****	-.025 & -0.05: .52	****	-0.05 & -.075: .53	****	-.075 & -0.1: .54	****	-0.1 & -.126: .55
-.126 & -.151: .56	****	-.151 & -.176: .57	****	-.176 & -.202: .58	****	-.202 & -.228: .59	****	-.228 & -.253: .60
-.253 & -.279: .61	****	-.279 & -.305: .62	****	-.305 & -.332: .63	****	-.332 & -.358: .64	****	-.358 & -.385: .65
-.385 & -.412: .66	****	-.412 & -0.44: .67	****	-0.44 & -.468 : .68	****	-.468 & -.496: .69	****	-.496 & -.524: .70
-.524 & -.553: .71	****	-.553 & -.583: .72	****	-.583 & -.613: .73	****	-.613 & -.643: .74	****	-.643 & -.674: .75
-.674 & -.706: .76	****	-.706 & -.739: .77	****	-.739 & -.772: .78	****	-.772 & -.806: .79	****	-.806 & -.842: .80
-.842 & -.878: .81	****	-.878 & -.915: .82	****	-.915 & -.954: .83	****	-.954 & -.994: .84	****	-.994 & -1.04: .85
-1.04 & -1.08: .86	****	-1.08 & -1.13: .87	****	-1.13 & -1.17: .88	****	-1.17 & -1.23: .89	****	-1.23 & -1.28: .90
-1.28 & -1.34: .91	****	-1.34 & -1.41: .92	****	-1.41 & -1.48: .93	****	-1.48 & -1.55: .94	****	-1.55 & -1.64: .95
-1.64 & -1.75: .96	****	-1.75 & -1.88: .97	****	-1.88 & -2.05: .98	****	-2.05 & -2.33: .99	****	-2.33 > : 1

FIGURE 7. Two-sided P-values for normal curves; ignore sign of Z + VS -.

IF Z WITHIN: P=	****	IF Z WITHIN: P=	****	IF Z WITHIN: P=	****	IF Z WITHIN: P=	****	IF Z WITHIN: P=
>3.291: .001	****	3.291 & 3.090: .002	****	3.090 & 2.968: .003	****	2.968 & 2.878: .004	****	2.878 & 2.807: .005
2.807 & 2.748: .006	****	2.748 & 2.697: .007	****	2.697 & 2.652: .008	****	2.652 & 2.612: .009	****	2.612 & 2.576: .010
2.576 & 2.512: .012	****	2.512 & 2.457: .014	****	2.457 & 2.409: .016	****	2.409 & 2.366: .018	****	2.366 & 2.326: .020
2.326 & 2.290: .022	****	2.290 & 2.257: .024	****	2.257 & 2.226: .026	****	2.226 & 2.197: .028	****	2.197 & 2.170: .030
2.170 & 2.144: .032	****	2.144 & 2.120: .034	****	2.120 & 2.097: .036	****	2.097 & 2.075: .038	****	2.075 & 2.054: .040
2.054 & 2.034: .042	****	2.034 & 2.014: .044	****	2.014 & 1.995: .046	****	1.995 & 1.977: .048	****	1.977 & 1.960: .050
1.960 & 1.919: .055	****	1.919 & 1.881: .060	****	1.881 & 1.845: .065	****	1.845 & 1.812: .070	****	1.812 & 1.780: .075
1.780 & 1.751: .080	****	1.751 & 1.722: .085	****	1.722 & 1.695: .090	****	1.695 & 1.670: .095	****	1.670 & 1.645: .100
1.645 & 1.598: 0.11	****	1.598 & 1.555: 0.12	****	1.555 & 1.514: 0.13	****	1.514 & 1.476: 0.14	****	1.476 & 1.440: 0.15
1.440 & 1.405: 0.16	****	1.405 & 1.372: 0.17	****	1.372 & 1.341: 0.18	****	1.341 & 1.311: 0.19	****	1.311 & 1.282: 0.20
1.282 & 1.254: 0.21	****	1.254 & 1.227: 0.22	****	1.227 & 1.200: 0.23	****	1.200 & 1.175: 0.24	****	1.175 & 1.150: 0.25
1.150 & 1.126: 0.26	****	1.126 & 1.103: 0.27	****	1.103 & 1.080: 0.28	****	1.080 & 1.058: 0.29	****	1.058 & 1.036: 0.30
1.036 & 1.015: 0.31	****	1.015 & 0.994: 0.32	****	0.994 & 0.974: 0.33	****	0.974 & 0.954: 0.34	****	0.954 & 0.935: 0.35
0.935 & 0.915: 0.36	****	0.915 & 0.896: .037	****	0.896 & 0.878: 0.38	****	0.878 & 0.860: 0.39	****	0.860 & 0.842: 0.40
0.842 & 0.824: 0.41	****	0.824 & 0.806: 0.42	****	0.806 & 0.789: 0.43	****	0.789 & 0.772: 0.44	****	0.772 & 0.755: 0.45
0.755 & 0.739: 0.46	****	0.739 & 0.722: 0.47	****	0.772 & 0.706: 0.48	****	0.706 & 0.690: 0.49	****	0.690 & 0.674: 0.50
0.674 & 0.659: 0.51	****	0.659 & 0.643: 0.52	****	0.643 & 0.628: 0.53	****	0.628 & 0.613: 0.54	****	0.613 & 0.598: 0.55
0.598 & 0.583: 0.56	****	0.583 & 0.568: 0.57	****	0.568 & 0.553: 0.58	****	0.553 & 0.539: 0.59	****	0.539 & 0.524: 0.60
0.524 & 0.510: 0.61	****	0.510 & 0.496: 0.62	****	0.496 & 0.482: 0.63	****	0.482 & 0.468: 0.64	****	0.468 & 0.454: 0.65
0.454 & 0.440: 0.66	****	0.440 & 0.426: 0.67	****	0.426 & 0.412: 0.68	****	0.412 & 0.399: 0.69	****	0.399 & 0.385: 0.70
0.385 & 0.372: 0.71	****	0.372 & 0.358: 0.72	****	0.358 & 0.345: 0.73	****	0.345 & 0.332: 0.74	****	0.332 & 0.319: 0.75
0.319 & 0.305: 0.76	****	0.305 & 0.292: 0.77	****	0.292 & 0.279: 0.78	****	0.279 & 0.266: 0.79	****	0.266 & 0.253: 0.80
0.253 & 0.240: 0.81	****	0.240 & 0.228: 0.82	****	0.228 & 0.215: 0.83	****	0.215 & 0.202: 0.84	****	0.202 & 0.189: 0.85
0.189 & 0.176: 0.86	****	0.176 & 0.164: 0.87	****	0.164 & 0.151: 0.88	****	0.151 & 0.138: 0.89	****	0.138 & 0.126: 0.90
0.126 & 0.113: 0.91	****	0.113 & 0.100: 0.92	****	0.100 & 0.088: 0.93	****	0.088 & 0.075: 0.94	****	0.075 & 0.063: 0.95
0.063 & 0.050: 0.96	****	0.050 & 0.038: 0.97	****	0.038 & 0.025: 0.98	****	0.025 & 0.013: 0.99	****	0.013 & 0.000: 1.00

FIGURE 8. Values for W in one-sided confidence intervals for approximately normally distributed estimates.

CONFIDENCE INTERVAL:	ABOVE	ESTIMATE - W SE	(UPPER CI)
CONFIDENCE INTERVAL:	BELOW	ESTIMATE + W SE	(LOWER CI)

W	Confidence (%)
0.84	80
1.282	90
1.645	95
2.33	99

Note: SE = Estimated standard deviation of the estimate, = "Standard Error"

Interpretation: Before the study is conducted, the probability that the true value of the parameter being estimated falls below the estimate + 1.654 standard error is about 95%. Thus, after the study is over, we have 95% confidence in the confidence statement. Note that once the study has been completed, the "game is over" and probability is no longer applicable.

FIGURE 9. Values for W in two-sided confidence intervals for approximately normally distributed estimates.

CONFIDENCE INTERVAL:	ESTIMATE - W SE	(LOWER LIMIT)
TO	ESTIMATE + W SE	(UPPER LIMIT)

W	Confidence (%)
1.282	80
1.654	90
1.96	95
2.58	99

Note: SE = Estimated standard deviation of the estimate, = "Standard Error"

Interpretation: Before the study is conducted, the probability that the true value of the parameter being estimated falls between the estimate ± 1.96 standard error is about 95%. Thus, after the study is over, we have 95% confidence in the confidence statement. Note that once the study has been completed, the "game is over" and probability is no longer applicable.

FIGURE 10. Survival in randomized pediatric cancer trial entries are in days with + to indicate survivors.

TREATMENT 1 — CONTROL THERAPY

1	1	2	7	12+	16	21+	43	47+	48
53	74	105	120	120	131	140	155+	172	195
195	209	237	252	256	265	287	309+	348	351
353	361	364	365	371	377	377	391	396	398
398	404	413	413	414	431	436	445	449	456
462	464	495	506	514	523	553	571	571	595
605+	605	617	629	641	650	671	700	766	769
772	795	801	802	804	828	844	870	874	898
908	932	949	964	973	977	984	1004	1028	1035
1047	1048	1062	1068	1084+	1158+	1173+	1195+	1321	1330
1367	1372	1442	1476	1514+	1559	1602	1639	1671	1678
1680	1690	1758	1777	1793	1797	1807+	1824	1861+	1922
1923+	1929	1935+	1946+	1952+	1984+	1987+	1992+	1993+	2019
2028+	2042+	2064+	2082+	2088+	2090	2110+	2122+	2129+	2166
2186+	2236	2245	2247+	2249+	2277+	2293+	2306+	2308+	2352
2353+	2353+	2405+	2436+	2464+	2465+	2491+	2501	2544+	2546
2580+	2605+	2630+	2640+	2667+	2674+	2678+	2684+	2693+	2733
2763+	2764+	2766+	2772+	2843+	2847+	2878+	2897+	2919+	2919
3018+	3037+	3047+	3070+	3074+	3454+				

TREATMENT 2 — TEST THERAPY

2	4	12	14	15	21	28	32	41	44
75	75	83+	98	98	104	104	133	139	206
213	215	223+	234	244	285+	285	290	313	318
319+	322+	349	357	371	372	375	381	390	394
424	436	443+	443	443	444	466+	475	479	514
531+	537	560	564	586	609	623	633	646	648
656	662	669	704	709	716	720	721	723	748
750	753	758	764	772	772	774	779	854+	888
893	912	952	959	959	960	980+	996	1000	1003
1039	1049	1053	1094	1176	1196	1213	1227+	1231	1245
1248+	1255+	1256	1274	1290	1331	1415	1433	1554	1572
1665+	1690	1700	1714+	1727+	1727	1767+	1772+	1779+	1779
1829	1835+	1842	1847	1848	1871+	1880+	1886+	1905+	1912
1914	1922+	1939	1961+	1983+	1987+	1992+	2003	2004+	2009
2029+	2051+	2051+	2058	2063+	2067+	2072+	2079+	2087+	2089
2121+	2145+	2154+	2155+	2174+	2195+	2196+	2213+	2241+	2254
2310+	2329+	2381+	2414+	2418+	2423+	2441	2447+	2453+	2497
2523+	2526+	2545+	2548+	2554+	2560+	2583+	2589+	2605+	2657
2671+	2697+	2714+	2714+	2719+	2723+	2730+	2736+	2744+	2746
2771	2772+	2815+	2841+	2863+	2863+	2876+	2928+	2957+	2959
2972+	3010+	3017+	3139+	3244+	3310+	3337+			

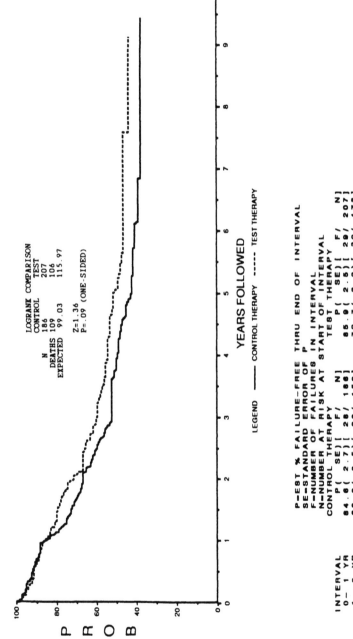

FIGURE 11. Survival comparison. See Figure 10 for raw data.

Chapter 3

DERIVATION OF THE STATISTICAL RESULTS

In this section, we shall demonstrate the plausibility of the large sample results, while pointing to the gap in the derivation. The mathematical sophistication of a formal proof is well beyond the scope of this Handbook. The aim of this partial derivation is to keep the mathematical level at that of a reader who has completed one year of Mathematical Statistics. More advanced readers are referred to Crowley (1974) and Breslow and Crowley (1974) for complete derivations.

To avoid complicated "stochastic processes", we shall initially make the following artificial assumption. Imagine that for each treatment there is a potentially infinite source of patients. Imagine further that for treatment 1, we shall pick sample sizes of N_1 day #1's, N_2 day #2's, N_3 day #3's, etc. Similarly, for treatment 2, we shall pick sample sizes of M_1 day#1's, M_2 day #2's, M_3 day#3's, etc. For each day, j, we shall restrict our analysis to the first N_j patients on treatment 1 at risk on day j and the first M_j patients on treatment 2 at risk on day j. We assume that the M_j and N_j are selected in advance.

Of course, this not a practical way to conduct a trial. The best we usually can do in practice is to pick N_1 and M_1. The fate of the patient forces the actual values of the remaining N's and M's to be random. Further, since the recruitment process has staggered entry, the actual N's an M's are hard to control. Nonetheless, our artificial method of restricting the analysis allows us to treat the sample sizes (N's and M's) as fixed, and the outcomes on each treatment by day combination as independent binomial variables. The gap in applying the large sample results, as if the actual day-specific sample sizes were in fact planned, can be closed by more advanced "stochastic process" methodology. Intuitively, the percent error caused by using the actual sample sizes versus the non-random sample sizes equated to the expected sample sizes is negligible.

In summary, we shall prove our results under this artificial assumption. We shall argue that the results hold in general on the basis of intuition.

The attached notation chart will be useful in the derivations. Further, Appendix I gives a review of mathematical statistics topics used in the derivation of our results.

	NOTATION CHART	
	TREATMENT 1	TREATMENT 2
1. j	Patient Day #	Patient Day #
2. Patients at Risk at Start of Day j	N_j	M_j
3. Failures on Day j	F_j	G_j
4. Probability of Surviving Day j given alive at start of Day j	P_j	R_j
5. Probability of Failing on Day j given alive at start of Day j	Q_j	T_j
6. Estimate of Parameter in 4	$\hat{P}_j = (N_j\text{-}F_j)/N_j$	$\hat{R}_j = (M_j\text{-}G_j)/M_j$
7. Estimate of Parameter in 5	$\hat{Q}_j = F_j/N_j$	$\hat{T}_j = G_j/M_j$
8. Cumulative Probability of Surviving at Least k days	$S_k = \prod_{j=1}^{k} P_j$	N/A
9. Estimate of S_k (Kaplan-Meier)	$\hat{S}_k = \prod_{j=1}^{k} \hat{P}_j$	N/A
10. D_k (Estimate Error of $\hat{S}_k$)	$D_k^2 = S_k^2 \sum_{j=1}^{k} \left\{ Q_j/(P_j N_j) \right\}$	N/A
11. $\hat{D}_k$ (Estimate of D_k)(Greenwood)	$\hat{D}_k^2 = \hat{S}_k^2 \sum_{j=1}^{k} \left\{ \hat{Q}_j/(\hat{P}_j N_j) \right\}$	N/A
12. E_j = Expected Failures (Logrank)	$E_j = (F_j + G_j) \left\{ N_j/(N_j + M_j) \right\}$	

3.1 Derivation of the Large Sample Distribution of the Logrank Statistic

Consider the following null Hypothesis:

H_0 : Treatment 1 has the same survival curve as Treatment 2.

In terms of the parameters in item 4 of the Notation Chart, this is equivalent to

$$H_0 : P_j = R_j \text{ for all } j$$

The "logrank" test is sensitive to

$$H_0 : P_j = R_j \text{ for all } j$$

against

$$H_A^1 : P_j > R_j \text{ for all } j \qquad \text{(Treatment 2 Superior)}$$

or

$$H_A^2 : P_j < R_j \text{ for all } j \qquad \text{(Treatment 1 Superior)}$$

or a two sided alternative

$$H_A^3 : H_A^1 \text{ or } H_A^2.$$

When mixed behavior occurs under H_A, some $P_j < R_j$ while other $P_j > R_j$, the logrank test may not have good performance.

Structurally, we compute the following quantities relevant to Treatment 1:

$$O = \sum_j F_j \qquad \text{(observed failures on Treatment 1)}, \qquad (1.1)$$

$$X = \sum_j E_j \qquad \text{(expected failures on Treatment 1)}, \qquad (1.2)$$

where

$$E_j = \frac{(F_j + G_j)N_j}{N_j + M_j} \qquad (1.3)$$

and

$$SE^2 = \sum_j \frac{N_j M_j (F_j + G_j)(N_j + M_j - F_j - G_j)}{(N_j + M_j)^2 (N_j + M_j - 1)}. \qquad (1.4)$$

If $N_j + M_j = 1$, then the jth term in (1.4) is defined as zero. The logrank statistic is

$$Z = \frac{O - X}{SE} \tag{1.5}$$

which under mild regularity conditions is asymptotically standard normal under H_0.

Note: If we interchange the definition of Treatment 1 and Treatment 2, all we do is change the sign of Z.

The P-value is obtained from (1.5) and Figure 6 or Figure 7,

depending on the Alternative Hypothesis.

Large Sample Result (Logrank)

If the assumptions below hold, then Z given in (1.5) is asymptotically standard normal.

Assumptions:

(a) $\quad \sum_j \frac{N_j M_j P_j Q_j}{N_j + M_j} \rightarrow \infty.$

(b) $\quad$ No day is a "wipeout." There exists a positive number $\delta < 1$ such that

$\quad P_j > 1 - \delta \quad$ for all j.

This is equivalent to bounding the failure rates away from one,

$\quad$ i.e., $Q_j < \delta \quad$ for all j. $\quad (Q_j = 1 - P_j)$

Plan of the Proof:

Step 1: Representation of Numerator of Z.

Prove
$$O - X = \sum_j \frac{(M_j F_j - N_j G_j)}{N_j + M_j}$$
(1.6)

with

F_j Binomial with Parameters N_j and Q_j

G_j Binomial with Parameters N_j and T_j

and $\{F_j\}, \{G_j\}$ mutually independent.

Step 2:

$$\frac{O - X}{V} \xrightarrow{D} N(0, 1)$$

where

$$V^2 = \sum_j \frac{N_j M_j P_j Q_j}{N_j + M_j}.$$
(1.7)

See A93 for a definition of convergence in distribution ($\xrightarrow{D}$).

Step 3: If Y is binomial with parameter Q and sample size N, then for $P = 1 - Q$

$$E\{Y(N-Y)\} = N(N-1)PQ$$

and

$$Var\{Y(N-Y)\} \le N(N-1)^2 PQ.$$

Step 4: Show

$$\frac{SE^2}{V^2} \xrightarrow{P} 1.$$

Step 5: Complete the proof.

Show that Z, defined in (1.5) converges in distribution to the standard normal.

Proof of Step 1:

$$O - X = \sum_j F_j - \sum_j E_j$$

$$= \sum_j F_j - \sum_j \frac{(F_j + G_j)N_j}{N_j + M_j}$$

$$= \sum_j \left\{ F_j - \frac{(F_j + G_j)N_j}{N_j + M_j} \right\}$$

$$= \sum_j \left\{ \frac{(N_j + M_j)F_j - (F_j + G_j)N_j}{N_j + M_j} \right\}$$

$$= \sum_j \frac{(M_j F_j - N_j G_j)}{N_j + M_j}. \tag{1.8}$$

Now F_j is Binomial with parameters N_j and Q_j

G_j is Binomial with parameters M_j and T_j

from the model assumed in the introduction to Chapter 3. This completes Step 1.

Proof of Step 2:

Let us represent the F_j and G_j as Bernoulli Random Variables, and apply the central limit theorem for Bernoulli variables.

$$F_j = \sum_{i=1}^{N_j} Y_{ij_1}$$

$$G_j = \sum_{i=1}^{M_j} Y_{ij_2}$$

with $Y_{ij_\ell} = 1$ or 0 depending respectively on whether for treatment ℓ, the ith patient trial on day j results in a failure or survival.

From (1.8),

$$O - X = \sum_j \sum_{i=1}^{N_j} \frac{M_j Y_{ij_1}}{N_j + M_j} - \sum_j \sum_{i=1}^{M_j} \frac{N_j Y_{ij_2}}{N_j + M_j} \tag{1.9}$$

Using (1.9) and the null hypothesis $Q_j = T_j$ we have

$$E(O-X) = \sum_j \sum_{i=1}^{N_j} \frac{M_j Q_j}{N_j + M_j} - \sum_j \sum_{i=1}^{M_j} \frac{N_j T_j}{N_j + M_j}$$

$$= \sum_j \frac{N_j M_j}{N_j + M_j} (Q_j - T_j) = 0;$$

$$Var(O-X) = \sum_j \sum_{i=1}^{N_j} \frac{M_j^2}{(N_j + M_j)^2} P_j Q_j + \sum_j \sum_{i=1}^{M_j} \frac{N_j^2}{(N_j + M_j)^2} R_j T_j$$

$$= \sum_j \left\{ \frac{M_j^2 N_j + N_j^2 M_j}{(N_j + M_j)^2} \right\} P_j Q_j \qquad \left(P_j Q_j = R_j T_j \right)$$

$$= \sum_j \frac{N_j M_j}{N_j + M_j} P_j Q_j$$

$$= V^2. \qquad \left(\text{From } (1.7) \right)$$

We wish to apply the Central Limit Theorem for Bernoulli Variables: (Condition 2), See A107

Let

$$b_{ijk} = \frac{M_j}{N_j + M_j} \qquad k = 1$$

$$= \frac{-N_j}{N_j + M_j} \qquad k = 2$$

From (1.9)

$$O - X = \sum_j \sum_{i=1}^{N_j} b_{ij1} Y_{ij1} + \sum_j \sum_{i=1}^{M_j} b_{ij2} Y_{ij2}$$

with

$$\sum_j \sum_{i=1}^{N_j} b_{ij1}^2 P_j Q_j + \sum_j \sum_{i=1}^{M_j} b_{ij2}^2 P_j Q_j = V^2.$$

Now $V^2 \to \infty.$ $\left(\text{Assumption (a)}\right)$

To apply condition 2 of A107 we must show

$$\frac{W}{V^3} \to 0 \text{ where}$$

$$W = \sum_j \sum_{i=1}^{N_j} |b_{ij1}|^3 \, Q_j + \sum_j \sum_{i=1}^{M_j} |b_{ij2}|^3 \, Q_j. \tag{1.10}$$

Now $$\sum_j \sum_{i=1}^{N_j} |b_{ij1}|^3 Q_j = \sum_j \frac{N_j M_j^3}{(N_j + M_j)^3} \, Q_j$$

and $$\sum_j \sum_{i=1}^{N_j} |b_{ij2}|^3 Q_j = \sum_j \frac{M_j N_j^3}{(N_j + M_j)^3} \, Q_j.$$

Thus

$$W = \sum_j \sum_{i=1}^{N_j} |b_{ij1}|^3 Q_j + \sum_j \sum_{i=1}^{M_j} |b_{ij2}|^3 Q_j = \sum_j \frac{N_j M_j (N_j^2 + M_j^2) Q_j}{(N_j + M_j)^3}.$$

Now since $(N_j + M_j)^2 \geq N_j^2 + M_j^2$

$$W \leq \sum_j \frac{N_j M_j Q_j}{(N_j + M_j)}.$$

From assumption (b), $P_j > 1 - \delta$ with δ fixed. Hence

$$W \leq \frac{1}{1-\delta} \sum_j \frac{N_j M_j Q_j P_j}{(N_j + M_j)} = \frac{V^2}{1 - \delta}$$

That is, since $V^2 \to \infty$

$$\frac{W}{V^3} < \frac{1}{(1-\delta)V} \to 0.$$

$\Big($The other condition 2 postulates are clearly satisfied since $Q_j < \delta$ (bounded away from one) and

$|b_{ijk}| \leq 1.$ $\Big)$

This completes the proof of step 2.

Proof of Step 3:

Let Y be binomial with sample size N and parameter Q. Let P = 1-Q.

For now, let N > 1.

$$E\Big(Y(N\text{-}Y)\Big) \;=\; \sum_{y=0}^{N} y(N\text{-}y)\binom{N}{y}Q^y P^{N\text{-}y}$$

$$= N(N\text{-}1)PQ \sum_{y=1}^{N\text{-}1} \binom{N\text{-}2}{y\text{-}1}Q^{y\text{-}1}P^{(N\text{-}2)\text{-}(y\text{-}1)}$$

$$= N(N\text{-}1)PQ \sum_{j=0}^{N\text{-}2} \binom{N\text{-}2}{j}Q^j P^{N\text{-}2\text{-}j}$$

$$= N(N\text{-}1)PQ. \tag{1.11}$$

Note: if N=1, formula (1.11) holds since both sides equal zero.

Similarly, if for now N > 2,

$$E\Big(Y(Y\text{-}1)(N\text{-}Y)\Big) \;=\; N(N\text{-}1)(N\text{-}2)PQ^2 \sum_{j=0}^{N\text{-}3} \binom{N\text{-}3}{j}Q^j P^{N\text{-}3\text{-}j}$$

$$= N(N\text{-}1)(N\text{-}2)PQ^2. \tag{1.12}$$

$$E\Big(Y(N\text{-}Y)(N\text{-}Y\text{-}1)\Big) = N(N\text{-}1)(N\text{-}2)P^2Q \sum_{j=0}^{N\text{-}3} \binom{N\text{-}3}{j}Q^j P^{N\text{-}3\text{-}j}$$

$$= N(N\text{-}1)(N\text{-}2)P^2Q. \tag{1.13}$$

If $N = 1$ or 2 (1.12) and (1.13) hold since both sides are zero.

If for now $N > 3$,

$$E\Big(Y(Y\text{-}1)(N\text{-}Y)(N\text{-}Y\text{-}1)\Big) = N(N\text{-}1)(N\text{-}2)(N\text{-}3)P^2Q^2 \sum_{j=0}^{N-4} \binom{N\text{-}4}{j} Q^j P^{N\text{-}4\text{-}j}$$

$$= N(N\text{-}1)(N\text{-}2)(N\text{-}3)P^2Q^2. \tag{1.14}$$

If $N = 1$, 2, or 3 (1.14) holds since both sides equal zero.

$$E\Big(Y^2(N\text{-}Y)^2\Big) = E\big\{Y(Y\text{-}1)(N\text{-}Y)(N\text{-}Y\text{-}1)\big\} + E\big\{Y(Y\text{-}1)(N\text{-}Y)\big\}$$

$$+ E\big\{Y(N\text{-}Y)(N\text{-}Y\text{-}1)\big\} + E\big\{Y(N\text{-}Y)\big\}. \tag{1.15}$$

The above is obtained by writing

$$Y^2(N\text{-}Y)^2 = Y(N\text{-}Y)\big\{(Y\text{-}1)+1\big\}\big\{(N\text{-}Y\text{-}1)+1\big\}.$$

Substitution of (1.11) - (1.14) into (1.15)

$$E\big\{Y^2(N\text{-}Y)^2\big\} = N(N\text{-}1)(N\text{-}2)(N\text{-}3)P^2Q^2 + N(N\text{-}1)(N\text{-}2)\big\{PQ^2 + Q^2P\big\} + N(N\text{-}1)PQ$$

$$= N(N\text{-}1)(N\text{-}2)(N\text{-}3)P^2Q^2 + N(N\text{-}1)(N\text{-}2)PQ(P+Q) + N(N\text{-}1)PQ$$

$$= N(N\text{-}1)(N\text{-}2)(N\text{-}3)P^2Q^2 + N(N\text{-}1)^2PQ. \tag{1.16}$$

$$\Big[E\big\{Y(N\text{-}Y)\big\}\Big]^2 = N^2(N\text{-}1)^2P^2Q^2. \tag{1.17}$$

Hence subtracting (1.17) from (1.16) yields:

$$\text{Var}\Big\{Y(N\text{-}Y)\Big\} = N(N\text{-}1)\Big\{(N\text{-}2)(N\text{-}3) - N(N\text{-}1)\Big\}P^2Q^2 + N(N\text{-}1)^2PQ$$

$$\leq N(N\text{-}1)^2PQ. \tag{1.18}$$

This competes the proof of step 3.

Proof of Step 4:

Under the null hypothesis $(F_j + G_j)$ are independent binomials with respective sample sizes $(N_j + M_j)$ and respective parameters Q_j.

From (1.4) and (1.11) successively,

$$
\begin{aligned}
E(SE^2) &= \sum_j \frac{N_j M_j E\Big\{(F_j + G_j)(N_j + M_j - F_j - G_j)\Big\}}{(N_j + M_j)^2 (N_j + M_j - 1)} \\
&= \sum_j \frac{N_j M_j(N_j + M_j)(N_j + M_j -1)P_j Q_j}{(N_j + M_j)^2(N_j + M_j -1)} \\
&= \sum_j \frac{N_j M_j P_j Q_j}{N_j + M_j} = V^2. \tag{1.19}
\end{aligned}
$$

from (1.4) and (1.18) successively,

$$
\begin{aligned}
\text{Var}(SE^2) &= \sum_j \frac{N_j^2 M_j^2 \text{Var}\Big\{(F_j + G_j)(N_j + M_j - F_j - G_j)\Big\}}{(N_j + M_j)^4 (N_j + M_j - 1)^2} \\
&\leq \sum_j \frac{N_j^2 M_j^2(N_j + M_j)(N_j + M_j -1)^2 P_j Q_j}{(N_j + M_j)^4(N_j + M_j -1)^2} \\
&\leq \sum_j \frac{N_j^2 M_j^2}{(N_j + M_j)^3} P_j Q_j
\end{aligned}
$$

Now since $\dfrac{N_j M_j}{(N_j + M_j)^2} \leq .25,$

$$\text{Var}(\text{SE}^2) \leq .25 \sum_j \frac{N_j M_j P_j Q_j}{N_j + M_j} = .25 V^2. \tag{1.20}$$

From Chebyshev's Theorem A51, (1.19) and (1.20), we have

$$P\left[\left|\frac{\text{SE}^2}{V^2} - 1\right| > \epsilon\right] = P\left[|\text{SE}^2 - V^2| > \epsilon V^2\right]$$

$$\leq \frac{\text{Var}(\text{SE}^2)}{\epsilon^2 V^4} \leq \frac{1}{4\epsilon^2 V^2} \to 0.$$

$V \to \infty$ by (1.7) and assumption (a).

That is, from the definition of covergence in probability

$$\frac{\text{SE}^2}{V^2} \xrightarrow{P} 1.$$

This complete the proof of step 4.

Proof of Step 5:

From step 2, we have

$$\frac{O - E}{V} \xrightarrow{D} N(0,1).$$

From step 4 and Appendix A98,

$$\frac{V}{\text{SE}} \xrightarrow{P} 1.$$

Hence by Slutzky's multiplicative theorem, Appendix A105,

$$\frac{O - E}{\text{SE}} = \left(\frac{O - E}{V}\right) \frac{V}{\text{SE}} \xrightarrow{D} N(0,1).$$

This completes the proof.

3.2 Derivation of the Large Sample Distribution of the Kaplan-Meier Statistic

Refer to the Treatment 1 column of the Notation Chart.

Large Sample Results

Assumptions:

(a)　The sample size on day j, N_j are non-increasing:

$$N_1 \geq N_2 \geq N_3 \geq \cdots$$

(b)　No day is a "wipeout". There exists a positive constant $\delta < 1$ such that P_j, the probability of surviving day j is:

$$P_j \geq 1 - \delta \quad \text{for all j.}$$

(c)　There is a minimum hazard on any given day. There exists a constant $c > 0$ such tha

$$Q_j > c \quad \text{for all j.}$$

Note that c can be small.

(d)　$D_k \rightarrow 0.$　　　　　　　　　(See Notation chart item 10)

(e)　$N_k D_k \rightarrow \infty.$

Conclusions:

(1)　$\dfrac{\hat{S}_k - S_k}{D_k} \xrightarrow{D} N(0,1).$　　　　(Standard Normal)

(2)　$\dfrac{\hat{S}_k - S_k}{\hat{D}_k} \xrightarrow{D} N(0,1).$

Note: The limiting process can involve Q_j (the probability of failure on day j given patient was alive at the start of day j), k (the day number at which survival is evaluated), as well as N_j (the number of patients at risk at the start of day j.)

$\hat{S}_k$ is given in item 9 of the Notation Chart, S_k in item 8, D_k in item 10, and $\hat{D}_k$ in item 11. The arrow represents the limiting distribution (approximate distribution). See Appendix A93.

Plan of Proof:

Step 1: Show that

$$\hat{S}_k - S_k = H_k + I_k,$$

where

$$H_k = S_k \sum_{j=1}^{k} (\hat{P}_j - P_j)/P_j,$$

and

$$I_k = S_k \sum_{j=1}^{k} (\hat{S}_{j-1} - S_{j-1})(\hat{P}_j - P_j)/(S_{j-1}P_j),$$

with

$$\hat{S}_0 = S_0 = 1.$$

Step 2: Show

$$\frac{H_k}{D_k} \xrightarrow{D} N(0, 1). \qquad \text{(Standard Normal)}$$

Step 3: Show

$$\frac{I_k}{D_k} \xrightarrow{P} 0.$$

Step 4: Show

$$\frac{\hat{S}_k - S_k}{D_k} \xrightarrow{D} N(0,1). \qquad \text{(Standard Normal)}$$

Step 5: Show

$$\frac{\hat{D}_k}{D_k} \xrightarrow{P} 1.$$

Step 6: Show

$$\frac{\hat{S}_k - S_k}{\hat{D}_k} \xrightarrow{D} N(0,1). \qquad \text{(Standard Normal)}$$

Before formally conducting the proof, some rationalization might be in order. First, Central Limit Theorems apply to sums. Hence, we would like to write $\left(\hat{S}_k - S_k\right)$ as a sum. The astute reader might ask, "Why not take logs?" This would work if each $\hat{P}_j$ was asymptotically normal. However, most of the applications will have a large N_j and a small Q_j (P_j near one). In these situations, the $\hat{P}_j$ may not be well approximated by normality.

The decomposition in Step 1 is motivated as follows.

Lemma (Product Difference to Summation):

Let

$$A_0 = 1 \qquad \text{and} \qquad A_j = a_1 a_2 \cdots a_j = \prod_{i=1}^{j} a_i,$$

$$B_0 = 1 \qquad \text{and} \qquad B_j = b_1 b_2 \cdots b_j = \prod_{i=1}^{j} b_i$$

with each $b_i \neq 0$. Then

$$A_k - B_k = B_k \sum_{j=1}^{k} \frac{A_{j-1}(a_j - b_j)}{B_j}.$$

Proof: By adding and subtracting common terms,

$$A_k - B_k = (a_1 b_2 \cdots b_k - b_1 b_2 \cdots b_k) + (a_1 a_2 b_3 \cdots b_k - a_1 b_2 \cdots b_k)$$

$$+ (a_1 a_2 a_3 b_4 \cdots b_k - a_1 a_2 b_3 \cdots b_k) + \cdots + (a_1 a_2 \cdots a_k - a_1 a_2 \cdots a_{k-1} b_k)$$

$$= \sum_{j=1}^{k} \left(\frac{A_j B_k}{B_j} - \frac{A_{j-1} B_k}{B_{j-1}} \right).$$

The term
$$b_{j+1} \cdots b_k = \frac{b_1 \cdots b_k}{b_1 \cdots b_j} = \frac{B_k}{B_j} \, ,$$

which produces the above sum.

Now
$$A_j = a_j A_{j-1}$$

and
$$B_{j-1} = \frac{B_j}{b_j}.$$

Substitution in the sum yields
$$A_k - B_k = B_k \sum \frac{A_{j-1}}{B_j} (a_j - b_j).$$

The completes the proof of the Lemma.

Proof of Step 1:

First, we apply the Lemma with

$$a_j = \hat{P}_j \qquad\qquad b_j = P_j$$

$$A_j = \hat{S}_j = \prod_{i=1}^{j} \hat{P}_j \qquad\qquad B_j = S_j = \prod_{i=1}^{j} P_j.$$

This yields
$$\hat{S}_k - S_k = S_k \sum_{j=1}^{k} \frac{\hat{S}_{j-1}}{S_j} (\hat{P}_j - P_j).$$

But since $S_j = S_{j-1} P_j$, this further reduces to

$$\hat{S}_k - S_k = S_k \sum_{j=1}^{k} \frac{\hat{S}_{j-1}(\hat{P}_j - P_j)}{S_{j-1} P_j}. \qquad\qquad (2.1)$$

Now since $\hat{S}_{j-1} = S_{j-1} + (\hat{S}_{j-1} - S_{j-1})$, equation (2.1) can be rewritten as

$$\hat{S}_k - S_k = S_k \sum_{j=1}^{k} \frac{(\hat{P}_j - P_j)}{P_j} + S_k \sum_{j=1}^{k} \frac{(\hat{S}_{j-1} - S_{j-1})(\hat{P}_j - P_j)}{S_{j-1} P_j}$$

$$= H_k + I_k. \qquad \qquad \text{(See Statement of Step 1).}$$

This complete the proof of Step 1.

Proof of Step 2: To prove normality, we shall apply the Central Limit Theorem for Bernoulli variables, See **A107**. Note that

$$(\hat{P}_j - P_j) = -(\hat{Q}_j - Q_j)$$

where $\qquad \qquad \hat{Q}_j = 1 - \hat{P}_j$ and $Q_j = 1 - P_j$.

Now $\qquad \qquad$ $H_k \quad = - S_k \sum_{j=1}^{k} \frac{(\hat{Q}_j - Q_j)}{P_j}$

$$= - S_k \sum_{j=1}^{k} \sum_{i=1}^{N_j} \frac{(Y_{ij} - Q_j)}{(N_j P_j)}, \qquad \qquad (2.2)$$

where Y_{ij} $i=1...N_j$, $j = 1...k$ are independent Bernoulli random variables with

$$P(Y_{ij} = 1) = Q_j$$

$$P(Y_{ij} = 0) = P_j = 1 - Q_j.$$

Let

$$V_k^2 = \text{Var}\Big(\frac{\sqrt{N} \, H_k}{D_k}\Big),$$

where

$$N = N_k D_k.$$

From (2.2),

$$V_k^2 = \sum_{j=1}^{k} \sum_{i=1}^{N_j} b_{ij}^2 P_j Q_j$$

where

$$b_{ij} = -\frac{\sqrt{N} S_k}{N_j P_j D_k}.$$

Hence

$$V_k^2 = \frac{N S_k^2}{D_k^2} \sum_{j=1}^{k} \sum_{i=1}^{N_j} \frac{Q_j}{N_j^2 P_j}$$

$$= \frac{N S_k^2}{D_k^2} \sum_{j=1}^{k} \frac{Q_j}{N_j P_j} = \frac{N D_k^2}{D_k^2}$$

$$= N.$$

Note that under our assumption (e), $N = N_k D_k \to \infty$ and hence $V_k^2 \to \infty$.

We next show that

$$|b_{ij}| \text{ is bounded.}$$

$$|b_{ij}| = \frac{\sqrt{N} S_k}{N_j P_j D_k}$$

$$= \frac{\sqrt{N_k D_k} S_k}{N_j P_j D_k}$$

$$\leq \frac{S_k}{P_j \sqrt{N_k D_k}} \qquad\qquad N_k \leq N_j$$

$$\leq \frac{S_k}{(1-\delta)\sqrt{N_k D_k}}. \qquad\qquad P_j \geq 1 - \delta \text{ by assumption (b)}$$

Now since $N_k D_k \to \infty$

$$\frac{1}{\sqrt{N_k D_k}} \to 0$$

and hence there exists $B \geq 0$:

$$\frac{1}{\sqrt{N_k D_k}} < B.$$

Thus,

$$|b_{ij}| < \frac{S_k}{(1-\delta)B} \qquad\qquad \text{throughout the limiting process.}$$

Further, the $Q_j = 1 - P_j \leq 1 - \delta < 1$ are bounded away from one. $\left(\text{Assumption (b)}\right)$. To obtain condition 2 of the Central Limit Theorem, (A107) we must show

$$\frac{\sum_{j=1}^{k} \sum_{i=1}^{N_j} |b_{ij}|^3 Q_j}{V_k^3} \to 0.$$

Now

$$\sum_{j=1}^{k} \sum_{i=1}^{N_j} \frac{|b_{ij}|^3 Q_j}{V_k^3} = \frac{N^{3/2} S_k^3}{N^{3/2} D_k^3} \sum_{j=1}^{k} \frac{Q_j}{N_j^2 P_j^3} \qquad \left(\text{uses } V_k^2 = N\right)$$

$$\leq \frac{S_k^3}{(1-\delta)^2 N_k D_k^3} \sum_{j=1}^{k} \frac{Q_j}{N_j P_j} \qquad (\text{Since } N_j \geq N_k, \ P_j \geq 1 - \delta)$$

$$\leq \frac{S_k D_k^2}{(1-\delta)^2 N_k D_k^3} = \frac{S_k}{(1-\delta)N_k D_k}$$

$$\to 0. \qquad \left(\text{Since } N_k D_k \to \infty \text{ per assumption (e)}\right)$$

Hence, from Appendix A107

$$\frac{\sqrt{N} H_k}{D_k V_k} \xrightarrow{D} N(0,1).$$

However, since $V_k^2 = N$, (as shown above)

$$\sqrt{N} \, \frac{H_k}{D_k V_k} = \frac{H_k}{D_k}$$

and hence

$$\frac{H_k}{D_k} \xrightarrow{D} N(0,1).$$

This completes step 2.

Proof of Step 3:   First note that I_k is a linear combination of terms

$$B_j = \left(\hat{S}_{j-1} - S_{j-1}\right)\left(\hat{P}_j - P_j\right). \tag{2.3}$$

We shall use the independence of $\hat{P}_1, \ldots, \hat{P}_k$ to obtain the means, variances, and covariances of the B's.   See A45 and A46.

(a) $E(B_j) = 0$

Proof: $E(B_j) = E\left(\hat{S}_{j-1} - S_{j-1}\right)E\left(\hat{P}_j - P_j\right).$ ($\hat{S}_{j-1}$ and $\hat{P}_j$ are independent)

Since $\hat{P}_j$ is the binomial estimate of P_j, (a) is clear.

(b) $Var(B_j) = Var(\hat{S}_{j-1})Var(\hat{P}_j)$

$$= \left[\prod_{i=1}^{j-1} \left\{P_i^2 + \frac{P_i Q_i}{N_i}\right\} - S_{j-1}^2\right] \frac{P_j Q_j}{N_j}$$

Proof: Note that $E(\hat{S}_{j-1}) = E(\hat{P}_1 \ldots \hat{P}_{j-1}) = E(\hat{P}_1) \cdots E(\hat{P}_{j-1}) = P_1 \cdots P_{j-1} = S_{j-1}.$

$$Var(B_j) = E(B_j^2) \qquad \left(\text{Since } E(B_j) = 0\right)$$

$$= E\left[\left(\hat{S}_{j-1} - S_{j-1}\right)^2 \left(\hat{P}_j - P_j\right)^2\right]$$

$$= E\left[\left(\hat{S}_{j-1} - S_{j-1}\right)^2\right] E\left[\left(\hat{P}_j - P_j\right)^2\right] \qquad \text{(Independence)}$$

$$= \mathrm{Var}\left(\hat{S}_{j-1}\right)\mathrm{Var}\left(\hat{P}_j\right). \tag{2.4}$$

Now
$$\mathrm{Var}\left(\hat{S}_{j-1}\right) = \mathrm{E}\left(\hat{P}_1^2,\ \hat{P}_2^2,\ ...,\ \hat{P}_{j-1}^2\right) - \left\{\mathrm{E}\left(\hat{S}_{j-1}\right)\right\}^2$$

$$= \mathrm{E}\left(\hat{P}_1^2\right) \cdots \mathrm{E}\left(\hat{P}_{j-1}^2\right) - S_{j-1}^2$$

$$= \prod_{i=1}^{j-1} \left\{P_i^2 + \frac{P_i Q_i}{N_i}\right\} - S_{j-1}^2, \tag{2.5}$$

since
$$\mathrm{Var}\left(\hat{P}_i\right) = \mathrm{E}\left(\hat{P}_i^2\right) - \left[\mathrm{E}\left(\hat{P}_i\right)\right]^2 \qquad \text{implies}$$

$$\mathrm{E}\left(\hat{P}_i^2\right) = \left[\mathrm{E}\left(\hat{P}_i\right)\right]^2 + \mathrm{Var}\left(\hat{P}_i\right) = P_i^2 + \frac{P_i Q_i}{N_i}.$$

Also
$$\mathrm{Var}\left(\hat{P}_j\right) = \frac{P_j Q_j}{N_j}. \tag{2.6}$$

Upon substituting (2.5) and (2.6) in (2.4), the result (b) holds.

(c)   Cov $(B_i,\ B_j) = 0.$ $i \neq j$

Proof:

Since $\mathrm{E}(B_i) = \mathrm{E}(B_j) = 0$

$$\mathrm{Cov}(B_i, B_j) = \mathrm{E}(B_i B_j).$$

Now if $j < i$,

$$\mathrm{Cov}(B_i,\ B_j) = \mathrm{E}\left[\left(\hat{S}_{i-1} - S_{i-1}\right)\left(\hat{P}_i - P_i\right)\left(\hat{S}_{j-1} - S_{j-1}\right)\left(\hat{P}_j - P_j\right)\right]$$

$$= \mathrm{E}\left(\hat{P}_i - P_i\right)\mathrm{E}\left[\left(\hat{S}_{i-1} - S_{i-1}\right)\left(\hat{S}_{j-1} - S_{j-1}\right)\left(\hat{P}_j - P_j\right)\right],$$

since $\left(\hat{S}_{i-1} - S_{i-1}\right)\left(\hat{S}_{j-1} - S_{j-1}\right)\left(\hat{P}_j - P_j\right)$ depends only on $\hat{P}_1, ..., \hat{P}_{i-1}$ and is therefore independent of $\hat{P}_i$. Now $E\left(\hat{P}_i - P_i\right) = 0$ and hence $Cov(B_i, B_j) = 0$.

The result follows in a similar fashion if $i < j$.

Next, we shall evaluate the mean and variance of I_k.

$$I_k = S_k \sum_{j=1}^{k} \frac{\left(\hat{S}_{j-1} - S_{j-1}\right)\left(\hat{P}_j - P_j\right)}{S_{j-1}P_j}$$

$$= S_k \sum_{j=2}^{k} \frac{B_j}{S_{j-1}P_j} \quad,$$

since $\hat{S}_0 = S_0 = 1$ and $B_j = \left(\hat{S}_{j-1} - S_{j-1}\right)\left(\hat{P}_j - P_j\right)$.

Using linear combinations per A45 and the fact that $Cov(B_i, B_j) = 0$

$$E(I_k) = S_k \sum_{j=2}^{k} \frac{E(B_j)}{S_{j-1}P_j} = 0$$

$$Var(I_k) = S_k^2 \sum_{j=2}^{k} \frac{Var(B_j)}{S_{j-1}^2 P_j^2}.$$

$$= S_k^2 \sum_{j=2}^{k} \frac{\left[\prod_{i=1}^{j-1}\left\{P_i^2 + \frac{P_iQ_i}{N_i}\right\} - S_{j-1}^2\right]}{S_{j-1}^2} \frac{Q_j}{N_jP_j}.$$

Since

$$S_{j-1} = \prod_{i=1}^{j-1} P_i,$$

$$Var(I_k) = S_k^2 \sum_{j=2}^{k} \left[\prod_{i=1}^{j-1}\left\{1 + \frac{Q_i}{N_iP_i}\right\} - 1\right] \frac{Q_j}{N_jP_j}. \tag{2.8}$$

Now applying the Lemma on converting products to sums, we see

$$\prod_{i=1}^{j-1}\left\{1+\frac{Q_i}{N_iP_i}\right\}-1=\sum_{i=1}^{j-1}A_{i-1}\frac{Q_i}{N_iP_i} \tag{2.9}$$

where

$$A_0 = 1$$

$$A_i = \prod_{\ell=1}^{i}\left\{1+\frac{Q_\ell}{N_\ell P_\ell}\right\} < A_j \quad i < j$$

$$\text{Var}(I_k) < S_k^2 \sum_{j=2}^{k}A_j\left(\sum_{i=1}^{j-1}\frac{Q_i}{N_iP_i}\right)\frac{Q_j}{N_jP_j} \leq \frac{A_kD_k^4}{S_k^2}. \tag{2.10}$$

The above follows since $A_j \leq A_k$ for $j \leq k$ and the definition of D_k^2 in the Notation Chart.

Consider (2.9) with $j = k$.

$$A_k - 1 = \sum_{i=1}^{k}A_{i-1}\frac{Q_i}{N_iP_i}$$

$$< A_k\sum_{i=1}^{k}\frac{Q_i}{N_iP_i} = \frac{A_k}{S_k^2}D_k^2.$$

From assumption (d), $D_k \to 0$ and hence if we are sufficiently far in the limiting process,

$$1 \leq A_k \leq \frac{S_k^2}{S_k^2 - D_k^2}.$$

Hence from (2.10):

$$\text{Var}(I_k) < \frac{D_k^4}{S_k^2 - D_k^2}. \tag{2.11}$$

By Chebyshev's Theorem, Appendix A51, for any small challenge number $\epsilon > 0$,

$$P\left[|I_k| > \epsilon D_k\right] < \frac{\text{Var}(I_k)}{\epsilon^2 D_k^2}$$

$$< \frac{D_k^4}{\epsilon^2 D_k^2\left(S_k^2 - D_k^2\right)} \qquad \left(\text{by } (2.11)\right)$$

$$< \frac{D_k^2}{\epsilon^2 \left(S_k^2 - D_k^2\right)}$$

$$\to 0. \qquad\qquad \text{Since } D_k^2 \to 0 \left(\text{Assumption (e)}\right)$$

That is,

$$P\left[|\frac{I_k}{D_k}| > \epsilon\right] \to 0$$

for any arbitrary $\epsilon > 0$. From the definition of convergence in probability (Appendix A92), this completes step 3.

Proof of Step 4: From steps 1, 2, 3

$$\frac{\hat{S}_k - S_k}{D_k} = \frac{H_k + I_k}{D_k} = \frac{H_k}{D_k} + \frac{I_k}{D_k}$$

$$\frac{H_k}{D_k} \xrightarrow{D} N(0,1) \qquad \frac{I_k}{D_k} \xrightarrow{P} 0.$$

By Slutsky's Theorem, Appendix A104, $\dfrac{\hat{S}_k - S_k}{D_k}$ has some limiting distribution as $\dfrac{H_k}{D_k}$, that is

$$\frac{\hat{S}_k - S_k}{D_k} \xrightarrow{D} N(0, 1).$$

This completes step 4.

Proof of Step 5: This is similar to Step 3, but considerably more involved. This itself is proved by a series of steps, which we shall call "tasks". The ultimate goal is to show that

$$\frac{\dot{D}_k}{D_k} \xrightarrow{P} 1$$

To avoid working with random variables in the denominator we shall consider

$$\hat{U}_k = \frac{\hat{D}_k^2}{\hat{S}_k}$$

$$= \hat{S}_k \sum_{j=1}^{k} \left\{ \frac{\hat{Q}_j}{(N_j \hat{P}_j)} \right\} \tag{2.12}$$

and

$$U_k = \frac{D_k^2}{S_k}$$

$$= S_k \sum_{j=1}^{k} \left\{ \frac{Q_j}{(N_j P_j)} \right\}.$$

The large picture of Step 5 is to prove that

$$\frac{\hat{U}_k}{U_k} \xrightarrow{P} 1.$$

Tasks are as follows:

Task 5.1: (See Notation Chart)

Let
$$Y_j = \frac{\hat{S}_k \hat{Q}_j}{\hat{P}_j} = \hat{P}_1 \hat{P}_2 \cdots \hat{P}_{j-1} \, \hat{Q}_j \, \hat{P}_{j+1} \cdots \hat{P}_k. \tag{2.13}$$

Show
$$E(Y_j) = \frac{S_k Q_j}{P_j}$$

and hence

$$E(\hat{U}_k) = U_k. \tag{2.14}$$

Task 5.2: Show that

$$\text{Var}(Y_j) = \frac{S_k^2}{P_j^2} \left[\left\{ Q_j^2 + \frac{P_j Q_j}{N_j} \right\} G_j + \frac{P_j Q_j}{N_j} \right] \tag{2.15}$$

where

$$G_j = \left\{ \prod_{\substack{i=1 \\ i \neq j}}^{k} \left(1 + \frac{Q_i}{N_i P_i} \right) \right\} - 1. \tag{2.16}$$

Task 5.3:

Show that

$$G_j \leq \left(1 + G_j \right) \sum_{\substack{i=1 \\ i \neq j}}^{k} \frac{Q_i}{N_i P_i}. \tag{2.17}$$

Task 5.4:

Show that if we are sufficiently far enough in the limiting process, there exists a constant B_0 such that for all $j \leq k$

$$\mathrm{Var}(Y_j) < B_0 D_k^2. \tag{2.18}$$

Task 5.5

Show that if we are sufficiently far enough in the limiting process, there exists a constant B_1 such that

$$\mathrm{Var}(\hat{U}_k) < B_1 D_k^6.$$

Task 5.6

Show

$$\frac{\hat{U}_k}{U_k} \xrightarrow{P} 1.$$

Task 5.7

Show

$$\frac{\hat{S}_k}{S_k} \xrightarrow{P} 1.$$

Task 5.8

Show

$$\frac{\hat{D}_k}{D_k} \xrightarrow{P} 1.$$

Proof of Task 5.1:

From (2.13) and the independence of the members of the product,

$$E(Y_j) = E\left(\hat{P}_1\hat{P}_2\cdots\hat{P}_{j-1}\hat{Q}_j\hat{P}_{j+1}\cdots\hat{P}_k\right)$$

$$= E(\hat{P}_1)E(\hat{P}_2)\cdots E(\hat{P}_{j-1})E(\hat{Q}_j)E(\hat{P}_{j+1})\cdots E(\hat{P}_k)$$

$$= P_1P_2\cdots P_{j-1}Q_jP_{j+1}\cdots P_k$$

$$= \frac{S_kQ_j}{P_j}.$$

$$E(\hat{U}_k) = E(\Sigma Y_j/N_j)$$

$$= \Sigma E(Y_j/N_j)$$

$$= S_k\Sigma \frac{Q_j}{N_jP_j} = U_k.$$

This completes task 5.1.

Proof of Task 5.2:

$$Y_j - E(Y_j) = \hat{Q}_j\prod_{\substack{i=1\\i\neq j}}^{k}\hat{P}_i - Q_j\prod_{\substack{i=1\\i\neq j}}^{k}P_i$$

$$= \hat{Q}_j\left\{\prod_{\substack{i=1\\i\neq j}}^{k}\hat{P}_i - \prod_{\substack{i=1\\i\neq j}}^{k}P_i\right\} + (\hat{Q}_j - Q_j)\prod_{\substack{i=1\\i\neq j}}^{k}P_i$$

$$= \text{Term 1} \qquad\qquad + \text{Term 2.}$$

Note that from independence $E(\text{Term 1}) = 0$, $E\{(\text{Term 1})(\text{Term 2})\} = 0$, and $E(\text{Term 2}) = 0$.

From A45 and A44,

$$\begin{aligned} \text{Var}(Y_j) &= \text{Var}(\text{Term 1}) + \text{Var}(\text{Term 2}) + 2\,\text{Cov}(\text{Term 1, Term 2}) \\ &= E(\text{Term } 1^2) + E(\text{Term } 2^2). \end{aligned} \tag{2.19}$$

By independence and the fact that from A43,

$$\text{Var}(T) = E\left[\left\{T - E(T)\right\}^2\right] = E(T^2) - \left\{E(T)\right\}^2,$$

$$\begin{aligned} E(\text{Term } 1^2) &= \left(Q_j^2 + \frac{P_j Q_j}{N_j}\right)\left\{ \prod_{\substack{i=1 \\ i \neq j}}^{k}\left(P_i^2 + \frac{P_i Q_i}{N_i}\right) - \prod_{\substack{i=1 \\ i \neq j}}^{k} P_i^2 \right\} \\ &= \frac{S_k^2}{P_j^2}\left(Q_j^2 + \frac{P_j Q_j}{N_j}\right)\left\{ \prod_{\substack{i=1 \\ i \neq j}}^{k}\left(1 + \frac{Q_i}{N_i P_i}\right) - 1 \right\} \\ &= \frac{S_k^2}{P_j^2}\left(Q_j^2 + \frac{P_j Q_j}{N_j}\right)G_j \qquad \left(\text{see } (2.16)\right) \end{aligned}$$

$$\begin{aligned} E(\text{Term } 2^2) &= \frac{P_j Q_j}{N_j}\prod_{\substack{i=1 \\ i \neq j}}^{k} P_i^2 \\ &= \frac{S_k^2}{P_j^2}\frac{P_j Q_j}{N_j}. \end{aligned}$$

From (2.19).

$$\text{Var}(Y_j) = \frac{S_k^2}{P_j^2}\left\{\left(Q_j^2 + \frac{P_j Q_j}{N_j}\right)G_j + \frac{P_j Q_j}{N_j}\right\}$$

This completes Task 5.2

Proof of Task 5.3:

$$G_j = \prod_{\substack{i=1 \\ i \neq j}}^{k} \left(1 + \frac{Q_i}{N_i P_i}\right) - 1.$$

This is a difference of products, to which we shall apply the product to summation lemma with

$$a_i = 1 + \frac{Q_i}{N_i P_i} \qquad\qquad b_i = 1$$

$$A_0 = 1 \qquad\qquad A_i = \prod_{\substack{\ell \leq i \\ \ell \neq j}} a_\ell \quad i \neq j \qquad A_j = A_{j-i}$$

$$B_0 = 1 \qquad\qquad B_i = b_1 b_2 \cdots b_i = 1.$$

$$G_j = \sum_{\substack{i=1 \\ i \neq j}}^{k} A_{i-1} \frac{Q_i}{N_i P_i}. \tag{2.20}$$

Now since $a_i \geq 1$

$$A_{i-1} \quad \leq \prod_{\substack{i=1 \\ i \neq j}}^{k} a_i = \prod_{\substack{i=1 \\ i \neq j}}^{k} \left\{1 + \frac{Q_i}{N_i P_i}\right\}$$

$$\leq 1 + G_j \qquad\qquad \left(\text{see } (2.16)\right).$$

Substituting in (2.20), we have

$$G_j \leq \left(1 + G_j\right) \sum_{\substack{i=1 \\ i \neq j}}^{k} \frac{Q_i}{N_i P_i}.$$

This completes Task 5.3

Proof of Task 5.4:

First note that since $0 \leq \hat{Q}_j \leq 1$

$$E(\hat{Q}_j^2) = Q_j^2 + \frac{P_j Q_j}{N_j} \leq 1. \tag{2.21}$$

Substitution of (2.17) and (2.21) into (2.15) yields

$$\text{Var}(Y_j) \leq \frac{S_k^2}{P_j^2}\left[\left(1 + G_j\right)\sum_{\substack{i=1 \\ i \neq j}}^{k}\frac{Q_i}{N_iP_i} + \frac{P_jQ_j}{N_j}\right]. \tag{2.22}$$

Now since $P_j \leq 1$, $\dfrac{P_jQ_j}{N_j} \leq \dfrac{Q_j}{P_jN_j}$.

Substitution into (2.22) yields

$$\text{Var}(Y_j) \leq \frac{S_k^2}{P_j^2}\left[\sum_{i=1}^{k}\frac{Q_i}{N_iP_i} + G_j\sum_{\substack{i=1 \\ i \neq j}}^{k}\frac{Q_i}{N_iP_i}\right]$$

$$\leq \frac{S_k^2}{P_j^2}\left[\left(1 + G_j\right)\sum_{i=1}^{k}\frac{Q_i}{N_iP_i}\right] \tag{2.23}$$

Returning once again to (2.17), we note

$$G_j \quad \leq \left(1 + G_j\right)\sum_{\substack{i=1 \\ i \neq j}}^{k}\frac{Q_i}{N_iP_i}$$

$$\leq \left(1 + G_j\right)\sum_{i=1}^{k}\frac{Q_i}{N_iP_i}. \tag{2.24}$$

Now since from assumption (d),

$$D_k^2 = S_k^2\sum_{i=1}^{k}\frac{Q_i}{N_iP_i} \rightarrow 0$$

and hence

$$\sum_{i=1}^{k}\frac{Q_i}{N_iP_i} \rightarrow 0. \qquad \left(S_k \text{ is fixed}\right)$$

Hence we can find for any constant ϵ, a critical point along the limiting process such that once we have passed this point,

$$\sum_{i=1}^{k} \frac{Q_i}{N_i P_i} < \epsilon.$$

That is, from (2.24)

$$G_j < (1 + G_j)\epsilon \qquad (2.25)$$

Since ϵ is arbitrary, let us select its value as

$$\epsilon = .5. \qquad (2.26)$$

This implies

$$G_j < \frac{\epsilon}{1 - \epsilon} = 1, \qquad (2.27)$$

as long as we are sufficiently far enough in the limiting process.

Substituting (2.27) into (2.23), we have

$$\text{Var}(Y_j) \leq \frac{2S_k^2}{P_j^2} \sum_{i=1}^{k} \frac{Q_i}{N_i P_i}.$$

Finally, from assumption (b), $P_j \geq 1 - \delta > 0$

$$\text{Var}(Y_j) \leq \frac{2S_k^2}{(1-\delta)^2} \sum_{i=1}^{k} \frac{Q_i}{N_i P_i} = \frac{2D_k^2}{(1-\delta)^2}.$$

As long as we are sufficiently far enough along the limiting process, we can use

$$\text{Var}(Y_j) \leq B_0 D_k^2,$$

where

$$B_0 = \frac{2}{(1-\delta)^2}.$$

This bound does not depend on j. This completes Task 5.4.

<u>Proof of Task 5.5</u>:

From (2.12) and (2.13)

$$\hat{U}_k = \sum_{j=1}^{k} Y_j/N_j \, .$$

From Appendix A45,

$$Var(\hat{U}_k) = \sum_{j=1}^{k} \frac{Var(Y_j)}{N_j^2} + 2 \sum_i \sum_{i<j} \frac{Cov(Y_i, Y_j)}{N_i N_j} \, . \tag{2.28}$$

Now from Appendix A50,

$$Cov(Y_i, Y_j) \leq \sqrt{Var(Y_i)Var(Y_j)}. \tag{2.29}$$

Substitution of (2.29) into (2.28) yields

$$Var(\hat{U}_k) \leq \sum_{i=1}^{k} \frac{Var(Y_j)}{N_j^2} + 2 \sum_{i<j} \frac{\sqrt{Var(Y_i)Var(Y_j)}}{N_i N_j}$$

$$\leq \left\{ \sum_{j=1}^{k} \frac{\sqrt{Var(Y_j)}}{N_j} \right\}^2 . \tag{2.30}$$

Substitution of (2.18) into (2.30) yields

$$Var(\hat{U}_k) \leq B_0 D_k^2 \left[\sum_{j=1}^{k} \left\{ \frac{1}{N_j} \right\} \right]^2 \tag{2.31}$$

From assumption (c), there exists a constant c > 0:

$$Q_j > c$$

That is,

$$\frac{Q_j}{P_j} = \frac{Q_j}{1 - Q_j} > \frac{c}{1 - c},$$

and hence

$$\frac{1-c}{c} \left(\frac{Q_j}{P_j} \right) > 1. \tag{2.32}$$

Substitution into (2.31) yields

$$Var(\hat{U}_k) \le \frac{B_0 D_k^2 (1-c)^2}{c^2} \left\{ \sum_{j=1}^{k} \frac{Q_i}{N_i P_i} \right\}^2$$

$$< \frac{B_0 D_k^2 (1-c)^2}{c^2} \frac{D_k^4}{S_k^4} \qquad \text{(See Notation Chart-10)}$$

$$< B_1 D_k^6,$$

where

$$B_1 = \frac{B_0 (1-c)^2}{c^2 S_k^4}.$$

This completes Task 5.5.

Proof of Task 5.6:

From (2.14) and Chebyshev's Theorem A51,

$$P\left[\left| \frac{\hat{U}_k}{U_k} - 1 \right| > \epsilon \right] = P\left[|\hat{U}_k - U_k| > \epsilon U_k \right]$$

$$< \frac{Var(\hat{U}_k)}{\epsilon^2 U_k^2}$$

$$< \frac{S_k^2 B_1 D_k^6}{\epsilon^2 D_k^4} \qquad \text{(Task 5.5 and Def of } U_k \text{)}$$

$$< \frac{S_k^2 B_1 D_k^2}{\epsilon^2} \to 0. \qquad \left(D_k^2 \to 0 \text{ by assumption (b)} \right)$$

This completes Task 5.6.

Proof of Task 5.7:

    From Step 4,

$$\frac{\hat{S}_k - S_k}{D_k} \xrightarrow{D} N(0,1).$$

    Since from Assumption (d), the asymptotic standard deviation, $D_k \rightarrow 0$, it follows that

$$\hat{S}_k \xrightarrow{P} S_k.$$

This completes Task 5.7.

Proof of Task 5.8:

    By definition

$$\hat{D}_k^2 = \hat{S}_k \hat{U}_k \quad \text{and} \quad D_k^2 = S_k U_k.$$

Hence from Tasks 5.6 and 5.7 and Appendix A101,

$$\frac{\hat{D}_k^2}{D_k^2} = \left(\frac{\hat{S}_k}{S_k}\right)\left(\frac{\hat{U}_k}{U_k}\right) \xrightarrow{P} 1.$$

Hence from A98,

$$\frac{\hat{D}_k}{D_k} \xrightarrow{P} 1.$$

This completes Step 5.

Proof of Step 6:

$$\frac{\hat{S}_k - S_k}{\hat{D}_k} = \frac{\hat{S}_k - S_k}{D_k}\left(\frac{D_k}{\hat{D}_k}\right).$$

Now since
$$\frac{\hat{S}_k - S_k}{D_k} \rightarrow N(0,1) \qquad \text{(Step 4)}$$

and
$$\frac{D_k}{\hat{D}_k} \xrightarrow{P} 1 \qquad \text{(Step 5 and Appendix A98)}$$

we apply the product version of Slutsky's Theorem (Appendix A105) to conclude that

$$\frac{\hat{S}_k - S_k}{\hat{D}_k} \rightarrow N(0,1).$$

This completes the proof. (You thought it would last forever.)

3.2.1 *Difference Between Kaplan-Meier Curves*

Suppose we have two independent Kaplan-Meier Estimators

Treatment 1:

$$\frac{\hat{S}_k - S_k}{D_k} \xrightarrow{D} N(0,1),$$

$$\frac{\hat{S}_k - S_k}{\hat{D}_k} \xrightarrow{D} N(0,1),$$

$$\frac{\hat{D}_k}{D_k} \xrightarrow{P} 1,$$

$$D_k \xrightarrow{P} 0.$$

Treatment 2:

$$\frac{\hat{S}_k^* - S_k^*}{\hat{D}_k^*} \xrightarrow{D} N(0,1),$$

$$\frac{\hat{S}_k^* - S_k^*}{\hat{D}_k^*} \xrightarrow{D} N(0,1),$$

$$\frac{\hat{D}_k^*}{D_k^*} \xrightarrow{P} 1,$$

$$D_k^* \xrightarrow{P} 0.$$

In addition, assume

$$\frac{D_k}{D_k^*} \to c. \qquad \text{(a constant)}$$

Then from A108 and A109,

$$\frac{\left(\hat{S}_k - \hat{S}_k^*\right) - \left(S_k - S_k^*\right)}{\sqrt{D_k^2 + D_k^{*2}}} \xrightarrow{D} N(0,1) \qquad (2.33)$$

and

$$\frac{\left(\hat{S}_k - \hat{S}_k^*\right) - \left(S_k - S_k^*\right)}{\sqrt{\hat{D}_k^2 + \hat{D}_k^{*2}}} \xrightarrow{D} N(0,1). \qquad (2.34)$$

3.3   Exponential Survival

In this section, we shall introduce methods for analysis of exponential survival data. Curiously, we discuss these methods despite the practical fact that we would rarely be in a position to assume exponentiality in the analysis of a clinical trial.

We introduce the methods for the following reasons. (1) We shall be able to show that the logrank and exponential comparisons of the treatments have nearly the same power when the data are in fact exponentially distributed; (2) Planning of clinical trials is generally done under "Proportional Hazards Models", which implies that by a single monotonic increasing transformation, the failure distribution of both therapies can be converted to exponential. The logrank test remains the same under monotonic increasing transformations; (3) Derivation of power and sample size for trials can be done in a straight forward manner for exponential survival. Because of (1), these calculations will apply to the logrank as well. Due to (2), the calculations can be extended to proportional hazards models.

This section will cover the following topics:

(A) Introduction to the Exponential Distribution

The key element is that

P_j = Probability of Surviving Day j, given patient is on Treatment 1 and is alive at the start of day j, is the same for all j. That is, P_j does not depend on j.

Similarly R_j the corresponding value for Treatment 2 does not depend on j.

(B) Estimation of the difference in daily failure rates.

Q - T = (1-P) - (1-R)

where $P_j \equiv P$ and $R_j \equiv R$.

Because of (A),

$$\hat{Q} = \frac{\text{Total failures on Treatment } \#1}{\text{Total days on Test for Treatment } \#1}$$

Similarly,

$$\hat{T} = \frac{\text{Total failures on Treatment } \#2}{\text{Total days on Test for Treatment } \#2}$$

Next, we shall obtain the large sample distribution properties of $(\hat{Q} - \hat{T})$. Unlike the logrank test, this will be done under the "alternate and null hypotheses." The logrank properties were restricted to the null hypothesis.

(C) Relate the logrank test to the exponential test when exponentiality holds. We shall see that the logrank test will have only slightly less power than the exponential test. The power loss will have no practical impact. (Section 3.4)

(D) Derive the sample size for exponential survival under given "planning parameters." (Sections 3.5, 3.6, and 3.7)

(E) Extend the derivation to "proportional hazard models." (Sections 3.8 and 3.9)

(F) Relate the derivation to the sample size tables. (Section 3.11)

(A) Introduction to the Exponential Distribution

The reader is first referred to the Appendix A70 - A76.

Note that if the "hazard rate" for Treatment #1 is λ_1, then under Treatment #1

$$P_j = P\left[\text{Survive day j}|\text{alive at start of day j}\right]$$

$$= \frac{P\left[\text{Survive at least j+1 days}\right]}{P\left[\text{Survive at least j days}\right]}$$

$$= \frac{\exp\left[-\lambda_1(j+1)\right]}{\exp\left[-\lambda_1 j\right]} = e^{-\lambda_1},$$

i.e., P_j does not depend on j.

Similarly, R_j does not depend on j, where R_j is defined in a similar manner to P_j, except for Treatment #2.

(B) Estimates of Failure Rates and Their Asymptotic Properties

If the survival distribution for each treatment is exponential, then a significant simplification of the survival analysis can be achieved. Under exponential survival, each day carries the same risk, regardless of how long the patient has survived.

Let us consider the following (See Notation Chart)

$$F = \sum_j F_j = \text{Total failures on Treatment \#1.}$$

$$N = \sum_j N_j = \text{Total days on test for Treatment \#1.}$$

$$G = \sum_j G_j = \text{Total failures on Treatment \#2.}$$

$$M = \sum_j M_j = \text{Total days on test for Treatment \#2.}$$

$$Q = Q_j \text{ (Same for all j)} = \text{Probability of failing on Treatment \#1 on any}$$
$$\text{day given alive at start of day.}$$

$T = T_j$ (Same for all j) = Probability of failing on Treatment #2 on any

day given alive at start of day.

$$\hat{Q} = F/N = \frac{\text{Total Failures on Treatment \#1}}{\text{Time on test for Treatment \#1}}$$

$$\hat{T} = G/M = \frac{\text{Total failures on Treatment \#2}}{\text{Time on test for Treatment \#2}}$$

Provided that $0 < Q < 1$ and $0 < T < 1$ are fixed, $N \to \infty$, $M \to \infty$ with $\dfrac{NQ(1-Q)}{MT(1-T)} \to c$ then

$$\frac{(\hat{Q} - \hat{T}) - (Q - T)}{\sqrt{\dfrac{Q(1-Q)}{N} + \dfrac{T(1-T)}{M}}} \to N(0, 1) \tag{3.1}$$

and

$$\frac{(\hat{Q} - \hat{T}) - (Q - T)}{\sqrt{\dfrac{\hat{Q}(1-\hat{Q})}{N} + \dfrac{\hat{T}(1-\hat{T})}{M}}} \to N(0, 1). \tag{3.2}$$

Further, if the probability of failure on any day is small, but NQ and MT are large,

$$\frac{(\hat{Q} - \hat{T}) - (Q - T)}{\sqrt{\dfrac{Q}{N} + \dfrac{T}{M}}} \text{ is approximately } N(0,1)$$

and

$$\frac{(\hat{Q} - \hat{T}) - (Q - T)}{\sqrt{\dfrac{\hat{Q}}{N} + \dfrac{\hat{T}}{M}}} \text{ is approximately } N(0, 1).$$

Proof of (3.1) and (3.2):

We shall prove normality of $\hat{Q}$ and $\hat{T}$ and apply **APPENDIX A108 and A109**.

We shall appeal to the Central Limit Theorem for Bernoulli random variables using subscripts

i = trial number

j = day number

k = treatment number

$Y_{ijk} = 1$ or 0 depending upon whether the ith trial on treatment k - day j

results in a failure or not.

$$N\hat{Q} = \sum_{j} \sum_{i=1}^{N_j} Y_{ij1} \tag{3.3}$$

$$M\hat{T} = \sum_{j} \sum_{i=1}^{M_j} Y_{ij2} \tag{3.4}$$

with $N = \Sigma N_j \quad$ and $\quad M = \Sigma M_j.$

To prove normality of $\hat{Q}$, we apply A107-1. If we let $b_{ij} = 1$ in (3.3), and note that

$$E(Y_{ij1}) = Q.$$

$$Var(N\hat{Q}) = \sum_{j} \sum_{i=1}^{N_j} Q(1\text{-}Q)$$

$$= NQ(1\text{-}Q).$$

and

$$\sum_{j} \sum_{i=1}^{N_j} b_{ij}^3 = \sum_{j} \sum_{i=1}^{N_j} 1 = N.$$

$$\frac{\Sigma b_{ij}^3}{\left[Var(N\hat{Q})\right]^{3/2}} = \frac{\left[Q(1\text{-}Q)\right]^{-\frac{3}{2}}}{\sqrt{N}} \to 0.$$

This yields all conditions of A107-1, that is

$$\frac{N\hat{Q} - NQ}{\sqrt{NQ(1-Q)}} = \sqrt{N} \frac{(\hat{Q} - Q)}{\sqrt{Q(1-Q)}} \xrightarrow{D} N(0,1). \tag{3.5}$$

Similarly,

$$\sqrt{M} \frac{(\hat{T} - T)}{\sqrt{T(1-T)}} \xrightarrow{D} N(0,1). \tag{3.6}$$

We apply Appendix A108 directly to obtain (3.1).

To prove (3.2), note that (3.5) implies

$$\hat{Q} \xrightarrow{P} Q$$

and hence by A98

$$\sqrt{\hat{Q}(1-\hat{Q})} \xrightarrow{P} \sqrt{Q(1-Q)}.$$

By A95,

$$\frac{\sqrt{\hat{Q}(1-\hat{Q})}}{\sqrt{Q(1-Q)}} \xrightarrow{P} 1.$$

That is,

$$\frac{\sqrt{\left\{\hat{Q}(1-\hat{Q})/N\right\}}}{\sqrt{\left\{Q(1-Q)/N\right\}}} \xrightarrow{P} 1.$$

Similarly

$$\frac{\sqrt{\left\{\hat{T}(1-\hat{T})/M\right\}}}{\sqrt{\left\{T(1-T)/M\right\}}} \xrightarrow{P} 1.$$

The conditions for A109 have been established and from A109, (3.2) follows.
This proves the major results.

The final findings are obtained by replacing $1 - Q$, $1 - T$, $1 - \hat{Q}$, and $1 - \hat{T}$ by one, in the standard error terms.

As is illustrated in the example below, the percentage error in doing so is extremely small.

Suppose that the one year survival rates are 50% on treatment 1 and 75% on treatment 2. That is,

$$(1-Q)^{365} = .50 \quad \text{and} \quad (1-T)^{365} = .75.$$

Hence

$$Q = .0019 \quad \text{and} \quad T = .00079.$$

$$\frac{Q}{N}(1-Q) = \frac{Q}{N}(.9981).$$

$$\frac{T}{M}(1-T) = \frac{T}{M}(.99921).$$

$$\frac{Q}{N} + \frac{T}{M} > \frac{Q(1-Q)}{N} + \frac{T(1-T)}{M}.$$

Thus, the simplified standard error $\sqrt{\frac{Q}{N} + \frac{T}{M}}$ overestimates $\sqrt{\frac{Q(1-Q)}{N} + \frac{T(1-T)}{M}}$.

Also,

$$.9981\left(\frac{Q}{N} + \frac{T}{M}\right) < \frac{Q(1-Q)}{N} + \frac{T(1-T)}{M}$$

and hence

$$\sqrt{.9981}\sqrt{\frac{Q}{N} + \frac{T}{M}} < \sqrt{\frac{Q(1-Q)}{N} + \frac{T(1-T)}{M}}.$$

That is, the percentage error is less than

$$100\left(1 - \sqrt{.9981}\right)\% = 0.095\%$$

In general, by the above argument, the percent error is less than

$$B = 100\left\{1 - \sqrt{1 - \text{MAX}(Q, T)}\right\}.$$

The following chart illustrates the bound on the percent error:

CHART
Bound on Error by Removing (1-Q) and (1-T) terms from Standard Error of
$$(\hat{Q} - \hat{T})$$

Inferior one year Survival	Bound on Error of Standard Normal Statistic
1%	0.63%
2%	0.54%
5%	0.41%
10%	0.31%
25%	0.19%
50%	0.095%
75%	0.039%

Each entry is less than 1%.

3.4   Application of the Logrank Test When Survival is Exponential

If we knew that survival was indeed exponential we would compare the survival by the exponential test, rather than the logrank test. In practice, we can never be certain that survival is exponential.

A key question is how much better is the exponential test than the logrank test when we do have exponential data. It will turn out that the logrank test does almost as well as the exponential test. The margin of superiority of the exponential test is so slight, that the logrank test should be preferred even when the exponential assumption seems reasonable. The reason is that power loss is minimal, while the logrank analysis gives a valid comparison, irrespective of the underlying survival curve.

First, we shall study the numerator of the logrank Statistic: (See Notation Chart)

$$O - X = \sum_j F_j - \sum_j E_j$$

$$= \sum_j \left(M_j F_j - N_j G_j \right) / \left(N_j + M_j \right) \qquad \left(\text{See } (1.6) \right)$$

$$= \sum_j M_j N_j \left(\hat{Q}_j - \hat{T}_j \right) / \left(N_j + M_j \right), \qquad (4.1)$$

where

$$\hat{Q}_j = F_j / N_j \quad \text{and} \quad \hat{T}_j = G_j / M_j.$$

Let

$$U_j = M_j N_j / \left(N_j + M_j \right) \qquad (4.2)$$

and

$$W_j = U_j / \sum_i U_i. \qquad (4.3)$$

That is,

$$U_j = \left(\sum_i U_i \right) W_j.$$

$$O - X = \left[\sum U_i \right] \sum_j W_j \left(\hat{Q}_j - \hat{T}_j \right).$$

$$\text{Var}(O\text{-}X) = \left[\sum U_i \right]^2 \sum_j W_j^2 \left\{ \frac{Q_j(1\text{-}Q_j)}{N_j} + \frac{T_j(1\text{-}T_j)}{M_j} \right\}.$$

The logrank statistic can therefore be rewritten as

$$Z = \frac{\sum W_j \left(\hat{Q}_j - \hat{T}_j \right)}{\sqrt{\hat{V}\text{ar}(O\text{-}X) / \left[\sum U_i \right]^2}}. \qquad (4.4)$$

The denominator is simply a standard error, and hence the key is the numerator in determining precision. (The null hypothesis is used to estimate the variance in the denominator.)

Note the numerator is a weighted average: (to estimate Q - T)

$$\text{Num} = \sum_j W_j \left(\hat{Q}_j - \hat{T}_j \right) = \sum_j W_j \hat{Q}_j - \sum_j W_j \hat{T}_j.$$

From (4.3) the weights, W_j, sum to unity.

The optimal exponential estimator would employ

$$\hat{Q} - \hat{T} = \sum_j \frac{N_j \hat{Q}_j}{N} - \sum_j \frac{M_j \hat{T}_j)}{M},$$

(minimizing the variance of $\hat{Q} - \hat{T}$).

Let us consider

$$W_j^* = \frac{N_j}{N} \text{ with } N = \Sigma N_j. \qquad (4.5)$$

$$\text{Var}(\sum_j W_j \hat{Q}_j) = \text{Var}\left(\sum_j \left\{ W_j^* + \left(W_j - W_j^* \right) \right\} \hat{Q}_j \right)$$

$$= \sum_j W_j^{*2} \text{Var}(\hat{Q}_j) + 2 \sum_j W_j^* (W_j - W_j^*) \, \text{Var}(\hat{Q}_j)$$
$$+ \sum_j (W_j - W_j^*)^2 \text{Var}(\hat{Q}_j)$$

$$= \sum_j \frac{N_j^2}{N^2} \frac{Q(1-Q)}{N_j} + 2 \sum_j \frac{N_j (W_j - W_j^*) Q(1-Q)}{N N_j}$$

$$+ \sum_j (W_j - W_j^*)^2 \frac{Q(1-Q)}{N_j}$$

$$= \frac{Q(1-Q)}{N^2} \Sigma N_j + 2 \frac{Q(1-Q)}{N} \Sigma (W_j - W_j^*) + Q(1-Q) \Sigma \frac{(W_j - W_j^*)^2}{N_j}.$$

Since $\Sigma N_j = N$, $\Sigma W_j = \Sigma W_j^* = 1$, and $N_j = N W_j^*$,

$$\text{Var}\left(\sum W_j \hat{Q}_j\right) = \frac{Q(1-Q)}{N} \left[1 + 0 + \frac{\Sigma \left(W_j - W_j^* \right)^2}{W_j^*} \right]$$

$$= \text{Var}(\hat{Q}) \left\{ 1 + \frac{\Sigma \left(W_j - W_j^* \right)^2}{W_j^*} \right\}. \qquad (4.6)$$

Case 1: If $N_j = M_j$ for all j then

$$U_j \quad = \frac{N_j M_j}{N_j + M_j} = \frac{N_j}{2}. \qquad \left(\text{From (4.2)}\right)$$

$$\sum U_i \; = \frac{\sum N_j}{2} = \frac{N}{2}.$$

$$W_j \quad = \frac{U_j}{\sum U_i} = \frac{N_j/2}{N/2} = \frac{N_j}{N} = W_j^*. \qquad \left(\text{See (4.3) and (4.5)}\right)$$

$$\text{Var}\left(\sum W_j \hat{Q}_j\right) = \text{Var}(\hat{Q}). \qquad \left(\text{See (4.6)}\right)$$

Similarly,

$$\text{Var}\left(\sum W_j \hat{T}_j\right) = \text{Var}(\hat{T}).$$

Case 2: If $N_j = aM_j$ for all j and a constant $a > 0$,

then

$$U_j \quad = \frac{N_j M_j}{N_j + M_j} = \frac{aM_j^2}{aM_j + M_j}.$$

$$= \frac{a}{1+a} M_j = \frac{N_j}{1+a}$$

$$\sum U_j = \frac{\sum N_j}{1+a} = \frac{N}{1+a}.$$

Hence

$$W_j = \frac{U_j}{\sum U_i} = \frac{N_j}{N}.$$

Hence

$$\text{Var}\left(\sum_j W_j \hat{Q}_j\right) = \text{Var}(\hat{Q})$$

and similarly $\text{Var}\left(\sum W_j \hat{T}_j\right) = \text{Var}(\hat{T}).$

This implies full efficiency of the logrank test, in case 2.

In a randomized trial, it turns out that the N_j and M_j stay close enough together so as not to make a practical difference in terms of efficiency.

This is illustrated in the following abbreviated example where Treatment #2 seems to have a decided edge in terms of sample size. See (4.2), (4.3), (4.5), (4.6).

Day	N_j	M_j	U_j	W_j	W_j^*	$(W_j\text{-}W_j^*)^2/W_j^*$
1	25	25	12.50	.1694	.1894	.0021
2	20	23	10.70	.1450	.1515	.0003
3	19	23	10.40	.1409	.1439	.0001
4	17	22	9.59	.1299	.1288	.0000+
5	16	22	9.26	.1255	.1212	.0002
6	14	20	8.24	.1116	.1061	.0003
7	11	18	6.82	.0924	.0833	.0010
8	10	17	6.30	.0853	.0756	.0012
Total	132	170	73.81			.0052

$$U_j = \frac{N_j M_j}{N_j + M_j}.$$

$$W_j = \frac{U_j}{\sum U_i}.$$

$$\mathrm{Var}\left(\Sigma W_j \hat{Q}_j\right) = 1.0052\ \mathrm{Var}(\hat{Q}).$$

Similarly, it can be shown that by switching rolls of Treatments #1 and #2:

$$\mathrm{Var}\left(\Sigma W_j \hat{T}_j\right) = 1.0086\ \mathrm{Var}(\hat{T}).$$

$$\text{Var}\left[\Sigma W_j(\hat{Q}_j - T_j)\right] = \text{Var}\left[\Sigma W_j\hat{Q}_j\right] + \text{Var}\left[\Sigma W_j\hat{T}_j\right]$$

$$= 1.0052 \text{ Var}(\hat{Q}) + 1.0086 \text{ Var}(\hat{T})$$

$$< 1.0086\left\{\text{Var}(\hat{Q}) + \text{Var}(\hat{T})\right\}$$

$$< 1.0086 \text{ Var } (\hat{Q} - \hat{T}).$$

The "relative efficiency" of the logrank test compared to the binomial test, the ratio of variances, is greater than

$$(1.0086)^{-1} = .9915.$$

This implies from A130 that the Exponential Test with 99.15% of the patients in each time frame is less efficient than the logrank test with the actual arrival pattern.

3.5   Exponential Survival with a Poisson Accrual Process

Up to now, we have treated the number at risk at the start of each day as fixed. In this section, we take another look at exponential survival under a Poisson accrual process. The results will be almost identical to the fixed daily sample size, but considerable added insight will be obtained. Initially, consider Treatment #1 only. The notation will be as follows:

Number of calendar days of patient accrual	s
Days from closure to accrual to analysis	t
Total duration of the trial in days	s+t
Expected daily accrual rate (Treatment #1)*	A (For Now)
Daily population failure rate (Treatment #1)	Q
Total time on test (Treatment #1)	N
Total failures on Treatment #1	F
$\hat{Q}$	F/N

Note: s = number of days from the opening of the trial until closure to accrual.

*Later, we shall set the accrual rate at (.5A) for each treatment.

Results for the one sample problem Poisson Arrival, Exponential Survival

(1) $\quad \frac{\sqrt{As}}{B}\left[\frac{F}{N} - Q\right] \xrightarrow{D} N(0, 1)$

where $\quad B^2 = Q^2\left[1 - \frac{1}{Qs}\left\{e^{-Qt} - e^{-Q(s+t)}\right\}\right]^{-1}$ $\qquad\qquad$ (5.1)

(2) $\quad \frac{N}{\sqrt{F}}\left[\frac{F}{N} - Q\right] \xrightarrow{D} N(0, 1)$

Plan of the Proof

Step 1.

First, we shall condition on N(s), the number of patients accrued to time s.

We shall represent, conditional that N(s) = n,

$$\frac{F}{n} - \frac{QN}{n}$$

as a sample mean of independent, identically distributed random variables

$$U_j = V_j - QW_j,$$

where

$$V_j = 1 \text{ if patient j dies during study}$$

and $\qquad\qquad\qquad\qquad = 0 \text{ otherwise,}$

$$W_j = \text{time on test for patient j}$$

$\qquad\qquad\qquad = \text{time from arrival to death or closure (earlier of two)}.$

The index j represents a randomly selected patient from the n.

$$F \quad = \Sigma V_j.$$

$$N \quad = \Sigma W_j.$$

$$F - NQ = \Sigma U_j = \Sigma(V_j - QW_j).$$

Step 2: Show that given $N(s) = n$,

$$E(U_j) = 0.$$
$$Var(U_j) = 1 - \frac{1}{Qs}\left\{e^{Qt} - e^{-Q(s+t)}\right\} = \sigma^2. \tag{5.2}$$

Step 3: Show that given $N(s) = n \to \infty$,

$$\frac{\Sigma U_j}{\sqrt{n}} \xrightarrow{D} N(0, \sigma^2)$$

where σ^2 is given in (5.2).

Step 4: Show that conditionally , as $n \to \infty$

$$\sqrt{n}\left[\frac{F}{N} - Q\right] \xrightarrow{D} N(0, B^2).$$

B^2 is defined in (5.1).

Step 5: Show that as $As \to \infty$,

$$\frac{n}{As} \xrightarrow{P} 1,$$

and hence conditionally as $n \to \infty$,

$$\sqrt{As} \left[\frac{F}{N} - Q \right] \xrightarrow{D} N(0, B^2).$$

Step 6: Show that

$$\hat{B}^2 = \frac{nF}{N^2}$$

satisfies

$$\frac{\hat{B}^2}{B^2} \xrightarrow{P} 1 \qquad \text{(conditionally as } n \to \infty),$$

and hence as $n \to \infty$,

$$\frac{N}{\sqrt{F}} \left[\frac{F}{N} - Q \right] \xrightarrow{D} N(0, 1).$$

Note that the conditional distributions and statistics in Steps 5 and 6 do not depend on n, and hence they are valid asymptotic unconditional distributions.

Proof of Step 1:

From our study of the Poisson Process, A143 we know that given $N(s)$, the number of patients accrued to time s, is equal to n, the arrival times form a random sample from the uniform distribution over the interval from 0 to s.

We need to study the variable

$$\frac{F}{n} - \frac{QN}{n} \tag{5.3}$$

That is,

$$\frac{F - QN}{n}.$$

Now, this can be viewed as the mean of the following independent identically distributed random variables: U_j, $j = 1, 2, \ldots, n$ with

$$U_j = V_j - QW_j.$$

$V_j = 1$ (death on study), 0 (survival) and W_j = time on test.

For each j, independently generate the following two independent random variables:

$$X_j = \text{Arrival time, uniform on the interval 0 to s}$$

$$Y_j = \text{Survival, exponential with hazard Q (Mean 1/Q)}.$$

Note that since the trial ends at s+t, the time at risk is $(s+t - X_j)$.

	If	$Y_j \leq s+t-X_j$, then $V_j = 1$	(patient dies during trial)
and		$W_j = Y_j$	(time on test of patient j is Y_j).

	If	$Y_j > s+t - X_j$ then	
		$V_j = 0$	(patient lives to end of trial)
and		$W_j = s+t - X_j$	(time on test of patient j is $s+t - X_j$).

Prior to reading the next body of material, it will be helpful to review Gamma integrals (Appendix A82 and A77).

By independence, and the fact that V_j and W_j are functions of X_j and Y_j, we have

$$E(V_j) = 0\, P(V_j = 0) + 1\, P(V_j = 1)$$

$$= \tfrac{1}{s} \int_0^s \int_0^{s+t-x} Q e^{-Qy} dy dx$$

$$= \tfrac{1}{s} \int_0^s \left[1 - e^{-Q(s+t-x)}\right] dx$$

$$= 1 - \tfrac{1}{Qs} \left\{ e^{-Qt} - e^{-Q(s+t)} \right\}. \tag{5.4}$$

$$E(W_j) = \tfrac{1}{s} \int_0^s \left[\int_0^{s+t-x} y Q e^{-Qy}\, dy + \int_{s+t-x}^{\infty} (s+t-x) Q e^{-Qy} dy \right] dx$$

$$= \tfrac{1}{s} \int_0^s \left[\tfrac{1}{Q} \left\{ 1 - e^{-Q(s+t-x)} - Q(s+t-x)e^{-Q(s+t-x)} \right\} + (s+t-x)e^{-Q(s+t-x)} \right] dx$$

$$= \tfrac{1}{Qs} \int_0^s \left\{ 1 - e^{-Q(s+t-x)} \right\} dx$$

$$= \tfrac{1}{Q} \left[1 - \tfrac{1}{Qs} \left\{ e^{-Qt} - e^{-Q(s+t)} \right\} \right] \tag{5.5}$$

For inference purposes, we shall need the second moments.

$$E\left((V_j - QW_j)^2 \right) = E(V_j^2) - 2QE\left(V_j W_j \right) + Q^2 E\left(W_j^2 \right). \tag{5.6}$$

This will be greatly simplified when we show

$$Q^2 E\left(W_j^2 \right) = 2QE\left(V_j W_j \right). \tag{5.7}$$

$$E\left(V_j W_j \right) = \tfrac{1}{s} \int_0^s \left[\int_0^{s+t-x} y Q e^{-Qy} dy \right] dx$$

$$= \tfrac{1}{Qs} \int_0^s \left\{ 1 - e^{-Q(s+t-x)} - Q(s+t-x)e^{-Q(s+t-x)} \right\} dx. \tag{5.8}$$

$$E\left(W_j^2 \right) = \tfrac{1}{s} \int_0^s \left\{ \int_0^{s+t-x} y^2 Q e^{-Qy} dy + \int_{s+t-x}^{\infty} (s+t-x)^2 Q e^{-Qy} dy \right\} dx. \tag{5.9}$$

We study

$$\int_0^{s+t-x} y^2 Q e^{-Qy} dy = \frac{2}{Q^2} \int_0^{s+t-x} \frac{Q^3}{2} y^2 e^{-Qy} dy$$

$$= \frac{2}{Q^2} \left\{ 1 - e^{Q(s+t-x)} - (s+t-x)Q e^{-Q(s+t-x)} - \frac{(s+t-x)^2}{2} Q^2 e^{-Q(s+t-x)} \right\}.$$

Also,

$$\int_{s+t-x}^{\infty} (s+t-x)^2 Q e^{-Qy} dy = (s+t-x)^2 e^{-Q(s+t-x)}.$$

Substituting these into (5.9), the first expression for $E\left(W_j^2\right)$ yields (after cancellation)

$$E\left(W_j^2\right) = \frac{2}{Q^2 s} \int_0^s \left\{ 1 - e^{-Q(s+t-x)} - (s+t-x) Q e^{-Q(s+t-x)} \right\} dx$$

$$= \frac{2}{Q} E\left(V_j W_j\right), \qquad \text{by (5.8)}.$$

Hence,

$$E\left(\left(V_j - Q_j W_j\right)^2\right) = E\left(V_j^2\right) - 2QE\left(V_j W_j\right) + Q^2 E\left(W_j^2\right)$$

$$= E\left(V_j^2\right) - 2QE\left(V_j W_j\right) + \frac{2Q^2 E\left(V_j W_j\right)}{Q}$$

$$= E\left(V_j^2\right).$$

However, V_j takes on only values of 0 or 1. Hence $V_j^2 = V_j$.

That is, from (5.6) and (5.4)

$$E\left(\left(V_j - Q_j W_j\right)^2\right) = E\left(V_j\right) = 1 - \frac{1}{Qs} \left\{ e^{Qt} - e^{-Q(s+t)} \right\}.$$

Summary for Step 2.

We have shown the following properties of the V_j and W_j given the number of arrivals in the time interval from 0 to s

$$E(V_j) = QE(W_j) = 1 - \frac{1}{Qs} \left\{ e^{Qt} - e^{-Q(s+t)} \right\}. \tag{5.10}$$

$$E\left(\left\{V_j - QW_j\right\}^2\right) = 1 - \frac{1}{Qs} \left\{ e^{Qt} - e^{-Q(s+t)} \right\}. \tag{5.11}$$

Hence $U_j = V_j - QW_j$ satisfies

$$E(U_j) = 0,$$

$$Var(U_j) = 1 - \frac{1}{Qs}\left\{e^{Qt} - e^{-Q(s+t)}\right\} = \sigma^2.$$

This completes Step 2.

Proof of Step 3.

Suppose that the number of arrivals, n, is sufficiently large to apply the central limit theorem on the V_j's:

$$\sqrt{n}\;\frac{\Sigma\left\{U_j - E(U_j)\right\}}{n}$$

$$= \frac{\Sigma U_j}{\sqrt{n}} \qquad \text{(From Step 2)}$$

$$\xrightarrow{D} N(0, \sigma^2). \qquad \text{(Central Limit Theorem A106 and Step 2)}.$$

This completes Step 3.

Proof of Step 4.

$$\sqrt{n}\left[\frac{F}{N} - Q\right] = \sqrt{n}\left\{\frac{\displaystyle\sum_{j=1}^{n} V_j/n}{\displaystyle\sum_{j=1}^{n} W_j/n} - Q\right\}$$

$$= \frac{1}{\Sigma(W_j/n)}\left\{\Sigma U_j/\sqrt{n}\right\}$$

$$= \frac{1}{E(W_j)}\left\{\sum_j U_j/\sqrt{n}\right\} + \left[\frac{1}{\Sigma W_j/n} - \frac{1}{E(W_j)}\right]\sum_j U_j/\sqrt{n}. \qquad (5.12)$$

By the Central Limit Theorem, A106,

$$\sqrt{n} \left\{ \Sigma \frac{W_j}{n} - E(W_j) \right\} \xrightarrow{D} N\left(0, \, Var(W_j)\right)$$

and hence

$$\Sigma \frac{W_j}{n} - E(W_j) \xrightarrow{P} 0.$$

$\sqrt{n} \left[\frac{F}{N} - Q \right]$ is the sum of two components: $\left(\text{See } (5.12) \right)$

a. $\dfrac{1}{E(W_j)} \left\{ \sum_j U_j / \sqrt{n} \right\} \xrightarrow{D} N\left(0, \, \dfrac{\sigma^2}{\left[E(W_j)\right]^2}\right)$

and

b. $\left[\dfrac{1}{\left[\Sigma W_j / n\right]} - \dfrac{1}{E(W_j)} \right] \sum_j U_j / \sqrt{n} \xrightarrow{P} 0.$

Hence, by Slutzky's Theorem **A104**,

$$\sqrt{n} \left[\frac{F}{N} - Q \right] \xrightarrow{D} N\left(0, \, \frac{\sigma^2}{\left[E(W_j)\right]^2}\right) \tag{5.13}$$

By (5.10),

$$QE(W_j) = 1 - \frac{1}{Qs} \left\{ e^{-Qt} - e^{-Q(s+t)} \right\}$$

$$= \sigma^2, \qquad\qquad \left(\text{See } (5.2) \right)$$

and hence

$$E(W_j) = \frac{\sigma^2}{Q}. \tag{5.14}$$

Substitution (5.14) into (5.13) yields

$$\sqrt{n}\left[\frac{F}{N} - Q\right] \xrightarrow{D} N\left(0, \frac{Q^2}{\sigma^2}\right). \tag{5.15}$$

Since

$$\sigma^2 = 1 - \frac{1}{Qs}\left\{e^{Qt} - e^{-Q(s+t)}\right\},$$

we see from (5.1), that

$$B^2 = \frac{Q^2}{\sigma^2}.$$

and hence from (5.15), the proof of Step 4 is complete.

Proof of Step 5.

Since n is Poisson with parameter As, we have from A62,

$$E(n) = As \quad \text{and} \quad Var(n) = As.$$

$$P\left[\left|\frac{n}{As} - 1\right| > \epsilon\right] = P\left[|n - As| > \epsilon As\right]$$

$$< \frac{Var(n)}{\epsilon^2 A^2 s^2} \qquad \text{(Chebyshev Theorem A51)}$$

$$< \frac{As}{\epsilon A^2 s^2} = \frac{1}{\epsilon As}$$

$$\rightarrow 0 \qquad \text{as } As \rightarrow \infty.$$

(It does not matter whether A is fixed and $s \rightarrow \infty$, or s is fixed and $A \rightarrow \infty$, or whether simply $As \rightarrow \infty$ on any path.) Thus

$$\frac{n}{As} \xrightarrow{P} 1.$$

Since

$$\sqrt{As}\left[\frac{F}{N} - Q\right] = \left\{\sqrt{n}\left[\frac{F}{N} - Q\right]\right\}\sqrt{\frac{As}{n}},$$

and since

$$\sqrt{\frac{As}{n}} \xrightarrow{P} 1, \qquad\qquad \text{(Appendix A98)},$$

Slutzky's Multiplicative Theorem, A105, and step 4 yield

$$\sqrt{As}\left[\frac{F}{N} - Q\right] \xrightarrow{D} N(0, B^2).$$

This completes Step 5.

Proof of Step 6:

Let

$$\hat{B}^2 = \frac{nF}{N^2}$$

$$= \frac{n}{F}\left(\frac{F}{N}\right)^2 \qquad\qquad (5.16)$$

Now since from Step 5:

$$\sqrt{As}\left(\frac{F}{N} - Q\right) \xrightarrow{D} N(0, B^2),$$

as $As \rightarrow \infty$

$$\frac{F}{N} \xrightarrow{P} Q. \qquad\qquad (5.17)$$

Now given n,

$$F = \Sigma V_j \qquad\qquad \text{(See Definition of } V_j \text{ in Step 1)}$$

is a sum independent identically distributed random variables. From (5.10), (5.1), and (5.2)

$$E(V_j) = 1 - \frac{1}{Qs}\left\{e^{-Qt} - e^{-Q(s+t)}\right\} = \frac{Q^2}{B^2}. \tag{5.18}$$

Since V_j is Bernoulli,

$$Var(V_j) = E(V_j)\left\{1 - E(V_j)\right\}.$$

Given n is sufficiently large,

$$\sqrt{n}\left\{\frac{F}{n} - E(V_j)\right\} \xrightarrow{D} N\left(0, Var(V_j)\right),$$

and hence

$$\frac{F}{n} - E(V_j) \xrightarrow{P} 0.$$

That is, from A95,

$$\frac{F}{nE(V_j)} \xrightarrow{P} 1.$$

from (5.18),

$$\frac{B^2F}{nQ^2} \xrightarrow{P} 1. \tag{5.19}$$

From Step 4, $F/(NQ) \xrightarrow{P} 1$. Application of A98 and A101 to (5.17) and (5.19) yields,

$$\frac{nF}{N^2B^2} = \left(\frac{nQ^2}{B^2F}\right)\left(\frac{F}{NQ}\right)^2 \xrightarrow{P} 1. \tag{5.20}$$

Now

$$\frac{N}{\sqrt{F}}\left[\frac{F}{N} - Q\right] = \frac{\sqrt{n}}{B}\left[\frac{F}{N} - Q\right]\left\{\frac{NB}{\sqrt{nF}}\right\}.$$

From Slutzky's Multiplicative Theorem A105 and (5.20)

$\frac{N}{\sqrt{F}}\left[\frac{F}{N} - Q\right]$ has the same asymptotic distribution as

$$\frac{\sqrt{n}}{B}\left[\frac{F}{N} - Q\right]$$

which is N(0, 1) from Step 4.

This completes the proof.

3.6   Extension to the Two-Sample Problem

Here the accrual rate is A/2 to each treatment, G is the number of failures on Treatment #2, M is the total time on test for Treatment #2. T is the daily failure rate on Treatment #2. We have by the one sample results (halving the accrual rate):

$$\sqrt{(.5A)s}\left[\frac{F}{N} - Q\right] \xrightarrow{D} N(0, B^2)$$

with

$$B^2 = Q^2\left[1 - \frac{1}{Qs}\left\{e^{-Qt} - e^{-Q(s+t)}\right\}\right]^{-1}.$$

$$\sqrt{(.5A)s}\left[\frac{G}{M} - T\right] \xrightarrow{D} N(0, C^2)$$

with

$$C^2 = T^2\left[1 - \frac{1}{Ts}\left\{e^{-Tt} - e^{-T(s+t)}\right\}\right]^{-1}$$

Hence

$$\sqrt{(.5A)s}\left[\left(\frac{F}{N} - \frac{G}{M}\right) - (Q - T)\right] \xrightarrow{D} N(0, B^2 + C^2),$$

Since from A142, each accrual process is independent, and A108 can be applied.

Now from the previous section, Steps 5 and 6

$$\frac{B^2/(.5As)}{F/N^2} \xrightarrow{P} 1$$

and

$$\frac{C^2/(.5As)}{G/M^2} \xrightarrow{P} 1.$$

Hence from the Appendix A103,

$$\frac{(B^2 + C^2)/(.5As)}{(F/N^2) + (G/M^2)} \xrightarrow{P} 1.$$

By applying A109, the following "Block" holds:

Results for Poisson Entry, Exponential Survival

(1) $\quad \dfrac{\sqrt{(.5As)}}{\sqrt{B^2 + C^2}} \left[\dfrac{F}{N} - \dfrac{G}{M} - (Q - T) \right] \xrightarrow{D} N(0, 1).$

(2) $\quad \left[\dfrac{F}{N} - \dfrac{G}{M} - (Q - T) \right] \dfrac{1}{\sqrt{(F/N^2) + (G/M^2)}} \xrightarrow{D} N(0, 1).$

(3) $\quad \dfrac{(B^2 + C^2)/(.5As)}{\sqrt{(F/N^2) + (G/M^2)}} \xrightarrow{P} 1.$

A = total accrual rate (A/2 to each treatment).

$$B^2 = Q^2 \left[1 - \frac{1}{Qs} \left\{ e^{-Qt} - e^{-Q(s+t)} \right\} \right]^{-1}.$$

$$C^2 = T^2 \left[1 - \frac{1}{Ts} \left\{ e^{-Tt} - e^{-T(s+t)} \right\} \right]^{-1}.$$

s = Accrual period. (Calendar time)

t = Time from end of accrual to analysis.

Q = Population daily failure rate on Treatment #1.

T = Population daily failure rate on Treatment #2.

F = Failures on Treatment #1 (observed).

G = Failures on Treatment #2 (observed).

N = Total time on test for Treatment #1 (total patient days).

M = Total time on test for Treatment #2.

Comparison with the fixed daily sample size version. The asymptotic result of section 3.3 was:

$$\frac{\hat{Q} - \hat{T} - (Q - T)}{\sqrt{\frac{\hat{Q}}{N} + \frac{\hat{T}}{M}}} \xrightarrow{D} N(0, 1). \qquad \text{(Approximately)}$$

Since $\hat{Q} = F/N$ and $\hat{T} = G/M$, this result is identical to (2) above.

<u>Sample size derivation</u> for exponential survival with Poisson Accrual.

Please consult Appendix A130(1) and A135, as well as template on final conclusions for Poisson entry above. We shall identify each term in A130:

$$H_0: \theta = Q - T = 0. \qquad\qquad (\theta_0 = 0)$$

$$H_A: \theta = Q - T = \theta_1 > 0. \qquad\qquad \text{(One-sided)}$$

$$\hat{\theta} = (F/N) - (G/M).$$

$$SE = \sqrt{(F/N^2) + (G/M^2)}.$$

$$S = \sqrt{(B^2 + C^2)/(.5As)}.$$

Hence the necessary sample size for Type I error α and Type II error β (Power 1 - β) satisfies

$$S = \frac{Q - T}{(Z_\alpha + Z_\beta)}.$$

This is solved by bisection. See block below for full definition of terms.

Sample Size Equation for Poisson Accrual and Exponential Survival:

Two sample trial with 50% allocation to each treatment. One-sided Test Calculation. (Replace α by $\alpha/2$ for two-sided test.)

$$\frac{B^2 + C^2}{(.5As)} = \frac{(Q - T)^2}{(Z_\alpha + Z_\beta)^2} \tag{6.1}$$

A = Total accrual rate.

s = Total calendar time of accrual.

t = Time from end of accrual to analysis (minimum follow-up time).

Q = Daily failure rate on Treatment #1 (under H_A).

T = Daily Failure Rate on Treatment #2 (under H_A).

α = Probability of a Type I error (Reject H_0 when true).

Z_α = Upper $100\alpha\%$ point of the standard normal distribution.

β = 1 - Power = Probability of a Type II error (Reject H_A when true).

Z_β = Upper $100\beta\%$ point of the standard normal

$$B^2 = Q^2 \left[1 - \frac{1}{Qs} \left\{ e^{-Qt} - e^{-Q(s+t)} \right\} \right]^{-1}. \tag{6.2}$$

$$C^2 = T^2 \left[1 - \frac{1}{Ts} \left\{ e^{-Tt} - e^{-T(s+t)} \right\} \right]^{-1}. \tag{6.3}$$

See Section 3.10 regarding losses to follow-up.

Tables for exponential survival appear in this volume under FACT = 1.0.

3.7 Exponential Survival with "Up-Front" Accrual

In rare instances, it is possible to accrue all patients at day 1 and follow them for t years.

We derive the distributional properties from the Poisson Arrival as follows:

Let the accrual period $s \rightarrow 0$.

Let the accrual rate $A \rightarrow \infty$,

with $As \rightarrow n$ (large).

From this, we shall obtain the analogous one sample and two sample results. Recall that

$$B^2 = Q^2 \left[1 - \frac{1}{Qs} \left\{ e^{-Qt} - e^{-Q(s+t)} \right\} \right]^{-1}.$$

By L'Hospital's rule, letting the accrual period $s \rightarrow 0$

$$\lim_{s \rightarrow 0} \frac{e^{-Qt} - e^{-Q(s+t)}}{Qs} = \lim_{s \rightarrow 0} \frac{Q e^{-Q(s+t)}}{Q} = e^{-Qt}.$$

Hence

$$B^2 \rightarrow Q^2 \left[1 - e^{-Qt} \right]^{-1} \quad \text{as } s \rightarrow 0.$$

This limit is used as the B^2 value in the variance formula.

Results for the One Sample Problem with Exponential Survival and Up Front Accrual:

(1) $\frac{\sqrt{n}}{B}\left[\frac{F}{N} - Q\right] \overset{D}{\to} N(0,\ 1)$,

where Q = Daily failure rate.

n = Number of patients treated.

F = Observed failures.

N = Total time on test.

$B^2 = Q^2\left[1 - e^{-Qt}\right]^{-1}$.

t = Time from accrual to analysis (follow-up time).

(2) $\frac{N}{\sqrt{F}}\left[\frac{F}{N} - Q\right] \overset{D}{\to} N(0,\ 1)$.

(3) $\frac{N/\sqrt{F}}{\sqrt{n}/B} \overset{P}{\to} 1$.

Results for the Two Sample Problem with Exponential Survival and Up Front Entry:

(1) $\quad \dfrac{\sqrt{.5n}}{\sqrt{B^2 + C^2}} \left[\dfrac{F}{N} - \dfrac{G}{M} - (Q - T) \right] \xrightarrow{D} N(0, 1)$

(2) $\quad \left[\dfrac{F}{N} - \dfrac{G}{M} - (Q - T) \right] \dfrac{1}{\sqrt{(F/N^2) + (G/M^2)}} \xrightarrow{D} N(0, 1)$

(3) $\quad \dfrac{(B^2 + C^2)/(.5n)}{\left[(F/N^2) + (G/M^2) \right]} \xrightarrow{P} 1$

with $\quad$ n = Total patient accrual (n/2 to each treatment).

$B^2 = Q^2 \left[1 - e^{-Qt} \right]^{-1}$

$C^2 = T^2 \left[1 - e^{-Tt} \right]^{-1}$.

t = Time from start of study to analysis (follow-up time)

Q = Population failure rate on Treatment #1.

T = Population Failure rate on Treatment #2.

117

Sample Size Equation for Up Front Accrual and Exponential Survival: Two sample trial with 50% allocation to each treatment. One-sided Test Calculation. (Replace α by $\alpha/2$ for two-sided test.)

$$\frac{B^2 + C^2}{.5n} = \frac{(Q - T)^2}{(Z_\alpha + Z_\beta)^2}. \tag{7.1}$$

n = Total number of patients.

t = Time from start of study to analysis (follow-up time).

Q = Daily failure rate (Treatment #1).

T = Daily failure rate (Treatment #2).

α = Probability of a Type I error (Reject H_0 when true).

β = Probability of a Type II error (Reject H_A when true).

Z_α = Upper (100α)% point of the standard normal distribution.

Z_β = Upper (100β)% point of the standard normal distribution.

$$B^2 = Q^2\left[1 - e^{-Qt}\right]^{-1} \tag{7.2}$$

$$C^2 = T^2\left[1 - e^{-Tt}\right]^{-1}. \tag{7.3}$$

See Section 3.10 regarding losses to follow-up

This formula also can be used when patients are entered in any fashion, but followed for a period t, not longer.

The tables for up front accrual are in this volume under FACT = 0.

3.8 Proportional Hazard Models and the Exponential Distribution

Let $F_j(x)$ be the cumulative distribution of survival for treatment j, j = 1, 2 respectively.

$$F_j(x) = P\left[X_j \leq x\right] \qquad \text{(See A20)}$$

The logrank test is defined for testing the null hypothesis

$$H_0: F_1(x) = F_2(x) \qquad \text{for all x.}$$

The alternative is typically for the one-sided case:

$$H_A: F_1(x) \leq F_2(x) \qquad \text{for all x} \qquad \text{(Treatment \#1 Superior)}$$

The proportional hazard formulation adds structure to H_A as follows:

$$H_A: \left\{1 - F_2(x)\right\} = \left\{1 - F_1(x)\right\}^\theta. \tag{8.1}$$

If the distributions have continuous densities represented by f_1 and f_2 respectively, then (8.1) can be considered as follows

$$F_2(x) = 1 - \left[\int_x^\infty f_1(t)dt\right]^\theta. \tag{8.2}$$

Differentiating both sides with respect to x yields for $\theta \neq 1$ (Chain rule)

$$F_2'(x) = f_2(x) = \theta f_1(x)\left[\int_x^\infty f_1(t)dt\right]^{\theta-1},$$

i.e.

$$f_2(x) = \theta f_1(x)\left\{1 - F_1(x)\right\}^{\theta-1}.$$

$$\frac{f_2(x)}{\left[1 - F_1(x)\right]^\theta} = \frac{\theta f_1(x)}{1 - F_1(x)}. \tag{8.3}$$

That is, from (8.1)

$$\frac{f_2(x)}{1 - F_2(x)} = \frac{\theta f_1(x)}{1 - F_1(x)} \tag{8.4}$$

Now if $\theta = 1$, (8.4), still holds.

From Appendix A75 which defines the Hazard Rate, the alternate hypothesis (8.1) states that the hazard rates are proportional. That is, the ratio of hazards is independent of time. If and only if $\theta < 1$, then Treatment #2 has a lower failure rate than Treatment #1.

The following result relates proportional hazards in a rather remarkable way to the exponential.

Theorem 8.1:

   Let

$$1 - F_1(x) = \left[1 - G(x)\right]^{\gamma} \tag{8.5}$$

and

$$1 - F_2(x) = \left[1 - G(x)\right]^{\gamma + \theta}, \tag{8.6}$$

where $G(x)$ is a known distribution but γ and θ are unknown.

   Let us transform the time scale x by

$$y = -\log\left\{1 - G(x)\right\}. \tag{8.7}$$

In this time scale, let Y_1 and Y_2 represent random variables of survival times on Treatments #1 and #2 respectively.

   Then Y_1 and Y_2 are respectively exponential with hazards γ and $(\gamma + \theta)$.

Proof:

   From the Appendix A89,

$$Y_1 \text{ has density } \frac{f_1(x)}{(dy/dx)}.$$

Now from (8.7),

$$\frac{dy}{dx} = \frac{G'(x)}{1 - G(x)}.$$

From Appendix A21 and (8.5)

$$f_1(x) = \gamma \left\{ 1 - G(x) \right\}^{\gamma-1} G'(x).$$

That is, Y_1 has density

$$h_1(y) = \gamma \left\{ 1 - G(x) \right\}^{\gamma} \qquad 0 < G(x) < 1.$$

But $1 - G(x) = e^{-y}$ $\left(\text{See } (8.7) \right)$ and hence Y_1 has density

$$h_1(y) = \gamma e^{-\gamma y} \qquad y > 0.$$

Similarly, replacing γ by $(\gamma + \theta)$ in the above yields the fact that Y_2 has density

$$h_2(y) = (\gamma + \theta) e^{-(\gamma+\theta)y} \qquad y > 0.$$

This completes the proof.

This transformation converts a proportional hazard problem to an exponential problem.

3.9 Considerations in Planning a Trial Under Proportional Hazards: Putting It All Together

Suppose you wished to plan a trial with the following endpoints:

Annual Accrual: 150 patients per year.

Two Year Survival: 40% Control (Treatment #1).

                       50% Experimental (Treatment #2).

Alpha: .05 (one-sided).

Power: 80%.

Distribution: Exponential (For Now).

A total of 300 patients are expected to be accrued through the two year endpoint, and hence from the sample size Tables (FACT = 1.0), 541 patients (3.6 years of accrual) are required. The trial would require an additional two years of follow-up.

As we have seen in Section 3.4, the power loss in the use of the logrank test as opposed to the exponential test would be of no practical concern.

However, the key issue one faces is non-exponentiality.

In answering this question, let us first consider an alternate study where the accrual is up front (or similarly if we only use the first two years of patient data.) Suppose further that you were willing to assume proportional hazards. Then the power loss caused by lack of knowledge of G would be minimal, since

(a)   There exists a monotonic transformation to convert the data to exponential.

(b)   The logrank test is invariant under monotonic transformations.

(c)   From Theorem 8.1 the exponential parameters do not involve G.

(d)   The logrank and exponential tests have virtually the same power, when exponentiality holds.

* The distribution G plays no role in planning this alternate study under proportional hazards and the planning endpoints.

In the actual setting of follow-up to the end of the study, the key sample size planning issue is what happens to patients after surviving to the endpoint of the study.

In the above example, the study is planned for 3.6 years of accrual plus 2 years of follow-up for a total of 5.6 years. Patients are at risk from 2.0 to 5.6 years.

To accommodate this in a quantitative way, we shall consider attaching exponential survival past the endpoint, but with a lower hazard. Proportional hazards are still assumed. More sophisticated assumptions would result in contingencies too numerous to tabulate.

The derivations will be conducted as follows:

	Daily Hazard Before t	Daily Hazard After t ($d \geq 0$)
Treatment #1	Q	dQ
Treatment #2	T	dT

Non-Exponentiality before the endpoint t will have virtually no impact on the power.

NOTE: If d=0, no failures will occur after t, and we can use up front accrual

results for design and analysis.

If $d > 0$, then the following transformation produces exponentiality under both treatments, with respective hazards Q and T. (Y is transformed time, X is original time.)

$$Y = X \qquad\qquad \text{if } X \leq t$$
$$= t + d(X\text{-}t) \qquad \text{if } X > t,$$

since for $x > t$, and treatment 1

$$P[X > x] \quad = P[X > t]P[X > x|X > t]$$

$$= e^{-Qt}e^{-dQ(x\text{-}t)}$$

$$= e^{-Q[t+d(x\text{-}t)]}$$

Setting $Y = t + d(X\text{-}t)$ and $y = t+d(x\text{-}t)$ yields exponentiality with hazard Q for $Y > t$.

Since it was assumed for $Y \leq t$, exponentiality holds for Y.

Based on exponential considerations, we would estimate Q by

$$\hat{Q} = \frac{\text{Total Failures}}{\text{Total accumulated time on test in Y-scale}}$$

Following the same derivation as seen in Section 3.5, "Exponential Survival with a Poisson Accrual

Process," the following results are obtained:

Let

F_1 = Number of failures occurring prior to patient time on test t.

N_1 = total accumulated time on test accrued by patients prior to t. (Clock only is on while patient is at risk less than t.)

F_2 = Number of failures occurring after patient time on test t.

N_2 = total accumulated time on test accrued by patients after t. (Clock starts for patient after t.)

with other notation per page 96.

Let

$$\hat{Q} = \frac{F_1 + F_2}{N_1 + dN_2}$$

be the exponential estimate of Q on the basis of the transformation to exponential. Then

$$\frac{\hat{Q} - Q}{V} \xrightarrow{D} N(0, 1), \tag{9.1}$$

where

$$V^2 = \frac{Q^2}{As}\left[1 - \frac{e^{-Qt}}{dQs}\left\{1 - e^{-dQs}\right\}\right]^{-1}. \tag{9.2}$$

Before proving this result, note that

(a)   If $d = 1$, we obtain the result for exponentials (5.1).

(b)   If $d \to 0$, by L'Hospital's rule

$$V^2 \to \frac{Q^2}{As}\left[1 - e^{-Qt}\right], \tag{9.3}$$

which agrees with the Up Front Accrual results in Section 3.7.

Proof: Although the proof can be derived exactly as in the exponential case, once $E(U_j)$ and $Var(U_j)$ are obtained, we shall outline an alternate proof.

Step 1: Show

$$\hat{Q} = \frac{F_1}{N_1} W_1 + \frac{F_2}{dN_2} W_2$$

where

$$W_1 = \frac{N_1}{N_1 + dN_2}$$

and

$$W_2 = \frac{dN_2}{N_1 + dN_2} = 1 - W_1.$$

Step 2: Apply results for Up Front Accrual to show

$$\frac{\left(\frac{F_1}{N_1} - Q\right)}{V_1} \xrightarrow{D} N(0, 1),$$

where

$$V_1^2 = \frac{Q^2}{As(1 - e^{-Qt})}$$

$$= \frac{Q}{E(N_1)}. \tag{9.4}$$

Step 3: Apply results for Poisson Accrual with minimum follow-up $t = 0$ to show that

$$\frac{\left(\frac{F_2}{dN_2} - Q\right)}{V_2} \xrightarrow{D} N(0, 1),$$

where

$$V_2^2 = \frac{Q^2}{As\ e^{-Qt}} \left[1 - \frac{1}{(dQs)}\left\{1 - e^{-dQs}\right\}\right]^{-1}$$

$$= \frac{Q}{dE(N_2)}. \tag{9.5}$$

This distribution holds independently of that of Step 2.

> NOTE ON STEP 3: From the appendix on Poisson Processes, since patients
>
> must survive at least t to accrue time past t, arrivals past time t is a
>
> Poisson process with arrival rate
>
> $$Ae^{-Qt} = A \ P[\text{Survival at least } t].$$
>
> The "clock" for this process starts at calendar time t.

Step 4: Based on the already established results,

$$\frac{N_1}{E(N_1)} \xrightarrow{P} 1$$

and

$$\frac{N_2}{E(N_2)} \xrightarrow{P} 1.$$

Step 5: Show that

$$\frac{Q^* - Q}{V} \xrightarrow{D} N(0, 1)$$

where

$$Q^* = \frac{V_2^2}{V_1^2 + V_2^2}\left(\frac{F_1}{N_1}\right) + \frac{V_1^2}{\left(V_1^2 + V_2^2\right)}\left(\frac{F_2}{dN_2}\right)$$

and V^2, V_1^2, and V_2^2 are given in (9.2), (9.4), and (9.5) respectively.

Step 6: Apply Step 4 and Slutzky's Theorem to show

$$\frac{\hat{Q} - Q}{V} \xrightarrow{D} N(0, 1).$$

Footnote:

It is readily seen that

$$V^2 = \left(\frac{1}{V_1^2} + \frac{1}{V_2^2} \right)^{-1}$$

$$= \left(\frac{E(N_1)}{Q} + \frac{dE(N_2)}{Q} \right)^{-1} \tag{9.6}$$

Now from (5.10),and Step 1 of Section 3.5

$$E(N_1) = \frac{E(F_1)}{Q}$$

and

$$E(N_2) = \frac{E(F_2)}{dQ}$$

Thus,

$$V^2 = \left(\frac{E(F_1) + E(F_2)}{Q^2} \right)^{-1} = \frac{Q^2}{\text{Expected Failures}}$$

for this reason, power is connected by many authors to the expected number of failures. If one has proportional hazards, and the expected number of failures per group is fixed, exponentiality is not a factor in the power of the logrank test.

Sample Size Equation for Poisson Accrual and Piecewise Exponential Survival:

Two sample trial with 50% allocation to each treatment. One-Sided Test Calculation (Replace α by $\alpha/2$ for Two-Sided Test.)

$$\frac{B^2 + C^2}{(.5As)} = \frac{(Q - T)^2}{(Z_\alpha + Z_\beta)^2} \tag{9.7}$$

A = Total accrual rate. (.5A to each treatment).

s = Total calendar time of accrual

t = Time from end of accrual to analysis (minimum follow-up time).

d = Hazard rate after t: Hazard Rate Before t (ratio) (FACT in Sample Size Tables)

Q = Daily failure rate on Treatment #1 (under H_A) to time t.

T = Daily failure rate on Treatment #2 (under H_A) to time t.

α = Probability of a Type I error (Reject H_0 when true).

Z_α = Upper $100\alpha\%$ point of the standard normal distribution.

β = 1 - Power = Probability of a Type II error (Reject H_A when true).

Z_β = Upper $100\beta\%$ point of the standard normal

$$B^2 = \frac{Q^2}{As}\left[1 - \frac{e^{-Qt}}{(dQs)}\left\{1 - e^{-dQs}\right\}\right]^{-1}. \qquad (9.8)$$

$$C^2 = \frac{T^2}{As}\left[1 - \frac{e^{-Tt}}{(dTs)}\left\{1 - e^{-dTs}\right\}\right]^{-1}. \qquad (9.9)$$

See Section 3.10 regarding losses to follow-up.

Tables for this Piecewise Exponential Model appear for

$$d = \text{``FACT''} = 0 \text{ (up front accrual) } \left(\lim_{d \to 0}\right)^*$$

$$= .25 \text{ (75\% hazard reduction)}$$

$$= .50 \text{ (50\% hazard reduction)}$$

$$= .75 \text{ (25\% hazard reduction)}$$

$$= 1.00 \text{ exponential.}$$

Increased hazard after t is not expected in any real clinical trial.

$\left(* \text{ See } (9.2) \text{ and } (9.3)\right)$

3.10   Losses to Follow-Up and Sample Size Adjustment

The calculations of our sample size formulas assume no losses to follow-up (completing losses). (Patients quitting the study or failing due to non-study related cause are "lost to follow-up".)

The validity of a trial can be compromised by such losses, unless that are "uninformative." $\Big($Rubinstein et al. (1981) provide a formal method to adjust for such uninformative losses to follow-up.$\Big)$ We shall employ a less formal approach.

If a loss is <u>uninformative</u>, then the prognosis of the patient, given the treatment and time of removal from follow-up is presumed to be identical to patients, who stay on study and who are on that treatment, from the corresponding time onward.

EXAMPLE 10.1

A patient, who has completed therapy, moves with her parents to a region where follow-up cannot be maintained. This patient had been followed for 27 months, and was on Treatment #2. This seems to be an uninformative loss to follow-up. It seems reasonable to believe that the patient's prognosis has not been altered by the move.

EXAMPLE 10.2

In a study of coronary heart disease (CHD) a patient on Treatment #2 died at 25 months from a Non-CHD cause. If the major analysis is to be based on CHD events, then this patient is censored at 25 months. The patient would be counted in all days prior to death and would not be considered a CHD failure. We are forced to assume that the loss is uninformative, but this might not be a valid assumption. For example, if the patient died of renal failure, it might be indicative of poor vascular function. The CHD outlook of such a patient, assuming that the patient had survived the renal failure, might be much worse than other patients.

In clinical trials, site specific endpoints are especially prone to bias caused by competing losses. If one treatment has a greater tendency for such losses than the other, it might be difficult or impossible to interpret the results of the trial. Even in studying mortality, if one treatment is more toxic than the other, that treatment may lose more patients to follow-up than the other, compromising the trial.

Aggressive follow-up is a requirement of all clinical trials. External references can help keep track of participants. The National Death Index (National Center for Health Statistics) can help identify which losses have survived or died, provided that proper identifiers are collected.

For planning a study, although losses to follow-up are included in the analysis up to the time they leave the study, and therefore in theory reduce the variance of the statistics, it is recommended that the study be planned as if these patients contribute nothing. This is a bit conservative if losses are indeed uninformative.

Adjustment for Losses to Follow-Up

Let

N_u = Sample Size obtained in the Table (Unadjusted).

Based on the length of time needed to accrue N_u patients let

L = Fraction of patients expected to be lost to follow-up.

Then the adjusted sample size is

$$N_A = \frac{N_u}{1 - L}.$$

For example, if

$$N_u = 513$$

$$L = .1 \qquad\qquad \text{(10\% to be lost to follow-up)}$$

$$N_A = \frac{513}{1 - .1} = \frac{513}{.9} = 570$$

The recommended trial size is 570.

3.11   Interpretation of the Tables

The sample size tables are obtained from (9.7) for Poisson Accrual, (7.1) for Up Front Accrual, and A133 for the Kaplan-Meier (Binomial). A computer routine readily solves (9.7) and (7.1) by bisection.

The parameters for the Table are identified with (9.7) and (7.1) as follows (the term minimum follow-up is t, the planned endpoint).

(a)   Expected Accrual Through Minimum Follow-up $= At$

    $= $ (Accrual Rate)(Minimum Follow-up).

(b)   PCONT $= e^{-tQ}$ ($t = $ endpoint duration in days) (Q under H_A)

        $= $ Control Anticipated Survival Rate at time t, under H_A.

(c)   DEL $= e^{-tT} - e^{-tQ}$ (T, Q under H_A)

        $= $ Difference between anticipated survival rates at time t

(d)   ALPHA $= \alpha$. (If two-sided, replace α by $\alpha/2$.)

(e)   POWER $= 1 - \beta$.

(f)   FACT$= d$

        $= $ Hazard rate after t: Hazard rate before t.

3.12   Multi-Treatment Trials (3+ Treatments)

NOTE: This section reflects the majority position among statisticians, but is not universally accepted.

Generally, there are several types of multi-treatment trials, with different objectives. We shall discuss the four most common. It is rarely practical to use more than three or four treatments.

3.12.1 Type A

The treatments in this type of trial are generally as follows: (3 treatment case)

Control

Control + X

Control + Y

The experimental questions are:

(1)   Is X efficacious?

(2)   Is Y efficacious?

In this trial, assume that comparison of (Control + X) versus (Control + Y) is not a primary

objective.

This type of trial would be conducted to learn what component of therapy may have efficacy.

Hopefully, this experience will lead to a future trial, incorporating one or both of X and Y.

This type of trial should be designed exactly as a two treatment trial, except that the accrual

per comparison is only two thirds that of a two treatment study.

$$\text{ACCRUAL THROUGH minimum follow-up} = \text{Atx} \qquad (12.1)$$

A = Annual Accrual Rate.

t = Minimum Follow-up (Endpoint).

$x = \dfrac{2}{\text{Number of Treatments}}.$ (2/3 in our case)

The analysis would compare Control versus Control + X by a one-sided logrank test and

Control versus Control + Y by a one-sided logrank test.

Controversy: Some statisticians insist on controlling the overall error rate of the study, by adjusting the P-Value through a formal "multiple comparison" procedure. We believe that such adjustment is not appropriate for the stated objectives of the trial, however.

- Had you done the two trials sequentially

Control	versus	Control + X	(Trial 1)
Control	versus	Control + Y	(Trial 2)

you would not have made such an adjustment.

- Why should data on the Control + Y group influence the efficacy comparison involving X?

3.12.2 Type B

The treatments for this type of trial are generally as follows: (3 treatment case).

Control

Control + X

Control + X + Y

The experimental questions are

(1)   Is X efficacious?

(2)   Is Y efficacious over and above X?

Here, the only adjustment centers around the accrual rate, which is handled as in Type A.

The following numerical example might be helpful.

Accrual Rate:	150 patients per year
Planning two year survival:	40% (Control)
	55% (Control + X)
	65% (Control + X + Y)

FACT = .50 (Post two year hazard = 50% of Pre two-year hazard)

ALPHA = .05 (one-sided)

POWER = .80

MINIMUM FOLLOW-UP = 2 years

Effective Accrual Rate = 100 per comparison per year (200 through minimum

follow-up)

From the tables, assuming negligible losses to follow-up we have:

Sample Size: 231 (Control versus Control + X)

437 (Control + X versus Control + X + Y)

In practice, we would run the trial as follows:

- Three-way randomization until 231 patients on Control or Control + X.

- Two-way randomization thereafter, with no controls. The first question will only utilize the 231 Control versus Control + X patients accrued in the three-way randomization.

In terms of total accrual time, at 150 patients per year, the three-way randomization would take about 2.3 years to accrue 346 patients (231 to question 1). It would take about 1.4 years to accrue the remaining 206 patients to answer the second question.

Technical Note: The accrual to the second experimental question was not Poisson, but skewed slightly toward later accrual. Thus it might be instructive to look at the second question assuming a two-way randomization throughout. (Accrual of 300 patients through the minimum follow-up.) This would require a sample of 467 (rather than 437). The actual trial has a greater expected follow-up than this hypothetical trial. Purists can either extend accrual by 30 in the two-way randomization, or extend follow-up about three additional months on the actual setup to achieve the power objective.

3.12.3   Type C

(All Two-Way Comparisons) The experimental questions here are

(1)   Is there a difference between X and Y?

(2)   Is there a difference between X and Z?

(3)   Is there a difference between Y and Z?

If these are your questions and you are willing to treat these as three distinct trials, then the only adjustments that would be made would be

(a)   Replace ALPHA by ALPHA/2. (two-sided tests)

(b)   Effective accrual per comparison through minimum follow-up is two-thirds of the actual accrual through minimum follow-up.

However if you wish to find the best treatment, then multiple comparisons are needed. In fact, the concept of power changes. Our sample size tables cannot be used to handle this situation, but can give conservative approximations.

The following example gives conservative sample sizes.

Annual Accrual:	225 per year (150 per comparison).
Minimum Follow-up:	2 years.
Two Year Survival:	50% versus 50% versus 70%.
FACT:	1.0 Exponential.
ALPHA:	(.05)/3 = .0167 (two sided).
POWER:	90% (two specific comparisons with the best treatment).

You are going to infer that two treatments differ if the P-Value for comparing these treatments is less than .0167 (two-sided).

When there is in fact no difference between the treatments there is less than a 5% chance of falsely declaring a difference. (3 comparisons × .0167).

Your study will be successful if each comparison with the truly best treatment has a P-Value of below .0167 favoring the truly best treatment.

Since for each of the two comparisons with the truly best treatments there is a 10% chance of failing to declare superiority of the truly best treatment is at most a 20% chance of failing to declare the truly best treatment superior to one or both competitors. (At least 80% overall power to achieve your goal.)

Now the sample size is based on these parameters

$$Z_\alpha = 2.40 \qquad \text{(Figure 7)}$$

$(Z_\alpha = 2.40$ corresponds to two-sided P-Value $= .0167)$ and

$$Z_\beta = 1.28. \qquad \text{(Figure 6)}$$

Note that the sample size equation (9.7) depends on Z_α and Z_β only through their sum $Z_\alpha + Z_\beta$.

The following approximation greatly increases the use of the tables:

Non-Tabulated Z_α and Z_β

Actual $(Z_\alpha + Z_\beta) = U_A$

Table $(Z_\alpha + Z_\beta) = U_T$ $\qquad$ (<u>Fairly Close</u> to U_A)

Sample Size from Table $= N_T$

Recommended Sample Size $= N_A$

$$N_A = N_T \frac{U_A^2}{U_T^2} \qquad\qquad (12.2)$$

This approximation is evident in (7.1) and (9.7).

For our tables we have

ALPHA	POWER	BETA	$(Z_\alpha + Z_\beta) = U_T$
.005	.8	.20	3.42
.005	.9	.10	3.86
.01	.8	.20	3.17
.01	.9	.10	3.61
.025	.8	.20	2.80
.025	.9	.10	3.24
.05	.8	.20	2.48
.05	.9	.10	2.92

In the actual application,

$$U_A = 2.40 + 1.28 = 3.68.$$

This yields for ACCRUAL THROUGH MINIMUM FOLLOW-UP of 300, PCONT = .50, DEL = .20, FACT = 1.0, ALPHA = .01, and POWER = .9, a sample size of N_T = 252 per comparison (126 per treatment). From (12.2)

$$N_A = \frac{(3.68)^2 \ 252}{(3.61)^2} = 262.$$

That is, 131 patients per treatment are required for a total of 393.

3.12.4 Type D

(2×2 Factorial Design) This is designed as follows:

Control

Control + X

Control + Y

Control + X + Y

The major therapeutic questions are usually

(1)   Is X efficacious?

(2)   Is Y efficacious?

Generally speaking, if one is willing to assume that no "qualitative interaction" between X and Y exists, then you can collapse this to <u>two</u> two-treatment questions. That is, if you believe that if X is efficacious, it will be efficacious whether or not Y is used, you may pool the groups without X and the groups with X for the comparison, (assuming that the randomization is about 25% to each treatment). Most statisticians would "stratify" the analysis as discussed in Section 3.13.

X Comparison: (Control) and (Control + Y) versus (Control + X) and (Control + X + Y)

Y Comparison: (Control) and (Control + X) versus (Control + Y) and Control + X + Y)

In the absence of qualitative interaction, the Type D trial allows you to answer two questions for the price of one.

A test for Qualitative Interaction appears in Gail and Simon(1985).

3.13   Stratified Logrank Test

Often in clinical trials, investigators worry about imbalance in important prognostic groups. For example, it may be known that the extent of disease at baseline, as measured by lesion size, is highly predictive of outcome. Lesions below 5 cm in diameter have excellent prognosis, while lesions over 10 cm in diameter have a terrible prognosis.

The trial can be conducted in two ways: Prospective Stratification or Retrospective Stratification.

(A)   Prospective Stratification

Conduct the study as if three trials were going on. Each trial will assign about equal numbers of patients to each treatment. Populations for Trials 1, 2, and 3 are respectively.

Stratum 1:   Patients with lesions less than 5 cm in diameter.

Stratum 2:   Patients with lesions 5 - 10 cm in diameter.

Stratum 3:   Patients with lesions over 10 cm in diameter.

The analysis for comparing outcomes in a two treatment trial is as follows: (Illustration for three strata.)

Stratum	Observed Failures on Treatment #1	Expected Failures on Treatment #1	Variance
1	O_1	X_1	V_1
2	O_2	X_2	V_2
3	O_3	X_3	V_3
Total	O_T	X_T	V_T

$$Z = \frac{O_T - X_T}{\sqrt{V_T}}$$
(13.1)

The P-Value is obtained from the standard normal table (Figure 6 or Figure 7) depending upon whether the comparison is one-sided or two-sided.

In summary, add the observed failures on Treatment #1 (O_T), the expected failures on Treatment #1 (X_T), and the variances (V_T). Compute Z as per (13.1) and determine the approximate P-Value from Figure 6 or Figure 7.

NOTE: As long as the number of strata is fairly small relative to the number of patients, prospective stratification has minimal impact on sample size requirements.

The sample size tables of this volume should be viewed as slightly conservative for use in prospectively stratified trials. That is, the sample size calculation slightly overestimates the true needs, since prospective stratification creates greater risk factor homogeneity between the groups than could be expected by chance. However, if the trial was conducted as a randomized study without using prospective stratification, the stratum specific percent of patients assigned to each treatment will still approach 50% as the sample size increases to infinity. In other words, for large trials with few strata, the benefit of stratification is more psychological than real.

(B) Retrospective Stratification

Patients would be randomized as if the trial was not stratified, but then the results would be analyzed as if the trial was stratified. This is a perfectly valid procedure. In practice, with a large sample size and few retrospective strata, the sample size requirements for the post stratified analysis are slightly less than those obtained in the tables, and slightly more than the Prospectively Stratified Trial. The sample size tables are therefore conservative, but only slightly so.

Retrospective Stratification is very useful for exploratory analysis. For example, if you conduct a trial and discover that the two treatment groups have a serious imbalance in an important prognostic category, you may wish to rerun the analysis, adjusting for this imbalance via retrospective stratification. If this analysis qualitatively agrees with the planned analysis, then the final results are supported. However, in the rare case that they disagree, the analysis still provides important insight as to an explanation of the difference or lack of a difference between the treatments.

3.13.1 Advice on Stratification

1. If used at all, limit the number of strata to a small number. When you stratify, the efficacy comparison only compares patients within the same strata.

    The following is illustrative

    Stratum 1: 10 patients assigned to Treatment #1

    10 patients assigned to Treatment #2

    Stratum 2: 10 patients assigned to Treatment #1

    10 patients assigned to Treatment #2

    In the stratified trial, the outcome of each Treatment #1 patient is compared to only 10 Treatment #2 patients, while the unstratified trial has 20 Treatment #2 patients to compare with each Treatment #1 patient. If you overstratify, the lack of ability to compare more than offsets the homogeneity created.

2. If you have stratum specific therapeutic questions, you must calculate sample size needs for each stratum separately. This could lead to a very much larger trial, than one where stratification is used only as a balancing tool.

3. If limited stratification is used to balance the randomization, the sample size requirements will be well approximated by the tables of this volume, derived for non-stratified studies.

3.14 Intuitive Justification Why the Logrank Test and Kaplan-Meier Estimation for Actual Accrual Process Behave in the Limit in the Same Way as the Fixed Binomial Assumption

Consider the distribution on Day j given the history of the trial up to day j. In other words, if we had the trial results for all patients' day 1, day 2, ..., day j-1, we would know N_j and M_j the number of patients at risk at the start of day j. Given this information, F_j and G_j are binomial with respective sample sizes N_j and M_j and respective failure rates Q_j and T_j. This <u>conditional</u> distribution is identical to the unconditional distribution of the fixed binomial case.

$$E\left(\frac{F_j}{N_j} \mid \text{Past History of Trial}\right) = Q_j \qquad (N_j \neq 0)$$

$$\text{Var}\left(\frac{F_j}{N_j} \mid \text{Past History of Trial}\right) = \frac{Q_j P_j}{N_j} \qquad (N_j \neq 0)$$

As an estimate of Q_j, if we can ignore the possibility that $N_j = 0$, we can say, F_j/N_j is unbiased for Q_j and is uncorrelated with its past history.

Consider an alternate trial where we shall preselect N_j as

$$N_j^* = E(N_j) \qquad \text{(rounded to an integer)}$$

If $N_j \geq N_j^*$ we shall use the actual first N_j^* day j's of the trial.

If $N_j \leq N_j^*$ we shall add an additional $(N_j^* - N_j)$ day j's (independent trials with failure rate Q_j).

Although it is difficult to prove at the mathematical level of this text, it is intuitively clear that $N_j/N_j^* \xrightarrow{P} 1$ as $N_j^* \rightarrow \infty$ and hence the <u>fraction</u> of overlapping data coverages to 100% as the $N_j^* \rightarrow \infty$.

In other words, it can be shown that the large sample distributions under the actual process and artificial fixed binomial accrual are identical. The non-overlapping data is an insignificant percentage.

3.15   Alternate Standard Error for the Kaplan-Meier Estimator

$\big($Peto et al (1977)$\big)$

The following estimator for D_k in 10 of the Notation chart of Chapter 3 is often used to obtain the standard error of the Kaplan-Meier estimator, $\hat{S}_k$.

$$D_k^* = \hat{S}_k\left\{(1 - \hat{S}_k)/N_{k+1}\right\}^{\frac{1}{2}} \qquad\qquad (15.1)$$

This is not completely rigorous but has a number of desirable properties. Some intuitive justification will be shown below.

Advantages over Greenwood Formula (Notation Chart item 11).

(a)   Can be computed from the Kaplan-Meier estimate at time k and the number of patients at risk at time k+1. It does not require a summation.

(b)   It agrees well with the Greenwood Formula except in the right hand portion of the curve where the fraction censored is high.

(c)   The standard error increases in the plateau portion of a curve as patients are censored. The Greenwood Formula has a standard error that changes only at times where failures occur. This leads to unrealistically low estimates of standard error by Greenwood's Formula for situations where few patients are at risk.

Disadvantages over Greenwood's Formula

(d) $\dfrac{\hat{S}_k - S_k}{D_k^*}$ is not in theory asymptotically standard normal. It is close to standard normal where the sample size is large and the fraction censored before k is small. We shall see in Theorem 15.1 that the use of D_k^* is conservative ($\hat{D}_k \leq D_k^*$).

(e)   If one wishes to make statements such as "We are 95% confident that the three year survival is <u>less</u> than x", then the Greenwood standard error is superior, even in plateaus. This is because survival curves are monotonically decreasing functions. Hence, if you are 95% confident that the two year survival is below 83%, you are certainly 95% confident that the three year survival is below 83%, whether or not there is a large fraction of patients censored between two and three years. If the two to three year interval is a plateau, for example, the upper limit of the 95% confidence interval for the alternate standard error will be higher at three years than at two years, while that of Greenwood's formula will remain fixed.

Intuitive justification of the alternative standard error:

<u>Theorem 15.1</u>: $D_k^* \geq \hat{D}_k$

<u>Proof</u>: From item 9 of the Notation Chart at the beginning of Chapter 3,

$$1 - \hat{S}_k = 1 - \prod_{j=1}^{k} \hat{P}_j. \tag{15.2}$$

Now, by applying the lemma on converting a difference of products to a sum (Section 3.2) with

$$a_j = 1 \quad \text{and } b_j = \hat{P}_j,$$

we have

$$1 - \hat{S}_k = \hat{S}_k \sum \frac{(1-\hat{P}_j)}{\hat{S}_j}$$

$$= \hat{S}_k \sum \frac{\hat{Q}_j}{\hat{S}_{j-1} \hat{P}_j}. \tag{15.3}$$

Now,

$$\frac{\hat{S}_k}{\hat{S}_{j-1}} = \frac{\hat{P}_1 \cdots \hat{P}_k}{\hat{P}_1 \cdots \hat{P}_{j-1}} = \hat{P}_j \cdots \hat{P}_k. \tag{15.4}$$

Now from the actual accrual process,

$$N_{j+1} = N_j - F_j - W_j$$

where

N_j = Number of Patients at risk at start of day j,

F_j = Failures on Day j,

W_j= Competing "withdrawals" on Day j.

The W_j include patients who are alive with time of risk j, as well as those who quit alive on day j.

$$\hat{P}_j = \frac{N_j - F_j}{N_j} \geq \frac{N_j - F_j - W_j}{N_j} = \frac{N_{j+1}}{N_j}. \tag{15.5}$$

As long as the W_j are small percentages of N_j, approximate equality holds.

By substituting (15.5) into (15.4) we have

$$\frac{\hat{S}_k}{\hat{S}_{j-1}} \geq \prod_{i=j}^{k} \frac{N_{i+1}}{N_i} = \frac{N_{k+1}}{N_j}. \tag{15.6}$$

Substituting (15.6) in (15.3) yields

$$1 - \hat{S}_k \geq N_{k+1} \sum_j \frac{\hat{Q}_j}{N_j \hat{P}_j}. \tag{15.7}$$

Hence

$$\frac{(1 - \hat{S}_k)}{N_{k+1}} \geq \sum_j \frac{\hat{Q}_j}{N_j \hat{P}_j}. \tag{15.8}$$

That is,

$$D_k^* = \hat{S}_k \left\{ \frac{(1 - \hat{S}_k)}{N_{k+1}} \right\}^{\frac{1}{2}}$$

$$\geq \hat{D}_k = \hat{S}_k \left\{ \sum \frac{\hat{Q}_j}{N_j \hat{P}_j} \right\}^{\frac{1}{2}} \qquad (15.9)$$

and approximate equality holds as long as the withdrawal rates are small fractions of the N_j.

Summary:

The use of D_k^* instead of $\hat{D}_k$ is <u>conservative</u>. The formula will underestimate the precision of the estimator $\hat{S}_k$, but only slightly if the fraction of competing withdrawals is small.

3.16   Connection Between Kaplan-Meier and Binomial

Note that if there are no censored patients to day k, under <u>actual accrual</u>, $N_{j+1} = N_j - F_j$ and

$$\hat{P}_j = \frac{N_{j+1}}{N_j}.$$

This yields from the Notation Chart:

$$\hat{S}_k = \prod_{j=1}^{k} \frac{N_{j+1}}{N_j}$$

$$= \frac{N_2}{N_1} \frac{N_3}{N_2} \cdots \frac{N_{k+1}}{N_k} = \frac{N_{k+1}}{N_1}. \qquad (16.1)$$

To evaluate $\hat{D}_k^2$, note hat

$$\hat{Q}_j = 1 - \hat{P}_j = \frac{N_j - N_{j+1}}{N_j}$$

and hence

$$\frac{\hat{Q}_j}{\hat{P}_j N_j} = \frac{N_j - N_{j+1}}{N_{j+1} N_j} = \frac{1}{N_{j+1}} - \frac{1}{N_j}$$

$$\sum_{j=1}^{k} \frac{\hat{Q}_j}{\hat{P}_j N_j} = \sum_{j=1}^{k} \left(\frac{1}{N_{j+1}} - \frac{1}{N_j} \right)$$

$$= \left(\frac{1}{N_2} - \frac{1}{N_1} \right) + \left(\frac{1}{N_3} - \frac{1}{N_2} \right) + \cdots + \left(\frac{1}{N_{k+1}} - \frac{1}{N_k} \right)$$

$$= \frac{1}{N_{k+1}} - \frac{1}{N_1} = \frac{N_1 - N_{k+1}}{N_1 N_{k+1}}$$

Hence

$$\hat{D}_k^2 = \hat{S}_k^2 \sum_{j=1}^{k} \frac{\hat{Q}_j}{\hat{P}_j N_j}$$

$$= \frac{N_{k+1}^2}{N_1^2} \frac{(N_1 - N_{k+1})}{N_1 N_{k+1}}$$

$$= \frac{N_{k+1}(N_1 - N_{k+1})}{N_1^3}$$

$$= \frac{\hat{S}_k (1 - \hat{S}_k)}{N_1} \qquad (16.2)$$

$\hat{D}_k^2$ is the binomial estimate of variance when there are no competing losses.

Studies based on the Kaplan-Meier are planned using the Tables with Binomial (BIN) value for FACT. Adjustments can be made for anticipated losses to follow-up. (See A133 for details).

REFERENCES

1. **Crowley, J.**, Asymptotic normality of a new nonparametric statistic for use in organ transplant studies, *J. Am. Statist. Assoc.*, 69, 1006, 1974.

2. **Breslow, N. E. and Crowley, J.**, A large sample study of the life table and product limit estimates under random censorship, *Ann. Statist.* 2, 437, 1974.

3. **Gail, M. and Simon, R.**, Testing for qualitative interactions between treatment effects and patient subsets, *Biometrics*, 41, 361, 1985.

4. **Greenwood, M.,** *The Errors of Sampling of the Survivorship Tables, Reports on Public Health and Statistical Subjects,* Her Majesty's Stationery Office, London, 1926, 33.

5. **Kaplan, E. L. and Meier, P.,** Nonparametric estimation from incomplete observations, *J. Am. Statist. Assoc.,* 53, 447, 1958.

6. **Peto, R., Pike, M. C., Armitage, P., Breslow, N. E., Cox, D. R., Howard, S. V., Mantel, N., McPherson, K., Peto, J., and Smith, P. G.,** Design and analysis of randomized clinical trials requiring prolonged observation of each patient, II. Analysis and example, *Br. J. Canc.,* 35, 1, 1977.

7. **Rubinstein, L. V., Gail, M. H., and Santner, T. J.,** Planning the duration of a comparitive clinical trial with loss to follow-up and a period of continued observation, *J. Chron. Dis.,* 34, 469, 1981.

APPENDIX I

A REVIEW OF MATHEMATICAL STATISTICS

In order to follow the material in Chapter 3, it is assumed that the reader has a working knowledge of undergraduate mathematical statistics at a level of one of the following texts: Mendenhall, Scheaffer, and Wackerly (1986), Fraser (1976), Mood, Graybill, and Boes (1974), or Hogg and Craig (1978).

In order that this book be self contained, this appendix will list definitions and results. For convenient reference each is assigned a number.

A.1 Sample Space(s) (Definition): The set of all possible outcomes of an experiment (study or observation process). This is denoted by S.

A.2 Event (Definition): A subset of the Sample Space.

A.3 Union of Events: $E_1 \cup E_2$ (Definition) An event consisting of outcomes in one or both of E_1, E_2. (Unites the two events into a larger event).

This definition extends to multi-events $E_1 \cup E_2 \cup E_3$ etc.

A.4 Intersection of Events: $E_1 E_2$ (Definition) An event consisting of outcomes common to E_1 and E_2. To qualify for membership in $E_1 E_2$, the outcome must be in both event E_1 and event E_2.

This definition extends to multi-events $E_1 E_2 E_3$ etc.

A.5 Complement of Event: ($\bar{E}$) (Definition) The complement of the event E, denoted by $\bar{E}$, consists of all outcomes in the sample space that are not in the event E. (E and $\bar{E}$ have no outcomes in common, and together make up the sample space $E \cup \bar{E} = S$).

A.6 Null Event: ($\emptyset$) (Definition) The Event which contains no outcomes in the sample space is called the null Event ($\emptyset = \bar{S}$).

A.7 Mutually Exclusive Events (Definition): The events E_1 and E_2 are mutually exclusive if they have no outcomes in common. $E_1 E_2 = \emptyset$.

This definition extends to multi-events E_1, E_2, ..., E_n as $E_i E_j = \emptyset$ for every i,j $1 \le i < j \le n$.

A.8 Probability (Definition) P is called a probability if

    (a)   P assigns a number $P(E)$ to every event in S with $0 \leq P(E) \leq 1$.

    (b)   If $E_1, E_2, \ldots, E_n$ are mutually exclusive events $P(E_1 \cup E_2 \cup \cdots \cup E_n) =$

           $P(E_1) + P(E_2) + \cdots + P(E_n)$

    (c) $P(S) = 1$ where S = Sample Space

Probability Results (Consequences)

A.9 (a) $P(\emptyset) = 0$

A.10 (b) $P(\bar{E}) = 1 - P(E)$

A.11 (c) $P(E_1 \cup E_2) = P(E_1) + P(E_2) - P(E_1 E_2)$

A.12 Conditional Probability $\left(P(E_1|E_2) \right)$ (Definition)

    The probability that E_1 occurs <u>given</u> that E_2 has occurred is denoted by

$$P(E_1|E_2) = \frac{P(E_1 E_2)}{P(E_2)} \qquad \left(P(E_2) > 0 \right)$$

A.13 Independence of two Events (Definition)

    Events E_1 and E_2 are independent if $P(E_1 E_2) = P(E_1)P(E_2)$.

A.14 Independence of two events (Consequence)

    Provided that $P(E_2) > 0$, then events E_1 and E_2 are independent if and only if

$$P(E_1|E_2) = P(E_1)$$

i.e. conditional and unconditional probability agree.

The proof is straight forward and is therefore omitted.

A.15 Mutual Independence of Several Events (Definition)

    Events E_1, E_2, ..., E_n are mutually independent if for every subset $\left\{ j_1, j_2, \ldots, j_k \right\}$ of the integers $\{1, 2, \ldots, n\}$

$$P(E_{j_1} E_{j_2} \ldots E_{j_k}) = P(E_{j_1}) P(E_{j_2}) \cdots P(E_{j_k})$$

A.16 <u>Random Variable</u> (Definition)

A random variable assigns a number to each outcome in the sample space S. That is, if X is a random variable, then for each sample point s in S, X assigns a number X(s).

A.17 <u>Random Vector</u> (Vector of Random Variables) (Definition)

A random vector, Y, of dimension k is a k-dimensional vector of random variables

$$(X_1, X_2, \ldots, X_k) = Y$$

A.18 <u>Cumulative Distribution Function of a Random Variable</u> (cdf) (Definition)

The cumulative distribution function of the random variable X, denoted by $F_X(x)$, is

$$F_X(x) = P[X \leq x] \quad -\infty < x < \infty$$

The cumulative distribution function is a plot of the probability that the random variable X takes on numerical value less than or equal to the number x. This is plotted over all real numbers, x.

A.19 <u>Discrete Random Variables</u> and <u>Probability Mass Function</u> (Definition)

A random variable X is "discrete" if the cumulative distribution is of the form

$$F_X(x) = \sum_{y:\, y \leq x} f_X(y)$$

y: y≤x refers to a <u>finite</u> set of <u>possible</u> values of the random variable X that assign values x or less and $F_X(x) \rightarrow 1$ as $x \rightarrow \infty$.

The function $f_X(y) = P[X=y]$ is called the <u>probability mass function</u>. This mass function will be zero except at a discrete set of values. (*This is not the most general definition, but it will meet our needs.*)

A.20 <u>Continuous Random Variables</u> and <u>Probability Density Function</u> (Definition)

A random variable X is "continuous" if the cumulative distribution is of the form

$$F_X(x) = \int_{-\infty}^{x} f_X(y)dy$$

with $F_X(x) \rightarrow 1$ as $x \rightarrow \infty$.

For the purpose of this volume, we shall assume that $f_X(y)$ has at most a finite number of discontinuities.

$f_X(y)$ is called the <u>probability density function</u>.

A.21 <u>Relation between Cumulative Distribution and Probability Density Function</u> (Consequence)

At each continuity point of $F_X(x)$,

$$f_X(x) = F'_X(x)$$

That is, the probability density is the derivative of the cumulative distribution function.

Proof:
$$F_X(x+h) - F_X(x) = \int_x^{x+h} f_X(y)dy$$

$$= hf_X(\xi) \quad \text{for some } \xi: \; x \leq \xi \leq x+h$$

The above follows from the Mean Value Theorem of Calculus.

If we divide both sides by h and let $h \to 0$, the result follows.

A.22 <u>Cumulative Distribution Function for Random Vectors</u> (Definition)

We shall write this in terms of two random vectors, but the extension to several dimensions is clear.

A random vector $X = (X_1, X_2)$ has cumulative distribution denoted by F_X:

$$F_X(x_1, x_2) = P(X_1 \leq x_1, X_2 \leq x_2)$$

A.23 <u>Multivariate Analogues of Mass Function and Density</u> (Definition)

(a) X_1, X_2 discrete

$$f_X(x_1, x_2) = P\left[X_1 = x_1, X_2 = x_2\right]$$
$$F_X(x_1, x_2) = \sum_{\substack{y_1 \leq x_1 \\ y_2 \leq x_2}} f_X(y_1, y_2)$$

$f_X(x_1, x_2)$ is the <u>joint probability mass function</u> of X_1, X_2

(b) X_1, X_2 continuous

$$F_X(x_1, x_2) = \int_{-\infty}^{x_1} \int_{-\infty}^{x_2} f_X(y_1, y_2) dy_2 dy_1$$

$f_X(x_1, x_2)$ is the <u>joint probability density</u> of X_1, X_2

Note: at continuity points of (x_1, x_2),

$$f_X(x_1, x_2) = \frac{\partial^2 F_X(x_1, x_2)}{\partial x_1 \partial x_2}$$

(c) X_1 discrete, X_2 continuous

$$F_X(x_1,x_2) = \sum_{y_1 \leq x_1} \int_{-\infty}^{x_2} f_X(y_1,y_2)dy_2$$

$f_X(x_1,x_2)$ is the joint probability mass-density of X_1, X_2

A.24 Marginal Distributions (Definition and Consequences)

Here, we also study two variables X_1, X_2 but the extension to several variables is clear.

$$P[X_1 \leq x_1] = \lim_{x_2 \to \infty} P[X_1 \leq x_1, X_2 \leq x_2]$$

$$= \lim_{x_2 \to \infty} F_X(x_1,x_2)$$

We denote this by the Marginal Cumulative Distribution of X_1:

$$F_{X_1}(x_1) = \lim_{x_2 \to \infty} F_X(x_1,x_2)$$

$$= \int_{-\infty}^{x_1} \left\{ \int_{-\infty}^{\infty} f_X(y_1,y_2)dy_2 \right\} dy_1 \quad X_1,X_2 \text{ continuous}$$

$$= \sum_{y_1 \leq x_1} \left\{ \sum_{\text{all } y_2} f_X(y_1,y_2) \right\} \quad X_1,X_2 \text{ discrete}$$

$$= \sum_{y_1 \leq x_1} \left\{ \int_{-\infty}^{\infty} f_X(y_1, y_2)dy_2 \right\} \quad X_1 \text{ discrete, } X_2 \text{ continuous}$$

$$= \int_{-\infty}^{x_1} \left\{ \sum_{\text{all } y_2} f_X(y_1, y_2) \right\} dy_1 \quad X_1 \text{ continuous, } X_2 \text{ discrete}$$

A25. If X_2 is discrete,

$$f_{X_1}(x_1) = \sum_{\text{all } y_2} f_X(x_1, y_2) \quad \text{is}$$

called the Marginal density of X_1 if X_1 is continuous, or Marginal Mass function of X_1 if X_1 is discrete.

A26. If X_2 is continuous

$$f_{X_1}(x_1) = \int_{-\infty}^{\infty} f_X(x_1, y_2) dy_2 \quad \text{is}$$

called the <u>Marginal density of X_1</u> if X_1 is <u>continuous</u>, or <u>Marginal Mass function</u> of X_1 if X_1 is <u>discrete</u>.

A27. <u>Conditional Distributions</u> (Definition and Consequences).

The <u>Conditional Mass function</u> (X_2 discrete) or <u>conditional density</u> (X_2 continuous) of X_2 given $X_1 = x_1$ is defined by

$$f_{X_2|X_1}(x_2|x_1) = \frac{f_X(x_1, x_2)}{f_{X_1}(x_1)}$$

If X_1 is continuous, the above definition only applies to continuity points of $f_{X_1}(x_1)$ and $f_X(x_1, x_2)$.

Notes: (a)   If X_1 and X_2 are discrete, then by the definitions of conditional probability and marginal probability mass function (See A12 and A25)

$$f_{X_2|X_1}(x_2|x_1) = P\left[X_2 = x_2 | X_1 = x_1\right]$$

(b)   If X_1 and X_2 are continuous, (x_1, x_2) is a continuity point of f_X, and Δ_1 and Δ_2 are small:

$$P\left[x_1 < X_1 < x_1 + \Delta_1, x_2 < X_2 < x_2 + \Delta_2\right] \simeq \Delta_1 \Delta_2 f_X(x_1, x_2)$$

$$P\left[x_1 < X_1 < x_1 + \Delta_1\right] \simeq \Delta_1 f_{X_1}(x_1)$$

$$P\left[x_2 < X_2 < x_2 + \Delta_2 | x_1 < X_1 < x_1 + \Delta_1\right] \simeq \frac{\Delta_2 f_X(x_1, x_2)}{f_{X_1}(x_1)}$$

That is, in the limit, as $\Delta_1 \to 0$ and $\Delta_2 \to 0$,

$$f_{X_1|X_2}(x_2|x_1) = \frac{f_X(x_1, x_2)}{f_{X_1}(x_1)}$$

behaves as a density, reflecting the appropriate conditional probability.

(c)   Similar analogues can be developed for mixed continuous-discrete random variables.

A28. <u>Conditioning on an Event</u> (consequence)

In the case of a continuous $X = (X_1, X_2)$ one may wish to find the conditional distribution of

X_2 given $X_1 < x_1$ rather than specifying an exact value of x_1.

$$P\left(X_2 < x_2 | X_1 < x_1\right) \quad = \frac{P\left[X_1 < x_1, X_2 < x_2\right]}{P\left[X_1 < x_1\right]}$$

$$= \frac{F_{X_1,X_2}(x_1, x_2)}{F_{X_1}(x_1)}$$

$$P\left[X_2 < x_2 | X_1 < x_1\right] \quad = \frac{\int_{-\infty}^{x_2}\left\{\int_{-\infty}^{x_1} f_X(y_1, y_2)dy_1\right\}dy_2}{F_{X_1}(x_1)}$$

This is the conditional cumulative distribution of X_2 given that $X_1 < x_1$.

Differentiating both sides with respect to x_2 yields the conditional density at continuity points:

$$f_{X_2|X_1}\left(x_2 | X_1 < x_1\right) = \frac{\int_{-\infty}^{x_1} f_X(y_1, x_2)dy_1}{F_{X_1}(x_1)}$$

A29. Independence of Random Variables (Definition)

Random variables X_1 and X_2 are independent if for every x_1, x_2

$$F_X(x_1, x_2) = F_{X_1}(x_1)F_{X_2}(x_2)$$

That is, for every x_1, x_2 the events

E_1 that $X_1 \leq x_1$ and

E_2 that $X_2 \leq x_2$ are independent.

A30. Independence of Random Variables (Consequence)

Independence is equivalent to

$$f_X(x_1, x_2) = f_{X_1}(x_1)f_{X_2}(x_2)$$

with f the mass or density function depending on whether the random variable is discrete or continuous.

The proofs are straight forward.

A31. Independence of Several Random Variables (Definition and Consequence)

Either of these definitions imply independence

$$F_X(x_1, x_2, ..., x_k) = P\big[X_1 \leq x_1, X_2 \leq x_2, ..., X_k \leq x_k\big]$$
$$= F_{X_1}(x_1)\, F_{X_2}(x_2)\, ...\, F_{X_k}(x_k) \quad \text{for all } (x_1, ..., x_k)$$

or

$$f_X(x_1, ..., x_k) = f_{X_1}(x_1)\, f_{X_2}(x_2)\, ...\, f_{X_k}(x_k) \quad \text{for all } (x_1, ..., x_k)$$

A32. <u>Random Samples</u> (Independently Identically Distributed <u>Random Variables</u>)

(<u>IID random variables</u>) (Definition)

Random variables X_1, X_2, ..., X_k are independent identically distributed if they are independent and each X_j has the same cumulative distribution:

$$F_{X_j}(x) = F(x) \quad \text{Same for all j.}$$

$$F_X(x_1, ..., x_k) = F(x_1) \, F(x_2) \, ... \, F(x_k)$$

or sufficiently

$$f_X(x_1, ..., x_k) = f(x_1) \, f(x_2) \, ... \, f(x_k).$$

EXPECTED VALUE OF A FUNCTION OF A RANDOM VECTOR
(Definition)

A33. (a)   Univariate

Let g(X) be a function of the random variable X. g(X) has <u>expectation</u> denoted by $E\big(g(X)\big)$

$$E\big(g(X)\big) \quad = \int_{-\infty}^{\infty} g(x)f_X(x)dx \qquad \text{(Continuous Case)}$$

$$= \sum_{\text{all x}} g(x)f_X(x) \qquad \text{(Discrete case)}$$

f_X is the probability density for continuous X or the probability mass function for discrete X.

A34. (b)   Multivariate

Let g(X) be a function of the random vector $X = (X_1, X_2)$. g(X) has expectation

$$E\big(g(X)\big) = \int_{-\infty}^{\infty}\int_{-\infty}^{\infty} g(x_1, x_2)f_X(x_1, x_2)dx_2dx_1 \quad \text{(Continuous case)}$$

with $f_X(x_1, x_2)$ the joint density of X_1. X_2.

Discrete components will involve sum instead of integral.

SPECIAL UNIVARIATE FUNCTIONS (Definitions)

We shall only write these for the continuous case, but they apply in similar fashion to the discrete case too.

A35. (a) $g(X) = X$ (The <u>population mean</u> of X)

We generally denote the mean by μ

$$\mu = E(X) = \int_{-\infty}^{\infty} xf_X(x)dx$$

A36. (b) $g(X) = (X-\mu)^2$ (The <u>population variance</u> of X)

We generally denote the population variance, the average squared deviation about the mean as σ^2

$$\begin{aligned}
Var(X) \quad &= E\big[(X-\mu)^2\big] \\
&= \int_{-\infty}^{\infty} (x-\mu)^2 f_X(x)dx
\end{aligned}$$

where $\mu = E(X)$

A37. (c) $g(X) = e^{tX}$ (The <u>moment generating function</u> of X)

We generally denote this by $M_X(t)$

$$M_X(t) = \int_{-\infty}^{\infty} e^{tx} f_X(x)dx$$

A38. SPECIAL MULTIVARIATE FUNCTION

$$g(X) = (X_1 - \mu_1)(X_2 - \mu_2) \qquad \text{(Population Covariance)}$$

where $\mu_1 = E(X_1) \quad \mu_2 = E(X_2)$

$$\text{Cov}(X_1, X_2) = \int_{-\infty}^{\infty}\int_{-\infty}^{\infty} (x_1 - \mu_1)(x_2 - \mu_2) f_X(x_1, x_2)dx_2\, dx_1$$

RESULTS ON EXPECTATIONS (Consequences)

A39. (a) Expectation of a constant is a constant.

If $g(X) = C$ for all X,

$E\big(g(X)\big) = C$

A40. (b) If C is a constant,

$E\big(Cg(X)\big) = CE\big(g(X)\big)$

A41. (c) Additivity of Expectation

$E\big\{g_1(X) + g_2(X)\big\} = E\big\{g_1(X)\big\} + E\big\{g_2(X)\big\}$

A42. (d) $E\big\{ \sum_{j=1}^{k} a_j g_j(X)\big\} = \Sigma a_j E\big\{g_j(X)\big\}$

A43. (e) Shortcut formula for Variance

$\sigma^2 = E\big((X-\mu)^2\big) = E(X^2) - \mu^2$

where $\mu = E(X)$

A44. (f)   Shortcut formula for covariance

$$\text{Cov}(X_1, X_2) = E\{(X_1 - \mu_1)(X_2 - \mu_2)\} \qquad \text{(Definition)}$$

$$= E(X_1 X_2) - \mu_1 \mu_2 \qquad \text{(Consequence)}$$

Proof (f)

$$(X_1 - \mu_1)(X_2 - \mu_2) = X_1 X_2 - \mu_1 X_2 - \mu_2 X_1 + \mu_1 \mu_2$$

Take expectation on both sides and apply (d) and (a) above.

A45. (g)   Linear Combinations (Consequence)

Let $Y = \sum_{j=1}^{k} a_j X_j$

Then $E(Y) = \sum_{j=1}^{k} a_j \mu_j$ where $\mu_j = E(X_j)$

and $\text{Var}(Y) = \sum_{j=1}^{k} a_j^2 \text{Var}(X_j) + 2 \sum_{j=2}^{k} \sum_{i=1}^{j-1} a_i a_j \text{Cov}(X_i, X_j)$

Proof (g):   The expectation of Y is obtained from (d) above.

$$\text{Var}(Y) = E\{Y - E(Y)\}^2 = E\left(\left\{\sum_{j=1}^{k} a_j (X_j - \mu_j)\right\}^2\right)$$

$$= \text{Desired Result by (d) above.}$$

A46. (h)   If X_1, X_2 are independent, then

$$\text{Cov}(X_1, X_2) = 0$$

Proof (h):

$$E(X_1 X_2) = \int_{-\infty}^{\infty} \int_{-\infty}^{\infty} x_1 x_2 f_X(x_1 x_2) dx_2 dx_1$$

$$= \int_{-\infty}^{\infty} \int_{-\infty}^{\infty} x_1 x_2 f_{X_1}(x_1) f_{X_2}(x_2) dx_2 dx_1 \qquad \text{(independence)}$$

$$= \int_{-\infty}^{\infty} x_1 f_{X_1}(x_1) dx_1 \int_{-\infty}^{\infty} x_2 f_{X_2}(x_2) dx_2$$

$$= \mu_1 \mu_2$$

From (f) above, $\text{Cov}(X_1, X_2) = 0$

(i) Expectation and Variance for classical statistics for independent identically distributed random variables.

Let $X_1, X_2, \ldots, X_n$ be independent identically distributed random variables with

$$E(X_j) = \mu \quad \text{and} \quad \text{Var}(X_j) = \sigma^2$$

Let $\bar{X} = \dfrac{\sum\limits_{j=1}^{n} X_j}{n}$ (Sample Mean)

$S^2 = \dfrac{\sum\limits_{j=1}^{n} (X_j - \bar{X})^2}{(n-1)}$ (Sample Variance)

A47. Then $E(\bar{X}) = \mu$

A48. $\text{Var}(\bar{X}) = \dfrac{\sigma^2}{n}$

A49. $E(S^2) = \sigma^2$

Proof (i): To prove the results for $\bar{X}$, we apply (g) above with $a_j = (\frac{1}{n})$. By independence $\text{Cov}(X_i, X_j) = 0$. This yields $E(\bar{X}) = \mu$ and $\text{Var}(\bar{X}) = \dfrac{\sigma^2}{n}$

$$S^2 = \dfrac{\left\{ \Sigma X_j^2 - n\bar{X}^2 \right\}}{(n-1)} \qquad (*)$$

From the shortcut formula

$$\sigma^2 = E(X_j^2) - \mu^2 \text{ implies } E(X_j^2) = \sigma^2 + \mu^2$$

and $\text{Var}(\bar{X}) = E(\bar{X})^2 - \left\{ E(\bar{X}) \right\}^2$

i.e. $\dfrac{\sigma^2}{n} = E(\bar{X})^2 - \mu^2.$

Hence $E(\bar{X})^2 = \mu^2 + \dfrac{\sigma^2}{n}.$

Taking expectation in (∗) yields:

$$E(S^2) = \frac{\left\{n\sigma^2 + n\mu^2 - n\mu^2 - \sigma^2\right\}}{(n-1)} = \sigma^2.$$

A50. (j) $\left|\text{Cov}\left(X_1, X_2\right)\right| \leq \left\{\text{Var}(X_1)\,\text{Var}(X_2)\right\}^{\frac{1}{2}}.$

Proof (j): $\text{Var}\left(X_1 + tX_2\right) = \text{Var}(X_1) + 2t\,\text{Cov}\left(X_1, X_2\right) + t^2\text{Var}(X_2).$

Since Var ( ) ≥ 0, we obtain the result by successively letting $t = \pm \sqrt{\dfrac{\text{Var}(X_1)}{\text{Var}(X_2)}}.$

Note that this implies

$$-1 \leq \rho = \frac{\text{Cov}\left(X_1, X_2\right)}{\left\{\text{Var}(X_1)\,\text{Var}(X_2)\right\}^{\frac{1}{2}}} \leq 1.$$

A51. (k)   Chebyshev Theorem (Consequence)

Let X be a random variable with mean μ and variance σ^2. Then for any $\epsilon > 0$

$$P\left[|X-\mu| \geq \epsilon\right] \leq \frac{\sigma^2}{\epsilon^2}.$$

Proof: We shall prove the result in the continuous case

$$\sigma^2 \quad = \int_{-\infty}^{\infty} (x-\mu)^2 f_X(x)dx$$

$$\geq \int_{|x-\mu|\geq\epsilon} (x-\mu)^2 f_X(x)dx$$

$$\geq \int_{|x-\mu|\geq\epsilon} \epsilon^2 f_X(x)dx$$

$$\geq \epsilon^2 P\left[|X-\mu|\geq\epsilon\right].$$

Divide both sides by ϵ^2 to obtain the result.

A52. (l)   "Consistency of $\bar{X}$" (Consequence)

If X_1, X_2, ... are independent, identically distributed random variables with mean μ and variance σ^2, then for any $\epsilon > 0$

$$P\left[|\bar{X}_n - \mu| > \epsilon\right] \to 0 \quad \text{as } n \to \infty$$

where $\bar{X}_n = \dfrac{\left(X_1 + \cdots + X_n\right)}{n}$.

In lay terms, this is the "Law of Averages".

Proof: From A47 and A48 above,

$$E(\bar{X}) = \mu \quad \text{and} \quad Var(\bar{X}) = \frac{\sigma^2}{n}.$$

By Chebyshev's Theorem (k)

$$P\left[|\bar{X} - \mu| > \epsilon\right] \leq \frac{Var(\bar{X})}{\epsilon^2} = \frac{\sigma^2}{n\epsilon^2} \to 0 \quad \text{as } n \to \infty.$$

MOMENT GENERATING FUNCTIONS (Consequences)

$$M_X(t) = E\left(e^{tX}\right)$$

A53. (a)   If two cumulative distributions differ, then they have different moment generating functions. (The moment generating function uniquely identifies the distribution. In fact, we only need to identify the moment generating function in an open interval about t=0.)

See Fraser (1976, pages 544-546).

A54. (b)   Provided the moment generating function is defined in a neighborhood of $t = 0$, then

$$E(X^n) = \lim_{t \to 0} M_X^{(n)}(t)$$

(The nth derivative of $M_X(t)$ with respect to t is taken, and then $t \to 0$.)

See Fraser (1976, page 233).

A55. (c) $M_X(t) = 1 + \sum\limits_{j=1}^{\infty} \dfrac{E(X^j)t^j}{j!}$

See Fraser (1976, page 233).

A56. (d)   If X_1, X_2, ..., X_n are independent random variables with respective moment generating

functions

$$M_{X_j}(t) \quad j = 1, ...,n$$

and

$$Y = \sum_{j=1}^{n} a_j X_j$$

with a_j's constant. Then Y has moment generating function

$$M_Y(t) = \prod_{j=1}^{n} M_{X_j}(a_j t)$$

Proof (d):

$$e^{tY} \qquad = e^{t\Sigma a_j X_j} = \prod_{j=1}^{n} e^{t a_j X_j}$$

$$E(e^{tY}) \qquad = \prod_{j=1}^{n} E\left(e^{t a_j X_j}\right) \qquad \text{by independence}$$

$$= \prod_{j=1}^{n} M_{X_j}(a_j t) \qquad \text{by definition.}$$

SPECIAL DISTRIBUTIONS

BINOMIAL DISTRIBUTION (Discrete)

A57. (a)   Probability Mass Function of the binomial distribution

$$f(x) = \binom{n}{x} p^x (1-p)^{n-x} \quad x = 0, 1, ..., n \quad 0 < p < 1$$

where

$$\binom{n}{x} = \frac{n!}{x!(n-x)!}$$

$$a! = a(a-1)...1$$

$$0! = 1$$

A58. (b)   Mean and Variance

$$\mu = np$$

$$\sigma^2 = np(1-p)$$

A59. (c)   Moment Generating Function

$$M_X(t) = \left\{pe^t + (1-p)\right\}^n$$

A60. (d)   How it arises in the real world.

If X is the number of observed failures, in a sample of n <u>independent</u> trials, each with probability of failure, p, then X has the Binomial Probability Mass Function above.

POISSON DISTRIBUTION (Discrete)

A61. (a)   Probability Mass Function of the Poisson distribution.

$$f(x) = \frac{\lambda^x e^{-\lambda}}{x!} \quad x = 0,1,2,\ldots \quad \lambda > 0$$

A62. (b)   Mean and Variance

$$\mu = \lambda$$

$$\sigma^2 = \lambda$$

A63. (c)   Moment Generating Function

$$M_X(t) = \exp\left\{\lambda(e^t - 1)\right\}$$

A64. (d)   How it arises in the real world.

Approximation to Binomial

If in the binomial, n is large and p is small with $np = \lambda$, then the Poisson distribution will provide an excellent approximation to the binomial distribution.

More formally, if $n \to \infty$ and $p \to 0$ with $np \to \lambda > 0$

$$\binom{n}{x} p^x (1-p)^{n-x} \rightarrow \frac{\lambda^x e^{-\lambda}}{x!} \quad x = 0,1,2,\ldots$$

UNIFORM DISTRIBUTION (Continuous)

A65. (a)   Probability Density Function of the uniform distribution.

$$f(x) = \frac{1}{B-A} \quad A < x < B$$

A66. (b)   Mean and Variance

$$\mu = (A+B)/2$$
$$\sigma^2 = (B-A)^2/12$$

A67. (c)   Moment Generating Function

$$M_x(t) = \frac{1}{\{(B-A)t\}} \left\{ e^{Bt} - e^{At} \right\}$$

A68. (d)   How it arises in the real world

Roundoff error

A69. (e)   Cumulative Distribution

$$F(x) = \frac{x-A}{B-A} \quad A < x < B$$

EXPONENTIAL DISTRIBUTION (Continuous)

A70. (a)   Probability Density Function of the exponential distribution

$$f(x) = \lambda e^{-\lambda x} \quad x > 0$$

A71. (b)   Mean and Variance

$$\mu = (1/\lambda)$$
$$\sigma^2 = (1/\lambda)^2$$

A72. (c)   Moment Generating Function

$$M_x(t) = \frac{\lambda}{\lambda - t} \quad t < \lambda$$

A73. (d) How it arises in the real world

- As survival time data - see (f) below

A74. (e) Cumulative Distribution

$$F(x) = 1 - e^{-\lambda x} \qquad x > 0$$

A75. (f) Memoriless Property (Constant Hazard)

Let $s > 0$ and $t > 0$ and let X have the exponential distribution.

$$P[X > t] = 1 - F(t) = e^{-\lambda t}$$

$$P[X > t+s | X > s] = \frac{P[X > t+s, \; X > s]}{P[X > s]}$$

$$= \frac{P[X > t+s]}{P[X > s]} = \frac{e^{-\lambda(t+s)}}{e^{-\lambda s}}$$

$$= e^{-\lambda t}$$

The exponential distribution would imply that an s-year old individual has the same outlook as a newborn, in terms of future life.

The "Hazard Rate" defined as

$$H(x) = \frac{f(x)}{1 - F(x)} \qquad x > 0$$

$$= \frac{\lambda e^{-\lambda x}}{e^{-\lambda x}} = \lambda$$

$H(x)dx$ is the instantaneous probability of failure between x and x+dx, given survival of at least x.

A76. Constant Hazard implies exponentiality, since if

$$\frac{f(x)}{1 - F(x)} = \lambda \qquad\qquad x > 0$$

$$\int_0^t \frac{f(x)dx}{1 - F(x)} = \lambda t$$

$$\log\left\{1-F(0)\right\} - \log\left\{1-F(t)\right\} = \lambda t$$

Now $F(0) = 0$ for a survival distribution, and hence

$$1-F(t) = e^{-\lambda t}$$

GAMMA DISTRIBUTION (Continuous)

A77. (a)   Probability Density of gamma distribution.

$$f(x) = \frac{\lambda^{\alpha}x^{\alpha-1}e^{-\lambda x}}{\Gamma(\alpha)} \qquad\qquad x>0,\ \alpha>0,\ \lambda>0$$

$$\Gamma(\alpha) = \int_0^{\infty}x^{\alpha-1}e^{-x}dx$$

Note: if α is an integer, n

$$\Gamma(n) = (n-1)!$$

A78. (b)   Mean and Variance

$$\mu = (\alpha/\lambda)$$
$$\sigma^2 = (\alpha/\lambda^2)$$

A79. (c)   Moment Generating Function

$$M_X(t) = \left\{\lambda/(\lambda-t)\right\}^{\alpha} \qquad\qquad t<\lambda$$

A80. (d)   How it arises in the real world

- If $\alpha=1$, note that we obtain the exponential distribution

- Let X_1, X_2, ..., X_n be independent, identically distributed exponential random variables with parameter λ.

Let $S = X_1 + X_2 + \cdots + X_n$

Then S has the Gamma Distribution with parameters $\alpha=n$ and λ.

Proof From A72 and A56, S has moment generating function

$$M_S(t) = M_{X_1}(t) \, M_{X_2}(t) \ldots M_{X_n}(t)$$

$$= \left\{ \frac{\lambda}{(\lambda - t)} \right\}^n.$$

From the uniqueness property of Moment Generating Functions A53, and by A79, the result holds.

A81.   The "chi-square" Distributions are special Gamma Distributions with

$$\alpha = (\text{Degrees of Freedom})/2$$

$$\lambda = \tfrac{1}{2}$$

A82. (e)   Cumulative Distribution (Integer α only)

Let $\alpha = n$ (an integer)

$$F(x) = 1 - \sum_{j=0}^{n-1} \frac{(\lambda x)^j e^{-\lambda x}}{j!}$$

Proof Differentiation of both sides yields result. (See A21.).

Note: Clear cut connection between Poisson Distribution and Gamma Distribution.

NORMAL DISTRIBUTION (Continuous)

A83. (a)   Probability Density of normal distribution

$$f(x) = \frac{1}{\sigma \sqrt{2\pi}} \exp\left\{ -(x-\mu)^2 / 2\sigma^2 \right\} \qquad\qquad -\infty < x < \infty$$

A84. (b)   Mean and Variance

We have so denoted the two parameters μ and σ^2 ($\pi = 3.1416$).

A85. (c)   Moment Generating Function

$$M_X(t) = \exp\left(\mu t + \tfrac{1}{2} \sigma^2 t^2 \right)$$

A86. (d)   Cumulative Distribution

No closed form, but the usual notation is

$F(x) = \Phi\left(\frac{x-\mu}{\sigma}\right)$ with

$\Phi(t) = \int_{-\infty}^{t} \frac{1}{\sqrt{2\pi}} \exp(-t^2/2)dt$

Φ is called the "Standard Normal" cumulative distribution, often denoted by N(0,1).

A87. (e)   Difference of independent normal random variables is normal.

Let X_1, X_2, be independent normally disributed random variables with respective means μ_1 and μ_2 and respective variances σ_1^2 and σ_2^2. Then $(X_1 - X_2)$ is normal with mean μ_1-μ_2 and variance $\sigma_1^2 + \sigma_2^2$.

Proof:

$$E\left(e^{t(X_1-X_2)}\right) = E(e^{tX_1})E(e^{-tX_2}) \qquad \text{Independence}$$

$$= \exp\left(\mu_1 t + \tfrac{1}{2}\sigma_1^2 t^2\right) \exp\left(-\mu_2 t + \tfrac{1}{2}\sigma_2^2 t^2\right) \qquad \text{by (c)}$$

$$= \exp\left\{(\mu_1-\mu_2)t + \tfrac{1}{2}(\sigma_1^2+\sigma_2^2)t^2\right\}$$

X_1-X_2 has the moment generating function of a normal random variable with mean $(\mu_1-\mu_2)$ and variance $\sigma_1^2 + \sigma_1^2$. The result follows from A53.

A88 (f)   How it arises in the real world.

-The Central Limit Theorem

Let X_1, X_2, ..., X_n, be a sequence of independent, identically distributed random variables with mean μ and variance σ^2. Let $\bar{X}_n = (X_1 + \cdots + X_n)/n$. then

$$P\left[\sqrt{n}\ (\bar{X}_n - \mu)/\sigma < x\right] \rightarrow \Phi(x)$$

This result is proved for the special case where the moment generating function exists in A106.

A89. TRANSFORMATIONS (CONTINUOUS VARIABLES) (Consequence)

Let $X = (X_1, ..., X_n)$ be a vector of continuous random variables with joint probability density

$$f_X(x_1, ..., x_n)$$

Let $Y = (Y_1, ..., Y_n)$ where $Y_j = g_j(X_1, ..., X_n)$ $j=1,2,...,n$ be a $1-1$ function of X. That is, no two different X vectors produce the same Y vector. Let $K = K(X)$ be the matrix function whose i^{th} row, j^{th} column is

$$K_{ij}(X) = \frac{\partial g_j}{\partial x_i}(X) \qquad 1 \leq i \leq n, \quad 1 \leq j \leq n$$

Then Y has joint density

$$f_Y(y_1, ..., y_n) = \frac{f_X(x_1, ..., x_n)}{|\det(K)|}$$

with

$$\det (K) = \text{Determinant}$$

$$|\ | = \text{Absolute value}$$

$$y_j = g_j(x_1, ..., x_n)$$

The reader is referred to Hogg and Craig (1978) for further details, but an example might be helpful.

A90. Example

Let X_1, X_2, ..., X_n be independent exponential random variables with common parameter λ.

$$f_X(x_1, ..., x_n) = \lambda^n \exp\left\{-\lambda(x_1 + \cdots + x_n)\right\} \qquad x_j > 0$$

Let

$$Y_1 = X_1$$

$$Y_2 = X_1 + X_2$$

$$Y_j = X_1 + \cdots + X_j$$

$$\vdots$$

$$Y_n = X_1 + \cdots + X_n$$

Note

$$X_1 = Y_1$$

$$X_2 = Y_2 - Y_1$$

$$X_3 = Y_3 - Y_2$$

$$\vdots$$

$$X_n = Y_n - Y_{n-1}$$

Hence, the value of the X's is uniquely determined by the Y's.

$$y_j = g_j(x_1, \ldots, x_n) = x_1 + \cdots + x_j$$

$$\frac{\partial g_j}{\partial x_i} = 1 \qquad\qquad i \leq j$$

$$= 0 \qquad\qquad i > j$$

$$K = \begin{bmatrix} 1 & 0 & 0 & \cdots & 0 \\ 1 & 1 & 0 & \cdots & 0 \\ 1 & 1 & 1 & 0 & 0 \\ \vdots & \vdots & \vdots & \vdots & \vdots \\ 1 & \cdots & \cdots & \cdots & 1 \end{bmatrix}$$

$$\det(K) = 1$$

$$f_y(y_1, \ldots, y_n) = \lambda^n \exp\left\{-\lambda(x_1 + \cdots + x_n)\right\}$$

$$= \lambda^n \exp\left\{-\lambda y_n\right\}$$

Now the range of the y's is obtained from the range of x's as

$$0 < x_1 < x_1 + x_2 < x_1 + x_2 + x_3 \cdots$$

$$0 < y_1 < y_2 < \cdots < y_n < \infty.$$

CONVERGENCE

A91. <u>Limit of a Sequence</u> (Definition)

Let A_1, A_2, ..., A_n, ... be a sequence of constants. We say that A_n converges to the constant A,

$(A_n \to A)$ if for every $\epsilon > 0$, there exists a number $N = N(\epsilon)$ such that whenever

$$n > N, \quad |A_n - A| < \epsilon.$$

In other words, you challenge me with a tiny number ϵ. I find a large N such that

$$|A_n - A| < \epsilon \text{ for all } n > N.$$

This can be done for every $\epsilon > 0$.

Example: $\quad A_n = 1/n \qquad\qquad\qquad A=0$

$\qquad\qquad\quad N(\epsilon) = 1/\epsilon$

A92. <u>Convergence in Probability</u> (Definition)

Let X_1, X_2, ..., X_n... be a sequence of random variables. We say that X_n converges in probability to the constant C

$$X_n \xrightarrow{P} C$$

If for every $\epsilon > 0$

$$P\left(|X_n - C| > \epsilon\right) \to 0 \text{ as } n \to \infty$$

In other words, regardless of what small "challenge number" ϵ you provide, the probability that X_n is within ϵ of C goes to one.

A93. <u>Convergence in Distribution</u> (Definition)

Let X_1, X_2, ... be a sequence of random variables with respective cumulative distributions

$$F_1(x), F_2(x), ..., F_n(x), ...$$

We say that

$$F_n \to F \text{ in distribution}$$

if

$$F_n(x) \rightarrow F(x) \qquad \text{at every continuity point of x.}$$

This is denoted by $F_n \xrightarrow{D} F$ or $X_n \xrightarrow{D} F$.

A94. Convergence in Distribution (Consequence)

Let X_1, X_2, ... be a sequence of random variables with respective cumulative distributions

$F_1(x)$, $F_2(x)$, ... and respective Moment Generating Functions

$$M_1(t), M_2(t), \ldots$$

Suppose that

$$M_n(t) \rightarrow M(t)$$

in a neighborhood of t=0 and M(t) is the Moment Generating Function of F(x). Then $F_n \xrightarrow{D} F$.

The proof is beyond the intended scope of the book. Interested readers might refer to

Fraser (1976, pages 248 and 558-560.)

Convergence in Probability (Consequence)

A95. (i)   If $X_n \xrightarrow{P} x$

and $Y_n = cX_n$ c constant

Then $Y_n \xrightarrow{P} cx$

A96. (ii)   If $X_n \xrightarrow{P} x$

and $Y_n \xrightarrow{P} 0$

then $X_n Y_n \xrightarrow{P} 0$

A97. (iii)   If $X_n \xrightarrow{P} x$ and $x \neq 0$,

then   if $Y_n = 1/X_n$

$Y_n \xrightarrow{P} (1/x)$

A98. (iv)   If $X_n \xrightarrow{P} x$ and g is continuous at x

then $g(X_n) \xrightarrow{P} g(x)$

Proofs:

(i)   If $c = 0$, then $Y_n = 0 = cx$ for all n and the result is immediate

If $c \neq 0$

$$P\big[|Y_n - cx| > \epsilon\big] = P\Big[|X_n - x| > \tfrac{\epsilon}{|c|}\Big] \to 0 \quad \text{since } X_n \xrightarrow{P} x$$

That is, $Y_n \xrightarrow{P} cx$

(ii)   The event that $|X_n Y_n| < \epsilon$ contains the intersection of the events

$|X_n - x| < \epsilon$ (event A_n)

and $|Y_n| < \frac{\epsilon}{|x| + \epsilon}$ (event B_n)

$\Big($Event A_n implies that $|X_n| < |x| + \epsilon$ and hence A_n and B_n together imply $|X_n Y_n| < \epsilon.\Big)$.

$$P(A_n B_n) \quad = P(A_n) - P(A_n \bar{B}_n)$$

$$\geq P(A_n) - P(\bar{B}_n)$$

Now $P(A_n) \to 1$ since $X_n \xrightarrow{P} x$

and $P(\bar{B}_n) \to 0$ since $Y_n \xrightarrow{P} 0$.

Hence $P(A_n B_n) \to 1$.

$$P\big[|X_n Y_n| < \epsilon\big] \geq P\big(A_n B_n\big) \to 1 \text{ as } n \to \infty.$$

(iii)   This is an important special case of (iv).

(iv)   Given $\epsilon > 0$, there exists a $\delta > 0$ such that whenever $|y - x| < \delta$, $|g(y) - g(x)| < \epsilon$.

Above is the definition of continuity.

Therefore, the event that $|g(X_n) - g(x)| < \epsilon$ contains the event A_n, that $|X_n - x| < \delta$. But

$$P\big[A_n\big] = P\big[|X_n - x| < \delta\big] \to 1 \qquad \text{(Convergence in P)}$$

Convergence in Probability (Consequence)

Let $X_n \xrightarrow{P} x$ and $Y_n \xrightarrow{P} y$. Then

A99. (v) $\quad X_n + Y_n \xrightarrow{P} x+y$

A100. (vi) $X_n - Y_n \xrightarrow{P} x-y$

A101. (vii) $\quad X_n Y_n \xrightarrow{P} xy$

A102. (viii) $X_n/Y_n \xrightarrow{P} x/y$ $\qquad\qquad$ if $y \neq 0$

Proofs:

(v) $\quad$ The event that $|X_n + Y_n - x - y| < \epsilon$ contains the event that both $|X_n - x| < \frac{\epsilon}{2}$ and $|Y_n - y| < \frac{\epsilon}{2}$.

$$P\left[|X_n - x| < \tfrac{\epsilon}{2}, \ |Y_n - y| < \tfrac{\epsilon}{2}\right] = P\left[|X_n - x| < \tfrac{\epsilon}{2}\right] - P\left[|X_n - x| < \tfrac{\epsilon}{2}, \ |Y_n - y| \geq \tfrac{\epsilon}{2}\right]$$

$$\geq P\left[|X_n - x| < \tfrac{\epsilon}{2}\right] - P\left[|Y_n - y| \geq \tfrac{\epsilon}{2}\right] \rightarrow 1 - 0 = 1 \text{ as } n \rightarrow \infty.$$

i.e., $\quad P\left[|X_n + Y_n - x - y| < \epsilon\right] \rightarrow 1$

(vi) $\quad$ By (i), $-Y_n \xrightarrow{P} -y$. Apply (v) to get desired result.

(vii) $\quad X_n Y_n - xy = X_n(Y_n - y) + (X_n - x)y$

$\qquad$ Now by (ii), $\qquad X_n(Y_n - y) \xrightarrow{P} 0$

$\qquad$ and by (i), $\qquad (X_n - x)y \xrightarrow{P} 0$

$\qquad$ Hence, by (v) $\qquad X_n Y_n - xy \xrightarrow{P} 0$

(viii) By (iii) $\frac{1}{Y_n} \xrightarrow{P} \frac{1}{y}$. Results hold by (vii) above.

A103. Convergence in Probability (Consequence)

If $\frac{A_n}{B_n} \xrightarrow{P} 1$

$\frac{C_n}{D_n} \xrightarrow{P} 1$

with $A_n, B_n, C_n, D_n > 0$

then $\frac{A_n + C_n}{B_n + D_n} \xrightarrow{P} 1$

Proof: Let $\epsilon > 0$, $\delta > 0$. Then there exists $N_1 > 0$ and $N_2 > 0$,

$$n > N_1 \Rightarrow P\left[1 - \epsilon < \left(\frac{A_n}{B_n}\right) < 1 + \epsilon\right] > 1 - \frac{\delta}{2}$$

$$n > N_2 \Rightarrow P\left[1 - \epsilon < \left(\frac{C_n}{D_n}\right) < 1 + \epsilon\right] > 1 - \frac{\delta}{2}$$

$$n > \max(N_1, N_2) \Rightarrow P\left[B_n(1-\epsilon) < A_n < B_n(1+\epsilon)\right] > 1 - \frac{\delta}{2}$$
$$P\left[D_n(1-\epsilon) < C_n < D_n(1+\epsilon)\right] > 1 - \frac{\delta}{2}.$$

The event

$$1 - \epsilon < \frac{A_n + C_n}{B_n + D_n} < 1 + \epsilon$$

includes the intersection of the events

$$B_n(1-\epsilon) < A_n < B_n(1+\epsilon)$$

with $D_n(1-\epsilon) < C_n < D_n(1+\epsilon)$ and hence

$$P\left[1 - \epsilon < \frac{A_n + C_n}{B_n + D_n} < 1 + \epsilon\right] > 1 - \delta.$$

Since ϵ and δ are arbitrary, the desired result holds.

A104. <u>Convergence in Distribution (Consequence)</u>

Slutzky's Theorem

Let $X_1, X_2, \ldots$ be a sequence of random variables with cumulative distributions

F_1, F_2, ..., with $X_n \xrightarrow{D} F$.

Let Y_n be a sequence of random variables with $Y_n \xrightarrow{P} 0$.

Then if $Z_n = X_n + Y_n$, and if we denote the cumulative distribution of Z_n by G_n, then

$$Z_n \xrightarrow{D} F \qquad (G_n \xrightarrow{D} F)$$

Note: We shall prove the result under the mild restriction that $F(x)$ has "isolated discontinuities." That is, at each point x_0 such that F has a jump at x_0, there exists an $h > 0$ such that F is continuous in the intervals $(x_0 - h, x_0)$ and $(x_0, x_0 + h)$. Continuous distributions and discrete distributions over the integers have this property. [A discrete distribution over the rational numbers is pathological, and does not have this property.]

Proof: Let x be a continuity point of F and let

A_n be the event that $X_n \leq x - \delta$

B_n be the event that $Y_n \leq \delta$

C_n be the event that $X_n + Y_n \leq x$.

Since C_n contains the intersection of A_n and B_n

$$G_n(x) = P(C_n) \geq P(A_n B_n)$$

$$\geq P(A_n) - P(A_n \bar{B}_n)$$

$$\geq P(A_n) - P(\bar{B}_n)$$

$$G_n(x) \geq F_n(x - \delta) - P(Y_n > \delta) \tag{1}$$

Similarly, if A_n^* is the event that $X_n > x + \delta$

B_n^* is the event that $Y_n > -\delta$

C_n^* is the event that $X_n + Y_n > x$

C_n^* contains the intersection of A_n^* and B_n^* and hence

$$1 - G_n(x) \geq 1 - F_n(x + \delta) - P(Y_n \leq -\delta)$$

i.e., $$G_n(x) \leq F_n(x + \delta) + P(Y_n \leq -\delta) \tag{2}$$

Combining (1) and (2) we have

$$F_n(x\text{-}\delta) - P(Y_n > \delta) \leq G_n(x) \leq F_n(x+\delta) + P(Y_n \leq -\delta). \qquad (3)$$

Since discontinuities are isolated, we can restrict our attention to $\delta < h$, where F is continuous in the interval $(x-h, x+h)$.

In (3), as $n \rightarrow \infty$ 　　　$F_n(x+\delta) \rightarrow F(x+\delta)$

$$F_n(x-\delta) \rightarrow F(x-\delta)$$

by convergence in distribution.

Also, 　　　$P(Y_n > \delta)$ and $P(Y_n \leq -\delta) \rightarrow 0$

since 　　　$Y_n \xrightarrow{P} 0.$

Hence: 　$\lim_{n \rightarrow \infty} \left\{ F_n(x\text{-}\delta) - P(Y_n > \delta) \right\} = F(x\text{-}\delta)$

and 　　$\lim_{n \rightarrow \infty} \left\{ F_n(x+\delta) + P(Y_n \leq \text{-}\delta) \right\} = F(x+\delta).$

Let $\epsilon > 0$ be arbitrary. Then there exists $N_1 = N_1(\epsilon, \delta)$ and $N_2 = N_2(\epsilon, \delta)$ such that whenever $n > N_1$

$$G_n(x) \geq F_n(x\text{-}\delta) - P(Y_n > \delta) > F(x\text{-}\delta) - \epsilon$$

and whenever $n > N_2$

$$G_n(x) \leq F_n(x+\delta) + P(Y_n \leq -\delta) < F(x+\delta) + \epsilon.$$

That is, whenever $n > \text{Max}(N_1, N_2)$

$$F(x-\delta) - \epsilon < G_n(x) < F(x+\delta) + \epsilon. \qquad (4)$$

Finally, from the continuity of F at x, there exists δ_0 such that

$$|F(x) - F(y)| \leq \epsilon \qquad\qquad \text{whenever}$$

$$|x-y| < \delta_0.$$

From (4), we have whenever

$$n > \text{Max}\left\{ N_1(\epsilon, \delta_0), N_2(\epsilon, \delta_0) \right\}$$

$$F(x) - 2\epsilon < G_n(x) < F(x) + 2\epsilon.$$

Since ϵ is arbitrary, we conclude that

$$G_n(x) \rightarrow F(x) \qquad \text{as } n \rightarrow \infty.$$

This completes the proof.

A105.   <u>Slutzky's Theorem for Products</u> (Consequence)

Let $X_n \xrightarrow{D} F$ and $Y_n \xrightarrow{P} 1$

then $Z_n = X_n Y_n \xrightarrow{D} F$

Note: We shall prove the result under the same isolated discontinuity assumption as the previous result.

Proof: It is clear that

$$X_n Y_n = X_n + X_n(Y_n\text{-}1)$$

If we can show

$$X_n(Y_n\text{-}1) \xrightarrow{P} 0$$

the desired result is immediate from the previous result.

Suppose we are challenged with two small number, δ and ϵ.

First, since F is a proper distribution we can find a real number B such that both B and -B are continuity points of F and

$$F(B) \text{ - } F(\text{-}B) > 1\text{-}\delta$$

Let A_n be the event that $\text{-}B < X_n \leq B$ and B_n be the event that $|Y_n\text{-}1| < \epsilon/B$.

The event that $|X_n(Y_n\text{-}1)| < \epsilon$ contains the intersection of A_n and B_n.

$$P(A_n B_n) = P(A_n) \text{ - } P(A_n \bar{B}_n)$$

$$\geq P(A_n) \text{ - } P(\bar{B}_n)$$

$$\geq F_n(B) \text{ - } F_n(\text{-}B) \text{ - } P(\bar{B}_n)$$

$$\rightarrow F(B) \text{ - } F(\text{-}B) \qquad \text{as } n \rightarrow \infty$$

$$> 1\text{-}\delta.$$

i.e., since we can make δ as close to zero as we wish, and since $|X_n(Y_n\text{-}1)| < \epsilon$ whenever $A_n B_n$ holds,

$$|X_n(Y_n\text{-}1)| \xrightarrow{P} 0.$$

This completes the proof.

A106.   Central Limit Theorem for Independent, Identically Distributed Random Variables

Let X_1, X_2, ..., X_n $\cdots$ be independent, identically distributed random variables with mean μ, variance σ^2, and moment generating function $M_X(t)$.

Then $Z_n = \dfrac{n^{\frac{1}{2}}(\bar{X}_n - \mu)}{\sigma} \xrightarrow{D} N(0,1)$ where $\bar{X}_n = \dfrac{(X_1 + \cdots + X_n)}{n}$ and $N(0,1)$ is the standard normal distribution.

Proof. It is seen from A56 and A40 that Z_n has moment generating function

$$M_{Z_n}(\theta) = \left[M_X\left(\frac{\theta}{\sigma\sqrt{n}}\right) \right]^n \exp\left[\frac{-\sqrt{n}\,\mu\theta}{\sigma}\right].$$

We need to show (See A94 and A85) that

$$M_{Z_n}(\theta) \rightarrow e^{\theta^2/2} \qquad \text{in a neighborhood of } \theta = 0.$$

That is, we must show

$$K_n(\theta) = \log M_{Z_n}(\theta) \rightarrow \frac{\theta^2}{2} \qquad \text{in a neighborhood of } \theta = 0. \tag{0}$$

$$K_n(\theta) = n \log M_X\left(\frac{\theta}{\sigma\sqrt{n}}\right) - \frac{\sqrt{n}\,\mu\theta}{\sigma} \tag{1}$$

By Taylor's Theorem, there exists θ^*: $|\theta^*| < |\theta|$

$$K_n(\theta) \doteq K_n(0) + \theta K_n'(0) + \frac{\theta^2}{2} K_n''(0) + \frac{\theta^3}{6} K_n'''(\theta^*). \tag{2}$$

From (1), and the fact that $M_X(0) = E\left(e^{0 X_n}\right) = 1$, we see that $K_n(0) = 0$

$$K_n'(\theta) = \frac{\sqrt{n}}{\sigma}\, \frac{M_X'\left\{\theta/(\sigma\sqrt{n})\right\}}{M_X\left\{\theta/(\sigma\sqrt{n})\right\}} - \frac{\sqrt{n}\,\mu}{\sigma} \tag{3}$$

$$K_n''(\theta) = \frac{1}{\sigma^2}\left[\frac{M_X''\{\theta/(\sigma\sqrt{n})\}}{M_X\{\theta/(\sigma\sqrt{n})\}} - \frac{\left[M_X'\{\theta/(\sigma\sqrt{n})\}\right]^2}{\left[M_X\{\theta/(\sigma\sqrt{n})\}\right]^2}\right] \tag{4}$$

$$K_n'''(\theta) = \frac{1}{\sqrt{n}\sigma^3}\left[\frac{M_X'''\{\theta/(\sigma\sqrt{n})\}}{M_X\{\theta/(\sigma\sqrt{n})\}} - \frac{3M_X''\{\theta/(\sigma\sqrt{n})\}M_X'\{\theta/(\sigma\sqrt{n})\}}{\left[M_X\{\theta/(\sigma\sqrt{n})\}\right]^2}\right.$$
$$\left. + \frac{2\left[M_X'\{\theta/(\sigma\sqrt{n})\}\right]^3}{\left[M_X\{\theta/(\sigma\sqrt{n})\}\right]^3}\right] \tag{5}$$

$$K_n(0) = 0 \tag{6}$$

$$K_n'(0) = \frac{\sqrt{n}}{\sigma}\frac{M_X'(0)}{M_X(0)} - \frac{\sqrt{n}}{\sigma}\mu = 0 \tag{7}$$

$$K_n''(0) = \frac{1}{\sigma^2}\left[E(X_j^2) - \mu^2\right] = \frac{\sigma^2}{\sigma^2} = 1 \tag{8}$$

$$K_n'''(\theta^*) \to 0 \text{ as shown below.} \tag{9}$$

Let

$$L_n(y) = \frac{M_X'''(y)}{M_X(y)} - \frac{3M_X''(y)M_X'(y)}{\left[M_X(y)\right]^2} + \frac{2\left[M_X'(y)\right]^3}{\left[M_X(y)\right]^3}.$$

Now

$$K_n'''(\theta^*) = \frac{1}{\sqrt{n}\sigma^3}L_n\{\theta^*/(\sigma\sqrt{n})\}$$

and $L_n(y) \to E(X_j^3) - 3\mu E(X_j^2) + 2\mu^3$ as $y \to 0$, a finite quantity. Hence

$$\lim_{n\to\infty}K_n'''(\theta^*) = \lim_{n\to\infty}\left(\frac{1}{\sqrt{n}\sigma^3}\right)\lim_{n\to\infty}L_n\{\theta^*/(\sigma\sqrt{n})\} = 0. \tag{10}$$

From (2), (6), (7), (8) and (10) we have

$$K_n(\theta) \rightarrow \theta^2/2 \quad \text{as } n \rightarrow \infty.$$

This completes the proof, as the requirement of equation (0) has been met.

A107.   A Central Limit Theorem for Sums of Bernoulli Random Variables

Let X_j, $j = 1, 2, ..., n$ be independent Bernoulli Random Variables (i.e., Binomial with sample size 1) with mass function

$$
\begin{aligned}
f_j(x) \quad &= 1 - p_j \quad && x = 0 \\
&= p_j \quad && x = 1 \\
\\
&= 0 \quad && \text{elsewhere.}
\end{aligned}
$$

Let b_j, $j = 1, 2, ...$ be constants and

$$S_n = \sum_{j=1}^{n} b_j X_j$$

$$M_n = \sum_{j=1}^{n} b_j p_j$$

$$V_n^2 = \sum_{j=1}^{n} b_j^2 p_j (1 - p_j)$$

and
$$Z_n = \frac{S_n - M_n}{V_n}$$

Without loss of generality in what follows, the b_j, p_j (and hence X_j) may be functions of n. We dropped the subscript n for ease of notation.

A107-1 Condition 1:

If $V_n \to \infty$

and

$$\frac{\sum_{j=1}^{n} |b_j|^3}{V_n^3} \to 0 \text{ as } n \to \infty.$$

then

$$Z_n \xrightarrow{D} N(0,1) \quad \text{(Standard Normal)}.$$

A107-2 Condition 2:

If the b_j and p_j are bounded:

$$p_j < p_0 < 1 \text{ and } |b_j| < B$$

and $V_n \to \infty$

and if

$$\frac{\sum_{j=1}^{n} |b_j|^3 p_j}{V_n^3} \to 0 \quad \text{as } n \to \infty$$

then

$$Z_n \xrightarrow{D} N(0,1).$$

This is designed to cover cases where $p_j \to 0$.

Proof. Let

$$K_n(\theta) = \log E\left(e^{\theta Z_n}\right) \tag{0}$$

the log of the moment generating function of Z_n.

It is readily seen by independence that

$$K_n(\theta) = \sum_{j=1}^{n} \left\{ \log \left[(1-p_j) + p_j e^{b_j \theta / V_n} \right] - \frac{b_j p_j \theta}{V_n} \right\} \tag{1}$$

To prove normality, we need to show that

$$K_n(\theta) \rightarrow \frac{\theta^2}{2} \quad \text{as } n \rightarrow \infty$$

in a neighborhood of $\theta = 0$.

By Taylor's Theorem, there exists a θ^* such that $|\theta^*| \leq |\theta|$ with

$$K_n(\theta) = K_n(0) + \theta K_n'(0) + \frac{\theta^2}{2}K_n''(0) + \frac{\theta^3}{6}K_n'''(\theta^*). \tag{2}$$

$$K_n'(\theta) = \sum_{j=1}^{n} \frac{b_j p_j}{V_n} \left\{ \frac{1}{(1-p_j)e^{-b_j\theta/V_n} + p_j} - 1 \right\}. \tag{3}$$

$$K_n''(\theta) = \sum_{j=1}^{n} \frac{b_j^2 p_j(1-p_j)}{V_n^2} \frac{e^{-b_j\theta/V_n}}{\left\{(1-p_j)e^{-b_j\theta/V_n} + p_j\right\}^2}. \tag{4}$$

$$K_n'''(\theta) = 2\sum_{j=1}^{n} \frac{b_j^3 p_j(1-p_j)^2 e^{-2b_j\theta/V_n}}{V_n^3\left\{(1-p_j)e^{-b_j\theta/V_n} + p_j\right\}^3} - \sum_{j=1}^{n} \frac{b_j^3 p_j(1-p_j)e^{-b_j\theta/V_n}}{V_n^3\left\{(1-p_j)e^{-b_j\theta/V_n} + p_j\right\}^2}. \tag{5}$$

Clearly, $K_n(0) = 0$ $\left(\text{Substitute } \theta = 0 \text{ in (1)}\right)$.

By substituting $\theta = 0$ in (3), we see that

$$(1 - p_j)e^{-b_j\theta/V_n} + p_j = (1 - p_j)e^0 + p_j$$
$$= 1.$$

Hence $\qquad K_n'(0) = 0.$

Similarly, from (4),

$$K_n''(0) = \sum_{j=1}^{n} \frac{b_j^2 p_j(1 - p_j)}{V_n^2} = \frac{V_n^2}{V_n^2} = 1.$$

To prove <u>Condition 1</u>, consider in (5)

$$\frac{p_j}{(1 - p_j)e^{-b_j\theta/V_n} + p_j} \leq 1,$$

and

$$\frac{(1 - p_j)e^{-b_j\theta/V_n}}{(1 - p_j)e^{-b_j\theta/V_n} + p_j} \leq 1.$$

Hence

$$|K_n'''(\theta)| < 3\,\frac{\Sigma|b_j|^3}{V_n^3} \tag{6}$$

Since the right hand side of (6) converges to zero, convergence to zero of $K_n'''(\theta)$ is assured.

From (2)

$$K_n(\theta) \rightarrow \frac{\theta^2}{2}$$

and hence

$$E\!\left(e^{\theta Z_n}\right) \rightarrow e^{\theta^2/2} \quad n \rightarrow \infty.$$

That is from **A94**,

$$Z_n \xrightarrow{D} N(0,1) \text{ under condition 1.}$$

For <u>Condition 2</u>, note that since $V_n \rightarrow \infty$ there exists for given $V_0 > 0$, a number N such that whenever

$$n > N, \quad V_n > V_0.$$

Now working with (5) we see

$$\frac{1}{(1 - p_j)e^{-b_j\theta/V_n} + p_j} \leq \frac{1}{(1 - p_j)e^{-b_j\theta/V_n}} \leq \frac{e^{|b_j\theta|/V_n}}{1 - p_j} \leq \frac{e^{|B\theta|/V_0}}{1 - p_0} \quad n > N,$$

also

$$\frac{(1 - p_j)e^{-b_j\theta/V_n}}{(1 - p_j)e^{-b_j\theta/V_n} + p_j} \leq 1.$$

Hence for $n > N$, Condition 2 and (5) yield

$$|K_n'''(\theta)| < \frac{3e^{|B\theta|/V_0}}{1 - p_0} \left\{ \frac{\sum_{j=1}^{n} |b_j|^3 p_j}{V_n^3} \right\} \qquad \text{for } n > N$$

$$\to 0.$$

Hence $\qquad K_n(\theta) \to \dfrac{\theta^2}{2}$

and hence $\qquad E\!\left(e^{\theta Z_n}\right) \to e^{\theta^2/2},$

and hence from A94,

$$Z_n \overset{D}{\to} N(0,1) \quad \text{under Condition 2.}$$

A107-EX   Example:

$$p_j = p_j(n) = \frac{1}{\sqrt[4]{n}}$$

$$b_j = 1$$

$$V_n^2 = n\left(\frac{1}{\sqrt[4]{n}}\right)\left(1 - \frac{1}{\sqrt[4]{n}}\right) = \sqrt{n} - 1$$

$$b_j^3 = 1 \text{ and } \Sigma b_j^3 = n$$

$$\frac{\Sigma b_j^3}{V_n^3} = \frac{n}{\left\{\sqrt{n}-1\right\}^{\frac{3}{2}}} \sim n^{\frac{1}{4}} \quad \text{(Condition 1 fails, since this does not go to zero).}$$

$$\frac{\Sigma b_j^3 p_j}{V_n^3} = \frac{\sqrt{n}}{\left\{\sqrt{n}-1\right\}^{\frac{3}{2}}} \rightarrow 0 \quad \text{(Condition 2 succeeds).}$$

A108 Asymptotic Normality of a Difference

Suppose X_n and Y_n are independent with

$$\frac{X_n - \mu_n}{\sigma_n} \xrightarrow{D} N(0, 1)$$

$$\frac{Y_n - \nu_n}{\lambda_n} \xrightarrow{D} N(0, 1)$$

$$\frac{\sigma_n}{\lambda_n} \rightarrow c$$

where μ_n, ν_n, σ_n, λ_n are constants.

Then
$$\frac{(X_n - Y_n) - (\mu_n - \nu_n)}{\sqrt{\sigma_n^2 + \lambda_n^2}} \xrightarrow{D} N(0, 1)$$

Proof:

$$\frac{X_n - \mu_n}{\sqrt{\sigma_n^2 + \lambda_n^2}} = \frac{X_n - \mu_n}{\sigma_n \sqrt{\left(1 + \frac{\lambda_n^2}{\sigma_n^2}\right)}} \cdot$$

Now since $\dfrac{\lambda_n^2}{\sigma_n^2} \rightarrow c^{-2}$,

Slutzky's Multiplicative Theorem A105 implies

$$\frac{X_n - \mu_n}{\sqrt{\sigma_n^2 + \lambda_n^2}} \quad \text{and} \quad \frac{X_n - \mu_n}{\sigma_n \sqrt{1 + c^{-2}}}$$

have the same asymptotic distribution. That is

$$\frac{X_n - \mu_n}{\sqrt{\sigma_n^2 + \lambda_n^2}} \xrightarrow{D} N\left(0, \frac{1}{1+c^{-2}}\right).$$

Similarly,

$$\frac{Y_n - \nu_n}{\sqrt{\sigma_n^2 + \lambda_n^2}} = \frac{Y_n - \nu_n}{\lambda_n \sqrt{1 + \frac{\sigma_n^2}{\lambda_n^2}}} \xrightarrow{D} N\left(0, \frac{1}{1+c^2}\right)$$

From A87,

$$\frac{(X_n - Y_n) - (\mu_n - \nu_n)}{\sqrt{\sigma_n^2 + \lambda_n^2}} \xrightarrow{D} N\left(0, \frac{1}{1+c^2} + \frac{1}{1+c^{-2}}\right)$$

But

$$\frac{1}{1+c^2} + \frac{1}{1+c^{-2}} = \frac{1}{1+c^2} + \frac{c^2}{1+c^2} = 1$$

Therefore

$$\frac{(X_n - Y_n) - (\mu_n - \nu_n)}{\sqrt{\sigma_n^2 + \lambda_n^2}} \xrightarrow{D} N\left(0, 1\right)$$

A109. Asymptotic Normality of a Difference

If in addition to the conditions of A108, we have consistent estimators

$$\frac{\hat{\sigma}_n}{\sigma_n} \xrightarrow{P} 1$$

$$\frac{\hat{\lambda}_n}{\lambda_n} \xrightarrow{P} 1$$

Then

$$\frac{(X_n - Y_n) - (\mu_n - \nu_n)}{\sqrt{\hat{\sigma}_n^2 + \hat{\lambda}_n^2}} \xrightarrow{D} N\left(0, 1\right)$$

Proof: By **A98**,

$$\frac{\hat{\sigma}_n^2}{\sigma_n^2} \xrightarrow{P} 1$$

$$\frac{\hat{\lambda}_n^2}{\lambda_n^2} \xrightarrow{P} 1.$$

Hence by **A103**,

$$\frac{\hat{\sigma}_n^2 + \hat{\lambda}_n^2}{\sigma_n^2 + \lambda_n^2} \xrightarrow{P} 1.$$

Again using **A98**, this yields

$$\frac{\sqrt{\hat{\sigma}_n^2 + \hat{\lambda}_n^2}}{\sqrt{\sigma_n^2 + \lambda_n^2}} \xrightarrow{P} 1.$$

Finally, from **A108** snd Slutsky's Theorem (Products), **A105**, the desired result holds.

STATISTICAL INFERENCE

A110   Estimator of a Parameter (Definition)

A random variable, whose observed value will estimate a parameter.

Example: The sample mean is an estimator of the population mean.

If the parameter is θ, then the estimator is denoted by $\hat\theta$

A111   Unbiased Estimator (Definition)

$\hat\theta$ is "unbiased" for θ if $E(\hat\theta) = \theta$

A112   Consistent Estimator(s) (Definition)

Let $\hat\theta_1, \hat\theta_2, \ldots$ be a sequence of estimators of θ. $\hat\theta_n$ is "consistent" for θ if

$$\hat\theta_n \xrightarrow{P} \theta.$$

(More formally, this is called "weakly consistent".)

A113   Maximum Likelihood Estimator

$\hat\theta$ is said to be the Maximum Likelihood Estimator of θ if the probability of the observed sample (discrete case) or joint density of the observed sample (continuous case) is maximized at $\theta = \hat\theta$.

This is a technique to obtain estimators of θ. In many applications, they are consistent, asymptotically normal, with optimal asymptotic variance.

We shall not make use of these estimators directly, but we include the definition for the sake of completeness.

A114   <u>Confidence Interval</u> (Definition)

Consider two estimators $\hat{\theta}_L < \hat{\theta}_U$ such that the <u>pre-experiment</u> probability

$$P\left[\hat{\theta}_L < \theta < \hat{\theta}_U\right] = (1 - \alpha).$$

We say that

$$\left(\hat{\theta}_L, \hat{\theta}_U\right)$$

form a $100(1-\alpha)\%$ confidence interval for θ.

After the data has been collected, probability no longer applies. $\hat{\theta}_L$, θ, and $\hat{\theta}_U$ are simply three numbers. We do have $100(1-\alpha)\%$ confidence in our procedure to capture the true θ. [The analogy is the following. Before playing a key ball game, Team A was believed to have a 90% chance of defeating team B. You might consider betting on the outcome. Even if you did not know the true result, but you did know that the game was over, you would not bet on the game.]

HYPOTHESIS TESTING (*P*-values)

A115   <u>Null Hypothesis</u> (Definition)

A statement corresponding to an <u>unsuccessful</u> outcome of an experiment or study. This is generally expressed in terms of parameters. The objective of the study will be achieved if we can reject the null hypothesis in favor of the alternate hypothesis below. We denote this by H_0.

A116   <u>Alternate Hypothesis</u> (Definition)

A statement corresponding to a successful outcome of an experiment or study. We denote this by H_A.

A117   Example (i):

Suppose you wish to determine which of two therapies is better in terms of five year survival. Since it is inconceivable that the five year survival under each therapy is exactly equal, the goal of the study is to find out which is better. The null hypothesis would be that the five year survival of each therapy is exactly the same. The alternative hypothesis is that there is a difference.

$$H_0 : P_5 = Q_5 \quad \text{versus} \quad H_A : P_5 \neq Q_5$$

with P_5 and Q_5 the five year survival under treatment A and treatment B respectively. If you can conclude beyond a reasonable doubt that one of the treatments is superior to the other (reject H_0), your research question will have been answered. If you fail to reject H_0, you <u>do not</u> conclude that the treatments are equal in terms of five year survival, but rather, that the data are <u>inconclusive</u>.

This is called a <u>two tailed test</u>. The burden of evidence is always placed on rejecting H_0 in favor of H_A.

A118   Example (ii):

Suppose that you wish to determine if the addition of a drug to a "standard" treatment improves five year survival, over the standard treatment alone. You wish to place the burden of proof on showing that the addition of the drug has efficacy. In this situation, we have

$$H_0 : P_5 \leq Q_5 \quad \text{versus} \quad H_A : P_5 > Q_5$$

In such situations, if you cannot demonstrate beyond a reasonable doubt that P_5, the five year survival of the "experimental" therapy (standard plus drug) is greater than Q_5, the five year survival of the "control" therapy (standard alone), then the drug is considered unproven in terms of efficacy, over and above the standard.

This is called a <u>one tailed test</u>. The burden of proof is to show a difference in a specified

direction.

Note: Some authors will write the one-tailed test as

$$H_0 : P_5 = Q_5 \text{ (instead of } H_0 : P_5 \leq Q_5) \quad \text{versus} \quad H_A : P_5 > Q_5$$

This is the most difficult situation among the candidates, where the more general H_0 is true, to

distinguish H_0 from H_A. For example, it is harder to distinguish between hypotheses for situation (A)

below than situation (B):

   Situation (A): $P_5 = Q_5 = .20$ versus $P_5 = .20 \quad Q_5 = .10$

   Situation (B): $P_5 = .10 \quad Q_5 = .20$ versus $P_5 = .20 \quad Q_5 = .10$

A119   <u>Test Statistic</u> (Definition)

   A quantity, calculated from the data, upon which the hypothesis test will be based. (The test

statistic is simply the value of a random variable.)

A120   <u>P-Value</u> (Definition)

   The maximum probability that in a replication of the study where the <u>null hypothesis is</u>

<u>true</u>, that the test statistic would be at <u>least as "extreme"</u> as the observed test statistic (from the

actual study.)

A121   <u>Two-Tailed Example</u> :

   Suppose you wanted to determine which of two candidates is preferred. We might sample N =

400 individuals and assume that p = true proportion favoring A and 1-p = true proportion favoring B.

Our hypotheses are

$$H_0 : p = .50 \quad \text{versus} \quad H_A ; p \neq .50$$

The test statistic generally used is

$$Z = 2\sqrt{n}(\hat{p} - .50)$$

where $\hat{p}$ is the fraction of the sample favoring A.

"Extreme" values for Z simply means large absolute values, since the larger $|Z|$ is, the stronger the evidence against H_0.

Under H_0, Z has an approximate standard normal distribution (see A107).

If for example, $\quad \hat{p} = \dfrac{176}{400} = .440,$

$$Z = -2.40.$$

The p-value, as obtained from Figure 7, is p = .018. This provides strong evidence against H_0, and in favor of Candidate B.

A122   One-Tailed Example:

Via a randomized trial, a drug company wishes to show that their experimental agent combination can place a higher proportion of relapsed acute lymphocytic leukemia patients into remission than can the standard agent combination. If P_E and P_S are the experimental and standard remission rates respectively, we wish to test

$$H_0 : P_E \leq P_S \quad \text{versus} \quad H_A : P_E > P_S.$$

The test statistic used will be

$$Z = \frac{\hat{P}_E - \hat{P}_S}{\sqrt{\left\{ \dfrac{\hat{P}_E(1-\hat{P}_E)}{N_E} + \dfrac{\hat{P}_S(1-\hat{P}_S)}{N_S} \right\}}}$$

where $\hat{P}_E$ and $\hat{P}_S$ are the fraction of patients achieving a remission on the experimental and standard agents respectively, and N_E and N_S are the corresponding sample sizes.

It is readily shown from results taken in our Appendix A107, A108, and A109, that if $P_E = P_S$, the hardest case in H_0 to distinguish from H_A, then Z is asymptotically standard normal.

Suppose that the observed value of Z is

$$Z = 0.37.$$

From Figure 6, this yields a p-value of $P = 0.36$. This value is certainly in the reasonable doubt range, and hence, we cannot reject H_0 in favor of H_A.

The study should be viewed as <u>inconclusive</u> as far as whether or not the experimental combination is superior.

A123   <u>Standard Reasonable Doubt Thresholds</u>:

The most widely used standard of doubt is a p-value of $P = .05$. In many fields, values below .05 are considered to be conclusive while values above are considered inconclusive.

Other than standardization, there is no logical basis for the 5% figure.

A124   <u>Planning a Study</u> (Sample Size Determination)

In planning a study, there are several ingredients needed in order to evaluate the sample size.

These include   - Null Hypothesis

- Alternative Hypothesis

- Test Statistic

A125   - "Type I Error." (α) Threshold p-value where you consider the dividing line between reasonable doubt and beyond reasonable doubt lies.) (This is usually denoted by α.)

A126   - "Planning Parameters." Under the alternate hypothesis, specify the parameters so that distributional properties of the test statistic can be evaluated. Generally, this involves specifying the smallest detectable difference of interest to the investigators, and other quantities such as accrual rate, "nuisance" parameters, and minimum follow-up time.

A127   - "Type II Error."(β) Specify the probability of failing to reject the null hypothesis, when the planning parameters represent the true target population conditions. (This is usually denoted by β.)

A128   - "Power" = 1 - Type II Error = probability of rejecting the null hypothesis when the planning parameters represent the true target population conditions. (This is usually denote by $\prod$.)

A129.   The most common situation is the following: We shall show this as one-sided.

$$H_0: \theta = \theta_0 \text{ (or } H_0 : \theta \leq \theta_0) \quad \text{versus} \quad H_A : \theta > \theta_0$$

Test Statistic: $Z = \dfrac{\hat{\theta} - \theta_0}{SE}$

with SE = "Standard Error" of $\hat{\theta}$, (estimated standard deviation of $\hat{\theta}$), calculated from the data and $Z \sim N(0,1)$ approximately, when H_0 is true.

To plan a study, enough information must be supplied to obtain an approximate distribution of Z under H_A.

This may involve specification of nuisance parameters.

For example, suppose that the planner of the acute lymphocytic leukemia study had come to you prior to conducting the study saying $\left(\text{see Chapter 2, Section 2.1.1}\right)$

"The study needs to be sensitive to a 10% difference. Type I Error should be $\alpha =$

.05 and Type II Error should be $\beta = .20$. I want equal sample sizes in each group."

$$H_0 : P_E - P_S = 0 \quad \text{versus} \quad H_A : P_E - P_S = .10$$

Test Statistic:

$$Z = \frac{\hat{P}_E - \hat{P}_S}{\sqrt{\left\{ \dfrac{\hat{P}_E(1-\hat{P}_E)}{N_E} + \dfrac{\hat{P}_S(1-\hat{P}_S)}{N_S} \right\}}}$$

where
$N = N_E + N_S = $ Total Sample Size

$N_E = $ Number assigned to Experimental Therapy

$N_S = $ Number assigned to Standard Therapy.

Z is approximately standard normal under H_0, but the approximate distribution of Z under H_A depends heavily on the actual values of P_E and P_S, not just the difference. The planner needs to provide values for both P_E and P_S under H_A to compute the sample size.

"Jumping ahead" somewhat, the required sample size in each group is: $\Big($See formal derivation in A132$\Big)$.

$$N_E = N_S = 617.5 \Big\{ P_E(1-P_E) + P_S(1-P_S) \Big\}$$

$$= 56 \quad P_S = 0, \ P_E = .1 \quad \text{(lowest possible value)}$$

$$= 306 \quad P_S = .45, \ P_E = .55 \quad \text{(highest possible value)}$$

It is much easier to distinguish 0 from 10% than it is to distinguish 45% from 55%.

A130   Sample Size Determination

The following derivation has not been developed in any text, yet is the basic background to nearly every sample size derivation. The ideas will look peculiar at first glance, but will make perfect sense once some examples are considered.

Consider the hypothesis:

$$H_0 : \theta = \theta_0 \quad \text{versus} \quad H_A : \theta = \theta_1 > \theta_0.$$

Let us set a Type I Error α and Power $(1-\beta)$ (Type II Error β). Suppose that $\alpha + \beta < 1$.

Let us assume that we have statistics $\hat{\theta}$ and SE such that SE > 0,

$$\text{(a)} \quad P\left[\frac{\hat{\theta} - \theta_0}{\text{SE}} > Z_\alpha\right] \simeq \alpha \text{ under } H_0$$

and

$$\text{(b)} \quad P\left[\frac{\hat{\theta} - \theta_1}{\text{SE}} - \frac{(Z_\alpha + Z_\beta)(\text{SE} - \text{S})}{\text{SE}} > - Z_\beta\right] \simeq 1 - \beta \text{ under } H_A,$$

with S, a function of the parameters of the distribution under H_A satisfying the implicit sample size equation:

$$S = \frac{\theta_1 - \theta_0}{Z_\alpha + Z_\beta} \qquad\qquad \text{(1). [KEY EQUATION]}$$

(Z_α and Z_β are generally the upper $(100\alpha)\%$ and $(100\beta)\%$ points of the standard normal distribution, but this derivation holds in general.)

Then

$$P\left[\frac{\hat{\theta} - \theta_0}{\text{SE}} > Z_\alpha\right] \simeq 1 - \beta \text{ under } H_A.$$

That is, if S is given as per the (1) [key equation], then the rejection region

$$\hat{\theta} > \theta_0 + Z_\alpha \text{SE}$$

has approximate Type I Error α and Power $1 - \beta$.

Proof:

This is simply a straight forward substitution for S in (b):

$$(1-\beta) \simeq P\left[\frac{\hat{\theta} - \theta_1}{SE} - \frac{(Z_\alpha + Z_\beta)(SE - S)}{SE} > -Z_\beta\right] \quad \text{(under } H_A)$$

$$= P\left[\frac{\hat{\theta} - \theta_1 + (Z_\alpha + Z_\beta)S}{SE} - (Z_\alpha + Z_\beta) > -Z_\beta\right]$$

$$= P\left[\frac{\hat{\theta} - \theta_1 + (\theta_1 - \theta_0)}{SE} > Z_\alpha\right] \quad (Z_\alpha + Z_\beta)S = \theta_1 - \theta_0$$

$$= P\left[\frac{\hat{\theta} - \theta_0}{SE} > Z_\alpha\right] \quad \text{(under } H_A)$$

That is, the Type I and Power requirements are approximately met when equation (1) holds.

Note: S is the asymptotic standard deviation of the statistic under H_A, and may contain nuisance parameters. [SE usually has the same form as S, except population values under H_A (from S) are replaced by consistent estimators to obtain SE.]

A131   Extension to the two-sided test

For the two-sided test, we can replace α by $\alpha/2$ can utilize the equation

$$S = \frac{|\theta_1 - \theta_0|}{Z_{\alpha/2} + Z_\beta}$$

This implies two calculations, one for each alternative

$$H_A : \theta = \theta_1 = \theta_0 + \Delta \quad (\Delta > 0) \quad \text{and} \quad H_A : \theta = \theta_1 = \theta_0 - \Delta \quad (\Delta > 0)$$

The larger resulting sample size would be used.

A132   Example 1: Two Sample Independent Binomial with Equal Sample Sizes

The following notation will be used:

	Treatment 1	Treatment 2
Failure Rate	Q	T
Sample Size	N	N
Failures in Sample	F	G
Estimated Failure Rate	$\hat{Q} = F/N$	$\hat{T} = G/N$

$\theta = Q - T$

$H_0 : \theta = 0$

$H_A : \theta = \theta_1 > 0$

$\hat{\theta} = \hat{Q} - \hat{T}$

$$SE = \sqrt{\frac{\hat{Q}(1-\hat{Q})}{N} + \frac{\hat{T}(1-\hat{T})}{N}}$$

$$S = \sqrt{\frac{Q(1-Q)}{N} + \frac{T(1-T)}{N}}$$

For large N, we know

(i) $\dfrac{\hat{\theta} - 0}{SE} \rightarrow N(0,1)$ under H_0

(ii) $\dfrac{S}{SE} \xrightarrow{P} 1$ under H_A

(iii) $\dfrac{\hat{\theta} - \theta_1}{SE} \rightarrow N(0,1)$ under H_A

The approximation (a) of A130 follows from (i), while from (ii), (iii) and Slutsky's theorem (A104), we have the approximation (b) of A130.

To obtain the sample size, we use

$$S = \frac{\theta_1 - \theta_0}{Z_\alpha + Z_\beta}.$$

Squaring both sides yields

$$\frac{Q(1\text{-}Q)}{N} + \frac{T(1\text{-}T)}{N} = \frac{(\theta_1 - \theta_0)^2}{(Z_\alpha + Z_\beta)^2} = \frac{(Q\text{-}T)^2}{(Z_\alpha + Z_\beta)^2}$$

That is, <u>each</u> sample size must be

A133
$$N = \frac{\left[Q(1\text{-}Q) + T(1\text{-}T)\right](Z_\alpha + Z_\beta)^2}{(Q\text{-}T)^2}$$

The two-tailed version of this formula appears in Pocock 1983, page 125.

A134   Example 2:

Let Q = .7 T = .5 in example 1, with α = .05 and β = .20 (80% power), then

$$N = \frac{[.21 + .25](1.645 + .84)^2}{(.2)^2} = 71 \qquad\qquad \text{(each treatment)}$$

How good are these approximations? The exact binomial probability that the left hand term in assumption (a) exceeds 1.645 is 5.2% under H_0 (common Q = T = .7), while the exact binomial probability that the left hand term in assumption (b) exceeds -.84 is exactly 79.5% under H_A. These are very close to the asymptotic values of 5% and 80% respectively. For an exact approach to the two sample binomial problem see Suissa and Shuster (1985).

In the binomial example, we used a standard error valid under H_0 and H_A. We did not pool the proportions in the calculation.

A135   <u>Summary of Sufficient Conditions to Apply the Sample Size Methodology</u>

(1) $\dfrac{(\hat{\theta} - \theta)}{SE}$ is asymptotically normal under H_0, $\theta = \theta_0$.

(2) $\dfrac{(\hat{\theta} - \theta)}{SE}$ is asymptotically normal under H_A, $\theta = \theta_1$.

(3) $\dfrac{S}{SE} \xrightarrow{P} 1$ under H_A.

(4)   SE   is calculated from the data.

(5)   S   is calculated from the parameters under H_A.

Note: This methodology will be employed to derive the sample size requirements for clinical trials, analyzed via censored survival analysis.

A136. THE POISSON PROCESS

INTRODUCTION

The Poisson Arrival Process is a logical model for patient accrual. The arrival rate (accrual rate), A, represents the expected number of patients to be accrued per unit time. The number of patients accrued in t units of time has a Poisson Distribution, with mean (At). Finally, conditional on the accrual of exactly n patients to time t, the n accrual times are uniformly distributed over the interval from 0 to t.

This will be an appropriate model for the accrual process of many clinical trials. First, there is a time scale, denoted by t. The Poisson process is simply a count N(t) of the number of patient entries up to time t.

Let us assume that we have a very large population of K patients. Assume that for a small interval of time of length Δt that the number of entries is binomial with sample size K and probability of incidence of disease

$$\frac{A\Delta t}{K}.$$

That is, the expected number of entries is $A\Delta t$, which is proportional to the length of Δt.

A137   A is called the arrival rate or accrual rate

We shall further assume that the number of entries in non-overlapping intervals are independent events. This may seem artificial, but is generally a reasonable assumption. The number of patients who develop the disease and leave the population will be a negligible fraction of K. The net effect of other departures and arrivals will also generally be a negligible fraction of K.

PLAN OF DERIVATION OF PROPERTIES

A138   (1) Divide the time scale from 0 to t into M intervals

$$0, \Delta t, 2\Delta t, ..., M\Delta t = T \quad M = T/\Delta t.$$

A139   (2) Prove that as $K \rightarrow \infty$ and $\Delta t \rightarrow 0$ the probability that every interval has no more than one arrival goes to one. Hence, we can consider (in the limit) that each interval will have either 0 or 1 arrival.

A140   (3) Prove that as $K \rightarrow \infty$ and $\Delta t \rightarrow 0$

$$P[N(t) = x] \rightarrow (At)^x e^{-At}/x! \quad x = 0, 1, ...$$

that is, N(t) is Poisson with mean At.

A141   (4) Prove that the waiting time between arrivals is exponential with parameter A.

A142   (5) Suppose we have a Poisson Process (with arrival rate A) where arrivals are assigned by independent flips of a fair coin to treatment 1 and treatment 2. Then the two processes $N_1(t)$, the number of patients assigned to Treatment 1 by time t and $N_2(t)$, the number of patients assigned to Treatment 2 by time t are independent Poisson Processes with common arrival rates A/2.

A143   (6) In a Poisson process, given that r entries occur by time t, the entry times are uniformly distributed over the time interval from zero to t.

Proofs:

Prior to proving the results, two lemmas will be derived for later use.

A144   <u>Lemma A:</u>

Let $0 < |y| < 1$. Then there exists $z : |z| \leq |y|$ and $\log(1+y) = y - \dfrac{y^2}{2(1+z)^2}$

Proof: This is simply Taylor's Theorem applied to the second derivative

$$f(y) = \log(1+y) \qquad\qquad f(0) = 0$$

$$f'(y) = \frac{1}{1+y} \qquad\qquad f'(0) = 1$$

$$f''(y) = \frac{-1}{(1+y)^2}$$

Taylor's theorem states that there exists a $z : |z| \leq |y|$ with
$$f(y) = f(0) + yf'(0) + \frac{y^2}{2}f''(z)$$

A145   Lemma B:

Let $A_n \to \infty$ and $B_n \to \infty$ with $\dfrac{A_n}{B_n} \to 1$ as $n \to \infty$

$$W_n = \left(1 + \frac{x}{B_n}\right)^{A_n} \to e^x \text{ as } n \to \infty.$$

Proof: Since $B_n \to \infty$, there exists N such that whenever $n > N$, $|x/B_n| < \frac{1}{2}$. We shall work exclusively with $n > N$ so that logs are defined, and x/B_n is bounded away from -1.

Let $Y_n = \log W_n = A_n \log(1 + \frac{x}{B_n})$.

From Lemma A (A144), there exists $z_n : |z_n| \leq |x/B_n|$

$$Y_n = \frac{A_n x}{B_n} - \frac{A_n x^2}{2B_n^2(1+z_n)^2}. \qquad (*)$$

$$\frac{A_n x}{B_n} \to x \qquad \text{since } A_n/B_n \to 1$$

We show that second term of $(*)$ goes to zero.

Recall that $B_n \to \infty$ and $A_n/B_n \to 1$. Hence

$$0 < \frac{A_n x^2}{B_n^2(1+z_n)^2} \leq \frac{4}{B_n}\left(\frac{A_n}{B_n}\right) x^2 \to 0,$$

which implies

$$\frac{A_n x^2}{2B_n^2(1+z_n)^2} \to 0.$$

That is,

$$Y_n = \log(W_n) \to x.$$

That is

$$W_n = e^{Y_n} \to e^x.$$

We now return to the major derivation of results.

Proofs: (Main results)

First, we shall prove (2), that in the limit, multiple entries in any interval become impossible.

For a given interval, the probability of zero or one entry is obtained from the binomial distribution as

$$C = \left(1 - \frac{A\Delta t}{K}\right)^K + \frac{KA\Delta t}{K}\left(1 - \frac{A\Delta t}{K}\right)^{K-1} \qquad \text{(See A136)}$$

$$= \left(1 - \frac{A\Delta t}{K}\right)^{K-1}\left\{1 - \frac{A\Delta t}{K} + \frac{KA\Delta t}{K}\right\}$$

$$= \left(1 - \frac{A\Delta t}{K}\right)^{K-1}\left\{1 + \left(\frac{K-1}{K}\right)A\Delta t\right\}.$$

By independence, the probability of 0 or 1 entry in all M intervals is

$$C^M = \left(1 - \frac{A\Delta t}{K}\right)^{M(K-1)}\left\{1 + \left(\frac{K-1}{K}\right)A\Delta t\right\}^M.$$

Now $\Delta t = T/M$ and hence

$$C^M = \left(1 - \frac{AT}{MK}\right)^{M(K-1)}\left\{1 + \frac{(K-1)}{KM}AT\right\}^M$$

We apply Lemma B to each term

$$\left(1 - \frac{AT}{KM}\right)^{M(K-1)} \to e^{-AT}$$

Since $M(K-1)$ and $MK \to \infty$ with $\dfrac{M(K-1)}{MK} \to 1$.

Also

$$\left\{ 1 + \frac{(K-1)AT}{KM} \right\}^M \to e^{AT}$$

since $M \to \infty$, $\dfrac{KM}{(K-1)} \to \infty$ with $\dfrac{M(K-1)}{KM} \to 1$.

That is, $C^M \to e^{-AT} e^{AT} = 1$. Thus (2) holds.

In the limit, we can ignore the possibility of multiple arrivals. This will help prove (3).

The probability of zero and one arrivals respectively in a single interval are

$$P_0 = \left(1 - \frac{A\Delta t}{K} \right)^K \text{ and } P_1 = A\Delta t \left(1 - \frac{A\Delta t}{K} \right)^{K-1}.$$

The probability of r arrivals to time t is simply binomial

$$D_r = \binom{M}{r} P_0^{M-r} P_1^r$$

$$= \binom{M}{r} \left(1 - \frac{A\Delta t}{K} \right)^{K(M-r)} (A\Delta t)^r \left(1 - \frac{A\Delta t}{K} \right)^{(K-1)r}$$

$$= \binom{M}{r} \left(1 - \frac{AT}{MK} \right)^{MK-r} \left(\frac{AT}{M} \right)^r \qquad (\Delta t = T/M)$$

$$= \left[\frac{M(M-1)(M-2)\cdots(M-r+1)}{M^r} \right] \frac{(At)^r}{r!} \left[1 - \frac{AT}{MK} \right]^{MK-r} \qquad (\ast\ast)$$

Now as $M \to \infty$, $K \to \infty$

$$\frac{M(M-1)\cdots(M-r+1)}{M^r} \to 1$$

and by Lemma B, and the fact that $MK \rightarrow \infty$ implies $MK\text{-}r \rightarrow \infty$ with $(MK\text{-}r)/MK \rightarrow 1$,

$$\left[1 - \frac{AT}{MK}\right]^{MK\text{-}r} \rightarrow e^{-AT}.$$

That is, substituting in (**) the probability of r arrivals to time t,

$$D_r \rightarrow \frac{(AT)^r e^{-AT}}{r!}$$

This completes the proof of (3).

To prove (4), let us first look at the first arrival and show the time is exponential.

In order to have an arrival before time t, we must have $N(t) \geq 1$, that is, at least one arrival prior to time t.

$$
\begin{aligned}
P(\text{Arrival prior to time t}) &= P\big((N(t) \geq 1\big) \\
&= 1 - P\big(N(t) = 0\big) \\
&= 1 - (AT)^0 e^{-AT}/0! \\
&= 1 - e^{-At}
\end{aligned}
$$

The cumulative distribution of T_1, the time to the first arrival satisfies (see A74)

$$P\big[T_1 < t\big] = 1 - e^{-At} \qquad \text{(exponential)}$$

Now let T_r be the time of the rth arrival. From the independent interval binomial construction, the number of arrivals in the next t units of time, given T_r is Poisson with parameter At. $\left(T_{r+1} - T_r \text{ is the time between the } r^{\text{th}} \text{ and } (r+1)^{\text{th}} \text{ arrival.}\right)$

Hence

$$P\big(T_{r+1} - T_r < t | T_r\big) = P(\text{at least one arrival in t units of time})$$

$$= 1 - P(\text{no arrivals in t units of time})$$

$$= 1 - e^{At} \qquad (\text{exponential})$$

This does not depend on T_r.

Hence, successive arrival times are independent identically distributed exponential random variables with parameter A.

To prove (5), it is clear from the structure discussed in A136 - A138 that each of $N_1(t)$ and $N_2(t)$ is a Poisson Process with accrual rate (A/2). Further, the number of arrivals in a given interval to Treatment 1 and number of arrivals to a second interval (mutually exclusive of the first) are mutually independent events, since the overall process

$$N(t) = N_1(t) + N_2(t)$$

is constructed that way.

To complete the proofs, we need to show the somewhat surprising result that X_1, the number of patients assigned to Treatment 1 in a given interval is statistically independent of X_2, the number of patients assigned to Treatment 2 in the same interval.

Assuming the interval is of length t_1

$$P\big[X_1 = x_1, X_2 = x_2\big] = \frac{(At)^{x_1+x_2}e^{-At}}{(x_1+x_2)!} \binom{x_1+x_2}{x_1}(.5)^{x_1+x_2}$$

$$= \left[\frac{(At/2)^{x_1}e^{-At/2}}{x_1!}\right]\left[\frac{(At/2)^{x_2}e^{-At/2}}{x_2!}\right]$$

$$= P\big[X_1 = x_2\big]P\big[X_2 = x_2\big]$$

Hence X_1 and X_2 are independent as required.

Proof of (6)

Since successive arrival times are independent, identically distributed, exponential random variables, the joint distribution of the successive times between arrivals of the first $(r+1)$ patients is

$$f(x_1, \ldots, x_{r+1}) = \prod_{j=1}^{r+1} \left(Ae^{-Ax_j} \right) \qquad \text{(exponential densities)}$$

$$= A^{r+1} e^{-A\sum_{j=1}^{r+1} x_j} \qquad x_1 > 0, \; x_2 > 0, \ldots$$

Note that this is only a function of

$$T_{r+1} = \sum_{j=1}^{r+1} x_j \quad \text{the total elapsed time until the } (r+1)^{st} \text{ arrival.}$$

Now we make a transformation (1-1)

$$T_1 = x_1$$

$$T_2 = x_1 + x_2$$

$$\vdots$$

$$T_r = x_1 + \cdots + x_r$$

$$T_{r+1} = (x_1 + x_2 + \cdots + x_{r+1}).$$

Since the Jacobian (see A89 and A90), is unity,

the joint distribution of

$T_1, \ldots, T_{r+1}$ is

$$g(T_1, \ldots, T_{r+1}) = A^{r+1} e^{-AT_{r+1}} \qquad 0 < T_1 < T_2 < \cdots < T_{r+1}.$$

Let U be the event that exactly r arrivals have occurred by time t.

The conditional density of $(T_1, \ldots, T_r)$ given U is

$$g(T_1, \ldots, T_r | U) = \frac{\int_t^\infty g(T_1, \ldots, T_{r+1}) dT_{r+1}}{P(U)}.$$

Once the $T_1, \ldots, T_r$ with $T_r < t$ are fixed, U occurs only if $T_{r+1} > t$.

$$P(U) = P(\text{Exactly r arrivals to time t})$$

$$= \frac{(At)^r e^{-At}}{r!} \qquad \text{(Poisson Process)}.$$

$$\int_t^\infty g(T_1, \ldots, T_{r+1}) dT_{r+1} = A^{r+1} \int_t^\infty e^{-AT_{r+1}} dT_{r+1}$$

$$= A^r e^{-At}$$

Hence, the conditional density of $T_1, \ldots, T_r$ given U is

$$g(T_1, \ldots, T_r | U) = r! t^r \qquad 0 < T_1 < \cdots < T_r < t$$

That is, the ordered arrival times are uniformly distributed from zero to t. (This conditional density does not depend on the values of the T's provided they are ordered as above.)

This completes the discussion of the Poisson Process.

REFERENCES

1. **Fraser, D. A. S.,** *Probability and Statistics: Theory and Applications,* Duxbury Press, Boston, 1976.

2. **Hogg, R. V. and Craig, A. T.,** *Introduction to Mathematical Statistics,* Macmillan, New York, 1978.

3. **Mendenhall, W., Scheaffer, R. L., and Wackerly, D. D.,** *Mathematical Statistics with Applications,* Duxbury Press, Boston, 1986.

4. **Mood, A. M., Graybill, F. A., and Boes, D. C.,** *Introduction to the Theory of Statistics,* McGraw-Hill, New York, 1974.

5. **Suissa, S. and Shuster, J. J.,** Exact unconditional sample sizes for the 2 × 2 binomial trial, *J. R. Statist. Soc., Ser. A,* 148, 315, 1985.

APPENDIX II

TABLE OF CONTENTS

TABLE 1: ALPHA= 0.005 POWER= 0.8   EXPECTED ACCRUAL THRU MINIMUM FOLLOW-UP= 30

PCONT=•••   REQUIRED NUMBER OF PATIENTS

	DEL=.10			DEL=.15			DEL=.20			DEL=.25			DEL=.30		
FACT=	1.0 / .75	.50 / .25	.00 / BIN	1.0 / .75	.50 / .25	.00 / BIN	1.0 / .75	.50 / .25	.00 / BIN	1.0 / 75	.50 / .25	.00 / BIN	1.0 / .75	.50 / .25	.00 / BIN
0.05 •••	245	246	265	142	144	155	100	101	109	78	79	84	64	65	69
•••	245	248	409	143	146	216	101	103	138	78	81	97	65	66	72
0.1 •••	386	388	444	204	206	236	134	136	155	99	101	113	78	80	89
•••	387	393	584	205	211	289	135	140	176	100	104	119	79	83	86
0.15 •••	498	502	609	250	254	307	158	162	193	113	117	137	87	90	105
•••	499	508	736	251	261	351	159	168	208	114	122	138	88	95	98
0.2 •••	580	585	755	282	287	368	173	179	226	122	127	157	92	97	117
•••	581	595	865	283	298	403	175	188	234	124	134	153	94	103	107
0.25 •••	631	639	879	300	308	419	182	190	252	126	133	172	95	101	127
•••	634	653	970	303	322	444	185	202	255	129	142	164	97	108	113
0.3 •••	656	666	980	307	318	459	184	195	272	127	136	183	95	102	133
•••	659	685	1052	311	336	475	188	209	269	130	147	171	98	111	117
0.35 •••	656	669	1056	305	318	488	182	194	285	125	135	189	93	102	136
•••	661	694	1110	309	341	496	186	212	278	129	149	175	97	111	119
0.4 •••	636	652	1108	294	311	506	175	190	292	120	133	192	90	100	136
•••	641	684	1145	300	338	507	181	211	281	125	147	175	94	110	117
0.45 •••	598	619	1134	276	297	512	166	183	292	114	128	190	85	96	134
•••	605	659	1157	284	329	507	172	206	278	119	143	171	89	107	113
0.5 •••	548	574	1134	255	280	506	154	174	286	107	121	184	80	91	128
•••	557	623	1145	264	315	496	161	198	269	112	137	164	84	102	107
0.55 •••	489	522	1109	231	259	490	141	162	274	98	113	174	73	84	120
•••	500	579	1110	241	297	475	149	187	255	104	129	153	78	95	98
0.6 •••	425	466	1059	205	236	462	127	149	255	89	104	160	66	77	109
•••	439	530	1052	217	276	444	136	173	234	95	119	138	71	87	86
0.65 •••	362	409	983	180	212	423	113	134	230	79	93	142	59	68	95
•••	379	478	970	192	251	403	121	157	208	85	107	119	63	77	72
0.7 •••	302	354	882	155	186	372	98	117	198	68	80	120	.	.	.
•••	321	424	865	167	223	351	106	138	176	73	92	97	.	.	.
0.75 •••	248	300	757	130	159	311	82	99	161	.	.	.	.	.	.
•••	267	366	736	142	192	289	89	116	138	.	.	.	.	.	.
0.8 •••	198	246	607	105	129	239	.	.	.	.	.	.	.	.	.
•••	217	305	584	115	155	216	.	.	.	.	.	.	.	.	.
0.85 •••	150	190	432	.	.	.	.	.	.	.	.	.	.	.	.
•••	166	235	409	.	.	.	.	.	.	.	.	.	.	.	.

TABLE 1: ALPHA= 0.005 POWER= 0.8 EXPECTED ACCRUAL THRU MINIMUM FOLLOW-UP= 40

		DEL=.10			DEL=.15			DEL=.20			DEL=.25			DEL=.30		
FACT=		1.0 .75	.50 .25	.00 BIN	1.0 .75	.50 .25	.00 BIN	1.0 .75	.50 .25	.00 BIN	1.0 75	.50 .25	.00 BIN	1.0 .75	.50 .25	.00 BIN
PCONT=***							REQUIRED	NUMBER OF	PATIENTS							
0.05	***	245	247	265	143	144	155	101	102	109	78	80	84	65	66	69
	***	246	249	409	143	147	216	101	104	138	79	81	97	65	67	72
0.1	***	387	390	444	205	208	236	135	138	155	100	102	113	79	81	89
	***	388	395	584	206	213	289	136	142	176	101	106	119	80	84	86
0.15	***	499	504	609	251	256	307	159	164	193	114	119	137	88	92	105
	***	501	513	736	253	265	351	161	171	208	116	124	138	90	96	98
0.2	***	581	588	755	283	291	368	175	183	226	124	130	157	94	100	117
	***	584	601	865	286	304	403	178	193	234	126	137	153	96	105	107
0.25	***	634	643	879	303	313	419	185	194	252	129	137	172	97	104	127
	***	637	662	970	306	330	444	188	207	255	132	146	164	100	111	113
0.3	***	659	672	980	311	324	459	188	200	272	130	140	183	98	106	133
	***	664	697	1052	316	346	475	192	217	269	134	152	171	101	114	117
0.35	***	661	677	1056	309	326	488	186	201	285	129	141	189	97	106	136
	***	666	710	1110	315	353	496	192	221	278	133	154	175	100	115	119
0.4	***	641	663	1108	300	321	506	181	198	292	125	139	192	94	104	136
	***	649	704	1145	307	353	507	187	220	281	130	153	175	98	114	117
0.45	***	605	633	1134	284	309	512	172	192	292	119	134	190	89	100	134
	***	615	683	1157	293	345	507	180	216	278	125	150	171	94	111	113
0.5	***	557	591	1134	264	293	506	161	183	286	112	128	184	84	95	128
	***	569	651	1145	275	333	496	170	208	269	119	144	164	89	106	107
0.55	***	500	542	1109	241	274	490	149	172	274	104	120	174	78	89	120
	***	515	611	1110	253	315	475	158	197	255	111	136	153	82	99	98
0.6	***	439	490	1059	217	252	462	136	159	255	95	110	160	71	81	109
	***	457	564	1052	230	294	444	145	183	234	101	125	138	75	90	86
0.65	***	379	435	983	192	228	423	121	143	230	85	99	142	63	72	95
	***	399	513	970	206	269	403	130	166	208	91	112	119	67	80	72
0.7	***	321	381	882	167	201	372	106	126	198	73	86	120	.	.	.
	***	344	457	865	181	240	351	114	146	176	78	97	97	.	.	.
0.75	***	267	326	757	142	173	311	89	106	161	.	.	.	.	.	.
	***	290	398	736	154	206	289	96	123	138	.	.	.	.	.	.
0.8	***	217	269	607	115	140	239	.	.	.	.	.	.	.	.	.
	***	237	332	584	125	166	216	.	.	.	.	.	.	.	.	.
0.85	***	166	208	432	.	.	.	.	.	.	.	.	.	.	.	.
	***	183	256	409	.	.	.	.	.	.	.	.	.	.	.	.

TABLE 1: ALPHA= 0.005 POWER= 0.8   EXPECTED ACCRUAL THRU MINIMUM FOLLOW-UP= 50

		DEL=.10			DEL=.15			DEL=.20			DEL=.25			DEL=.30		
FACT=		1.0 .75	.50 .25	.00 BIN	1.0 .75	.50 .25	.00 BIN	1.0 .75	.50 .25	.00 BIN	1.0 75	.50 .25	.00 BIN	1.0 .75	.50 .25	.00 BIN
PCONT=***							REQUIRED NUMBER OF PATIENTS								
0.05	***	246	247	265	143	145	155	101	103	109	79	80	84	65	66	69
	***	246	250	409	144	147	216	102	105	138	79	82	97	65	67	72
0.1	***	388	391	444	205	209	236	136	139	155	100	103	113	80	82	89
	***	389	398	584	207	215	289	137	144	176	102	107	119	81	85	86
0.15	***	500	506	609	252	259	307	160	166	193	115	120	137	90	94	105
	***	502	517	736	255	268	351	163	173	208	117	126	138	91	97	98
0.2	***	583	591	755	285	294	368	177	186	226	125	132	157	96	101	117
	***	586	608	865	288	309	403	180	196	234	128	140	153	98	107	107
0.25	***	636	648	879	305	318	419	187	198	252	131	140	172	99	106	127
	***	640	671	970	310	337	444	191	212	255	135	150	164	102	113	113
0.3	***	662	678	980	314	331	459	191	205	272	133	144	183	100	109	133
	***	668	709	1052	320	355	475	197	222	269	137	156	171	104	117	117
0.35	***	665	686	1056	314	334	488	191	207	285	132	145	189	99	109	136
	***	672	725	1110	321	364	496	197	227	278	137	158	175	103	118	119
0.4	***	647	674	1108	305	330	506	186	205	292	129	143	192	97	107	136
	***	656	722	1145	314	364	507	193	227	281	135	158	175	101	117	117
0.45	***	612	646	1134	291	320	512	178	200	292	124	139	190	93	104	134
	***	624	705	1157	302	358	507	186	224	278	130	155	171	97	114	113
0.5	***	566	608	1134	272	305	506	168	191	286	117	133	184	88	99	128
	***	580	676	1145	284	347	496	177	216	269	124	149	164	92	109	107
0.55	***	511	562	1109	250	286	490	156	180	274	109	125	174	81	92	120
	***	529	637	1110	264	330	475	166	205	255	116	140	153	86	102	98
0.6	***	453	511	1059	227	265	462	143	167	255	100	115	160	74	84	109
	***	474	592	1052	242	308	444	152	191	234	106	130	138	78	93	86
0.65	***	394	458	983	203	240	423	128	151	230	89	103	142	66	75	95
	***	418	541	970	218	283	403	137	174	208	95	116	119	69	82	72
0.7	***	338	404	882	178	213	372	112	133	198	77	89	120	.	.	.
	***	363	485	865	192	252	351	121	153	176	82	100	97	.	.	.
0.75	***	285	348	757	151	183	311	94	112	161	.	.	.	.	.	.
	***	309	423	736	164	216	289	102	127	138	.	.	.	.	.	.
0.8	***	232	289	607	123	149	239	.	.	.	.	.	.	.	.	.
	***	255	353	584	133	174	216	.	.	.	.	.	.	.	.	.
0.85	***	179	223	432	.	.	.	.	.	.	.	.	.	.	.	.
	***	196	271	409	.	.	.	.	.	.	.	.	.	.	.	.

TABLE 1: ALPHA= 0.005 POWER= 0.8   EXPECTED ACCRUAL THRU MINIMUM FOLLOW-UP= 60

		DEL=.10			DEL=.15			DEL=.20			DEL=.25			DEL=.30		
FACT=		1.0 .75	.50 .25	.00 BIN	1.0 .75	.50 .25	.00 BIN	1.0 .75	.50 .25	.00 BIN	1.0 75	.50 .25	.00 BIN	1.0 .75	.50 .25	.00 BIN
PCONT=***		REQUIRED NUMBER OF PATIENTS														
0.05	***	246	248	265	143	146	155	101	103	109	79	81	84	65	67	69
	•••	247	251	409	144	148	216	102	105	138	80	82	97	66	68	72
0.1	***	388	393	444	206	211	236	136	140	155	101	104	113	80	83	89
	•••	390	400	584	208	217	289	138	145	176	102	107	119	81	85	86
0.15	***	502	508	609	254	261	307	162	168	193	117	122	137	91	95	105
	•••	504	521	736	256	271	351	164	175	208	119	127	138	92	98	98
0.2	***	585	595	755	287	298	368	179	188	226	127	134	157	97	103	117
	•••	588	614	865	291	313	403	183	199	234	130	142	153	100	108	107
0.25	***	639	653	879	308	322	419	190	202	252	133	142	172	101	108	127
	•••	643	679	970	313	343	444	194	216	255	137	152	164	104	115	113
0.3	***	666	685	980	318	336	459	195	209	272	136	147	183	102	111	133
	•••	672	720	1052	324	362	475	200	227	269	140	158	171	106	119	117
0.35	***	669	694	1056	318	341	488	194	212	285	135	149	189	102	111	136
	•••	677	739	1110	326	372	496	201	232	278	141	161	175	106	120	119
0.4	***	652	684	1108	311	338	506	190	211	292	133	147	192	100	110	136
	•••	663	739	1145	321	374	507	198	233	281	139	161	175	104	119	117
0.45	***	619	659	1134	297	329	512	183	206	292	128	143	190	96	107	134
	•••	633	724	1157	309	369	507	192	230	278	134	158	171	100	116	113
0.5	***	574	623	1134	280	315	506	174	198	286	121	137	184	91	102	128
	•••	591	697	1145	293	358	496	183	223	269	128	153	164	95	111	107
0.55	***	522	579	1109	259	297	490	162	187	274	113	129	174	84	95	120
	•••	542	661	1110	274	342	475	172	212	255	120	144	153	89	104	98
0.6	***	466	530	1059	236	276	462	149	173	255	104	119	160	77	87	109
	•••	490	617	1052	252	320	444	159	197	234	110	133	138	81	95	86
0.65	***	409	478	983	212	251	423	134	157	230	93	107	142	68	77	95
	•••	435	566	970	228	294	403	143	179	208	99	119	119	72	84	72
0.7	***	354	424	882	187	223	372	117	138	198	80	92	120	.	.	.
	•••	381	508	865	201	262	351	126	157	176	86	102	97	.	.	.
0.75	***	300	366	757	159	192	311	99	116	161	.	.	.	.	.	.
	•••	326	444	736	173	225	289	106	131	138	.	.	.	.	.	.
0.8	***	246	305	607	129	155	239	.	.	.	.	.	.	.	.	.
	•••	269	370	584	140	181	216	.	.	.	.	.	.	.	.	.
0.85	***	190	235	432	.	.	.	.	.	.	.	.	.	.	.	.
	•••	208	284	409	.	.	.	.	.	.	.	.	.	.	.	.

TABLE 1: ALPHA= 0.005 POWER= 0.8 EXPECTED ACCRUAL THRU MINIMUM FOLLOW-UP= 70

		DEL=.10			DEL=.15			DEL=.20			DEL=.25			DEL=.30	
FACT=	1.0 .75	.50 .25	.00 BIN	1.0 .75	.50 .25	.00 BIN	1.0 .75	.50 .25	.00 BIN	1.0 75	.50 .25	.00 BIN	1.0 .75	.50 .25	.00 BIN

PCONT=•••   REQUIRED NUMBER OF PATIENTS

PCONT		DEL=.10			DEL=.15			DEL=.20			DEL=.25			DEL=.30		
		1.0/.75	.50/.25	.00/BIN	1.0/.75	.50/.25	.00/BIN	1.0/.75	.50/.25	.00/BIN	1.0/75	.50/.25	.00/BIN	1.0/.75	.50/.25	.00/BIN
0.05	***	246	249	265	144	146	155	102	104	109	79	81	84	66	67	69
	•••	247	252	409	145	149	216	102	105	138	80	82	97	66	68	72
0.1	***	389	394	444	207	212	236	137	141	155	102	105	113	81	83	89
	•••	391	403	584	209	218	289	139	146	176	103	108	119	82	86	86
0.15	***	503	511	609	255	263	307	163	170	193	118	123	137	91	95	105
	•••	505	525	736	258	274	351	165	177	208	120	128	138	93	99	98
0.2	***	586	598	755	289	301	368	181	190	226	128	136	157	98	104	117
	•••	590	620	865	293	317	403	185	201	234	131	143	153	101	109	107
0.25	***	641	657	879	311	326	419	192	205	252	135	145	172	103	110	127
	•••	646	687	970	316	348	444	197	219	255	139	154	164	106	116	113
0.3	***	669	691	980	321	341	459	198	213	272	138	150	183	104	113	133
	•••	676	730	1052	329	369	475	204	230	269	143	161	171	108	120	117
0.35	***	673	702	1056	322	348	488	198	217	285	138	152	189	104	114	136
	•••	683	752	1110	332	380	496	205	237	278	144	164	175	108	122	119
0.4	***	658	694	1108	316	346	506	195	216	292	136	150	192	102	112	136
	•••	670	754	1145	327	383	507	203	238	281	142	164	175	106	121	117
0.45	***	626	672	1134	304	338	512	188	211	292	131	147	190	98	109	134
	•••	642	742	1157	317	379	507	197	235	278	138	161	171	103	118	113
0.5	***	583	638	1134	287	325	506	179	203	286	125	141	184	93	104	128
	•••	602	717	1145	301	368	496	189	228	269	132	156	164	98	113	107
0.55	***	532	595	1109	267	307	490	167	192	274	117	133	174	87	97	120
	•••	555	682	1110	282	352	475	177	217	255	123	147	153	91	106	98
0.6	***	478	548	1059	244	285	462	154	179	255	107	122	160	79	89	109
	•••	504	638	1052	261	330	444	164	202	234	114	136	138	83	97	86
0.65	***	423	496	983	220	261	423	139	162	230	96	110	142	70	79	95
	•••	451	587	970	236	303	403	149	184	208	102	121	119	74	85	72
0.7	***	368	441	882	194	232	372	122	143	198	83	95	120	.	.	.
	•••	396	528	865	210	271	351	131	161	176	88	104	97	.	.	.
0.75	***	313	383	757	166	199	311	103	119	161	.	.	.	.	.	.
	•••	341	462	736	180	232	289	110	134	138	.	.	.	.	.	.
0.8	***	258	319	607	135	161	239	.	.	.	.	.	.	.	.	.
	•••	283	385	584	146	186	216	.	.	.	.	.	.	.	.	.
0.85	***	199	246	432	.	.	.	.	.	.	.	.	.	.	.	.
	•••	218	295	409	.	.	.	.	.	.	.	.	.	.	.	.

TABLE 1: ALPHA= 0.005 POWER= 0.8   EXPECTED ACCRUAL THRU MINIMUM FOLLOW-UP= 80

		DEL=.10			DEL=.15			DEL=.20			DEL=.25			DEL=.30		
FACT=		1.0 .75	.50 .25	.00 BIN	1.0 .75	.50 .25	.00 BIN	1.0 .75	.50 .25	.00 BIN	1.0 75	.50 .25	.00 BIN	1.0 .75	.50 .25	.00 BIN
PCONT=***					REQUIRED NUMBER OF PATIENTS											
0.05	***	247	249	265	144	147	155	102	104	109	80	81	84	66	67	69
	***	248	253	409	145	149	216	103	106	138	80	82	97	66	68	72
0.1	***	390	395	444	208	213	236	138	142	155	102	106	113	81	84	89
	***	392	405	584	210	220	289	140	146	176	104	109	119	82	86	86
0.15	***	504	513	609	256	265	307	164	171	193	119	124	137	92	96	105
	***	507	529	736	259	276	351	167	178	208	121	129	138	94	100	98
0.2	***	588	601	755	291	304	368	183	193	226	130	137	157	100	105	117
	***	593	625	865	295	320	403	186	203	234	133	144	153	102	110	107
0.25	***	643	662	879	313	330	419	194	207	252	137	146	172	104	111	127
	***	650	695	970	319	352	444	199	221	255	141	155	164	107	117	113
0.3	***	672	697	980	324	346	459	200	217	272	140	152	183	106	114	133
	***	681	740	1052	333	374	475	207	233	269	145	163	171	109	121	117
0.35	***	677	710	1056	326	353	488	201	221	285	141	154	189	106	115	136
	***	689	763	1110	337	386	496	209	240	278	146	166	175	110	123	119
0.4	***	663	704	1108	321	353	506	198	220	292	139	153	192	104	114	136
	***	677	768	1145	333	390	507	207	242	281	145	167	175	108	123	117
0.45	***	633	683	1134	309	345	512	192	216	292	134	150	190	100	111	134
	***	651	758	1157	323	387	507	202	239	278	141	164	171	105	120	113
0.5	***	591	651	1134	293	333	506	183	208	286	128	144	184	95	106	128
	***	613	734	1145	309	377	496	193	232	269	135	158	164	100	115	107
0.55	***	542	611	1109	274	315	490	172	197	274	120	136	174	89	99	120
	***	568	700	1110	290	361	475	182	221	255	127	149	153	93	107	98
0.6	***	490	564	1059	252	294	462	159	183	255	110	125	160	81	90	109
	***	518	657	1052	269	339	444	169	206	234	117	138	138	85	98	86
0.65	***	435	513	983	228	269	423	143	166	230	99	112	142	72	80	95
	***	465	606	970	244	311	403	153	187	208	105	123	119	75	86	72
0.7	***	381	457	882	201	240	372	126	146	198	86	97	120	.	.	.
	***	411	546	865	217	278	351	135	164	176	90	106	97	.	.	.
0.75	***	326	398	757	173	206	311	106	123	161	.	.	.	.	.	.
	***	354	477	736	186	238	289	113	136	138	.	.	.	.	.	.
0.8	***	269	332	607	140	166	239	.	.	.	.	.	.	.	.	.
	***	294	398	584	151	190	216	.	.	.	.	.	.	.	.	.
0.85	***	208	256	432	.	.	.	.	.	.	.	.	.	.	.	.
	***	227	304	409	.	.	.	.	.	.	.	.	.	.	.	.

TABLE 1: ALPHA= 0.005 POWER= 0.8   EXPECTED ACCRUAL THRU MINIMUM FOLLOW-UP= 90

		DEL=.10			DEL=.15			DEL=.20			DEL=.25			DEL=.30		
FACT=		1.0 .75	.50 .25	.00 BIN	1.0 .75	.50 .25	.00 BIN	1.0 .75	.50 .25	.00 BIN	1.0 75	.50 .25	.00 BIN	1.0 .75	.50 .25	.00 BIN
PCONT=***					REQUIRED NUMBER OF PATIENTS											
0.05	***	247	250	265	145	147	155	102	104	109	80	81	84	66	67	69
	***	248	254	409	146	150	216	103	106	138	81	83	97	66	68	72
0.1	***	390	397	444	209	214	236	138	143	155	103	106	113	82	84	89
	***	393	407	584	211	221	289	140	147	176	104	109	119	83	86	86
0.15	***	505	515	609	257	267	307	165	172	193	119	125	137	93	97	105
	***	508	532	736	261	278	351	168	180	208	122	129	138	95	100	98
0.2	***	590	605	755	293	306	368	184	194	226	131	139	157	100	106	117
	***	595	630	865	298	323	403	188	205	234	134	145	153	103	110	107
0.25	***	646	667	879	316	334	419	196	210	252	138	148	172	105	112	127
	***	653	702	970	322	356	444	202	223	255	142	157	164	108	118	113
0.3	***	675	703	980	328	351	459	203	219	272	142	154	183	107	116	133
	***	685	749	1052	336	379	475	209	236	269	147	164	171	111	122	117
0.35	***	682	717	1056	330	359	488	204	224	285	143	156	189	108	117	136
	***	694	774	1110	341	392	496	212	243	278	149	168	175	111	124	119
0.4	***	668	713	1108	326	359	506	202	224	292	141	156	192	106	116	136
	***	684	781	1145	338	397	507	211	245	281	147	169	175	110	124	117
0.45	***	640	694	1134	315	352	512	196	220	292	137	152	190	102	113	134
	***	659	772	1157	329	394	507	206	243	278	143	166	171	107	121	113
0.5	***	600	664	1134	299	340	506	187	212	286	131	147	184	97	108	128
	***	623	750	1145	315	384	496	198	236	269	137	160	164	102	116	107
0.55	***	552	624	1109	280	323	490	176	201	274	123	138	174	91	101	120
	***	579	717	1110	297	368	475	187	225	255	129	151	153	95	108	98
0.6	***	501	579	1059	259	302	462	163	187	255	113	128	160	83	92	109
	***	530	674	1052	276	346	444	173	210	234	119	140	138	87	99	86
0.65	***	447	528	983	234	276	423	147	170	230	101	114	142	73	81	95
	***	478	622	970	251	318	403	157	191	208	107	125	119	77	87	72
0.7	***	393	472	882	208	246	372	129	150	198	88	99	120	.	.	.
	***	424	562	865	223	284	351	138	167	176	92	107	97	.	.	.
0.75	***	337	411	757	178	211	311	109	125	161	.	.	.	.	.	.
	***	366	491	736	192	243	289	116	138	138	.	.	.	.	.	.
0.8	***	279	343	607	145	170	239	.	.	.	.	.	.	.	.	.
	***	305	409	584	155	194	216	.	.	.	.	.	.	.	.	.
0.85	***	216	264	432	.	.	.	.	.	.	.	.	.	.	.	.
	***	235	312	409	.	.	.	.	.	.	.	.	.	.	.	.

TABLE 1: ALPHA= 0.005 POWER= 0.8   EXPECTED ACCRUAL THRU MINIMUM FOLLOW-UP= 100

		DEL=.10			DEL=.15			DEL=.20			DEL=.25			DEL=.30		
FACT=		1.0 .75	.50 .25	.00 BIN	1.0 .75	.50 .25	.00 BIN	1.0 .75	.50 .25	.00 BIN	1.0 75	.50 .25	.00 BIN	1.0 .75	.50 .25	.00 BIN
PCONT=***				REQUIRED NUMBER OF PATIENTS												
0.05	***	247	250	265	145	147	155	103	105	109	80	82	84	66	67	69
	***	248	254	409	146	150	216	104	106	138	81	83	97	67	68	72
0.1	***	391	398	444	209	215	236	139	144	155	103	107	113	82	85	89
	***	394	408	584	212	222	289	141	148	176	105	109	119	83	86	86
0.15	***	506	517	609	259	268	307	166	173	193	120	126	137	94	97	105
	***	510	535	736	262	280	351	169	180	208	123	130	138	95	100	98
0.2	***	591	608	755	294	309	368	186	196	226	132	140	157	101	107	117
	***	597	635	865	300	325	403	190	206	234	135	146	153	104	111	107
0.25	***	648	671	879	318	337	419	198	212	252	140	150	172	106	113	127
	***	656	708	970	325	360	444	204	225	255	144	158	164	109	118	113
0.3	***	678	709	980	331	355	459	205	222	272	144	156	183	109	117	133
	***	689	757	1052	340	383	475	212	238	269	149	166	171	112	123	117
0.35	***	686	725	1056	334	364	488	207	227	285	145	158	189	109	118	136
	***	699	784	1110	345	397	496	215	246	278	151	170	175	113	125	119
0.4	***	674	722	1108	330	364	506	205	227	292	143	158	192	107	117	136
	***	691	793	1145	343	403	507	214	249	281	149	170	175	112	125	117
0.45	***	646	705	1134	320	358	512	200	224	292	139	155	190	104	114	134
	***	668	786	1157	335	400	507	209	246	278	146	168	171	108	122	113
0.5	***	608	676	1134	305	347	506	191	216	286	133	149	184	99	109	128
	***	633	765	1145	322	391	496	201	239	269	140	162	164	103	117	107
0.55	***	562	637	1109	286	330	490	180	205	274	125	140	174	92	102	120
	***	590	732	1110	304	375	475	191	228	255	132	153	153	97	109	98
0.6	***	511	592	1059	265	308	462	167	191	255	115	130	160	84	93	109
	***	542	689	1052	282	352	444	177	213	234	121	141	138	88	100	86
0.65	***	458	541	983	240	283	423	151	174	230	103	116	142	75	82	95
	***	490	637	970	258	324	403	160	193	208	109	126	119	78	88	72
0.7	***	404	485	882	213	252	372	133	153	198	89	100	120	.	.	.
	***	436	576	865	229	289	351	141	169	176	94	108	97	.	.	.
0.75	***	348	423	757	183	216	311	112	127	161	.	.	.	.	.	.
	***	378	504	736	197	247	289	118	140	138	.	.	.	.	.	.
0.8	***	289	353	607	149	174	239	.	.	.	.	.	.	.	.	.
	***	315	419	584	159	197	216	.	.	.	.	.	.	.	.	.
0.85	***	223	271	432	.	.	.	.	.	.	.	.	.	.	.	.
	***	243	319	409	.	.	.	.	.	.	.	.	.	.	.	.

TABLE 1: ALPHA= 0.005 POWER= 0.8 EXPECTED ACCRUAL THRU MINIMUM FOLLOW-UP= 110

	DEL=.05			DEL=.10			DEL=.15			DEL=.20			DEL=.25		
FACT=	1.0 .75	.50 .25	.00 BIN	1.0 .75	.50 .25	.00 BIN	1.0 .75	.50 .25	.00 BIN	1.0 75	.50 .25	.00 BIN	1.0 .75	.50 .25	.00 BIN
PCONT=***				REQUIRED NUMBER OF PATIENTS											
0.05 ***	698	700	746	248	251	265	145	148	155	103	105	109	80	82	84
***	698	706	1285	249	255	409	146	150	216	104	106	138	81	83	97
0.1 ***	1271	1278	1439	392	399	444	210	216	236	140	144	155	104	107	113
***	1274	1291	2033	394	410	584	212	222	289	142	148	176	105	110	119
0.15 ***	1758	1769	2109	507	519	609	260	270	307	167	174	193	121	126	137
***	1762	1790	2687	511	538	736	264	281	351	170	181	208	123	131	138
0.2 ***	2133	2150	2721	593	611	755	296	311	368	187	197	226	133	141	157
***	2139	2182	3247	599	639	865	302	328	403	191	207	234	136	147	153
0.25 ***	2393	2416	3256	650	675	879	320	340	419	200	214	252	141	151	172
***	2400	2462	3714	659	714	970	328	363	444	206	227	255	145	159	164
0.3 ***	2540	2572	3703	682	714	980	333	359	459	207	224	272	146	157	183
***	2551	2634	4088	693	765	1052	343	387	475	214	241	269	150	167	171
0.35 ***	2586	2627	4055	690	732	1056	338	368	488	210	230	285	147	160	189
***	2600	2710	4368	705	794	1110	350	402	496	218	248	278	152	171	175
0.4 ***	2542	2596	4308	679	731	1108	334	370	506	208	230	292	145	160	192
***	2560	2703	4555	698	804	1145	348	408	507	217	251	281	152	172	175
0.45 ***	2423	2492	4462	653	715	1134	325	364	512	203	227	292	141	157	190
***	2446	2628	4649	676	798	1157	340	406	507	213	249	278	148	169	171
0.5 ***	2244	2332	4514	616	687	1134	310	353	506	194	220	286	135	151	184
***	2273	2501	4649	642	778	1145	327	397	496	205	242	269	142	163	164
0.55 ***	2021	2133	4464	571	650	1109	292	336	490	183	209	274	127	142	174
***	2059	2337	4555	601	746	1110	310	381	475	194	231	255	134	154	153
0.6 ***	1773	1913	4312	521	605	1059	270	315	462	170	194	255	117	131	160
***	1821	2150	4368	553	703	1052	288	358	444	180	216	234	123	142	138
0.65 ***	1517	1686	4059	468	554	983	246	288	423	154	177	230	105	118	142
***	1576	1948	4088	502	651	970	263	329	403	164	196	208	111	127	119
0.7 ***	1269	1463	3706	414	497	882	219	257	372	135	155	198	91	101	120
***	1340	1737	3714	447	588	865	235	294	351	144	171	176	96	109	97
0.75 ***	1043	1248	3253	357	434	757	188	221	311	114	129	161	.	.	.
***	1120	1518	3247	388	515	736	202	251	289	121	142	138	.	.	.
0.8 ***	841	1039	2702	297	362	607	152	178	239	.	.	.	.	.	.
***	917	1287	2687	323	428	584	163	199	216	.	.	.	.	.	.
0.85 ***	655	828	2053	230	278	432	.	.	.	.	.	.	.	.	.
***	722	1035	2033	249	325	409	.	.	.	.	.	.	.	.	.
0.9 ***	467	596	1308	.	.	.	.	.	.	.	.	.	.	.	.
***	518	741	1285	.	.	.	.	.	.	.	.	.	.	.	.

TABLE 1: ALPHA= 0.005 POWER= 0.8   EXPECTED ACCRUAL THRU MINIMUM FOLLOW-UP= 120

		DEL=.05			DEL=.10			DEL=.15			DEL=.20			DEL=.25		
FACT=		1.0 .75	.50 .25	.00 BIN	1.0 .75	.50 .25	.00 BIN	1.0 .75	.50 .25	.00 BIN	1.0 75	.50 .25	.00 BIN	1.0 .75	.50 .25	.00 BIN
PCONT=***							REQUIRED NUMBER OF PATIENTS								
0.05 ***		698	701	746	248	251	265	146	148	155	103	105	109	81	82	84
•••		699	707	1285	249	256	409	147	151	216	104	107	138	81	83	97
0.1 ***		1272	1279	1439	393	400	444	211	217	236	140	145	155	104	107	113
•••		1274	1293	2033	395	411	584	213	223	289	142	148	176	106	110	119
0.15 ***		1759	1771	2109	508	521	609	261	271	307	168	175	193	122	127	137
•••		1763	1794	2687	513	541	736	265	283	351	171	182	208	124	131	138
0.2 ***		2135	2153	2721	595	614	755	298	313	368	188	199	226	134	142	157
•••		2141	2188	3247	601	643	865	304	330	403	193	208	234	137	147	153
0.25 ***		2395	2420	3256	653	679	879	322	343	419	202	216	252	142	152	172
•••		2403	2470	3714	662	720	970	330	366	444	207	228	255	146	160	164
0.3 ***		2543	2577	3703	685	720	980	336	362	459	209	227	272	147	158	183
•••		2554	2646	4088	697	772	1052	346	391	475	217	242	269	152	168	171
0.35 ***		2589	2635	4055	694	739	1056	341	372	488	212	232	285	149	161	189
•••		2605	2725	4368	710	802	1110	353	406	496	220	250	278	154	172	175
0.4 ***		2547	2606	4308	684	739	1108	338	374	506	211	233	292	147	161	192
•••		2566	2722	4555	704	814	1145	353	412	507	220	253	281	153	173	175
0.45 ***		2429	2505	4462	659	724	1134	329	369	512	206	230	292	143	158	190
•••		2454	2652	4649	683	809	1157	345	411	507	216	251	278	150	170	171
0.5 ***		2252	2348	4513	623	697	1134	315	358	506	198	223	286	137	153	184
•••		2284	2529	4555	651	790	1145	333	402	496	208	245	269	144	165	164
0.55 ***		2032	2153	4464	579	661	1109	297	342	490	187	212	274	129	144	174
•••		2073	2370	4555	610	759	1110	315	386	475	197	234	255	136	156	153
0.6 ***		1786	1937	4312	530	617	1059	276	320	462	173	197	255	119	133	160
•••		1838	2186	4368	564	716	1052	294	363	444	183	218	234	125	144	138
0.65 ***		1533	1714	4059	478	566	983	251	294	423	157	179	230	107	119	142
•••		1597	1987	4088	513	663	970	269	334	403	166	198	208	112	128	119
0.7 ***		1289	1493	3706	424	508	882	223	262	372	138	157	198	92	102	120
•••		1364	1776	3714	457	600	865	240	298	351	146	173	176	97	110	97
0.75 ***		1065	1279	3253	366	444	757	192	225	311	116	131	161	.	.	.
•••		1146	1555	3247	397	525	736	206	254	289	122	143	138	.	.	.
0.8 ***		863	1068	2702	305	370	607	155	181	239	.	.	.	.	.	.
•••		942	1321	2687	331	436	584	166	202	216	.	.	.	.	.	.
0.85 ***		675	852	2053	235	284	432	.	.	.	.	.	.	.	.	.
•••		744	1063	2033	256	330	409	.	.	.	.	.	.	.	.	.
0.9 ***		482	613	1308	.	.	.	.	.	.	.	.	.	.	.	.
•••		534	760	1285	.	.	.	.	.	.	.	.	.	.	.	.

TABLE 1: ALPHA= 0.005 POWER= 0.8   EXPECTED ACCRUAL THRU MINIMUM FOLLOW-UP= 130

	DEL=.05			DEL=.10			DEL=.15			DEL=.20			DEL=.25		
FACT=	1.0 .75	.50 .25	.00 BIN	1.0 .75	.50 .25	.00 BIN	1.0 .75	.50 .25	.00 BIN	1.0 75	.50 .25	.00 BIN	1.0 .75	.50 .25	.00 BIN

PCONT=***   REQUIRED NUMBER OF PATIENTS

PCONT	1.0/.75	.50/.25	.00/BIN	1.0/.75	.50/.25	.00/BIN	1.0/.75	.50/.25	.00/BIN	1.0/.75	.50/.25	.00/BIN	1.0/.75	.50/.25	.00/BIN
0.05	698	701	746	248	252	265	146	148	155	103	105	109	81	82	84
	699	708	1285	250	256	409	147	151	216	104	107	138	81	83	97
0.1	1273	1280	1439	393	401	444	211	218	236	141	145	155	105	108	113
	1275	1295	2033	396	413	584	214	224	289	143	149	176	106	110	119
0.15	1760	1773	2109	510	523	609	262	273	307	169	176	193	122	127	137
	1764	1798	2687	514	543	736	266	284	351	172	183	208	125	131	138
0.2	2136	2156	2721	596	617	755	299	315	368	189	200	226	135	142	157
	2143	2194	3247	604	647	865	305	332	403	194	210	234	138	148	153
0.25	2397	2424	3256	655	683	879	324	345	419	203	217	252	144	153	172
	2406	2479	3714	665	725	970	333	368	444	209	230	255	148	160	164
0.3	2546	2583	3703	688	725	980	339	366	459	211	229	272	148	160	183
	2558	2657	4088	701	779	1052	349	394	475	219	244	269	153	169	171
0.35	2593	2642	4055	698	745	1056	344	376	488	215	235	285	150	163	189
	2610	2740	4368	715	811	1110	357	410	496	223	252	278	156	173	175
0.4	2552	2615	4308	689	747	1108	342	379	506	213	236	292	149	163	192
	2573	2741	4555	710	823	1145	357	417	507	223	256	281	155	174	175
0.45	2435	2517	4462	665	733	1134	334	374	512	209	232	292	145	160	190
	2463	2675	4649	691	820	1157	350	415	507	219	254	278	152	172	171
0.5	2260	2364	4514	630	707	1134	320	363	506	200	225	286	139	154	184
	2295	2557	4649	660	802	1145	338	407	496	211	247	269	146	166	164
0.55	2042	2173	4464	587	672	1109	302	347	490	190	214	274	131	146	174
	2086	2402	4555	620	771	1110	321	391	475	200	236	255	137	157	153
0.6	1799	1961	4312	539	628	1059	281	325	462	176	200	255	121	134	160
	1855	2221	4368	574	728	1052	299	368	444	186	220	234	127	145	138
0.65	1549	1741	4059	487	577	983	256	299	423	160	182	230	108	120	142
	1617	2024	4088	523	675	970	274	338	403	169	200	208	114	129	119
0.7	1309	1522	3706	433	519	882	228	267	372	140	159	198	94	103	120
	1387	1813	3714	467	610	865	244	302	351	149	174	176	98	110	97
0.75	1087	1307	3253	375	453	757	196	229	311	118	133	161	.	.	.
	1170	1591	3247	406	534	736	210	257	289	124	144	138	.	.	.
0.8	884	1095	2702	312	378	607	158	183	239	.	.	.	.	.	.
	965	1353	2687	339	444	584	169	204	216	.	.	.	.	.	.
0.85	693	875	2053	241	290	432	.	.	.	.	.	.	.	.	.
	764	1089	2033	261	335	409	.	.	.	.	.	.	.	.	.
0.9	496	629	1308	.	.	.	.	.	.	.	.	.	.	.	.
	549	777	1285	.	.	.	.	.	.	.	.	.	.	.	.

TABLE 1: ALPHA= 0.005 POWER= 0.8   EXPECTED ACCRUAL THRU MINIMUM FOLLOW-UP= 140

		DEL=.05			DEL=.10			DEL=.15			DEL=.20			DEL=.25		
FACT=		1.0	.50	.00	1.0	.50	.00	1.0	.50	.00	1.0	.50	.00	1.0	.50	.00
		.75	.25	BIN	.75	.25	BIN	.75	.25	BIN	75	.25	BIN	.75	.25	BIN
PCONT=•••					REQUIRED NUMBER OF PATIENTS											
0.05	•••	698	702	746	249	252	265	146	149	155	104	105	109	81	82	84
	•••	700	709	1285	250	256	409	147	151	216	104	107	138	81	83	97
0.1	•••	1273	1281	1439	394	403	444	212	218	236	141	146	155	105	108	113
	•••	1276	1297	2033	397	414	584	214	224	289	143	149	176	106	110	119
0.15	•••	1761	1775	2109	511	525	609	263	274	307	170	177	193	123	128	137
	•••	1766	1802	2687	516	546	736	267	285	351	173	183	208	125	132	138
0.2	•••	2138	2159	2721	598	620	755	301	317	368	190	201	226	136	143	157
	•••	2145	2200	3247	606	651	865	307	333	403	195	210	234	139	148	153
0.25	•••	2399	2428	3256	657	687	879	326	348	419	205	219	252	145	154	172
	•••	2409	2487	3714	668	730	970	335	370	444	211	231	255	149	161	164
0.3	•••	2549	2589	3703	691	730	980	342	369	459	213	230	272	150	161	183
	•••	2562	2668	4088	705	785	1052	352	397	475	220	245	269	154	169	171
0.35	•••	2597	2650	4055	702	752	1056	348	380	488	217	237	285	152	164	189
	•••	2615	2754	4368	720	818	1110	360	413	496	225	254	278	157	174	175
0.4	•••	2556	2625	4309	694	754	1108	346	383	506	216	238	292	151	164	192
	•••	2579	2759	4555	716	832	1145	361	420	507	225	257	281	156	175	175
0.45	•••	2442	2530	4462	672	742	1134	338	379	512	211	235	292	147	161	190
	•••	2471	2697	4649	698	829	1157	354	420	507	221	256	278	153	173	171
0.5	•••	2268	2380	4514	638	717	1134	325	368	506	203	228	286	141	156	184
	•••	2305	2584	4649	668	812	1145	342	411	496	214	249	269	147	167	164
0.55	•••	2052	2193	4464	595	682	1109	307	352	490	192	217	274	133	147	174
	•••	2100	2433	4555	629	781	1110	325	395	475	203	238	255	139	158	153
0.6	•••	1812	1984	4312	548	638	1059	286	330	462	179	202	255	122	136	160
	•••	1872	2255	4368	583	739	1052	304	372	444	189	222	234	128	146	138
0.65	•••	1565	1767	4059	496	587	983	261	303	423	162	184	230	110	121	142
	•••	1638	2059	4088	532	685	970	278	342	403	171	201	208	115	130	119
0.7	•••	1327	1549	3706	441	528	882	232	271	372	143	161	198	95	104	120
	•••	1410	1849	3714	476	620	865	248	305	351	151	176	176	99	111	97
0.75	•••	1107	1335	3253	383	462	757	199	232	311	119	134	161	.	.	.
	•••	1194	1624	3247	415	542	736	213	260	289	126	145	138	.	.	.
0.8	•••	904	1120	2702	319	385	607	161	186	239	.	.	.	.	.	.
	•••	988	1383	2687	346	450	584	172	206	216	.	.	.	.	.	.
0.85	•••	711	897	2053	246	295	432	.	.	.	.	.	.	.	.	.
	•••	784	1113	2033	266	339	409	.	.	.	.	.	.	.	.	.
0.9	•••	510	645	1308	.	.	.	.	.	.	.	.	.	.	.	.
	•••	563	793	1285	.	.	.	.	.	.	.	.	.	.	.	.

TABLE 1: ALPHA= 0.005 POWER= 0.8   EXPECTED ACCRUAL THRU MINIMUM FOLLOW-UP= 150

		DEL=.05			DEL=.10			DEL=.15			DEL=.20			DEL=.25		
FACT=		1.0 .75	.50 .25	.00 BIN	1.0 .75	.50 .25	.00 BIN	1.0 .75	.50 .25	.00 BIN	1.0 75	.50 .25	.00 BIN	1.0 .75	.50 .25	.00 BIN
PCONT=***					REQUIRED NUMBER OF PATIENTS											
0.05	***	698	703	746	249	253	265	146	149	155	104	106	109	81	82	84
	•••	700	710	1285	250	257	409	147	151	216	105	107	138	82	83	97
0.1	***	1274	1282	1439	395	404	444	212	219	236	142	146	155	105	108	113
	•••	1277	1299	2033	398	415	584	215	225	289	143	149	176	107	110	119
0.15	***	1762	1777	2109	512	527	609	264	275	307	170	178	193	123	128	137
	•••	1767	1806	2687	517	548	736	268	286	351	173	184	208	126	132	138
0.2	***	2139	2162	2721	600	623	755	302	318	368	191	202	226	137	144	157
	•••	2147	2206	3247	608	654	865	309	335	403	196	211	234	140	149	153
0.25	***	2401	2432	3256	660	691	879	328	350	419	206	220	252	146	155	172
	•••	2411	2495	3714	671	734	970	337	372	444	212	232	255	149	161	164
0.3	***	2552	2594	3703	694	735	980	344	371	459	215	232	272	151	162	183
	•••	2566	2679	4088	709	791	1052	355	399	475	222	247	269	156	170	171
0.35	***	2601	2658	4055	706	758	1056	350	383	488	219	238	285	153	165	189
	•••	2620	2769	4368	725	825	1110	364	416	496	227	256	278	158	175	175
0.4	***	2561	2635	4308	699	761	1108	349	387	506	218	240	292	152	165	192
	•••	2586	2777	4555	722	840	1145	364	424	507	227	259	281	158	176	175
0.45	***	2448	2542	4462	677	750	1134	342	383	512	214	237	292	148	163	190
	•••	2479	2719	4649	705	839	1157	358	423	507	224	257	278	155	173	171
0.5	***	2276	2396	4514	644	726	1134	329	373	506	206	230	286	142	157	184
	•••	2316	2610	4649	676	822	1145	347	415	496	216	251	269	149	168	164
0.55	***	2062	2212	4463	603	691	1109	311	356	490	195	219	274	134	148	174
	•••	2113	2462	4555	637	791	1110	330	399	475	205	239	255	140	159	153
0.6	***	1825	2006	4312	556	648	1059	290	335	462	181	204	255	124	137	160
	•••	1888	2287	4368	592	749	1052	308	376	444	191	223	234	130	146	138
0.65	***	1581	1792	4059	505	596	983	265	307	423	164	186	230	111	122	142
	•••	1657	2092	4088	541	695	970	283	346	403	174	203	208	116	131	119
0.7	***	1346	1576	3706	450	537	882	236	274	372	145	163	198	96	105	120
	•••	1431	1882	3714	485	629	865	252	308	351	152	177	176	100	111	97
0.75	***	1127	1361	3253	390	470	757	203	235	311	121	135	161	.	.	.
	•••	1216	1656	3247	423	550	736	216	262	289	127	146	138	.	.	.
0.8	***	923	1144	2702	325	392	607	164	188	239	.	.	.	.	.	.
	•••	1009	1411	2687	353	456	584	174	207	216	.	.	.	.	.	.
0.85	***	728	917	2053	251	299	432	.	.	.	.	.	.	.	.	.
	•••	802	1136	2033	271	343	409	.	.	.	.	.	.	.	.	.
0.9	***	522	659	1308	.	.	.	.	.	.	.	.	.	.	.	.
	•••	577	808	1285	.	.	.	.	.	.	.	.	.	.	.	.

TABLE 1: ALPHA= 0.005 POWER= 0.8   EXPECTED ACCRUAL THRU MINIMUM FOLLOW-UP= 160

	DEL=.05			DEL=.10			DEL=.15			DEL=.20			DEL=.25		
FACT=	1.0 .75	.50 .25	.00 BIN	1.0 .75	.50 .25	.00 BIN	1.0 .75	.50 .25	.00 BIN	1.0 75	.50 .25	.00 BIN	1.0 .75	.50 .25	.00 BIN
PCONT=***				REQUIRED NUMBER OF PATIENTS											
0.05	699	703	746	249	253	265	147	149	155	104	106	109	81	82	84
	700	711	1285	251	257	409	148	151	216	105	107	138	82	83	97
0.1	1274	1284	1439	395	405	444	213	220	236	142	146	155	106	109	113
	1277	1302	2033	399	416	584	216	226	289	144	150	176	107	111	119
0.15	1763	1779	2109	513	529	609	265	276	307	171	178	193	124	129	137
	1768	1810	2687	519	550	736	269	287	351	174	184	208	126	132	138
0.2	2141	2165	2721	601	625	755	304	320	368	193	203	226	137	144	157
	2149	2212	3247	610	657	865	310	336	403	197	212	234	140	149	153
0.25	2403	2437	3256	662	695	879	330	352	419	207	221	252	146	155	172
	2414	2503	3714	674	739	970	339	374	444	213	233	255	150	162	164
0.3	2554	2600	3703	697	740	980	346	374	459	217	233	272	152	163	183
	2570	2690	4068	713	796	1052	357	402	475	224	248	269	157	171	171
0.35	2605	2665	4055	710	763	1056	353	386	488	221	240	285	154	166	189
	2625	2783	4368	730	832	1110	367	419	496	229	257	278	159	176	175
0.4	2566	2645	4308	704	768	1108	353	390	506	220	242	292	153	167	192
	2593	2795	4555	728	848	1145	368	427	507	229	261	281	159	177	175
0.45	2454	2555	4462	683	758	1134	345	387	512	216	239	292	150	164	190
	2488	2741	4649	712	847	1157	362	427	507	226	259	278	156	174	171
0.5	2284	2411	4514	651	734	1134	333	377	506	208	232	286	144	158	184
	2327	2635	4649	683	831	1145	351	418	496	218	252	269	150	169	164
0.55	2073	2231	4464	611	700	1109	315	361	490	197	221	274	136	149	174
	2127	2491	4555	646	801	1110	334	403	475	208	241	255	142	159	153
0.6	1838	2028	4312	564	657	1059	294	339	462	183	206	255	125	138	160
	1905	2318	4368	601	758	1052	313	379	444	193	225	234	131	147	138
0.65	1597	1816	4059	513	606	983	269	311	423	166	187	230	112	123	142
	1677	2125	4088	550	704	970	287	349	403	176	204	208	117	131	119
0.7	1364	1601	3706	457	546	882	240	278	372	146	164	198	97	106	120
	1453	1914	3714	493	637	865	256	311	351	154	178	176	101	112	97
0.75	1146	1386	3253	398	477	757	206	238	311	123	136	161	.	.	.
	1238	1686	3247	430	557	736	219	265	289	129	147	138	.	.	.
0.8	942	1168	2702	332	398	607	166	190	239	.	.	.	.	.	.
	1030	1437	2687	359	462	584	177	209	216	.	.	.	.	.	.
0.85	744	936	2053	256	304	432	.	.	.	.	.	.	.	.	.
	820	1157	2033	276	347	409	.	.	.	.	.	.	.	.	.
0.9	534	673	1308	.	.	.	.	.	.	.	.	.	.	.	.
	589	822	1285	.	.	.	.	.	.	.	.	.	.	.	.

TABLE 1: ALPHA= 0.005 POWER= 0.8   EXPECTED ACCRUAL THRU MINIMUM FOLLOW-UP= 170

	DEL=.05			DEL=.10			DEL=.15			DEL=.20			DEL=.25		
FACT=	1.0 .75	.50 .25	.00 BIN	1.0 .75	.50 .25	.00 BIN	1.0 .75	.50 .25	.00 BIN	1.0 75	.50 .25	.00 BIN	1.0 .75	.50 .25	.00 BIN
PCONT=***			REQUIRED NUMBER OF PATIENTS												
0.05 ***	699	704	746	249	253	265	147	149	155	104	106	109	81	82	84
•••	701	711	1285	251	258	409	148	152	216	105	107	138	82	83	97
0.1 ***	1275	1285	1439	396	406	444	214	220	236	143	147	155	106	109	113
•••	1278	1304	2033	399	417	584	216	226	289	144	150	176	107	111	119
0.15 ***	1764	1781	2109	514	531	609	266	277	307	172	179	193	124	129	137
•••	1770	1813	2687	520	552	736	270	288	351	175	185	208	126	133	138
0.2 ***	2142	2167	2721	603	628	755	305	322	368	194	204	226	138	145	157
•••	2151	2217	3247	612	660	865	312	338	403	198	212	234	141	150	153
0.25 ***	2405	2441	3256	664	698	879	332	354	419	209	222	252	147	156	172
•••	2417	2511	3714	677	743	970	341	376	444	215	234	255	151	163	164
0.3 ***	2557	2606	3703	700	744	980	349	377	459	218	235	272	153	163	183
•••	2574	2701	4088	716	801	1052	360	404	475	225	249	269	157	171	171
0.35 ***	2608	2673	4055	714	769	1056	356	389	488	222	242	285	155	167	189
•••	2630	2797	4368	734	838	1110	370	422	496	231	258	278	160	176	175
0.4 ***	2571	2655	4308	709	775	1108	356	394	506	222	244	292	154	168	192
•••	2599	2813	4555	734	855	1145	371	430	507	231	262	281	160	177	175
0.45 ***	2460	2567	4462	689	765	1134	349	390	512	218	241	292	151	165	190
•••	2496	2762	4649	718	855	1157	366	430	507	228	260	278	157	175	171
0.5 ***	2292	2427	4513	657	743	1134	337	381	507	210	234	286	145	159	184
•••	2338	2660	4649	690	839	1145	355	422	496	221	254	269	151	169	164
0.55 ***	2083	2249	4464	618	709	1109	319	364	490	199	223	274	137	150	174
•••	2140	2518	4555	653	810	1110	338	406	475	210	242	255	143	160	153
0.6 ***	1851	2050	4312	571	666	1059	298	342	462	185	208	255	126	139	160
•••	1921	2348	4368	609	767	1052	317	383	444	195	226	234	132	148	138
0.65 ***	1612	1840	4059	520	614	983	273	315	423	168	189	230	113	124	142
•••	1695	2155	4088	558	712	970	290	352	403	177	205	208	118	132	119
0.7 ***	1381	1626	3706	465	554	882	243	281	372	148	166	198	98	106	120
•••	1473	1944	3714	501	645	865	259	313	351	156	179	176	102	112	97
0.75 ***	1164	1410	3253	404	485	757	209	240	311	124	137	161	.	.	.
•••	1258	1715	3247	437	564	736	222	266	289	130	147	138	.	.	.
0.8 ***	960	1190	2702	337	404	607	168	19?	239	.	.	.	.	.	.
•••	1049	1462	2687	365	467	584	179	210	216	.	.	.	.	.	.
0.85 ***	759	955	2053	260	308	432	.	.	.	.	.	.	.	.	.
•••	836	1177	2033	280	350	409	.	.	.	.	.	.	.	.	.
0.9 ***	545	685	1308	.	.	.	.	.	.	.	.	.	.	.	.
•••	602	835	1285	.	.	.	.	.	.	.	.	.	.	.	.

TABLE 1: ALPHA= 0.005 POWER= 0.8   EXPECTED ACCRUAL THRU MINIMUM FOLLOW-UP= 180

		DEL=.05			DEL=.10			DEL=.15			DEL=.20			DEL=.25		
FACT=		1.0 .75	.50 .25	.00 BIN	1.0 .75	.50 .25	.00 BIN	1.0 .75	.50 .25	.00 BIN	1.0 75	.50 .25	.00 BIN	1.0 .75	.50 .25	.00 BIN
PCONT=***				REQUIRED NUMBER OF PATIENTS												
0.05	***	699	704	746	250	254	265	147	150	155	104	106	109	81	83	84
	•••	701	712	1285	251	258	409	148	152	216	105	107	138	82	83	97
0.1	***	1275	1286	1439	397	407	444	214	221	236	143	147	155	106	109	113
	•••	1279	1306	2033	400	418	584	217	226	289	145	150	176	107	111	119
0.15	***	1765	1783	2109	515	532	609	267	278	307	172	180	193	125	129	137
	•••	1771	1817	2687	521	554	736	271	289	351	175	185	208	127	133	138
0.2	***	2144	2170	2721	605	630	755	306	323	368	194	205	226	139	145	157
	•••	2153	2223	3247	614	663	865	313	339	403	199	213	234	141	150	153
0.25	***	2407	2445	3256	667	702	879	334	356	419	210	223	252	148	157	172
	•••	2420	2519	3714	679	747	970	343	378	444	216	235	255	152	163	164
0.3	***	2560	2612	3703	703	749	980	351	379	459	219	236	272	154	164	183
	•••	2577	2712	4088	720	806	1052	362	406	475	227	250	269	158	172	171
0.35	***	2612	2680	4055	717	774	1056	359	392	488	224	243	285	156	168	189
	•••	2635	2811	4368	739	844	1110	372	424	496	232	259	278	162	177	175
0.4	***	2576	2664	4308	713	781	1108	359	397	506	224	245	292	156	168	192
	•••	2606	2830	4555	739	862	1145	374	433	507	233	263	281	161	178	175
0.45	***	2467	2580	4462	694	772	1134	352	394	512	220	243	292	152	166	190
	•••	2505	2783	4649	724	863	1157	369	433	507	230	262	278	158	176	171
0.5	***	2300	2442	4513	664	750	1134	340	384	506	212	236	286	147	160	184
	•••	2348	2684	4555	697	847	1145	358	425	496	223	255	269	153	170	164
0.55	***	2093	2267	4464	624	717	1109	323	368	490	201	225	274	138	151	174
	•••	2153	2545	4555	661	818	1110	342	409	475	212	244	255	144	161	153
0.6	***	1863	2071	4312	579	674	1059	302	346	462	187	210	255	128	140	160
	•••	1937	2376	4368	617	776	1052	320	386	444	197	227	234	133	148	138
0.65	***	1628	1863	4059	528	622	983	276	318	423	170	191	230	114	125	142
	•••	1714	2184	4088	566	720	970	294	354	403	179	206	208	119	132	119
0.7	***	1398	1650	3706	472	562	882	246	284	372	150	167	198	99	107	120
	•••	1493	1973	3714	508	652	865	262	316	351	157	180	176	102	113	97
0.75	***	1182	1433	3253	411	491	757	211	243	311	125	138	161	.	.	.
	•••	1279	1742	3247	444	570	736	225	268	289	131	148	138	.	.	.
0.8	***	977	1211	2702	343	409	607	170	194	239	.	.	.	.	.	.
	•••	1068	1486	2687	370	472	584	181	211	216	.	.	.	.	.	.
0.85	***	774	972	2053	264	312	432	.	.	.	.	.	.	.	.	.
	•••	852	1196	2033	284	353	409	.	.	.	.	.	.	.	.	.
0.9	***	556	697	1308	.	.	.	.	.	.	.	.	.	.	.	.
	•••	613	847	1285	.	.	.	.	.	.	.	.	.	.	.	.

TABLE 1: ALPHA= 0.005 POWER= 0.8   EXPECTED ACCRUAL THRU MINIMUM FOLLOW-UP= 190

PCONT=	DEL=.05			DEL=.10			DEL=.15			DEL=.20			DEL=.25		
FACT=	1.0 / .75	.50 / .25	.00 BIN	1.0 / .75	.50 / .25	.00 BIN	1.0 / .75	.50 / .25	.00 BIN	1.0 / 75	.50 / .25	.00 BIN	1.0 / .75	.50 / .25	.00 BIN
	REQUIRED NUMBER OF PATIENTS														
0.05 ***	700	705	746	250	254	265	147	150	155	104	106	109	81	83	84
•••	701	713	1285	252	258	409	148	152	216	105	107	138	82	83	97
0.1 ***	1276	1287	1439	397	407	444	215	221	236	143	147	155	106	109	113
•••	1280	1308	2033	401	419	584	217	227	289	145	150	176	108	111	119
0.15 ***	1766	1785	2109	516	534	609	268	279	307	173	180	193	125	130	137
•••	1772	1821	2687	523	555	736	272	289	351	176	185	208	127	133	138
0.2 ***	2145	2173	2721	606	633	755	307	324	368	195	206	226	139	146	157
•••	2155	2228	3247	616	666	865	314	340	403	200	214	234	142	150	153
0.25 ***	2409	2449	3256	669	705	879	335	358	419	211	224	252	149	157	172
•••	2423	2527	3714	682	750	970	345	379	444	217	235	255	153	163	164
0.3 ***	2563	2617	3703	706	753	980	353	381	459	221	237	272	155	165	183
•••	2581	2722	4088	724	811	1052	364	408	475	228	251	269	159	172	171
0.35 ***	2616	2688	4055	721	779	1056	361	395	488	225	245	285	157	169	189
•••	2640	2824	4368	743	850	1110	375	426	496	234	260	278	162	177	175
0.4 ***	2581	2674	4308	718	787	1108	362	400	506	226	247	292	157	169	192
•••	2612	2846	4555	744	868	1145	377	435	507	235	264	281	162	179	175
0.45 ***	2473	2592	4462	700	779	1134	355	397	512	222	245	292	154	167	190
•••	2513	2803	4649	730	870	1157	373	436	507	232	263	278	159	176	171
0.5 ***	2308	2457	4514	670	758	1134	344	388	506	214	238	286	148	161	184
•••	2359	2707	4649	704	855	1145	362	428	496	225	256	269	154	171	164
0.55 ***	2103	2285	4464	631	725	1109	326	372	490	203	227	274	139	152	174
•••	2167	2571	4555	668	826	1110	345	412	475	214	245	255	145	161	153
0.6 ***	1876	2091	4312	586	682	1059	305	349	462	189	211	255	129	140	160
•••	1953	2404	4368	624	783	1052	324	388	444	199	229	234	134	149	138
0.65 ***	1642	1885	4059	535	630	983	279	321	423	172	192	230	115	126	142
•••	1732	2213	4088	573	728	970	297	357	403	181	207	208	120	133	119
0.7 ***	1415	1672	3706	478	569	882	249	287	372	151	168	198	99	108	120
•••	1512	2001	3714	515	659	865	265	318	351	159	181	176	103	113	97
0.75 ***	1199	1456	3253	417	498	757	214	245	311	126	139	161	.	.	.
•••	1298	1768	3247	450	576	736	227	270	289	132	148	138	.	.	.
0.8 ***	993	1231	2702	348	415	607	172	195	239	.	.	.	.	.	.
•••	1086	1509	2687	376	476	584	182	213	216	.	.	.	.	.	.
0.85 ***	788	989	2053	268	315	432	.	.	.	.	.	.	.	.	.
•••	868	1214	2033	288	356	409	.	.	.	.	.	.	.	.	.
0.9 ***	567	709	1308	.	.	.	.	.	.	.	.	.	.	.	.
•••	624	859	1285	.	.	.	.	.	.	.	.	.	.	.	.

TABLE 1: ALPHA= 0.005 POWER= 0.8   EXPECTED ACCRUAL THRU MINIMUM FOLLOW-UP= 200

	DEL=.05			DEL=.10			DEL=.15			DEL=.20			DEL=.25		
FACT=	1.0 .75	.50 .25	.00 BIN	1.0 .75	.50 .25	.00 BIN	1.0 .75	.50 .25	.00 BIN	1.0 75	.50 .25	.00 BIN	1.0 .75	.50 .25	.00 BIN
PCONT=***				REQUIRED	NUMBER OF	PATIENTS									
0.05 ***	700	705	746	250	254	265	147	150	155	105	106	109	82	83	84
•••	702	714	1285	252	258	409	149	152	216	105	107	138	82	83	97
0.1 ***	1277	1288	1439	398	408	444	215	222	236	144	148	155	107	109	113
•••	1281	1310	2033	402	420	584	218	227	289	145	151	176	108	111	119
0.15 ***	1767	1787	2109	517	535	609	268	280	307	173	180	193	126	130	137
•••	1773	1824	2687	524	557	736	273	290	351	177	186	208	128	133	138
0.2 ***	2147	2176	2721	608	635	755	309	325	368	196	206	226	140	146	157
•••	2157	2234	3247	618	668	865	316	341	403	200	214	234	143	151	153
0.25 ***	2412	2454	3256	671	708	879	337	360	419	212	225	252	150	158	172
•••	2426	2535	3714	685	754	970	346	381	444	218	236	255	153	164	164
0.3 ***	2566	2623	3703	709	757	980	355	383	459	222	238	272	156	166	183
•••	2585	2733	4088	727	815	1052	367	410	475	229	252	269	160	173	171
0.35 ***	2620	2695	4055	725	784	1056	364	397	488	227	246	285	158	170	189
•••	2645	2838	4368	747	855	1110	377	428	496	235	261	278	163	178	175
0.4 ***	2586	2684	4308	722	793	1108	364	403	506	227	249	292	158	170	192
•••	2619	2863	4555	749	875	1145	380	438	507	236	265	281	163	179	175
0.45 ***	2479	2604	4462	705	786	1134	358	400	512	224	246	292	155	168	190
•••	2521	2822	4649	736	876	1157	376	438	507	233	264	278	160	177	171
0.5 ***	2316	2472	4514	676	765	1134	347	391	506	216	239	286	149	162	184
•••	2369	2729	4649	711	862	1145	365	430	496	226	258	269	155	171	164
0.55 ***	2113	2303	4464	637	732	1109	330	375	490	205	228	274	140	153	174
•••	2180	2596	4555	675	833	1110	349	415	475	215	246	255	146	162	153
0.6 ***	1888	2111	4312	592	689	1059	308	352	462	191	213	255	130	141	160
•••	1968	2430	4368	631	791	1052	327	391	444	201	230	234	135	149	138
0.65 ***	1657	1906	4059	541	637	983	283	324	423	174	193	230	116	126	142
•••	1749	2240	4088	580	735	970	300	359	403	182	208	208	121	133	119
0.7 ***	1431	1695	3706	485	576	882	252	289	372	153	169	198	100	108	120
•••	1531	2027	3714	522	665	865	268	320	351	160	181	176	104	113	97
0.75 ***	1216	1477	3253	423	504	757	216	247	311	127	140	161	.	.	.
•••	1317	1792	3247	456	581	736	230	272	289	133	149	138	.	.	.
0.8 ***	1009	1250	2702	353	419	607	174	197	239	.	.	.	.	.	.
•••	1103	1531	2687	381	480	584	184	214	216	.	.	.	.	.	.
0.85 ***	802	1005	2053	271	319	432	.	.	.	.	.	.	.	.	.
•••	882	1231	2033	291	359	409	.	.	.	.	.	.	.	.	.
0.9 ***	577	720	1308	.	.	.	.	.	.	.	.	.	.	.	.
•••	635	870	1285	.	.	.	.	.	.	.	.	.	.	.	.

TABLE 1: ALPHA= 0.005 POWER= 0.8   EXPECTED ACCRUAL THRU MINIMUM FOLLOW-UP= 225

		DEL=.05			DEL=.10			DEL=.15			DEL=.20			DEL=.25		
FACT=		1.0 .75	.50 .25	.00 BIN	1.0 .75	.50 .25	.00 BIN	1.0 .75	.50 .25	.00 BIN	1.0 75	.50 .25	.00 BIN	1.0 .75	.50 .25	.00 BIN
PCONT=•••					REQUIRED NUMBER OF PATIENTS											
0.05	•••	701	706	746	251	255	265	148	150	155	105	106	109	82	83	84
	•••	703	716	1285	253	259	409	149	152	216	106	107	138	82	84	97
0.1	•••	1278	1291	1439	399	410	444	216	223	236	144	148	155	107	110	113
	•••	1283	1315	2033	404	422	584	219	228	289	146	151	176	108	111	119
0.15	•••	1769	1791	2109	520	539	609	270	282	307	175	182	193	126	131	137
	•••	1777	1833	2687	527	560	736	275	291	351	178	186	208	128	134	138
0.2	•••	2150	2184	2721	612	640	755	311	328	368	198	208	226	141	147	157
	•••	2162	2247	3247	623	674	865	318	343	403	202	215	234	144	151	153
0.25	•••	2417	2464	3256	676	716	879	341	363	419	214	227	252	151	159	172
	•••	2433	2554	3714	691	761	970	350	384	444	220	237	255	155	165	164
0.3	•••	2573	2637	3703	716	767	980	360	388	459	225	241	272	157	167	183
	•••	2594	2758	4088	735	825	1052	371	413	475	232	253	269	162	174	171
0.35	•••	2629	2714	4055	734	796	1056	369	403	488	230	249	285	160	171	189
	•••	2658	2870	4368	758	867	1110	383	433	496	239	263	278	165	179	175
0.4	•••	2598	2708	4308	733	806	1108	371	409	506	231	252	292	160	172	192
	•••	2635	2903	4555	761	888	1145	387	443	507	240	268	281	165	180	175
0.45	•••	2495	2634	4462	717	801	1134	365	407	512	228	250	292	157	169	190
	•••	2542	2869	4649	750	892	1157	383	444	507	237	267	278	163	178	171
0.5	•••	2336	2508	4514	690	781	1134	354	398	507	220	243	286	151	164	184
	•••	2396	2783	4649	726	878	1145	373	436	496	230	260	269	157	172	164
0.55	•••	2138	2346	4464	653	749	1109	338	382	490	209	232	274	143	155	174
	•••	2212	2654	4555	691	850	1110	356	420	475	219	249	255	148	163	153
0.6	•••	1919	2159	4312	608	707	1059	316	360	462	195	216	255	132	143	160
	•••	2006	2492	4368	648	807	1052	335	396	444	204	232	234	137	150	138
0.65	•••	1693	1958	4059	557	654	983	290	331	423	177	196	230	118	128	142
	•••	1792	2303	4088	597	750	970	307	365	403	186	210	208	122	134	119
0.7	•••	1470	1747	3706	500	591	882	259	295	372	156	172	198	102	109	120
	•••	1576	2089	3714	537	679	865	274	324	351	163	183	176	105	114	97
0.75	•••	1256	1527	3253	436	517	757	222	252	311	130	142	161	.	.	.
	•••	1361	1850	3247	470	593	736	235	275	289	135	150	138	.	.	.
0.8	•••	1047	1296	2702	364	430	606	178	200	239	.	.	.	.	.	.
	•••	1145	1581	2687	392	490	584	188	216	216	.	.	.	.	.	.
0.85	•••	834	1042	2053	280	326	432	.	.	.	.	.	.	.	.	.
	•••	917	1271	2033	299	365	409	.	.	.	.	.	.	.	.	.
0.9	•••	600	745	1308	.	.	.	.	.	.	.	.	.	.	.	.
	•••	659	895	1285	.	.	.	.	.	.	.	.	.	.	.	.

TABLE 1: ALPHA= 0.005 POWER= 0.8   EXPECTED ACCRUAL THRU MINIMUM FOLLOW-UP= 250

PCONT=***	DEL=.05			DEL=.10			DEL=.15			DEL=.20			DEL=.25		
FACT=	1.0 .75	.50 .25	.00 BIN	1.0 .75	.50 .25	.00 BIN	1.0 .75	.50 .25	.00 BIN	1.0 75	.50 .25	.00 BIN	1.0 .75	.50 .25	.00 BIN
				REQUIRED NUMBER OF PATIENTS											
0.05 ***	701	708	746	252	256	265	148	151	155	105	107	109	82	83	84
***	703	717	1285	253	259	409	149	152	216	106	108	138	82	84	97
0.1 ***	1280	1294	1439	401	412	444	217	224	236	145	149	155	108	110	113
***	1284	1319	2033	405	423	584	220	229	289	147	151	176	109	112	119
0.15 ***	1772	1796	2109	522	542	609	272	283	307	176	182	193	127	131	137
***	1780	1842	2687	530	564	736	277	293	351	179	187	208	129	134	138
0.2 ***	2154	2191	2721	616	645	755	314	331	368	199	209	226	142	148	157
***	2167	2260	3247	627	679	865	321	345	403	204	216	234	144	152	153
0.25 ***	2422	2474	3256	682	722	879	344	367	419	217	229	252	152	160	172
***	2439	2572	3714	697	768	970	354	386	444	222	239	255	156	165	164
0.3 ***	2580	2651	3703	723	775	980	364	392	459	228	243	272	159	168	183
***	2604	2782	4088	743	834	1052	376	417	475	234	255	269	163	174	171
0.35 ***	2639	2732	4055	742	807	1056	374	408	488	233	252	285	162	173	189
***	2670	2901	4368	767	878	1110	388	437	496	241	265	278	167	180	175
0.4 ***	2611	2731	4309	743	819	1108	377	415	506	234	254	292	162	174	192
***	2652	2940	4555	773	901	1145	393	447	507	243	270	281	167	181	175
0.45 ***	2511	2663	4462	729	814	1134	372	413	512	231	253	292	159	171	190
***	2563	2913	4649	763	905	1157	389	449	507	241	269	278	164	179	171
0.5 ***	2356	2543	4514	703	796	1134	361	404	507	224	246	286	153	165	184
***	2422	2833	4649	740	893	1145	379	441	496	234	262	269	159	173	164
0.55 ***	2163	2386	4464	666	765	1109	344	388	490	213	235	274	145	156	174
***	2243	2709	4555	706	865	1110	363	425	475	223	251	255	150	164	153
0.6 ***	1949	2204	4312	622	722	1059	323	366	462	199	219	255	134	144	160
***	2043	2549	4368	663	822	1052	341	401	444	208	234	234	139	151	138
0.65 ***	1727	2005	4059	571	669	983	296	336	423	180	199	230	120	129	142
***	1832	2361	4088	611	764	970	314	369	403	189	212	208	124	135	119
0.7 ***	1508	1795	3706	514	605	882	264	300	372	158	174	198	103	110	120
***	1618	2145	3714	551	692	865	280	328	351	165	184	176	106	114	97
0.75 ***	1293	1574	3253	449	529	757	227	256	311	132	143	161	.	.	.
***	1402	1902	3247	482	604	736	239	278	289	137	151	138	.	.	.
0.8 ***	1082	1337	2702	374	440	607	182	203	239	.	.	.	.	.	.
***	1182	1626	2687	402	498	584	191	218	216	.	.	.	.	.	.
0.85 ***	864	1076	2053	287	332	432	.	.	.	.	.	.	.	.	.
***	949	1306	2033	307	370	409	.	.	.	.	.	.	.	.	.
0.9 ***	621	768	1308	.	.	.	.	.	.	.	.	.	.	.	.
***	681	917	1285	.	.	.	.	.	.	.	.	.	.	.	.

TABLE 1: ALPHA= 0.005 POWER= 0.8   EXPECTED ACCRUAL THRU MINIMUM FOLLOW-UP= 275

		DEL=.05			DEL=.10			DEL=.15			DEL=.20			DEL=.25		
FACT=		1.0 .75	.50 .25	.00 BIN	1.0 .75	.50 .25	.00 BIN	1.0 .75	.50 .25	.00 BIN	1.0 75	.50 .25	.00 BIN	1.0 .75	.50 .25	.00 BIN
PCONT=•••		REQUIRED NUMBER OF PATIENTS														
0.05	•••	702	709	746	252	256	265	149	151	155	105	107	109	82	83	84
	•••	704	718	1285	254	260	409	150	153	216	106	108	138	83	84	97
0.1	•••	1281	1297	1439	402	414	444	218	224	236	145	149	155	108	110	113
	•••	1286	1324	2033	407	425	584	221	229	289	147	151	176	109	112	119
0.15	•••	1774	1801	2109	525	545	609	274	285	307	177	183	193	128	132	137
	•••	1783	1849	2687	533	566	736	278	294	351	180	188	208	130	134	138
0.2	•••	2158	2199	2721	619	650	755	316	333	368	201	210	226	143	148	157
	•••	2171	2273	3247	631	683	865	323	347	403	205	217	234	145	152	153
0.25	•••	2427	2485	3256	686	729	879	347	370	419	218	231	252	154	161	172
	•••	2446	2590	3714	703	774	970	357	389	444	224	240	255	157	166	164
0.3	•••	2587	2665	3703	729	783	980	368	396	459	230	245	272	160	169	183
	•••	2613	2805	4088	750	842	1052	380	420	475	237	256	269	164	175	171
0.35	•••	2648	2751	4055	750	816	1056	379	412	488	236	254	285	164	174	189
	•••	2683	2930	4368	776	887	1110	393	440	496	244	267	278	168	180	175
0.4	•••	2623	2754	4308	752	830	1108	382	420	506	237	257	292	164	175	192
	•••	2668	2976	4555	783	912	1145	398	451	507	246	271	281	169	182	175
0.45	•••	2527	2692	4462	740	827	1134	378	419	512	234	255	292	161	172	190
	•••	2584	2955	4649	775	917	1157	395	453	507	244	270	278	166	180	171
0.5	•••	2376	2577	4513	715	809	1134	367	410	506	227	248	286	155	167	184
	•••	2447	2879	4649	753	905	1145	385	446	496	237	264	269	160	174	164
0.55	•••	2188	2425	4463	679	779	1109	351	394	490	216	237	274	147	158	174
	•••	2273	2759	4555	720	877	1110	369	430	475	226	252	255	152	165	153
0.6	•••	1978	2247	4312	635	736	1059	329	371	462	202	221	255	135	145	160
	•••	2078	2602	4368	677	834	1052	347	405	444	210	235	234	140	152	138
0.65	•••	1760	2050	4059	584	683	983	302	341	423	183	201	230	121	130	142
	•••	1870	2414	4088	625	776	970	319	373	403	191	213	208	125	135	119
0.7	•••	1543	1840	3706	526	618	882	270	304	372	161	175	198	104	111	120
	•••	1657	2197	3714	564	703	865	285	331	351	167	185	176	107	115	97
0.75	•••	1328	1616	3253	460	540	757	231	259	311	134	145	161	.	.	.
	•••	1441	1949	3247	494	613	736	244	280	289	139	152	138	.	.	.
0.8	•••	1114	1375	2702	384	449	607	185	205	239	.	.	.	.	.	.
	•••	1217	1667	2687	411	505	584	194	220	216	.	.	.	.	.	.
0.85	•••	891	1107	2053	293	338	432	.	.	.	.	.	.	.	.	.
	•••	978	1337	2033	313	374	409	.	.	.	.	.	.	.	.	.
0.9	•••	641	789	1308	.	.	.	.	.	.	.	.	.	.	.	.
	•••	701	936	1285	.	.	.	.	.	.	.	.	.	.	.	.

TABLE 1: ALPHA= 0.005 POWER= 0.8   EXPECTED ACCRUAL THRU MINIMUM FOLLOW-UP= 300

		DEL=.05			DEL=.10			DEL=.15			DEL=.20			DEL=.25		
FACT=		1.0 .75	.50 .25	.00 BIN	1.0 .75	.50 .25	.00 BIN	1.0 .75	.50 .25	.00 BIN	1.0 75	.50 .25	.00 BIN	1.0 .75	.50 .25	.00 BIN
PCONT=•••							REQUIRED	NUMBER	OF PATIENTS						
0.05	•••	703	710	746	253	257	265	149	151	155	106	107	109	82	83	84
	•••	705	720	1285	254	260	409	150	153	216	106	108	138	83	84	97
0.1	•••	1282	1299	1439	403	415	444	219	225	236	146	149	154	108	110	113
	•••	1288	1328	2033	408	426	584	222	229	289	148	152	176	109	112	119
0.15	•••	1777	1806	2109	527	548	609	275	286	307	178	184	193	128	132	137
	•••	1786	1857	2687	535	569	736	280	295	351	180	188	208	130	134	138
0.2	•••	2161	2206	2721	622	654	755	318	335	368	202	211	226	144	149	157
	•••	2176	2285	3247	635	687	865	325	348	403	206	217	234	146	152	153
0.25	•••	2432	2495	3256	691	734	879	350	372	419	220	232	252	154	161	172
	•••	2453	2606	3714	708	780	970	360	391	444	225	241	255	158	166	164
0.3	•••	2594	2679	3703	735	791	980	371	399	459	232	247	272	162	170	183
	•••	2623	2827	4088	757	849	1052	383	422	475	238	257	269	166	176	171
0.35	•••	2658	2769	4055	757	825	1056	383	416	488	238	256	285	165	175	189
	•••	2695	2958	4368	784	895	1110	397	443	496	246	268	278	169	181	175
0.4	•••	2635	2777	4308	761	840	1108	387	424	506	240	259	292	165	176	192
	•••	2684	3009	4555	793	921	1145	403	455	507	248	273	281	170	183	175
0.45	•••	2542	2719	4462	750	838	1134	383	423	512	237	257	292	163	173	190
	•••	2604	2994	4649	786	928	1157	400	457	507	246	272	278	168	181	171
0.5	•••	2396	2610	4513	726	822	1134	373	415	506	230	251	286	157	168	184
	•••	2472	2923	4555	765	917	1145	391	449	496	239	265	269	162	175	164
0.55	•••	2212	2462	4463	691	791	1109	356	399	490	219	239	274	148	159	174
	•••	2303	2806	4555	732	889	1110	375	433	475	228	254	255	153	166	153
0.6	•••	2006	2287	4312	648	749	1059	334	376	462	204	223	255	137	146	160
	•••	2111	2651	4368	689	846	1052	352	409	444	213	237	234	141	152	138
0.65	•••	1792	2092	4059	596	695	983	307	346	423	186	202	230	122	130	142
	•••	1906	2463	4088	637	787	970	324	376	403	193	214	208	126	136	119
0.7	•••	1576	1882	3706	537	629	882	274	308	372	163	177	198	105	111	120
	•••	1694	2244	3714	576	712	865	289	334	351	169	186	176	108	115	97
0.75	•••	1361	1656	3253	470	550	757	235	262	311	135	146	161	.	.	.
	•••	1477	1993	3247	504	621	736	247	282	289	140	152	138	.	.	.
0.8	•••	1144	1411	2702	392	456	607	188	207	239	.	.	.	.	.	.
	•••	1250	1704	2687	419	511	584	197	221	216	.	.	.	.	.	.
0.85	•••	917	1136	2053	299	343	432	.	.	.	.	.	.	.	.	.
	•••	1005	1366	2033	319	378	409	.	.	.	.	.	.	.	.	.
0.9	•••	659	808	1308	.	.	.	.	.	.	.	.	.	.	.	.
	•••	720	954	1285	.	.	.	.	.	.	.	.	.	.	.	.

TABLE 1: ALPHA= 0.005 POWER= 0.8   EXPECTED ACCRUAL THRU MINIMUM FOLLOW-UP= 325

		DEL=.05			DEL=.10			DEL=.15			DEL=.20			DEL=.25		
FACT=		1.0 .75	.50 .25	.00 BIN	1.0 .75	.50 .25	.00 BIN	1.0 .75	.50 .25	.00 BIN	1.0 75	.50 .25	.00 BIN	1.0 .75	.50 .25	.00 BIN
PCONT=***		REQUIRED NUMBER OF PATIENTS														
0.05	***	703	711	746	253	257	265	149	151	155	106	107	109	82	83	84
	•••	706	721	1285	255	261	409	150	153	216	106	108	138	83	84	97
0.1	***	1284	1302	1439	405	416	444	220	226	236	146	150	155	109	111	113
	•••	1290	1331	2033	409	427	584	222	230	289	148	152	176	110	112	119
0.15	***	1779	1811	2109	529	550	609	276	287	307	179	184	193	129	132	137
	•••	1790	1864	2687	538	571	736	281	295	351	181	188	208	131	135	138
0.2	***	2165	2213	2721	626	658	755	320	337	368	203	212	226	144	149	157
	•••	2181	2296	3247	639	690	865	327	349	403	207	218	234	147	153	153
0.25	***	2438	2505	3256	696	740	879	353	375	419	222	233	252	156	162	172
	•••	2460	2623	3714	713	785	970	362	392	444	227	241	255	159	167	164
0.3	***	2602	2693	3703	741	798	980	375	402	459	234	248	272	163	171	183
	•••	2632	2849	4088	764	855	1052	387	424	475	240	258	269	167	176	171
0.35	***	2667	2786	4055	765	833	1056	387	420	488.	241	257	285	167	176	189
	•••	2708	2985	4368	792	903	1110	401	446	496	248	269	278	171	182	175
0.4	***	2647	2799	4308	770	850	1108	391	428	506	242	261	292	167	177	192
	•••	2700	3041	4555	802	930	1145	407	458	507	251	274	281	171	183	175
0.45	***	2558	2746	4462	760	849	1134	388	428	512	240	259	292	164	174	190
	•••	2624	3031	4649	796	937	1157	405	460	507	249	273	278	169	181	171
0.5	***	2415	2642	4514	736	833	1134	378	419	506	233	253	286	158	169	184
	•••	2496	2964	4649	776	927	1145	396	453	496	242	267	269	163	176	164
0.55	***	2235	2498	4463	703	803	1109	362	403	490	222	242	274	150	160	174
	•••	2332	2850	4555	744	900	1110	380	437	475	231	255	255	154	166	153
0.6	***	2034	2326	4312	659	761	1059	340	380	462	207	225	255	138	147	160
	•••	2143	2697	4368	701	856	1052	357	412	444	215	238	234	142	153	138
0.65	***	1822	2132	4059	608	706	983	312	350	423	188	204	230	123	131	142
	•••	1941	2509	4088	649	796	970	328	379	403	195	215	208	127	136	119
0.7	***	1608	1922	3706	548	639	882	278	311	372	165	178	198	106	112	120
	•••	1730	2288	3714	586	721	865	293	336	351	171	187	176	109	116	97
0.75	***	1392	1693	3253	479	559	757	238	265	311	137	147	161	.	.	.
	•••	1511	2033	3247	513	628	736	250	284	289	141	153	138	.	.	.
0.8	***	1173	1444	2702	400	463	606	190	209	239	.	.	.	.	.	.
	•••	1281	1738	2687	427	516	584	199	222	216	.	.	.	.	.	.
0.85	***	941	1162	2053	305	348	432	.	.	.	.	.	.	.	.	.
	•••	1030	1392	2033	324	381	409	.	.	.	.	.	.	.	.	.
0.9	***	676	825	1308	.	.	.	.	.	.	.	.	.	.	.	.
	•••	737	970	1285	.	.	.	.	.	.	.	.	.	.	.	.

TABLE 1: ALPHA= 0.005 POWER= 0.8   EXPECTED ACCRUAL THRU MINIMUM FOLLOW-UP= 350

	DEL=.05			DEL=.10			DEL=.15			DEL=.20			DEL=.25		
FACT=	1.0 .75	.50 .25	.00 BIN	1.0 .75	.50 .25	.00 BIN	1.0 .75	.50 .25	.00 BIN	1.0 75	.50 .25	.00 BIN	1.0 .75	.50 .25	.00 BIN
PCONT=***	REQUIRED NUMBER OF PATIENTS														
0.05 ***	704	712	746	254	258	265	150	152	155	106	107	109	82	83	84
•••	707	722	1285	255	261	409	150	153	216	106	108	138	83	84	97
0.1 ***	1285	1305	1439	406	418	444	220	226	236	147	150	155	109	111	113
•••	1292	1335	2033	411	428	584	223	230	289	148	152	176	110	112	119
0.15 ***	1782	1815	2109	531	553	609	278	288	307	179	185	193	129	133	137
•••	1793	1871	2687	540	573	736	282	296	351	182	189	208	131	135	138
0.2 ***	2169	2220	2721	629	662	755	322	338	369	204	213	226	145	150	157
•••	2186	2307	3247	642	693	865	329	351	403	208	218	234	147	153	153
0.25 ***	2443	2515	3256	700	745	879	355	377	419	223	234	252	156	163	172
•••	2467	2638	3714	718	789	970	365	394	444	228	242	255	159	167	164
0.3 ***	2609	2706	3703	747	804	980	378	405	459	236	249	272	164	171	183
•••	2642	2869	4088	769	861	1052	390	426	475	242	259	269	167	176	171
0.35 ***	2676	2804	4055	772	841	1056	391	423	488	243	259	285	168	176	189
•••	2720	3010	4368	799	910	1110	405	448	496	250	270	278	172	182	175
0.4 ***	2659	2821	4308	778	859	1108	395	432	506	245	262	292	168	178	192
•••	2716	3071	4555	811	938	1145	411	460	507	253	275	281	173	184	175
0.45 ***	2573	2773	4462	769	859	1134	392	432	512	242	261	292	165	175	190
•••	2644	3065	4649	806	946	1157	409	463	507	251	274	278	170	182	171
0.5 ***	2434	2672	4514	747	843	1134	382	423	507	235	255	286	160	170	184
•••	2520	3002	4649	786	936	1145	400	456	496	244	268	269	164	176	164
0.55 ***	2258	2532	4464	713	814	1109	366	407	490	224	243	274	151	160	174
•••	2359	2892	4555	755	909	1110	384	440	475	233	256	255	155	167	153
0.6 ***	2060	2362	4312	670	771	1058	344	384	462	209	227	255	139	148	160
•••	2174	2740	4368	712	865	1052	362	415	444	217	239	234	143	153	138
0.65 ***	1851	2170	4059	618	716	983	316	353	423	190	206	230	125	132	142
•••	1974	2551	4088	659	805	970	332	381	403	197	216	208	128	136	119
0.7 ***	1638	1959	3706	558	649	882	282	314	372	166	179	198	107	112	120
•••	1763	2329	3714	596	729	865	297	338	351	172	188	176	109	116	97
0.75 ***	1422	1728	3253	488	567	757	242	267	311	138	148	161	.	.	.
•••	1543	2070	3247	521	634	736	253	286	289	142	153	138	.	.	.
0.8 ***	1200	1475	2702	407	469	607	193	211	239	.	.	.	.	.	.
•••	1310	1770	2687	434	521	584	201	223	216	.	.	.	.	.	.
0.85 ***	964	1187	2053	310	352	432	.	.	.	.	.	.	.	.	.
•••	1054	1417	2033	328	384	409	.	.	.	.	.	.	.	.	.
0.9 ***	691	841	1308	.	.	.	.	.	.	.	.	.	.	.	.
•••	753	984	1285	.	.	.	.	.	.	.	.	.	.	.	.

TABLE 1: ALPHA= 0.005 POWER= 0.8   EXPECTED ACCRUAL THRU MINIMUM FOLLOW-UP= 375

		DEL=.05			DEL=.10			DEL=.15			DEL=.20			DEL=.25		
FACT=		1.0 .75	.50 .25	.00 BIN	1.0 .75	.50 .25	.00 BIN	1.0 .75	.50 .25	.00 BIN	1.0 75	.50 .25	.00 BIN	1.0 .75	.50 .25	.00 BIN
PCONT=•••					REQUIRED NUMBER OF PATIENTS											
0.05	***	704 707	713 723	746 1285	254 256	258 261	265 409	150 151	152 153	155 216	106 107	107 108	109 138	83 83	83 84	84 97
0.1	***	1287 1294	1307 1338	1439 2033	407 412	419 429	444 584	221 223	227 230	236 289	147 149	150 152	155 176	109 110	111 112	113 119
0.15	***	1784 1796	1820 1878	2109 2687	533 542	555 575	609 736	279 283	289 297	307 351	180 182	185 189	193 208	130 131	133 135	137 138
0.2	***	2173 2191	2227 2317	2721 3247	632 645	665 696	755 865	324 331	340 351	368 403	205 209	213 219	226 234	145 148	150 153	157 153
0.25	***	2448 2474	2525 2653	3256 3714	704 722	749 793	879 970	358 367	379 395	419 444	224 229	235 242	252 255	157 160	163 167	172 164
0.3	***	2616 2651	2720 2888	3703 4088	752 775	810 866	980 1052	380 392	407 428	460 475	237 243	250 260	272 269	165 168	172 177	183 171
0.35	***	2686 2732	2821 3034	4055 4368	778 807	848 916	1056 1110	394 408	426 450	488 496	244 251	260 271	285 278	169 173	177 183	189 175
0.4	***	2672 2731	2842 3100	4308 4555	786 819	867 945	1108 1145	399 415	435 463	506 507	247 254	264 276	292 281	169 174	178 184	192 175
0.45	***	2589 2663	2798 3098	4462 4649	778 815	868 954	1134 1157	396 413	435 465	512 507	244 253	263 275	292 278	167 171	176 182	190 171
0.5	***	2453 2543	2701 3039	4513 4649	756 796	853 945	1134 1145	387 404	427 458	506 496	237 246	256 269	286 269	161 165	170 177	184 164
0.55	***	2281 2386	2564 2930	4464 4555	723 765	824 918	1109 1110	371 388	411 442	490 475	226 235	245 257	274 255	152 156	161 167	174 153
0.6	***	2086 2204	2397 2780	4312 4368	680 722	781 874	1059 1052	348 366	388 417	462 444	211 219	228 240	255 234	140 144	149 154	160 138
0.65	***	1879 2005	2206 2591	4059 4088	628 669	726 813	983 970	320 336	356 383	423 403	192 199	207 217	230 208	125 129	133 137	142 119
0.7	***	1667 1795	1994 2367	3706 3714	567 605	657 736	882 865	286 300	317 340	372 351	168 174	180 188	198 176	107 110	113 116	120 97
0.75	***	1450 1573	1761 2105	3253 3247	496 529	574 640	757 736	244 256	270 287	311 289	139 143	148 154	161 138	. .	. .	. .
0.8	***	1226 1337	1503 1799	2702 2687	413 440	475 525	606 584	195 203	212 224	239 216	. .	. .	. .	. .	. .	. .
0.85	***	985 1076	1210 1439	2053 2033	314 332	355 386	432 409	. .	. .	. .	. .	. .	. .	. .	. .	. .
0.9	***	706 768	856 997	1308 1285	. .	. .	. .	. .	. .	. .	. .	. .	. .	. .	. .	. .

TABLE 1: ALPHA= 0.005 POWER= 0.8   EXPECTED ACCRUAL THRU MINIMUM FOLLOW-UP= 400

		DEL=.05			DEL=.10			DEL=.15			DEL=.20			DEL=.25		
FACT=		1.0 .75	.50 .25	.00 BIN	1.0 .75	.50 .25	.00 BIN	1.0 .75	.50 .25	.00 BIN	1.0 75	.50 .25	.00 BIN	1.0 .75	.50 .25	.00 BIN
PCONT=•••		REQUIRED NUMBER OF PATIENTS														
0.05	***	705	714	746	254	258	265	150	152	155	106	107	109	83	83	84
	•••	708	724	1285	256	261	409	151	153	216	107	108	138	83	84	97
0.1	***	1288	1310	1439	408	420	444	222	227	236	148	151	155	109	111	113
	•••	1296	1342	2033	413	429	584	224	231	289	149	152	176	110	112	119
0.15	***	1787	1824	2109	535	557	609	280	290	307	180	186	193	130	133	137
	•••	1800	1884	2687	544	576	736	284	297	351	183	189	208	132	135	138
0.2	***	2176	2234	2721	635	668	755	325	341	368	206	214	226	146	151	157
	•••	2196	2327	3247	648	699	865	332	352	403	210	219	234	148	153	153
0.25	***	2454	2535	3256	708	754	879	360	381	419	225	236	252	158	164	172
	•••	2481	2667	3714	727	797	970	369	397	444	230	243	255	161	167	164
0.3	***	2623	2733	3703	757	815	980	383	410	459	238	252	272	166	173	183
	•••	2661	2907	4088	781	871	1052	395	430	475	244	260	269	169	177	171
0.35	***	2695	2838	4055	784	855	1056	397	428	488	246	261	285	170	178	189
	•••	2745	3057	4368	813	922	1110	411	452	496	253	272	278	173	183	175
0.4	***	2684	2863	4308	793	875	1108	403	438	506	249	265	292	170	179	192
	•••	2747	3128	4555	826	952	1145	418	465	507	256	277	281	174	185	175
0.45	***	2604	2822	4462	786	876	1134	400	438	512	246	264	292	168	177	190
	•••	2682	3130	4649	823	962	1157	417	468	507	254	276	278	172	183	171
0.5	***	2472	2729	4514	765	862	1134	391	430	506	239	258	286	162	171	184
	•••	2566	3073	4649	805	952	1145	408	461	496	248	270	269	166	177	164
0.55	***	2303	2596	4464	732	833	1109	375	415	490	228	246	274	153	162	174
	•••	2412	2967	4555	774	925	1110	392	445	475	236	258	255	157	167	153
0.6	***	2111	2430	4312	689	791	1059	352	391	462	213	230	255	141	149	160
	•••	2233	2818	4368	732	881	1052	369	420	444	221	241	234	145	154	138
0.65	***	1906	2240	4059	637	735	983	324	359	423	193	208	230	126	133	142
	•••	2036	2629	4088	678	820	970	340	385	403	200	218	208	129	137	119
0.7	***	1695	2027	3706	576	665	882	289	320	372	169	181	198	108	113	120
	•••	1825	2402	3714	614	742	865	303	342	351	175	189	176	110	116	97
0.75	***	1477	1792	3253	504	581	757	247	272	311	140	149	161	.	.	.
	•••	1602	2137	3247	537	645	736	258	289	289	144	154	138	.	.	.
0.8	***	1250	1531	2702	419	480	607	197	214	239	.	.	.	.	.	.
	•••	1363	1827	2687	446	529	584	205	225	216	.	.	.	.	.	.
0.85	***	1005	1231	2053	319	359	432	.	.	.	.	.	.	.	.	.
	•••	1097	1459	2033	336	389	409	.	.	.	.	.	.	.	.	.
0.9	***	720	870	1308	.	.	.	.	.	.	.	.	.	.	.	.
	•••	782	1009	1285	.	.	.	.	.	.	.	.	.	.	.	.

TABLE 1: ALPHA= 0.005 POWER= 0.8   EXPECTED ACCRUAL THRU MINIMUM FOLLOW-UP= 425

		DEL=.05			DEL=.10			DEL=.15			DEL=.20			DEL=.25		
FACT=		1.0 .75	.50 .25	.00 BIN	1.0 .75	.50 .25	.00 BIN	1.0 .75	.50 .25	.00 BIN	1.0 75	.50 .25	.00 BIN	1.0 .75	.50 .25	.00 BIN
PCONT=•••		REQUIRED NUMBER OF PATIENTS														
0.05	•••	706	715	746	255	259	265	150	152	155	106	107	109	83	84	84
	•••	709	725	1285	256	261	409	151	153	216	107	108	138	83	84	97
0.1	•••	1290	1312	1439	409	421	444	222	227	236	148	151	155	110	111	113
	•••	1298	1344	2033	414	430	584	224	231	289	149	153	176	110	112	119
0.15	•••	1789	1829	2109	537	559	609	281	291	307	181	186	193	130	133	137
	•••	1803	1890	2687	546	578	736	285	298	351	183	189	208	132	135	138
0.2	•••	2180	2241	2721	638	671	755	327	342	368	207	215	226	146	151	156
	•••	2201	2337	3247	651	701	865	334	353	403	210	220	234	149	154	153
0.25	•••	2459	2544	3256	712	758	879	362	382	419	226	237	252	158	164	172
	•••	2488	2680	3714	731	800	970	371	398	444	231	243	255	161	168	164
0.3	•••	2630	2745	3703	762	821	980	386	411	459	240	252	272	166	173	183
	•••	2670	2925	4088	786	875	1052	397	431	475	246	261	269	170	178	171
0.35	•••	2705	2854	4055	790	861	1056	400	431	488	248	262	285	170	178	189
	•••	2757	3079	4368	819	927	1110	414	454	496	254	272	278	174	183	175
0.4	•••	2696	2883	4309	800	882	1108	406	441	506	250	266	292	171	180	192
	•••	2762	3154	4555	834	958	1145	421	467	507	258	278	281	175	185	175
0.45	•••	2619	2846	4462	793	884	1133	404	441	512	248	265	292	169	178	190
	•••	2701	3159	4649	831	968	1157	420	470	507	256	277	278	173	183	171
0.5	•••	2490	2756	4514	773	870	1134	394	433	506	241	259	286	163	172	184
	•••	2588	3106	4649	814	960	1145	411	463	496	249	271	269	167	178	164
0.55	•••	2325	2626	4464	741	842	1109	379	418	490	230	247	274	154	163	174
	•••	2438	3002	4555	783	933	1110	396	447	475	238	259	255	158	168	153
0.6	•••	2135	2462	4312	698	799	1059	356	394	462	215	231	255	142	150	160
	•••	2261	2853	4368	741	888	1052	373	422	444	222	241	234	146	154	138
0.65	•••	1932	2272	4059	646	743	983	327	362	423	195	209	230	127	133	142
	•••	2065	2664	4088	687	827	970	343	387	403	201	218	208	130	137	119
0.7	•••	1721	2059	3706	584	672	882	292	322	372	170	182	198	108	113	120
	•••	1854	2436	3714	621	748	865	306	343	351	176	189	176	111	116	97
0.75	•••	1503	1822	3253	511	587	757	249	273	311	141	150	161	.	.	.
	•••	1630	2167	3247	544	650	736	260	290	289	145	155	138	.	.	.
0.8	•••	1274	1556	2702	425	485	606	198	215	239	.	.	.	.	.	.
	•••	1387	1852	2687	451	532	584	206	226	216	.	.	.	.	.	.
0.85	•••	1024	1251	2053	322	362	432	.	.	.	.	.	.	.	.	.
	•••	1117	1478	2033	340	391	409	.	.	.	.	.	.	.	.	.
0.9	•••	733	883	1308	.	.	.	.	.	.	.	.	.	.	.	.
	•••	796	1020	1285	.	.	.	.	.	.	.	.	.	.	.	.

TABLE 1: ALPHA= 0.005 POWER= 0.8   EXPECTED ACCRUAL THRU MINIMUM FOLLOW-UP= 450

		DEL=.05			DEL=.10			DEL=.15			DEL=.20			DEL=.25		
FACT=		1.0 .75	.50 .25	.00 BIN	1.0 .75	.50 .25	.00 BIN	1.0 .75	.50 .25	.00 BIN	1.0 75	.50 .25	.00 BIN	1.0 .75	.50 .25	.00 BIN
PCONT=***					REQUIRED NUMBER OF PATIENTS											
0.05	***	706	716	746	255	259	265	150	152	155	106	108	109	83	84	84
	•••	710	726	1285	257	262	409	151	153	216	107	108	138	83	84	97
0.1	***	1291	1315	1439	410	422	444	222	228	236	148	151	154	110	111	113
	•••	1299	1347	2033	415	431	584	225	231	289	149	153	176	111	112	119
0.15	***	1791	1833	2109	539	560	609	282	291	307	181	186	193	131	134	137
	•••	1806	1896	2687	548	579	736	286	298	351	184	190	208	132	135	138
0.2	***	2184	2247	2721	640	674	755	328	343	369	208	215	226	147	151	157
	•••	2206	2346	3247	654	703	865	335	354	403	211	220	234	149	154	153
0.25	***	2464	2554	3256	716	761	879	363	384	419	227	237	252	159	165	172
	•••	2495	2693	3714	734	803	970	372	399	444	232	244	255	162	168	164
0.3	***	2637	2758	3703	766	825	980	388	414	459	241	253	272	167	174	183
	•••	2679	2942	4088	791	879	1052	399	432	475	247	262	269	170	178	171
0.35	***	2714	2870	4055	796	867	1056	403	433	488	249	263	285	171	179	189
	•••	2769	3100	4368	825	932	1110	416	456	496	255	273	278	175	184	175
0.4	***	2708	2903	4308	806	888	1108	409	443	506	252	268	292	172	180	192
	•••	2777	3178	4555	840	964	1145	424	469	507	259	279	281	176	185	175
0.45	***	2634	2869	4462	801	892	1134	407	444	512	249	267	292	169	178	190
	•••	2719	3188	4649	838	974	1157	423	471	507	257	278	278	174	183	171
0.5	***	2508	2783	4514	781	878	1134	398	436	507	243	260	286	164	172	184
	•••	2610	3137	4649	822	966	1145	415	465	496	251	272	269	168	178	164
0.55	***	2346	2654	4464	749	850	1109	382	420	490	232	249	274	155	163	174
	•••	2462	3035	4555	792	939	1110	399	449	475	239	260	255	159	168	153
0.6	***	2159	2493	4312	707	807	1059	360	396	462	216	232	255	143	150	160
	•••	2287	2887	4368	749	895	1052	376	423	444	223	242	234	146	155	138
0.65	***	1958	2303	4059	654	750	983	330	365	423	196	210	230	127	134	142
	•••	2093	2697	4088	695	833	970	346	389	403	203	219	208	131	138	119
0.7	***	1747	2089	3706	591	679	882	295	324	372	172	183	198	109	114	120
	•••	1882	2467	3714	629	753	865	308	345	351	177	190	176	111	117	97
0.75	***	1527	1850	3253	517	593	757	252	275	311	142	150	161	.	.	.
	•••	1656	2196	3247	550	654	736	262	291	289	146	155	138	.	.	.
0.8	***	1296	1581	2702	430	490	606	200	216	239	.	.	.	.	.	.
	•••	1411	1876	2687	456	536	584	207	226	216	.	.	.	.	.	.
0.85	***	1042	1271	2053	326	365	432	.	.	.	.	.	.	.	.	.
	•••	1136	1496	2033	343	393	409	.	.	.	.	.	.	.	.	.
0.9	***	745	895	1308	.	.	.	.	.	.	.	.	.	.	.	.
	•••	808	1031	1285	.	.	.	.	.	.	.	.	.	.	.	.

TABLE 1: ALPHA= 0.005 POWER= 0.8   EXPECTED ACCRUAL THRU MINIMUM FOLLOW-UP= 475

PCONT	DEL=.05 1.0/.75	.50/.25	.00/BIN	DEL=.10 1.0/.75	.50/.25	.00/BIN	DEL=.15 1.0/.75	.50/.25	.00/BIN	DEL=.20 1.0/75	.50/.25	.00/BIN	DEL=.25 1.0/.75	.50/.25	.00/BIN
						REQUIRED NUMBER OF PATIENTS								
0.05 ***	707	716	746	255	259	265	150	152	155	107	108	109	83	83	84
•••	710	726	1285	257	262	409	151	153	216	107	108	138	83	84	97
0.1 ***	1293	1317	1439	411	422	444	223	228	236	148	151	155	110	111	113
•••	1301	1350	2033	416	431	584	225	232	289	150	153	176	110	112	119
0.15 ***	1794	1837	2109	541	562	609	282	292	307	182	187	193	131	134	137
•••	1809	1901	2687	549	580	736	287	298	351	184	190	208	132	135	138
0.2 ***	2187	2254	2721	643	676	755	330	344	368	208	216	226	147	151	157
•••	2211	2354	3247	657	705	865	336	355	403	212	220	234	149	154	153
0.25 ***	2469	2563	3256	719	765	879	365	385	419	228	238	252	159	165	172
•••	2502	2705	3714	738	806	970	374	400	444	233	244	255	162	168	164
0.3 ***	2644	2770	3703	771	830	980	390	415	459	242	254	272	167	174	183
•••	2688	2958	4088	795	883	1052	401	433	475	248	262	269	171	178	171
0.35 ***	2723	2885	4055	801	872	1056	406	435	488	250	264	285	172	179	189
•••	2781	3120	4368	831	937	1110	419	457	496	257	273	278	175	184	175
0.4 ***	2720	2922	4309	813	895	1108	412	445	506	253	269	292	173	181	192
•••	2792	3202	4555	847	969	1145	427	470	507	260	279	281	177	186	175
0.45 ***	2649	2892	4462	808	899	1134	410	447	512	251	267	292	170	179	190
•••	2737	3215	4649	846	980	1157	426	473	507	259	279	278	174	184	171
0.5 ***	2526	2808	4513	789	886	1134	401	439	507	244	261	286	165	173	184
•••	2631	3167	4649	830	972	1145	418	467	496	252	272	269	168	178	164
0.55 ***	2366	2682	4463	757	857	1109	385	423	490	233	250	274	156	164	174
•••	2486	3066	4555	799	946	1110	402	450	475	241	260	255	159	168	153
0.6 ***	2182	2521	4312	715	815	1059	363	399	462	218	233	255	143	151	160
•••	2313	2919	4368	757	901	1052	379	425	444	225	243	234	147	155	138
0.65 ***	1982	2333	4059	662	757	983	333	367	422	197	211	230	128	134	142
•••	2119	2729	4088	702	839	970	348	390	403	204	219	208	131	138	119
0.7 ***	1771	2118	3706	598	686	882	298	326	372	172	184	198	110	114	120
•••	1909	2497	3714	636	758	865	310	346	351	178	190	176	112	117	97
0.75 ***	1551	1876	3253	523	599	756	254	276	311	143	150	161	.	.	.
•••	1681	2222	3247	556	659	736	264	292	289	146	155	138	.	.	.
0.8 ***	1317	1604	2702	435	494	606	202	217	239	.	.	.	.	.	.
•••	1433	1898	2687	461	539	584	209	227	216	.	.	.	.	.	.
0.85 ***	1060	1289	2053	329	367	432	.	.	.	.	.	.	.	.	.
•••	1153	1513	2033	346	395	409	.	.	.	.	.	.	.	.	.
0.9 ***	757	906	1308	.	.	.	.	.	.	.	.	.	.	.	.
•••	820	1040	1285	.	.	.	.	.	.	.	.	.	.	.	.

TABLE 1: ALPHA= 0.005 POWER= 0.8   EXPECTED ACCRUAL THRU MINIMUM FOLLOW-UP= 500

PCONT=***		DEL=.05			DEL=.10			DEL=.15			DEL=.20			DEL=.25		
FACT=		1.0 .75	.50 .25	.00 BIN	1.0 .75	.50 .25	.00 BIN	1.0 .75	.50 .25	.00 BIN	1.0 75	.50 .25	.00 BIN	1.0 .75	.50 .25	.00 BIN
					REQUIRED NUMBER OF PATIENTS											
0.05	***	708	717	746	256	259	265	151	153	155	107	108	109	83	83	84
	•••	711	727	1285	258	262	409	152	153	216	107	108	138	83	84	97
0.1	***	1294	1319	1439	412	423	444	223	228	236	149	151	155	110	112	113
	•••	1303	1353	2033	417	432	584	226	232	289	150	153	176	111	113	119
0.15	***	1796	1842	2109	542	563	609	283	293	307	183	187	193	131	134	137
	•••	1812	1906	2687	551	581	736	288	299	351	185	190	208	133	136	138
0.2	***	2191	2260	2721	645	678	755	331	345	368	209	216	226	148	152	157
	•••	2215	2363	3247	659	707	865	337	355	403	212	221	234	150	154	153
0.25	***	2474	2572	3256	723	768	879	367	386	419	229	238	252	160	165	172
	•••	2508	2717	3714	742	809	970	376	400	444	233	245	255	163	168	164
0.3	***	2651	2782	3703	775	834	980	392	417	459	243	255	272	168	174	183
	•••	2697	2973	4088	800	887	1052	403	435	475	248	263	269	171	178	171
0.35	***	2733	2901	4055	807	878	1056	408	437	488	252	265	285	173	180	189
	•••	2792	3139	4368	836	941	1110	421	458	496	258	274	278	176	184	175
0.4	***	2731	2940	4308	819	901	1108	415	448	506	254	270	292	173	181	192
	•••	2807	3225	4555	853	973	1145	429	472	507	262	280	281	177	186	175
0.45	***	2663	2913	4462	814	905	1134	413	449	512	253	269	292	171	179	190
	•••	2755	3240	4649	853	985	1157	429	475	507	260	279	278	175	184	171
0.5	***	2543	2833	4513	796	893	1134	404	441	507	246	262	286	165	173	184
	•••	2652	3195	4649	837	978	1145	421	468	496	253	273	269	169	178	164
0.55	***	2386	2709	4463	765	865	1109	388	425	490	235	251	274	156	164	174
	•••	2509	3096	4555	807	951	1110	405	452	475	242	261	255	160	169	153
0.6	***	2204	2549	4312	722	822	1058	366	401	462	219	234	255	144	151	160
	•••	2338	2949	4368	764	907	1052	382	427	444	226	243	234	148	155	138
0.65	***	2005	2361	4059	669	764	983	336	369	423	199	212	230	129	135	142
	•••	2145	2758	4088	710	844	970	351	392	403	205	220	208	132	138	119
0.7	***	1795	2145	3706	605	692	882	300	328	372	173	184	198	110	114	120
	•••	1934	2525	3714	642	763	865	313	347	351	178	191	176	112	117	97
0.75	***	1573	1902	3253	529	603	757	256	278	311	143	151	161	.	.	.
	•••	1705	2248	3247	562	663	736	266	293	289	147	156	138	.	.	.
0.8	***	1337	1626	2702	440	498	607	203	218	239	.	.	.	.	.	.
	•••	1454	1919	2687	465	541	584	210	228	216	.	.	.	.	.	.
0.85	***	1076	1306	2053	333	370	432	.	.	.	.	.	.	.	.	.
	•••	1171	1528	2033	349	396	409	.	.	.	.	.	.	.	.	.
0.9	***	768	917	1308	.	.	.	.	.	.	.	.	.	.	.	.
	•••	831	1049	1285	.	.	.	.	.	.	.	.	.	.	.	.

TABLE 1: ALPHA= 0.005 POWER= 0.8   EXPECTED ACCRUAL THRU MINIMUM FOLLOW-UP= 550

		DEL=.02			DEL=.05			DEL=.10			DEL=.15			DEL=.20	
FACT=		1.0 .75	.50 .25	.00 BIN	1.0 .75	.50 .25	.00 BIN	1.0 .75	.50 .25	.00 BIN	1.0 75	.50 .25	.00 BIN	1.0 .75	.50 .25	.00 BIN
PCONT=***					REQUIRED NUMBER OF PATIENTS											
0.05	***	3323	3334	3518	709	718	746	256	260	265	151	153	155	107	107	109
	***	3326	3356	6576	712	728	1285	258	262	409	151	154	216	107	108	138
0.1	***	6912	6939	7730	1297	1324	1439	413	424	444	224	229	236	149	151	155
	***	6921	6993	11423	1306	1357	2033	418	433	584	226	232	289	150	153	176
0.15	***	10096	10144	11954	1801	1849	2109	545	566	609	285	294	307	183	188	193
	***	10112	10240	15685	1818	1916	2687	554	583	736	289	300	351	185	190	208
0.2	***	12630	12704	15884	2199	2273	2721	650	683	755	333	347	368	210	217	226
	***	12654	12851	19364	2225	2378	3247	664	710	865	339	356	403	213	221	234
0.25	***	14446	14552	19367	2485	2590	3256	729	774	879	370	389	419	231	239	252
	***	14481	14764	22459	2522	2739	3714	748	814	970	378	402	444	235	245	255
0.3	***	15555	15701	22320	2665	2805	3703	783	842	980	396	420	459	245	256	272
	***	15604	15991	24970	2715	3002	4088	808	893	1052	407	437	475	250	263	269
0.35	***	16008	16201	24695	2751	2930	4055	816	887	1056	412	440	488	254	267	285
	***	16072	16588	26897	2815	3175	4368	846	949	1110	425	461	496	259	275	278
0.4	***	15876	16128	26463	2754	2975	4309	830	912	1108	420	451	506	257	271	292
	***	15960	16632	28240	2835	3267	4555	864	982	1145	434	474	507	264	280	281
0.45	***	15243	15569	27607	2692	2955	4462	827	917	1134	419	453	512	255	270	292
	***	15352	16215	28999	2789	3288	4649	865	995	1157	434	478	507	262	280	278
0.5	***	14205	14621	28119	2577	2879	4513	809	905	1134	410	445	506	248	264	286
	***	14344	15438	29175	2692	3247	4649	850	988	1145	426	471	496	256	274	269
0.55	***	12860	13389	27996	2425	2759	4463	779	877	1109	394	430	490	237	252	274
	***	13036	14403	28766	2554	3151	4555	821	961	1110	410	455	475	244	262	255
0.6	***	11311	11983	27237	2247	2602	4312	736	834	1059	371	405	462	221	235	255
	***	11537	13205	27773	2386	3005	4368	778	916	1052	386	429	444	228	244	234
0.65	***	9670	10510	25846	2050	2414	4059	683	776	983	341	373	423	201	213	230
	***	9958	11921	26196	2194	2814	4088	723	853	970	355	394	403	206	221	208
0.7	***	8051	9061	23826	1840	2197	3706	618	703	882	304	331	372	175	185	198
	***	8408	10599	24036	1982	2577	3714	654	771	865	316	349	351	180	191	176
0.75	***	6557	7688	21181	1616	1949	3253	540	613	756	259	280	311	145	152	161
	***	6972	9261	21291	1750	2294	3247	572	669	736	269	294	289	148	156	138
0.8	***	5241	6395	17920	1375	1667	2702	448	505	607	205	220	239	.	.	.
	***	5677	7894	17963	1494	1958	2687	473	546	584	212	228	216	.	.	.
0.85	***	4083	5143	14047	1107	1337	2053	338	374	432	.	.	.	.	.	.
	***	4491	6453	14050	1202	1557	2033	354	399	409	.	.	.	.	.	.
0.9	***	2994	3840	9570	789	936	1308	.	.	.	.	.	.	.	.	.
	***	3325	4839	9554	851	1065	1285	.	.	.	.	.	.	.	.	.
0.95	***	1787	2271	4496	.	.	.	.	.	.	.	.	.	.	.	.
	***	1981	2793	4474	.	.	.	.	.	.	.	.	.	.	.	.

TABLE 1: ALPHA= 0.005 POWER= 0.8   EXPECTED ACCRUAL THRU MINIMUM FOLLOW-UP= 600

		DEL=.02			DEL=.05			DEL=.10			DEL=.15			DEL=.20		
FACT=		1.0 .75	.50 .25	.00 BIN	1.0 .75	.50 .25	.00 BIN	1.0 .75	.50 .25	.00 BIN	1.0 75	.50 .25	.00 BIN	1.0 .75	.50 .25	.00 BIN
PCONT=***					REQUIRED	NUMBER OF	PATIENTS									
0.05	***	3323	3336	3518	710	719	746	257	260	265	151	152	155	107	108	109
	***	3328	3360	6576	713	729	1285	258	262	409	152	154	216	107	108	138
0.1	***	6914	6944	7730	1299	1328	1439	415	426	444	225	229	236	149	152	154
	***	6925	7003	11423	1310	1361	2033	420	433	584	227	232	289	151	153	176
0.15	***	10100	10153	11954	1806	1857	2109	548	569	609	286	295	307	184	188	193
	***	10118	10257	15685	1824	1925	2687	557	585	736	290	300	351	186	190	208
0.2	***	12636	12717	15884	2206	2285	2721	654	687	755	335	348	368	211	217	226
	***	12663	12878	19364	2234	2393	3247	668	713	865	341	357	403	214	221	234
0.25	***	14455	14571	19367	2495	2606	3256	734	780	879	372	391	419	232	241	252
	***	14494	14802	22459	2535	2760	3714	754	818	970	380	403	444	236	246	255
0.3	***	15568	15727	22320	2679	2827	3703	791	849	980	399	422	459	247	257	272
	***	15621	16044	24970	2732	3029	4088	815	898	1052	409	438	475	251	264	269
0.35	***	16025	16237	24695	2768	2958	4055	825	895	1056	416	443	488	256	268	285
	***	16096	16658	26897	2837	3208	4368	855	955	1110	428	463	496	261	275	278
0.4	***	15898	16174	26462	2777	3009	4308	840	921	1108	424	455	506	259	273	292
	***	15990	16723	28240	2863	3306	4555	874	990	1145	437	476	507	265	281	281
0.45	***	15273	15628	27607	2719	2994	4462	838	928	1133	423	457	512	257	272	292
	***	15391	16330	28999	2822	3332	4649	876	1003	1157	438	480	507	264	281	278
0.5	***	14243	14696	28119	2610	2923	4513	822	917	1134	415	449	506	251	265	286
	***	14394	15581	29175	2729	3295	4649	862	997	1145	430	473	496	257	275	269
0.55	***	12908	13485	27996	2462	2806	4463	791	889	1109	399	433	490	239	254	274
	***	13100	14575	28766	2596	3201	4555	833	970	1110	415	457	475	246	263	255
0.6	***	11373	12102	27237	2287	2651	4312	749	846	1058	376	409	462	223	237	255
	***	11619	13404	27773	2430	3056	4368	790	925	1052	391	431	444	230	245	234
0.65	***	9749	10654	25846	2092	2463	4059	695	787	983	346	376	422	202	214	230
	***	10061	12140	26196	2240	2864	4088	734	861	970	359	396	403	208	221	208
0.7	***	8150	9225	23825	1882	2244	3706	629	712	882	308	334	372	177	186	198
	***	8533	10829	24036	2027	2624	3714	665	778	865	320	350	351	181	192	176
0.75	***	6674	7862	21181	1656	1993	3253	550	621	757	262	282	311	146	152	161
	***	7113	9488	21291	1792	2336	3247	581	675	736	271	295	289	149	156	138
0.8	***	5367	6565	17920	1411	1704	2702	456	511	607	207	221	239	.	.	.
	***	5821	8104	17963	1531	1993	2687	480	550	584	214	229	216	.	.	.
0.85	***	4202	5294	14047	1136	1366	2053	343	378	432	.	.	.	.	.	.
	***	4624	6633	14050	1231	1582	2033	359	401	409	.	.	.	.	.	.
0.9	***	3091	3958	9570	808	954	1308	.	.	.	.	.	.	.	.	.
	***	3431	4972	9554	870	1079	1285	.	.	.	.	.	.	.	.	.
0.95	***	1844	2335	4496	.	.	.	.	.	.	.	.	.	.	.	.
	***	2041	2859	4474	.	.	.	.	.	.	.	.	.	.	.	.

TABLE 1: ALPHA= 0.005 POWER= 0.8 EXPECTED ACCRUAL THRU MINIMUM FOLLOW-UP= 650

	DEL=.02			DEL=.05			DEL=.10			DEL=.15			DEL=.20		
FACT=	1.0 .75	.50 .25	.00 BIN	1.0 .75	.50 .25	.00 BIN	1.0 .75	.50 .25	.00 BIN	1.0 75	.50 .25	.00 BIN	1.0 .75	.50 .25	.00 BIN

PCONT=*** REQUIRED NUMBER OF PATIENTS

PCONT	1.0 .75	.50 .25	.00 BIN	1.0 .75	.50 .25	.00 BIN	1.0 .75	.50 .25	.00 BIN	1.0 .75	.50 .25	.00 BIN	1.0 .75	.50 .25	.00 BIN
0.05 ***	3325	3338	3518	711	721	746	257	261	265	151	153	155	107	108	109
***	3329	3364	6576	715	730	1285	259	263	409	152	154	216	107	108	138
0.1 ***	6917	6949	7730	1302	1332	1439	416	427	444	226	230	236	150	152	155
***	6928	7013	11423	1313	1365	2033	421	434	584	228	233	289	151	153	176
0.15 ***	10105	10161	11954	1810	1865	2109	550	571	609	287	295	307	184	188	193
***	10124	10274	15685	1830	1933	2687	559	587	736	291	301	351	186	191	208
0.2 ***	12643	12730	15884	2213	2296	2721	658	691	755	337	349	368	212	218	226
***	12672	12904	19364	2243	2406	3247	672	716	865	342	358	403	215	222	234
0.25 ***	14465	14590	19367	2505	2623	3256	740	785	879	375	392	419	233	241	252
***	14507	14841	22459	2547	2779	3714	759	821	970	383	404	444	237	246	255
0.3 ***	15582	15753	22320	2693	2849	3703	798	855	980	402	424	459	248	258	272
***	15639	16096	24970	2749	3054	4088	822	903	1052	412	439	475	253	264	269
0.35 ***	16043	16272	24695	2786	2985	4055	834	903	1056	419	446	488	257	269	285
***	16119	16728	26897	2859	3239	4368	863	961	1110	432	464	496	263	276	278
0.4 ***	15922	16220	26463	2799	3041	4309	850	930	1108	428	458	506	261	274	292
***	16021	16813	28240	2889	3343	4555	884	997	1145	441	478	507	267	282	281
0.45 ***	15303	15687	27607	2746	3031	4462	849	938	1134	428	460	512	259	273	292
***	15431	16445	28999	2854	3373	4649	887	1011	1157	442	482	507	266	282	278
0.5 ***	14281	14772	28119	2642	2964	4514	833	927	1134	419	453	506	253	267	286
***	14445	15722	29175	2766	3339	4649	873	1004	1145	435	476	496	259	276	269
0.55 ***	12956	13580	27996	2498	2850	4463	803	900	1109	403	437	490	242	255	274
***	13165	14743	28766	2635	3247	4555	844	978	1110	419	460	475	248	263	255
0.6 ***	11435	12220	27237	2326	2697	4312	761	856	1059	380	412	462	225	238	255
***	11700	13596	27773	2472	3104	4368	802	933	1052	395	434	444	231	246	234
0.65 ***	9828	10795	25846	2132	2509	4059	706	796	983	350	379	423	204	215	230
***	10163	12349	26196	2283	2910	4088	745	868	970	363	398	403	210	222	208
0.7 ***	8248	9384	23826	1921	2288	3706	639	721	882	311	336	372	178	187	198
***	8655	11046	24036	2069	2668	3714	675	784	865	323	352	351	182	192	176
0.75 ***	6789	8028	21181	1693	2033	3253	559	628	757	265	284	311	146	153	161
***	7249	9701	21291	1831	2375	3247	589	680	736	274	296	289	150	157	138
0.8 ***	5487	6726	17920	1444	1738	2702	463	516	606	209	222	239	.	.	.
***	5959	8302	17963	1565	2025	2687	487	554	584	215	230	216	.	.	.
0.85 ***	4315	5437	14047	1162	1393	2053	348	381	432	.	.	.	.	.	.
***	4749	6801	14050	1258	1605	2033	363	403	409	.	.	.	.	.	.
0.9 ***	3183	4068	9570	825	970	1308	.	.	.	.	.	.	.	.	.
***	3530	5096	9554	887	1091	1285	.	.	.	.	.	.	.	.	.
0.95 ***	1898	2395	4496	.	.	.	.	.	.	.	.	.	.	.	.
***	2098	2919	4474	.	.	.	.	.	.	.	.	.	.	.	.

TABLE 1: ALPHA= 0.005 POWER= 0.8   EXPECTED ACCRUAL THRU MINIMUM FOLLOW-UP= 700

		DEL=.02			DEL=.05			DEL=.10			DEL=.15			DEL=.20	
FACT=	1.0 .75	.50 .25	.00 BIN	1.0 .75	.50 .25	.00 BIN	1.0 .75	.50 .25	.00 BIN	1.0 75	.50 .25	.00 BIN	1.0 .75	.50 .25	.00 BIN
PCONT=•••			REQUIRED NUMBER OF PATIENTS												
0.05 ***	3326	3340	3518	712	722	746	257	261	265	152	153	155	107	108	109
•••	3330	3368	6576	716	731	1285	259	263	409	152	154	216	107	108	138
0.1 ***	6920	6954	7730	1305	1335	1439	418	428	444	226	230	236	150	152	155
•••	6931	7023	11423	1316	1369	2033	422	435	584	228	233	289	151	153	176
0.15 ***	10109	10170	11954	1815	1871	2109	553	573	609	288	296	307	185	189	193
•••	10129	10292	15685	1836	1940	2687	562	588	736	292	301	351	187	191	208
0.2 ***	12650	12744	15884	2220	2307	2721	662	693	754	338	351	369	213	219	226
•••	12681	12931	19364	2252	2418	3247	675	718	865	344	359	403	215	222	234
0.25 ***	14475	14610	19367	2515	2638	3256	745	789	879	377	394	419	234	242	252
•••	14520	14879	22459	2560	2796	3714	764	825	970	385	405	444	238	247	255
0.3 ***	15595	15780	22320	2706	2869	3703	804	861	980	405	426	459	249	259	272
•••	15657	16149	24970	2766	3077	4088	828	907	1052	415	441	475	254	265	269
0.35 ***	16060	16307	24695	2804	3010	4055	841	910	1056	423	448	488	259	270	285
•••	16143	16797	26897	2880	3267	4368	870	966	1110	434	466	496	264	277	278
0.4 ***	15944	16266	26462	2821	3071	4308	859	938	1108	432	460	506	262	275	292
•••	16052	16903	28240	2915	3376	4555	893	1003	1145	445	480	507	268	283	281
0.45 ***	15332	15746	27607	2773	3065	4462	859	946	1134	432	463	512	261	275	292
•••	15470	16557	28999	2884	3411	4649	896	1017	1157	446	484	507	267	282	278
0.5 ***	14318	14847	28119	2672	3002	4513	843	936	1134	423	456	506	254	268	286
•••	14495	15860	29175	2800	3380	4649	884	1011	1145	438	478	496	261	276	269
0.55 ***	13004	13675	27996	2532	2892	4464	814	909	1109	408	439	490	243	256	274
•••	13229	14906	28766	2673	3290	4555	855	985	1110	422	461	475	249	264	255
0.6 ***	11496	12337	27237	2362	2740	4312	772	865	1059	384	415	462	227	239	255
•••	11782	13781	27773	2512	3147	4368	812	940	1052	398	436	444	233	246	234
0.65 ***	9906	10932	25846	2170	2551	4059	716	805	983	353	381	422	205	216	229
•••	10264	12550	26196	2323	2952	4088	755	875	970	366	399	403	211	222	208
0.7 ***	8345	9537	23826	1958	2329	3706	649	729	882	314	338	372	179	188	198
•••	8775	11253	24036	2108	2707	3714	684	790	865	325	353	351	183	193	176
0.75 ***	6900	8187	21181	1728	2070	3253	567	634	757	268	286	311	148	153	161
•••	7380	9903	21291	1867	2410	3247	597	684	736	276	297	289	150	157	138
0.8 ***	5603	6879	17920	1475	1770	2702	469	520	607	211	223	239	.	.	.
•••	6090	8488	17963	1596	2053	2687	492	557	584	217	230	216	.	.	.
0.85 ***	4422	5572	14047	1187	1416	2053	352	384	432	.	.	.	.	.	.
•••	4868	6959	14050	1283	1626	2033	366	405	409	.	.	.	.	.	.
0.9 ***	3269	4172	9570	841	984	1308	.	.	.	.	.	.	.	.	.
•••	3624	5212	9554	902	1103	1285	.	.	.	.	.	.	.	.	.
0.95 ***	1948	2451	4496	.	.	.	.	.	.	.	.	.	.	.	.
•••	2151	2974	4474	.	.	.	.	.	.	.	.	.	.	.	.

TABLE 1: ALPHA= 0.005 POWER= 0.8   EXPECTED ACCRUAL THRU MINIMUM FOLLOW-UP= 750

PCONT=•••		DEL=.02 1.0 .75	DEL=.02 .50 .25	DEL=.02 .00 BIN	DEL=.05 1.0 .75	DEL=.05 .50 .25	DEL=.05 .00 BIN	DEL=.10 1.0 .75	DEL=.10 .50 .25	DEL=.10 .00 BIN	DEL=.15 1.0 75	DEL=.15 .50 .25	DEL=.15 .00 BIN	DEL=.20 1.0 .75	DEL=.20 .50 .25	DEL=.20 .00 BIN
								REQUIRED NUMBER OF PATIENTS								
0.05	***	3327	3342	3518	713	723	746	258	261	265	152	153	154	107	108	109
	•••	3332	3371	6576	717	732	1285	259	263	409	153	154	216	108	109	138
0.1	***	6922	6959	7729	1308	1339	1439	419	429	444	227	230	235	150	152	154
	•••	6934	7032	11423	1319	1372	2033	423	435	584	229	233	289	151	154	176
0.15	***	10114	10179	11954	1820	1878	2109	555	574	609	289	296	307	185	189	193
	•••	10135	10309	15685	1841	1947	2687	564	589	736	293	301	351	187	191	208
0.2	***	12656	12757	15884	2227	2317	2721	665	696	755	340	351	368	214	219	226
	•••	12690	12958	19364	2260	2429	3247	679	720	865	345	359	403	216	222	234
0.25	***	14485	14629	19367	2525	2652	3256	749	793	879	379	395	419	235	243	252
	•••	14532	14917	22459	2572	2813	3714	769	828	970	386	406	444	238	247	255
0.3	***	15608	15806	22320	2719	2888	3703	810	866	980	407	428	460	250	260	272
	•••	15674	16201	24970	2782	3098	4088	834	911	1052	417	442	475	255	265	269
0.35	***	16078	16342	24695	2821	3034	4055	848	916	1056	426	450	488	260	271	285
	•••	16166	16866	26897	2900	3294	4368	878	971	1110	437	467	496	265	277	278
0.4	***	15967	16312	26463	2842	3100	4309	867	945	1107	435	463	505	264	276	292
	•••	16082	16991	28240	2940	3407	4555	901	1008	1145	447	482	507	270	283	281
0.45	***	15362	15805	27607	2798	3098	4462	868	954	1134	435	465	512	263	275	292
	•••	15510	16669	28999	2914	3445	4649	905	1024	1157	449	485	507	269	283	278
0.5	***	14356	14922	28119	2701	3039	4513	853	945	1134	427	458	506	256	269	286
	•••	14545	15995	29175	2832	3417	4649	893	1018	1145	441	479	496	262	277	269
0.55	***	13053	13769	27996	2564	2930	4464	824	918	1109	411	442	490	245	257	274
	•••	13293	15065	28766	2709	3330	4555	865	992	1110	425	463	475	250	265	255
0.6	***	11557	12452	27237	2397	2780	4312	781	874	1059	387	417	462	229	240	255
	•••	11862	13960	27773	2549	3187	4368	821	946	1052	401	437	444	234	247	234
0.65	***	9984	11066	25846	2206	2591	4059	726	813	983	356	383	423	207	217	230
	•••	10363	12742	26196	2361	2991	4088	764	880	970	369	401	403	212	223	208
0.7	***	8440	9685	23825	1994	2366	3706	657	736	882	317	340	372	180	188	199
	•••	8891	11450	24036	2145	2743	3714	692	795	865	328	355	351	184	193	176
0.75	***	7008	8339	21182	1761	2104	3253	574	640	757	270	287	311	148	154	161
	•••	7506	10096	21291	1902	2441	3247	604	688	736	278	298	289	151	157	138
0.8	***	5714	7025	17920	1504	1799	2702	475	525	606	213	224	239	.	.	.
	•••	6216	8664	17963	1626	2079	2687	498	560	584	218	231	216	.	.	.
0.85	***	4525	5700	14047	1210	1439	2053	355	386	432	.	.	.	.	.	.
	•••	4982	7108	14050	1306	1644	2033	370	407	409	.	.	.	.	.	.
0.9	***	3352	4270	9570	856	997	1308	.	.	.	.	.	.	.	.	.
	•••	3714	5320	9554	917	1113	1285	.	.	.	.	.	.	.	.	.
0.95	***	1996	2503	4496	.	.	.	.	.	.	.	.	.	.	.	.
	•••	2201	3025	4474	.	.	.	.	.	.	.	.	.	.	.	.

TABLE 1: ALPHA= 0.005 POWER= 0.8   EXPECTED ACCRUAL THRU MINIMUM FOLLOW-UP= 800

PCONT=***	DEL=.02			DEL=.05			DEL=.10			DEL=.15			DEL=.20		
FACT=	1.0 / .75	.50 / .25	.00 / BIN	1.0 / .75	.50 / .25	.00 / BIN	1.0 / .75	.50 / .25	.00 / BIN	1.0 / 75	.50 / .25	.00 / BIN	1.0 / .75	.50 / .25	.00 / BIN
	REQUIRED NUMBER OF PATIENTS														
0.05	3328	3345	3518	714	724	746	258	261	265	152	153	155	107	108	109
	3333	3375	6576	718	733	1285	260	263	409	153	154	216	108	108	138
0.1	6925	6964	7730	1310	1342	1439	420	429	444	227	231	236	151	152	155
	6938	7042	11423	1322	1375	2033	424	436	584	229	233	289	151	153	176
0.15	10118	10187	11954	1824	1884	2109	557	576	609	290	297	307	186	189	193
	10141	10326	15685	1847	1953	2687	565	590	736	293	302	351	187	191	208
0.2	12663	12771	15884	2234	2327	2721	668	699	755	341	352	368	214	219	226
	12699	12984	19364	2269	2440	3247	682	722	865	346	360	403	217	222	234
0.25	14494	14648	19367	2535	2667	3256	754	797	879	381	397	419	236	243	252
	14545	14955	22459	2584	2828	3714	773	830	970	388	407	444	239	247	255
0.3	15621	15833	22320	2733	2907	3703	815	871	980	410	430	459	252	260	272
	15692	16253	24970	2798	3118	4088	839	914	1052	419	443	475	256	266	269
0.35	16096	16377	24695	2838	3057	4055	855	922	1056	428	452	488	261	272	285
	16190	16935	26897	2920	3318	4368	884	975	1110	439	468	496	266	278	278
0.4	15990	16358	26463	2863	3128	4308	875	952	1108	438	465	506	265	277	292
	16113	17080	28240	2964	3435	4555	908	1013	1145	450	483	507	271	284	281
0.45	15391	15864	27607	2822	3130	4462	876	962	1134	438	468	512	264	276	292
	15549	16779	28999	2941	3478	4649	913	1029	1157	452	487	507	270	284	278
0.5	14394	14997	28119	2729	3073	4514	862	952	1134	430	461	506	258	270	286
	14596	16128	29175	2864	3453	4649	901	1023	1145	444	481	496	263	278	269
0.55	13101	13862	27996	2596	2967	4464	833	925	1108	415	445	490	246	258	274
	13357	15220	28766	2743	3366	4555	873	997	1110	428	465	475	252	265	255
0.6	11619	12565	27237	2430	2818	4312	791	881	1059	391	420	462	230	241	255
	11943	14133	27773	2585	3224	4368	830	951	1052	404	438	444	235	247	234
0.65	10061	11198	25846	2240	2629	4059	735	820	983	359	385	423	208	218	230
	10462	12927	26196	2397	3027	4088	772	885	970	371	402	403	213	223	208
0.7	8533	9828	23826	2027	2402	3706	665	742	882	320	342	372	181	189	198
	9005	11638	24036	2180	2777	3714	699	799	865	330	356	351	185	193	176
0.75	7113	8486	21181	1792	2137	3253	581	645	757	272	289	311	149	154	161
	7629	10278	21291	1934	2471	3247	610	692	736	280	299	289	152	158	138
0.8	5821	7165	17920	1531	1827	2702	480	529	607	214	225	239	.	.	.
	6337	8831	17963	1653	2103	2687	502	562	584	219	232	216	.	.	.
0.85	4624	5822	14047	1231	1459	2053	359	389	432	.	.	.	.	.	.
	5090	7248	14050	1327	1661	2033	373	408	409	.	.	.	.	.	.
0.9	3431	4364	9570	870	1009	1308	.	.	.	.	.	.	.	.	.
	3799	5422	9554	930	1122	1285	.	.	.	.	.	.	.	.	.
0.95	2041	2552	4496	.	.	.	.	.	.	.	.	.	.	.	.
	2248	3073	4474	.	.	.	.	.	.	.	.	.	.	.	.

TABLE 1: ALPHA= 0.005 POWER= 0.8   EXPECTED ACCRUAL THRU MINIMUM FOLLOW-UP= 850

		DEL=.02			DEL=.05			DEL=.10			DEL=.15			DEL=.20		
FACT=		1.0 .75	.50 .25	.00 BIN	1.0 .75	.50 .25	.00 BIN	1.0 .75	.50 .25	.00 BIN	1.0 75	.50 .25	.00 BIN	1.0 .75	.50 .25	.00 BIN
PCONT=•••				REQUIRED NUMBER OF PATIENTS												
0.05	***	3329 3347 3518	715 725 746	259 261 265	152 153 155	107 108 108
	•••	3335 3378 6576	719 733 1285	260 263 409	153 154 216	107 108 138
0.1	***	6927 6969 7730	1312 1344 1439	420 430 444	227 231 236	151 153 155
	•••	6941 7052 11423	1325 1377 2033	425 436 584	229 233 289	152 154 176
0.15	***	10122 10196 11954	1829 1890 2109	559 578 609	291 298 307	186 189 193
	•••	10147 10342 15685	1852 1959 2687	567 591 736	294 302 351	188 191 208
0.2	***	12670 12784 15883	2241 2337 2721	671 701 754	342 353 368	215 220 226
	•••	12708 13011 19364	2277 2450 3247	684 724 865	347 360 403	217 223 234
0.25	***	14504 14668 19367	2544 2680 3256	758 800 879	382 397 419	237 243 252
	•••	14558 14993 22459	2595 2842 3714	776 833 970	389 408 444	240 248 255
0.3	***	15634 15859 22320	2745 2925 3703	820 875 980	411 431 459	253 261 272
	•••	15709 16305 24970	2813 3137 4088	844 918 1052	420 444 475	256 266 269
0.35	***	16114 16412 24695	2854 3079 4055	861 927 1056	430 454 488	262 272 285
	•••	16213 17003 26897	2940 3341 4368	890 979 1110	442 469 496	267 278 278
0.4	***	16013 16404 26462	2883 3154 4309	882 958 1108	441 467 506	266 277 292
	•••	16143 17167 28240	2987 3462 4555	915 1017 1145	452 484 507	272 284 281
0.45	***	15421 15923 27607	2846 3160 4462	884 968 1133	441 470 512	265 277 292
	•••	15589 16887 28999	2968 3508 4649	921 1034 1157	454 488 507	271 284 278
0.5	***	14432 15072 28119	2756 3106 4514	870 960 1134	433 463 507	259 271 286
	•••	14646 16258 29175	2894 3485 4649	909 1029 1145	447 482 496	265 278 269
0.55	***	13149 13955 27996	2626 3002 4464	842 933 1109	418 447 490	247 259 274
	•••	13421 15370 28766	2775 3401 4555	882 1003 1110	431 466 475	253 266 255
0.6	***	11680 12676 27237	2462 2853 4312	799 888 1058	394 422 462	231 241 255
	•••	12023 14300 27773	2619 3258 4368	838 956 1052	407 440 444	236 248 234
0.65	***	10137 11325 25846	2272 2664 4059	743 827 983	362 387 423	209 219 230
	•••	10559 13105 26196	2431 3060 4088	780 890 970	374 403 403	213 224 208
0.7	***	8625 9966 23826	2059 2436 3706	672 748 882	322 343 372	182 189 198
	•••	9116 11819 24036	2213 2808 3714	706 803 865	332 357 351	186 193 176
0.75	***	7215 8627 21182	1822 2167 3253	587 650 757	273 290 311	150 155 161
	•••	7747 10453 21291	1964 2498 3247	615 695 736	281 299 289	152 157 138
0.8	***	5925 7298 17920	1556 1852 2702	485 532 606	215 225 239	. . .
	•••	6453 8990 17963	1680 2125 2687	507 564 584	220 232 216	. . .
0.85	***	4718 5938 14047	1251 1478 2053	362 391 432	. . .	. . .
	•••	5194 7381 14050	1347 1677 2033	375 409 409	. . .	. . .
0.9	***	3506 4452 9570	883 1020 1308	. . .	. . .	. . .
	•••	3880 5517 9554	943 1130 1285	. . .	. . .	. . .
0.95	***	2084 2598 4496	. . .	. . .	. . .	. . .
	•••	2293 3117 4474	. . .	. . .	. . .	. . .

TABLE 1: ALPHA= 0.005 POWER= 0.8   EXPECTED ACCRUAL THRU MINIMUM FOLLOW-UP= 900

		DEL=.02			DEL=.05			DEL=.10			DEL=.15			DEL=.20		
FACT=		1.0 / .75	.50 / .25	.00 / BIN	1.0 / .75	.50 / .25	.00 / BIN	1.0 / .75	.50 / .25	.00 / BIN	1.0 / 75	.50 / .25	.00 / BIN	1.0 / .75	.50 / .25	.00 / BIN
PCONT=•••						REQUIRED	NUMBER	OF	PATIENTS							
0.05	***	3330	3349	3518	716	726	746	259	262	265	152	153	155	108	108	109
	•••	3336	3381	6576	720	734	1285	260	263	409	153	154	216	108	108	138
0.1	***	6930	6974	7730	1315	1347	1439	422	431	444	228	231	236	151	153	154
	•••	6944	7061	11423	1328	1380	2033	426	437	584	230	234	289	152	154	176
0.15	***	10126	10205	11954	1833	1896	2109	560	579	609	291	298	307	186	190	193
	•••	10153	10360	15685	1857	1964	2687	569	592	736	294	302	351	188	191	208
0.2	***	12677	12798	15884	2247	2346	2721	674	703	755	343	354	369	215	220	226
	•••	12717	13037	19364	2285	2459	3247	687	725	865	348	361	403	217	223	234
0.25	***	14513	14687	19367	2554	2693	3256	761	803	879	384	398	419	237	244	252
	•••	14571	15031	22459	2606	2855	3714	780	835	970	390	408	444	240	248	255
0.3	***	15648	15886	22320	2757	2942	3702	825	879	980	414	432	459	253	262	272
	•••	15727	16356	24970	2827	3155	4088	849	920	1052	422	444	475	257	266	269
0.35	***	16131	16448	24695	2870	3100	4055	867	932	1056	433	456	488	263	273	285
	•••	16237	17071	26897	2958	3363	4368	896	982	1110	443	470	496	268	279	278
0.4	***	16037	16449	26462	2903	3178	4308	888	964	1108	443	469	506	268	279	292
	•••	16174	17253	28240	3009	3488	4555	921	1021	1145	455	485	507	273	285	281
0.45	***	15450	15982	27606	2870	3188	4462	892	974	1134	444	471	512	267	278	292
	•••	15628	16993	28999	2994	3537	4649	928	1038	1157	456	489	507	272	285	278
0.5	***	14470	15146	28119	2783	3137	4514	878	966	1134	436	465	506	260	272	286
	•••	14697	16385	29175	2923	3516	4649	917	1033	1145	450	483	496	266	279	269
0.55	***	13197	14046	27996	2654	3035	4464	850	939	1109	420	449	489	249	260	273
	•••	13485	15516	28766	2806	3432	4555	889	1007	1110	433	467	475	254	266	255
0.6	***	11741	12785	27237	2493	2887	4312	807	895	1059	396	423	462	232	242	255
	•••	12102	14461	27773	2651	3290	4368	846	961	1052	409	441	444	237	248	234
0.65	***	10214	11450	25846	2303	2697	4059	750	833	983	365	389	423	210	219	230
	•••	10654	13275	26196	2463	3092	4088	787	894	970	376	404	403	215	224	208
0.7	***	8715	10100	23825	2089	2467	3706	679	753	882	324	344	372	183	190	198
	•••	9225	11991	24036	2244	2837	3714	712	807	865	334	357	351	186	194	176
0.75	***	7314	8762	21182	1850	2196	3253	593	654	757	275	290	311	150	155	161
	•••	7862	10619	21291	1992	2523	3247	621	698	736	282	300	289	153	158	138
0.8	***	6025	7426	17920	1581	1876	2702	489	536	606	216	226	239	.	.	.
	•••	6565	9141	17963	1704	2145	2687	510	567	584	221	233	216	.	.	.
0.85	***	4809	6050	14047	1271	1496	2053	365	393	432	.	.	.	.	.	.
	•••	5294	7507	14050	1366	1691	2033	378	411	409	.	.	.	.	.	.
0.9	***	3578	4537	9570	895	1031	1308	.	.	.	.	.	.	.	.	.
	•••	3958	5608	9554	954	1137	1285	.	.	.	.	.	.	.	.	.
0.95	***	2125	2641	4496	.	.	.	.	.	.	.	.	.	.	.	.
	•••	2335	3158	4474	.	.	.	.	.	.	.	.	.	.	.	.

TABLE 1: ALPHA= 0.005 POWER= 0.8   EXPECTED ACCRUAL THRU MINIMUM FOLLOW-UP= 950

		DEL=.02			DEL=.05			DEL=.10			DEL=.15			DEL=.20		
FACT=		1.0 .75	.50 .25	.00 BIN	1.0 .75	.50 .25	.00 BIN	1.0 .75	.50 .25	.00 BIN	1.0 75	.50 .25	.00 BIN	1.0 .75	.50 .25	.00 BIN
PCONT=***				REQUIRED NUMBER OF PATIENTS												
0.05	***	3331	3351	3518	716	726	746	259	262	265	152	153	154	108	108	109
	***	3338	3384	6576	720	735	1285	260	263	409	153	154	216	108	108	138
0.1	***	6932	6979	7729	1317	1350	1439	422	431	444	228	232	236	151	153	154
	***	6948	7070	11423	1330	1382	2033	426	437	584	230	234	289	152	153	176
0.15	***	10131	10213	11954	1837	1901	2109	562	580	609	292	298	307	186	190	193
	***	10159	10376	15685	1862	1969	2687	570	593	736	295	302	351	188	191	208
0.2	***	12683	12811	15884	2254	2354	2721	676	705	755	344	355	368	216	220	226
	***	12726	13063	19364	2293	2467	3247	689	726	865	349	361	403	218	223	234
0.25	***	14523	14706	19367	2563	2705	3256	765	806	879	385	400	419	238	244	252
	***	14584	15068	22459	2617	2867	3714	783	837	970	392	409	444	241	248	255
0.3	***	15661	15912	22320	2770	2958	3703	830	883	980	415	433	459	254	262	272
	***	15745	16408	24970	2842	3171	4088	853	923	1052	424	445	475	258	267	269
0.35	***	16148	16483	24695	2885	3120	4055	872	937	1056	435	457	488	264	273	285
	***	16260	17138	26897	2976	3383	4368	901	985	1110	445	471	496	268	279	278
0.4	***	16059	16495	26462	2921	3202	4308	895	969	1108	445	470	505	268	279	292
	***	16205	17338	28240	3031	3511	4555	927	1025	1145	457	486	507	273	285	281
0.45	***	15480	16040	27607	2892	3215	4461	899	980	1134	447	473	512	267	279	292
	***	15667	17098	28999	3019	3563	4649	934	1042	1157	458	490	507	273	285	278
0.5	***	14508	15220	28119	2808	3167	4513	886	972	1134	439	467	507	261	272	286
	***	14746	16509	29175	2951	3545	4649	923	1037	1145	451	485	496	267	279	269
0.55	***	13245	14137	27996	2682	3066	4463	857	945	1109	423	450	490	249	260	274
	***	13548	15658	28766	2836	3462	4555	896	1011	1110	436	468	475	255	267	255
0.6	***	11801	12893	27237	2521	2918	4312	815	901	1059	399	425	462	233	242	255
	***	12181	14617	27773	2682	3320	4368	853	965	1052	411	442	444	238	248	234
0.65	***	10289	11572	25846	2332	2728	4059	757	838	983	367	390	422	211	219	230
	***	10748	13440	26196	2494	3121	4088	793	898	970	378	405	403	215	224	208
0.7	***	8804	10231	23826	2117	2497	3706	686	758	882	326	346	372	184	190	198
	***	9331	12157	24036	2274	2864	3714	718	810	865	335	358	351	186	194	176
0.75	***	7411	8893	21181	1876	2223	3253	599	659	756	276	292	311	150	155	161
	***	7974	10778	21291	2020	2547	3247	625	700	736	283	300	289	153	158	138
0.8	***	6122	7550	17919	1604	1898	2702	494	539	606	217	227	239	.	.	.
	***	6673	9285	17963	1727	2165	2687	514	568	584	222	232	216	.	.	.
0.85	***	4897	6157	14047	1288	1512	2053	367	394	432	.	.	.	.	.	.
	***	5390	7627	14050	1383	1704	2033	380	412	409	.	.	.	.	.	.
0.9	***	3647	4617	9570	906	1040	1308	.	.	.	.	.	.	.	.	.
	***	4032	5694	9554	964	1144	1285	.	.	.	.	.	.	.	.	.
0.95	***	2164	2682	4496	.	.	.	.	.	.	.	.	.	.	.	.
	***	2376	3197	4474	.	.	.	.	.	.	.	.	.	.	.	.

TABLE 1: ALPHA= 0.005 POWER= 0.8   EXPECTED ACCRUAL THRU MINIMUM FOLLOW-UP= 1000

PCONT=		DEL=.02			DEL=.05			DEL=.10			DEL=.15			DEL=.20		
FACT=		1.0 .75	.50 .25	.00 BIN	1.0 .75	.50 .25	.00 BIN	1.0 .75	.50 .25	.00 BIN	1.0 75	.50 .25	.00 BIN	1.0 .75	.50 .25	.00 BIN
					REQUIRED NUMBER OF PATIENTS											
0.05	•••	3332	3353	3518	717	727	746	260	262	265	153	153	155	108	108	109
	•••	3339	3387	6576	721	735	1285	261	263	409	153	154	216	108	108	138
0.1	•••	6935	6984	7730	1320	1353	1439	423	432	445	228	231	236	151	153	155
	•••	6951	7079	11423	1333	1384	2033	427	438	584	230	233	289	152	154	176
0.15	•••	10135	10222	11954	1841	1906	2110	563	581	609	293	299	307	187	190	193
	•••	10164	10393	15685	1866	1974	2687	571	593	736	296	303	351	188	191	208
0.2	•••	12690	12824	15884	2260	2363	2721	678	707	755	345	355	368	216	221	226
	•••	12735	13090	19364	2300	2475	3247	691	728	865	350	361	403	218	223	234
0.25	•••	14533	14725	19367	2572	2717	3256	768	809	879	386	400	420	238	245	252
	•••	14597	15106	22459	2628	2879	3714	786	838	970	393	409	444	241	248	255
0.3	•••	15674	15938	22320	2782	2973	3703	834	886	980	417	435	460	255	263	272
	•••	15762	16458	24970	2856	3186	4088	857	925	1052	425	446	475	258	267	269
0.35	•••	16166	16518	24695	2901	3139	4055	878	941	1056	437	458	488	265	274	285
	•••	16283	17204	26897	2993	3402	4368	906	988	1110	447	472	496	270	280	278
0.4	•••	16082	16541	26463	2940	3225	4308	901	973	1108	448	471	506	270	280	292
	•••	16235	17422	28240	3051	3534	4555	933	1028	1145	458	487	507	275	285	281
0.45	•••	15510	16099	27607	2913	3240	4462	905	985	1134	449	475	512	269	279	292
	•••	15707	17200	28999	3043	3588	4649	940	1046	1157	461	491	507	274	285	278
0.5	•••	14545	15293	28119	2833	3195	4513	893	978	1134	441	468	506	262	273	286
	•••	14797	16630	29175	2977	3572	4649	930	1041	1145	454	485	496	267	279	269
0.55	•••	13293	14227	27996	2709	3096	4463	865	951	1109	425	452	490	251	261	274
	•••	13611	15796	28766	2865	3491	4555	903	1016	1110	438	469	475	256	267	255
0.6	•••	11863	12999	27237	2550	2949	4312	821	906	1058	401	426	462	234	243	255
	•••	12259	14768	27773	2711	3349	4368	859	969	1052	413	443	444	238	249	234
0.65	•••	10363	11691	25846	2361	2758	4059	764	844	983	369	391	423	211	220	230
	•••	10841	13598	26196	2523	3148	4088	800	901	970	380	406	403	216	225	208
0.7	•••	8891	10358	23826	2145	2525	3706	691	763	882	328	347	372	184	191	198
	•••	9435	12317	24036	2301	2889	3714	723	813	865	337	359	351	187	195	176
0.75	•••	7506	9020	21181	1902	2248	3253	603	663	756	278	293	311	151	156	161
	•••	8081	10931	21291	2045	2570	3247	630	703	736	285	301	289	153	158	138
0.8	•••	6216	7669	17920	1626	1920	2702	498	541	606	218	228	239	.	.	.
	•••	6778	9423	17963	1749	2183	2687	518	570	584	223	233	216	.	.	.
0.85	•••	4981	6260	14047	1306	1528	2053	370	396	432	.	.	.	.	.	.
	•••	5483	7742	14050	1401	1717	2033	382	413	409	.	.	.	.	.	.
0.9	•••	3714	4694	9570	917	1049	1308	.	.	.	.	.	.	.	.	.
	•••	4104	5775	9554	975	1151	1285	.	.	.	.	.	.	.	.	.
0.95	•••	2201	2721	4496	.	.	.	.	.	.	.	.	.	.	.	.
	•••	2414	3233	4474	.	.	.	.	.	.	.	.	.	.	.	.

TABLE 1: ALPHA= 0.005 POWER= 0.8   EXPECTED ACCRUAL THRU MINIMUM FOLLOW-UP= 1100

	DEL=.02			DEL=.05			DEL=.10			DEL=.15			DEL=.20		
FACT=	1.0	.50	.00	1.0	.50	.00	1.0	.50	.00	1.0	.50	.00	1.0	.50	.00
	.75	.25	BIN	.75	.25	BIN	.75	.25	BIN	75	.25	BIN	.75	.25	BIN
PCONT=•••					REQUIRED NUMBER OF PATIENTS										
0.05 ***	3334	3356	3518	719	728	746	260	262	265	153	153	155	107	108	109
•••	3342	3393	6576	723	736	1285	261	263	409	153	154	216	108	109	138
0.1 ***	6939	6993	7730	1324	1357	1439	424	433	444	229	232	236	151	153	155
•••	6957	7096	11423	1337	1387	2033	428	438	584	230	234	289	152	153	176
0.15 ***	10144	10240	11954	1849	1916	2109	566	583	609	294	300	307	188	191	193
•••	10176	10426	15685	1875	1982	2687	574	595	736	296	303	351	189	192	208
0.2 ***	12704	12851	15883	2273	2378	2721	683	710	755	346	356	368	217	221	225
•••	12753	13140	19364	2314	2490	3247	695	730	865	351	362	403	219	224	234
0.25 ***	14552	14763	19367	2590	2739	3256	774	813	879	389	401	419	239	246	252
•••	14623	15179	22459	2648	2900	3714	792	841	970	395	410	444	242	248	255
0.3 ***	15701	15991	22320	2805	3002	3703	841	892	980	419	437	459	256	263	272
•••	15797	16559	24970	2882	3214	4088	864	929	1052	428	447	475	259	268	269
0.35 ***	16201	16588	24695	2930	3175	4055	887	949	1056	440	461	488	267	275	285
•••	16330	17335	26897	3026	3437	4368	914	994	1110	450	473	496	270	280	278
0.4 ***	16128	16632	26463	2975	3267	4309	912	983	1108	451	474	505	271	280	291
•••	16296	17586	28240	3090	3574	4555	943	1034	1145	462	488	507	276	286	281
0.45 ***	15569	16215	27607	2955	3288	4462	917	995	1134	453	478	511	270	280	292
•••	15786	17400	28999	3088	3634	4649	951	1052	1157	464	493	507	275	286	278
0.5 ***	14621	15438	28119	2879	3247	4514	906	988	1134	445	471	506	264	274	286
•••	14897	16865	29175	3027	3621	4649	942	1048	1145	457	487	496	269	280	269
0.55 ***	13389	14403	27996	2759	3151	4463	877	961	1109	430	455	489	252	262	274
•••	13738	16062	28766	2918	3542	4555	914	1022	1110	441	471	475	257	268	255
0.6 ***	11983	13205	27237	2602	3005	4312	834	917	1059	405	429	462	235	244	255
•••	12413	15057	27773	2766	3400	4368	871	976	1052	417	444	444	239	249	234
0.65 ***	10510	11921	25846	2414	2814	4059	776	853	983	373	394	423	213	221	230
•••	11022	13900	26196	2578	3198	4088	811	907	970	382	407	403	217	225	208
0.7 ***	9061	10599	23825	2197	2578	3706	703	771	882	331	349	372	185	191	198
•••	9636	12618	24036	2354	2935	3714	733	818	865	340	360	351	188	195	176
0.75 ***	7688	9261	21181	1949	2294	3253	613	669	756	280	294	311	152	156	161
•••	8289	11219	21291	2094	2609	3247	638	707	736	287	302	289	154	158	138
0.8 ***	6395	7894	17920	1666	1958	2702	505	546	606	219	228	239	.	.	.
•••	6977	9682	17963	1790	2215	2687	523	573	584	224	233	216	.	.	.
0.85 ***	5142	6453	14047	1337	1556	2053	374	399	433	.	.	.	.	.	.
•••	5658	7957	14050	1431	1739	2033	386	414	409	.	.	.	.	.	.
0.9 ***	3840	4839	9570	936	1065	1308	.	.	.	.	.	.	.	.	.
•••	4239	5926	9554	993	1162	1285	.	.	.	.	.	.	.	.	.
0.95 ***	2271	2793	4496	.	.	.	.	.	.	.	.	.	.	.	.
•••	2486	3299	4474	.	.	.	.	.	.	.	.	.	.	.	.

TABLE 1: ALPHA= 0.005 POWER= 0.8   EXPECTED ACCRUAL THRU MINIMUM FOLLOW-UP= 1200

		DEL=.02			DEL=.05			DEL=.10			DEL=.15			DEL=.20		
FACT=		1.0 .75	.50 .25	.00 BIN	1.0 .75	.50 .25	.00 BIN	1.0 .75	.50 .25	.00 BIN	1.0 75	.50 .25	.00 BIN	1.0 .75	.50 .25	.00 BIN
PCONT=***					REQUIRED NUMBER OF PATIENTS											
0.05	***	3336	3360	3518	719	729	746	260	262	265	152	154	154	108	108	109
	***	3344	3397	6576	724	736	1285	261	264	409	153	154	216	108	109	138
0.1	***	6944	7003	7729	1327	1361	1439	426	433	444	229	232	235	151	153	154
	***	6964	7113	11423	1342	1391	2033	429	439	584	231	234	289	152	154	176
0.15	***	10153	10257	11954	1857	1924	2109	568	585	609	295	300	307	188	190	193
	***	10187	10458	15685	1884	1990	2687	576	595	736	297	303	351	189	192	208
0.2	***	12717	12877	15883	2284	2392	2721	687	713	754	348	357	368	217	221	226
	***	12770	13191	19364	2327	2503	3247	699	731	865	352	362	403	219	223	234
0.25	***	14571	14802	19366	2606	2760	3256	780	817	879	391	403	419	241	246	252
	***	14648	15251	22459	2667	2919	3714	796	844	970	397	410	444	243	249	255
0.3	***	15727	16044	22320	2827	3029	3703	849	898	979	422	438	459	257	264	271
	***	15832	16656	24970	2907	3239	4088	871	933	1052	430	448	475	260	268	269
0.35	***	16237	16657	24694	2958	3208	4054	895	955	1056	443	463	488	268	275	285
	***	16377	17461	26897	3057	3468	4368	922	997	1110	452	474	496	271	280	278
0.4	***	16174	16723	26462	3009	3306	4308	921	990	1108	454	476	505	273	281	292
	***	16357	17745	28240	3127	3611	4555	952	1039	1145	465	490	507	277	286	281
0.45	***	15628	16330	27607	2994	3332	4462	928	1003	1133	457	480	511	271	281	292
	***	15864	17593	28999	3130	3675	4649	961	1057	1157	467	494	507	276	286	278
0.5	***	14696	15581	28119	2923	3295	4513	916	997	1134	449	473	506	265	274	286
	***	14997	17089	29175	3073	3665	4649	952	1054	1145	460	488	496	270	280	269
0.55	***	13485	14575	27996	2806	3201	4463	889	970	1109	433	457	490	253	262	274
	***	13862	16313	28766	2967	3588	4555	925	1028	1110	445	472	475	258	268	255
0.6	***	12102	13404	27237	2651	3056	4312	846	925	1058	409	431	462	237	245	255
	***	12565	15329	27773	2818	3446	4368	881	982	1052	419	445	444	241	250	234
0.65	***	10654	12139	25846	2463	2864	4059	787	861	982	376	396	422	214	221	229
	***	11197	14182	26196	2629	3242	4088	820	913	970	385	408	403	217	225	208
0.7	***	9225	10828	23825	2244	2624	3706	712	778	882	334	350	372	186	192	198
	***	9828	12899	24036	2402	2976	3714	742	823	865	342	361	351	189	195	176
0.75	***	7861	9487	21181	1993	2336	3253	621	675	757	282	295	311	152	156	160
	***	8485	11485	21291	2137	2644	3247	645	711	736	289	302	289	154	158	138
0.8	***	6565	8104	17920	1704	1993	2701	511	550	607	220	229	239	.	.	.
	***	7165	9921	17963	1826	2242	2687	529	575	584	225	234	216	.	.	.
0.85	***	5294	6633	14047	1366	1582	2053	378	401	432	.	.	.	.	.	.
	***	5822	8153	14050	1459	1758	2033	388	415	409	.	.	.	.	.	.
0.9	***	3958	4972	9570	954	1079	1308	.	.	.	.	.	.	.	.	.
	***	4363	6064	9554	1009	1172	1285	.	.	.	.	.	.	.	.	.
0.95	***	2335	2859	4496	.	.	.	.	.	.	.	.	.	.	.	.
	***	2551	3358	4474	.	.	.	.	.	.	.	.	.	.	.	.

TABLE 1: ALPHA= 0.005 POWER= 0.8   EXPECTED ACCRUAL THRU MINIMUM FOLLOW-UP= 1300

	DEL=.02			DEL=.05			DEL=.10			DEL=.15			DEL=.20		
FACT=	1.0 .75	.50 .25	.00 BIN	1.0 .75	.50 .25	.00 BIN	1.0 .75	.50 .25	.00 BIN	1.0 75	.50 .25	.00 BIN	1.0 .75	.50 .25	.00 BIN

PCONT=*** REQUIRED NUMBER OF PATIENTS

PCONT	1.0/.75	.50/.25	.00/BIN	1.0/.75	.50/.25	.00/BIN	1.0/.75	.50/.25	.00/BIN	1.0/.75	.50/.25	.00/BIN	1.0/.75	.50/.25	.00/BIN
0.05	3339	3364	3518	721	731	746	261	263	265	153	154	154	108	108	109
	3347	3403	6576	725	737	1285	262	264	409	154	154	216	108	108	138
0.1	6949	7013	7729	1332	1365	1439	427	434	445	230	232	236	152	153	154
	6971	7130	11423	1345	1394	2033	431	439	584	231	234	289	153	154	176
0.15	10161	10274	11954	1865	1933	2109	570	587	609	295	301	307	188	191	193
	10198	10489	15685	1892	1996	2687	578	596	736	297	304	351	189	192	208
0.2	12730	12904	15884	2296	2406	2721	691	716	755	349	357	368	218	222	226
	12789	13240	19364	2340	2514	3247	702	733	865	354	362	403	219	224	234
0.25	14590	14840	19367	2623	2779	3256	785	822	879	393	404	419	241	246	252
	14674	15322	22459	2684	2936	3714	801	847	970	398	411	444	244	249	255
0.3	15754	16097	22320	2849	3054	3703	856	903	980	424	440	459	258	264	272
	15868	16752	24970	2931	3262	4088	877	936	1052	432	448	475	261	268	269
0.35	16272	16728	24695	2984	3239	4054	903	961	1056	446	464	488	269	276	285
	16424	17586	26897	3086	3496	4368	929	1002	1110	454	475	496	272	280	278
0.4	16220	16813	26462	3041	3343	4309	930	997	1108	458	479	505	274	282	292
	16419	17899	28240	3162	3644	4555	960	1044	1145	467	491	507	278	287	281
0.45	15687	16445	27606	3031	3373	4462	938	1011	1134	460	482	512	273	282	292
	15943	17777	28999	3169	3712	4649	970	1063	1157	471	496	507	277	287	278
0.5	14772	15722	28119	2964	3339	4514	927	1004	1134	453	476	506	266	276	286
	15096	17304	29175	3117	3706	4649	962	1059	1145	463	490	496	271	280	269
0.55	13580	14743	27995	2850	3248	4463	900	978	1109	436	460	490	255	263	274
	13986	16553	28766	3013	3630	4555	935	1033	1110	448	474	475	259	268	255
0.6	12220	13596	27238	2697	3104	4312	856	933	1059	412	434	461	238	245	255
	12712	15586	27773	2865	3488	4368	890	986	1052	422	447	444	241	250	234
0.65	10795	12349	25846	2509	2910	4059	796	869	983	379	398	422	215	222	230
	11368	14447	26196	2676	3282	4088	829	917	970	388	409	403	219	226	208
0.7	9384	11046	23826	2288	2668	3706	721	784	882	336	352	372	187	193	198
	10012	13162	24036	2447	3013	3714	750	827	865	344	362	351	189	195	176
0.75	8027	9701	21181	2033	2375	3253	628	680	757	284	297	311	153	157	161
	8672	11734	21291	2177	2676	3247	652	714	736	290	303	289	155	159	138
0.8	6726	8302	17920	1738	2025	2702	516	553	606	222	230	239	.	.	.
	7342	10143	17963	1860	2268	2687	533	578	584	226	234	216	.	.	.
0.85	5437	6802	14047	1393	1605	2053	381	403	432	.	.	.	.	.	.
	5976	8335	14050	1485	1775	2033	392	417	409	.	.	.	.	.	.
0.9	4068	5096	9571	969	1091	1308	.	.	.	.	.	.	.	.	.
	4481	6190	9554	1024	1181	1285	.	.	.	.	.	.	.	.	.
0.95	2395	2919	4496	.	.	.	.	.	.	.	.	.	.	.	.
	2612	3412	4474	.	.	.	.	.	.	.	.	.	.	.	.

TABLE 1: ALPHA= 0.005 POWER= 0.8   EXPECTED ACCRUAL THRU MINIMUM FOLLOW-UP= 1400

	DEL=.02			DEL=.05			DEL=.10			DEL=.15			DEL=.20		
FACT=	1.0 .75	.50 .25	.00 BIN	1.0 .75	.50 .25	.00 BIN	1.0 .75	.50 .25	.00 BIN	1.0 75	.50 .25	.00 BIN	1.0 .75	.50 .25	.00 BIN

PCONT=••• REQUIRED NUMBER OF PATIENTS

PCONT	DEL=.02 1.0/.75	.50/.25	.00/BIN	DEL=.05 1.0/.75	.50/.25	.00/BIN	DEL=.10 1.0/.75	.50/.25	.00/BIN	DEL=.15 1.0/75	.50/.25	.00/BIN	DEL=.20 1.0/.75	.50/.25	.00/BIN
0.05 ***	3340	3368	3518	722	731	746	261	262	265	153	154	155	108	108	108
•••	3349	3407	6576	726	738	1285	262	264	409	153	154	216	108	108	138
0.1 ***	6954	7023	7730	1335	1368	1439	428	435	444	230	233	235	152	153	155
•••	6977	7145	11423	1349	1396	2033	431	439	584	231	234	289	152	154	176
0.15 ***	10170	10292	11954	1872	1940	2110	573	588	609	296	301	307	189	191	193
•••	10210	10519	15685	1900	2003	2687	580	598	736	298	304	351	190	192	208
0.2 ***	12743	12932	15884	2307	2418	2721	693	718	754	351	359	368	219	222	226
•••	12806	13289	19364	2352	2524	3247	704	734	865	354	363	403	220	224	234
0.25 ***	14610	14878	19367	2638	2796	3256	789	825	878	394	405	419	241	247	252
•••	14699	15392	22459	2701	2951	3714	805	849	970	399	412	444	244	249	255
0.3 ***	15780	16149	22320	2869	3077	3703	861	907	980	426	441	459	259	265	272
•••	15903	16845	24970	2952	3283	4088	882	939	1052	433	450	475	262	269	269
0.35 ***	16306	16797	24695	3010	3267	4055	910	966	1056	448	465	488	270	276	285
•••	16471	17706	26897	3113	3522	4368	935	1005	1110	457	476	496	273	281	278
0.4 ***	16265	16902	26463	3071	3376	4308	938	1003	1108	460	480	506	275	283	292
•••	16480	18049	28240	3195	3674	4555	967	1047	1145	470	492	507	279	287	281
0.45 ***	15746	16558	27607	3065	3411	4462	946	1018	1134	463	484	512	275	283	292
•••	16021	17956	28999	3206	3746	4649	978	1067	1157	472	497	507	278	287	278
0.5 ***	14847	15860	28119	3002	3380	4513	936	1011	1134	456	478	507	268	276	286
•••	15195	17509	29175	3157	3741	4649	970	1064	1145	466	491	496	272	281	269
0.55 ***	13674	14906	27996	2892	3290	4463	909	985	1109	439	461	490	256	264	274
•••	14107	16780	28766	3055	3667	4555	943	1039	1110	450	474	475	260	269	255
0.6 ***	12337	13781	27237	2740	3146	4312	865	940	1059	415	436	462	239	246	255
•••	12857	15829	27773	2908	3525	4368	899	990	1052	424	448	444	242	250	234
0.65 ***	10932	12550	25846	2551	2952	4059	805	875	983	381	399	423	216	222	229
•••	11532	14697	26196	2719	3318	4088	836	921	970	389	410	403	220	226	208
0.7 ***	9537	11253	23825	2328	2707	3706	729	790	882	338	353	372	188	192	199
•••	10188	13408	24036	2488	3045	3714	757	830	865	346	362	351	190	195	176
0.75 ***	8186	9903	21181	2070	2410	3253	634	684	757	286	297	311	153	157	161
•••	8850	11966	21291	2214	2705	3247	657	717	736	291	304	289	155	159	138
0.8 ***	6879	8488	17920	1770	2053	2702	521	556	606	223	230	239	.	.	.
•••	7509	10350	17963	1891	2290	2687	537	579	584	227	234	216	.	.	.
0.85 ***	5572	6959	14047	1417	1626	2053	384	405	432	.	.	.	.	.	.
•••	6121	8503	14050	1508	1790	2033	394	417	409	.	.	.	.	.	.
0.9 ***	4172	5211	9570	984	1102	1308	.	.	.	.	.	.	.	.	.
•••	4590	6306	9554	1037	1188	1285	.	.	.	.	.	.	.	.	.
0.95 ***	2451	2974	4497	.	.	.	.	.	.	.	.	.	.	.	.
•••	2669	3460	4474	.	.	.	.	.	.	.	.	.	.	.	.

TABLE 1: ALPHA= 0.005 POWER= 0.8 EXPECTED ACCRUAL THRU MINIMUM FOLLOW-UP= 1500

	DEL=.02			DEL=.05			DEL=.10			DEL=.15			DEL=.20		
FACT=	1.0 .75	.50 .25	.00 BIN	1.0 .75	.50 .25	.00 BIN	1.0 .75	.50 .25	.00 BIN	1.0 75	.50 .25	.00 BIN	1.0 .75	.50 .25	.00 BIN
PCONT=***						REQUIRED NUMBER OF PATIENTS									
0.05 ***	3342 3371		3518	723 732		746	260 262		265	153 154		155	108 109		109
***	3352 3412		6576	727 738		1285	262 264		409	154 154		216	108 109		138
0.1 ***	6959 7032		7730	1339 1372		1439	428 435		444	230 232		235	152 154		155
***	6983 7160		11423	1353 1399		2033	432 440		584	232 234		289	153 154		176
0.15 ***	10178 10309		11954	1878 1947		2109	575 589		608	296 301		307	189 191		193
***	10222 10549		15685	1906 2008		2687	581 598		736	299 304		351	190 192		208
0.2 ***	12757 12958		15884	2317 2429		2722	697 720		755	352 359		368	219 222		226
***	12824 13336		19364	2362 2534		3247	707 736		865	355 364		403	220 224		234
0.25 ***	14629 14917		19367	2652 2812		3256	793 828		879	395 406		419	243 247		252
***	14725 15459		22459	2717 2965		3714	809 850		970	400 412		444	245 249		255
0.3 ***	15806 16201		22320	2888 3098		3703	866 911		980	428 442		459	260 265		272
***	15938 16937		24970	2973 3302		4088	887 941		1052	435 450		475	262 268		269
0.35 ***	16342 16867		24695	3034 3293		4055	916 971		1057	450 467		488	271 277		285
***	16518 17824		26897	3139 3545		4368	941 1008		1110	458 477		496	274 281		278
0.4 ***	16312 16991		26463	3100 3407		4309	945 1008		1107	463 482		505	275 283		292
***	16540 18193		28240	3225 3701		4555	973 1051		1145	472 493		507	279 288		281
0.45 ***	15805 16669		27607	3098 3445		4462	954 1024		1133	465 485		512	275 283		292
***	16099 18127		28999	3240 3776		4649	985 1070		1157	474 498		507	279 288		278
0.5 ***	14922 15995		28119	3039 3417		4513	945 1018		1134	458 479		506	269 277		286
***	15292 17705		29175	3195 3774		4649	978 1068		1145	468 492		496	273 281		269
0.55 ***	13769 15065		27995	2930 3330		4463	918 992		1109	442 463		489	257 265		274
***	14227 16997		28766	3095 3701		4555	952 1042		1110	452 475		475	260 269		255
0.6 ***	12452 13960		27237	2780 3187		4312	874 946		1058	417 437		462	240 247		255
***	12999 16059		27773	2949 3560		4368	907 995		1052	427 448		444	243 250		234
0.65 ***	11066 12742		25846	2591 2990		4059	813 880		982	383 401		423	217 223		230
***	11690 14933		26196	2758 3350		4088	844 925		970	392 410		403	220 226		208
0.7 ***	9684 11450		23825	2366 2743		3706	736 795		882	340 354		372	188 193		199
***	10357 13640		24036	2525 3075		3714	763 833		865	347 363		351	190 196		176
0.75 ***	8339 10096		21182	2105 2441		3253	640 688		757	287 298		311	154 157		161
***	9020 12185		21291	2248 2730		3247	663 719		736	292 305		289	155 159		138
0.8 ***	7025 8664		17919	1799 2079		2702	525 560		607	224 230		239	. .		.
***	7669 10544		17963	1919 2310		2687	541 581		584	228 235		216	. .		.
0.85 ***	5700 7108		14047	1439 1644		2053	386 407		432	. .		.	. .		.
***	6260 8660		14050	1528 1804		2033	395 419		409	. .		.	. .		.
0.9 ***	4270 5320		9570	997 1113		1308	. .		.	. .		.	. .		.
***	4694 6413		9554	1049 1195		1285	. .		.	. .		.	. .		.
0.95 ***	2503 3025		4496	. .		.	. .		.	. .		.	. .		.
***	2722 3504		4474	. .		.	. .		.	. .		.	. .		.

TABLE 1: ALPHA= 0.005 POWER= 0.8   EXPECTED ACCRUAL THRU MINIMUM FOLLOW-UP= 1600

	DEL=.02			DEL=.05			DEL=.10			DEL=.15			DEL=.20		
FACT=	1.0 .75	.50 .25	.00 BIN	1.0 .75	.50 .25	.00 BIN	1.0 .75	.50 .25	.00 BIN	1.0 75	.50 .25	.00 BIN	1.0 .75	.50 .25	.00 BIN
PCONT=***	REQUIRED NUMBER OF PATIENTS														
0.05 ***	3345 3355	3375 3415	3518 6576	724 728	733 739	746 1285	261 262	263 264	265 409	153 154	154 154	155 216	108 108	108 109	109 138
0.1 ***	6964 6990	7042 7175	7730 11423	1342 1356	1375 1401	1439 2033	429 432	436 440	444 584	231 232	233 234	236 289	152 153	153 154	155 176
0.15 ***	10187 10234	10326 10578	11954 15685	1884 1913	1953 2013	2109 2687	576 583	590 599	609 736	297 299	302 304	307 351	189 190	191 192	193 208
0.2 ***	12771 12842	12984 13382	15884 19364	2327 2373	2440 2542	2721 3247	699 710	722 737	755 865	352 356	360 364	368 403	219 221	222 224	226 234
0.25 ***	14648 14751	14955 15526	19367 22459	2667 2732	2828 2978	3256 3714	797 812	830 852	879 970	397 401	407 413	419 444	243 245	247 250	252 255
0.3 ***	15833 15974	16253 17026	22320 24970	2907 2993	3118 3319	3703 4088	871 891	914 943	980 1052	430 436	443 451	459 475	260 263	266 269	272 269
0.35 ***	16377 16565	16935 17937	24695 26897	3057 3164	3318 3566	4055 4368	922 946	975 1011	1056 1110	452 460	468 478	488 496	272 275	278 281	285 278
0.4 ***	16358 16602	17080 18333	26463 28240	3128 3254	3435 3726	4308 4555	952 980	1013 1054	1108 1145	465 473	483 494	506 507	277 280	284 288	292 281
0.45 ***	15864 16177	16779 18293	27607 28999	3130 3273	3478 3804	4462 4649	962 992	1029 1074	1134 1157	468 477	487 499	512 507	276 280	284 288	292 278
0.5 ***	14997 15390	16128 17893	28119 29175	3073 3230	3453 3805	4514 4649	952 985	1023 1072	1134 1145	461 470	481 493	506 496	270 274	278 282	286 269
0.55 ***	13862 14345	15220 17203	27996 28766	2967 3133	3366 3732	4464 4555	925 958	997 1046	1109 1110	445 454	465 476	490 475	258 262	265 269	274 255
0.6 ***	12565 13137	14133 16278	27237 27773	2818 2987	3224 3591	4312 4368	881 913	951 998	1059 1052	420 428	438 449	462 444	241 244	247 251	255 234
0.65 ***	11198 11845	12927 15157	25846 26196	2629 2796	3027 3381	4059 4088	820 850	885 928	983 970	385 393	402 412	423 403	218 220	223 226	230 208
0.7 ***	9828 10520	11638 13860	23826 24036	2402 2561	2777 3102	3706 3714	742 768	799 836	882 865	342 348	356 364	372 351	189 191	193 196	198 176
0.75 ***	8486 9182	10278 12391	21181 21291	2137 2279	2471 2753	3253 3247	645 667	692 721	757 736	289 294	299 305	311 289	154 156	158 159	161 138
0.8 ***	7165 7821	8831 10725	17920 17963	1827 1946	2103 2328	2702 2687	529 544	562 583	607 584	225 228	232 235	239 216	. .	. .	. .
0.85 ***	5822 6390	7248 8807	14047 14050	1459 1548	1661 1816	2053 2033	389 398	408 420	432 409	. .	. .	. .	. .	. .	. .
0.9 ***	4364 4792	5422 6513	9570 9554	1009 1060	1122 1201	1308 1285	. .	. .	. .	. .	. .	. .	. .	. .	. .
0.95 ***	2552 2770	3073 3544	4496 4474	. .	. .	. .	. .	. .	. .	. .	. .	. .	. .	. .	. .

TABLE 1: ALPHA= 0.005 POWER= 0.8   EXPECTED ACCRUAL THRU MINIMUM FOLLOW-UP= 1700

		DEL=.02			DEL=.05			DEL=.10			DEL=.15			DEL=.20		
FACT=		1.0 .75	.50 .25	.00 BIN	1.0 .75	.50 .25	.00 BIN	1.0 .75	.50 .25	.00 BIN	1.0 75	.50 .25	.00 BIN	1.0 .75	.50 .25	.00 BIN
PCONT=***							REQUIRED NUMBER OF PATIENTS								
0.05	***	3347	3378	3518	724	733	746	261	263	265	153	154	155	108	108	108
	•••	3357	3419	6576	729	739	1285	262	265	409	154	154	216	108	108	138
0.1	***	6969	7052	7730	1344	1377	1438	430	437	444	231	233	236	153	154	155
	•••	6997	7189	11423	1359	1402	2033	433	440	584	233	235	289	153	154	176
0.15	***	10195	10342	11954	1890	1959	2109	578	590	609	297	301	307	189	191	193
	•••	10246	10606	15685	1919	2018	2687	584	599	736	299	305	351	190	192	208
0.2	***	12784	13011	15883	2336	2450	2721	701	724	754	352	360	369	220	223	226
	•••	12860	13427	19364	2383	2550	3247	712	737	865	357	364	403	221	224	234
0.25	***	14668	14993	19367	2681	2842	3257	800	833	879	397	408	420	243	248	252
	•••	14776	15590	22459	2747	2990	3714	815	853	970	403	413	444	245	250	255
0.3	***	15859	16305	22320	2925	3138	3703	875	918	979	431	444	459	261	265	272
	•••	16009	17112	24970	3011	3334	4088	894	945	1052	437	452	475	263	269	269
0.35	***	16412	17003	24694	3079	3342	4054	928	979	1056	454	469	488	272	278	284
	•••	16611	18047	26897	3186	3586	4368	951	1013	1110	461	478	496	275	282	278
0.4	***	16404	17167	26462	3153	3463	4308	958	1017	1108	466	484	505	277	284	292
	•••	16662	18468	28240	3281	3750	4555	985	1057	1145	475	494	507	280	288	281
0.45	***	15922	16887	27607	3160	3508	4461	968	1034	1134	469	488	512	277	284	292
	•••	16254	18451	28999	3303	3830	4649	998	1077	1157	478	499	507	280	288	278
0.5	***	15072	16258	28119	3106	3485	4513	959	1028	1134	463	482	507	271	278	286
	•••	15486	18074	29175	3264	3832	4649	991	1075	1145	471	494	496	274	282	269
0.55	***	13955	15370	27996	3002	3401	4464	933	1003	1109	447	466	490	259	265	274
	•••	14462	17401	28766	3168	3761	4555	964	1049	1110	456	477	475	262	270	255
0.6	***	12675	14300	27237	2853	3257	4311	888	956	1058	422	440	462	241	248	255
	•••	13272	16487	27773	3023	3619	4368	919	1002	1052	430	450	444	244	250	234
0.65	***	11325	13105	25846	2664	3060	4058	826	890	983	386	403	423	219	224	229
	•••	11995	15369	26196	2832	3407	4088	856	932	970	395	412	403	221	226	208
0.7	***	9966	11819	23825	2436	2808	3706	748	803	882	343	357	372	189	193	199
	•••	10677	14069	24036	2594	3127	3714	773	838	865	350	364	351	191	195	176
0.75	***	8626	10453	21182	2167	2498	3253	650	695	756	290	299	311	155	157	160
	•••	9338	12585	21291	2309	2774	3247	671	724	736	294	305	289	156	159	138
0.8	***	7298	8990	17920	1852	2125	2702	532	564	607	225	231	239	.	.	.
	•••	7966	10897	17963	1970	2345	2687	547	584	584	228	236	216	.	.	.
0.85	***	5938	7381	14047	1478	1676	2052	391	409	432	.	.	.	.	.	.
	•••	6514	8945	14050	1566	1827	2033	399	420	409	.	.	.	.	.	.
0.9	***	4452	5518	9570	1020	1130	1308	.	.	.	.	.	.	.	.	.
	•••	4884	6606	9554	1070	1206	1285	.	.	.	.	.	.	.	.	.
0.95	***	2598	3117	4496	.	.	.	.	.	.	.	.	.	.	.	.
	•••	2815	3580	4474	.	.	.	.	.	.	.	.	.	.	.	.

TABLE 1: ALPHA= 0.005 POWER= 0.8   EXPECTED ACCRUAL THRU MINIMUM FOLLOW-UP= 1800

	DEL=.02			DEL=.05			DEL=.10			DEL=.15			DEL=.20		
FACT=	1.0 .75	.50 .25	.00 BIN	1.0 .75	.50 .25	.00 BIN	1.0 .75	.50 .25	.00 BIN	1.0 75	.50 .25	.00 BIN	1.0 .75	.50 .25	.00 BIN
PCONT=***	REQUIRED NUMBER OF PATIENTS														
0.05 ***	3349	3381	3518	726	733	746	262	263	265	153	154	155	108	108	109
•••	3360	3422	6576	729	739	1285	262	264	409	154	154	216	108	109	138
0.1 ***	6974	7060	7730	1347	1380	1439	431	436	444	231	234	236	153	154	154
•••	7003	7202	11423	1361	1404	2033	433	440	584	233	234	289	153	154	176
0.15 ***	10205	10360	11954	1896	1964	2109	579	591	609	298	303	307	190	191	193
•••	10257	10632	15685	1925	2022	2687	585	600	736	300	305	351	190	192	208
0.2 ***	12798	13038	15884	2346	2459	2721	703	726	755	354	361	369	220	222	226
•••	12878	13470	19364	2393	2557	3247	713	739	865	357	364	403	222	224	234
0.25 ***	14687	15031	19367	2693	2855	3255	803	834	879	398	408	420	244	247	252
•••	14802	15652	22459	2760	3000	3714	818	855	970	402	414	444	246	249	255
0.3 ***	15886	16356	22320	2942	3154	3702	879	920	980	432	444	459	262	267	272
•••	16044	17196	24970	3030	3349	4088	897	947	1052	438	451	475	264	269	269
0.35 ***	16447	17070	24695	3100	3363	4054	933	982	1056	456	470	488	273	279	285
•••	16658	18155	26897	3208	3603	4368	955	1014	1110	462	479	496	276	282	278
0.4 ***	16449	17253	26462	3178	3487	4308	964	1021	1108	469	485	506	279	285	292
•••	16723	18598	28240	3306	3771	4555	990	1059	1145	477	495	507	281	288	281
0.45 ***	15981	16993	27606	3188	3537	4461	974	1038	1134	471	489	512	278	285	292
•••	16330	18605	28999	3332	3853	4649	1003	1080	1157	480	499	507	281	288	278
0.5 ***	15146	16384	28119	3138	3516	4513	966	1032	1134	465	483	506	272	279	285
•••	15581	18247	29175	3295	3858	4649	996	1077	1145	474	494	496	274	282	269
0.55 ***	14046	15516	27996	3035	3432	4464	939	1007	1109	449	467	489	260	267	273
•••	14575	17589	28766	3201	3788	4555	971	1052	1110	458	478	475	263	270	255
0.6 ***	12786	14460	27237	2886	3291	4312	895	960	1059	423	441	462	242	249	255
•••	13404	16686	27773	3057	3645	4368	924	1005	1052	432	451	444	245	252	234
0.65 ***	11450	13275	25845	2697	3091	4059	832	894	983	389	404	423	219	224	229
•••	12140	15571	26196	2864	3433	4088	861	933	970	396	413	403	222	227	208
0.7 ***	10100	11991	23825	2467	2837	3705	753	807	882	344	357	372	190	193	198
•••	10829	14265	24036	2625	3150	3714	778	840	865	351	364	351	192	195	176
0.75 ***	8763	10619	21181	2196	2523	3253	654	697	757	290	300	312	155	157	161
•••	9488	12768	21291	2337	2794	3247	675	724	736	296	306	289	156	159	138
0.8 ***	7426	9141	17920	1876	2145	2702	535	567	606	226	233	240	.	.	.
•••	8104	11058	17963	1993	2360	2687	550	585	584	229	235	216	.	.	.
0.85 ***	6050	7507	14046	1496	1691	2053	393	411	432	.	.	.	.	.	.
•••	6633	9074	14050	1581	1837	2033	402	420	409	.	.	.	.	.	.
0.9 ***	4537	5608	9570	1030	1137	1308	.	.	.	.	.	.	.	.	.
•••	4972	6693	9554	1079	1212	1285	.	.	.	.	.	.	.	.	.
0.95 ***	2641	3158	4497	.	.	.	.	.	.	.	.	.	.	.	.
•••	2859	3615	4474	.	.	.	.	.	.	.	.	.	.	.	.

TABLE 1: ALPHA= 0.005 POWER= 0.8   EXPECTED ACCRUAL THRU MINIMUM FOLLOW-UP= 1900

		DEL=.02			DEL=.05			DEL=.10			DEL=.15			DEL=.20		
FACT=		1.0 .75	.50 .25	.00 BIN	1.0 .75	.50 .25	.00 BIN	1.0 .75	.50 .25	.00 BIN	1.0 75	.50 .25	.00 BIN	1.0 .75	.50 .25	.00 BIN
PCONT=•••					REQUIRED NUMBER OF PATIENTS											
0.05	•••	3351 3363	3384 3426	3518 6576	726 730	735 740	745 1285	262 262	263 265	265 409	153 154	154 154	154 216	108 108	108 109	109 138
0.1	•••	6979 7010	7070 7215	7729 11423	1350 1364	1382 1406	1439 2033	431 434	437 440	444 584	231 232	234 235	236 289	153 153	153 154	154 176
0.15	•••	10214 10268	10376 10659	11954 15685	1901 1930	1968 2025	2109 2687	581 586	592 600	609 736	298 300	303 305	307 351	190 191	191 192	193 208
0.2	•••	12810 12896	13064 13512	15884 19364	2355 2402	2467 2564	2721 3247	705 714	726 740	755 865	355 357	361 364	368 403	220 222	223 224	225 234
0.25	•••	14706 14828	15068 15713	19367 22459	2705 2773	2868 3010	3256 3714	806 820	837 856	878 970	400 403	408 413	419 444	244 246	248 250	251 255
0.3	•••	15912 16078	16408 17278	22320 24970	2958 3046	3170 3362	3702 4088	883 901	923 949	980 1052	433 439	445 452	459 475	262 265	267 269	272 269
0.35	•••	16482 16704	17138 18259	24695 26897	3119 3229	3383 3621	4055 4368	937 959	985 1018	1056 1110	457 464	471 479	488 496	273 277	279 282	285 278
0.4	•••	16495 16783	17337 18724	26462 28240	3203 3331	3511 3790	4308 4555	969 995	1025 1061	1108 1145	470 477	486 495	505 507	279 282	285 288	292 281
0.45	•••	16040 16406	17097 18753	27607 28999	3214 3359	3562 3876	4461 4649	980 1008	1042 1083	1134 1157	474 482	490 501	512 507	279 281	285 288	292 278
0.5	•••	15220 15675	16508 18413	28119 29175	3167 3325	3545 3882	4514 4649	972 1002	1037 1080	1134 1145	467 475	484 495	507 496	272 275	279 282	286 269
0.55	•••	14137 14688	15658 17771	27996 28766	3066 3232	3462 3811	4464 4555	945 976	1011 1054	1109 1110	450 459	467 478	490 475	260 263	267 270	274 255
0.6	•••	12893 13532	14617 16875	27238 27773	2918 3089	3320 3669	4312 4368	901 931	965 1007	1059 1052	425 433	441 451	462 444	242 246	248 251	255 234
0.65	•••	11572 12281	13440 15764	25846 26196	2728 2895	3120 3457	4058 4088	838 866	897 935	983 970	391 398	405 413	422 403	220 222	224 227	230 208
0.7	•••	10231 10974	12157 14453	23826 24036	2497 2654	2864 3170	3706 3714	759 782	809 843	882 865	345 351	358 364	372 351	190 192	194 196	198 176
0.75	•••	8893 9632	10777 12943	21181 21291	2223 2362	2547 2812	3253 3247	659 678	700 726	756 736	292 296	300 306	311 289	155 156	158 159	161 138
0.8	•••	7550 8238	9285 11211	17919 17963	1899 2014	2165 2374	2701 2687	539 552	569 586	607 584	227 229	232 236	239 216	. .	. .	. .
0.85	•••	6157 6746	7627 9197	14047 14050	1512 1597	1704 1846	2053 2033	394 402	412 421	432 409	. .	. .	. .	. .	. .	. .
0.9	•••	4617 5056	5694 6773	9570 9554	1040 1087	1144 1216	1308 1285	. .	. .	. .	. .	. .	. .	. .	. .	. .
0.95	•••	2682 2899	3196 3647	4495 4474	. .	. .	. .	. .	. .	. .	. .	. .	. .	. .	. .	. .

TABLE 1: ALPHA= 0.005 POWER= 0.8   EXPECTED ACCRUAL THRU MINIMUM FOLLOW-UP= 2000

		DEL=.02			DEL=.05			DEL=.10			DEL=.15			DEL=.20		
FACT=		1.0 .75	.50 .25	.00 BIN	1.0 .75	.50 .25	.00 BIN	1.0 .75	.50 .25	.00 BIN	1.0 75	.50 .25	.00 BIN	1.0 .75	.50 .25	.00 BIN
PCONT=•••					REQUIRED NUMBER OF PATIENTS											
0.05	***	3352	3387	3519	727	735	746	262	264	265	154	154	155	109	109	109
	•••	3365	3429	6576	731	740	1285	262	265	409	154	155	216	109	109	138
0.1	***	6984	7079	7730	1352	1384	1439	432	437	445	231	234	236	152	154	155
	•••	7016	7227	11423	1366	1407	2033	435	441	584	232	235	289	154	154	176
0.15	***	10222	10394	11954	1906	1974	2110	581	594	609	299	302	307	190	191	194
	•••	10280	10684	15685	1935	2029	2687	587	601	736	301	305	351	191	192	208
0.2	***	12824	13090	15884	2362	2475	2721	707	727	755	355	361	369	221	224	226
	•••	12914	13552	19364	2410	2570	3247	717	740	865	359	365	403	222	225	234
0.25	***	14725	15106	19367	2717	2879	3256	809	839	879	400	409	420	245	249	252
	•••	14854	15772	22459	2785	3020	3714	822	857	970	405	414	444	246	250	255
0.3	***	15939	16459	22320	2974	3186	3702	886	925	980	435	446	460	262	267	272
	•••	16114	17357	24970	3062	3374	4088	905	950	1052	440	452	475	265	270	269
0.35	***	16517	17204	24695	3139	3402	4055	941	989	1056	459	472	489	274	280	285
	•••	16751	18359	26897	3249	3636	4368	962	1019	1110	465	480	496	276	282	278
0.4	***	16541	17422	26462	3225	3534	4309	974	1029	1107	471	487	506	280	285	292
	•••	16844	18847	28240	3354	3809	4555	999	1064	1145	479	496	507	282	289	281
0.45	***	16099	17200	27607	3240	3589	4462	985	1046	1134	475	491	512	279	285	292
	•••	16482	18895	28999	3386	3896	4649	1014	1085	1157	482	501	507	282	289	278
0.5	***	15292	16630	28119	3195	3572	4514	977	1041	1134	469	485	506	272	279	286
	•••	15769	18574	29175	3354	3904	4649	1007	1082	1145	476	495	496	276	282	269
0.55	***	14227	15796	27996	3096	3491	4464	951	1016	1109	452	469	490	261	267	274
	•••	14799	17945	28766	3262	3835	4555	981	1057	1110	460	479	475	264	270	255
0.6	***	12999	14769	27237	2949	3349	4312	906	969	1059	426	442	462	244	249	255
	•••	13659	17057	27773	3119	3692	4368	935	1009	1052	435	451	444	246	251	234
0.65	***	11691	13599	25846	2759	3149	4059	844	901	982	391	406	422	220	225	230
	•••	12417	15947	26196	2925	3477	4088	871	939	970	399	414	403	222	227	208
0.7	***	10357	12317	23826	2525	2889	3706	762	814	882	347	359	372	191	195	199
	•••	11116	14631	24036	2681	3191	3714	786	845	865	352	365	351	192	196	176
0.75	***	9020	10931	21181	2247	2570	3254	662	702	756	292	301	311	156	159	161
	•••	9770	13110	21291	2386	2829	3247	681	727	736	296	306	289	157	160	138
0.8	***	7669	9422	17920	1920	2182	2702	541	570	606	227	232	239	.	.	.
	•••	8365	11356	17963	2035	2386	2687	555	587	584	230	236	216	.	.	.
0.85	***	6260	7742	14047	1529	1717	2054	396	412	432	.	.	.	.	.	.
	•••	6855	9312	14050	1612	1855	2033	404	422	409	.	.	.	.	.	.
0.9	***	4694	5775	9570	1049	1151	1307	.	.	.	.	.	.	.	.	.
	•••	5136	6850	9554	1095	1220	1285	.	.	.	.	.	.	.	.	.
0.95	***	2721	3234	4496	.	.	.	.	.	.	.	.	.	.	.	.
	•••	2939	3676	4474	.	.	.	.	.	.	.	.	.	.	.	.

TABLE 1: ALPHA= 0.005 POWER= 0.8   EXPECTED ACCRUAL THRU MINIMUM FOLLOW-UP= 2250

		DEL=.02			DEL=.05			DEL=.10			DEL=.15			DEL=.20		
FACT=		1.0 .75	.50 .25	.00 BIN	1.0 .75	.50 .25	.00 BIN	1.0 .75	.50 .25	.00 BIN	1.0 75	.50 .25	.00 BIN	1.0 .75	.50 .25	.00 BIN
PCONT=***				REQUIRED NUMBER OF PATIENTS												
0.05	***	3358	3394	3518	728	735	747	263	264	266	153	154	154	108	108	109
	***	3371	3435	6576	732	741	1285	263	264	409	154	154	216	108	108	138
0.1	***	6996	7101	7730	1358	1389	1438	433	438	444	232	233	236	153	153	154
	***	7032	7255	11423	1372	1410	2033	435	441	584	233	235	289	153	154	176
0.15	***	10244	10434	11954	1918	1984	2109	583	595	609	299	303	306	191	192	194
	***	10309	10743	15685	1947	2036	2687	589	601	736	300	305	351	191	192	208
0.2	***	12857	13154	15883	2382	2493	2721	711	730	755	357	362	368	221	223	226
	***	12958	13650	19364	2429	2583	3247	719	742	865	359	365	403	222	224	234
0.25	***	14774	15197	19366	2744	2905	3256	815	842	879	402	410	419	246	249	252
	***	14917	15913	22459	2812	3040	3714	828	859	970	406	415	444	247	250	255
0.3	***	16004	16584	22319	3009	3222	3702	894	930	980	437	447	460	263	267	271
	***	16201	17546	24970	3098	3402	4088	911	953	1052	441	452	475	266	269	269
0.35	***	16604	17367	24695	3183	3445	4055	950	994	1056	461	474	488	275	280	285
	***	16866	18597	26897	3293	3670	4368	972	1022	1110	466	480	496	277	282	278
0.4	***	16654	17625	26462	3278	3584	4308	984	1036	1108	475	489	506	281	286	292
	***	16992	19135	28240	3407	3849	4555	1008	1068	1145	482	496	507	284	289	281
0.45	***	16244	17448	27607	3300	3644	4462	997	1053	1133	478	493	511	281	286	292
	***	16669	19229	28999	3445	3942	4649	1023	1090	1157	485	502	507	284	289	278
0.5	***	15474	16922	28119	3259	3633	4514	989	1050	1135	472	488	506	274	280	286
	***	15996	18947	29175	3417	3953	4649	1018	1088	1145	479	496	496	277	282	269
0.55	***	14446	16126	27995	3164	3554	4463	964	1023	1109	455	471	489	263	268	274
	***	15065	18348	28766	3329	3885	4555	991	1063	1110	464	480	475	266	271	255
0.6	***	13255	15127	27237	3019	3413	4311	919	977	1059	430	444	462	244	250	254
	***	13960	17479	27773	3186	3742	4368	946	1014	1052	437	452	444	247	252	234
0.65	***	11977	13972	25845	2826	3209	4060	854	910	983	395	407	423	221	224	230
	***	12742	16373	26196	2991	3525	4088	880	943	970	401	415	403	223	227	208
0.7	***	10658	12689	23826	2590	2946	3705	773	820	882	350	359	372	191	195	198
	***	11449	15044	24036	2743	3233	3714	794	848	865	354	365	351	192	196	176
0.75	***	9319	11288	21182	2305	2618	3252	671	708	756	294	302	311	156	159	162
	***	10095	13491	21291	2441	2864	3247	689	731	736	298	306	289	157	160	138
0.8	***	7948	9744	17920	1967	2221	2701	547	573	606	229	233	239	.	.	.
	***	8664	11688	17963	2080	2414	2687	559	589	584	230	236	216	.	.	.
0.85	***	6499	8007	14047	1563	1743	2053	399	415	433	.	.	.	.	.	.
	***	7108	9575	14050	1644	1873	2033	406	423	409	.	.	.	.	.	.
0.9	***	4872	5962	9570	1068	1166	1307	.	.	.	.	.	.	.	.	.
	***	5320	7022	9554	1112	1229	1285	.	.	.	.	.	.	.	.	.
0.95	***	2811	3314	4497	.	.	.	.	.	.	.	.	.	.	.	.
	***	3026	3740	4474	.	.	.	.	.	.	.	.	.	.	.	.

TABLE 1: ALPHA= 0.005 POWER= 0.8   EXPECTED ACCRUAL THRU MINIMUM FOLLOW-UP= 2500

		DEL=.02			DEL=.05			DEL=.10			DEL=.15			DEL=.20		
FACT=		1.0 .75	.50 .25	.00 BIN	1.0 .75	.50 .25	.00 BIN	1.0 .75	.50 .25	.00 BIN	1.0 75	.50 .25	.00 BIN	1.0 .75	.50 .25	.00 BIN
PCONT=***		REQUIRED NUMBER OF PATIENTS														
0.05	***	3362	3399	3518	729	737	746	262	264	265	154	154	154	108	109	109
	•••	3376	3440	6576	733	742	1285	264	264	409	154	154	216	109	109	138
0.1	***	7009	7121	7729	1364	1392	1439	434	439	445	233	234	236	152	154	154
	•••	7048	7281	11423	1376	1412	2033	436	442	584	233	234	289	152	154	176
0.15	***	10265	10473	11954	1929	1993	2109	586	596	609	299	302	308	190	192	193
	•••	10337	10798	15685	1958	2042	2687	590	602	736	301	306	351	192	193	208
0.2	***	12892	13215	15884	2399	2509	2721	715	733	754	358	362	368	221	223	226
	•••	13002	13740	19364	2446	2595	3247	723	743	865	361	365	403	223	224	234
0.25	***	14821	15287	19367	2770	2927	3256	820	845	879	404	411	420	246	249	252
	•••	14981	16043	22459	2837	3056	3714	833	861	970	408	415	444	248	251	255
0.3	***	16070	16704	22320	3042	3251	3702	901	934	979	439	448	459	264	268	271
	•••	16287	17720	24970	3131	3424	4088	917	956	1052	443	454	475	265	270	269
0.35	***	16693	17524	24695	3224	3483	4054	958	999	1056	464	474	489	276	281	286
	•••	16981	18818	26897	3334	3698	4368	977	1026	1110	468	481	496	277	283	278
0.4	***	16768	17823	26462	3324	3627	4309	993	1042	1108	477	490	506	283	287	292
	•••	17137	19401	28240	3454	3883	4555	1015	1071	1145	484	498	507	284	289	281
0.45	***	16387	17686	27608	3352	3693	4462	1008	1061	1134	481	495	512	281	287	292
	•••	16851	19537	28999	3498	3979	4649	1033	1093	1157	487	502	507	284	289	278
0.5	***	15651	17198	28118	3317	3686	4514	1001	1058	1134	474	489	506	274	281	286
	•••	16215	19290	29175	3474	3993	4649	1026	1092	1145	483	498	496	277	284	269
0.55	***	14659	16434	27996	3224	3609	4464	974	1031	1109	459	473	490	264	268	273
	•••	15320	18717	28766	3389	3927	4555	1001	1067	1110	465	481	475	265	271	255
0.6	***	13501	15459	27237	3081	3467	4312	929	984	1059	433	446	462	245	249	254
	•••	14245	17862	27773	3246	3784	4368	954	1018	1052	439	454	444	248	252	234
0.65	***	12245	14317	25846	2887	3262	4059	865	915	983	396	409	423	221	226	229
	•••	13046	16758	26196	3049	3565	4088	889	946	970	402	415	403	223	227	208
0.7	***	10939	13033	23826	2646	2995	3706	781	824	883	351	361	371	192	195	198
	•••	11759	15414	24036	2798	3270	3714	801	851	865	356	367	351	193	196	176
0.75	***	9596	11612	21181	2356	2661	3252	677	712	756	296	302	311	156	159	161
	•••	10395	13833	21291	2489	2895	3247	693	733	736	299	308	289	158	159	138
0.8	***	8204	10034	17920	2009	2256	2701	551	576	606	229	234	239	.	.	.
	•••	8937	11984	17963	2118	2437	2687	564	590	584	233	237	216	.	.	.
0.85	***	6718	8245	14046	1593	1767	2052	402	417	433	.	.	.	.	.	.
	•••	7337	9808	14050	1671	1889	2033	409	424	409	.	.	.	.	.	.
0.9	***	5036	6127	9570	1086	1176	1308	.	.	.	.	.	.	.	.	.
	•••	5486	7173	9554	1127	1236	1285	.	.	.	.	.	.	.	.	.
0.95	***	2890	3386	4496	.	.	.	.	.	.	.	.	.	.	.	.
	•••	3102	3795	4474	.	.	.	.	.	.	.	.	.	.	.	.

TABLE 1: ALPHA= 0.005 POWER= 0.8   EXPECTED ACCRUAL THRU MINIMUM FOLLOW-UP= 2750

	DEL=.02			DEL=.05			DEL=.10			DEL=.15			DEL=.20		
FACT=	1.0 .75	.50 .25	.00 BIN	1.0 .75	.50 .25	.00 BIN	1.0 .75	.50 .25	.00 BIN	1.0 75	.50 .25	.00 BIN	1.0 .75	.50 .25	.00 BIN
PCONT=***						REQUIRED NUMBER OF PATIENTS									
0.05 ***	3367	3406	3518	731	737	745	263	264	266	154	154	154	108	108	109
***	3382	3446	6576	734	742	1285	263	264	409	154	154	216	108	108	138
0.1 ***	7021	7141	7729	1367	1395	1438	434	439	445	233	233	235	153	154	154
***	7063	7304	11423	1380	1414	2033	436	441	584	233	235	289	154	154	176
0.15 ***	10286	10511	11954	1938	2002	2109	587	597	608	300	304	307	190	192	194
***	10365	10848	15685	1965	2047	2687	593	603	736	302	305	351	192	192	208
0.2 ***	12924	13277	15884	2414	2522	2722	718	734	754	359	362	369	221	223	226
***	13047	13824	19364	2460	2604	3247	725	744	865	360	366	403	223	225	234
0.25 ***	14869	15374	19366	2793	2947	3257	824	848	879	405	412	419	247	249	252
***	15044	16166	22459	2859	3071	3714	835	862	970	408	415	444	247	250	255
0.3 ***	16135	16822	22319	3071	3277	3704	905	938	979	441	450	460	264	267	271
***	16374	17883	24970	3160	3444	4088	920	957	1052	445	453	475	266	269	269
0.35 ***	16780	17677	24694	3260	3516	4054	965	1005	1057	465	476	487	276	281	285
***	17093	19022	26897	3370	3724	4368	982	1027	1110	470	483	496	278	283	278
0.4 ***	16881	18012	26463	3368	3667	4309	1002	1046	1108	479	491	505	283	287	291
***	17282	19647	28240	3495	3911	4555	1022	1075	1145	486	498	507	285	290	281
0.45 ***	16529	17912	27606	3401	3738	4461	1015	1065	1134	484	496	511	283	287	291
***	17029	19820	28999	3545	4013	4649	1039	1098	1157	489	503	507	285	290	278
0.5 ***	15825	17458	28118	3370	3732	4513	1010	1064	1134	477	491	507	276	281	287
***	16427	19603	29175	3526	4028	4649	1034	1096	1145	484	498	496	278	283	269
0.55 ***	14866	16725	27996	3281	3659	4463	984	1037	1108	460	474	489	264	269	273
***	15564	19053	28766	3442	3965	4555	1009	1070	1110	467	483	475	266	271	255
0.6 ***	13735	15769	27237	3136	3516	4312	938	989	1058	434	448	462	245	250	256
***	14514	18211	27773	3301	3820	4368	962	1022	1052	441	455	444	249	252	234
0.65 ***	12500	14636	25846	2942	3310	4059	872	920	982	398	410	422	223	226	230
***	13330	17106	26196	3102	3600	4088	895	950	970	405	415	403	225	228	208
0.7 ***	11202	13349	23825	2698	3037	3705	789	830	881	353	362	373	192	195	199
***	12048	15750	24036	2845	3301	3714	807	854	865	357	367	351	194	197	176
0.75 ***	9855	11910	21181	2401	2698	3253	683	716	755	297	304	311	157	159	161
***	10673	14140	21291	2531	2921	3247	699	735	736	300	307	289	157	159	138
0.8 ***	8442	10300	17919	2047	2284	2701	556	579	606	230	235	239	.	.	.
***	9190	12249	17963	2151	2457	2687	566	593	584	232	236	216	.	.	.
0.85 ***	6921	8462	14047	1620	1787	2054	405	417	432	.	.	.	.	.	.
***	7548	10015	14050	1696	1900	2033	410	424	409	.	.	.	.	.	.
0.9 ***	5183	6278	9569	1099	1185	1308	.	.	.	.	.	.	.	.	.
***	5637	7306	9554	1140	1240	1285	.	.	.	.	.	.	.	.	.
0.95 ***	2961	3449	4495	.	.	.	.	.	.	.	.	.	.	.	.
***	3171	3842	4474	.	.	.	.	.	.	.	.	.	.	.	.

TABLE 1: ALPHA= 0.005 POWER= 0.8   EXPECTED ACCRUAL THRU MINIMUM FOLLOW-UP= 3000

		DEL=.01			DEL=.02			DEL=.05			DEL=.10			DEL=.15		
FACT=		1.0 .75	.50 .25	.00 BIN	1.0 .75	.50 .25	.00 BIN	1.0 .75	.50 .25	.00 BIN	1.0 75	.50 .25	.00 BIN	1.0 .75	.50 .25	.00 BIN
PCONT=***					REQUIRED NUMBER OF PATIENTS											
0.01	***	1783 1790	1802		655 658	658		215 215	215		110 110	110		77 77	77	
	•••	1787 1795	6891		658 658	2278		215 215	620		110 110	252		77 77	150	
0.02	***	3940 3958	4018		1265 1273	1285		343 343	347		152 152	152		100 100	100	
	•••	3947 3977	11376		1270 1277	3387		343 347	792		152 152	293		100 100	167	
0.05	***	11923 11980	12553		3370 3410	3520		733 737	745		260 265	265		152 152	155	
	•••	11942 12080	24269		3388 3448	6576		733 740	1285		265 265	409		152 155	216	
0.1	***	26300 26440	29222		7033 7160	7730		1370 1397	1438		433 440	445		230 235	235	
	•••	26345 26713	43890		7078 7325	11423		1382 1415	2033		437 440	584		235 235	289	
0.15	***	39265 39515	46165		10307 10547	11953		1948 2008	2110		587 598	610		302 302	305	
	•••	39347 40010	61175		10393 10892	15685		1975 2050	2687		595 602	736		302 305	351	
0.2	***	49705 50090	62023		12958 13337	15883		2428 2533	2720		718 737	755		358 362	370	
	•••	49832 50863	76124		13090 13903	19364		2473 2612	3247		725 745	865		362 365	403	
0.25	***	57295 57857	76145		14915 15460	19367		2810 2965	3257		827 850	880		407 410	418	
	•••	57482 58970	88737		15107 16280	22459		2878 3085	3714		838 865	970		407 415	444	
0.3	***	62060 62830	88175		16202 16937	22318		3100 3302	3703		910 940	980		440 448	460	
	•••	62315 64367	99015		16457 18035	24970		3185 3460	4088		925 958	1052		445 455	475	
0.35	***	64187 65215	97907		16865 17822	24695		3295 3545	4055		970 1007	1055		467 478	490	
	•••	64532 67258	106956		17203 19213	26897		3403 3745	4368		988 1030	1110		470 482	496	
0.4	***	63962 65305	105223		16990 18193	26462		3407 3700	4307		1007 1052	1108		482 493	505	
	•••	64408 67960	112562		17420 19873	28240		3535 3935	4555		1030 1078	1145		485 497	507	
0.45	***	61720 63452	110050		16667 18125	27605		3445 3775	4460		1022 1070	1135		485 497	512	
	•••	62297 66830	115833		17200 20083	28999		3587 4040	4649		1045 1100	1157		490 505	507	
0.5	***	57842 60055	112352		15995 17705	28120		3418 3775	4513		1018 1067	1135		478 493	505	
	•••	58580 64262	116767		16630 19895	29175		3572 4060	4649		1040 1097	1145		485 497	496	
0.55	***	52727 55528	112112		15065 16997	27995		3328 3700	4465		992 1040	1108		463 475	490	
	•••	53668 60632	115365		15797 19363	28766		3490 3995	4555		1015 1075	1110		467 482	475	
0.6	***	46805 50312	109330		13960 16060	27238		3185 3560	4310		947 995	1060		437 448	463	
	•••	47998 56270	111628		14770 18530	27773		3347 3853	4368		970 1025	1052		440 455	444	
0.65	***	40525 44788	104015		12740 14935	25847		2990 3350	4060		880 925	980		400 410	422	
	•••	42013 51437	105555		13600 17425	26196		3148 3628	4088		902 950	970		407 415	403	
0.7	***	34322 39253	96185		11450 13640	23825		2743 3073	3707		793 835	883		355 362	373	
	•••	36107 46300	97146		12317 16055	24036		2890 3328	3714		812 857	865		358 365	351	
0.75	***	28547 33857	85858		10097 12185	21182		2440 2728	3253		688 718	755		298 305	310	
	•••	30530 40907	86402		10930 14420	21291		2570 2942	3247		703 737	736		302 305	289	
0.8	***	23323 28600	73055		8665 10543	17920		2080 2308	2702		560 580	605		230 235	238	
	•••	25337 35222	73321		9422 12490	17963		2180 2473	2687		568 595	584		230 238	216	
0.85	***	18542 23315	57808		7108 8660	14045		1645 1802	2053		407 418	433		. .	.	
	•••	20395 29060	57905		7742 10198	14050		1715 1910	2033		410 425	409		. .	.	
0.9	***	13870 17668	40142		5320 6415	9568		1112 1195	1307		. .	.		. .	.	
	•••	15365 22033	40153		5773 7423	9554		1150 1247	1285		. .	.		. .	.	
0.95	***	8590 10817	20083		3025 3505	4495		. .	.		. .	.		. .	.	
	•••	9490 13157	20065		3235 3883	4474		. .	.		. .	.		. .	.	
0.98	***	4123 4922	6913		. .	.		. .	.		. .	.		. .	.	
	•••	4460 5615	6891		. .	.		. .	.		. .	.		. .	.	

TABLE 1: ALPHA= 0.005 POWER= 0.8   EXPECTED ACCRUAL THRU MINIMUM FOLLOW-UP= 3250

		DEL=.01			DEL=.02			DEL=.05			DEL=.10			DEL=.15		
FACT=		1.0 .75	.50 .25	.00 BIN	1.0 .75	.50 .25	.00 BIN	1.0 .75	.50 .25	.00 BIN	1.0 75	.50 .25	.00 BIN	1.0 .75	.50 .25	.00 BIN
PCONT=***				REQUIRED NUMBER OF PATIENTS												
0.01	***	1785	1790	1801	656	656	661	217	217	217	108	108	108	76	79	79
	***	1785	1793	6891	656	661	2278	217	217	620	108	108	252	76	79	150
0.02	***	3943	3959	4019	1265	1273	1286	344	344	347	152	152	152	100	100	100
	***	3951	3979	11376	1270	1278	3387	344	344	792	152	152	293	100	100	167
0.05	***	11930	11990	12556	3374	3415	3516	734	737	745	263	263	266	152	152	152
	***	11949	12096	24269	3391	3456	6576	734	742	1285	263	266	409	152	152	216
0.1	***	26311	26461	29223	7042	7177	7729	1376	1400	1441	436	441	444	233	233	233
	***	26363	26758	43890	7091	7342	11423	1387	1419	2033	436	441	584	233	233	289
0.15	***	39287	39555	46164	10329	10584	11954	1956	2012	2110	591	599	607	303	303	306
	***	39376	40091	61175	10419	10934	15685	1980	2053	2687	596	604	736	303	306	351
0.2	***	49736	50158	62024	12989	13393	15882	2443	2546	2719	721	737	753	360	363	368
	***	49877	50995	76124	13133	13972	19364	2489	2619	3247	729	745	865	360	368	403
0.25	***	57341	57950	76145	14964	15541	19367	2830	2979	3256	831	851	880	409	412	420
	***	57544	59156	88737	15167	16386	22459	2895	3093	3714	839	864	970	409	417	444
0.3	***	62126	62958	88175	16264	17049	22322	3123	3321	3702	916	945	981	441	449	458
	***	62403	64621	99015	16540	18178	24970	3212	3475	4088	929	961	1052	444	452	475
0.35	***	64274	65387	97909	16951	17967	24694	3326	3569	4052	978	1010	1054	469	477	490
	***	64645	67594	106956	17312	19392	26897	3431	3764	4368	994	1029	1110	474	482	496
0.4	***	64071	65531	105221	17101	18368	26461	3442	3732	4309	1013	1054	1108	482	493	506
	***	64559	68391	112562	17561	20088	28240	3569	3959	4555	1034	1078	1145	490	498	507
0.45	***	61866	63738	110051	16805	18333	27606	3483	3808	4463	1029	1075	1132	485	498	509
	***	62492	67370	115833	17366	20326	28999	3626	4065	4649	1051	1102	1157	493	506	507
0.5	***	58026	60420	112350	16158	17937	28118	3459	3813	4512	1026	1070	1135	482	493	506
	***	58828	64913	116767	16824	20164	29175	3613	4084	4555	1046	1099	1145	485	498	496
0.55	***	52961	55987	112111	15256	17252	27996	3374	3740	4463	997	1046	1108	466	477	490
	***	53980	61390	115365	16017	19649	28766	3532	4024	4555	1021	1075	1110	469	482	475
0.6	***	47106	50865	109333	14176	16329	27238	3231	3597	4312	953	997	1059	436	449	461
	***	48390	57121	111628	15009	18828	27773	3391	3878	4368	972	1026	1052	444	452	444
0.65	***	40904	45432	104019	12970	15213	25843	3036	3386	4060	888	929	981	401	412	420
	***	42496	52352	105555	13851	17718	26196	3191	3654	4088	907	953	970	409	417	403
0.7	***	34781	39961	96186	11686	13913	23825	2784	3109	3708	799	834	883	355	363	371
	***	36666	47236	97146	12567	16337	24036	2928	3350	3714	818	859	865	360	368	351
0.75	***	29066	34586	85859	10321	12442	21179	2476	2757	3253	693	721	758	298	303	311
	***	31138	41822	86402	11169	14676	21291	2603	2963	3247	704	737	736	303	306	289
0.8	***	23858	29296	73057	8870	10768	17918	2110	2329	2700	563	582	607	230	233	238
	***	25946	36061	73321	9639	12710	17963	2207	2489	2687	571	596	584	233	238	216
0.85	***	19039	23934	57812	7283	8843	14046	1663	1817	2053	409	420	433	.	.	.
	***	20944	29776	57905	7919	10370	14050	1736	1920	2033	412	425	409	.	.	.
0.9	***	14273	18149	40143	5446	6535	9569	1124	1200	1306	.	.	.	.	.	.
	***	15801	22561	40153	5901	7529	9554	1159	1249	1285	.	.	.	.	.	.
0.95	***	8834	11088	20082	3085	3553	4496	.	.	.	.	.	.	.	.	.
	***	9744	13425	20065	3288	3919	4474	.	.	.	.	.	.	.	.	.
0.98	***	4219	5008	6912	.	.	.	.	.	.	.	.	.	.	.	.
	***	4556	5685	6891	.	.	.	.	.	.	.	.	.	.	.	.

TABLE 1: ALPHA= 0.005 POWER= 0.8   EXPECTED ACCRUAL THRU MINIMUM FOLLOW-UP= 3500

		DEL=.01			DEL=.02			DEL=.05			DEL=.10			DEL=.15		
FACT=		1.0 .75	.50 .25	.00 BIN	1.0 .75	.50 .25	.00 BIN	1.0 .75	.50 .25	.00 BIN	1.0 75	.50 .25	.00 BIN	1.0 .75	.50 .25	.00 BIN
PCONT=***							REQUIRED NUMBER OF PATIENTS								
0.01	***	1788	1791	1800	659	659	659	216	216	216	108	108	108	76	76	76
	•••	1788	1796	6891	659	659	2278	216	216	620	108	108	252	76	76	150
0.02	***	3943	3961	4019	1266	1275	1283	344	344	344	151	151	151	99	99	99
	•••	3952	3984	11376	1271	1280	3387	344	344	792	151	151	293	99	99	167
0.05	***	11932	11999	12553	3380	3418	3520	732	738	746	265	265	265	155	155	155
	•••	11955	12113	24269	3398	3459	6576	738	741	1285	265	265	409	155	155	216
0.1	***	26323	26483	29222	7055	7195	7729	1376	1403	1438	435	440	443	233	233	233
	•••	26378	26804	43890	7108	7361	11423	1388	1420	2033	440	440	584	233	233	289
0.15	***	39308	39596	46168	10348	10620	11955	1963	2018	2111	592	598	610	300	303	309
	•••	39404	40174	61175	10445	10975	15685	1989	2059	2687	592	601	736	303	303	351
0.2	***	49767	50219	62023	13023	13446	15884	2453	2552	2718	723	738	755	361	365	370
	•••	49921	51120	76124	13175	14041	19364	2496	2622	3247	729	746	865	361	365	403
0.25	***	57388	58041	76145	15012	15621	19366	2846	2995	3258	834	855	878	408	414	417
	•••	57607	59340	88737	15228	16482	22459	2911	3103	3714	843	863	970	408	417	444
0.3	***	62189	63090	88176	16330	17153	22318	3144	3340	3704	916	948	977	443	452	458
	•••	62490	64870	99015	16622	18308	24970	3231	3488	4088	930	960	1052	449	452	475
0.35	***	64359	65558	97906	17039	18101	24695	3354	3593	4054	983	1012	1056	470	478	487
	•••	64761	67923	106956	17418	19559	26897	3459	3783	4363	995	1035	1110	475	484	496
0.4	***	64184	65753	105221	17208	18535	26463	3476	3760	4308	1018	1056	1108	484	493	505
	•••	64709	68813	112562	17690	20285	28240	3599	3978	4555	1038	1082	1145	487	501	507
0.45	***	62008	64026	110051	16937	18530	27609	3523	3844	4460	1035	1079	1135	487	501	510
	•••	62682	67897	115833	17529	20551	28999	3660	4089	4649	1056	1105	1157	493	505	507
0.5	***	58211	60783	112353	16321	18162	28120	3503	3844	4512	1030	1073	1135	484	493	505
	•••	59074	65549	116767	17016	20411	29175	3651	4110	4649	1053	1105	1145	487	501	496
0.55	***	53197	56440	112111	15441	17497	27994	3415	3774	4465	1003	1053	1108	466	478	487
	•••	54291	62122	115365	16228	19918	28766	3573	4048	4555	1026	1079	1110	470	484	475
0.6	***	47405	51409	109328	14379	16587	27236	3275	3634	4311	960	1003	1056	440	449	461
	•••	48784	57940	111628	15240	19098	27773	3433	3905	4368	977	1030	1052	443	458	444
0.65	***	41276	46057	104017	13189	15473	25845	3077	3418	4057	890	933	983	405	414	423
	•••	42968	53223	105555	14090	17987	26196	3226	3678	4088	913	956	970	408	417	403
0.7	***	35233	40641	96186	11906	14169	23823	2823	3138	3704	802	837	881	356	365	370
	•••	37216	48122	97146	12807	16596	24036	2963	3371	3714	820	860	865	361	370	351
0.75	***	29569	35283	85858	10538	12676	21181	2508	2785	3252	697	723	755	300	303	309
	•••	31721	42680	86402	11398	14913	21291	2631	2981	3247	711	738	736	303	309	289
0.8	***	24371	29963	73056	9068	10978	17917	2138	2351	2701	566	583	606	230	233	239
	•••	26518	36849	73321	9841	12909	17963	2234	2500	2687	575	592	584	233	239	216
0.85	***	19510	24520	57808	7443	9010	14046	1683	1831	2053	408	423	431	.	.	.
	•••	21461	30444	57905	8088	10520	14050	1753	1931	2033	414	426	409	.	.	.
0.9	***	14653	18600	40142	5562	6647	9570	1135	1210	1306	.	.	.	.	.	.
	•••	16216	23050	40153	6017	7624	9554	1170	1254	1285	.	.	.	.	.	.
0.95	***	9068	11337	20084	3138	3599	4495	.	.	.	.	.	.	.	.	.
	•••	9986	13665	20065	3340	3952	4474	.	.	.	.	.	.	.	.	.
0.98	***	4308	5086	6915	.	.	.	.	.	.	.	.	.	.	.	.
	•••	4640	5746	6891	.	.	.	.	.	.	.	.	.	.	.	.

TABLE 1: ALPHA= 0.005 POWER= 0.8   EXPECTED ACCRUAL THRU MINIMUM FOLLOW-UP= 3750

	DEL=.01			DEL=.02			DEL=.05			DEL=.10			DEL=.15		
FACT=	1.0 .75	.50 .25	.00 BIN	1.0 .75	.50 .25	.00 BIN	1.0 .75	.50 .25	.00 BIN	1.0 75	.50 .25	.00 BIN	1.0 .75	.50 .25	.00 BIN
PCONT=***						REQUIRED NUMBER OF PATIENTS									
0.01 ***	1788 1788	1793 1793	1803 6891	659 659	659 659	659 2278	218 218	218 218	218 620	109 109	109 109	109 252	78 78	78 78	78 150
0.02 ***	3944 3953	3963 3981	4019 11376	1268 1272	1272 1278	1281 3387	344 344	344 344	344 792	153 153	153 153	153 293	100 100	100 100	100 167
0.05 ***	11937 11959	12006 12125	12556 24269	3381 3400	3425 3462	3518 6576	734 738	738 743	747 1285	265 265	265 265	265 409	153 153	153 153	153 216
0.1 ***	26337 26393	26509 26847	29225 43890	7066 7122	7212 7375	7728 11423	1381 1390	1403 1422	1437 2033	438 438	438 443	443 584	231 231	231 237	237 289
0.15 ***	39325 39434	39640 40253	46165 61175	10372 10475	10653 11013	11956 15685	1966 1994	2022 2059	2106 2687	593 597	597 603	606 736	303 303	303 306	306 351
0.2 ***	49803 49962	50284 51247	62022 76124	13056 13216	13503 14103	15884 19364	2463 2509	2562 2631	2722 3247	725 734	738 747	753 865	359 363	363 368	368 403
0.25 ***	57434 57668	58137 59519	76147 88737	15059 15288	15697 16578	19366 22459	2862 2928	3006 3109	3256 3714	837 847	856 865	878 970	409 409	415 415	419 444
0.3 ***	62253 62575	63218 65122	88175 99015	16394 16703	17256 18434	22319 24970	3166 3250	3359 3500	3700 4088	922 934	950 963	978 1052	443 447	453 456	456 475
0.35 ***	64447 64872	65731 68253	97909 106956	17122 17525	18231 19713	24697 26897	3378 3481	3616 3794	4053 4368	981 1000	1015 1034	1056 1110	472 475	481 484	490 496
0.4 ***	64297 64859	65975 69231	105222 112562	17318 17825	18691 20472	26463 28240	3503 3625	3784 3997	4309 4555	1025 1043	1062 1081	1109 1145	484 490	494 500	503 507
0.45 ***	62153 62875	64315 68412	110050 115833	17069 17684	18715 20763	27606 28999	3556 3691	3869 4109	4459 4649	1038 1062	1081 1103	1131 1157	490 494	500 503	513 507
0.5 ***	58394 59318	61147 66162	112353 116767	16478 17197	18372 20647	28118 29175	3537 3687	3875 4131	4512 4649	1034 1056	1081 1103	1131 1145	484 490	494 500	503 496
0.55 ***	53434 54603	56884 62834	112113 115365	15622 16431	17725 20163	27997 28766	3456 3606	3803 4072	4465 4555	1009 1028	1053 1081	1109 1110	466 472	475 484	490 475
0.6 ***	47703 49169	51944 58722	109328 111628	14581 15456	16831 19353	27237 27773	3312 3466	3663 3925	4309 4368	963 981	1006 1028	1056 1052	443 447	453 456	462 444
0.65 ***	41647 43437	46666 54059	104018 105555	13400 14318	15715 18241	25844 26196	3115 3259	3453 3700	4056 4088	897 916	934 959	981 970	406 409	415 419	425 403
0.7 ***	35675 37747	41300 48968	96184 97146	12115 13034	14406 16834	23825 24036	2856 2993	3166 3391	3706 3714	809 822	841 859	884 865	359 359	363 368	372 351
0.75 ***	30059 32284	35950 43497	85859 86402	10737 11613	12903 15128	21181 21291	2543 2659	2806 2993	3250 3247	700 709	725 738	756 736	303 303	306 306	312 289
0.8 ***	24865 27072	30593 37591	73056 73321	9250 10034	11172 13090	17918 17963	2159 2256	2369 2515	2703 2687	569 575	584 597	606 584	231 231	237 237	237 216
0.85 ***	19962 21959	25072 31066	57809 57905	7597 8243	9166 10662	14047 14050	1700 1765	1844 1938	2050 2033	409 415	419 425	434 409	. .	. .	. .
0.9 ***	15016 16606	19025 23509	40141 40153	5675 6128	6753 7713	9569 9554	1141 1175	1212 1259	1306 1285	. .	. .	. .	. .	. .	. .
0.95 ***	9284 10212	11572 13891	20084 20065	3184 3387	3640 3981	4497 4474	. .	. .	. .	. .	. .	. .	. .	. .	. .
0.98 ***	4384 4718	5159 5806	6912 6891	. .	. .	. .	. .	. .	. .	. .	. .	. .	. .	. .	. .

TABLE 1: ALPHA= 0.005 POWER= 0.8   EXPECTED ACCRUAL THRU MINIMUM FOLLOW-UP= 4000

	DEL=.01			DEL=.02			DEL=.05			DEL=.10			DEL=.15		
FACT=	1.0 .75	.50 .25	.00 BIN	1.0 .75	.50 .25	.00 BIN	1.0 .75	.50 .25	.00 BIN	1.0 75	.50 .25	.00 BIN	1.0 .75	.50 .25	.00 BIN

PCONT=***   REQUIRED NUMBER OF PATIENTS

PCONT	1.0 .75	.50 .25	.00 BIN	1.0 .75	.50 .25	.00 BIN	1.0 .75	.50 .25	.00 BIN	1.0 .75	.50 .25	.00 BIN	1.0 .75	.50 .25	.00 BIN
0.01	1787	1793	1803	657	657	657	217	217	217	107	107	107	77	77	77
	1787	1797	6891	657	657	2278	217	217	620	107	107	252	77	77	150
0.02	3947	3967	4017	1267	1277	1283	343	343	347	153	153	153	97	97	97
	3953	3987	11376	1273	1277	3387	343	347	792	153	153	293	97	97	167
0.05	11943	12017	12553	3387	3427	3517	733	737	747	263	263	263	153	153	153
	11967	12137	24269	3403	3463	6576	737	743	1285	263	263	409	153	153	216
0.1	26347	26533	29223	7077	7227	7727	1383	1407	1437	437	443	443	233	233	237
	26407	26887	43890	7133	7393	11423	1393	1423	2033	437	443	584	233	233	289
0.15	39347	39683	46167	10393	10683	11953	1973	2027	2107	593	603	607	303	303	307
	39457	40333	61175	10497	11047	15685	1997	2063	2687	597	603	736	303	307	351
0.2	49833	50347	62023	13087	13553	15883	2473	2567	2723	727	737	753	363	363	367
	50007	51373	76124	13257	14163	19364	2517	2637	3247	733	747	865	363	367	403
0.25	57483	58227	76163	15107	15773	19367	2877	3017	3257	837	857	877	407	413	417
	57733	59703	88737	15347	16663	22459	2943	3117	3714	847	867	970	413	417	444
0.3	62317	63343	88173	16457	17357	22317	3187	3373	3703	923	947	977	447	453	457
	62657	65367	99015	16783	18547	24970	3267	3513	4088	937	963	1052	447	457	475
0.35	64533	65903	97907	17203	18357	24693	3403	3637	4053	987	1017	1057	473	477	487
	64987	68577	106956	17627	19857	26897	3503	3807	4368	1003	1037	1110	477	483	496
0.4	64407	66197	105223	17423	18847	26463	3533	3807	4307	1027	1063	1107	487	497	507
	65007	69643	112562	17947	20643	28240	3653	4013	4555	1043	1083	1145	493	503	507
0.45	62297	64597	110047	17197	18893	27607	3587	3897	4463	1047	1083	1133	493	503	513
	63067	68917	115833	17837	20963	28999	3723	4127	4649	1063	1107	1157	497	507	507
0.5	58577	61503	112353	16627	18573	28117	3573	3903	4513	1043	1083	1133	483	493	507
	59563	66763	116767	17373	20863	29175	3717	4153	4649	1063	1107	1145	487	503	496
0.55	53667	57327	112113	15797	17943	27997	3493	3833	4463	1017	1057	1107	467	477	487
	54913	63523	115365	16627	20393	28766	3643	4093	4555	1033	1083	1110	473	483	475
0.6	47997	52467	109333	14767	17057	27237	3347	3693	4313	967	1007	1057	443	453	463
	49553	59477	111628	15667	19587	27773	3503	3947	4368	987	1033	1052	447	457	444
0.65	42013	47257	104017	13597	15947	25847	3147	3477	4057	903	937	983	407	413	423
	43897	54853	105555	14533	18473	26196	3293	3717	4088	917	957	970	407	417	403
0.7	36107	41933	96187	12317	14633	23827	2887	3193	3707	813	843	883	357	363	373
	38263	49767	97146	13247	17057	24036	3023	3407	3714	827	863	865	363	367	351
0.75	30527	36587	85857	10933	13107	21183	2567	2827	3253	703	727	757	303	307	313
	32827	44273	86402	11813	15327	21291	2687	3007	3247	713	743	736	303	307	289
0.8	25337	31197	73057	9423	11357	17917	2183	2387	2703	567	587	607	233	237	237
	27603	38297	73321	10213	13263	17963	2273	2523	2687	577	597	584	233	237	216
0.85	20393	25597	57807	7743	9313	14047	1717	1853	2053	413	423	433	.	.	.
	22433	31657	57905	8393	10787	14050	1783	1943	2033	417	427	409	.	.	.
0.9	15363	19427	40143	5773	6847	9567	1153	1217	1307	.	.	.	.	.	.
	16977	23937	40153	6227	7793	9554	1183	1263	1285	.	.	.	.	.	.
0.95	9487	11787	20083	3233	3677	4497	.	.	.	.	.	.	.	.	.
	10427	14097	20065	3427	4007	4474	.	.	.	.	.	.	.	.	.
0.98	4463	5227	6913	.	.	.	.	.	.	.	.	.	.	.	.
	4793	5853	6891	.	.	.	.	.	.	.	.	.	.	.	.

TABLE 1: ALPHA= 0.005 POWER= 0.8   EXPECTED ACCRUAL THRU MINIMUM FOLLOW-UP= 4250

	DEL=.01			DEL=.02			DEL=.05			DEL=.10			DEL=.15		
FACT=	1.0 .75	.50 .25	.00 BIN	1.0 .75	.50 .25	.00 BIN	1.0 .75	.50 .25	.00 BIN	1.0 75	.50 .25	.00 BIN	1.0 .75	.50 .25	.00 BIN
PCONT=***					REQUIRED NUMBER OF PATIENTS										
0.01 ***	1788	1792	1799	655	655	662	216	216	216	109	109	109	78	78	78
•••	1788	1799	6891	655	662	2278	216	216	620	109	109	252	78	78	150
0.02 ***	3949	3966	4019	1272	1278	1282	343	343	347	152	152	152	99	99	99
•••	3956	3988	11376	1272	1278	3387	343	343	792	152	152	293	99	99	167
0.05 ***	11950	12024	12555	3393	3428	3520	736	740	747	262	262	262	152	152	156
•••	11978	12148	24269	3407	3467	6576	736	740	1285	262	262	409	152	152	216
0.1 ***	26357	26555	29222	7090	7239	7728	1384	1410	1438	439	439	443	230	237	237
•••	26421	26927	43890	7147	7402	11423	1395	1420	2033	439	443	584	230	237	289
0.15 ***	39369	39723	46162	10416	10713	11956	1979	2033	2107	592	602	609	301	305	305
•••	39486	40410	61175	10526	11078	15685	2005	2064	2687	598	602	736	305	305	351
0.2 ***	49866	50412	62021	13118	13603	15881	2483	2574	2723	730	740	751	358	364	368
•••	50047	51496	76124	13295	14219	19364	2525	2638	3247	736	747	865	364	364	403
0.25 ***	57527	58324	76146	15154	15845	19366	2893	3031	3254	843	857	878	411	411	418
•••	57793	59879	88737	15403	16748	22459	2950	3127	3714	847	868	970	411	418	444
0.3 ***	62383	63470	88173	16518	17453	22320	3205	3386	3701	928	953	981	443	453	460
•••	62744	65613	99015	16858	18661	24970	3286	3520	4088	938	963	1052	449	453	475
0.35 ***	64618	66069	97906	17283	18480	24696	3424	3652	4055	991	1023	1055	471	481	485
•••	65103	68896	106956	17726	19999	26897	3524	3822	4368	1006	1038	1110	475	485	496
0.4 ***	64522	66420	105223	17524	18994	26463	3556	3828	4306	1034	1066	1108	485	496	507
•••	65156	70047	112562	18070	20807	28240	3679	4030	4555	1048	1087	1145	492	503	507
0.45 ***	62440	64883	110050	17326	19064	27607	3616	3917	4459	1048	1087	1133	492	503	513
•••	63258	69410	115833	17985	21147	28999	3747	4140	4649	1066	1108	1157	496	507	507
0.5 ***	58763	61862	112352	16780	18767	28117	3605	3928	4512	1044	1087	1133	485	496	507
•••	59811	67348	116767	17538	21066	29175	3747	4168	4649	1066	1108	1145	492	503	496
0.55 ***	53904	57765	112112	15962	18151	27993	3524	3860	4466	1017	1059	1108	471	481	492
•••	55221	64182	115365	16816	20609	28766	3673	4108	4555	1038	1080	1110	475	485	475
0.6 ***	48294	52979	109328	14953	17273	27235	3382	3715	4310	974	1013	1059	443	453	460
•••	49934	60204	111628	15866	19812	27773	3531	3966	4368	991	1034	1052	449	453	444
0.65 ***	42376	47837	104015	13788	16164	25847	3180	3503	4055	906	938	981	407	411	422
•••	44345	55614	105555	14740	18693	26196	3322	3733	4088	921	959	970	411	418	403
0.7 ***	36525	42546	96185	12509	14840	23824	2918	3212	3705	815	847	878	358	364	368
•••	38763	50536	97146	13448	17262	24036	3053	3418	3714	832	864	865	364	368	351
0.75 ***	30986	37201	85857	11110	13306	21183	2596	2844	3254	704	730	758	301	305	311
•••	33348	45004	86402	12003	15516	21291	2706	3021	3247	715	740	736	305	311	289
0.8 ***	25794	31772	73054	9587	11525	17921	2203	2398	2702	570	588	609	230	237	237
•••	28110	38965	73321	10384	13416	17963	2292	2532	2687	581	598	584	237	237	216
0.85 ***	20807	26098	57807	7876	9449	14049	1728	1863	2054	411	422	432	.	.	.
•••	22883	32218	57905	8528	10908	14050	1792	1948	2033	418	428	409	.	.	.
0.9 ***	15696	19808	40144	5872	6941	9570	1155	1225	1310	.	.	.	.	.	.
•••	17333	24338	40153	6325	7859	9554	1187	1261	1285	.	.	.	.	.	.
0.95 ***	9683	11992	20084	3276	3711	4498	.	.	.	.	.	.	.	.	.
•••	10628	14294	20065	3467	4030	4474	.	.	.	.	.	.	.	.	.
0.98 ***	4533	5288	6913	.	.	.	.	.	.	.	.	.	.	.	.
•••	4859	5900	6891	.	.	.	.	.	.	.	.	.	.	.	.

TABLE 1: ALPHA= 0.005 POWER= 0.8   EXPECTED ACCRUAL THRU MINIMUM FOLLOW-UP= 4500

		DEL=.01			DEL=.02			DEL=.05			DEL=.10			DEL=.15		
FACT=		1.0 .75	.50 .25	.00 BIN	1.0 .75	.50 .25	.00 BIN	1.0 .75	.50 .25	.00 BIN	1.0 75	.50 .25	.00 BIN	1.0 .75	.50 .25	.00 BIN
PCONT=•••					REQUIRED NUMBER OF PATIENTS											
0.01	•••	1785 1792	1803	656 660	660	217 217	217	109 109	109	75 75	75
	•••	1792 1796	6891	656 660	2278	217 217	620	109 109	252	75 75	150
0.02	•••	3952 3968	4020	1268 1275	1286	345 345	345	150 150	150	98 98	98
	•••	3956 3986	11376	1275 1279	3387	345 345	792	150 150	293	98 98	167
0.05	•••	11951 12034	12551	3394 3435	3518	735 739	746	262 262	266	154 154	154
	•••	11978 12158	24269	3412 3468	6576	739 746	1285	262 262	409	154 154	216
0.1	•••	26366 26580	29224	7102 7253	7732	1387 1410	1436	435 442	442	233 233	233
	•••	26441 26970	43890	7158 7417	11423	1398 1425	2033	442 442	584	233 233	289
0.15	•••	39390 39765	46166	10432 10740	11955	1983 2033	2107	593 600	611	300 307	307
	•••	39513 40485	61175	10549 11107	15685	2006 2066	2687	600 604	736	300 307	351
0.2	•••	49897 50482	62025	13155 13650	15881	2494 2584	2719	728 739	757	363 363	368
	•••	50093 51618	76124	13335 14268	19364	2535 2640	3247	735 746	865	363 368	403
0.25	•••	57574 58413	76143	15195 15911	19365	2906 3041	3255	840 858	881	408 413	420
	•••	57855 60056	88737	15461 16822	22459	2966 3131	3714	851 870	970	413 420	444
0.3	•••	62445 63600	88174	16586 17546	22316	3221 3401	3705	930 953	982	446 453	458
	•••	62828 65850	99015	16935 18761	24970	3300 3529	4088	941 964	1052	446 458	475
0.35	•••	64702 66243	97905	17366 18600	24697	3446 3671	4053	993 1020	1054	476 480	487
	•••	65213 69213	106956	17823 20130	26897	3547 3833	4368	1009 1038	1110	476 487	496
0.4	•••	64635 66641	105225	17625 19133	26463	3585 3851	4305	1038 1065	1106	487 498	503
	•••	65303 70444	112562	18195 20962	28240	3698 4042	4555	1050 1088	1145	491 503	507
0.45	•••	62587 65168	110051	17445 19230	27604	3641 3941	4463	1054 1088	1133	491 503	510
	•••	63453 69893	115833	18127 21322	28999	3776 4155	4649	1072 1110	1157	498 510	507
0.5	•••	58946 62216	112350	16923 18948	28117	3630 3952	4515	1050 1088	1133	487 498	503
	•••	60056 67913	116767	17704 21259	29175	3772 4181	4649	1065 1110	1145	491 503	496
0.55	•••	54138 58193	112110	16125 18345	27993	3551 3885	4463	1020 1061	1110	469 480	487
	•••	55526 64826	115365	16995 20816	28766	3698 4125	4555	1043 1083	1110	476 487	475
0.6	•••	48585 53479	109331	15128 17479	27240	3412 3743	4312	975 1016	1061	442 453	465
	•••	50313 60904	111628	16057 20017	27773	3558 3979	4368	993 1031	1052	446 458	444
0.65	•••	42735 48394	104014	13969 16372	25845	3210 3525	4058	908 941	982	408 413	424
	•••	44790 56343	105555	14932 18896	26196	3349 3750	4088	926 960	970	408 420	403
0.7	•••	36941 43129	96184	12686 15045	23824	2944 3232	3705	818 847	881	356 363	375
	•••	39255 51270	97146	13638 17456	24036	3075 3435	3714	836 863	865	363 368	351
0.75	•••	31429 37785	85856	11287 13492	21180	2618 2865	3255	705 728	757	300 307	311
	•••	33859 45705	86402	12187 15690	21291	2730 3034	3247	716 746	736	307 307	289
0.8	•••	26238 32325	73054	9746 11685	17918	2220 2415	2703	570 588	604	233 233	240
	•••	28601 39596	73321	10545 13564	17963	2310 2539	2687	581 600	584	233 240	216
0.85	•••	21210 26576	57810	8006 9577	14048	1740 1871	2051	413 424	431	. .	.
	•••	23318 32745	57905	8659 11017	14050	1803 1954	2033	420 424	409	. .	.
0.9	•••	16012 20175	40143	5959 7023	9570	1166 1230	1308	. .	.	. .	.
	•••	17670 24720	40153	6416 7928	9554	1196 1263	1285	. .	.	. .	.
0.95	•••	9870 12187	20085	3315 3738	4496	. .	.	. .	.	. .	.
	•••	10819 14471	20065	3502 4053	4474	. .	.	. .	.	. .	.
0.98	•••	4598 5347	6915	. .	.	. .	.	. .	.	. .	.
	•••	4920 5943	6891	. .	.	. .	.	. .	.	. .	.

TABLE 1: ALPHA= 0.005 POWER= 0.8   EXPECTED ACCRUAL THRU MINIMUM FOLLOW-UP= 4750

	DEL=.01			DEL=.02			DEL=.05			DEL=.10			DEL=.15		
FACT=	1.0 .75	.50 .25	.00 BIN	1.0 .75	.50 .25	.00 BIN	1.0 .75	.50 .25	.00 BIN	1.0 75	.50 .25	.00 BIN	1.0 .75	.50 .25	.00 BIN

PCONT=***   REQUIRED NUMBER OF PATIENTS

PCONT	.01	.01	.01	.02	.02	.02	.05	.05	.05	.10	.10	.10	.15	.15	.15
0.01 ***	1789	1789	1801	657	657	661	217	217	217	110	110	110	75	79	79
***	1789	1797	6891	657	661	2278	217	217	620	110	110	252	79	79	150
0.02 ***	3950	3970	4022	1274	1274	1286	340	348	348	150	150	150	98	98	98
***	3958	3986	11376	1274	1279	3387	348	348	792	150	150	293	98	98	167
0.05 ***	11954	12045	12555	3400	3435	3518	732	740	744	265	265	265	150	150	150
***	11985	12168	24269	3412	3471	6576	740	744	1285	265	265	409	150	150	216
0.1 ***	26382	26603	29220	7109	7271	7727	1393	1409	1440	435	443	443	234	234	234
***	26454	27007	43890	7169	7430	11423	1397	1421	2033	435	443	584	234	234	289
0.15 ***	39409	39801	46166	10453	10767	11954	1987	2039	2110	597	602	609	305	305	305
***	39540	40561	61175	10572	11135	15685	2010	2070	2687	597	602	736	305	305	351
0.2 ***	49930	50543	62019	13185	13695	15885	2502	2585	2723	732	740	756	360	364	364
***	50132	51735	76124	13375	14317	19364	2537	2644	3247	732	744	865	364	364	403
0.25 ***	57621	58504	76146	15244	15980	19364	2917	3048	3257	847	858	875	412	412	419
***	57918	60233	88737	15512	16894	22459	2977	3138	3714	851	870	970	412	419	444
0.3 ***	62506	63729	88175	16645	17630	22321	3238	3412	3701	930	953	977	447	455	459
***	62917	66088	99015	17008	18861	24970	3317	3535	4088	942	965	1052	447	455	475
0.35 ***	64786	66413	97905	17448	18711	24696	3464	3685	4053	994	1025	1053	471	478	490
***	65328	69524	106956	17915	20250	26897	3559	3839	4368	1013	1037	1110	478	483	496
0.4 ***	64746	66864	105220	17725	19269	26461	3606	3867	4307	1037	1072	1108	490	495	507
***	65451	70838	112562	18307	21110	28240	3720	4053	4555	1053	1089	1145	495	502	507
0.45 ***	62732	65451	110049	17567	19383	27605	3673	3962	4461	1053	1089	1132	495	502	514
***	63642	70367	115833	18267	21490	28999	3796	4172	4649	1072	1112	1157	495	507	507
0.5 ***	59134	62565	112353	17060	19122	28116	3661	3974	4516	1053	1089	1132	490	495	507
***	60297	68467	116767	17863	21438	29175	3796	4200	4649	1072	1112	1145	490	502	496
0.55 ***	54372	58618	112108	16284	18540	27997	3582	3903	4461	1029	1065	1108	471	478	490
***	55832	65447	115365	17167	21003	28766	3725	4140	4555	1048	1084	1110	478	483	475
0.6 ***	48881	53968	109329	15298	17673	27237	3440	3760	4314	982	1017	1060	443	455	459
***	50679	61575	111628	16241	20215	27773	3582	3993	4368	994	1037	1052	447	455	444
0.65 ***	43090	48933	104017	14147	16569	25843	3238	3547	4057	910	946	982	407	412	424
***	45216	57044	105555	15120	19087	26196	3376	3760	4088	930	965	970	412	419	403
0.7 ***	37350	43703	96184	12864	15232	23825	2972	3250	3708	823	851	882	360	364	372
***	39730	51968	97146	13826	17638	24036	3095	3447	3714	835	863	865	364	372	351
0.75 ***	31864	38352	85860	11451	13664	21181	2640	2882	3250	709	732	756	300	305	312
***	34350	46375	86402	12358	15849	21291	2751	3043	3247	720	744	736	305	305	289
0.8 ***	26663	32854	73058	9888	11843	17915	2240	2426	2699	573	590	609	234	234	241
***	29066	40200	73321	10695	13700	17963	2324	2549	2687	585	597	584	234	234	216
0.85 ***	21592	27031	57811	8130	9693	14044	1754	1880	2050	412	424	431	.	.	.
***	23734	33246	57905	8784	11118	14050	1813	1955	2033	419	424	409	.	.	.
0.9 ***	16312	20516	40141	6048	7097	9567	1172	1231	1310	.	.	.	.	.	.
***	17994	25076	40153	6499	7988	9554	1203	1267	1285	.	.	.	.	.	.
0.95 ***	10042	12370	20084	3352	3768	4497	.	.	.	.	.	.	.	.	.
***	11000	14638	20065	3535	4069	4474	.	.	.	.	.	.	.	.	.
0.98 ***	4658	5399	6915	.	.	.	.	.	.	.	.	.	.	.	.
***	4979	5981	6891	.	.	.	.	.	.	.	.	.	.	.	.

TABLE 1: ALPHA= 0.005 POWER= 0.8   EXPECTED ACCRUAL THRU MINIMUM FOLLOW-UP= 5000

PCONT	FACT	DEL=.01 1.0/.75	.50/.25	.00/BIN	DEL=.02 1.0/.75	.50/.25	.00/BIN	DEL=.05 1.0/.75	.50/.25	.00/BIN	DEL=.10 1.0/75	.50/.25	.00/BIN	DEL=.15 1.0/.75	.50/.25	.00/BIN
						REQUIRED NUMBER OF PATIENTS									
0.01	***	1791	1791	1804	658	658	658	216	216	216	108	108	108	79	79	79
	•••	1791	1796	6891	658	658	2278	216	216	620	108	108	252	79	79	150
0.02	***	3954	3971	4021	1271	1279	1283	346	346	346	154	154	154	96	96	96
	•••	3958	3991	11376	1271	1279	3387	346	346	792	154	154	293	96	96	167
0.05	***	11958	12054	12554	3396	3441	3516	733	741	746	266	266	266	154	154	154
	•••	11991	12179	24269	3416	3471	6576	741	741	1285	266	266	409	154	154	216
0.1	***	26391	26621	29221	7121	7279	7729	1391	1408	1441	441	441	446	233	233	233
	•••	26471	27046	43890	7183	7441	11423	1404	1421	2033	441	441	584	233	233	289
0.15	***	39433	39846	46166	10471	10796	11954	1991	2041	2108	596	604	608	304	304	308
	•••	39571	40641	61175	10596	11158	15685	2016	2071	2687	596	604	736	304	304	351
0.2	***	49958	50608	62021	13216	13741	15883	2508	2596	2721	733	741	754	358	366	366
	•••	50179	51858	76124	13408	14366	19364	2546	2654	3247	733	746	865	366	366	403
0.25	***	57666	58604	76146	15283	16041	19366	2929	3054	3254	846	858	879	408	416	421
	•••	57979	60404	88737	15566	16966	22459	2983	3141	3714	854	871	970	408	416	444
0.3	***	62571	63858	88171	16704	17721	22321	3254	3421	3704	933	954	979	446	454	458
	•••	63004	66329	99015	17083	18954	24970	3329	3546	4088	946	966	1052	454	454	475
0.35	***	64871	66583	97908	17521	18816	24696	3483	3696	4054	996	1029	1054	471	479	491
	•••	65446	69829	106956	18008	20366	26897	3579	3854	4368	1008	1041	1110	479	483	496
0.4	***	64858	67083	105221	17821	19404	26458	3629	3883	4308	1041	1071	1108	491	496	504
	•••	65604	71216	112562	18421	21246	28240	3741	4066	4555	1054	1091	1145	491	504	507
0.45	***	62871	65729	110046	17683	19533	27608	3691	3979	4458	1058	1091	1133	496	504	504
	•••	63833	70829	115833	18396	21646	28999	3821	4183	4649	1079	1108	1157	496	508	507
0.5	***	59316	62916	112354	17196	19291	28116	3683	3991	4516	1058	1091	1133	491	496	504
	•••	60541	69004	116767	18016	21608	29175	3821	4208	4649	1071	1108	1145	491	504	496
0.55	***	54604	59033	112108	16433	18716	27996	3608	3929	4466	1029	1066	1108	471	479	491
	•••	56141	66054	115365	17333	21183	28766	3754	4154	4555	1046	1083	1110	479	483	475
0.6	***	49171	54454	109329	15458	17858	27233	3466	3783	4308	983	1016	1058	446	454	454
	•••	51046	62229	111628	16416	20396	27773	3608	4008	4368	996	1033	1052	446	458	444
0.65	***	43441	49466	104016	14316	16758	25846	3258	3566	4058	916	946	983	408	416	421
	•••	45641	57721	105555	15296	19271	26196	3396	3779	4088	929	966	970	408	416	403
0.7	***	37746	44254	96183	13033	15416	23829	2996	3271	3704	821	854	883	358	366	371
	•••	40191	52641	97146	13996	17808	24036	3116	3458	3714	833	866	865	366	371	351
0.75	***	32283	38904	85858	11608	13833	21179	2658	2896	3254	708	733	754	304	308	308
	•••	34821	47016	86402	12521	16004	21291	2766	3054	3247	721	746	736	304	308	289
0.8	***	27071	33358	73058	10033	11983	17921	2254	2433	2704	579	591	604	233	233	241
	•••	29521	40779	73321	10841	13829	17963	2341	2554	2687	583	596	584	233	233	216
0.85	***	21958	27471	57808	8246	9808	14046	1766	1891	2054	416	421	433	.	.	.
	•••	24133	33721	57905	8896	11216	14050	1821	1966	2033	421	429	409	.	.	.
0.9	***	16604	20846	40141	6129	7171	9571	1179	1233	1308	.	.	.	.	.	.
	•••	18304	25416	40153	6579	8046	9554	1204	1271	1285	.	.	.	.	.	.
0.95	***	10208	12546	20083	3383	3796	4496	.	.	.	.	.	.	.	.	.
	•••	11171	14796	20065	3566	4091	4474	.	.	.	.	.	.	.	.	.
0.98	***	4716	5446	6916	.	.	.	.	.	.	.	.	.	.	.	.
	•••	5033	6016	6891	.	.	.	.	.	.	.	.	.	.	.	.

TABLE 1: ALPHA= 0.005 POWER= 0.8   EXPECTED ACCRUAL THRU MINIMUM FOLLOW-UP= 5500

	DEL=.01			DEL=.02			DEL=.05			DEL=.10			DEL=.15		
FACT=	1.0 / .75	.50 / .25	.00 / BIN	1.0 / .75	.50 / .25	.00 / BIN	1.0 / .75	.50 / .25	.00 / BIN	1.0 / 75	.50 / .25	.00 / BIN	1.0 / .75	.50 / .25	.00 / BIN

PCONT=***   REQUIRED NUMBER OF PATIENTS

PCONT	1.0	.50	.00	1.0	.50	.00	1.0	.50	.00	1.0	.50	.00	1.0	.50	.00
0.01	1791	1791	1797	655	655	655	215	215	215	105	105	105	78	78	78
	1791	1797	6891	655	655	2278	215	215	620	105	105	252	78	78	150
0.02	3955	3978	4019	1274	1274	1283	348	348	348	147	147	147	100	100	100
	3964	3991	11376	1274	1283	3387	348	348	792	147	147	293	100	100	167
0.05	11972	12068	12558	3405	3447	3515	738	738	746	265	265	265	155	155	155
	12008	12200	24269	3419	3474	6576	738	746	1285	265	265	409	155	155	216
0.1	26418	26670	29223	7140	7305	7731	1393	1412	1439	435	444	444	229	238	238
	26500	27119	43890	7209	7456	11423	1406	1425	2033	435	444	584	238	238	289
0.15	39472	39925	46163	10509	10844	11953	2003	2044	2108	595	600	609	306	306	306
	39623	40786	61175	10638	11210	15685	2025	2072	2687	600	609	736	306	306	351
0.2	50026	50733	62022	13278	13823	15885	2520	2603	2718	733	746	752	361	367	367
	50265	52094	76124	13484	14447	19364	2561	2658	3247	738	746	865	361	367	403
0.25	57759	58785	76143	15376	16165	19364	2946	3070	3254	848	862	875	408	416	416
	58103	60743	88737	15670	17087	22459	3001	3153	3714	856	870	970	416	416	444
0.3	62704	64112	88174	16820	17884	22320	3276	3441	3703	939	958	980	449	449	458
	63171	66788	99015	17224	19122	24970	3350	3557	4088	944	966	1052	449	458	475
0.35	65047	66925	97909	17678	19020	24690	3515	3722	4052	1008	1027	1054	477	485	485
	65674	70423	106956	18187	20579	26897	3606	3868	4368	1013	1040	1110	477	485	496
0.4	65083	67522	105224	18008	19644	26464	3667	3909	4308	1049	1076	1109	490	499	504
	65899	71958	112562	18640	21500	28240	3777	4079	4555	1063	1090	1145	490	499	507
0.45	63163	66284	110050	17912	19818	27605	3735	4010	4459	1063	1095	1131	499	504	513
	64216	71715	115833	18654	21935	28999	3859	4203	4649	1082	1118	1157	499	504	507
0.5	59684	63598	112347	17458	19603	28114	3730	4024	4514	1063	1095	1131	490	499	504
	61026	70033	116767	18305	21927	29175	3868	4239	4649	1076	1109	1145	490	499	496
0.55	55064	59844	112113	16724	19053	27999	3661	3964	4464	1035	1068	1109	471	485	490
	56736	67200	115365	17650	21514	28766	3799	4175	4555	1054	1090	1110	477	485	475
0.6	49743	55380	109330	15767	18209	27234	3515	3818	4313	985	1021	1054	449	458	463
	51764	63460	111628	16751	20730	27773	3653	4033	4368	1008	1040	1052	449	458	444
0.65	44119	50480	104014	14634	17109	25845	3309	3598	4060	917	953	980	408	416	422
	46465	58997	105555	15638	19603	26196	3441	3799	4088	930	966	970	416	416	403
0.7	38518	45310	96185	13347	15748	23824	3034	3304	3703	829	856	884	361	367	375
	41080	53904	97146	14331	18118	24036	3158	3474	3714	843	870	865	361	367	351
0.75	33092	39939	85859	11911	14139	21179	2699	2919	3254	719	733	752	306	306	312
	35732	48217	86402	12824	16284	21291	2800	3070	3247	724	746	736	306	306	289
0.8	27861	34324	73058	10303	12247	17920	2286	2457	2699	581	595	609	238	238	238
	30383	41850	73321	11105	14062	17963	2360	2567	2687	587	600	584	238	238	216
0.85	22664	28293	57809	8460	10014	14048	1783	1901	2053	416	422	430	.	.	.
	24891	34613	57905	9112	11389	14050	1838	1970	2033	422	430	409	.	.	.
0.9	17155	21468	40140	6279	7305	9565	1186	1241	1310	.	.	.	.	.	.
	18888	26046	40153	6719	8149	9554	1214	1274	1285	.	.	.	.	.	.
0.95	10528	12865	20084	3447	3840	4492	.	.	.	.	.	.	.	.	.
	11490	15088	20065	3625	4120	4474	.	.	.	.	.	.	.	.	.
0.98	4822	5537	6912	.	.	.	.	.	.	.	.	.	.	.	.
	5138	6081	6891	.	.	.	.	.	.	.	.	.	.	.	.

TABLE 1: ALPHA= 0.005 POWER= 0.8   EXPECTED ACCRUAL THRU MINIMUM FOLLOW-UP= 6000

		DEL=.01			DEL=.02			DEL=.05			DEL=.10			DEL=.15		
FACT=		1.0 .75	.50 .25	.00 BIN	1.0 .75	.50 .25	.00 BIN	1.0 .75	.50 .25	.00 BIN	1.0 75	.50 .25	.00 BIN	1.0 .75	.50 .25	.00 BIN
PCONT=***					REQUIRED NUMBER OF PATIENTS											
0.01	***	1789	1795	1804	655	655	655	214	214	214	109	109	109	79	79	79
	***	1789	1795	6891	655	655	2278	214	214	620	109	109	252	79	79	150
0.02	***	3955	3979	4015	1270	1279	1285	340	349	349	154	154	154	100	100	100
	***	3964	3994	11376	1279	1279	3387	340	349	792	154	154	293	100	100	167
0.05	***	11980	12079	12550	3409	3445	3520	739	739	745	265	265	265	154	154	154
	***	12019	12214	24269	3430	3475	6576	739	745	1285	265	265	409	154	154	216
0.1	***	26440	26710	29224	7159	7324	7729	1399	1414	1435	439	439	445	235	235	235
	***	26530	27190	43890	7225	7474	11423	1405	1429	2033	439	445	584	235	235	289
0.15	***	39514	40009	46165	10549	10894	11950	2005	2050	2110	595	604	610	304	304	304
	***	39679	40924	61175	10684	11254	15685	2029	2080	2687	604	604	736	304	304	351
0.2	***	50089	50860	62020	13339	13900	15880	2530	2614	2719	739	745	754	364	364	370
	***	50350	52315	76124	13549	14524	19364	2569	2659	3247	739	745	865	364	364	403
0.25	***	57859	58969	76144	15460	16279	19369	2965	3085	3259	850	865	880	409	415	415
	***	58225	61075	88737	15769	17200	22459	3019	3160	3714	859	874	970	415	415	444
0.3	***	62830	64369	88174	16939	18034	22315	3304	3460	3700	940	955	979	445	454	460
	***	63340	67234	99015	17359	19279	24970	3370	3565	4088	949	970	1052	454	454	475
0.35	***	65215	67255	97909	17824	19210	24694	3544	3745	4054	1009	1030	1054	475	484	490
	***	65899	70999	106956	18355	20770	26897	3634	3880	4368	1015	1039	1110	475	484	496
0.4	***	65305	67960	105220	18190	19870	26464	3700	3934	4309	1054	1075	1105	490	499	505
	***	66199	72670	112562	18844	21730	28240	3805	4099	4555	1060	1090	1145	499	499	507
0.45	***	63454	66829	110050	18124	20080	27604	3775	4039	4459	1069	1099	1135	499	505	514
	***	64600	72565	115833	18895	22195	28999	3895	4219	4649	1084	1114	1157	499	505	507
0.5	***	60055	64264	112354	17704	19894	28120	3775	4060	4510	1069	1099	1135	490	499	505
	***	61504	71005	116767	18574	22210	29175	3904	4255	4649	1084	1114	1145	490	499	496
0.55	***	55525	60634	112114	16999	19360	27994	3700	3994	4465	1039	1075	1105	475	484	490
	***	57325	68284	115365	17944	21814	28766	3835	4195	4555	1054	1090	1110	475	484	475
0.6	***	50314	56269	109330	16060	18529	27235	3559	3850	4309	994	1024	1060	445	454	460
	***	52465	64615	111628	17059	21040	27773	3694	4054	4368	1009	1039	1052	454	460	444
0.65	***	44785	51439	104014	14935	17425	25849	3349	3625	4060	925	949	979	409	415	424
	***	47260	60184	105555	15949	19900	26196	3475	3814	4088	940	964	970	415	415	403
0.7	***	39250	46300	96184	13639	16054	23824	3070	3325	3709	835	859	880	364	364	370
	***	41929	55084	97144	14629	18400	24036	3190	3490	3714	844	865	865	364	370	351
0.75	***	33859	40909	85855	12184	14419	21184	2725	2944	3250	715	739	754	304	304	310
	***	36589	49324	86402	13105	16534	21291	2830	3079	3247	724	745	736	304	310	289
0.8	***	28600	35224	73054	10540	12490	17920	2305	2470	2704	580	595	604	235	235	235
	***	31195	42844	73321	11359	14269	17963	2389	2575	2687	589	595	584	235	235	216
0.85	***	23314	29059	57805	8659	10195	14044	1804	1909	2050	415	424	430	.	.	.
	***	25594	35425	57905	9310	11539	14050	1855	1975	2033	424	430	409	.	.	.
0.9	***	17665	22030	40144	6415	7420	9565	1195	1249	1309	.	.	.	.	.	.
	***	19429	26614	40153	6850	8239	9554	1219	1279	1285	.	.	.	.	.	.
0.95	***	10819	13159	20080	3505	3880	4495	.	.	.	.	.	.	.	.	.
	***	11785	15340	20065	3679	4144	4474	.	.	.	.	.	.	.	.	.
0.98	***	4924	5614	6910	.	.	.	.	.	.	.	.	.	.	.	.
	***	5224	6139	6891	.	.	.	.	.	.	.	.	.	.	.	.

| TABLE 1: ALPHA= 0.005 POWER= 0.8 | | | EXPECTED ACCRUAL THRU MINIMUM FOLLOW-UP= 6500 | | | | | | | | | | | |

	DEL=.01			DEL=.02			DEL=.05			DEL=.10			DEL=.15		
FACT=	1.0 .75	.50 .25	.00 BIN	1.0 .75	.50 .25	.00 BIN	1.0 .75	.50 .25	.00 BIN	1.0 75	.50 .25	.00 BIN	1.0 .75	.50 .25	.00 BIN

PCONT=*** REQUIRED NUMBER OF PATIENTS

PCONT	DEL=.01 1.0/.75	DEL=.01 .50/.25	DEL=.01 .00/BIN	DEL=.02 1.0/.75	DEL=.02 .50/.25	DEL=.02 .00/BIN	DEL=.05 1.0/.75	DEL=.05 .50/.25	DEL=.05 .00/BIN	DEL=.10 1.0/.75	DEL=.10 .50/.25	DEL=.10 .00/BIN	DEL=.15 1.0/.75	DEL=.15 .50/.25	DEL=.15 .00/BIN
0.01 ***	1792	1792	1798	655	661	661	216	216	216	108	108	108	76	76	76
•••	1792	1798	6891	661	661	2278	216	216	620	108	108	252	76	76	150
0.02 ***	3959	3976	4018	1272	1278	1288	346	346	346	151	151	151	102	102	102
•••	3970	3992	11376	1278	1278	3387	346	346	792	151	151	293	102	102	167
0.05 ***	11987	12095	12556	3417	3456	3515	736	742	742	265	265	265	151	151	151
•••	12030	12231	24269	3433	3482	6576	742	742	1285	265	265	409	151	151	216
0.1 ***	26460	26758	29222	7177	7339	7729	1402	1418	1441	443	443	443	232	232	232
•••	26563	27256	43890	7242	7486	11423	1408	1424	2033	443	443	584	232	232	289
0.15 ***	39557	40093	46161	10583	10931	11954	2009	2052	2107	596	606	606	303	303	303
•••	39736	41058	61175	10719	11288	15685	2036	2074	2687	596	606	736	303	303	351
0.2 ***	50158	50997	62021	13395	13997	15881	2546	2621	2718	736	742	752	362	368	368
•••	50434	52541	76124	13622	14587	19364	2578	2670	3247	742	752	865	362	368	403
0.25 ***	57952	59155	76142	15540	16385	19364	2978	3092	3255	850	866	882	411	417	417
•••	58348	61397	88737	15865	17301	22459	3033	3163	3714	856	872	970	411	417	444
0.3 ***	62957	64621	88177	17051	18178	22322	3320	3472	3699	947	963	980	449	449	460
•••	63516	67659	99015	17479	19413	24970	3391	3580	4088	953	969	1052	449	460	475
0.35 ***	65384	67594	97911	17967	19391	24694	3569	3764	4051	1012	1028	1051	476	482	492
•••	66126	71553	106956	18519	20941	26897	3661	3888	4368	1018	1045	1110	482	482	496
0.4 ***	65531	68391	105223	18367	20090	26460	3732	3959	4311	1051	1077	1110	492	498	508
•••	66500	73357	112562	19039	21932	28240	3840	4116	4555	1067	1093	1145	498	498	507
0.45 ***	63737	67367	110050	18335	20323	27603	3807	4067	4463	1077	1099	1132	498	508	508
•••	64978	73379	115833	19121	22430	28999	3927	4236	4649	1083	1116	1157	498	508	507
0.5 ***	60422	64913	112347	17934	20161	28117	3813	4083	4512	1067	1099	1132	492	498	508
•••	61982	71933	116767	18828	22468	29175	3937	4268	4649	1083	1116	1145	498	498	496
0.55 ***	55986	61387	112113	17252	19651	27993	3742	4024	4463	1045	1077	1110	476	482	492
•••	57909	69301	115365	18221	22088	28766	3872	4213	4555	1061	1093	1110	476	482	475
0.6 ***	50867	57123	109335	16326	18828	27240	3596	3878	4311	996	1028	1061	449	449	460
•••	53148	65703	111628	17343	21314	27773	3726	4067	4368	1012	1045	1052	449	460	444
0.65 ***	45429	52352	104021	15215	17717	25842	3385	3651	4057	931	953	980	411	417	417
•••	48023	61289	105555	16238	20171	26196	3504	3829	4088	937	969	970	411	417	403
0.7 ***	39963	47233	96188	13915	16336	23827	3108	3352	3710	833	856	882	362	368	368
•••	42742	56171	97146	14912	18649	24036	3222	3504	3714	850	866	865	362	368	351
0.75 ***	34585	41822	85859	12442	14678	21178	2757	2962	3255	720	736	758	303	303	313
•••	37396	50353	86402	13368	16758	21291	2854	3092	3247	726	742	736	303	313	289
0.8 ***	29293	36063	73054	10768	12712	17918	2328	2491	2702	579	596	606	232	238	238
•••	31958	43756	73321	11581	14451	17963	2399	2588	2687	590	596	584	238	238	216
0.85 ***	23931	29775	57812	8845	10372	14045	1814	1922	2052	417	427	433	.	.	.
•••	26254	36167	57905	9495	11672	14050	1863	1977	2033	417	427	409	.	.	.
0.9 ***	18146	22560	40142	6537	7528	9566	1197	1246	1305	.	.	.	.	.	.
•••	19933	27132	40153	6966	8314	9554	1223	1278	1285	.	.	.	.	.	.
0.95 ***	11087	13427	20079	3553	3921	4496	.	.	.	.	.	.	.	.	.
•••	12062	15572	20065	3716	4171	4474	.	.	.	.	.	.	.	.	.
0.98 ***	5010	5682	6911	.	.	.	.	.	.	.	.	.	.	.	.
•••	5308	6186	6891	.	.	.	.	.	.	.	.	.	.	.	.

TABLE 1: ALPHA= 0.005   POWER= 0.8   EXPECTED ACCRUAL THRU MINIMUM FOLLOW-UP= 7000

		DEL=.01			DEL=.02			DEL=.05			DEL=.10			DEL=.15	
FACT=	1.0 .75	.50 .25	.00 BIN	1.0 .75	.50 .25	.00 BIN	1.0 .75	.50 .25	.00 BIN	1.0 75	.50 .25	.00 BIN	1.0 .75	.50 .25	.00 BIN
PCONT=***				REQUIRED NUMBER OF PATIENTS											
0.01 ***	1790	1796	1796	659	659	659	215	215	215	110	110	110	75	75	75
•••	1790	1796	6891	659	659	2278	215	215	620	110	110	252	75	75	150
0.02 ***	3960	3984	4019	1271	1282	1282	344	344	344	151	151	151	99	99	99
•••	3966	3995	11376	1271	1282	3387	344	344	792	151	151	293	99	99	167
0.05 ***	11999	12115	12552	3417	3459	3522	740	740	746	267	267	267	151	151	151
•••	12034	12244	24269	3435	3487	6576	740	746	1285	267	267	409	151	151	216
0.1 ***	26482	26804	29219	7197	7361	7729	1405	1422	1440	442	442	442	232	232	232
•••	26594	27322	43890	7267	7501	11423	1411	1429	2033	442	442	584	232	232	289
0.15 ***	39596	40174	46170	10616	10977	11957	2017	2059	2111	600	600	606	302	302	309
•••	39789	41189	61175	10756	11316	15685	2035	2076	2687	600	606	736	302	302	351
0.2 ***	50219	51122	62025	13445	14040	15884	2549	2619	2717	740	746	757	361	361	372
•••	50516	52750	76124	13679	14652	19364	2584	2671	3247	740	746	865	361	361	403
0.25 ***	58041	59336	76147	15621	16479	19366	2997	3102	3260	851	862	880	414	414	414
•••	58479	61699	88737	15954	17400	22459	3050	3172	3714	862	869	970	414	414	444
0.3 ***	63092	64866	88176	17155	18310	22317	3336	3487	3704	950	956	974	449	449	460
•••	63687	68080	99015	17599	19541	24970	3406	3581	4088	950	967	1052	449	460	475
0.35 ***	65560	67922	97906	18100	19559	24697	3592	3785	4054	1009	1037	1055	477	484	484
•••	66354	72087	106956	18666	21110	26897	3680	3896	4368	1020	1044	1110	477	484	496
0.4 ***	65752	68815	105221	18537	20287	26465	3756	3977	4310	1055	1079	1107	495	501	501
•••	66791	74012	112562	19226	22125	28240	3861	4124	4555	1072	1090	1145	495	501	507
0.45 ***	64026	67894	110051	18526	20550	27609	3844	4089	4456	1079	1107	1131	501	501	512
•••	65356	74152	115833	19331	22639	28999	3949	4246	4649	1090	1114	1157	501	512	507
0.5 ***	60782	65549	112355	18159	20410	28116	3844	4106	4509	1072	1107	1131	495	501	501
•••	62451	72811	116767	19062	22702	29175	3966	4281	4649	1090	1114	1145	495	501	496
0.55 ***	56442	62119	112110	17494	19920	27994	3774	4047	4467	1055	1079	1107	477	484	484
•••	58472	70267	115365	18474	22335	28766	3896	4229	4555	1061	1090	1110	477	484	475
0.6 ***	51409	57936	109327	16584	19097	27235	3634	3907	4310	1002	1026	1055	449	460	460
•••	53806	66721	111628	17610	21565	27773	3756	4082	4368	1020	1044	1052	449	460	444
0.65 ***	46054	53222	104014	15475	17984	25841	3417	3680	4054	932	956	985	414	414	425
•••	48760	62340	105555	16507	20410	26196	3540	3844	4088	939	967	970	414	414	403
0.7 ***	40640	48119	96185	14169	16595	23822	3137	3371	3704	834	862	880	361	372	372
•••	43516	57195	97146	15166	18876	24036	3242	3522	3714	851	869	865	361	372	351
0.75 ***	35285	42676	85860	12675	14915	21180	2787	2980	3249	722	740	757	302	309	309
•••	38172	51315	86402	13609	16962	21291	2875	3102	3247	729	746	736	302	309	289
0.8 ***	29965	36849	73056	10977	12909	17914	2350	2496	2700	582	589	606	232	239	239
•••	32677	44612	73321	11789	14617	17963	2420	2595	2687	589	600	584	232	239	216
0.85 ***	24522	30444	57807	9006	10522	14046	1831	1930	2052	425	425	431	.	.	.
•••	26885	36860	57905	9654	11789	14050	1877	1989	2033	425	431	409	.	.	.
0.9 ***	18596	23052	40139	6644	7624	9566	1212	1254	1306	.	.	.	.	.	.
•••	20399	27609	40153	7075	8387	9554	1230	1282	1285	.	.	.	.	.	.
0.95 ***	11334	13661	20084	3599	3949	4491	.	.	.	.	.	.	.	.	.
•••	12314	15779	20065	3756	4187	4474	.	.	.	.	.	.	.	.	.
0.98 ***	5086	5745	6917	.	.	.	.	.	.	.	.	.	.	.	.
•••	5377	6224	6891	.	.	.	.	.	.	.	.	.	.	.	.

TABLE 1: ALPHA= 0.005 POWER= 0.8   EXPECTED ACCRUAL THRU MINIMUM FOLLOW-UP= 7500

		DEL=.01			DEL=.02			DEL=.05			DEL=.10			DEL=.15		
FACT=		1.0 .75	.50 .25	.00 BIN	1.0 .75	.50 .25	.00 BIN	1.0 .75	.50 .25	.00 BIN	1.0 75	.50 .25	.00 BIN	1.0 .75	.50 .25	.00 BIN
PCONT=***					REQUIRED NUMBER OF PATIENTS											
0.01	***	1793	1793	1805	661	661	661	218	218	218	106	106	106	80	80	80
	***	1793	1793	6891	661	661	2278	218	218	620	106	106	252	80	80	150
0.02	***	3961	3980	4018	1268	1280	1280	343	343	343	155	155	155	99	99	99
	***	3968	3999	11376	1280	1280	3387	343	343	792	155	155	293	99	99	167
0.05	***	12005	12125	12556	3425	3462	3518	736	743	743	268	268	268	155	155	155
	***	12050	12256	24269	3436	3481	6576	743	743	1285	268	268	409	155	155	216
0.1	***	26506	26843	29225	7212	7374	7730	1400	1418	1437	436	443	443	230	237	237
	***	26618	27380	43890	7280	7512	11423	1411	1430	2033	443	443	584	230	237	289
0.15	***	39643	40250	46168	10655	11011	11956	2018	2056	2105	593	605	605	305	305	305
	***	39849	41318	61175	10793	11349	15685	2037	2086	2687	605	605	736	305	305	351
0.2	***	50281	51249	62018	13505	14105	15886	2562	2630	2724	736	743	755	361	368	368
	***	50611	52955	76124	13737	14705	19364	2593	2675	3247	743	755	865	361	368	403
0.25	***	58137	59518	76149	15699	16580	19362	3005	3106	3256	856	868	875	418	418	418
	***	58599	62000	88737	16043	17480	22459	3050	3174	3714	856	875	970	418	418	444
0.3	***	63218	65124	88175	17255	18436	22318	3361	3500	3699	950	961	980	455	455	455
	***	63856	68480	99015	17724	19655	24970	3425	3593	4088	950	968	1052	455	455	475
0.35	***	65731	68255	97906	18230	19711	24699	3612	3793	4055	1018	1036	1055	481	481	493
	***	66586	72605	106956	18818	21249	26897	3699	3912	4368	1025	1043	1110	481	481	496
0.4	***	65975	69230	105218	18687	20468	26461	3781	3999	4306	1062	1081	1111	493	500	500
	***	67081	74637	112562	19400	22299	28240	3886	4137	4555	1074	1093	1145	500	500	507
0.45	***	64318	68412	110049	18718	20761	27605	3868	4111	4456	1081	1100	1130	500	500	511
	***	65731	74893	115833	19531	22843	28999	3980	4261	4649	1093	1118	1157	500	511	507
0.5	***	61149	66162	112355	18368	20649	28118	3875	4130	4512	1081	1100	1130	493	500	500
	***	62911	73643	116767	19287	22918	29175	3987	4299	4555	1093	1118	1145	500	500	496
0.55	***	56881	62836	112111	17724	20161	27999	3800	4074	4468	1055	1081	1111	474	481	493
	***	59030	71180	115365	18718	22561	28766	3924	4243	4555	1062	1093	1110	481	481	475
0.6	***	51943	58718	109325	16831	19355	27237	3661	3924	4306	1006	1025	1055	455	455	462
	***	54455	67681	111628	17862	21793	27773	3781	4100	4368	1018	1043	1052	455	455	444
0.65	***	46662	54061	104018	15718	18237	25843	3455	3699	4055	931	961	980	418	418	425
	***	49468	63312	105555	16756	20630	26196	3568	3856	4088	943	968	970	418	418	403
0.7	***	41300	48968	96181	14405	16831	23825	3162	3387	3706	837	856	886	361	368	368
	***	44255	58156	97146	15418	19081	24036	3268	3530	3714	849	868	865	368	368	351
0.75	***	35949	43493	85861	12905	15125	21181	2806	2993	3249	725	736	755	305	305	312
	***	38900	52212	86402	13831	17143	21291	2893	3106	3247	736	743	736	305	305	289
0.8	***	30593	37587	73055	11168	13093	17918	2368	2518	2705	586	593	605	237	237	237
	***	33361	45406	73321	11986	14768	17963	2431	2600	2687	586	605	584	237	237	216
0.85	***	25074	31062	57811	9162	10662	14049	1843	1936	2049	418	425	436	.	.	.
	***	27474	37505	57905	9811	11900	14050	1887	1993	2033	425	425	409	.	.	.
0.9	***	19025	23506	40137	6755	7711	9568	1212	1261	1306	.	.	.	.	.	.
	***	20843	28055	40153	7175	8443	9554	1231	1280	1285	.	.	.	.	.	.
0.95	***	11574	13887	20086	3643	3980	4493	.	.	.	.	.	.	.	.	.
	***	12549	15968	20065	3793	4205	4474	.	.	.	.	.	.	.	.	.
0.98	***	5161	5806	6912	.	.	.	.	.	.	.	.	.	.	.	.
	***	5450	6268	6891	.	.	.	.	.	.	.	.	.	.	.	.

TABLE 1: ALPHA= 0.005 POWER= 0.8   EXPECTED ACCRUAL THRU MINIMUM FOLLOW-UP= 8000

		DEL=.01			DEL=.02			DEL=.05			DEL=.10			DEL=.15		
FACT=		1.0 .75	.50 .25	.00 BIN	1.0 .75	.50 .25	.00 BIN	1.0 .75	.50 .25	.00 BIN	1.0 75	.50 .25	.00 BIN	1.0 .75	.50 .25	.00 BIN
PCONT=•••					REQUIRED NUMBER OF PATIENTS											
0.01	•••	1793	1793	1805	653	653	653	213	213	213	105	105	105	73	73	73
	•••	1793	1793	6891	653	653	2278	213	213	620	105	105	252	73	73	150
0.02	•••	3965	3985	4013	1273	1273	1285	345	345	345	153	153	153	93	93	93
	•••	3973	3993	11376	1273	1285	3387	345	345	792	153	153	293	93	93	167
0.05	•••	12013	12133	12553	3425	3465	3513	733	745	745	265	265	265	153	153	153
	•••	12065	12273	24269	3445	3485	6576	745	745	1285	265	265	409	153	153	216
0.1	•••	26533	26885	29225	7225	7393	7725	1405	1425	1433	445	445	445	233	233	233
	•••	26653	27433	43890	7293	7525	11423	1413	1425	2033	445	445	584	233	233	289
0.15	•••	39685	40333	46165	10685	11045	11953	2025	2065	2105	605	605	605	305	305	305
	•••	39905	41433	61175	10833	11385	15685	2045	2085	2687	605	605	736	305	305	351
0.2	•••	50345	51373	62025	13553	14165	15885	2565	2633	2725	733	745	753	365	365	365
	•••	50693	53153	76124	13793	14753	19364	2605	2673	3247	745	753	865	365	365	403
0.25	•••	58225	59705	76145	15773	16665	19365	3013	3113	3253	853	865	873	413	413	413
	•••	58725	62285	88737	16125	17565	22459	3065	3185	3714	865	873	970	413	413	444
0.3	•••	63345	65365	88173	17353	18545	22313	3373	3513	3705	945	965	973	453	453	453
	•••	64025	68865	99015	17825	19765	24970	3433	3593	4088	953	973	1052	453	453	475
0.35	•••	65905	68573	97905	18353	19853	24693	3633	3805	4053	1013	1033	1053	473	485	485
	•••	66805	73093	106956	18953	21385	26897	3713	3913	4368	1025	1045	1110	485	485	496
0.4	•••	66193	69645	105225	18845	20645	26465	3805	4013	4305	1065	1085	1105	493	505	505
	•••	67373	75233	112562	19565	22453	28240	3905	4145	4555	1073	1093	1145	493	505	507
0.45	•••	64593	68913	110045	18893	20965	27605	3893	4125	4465	1085	1105	1133	505	505	513
	•••	66105	75593	115833	19725	23025	28999	4005	4273	4649	1093	1113	1157	505	505	507
0.5	•••	61505	66765	112353	18573	20865	28113	3905	4153	4513	1085	1105	1133	493	505	505
	•••	63365	74445	116767	19505	23113	29175	4013	4313	4649	1093	1113	1145	493	505	496
0.55	•••	57325	63525	112113	17945	20393	27993	3833	4093	4465	1053	1085	1105	473	485	485
	•••	59573	72045	115365	18945	22765	28766	3953	4253	4555	1065	1093	1110	485	485	475
0.6	•••	52465	59473	109333	17053	19585	27233	3693	3945	4313	1005	1033	1053	453	453	465
	•••	55073	68593	111628	18093	22005	27773	3805	4105	4368	1025	1045	1052	453	453	444
0.65	•••	47253	54853	104013	15945	18473	25845	3473	3713	4053	933	953	985	413	413	425
	•••	50145	64245	105555	16993	20833	26196	3585	3873	4088	945	973	970	413	413	403
0.7	•••	41933	49765	96185	14633	17053	23825	3193	3405	3705	845	865	885	365	365	373
	•••	44965	59053	97197	15645	19273	24036	3293	3545	3714	853	873	865	365	373	351
0.75	•••	36585	44273	85853	13105	15325	21185	2825	3005	3253	725	745	753	305	305	313
	•••	39593	53053	86402	14045	17305	21291	2913	3125	3247	733	745	736	305	305	289
0.8	•••	31193	38293	73053	11353	13265	17913	2385	2525	2705	585	593	605	233	233	233
	•••	34013	46145	73321	12165	14905	17963	2453	2605	2687	593	605	584	233	233	216
0.85	•••	25593	31653	57805	9313	10785	14045	1853	1945	2053	425	425	433	.	.	.
	•••	28025	38105	57905	9945	12005	14050	1893	1993	2033	425	425	409	.	.	.
0.9	•••	19425	23933	40145	6845	7793	9565	1213	1265	1305	.	.	.	.	.	.
	•••	21265	28453	40153	7265	8505	9554	1233	1285	1285	.	.	.	.	.	.
0.95	•••	11785	14093	20085	3673	4005	4493	.	.	.	.	.	.	.	.	.
	•••	12765	16133	20065	3825	4225	4474	.	.	.	.	.	.	.	.	.
0.98	•••	5225	5853	6913	.	.	.	.	.	.	.	.	.	.	.	.
	•••	5505	6293	6891	.	.	.	.	.	.	.	.	.	.	.	.

TABLE 1: ALPHA= 0.005 POWER= 0.8   EXPECTED ACCRUAL THRU MINIMUM FOLLOW-UP= 8500

	DEL=.01			DEL=.02			DEL=.05			DEL=.10			DEL=.15		
FACT=	1.0 .75	.50 .25	.00 BIN	1.0 .75	.50 .25	.00 BIN	1.0 .75	.50 .25	.00 BIN	1.0 75	.50 .25	.00 BIN	1.0 .75	.50 .25	.00 BIN
PCONT=•••					REQUIRED NUMBER OF PATIENTS										
0.01 •••	1791	1799	1799	651	665	665	218	218	218	112	112	112	78	78	78
•••	1791	1799	6891	651	665	2278	218	218	620	112	112	252	78	78	150
0.02 •••	3966	3988	4022	1281	1281	1281	346	346	346	155	155	155	99	99	99
•••	3980	4001	11376	1281	1281	3387	346	346	792	155	155	293	99	99	167
0.05 •••	12020	12148	12551	3427	3470	3520	736	736	750	261	261	261	155	155	155
•••	12076	12288	24269	3448	3491	6576	736	750	1285	261	261	409	155	155	216
0.1 •••	26555	26930	29225	7239	7401	7728	1408	1416	1438	439	439	439	240	240	240
•••	26683	27482	43890	7316	7536	11423	1416	1430	2033	439	439	584	240	240	289
0.15 •••	39722	40410	46161	10716	11077	11956	2033	2067	2110	601	601	609	303	303	303
•••	39956	41550	61175	10865	11404	15685	2046	2088	2687	601	609	736	303	303	351
0.2 •••	50411	51495	62021	13606	14222	15880	2577	2641	2726	736	750	750	367	367	367
•••	50780	53343	76124	13848	14804	19364	2606	2670	3247	736	750	865	367	367	403
0.25 •••	58324	59875	76145	15845	16751	19365	3031	3130	3257	856	864	877	410	418	418
•••	58847	62566	88737	16206	17630	22459	3074	3180	3714	864	877	970	418	418	444
0.3 •••	63466	65613	88172	17452	18663	22318	3385	3520	3703	949	962	983	452	452	460
•••	64202	69238	99015	17928	19861	24970	3448	3605	4088	962	970	1052	452	460	475
0.35 •••	66072	68898	97905	18480	20002	24698	3648	3818	4051	1026	1034	1055	481	481	481
•••	67036	73573	106956	19088	21511	26897	3733	3924	4368	1026	1047	1110	481	481	496
0.4 •••	66420	70046	105223	18990	20810	26462	3831	4030	4306	1068	1090	1111	495	503	503
•••	67666	75813	112562	19726	22603	28240	3916	4150	4555	1076	1098	1145	495	503	507
0.45 •••	64882	69408	110046	19067	21150	27610	3916	4136	4455	1090	1111	1132	503	503	516
•••	66463	76272	115833	19904	23190	28999	4022	4285	4649	1098	1119	1157	503	503	507
0.5 •••	61865	67347	112355	18770	21065	28120	3924	4171	4511	1090	1111	1132	495	503	503
•••	63820	75210	116767	19705	23296	29175	4043	4320	4649	1098	1119	1145	495	503	496
0.55 •••	57763	64181	112108	18153	20605	27992	3860	4107	4468	1055	1076	1111	481	481	495
•••	60109	72872	115365	19160	22956	28766	3980	4264	4555	1068	1098	1110	481	481	475
0.6 •••	52982	60207	109324	17269	19811	27235	3711	3966	4306	1013	1034	1055	452	452	460
•••	55681	69459	111628	18323	22191	27773	3831	4115	4368	1026	1047	1052	452	460	444
0.65 •••	47840	55617	104011	16164	18693	25846	3499	3733	4051	941	962	983	410	418	418
•••	50801	65116	105555	17218	21022	26196	3605	3881	4088	949	970	970	418	418	403
0.7 •••	42548	50538	96183	14838	17261	23827	3215	3414	3703	843	864	877	367	367	367
•••	45638	59910	97146	15858	19450	24036	3308	3555	3714	856	877	865	367	367	351
0.75 •••	37201	45000	85856	13308	15518	21179	2840	3023	3257	728	736	758	303	311	311
•••	40261	53853	86402	14230	17460	21291	2925	3130	3247	736	750	736	303	311	289
0.8 •••	31775	38965	73050	11523	13415	17920	2394	2535	2705	588	601	609	240	240	240
•••	34630	46841	73321	12331	15030	17963	2458	2606	2687	588	601	584	240	240	216
0.85 •••	26101	32221	57806	9449	10907	14052	1863	1948	2054	418	431	431	.	.	.
•••	28553	38660	57905	10078	12084	14050	1905	2003	2033	418	431	409	.	.	.
0.9 •••	19811	24337	40147	6941	7855	9568	1225	1260	1310	.	.	.	.	.	.
•••	21660	28842	40153	7345	8548	9554	1238	1281	1285	.	.	.	.	.	.
0.95 •••	11991	14294	20087	3711	4030	4498	.	.	.	.	.	.	.	.	.
•••	12968	16291	20065	3860	4235	4474	.	.	.	.	.	.	.	.	.
0.98 •••	5284	5900	6912	.	.	.	.	.	.	.	.	.	.	.	.
•••	5560	6325	6891	.	.	.	.	.	.	.	.	.	.	.	.

TABLE 1: ALPHA= 0.005 POWER= 0.8   EXPECTED ACCRUAL THRU MINIMUM FOLLOW-UP= 9000

PCONT		DEL=.01 1.0 / .75	DEL=.01 .50 / .25	DEL=.01 .00 / BIN	DEL=.02 1.0 / .75	DEL=.02 .50 / .25	DEL=.02 .00 / BIN	DEL=.05 1.0 / .75	DEL=.05 .50 / .25	DEL=.05 .00 / BIN	DEL=.10 1.0 / 75	DEL=.10 .50 / .25	DEL=.10 .00 / BIN	DEL=.15 1.0 / .75	DEL=.15 .50 / .25	DEL=.15 .00 / BIN
					REQUIRED NUMBER OF PATIENTS											
0.01	***	1792	1792	1806	659	659	659	217	217	217	105	105	105	74	74	74
	•••	1792	1792	6891	659	659	2278	217	217	620	105	105	252	74	74	150
0.02	***	3966	3989	4020	1275	1275	1289	344	344	344	150	150	150	96	96	96
	•••	3975	3997	11376	1275	1275	3387	344	344	792	150	150	293	96	96	167
0.05	***	12030	12156	12547	3435	3471	3516	735	749	749	262	262	262	150	150	150
	•••	12075	12291	24269	3449	3494	6576	735	749	1285	262	262	409	150	150	216
0.1	***	26579	26970	29220	7251	7417	7732	1410	1424	1432	442	442	442	231	231	231
	•••	26714	27532	43890	7327	7544	11423	1410	1432	2033	442	442	584	231	231	289
0.15	***	39764	40484	46162	10739	11107	11954	2031	2062	2107	600	600	614	307	307	307
	•••	40011	41662	61175	10896	11422	15685	2054	2085	2687	600	600	736	307	307	351
0.2	***	50482	51621	62025	13650	14271	15877	2580	2639	2715	735	749	757	366	366	366
	•••	50865	53534	76124	13897	14842	19364	2616	2684	3247	749	749	865	366	366	403
0.25	***	58416	60059	76146	15914	16822	19365	3044	3134	3255	861	870	884	411	420	420
	•••	58965	62835	88737	16282	17700	22459	3089	3187	3714	861	870	970	411	420	444
0.3	***	63600	65850	88170	17542	18757	22312	3404	3525	3705	951	960	982	456	456	456
	•••	64365	69599	99015	18037	19950	24970	3457	3606	4088	960	974	1052	456	456	475
0.35	***	66246	69216	97904	18600	20130	24697	3674	3831	4056	1019	1041	1050	479	487	487
	•••	67259	74031	106956	19207	21629	26807	3741	3930	4368	1027	1050	1110	479	487	496
0.4	***	66637	70440	105225	19131	20962	26466	3854	4042	4304	1064	1086	1109	501	501	501
	•••	67956	76371	112562	19874	22740	28240	3930	4155	4555	1072	1095	1145	501	501	507
0.45	***	65166	69891	110054	19230	21322	27600	3944	4155	4461	1086	1109	1131	501	510	510
	•••	66831	76911	115833	20085	23347	28999	4042	4290	4649	1095	1117	1157	501	510	507
0.5	***	62219	67911	112349	18951	21255	28117	3952	4177	4515	1086	1109	1131	501	501	501
	•••	64266	75930	116767	19896	23460	29175	4056	4335	4649	1095	1117	1145	501	501	496
0.55	***	58191	64829	112110	18344	20819	27996	3885	4124	4461	1064	1086	1109	479	487	487
	•••	60630	73657	115365	19365	23136	28766	3997	4281	4555	1072	1095	1110	479	487	475
0.6	***	53475	60900	109334	17475	20017	27240	3741	3975	4312	1019	1027	1064	456	456	465
	•••	56265	70282	111628	18532	22371	27773	3854	4132	4368	1027	1041	1052	456	456	444
0.65	***	48390	56346	104010	16372	18892	25845	3525	3750	4056	937	960	982	411	420	420
	•••	51441	65940	105555	17421	21201	26196	3629	3885	4088	951	974	970	411	420	403
0.7	***	43125	51270	96180	15045	17452	23820	3232	3435	3705	847	861	884	366	366	375
	•••	46297	60720	97146	16057	19612	24036	3322	3561	3714	861	870	865	366	366	351
0.75	***	37784	45704	85852	13492	15689	21179	2864	3030	3255	726	749	757	307	307	307
	•••	40911	54600	86402	14415	17601	21291	2940	3134	3247	735	749	736	307	307	289
0.8	***	32325	39592	73050	11684	13560	17916	2414	2535	2706	591	600	600	231	240	240
	•••	35219	47504	73321	12494	15149	17963	2467	2616	2687	591	600	584	240	240	216
0.85	***	26579	32744	57809	9577	11017	14046	1874	1950	2054	420	420	434	.	.	.
	•••	29062	39187	57905	10199	12179	14050	1905	1995	2033	420	434	409	.	.	.
0.9	***	20175	24720	40146	7026	7926	9569	1230	1266	1311	.	.	.	.	.	.
	•••	22034	29189	40153	7417	8601	9554	1244	1289	1285	.	.	.	.	.	.
0.95	***	12187	14474	20085	3741	4056	4492	.	.	.	.	.	.	.	.	.
	•••	13155	16440	20065	3885	4245	4474	.	.	.	.	.	.	.	.	.
0.98	***	5347	5946	6914	.	.	.	.	.	.	.	.	.	.	.	.
	•••	5617	6360	6891	.	.	.	.	.	.	.	.	.	.	.	.

TABLE 1: ALPHA= 0.005 POWER= 0.8   EXPECTED ACCRUAL THRU MINIMUM FOLLOW-UP= 9500

	DEL=.01			DEL=.02			DEL=.05			DEL=.10			DEL=.15		
FACT=	1.0 .75	.50 .25	.00 BIN	1.0 .75	.50 .25	.00 BIN	1.0 .75	.50 .25	.00 BIN	1.0 75	.50 .25	.00 BIN	1.0 .75	.50 .25	.00 BIN
PCONT=***					REQUIRED NUMBER OF PATIENTS										
0.01 ***	1788	1797	1797	657	657	657	220	220	220	110	110	110	78	78	78
***	1797	1797	6891	657	657	2278	220	220	620	110	110	252	78	78	150
0.02 ***	3973	3982	4020	1274	1274	1289	348	348	348	149	149	149	101	101	101
***	3982	4005	11376	1274	1274	3387	348	348	792	149	149	293	101	101	167
0.05 ***	12048	12166	12555	3435	3474	3521	743	743	743	268	268	268	149	149	149
***	12095	12309	24269	3450	3498	6576	743	743	1285	268	268	409	149	149	216
0.1 ***	26606	27010	29219	7274	7425	7725	1408	1417	1440	443	443	443	229	229	229
***	26749	27580	43890	7330	7544	11423	1417	1431	2033	443	443	584	229	229	289
0.15 ***	39797	40557	46162	10765	11130	11953	2034	2073	2105	600	600	609	300	300	300
***	40058	41768	61175	10917	11454	15685	2049	2082	2687	600	609	736	300	300	351
0.2 ***	50546	51734	62018	13695	14313	15880	2580	2643	2723	743	743	752	363	363	363
***	50950	53705	76124	13948	14883	19364	2619	2675	3247	743	752	865	363	363	403
0.25 ***	58503	60236	76149	15975	16893	19363	3046	3141	3260	861	870	870	410	419	419
***	59096	63086	88737	16346	17757	22459	3094	3189	3714	861	870	970	410	419	444
0.3 ***	63728	66088	88175	17629	18864	22317	3412	3530	3697	956	965	980	458	458	458
***	64535	69950	99015	18128	20037	24970	3474	3616	4088	956	965	1052	458	458	475
0.35 ***	66411	69523	97904	18707	20250	24692	3688	3839	4053	1028	1037	1051	481	481	490
***	67480	74472	106956	19339	21723	26897	3759	3934	4368	1028	1051	1110	481	481	496
0.4 ***	66863	70838	105219	19268	21105	26464	3863	4053	4305	1075	1084	1108	490	505	505
***	68249	76909	112562	20013	22863	28240	3949	4172	4555	1075	1099	1145	505	505	507
0.45 ***	65447	70363	110049	19386	21485	27604	3958	4172	4457	1084	1108	1132	505	505	514
***	67195	77535	115833	20241	23495	28999	4053	4305	4649	1099	1123	1157	505	505	507
0.5 ***	62564	68463	112353	19125	21438	28112	3973	4195	4519	1084	1108	1132	490	505	505
***	64701	76624	116767	20075	23623	29175	4077	4338	4649	1099	1123	1145	505	505	496
0.55 ***	58621	65447	112106	18540	21001	27993	3901	4139	4457	1060	1084	1108	481	481	490
***	61139	74400	115365	19553	23305	28766	4020	4281	4555	1075	1099	1110	481	490	475
0.6 ***	53966	61575	109328	17676	20218	27233	3759	3996	4314	1013	1037	1060	458	458	458
***	56840	71066	111628	18730	22545	27773	3863	4139	4368	1028	1051	1052	458	458	444
0.65 ***	48931	57039	104017	16569	19087	25846	3545	3759	4053	942	965	980	410	419	419
***	52052	66729	105555	17614	21358	26196	3649	3901	4088	956	965	970	419	419	403
0.7 ***	43706	51971	96179	15230	17638	23828	3245	3450	3711	847	861	885	363	372	372
***	46922	61480	97146	16251	19766	24036	3340	3569	3714	861	870	865	363	372	351
0.75 ***	38348	46375	85863	13663	15848	21177	2880	3046	3245	728	743	752	300	300	315
***	41521	55305	86402	14589	17733	21291	2951	3141	3247	743	752	736	300	315	289
0.8 ***	32853	40200	73061	11843	13695	17914	2429	2548	2699	585	600	609	229,	229	244
***	35783	48109	73321	12641	15254	17963	2485	2619	2687	585	600	584	229	244	216
0.85 ***	27034	33242	57814	9696	11121	14043	1883	1954	2049	419	419	434	.	.	.
***	29537	39678	57905	10314	12247	14050	1915	2001	2033	419	434	409	.	.	.
0.9 ***	20512	25072	40144	7093	7986	9563	1227	1265	1313	.	.	.	.	.	.
***	22388	29513	40153	7497	8637	9554	1250	1289	1285	.	.	.	.	.	.
0.95 ***	12365	14636	20084	3768	4068	4495	.	.	.	.	.	.	.	.	.
***	13339	16569	20065	3910	4258	4474	.	.	.	.	.	.	.	.	.
0.98 ***	5398	5977	6918	.	.	.	.	.	.	.	.	.	.	.	.
***	5659	6380	6891	.	.	.	.	.	.	.	.	.	.	.	.

TABLE 1: ALPHA= 0.005 POWER= 0.8   EXPECTED ACCRUAL THRU MINIMUM FOLLOW-UP= 10000

		DEL=.01			DEL=.02			DEL=.05			DEL=.10			DEL=.15		
FACT=		1.0 .75	.50 .25	.00 BIN	1.0 .75	.50 .25	.00 BIN	1.0 .75	.50 .25	.00 BIN	1.0 75	.50 .25	.00 BIN	1.0 .75	.50 .25	.00 BIN
PCONT=***						REQUIRED NUMBER OF PATIENTS										
0.01	***	1791 1791	1791 1791	1807 6891	657 657	657 657	657 2278	216 216	216 216	216 620	107 107	107 107	107 252	82 82	82 82	82 150
0.02	***	3966 3982	3991 4007	4016 11376	1282 1282	1282 1282	1282 3387	341 341	341 341	341 792	157 157	157 157	157 293	91 91	91 91	91 167
0.05	***	12057 12107	12182 12316	12557 24269	3441 3457	3466 3491	3516 6576	741 741	741 741	741 1285	266 266	266 266	266 409	157 157	157 157	157 216
0.1	***	26616 26766	27041 27632	29216 43890	7282 7341	7441 7557	7732 11423	1407 1416	1416 1432	1441 2033	441 441	441 441	441 584	232 232	232 232	232 289
0.15	***	39841 40116	40641 41866	46166 61175	10791 10941	11157 11466	11957 15685	2041 2057	2066 2091	2107 2687	607 607	607 607	607 736	307 307	307 307	307 351
0.2	***	50607 51032	51857 53866	62016 76124	13741 13991	14366 14932	15882 19364	2591 2616	2657 2682	2716 3247	741 741	741 757	757 865	366 366	366 366	366 403
0.25	***	58607 59216	60407 63341	76141 88737	16041 16416	16966 17816	19366 22459	3057 3091	3141 3191	3257 3714	857 866	866 866	882 970	416 416	416 416	416 444
0.3	***	63857 64707	66332 70282	88166 99015	17716 18216	18957 20116	22316 24970	3416 3482	3541 3616	3707 4088	957 957	966 966	982 1052	457 457	457 457	457 475
0.35	***	66582 67707	69832 74891	97907 106956	18816 19441	20366 21832	24691 26897	3691 3766	3857 3941	4057 4368	1032 1032	1041 1041	1057 1110	482 482	482 482	491 496
0.4	***	67082 68532	71216 77416	105216 112562	19407 20157	21241 22991	26457 28240	3882 3966	4066 4166	4307 4555	1066 1082	1091 1091	1107 1145	491 491	507 507	507 507
0.45	***	65732 67541	70832 78132	110041 115833	19532 20407	21641 23632	27607 28999	3982 4066	4182 4307	4457 4649	1091 1107	1107 1116	1132 1157	507 507	507 507	507 507
0.5	***	62916 65132	69007 77282	112357 116767	19291 20241	21607 23766	28116 29175	3991 4091	4207 4341	4516 4649	1091 1107	1107 1116	1132 1145	491 491	507 507	507 496
0.55	***	59032 61641	66057 75116	112107 115365	18716 19741	21182 23457	27991 28766	3932 4032	4157 4291	4466 4555	1066 1082	1082 1091	1107 1110	482 482	482 482	491 475
0.6	***	54457 57391	62232 71807	109332 111628	17857 18916	20391 22691	27232 27773	3782 3891	4007 4141	4307 4368	1016 1032	1032 1041	1057 1052	457 457	457 457	457 444
0.65	***	49466 52641	57716 67482	104016 105555	16757 17807	19266 21507	25841 26196	3566 3666	3782 3907	4057 4368	941 957	966 966	982 970	416 416	416 416	416 403
0.7	***	44257 47532	52641 62207	96182 97146	15416 16416	17807 19907	23832 24036	3266 3357	3457 3566	3707 3714	857 857	866 866	882 865	366 366	366 366	366 351
0.75	***	38907 42116	47016 55982	85857 86402	13832 14757	16007 17857	21182 21291	2891 2966	3057 3141	3257 3247	732 741	741 741	757 736	307 307	307 307	307 289
0.8	***	33357 36332	40782 48707	73057 73321	11982 12782	13832 15341	17916 17963	2432 2491	2557 2616	2707 2687	591 591	591 607	607 584	232 232	232 241	241 216
0.85	***	27466 30007	33716 40141	57807 57905	9807 10416	11216 12316	14041 14050	1891 1916	1966 2007	2107 2033	416 432	432 432	432 409	. .	. .	. .
0.9	***	20841 22732	25416 29832	40141 40153	7166 7566	8041 8682	9566 9554	1232 1257	1266 1291	1307 1285	. .	. .	. .	. .	. .	. .
0.95	***	12541 13507	14791 16707	20082 20065	3791 3932	4091 4266	4491 4474	. .	. .	. .	. .	. .	. .	. .	. .	. .
0.98	***	5441 5707	6016 6407	6916 6891	. .	. .	. .	. .	. .	. .	. .	. .	. .	. .	. .	. .

TABLE 1: ALPHA= 0.005 POWER= 0.8 EXPECTED ACCRUAL THRU MINIMUM FOLLOW-UP= 11000

		DEL=.01			DEL=.02			DEL=.05			DEL=.10			DEL=.15		
FACT=		1.0 .75	.50 .25	.00 BIN	1.0 .75	.50 .25	.00 BIN	1.0 .75	.50 .25	.00 BIN	1.0 75	.50 .25	.00 BIN	1.0 .75	.50 .25	.00 BIN
PCONT=***							REQUIRED NUMBER OF PATIENTS								
0.01	***	1795	1795	1795	650	650	650	210	210	210	100	100	100	73	73	73
	•••	1795	1795	6891	650	650	2278	210	210	620	100	100	252	73	73	150
0.02	***	3978	3995	4022	1272	1283	1283	348	348	348	145	145	145	100	100	100
	•••	3978	4005	11376	1283	1283	3387	348	348	792	145	145	293	100	100	167
0.05	***	12063	12200	12558	3445	3472	3510	733	750	750	265	265	265	155	155	155
	•••	12118	12327	24269	3455	3500	6576	733	750	1285	265	265	409	155	155	216
0.1	***	26665	27122	29223	7305	7460	7735	1410	1420	1437	447	447	447	238	238	238
	•••	26830	27710	43890	7377	7570	11423	1420	1437	2033	447	447	584	238	238	289
0.15	***	39920	40790	46163	10842	11210	11953	2042	2070	2108	595	612	612	310	310	310
	•••	40223	42055	61175	11007	11502	15685	2053	2097	2687	595	612	736	310	310	351
0.2	***	50728	52092	62020	13823	14445	15885	2603	2658	2713	750	750	750	365	365	365
	•••	51195	54193	76124	14087	14995	19364	2630	2685	3247	750	750	865	365	365	403
0.25	***	58785	60738	76138	16160	17085	19367	3070	3153	3252	860	870	870	420	420	420
	•••	59462	63807	88737	16545	17920	22459	3115	3197	3714	860	870	970	420	420	444
0.3	***	64110	66788	88172	17882	19120	22320	3445	3555	3703	953	970	980	447	458	458
	•••	65028	70913	99015	18388	20258	24970	3500	3620	4088	953	970	1052	458	458	475
0.35	***	66925	70418	97907	19020	20577	24685	3720	3868	4050	1025	1035	1052	485	485	485
	•••	68135	75698	106956	19670	22007	26897	3785	3950	4368	1035	1052	1110	485	485	496
0.4	***	67520	71958	105222	19642	21495	26462	3912	4077	4308	1080	1090	1107	502	502	502
	•••	69087	78382	112562	20412	23200	28240	3995	4187	4555	1080	1090	1145	502	502	507
0.45	***	66282	71710	110045	19818	21935	27600	4005	4198	4462	1090	1118	1135	502	502	513
	•••	68245	79245	115833	20698	23860	28999	4105	4325	4649	1107	1118	1157	502	513	507
0.5	***	63598	70033	112345	19598	21925	28112	4022	4242	4517	1090	1107	1135	502	502	502
	•••	65963	78530	116767	20560	24042	29175	4132	4363	4649	1107	1118	1145	502	502	496
0.55	***	59847	67200	112108	19048	21512	28002	3967	4170	4462	1063	1090	1107	485	485	485
	•••	62597	76457	115365	20082	23723	28766	4060	4308	4555	1080	1090	1110	485	485	475
0.6	***	55375	63460	109330	18212	20725	27232	3813	4033	4308	1025	1035	1052	458	458	458
	•••	58472	73195	111628	19268	22970	27773	3923	4160	4368	1025	1052	1052	458	458	444
0.65	***	50480	58995	104012	17112	19598	25840	3593	3802	4060	953	970	980	420	420	420
	•••	53780	68878	105555	18157	21787	26196	3692	3912	4088	953	970	970	420	420	403
0.7	***	45310	53907	96185	15748	18113	23822	3307	3472	3703	860	870	887	365	365	375
	•••	48693	63560	97146	16755	20148	24036	3390	3582	3714	860	870	865	365	365	351
0.75	***	39937	48215	85862	14142	16287	21182	2922	3070	3252	733	750	750	310	310	310
	•••	43237	57218	86402	15060	18075	21291	2988	3153	3247	733	750	736	310	310	289
0.8	***	34327	41845	73058	12245	14060	17920	2455	2565	2702	595	595	612	238	238	238
	•••	37352	49782	73321	13025	15517	17963	2510	2630	2687	595	595	584	238	238	216
0.85	***	28288	34613	57812	10017	11392	14043	1905	1970	2053	420	430	430	.	.	.
	•••	30862	40993	57905	10622	12437	14050	1932	2015	2033	420	430	409	.	.	.
0.9	***	21468	26050	40140	7305	8147	9560	1245	1272	1310	.	.	.	.	.	.
	•••	23355	30395	40153	7680	8752	9554	1255	1283	1285	.	.	.	.	.	.
0.95	***	12860	15088	20082	3840	4115	4490	.	.	.	.	.	.	.	.	.
	•••	13812	16920	20065	3967	4297	4474	.	.	.	.	.	.	.	.	.
0.98	***	5535	6085	6910	.	.	.	.	.	.	.	.	.	.	.	.
	•••	5782	6442	6891	.	.	.	.	.	.	.	.	.	.	.	.

TABLE 1: ALPHA= 0.005 POWER= 0.8   EXPECTED ACCRUAL THRU MINIMUM FOLLOW-UP= 12000

	DEL=.01			DEL=.02			DEL=.05			DEL=.10			DEL=.15		
FACT=	1.0 .75	.50 .25	.00 BIN	1.0 .75	.50 .25	.00 BIN	1.0 .75	.50 .25	.00 BIN	1.0 75	.50 .25	.00 BIN	1.0 .75	.50 .25	.00 BIN

PCONT=•••   REQUIRED NUMBER OF PATIENTS

PCONT	DEL=.01 1.0/.75	.50/.25	.00/BIN	DEL=.02 1.0/.75	.50/.25	.00/BIN	DEL=.05 1.0/.75	.50/.25	.00/BIN	DEL=.10 1.0/.75	.50/.25	.00/BIN	DEL=.15 1.0/.75	.50/.25	.00/BIN
0.01 •••	1789	1789	1808	649	649	649	218	218	218	109	109	109	79	79	79
•••	1789	1789	6891	649	649	2278	218	218	620	109	109	252	79	79	150
0.02 •••	3979	3998	4009	1279	1279	1279	349	349	349	158	158	158	98	98	98
•••	3979	4009	11376	1279	1279	3387	349	349	792	158	158	293	98	98	167
0.05 •••	12079	12218	12548	3439	3469	3518	739	739	739	259	259	259	158	158	158
•••	12139	12338	24269	3458	3499	6576	739	739	1285	259	259	409	158	158	216
0.1 •••	26708	27188	29228	7328	7478	7729	1418	1429	1429	439	439	439	229	229	229
•••	26888	27788	43890	7388	7579	11423	1418	1429	2033	439	439	584	229	229	289
0.15 •••	40009	40928	46159	10898	11258	11948	2048	2078	2108	608	608	608	308	308	308
•••	40328	42229	61175	11048	11528	15685	2059	2089	2687	608	608	736	308	308	351
0.2 •••	50858	52309	62018	13898	14528	15878	2618	2659	2719	739	739	758	368	368	368
•••	51368	54488	76124	14168	15049	19364	2629	2689	3247	739	758	865	368	368	403
0.25 •••	58969	61069	76148	16279	17198	19369	3079	3158	3259	859	878	878	409	409	409
•••	59708	64249	88737	16658	18008	22459	3109	3199	3714	859	878	970	409	409	444
0.3 •••	64369	67238	88178	18038	19279	22309	3458	3559	3698	949	968	979	458	458	458
•••	65359	71509	99015	18548	20378	24970	3518	3638	4088	968	968	1052	458	458	475
0.35 •••	67249	70999	97909	19208	20768	24698	3739	3878	4058	1028	1039	1058	488	488	488
•••	68569	76448	106956	19849	22159	26897	3799	3968	4368	1039	1039	1110	488	488	496
0.4 •••	67958	72668	105218	19868	21728	26468	3938	4099	4309	1069	1088	1099	499	499	499
•••	69638	79279	112562	20648	23389	28240	4009	4189	4555	1088	1099	1145	499	499	507
0.45 •••	66829	72559	110048	20078	22189	27608	4039	4219	4459	1099	1118	1129	499	499	518
•••	68918	80288	115833	20959	24079	28999	4129	4328	4649	1099	1118	1157	499	499	507
0.5 •••	64268	70999	112358	19898	22208	28118	4058	4249	4508	1099	1118	1129	499	499	499
•••	66758	79688	116767	20858	24259	29175	4148	4369	4649	1099	1118	1145	499	499	496
0.55 •••	60638	68288	112118	19358	21818	27998	3998	4189	4459	1069	1088	1099	488	488	488
•••	63518	77689	115365	20389	23978	28766	4088	4328	4555	1088	1099	1110	488	488	475
0.6 •••	56269	64609	109328	18529	21038	27229	3848	4058	4309	1028	1039	1058	458	458	458
•••	59479	74479	111628	19579	23228	27773	3949	4178	4368	1028	1039	1052	458	458	444
0.65 •••	51439	60188	104018	17419	19898	25849	3619	3818	4058	949	968	979	409	409	428
•••	54848	70159	105555	18469	22028	26196	3709	3919	4088	949	968	970	409	428	403
0.7 •••	46298	55088	96188	16058	18398	23828	3319	3488	3709	859	859	878	368	368	368
•••	49759	64789	97146	17059	20378	24036	3409	3589	3714	859	878	865	368	368	351
0.75 •••	40909	49328	85849	14419	16538	21188	2948	3079	3248	739	739	758	308	308	308
•••	44269	58358	86402	15319	18259	21291	3008	3158	3247	739	758	736	308	308	289
0.8 •••	35228	42848	73058	12488	14269	17918	2468	2569	2708	589	589	608	229	229	229
•••	38299	50768	73321	13268	15668	17963	2528	2629	2687	589	608	584	229	229	216
0.85 •••	29059	35419	57799	10189	11539	14048	1909	1969	2048	428	428	428	.	.	.
•••	31658	41768	57905	10789	12548	14050	1939	2018	2033	428	428	409	.	.	.
0.9 •••	22028	26618	40148	7418	8239	9559	1249	1279	1309	.	.	.	.	.	.
•••	23929	30889	40153	7789	8798	9554	1268	1298	1285	.	.	.	.	.	.
0.95 •••	13159	15338	20078	3878	4148	4489	.	.	.	.	.	.	.	.	.
•••	14089	17119	20065	4009	4309	4474	.	.	.	.	.	.	.	.	.
0.98 •••	5618	6139	6908	.	.	.	.	.	.	.	.	.	.	.	.
•••	5858	6488	6891	.	.	.	.	.	.	.	.	.	.	.	.

TABLE 1: ALPHA= 0.005 POWER= 0.8   EXPECTED ACCRUAL THRU MINIMUM FOLLOW-UP= 13000

	DEL=.01			DEL=.02			DEL=.05			DEL=.10			DEL=.15		
FACT=	1.0 .75	.50 .25	.00 BIN	1.0 .75	.50 .25	.00 BIN	1.0 .75	.50 .25	.00 BIN	1.0 75	.50 .25	.00 BIN	1.0 .75	.50 .25	.00 BIN

PCONT=***   REQUIRED NUMBER OF PATIENTS

PCONT	D.01 (1.0/.75)	D.01 (.50/.25)	D.01 (.00/BIN)	D.02 (1.0/.75)	D.02 (.50/.25)	D.02 (.00/BIN)	D.05 (1.0/.75)	D.05 (.50/.25)	D.05 (.00/BIN)	D.10 (1.0/.75)	D.10 (.50/.25)	D.10 (.00/BIN)	D.15 (1.0/.75)	D.15 (.50/.25)	D.15 (.00/BIN)
0.01 ***	1796	1796	1796	659	659	659	216	216	216	106	106	106	74	74	74
***	1796	1796	6891	659	659	2278	216	216	620	106	106	252	74	74	150
0.02 ***	3974	3986	4018	1276	1276	1288	346	346	346	151	151	151	106	106	106
***	3986	4006	11376	1276	1276	3387	346	346	792	151	151	293	106	106	167
0.05 ***	12099	12229	12554	3454	3486	3519	736	736	736	269	269	269	151	151	151
***	12143	12359	24269	3466	3498	6576	736	736	1285	269	269	409	151	151	216
0.1 ***	26756	27256	29226	7333	7484	7723	1418	1418	1439	443	443	443	236	236	236
***	26931	27861	43890	7398	7593	11423	1418	1439	2033	443	443	584	236	236	289
0.15 ***	40093	41056	46159	10929	11286	11948	2056	2068	2101	606	606	606	301	301	301
***	40439	42389	61175	11091	11558	15685	2068	2089	2687	606	606	736	301	301	351
0.2 ***	51001	52541	62019	13963	14581	15881	2621	2674	2718	736	756	756	366	366	366
***	51533	54771	76124	14244	15101	19364	2641	2686	3247	736	756	865	366	366	403
0.25 ***	59159	61401	76136	16389	17299	19358	3096	3161	3259	866	866	886	411	411	411
***	59939	64663	88737	16779	18091	22459	3129	3206	3714	866	866	970	411	411	444
0.3 ***	64619	67653	88181	18176	19411	22316	3466	3584	3693	963	963	984	443	464	464
***	65691	72061	99015	18696	20484	24970	3519	3628	4088	963	963	1052	464	464	475
0.35 ***	67588	71553	97911	19391	20939	24688	3758	3888	4051	1028	1049	1049	476	476	496
***	69006	77143	106956	20041	22304	26897	3823	3974	4368	1028	1049	1110	476	476	496
0.4 ***	68389	73361	105223	20094	21926	26464	3953	4116	4311	1081	1093	1114	496	496	508
***	70176	80121	112562	20853	23551	28240	4039	4201	4555	1081	1093	1145	496	496	507
0.45 ***	67361	73373	110054	20321	22434	27601	4071	4234	4461	1093	1114	1126	508	508	508
***	69571	81238	115833	21211	24266	28999	4148	4343	4649	1114	1126	1157	508	508	507
0.5 ***	64911	71931	112341	20159	22466	28121	4083	4266	4506	1093	1114	1126	496	496	508
***	67544	80751	116767	21134	24481	29175	4169	4376	4649	1114	1126	1145	496	508	496
0.55 ***	61381	69299	112113	19651	22088	27991	4018	4213	4461	1081	1093	1114	476	476	496
***	64403	78833	115365	20679	24189	28766	4116	4331	4555	1081	1093	1110	476	496	475
0.6 ***	57123	65703	109339	18826	21308	27244	3876	4071	4311	1028	1049	1061	443	464	464
***	60438	75648	111628	19878	23441	27773	3974	4181	4368	1028	1049	1052	464	464	444
0.65 ***	52346	61283	104021	17721	20171	25846	3649	3823	4051	951	963	984	411	411	411
***	55856	71326	105555	18761	22239	26196	3746	3941	4088	963	963	970	411	411	403
0.7 ***	47231	56169	96188	16336	18643	23831	3356	3498	3714	854	866	886	366	366	366
***	50786	65919	97146	17331	20561	24036	3421	3596	3714	866	866	865	366	366	351
0.75 ***	41816	50351	85853	14678	16758	21178	2966	3096	3259	736	736	756	301	313	313
***	45249	59386	86402	15576	18436	21291	3031	3161	3247	736	756	736	301	313	289
0.8 ***	36063	43754	73048	12716	14451	17916	2491	2588	2706	594	594	606	236	236	236
***	39171	51651	73321	13464	15804	17963	2544	2641	2687	594	606	584	236	236	216
0.85 ***	29779	36161	57806	10376	11676	14049	1926	1971	2056	431	431	431	.	.	.
***	32391	42454	57905	10941	12631	14050	1959	2024	2033	431	431	409	.	.	.
0.9 ***	22564	27126	40146	7528	8308	9564	1244	1276	1309	.	.	.	.	.	.
***	24461	31339	40153	7886	8849	9554	1256	1288	1285	.	.	.	.	.	.
0.95 ***	13431	15576	20073	3921	4169	4494	.	.	.	.	.	.	.	.	.
***	14353	17299	20065	4039	4311	4474	.	.	.	.	.	.	.	.	.
0.98 ***	5676	6184	6911	.	.	.	.	.	.	.	.	.	.	.	.
***	5903	6509	6891	.	.	.	.	.	.	.	.	.	.	.	.

TABLE 1: ALPHA= 0.005 POWER= 0.8   EXPECTED ACCRUAL THRU MINIMUM FOLLOW-UP= 14000

PCONT=	FACT=	DEL=.01 1.0 .75	.50 .25	.00 BIN	DEL=.02 1.0 .75	.50 .25	.00 BIN	DEL=.05 1.0 .75	.50 .25	.00 BIN	DEL=.10 1.0 75	.50 .25	.00 BIN	DEL=.15 1.0 .75	.50 .25	.00 BIN
						REQUIRED NUMBER OF PATIENTS									
0.01	***	1794	1794	1794	652	652	652	219	219	219	114	114	114	79	79	79
	***	1794	1794	6891	652	652	2278	219	219	620	114	114	252	79	79	150
0.02	***	3977	3999	4012	1282	1282	1282	337	337	337	149	149	149	92	92	92
	***	3977	4012	11376	1282	1282	3387	337	337	792	149	149	293	92	92	167
0.05	***	12119	12237	12552	3452	3487	3522	744	744	744	267	267	267	149	149	149
	***	12167	12364	24269	3474	3487	6576	744	744	1285	267	267	409	149	149	216
0.1	***	26797	27322	29212	7359	7499	7722	1422	1422	1444	442	442	442	232	232	232
	***	26994	27917	43890	7429	7604	11423	1422	1422	2033	442	442	584	232	232	289
0.15	***	40167	41182	46174	10977	11314	11957	2052	2074	2109	604	604	604	302	302	302
	***	40539	42534	61175	11117	11572	15685	2074	2087	2687	604	604	736	302	302	351
0.2	***	51122	52754	62029	14044	14652	15877	2612	2669	2717	744	744	757	359	359	372
	***	51704	55029	76124	14302	15142	19364	2647	2682	3247	744	744	865	359	359	403
0.25	***	59334	61692	76147	16472	17404	19364	3102	3172	3264	862	862	884	407	407	407
	***	60174	65039	88737	16879	18152	22459	3137	3207	3714	862	884	970	407	407	444
0.3	***	64864	68084	88174	18314	19539	22317	3487	3579	3697	954	967	967	442	464	464
	***	66019	72577	99015	18817	20589	24970	3522	3649	4088	967	967	1052	464	464	475
0.35	***	67922	72087	97904	19552	21114	24697	3789	3894	4047	1037	1037	1059	477	477	477
	***	69414	77792	106956	20204	22422	26897	3837	3977	4368	1037	1037	1110	477	477	496
0.4	***	68819	74012	105219	20287	22129	26469	3977	4117	4314	1072	1094	1107	499	499	499
	***	70709	80907	112562	21057	23704	28240	4047	4209	4555	1094	1094	1145	499	499	507
0.45	***	67887	74152	110049	20554	22632	27602	4082	4244	4454	1107	1107	1129	499	512	512
	***	70219	82132	115833	21429	24439	28999	4174	4349	4649	1107	1129	1157	499	512	507
0.5	***	65542	72809	112359	20414	22702	28114	4104	4279	4502	1107	1107	1129	499	499	499
	***	68272	81734	116767	21372	24662	29175	4187	4397	4649	1107	1129	1145	499	499	496
0.55	***	62112	70267	112114	19924	22339	27987	4047	4222	4467	1072	1094	1107	477	477	477
	***	65249	79892	115365	20939	24382	28766	4139	4327	4555	1072	1094	1110	477	477	475
0.6	***	57934	66719	109327	19097	21569	27239	3907	4082	4314	1024	1037	1059	464	464	464
	***	61364	76742	111628	20147	23634	27773	3977	4187	4368	1037	1059	1052	464	464	444
0.65	***	53222	62344	104007	17977	20414	25839	3684	3837	4047	954	967	989	407	407	429
	***	56814	72424	105555	19027	22422	26196	3754	3942	4088	954	967	970	407	429	403
0.7	***	48112	57199	96189	16599	18874	23822	3369	3522	3697	862	862	884	372	372	372
	***	51739	66964	97146	17579	20742	24036	3439	3614	3714	862	884	865	372	372	351
0.75	***	42674	51319	85864	14919	16962	21184	2984	3102	3242	744	744	757	302	302	302
	***	46152	60349	86402	15794	18572	21291	3032	3172	3247	744	744	736	302	302	289
0.8	***	36842	44612	73054	12902	14617	17907	2494	2599	2704	582	604	604	232	232	232
	***	39992	52474	73321	13659	15934	17963	2542	2647	2687	604	604	584	232	232	216
0.85	***	30437	36864	57807	10522	11782	14044	1934	1982	2052	429	429	429	.	.	.
	***	33084	43094	57905	11082	12727	14050	1947	2017	2033	429	429	409	.	.	.
0.9	***	23052	27602	40132	7617	8387	9564	1247	1282	1304	.	.	.	.	.	.
	***	24964	31754	40153	7967	8899	9554	1269	1282	1285	.	.	.	.	.	.
0.95	***	13659	15772	20077	3942	4187	4489	.	.	.	.	.	.	.	.	.
	***	14582	17439	20065	4069	4327	4474	.	.	.	.	.	.	.	.	.
0.98	***	5749	6217	6917	.	.	.	.	.	.	.	.	.	.	.	.
	***	5972	6532	6891	.	.	.	.	.	.	.	.	.	.	.	.

TABLE 1: ALPHA= 0.005 POWER= 0.8   EXPECTED ACCRUAL THRU MINIMUM FOLLOW-UP= 15000

	DEL=.01			DEL=.02			DEL=.05			DEL=.10			DEL=.15		
FACT=	1.0 .75	.50 .25	.00 BIN	1.0 .75	.50 .25	.00 BIN	1.0 .75	.50 .25	.00 BIN	1.0 75	.50 .25	.00 BIN	1.0 .75	.50 .25	.00 BIN

PCONT=***   REQUIRED NUMBER OF PATIENTS

PCONT	DEL=.01 1.0/.75	.50/.25	.00/BIN	DEL=.02 1.0/.75	.50/.25	.00/BIN	DEL=.05 1.0/.75	.50/.25	.00/BIN	DEL=.10 1.0/.75	.50/.25	.00/BIN	DEL=.15 1.0/.75	.50/.25	.00/BIN
0.01	1786	1786	1810	661	661	661	211	211	211	99	99	99	85	85	85
	1786	1786	6891	661	661	2278	211	211	620	99	99	252	85	85	150
0.02	3985	3999	4022	1285	1285	1285	347	347	347	160	160	160	99	99	99
	3985	3999	11376	1285	1285	3387	347	347	792	160	160	293	99	99	167
0.05	12122	12249	12549	3460	3474	3511	736	736	736	272	272	272	160	160	160
	12174	12385	24269	3474	3497	6576	736	736	1285	272	272	409	160	160	216
0.1	26836	27385	29222	7374	7510	7735	1411	1435	1435	436	436	436	235	235	235
	27047	27961	43890	7435	7599	11423	1411	1435	2033	436	436	584	235	235	289
0.15	40247	41311	46172	11011	11349	11949	2049	2086	2110	610	610	610	310	310	310
	40636	42685	61175	11161	11597	15685	2072	2086	2687	610	610	736	310	310	351
0.2	51249	52960	62011	14110	14710	15886	2635	2672	2724	736	760	760	361	361	361
	51849	55261	76124	14372	15174	19364	2649	2686	3247	736	760	865	361	361	403
0.25	59522	61997	76149	16585	17485	19360	3099	3174	3249	872	872	872	422	422	422
	60399	65410	88737	16960	18211	22459	3136	3211	3714	872	872	970	422	422	444
0.3	65124	68485	88172	18436	19660	22322	3497	3586	3699	961	961	985	460	460	460
	66324	73060	99015	18947	20672	24970	3549	3647	4088	961	985	1052	460	460	475
0.35	68260	72610	97899	19711	21249	24699	3797	3910	4060	1036	1036	1060	474	474	497
	69835	78399	106956	20372	22547	26897	3849	3985	4368	1036	1060	1110	474	474	496
0.4	69235	74635	105211	20461	22299	26461	3999	4135	4299	1074	1097	1111	497	497	497
	71222	81624	112562	21249	23836	28240	4060	4210	4555	1097	1097	1145	497	497	507
0.45	68410	74897	110049	20761	22847	27610	4111	4261	4449	1097	1111	1135	497	511	511
	70824	82974	115833	21647	24586	28999	4186	4360	4649	1111	1135	1157	511	511	507
0.5	66160	73636	112360	20649	22922	28111	4135	4299	4510	1097	1111	1135	497	497	497
	69010	82660	116767	21610	24835	29175	4210	4397	4649	1111	1135	1145	497	497	496
0.55	62836	71185	112111	20161	22561	27999	4074	4247	4472	1074	1097	1111	474	474	497
	66047	80874	115365	21174	24572	28766	4149	4336	4555	1074	1097	1110	474	497	475
0.6	58711	67674	109322	19360	21797	27235	3924	4097	4299	1022	1036	1060	460	460	460
	62222	77761	111628	20386	23822	27773	3999	4186	4368	1036	1060	1052	460	460	444
0.65	54061	63310	104011	18235	20635	25847	3699	3849	4060	961	961	985	422	422	422
	57722	73435	105555	19261	22585	26196	3774	3947	4088	961	985	970	422	422	403
0.7	48961	58149	96174	16824	19074	23822	3385	3535	3699	849	872	886	361	361	361
	52636	67922	97146	17799	20897	24036	3460	3610	3714	872	872	865	361	361	351
0.75	43486	52210	85861	15122	17147	21174	2986	3099	3249	736	736	760	310	310	310
	47011	61224	86402	15999	18722	21291	3047	3174	3247	736	760	736	310	310	289
0.8	37585	45399	73060	13097	14761	17911	2522	2597	2710	586	610	610	235	235	235
	40772	53222	73321	13824	16036	17963	2560	2649	2687	586	610	584	235	235	216
0.85	31060	37510	57811	10660	11897	14049	1936	1997	2049	422	422	436	.	.	.
	33722	43674	57905	11222	12797	14050	1960	2011	2033	422	422	409	.	.	.
0.9	23499	28060	40135	7711	8447	9572	1261	1285	1299	.	.	.	.	.	.
	25411	32110	40153	8049	8935	9554	1261	1299	1285	.	.	.	.	.	.
0.95	13885	15961	20086	3985	4210	4486	.	.	.	.	.	.	.	.	.
	14799	17574	20065	4097	4336	4474	.	.	.	.	.	.	.	.	.
0.98	5799	6272	6910	.	.	.	.	.	.	.	.	.	.	.	.
	6010	6549	6891	.	.	.	.	.	.	.	.	.	.	.	.

TABLE 1: ALPHA= 0.005 POWER= 0.8   EXPECTED ACCRUAL THRU MINIMUM FOLLOW-UP= 17000

		DEL=.01			DEL=.02			DEL=.05			DEL=.10			DEL=.15		
FACT=		1.0 .75	.50 .25	.00 BIN	1.0 .75	.50 .25	.00 BIN	1.0 .75	.50 .25	.00 BIN	1.0 75	.50 .25	.00 BIN	1.0 .75	.50 .25	.00 BIN
PCONT=•••					REQUIRED NUMBER OF PATIENTS											
0.01	•••	1796	1796	1796	665	665	665	224	224	224	112	112	112	70	70	70
	•••	1796	1796	6891	665	665	2278	224	224	620	112	112	252	70	70	150
0.02	•••	3980	4006	4022	1286	1286	1286	351	351	351	155	155	155	96	96	96
	•••	3980	4006	11376	1286	1286	3387	351	351	792	155	155	293	96	96	167
0.05	•••	12140	12294	12549	3470	3496	3512	734	750	750	266	266	266	155	155	155
	•••	12209	12395	24269	3470	3496	6576	734	750	1285	266	266	409	155	155	216
0.1	•••	26930	27482	29225	7406	7534	7720	1414	1430	1430	436	436	436	240	240	240
	•••	27142	28077	43890	7465	7619	11423	1430	1430	2033	436	436	584	240	240	289
0.15	•••	40402	41550	46166	11077	11401	11954	2067	2094	2110	606	606	606	309	309	309
	•••	40827	42910	61175	11231	11630	15685	2067	2094	2687	606	606	736	309	309	351
0.2	•••	51495	53349	62019	14222	14801	15880	2646	2662	2731	750	750	750	367	367	367
	•••	52175	55702	76124	14477	15242	19364	2662	2689	3247	750	750	865	367	367	403
0.25	•••	59867	62571	76145	16756	17622	19365	3130	3172	3257	861	877	877	410	410	410
	•••	60845	66056	88737	17139	18329	22459	3156	3215	3714	861	877	970	410	410	444
0.3	•••	65605	69244	88172	18669	19859	22324	3512	3597	3709	962	962	989	452	452	452
	•••	66922	73935	99015	19179	20836	24970	3555	3640	4088	962	962	1052	452	452	475
0.35	•••	68904	73579	97905	20002	21516	24704	3810	3921	4049	1031	1047	1047	479	479	479
	•••	70620	79529	106956	20640	22749	26897	3879	3980	4368	1047	1047	1110	479	479	496
0.4	•••	70051	75805	105215	20810	22595	26462	4022	4150	4304	1090	1090	1116	495	495	495
	•••	72192	82971	112562	21575	24082	28240	4091	4219	4555	1090	1090	1145	495	495	507
0.45	•••	69414	76272	110044	21150	23190	27610	4134	4277	4447	1116	1116	1132	495	495	521
	•••	72006	84501	115833	22026	24874	28999	4219	4362	4649	1116	1132	1157	495	495	507
0.5	•••	67347	75210	112355	21065	23301	28120	4176	4320	4516	1116	1116	1132	495	495	495
	•••	70365	84331	116767	22026	25129	29175	4235	4405	4649	1116	1132	1145	495	495	496
0.55	•••	64186	72872	112100	20597	22961	27992	4107	4261	4474	1074	1090	1116	479	479	495
	•••	67560	82647	115365	21617	24874	28766	4176	4362	4555	1090	1090	1110	479	479	475
0.6	•••	60207	69456	109321	19816	22196	27227	3964	4107	4304	1031	1047	1047	452	452	452
	•••	63846	79587	111628	20836	24109	27773	4049	4219	4368	1031	1047	1052	452	452	444
0.65	•••	55617	65121	104009	18685	21022	25851	3725	3879	4049	962	962	989	410	410	410
	•••	59400	75236	105555	19705	22876	26196	3810	3964	4088	962	962	970	410	410	403
0.7	•••	50544	59910	96189	17266	19450	23827	3411	3555	3709	861	877	877	367	367	367
	•••	54300	69642	97146	18217	21150	24036	3470	3624	3714	861	877	865	367	367	351
0.75	•••	44992	53859	85861	15524	17452	21176	3029	3130	3257	734	750	750	309	309	309
	•••	48589	62784	86402	16374	18940	21291	3071	3172	3247	750	750	736	309	309	289
0.8	•••	38957	46846	73042	13415	15030	17920	2535	2604	2705	606	606	606	240	240	240
	•••	42187	54555	73321	14137	16204	17963	2561	2646	2687	606	606	584	240	240	216
0.85	•••	32226	38660	57811	10907	12081	14052	1940	2009	2051	436	436	436	.	.	.
	•••	34877	44695	57905	11444	12931	14050	1966	2025	2033	436	436	409	.	.	.
0.9	•••	24337	28842	40147	7847	8554	9574	1260	1286	1302	.	.	.	.	.	.
	•••	26234	32752	40153	8171	8995	9554	1260	1286	1285	.	.	.	.	.	.
0.95	•••	14291	16289	20087	4022	4235	4490	.	.	.	.	.	.	.	.	.
	•••	15184	17819	20065	4134	4362	4474	.	.	.	.	.	.	.	.	.
0.98	•••	5892	6317	6912	.	.	.	.	.	.	.	.	.	.	.	.
	•••	6105	6599	6891	.	.	.	.	.	.	.	.	.	.	.	.

TABLE 1: ALPHA= 0.005 POWER= 0.8   EXPECTED ACCRUAL THRU MINIMUM FOLLOW-UP= 20000

FACT=	DEL=.01			DEL=.02			DEL=.05			DEL=.10			DEL=.15		
	1.0 .75	.50 .25	.00 BIN	1.0 .75	.50 .25	.00 BIN	1.0 .75	.50 .25	.00 BIN	1.0 75	.50 .25	.00 BIN	1.0 .75	.50 .25	.00 BIN
PCONT=•••			REQUIRED NUMBER OF PATIENTS												
0.01 •••	1782	1782	1813	663	663	663	213	213	213	113	113	113	82	82	82
•••	1782	1782	6891	663	663	2278	213	213	620	113	113	252	82	82	150
0.02 •••	3982	4013	4013	1282	1282	1282	332	332	332	163	163	163	82	82	82
•••	3982	4013	11376	1282	1282	3387	332	332	792	163	163	293	82	82	167
0.05 •••	12182	12313	12563	3463	3482	3513	732	732	732	263	263	263	163	163	163
•••	12232	12413	24269	3482	3513	6576	732	732	1285	263	263	409	163	163	216
0.1 •••	27032	27632	29213	7432	7563	7732	1413	1432	1432	432	432	432	232	232	232
•••	27282	28182	43890	7482	7632	11423	1432	1432	2033	432	432	584	232	232	289
0.15 •••	40632	41863	46163	11163	11463	11963	2063	2082	2113	613	613	613	313	313	313
•••	41113	43232	61175	11313	11682	15685	2082	2082	2687	613	613	736	313	313	351
0.2 •••	51863	53863	62013	14363	14932	15882	2663	2682	2713	732	763	763	363	363	363
•••	52613	56263	76124	14613	15332	19364	2663	2713	3247	732	763	865	363	363	403
0.25 •••	60413	63332	76132	16963	17813	19363	3132	3182	3263	863	863	882	413	413	413
•••	61482	66882	88737	17332	18463	22459	3163	3213	3714	863	882	970	413	413	444
0.3 •••	66332	70282	88163	18963	20113	22313	3532	3613	3713	963	963	982	463	463	463
•••	67813	75082	99015	19463	21013	24970	3582	3663	4088	963	982	1052	463	463	475
0.35 •••	69832	74882	97913	20363	21832	24682	3863	3932	4063	1032	1032	1063	482	482	482
•••	71732	80963	106956	21013	22963	26897	3882	3982	4368	1032	1063	1110	482	482	496
0.4 •••	71213	77413	105213	21232	22982	26463	4063	4163	4313	1082	1082	1113	513	513	513
•••	73582	84682	112562	22013	24363	28240	4113	4232	4555	1082	1113	1145	513	513	507
0.45 •••	70832	78132	110032	21632	23632	27613	4182	4313	4463	1113	1113	1132	513	513	513
•••	73632	86463	115833	22513	25182	28999	4232	4382	4649	1113	1132	1157	513	513	507
0.5 •••	69013	77282	112363	21613	23763	28113	4213	4332	4513	1113	1113	1132	513	513	513
•••	72232	86482	116767	22532	25463	29175	4282	4432	4649	1113	1132	1145	513	513	496
0.55 •••	66063	75113	112113	21182	23463	27982	4163	4282	4463	1082	1082	1113	482	482	482
•••	69632	84932	115365	22163	25232	28766	4213	4363	4555	1082	1113	1110	482	482	475
0.6 •••	62232	71813	109332	20382	22682	27232	4013	4132	4313	1032	1032	1063	463	463	463
•••	66032	81932	111628	21382	24482	27773	4063	4213	4368	1032	1063	1052	463	463	444
0.65 •••	57713	67482	104013	19263	21513	25832	3782	3913	4063	963	963	982	413	413	413
•••	61632	77563	105555	20263	23232	26196	3832	3982	4088	963	982	970	413	413	403
0.7 •••	52632	62213	96182	17813	19913	23832	3463	3563	3713	863	863	882	363	363	363
•••	56513	71832	97146	18732	21482	24036	3513	3632	3714	863	882	865	363	363	351
0.75 •••	47013	55982	85863	16013	17863	21182	3063	3132	3263	732	732	763	313	313	313
•••	50682	64782	86402	16832	19213	21291	3082	3182	3247	732	763	736	313	313	289
0.8 •••	40782	48713	73063	13832	15332	17913	2563	2613	2713	582	613	613	232	232	232
•••	44063	56232	73321	14513	16432	17963	2582	2663	2687	582	613	584	232	232	216
0.85 •••	33713	40132	57813	11213	12313	14032	1963	2013	2063	432	432	432	.	.	.
•••	36413	45963	57905	11713	13063	14050	1982	2032	2033	432	432	409	.	.	.
0.9 •••	25413	29832	40132	8032	8682	9563	1263	1282	1313	.	.	.	.	.	.
•••	27282	33532	40153	8332	9082	9554	1282	1282	1285	.	.	.	.	.	.
0.95 •••	14782	16713	20082	4082	4263	4482	.	.	.	.	.	.	.	.	.
•••	15632	18082	20065	4182	4382	4474	.	.	.	.	.	.	.	.	.
0.98 •••	6013	6413	6913	.	.	.	.	.	.	.	.	.	.	.	.
•••	6213	6632	6891	.	.	.	.	.	.	.	.	.	.	.	.

TABLE 1: ALPHA= 0.005 POWER= 0.8   EXPECTED ACCRUAL THRU MINIMUM FOLLOW-UP= 25000

	DEL=.01			DEL=.02			DEL=.05			DEL=.10			DEL=.15		
FACT=	1.0 .75	.50 .25	.00 BIN	1.0 .75	.50 .25	.00 BIN	1.0 .75	.50 .25	.00 BIN	1.0 75	.50 .25	.00 BIN	1.0 .75	.50 .25	.00 BIN
PCONT=***				REQUIRED NUMBER OF PATIENTS											
0.01 ***	1790 1790	1790	665 665	665	204 204	204	102 102	102	79 79	79
•••	1790 1790	6891	665 665	2278	204 204	620	102 102	252	79 79	150
0.02 ***	3977 4016	4016	1266 1290	1290	352 352	352	141 141	141	102 102	102
•••	4016 4016	11376	1290 1290	3387	352 352	792	141 141	293	102 102	167
0.05 ***	12227 12352	12540	3477 3477	3516	727 727	727	266 266	266	141 141	141
•••	12266 12454	24269	3477 3516	6576	727 727	1285	266 266	409	141 141	216
0.1 ***	27227 27829	29227	7477 7579	7727	1415 1415	1454	454 454	454	227 227	227
•••	27454 28352	43890	7540 7641	11423	1415 1415	2033	454 454	584	227 227	289
0.15 ***	40977 42290	46165	11266 11540	11954	2079 2079	2102	602 602	602	290 290	290
•••	41516 43641	61175	11391 11727	15685	2079 2102	2687	602 602	736	290 290	351
0.2 ***	52415 54641	62016	14540 15079	15891	2665 2704	2727	727 766	766	352 352	352
•••	53290 57016	76124	14790 15415	19364	2665 2704	3247	766 766	865	352 352	403
0.25 ***	61227 64454	76141	17266 18040	19352	3165 3204	3266	852 852	891	415 415	415
•••	62477 68016	88737	17602 18602	22459	3165 3227	3714	852 891	970	415 415	444
0.3 ***	67454 71790	88165	19352 20415	22329	3579 3641	3704	977 977	977	454 454	454
•••	69102 76641	99015	19829 21227	24970	3602 3665	4088	977 977	1052	454 454	475
0.35 ***	71266 76790	97915	20852 22227	24704	3891 3954	4040	1040 1040	1040	477 477	477
•••	73415 82891	106956	21477 23266	26897	3915 4016	4368	1040 1040	1110	477 477	496
0.4 ***	73016 79704	105227	21829 23477	26454	4102 4204	4290	1079 1102	1102	516 516	516
•••	75641 87016	112562	22540 24704	28240	4141 4266	4555	1102 1102	1145	516 516	507
0.45 ***	72977 80766	110040	22329 24165	27602	4227 4329	4454	1102 1102	1141	516 516	516
•••	76040 89102	115833	23141 25579	28999	4290 4391	4649	1102 1141	1157	516 516	507
0.5 ***	71477 80227	112352	22329 24391	28102	4266 4391	4516	1102 1102	1141	516 516	516
•••	74954 89391	116767	23227 25891	29175	4329 4454	4649	1102 1141	1145	516 516	496
0.55 ***	68790 78266	112102	21954 24079	27977	4204 4329	4454	1079 1102	1102	477 477	477
•••	72602 88016	115365	22891 25665	28766	4266 4391	4555	1102 1102	1110	477 477	475
0.6 ***	65165 75079	109329	21165 23329	27227	4040 4165	4290	1040 1040	1040	454 454	454
•••	69165 85079	111628	22141 24915	27773	4102 4227	4368	1040 1040	1052	454 454	444
0.65 ***	60727 70766	104016	20040 22141	25852	3829 3915	4040	954 977	977	415 415	415
•••	64829 80641	105555	20954 23665	26196	3891 3977	4088	977 977	970	415 415	403
0.7 ***	55641 65352	96165	18516 20477	23829	3516 3602	3704	852 891	891	352 352	352
•••	59641 74766	97146	19391 21852	24036	3540 3641	3714	852 891	865	352 352	351
0.75 ***	49852 58891	85852	16641 18352	21165	3079 3165	3266	727 766	766	290 290	290
•••	53579 67391	86402	17415 19540	21291	3141 3204	3247	727 766	736	290 290	289
0.8 ***	43290 51204	73040	14352 15727	17915	2579 2641	2704	602 602	602	227 227	227
•••	46602 58415	73321	14977 16665	17963	2602 2665	2687	602 602	584	227 227	216
0.85 ***	35790 42102	57790	11602 12602	14040	1977 2016	2040	415 415	415	. .	.
•••	38477 47602	57905	12040 13227	14050	1977 2040	2033	415 415	409	. .	.
0.9 ***	26891 31102	40141	8266 8829	9579	1266 1290	1290	. .	.	. .	.
•••	28704 34540	40153	8540 9165	9554	1290 1290	1285	. .	.	. .	.
0.95 ***	15454 17204	20079	4165 4329	4477	. .	.	. .	.	. .	.
•••	16227 18415	20065	4227 4391	4474	. .	.	. .	.	. .	.
0.98 ***	6165 6477	6915	. .	.	. .	.	. .	.	. .	.
•••	6329 6704	6891	. .	.	. .	.	. .	.	. .	.

TABLE 2: ALPHA= 0.005 POWER= 0.9   EXPECTED ACCRUAL THRU MINIMUM FOLLOW-UP= 30

		DEL=.10			DEL=.15			DEL=.20			DEL=.25			DEL=.30		
FACT=		1.0 .75	.50 .25	.00 BIN	1.0 .75	.50 .25	.00 BIN	1.0 .75	.50 .25	.00 BIN	1.0 75	.50 .25	.00 BIN	1.0 .75	.50 .25	.00 BIN
PCONT=***						REQUIRED NUMBER OF PATIENTS										
0.05	***	312	313	338	181	182	197	127	128	138	99	100	107	81	83	88
	•••	312	315	521	181	184	275	128	131	175	99	102	123	82	84	91
0.1	***	491	494	566	259	261	300	170	172	197	125	127	144	99	101	113
	•••	492	498	744	260	266	368	171	177	224	126	131	152	99	104	110
0.15	***	634	637	776	317	321	391	200	204	246	142	146	175	110	113	133
	•••	635	644	938	318	328	447	201	211	265	144	153	175	111	119	124
0.2	***	737	742	961	357	363	469	219	225	288	153	159	199	116	121	149
	•••	739	752	1102	359	374	513	221	235	298	155	167	195	118	128	136
0.25	***	802	810	1120	380	388	534	229	238	321	158	166	219	119	126	161
	•••	805	824	1235	383	403	566	232	251	324	161	177	209	121	134	144
0.3	***	833	843	1248	389	399	585	232	242	346	158	168	233	118	127	169
	•••	836	862	1340	392	419	606	235	260	343	162	182	218	121	137	149
0.35	***	833	845	1346	384	398	622	228	241	363	155	167	241	115	125	173
	•••	837	870	1414	389	423	632	232	262	354	159	183	223	119	138	151
0.4	***	806	822	1411	369	387	644	218	235	372	149	163	244	110	122	174
	•••	811	854	1459	375	417	645	224	259	358	154	181	223	115	135	149
0.45	***	757	777	1444	346	368	652	205	225	372	140	156	242	104	117	170
	•••	764	818	1474	353	404	645	212	252	354	146	176	218	109	131	144
0.5	***	691	717	1445	317	343	645	189	211	365	130	147	234	97	110	163
	•••	700	768	1459	326	384	632	198	241	343	137	168	209	102	125	136
0.55	***	613	647	1413	284	315	624	172	196	349	119	137	222	89	102	153
	•••	625	708	1414	296	360	606	181	226	324	126	157	195	94	116	124
0.6	***	530	572	1349	251	285	588	154	179	325	107	125	204	80	93	139
	•••	544	643	1340	264	333	566	164	209	298	114	145	175	85	106	110
0.65	***	446	497	1252	217	254	538	136	161	293	95	112	181	71	82	121
	•••	464	575	1235	231	302	513	145	189	265	102	130	152	75	94	91
0.7	***	367	425	1124	186	222	474	117	141	253	82	97	153	.	.	.
	•••	388	506	1102	200	268	447	126	167	224	88	112	123	.	.	.
0.75	***	297	356	964	155	189	396	98	118	205	.	.	.	.	.	.
	•••	319	435	938	168	230	368	106	140	175	.	.	.	.	.	.
0.8	***	234	291	773	124	153	304	.	.	.	.	.	.	.	.	.
	•••	256	361	744	136	186	275	.	.	.	.	.	.	.	.	.
0.85	***	176	223	551	.	.	.	.	.	.	.	.	.	.	.	.
	•••	195	279	521	.	.	.	.	.	.	.	.	.	.	.	.

TABLE 2: ALPHA= 0.005 POWER= 0.9   EXPECTED ACCRUAL THRU MINIMUM FOLLOW-UP= 40

	DEL=.10			DEL=.15			DEL=.20			DEL=.25			DEL=.30		
FACT=	1.0 .75	.50 .25	.00 BIN	1.0 .75	.50 .25	.00 BIN	1.0 .75	.50 .25	.00 BIN	1.0 75	.50 .25	.00 BIN	1.0 .75	.50 .25	.00 BIN

PCONT=*** — REQUIRED NUMBER OF PATIENTS

PCONT	D.10 F1	D.10 F2	D.10 BIN	D.15 F1	D.15 F2	D.15 BIN	D.20 F1	D.20 F2	D.20 BIN	D.25 F1	D.25 F2	D.25 BIN	D.30 F1	D.30 F2	D.30 BIN
0.05 ***	312	313	338	181	183	197	128	129	138	99	101	107	82	83	88
•••	312	316	521	182	186	275	128	132	175	100	103	123	82	85	91
0.1 ***	492	495	566	260	263	300	171	174	197	126	129	144	100	102	113
•••	493	501	744	261	269	368	172	179	224	127	133	152	101	106	110
0.15 ***	635	639	776	318	324	391	201	206	246	144	149	175	111	116	133
•••	636	648	938	320	333	447	203	215	265	146	155	175	113	121	124
0.2 ***	739	745	961	359	366	469	221	229	288	155	162	199	118	124	149
•••	741	759	1102	361	380	513	224	240	298	158	171	195	120	131	136
0.25 ***	805	814	1120	383	393	534	232	243	321	161	170	219	121	129	161
•••	808	833	1235	386	412	566	236	258	324	164	182	209	124	138	144
0.3 ***	836	849	1248	392	406	585	235	249	346	162	174	233	121	131	169
•••	841	875	1340	397	430	606	240	268	343	166	188	218	125	142	149
0.35 ***	837	854	1346	389	407	622	232	249	363	160	173	241	119	130	173
•••	843	887	1414	395	437	632	238	272	354	165	190	223	124	143	151
0.4 ***	811	833	1411	375	398	644	224	244	372	154	170	244	115	128	174
•••	818	875	1459	383	434	645	231	270	358	160	189	223	120	141	149
0.45 ***	764	791	1444	353	381	652	212	235	372	146	164	242	109	123	170
•••	773	844	1474	363	423	645	221	264	354	153	184	218	115	137	144
0.5 ***	700	735	1445	326	359	645	198	223	365	137	156	234	102	116	163
•••	711	799	1459	338	405	632	207	254	343	144	176	209	108	130	136
0.55 ***	625	669	1413	296	333	624	181	208	349	126	145	222	94	108	153
•••	640	744	1414	309	382	606	192	240	324	134	166	195	100	122	124
0.6 ***	544	598	1349	264	304	588	164	191	325	114	133	204	85	99	139
•••	563	682	1340	279	355	566	175	222	298	122	153	175	91	111	110
0.65 ***	464	526	1252	231	273	538	145	172	293	102	119	181	75	87	121
•••	486	616	1235	247	323	513	156	202	265	109	137	152	80	98	91
0.7 ***	388	456	1124	200	240	474	126	151	253	88	103	153	.	.	.
•••	413	546	1102	216	288	447	136	177	224	94	118	123	.	.	.
0.75 ***	319	387	964	168	205	396	106	127	205	.	.	.	.	.	.
•••	345	473	938	183	247	368	115	149	175	.	.	.	.	.	.
0.8 ***	256	318	773	136	167	304	.	.	.	.	.	.	.	.	.
•••	280	394	744	148	200	275	.	.	.	.	.	.	.	.	.
0.85 ***	195	245	551	.	.	.	.	.	.	.	.	.	.	.	.
•••	214	304	521	.	.	.	.	.	.	.	.	.	.	.	.

TABLE 2: ALPHA= 0.005 POWER= 0.9   EXPECTED ACCRUAL THRU MINIMUM FOLLOW-UP= 50

	DEL=.10			DEL=.15			DEL=.20			DEL=.25			DEL=.30		
FACT=	1.0 .75	.50 .25	.00 BIN	1.0 .75	.50 .25	.00 BIN	1.0 .75	.50 .25	.00 BIN	1.0 75	.50 .25	.00 BIN	1.0 .75	.50 .25	.00 BIN
PCONT=•••	REQUIRED NUMBER OF PATIENTS														
0.05 •••	312	314	338	182	184	197	128	130	138	100	101	107	82	84	88
•••	313	317	521	182	187	275	129	132	175	100	103	123	83	85	91
0.1 •••	493	496	566	261	265	300	171	175	197	127	130	144	100	103	113
•••	494	503	744	262	271	368	173	181	224	128	134	152	102	107	110
0.15 •••	636	642	776	320	326	391	202	209	246	145	151	175	112	117	133
•••	638	653	938	322	337	447	205	218	265	147	158	175	114	122	124
0.2 •••	740	749	961	361	370	469	223	232	288	157	165	199	120	126	149
•••	743	766	1102	364	386	513	226	245	298	160	175	195	122	134	136
0.25 •••	807	819	1120	385	398	534	235	247	321	164	174	219	124	132	161
•••	811	842	1235	390	420	566	239	264	324	167	186	209	127	141	144
0.3 •••	840	855	1248	396	413	585	239	255	346	165	178	233	124	135	169
•••	845	887	1340	401	440	606	245	275	343	170	193	218	128	145	149
0.35 •••	841	862	1346	393	415	622	237	256	363	164	179	241	123	134	173
•••	848	903	1414	401	449	632	244	280	354	169	196	223	127	146	151
0.4 •••	816	844	1411	381	408	644	230	252	372	159	176	244	119	132	174
•••	826	895	1459	390	448	645	238	280	358	165	195	223	124	145	149
0.45 •••	771	805	1444	361	393	652	219	244	372	152	170	242	113	127	170
•••	782	868	1474	372	439	645	228	274	354	159	190	218	119	141	144
0.5 •••	708	752	1445	335	372	645	205	232	365	143	162	234	107	121	163
•••	723	827	1459	349	422	632	216	264	343	150	183	209	113	134	136
0.55 •••	636	689	1413	306	347	624	189	218	349	132	152	222	99	113	153
•••	654	775	1414	321	400	606	200	250	324	140	172	195	105	126	124
0.6 •••	558	621	1349	275	319	588	172	201	325	120	140	204	90	103	139
•••	581	715	1340	292	373	566	184	233	298	128	159	175	95	115	110
0.65 •••	481	552	1252	244	288	538	154	182	293	107	125	181	79	91	121
•••	507	650	1235	261	341	513	165	211	265	115	143	152	84	101	91
0.7 •••	407	482	1124	212	255	474	134	160	253	93	108	153	.	.	.
•••	436	580	1102	229	304	447	144	186	224	99	123	123	.	.	.
0.75 •••	339	413	964	180	219	396	113	134	205	.	.	.	.	.	.
•••	367	504	938	195	261	368	122	155	175	.	.	.	.	.	.
0.8 •••	274	341	773	145	177	304	.	.	.	.	.	.	.	.	.
•••	300	420	744	158	211	275	.	.	.	.	.	.	.	.	.
0.85 •••	210	263	551	.	.	.	.	.	.	.	.	.	.	.	.
•••	231	324	521	.	.	.	.	.	.	.	.	.	.	.	.

TABLE 2: ALPHA= 0.005 POWER= 0.9   EXPECTED ACCRUAL THRU MINIMUM FOLLOW-UP= 60

		DEL=.10			DEL=.15			DEL=.20			DEL=.25			DEL=.30		
FACT=		1.0 .75	.50 .25	.00 BIN	1.0 .75	.50 .25	.00 BIN	1.0 .75	.50 .25	.00 BIN	1.0 75	.50 .25	.00 BIN	1.0 .75	.50 .25	.00 BIN
PCONT=•••					REQUIRED NUMBER OF PATIENTS											
0.05	•••	313	315	338	182	184	197	128	131	138	100	102	107	83	84	88
	•••	313	318	521	183	187	275	129	133	175	101	104	123	83	85	91
0.1	•••	494	498	566	261	266	300	172	177	197	127	131	144	101	104	113
	•••	495	506	744	263	273	368	174	182	224	129	135	152	102	107	110
0.15	•••	637	644	776	321	328	391	204	211	246	146	153	175	113	119	133
	•••	639	657	938	323	341	447	206	220	265	149	159	175	116	124	124
0.2	•••	742	752	961	363	374	469	225	235	288	159	167	199	121	128	149
	•••	745	772	1102	366	391	513	229	248	298	162	177	195	124	135	136
0.25	•••	810	824	1120	388	403	534	238	251	321	166	177	219	126	134	161
	•••	814	851	1235	393	427	566	242	268	324	170	189	209	129	143	144
0.3	•••	843	862	1248	399	419	585	242	260	346	168	182	233	127	137	169
	•••	849	899	1340	406	449	606	249	281	343	174	197	218	131	148	149
0.35	•••	845	870	1346	398	423	622	241	262	363	167	183	241	125	138	173
	•••	854	918	1414	407	460	632	249	287	354	173	200	223	130	149	151
0.4	•••	822	854	1411	387	417	644	235	259	372	163	181	244	122	135	174
	•••	833	914	1459	398	460	645	244	287	358	170	199	223	128	148	149
0.45	•••	777	818	1444	368	404	652	225	252	372	156	176	242	117	131	170
	•••	791	890	1474	381	452	645	235	282	354	164	195	218	123	144	144
0.5	•••	717	768	1445	343	384	645	211	241	365	147	168	234	110	125	163
	•••	735	852	1459	359	437	632	223	273	343	156	188	209	116	138	136
0.55	•••	647	708	1413	315	360	624	196	226	349	137	157	222	102	116	153
	•••	669	803	1414	333	415	606	208	259	324	145	177	195	108	129	124
0.6	•••	572	643	1349	286	333	588	179	209	325	125	145	204	93	106	139
	•••	598	744	1340	304	388	566	191	241	298	133	164	175	99	118	110
0.65	•••	497	575	1252	254	302	538	161	189	293	112	130	181	82	94	121
	•••	526	679	1235	273	355	513	172	219	265	119	147	152	87	104	91
0.7	•••	425	506	1124	223	268	474	141	167	253	97	112	153	.	.	.
	•••	456	608	1102	240	317	447	151	192	224	103	126	123	.	.	.
0.75	•••	356	435	964	189	230	396	118	140	205	.	.	.	.	.	.
	•••	387	530	938	205	272	368	127	160	175	.	.	.	.	.	.
0.8	•••	290	361	773	153	186	304	.	.	.	.	.	.	.	.	.
	•••	318	442	744	167	219	275	.	.	.	.	.	.	.	.	.
0.85	•••	223	279	551	.	.	.	.	.	.	.	.	.	.	.	.
	•••	245	340	521	.	.	.	.	.	.	.	.	.	.	.	.

TABLE 2: ALPHA= 0.005 POWER= 0.9   EXPECTED ACCRUAL THRU MINIMUM FOLLOW-UP= 70

		DEL=.10			DEL=.15			DEL=.20			DEL=.25			DEL=.30		
FACT=		1.0 .75	.50 .25	.00 BIN	1.0 .75	.50 .25	.00 BIN	1.0 .75	.50 .25	.00 BIN	1.0 75	.50 .25	.00 BIN	1.0 .75	.50 .25	.00 BIN
PCONT=***					REQUIRED NUMBER OF PATIENTS											
0.05	***	313	315	338	182	185	197	129	131	138	100	102	107	83	84	88
	•••	314	319	521	183	188	275	130	133	175	101	104	123	83	86	91
0.1	***	494	499	566	262	267	300	173	178	197	128	132	144	102	105	113
	•••	496	508	744	264	275	368	175	183	224	130	136	152	103	108	110
0.15	***	638	646	776	322	331	391	205	213	246	148	154	175	115	120	133
	•••	641	662	938	325	344	447	208	222	265	150	161	175	117	125	124
0.2	***	744	755	961	365	377	469	227	238	288	161	170	199	123	130	149
	•••	748	778	1102	369	396	513	231	251	298	164	179	195	126	137	136
0.25	***	812	828	1120	391	408	534	240	255	321	168	180	219	127	136	161
	•••	817	860	1235	397	433	566	246	272	324	173	192	209	131	145	144
0.3	***	846	868	1248	403	425	585	246	264	346	171	185	233	129	140	169
	•••	853	910	1340	410	457	606	253	286	343	177	200	218	133	150	149
0.35	***	850	879	1346	402	430	622	245	267	363	170	187	241	128	140	173
	•••	859	932	1414	412	469	632	254	293	354	177	204	223	133	152	151
0.4	***	827	865	1411	392	426	644	240	265	372	167	185	244	125	138	174
	•••	840	931	1459	404	471	645	249	293	358	174	203	223	131	150	149
0.45	***	784	832	1444	374	414	652	230	258	372	160	180	242	120	134	170
	•••	800	910	1474	389	464	645	241	289	354	168	199	218	126	147	144
0.5	***	726	784	1445	351	395	645	217	248	365	152	172	234	114	128	163
	•••	746	875	1459	368	449	632	229	279	343	160	192	209	119	140	136
0.55	***	658	727	1413	324	372	624	202	234	349	142	162	222	106	119	153
	•••	682	827	1414	343	428	606	215	266	324	150	181	195	111	131	124
0.6	***	585	663	1349	295	344	588	186	216	325	130	149	204	96	109	139
	•••	614	770	1340	314	401	566	198	247	298	138	167	175	102	120	110
0.65	***	512	596	1252	264	313	538	167	196	293	116	134	181	85	96	121
	•••	544	705	1235	284	367	513	179	225	265	123	150	152	90	106	91
0.7	***	441	527	1124	232	278	474	146	172	253	100	116	153	.	.	.
	•••	474	633	1102	250	328	447	157	197	224	107	129	123	.	.	.
0.75	***	372	455	964	198	239	396	123	145	205	.	.	.	.	.	.
	•••	405	552	938	214	281	368	132	165	175	.	.	.	.	.	.
0.8	***	305	378	773	160	194	304	.	.	.	.	.	.	.	.	.
	•••	334	461	744	174	226	275	.	.	.	.	.	.	.	.	.
0.85	***	235	292	551	.	.	.	.	.	.	.	.	.	.	.	.
	•••	258	354	521	.	.	.	.	.	.	.	.	.	.	.	.

TABLE 2: ALPHA= 0.005 POWER= 0.9   EXPECTED ACCRUAL THRU MINIMUM FOLLOW-UP= 80

		DEL=.10			DEL=.15			DEL=.20			DEL=.25			DEL=.30	
FACT=	1.0 .75	.50 .25	.00 BIN	1.0 .75	.50 .25	.00 BIN	1.0 .75	.50 .25	.00 BIN	1.0 75	.50 .25	.00 BIN	1.0 .75	.50 .25	.00 BIN
PCONT=•••					REQUIRED	NUMBER	OF	PATIENTS							
0.05 ***	313	316	338	183	186	197	129	132	138	101	103	107	83	85	88
•••	314	320	521	184	189	275	130	134	175	101	104	123	84	86	91
0.1 ***	495	501	566	263	269	300	174	179	197	129	133	144	102	106	113
•••	497	511	744	265	277	368	176	185	224	131	137	152	104	109	110
0.15 ***	639	648	776	324	333	391	206	215	246	149	155	175	116	121	133
•••	642	666	938	327	346	447	209	224	265	151	162	175	118	125	124
0.2 ***	745	759	961	366	380	469	229	240	288	162	171	199	124	131	149
•••	750	784	1102	371	400	513	233	254	298	166	181	195	127	138	136
0.25 ***	814	833	1120	393	412	534	243	258	321	170	182	219	129	138	161
•••	821	868	1235	400	439	566	248	276	324	175	194	209	133	146	144
0.3 ***	849	875	1248	406	430	585	249	268	346	174	188	233	131	142	169
•••	858	921	1340	415	464	606	256	290	343	180	203	218	136	152	149
0.35 ***	854	887	1346	407	437	622	249	272	363	173	190	241	130	143	173
•••	865	946	1414	418	477	632	258	297	354	180	207	223	136	154	151
0.4 ***	833	875	1411	398	434	644	244	270	372	170	189	244	128	141	174
•••	847	947	1459	411	480	645	254	299	358	178	207	223	133	153	149
0.45 ***	791	844	1444	381	423	652	235	264	372	164	184	242	123	137	170
•••	810	929	1474	397	474	645	247	295	354	172	203	218	129	149	144
0.5 ***	735	799	1445	359	405	645	223	254	365	156	176	234	116	130	163
•••	757	896	1459	376	460	632	235	285	343	164	195	209	122	142	136
0.55 ***	669	744	1413	333	382	624	208	240	349	145	166	222	108	122	153
•••	696	850	1414	352	439	606	221	272	324	154	185	195	114	133	124
0.6 ***	598	682	1349	304	355	588	191	222	325	133	153	204	99	111	139
•••	629	794	1340	324	412	566	204	253	298	142	170	175	104	122	110
0.65 ***	526	616	1252	273	323	538	172	202	293	119	137	181	87	98	121
•••	560	728	1235	293	378	513	184	230	265	127	153	152	92	107	91
0.7 ***	456	546	1124	240	288	474	151	177	253	103	118	153	.	.	.
•••	491	655	1102	259	337	447	162	202	224	110	131	123	.	.	.
0.75 ***	387	473	964	205	247	396	127	149	205	.	.	.	.	.	.
•••	421	572	938	222	289	368	136	168	175	.	.	.	.	.	.
0.8 ***	318	394	773	167	200	304	.	.	.	.	.	.	.	.	.
•••	348	477	744	180	232	275	.	.	.	.	.	.	.	.	.
0.85 ***	245	304	551	.	.	.	.	.	.	.	.	.	.	.	.
•••	269	366	521	.	.	.	.	.	.	.	.	.	.	.	.

TABLE 2: ALPHA= 0.005 POWER= 0.9   EXPECTED ACCRUAL THRU MINIMUM FOLLOW-UP= 90

PCONT=*** REQUIRED NUMBER OF PATIENTS

PCONT		DEL=.10 1.0/.75	DEL=.10 .50/.25	DEL=.10 .00/BIN	DEL=.15 1.0/.75	DEL=.15 .50/.25	DEL=.15 .00/BIN	DEL=.20 1.0/.75	DEL=.20 .50/.25	DEL=.20 .00/BIN	DEL=.25 1.0/75	DEL=.25 .50/.25	DEL=.25 .00/BIN	DEL=.30 1.0/.75	DEL=.30 .50/.25	DEL=.30 .00/BIN
0.05	***	314	317	338	183	186	197	130	132	138	101	103	107	83	85	88
	***	315	321	521	184	189	275	130	134	175	102	105	123	84	86	91
0.1	***	496	502	566	264	270	300	175	180	197	130	134	144	103	106	113
	***	498	513	744	266	278	368	177	185	224	131	138	152	104	109	110
0.15	***	640	651	776	325	335	391	208	216	246	150	157	175	116	122	133
	***	644	669	938	328	349	447	211	226	265	153	163	175	119	126	124
0.2	***	747	762	961	368	383	469	230	243	288	164	173	199	125	133	149
	***	752	790	1102	374	404	513	235	256	298	167	182	195	128	139	136
0.25	***	817	838	1120	396	416	534	245	261	321	172	184	219	131	140	161
	***	824	876	1235	403	443	566	251	279	324	177	196	209	134	148	144
0.3	***	852	881	1248	409	435	585	252	272	346	176	191	233	133	144	169
	***	862	931	1340	419	470	606	260	294	343	182	205	218	137	153	149
0.35	***	858	895	1346	411	443	622	252	276	363	176	193	241	132	145	173
	***	870	958	1414	423	484	632	262	302	354	183	209	223	138	155	151
0.4	***	838	885	1411	403	441	644	248	275	372	173	192	244	130	143	174
	***	854	962	1459	417	488	645	259	303	358	181	209	223	135	154	149
0.45	***	798	856	1444	387	431	652	240	269	372	167	187	242	125	139	170
	***	819	946	1474	404	483	645	252	300	354	176	206	218	131	150	144
0.5	***	743	813	1445	366	414	645	228	259	364	159	180	234	119	133	163
	***	768	915	1459	384	470	632	241	291	343	168	198	209	125	144	136
0.55	***	679	760	1413	340	392	624	213	245	349	149	169	222	111	124	153
	***	708	870	1414	360	449	606	226	277	324	157	187	195	116	135	124
0.6	***	610	699	1349	312	364	588	196	228	325	137	156	204	101	113	139
	***	643	814	1340	333	421	566	209	258	298	145	173	175	106	123	110
0.65	***	539	634	1252	281	333	538	177	207	293	123	140	181	89	100	121
	***	575	749	1235	302	387	513	189	234	265	130	155	152	94	108	91
0.7	***	469	564	1124	248	296	474	156	182	253	106	121	153	.	.	.
	***	506	675	1102	268	345	447	167	205	224	112	133	123	.	.	.
0.75	***	400	489	964	212	254	396	131	152	205	.	.	.	.	.	.
	***	435	590	938	229	296	368	140	171	175	.	.	.	.	.	.
0.8	***	330	407	773	172	206	304	.	.	.	.	.	.	.	.	.
	***	361	492	744	186	237	275	.	.	.	.	.	.	.	.	.
0.85	***	255	314	551	.	.	.	.	.	.	.	.	.	.	.	.
	***	279	376	521	.	.	.	.	.	.	.	.	.	.	.	.

TABLE 2: ALPHA= 0.005 POWER= 0.9   EXPECTED ACCRUAL THRU MINIMUM FOLLOW-UP= 100

		DEL=.10			DEL=.15			DEL=.20			DEL=.25			DEL=.30		
FACT=		1.0 .75	.50 .25	.00 BIN	1.0 .75	.50 .25	.00 BIN	1.0 .75	.50 .25	.00 BIN	1.0 75	.50 .25	.00 BIN	1.0 .75	.50 .25	.00 BIN
PCONT=***				REQUIRED	NUMBER	OF	PATIENTS									
0.05	***	314	317	338	184	187	197	130	132	138	101	103	107	84	85	88
	•••	315	322	521	185	190	275	131	135	175	102	105	123	84	86	91
0.1	***	496	503	566	265	271	300	175	181	197	130	134	144	103	107	113
	•••	499	515	744	267	279	368	177	186	224	132	138	152	105	109	110
0.15	***	642	653	776	326	337	391	209	218	246	151	158	175	117	122	133
	•••	645	673	938	330	351	447	212	227	265	154	164	175	119	127	124
0.2	***	749	766	961	370	386	469	232	245	288	165	175	199	126	134	149
	•••	754	795	1102	376	407	513	237	258	298	169	183	195	129	140	136
0.25	***	819	842	1120	398	420	534	247	264	321	174	186	219	132	141	161
	•••	827	884	1235	406	448	566	254	281	324	179	198	209	136	149	144
0.3	***	855	887	1248	413	440	585	255	275	346	178	193	233	135	145	169
	•••	866	941	1340	423	475	606	263	297	343	184	207	218	139	154	149
0.35	***	862	903	1346	415	449	622	256	280	363	179	196	241	134	146	173
	•••	876	970	1414	428	491	632	265	305	354	186	211	223	139	156	151
0.4	***	844	895	1411	408	448	644	252	280	372	176	195	244	132	145	174
	•••	861	976	1459	423	496	645	263	308	358	184	212	223	137	156	149
0.45	***	805	868	1444	393	439	652	244	274	372	170	190	242	127	141	170
	•••	827	962	1474	410	491	645	256	304	354	179	208	218	133	152	144
0.5	***	752	827	1445	372	422	645	232	264	365	162	183	234	121	134	163
	•••	779	932	1459	392	478	632	245	295	343	171	201	209	127	145	136
0.55	***	689	775	1413	347	400	624	218	250	349	152	172	222	113	126	153
	•••	721	889	1414	368	458	606	231	281	324	161	190	195	118	136	124
0.6	***	621	715	1349	319	373	588	201	233	325	140	159	204	103	115	139
	•••	657	834	1340	341	430	566	214	262	298	148	175	175	108	124	110
0.65	***	552	650	1252	288	341	538	182	211	293	125	143	181	91	101	121
	•••	590	768	1235	310	395	513	194	238	265	133	157	152	96	109	91
0.7	***	482	580	1124	255	304	474	160	186	253	108	123	153	.	.	.
	•••	520	692	1102	275	352	447	171	209	224	115	134	123	.	.	.
0.75	***	413	504	964	219	261	396	134	155	205	.	.	.	.	.	.
	•••	449	605	938	236	302	368	143	173	175	.	.	.	.	.	.
0.8	***	341	420	773	177	211	304	.	.	.	.	.	.	.	.	.
	•••	373	505	744	191	241	275	.	.	.	.	.	.	.	.	.
0.85	***	263	324	551	.	.	.	.	.	.	.	.	.	.	.	.
	•••	288	385	521	.	.	.	.	.	.	.	.	.	.	.	.

TABLE 2: ALPHA= 0.005 POWER= 0.9   EXPECTED ACCRUAL THRU MINIMUM FOLLOW-UP= 110

		DEL=.05			DEL=.10			DEL=.15			DEL=.20			DEL=.25		
FACT=		1.0 .75	.50 .25	.00 BIN	1.0 .75	.50 .25	.00 BIN	1.0 .75	.50 .25	.00 BIN	1.0 75	.50 .25	.00 BIN	1.0 .75	.50 .25	.00 BIN
PCONT=•••				REQUIRED	NUMBER	OF	PATIENTS									
0.05	•••	888	891	951	314	318	338	184	187	197	130	133	138	102	103	107
	•••	889	896	1637	316	323	521	185	190	275	131	135	175	102	105	123
0.1	•••	1618	1624	1833	497	505	566	265	272	300	176	181	197	131	135	144
	•••	1620	1637	2590	500	517	744	268	280	368	178	187	224	132	138	152
0.15	•••	2237	2248	2687	643	655	776	327	339	391	210	219	246	152	159	175
	•••	2240	2269	3423	647	676	938	331	353	447	213	228	265	155	164	175
0.2	•••	2713	2730	3467	750	769	961	372	389	469	234	247	288	166	176	199
	•••	2719	2762	4137	757	801	1102	378	410	513	239	260	298	170	184	195
0.25	•••	3042	3065	4148	821	847	1120	401	423	534	249	266	321	176	188	219
	•••	3049	3111	4732	830	891	1235	409	452	566	256	284	324	181	199	209
0.3	•••	3228	3259	4717	859	893	1248	416	445	585	257	278	346	180	195	233
	•••	3238	3322	5208	870	950	1340	427	480	606	265	300	343	186	208	218
0.35	•••	3283	3324	5166	866	911	1346	419	454	622	259	284	363	181	198	241
	•••	3297	3408	5565	881	981	1414	432	497	632	269	309	354	188	213	223
0.4	•••	3223	3277	5489	849	904	1411	413	454	644	256	283	372	178	197	244
	•••	3241	3385	5803	868	989	1459	428	503	645	267	311	358	186	214	223
0.45	•••	3067	3137	5684	812	879	1444	398	446	652	248	278	372	173	193	242
	•••	3090	3275	5922	836	977	1474	417	499	645	260	308	354	181	210	218
0.5	•••	2834	2923	5750	760	840	1445	379	430	645	237	269	365	165	185	234
	•••	2864	3097	5922	789	949	1459	399	486	632	250	299	343	174	203	209
0.55	•••	2544	2657	5687	699	789	1413	354	408	624	222	255	349	155	175	222
	•••	2582	2872	5803	732	906	1414	375	465	606	236	285	324	163	192	195
0.6	•••	2219	2362	5493	633	730	1348	326	381	588	205	237	325	142	161	204
	•••	2267	2618	5565	670	851	1340	348	437	566	218	266	298	151	177	175
0.65	•••	1882	2059	5171	564	665	1252	295	348	538	186	215	293	128	145	181
	•••	1943	2352	5208	603	785	1235	317	402	513	198	241	265	135	158	152
0.7	•••	1555	1766	4721	495	594	1124	262	311	474	163	189	253	111	125	153
	•••	1630	2080	4732	534	708	1102	282	359	447	174	211	224	117	136	123
0.75	•••	1258	1489	4144	424	517	964	224	267	396	137	158	205	.	.	.
	•••	1344	1805	4137	461	620	938	242	307	368	146	175	175	.	.	.
0.8	•••	999	1228	3442	351	431	773	182	215	304	.	.	.	.	.	.
	•••	1086	1523	3423	383	516	744	196	245	275	.	.	.	.	.	.
0.85	•••	768	972	2616	271	332	551	.	.	.	.	.	.	.	.	.
	•••	847	1222	2590	296	393	521	.	.	.	.	.	.	.	.	.
0.9	•••	545	699	1666	.	.	.	.	.	.	.	.	.	.	.	.
	•••	605	877	1637	.	.	.	.	.	.	.	.	.	.	.	.

TABLE 2: ALPHA= 0.005 POWER= 0.9   EXPECTED ACCRUAL THRU MINIMUM FOLLOW-UP= 120

		DEL=.05			DEL=.10			DEL=.15			DEL=.20			DEL=.25		
FACT=		1.0 .75	.50 .25	.00 BIN	1.0 .75	.50 .25	.00 BIN	1.0 .75	.50 .25	.00 BIN	1.0 75	.50 .25	.00 BIN	1.0 .75	.50 .25	.00 BIN
PCONT=•••					REQUIRED	NUMBER OF	PATIENTS									
0.05	•••	888	891	950	315	318	338	184	187	197	130	133	138	102	104	107
	•••	889	897	1637	316	324	521	186	191	275	131	135	175	103	105	123
0.1	•••	1619	1626	1833	498	506	566	266	273	300	177	182	197	131	135	144
	•••	1621	1639	2590	501	519	744	269	282	368	179	187	224	133	139	152
0.15	•••	2238	2250	2687	644	657	776	328	340	391	211	220	246	153	159	175
	•••	2242	2273	3423	648	680	938	333	355	447	214	229	265	155	165	175
0.2	•••	2715	2733	3467	752	772	961	374	391	469	235	248	288	167	177	199
	•••	2721	2768	4137	759	805	1102	380	413	513	240	262	298	171	185	195
0.25	•••	3044	3069	4148	824	851	1120	403	427	534	251	268	321	177	189	219
	•••	3052	3119	4732	833	897	1235	412	456	566	258	286	324	182	200	209
0.3	•••	3230	3265	4717	862	899	1248	419	449	585	260	281	346	182	197	233
	•••	3242	3333	5208	874	958	1340	430	485	606	268	302	343	188	210	218
0.35	•••	3286	3332	5166	870	918	1346	423	460	622	262	287	363	183	200	241
	•••	3302	3423	5565	887	992	1414	437	502	632	272	311	354	190	215	223
0.4	•••	3228	3287	5489	854	914	1411	417	460	644	259	287	372	181	199	244
	•••	3248	3405	5803	875	1002	1459	434	509	645	270	314	358	189	216	223
0.45	•••	3074	3149	5684	818	890	1444	404	452	652	252	282	372	176	195	242
	•••	3099	3299	5922	844	991	1474	423	505	645	264	311	354	184	212	218
0.5	•••	2842	2939	5750	768	852	1445	384	437	645	241	273	364	168	188	234
	•••	2874	3127	5922	799	964	1459	405	493	632	254	302	343	176	205	209
0.55	•••	2554	2677	5687	708	802	1413	360	415	624	226	259	349	157	177	222
	•••	2596	2908	5803	744	922	1414	382	472	606	240	288	324	166	194	195
0.6	•••	2232	2387	5493	643	744	1348	333	388	588	209	241	325	145	163	204
	•••	2285	2660	5565	682	867	1340	355	444	566	222	269	298	153	179	175
0.65	•••	1899	2089	5171	575	679	1252	302	355	538	189	219	293	130	147	181
	•••	1965	2397	5208	616	801	1235	323	408	513	202	244	265	137	160	152
0.7	•••	1576	1799	4721	506	608	1124	268	317	474	167	192	253	112	126	153
	•••	1656	2126	4732	546	723	1102	288	364	447	177	214	224	118	137	123
0.75	•••	1283	1524	4144	435	530	964	229	272	396	140	160	205	.	.	.
	•••	1372	1850	4137	473	633	938	247	312	368	149	177	175	.	.	.
0.8	•••	1024	1262	3442	361	442	773	186	219	304	.	.	.	.	.	.
	•••	1115	1564	3423	394	527	744	200	248	275	.	.	.	.	.	.
0.85	•••	791	1001	2615	279	340	551	.	.	.	.	.	.	.	.	.
	•••	873	1256	2590	304	401	521	.	.	.	.	.	.	.	.	.
0.9	•••	563	720	1666	.	.	.	.	.	.	.	.	.	.	.	.
	•••	625	901	1637	.	.	.	.	.	.	.	.	.	.	.	.

TABLE 2: ALPHA= 0.005 POWER= 0.9 — EXPECTED ACCRUAL THRU MINIMUM FOLLOW-UP= 130

REQUIRED NUMBER OF PATIENTS

PCONT	FACT	DEL=.05 1.0/.75	DEL=.05 .50/.25	DEL=.05 .00/BIN	DEL=.10 1.0/.75	DEL=.10 .50/.25	DEL=.10 .00/BIN	DEL=.15 1.0/.75	DEL=.15 .50/.25	DEL=.15 .00/BIN	DEL=.20 1.0/75	DEL=.20 .50/.25	DEL=.20 .00/BIN	DEL=.25 1.0/.75	DEL=.25 .50/.25	DEL=.25 .00/BIN
0.05	***	888	892	951	315	319	338	185	188	197	131	133	139	102	104	107
	•••	889	898	1637	316	324	521	186	191	275	132	135	175	103	105	123
0.1	***	1619	1627	1833	498	507	566	267	274	300	177	183	197	132	136	144
	•••	1622	1642	2590	501	520	744	270	282	368	180	188	224	133	139	152
0.15	***	2239	2251	2687	645	659	776	330	342	391	212	221	246	153	160	175
	•••	2243	2277	3423	650	683	938	334	357	447	216	230	265	156	166	175
0.2	***	2716	2736	3467	754	775	961	375	394	469	237	250	288	169	178	200
	•••	2723	2774	4137	761	810	1102	382	415	513	242	263	298	173	186	195
0.25	***	3046	3073	4148	826	856	1120	405	430	534	253	270	321	178	191	219
	•••	3055	3128	4732	836	904	1235	415	459	566	260	287	324	184	201	209
0.3	***	3233	3270	4717	865	905	1248	422	453	585	262	283	346	184	198	233
	•••	3246	3345	5208	879	966	1340	434	489	606	271	304	343	190	211	218
0.35	***	3290	3339	5166	875	925	1346	427	464	622	265	290	363	185	202	241
	•••	3307	3438	5565	892	1002	1414	441	507	632	275	314	354	192	216	223
0.4	***	3233	3297	5489	860	922	1411	422	466	644	262	290	372	183	201	244
	•••	3254	3424	5803	882	1013	1459	439	514	645	274	317	358	191	217	223
0.45	***	3080	3162	5684	825	901	1444	409	458	652	255	286	372	178	197	242
	•••	3107	3324	5922	852	1004	1474	428	511	645	268	314	354	186	214	218
0.5	***	2850	2955	5750	776	864	1445	390	443	645	244	276	365	170	190	234
	•••	2885	3157	5922	809	978	1459	411	499	632	257	306	343	179	206	209
0.55	***	2565	2698	5687	718	815	1413	366	422	624	230	262	349	160	179	222
	•••	2609	2943	5803	755	936	1414	389	479	606	244	291	324	168	195	195
0.6	***	2245	2412	5493	653	758	1349	339	394	588	213	244	325	147	165	204
	•••	2302	2700	5565	694	882	1340	361	451	566	226	272	298	155	180	175
0.65	***	1915	2119	5171	586	693	1252	308	361	538	193	222	293	132	148	181
	•••	1986	2440	5208	628	815	1235	330	414	513	205	247	265	139	161	152
0.7	***	1597	1832	4721	517	621	1124	273	322	474	170	195	253	114	128	153
	•••	1682	2170	4732	558	737	1102	293	369	447	180	216	224	120	138	123
0.75	***	1306	1558	4144	445	541	964	234	277	396	143	163	205	.	.	.
	•••	1400	1892	4137	484	645	938	252	316	368	151	179	175	.	.	.
0.8	***	1048	1294	3442	370	452	773	190	223	304	.	.	.	.	.	.
	•••	1142	1603	3423	403	537	744	204	251	275	.	.	.	.	.	.
0.85	***	813	1029	2615	286	347	551	.	.	.	.	.	.	.	.	.
	•••	897	1288	2590	311	407	521	.	.	.	.	.	.	.	.	.
0.9	***	579	740	1666	.	.	.	.	.	.	.	.	.	.	.	.
	•••	643	923	1637	.	.	.	.	.	.	.	.	.	.	.	.

TABLE 2: ALPHA= 0.005 POWER= 0.9   EXPECTED ACCRUAL THRU MINIMUM FOLLOW-UP= 140

PCONT= ***		DEL=.05			DEL=.10			DEL=.15			DEL=.20			DEL=.25		
FACT=		1.0 .75	.50 .25	.00 BIN	1.0 .75	.50 .25	.00 BIN	1.0 .75	.50 .25	.00 BIN	1.0 75	.50 .25	.00 BIN	1.0 .75	.50 .25	.00 BIN
		REQUIRED NUMBER OF PATIENTS														
0.05	***	888	892	951	315	319	338	185	188	197	131	133	138	102	104	107
	•••	890	899	1637	317	325	521	186	191	275	132	135	175	103	105	123
0.1	***	1620	1628	1833	499	508	566	267	275	300	178	183	197	132	136	144
	•••	1622	1644	2590	502	522	744	270	283	368	180	188	224	134	139	152
0.15	***	2240	2253	2687	646	662	776	331	344	391	213	222	246	154	161	175
	•••	2244	2281	3423	651	685	938	336	358	447	217	231	265	157	166	175
0.2	***	2718	2739	3467	755	778	961	377	396	469	238	251	288	170	179	200
	•••	2725	2780	4137	763	814	1102	384	417	513	243	264	298	174	187	195
0.25	***	3048	3077	4148	828	860	1120	408	433	534	255	272	321	180	192	219
	•••	3058	3136	4732	839	910	1235	417	462	566	262	289	324	185	202	209
0.3	***	3236	3276	4717	868	910	1248	425	457	585	264	286	346	185	200	233
	•••	3249	3356	5208	883	974	1340	437	493	606	273	306	343	192	212	218
0.35	***	3294	3347	5166	879	932	1346	430	469	622	267	293	363	187	204	241
	•••	3312	3453	5565	898	1011	1414	445	512	632	278	316	354	194	218	223
0.4	***	3238	3307	5489	865	931	1411	426	471	644	265	293	372	185	203	244
	•••	3261	3443	5803	888	1024	1459	443	519	645	277	320	358	193	219	223
0.45	***	3086	3175	5684	832	910	1444	414	464	652	258	289	372	180	199	242
	•••	3116	3348	5922	860	1016	1474	433	517	645	271	317	354	188	215	218
0.5	***	2858	2971	5750	784	875	1445	395	449	645	248	279	365	172	192	234
	•••	2896	3186	5922	818	991	1459	417	505	632	261	308	343	181	208	209
0.55	***	2575	2718	5687	727	827	1413	372	428	624	234	266	349	162	181	222
	•••	2623	2977	5803	765	950	1414	395	485	606	247	294	324	170	197	195
0.6	***	2259	2437	5493	663	770	1349	344	401	588	216	247	325	149	167	204
	•••	2319	2738	5565	705	896	1340	367	456	566	229	274	298	157	181	175
0.65	***	1932	2147	5171	596	705	1252	313	367	538	196	225	293	134	150	181
	•••	2008	2481	5208	639	829	1235	335	419	513	208	249	265	141	162	152
0.7	***	1617	1863	4721	527	633	1124	278	328	474	172	197	253	116	129	153
	•••	1706	2212	4732	569	749	1102	299	374	447	183	218	224	122	139	123
0.75	***	1329	1590	4144	455	552	964	239	281	396	145	165	205	.	.	.
	•••	1427	1933	4137	494	655	938	257	319	368	153	180	175	.	.	.
0.8	***	1071	1324	3442	378	461	773	194	226	304	.	.	.	.	.	.
	•••	1168	1639	3423	412	545	744	207	254	275	.	.	.	.	.	.
0.85	***	834	1054	2615	292	354	551	.	.	.	.	.	.	.	.	.
	•••	920	1318	2590	318	413	521	.	.	.	.	.	.	.	.	.
0.9	***	595	758	1666	.	.	.	.	.	.	.	.	.	.	.	.
	•••	660	943	1637	.	.	.	.	.	.	.	.	.	.	.	.

TABLE 2: ALPHA= 0.005 POWER= 0.9   EXPECTED ACCRUAL THRU MINIMUM FOLLOW-UP= 150

		DEL=.05			DEL=.10			DEL=.15			DEL=.20			DEL=.25		
FACT=		1.0 .75	.50 .25	.00 BIN	1.0 .75	.50 .25	.00 BIN	1.0 .75	.50 .25	.00 BIN	1.0 75	.50 .25	.00 BIN	1.0 .75	.50 .25	.00 BIN
PCONT=***					REQUIRED NUMBER OF PATIENTS											
0.05	***	889	893	950	316	320	338	185	189	197	131	134	138	102	104	107
	•••	890	900	1637	317	325	521	187	192	275	132	136	175	103	105	123
0.1	***	1620	1629	1833	500	510	566	268	276	300	178	184	197	133	137	144
	•••	1623	1646	2590	503	523	744	271	284	368	181	189	224	134	140	152
0.15	***	2241	2255	2687	647	664	776	332	345	391	214	223	246	155	161	175
	•••	2246	2285	3423	653	688	938	337	360	447	218	232	265	158	167	175
0.2	***	2719	2742	3467	757	781	961	379	398	469	239	253	288	171	180	199
	•••	2727	2786	4137	766	818	1102	386	419	513	245	265	298	175	188	195
0.25	***	3050	3082	4148	831	864	1120	410	436	534	256	274	321	181	193	219
	•••	3061	3145	4732	842	915	1235	420	465	566	264	290	324	186	203	209
0.3	***	3239	3282	4717	871	916	1248	427	460	585	266	288	346	187	201	233
	•••	3253	3367	5208	887	981	1340	440	497	606	275	308	343	193	213	218
0.35	***	3298	3355	5166	883	939	1346	433	473	622	270	295	363	189	205	241
	•••	3317	3467	5565	903	1020	1414	449	516	632	280	318	354	196	219	223
0.4	***	3243	3317	5489	870	939	1411	430	475	644	268	296	372	187	205	244
	•••	3268	3462	5803	895	1034	1459	448	524	645	280	322	358	195	220	223
0.45	***	3092	3187	5684	838	920	1444	418	469	652	261	292	372	182	201	242
	•••	3124	3371	5922	868	1028	1474	439	522	645	274	319	354	190	217	218
0.5	***	2866	2987	5750	792	885	1445	400	455	645	251	283	365	174	194	234
	•••	2907	3214	5922	827	1003	1459	422	510	632	264	311	343	183	209	209
0.55	***	2585	2738	5687	735	839	1413	377	434	624	237	269	349	164	183	222
	•••	2637	3010	5803	775	963	1414	400	490	606	250	297	324	172	198	195
0.6	***	2272	2461	5493	673	782	1349	350	406	588	220	250	325	151	169	204
	•••	2336	2775	5565	715	909	1340	373	461	566	233	277	298	159	183	175
0.65	***	1948	2175	5171	606	717	1252	319	373	538	199	227	293	136	151	181
	•••	2029	2520	5208	650	841	1235	341	424	513	211	251	265	142	163	152
0.7	***	1637	1894	4721	537	644	1124	283	333	474	175	200	253	117	130	153
	•••	1731	2252	4732	580	761	1102	304	378	447	186	220	224	123	139	123
0.75	***	1351	1620	4144	464	562	964	243	285	396	147	166	205	.	.	.
	•••	1452	1971	4137	504	666	938	261	323	368	155	181	175	.	.	.
0.8	***	1093	1352	3442	386	469	773	197	229	304	.	.	.	.	.	.
	•••	1193	1673	3423	420	554	744	211	256	275	.	.	.	.	.	.
0.85	***	854	1079	2615	298	360	551	.	.	.	.	.	.	.	.	.
	•••	941	1346	2590	324	419	521	.	.	.	.	.	.	.	.	.
0.9	***	610	776	1666	.	.	.	.	.	.	.	.	.	.	.	.
	•••	676	962	1637	.	.	.	.	.	.	.	.	.	.	.	.

TABLE 2: ALPHA= 0.005 POWER= 0.9   EXPECTED ACCRUAL THRU MINIMUM FOLLOW-UP= 160

		DEL=.05			DEL=.10			DEL=.15			DEL=.20			DEL=.25		
FACT=		1.0 .75	.50 .25	.00 BIN	1.0 .75	.50 .25	.00 BIN	1.0 .75	.50 .25	.00 BIN	1.0 75	.50 .25	.00 BIN	1.0 .75	.50 .25	.00 BIN
PCONT=***					REQUIRED NUMBER OF PATIENTS											
0.05	***	889	893	951	316	320	338	186	189	197	132	134	138	103	104	107
	***	890	901	1637	318	326	521	187	192	275	133	136	175	103	106	123
0.1	***	1621	1630	1833	501	511	566	269	277	300	179	185	197	133	137	144
	***	1624	1649	2590	504	525	744	272	285	368	181	189	224	135	140	152
0.15	***	2242	2257	2687	648	666	776	333	346	391	215	224	246	155	162	175
	***	2247	2289	3423	654	691	938	338	361	447	218	232	265	158	167	175
0.2	***	2721	2745	3467	759	784	961	380	400	469	240	254	288	171	181	199
	***	2729	2792	4137	768	822	1102	388	421	513	246	266	298	175	188	195
0.25	***	3052	3086	4148	833	868	1120	412	439	534	258	276	321	182	194	219
	***	3063	3153	4732	846	921	1235	422	467	566	265	292	324	187	204	209
0.3	***	3242	3288	4717	875	921	1248	430	464	585	268	290	346	188	203	233
	***	3257	3379	5208	891	988	1340	443	500	606	277	310	343	194	214	218
0.35	***	3302	3362	5166	887	946	1346	437	477	622	272	297	363	190	207	241
	***	3322	3482	5565	908	1028	1414	453	520	632	283	320	354	197	220	223
0.4	***	3248	3327	5489	875	947	1411	434	480	644	270	299	372	189	207	244
	***	3274	3481	5803	901	1044	1459	452	528	645	282	324	358	196	221	223
0.45	***	3099	3200	5684	844	929	1444	423	474	652	264	295	372	184	203	242
	***	3133	3395	5922	876	1038	1474	443	527	645	277	322	354	192	218	218
0.5	***	2874	3003	5750	799	896	1445	405	460	645	254	285	365	176	195	234
	***	2917	3242	5922	835	1015	1459	427	515	632	267	313	343	185	211	209
0.55	***	2596	2757	5687	744	850	1413	382	439	624	240	272	349	166	185	222
	***	2650	3042	5803	784	975	1414	405	495	606	253	299	324	174	199	195
0.6	***	2285	2485	5493	682	794	1349	355	412	588	222	253	325	153	170	204
	***	2354	2811	5565	725	921	1340	378	466	566	235	279	298	161	184	175
0.65	***	1965	2202	5171	616	728	1252	323	378	538	202	230	293	137	153	181
	***	2049	2558	5208	660	853	1235	346	429	513	214	253	265	144	164	152
0.7	***	1656	1923	4721	546	655	1124	288	337	474	177	202	253	118	131	153
	***	1754	2290	4732	590	772	1102	308	382	447	188	221	224	124	140	123
0.75	***	1372	1650	4144	473	572	964	247	289	396	149	168	205	.	.	.
	***	1477	2007	4137	513	675	938	265	326	368	157	183	175	.	.	.
0.8	***	1115	1380	3442	394	477	773	200	232	304	.	.	.	.	.	.
	***	1217	1706	3423	428	561	744	214	258	275	.	.	.	.	.	.
0.85	***	873	1102	2616	304	366	551	.	.	.	.	.	.	.	.	.
	***	962	1373	2590	330	424	521	.	.	.	.	.	.	.	.	.
0.9	***	625	793	1666	.	.	.	.	.	.	.	.	.	.	.	.
	***	691	980	1637	.	.	.	.	.	.	.	.	.	.	.	.

TABLE 2: ALPHA= 0.005 POWER= 0.9   EXPECTED ACCRUAL THRU MINIMUM FOLLOW-UP= 170

		DEL=.05			DEL=.10			DEL=.15			DEL=.20			DEL=.25		
FACT=		1.0 .75	.50 .25	.00 BIN	1.0 .75	.50 .25	.00 BIN	1.0 .75	.50 .25	.00 BIN	1.0 75	.50 .25	.00 BIN	1.0 .75	.50 .25	.00 BIN
PCONT=•••				REQUIRED NUMBER OF PATIENTS												
0.05	•••	889	894	951	316	321	338	186	189	197	132	134	138	103	104	107
	•••	891	902	1637	318	326	521	187	192	275	133	136	175	103	106	123
0.1	•••	1622	1631	1833	501	512	566	269	277	300	179	185	197	133	137	144
	•••	1625	1651	2590	505	526	744	273	285	368	182	190	224	135	140	152
0.15	•••	2243	2259	2687	650	667	776	334	348	391	215	225	246	156	162	175
	•••	2248	2293	3423	656	693	938	339	362	447	219	233	265	159	167	175
0.2	•••	2722	2748	3467	761	787	961	382	402	469	242	255	288	172	181	199
	•••	2731	2798	4137	770	826	1102	390	423	513	247	267	298	176	189	195
0.25	•••	3054	3090	4148	836	872	1120	414	441	534	259	277	321	183	195	219
	•••	3066	3161	4732	849	926	1235	425	470	566	267	293	324	188	204	209
0.3	•••	3245	3293	4717	878	926	1248	433	467	585	270	292	346	190	204	233
	•••	3261	3390	5208	895	995	1340	446	503	606	279	311	343	196	215	218
0.35	•••	3305	3370	5166	891	952	1346	440	481	622	274	300	363	192	208	241
	•••	3327	3497	5565	913	1036	1414	456	523	632	285	322	354	199	221	223
0.4	•••	3253	3336	5489	880	955	1411	437	484	644	273	301	372	190	208	244
	•••	3281	3500	5803	908	1053	1459	456	532	645	285	326	358	198	222	223
0.45	•••	3105	3212	5684	850	938	1444	427	479	652	267	297	372	186	204	242
	•••	3141	3417	5922	883	1049	1474	448	531	645	280	324	354	194	219	218
0.5	•••	2882	3019	5750	806	905	1445	410	465	645	256	288	365	178	197	234
	•••	2928	3270	5922	844	1026	1459	432	520	632	270	315	343	186	212	209
0.55	•••	2606	2777	5686	752	860	1413	387	444	624	243	274	349	168	186	222
	•••	2664	3074	5803	794	986	1414	410	500	606	256	301	324	176	200	195
0.6	•••	2298	2508	5493	691	804	1349	360	416	588	225	256	325	155	172	204
	•••	2371	2845	5565	735	932	1340	383	470	566	238	281	298	162	185	175
0.65	•••	1981	2229	5171	625	739	1252	328	382	538	204	232	293	139	154	181
	•••	2069	2594	5208	670	864	1235	351	433	513	216	255	265	145	165	152
0.7	•••	1675	1951	4721	555	665	1124	292	341	474	180	204	253	120	132	153
	•••	1777	2326	4732	599	782	1102	313	386	447	190	222	224	125	141	123
0.75	•••	1393	1678	4144	481	581	964	251	293	396	151	169	205	.	.	.
	•••	1501	2042	4137	521	684	938	269	329	368	159	184	175	.	.	.
0.8	•••	1135	1406	3442	401	485	773	203	235	304	.	.	.	.	.	.
	•••	1240	1737	3423	435	568	744	217	260	275	.	.	.	.	.	.
0.85	•••	891	1124	2615	309	371	551	.	.	.	.	.	.	.	.	.
	•••	982	1398	2590	335	429	521	.	.	.	.	.	.	.	.	.
0.9	•••	638	808	1666	.	.	.	.	.	.	.	.	.	.	.	.
	•••	706	997	1637	.	.	.	.	.	.	.	.	.	.	.	.

TABLE 2: ALPHA= 0.005 POWER= 0.9   EXPECTED ACCRUAL THRU MINIMUM FOLLOW-UP= 180

	DEL=.05			DEL=.10			DEL=.15			DEL=.20			DEL=.25		
FACT=	1.0 / .75	.50 / .25	.00 / BIN	1.0 / .75	.50 / .25	.00 / BIN	1.0 / .75	.50 / .25	.00 / BIN	1.0 / 75	.50 / .25	.00 / BIN	1.0 / .75	.50 / .25	.00 / BIN
PCONT=***				REQUIRED NUMBER OF PATIENTS											
0.05 ***	890	894	951	317	321	338	186	189	197	132	134	138	103	105	107
•••	891	903	1637	318	327	521	188	192	275	133	136	175	104	106	123
0.1 ***	1622	1632	1833	502	513	566	270	278	300	180	185	197	134	138	144
•••	1626	1653	2590	506	527	744	273	286	368	182	190	224	135	140	152
0.15 ***	2244	2261	2687	651	669	776	335	349	391	216	226	246	157	163	175
•••	2250	2296	3423	657	695	938	341	363	447	220	233	265	159	168	175
0.2 ***	2724	2751	3467	762	790	961	383	404	469	243	256	288	173	182	199
•••	2733	2804	4137	772	830	1102	391	425	513	248	268	298	177	189	195
0.25 ***	3057	3094	4148	838	876	1120	416	443	534	261	279	321	184	196	219
•••	3069	3169	4732	852	930	1235	427	472	566	268	294	324	189	205	209
0.3 ***	3248	3299	4717	881	931	1248	435	470	585	272	294	346	191	205	233
•••	3265	3401	5208	899	1001	1340	449	506	606	281	313	343	197	216	218
0.35 ***	3309	3377	5166	895	958	1346	443	484	622	276	302	363	193	209	241
•••	3332	3511	5565	918	1043	1414	460	527	632	287	324	354	200	222	223
0.4 ***	3258	3346	5489	885	962	1411	441	488	644	275	303	372	192	209	244
•••	3287	3518	5803	914	1061	1459	460	536	645	287	328	358	199	223	223
0.45 ***	3111	3225	5684	856	946	1444	431	483	652	269	300	372	187	206	242
•••	3149	3440	5922	890	1058	1474	452	535	645	282	326	354	195	220	218
0.5 ***	2890	3035	5750	813	915	1445	414	470	645	259	291	364	180	198	234
•••	2939	3296	5922	852	1036	1459	437	524	632	273	317	343	188	213	209
0.55 ***	2616	2796	5687	760	870	1413	392	449	624	245	277	349	169	187	222
•••	2677	3104	5803	803	997	1414	415	504	606	259	303	324	177	201	195
0.6 ***	2311	2531	5493	699	814	1349	364	421	588	228	258	325	156	173	204
•••	2387	2878	5565	744	943	1340	388	475	566	241	283	298	164	185	175
0.65 ***	1997	2255	5171	634	749	1252	333	387	538	207	234	293	140	155	181
•••	2089	2629	5208	679	874	1235	355	436	513	219	256	265	147	165	152
0.7 ***	1694	1978	4721	564	675	1124	296	345	474	182	205	253	121	133	153
•••	1799	2361	4732	608	791	1102	317	389	447	192	224	224	126	141	123
0.75 ***	1414	1705	4144	489	590	964	254	296	396	152	171	205	.	.	.
•••	1524	2075	4137	530	692	938	272	331	368	161	185	175	.	.	.
0.8 ***	1155	1431	3442	407	492	773	206	237	304	.	.	.	.	.	.
•••	1262	1766	3423	442	575	744	219	262	275	.	.	.	.	.	.
0.85 ***	908	1146	2616	314	376	551	.	.	.	.	.	.	.	.	.
•••	1001	1422	2590	340	433	521	.	.	.	.	.	.	.	.	.
0.9 ***	651	823	1666	.	.	.	.	.	.	.	.	.	.	.	.
•••	720	1013	1637	.	.	.	.	.	.	.	.	.	.	.	.

TABLE 2: ALPHA= 0.005 POWER= 0.9   EXPECTED ACCRUAL THRU MINIMUM FOLLOW-UP= 190

PCONT=•••		DEL=.05			DEL=.10			DEL=.15			DEL=.20			DEL=.25		
FACT=		1.0 .75	.50 .25	.00 BIN	1.0 .75	.50 .25	.00 BIN	1.0 .75	.50 .25	.00 BIN	1.0 75	.50 .25	.00 BIN	1.0 .75	.50 .25	.00 BIN
				REQUIRED NUMBER OF PATIENTS												
0.05	•••	890	895	951	317	322	338	186	190	197	132	134	138	103	105	107
	•••	892	904	1637	319	327	521	188	192	275	133	136	175	104	106	123
0.1	•••	1623	1634	1833	503	514	566	271	279	300	180	186	197	134	138	144
	•••	1626	1655	2590	507	529	744	274	287	368	183	190	224	136	141	152
0.15	•••	2245	2263	2687	652	671	776	336	350	391	217	226	246	157	163	175
	•••	2251	2300	3423	659	698	938	342	364	447	221	234	265	160	168	175
0.2	•••	2725	2754	3467	764	793	961	385	405	469	244	257	288	174	183	199
	•••	2735	2809	4137	774	833	1102	393	426	513	249	269	298	178	190	195
0.25	•••	3059	3099	4148	840	880	1120	418	446	534	262	280	321	185	197	219
	•••	3072	3178	4732	854	935	1235	429	474	566	270	295	324	190	206	209
0.3	•••	3250	3305	4717	884	936	1248	438	473	585	273	295	346	192	206	233
	•••	3268	3412	5208	903	1007	1340	452	508	606	283	314	343	198	216	218
0.35	•••	3313	3385	5166	899	964	1346	446	488	622	278	303	363	194	210	241
	•••	3337	3526	5565	923	1050	1414	463	530	632	289	325	354	201	222	223
0.4	•••	3263	3356	5489	890	969	1411	445	492	644	277	306	372	193	211	244
	•••	3294	3536	5803	919	1070	1459	464	540	645	289	330	358	201	224	223
0.45	•••	3118	3238	5684	862	955	1444	435	487	652	272	302	372	189	207	242
	•••	3158	3462	5922	897	1067	1474	456	539	645	284	328	354	197	221	218
0.5	•••	2899	3050	5750	820	924	1445	418	474	645	262	293	364	181	200	234
	•••	2950	3322	5922	860	1046	1459	441	528	632	275	319	343	189	213	209
0.55	•••	2626	2815	5687	767	880	1413	396	453	624	248	279	349	171	189	222
	•••	2691	3134	5803	811	1007	1414	420	508	606	261	305	324	179	202	195
0.6	•••	2324	2553	5493	707	824	1348	369	426	588	230	260	325	158	174	204
	•••	2404	2910	5565	753	953	1340	392	478	566	243	284	298	165	186	175
0.65	•••	2013	2280	5171	642	759	1252	337	391	538	209	236	293	141	156	181
	•••	2109	2662	5208	688	884	1235	359	440	513	221	258	265	148	166	152
0.7	•••	1713	2005	4721	572	684	1124	300	349	474	184	207	253	122	134	153
	•••	1821	2394	4732	617	800	1102	321	392	447	194	225	224	127	142	123
0.75	•••	1433	1731	4144	496	598	964	258	299	396	154	172	205	.	.	.
	•••	1547	2106	4137	538	700	938	275	334	368	162	185	175	.	.	.
0.8	•••	1174	1455	3442	414	499	773	208	239	304	.	.	.	.	.	.
	•••	1283	1794	3423	449	581	744	222	264	275	.	.	.	.	.	.
0.85	•••	925	1166	2616	319	381	551	.	.	.	.	.	.	.	.	.
	•••	1020	1444	2590	345	437	521	.	.	.	.	.	.	.	.	.
0.9	•••	664	838	1666	.	.	.	.	.	.	.	.	.	.	.	.
	•••	733	1028	1637	.	.	.	.	.	.	.	.	.	.	.	.

TABLE 2: ALPHA= 0.005 POWER= 0.9   EXPECTED ACCRUAL THRU MINIMUM FOLLOW-UP= 200

		DEL=.05			DEL=.10			DEL=.15			DEL=.20			DEL=.25		
FACT=		1.0 .75	.50 .25	.00 BIN	1.0 .75	.50 .25	.00 BIN	1.0 .75	.50 .25	.00 BIN	1.0 75	.50 .25	.00 BIN	1.0 .75	.50 .25	.00 BIN
PCONT=***					REQUIRED	NUMBER OF	PATIENTS									
0.05	***	890	895	951	317	322	338	187	190	197	132	135	138	103	105	107
	•••	892	905	1637	319	327	521	188	193	275	133	136	175	104	106	123
0.1	***	1623	1635	1833	503	515	566	271	279	300	181	186	197	134	138	144
	•••	1627	1657	2590	508	530	744	274	287	368	183	190	224	136	141	152
0.15	***	2246	2265	2687	653	673	776	337	351	391	218	227	246	158	164	175
	•••	2252	2304	3423	660	700	938	343	365	447	222	234	265	160	168	175
0.2	***	2727	2756	3467	766	795	961	386	407	469	245	258	288	175	183	199
	•••	2737	2815	4137	776	836	1102	394	428	513	250	270	298	178	190	195
0.25	***	3061	3103	4148	842	884	1120	420	448	534	264	281	321	186	198	219
	•••	3075	3186	4732	857	939	1235	431	476	566	271	296	324	191	206	209
0.3	***	3253	3310	4717	887	941	1248	440	475	585	275	297	346	193	207	233
	•••	3272	3423	5208	906	1012	1340	454	511	606	284	315	343	199	217	218
0.35	***	3317	3392	5166	903	970	1346	449	491	622	280	305	363	196	211	241
	•••	3342	3540	5565	928	1057	1414	466	533	632	291	327	354	203	223	223
0.4	***	3268	3366	5489	895	976	1411	448	496	644	280	308	372	195	212	244
	•••	3300	3554	5803	925	1077	1459	467	543	645	291	331	358	202	225	223
0.45	***	3124	3250	5684	868	962	1444	439	491	652	274	304	372	190	208	242
	•••	3166	3484	5922	904	1076	1474	460	542	645	287	329	354	198	222	218
0.5	***	2907	3066	5750	827	932	1445	422	478	645	264	295	365	183	201	234
	•••	2960	3348	5922	867	1055	1459	445	532	632	277	321	343	191	214	209
0.55	***	2637	2834	5687	775	889	1413	400	458	624	250	281	349	172	190	222
	•••	2704	3162	5803	819	1017	1414	424	511	606	263	306	324	180	203	195
0.6	***	2337	2576	5493	715	834	1349	373	430	588	233	262	325	159	175	204
	•••	2420	2942	5565	762	963	1340	397	482	566	245	286	298	166	187	175
0.65	***	2029	2304	5171	650	768	1252	341	395	538	211	238	293	143	157	181
	•••	2128	2694	5208	697	893	1235	363	443	513	223	259	265	149	167	152
0.7	***	1731	2030	4721	580	692	1124	304	352	474	186	209	253	123	134	153
	•••	1842	2426	4732	625	808	1102	324	395	447	196	226	224	128	142	123
0.75	***	1452	1757	4144	504	605	964	261	302	396	155	173	205	.	.	.
	•••	1568	2137	4137	545	707	938	278	336	368	163	186	175	.	.	.
0.8	***	1193	1479	3442	420	505	773	211	241	304	.	.	.	.	.	.
	•••	1304	1821	3423	455	586	744	224	265	275	.	.	.	.	.	.
0.85	***	942	1185	2616	324	385	551	.	.	.	.	.	.	.	.	.
	•••	1037	1466	2590	350	440	521	.	.	.	.	.	.	.	.	.
0.9	***	676	852	1666	.	.	.	.	.	.	.	.	.	.	.	.
	•••	746	1042	1637	.	.	.	.	.	.	.	.	.	.	.	.

TABLE 2: ALPHA= 0.005 POWER= 0.9   EXPECTED ACCRUAL THRU MINIMUM FOLLOW-UP= 225

		DEL=.05			DEL=.10			DEL=.15			DEL=.20			DEL=.25		
PCONT=***	FACT=	1.0 .75	.50 .25	.00 BIN	1.0 .75	.50 .25	.00 BIN	1.0 .75	.50 .25	.00 BIN	1.0 75	.50 .25	.00 BIN	1.0 .75	.50 .25	.00 BIN
					REQUIRED NUMBER OF PATIENTS											
0.05	***	891	897	950	318	323	338	187	190	197	133	135	138	104	105	107
	***	893	907	1637	320	328	521	189	193	275	134	136	175	104	106	123
0.1	***	1625	1638	1833	505	517	566	273	281	300	182	187	197	135	139	144
	***	1629	1663	2590	510	532	744	276	288	368	184	191	224	137	141	152
0.15	***	2248	2270	2687	656	677	776	339	354	391	219	228	246	159	165	175
	***	2255	2313	3423	664	704	938	345	367	447	223	236	265	161	169	175
0.2	***	2730	2764	3467	770	802	961	389	411	469	247	260	288	176	185	200
	***	2742	2829	4137	781	843	1102	398	431	513	253	271	298	180	191	195
0.25	***	3066	3113	4148	848	892	1120	424	453	534	267	284	321	188	199	219
	***	3082	3206	4732	864	950	1235	436	480	566	274	298	324	193	207	209
0.3	***	3260	3325	4717	894	952	1248	446	482	585	279	300	346	195	209	233
	***	3282	3450	5208	916	1025	1340	460	516	606	288	318	343	201	219	218
0.35	***	3326	3411	5166	912	984	1346	456	498	622	284	309	363	199	214	241
	***	3355	3575	5565	939	1073	1414	473	539	632	295	330	354	205	225	223
0.4	***	3280	3390	5489	907	992	1411	456	504	644	284	312	372	198	214	244
	***	3317	3598	5803	939	1095	1459	475	550	645	296	335	358	205	226	223
0.45	***	3140	3281	5684	882	981	1444	447	500	652	279	309	372	194	211	242
	***	3187	3536	5922	920	1096	1474	469	550	645	292	333	354	201	223	218
0.5	***	2927	3104	5750	843	952	1445	432	488	645	270	300	365	186	203	234
	***	2987	3409	5922	885	1076	1459	455	540	632	283	324	343	194	216	209
0.55	***	2662	2881	5687	793	910	1413	410	467	624	256	286	349	176	192	222
	***	2738	3231	5803	839	1039	1414	434	520	606	269	310	324	183	204	195
0.6	***	2368	2629	5493	734	855	1349	383	439	588	238	267	325	162	177	204
	***	2461	3015	5565	782	984	1340	406	490	566	250	289	298	169	188	175
0.65	***	2067	2363	5171	669	789	1252	350	404	538	216	242	293	145	159	181
	***	2175	2770	5208	717	914	1235	373	450	513	227	262	265	151	168	152
0.7	***	1774	2091	4721	598	712	1124	312	360	474	190	212	253	125	136	153
	***	1893	2501	4732	644	827	1102	333	401	447	200	228	224	130	143	123
0.75	***	1498	1816	4144	520	623	964	268	308	396	159	176	205	.	.	.
	***	1620	2207	4137	562	723	938	285	341	368	166	188	175	.	.	.
0.8	***	1237	1533	3442	434	519	773	216	246	304	.	.	.	.	.	.
	***	1352	1883	3423	469	599	744	229	269	275	.	.	.	.	.	.
0.85	***	980	1231	2615	334	395	551	.	.	.	.	.	.	.	.	.
	***	1079	1516	2590	360	449	521	.	.	.	.	.	.	.	.	.
0.9	***	704	883	1666	.	.	.	.	.	.	.	.	.	.	.	.
	***	776	1074	1637	.	.	.	.	.	.	.	.	.	.	.	.

CRC Handbook of Sample Size Guidelines for Clinical Trials

TABLE 2: ALPHA= 0.005 POWER= 0.9   EXPECTED ACCRUAL THRU MINIMUM FOLLOW-UP= 250

PCONT=***	DEL=.05			DEL=.10			DEL=.15			DEL=.20			DEL=.25		
FACT=	1.0 / .75	.50 / .25	.00 BIN	1.0 / .75	.50 / .25	.00 BIN	1.0 / .75	.50 / .25	.00 BIN	1.0 / 75	.50 / .25	.00 BIN	1.0 / .75	.50 / .25	.00 BIN
						REQUIRED NUMBER OF PATIENTS									
0.05 ***	891	898	951	319	324	338	188	191	197	133	135	138	104	105	107
•••	894	909	1637	321	329	521	189	193	275	134	137	175	104	106	123
0.1 ***	1626	1641	1833	507	520	566	274	282	300	183	188	197	136	139	144
•••	1631	1668	2590	512	534	744	277	289	368	185	192	224	137	141	152
0.15 ***	2251	2275	2687	658	681	776	341	356	391	221	230	246	160	165	175
•••	2259	2322	3423	667	709	938	347	369	447	224	236	265	162	169	175
0.2 ***	2734	2771	3467	774	808	961	392	414	469	249	262	288	178	186	199
•••	2747	2843	4137	786	850	1102	401	434	513	255	272	298	181	192	195
0.25 ***	3071	3124	4148	854	901	1120	428	457	534	269	287	321	190	201	219
•••	3089	3226	4732	871	959	1235	440	484	566	277	300	324	195	208	209
0.3 ***	3267	3339	4717	902	962	1248	451	487	585	282	303	346	198	210	233
•••	3291	3476	5208	924	1037	1340	466	521	606	291	320	343	203	220	218
0.35 ***	3336	3430	5166	922	997	1346	462	505	622	288	313	363	201	216	241
•••	3367	3609	5565	950	1087	1414	480	545	632	299	332	354	207	226	223
0.4 ***	3292	3414	5489	918	1007	1411	463	512	644	289	316	372	200	216	244
•••	3333	3639	5803	952	1111	1459	483	557	645	300	337	358	208	228	223
0.45 ***	3156	3312	5684	895	998	1444	455	508	652	284	313	372	196	213	242
•••	3208	3586	5922	935	1113	1474	477	557	645	296	336	354	204	225	218
0.5 ***	2947	3142	5750	858	971	1445	440	496	645	274	304	364	189	206	234
•••	3014	3467	5922	902	1095	1459	464	547	632	287	328	343	196	218	209
0.55 ***	2688	2925	5687	809	929	1413	419	476	624	261	290	349	178	194	222
•••	2771	3295	5803	857	1058	1414	443	527	606	273	313	324	186	206	195
0.6 ***	2400	2680	5493	751	875	1349	391	447	588	242	270	325	164	179	204
•••	2500	3084	5565	801	1004	1340	415	497	566	255	292	298	171	190	175
0.65 ***	2104	2418	5171	686	808	1252	358	411	538	220	246	293	147	160	181
•••	2220	2840	5208	736	932	1235	381	457	513	231	264	265	153	169	152
0.7 ***	1816	2148	4721	615	730	1124	320	367	474	194	215	253	127	137	153
•••	1942	2570	4732	662	844	1102	340	406	447	203	230	224	132	144	123
0.75 ***	1541	1872	4144	536	639	964	274	314	396	162	178	205	.	.	.
•••	1669	2272	4137	578	738	938	291	345	368	169	189	175	.	.	.
0.8 ***	1278	1584	3442	447	532	773	221	250	304	.	.	.	.	.	.
•••	1397	1940	3423	482	610	744	234	272	275	.	.	.	.	.	.
0.85 ***	1015	1273	2616	344	404	551	.	.	.	.	.	.	.	.	.
•••	1117	1560	2590	369	456	521	.	.	.	.	.	.	.	.	.
0.9 ***	730	912	1666	.	.	.	.	.	.	.	.	.	.	.	.
•••	803	1103	1637	.	.	.	.	.	.	.	.	.	.	.	.

313

TABLE 2: ALPHA= 0.005 POWER= 0.9	EXPECTED ACCRUAL THRU MINIMUM FOLLOW-UP= 275

		DEL=.05			DEL=.10			DEL=.15			DEL=.20			DEL=.25		
FACT=		1.0 .75	.50 .25	.00 BIN	1.0 .75	.50 .25	.00 BIN	1.0 .75	.50 .25	.00 BIN	1.0 75	.50 .25	.00 BIN	1.0 .75	.50 .25	.00 BIN
PCONT=•••					REQUIRED NUMBER OF PATIENTS											
0.05	***	892	899	950	319	325	338	188	191	197	133	135	138	104	105	107
	•••	894	911	1637	321	330	521	189	194	275	134	137	175	105	106	123
0.1	***	1628	1643	1833	508	521	566	275	283	300	183	188	197	136	139	144
	•••	1633	1673	2590	513	536	744	278	290	368	186	192	224	138	142	152
0.15	***	2253	2280	2687	661	685	776	343	358	391	222	231	246	161	166	175
	•••	2262	2331	3423	670	712	938	349	370	447	226	237	265	163	170	175
0.2	***	2738	2779	3467	777	813	961	395	417	469	251	264	288	179	187	199
	•••	2751	2857	4137	791	856	1102	404	436	513	257	274	298	182	192	195
0.25	***	3076	3134	4148	859	908	1120	432	461	534	272	289	321	191	202	219
	•••	3096	3245	4732	877	967	1235	444	487	566	279	302	324	196	209	209
0.3	***	3275	3353	4717	909	972	1248	456	492	585	285	306	346	200	212	233
	•••	3301	3502	5208	933	1047	1340	471	525	606	294	322	343	205	221	218
0.35	***	3345	3449	5166	930	1009	1346	468	511	622	292	316	363	203	217	241
	•••	3380	3641	5565	960	1099	1414	486	550	632	302	334	354	210	227	223
0.4	***	3304	3439	5489	929	1021	1411	469	518	644	293	319	372	203	218	244
	•••	3350	3680	5803	964	1126	1459	490	562	645	304	340	358	210	229	223
0.45	***	3171	3342	5684	908	1013	1444	462	516	652	288	316	372	199	215	242
	•••	3229	3634	5922	949	1129	1474	485	563	645	300	338	354	206	226	218
0.5	***	2967	3179	5750	872	988	1445	448	504	645	279	308	365	191	208	234
	•••	3040	3521	5922	918	1112	1459	471	553	632	291	330	343	199	219	209
0.55	***	2713	2969	5687	824	947	1413	426	483	624	265	294	349	181	196	222
	•••	2803	3355	5803	873	1075	1414	450	533	606	277	315	324	188	207	195
0.6	***	2431	2728	5493	767	893	1349	399	455	588	247	274	325	167	181	204
	•••	2538	3147	5565	818	1021	1340	423	502	566	259	294	298	173	191	175
0.65	***	2140	2470	5171	702	826	1252	366	418	538	224	249	293	149	162	181
	•••	2263	2905	5208	752	949	1235	388	462	513	235	267	265	155	170	152
0.7	***	1855	2201	4721	630	746	1124	326	373	474	197	217	253	128	138	153
	•••	1987	2633	4732	678	859	1102	346	411	447	206	232	224	133	145	123
0.75	***	1582	1923	4144	550	653	964	280	319	396	164	180	205	.	.	.
	•••	1714	2331	4137	592	750	938	297	349	368	171	191	175	.	.	.
0.8	***	1316	1630	3442	459	543	773	225	253	304	.	.	.	.	.	.
	•••	1439	1991	3423	494	619	744	238	274	275	.	.	.	.	.	.
0.85	***	1048	1311	2615	352	412	551	.	.	.	.	.	.	.	.	.
	•••	1152	1601	2590	378	462	521	.	.	.	.	.	.	.	.	.
0.9	***	754	938	1666	.	.	.	.	.	.	.	.	.	.	.	.
	•••	828	1129	1637	.	.	.	.	.	.	.	.	.	.	.	.

TABLE 2: ALPHA= 0.005 POWER= 0.9   EXPECTED ACCRUAL THRU MINIMUM FOLLOW-UP= 300

	DEL=.05			DEL=.10			DEL=.15			DEL=.20			DEL=.25		
FACT=	1.0 .75	.50 .25	.00 BIN	1.0 .75	.50 .25	.00 BIN	1.0 .75	.50 .25	.00 BIN	1.0 75	.50 .25	.00 BIN	1.0 .75	.50 .25	.00 BIN
PCONT=***				REQUIRED NUMBER OF PATIENTS											
0.05 ***	893	900	950	320	325	338	189	192	197	134	136	138	104	105	107
•••	895	912	1637	322	330	521	190	194	275	134	137	175	105	106	123
0.1 ***	1629	1646	1833	510	523	566	276	284	300	184	189	197	137	140	144
•••	1635	1677	2590	515	538	744	279	290	368	186	192	224	138	142	152
0.15 ***	2255	2285	2687	664	688	775	345	360	391	223	232	246	161	167	175
•••	2265	2340	3423	673	715	938	351	372	447	227	238	265	164	170	175
0.2 ***	2742	2786	3467	781	818	961	398	419	469	253	265	288	180	187	199
•••	2756	2870	4137	795	861	1102	407	438	513	258	274	298	183	193	195
0.25 ***	3082	3145	4148	864	915	1120	436	465	534	274	290	321	193	203	219
•••	3103	3263	4732	884	974	1235	448	490	566	281	303	324	197	210	209
0.3 ***	3282	3367	4717	916	981	1248	460	496	585	288	308	346	201	213	232
•••	3310	3526	5208	941	1056	1340	475	529	606	297	323	343	207	221	218
0.35 ***	3355	3467	5166	939	1020	1346	473	516	622	295	318	363	205	219	241
•••	3392	3673	5565	970	1110	1414	491	554	632	305	336	354	211	228	223
0.4 ***	3317	3462	5489	939	1034	1411	475	524	644	296	322	372	205	220	244
•••	3366	3718	5803	976	1138	1459	496	567	645	307	342	358	212	230	223
0.45 ***	3187	3371	5684	920	1028	1444	469	522	652	292	319	372	201	217	242
•••	3250	3679	5922	962	1143	1474	491	568	645	304	341	354	208	227	218
0.5 ***	2987	3214	5750	885	1003	1445	455	510	645	283	311	364	194	209	234
•••	3066	3572	5922	932	1127	1459	478	558	632	295	332	343	201	220	209
0.55 ***	2738	3010	5686	839	963	1413	434	490	624	269	297	349	183	198	222
•••	2834	3411	5803	889	1091	1414	458	538	606	281	318	324	190	208	195
0.6 ***	2461	2775	5493	782	909	1348	406	461	588	250	277	325	169	183	204
•••	2575	3206	5565	833	1036	1340	430	508	566	262	296	298	175	192	175
0.65 ***	2175	2520	5171	717	841	1252	373	424	538	227	251	292	151	163	181
•••	2304	2965	5208	768	963	1235	395	466	513	238	268	265	157	171	152
0.7 ***	1894	2251	4721	644	761	1124	333	378	474	200	220	253	130	139	153
•••	2030	2692	4732	692	872	1102	352	415	447	208	234	224	134	145	123
0.75 ***	1620	1971	4144	562	666	964	285	323	396	166	181	205	.	.	.
•••	1756	2385	4137	605	761	938	302	352	368	173	192	175	.	.	.
0.8 ***	1352	1673	3442	469	553	773	229	256	304	.	.	.	.	.	.
•••	1479	2038	3423	505	628	744	241	276	275	.	.	.	.	.	.
0.85 ***	1079	1346	2615	360	419	551	.	.	.	.	.	.	.	.	.
•••	1185	1638	2590	385	467	521	.	.	.	.	.	.	.	.	.
0.9 ***	776	962	1666	.	.	.	.	.	.	.	.	.	.	.	.
•••	852	1152	1637	.	.	.	.	.	.	.	.	.	.	.	.

TABLE 2: ALPHA= 0.005 POWER= 0.9   EXPECTED ACCRUAL THRU MINIMUM FOLLOW-UP= 325

		DEL=.05			DEL=.10			DEL=.15			DEL=.20			DEL=.25		
FACT=		1.0 / .75	.50 / .25	.00 / BIN	1.0 / .75	.50 / .25	.00 / BIN	1.0 / .75	.50 / .25	.00 / BIN	1.0 / 75	.50 / .25	.00 / BIN	1.0 / .75	.50 / .25	.00 / BIN

PCONT=*** REQUIRED NUMBER OF PATIENTS

PCONT		DEL=.05			DEL=.10			DEL=.15			DEL=.20			DEL=.25		
0.05	***	893	902	950	320	326	338	189	192	197	134	136	138	104	106	107
	***	896	914	1637	323	330	521	190	194	275	135	137	175	105	106	123
0.1	***	1630	1649	1833	511	525	566	277	285	300	185	189	197	137	140	144
	***	1637	1682	2590	517	539	744	280	291	368	187	193	224	138	142	152
0.15	***	2258	2290	2687	666	691	775	347	361	391	224	232	246	162	167	175
	***	2268	2348	3423	676	719	938	353	373	447	228	238	265	164	171	175
0.2	***	2745	2793	3467	785	823	961	400	422	469	254	266	288	181	188	199
	***	2761	2883	4137	800	866	1102	409	440	513	260	275	298	184	193	195
0.25	***	3087	3155	4148	869	922	1120	439	468	534	276	292	321	194	204	219
	***	3110	3281	4732	889	980	1235	451	493	566	283	304	324	199	210	209
0.3	***	3289	3381	4717	922	990	1248	465	500	585	290	310	346	203	214	233
	***	3320	3550	5208	948	1065	1340	480	532	606	299	325	343	208	222	218
0.35	***	3364	3486	5166	947	1030	1346	478	521	622	298	321	363	207	220	241
	***	3405	3703	5565	979	1120	1414	496	558	632	308	338	354	213	229	223
0.4	***	3329	3486	5489	949	1046	1411	481	529	644	299	325	372	207	221	244
	***	3382	3755	5803	987	1150	1459	502	571	645	311	344	358	213	231	223
0.45	***	3203	3400	5684	931	1041	1444	475	528	652	295	322	372	203	218	242
	***	3271	3723	5922	975	1156	1474	497	573	645	307	343	354	210	228	218
0.5	***	3007	3249	5750	898	1018	1445	461	517	645	286	314	365	196	211	234
	***	3092	3621	5922	946	1141	1459	485	563	632	298	335	343	203	221	209
0.55	***	2762	3050	5686	852	978	1413	440	496	624	272	300	349	185	199	222
	***	2865	3464	5803	903	1105	1414	464	543	606	284	320	324	192	209	195
0.6	***	2491	2819	5493	796	924	1349	413	467	588	254	279	325	171	184	204
	***	2611	3262	5565	848	1050	1340	436	512	566	265	298	298	177	193	175
0.65	***	2209	2567	5171	731	856	1252	379	430	538	231	253	292	153	164	181
	***	2344	3022	5208	782	977	1235	401	471	513	241	270	265	158	172	152
0.7	***	1930	2299	4721	658	774	1124	338	383	474	202	222	253	131	140	153
	***	2072	2747	4732	706	884	1102	358	418	447	211	235	224	135	146	123
0.75	***	1657	2016	4144	574	677	964	290	327	396	168	183	205	.	.	.
	***	1797	2435	4137	617	771	938	306	354	368	175	193	175	.	.	.
0.8	***	1386	1714	3442	479	563	773	233	259	304	.	.	.	.	.	.
	***	1516	2082	3423	515	636	744	244	278	275	.	.	.	.	.	.
0.85	***	1108	1379	2615	367	425	551	.	.	.	.	.	.	.	.	.
	***	1216	1672	2590	392	472	521	.	.	.	.	.	.	.	.	.
0.9	***	797	984	1666	.	.	.	.	.	.	.	.	.	.	.	.
	***	873	1173	1637	.	.	.	.	.	.	.	.	.	.	.	.

TABLE 2: ALPHA= 0.005 POWER= 0.9   EXPECTED ACCRUAL THRU MINIMUM FOLLOW-UP= 350

		DEL=.05			DEL=.10			DEL=.15			DEL=.20			DEL=.25		
FACT=		1.0	.50	.00	1.0	.50	.00	1.0	.50	.00	1.0	.50	.00	1.0	.50	.00
		.75	.25	BIN	.75	.25	BIN	.75	.25	BIN	75	.25	BIN	.75	.25	BIN
PCONT=***					REQUIRED NUMBER OF PATIENTS											
0.05	***	894	903	950	321	327	338	189	192	197	134	136	138	104	106	107
	•••	897	915	1637	323	331	521	191	194	275	135	137	175	105	106	123
0.1	***	1632	1652	1833	512	527	566	278	286	300	185	190	197	138	140	144
	•••	1639	1686	2590	518	541	744	281	292	368	187	193	224	139	142	152
0.15	***	2260	2294	2687	668	694	776	348	363	391	225	233	246	163	168	175
	•••	2272	2356	3423	678	721	938	355	374	447	229	239	265	165	171	175
0.2	***	2749	2801	3467	789	828	961	403	424	469	256	268	288	182	189	199
	•••	2766	2895	4137	804	870	1102	412	442	513	261	276	298	185	194	195
0.25	***	3092	3165	4148	874	928	1120	442	471	534	278	294	321	195	205	219
	•••	3117	3299	4732	895	986	1235	454	495	566	285	305	324	200	211	209
0.3	***	3296	3395	4717	929	998	1248	468	504	585	293	312	346	204	215	232
	•••	3329	3574	5208	955	1072	1340	484	535	606	301	326	343	209	223	218
0.35	***	3373	3504	5166	955	1040	1346	483	525	622	300	323	363	209	221	241
	•••	3417	3732	5565	988	1130	1414	501	561	632	311	339	354	214	230	223
0.4	***	3341	3509	5489	958	1057	1411	486	534	644	302	327	372	209	223	244
	•••	3398	3790	5803	997	1161	1459	507	575	645	313	346	358	215	232	223
0.45	***	3219	3429	5684	942	1053	1444	481	533	652	299	325	372	205	219	242
	•••	3291	3764	5922	986	1168	1474	503	577	645	310	344	354	211	229	218
0.5	***	3027	3283	5750	910	1031	1445	467	522	645	289	316	365	198	212	234
	•••	3117	3667	5922	959	1153	1459	491	568	632	301	336	343	204	222	209
0.55	***	2787	3089	5687	865	992	1413	447	502	624	275	302	349	187	201	222
	•••	2896	3514	5803	916	1118	1414	470	547	606	287	321	324	193	210	195
0.6	***	2519	2862	5493	809	938	1349	419	473	588	257	282	325	172	185	204
	•••	2646	3315	5565	862	1063	1340	442	516	566	268	300	298	178	193	175
0.65	***	2242	2611	5171	744	869	1252	384	435	538	233	256	293	154	165	181
	•••	2382	3075	5208	796	988	1235	406	474	513	243	271	265	159	172	152
0.7	***	1965	2343	4721	670	787	1124	343	387	474	204	223	253	132	141	153
	•••	2111	2798	4732	718	895	1102	363	421	447	213	236	224	136	146	123
0.75	***	1692	2058	4144	585	688	964	294	330	396	170	184	205	.	.	.
	•••	1835	2483	4137	628	780	938	310	357	368	176	193	175	.	.	.
0.8	***	1419	1751	3442	488	571	773	236	261	304	.	.	.	.	.	.
	•••	1550	2123	3423	524	642	744	247	279	275	.	.	.	.	.	.
0.85	***	1135	1410	2616	374	431	551	.	.	.	.	.	.	.	.	.
	•••	1245	1703	2590	398	476	521	.	.	.	.	.	.	.	.	.
0.9	***	816	1005	1666	.	.	.	.	.	.	.	.	.	.	.	.
	•••	893	1193	1637	.	.	.	.	.	.	.	.	.	.	.	.

TABLE 2: ALPHA= 0.005 POWER= 0.9   EXPECTED ACCRUAL THRU MINIMUM FOLLOW-UP= 375

	DEL=.05			DEL=.10			DEL=.15			DEL=.20			DEL=.25		
FACT=	1.0 .75	.50 .25	.00 BIN	1.0 .75	.50 .25	.00 BIN	1.0 .75	.50 .25	.00 BIN	1.0 75	.50 .25	.00 BIN	1.0 .75	.50 .25	.00 BIN
PCONT=***					REQUIRED NUMBER OF PATIENTS										
0.05 ***	895	904	950	322	327	338	190	193	197	134	136	138	105	106	107
***	898	917	1637	324	331	521	191	194	275	135	137	175	105	107	123
0.1 ***	1633	1655	1833	514	528	566	279	286	300	186	190	197	138	140	144
***	1640	1690	2590	520	542	744	282	292	368	188	193	224	139	142	152
0.15 ***	2263	2299	2687	671	697	776	350	364	391	226	234	246	163	168	175
***	2275	2364	3423	681	724	938	356	375	447	230	239	265	165	171	175
0.2 ***	2753	2808	3467	792	832	961	405	426	469	257	269	288	183	190	200
***	2771	2907	4137	808	874	1102	414	443	513	262	277	298	186	194	195
0.25 ***	3097	3176	4148	879	934	1120	445	474	534	280	295	321	197	205	219
***	3124	3316	4732	901	992	1235	457	497	566	287	306	324	200	211	209
0.3 ***	3303	3409	4717	935	1005	1248	472	508	585	295	314	346	205	216	233
***	3339	3596	5208	962	1079	1340	487	537	606	303	327	343	210	223	218
0.35 ***	3383	3522	5166	963	1049	1346	487	529	622	303	325	363	210	222	241
***	3430	3760	5565	997	1138	1414	505	564	632	313	341	354	215	230	223
0.4 ***	3354	3532	5489	967	1068	1411	491	539	644	305	329	372	210	224	244
***	3415	3824	5803	1007	1171	1459	512	578	645	316	347	358	216	233	223
0.45 ***	3234	3457	5684	953	1065	1444	486	538	652	301	327	372	207	221	242
***	3312	3803	5922	998	1179	1474	508	580	645	313	346	354	213	230	218
0.5 ***	3046	3316	5750	921	1043	1445	473	527	645	292	319	365	199	213	234
***	3142	3711	5922	971	1165	1459	496	571	632	304	338	343	206	223	209
0.55 ***	2811	3126	5687	877	1005	1413	452	507	624	278	304	349	188	202	222
***	2925	3562	5803	929	1129	1414	476	551	606	290	323	324	194	211	195
0.6 ***	2548	2902	5493	822	950	1348	425	478	588	260	284	325	174	186	204
***	2680	3364	5565	875	1074	1340	447	520	566	270	301	298	179	194	175
0.65 ***	2274	2654	5171	756	882	1252	390	439	538	236	257	293	155	166	181
***	2418	3124	5208	808	999	1235	411	478	513	245	272	265	160	173	152
0.7 ***	1998	2386	4721	681	798	1124	348	391	474	207	225	253	133	141	152
***	2148	2845	4732	730	905	1102	367	424	447	215	237	224	137	147	123
0.75 ***	1725	2099	4144	596	698	964	298	333	396	172	185	205	.	.	.
***	1871	2526	4137	639	789	938	314	359	368	178	194	175	.	.	.
0.8 ***	1450	1787	3442	497	579	773	238	264	304	.	.	.	.	.	.
***	1584	2160	3423	532	649	744	250	281	275	.	.	.	.	.	.
0.85 ***	1161	1439	2615	380	436	551	.	.	.	.	.	.	.	.	.
***	1273	1732	2590	404	480	521	.	.	.	.	.	.	.	.	.
0.9 ***	834	1024	1666	.	.	.	.	.	.	.	.	.	.	.	.
***	912	1210	1637	.	.	.	.	.	.	.	.	.	.	.	.

TABLE 2: ALPHA= 0.005 POWER= 0.9   EXPECTED ACCRUAL THRU MINIMUM FOLLOW-UP= 400

		DEL=.05			DEL=.10			DEL=.15			DEL=.20			DEL=.25		
FACT=		1.0 .75	.50 .25	.00 BIN	1.0 .75	.50 .25	.00 BIN	1.0 .75	.50 .25	.00 BIN	1.0 75	.50 .25	.00 BIN	1.0 .75	.50 .25	.00 BIN
PCONT=***							REQUIRED NUMBER OF PATIENTS								
0.05	***	895	905	951	322	327	338	190	193	197	135	136	138	105	106	107
	***	899	918	1637	324	332	521	191	195	275	135	137	175	105	107	123
0.1	***	1635	1657	1833	515	530	566	279	287	300	186	190	197	138	141	144
	***	1643	1694	2590	521	543	744	283	293	368	188	193	224	139	142	152
0.15	***	2265	2304	2687	673	700	776	351	365	391	227	234	246	164	168	175
	***	2278	2371	3423	684	726	938	357	376	447	230	240	265	166	171	175
0.2	***	2756	2815	3467	795	836	961	407	428	469	258	270	288	183	190	199
	***	2776	2919	4137	811	877	1102	416	444	513	263	277	298	186	194	195
0.25	***	3103	3186	4148	884	939	1120	448	476	534	281	296	321	198	206	219
	***	3131	3332	4732	906	997	1235	460	499	566	288	307	324	201	212	209
0.3	***	3310	3423	4717	941	1012	1248	475	511	585	297	315	346	207	217	233
	***	3348	3618	5208	969	1086	1340	491	539	606	305	328	343	211	224	218
0.35	***	3392	3540	5166	970	1057	1346	491	533	622	305	327	363	211	223	241
	***	3443	3787	5565	1005	1146	1414	509	567	632	315	342	354	217	231	223
0.4	***	3366	3554	5489	976	1077	1411	496	543	644	308	331	372	212	225	244
	***	3431	3857	5803	1017	1180	1459	516	581	645	318	348	358	218	233	223
0.45	***	3250	3484	5684	962	1076	1444	491	542	652	304	329	372	208	222	242
	***	3332	3841	5922	1008	1189	1474	513	584	645	315	347	354	214	231	218
0.5	***	3066	3348	5750	932	1055	1445	478	532	645	295	321	365	201	214	234
	***	3166	3753	5922	982	1175	1459	501	575	632	307	339	343	207	223	209
0.55	***	2834	3162	5687	889	1017	1413	458	511	624	281	306	349	190	203	222
	***	2954	3607	5803	941	1140	1414	481	555	606	292	324	324	196	211	195
0.6	***	2576	2942	5493	834	963	1349	430	482	588	262	286	325	175	187	204
	***	2713	3411	5565	887	1085	1340	452	523	566	273	302	298	181	194	175
0.65	***	2304	2694	5171	768	893	1252	395	443	538	238	259	293	157	167	181
	***	2453	3171	5208	820	1009	1235	416	481	513	248	274	265	161	173	152
0.7	***	2030	2426	4721	692	808	1124	352	395	474	209	226	253	134	142	153
	***	2184	2890	4732	741	914	1102	371	427	447	217	238	224	138	147	123
0.75	***	1757	2137	4144	605	707	964	302	336	396	173	186	205	.	.	.
	***	1906	2568	4137	648	796	938	317	361	368	179	195	175	.	.	.
0.8	***	1479	1821	3442	505	586	773	241	265	304	.	.	.	.	.	.
	***	1615	2196	3423	540	654	744	252	282	275	.	.	.	.	.	.
0.85	***	1185	1466	2616	385	440	551	.	.	.	.	.	.	.	.	.
	***	1298	1759	2590	409	484	521	.	.	.	.	.	.	.	.	.
0.9	***	852	1042	1666	.	.	.	.	.	.	.	.	.	.	.	.
	***	930	1227	1637	.	.	.	.	.	.	.	.	.	.	.	.

TABLE 2: ALPHA= 0.005 POWER= 0.9   EXPECTED ACCRUAL THRU MINIMUM FOLLOW-UP= 425

		DEL=.05			DEL=.10			DEL=.15			DEL=.20			DEL=.25		
FACT=		1.0 .75	.50 .25	.00 BIN	1.0 .75	.50 .25	.00 BIN	1.0 .75	.50 .25	.00 BIN	1.0 75	.50 .25	.00 BIN	1.0 .75	.50 .25	.00 BIN
PCONT=***					REQUIRED NUMBER OF PATIENTS											
0.05	***	896	906	951	323	328	338	190	193	197	135	136	138	105	106	107
	•••	900	919	1637	325	332	521	191	195	275	136	137	175	105	107	123
0.1	***	1636	1660	1833	516	531	566	280	288	300	187	191	197	138	141	144
	•••	1644	1698	2590	522	544	744	283	293	368	188	193	224	139	142	152
0.15	***	2268	2309	2687	675	702	776	353	366	391	228	235	246	164	169	175
	•••	2281	2378	3423	686	728	938	358	376	447	231	240	265	166	171	175
0.2	***	2760	2822	3467	799	840	961	409	430	469	259	271	288	184	190	200
	•••	2781	2930	4137	815	881	1102	418	445	513	264	278	298	187	195	195
0.25	***	3108	3196	4148	888	945	1120	451	478	534	283	297	321	198	207	219
	•••	3138	3348	4732	911	1002	1235	462	501	566	289	308	324	202	212	209
0.3	***	3317	3436	4717	946	1019	1248	479	513	585	298	316	346	208	218	232
	•••	3358	3638	5208	975	1092	1340	494	541	606	307	329	343	212	224	218
0.35	***	3402	3558	5166	977	1065	1346	495	536	622	307	328	363	213	224	241
	•••	3455	3813	5565	1012	1154	1414	512	569	632	317	343	354	218	232	223
0.4	***	3378	3576	5489	985	1087	1411	500	547	644	310	333	372	213	226	244
	•••	3447	3888	5803	1026	1188	1459	520	584	645	320	349	358	219	234	223
0.45	***	3265	3510	5684	972	1086	1444	496	546	652	306	331	372	209	223	242
	•••	3351	3877	5922	1018	1198	1474	518	587	645	317	349	354	215	231	218
0.5	***	3085	3379	5750	943	1066	1445	483	536	645	298	323	365	202	215	234
	•••	3190	3793	5922	993	1185	1459	506	578	632	309	340	343	208	224	209
0.55	***	2858	3197	5687	900	1028	1413	462	516	624	283	308	349	191	204	222
	•••	2983	3650	5803	952	1150	1414	486	558	606	295	325	324	197	212	195
0.6	***	2602	2979	5493	844	974	1348	435	486	588	264	288	325	176	188	204
	•••	2744	3455	5565	898	1095	1340	457	526	566	275	303	298	181	195	175
0.65	***	2334	2733	5171	779	904	1252	399	447	538	240	261	292	158	167	181
	•••	2487	3215	5208	831	1019	1235	420	483	513	249	274	265	162	173	152
0.7	***	2061	2465	4721	702	818	1124	356	398	474	210	227	253	135	143	153
	•••	2219	2933	4732	751	922	1102	375	429	447	218	239	224	139	147	123
0.75	***	1787	2173	4144	614	715	964	305	339	396	175	187	205	.	.	.
	•••	1939	2606	4137	657	803	938	320	363	368	180	195	175	.	.	.
0.8	***	1507	1853	3442	512	593	773	244	267	305	.	.	.	.	.	.
	•••	1645	2229	3423	547	659	744	254	283	275	.	.	.	.	.	.
0.85	***	1209	1492	2615	391	445	551	.	.	.	.	.	.	.	.	.
	•••	1323	1785	2590	414	487	521	.	.	.	.	.	.	.	.	.
0.9	***	868	1059	1666	.	.	.	.	.	.	.	.	.	.	.	.
	•••	946	1242	1637	.	.	.	.	.	.	.	.	.	.	.	.

TABLE 2: ALPHA= 0.005 POWER= 0.9   EXPECTED ACCRUAL THRU MINIMUM FOLLOW-UP= 450

PCONT= •••	FACT=	DEL=.05 1.0 .75	.50 .25	.00 BIN	DEL=.10 1.0 .75	.50 .25	.00 BIN	DEL=.15 1.0 .75	.50 .25	.00 BIN	DEL=.20 1.0 75	.50 .25	.00 BIN	DEL=.25 1.0 .75	.50 .25	.00 BIN
							REQUIRED NUMBER OF PATIENTS								
0.05 •••		897	907	950	323	328	338	190	193	197	135	136	138	105	106	107
•••		900	920	1637	325	332	521	192	195	275	136	137	175	105	107	123
0.1 •••		1638	1662	1833	517	532	566	281	288	300	187	191	197	139	141	144
•••		1646	1701	2590	523	545	744	284	293	368	189	194	224	140	143	152
0.15 •••		2270	2313	2687	677	704	775	354	367	391	228	235	246	165	169	175
•••		2285	2385	3423	688	730	938	360	377	447	231	240	265	167	172	175
0.2 •••		2764	2829	3467	802	843	961	411	431	469	261	271	288	185	191	199
•••		2786	2941	4137	819	884	1102	419	447	513	265	279	298	188	195	195
0.25 •••		3113	3206	4148	892	950	1120	453	480	534	284	298	321	199	207	219
•••		3145	3363	4732	915	1006	1235	465	502	566	291	308	324	203	212	209
0.3 •••		3325	3450	4717	952	1025	1248	482	516	585	300	318	346	209	219	233
•••		3367	3658	5208	981	1098	1340	496	543	606	308	330	343	213	225	218
0.35 •••		3411	3575	5166	984	1073	1346	498	539	622	309	330	363	213	225	241
•••		3467	3839	5565	1020	1161	1414	516	572	632	318	344	354	219	232	223
0.4 •••		3390	3597	5489	992	1095	1411	504	550	644	312	334	372	214	226	244
•••		3462	3918	5803	1034	1196	1459	524	587	645	322	351	358	220	234	223
0.45 •••		3281	3536	5684	981	1096	1444	500	550	652	309	333	372	211	223	242
•••		3371	3911	5922	1028	1207	1474	522	590	645	320	350	354	217	232	218
0.5 •••		3104	3409	5750	952	1076	1445	488	540	645	300	324	365	203	216	234
•••		3214	3831	5922	1003	1194	1459	510	581	632	311	342	343	209	224	209
0.55 •••		2881	3231	5687	910	1039	1413	467	519	624	286	310	348	192	204	222
•••		3010	3690	5803	963	1159	1414	490	561	606	297	327	324	198	212	195
0.6 •••		2629	3015	5493	855	984	1349	439	490	588	267	289	325	177	189	204
•••		2775	3497	5565	909	1104	1340	461	529	566	277	304	298	183	195	175
0.65 •••		2363	2770	5171	789	914	1252	404	450	538	242	262	293	159	168	181
•••		2520	3257	5208	841	1027	1235	424	486	513	251	275	265	163	174	152
0.7 •••		2091	2501	4721	712	828	1124	360	401	474	212	228	253	136	143	153
•••		2251	2973	4732	761	929	1102	378	431	447	219	239	224	139	147	123
0.75 •••		1816	2207	4144	623	723	964	308	341	396	176	188	205	.	.	.
•••		1971	2643	4137	666	809	938	323	364	368	181	195	175	.	.	.
0.8 •••		1533	1883	3442	519	599	773	246	269	304	.	.	.	.	.	.
•••		1673	2260	3423	554	664	744	256	284	275	.	.	.	.	.	.
0.85 •••		1231	1516	2615	395	449	551	.	.	.	.	.	.	.	.	.
•••		1346	1808	2590	419	489	521	.	.	.	.	.	.	.	.	.
0.9 •••		883	1074	1666	.	.	.	.	.	.	.	.	.	.	.	.
•••		962	1256	1637	.	.	.	.	.	.	.	.	.	.	.	.

TABLE 2: ALPHA= 0.005 POWER= 0.9   EXPECTED ACCRUAL THRU MINIMUM FOLLOW-UP= 475

	DEL=.05			DEL=.10			DEL=.15			DEL=.20			DEL=.25		
FACT=	1.0 .75	.50 .25	.00 BIN	1.0 .75	.50 .25	.00 BIN	1.0 .75	.50 .25	.00 BIN	1.0 75	.50 .25	.00 BIN	1.0 .75	.50 .25	.00 BIN
PCONT=***				REQUIRED NUMBER OF PATIENTS											
0.05 ***	897	908	951	324	329	338	191	193	197	135	137	138	105	106	107
***	901	921	1637	326	333	521	192	195	275	136	137	175	105	107	123
0.1 ***	1639	1665	1833	518	533	566	281	289	300	187	191	197	139	141	144
***	1648	1705	2590	525	546	744	285	294	368	189	194	224	140	143	152
0.15 ***	2273	2318	2687	679	707	775	355	368	391	229	236	246	165	169	175
***	2288	2392	3423	690	732	938	361	378	447	232	241	265	167	172	175
0.2 ***	2768	2836	3467	805	847	961	412	432	469	261	272	288	185	191	200
***	2791	2951	4137	822	887	1102	421	448	513	266	279	298	188	195	195
0.25 ***	3118	3216	4148	897	954	1120	455	482	534	285	299	321	200	208	219
***	3152	3378	4732	920	1010	1235	467	503	566	292	309	324	204	213	209
0.3 ***	3332	3463	4717	957	1031	1248	485	519	585	302	319	346	210	219	232
***	3377	3678	5208	987	1103	1340	499	545	606	309	331	343	214	225	218
0.35 ***	3421	3592	5166	990	1080	1346	502	542	622	311	331	363	215	226	241
***	3480	3863	5565	1027	1167	1414	519	574	632	320	345	354	220	232	223
0.4 ***	3402	3619	5489	1000	1103	1411	508	553	644	314	336	372	215	227	244
***	3478	3946	5803	1042	1203	1459	528	589	645	324	352	358	221	235	223
0.45 ***	3296	3562	5684	989	1105	1444	504	554	652	311	334	372	212	224	242
***	3391	3944	5922	1037	1214	1474	526	592	645	321	351	354	218	232	218
0.5 ***	3123	3438	5750	962	1086	1445	492	544	645	302	326	365	205	217	234
***	3238	3868	5922	1013	1202	1459	514	583	632	313	343	343	210	225	209
0.55 ***	2903	3263	5687	920	1049	1413	472	523	624	288	311	349	193	205	222
***	3037	3729	5803	973	1168	1414	494	563	606	299	327	324	199	213	195
0.6 ***	2655	3050	5493	865	994	1348	443	493	588	269	291	324	178	189	204
***	2805	3537	5565	919	1112	1340	465	531	566	279	305	298	184	196	175
0.65 ***	2391	2806	5171	799	924	1252	408	454	538	244	263	292	159	169	181
***	2551	3297	5208	851	1035	1235	428	488	513	253	276	265	164	174	152
0.7 ***	2120	2536	4721	721	836	1124	364	404	474	213	229	253	137	144	153
***	2283	3011	4732	770	936	1102	381	433	447	221	240	224	140	148	123
0.75 ***	1844	2240	4144	631	731	964	311	343	396	177	188	205	.	.	.
***	2001	2677	4137	674	815	938	325	366	368	182	196	175	.	.	.
0.8 ***	1559	1912	3442	526	604	773	248	270	304	.	.	.	.	.	.
***	1701	2289	3423	560	668	744	258	285	275	.	.	.	.	.	.
0.85 ***	1252	1539	2616	400	452	551	.	.	.	.	.	.	.	.	.
***	1368	1830	2590	423	492	521	.	.	.	.	.	.	.	.	.
0.9 ***	898	1089	1666	.	.	.	.	.	.	.	.	.	.	.	.
***	977	1269	1637	.	.	.	.	.	.	.	.	.	.	.	.

TABLE 2: ALPHA= 0.005 POWER= 0.9   EXPECTED ACCRUAL THRU MINIMUM FOLLOW-UP= 500

	DEL=.05			DEL=.10			DEL=.15			DEL=.20			DEL=.25		
FACT=	1.0 .75	.50 .25	.00 BIN	1.0 .75	.50 .25	.00 BIN	1.0 .75	.50 .25	.00 BIN	1.0 75	.50 .25	.00 BIN	1.0 .75	.50 .25	.00 BIN
PCONT=•••				REQUIRED NUMBER OF PATIENTS											
0.05 ***	898	909	951	324	329	338	191	193	197	135	137	138	105	106	107
•••	902	922	1637	326	333	521	192	195	275	136	138	175	106	107	123
0.1 ***	1641	1668	1833	520	534	566	282	289	300	188	192	197	139	141	144
•••	1650	1708	2590	526	547	744	285	294	368	189	194	224	140	143	152
0.15 ***	2275	2323	2687	681	708	776	356	369	391	230	236	246	165	169	175
•••	2291	2398	3423	693	733	938	362	378	447	233	241	265	167	172	175
0.2 ***	2771	2843	3467	808	850	961	414	434	469	262	273	288	186	192	199
•••	2796	2961	4137	825	889	1102	423	448	513	267	279	298	188	195	195
0.25 ***	3124	3226	4148	901	958	1120	457	484	534	287	300	321	201	208	219
•••	3158	3392	4732	924	1014	1235	469	505	566	293	309	324	204	213	209
0.3 ***	3339	3476	4717	963	1037	1248	487	521	585	303	320	346	210	220	233
•••	3386	3696	5208	993	1108	1340	502	547	606	311	331	343	215	226	218
0.35 ***	3430	3609	5166	997	1087	1346	505	545	622	313	332	363	216	226	241
•••	3492	3886	5565	1033	1173	1414	522	576	632	322	346	354	221	233	223
0.4 ***	3414	3639	5489	1008	1111	1411	512	557	644	316	338	372	216	228	244
•••	3493	3974	5803	1050	1210	1459	531	591	645	326	352	358	222	235	223
0.45 ***	3312	3586	5684	998	1113	1444	508	557	652	313	336	372	213	225	242
•••	3410	3975	5922	1045	1222	1474	529	595	645	323	352	354	218	233	218
0.5 ***	3142	3467	5750	971	1095	1445	496	547	645	304	328	364	206	218	234
•••	3261	3903	5922	1022	1210	1459	518	586	632	315	343	343	211	225	209
0.55 ***	2925	3295	5687	929	1058	1413	476	527	624	290	313	348	194	206	222
•••	3063	3766	5803	983	1176	1414	498	566	606	300	328	324	200	213	195
0.6 ***	2680	3083	5493	875	1003	1348	448	497	588	270	292	325	179	190	204
•••	2834	3575	5565	928	1120	1340	469	533	566	280	306	298	184	196	175
0.65 ***	2418	2840	5171	808	933	1252	411	457	538	246	264	293	160	169	181
•••	2582	3335	5208	860	1042	1235	431	490	513	254	277	265	164	174	152
0.7 ***	2148	2570	4721	730	844	1124	367	406	474	215	230	253	137	144	153
•••	2314	3047	4732	778	943	1102	384	435	447	222	240	224	140	148	123
0.75 ***	1872	2272	4144	639	738	964	314	345	396	178	189	205	.	.	.
•••	2030	2710	4137	681	820	938	328	367	368	183	197	175	.	.	.
0.8 ***	1583	1940	3442	532	610	773	250	272	304	.	.	.	.	.	.
•••	1727	2317	3423	566	672	744	260	286	275	.	.	.	.	.	.
0.85 ***	1273	1560	2616	404	456	551	.	.	.	.	.	.	.	.	.
•••	1390	1851	2590	427	494	521	.	.	.	.	.	.	.	.	.
0.9 ***	912	1103	1666	.	.	.	.	.	.	.	.	.	.	.	.
•••	991	1281	1637	.	.	.	.	.	.	.	.	.	.	.	.

TABLE 2: ALPHA= 0.005 POWER= 0.9   EXPECTED ACCRUAL THRU MINIMUM FOLLOW-UP= 550

	DEL=.02			DEL=.05			DEL=.10			DEL=.15			DEL=.20		
FACT=	1.0	.50	.00	1.0	.50	.00	1.0	.50	.00	1.0	.50	.00	1.0	.50	.00
	.75	.25	BIN	.75	.25	BIN	.75	.25	BIN	75	.25	BIN	.75	.25	BIN
PCONT=***				REQUIRED NUMBER OF PATIENTS											
0.05 ***	4230	4241	4482	899	910	950	324	330	338	191	193	197	135	137	138
***	4234	4264	8378	903	924	1637	327	333	521	192	195	275	136	137	175
0.1 ***	8799	8826	9848	1643	1673	1833	521	536	566	283	290	300	188	192	197
***	8808	8880	14553	1654	1714	2590	528	548	744	286	294	368	190	194	224
0.15 ***	12849	12897	15230	2280	2331	2687	685	712	775	358	371	391	231	237	246
***	12865	12993	19984	2298	2410	3423	696	736	938	363	379	447	234	241	265
0.2 ***	16070	16144	20236	2778	2857	3467	813	855	961	417	436	469	264	274	288
***	16095	16292	24671	2806	2980	4137	831	894	1102	426	450	513	268	280	298
0.25 ***	18375	18481	24674	3134	3245	4148	908	967	1119	461	487	534	289	302	321
***	18411	18693	28614	3172	3419	4732	932	1020	1235	473	507	566	294	310	324
0.3 ***	19778	19923	28436	3353	3502	4717	972	1047	1248	492	525	585	306	322	346
***	19826	20214	31813	3404	3732	5208	1003	1116	1340	507	550	606	313	333	343
0.35 ***	20341	20535	31462	3449	3641	5166	1009	1099	1346	511	550	622	316	334	363
***	20406	20922	34268	3516	3930	5565	1046	1183	1414	528	579	632	324	347	354
0.4 ***	20157	20409	33714	3438	3679	5489	1021	1126	1411	518	562	644	319	340	372
***	20241	20914	35979	3524	4026	5803	1064	1222	1459	537	595	645	329	354	358
0.45 ***	19331	19657	35172	3341	3634	5684	1013	1129	1444	516	563	652	316	338	372
***	19440	20306	36946	3447	4035	5922	1061	1235	1474	536	599	645	327	353	354
0.5 ***	17984	18400	35824	3179	3521	5750	987	1112	1445	503	553	645	308	330	365
***	18122	19227	37169	3305	3968	5922	1039	1224	1459	525	590	632	318	345	343
0.55 ***	16238	16768	35667	2969	3355	5687	947	1075	1413	483	533	624	294	316	349
***	16415	17810	36648	3114	3836	5803	1001	1190	1414	505	570	606	303	330	324
0.6 ***	14225	14902	34701	2728	3147	5493	893	1020	1348	455	503	588	274	294	324
***	14451	16187	35384	2889	3646	5565	946	1134	1340	476	538	566	283	308	298
0.65 ***	12082	12941	32928	2470	2905	5171	826	949	1252	418	462	538	248	267	292
***	12372	14472	33375	2640	3405	5208	877	1056	1235	437	494	513	257	278	265
0.7 ***	9953	11019	30354	2201	2633	4721	746	859	1124	373	411	474	217	232	253
***	10322	12744	30622	2372	3114	4732	794	954	1102	390	437	447	224	241	224
0.75 ***	7982	9226	26986	1923	2331	4144	653	750	964	319	349	396	180	191	205
***	8431	11037	27126	2085	2771	4137	695	830	938	332	369	368	185	197	175
0.8 ***	6269	7583	22830	1630	1991	3442	543	619	773	253	274	305	.	.	.
***	6760	9342	22885	1776	2368	3423	576	679	744	263	288	275	.	.	.
0.85 ***	4808	6042	17896	1311	1601	2615	412	462	551	.	.	.	.	.	.
***	5280	7604	17900	1429	1889	2590	434	498	521	.	.	.	.	.	.
0.9 ***	3490	4491	12193	938	1129	1666	.	.	.	.	.	.	.	.	.
***	3879	5701	12172	1018	1303	1637	.	.	.	.	.	.	.	.	.
0.95 ***	2081	2670	5728	.	.	.	.	.	.	.	.	.	.	.	.
***	2314	3324	5699	.	.	.	.	.	.	.	.	.	.	.	.

TABLE 2: ALPHA= 0.005 POWER= 0.9   EXPECTED ACCRUAL THRU MINIMUM FOLLOW-UP= 600

		DEL=.02			DEL=.05			DEL=.10			DEL=.15			DEL=.20		
FACT=		1.0 .75	.50 .25	.00 BIN	1.0 .75	.50 .25	.00 BIN	1.0 .75	.50 .25	.00 BIN	1.0 75	.50 .25	.00 BIN	1.0 .75	.50 .25	.00 BIN
PCONT=***				REQUIRED NUMBER OF PATIENTS												
0.05	***	4231	4243	4482	900	912	950	325	330	338	191	194	197	136	137	138
	•••	4235	4268	8378	905	925	1637	327	333	521	193	195	275	136	137	175
0.1	***	8801	8831	9848	1646	1677	1833	523	538	566	284	290	300	189	192	197
	•••	8811	8890	14553	1657	1719	2590	530	549	744	287	295	368	190	194	224
0.15	***	12854	12906	15230	2285	2340	2687	688	715	775	359	371	391	232	238	246
	•••	12871	13010	19984	2304	2421	3423	700	739	938	365	380	447	234	242	265
0.2	***	16077	16157	20236	2786	2870	3467	818	861	961	419	438	469	265	274	287
	•••	16104	16318	24671	2815	2998	4137	836	898	1102	428	451	513	269	280	298
0.25	***	18385	18500	24674	3145	3263	4148	915	974	1120	465	490	534	290	303	321
	•••	18424	18731	28614	3186	3444	4732	939	1027	1235	476	509	566	296	311	324
0.3	***	19791	19950	28436	3367	3526	4717	981	1056	1248	496	529	585	308	323	346
	•••	19844	20267	31813	3423	3765	5208	1012	1124	1340	511	552	606	315	334	343
0.35	***	20359	20570	31462	3467	3673	5166	1019	1110	1346	516	554	622	318	336	363
	•••	20429	20992	34268	3540	3971	5565	1057	1192	1414	532	582	632	326	348	354
0.4	***	20179	20455	33714	3462	3718	5489	1034	1138	1411	524	566	644	322	342	372
	•••	20271	21005	35979	3554	4075	5803	1077	1233	1459	543	598	645	331	355	358
0.45	***	19361	19716	35172	3371	3679	5684	1028	1143	1444	522	568	652	319	341	372
	•••	19479	20424	36946	3484	4090	5922	1076	1247	1474	542	602	645	329	355	354
0.5	***	18021	18475	35824	3214	3572	5750	1003	1127	1445	510	558	645	311	332	364
	•••	18173	19375	37169	3347	4028	5922	1055	1237	1459	532	594	632	321	347	343
0.55	***	16286	16865	35668	3010	3411	5686	963	1091	1413	490	538	624	296	317	349
	•••	16479	17992	36648	3162	3899	5803	1017	1203	1414	511	574	606	306	331	324
0.6	***	14287	15023	34701	2775	3206	5493	909	1036	1348	461	508	588	277	296	325
	•••	14533	16404	35384	2941	3712	5565	962	1146	1340	482	541	566	286	309	298
0.65	***	12161	13092	32928	2520	2965	5171	841	963	1252	424	466	538	251	268	292
	•••	12476	14718	33375	2694	3470	5208	893	1067	1235	443	497	513	259	279	265
0.7	***	10055	11198	30354	2251	2692	4721	761	872	1124	378	415	474	220	233	253
	•••	10453	13008	30622	2426	3175	4732	808	965	1102	395	440	447	226	242	224
0.75	***	8108	9422	26986	1971	2385	4144	665	761	964	323	352	396	181	191	205
	•••	8585	11303	27126	2137	2826	4137	707	838	938	336	371	368	186	198	175
0.8	***	6410	7780	22830	1673	2038	3442	553	628	773	256	276	304	.	.	.
	•••	6924	9593	22885	1821	2414	3423	586	685	744	265	289	275	.	.	.
0.85	***	4945	6221	17896	1346	1638	2615	419	467	551	.	.	.	.	.	.
	•••	5434	7822	17900	1466	1924	2590	440	502	521	.	.	.	.	.	.
0.9	***	3604	4633	12193	962	1152	1666	.	.	.	.	.	.	.	.	.
	•••	4004	5866	12172	1042	1323	1637	.	.	.	.	.	.	.	.	.
0.95	***	2150	2749	5728	.	.	.	.	.	.	.	.	.	.	.	.
	•••	2388	3409	5699	.	.	.	.	.	.	.	.	.	.	.	.

TABLE 2: ALPHA= 0.005 POWER= 0.9   EXPECTED ACCRUAL THRU MINIMUM FOLLOW-UP= 650

		DEL=.02			DEL=.05			DEL=.10			DEL=.15			DEL=.20		
FACT=		1.0 .75	.50 .25	.00 BIN	1.0 .75	.50 .25	.00 BIN	1.0 .75	.50 .25	.00 BIN	1.0 75	.50 .25	.00 BIN	1.0 .75	.50 .25	.00 BIN
PCONT=***		REQUIRED NUMBER OF PATIENTS														
0.05	***	4232	4246	4482	902	913	951	326	331	338	192	194	197	136	137	138
	•••	4236	4272	8378	906	926	1637	328	334	521	193	195	275	136	138	175
0.1	***	8804	8836	9848	1649	1682	1833	525	539	566	285	291	300	189	192	197
	•••	8814	8900	14553	1661	1724	2590	531	550	744	288	295	368	191	194	224
0.15	***	12858	12915	15230	2290	2348	2687	691	718	775	361	373	391	232	238	246
	•••	12877	13028	19984	2310	2431	3423	703	741	938	366	381	447	235	242	265
0.2	***	16084	16171	20236	2793	2883	3467	823	866	961	422	440	469	266	275	288
	•••	16113	16345	24671	2824	3015	4137	841	902	1102	430	452	513	270	281	298
0.25	***	18395	18520	24674	3155	3282	4149	922	981	1120	468	493	534	292	304	321
	•••	18436	18770	28614	3199	3468	4732	946	1032	1235	479	510	566	298	312	324
0.3	***	19805	19976	28437	3382	3551	4717	990	1065	1248	500	532	585	310	325	346
	•••	19862	20320	31813	3441	3796	5208	1021	1131	1340	515	554	606	317	335	343
0.35	***	20377	20605	31462	3486	3703	5166	1030	1120	1346	521	558	622	321	338	363
	•••	20453	21062	34268	3564	4009	5565	1068	1201	1414	537	584	632	329	349	354
0.4	***	20203	20501	33714	3486	3755	5489	1046	1150	1411	529	571	644	325	344	372
	•••	20302	21097	35979	3583	4120	5803	1089	1243	1459	548	601	645	333	356	358
0.45	***	19391	19775	35172	3400	3723	5684	1041	1156	1444	528	573	652	322	343	372
	•••	19519	20541	36946	3519	4141	5922	1089	1258	1474	548	606	645	332	356	354
0.5	***	18059	18551	35825	3249	3621	5750	1017	1141	1445	517	563	645	314	335	365
	•••	18223	19521	37169	3389	4084	5922	1069	1248	1459	537	597	632	323	348	343
0.55	***	16335	16961	35667	3050	3464	5687	978	1105	1413	496	543	624	300	320	348
	•••	16544	18171	36648	3208	3959	5803	1032	1215	1414	517	577	606	309	333	324
0.6	***	14349	15144	34701	2819	3262	5493	924	1050	1349	467	512	588	279	298	324
	•••	14615	16615	35384	2991	3772	5565	977	1158	1340	487	544	566	288	310	298
0.65	***	12240	13242	32928	2567	3022	5171	856	977	1252	430	471	538	253	270	292
	•••	12581	14956	33375	2746	3529	5208	907	1077	1235	448	500	513	261	280	265
0.7	***	10156	11372	30354	2299	2746	4721	774	884	1124	383	418	474	222	235	253
	•••	10583	13260	30622	2477	3232	4732	821	974	1102	399	443	447	228	243	224
0.75	***	8231	9611	26986	2016	2435	4144	678	771	964	327	354	396	183	192	205
	•••	8735	11556	27126	2185	2876	4137	718	846	938	340	373	368	188	198	175
0.8	***	6545	7967	22830	1714	2082	3442	563	636	773	259	278	305	.	.	.
	•••	7080	9830	22885	1863	2456	3423	595	691	744	268	290	275	.	.	.
0.85	***	5076	6389	17896	1379	1672	2615	425	472	551	.	.	.	.	.	.
	•••	5580	8026	17900	1500	1955	2590	446	505	521	.	.	.	.	.	.
0.9	***	3711	4765	12193	984	1173	1666	.	.	.	.	.	.	.	.	.
	•••	4122	6019	12172	1064	1341	1637	.	.	.	.	.	.	.	.	.
0.95	***	2215	2823	5728	.	.	.	.	.	.	.	.	.	.	.	.
	•••	2457	3486	5699	.	.	.	.	.	.	.	.	.	.	.	.

TABLE 2: ALPHA= 0.005 POWER= 0.9   EXPECTED ACCRUAL THRU MINIMUM FOLLOW-UP= 700

	DEL=.02			DEL=.05			DEL=.10			DEL=.15			DEL=.20		
FACT=	1.0 .75	.50 .25	.00 BIN	1.0 .75	.50 .25	.00 BIN	1.0 .75	.50 .25	.00 BIN	1.0 75	.50 .25	.00 BIN	1.0 .75	.50 .25	.00 BIN

PCONT=•••   REQUIRED NUMBER OF PATIENTS

PCONT	1.0	.50	.00	1.0	.50	.00	1.0	.50	.00	1.0	.50	.00	1.0	.50	.00
0.05 •••	4233	4247	4482	903	915	950	327	331	338	192	194	197	136	137	138
•••	4238	4276	8378	908	928	1637	328	334	521	193	195	275	136	138	175
0.1 •••	8806	8841	9847	1652	1686	1833	527	541	566	285	292	300	190	193	197
•••	8818	8910	14553	1664	1729	2590	533	551	744	289	296	368	191	194	224
0.15 •••	12862	12924	15230	2294	2356	2687	694	721	775	362	374	391	233	239	246
•••	12883	13045	19984	2316	2441	3423	706	743	938	368	381	447	236	242	265
0.2 •••	16090	16184	20236	2801	2895	3467	828	870	961	424	442	469	268	276	288
•••	16122	16372	24671	2834	3030	4137	845	905	1102	432	453	513	271	282	298
0.25 •••	18404	18539	24674	3165	3299	4148	928	986	1120	471	495	534	294	305	321
•••	18449	18809	28614	3212	3490	4732	953	1037	1235	481	512	566	299	312	324
0.3 •••	19818	20003	28436	3395	3574	4717	998	1073	1248	504	534	585	312	326	346
•••	19879	20372	31813	3459	3824	5208	1029	1137	1340	518	556	606	318	335	343
0.35 •••	20394	20640	31462	3504	3732	5166	1040	1130	1346	525	561	622	323	339	363
•••	20476	21132	34268	3586	4044	5565	1077	1208	1414	541	586	632	331	350	354
0.4 •••	20225	20547	33714	3509	3790	5489	1057	1161	1411	534	575	644	327	345	372
•••	20333	21188	35979	3612	4162	5803	1101	1251	1459	552	604	645	335	357	358
0.45 •••	19420	19834	35172	3429	3764	5684	1053	1168	1444	533	577	652	325	344	372
•••	19558	20657	36946	3553	4188	5922	1102	1267	1474	552	608	645	334	357	354
0.5 •••	18097	18626	35824	3283	3667	5750	1031	1153	1445	522	568	645	317	336	365
•••	18274	19667	37169	3428	4136	5922	1083	1258	1459	542	600	632	326	349	343
0.55 •••	16383	17057	35667	3089	3514	5686	992	1118	1413	502	547	624	302	321	348
•••	16608	18348	36648	3253	4013	5803	1045	1225	1414	522	579	606	311	334	324
0.6 •••	14410	15265	34701	2862	3315	5493	938	1062	1349	473	516	588	282	299	324
•••	14697	16821	35384	3039	3828	5565	991	1167	1340	492	547	566	290	311	298
0.65 •••	12319	13388	32928	2611	3075	5171	869	989	1252	435	474	538	256	271	292
•••	12685	15184	33375	2795	3584	5208	920	1087	1235	453	502	513	263	281	265
0.7 •••	10256	11542	30354	2343	2798	4721	786	895	1124	387	422	474	223	236	253
•••	10710	13501	30622	2525	3283	4732	833	982	1102	403	444	447	229	243	224
0.75 •••	8352	9792	26986	2058	2483	4144	688	780	964	330	357	396	184	193	205
•••	8881	11795	27126	2229	2922	4137	728	852	938	342	374	368	188	199	175
0.8 •••	6676	8145	22830	1752	2123	3442	571	642	772	261	279	304	.	.	.
•••	7231	10054	22885	1903	2494	3423	603	695	744	270	291	275	.	.	.
0.85 •••	5200	6550	17896	1410	1703	2616	431	476	551	.	.	.	.	.	.
•••	5720	8218	17900	1531	1983	2590	451	508	521	.	.	.	.	.	.
0.9 •••	3813	4891	12192	1005	1192	1666	.	.	.	.	.	.	.	.	.
•••	4234	6162	12172	1084	1356	1637	.	.	.	.	.	.	.	.	.
0.95 •••	2276	2892	5728	.	.	.	.	.	.	.	.	.	.	.	.
•••	2522	3558	5699	.	.	.	.	.	.	.	.	.	.	.	.

TABLE 2: ALPHA= 0.005 POWER= 0.9   EXPECTED ACCRUAL THRU MINIMUM FOLLOW-UP= 750

	DEL=.02			DEL=.05			DEL=.10			DEL=.15			DEL=.20		
FACT=	1.0 .75	.50 .25	.00 BIN	1.0 .75	.50 .25	.00 BIN	1.0 .75	.50 .25	.00 BIN	1.0 75	.50 .25	.00 BIN	1.0 .75	.50 .25	.00 BIN

PCONT=•••   REQUIRED NUMBER OF PATIENTS

PCONT	DEL=.02 1.0/.75	.50/.25	.00/BIN	DEL=.05 1.0/.75	.50/.25	.00/BIN	DEL=.10 1.0/.75	.50/.25	.00/BIN	DEL=.15 1.0/.75	.50/.25	.00/BIN	DEL=.20 1.0/.75	.50/.25	.00/BIN
0.05 •••	4234	4249	4482	904	917	950	327	331	338	192	194	197	136	137	139
•••	4239	4280	8378	909	929	1637	329	334	521	193	196	275	137	138	175
0.1 •••	8809	8845	9848	1654	1690	1833	528	542	566	286	292	300	190	193	197
•••	8821	8920	14553	1668	1733	2590	534	552	744	289	295	368	191	195	224
0.15 •••	12867	12932	15230	2299	2364	2687	697	724	775	364	375	391	234	239	246
•••	12889	13062	19984	2322	2449	3423	709	745	938	369	382	447	236	243	265
0.2 •••	16097	16198	20236	2808	2907	3467	832	874	961	426	443	469	269	277	288
•••	16131	16399	24671	2843	3044	4137	850	908	1102	434	454	513	273	282	298
0.25 •••	18414	18559	24674	3175	3316	4148	934	992	1120	474	497	534	295	306	321
•••	18462	18847	28614	3226	3510	4732	958	1041	1235	484	513	566	300	313	324
0.3 •••	19831	20029	28436	3409	3596	4717	1005	1079	1248	507	537	585	314	327	346
•••	19897	20425	31813	3476	3851	5208	1037	1143	1340	521	558	606	320	336	343
0.35 •••	20412	20675	31462	3522	3760	5166	1049	1138	1345	529	564	622	325	341	363
•••	20500	21202	34268	3609	4078	5565	1086	1215	1414	544	589	632	332	351	354
0.4 •••	20248	20593	33714	3532	3824	5489	1068	1171	1411	539	578	644	329	347	372
•••	20363	21279	35979	3640	4201	5803	1111	1259	1459	557	606	645	337	358	358
0.45 •••	19450	19894	35172	3457	3803	5684	1065	1179	1444	538	580	652	327	346	372
•••	19598	20772	36946	3586	4233	5922	1114	1276	1474	557	610	645	336	358	354
0.5 •••	18135	18702	35824	3316	3711	5750	1043	1165	1445	527	571	645	319	338	364
•••	18324	19810	37169	3467	4184	5922	1095	1267	1459	547	603	632	327	350	343
0.55 •••	16431	17152	35667	3126	3562	5687	1005	1129	1413	507	551	624	304	323	349
•••	16672	18520	36648	3295	4064	5803	1058	1234	1414	527	582	606	313	334	324
0.6 •••	14471	15384	34701	2902	3364	5494	950	1074	1348	477	520	588	284	301	325
•••	14779	17020	35384	3084	3880	5565	1003	1176	1340	497	549	566	292	312	298
0.65 •••	12398	13533	32929	2654	3124	5171	881	999	1252	439	478	538	258	273	293
•••	12788	15404	33375	2840	3635	5208	933	1095	1235	457	504	513	265	281	265
0.7 •••	10355	11707	30354	2386	2845	4721	798	904	1124	391	424	474	225	237	252
•••	10836	13731	30622	2570	3331	4732	844	989	1102	406	446	447	230	244	224
0.75 •••	8470	9967	26986	2099	2526	4144	698	789	964	333	359	396	185	194	205
•••	9022	12024	27126	2272	2965	4137	738	858	938	345	376	368	189	199	175
0.8 •••	6802	8316	22830	1787	2160	3442	579	649	773	264	281	304	.	.	.
•••	7375	10266	22885	1940	2530	3423	610	700	744	272	292	275	.	.	.
0.85 •••	5320	6702	17896	1439	1732	2615	436	480	550	.	.	.	.	.	.
•••	5853	8400	17900	1560	2008	2590	456	510	521	.	.	.	.	.	.
0.9 •••	3911	5009	12192	1024	1210	1666	.	.	.	.	.	.	.	.	.
•••	4341	6297	12172	1103	1371	1637	.	.	.	.	.	.	.	.	.
0.95 •••	2333	2957	5728	.	.	.	.	.	.	.	.	.	.	.	.
•••	2584	3624	5699	.	.	.	.	.	.	.	.	.	.	.	.

TABLE 2: ALPHA= 0.005 POWER= 0.9   EXPECTED ACCRUAL THRU MINIMUM FOLLOW-UP= 800

		DEL=.02			DEL=.05			DEL=.10			DEL=.15			DEL=.20		
FACT=		1.0 .75	.50 .25	.00 BIN	1.0 .75	.50 .25	.00 BIN	1.0 .75	.50 .25	.00 BIN	1.0 75	.50 .25	.00 BIN	1.0 .75	.50 .25	.00 BIN
PCONT=•••		REQUIRED NUMBER OF PATIENTS														
0.05	•••	4235	4252	4482	905	918	951	327	332	338	193	195	197	136	137	138
	•••	4241	4284	8378	910	930	1637	329	334	521	194	196	275	137	138	175
0.1	•••	8811	8851	9848	1657	1694	1833	530	543	566	287	293	300	190	193	197
	•••	8824	8930	14553	1671	1737	2590	536	553	744	290	296	368	192	195	224
0.15	•••	12871	12941	15230	2304	2371	2687	700	726	776	365	376	391	234	240	246
	•••	12894	13080	19984	2328	2458	3423	711	746	938	370	383	447	237	243	265
0.2	•••	16104	16211	20236	2815	2919	3467	836	877	961	428	444	469	270	277	288
	•••	16140	16426	24671	2852	3058	4137	854	911	1102	435	455	513	273	282	298
0.25	•••	18424	18578	24674	3186	3332	4148	939	997	1120	476	499	534	296	307	321
	•••	18475	18886	28614	3238	3530	4732	964	1045	1235	486	514	566	301	313	324
0.3	•••	19844	20055	28436	3423	3618	4717	1012	1086	1248	511	539	585	315	328	346
	•••	19915	20478	31813	3493	3877	5208	1044	1148	1340	524	559	606	321	337	343
0.35	•••	20429	20711	31462	3540	3787	5166	1057	1146	1346	533	567	622	327	342	363
	•••	20523	21272	34268	3631	4110	5565	1095	1221	1414	548	590	632	334	352	354
0.4	•••	20271	20639	33714	3554	3857	5489	1077	1180	1411	543	581	644	331	348	372
	•••	20394	21370	35979	3666	4238	5803	1121	1266	1459	560	608	645	339	359	358
0.45	•••	19479	19952	35172	3484	3841	5684	1076	1189	1444	542	584	652	329	347	372
	•••	19637	20887	36946	3618	4275	5922	1124	1284	1474	561	613	645	338	359	354
0.5	•••	18173	18777	35824	3348	3753	5750	1055	1175	1445	532	575	645	321	339	365
	•••	18374	19952	37169	3503	4230	5922	1106	1275	1459	551	605	632	329	351	343
0.55	•••	16479	17248	35668	3162	3607	5687	1017	1140	1413	511	555	624	306	324	349
	•••	16736	18689	36648	3335	4112	5803	1070	1242	1414	531	584	606	315	335	324
0.6	•••	14533	15503	34701	2942	3411	5493	963	1085	1349	482	523	588	286	302	325
	•••	14861	17215	35384	3126	3928	5565	1015	1184	1340	501	551	566	294	313	298
0.65	•••	12477	13674	32928	2694	3171	5171	893	1009	1252	443	481	538	259	274	293
	•••	12890	15617	33375	2884	3682	5208	944	1102	1235	460	506	513	266	282	265
0.7	•••	10454	11867	30354	2426	2890	4721	808	914	1124	395	427	474	226	238	253
	•••	10959	13953	30622	2613	3375	4732	854	996	1102	409	448	447	232	245	224
0.75	•••	8585	10136	26986	2137	2568	4144	707	796	964	336	361	396	186	195	205
	•••	9159	12241	27126	2312	3004	4137	746	863	938	348	377	368	190	199	175
0.8	•••	6924	8480	22830	1821	2196	3442	586	654	773	265	282	304	.	.	.
	•••	7515	10468	22885	1975	2562	3423	617	703	744	273	292	275	.	.	.
0.85	•••	5434	6847	17896	1466	1759	2616	440	484	551	.	.	.	.	.	.
	•••	5981	8572	17900	1588	2032	2590	460	513	521	.	.	.	.	.	.
0.9	•••	4004	5122	12193	1042	1227	1666	.	.	.	.	.	.	.	.	.
	•••	4442	6424	12172	1121	1384	1637	.	.	.	.	.	.	.	.	.
0.95	•••	2388	3018	5728	.	.	.	.	.	.	.	.	.	.	.	.
	•••	2642	3686	5699	.	.	.	.	.	.	.	.	.	.	.	.

TABLE 2: ALPHA= 0.005 POWER= 0.9   EXPECTED ACCRUAL THRU MINIMUM FOLLOW-UP= 850

	DEL=.02			DEL=.05			DEL=.10			DEL=.15			DEL=.20		
FACT=	1.0 .75	.50 .25	.00 BIN	1.0 .75	.50 .25	.00 BIN	1.0 .75	.50 .25	.00 BIN	1.0 75	.50 .25	.00 BIN	1.0 .75	.50 .25	.00 BIN

PCONT=•••   REQUIRED NUMBER OF PATIENTS

PCONT	DEL=.02 1.0/.75	.50/.25	.00/BIN	DEL=.05 1.0/.75	.50/.25	.00/BIN	DEL=.10 1.0/.75	.50/.25	.00/BIN	DEL=.15 1.0/.75	.50/.25	.00/BIN	DEL=.20 1.0/.75	.50/.25	.00/BIN
0.05 ***	4236	4254	4482	906	919	951	328	332	338	193	195	197	136	137	138
•••	4242	4287	8378	911	930	1637	329	334	521	193	196	275	137	138	175
0.1 ***	8814	8856	9847	1660	1698	1833	531	544	566	288	293	300	191	193	197
•••	8827	8939	14553	1674	1740	2590	537	553	744	290	296	368	192	195	224
0.15 ***	12875	12949	15230	2309	2378	2687	702	728	776	366	376	391	235	240	246
•••	12900	13097	19984	2334	2466	3423	713	748	938	371	383	447	237	243	265
0.2 ***	16110	16224	20236	2822	2930	3467	839	881	961	429	445	469	271	278	288
•••	16149	16453	24671	2861	3071	4137	858	913	1102	437	456	513	274	282	298
0.25 ***	18433	18597	24674	3196	3348	4148	945	1002	1120	478	501	534	297	308	321
•••	18488	18924	28614	3251	3548	4732	969	1048	1235	488	515	566	302	314	324
0.3 ***	19857	20082	28436	3436	3638	4717	1019	1092	1248	513	542	585	316	329	346
•••	19932	20530	31813	3510	3901	5208	1050	1152	1340	526	561	606	323	337	343
0.35 ***	20447	20746	31462	3557	3813	5166	1065	1154	1346	536	569	622	328	343	363
•••	20547	21342	34268	3652	4139	5565	1103	1227	1414	551	592	632	335	352	354
0.4 ***	20295	20684	33714	3576	3888	5489	1087	1188	1411	547	584	644	333	349	372
•••	20425	21459	35979	3692	4273	5803	1130	1273	1459	563	610	645	341	360	358
0.45 ***	19509	20011	35172	3510	3877	5684	1086	1198	1444	546	587	652	331	349	372
•••	19677	21000	36946	3649	4314	5922	1134	1291	1474	565	615	645	339	359	354
0.5 ***	18210	18853	35824	3379	3793	5750	1066	1185	1445	536	578	645	323	340	365
•••	18425	20091	37169	3538	4272	5922	1117	1283	1459	555	607	632	331	351	343
0.55 ***	16527	17343	35667	3197	3650	5687	1028	1150	1413	515	558	624	308	325	349
•••	16800	18855	36648	3374	4157	5803	1081	1250	1414	535	586	606	316	336	324
0.6 ***	14595	15620	34701	2979	3455	5493	974	1094	1348	486	526	588	288	304	325
•••	14942	17403	35384	3167	3973	5565	1026	1192	1340	504	553	566	295	313	298
0.65 ***	12555	13814	32928	2734	3215	5171	904	1019	1252	447	483	538	260	274	292
•••	12992	15822	33375	2926	3726	5208	954	1109	1235	463	508	513	267	283	265
0.7 ***	10551	12023	30354	2465	2933	4721	818	922	1124	398	429	474	227	239	253
•••	11080	14165	30622	2653	3416	4732	864	1002	1102	412	449	447	233	245	224
0.75 ***	8698	10298	26985	2173	2606	4145	715	803	964	339	362	396	187	195	205
•••	9293	12450	27126	2349	3041	4137	754	868	938	350	378	368	191	200	175
0.8 ***	7042	8637	22830	1853	2229	3442	593	660	773	267	283	305	.	.	.
•••	7650	10660	22885	2007	2592	3423	622	707	744	275	293	275	.	.	.
0.85 ***	5545	6987	17896	1491	1785	2616	445	487	550	.	.	.	.	.	.
•••	6103	8736	17900	1614	2054	2590	464	514	521	.	.	.	.	.	.
0.9 ***	4094	5229	12192	1058	1242	1666	.	.	.	.	.	.	.	.	.
•••	4540	6544	12172	1137	1396	1637	.	.	.	.	.	.	.	.	.
0.95 ***	2440	3076	5728	.	.	.	.	.	.	.	.	.	.	.	.
•••	2696	3744	5699	.	.	.	.	.	.	.	.	.	.	.	.

TABLE 2: ALPHA= 0.005 POWER= 0.9   EXPECTED ACCRUAL THRU MINIMUM FOLLOW-UP= 900

	DEL=.02			DEL=.05			DEL=.10			DEL=.15			DEL=.20		
FACT=	1.0 .75	.50 .25	.00 BIN	1.0 .75	.50 .25	.00 BIN	1.0 .75	.50 .25	.00 BIN	1.0 75	.50 .25	.00 BIN	1.0 .75	.50 .25	.00 BIN
PCONT=•••				REQUIRED NUMBER OF PATIENTS											
0.05 ***	4237 4256 4482			907 920 950			328 332 338			193 195 197			136 137 138		
•••	4243 4291 8378			912 932 1637			330 335 521			194 196 275			137 138 175		
0.1 ***	8816 8861 9848			1662 1701 1833			532 545 566			288 293 300			191 194 197		
•••	8831 8949 14553			1677 1744 2590			538 554 744			290 297 368			192 195 224		
0.15 ***	12880 12958 15230			2313 2385 2687			704 730 775			367 377 391			235 240 246		
•••	12906 13114 19984			2340 2473 3423			716 749 938			371 383 447			237 243 265		
0.2 ***	16118 16238 20236			2829 2941 3467			843 884 962			431 447 469			271 279 288		
•••	16157 16479 24671			2870 3083 4137			861 915 1102			438 457 513			275 282 298		
0.25 ***	18443 18616 24674			3206 3363 4148			950 1006 1119			480 502 534			298 308 321		
•••	18501 18962 28614			3263 3565 4732			974 1051 1235			490 516 566			303 314 324		
0.3 ***	19871 20108 28436			3450 3658 4717			1025 1098 1248			516 543 585			318 330 346		
•••	19950 20583 31813			3527 3924 5208			1056 1156 1340			528 561 606			324 338 343		
0.35 ***	20465 20781 31462			3575 3839 5166			1073 1161 1346			539 572 622			330 344 363		
•••	20570 21411 34268			3673 4167 5565			1110 1232 1414			554 594 632			336 353 354		
0.4 ***	20317 20730 33714			3597 3918 5489			1095 1196 1411			550 587 644			334 351 372		
•••	20455 21549 35979			3718 4305 5803			1139 1278 1459			567 612 645			342 360 358		
0.45 ***	19539 20071 35172			3536 3911 5684			1096 1207 1444			550 590 652			333 350 372		
•••	19716 21113 36946			3680 4350 5922			1143 1297 1474			568 617 645			341 360 354		
0.5 ***	18248 18928 35825			3409 3831 5750			1076 1194 1445			540 581 645			324 342 365		
•••	18475 20228 37169			3573 4312 5922			1127 1289 1459			558 609 632			333 352 343		
0.55 ***	16575 17437 35668			3231 3690 5687			1038 1160 1413			519 560 624			310 326 348		
•••	16865 19017 36648			3411 4199 5803			1091 1256 1414			538 588 606			317 336 324		
0.6 ***	14656 15736 34701			3015 3497 5493			984 1104 1349			489 529 588			289 305 325		
•••	15023 17587 35384			3206 4015 5565			1036 1198 1340			507 555 566			296 314 298		
0.65 ***	12633 13950 32928			2771 3257 5171			914 1027 1252			450 486 539			262 275 293		
•••	13092 16021 33375			2966 3767 5208			963 1115 1235			467 509 513			269 283 265		
0.7 ***	10647 12174 30354			2501 2973 4721			828 929 1124			401 431 474			228 239 253		
•••	11198 14369 30622			2692 3455 4732			872 1007 1102			415 450 447			234 245 224		
0.75 ***	8808 10455 26985			2207 2643 4144			723 809 964			341 364 396			188 195 205		
•••	9423 12650 27126			2385 3075 4137			761 873 938			352 379 368			191 200 175		
0.8 ***	7156 8788 22830			1883 2260 3442			599 664 773			269 284 305			. . .		
•••	7779 10844 22885			2039 2620 3423			628 710 744			276 294 275			. . .		
0.85 ***	5651 7120 17896			1515 1808 2615			449 489 551			. . .			. . .		
•••	6221 8891 17900			1638 2074 2590			467 516 521			. . .			. . .		
0.9 ***	4179 5332 12192			1074 1256 1666			. . .			. . .			. . .		
•••	4632 6658 12172			1152 1406 1637			. . .			. . .			. . .		
0.95 ***	2490 3131 5728			. . .			. . .			. . .			. . .		
•••	2749 3798 5699			. . .			. . .			. . .			. . .		

TABLE 2: ALPHA= 0.005 POWER= 0.9   EXPECTED ACCRUAL THRU MINIMUM FOLLOW-UP= 950

	DEL=.02			DEL=.05			DEL=.10			DEL=.15			DEL=.20		
FACT=	1.0 .75	.50 .25	.00 BIN	1.0 .75	.50 .25	.00 BIN	1.0 .75	.50 .25	.00 BIN	1.0 75	.50 .25	.00 BIN	1.0 .75	.50 .25	.00 BIN
PCONT=***	REQUIRED NUMBER OF PATIENTS														
0.05 ***	4238	4258	4482	908	921	951	329	333	338	193	195	197	137	137	139
***	4245	4295	8378	913	932	1637	330	335	521	194	196	275	137	138	175
0.1 ***	8818	8865	9848	1665	1705	1833	533	546	566	289	293	300	191	194	197
***	8834	8959	14553	1680	1747	2590	539	555	744	291	296	368	192	195	224
0.15 ***	12884	12967	15230	2318	2392	2687	707	732	775	368	378	391	236	241	246
***	12912	13131	19984	2345	2480	3423	717	750	938	372	384	447	238	243	265
0.2 ***	16124	16251	20236	2837	2951	3467	847	887	961	432	448	469	272	279	287
***	16166	16506	24671	2879	3093	4137	864	918	1102	439	457	513	275	283	298
0.25 ***	18453	18635	24674	3216	3378	4148	954	1010	1120	482	503	534	299	309	321
***	18513	19000	28614	3275	3581	4732	978	1054	1235	492	517	566	304	314	324
0.3 ***	19884	20135	28437	3463	3678	4718	1031	1103	1248	519	545	585	319	331	346
***	19967	20635	31813	3542	3945	5208	1062	1160	1340	530	563	606	324	338	343
0.35 ***	20482	20816	31462	3592	3863	5166	1079	1167	1345	542	574	622	331	345	363
***	20593	21480	34268	3693	4193	5565	1117	1236	1414	557	595	632	337	353	354
0.4 ***	20340	20777	33714	3618	3946	5489	1103	1203	1411	553	589	644	336	352	372
***	20486	21638	35979	3743	4337	5803	1147	1283	1459	570	613	645	343	361	358
0.45 ***	19568	20129	35171	3561	3944	5684	1104	1214	1444	553	592	652	334	350	372
***	19755	21224	36946	3708	4385	5922	1152	1303	1474	571	618	645	342	361	354
0.5 ***	18286	19003	35825	3438	3868	5750	1085	1203	1445	543	583	645	326	343	365
***	18526	20364	37169	3605	4350	5922	1136	1296	1459	562	610	632	334	353	343
0.55 ***	16624	17531	35667	3263	3730	5687	1049	1168	1413	523	563	623	311	327	349
***	16929	19176	36648	3447	4238	5803	1100	1262	1414	542	590	606	319	337	324
0.6 ***	14717	15851	34701	3050	3537	5493	994	1112	1349	494	532	588	291	305	324
***	15104	17766	35384	3244	4055	5565	1046	1204	1340	511	557	566	298	314	298
0.65 ***	12711	14085	32928	2806	3297	5171	923	1035	1252	454	488	538	263	276	292
***	13192	16213	33375	3003	3806	5208	972	1121	1235	469	510	513	269	284	265
0.7 ***	10742	12322	30354	2536	3011	4721	836	937	1123	404	433	474	229	240	253
***	11315	14567	30622	2729	3491	4732	880	1012	1102	417	451	447	235	246	224
0.75 ***	8916	10608	26986	2240	2677	4144	730	815	964	343	366	396	188	196	205
***	9549	12841	27126	2419	3107	4137	768	876	938	353	380	368	192	200	175
0.8 ***	7268	8934	22830	1912	2289	3442	604	668	773	270	285	304	.	.	.
***	7905	11019	22885	2068	2646	3423	633	713	744	277	294	275	.	.	.
0.85 ***	5753	7248	17896	1539	1830	2616	452	492	551	.	.	.	.	.	.
***	6334	9039	17900	1661	2092	2590	470	518	521	.	.	.	.	.	.
0.9 ***	4262	5430	12193	1089	1269	1666	.	.	.	.	.	.	.	.	.
***	4722	6766	12172	1166	1416	1637	.	.	.	.	.	.	.	.	.
0.95 ***	2538	3183	5728	.	.	.	.	.	.	.	.	.	.	.	.
***	2799	3849	5699	.	.	.	.	.	.	.	.	.	.	.	.

TABLE 2: ALPHA= 0.005 POWER= 0.9   EXPECTED ACCRUAL THRU MINIMUM FOLLOW-UP= 1000

PCONT=***		DEL=.02			DEL=.05			DEL=.10			DEL=.15			DEL=.20		
FACT=		1.0 .75	.50 .25	.00 BIN	1.0 .75	.50 .25	.00 BIN	1.0 .75	.50 .25	.00 BIN	1.0 75	.50 .25	.00 BIN	1.0 .75	.50 .25	.00 BIN
0.05	***	4239	4260	4482	909	922	951	329	333	338	193	195	197	136	138	138
	•••	4246	4298	8378	914	933	1637	331	335	521	194	196	275	137	138	175
0.1	***	8821	8870	9848	1668	1708	1833	535	546	566	289	294	300	191	194	197
	•••	8838	8968	14553	1683	1750	2590	540	555	744	291	297	368	193	195	224
0.15	***	12889	12976	15230	2323	2398	2687	708	733	776	369	378	391	236	241	246
	•••	12918	13149	19984	2351	2486	3423	720	751	938	373	384	447	238	243	265
0.2	***	16131	16265	20236	2843	2961	3467	850	890	961	434	448	470	273	279	288
	•••	16175	16532	24671	2887	3105	4137	867	919	1102	441	458	513	276	283	298
0.25	***	18462	18655	24674	3226	3392	4148	958	1014	1120	484	505	534	300	310	321
	•••	18526	19038	28614	3287	3596	4732	983	1056	1235	493	518	566	305	315	324
0.3	***	19897	20161	28436	3476	3696	4717	1036	1108	1248	521	547	585	320	331	346
	•••	19985	20687	31813	3558	3965	5208	1068	1164	1340	533	564	606	325	338	343
0.35	***	20500	20851	31462	3609	3886	5166	1086	1173	1346	545	576	622	332	346	363
	•••	20617	21550	34268	3713	4218	5565	1124	1241	1414	559	596	632	338	354	354
0.4	***	20363	20822	33714	3640	3974	5489	1111	1210	1411	556	591	644	338	352	372
	•••	20516	21726	35979	3767	4366	5803	1154	1288	1459	572	615	645	345	361	358
0.45	***	19598	20189	35172	3586	3975	5685	1113	1222	1444	557	595	652	336	351	372
	•••	19795	21335	36946	3736	4418	5922	1160	1308	1474	574	620	645	343	361	354
0.5	***	18324	19078	35825	3466	3903	5750	1095	1210	1445	547	586	645	328	343	365
	•••	18576	20497	37169	3637	4386	5922	1145	1301	1459	565	612	632	335	353	343
0.55	***	16672	17625	35668	3295	3766	5686	1058	1176	1413	526	566	624	313	328	348
	•••	16993	19331	36648	3481	4275	5803	1110	1268	1414	545	591	606	320	338	324
0.6	***	14779	15965	34701	3083	3575	5493	1003	1120	1348	496	533	588	292	306	325
	•••	15185	17940	35384	3280	4093	5565	1055	1210	1340	513	558	566	299	315	298
0.65	***	12788	14216	32928	2840	3335	5171	933	1042	1252	456	490	538	265	277	293
	•••	13291	16400	33375	3040	3843	5208	981	1126	1235	472	511	513	270	285	265
0.7	***	10836	12466	30355	2570	3047	4721	844	943	1124	406	435	475	230	240	253
	•••	11430	14756	30622	2764	3525	4732	888	1016	1102	420	453	447	235	246	224
0.75	***	9022	10755	26986	2271	2710	4145	738	820	964	345	367	396	190	196	205
	•••	9672	13026	27126	2451	3137	4137	775	880	938	355	380	368	193	200	175
0.8	***	7376	9075	22830	1940	2316	3442	610	672	773	271	286	305	.	.	.
	•••	8027	11188	22885	2096	2671	3423	638	715	744	278	295	275	.	.	.
0.85	***	5853	7371	17896	1560	1851	2616	456	495	551	.	.	.	.	.	.
	•••	6444	9181	17900	1683	2110	2590	473	520	521	.	.	.	.	.	.
0.9	***	4341	5524	12193	1103	1281	1666	.	.	.	.	.	.	.	.	.
	•••	4808	6869	12172	1180	1426	1637	.	.	.	.	.	.	.	.	.
0.95	***	2583	3233	5728	.	.	.	.	.	.	.	.	.	.	.	.
	•••	2846	3897	5699	.	.	.	.	.	.	.	.	.	.	.	.

PCONT=***   REQUIRED NUMBER OF PATIENTS

TABLE 2: ALPHA= 0.005 POWER= 0.9 EXPECTED ACCRUAL THRU MINIMUM FOLLOW-UP= 1100

REQUIRED NUMBER OF PATIENTS

		DEL=.02			DEL=.05			DEL=.10			DEL=.15			DEL=.20		
	FACT=	1.0 .75	.50 .25	.00 BIN	1.0 .75	.50 .25	.00 BIN	1.0 .75	.50 .25	.00 BIN	1.0 75	.50 .25	.00 BIN	1.0 .75	.50 .25	.00 BIN
PCONT=***																
0.05	***	4241	4264	4482	910	923	950	329	333	338	193	195	197	137	137	138
	***	4249	4305	8378	916	934	1637	331	335	521	195	196	275	137	138	175
0.1	***	8826	8881	9848	1673	1714	1833	536	548	566	290	294	300	192	194	197
	***	8844	8987	14553	1688	1755	2590	541	555	744	291	297	368	193	195	224
0.15	***	12897	12993	15229	2331	2410	2688	712	736	775	371	379	391	237	241	246
	***	12929	13183	19984	2361	2497	3423	723	753	938	375	384	447	239	243	265
0.2	***	16144	16292	20236	2857	2980	3467	855	894	961	436	450	469	274	280	287
	***	16193	16585	24671	2903	3124	4137	873	922	1102	443	459	513	276	283	298
0.25	***	18481	18693	24674	3244	3419	4148	967	1020	1119	488	507	534	302	310	321
	***	18551	19115	28614	3310	3624	4732	990	1061	1235	496	519	566	306	315	324
0.3	***	19923	20214	28436	3502	3732	4717	1047	1116	1248	525	549	585	322	333	346
	***	20020	20791	31813	3589	4003	5208	1077	1170	1340	536	565	606	327	339	343
0.35	***	20535	20921	31462	3641	3930	5166	1099	1183	1346	549	579	621	334	347	363
	***	20663	21686	34268	3751	4265	5565	1136	1248	1414	563	598	632	340	355	354
0.4	***	20409	20914	33714	3679	4026	5489	1126	1222	1411	562	595	644	340	354	372
	***	20578	21900	35979	3813	4420	5803	1167	1297	1459	577	617	645	346	362	358
0.45	***	19656	20306	35172	3634	4035	5684	1129	1236	1444	563	599	652	338	353	372
	***	19874	21550	36946	3790	4479	5922	1176	1318	1474	580	622	645	345	362	354
0.5	***	18400	19227	35824	3521	3968	5750	1112	1225	1445	553	590	645	330	345	364
	***	18677	20756	37169	3697	4451	5922	1160	1312	1459	570	615	632	337	354	343
0.55	***	16768	17810	35667	3355	3836	5686	1075	1190	1413	533	570	624	316	330	349
	***	17120	19632	36648	3546	4344	5803	1126	1279	1414	550	594	606	323	338	324
0.6	***	14902	16187	34701	3147	3646	5493	1020	1134	1348	503	538	588	294	307	324
	***	15344	18274	35384	3348	4162	5565	1071	1220	1340	518	560	566	301	316	298
0.65	***	12941	14472	32928	2905	3405	5171	949	1055	1252	462	494	538	267	278	292
	***	13485	16754	33375	3108	3910	5208	995	1135	1235	477	514	513	272	285	265
0.7	***	11019	12743	30354	2633	3114	4721	859	954	1123	411	437	474	232	241	252
	***	11653	15117	30622	2830	3587	4732	901	1024	1102	423	455	447	236	247	224
0.75	***	9226	11037	26986	2331	2771	4144	750	830	964	349	369	396	191	197	205
	***	9910	13374	27126	2512	3191	4137	786	886	938	358	382	368	194	201	175
0.8	***	7583	9343	22830	1991	2368	3442	620	679	773	274	287	305	.	.	.
	***	8260	11505	22885	2148	2715	3423	646	720	744	280	296	275	.	.	.
0.85	***	6043	7605	17896	1601	1889	2615	462	499	551	.	.	.	.	.	.
	***	6652	9448	17900	1723	2141	2590	478	522	521	.	.	.	.	.	.
0.9	***	4492	5702	12193	1129	1303	1666	.	.	.	.	.	.	.	.	.
	***	4970	7061	12172	1204	1442	1637	.	.	.	.	.	.	.	.	.
0.95	***	2670	3324	5728	.	.	.	.	.	.	.	.	.	.	.	.
	***	2936	3986	5699	.	.	.	.	.	.	.	.	.	.	.	.

TABLE 2: ALPHA= 0.005 POWER= 0.9   EXPECTED ACCRUAL THRU MINIMUM FOLLOW-UP= 1200

	DEL=.02			DEL=.05			DEL=.10			DEL=.15			DEL=.20		
FACT=	1.0 .75	.50 .25	.00 BIN	1.0 .75	.50 .25	.00 BIN	1.0 .75	.50 .25	.00 BIN	1.0 75	.50 .25	.00 BIN	1.0 .75	.50 .25	.00 BIN

PCONT=•••   REQUIRED NUMBER OF PATIENTS

PCONT		1.0/.75	.50/.25	.00/BIN	1.0/.75	.50/.25	.00/BIN	1.0/.75	.50/.25	.00/BIN	1.0/75	.50/.25	.00/BIN	1.0/.75	.50/.25	.00/BIN
0.05	***	4243	4268	4482	912	925	950	330	333	337	193	195	196	136	137	138
	•••	4252	4310	8378	917	935	1637	331	335	521	194	196	275	137	138	175
0.1	***	8830	8890	9847	1677	1719	1833	538	549	565	290	295	300	192	194	196
	•••	8851	9005	14553	1694	1760	2590	543	556	744	292	297	368	193	196	224
0.15	***	12906	13010	15229	2340	2421	2687	715	739	775	371	380	391	238	241	246
	•••	12940	13216	19984	2371	2508	3423	726	754	938	376	385	447	239	244	265
0.2	***	16157	16318	20236	2870	2998	3467	861	898	961	438	451	469	274	280	287
	•••	16211	16637	24671	2919	3141	4137	877	925	1102	444	460	513	277	283	298
0.25	***	18500	18731	24673	3263	3444	4148	973	1027	1120	490	508	534	303	311	321
	•••	18577	19189	28614	3332	3650	4732	997	1065	1235	499	520	566	307	316	324
0.3	***	19950	20266	28436	3526	3765	4717	1056	1124	1248	529	552	585	323	334	346
	•••	20055	20893	31813	3617	4036	5208	1086	1175	1340	539	567	606	328	340	343
0.35	***	20570	20992	31462	3673	3970	5166	1110	1192	1345	553	582	622	336	348	363
	•••	20710	21820	34268	3787	4306	5565	1146	1255	1414	566	600	632	342	355	354
0.4	***	20455	21005	33714	3718	4075	5489	1138	1233	1411	566	598	644	342	355	372
	•••	20638	22071	35979	3856	4468	5803	1180	1305	1459	581	619	645	348	363	358
0.45	***	19716	20424	35172	3679	4090	5684	1143	1247	1444	568	602	652	340	355	372
	•••	19952	21761	36946	3841	4534	5922	1189	1327	1474	583	625	645	347	363	354
0.5	***	18475	19375	35824	3572	4028	5750	1126	1237	1444	558	594	645	332	346	364
	•••	18777	21007	37169	3753	4510	5922	1175	1321	1459	574	616	632	339	355	343
0.55	***	16864	17992	35668	3411	3899	5686	1090	1203	1413	538	574	624	317	331	349
	•••	17248	19920	36648	3607	4405	5803	1140	1288	1414	554	596	606	324	340	324
0.6	***	15023	16404	34701	3206	3712	5493	1036	1146	1348	508	541	588	296	309	325
	•••	15502	18591	35384	3411	4224	5565	1084	1228	1340	523	562	566	302	316	298
0.65	***	13092	14718	32928	2965	3470	5171	963	1067	1252	466	496	538	268	279	292
	•••	13674	17089	33375	3171	3970	5208	1009	1143	1235	481	515	513	274	286	265
0.7	***	11198	13008	30354	2692	3175	4721	872	964	1123	415	440	474	233	242	253
	•••	11866	15454	30622	2890	3643	4732	913	1031	1102	427	456	447	238	247	224
0.75	***	9422	11303	26986	2385	2826	4144	761	838	964	352	371	396	191	198	205
	•••	10135	13699	27126	2567	3240	4137	796	892	938	361	382	368	194	201	175
0.8	***	7780	9593	22830	2038	2413	3442	628	685	772	276	289	304	.	.	.
	•••	8479	11800	22885	2195	2754	3423	654	724	744	282	296	275	.	.	.
0.85	***	6220	7822	17896	1638	1924	2615	467	502	550	.	.	.	.	.	.
	•••	6847	9693	17900	1759	2169	2590	484	524	521	.	.	.	.	.	.
0.9	***	4633	5866	12193	1152	1323	1666	.	.	.	.	.	.	.	.	.
	•••	5122	7236	12172	1226	1456	1637	.	.	.	.	.	.	.	.	.
0.95	***	2749	3409	5728	.	.	.	.	.	.	.	.	.	.	.	.
	•••	3018	4065	5699	.	.	.	.	.	.	.	.	.	.	.	.

TABLE 2: ALPHA= 0.005 POWER= 0.9   EXPECTED ACCRUAL THRU MINIMUM FOLLOW-UP= 1300

		DEL=.02			DEL=.05			DEL=.10			DEL=.15			DEL=.20		
FACT=		1.0 .75	.50 .25	.00 BIN	1.0 .75	.50 .25	.00 BIN	1.0 .75	.50 .25	.00 BIN	1.0 75	.50 .25	.00 BIN	1.0 .75	.50 .25	.00 BIN
PCONT=***		REQUIRED NUMBER OF PATIENTS														
0.05	***	4245 4272 4483			913 926 951			331 334 338			194 195 197			136 137 138		
	•••	4254 4317 8378			919 936 1637			332 336 521			195 196 275			137 138 175		
0.1	***	8836 8900 9847			1682 1724 1833			539 550 565			291 295 300			193 194 197		
	•••	8857 9023 14553			1699 1764 2590			544 557 744			293 297 368			193 196 224		
0.15	***	12915 13028 15229			2348 2431 2687			718 741 775			373 381 391			238 242 246		
	•••	12952 13250 19984			2381 2517 3423			729 757 938			376 385 447			240 244 265		
0.2	***	16171 16345 20236			2883 3014 3467			865 902 961			440 453 469			276 281 288		
	•••	16229 16690 24671			2934 3157 4137			882 927 1102			446 460 513			278 284 298		
0.25	***	18520 18770 24674			3282 3468 4149			981 1032 1120			492 510 534			304 312 321		
	•••	18603 19264 28614			3353 3673 4732			1004 1069 1235			500 521 566			308 316 324		
0.3	***	19976 20320 28437			3551 3796 4717			1064 1131 1248			531 554 585			325 335 346		
	•••	20091 20994 31813			3645 4067 5208			1094 1180 1340			542 568 606			329 340 343		
0.35	***	20605 21063 31462			3703 4009 5166			1121 1201 1345			557 584 622			338 349 363		
	•••	20758 21953 34268			3822 4344 5565			1156 1260 1414			570 601 632			344 356 354		
0.4	***	20501 21097 33714			3755 4119 5489			1150 1242 1411			571 601 644			344 356 372		
	•••	20700 22239 35979			3898 4513 5803			1191 1311 1459			585 621 645			349 363 358		
0.45	***	19775 20541 35171			3723 4141 5684			1156 1258 1444			573 605 652			343 356 372		
	•••	20031 21966 36946			3888 4583 5922			1201 1334 1474			588 627 645			349 363 354		
0.5	***	18551 19521 35825			3621 4084 5750			1141 1248 1445			563 597 645			335 348 365		
	•••	18878 21249 37169			3806 4564 5922			1188 1329 1459			579 619 632			341 356 343		
0.55	***	16961 18172 35667			3464 3958 5687			1105 1215 1413			543 577 624			319 332 349		
	•••	17374 20195 36648			3664 4461 5803			1153 1296 1414			558 598 606			326 341 324		
0.6	***	15144 16615 34701			3262 3773 5493			1050 1158 1349			512 544 588			298 310 324		
	•••	15658 18892 35384			3469 4280 5565			1098 1236 1340			527 564 566			304 317 298		
0.65	***	13241 14956 32928			3022 3529 5172			977 1077 1252			471 500 538			270 280 292		
	•••	13860 17405 33375			3230 4024 5208			1021 1150 1235			484 518 513			275 286 265		
0.7	***	11373 13260 30354			2746 3231 4722			884 973 1124			419 443 474			235 243 253		
	•••	12074 15772 30622			2946 3693 4732			925 1037 1102			430 458 447			239 248 224		
0.75	***	9611 11555 26986			2435 2876 4145			771 846 964			354 373 396			193 198 205		
	•••	10351 14003 27126			2619 3284 4137			805 896 938			363 383 368			195 201 175		
0.8	***	7967 9830 22831			2083 2456 3442			635 691 773			278 290 305			. . .		
	•••	8688 12075 22885			2239 2789 3423			661 727 744			284 297 275			. . .		
0.85	***	6389 8026 17896			1672 1955 2616			472 505 551			. . .			. . .		
	•••	7031 9922 17900			1792 2193 2590			487 526 521			. . .			. . .		
0.9	***	4765 6019 12192			1173 1341 1667			. . .			. . .			. . .		
	•••	5264 7398 12172			1246 1469 1637			. . .			. . .			. . .		
0.95	***	2823 3487 5728			. . .			. . .			. . .			. . .		
	•••	3095 4137 5699			. . .			. . .			. . .			. . .		

TABLE 2: ALPHA= 0.005 POWER= 0.9   EXPECTED ACCRUAL THRU MINIMUM FOLLOW-UP= 1400

		DEL=.02			DEL=.05			DEL=.10			DEL=.15			DEL=.20		
FACT=		1.0 .75	.50 .25	.00 BIN	1.0 .75	.50 .25	.00 BIN	1.0 .75	.50 .25	.00 BIN	1.0 75	.50 .25	.00 BIN	1.0 .75	.50 .25	.00 BIN
PCONT=***					REQUIRED NUMBER OF PATIENTS											
0.05	***	4247	4276	4483	915	927	950	331	334	338	194	195	197	137	137	138
	•••	4257	4322	8378	920	937	1637	332	336	521	195	196	275	137	138	175
0.1	***	8841	8910	9847	1686	1729	1833	541	551	566	291	296	300	192	194	197
	•••	8864	9040	14553	1704	1767	2590	545	558	744	293	297	368	193	196	224
0.15	***	12924	13045	15229	2356	2440	2687	721	743	775	374	381	391	239	242	246
	•••	12964	13282	19984	2390	2525	3423	731	758	938	377	386	447	241	244	265
0.2	***	16184	16372	20236	2895	3030	3467	870	905	962	442	453	469	276	282	288
	•••	16247	16740	24671	2948	3172	4137	885	929	1102	447	461	513	278	284	298
0.25	***	18539	18809	24674	3299	3489	4148	986	1037	1120	495	512	534	305	312	321
	•••	18629	19337	28614	3373	3694	4732	1009	1072	1235	503	522	566	309	317	324
0.3	***	20002	20373	28437	3573	3825	4717	1073	1137	1249	535	556	585	326	335	346
	•••	20126	21094	31813	3671	4095	5208	1101	1184	1340	544	570	606	331	340	343
0.35	***	20640	21132	31461	3732	4044	5166	1130	1208	1346	561	586	622	339	350	363
	•••	20805	22083	34268	3854	4378	5565	1165	1265	1414	573	603	632	345	356	354
0.4	***	20547	21188	33714	3790	4161	5489	1161	1251	1411	575	604	644	346	357	372
	•••	20761	22403	35979	3937	4553	5803	1200	1318	1459	588	622	645	351	364	358
0.45	***	19834	20657	35171	3763	4189	5684	1168	1267	1444	577	608	652	344	357	372
	•••	20110	22165	36946	3933	4629	5922	1212	1340	1474	591	628	645	350	364	354
0.5	***	18626	19666	35824	3667	4136	5750	1153	1258	1445	568	600	645	336	349	365
	•••	18978	21483	37169	3855	4613	5922	1200	1335	1459	583	620	632	342	356	343
0.55	***	17056	18348	35668	3514	4014	5687	1117	1225	1413	547	579	624	321	333	348
	•••	17500	20459	36648	3716	4512	5803	1165	1303	1414	563	599	606	327	340	324
0.6	***	15264	16821	34701	3314	3828	5493	1062	1167	1348	516	547	588	299	311	325
	•••	15813	19179	35384	3524	4331	5565	1109	1242	1340	530	566	566	305	318	298
0.65	***	13388	15184	32928	3075	3584	5171	989	1087	1252	474	501	538	271	281	292
	•••	14040	17705	33375	3284	4073	5208	1032	1156	1235	487	519	513	276	286	265
0.7	***	11542	13500	30355	2797	3283	4721	895	982	1123	422	444	474	236	243	253
	•••	12274	16072	30622	2999	3739	4732	934	1042	1102	432	458	447	240	248	224
0.75	***	9792	11795	26986	2482	2922	4144	780	852	963	357	374	396	193	199	205
	•••	10558	14289	27126	2666	3323	4137	813	900	938	365	385	368	196	201	175
0.8	***	8145	10054	22830	2123	2495	3442	642	696	773	279	290	304	.	.	.
	•••	8886	12331	22885	2279	2821	3423	667	730	744	285	297	275	.	.	.
0.85	***	6549	8218	17896	1704	1983	2615	476	508	550	.	.	.	.	.	.
	•••	7206	10133	17900	1823	2215	2590	491	528	521	.	.	.	.	.	.
0.9	***	4890	6163	12192	1193	1356	1666	.	.	.	.	.	.	.	.	.
	•••	5398	7548	12172	1264	1480	1637	.	.	.	.	.	.	.	.	.
0.95	***	2892	3558	5729	.	.	.	.	.	.	.	.	.	.	.	.
	•••	3166	4203	5699	.	.	.	.	.	.	.	.	.	.	.	.

337

TABLE 2: ALPHA= 0.005 POWER= 0.9   EXPECTED ACCRUAL THRU MINIMUM FOLLOW-UP= 1500

		DEL=.02			DEL=.05			DEL=.10			DEL=.15			DEL=.20		
FACT=		1.0 .75	.50 .25	.00 BIN	1.0 .75	.50 .25	.00 BIN	1.0 .75	.50 .25	.00 BIN	1.0 75	.50 .25	.00 BIN	1.0 .75	.50 .25	.00 BIN
PCONT=***						REQUIRED	NUMBER	OF PATIENTS								
0.05	***	4250 4280	4482		917 929	950		331 334	337		194 196	197		137 138	139	
	•••	4260 4327	8378		922 937	1637		333 335	521		195 196	275		138 138	175	
0.1	***	8845 8919	9847		1690 1733	1833		542 552	566		292 295	300		193 195	197	
	•••	8870 9058	14553		1708 1771	2590		547 559	744		293 298	368		194 196	224	
0.15	***	12932 13062	15230		2363 2450	2687		724 744	775		375 382	391		239 243	247	
	•••	12976 13314	19984		2398 2533	3423		733 758	938		378 386	447		241 245	265	
0.2	***	16198 16399	20236		2907 3044	3467		874 908	961		443 455	470		277 282	288	
	•••	16265 16792	24671		2962 3185	4137		890 932	1102		448 461	513		279 285	298	
0.25	***	18559 18847	24674		3316 3510	4148		992 1040	1119		497 513	534		305 313	320	
	•••	18654 19410	28614		3392 3713	4732		1013 1074	1235		504 523	566		309 317	324	
0.3	***	20029 20425	28436		3596 3851	4717		1079 1143	1248		537 558	585		327 335	346	
	•••	20161 21192	31813		3697 4120	5208		1108 1188	1340		547 570	606		331 341	343	
0.35	***	20675 21202	31462		3760 4078	5165		1138 1215	1345		564 589	622		341 350	363	
	•••	20852 22210	34268		3886 4410	5565		1173 1270	1414		575 604	632		346 357	354	
0.4	***	20593 21279	33714		3824 4201	5489		1171 1259	1411		578 607	644		347 358	372	
	•••	20822 22563	35979		3974 4591	5803		1210 1323	1459		592 623	645		352 365	358	
0.45	***	19894 20772	35172		3803 4233	5684		1179 1276	1444		580 610	652		346 358	372	
	•••	20189 22358	36946		3975 4670	5922		1222 1346	1474		594 630	645		352 365	358	
0.5	***	18702 19810	35825		3712 4184	5750		1164 1267	1445		571 603	645		337 350	365	
	•••	19078 21709	37169		3903 4657	5922		1210 1342	1459		586 622	632		343 357	343	
0.55	***	17152 18520	35667		3562 4064	5687		1130 1234	1413		551 582	623		322 335	349	
	•••	17625 20713	36648		3766 4558	5803		1175 1309	1414		565 601	606		328 341	324	
0.6	***	15384 17020	34702		3364 3879	5494		1074 1177	1348		519 549	588		301 312	324	
	•••	15965 19453	35384		3575 4378	5565		1120 1249	1340		533 567	566		307 318	298	
0.65	***	13533 15404	32929		3125 3635	5171		999 1095	1252		478 504	538		273 281	292	
	•••	14216 17989	33375		3335 4118	5208		1042 1162	1235		490 519	513		277 287	265	
0.7	***	11707 13732	30354		2845 3331	4721		905 989	1124		425 446	474		237 244	252	
	•••	12466 16355	30622		3047 3780	4732		943 1047	1102		435 459	447		240 248	224	
0.75	***	9967 12023	26986		2527 2965	4145		788 858	964		359 376	397		194 199	204	
	•••	10755 14557	27126		2710 3359	4137		820 905	938		367 385	368		197 202	175	
0.8	***	8317 10265	22830		2160 2529	3442		649 699	772		281 292	305		. .	.	
	•••	9075 12573	22885		2317 2849	3423		672 732	744		286 298	275		. .	.	
0.85	***	6702 8400	17896		1732 2008	2615		480 510	550		. .	.		. .	.	
	•••	7372 10332	17900		1852 2235	2590		494 529	521		. .	.		. .	.	
0.9	***	5009 6297	12192		1210 1370	1666		. .	.		. .	.		. .	.	
	•••	5524 7687	12172		1282 1490	1637		. .	.		. .	.		. .	.	
0.95	***	2957 3624	5728		. .	.		. .	.		. .	.		. .	.	
	•••	3232 4263	5699		. .	.		. .	.		. .	.		. .	.	

TABLE 2: ALPHA= 0.005 POWER= 0.9   EXPECTED ACCRUAL THRU MINIMUM FOLLOW-UP= 1600

	DEL=.02			DEL=.05			DEL=.10			DEL=.15			DEL=.20		
FACT=	1.0 .75	.50 .25	.00 BIN	1.0 .75	.50 .25	.00 BIN	1.0 .75	.50 .25	.00 BIN	1.0 75	.50 .25	.00 BIN	1.0 .75	.50 .25	.00 BIN
PCONT=•••	REQUIRED NUMBER OF PATIENTS														
0.05 ***	4252 4284		4482	918 930		951	332 334		338	195 196		197	137 138		138
•••	4263 4332		8378	923 939		1637	333 336		521	195 196		275	138 138		175
0.1 ***	8851 8930		9848	1694 1737		1833	543 553		566	293 296		300	193 195		197
•••	8877 9074		14553	1712 1774		2590	548 559		744	294 298		368	194 196		224
0.15 ***	12941 13080		15230	2371 2458		2687	726 746		776	376 383		391	240 243		246
•••	12987 13346		19984	2406 2540		3423	735 759		938	379 387		447	241 244		265
0.2 ***	16211 16426		20236	2919 3058		3467	877 911		961	444 455		469	277 282		288
•••	16283 16841		24671	2974 3197		4137	893 933		1102	450 462		513	280 285		298
0.25 ***	18578 18886		24674	3332 3530		4148	997 1045		1120	499 514		534	307 313		321
•••	18680 19482		28614	3410 3731		5235	1018 1077		1235	506 524		566	310 317		324
0.3 ***	20055 20478		28436	3618 3877		4717	1086 1148		1248	539 559		585	328 337		346
•••	20196 21289		31813	3720 4144		5208	1114 1191		1340	549 571		606	332 341		343
0.35 ***	20711 21272		31462	3787 4110		5166	1146 1221		1346	567 590		622	342 352		363
•••	20898 22336		34268	3916 4439		5565	1180 1274		1414	578 605		632	347 357		354
0.4 ***	20639 21370		33714	3857 4238		5489	1180 1266		1411	581 608		644	348 359		372
•••	20883 22719		35979	4009 4625		5803	1218 1327		1459	594 625		645	353 365		358
0.45 ***	19952 20887		35172	3841 4275		5684	1189 1284		1444	584 613		652	347 359		372
•••	20267 22546		36946	4016 4709		5922	1231 1351		1474	598 631		645	353 365		354
0.5 ***	18777 19952		35824	3753 4230		5750	1175 1275		1445	575 605		645	339 351		365
•••	19177 21926		37169	3947 4699		5922	1220 1347		1459	589 623		632	345 357		343
0.55 ***	17248 18689		35668	3607 4112		5687	1140 1242		1413	555 584		624	324 335		349
•••	17749 20956		36648	3813 4601		5803	1186 1314		1414	568 603		606	330 342		324
0.6 ***	15503 17215		34701	3411 3928		5493	1085 1184		1349	523 551		588	302 313		325
•••	16113 19715		35384	3624 4421		5565	1130 1254		1340	536 568		566	307 318		298
0.65 ***	13674 15617		32928	3171 3682		5171	1009 1102		1252	481 506		538	274 282		293
•••	14388 18260		33375	3383 4160		5208	1051 1167		1235	493 521		513	278 287		265
0.7 ***	11867 13953		30354	2890 3375		4721	914 996		1124	427 448		474	238 245		253
•••	12653 16624		30622	3093 3818		4732	951 1051		1102	437 460		447	241 249		224
0.75 ***	10136 12241		26986	2568 3004		4144	796 863		964	361 377		396	195 199		205
•••	10945 14812		27126	2751 3392		4137	827 908		938	369 386		368	197 202		175
0.8 ***	8480 10468		22830	2196 2562		3442	654 703		773	282 292		304	.	.	.
•••	9255 12800		22885	2351 2875		3423	677 735		744	287 298		275	.	.	.
0.85 ***	6847 8572		17896	1759 2032		2616	484 513		551	.	.	.	.	.	.
•••	7528 10519		17900	1877 2252		2590	497 530		521	.	.	.	.	.	.
0.9 ***	5122 6424		12193	1227 1384		1666	.	.	.	.	.	.	.	.	.
•••	5644 7817		12172	1296 1499		1637	.	.	.	.	.	.	.	.	.
0.95 ***	3018 3686		5728	.	.	.	.	.	.	.	.	.	.	.	.
•••	3295 4318		5699	.	.	.	.	.	.	.	.	.	.	.	.

TABLE 2: ALPHA= 0.005 POWER= 0.9   EXPECTED ACCRUAL THRU MINIMUM FOLLOW-UP= 1700

	DEL=.02			DEL=.05			DEL=.10			DEL=.15			DEL=.20		
FACT=	1.0	.50	.00	1.0	.50	.00	1.0	.50	.00	1.0	.50	.00	1.0	.50	.00
	.75	.25	BIN	.75	.25	BIN	.75	.25	BIN	75	.25	BIN	.75	.25	BIN
PCONT=***				REQUIRED NUMBER OF PATIENTS											
0.05 ***	4254	4287	4483	919	930	951	331	335	338	194	195	197	137	138	138
***	4266	4337	8378	924	939	1637	333	335	521	195	197	275	138	138	175
0.1 ***	8856	8939	9847	1698	1740	1833	544	554	566	293	296	301	193	195	197
***	8884	9090	14553	1716	1778	2590	548	559	744	294	299	368	194	195	224
0.15 ***	12949	13097	15230	2378	2466	2687	728	748	775	376	384	391	240	243	246
***	12998	13377	19984	2414	2547	3423	737	760	938	379	386	447	241	244	265
0.2 ***	16224	16453	20236	2930	3070	3467	881	913	962	445	456	469	278	282	288
***	16301	16889	24671	2987	3208	4137	896	935	1102	450	462	513	280	284	298
0.25 ***	18597	18924	24674	3348	3548	4148	1002	1047	1120	500	515	534	308	313	321
***	18706	19552	28614	3427	3747	4732	1023	1079	1235	508	524	566	310	318	324
0.3 ***	20082	20531	28436	3638	3901	4717	1092	1151	1248	542	561	585	329	337	346
***	20232	21384	31813	3743	4165	5208	1119	1194	1340	550	571	606	333	341	343
0.35 ***	20746	21342	31462	3813	4139	5166	1154	1227	1346	569	592	622	343	352	363
***	20945	22458	34268	3944	4466	5565	1187	1278	1414	580	605	632	347	357	354
0.4 ***	20684	21459	33714	3888	4273	5489	1188	1273	1411	584	610	644	350	360	372
***	20945	22870	35979	4043	4656	5803	1226	1331	1459	596	626	645	355	365	358
0.45 ***	20011	21000	35172	3877	4313	5684	1198	1291	1444	586	615	652	348	359	372
***	20346	22729	36946	4053	4744	5922	1240	1355	1474	600	632	645	354	365	354
0.5 ***	18853	20091	35824	3793	4272	5750	1185	1282	1445	578	607	645	340	352	364
***	19277	22136	37169	3988	4738	5922	1229	1351	1459	592	624	632	345	358	343
0.55 ***	17343	18855	35667	3650	4156	5686	1151	1249	1413	558	586	624	325	335	348
***	17871	21189	36648	3858	4641	5803	1195	1319	1414	571	603	606	330	342	324
0.6 ***	15620	17403	34701	3455	3973	5493	1094	1192	1348	526	554	588	304	313	325
***	16260	19965	35384	3669	4460	5565	1138	1259	1340	539	569	566	308	318	298
0.65 ***	13814	15822	32928	3215	3726	5171	1019	1109	1251	483	508	539	274	282	292
***	14555	18518	33375	3427	4198	5208	1059	1171	1235	495	522	513	278	288	265
0.7 ***	12023	14165	30355	2932	3416	4721	922	1002	1124	429	449	474	239	245	253
***	12833	16879	30622	3135	3852	4732	958	1055	1102	439	461	447	242	248	224
0.75 ***	10297	12450	26985	2606	3041	4145	803	868	964	362	378	396	195	199	205
***	11128	15053	27126	2790	3422	4137	833	911	938	369	386	368	197	202	175
0.8 ***	8637	10660	22830	2229	2592	3442	660	707	772	284	293	305	.	.	.
***	9428	13015	22885	2384	2898	3423	681	736	744	288	299	275	.	.	.
0.85 ***	6987	8736	17896	1785	2054	2616	486	514	550	.	.	.	.	.	.
***	7679	10694	17900	1902	2268	2590	499	531	521	.	.	.	.	.	.
0.9 ***	5229	6544	12192	1242	1396	1666	.	.	.	.	.	.	.	.	.
***	5758	7938	12172	1310	1508	1637	.	.	.	.	.	.	.	.	.
0.95 ***	3076	3744	5728	.	.	.	.	.	.	.	.	.	.	.	.
***	3353	4369	5699	.	.	.	.	.	.	.	.	.	.	.	.

TABLE 2: ALPHA= 0.005 POWER= 0.9   EXPECTED ACCRUAL THRU MINIMUM FOLLOW-UP= 1800

	DEL=.02			DEL=.05			DEL=.10			DEL=.15			DEL=.20		
FACT=	1.0 .75	.50 .25	.00 BIN	1.0 .75	.50 .25	.00 BIN	1.0 .75	.50 .25	.00 BIN	1.0 75	.50 .25	.00 BIN	1.0 .75	.50 .25	.00 BIN
PCONT=***				REQUIRED NUMBER OF PATIENTS											
0.05 ***	4256	4290	4482	920	931	951	332	335	337	195	195	197	137	138	138
***	4268	4341	8378	924	939	1637	333	336	521	195	197	275	137	138	175
0.1 ***	8860	8949	9848	1701	1743	1833	546	555	566	294	297	300	193	195	197
***	8891	9105	14553	1719	1779	2590	549	559	744	294	298	368	195	195	224
0.15 ***	12957	13114	15230	2385	2472	2688	730	749	775	377	384	391	240	243	246
***	13011	13407	19984	2421	2553	3423	739	762	938	380	387	447	242	245	265
0.2 ***	16238	16479	20236	2940	3082	3467	884	915	962	447	456	469	279	282	288
***	16318	16938	24671	2998	3217	4137	897	936	1102	451	462	513	280	285	298
0.25 ***	18616	18962	24673	3363	3565	4148	1005	1050	1119	501	516	534	308	314	321
***	18731	19621	28614	3444	3763	4732	1027	1081	1235	508	524	566	312	317	324
0.3 ***	20108	20583	28437	3658	3924	4717	1098	1156	1248	543	561	585	330	337	346
***	20267	21477	31813	3765	4186	5208	1124	1196	1340	552	573	606	334	342	343
0.35 ***	20781	21411	31461	3838	4167	5166	1161	1232	1345	571	594	622	344	353	363
***	20992	22578	34268	3971	4491	5565	1192	1281	1414	582	606	632	348	357	354
0.4 ***	20730	21549	33714	3918	4305	5489	1196	1278	1410	587	612	645	351	360	372
***	21006	23019	35979	4074	4686	5803	1233	1335	1459	598	627	645	355	366	358
0.45 ***	20071	21113	35172	3910	4350	5685	1207	1297	1444	589	616	652	350	360	372
***	20424	22905	36946	4090	4776	5922	1248	1360	1474	603	633	645	354	366	354
0.5 ***	18928	20229	35824	3831	4312	5750	1194	1289	1444	580	609	645	342	352	364
***	19374	22339	37169	4029	4772	5922	1237	1356	1459	594	625	632	346	357	343
0.55 ***	17437	19017	35668	3690	4198	5687	1160	1257	1413	560	588	624	326	336	348
***	17992	21414	36648	3899	4677	5803	1203	1324	1414	573	605	606	332	342	324
0.6 ***	15736	17587	34701	3498	4015	5493	1104	1198	1349	528	555	588	305	314	325
***	16404	20205	35384	3712	4497	5565	1146	1263	1340	541	570	566	309	319	298
0.65 ***	13950	16021	32928	3257	3768	5172	1027	1115	1252	486	510	539	276	283	292
***	14718	18764	33375	3471	4233	5208	1068	1174	1235	497	523	513	279	288	265
0.7 ***	12174	14370	30354	2973	3455	4722	929	1007	1124	431	450	474	240	245	253
***	13008	17122	30622	3176	3885	4732	965	1059	1102	440	461	447	242	249	224
0.75 ***	10455	12649	26985	2643	3075	4144	809	873	964	364	379	396	195	200	204
***	11304	15282	27126	2826	3450	4137	838	913	938	371	387	368	198	202	175
0.8 ***	8788	10844	22830	2260	2620	3442	663	710	773	285	294	305	.	.	.
***	9594	13218	22885	2414	2920	3423	685	738	744	289	299	275	.	.	.
0.85 ***	7120	8891	17896	1808	2074	2616	489	516	551	.	.	.	.	.	.
***	7822	10860	17900	1923	2283	2590	501	532	521	.	.	.	.	.	.
0.9 ***	5332	6657	12192	1255	1406	1666	.	.	.	.	.	.	.	.	.
***	5865	8052	12172	1323	1515	1637	.	.	.	.	.	.	.	.	.
0.95 ***	3131	3798	5728	.	.	.	.	.	.	.	.	.	.	.	.
***	3408	4416	5699	.	.	.	.	.	.	.	.	.	.	.	.

TABLE 2: ALPHA= 0.005 POWER= 0.9   EXPECTED ACCRUAL THRU MINIMUM FOLLOW-UP= 1900

PCONT	DEL=.02			DEL=.05			DEL=.10			DEL=.15			DEL=.20		
FACT=	1.0 .75	.50 .25	.00 BIN	1.0 .75	.50 .25	.00 BIN	1.0 .75	.50 .25	.00 BIN	1.0 75	.50 .25	.00 BIN	1.0 .75	.50 .25	.00 BIN
PCONT=•••						REQUIRED NUMBER OF PATIENTS									
0.05 •••	4258	4295	4483	921	932	951	332	334	338	194	196	197	137	137	139
•••	4271	4345	8378	926	940	1637	334	336	521	196	196	275	137	137	175
0.1 •••	8865	8958	9848	1705	1747	1833	546	554	566	293	296	300	193	194	197
•••	8896	9121	14553	1723	1782	2590	550	560	744	296	298	368	194	196	224
0.15 •••	12967	13131	15230	2391	2479	2687	731	750	775	377	383	391	241	243	246
•••	13022	13436	19984	2428	2557	3423	740	762	938	381	387	447	242	244	265
0.2 •••	16251	16506	20236	2951	3093	3467	887	918	961	448	457	469	279	282	287
•••	16336	16985	24671	3009	3227	4137	901	937	1102	452	463	513	281	285	298
0.25 •••	18635	19000	24673	3378	3581	4148	1010	1054	1120	503	517	534	308	315	320
•••	18757	19688	28614	3460	3776	4732	1030	1083	1235	509	524	566	311	318	324
0.3 •••	20135	20635	28437	3678	3944	4718	1103	1160	1248	545	562	585	331	338	346
•••	20301	21568	31813	3785	4203	5208	1129	1199	1340	553	573	606	334	342	343
0.35 •••	20816	21481	31461	3863	4193	5165	1167	1236	1345	573	595	622	345	353	363
•••	21039	22695	34268	3997	4514	5565	1198	1284	1414	584	608	632	349	358	354
0.4 •••	20776	21638	33714	3946	4336	5488	1203	1284	1410	589	612	645	351	361	372
•••	21066	23163	35979	4105	4713	5803	1239	1338	1459	600	628	645	356	365	358
0.45 •••	20129	21224	35171	3944	4385	5684	1215	1303	1444	592	619	652	350	361	372
•••	20502	23077	36946	4124	4808	5922	1255	1364	1474	604	634	645	355	365	354
0.5 •••	19003	20364	35824	3868	4350	5749	1203	1295	1445	583	610	645	343	353	364
•••	19472	22536	37169	4066	4806	5922	1244	1360	1459	596	627	632	348	358	343
0.55 •••	17531	19175	35667	3730	4238	5687	1168	1262	1413	562	590	623	327	337	349
•••	18111	21631	36648	3940	4712	5803	1211	1327	1414	576	605	606	332	343	324
0.6 •••	15850	17766	34701	3537	4055	5493	1113	1204	1349	532	557	588	305	315	324
•••	16545	20436	35384	3752	4531	5565	1154	1267	1340	543	571	566	310	319	298
0.65 •••	14085	16213	32928	3298	3806	5171	1035	1121	1251	488	510	538	277	284	292
•••	14877	19000	33375	3510	4265	5208	1075	1178	1235	498	524	513	280	288	265
0.7 •••	12323	14567	30354	3011	3491	4721	937	1011	1123	433	451	474	239	246	253
•••	13178	17354	30622	3213	3915	4732	971	1061	1102	441	462	447	243	249	224
0.75 •••	10608	12841	26986	2678	3108	4144	814	876	964	365	380	396	196	201	205
•••	11473	15500	27126	2859	3476	4137	843	915	938	372	387	368	198	203	175
0.8 •••	8934	11019	22830	2289	2645	3442	668	712	773	285	294	304	.	.	.
•••	9753	13411	22885	2443	2940	3423	688	740	744	289	299	275	.	.	.
0.85 •••	7248	9039	17897	1830	2092	2616	491	517	551	.	.	.	.	.	.
•••	7960	11016	17900	1945	2296	2590	505	533	521	.	.	.	.	.	.
0.9 •••	5430	6766	12193	1269	1417	1666	.	.	.	.	.	.	.	.	.
•••	5969	8159	12172	1334	1521	1637	.	.	.	.	.	.	.	.	.
0.95 •••	3182	3849	5728	.	.	.	.	.	.	.	.	.	.	.	.
•••	3461	4459	5699	.	.	.	.	.	.	.	.	.	.	.	.

TABLE 2: ALPHA= 0.005 POWER= 0.9   EXPECTED ACCRUAL THRU MINIMUM FOLLOW-UP= 2000

	DEL=.02			DEL=.05			DEL=.10			DEL=.15			DEL=.20		
FACT=	1.0 .75	.50 .25	.00 BIN	1.0 .75	.50 .25	.00 BIN	1.0 .75	.50 .25	.00 BIN	1.0 75	.50 .25	.00 BIN	1.0 .75	.50 .25	.00 BIN
PCONT=***				REQUIRED NUMBER OF PATIENTS											
0.05 ***	4260	4297	4482	922	934	951	332	335	337	195	196	197	137	137	139
***	4274	4350	8378	927	941	1637	334	336	521	195	196	275	137	139	175
0.1 ***	8870	8969	9847	1707	1750	1834	546	555	566	294	297	300	194	195	197
***	8904	9136	14553	1726	1784	2590	551	560	744	295	299	368	195	196	224
0.15 ***	12976	13149	15230	2397	2486	2687	734	751	776	379	384	391	241	244	246
***	13034	13465	19984	2435	2562	3423	741	762	938	381	387	447	242	245	265
0.2 ***	16265	16532	20236	2961	3105	3467	890	919	961	449	457	470	279	284	287
***	16355	17031	24671	3020	3236	4137	904	939	1102	454	464	513	281	285	298
0.25 ***	18655	19039	24674	3392	3596	4149	1014	1056	1120	505	517	534	310	315	321
***	18784	19755	28614	3475	3790	4732	1034	1085	1235	511	526	566	312	317	324
0.3 ***	20161	20687	28436	3696	3965	4717	1107	1164	1249	547	564	585	331	339	346
***	20337	21657	31813	3806	4221	5208	1134	1201	1340	555	574	606	335	342	343
0.35 ***	20851	21550	31462	3886	4219	5166	1172	1241	1346	576	596	622	346	354	364
***	21086	22810	34268	4021	4536	5565	1204	1286	1414	585	609	632	350	359	354
0.4 ***	20822	21726	33714	3974	4366	5489	1210	1289	1411	591	615	644	352	361	372
***	21127	23304	35979	4134	4739	5803	1246	1342	1459	602	629	645	356	366	358
0.45 ***	20189	21335	35172	3975	4419	5685	1222	1309	1444	595	620	652	351	361	372
***	20580	23242	36946	4157	4836	5922	1261	1367	1474	606	635	645	356	366	354
0.5 ***	19077	20497	35825	3902	4386	5750	1210	1301	1445	586	612	645	344	354	365
***	19570	22725	37169	4102	4836	5922	1251	1364	1459	599	627	632	349	359	343
0.55 ***	17625	19331	35667	3766	4275	5686	1176	1269	1412	566	591	624	329	337	349
***	18231	21840	36648	3977	4744	5803	1219	1331	1414	577	606	606	332	344	324
0.6 ***	15965	17940	34701	3575	4094	5494	1120	1210	1349	534	557	589	306	315	325
***	16685	20656	35384	3791	4562	5565	1161	1271	1340	545	572	566	310	320	298
0.65 ***	14216	16400	32929	3335	3842	5171	1042	1126	1252	490	511	539	277	285	292
***	15032	19226	33375	3549	4296	5208	1081	1181	1235	500	524	513	281	289	265
0.7 ***	12466	14756	30355	3047	3525	4721	942	1016	1124	435	452	475	240	246	252
***	13341	17576	30622	3249	3942	4732	976	1064	1102	444	462	447	244	250	224
0.75 ***	10755	13026	26986	2710	3137	4145	820	880	964	367	380	396	196	200	205
***	11637	15709	27126	2892	3499	4137	847	917	938	374	387	368	199	202	175
0.8 ***	9075	11187	22830	2316	2671	3442	672	715	772	286	295	305	.	.	.
***	9906	13596	22885	2469	2959	3423	692	741	744	290	300	275	.	.	.
0.85 ***	7371	9181	17896	1851	2110	2616	495	520	551	.	.	.	.	.	.
***	8091	11165	17900	1965	2310	2590	506	534	521	.	.	.	.	.	.
0.9 ***	5524	6869	12192	1281	1426	1666	.	.	.	.	.	.	.	.	.
***	6067	8260	12172	1346	1527	1637	.	.	.	.	.	.	.	.	.
0.95 ***	3232	3897	5729	.	.	.	.	.	.	.	.	.	.	.	.
***	3511	4500	5699	.	.	.	.	.	.	.	.	.	.	.	.

TABLE 2: ALPHA= 0.005 POWER= 0.9   EXPECTED ACCRUAL THRU MINIMUM FOLLOW-UP= 2250

		DEL=.02			DEL=.05			DEL=.10			DEL=.15			DEL=.20		
FACT=		1.0 .75	.50 .25	.00 BIN	1.0 .75	.50 .25	.00 BIN	1.0 .75	.50 .25	.00 BIN	1.0 75	.50 .25	.00 BIN	1.0 .75	.50 .25	.00 BIN
PCONT=***		REQUIRED NUMBER OF PATIENTS														
0.05	***	4265	4305	4483	924	935	950	333	336	337	195	196	196	137	137	139
	***	4280	4358	8378	929	942	1637	334	336	521	195	196	275	137	137	175
0.1	***	8883	8991	9848	1715	1756	1833	548	556	567	295	297	300	194	195	196
	***	8920	9170	14553	1734	1788	2590	552	561	744	295	298	368	195	196	224
0.15	***	12998	13192	15229	2413	2500	2687	736	753	776	379	385	390	241	243	246
	***	13062	13535	19984	2449	2573	3423	745	763	938	382	388	447	243	244	265
0.2	***	16298	16598	20235	2985	3129	3467	896	924	961	449	460	469	280	284	288
	***	16399	17142	24671	3044	3255	4137	908	941	1102	454	464	513	282	285	298
0.25	***	18703	19133	24674	3425	3632	4148	1022	1063	1119	507	520	534	311	314	320
	***	18848	19915	28614	3509	3818	4732	1040	1088	1235	513	527	566	313	317	324
0.3	***	20227	20817	28437	3740	4012	4718	1119	1171	1248	550	567	584	333	339	345
	***	20426	21872	31813	3852	4259	5208	1143	1205	1340	558	575	606	336	343	343
0.35	***	20939	21719	31462	3940	4274	5166	1185	1250	1346	579	599	621	347	354	362
	***	21202	23083	34268	4078	4584	5565	1214	1292	1414	589	610	632	351	358	354
0.4	***	20937	21943	33715	4038	4432	5488	1225	1299	1411	596	617	644	354	362	372
	***	21279	23639	35979	4201	4795	5803	1258	1348	1459	606	629	645	358	367	358
0.45	***	20336	21604	35172	4048	4493	5684	1239	1320	1444	600	623	652	354	362	372
	***	20773	23637	36946	4232	4899	5922	1275	1375	1474	610	637	645	358	367	354
0.5	***	19264	20820	35824	3984	4466	5750	1227	1315	1445	592	615	645	345	354	364
	***	19810	23173	37169	4184	4904	5922	1267	1371	1459	603	629	632	350	359	343
0.55	***	17856	19706	35668	3853	4359	5686	1194	1281	1413	570	595	624	330	339	348
	***	18520	22330	36648	4064	4814	5803	1234	1340	1414	582	609	606	334	343	324
0.6	***	16242	18354	34702	3663	4178	5494	1137	1222	1348	538	561	587	308	316	325
	***	17021	21174	35384	3880	4633	5565	1177	1278	1340	550	573	566	312	320	298
0.65	***	14534	16840	32928	3422	3926	5171	1059	1137	1253	494	514	538	278	285	292
	***	15404	19752	33375	3635	4363	5208	1095	1188	1235	505	525	513	281	288	265
0.7	***	12810	15204	30354	3130	3602	4721	957	1026	1123	438	455	474	241	247	253
	***	13732	18089	30622	3331	4003	4732	989	1070	1102	446	464	447	244	250	224
0.75	***	11105	13457	26985	2786	3205	4144	832	887	964	370	382	396	196	201	205
	***	12023	16190	27126	2966	3552	4137	857	922	938	375	389	368	199	202	175
0.8	***	9406	11581	22830	2379	2725	3442	680	721	773	288	295	305	.	.	.
	***	10265	14020	22885	2530	2999	3423	700	745	744	292	299	275	.	.	.
0.85	***	7660	9510	17895	1898	2148	2615	499	523	551	.	.	.	.	.	.
	***	8399	11505	17900	2008	2337	2590	510	536	521	.	.	.	.	.	.
0.9	***	5744	7106	12192	1309	1447	1666	.	.	.	.	.	.	.	.	.
	***	6297	8489	12172	1371	1541	1637	.	.	.	.	.	.	.	.	.
0.95	***	3346	4006	5729	.	.	.	.	.	.	.	.	.	.	.	.
	***	3624	4591	5699	.	.	.	.	.	.	.	.	.	.	.	.

TABLE 2: ALPHA= 0.005 POWER= 0.9   EXPECTED ACCRUAL THRU MINIMUM FOLLOW-UP= 2500

		DEL=.02			DEL=.05			DEL=.10			DEL=.15			DEL=.20		
FACT=		1.0 .75	.50 .25	.00 BIN	1.0 .75	.50 .25	.00 BIN	1.0 .75	.50 .25	.00 BIN	1.0 75	.50 .25	.00 BIN	1.0 .75	.50 .25	.00 BIN
PCONT=***				REQUIRED NUMBER OF PATIENTS												
0.05	***	4270	4314	4483	926	936	951	334	336	337	195	196	196	137	139	139
	•••	4286	4367	8378	931	942	1637	334	337	521	195	196	275	137	139	175
0.1	***	8895	9014	9848	1721	1762	1833	549	558	565	295	298	299	195	195	196
	•••	8936	9201	14553	1739	1792	2590	552	561	744	296	298	368	195	196	224
0.15	***	13018	13233	15229	2426	2512	2687	740	756	776	381	386	390	242	243	246
	•••	13092	13599	19984	2464	2583	3423	746	765	938	383	389	447	243	245	265
0.2	***	16333	16664	20236	3006	3149	3467	899	926	961	452	461	470	281	284	287
	•••	16443	17246	24671	3067	3271	4137	912	942	1102	456	464	513	283	286	298
0.25	***	18751	19226	24673	3456	3662	4148	1029	1067	1120	509	521	534	312	315	321
	•••	18911	20067	28614	3542	3842	4732	1046	1090	1235	515	527	566	314	318	324
0.3	***	20293	20943	28436	3781	4051	4717	1127	1177	1248	552	568	586	334	340	346
	•••	20512	22076	31813	3893	4293	5208	1151	1209	1340	561	576	606	337	343	343
0.35	***	21027	21887	31462	3990	4324	5165	1196	1258	1345	583	601	621	348	356	364
	•••	21318	23342	34268	4129	4624	5565	1224	1296	1414	592	611	632	352	359	354
0.4	***	21051	22156	33714	4098	4492	5489	1237	1309	1411	599	620	643	356	364	371
	•••	21429	23952	35979	4262	4843	5803	1270	1354	1459	609	631	645	359	367	358
0.45	***	20483	21865	35171	4115	4559	5684	1252	1331	1443	604	626	651	356	364	371
	•••	20962	24004	36946	4301	4954	5922	1289	1381	1474	614	639	645	359	367	354
0.5	***	19448	21129	35824	4056	4537	5749	1243	1324	1445	595	618	645	346	356	364
	•••	20045	23587	37169	4259	4962	5922	1281	1377	1459	606	631	632	351	359	343
0.55	***	18083	20059	35667	3929	4434	5687	1209	1292	1412	574	596	623	333	340	348
	•••	18799	22779	36648	4142	4874	5803	1246	1345	1414	586	609	606	336	343	324
0.6	***	16511	18743	34701	3743	4252	5493	1152	1233	1348	543	564	589	309	317	324
	•••	17342	21645	35384	3959	4693	5565	1189	1284	1340	552	574	566	314	320	298
0.65	***	14837	17249	32927	3499	3998	5171	1073	1146	1252	498	517	539	279	286	292
	•••	15754	20229	33375	3712	4420	5208	1108	1193	1235	508	526	513	283	289	265
0.7	***	13136	15617	30354	3204	3668	4721	970	1034	1123	442	456	474	242	248	252
	•••	14095	18556	30622	3402	4056	4732	999	1074	1102	448	465	447	245	249	224
0.75	***	11431	13852	26986	2851	3262	4145	842	895	964	371	384	396	198	201	204
	•••	12381	16623	27126	3029	3595	4137	867	926	938	377	390	368	199	202	175
0.8	***	9714	11939	22831	2436	2771	3442	689	724	773	289	296	304	.	.	.
	•••	10596	14399	22885	2583	3034	3423	706	748	744	293	299	275	.	.	.
0.85	***	7926	9809	17896	1940	2181	2615	504	524	551	.	.	.	.	.	.
	•••	8683	11809	17900	2046	2359	2590	514	537	521	.	.	.	.	.	.
0.9	***	5943	7318	12192	1333	1464	1667	.	.	.	.	.	.	.	.	.
	•••	6504	8692	12172	1392	1551	1637	.	.	.	.	.	.	.	.	.
0.95	***	3448	4101	5727	.	.	.	.	.	.	.	.	.	.	.	.
	•••	3724	4668	5699	.	.	.	.	.	.	.	.	.	.	.	.

TABLE 2: ALPHA= 0.005 POWER= 0.9   EXPECTED ACCRUAL THRU MINIMUM FOLLOW-UP= 2750

		DEL=.02			DEL=.05			DEL=.10			DEL=.15			DEL=.20		
FACT=		1.0 .75	.50 .25	.00 BIN	1.0 .75	.50 .25	.00 BIN	1.0 .75	.50 .25	.00 BIN	1.0 75	.50 .25	.00 BIN	1.0 .75	.50 .25	.00 BIN
PCONT=•••		REQUIRED NUMBER OF PATIENTS														
0.05	***	4275	4320	4482	927	936	950	333	336	338	195	195	197	137	139	139
	•••	4291	4374	8378	931	943	1637	335	336	521	195	197	275	137	139	175
0.1	***	8908	9036	9848	1727	1766	1834	551	558	565	295	297	300	194	195	197
	•••	8953	9231	14553	1745	1796	2590	555	562	744	297	298	368	195	195	224
0.15	***	13041	13274	15229	2439	2522	2687	742	758	775	381	386	391	242	243	247
	•••	13120	13660	19984	2474	2590	3423	749	766	938	383	388	447	243	245	265
0.2	***	16365	16728	20236	3026	3169	3466	903	929	962	453	460	469	281	285	288
	•••	16488	17347	24671	3086	3285	4137	916	943	1102	456	465	513	283	287	298
0.25	***	18799	19320	24674	3484	3690	4148	1036	1070	1120	511	522	534	312	315	321
	•••	18974	20210	28614	3571	3863	4732	1051	1092	1235	517	527	566	314	319	324
0.3	***	20360	21070	28436	3817	4089	4718	1136	1182	1247	555	569	586	335	340	346
	•••	20600	22267	31813	3930	4320	5208	1158	1213	1340	562	577	606	338	343	343
0.35	***	21114	22051	31461	4035	4370	5166	1206	1264	1345	586	603	621	350	355	364
	•••	21434	23584	34268	4176	4660	5565	1233	1301	1414	594	611	632	353	359	354
0.4	***	21166	22362	33714	4152	4544	5489	1249	1316	1411	603	621	644	357	364	373
	•••	21579	24247	35979	4315	4884	5803	1280	1359	1459	611	632	645	360	367	358
0.45	***	20628	22116	35172	4178	4618	5684	1264	1339	1443	608	627	652	357	364	373
	•••	21150	24347	36946	4361	5001	5922	1299	1387	1474	617	639	645	360	367	354
0.5	***	19631	21425	35825	4123	4601	5750	1256	1333	1445	599	620	645	349	355	364
	•••	20274	23971	37169	4326	5013	5922	1292	1383	1459	610	632	632	352	360	343
0.55	***	18304	20394	35667	4000	4499	5687	1222	1301	1412	579	599	624	333	340	349
	•••	19070	23194	36648	4212	4927	5803	1257	1350	1414	589	611	606	336	345	324
0.6	***	16769	19108	34701	3815	4319	5492	1165	1240	1349	546	565	587	311	318	324
	•••	17647	22078	35384	4028	4745	5565	1201	1290	1340	555	577	566	314	321	298
0.65	***	15128	17632	32929	3571	4061	5171	1084	1154	1253	501	518	538	281	287	291
	•••	16085	20668	33375	3780	4470	5208	1116	1199	1235	510	527	513	283	290	265
0.7	***	13442	15999	30354	3270	3728	4721	979	1041	1123	445	459	474	243	247	252
	•••	14435	18981	30622	3468	4100	4732	1009	1079	1102	452	465	447	245	250	224
0.75	***	11737	14219	26986	2911	3313	4145	850	900	964	374	384	397	199	202	204
	•••	12715	17017	27126	3086	3633	4137	874	929	938	379	390	368	201	202	175
0.8	***	9999	12268	22830	2485	2813	3442	694	728	773	290	297	304	.	.	.
	•••	10903	14743	22885	2629	3064	3423	711	749	744	294	300	275	.	.	.
0.85	***	8172	10081	17897	1976	2210	2615	507	527	551	.	.	.	.	.	.
	•••	8940	12081	17900	2081	2378	2590	517	538	521	.	.	.	.	.	.
0.9	***	6127	7512	12192	1352	1477	1666	.	.	.	.	.	.	.	.	.
	•••	6694	8870	12172	1411	1560	1637	.	.	.	.	.	.	.	.	.
0.95	***	3540	4186	5728	.	.	.	.	.	.	.	.	.	.	.	.
	•••	3815	4736	5699	.	.	.	.	.	.	.	.	.	.	.	.

TABLE 2: ALPHA= 0.005 POWER= 0.9   EXPECTED ACCRUAL THRU MINIMUM FOLLOW-UP= 3000

	DEL=.01			DEL=.02			DEL=.05			DEL=.10			DEL=.15		
FACT=	1.0 .75	.50 .25	.00 BIN	1.0 .75	.50 .25	.00 BIN	1.0 .75	.50 .25	.00 BIN	1.0 75	.50 .25	.00 BIN	1.0 .75	.50 .25	.00 BIN
PCONT=***				REQUIRED NUMBER OF PATIENTS											
0.01 ***	2270 2278	2293	835 838	842	275 275	275	140 140	140	100 100	100
•••	2275 2285	8779	835 838	2902	275 275	790	140 140	321	100 100	191
0.02 ***	5015 5035	5120	1610 1618	1637	437 437	440	193 193	193	125 125	125
•••	5023 5057	14493	1615 1625	4316	437 440	1009	193 193	373	125 125	213
0.05 ***	15178 15235	15995	4280 4325	4483	928 935	950	332 335	335	197 197	197
•••	15197 15340	30920	4295 4378	8378	932 943	1637	335 335	521	197 197	275
0.1 ***	33470 33610	37232	8920 9058	9845	1735 1772	1832	553 557	565	295 298	298
•••	33515 33887	55917	8968 9257	14553	1750 1798	2590	553 560	744	298 298	368
0.15 ***	49955 50207	58817	13063 13315	15230	2450 2533	2687	745 760	775	380 385	392
•••	50042 50705	77939	13150 13715	19984	2485 2597	3423	752 767	938	385 388	447
0.2 ***	63220 63605	79018	16397 16790	20237	3043 3185	3467	910 932	962	455 460	470
•••	63347 64382	96984	16532 17440	24671	3103 3298	4137	917 943	1102	455 463	513
0.25 ***	72842 73405	97010	18845 19408	24673	3508 3715	4150	1040 1075	1120	512 523	535
•••	73030 74522	113054	19037 20345	28614	3595 3883	4732	1055 1097	1235	515 527	566
0.3 ***	78857 79625	112337	20425 21193	28435	3850 4120	4715	1142 1187	1247	557 568	583
•••	79112 81167	126148	20687 22450	31813	3965 4345	5208	1165 1213	1340	565 575	606
0.35 ***	81497 82525	124738	21200 22210	31460	4078 4408	5165	1213 1270	1345	587 602	620
•••	81838 84575	136266	21550 23815	34268	4217 4690	5565	1240 1303	1414	595 613	632
0.4 ***	81122 82465	134057	21280 22562	33715	4202 4592	5488	1258 1322	1412	605 625	643
•••	81568 85142	143408	21725 24523	35979	4367 4922	5803	1288 1363	1459	613 632	645
0.45 ***	78160 79892	140207	20773 22360	35173	4232 4670	5683	1277 1345	1442	610 628	650
•••	78737 83327	147574	21335 24665	36946	4420 5042	5922	1307 1390	1474	620 640	645
0.5 ***	73082 75298	143140	19810 21707	35825	4183 4655	5750	1265 1340	1445	602 620	643
•••	73825 79637	148765	20495 24328	37169	4385 5057	5922	1300 1390	1459	613 632	632
0.55 ***	66403 69223	142835	18520 20713	35668	4063 4558	5687	1232 1307	1412	583 602	625
•••	67345 74590	146979	19330 23578	36648	4273 4975	5803	1270 1355	1414	590 613	606
0.6 ***	58645 62215	139292	17020 19453	34700	3880 4378	5492	1175 1247	1348	550 568	587
•••	59845 68660	142218	17938 22480	35384	4093 4790	5565	1210 1292	1340	557 575	566
0.65 ***	50372 54823	132520	15403 17987	32927	3635 4120	5170	1093 1160	1250	505 520	538
•••	51898 62233	134480	16400 21070	33375	3842 4513	5208	1127 1202	1235	512 527	513
0.7 ***	42167 47500	122545	13730 16355	30355	3332 3778	4720	988 1048	1123	445 460	475
•••	44057 55562	123767	14755 19370	30622	3523 4138	4732	1015 1082	1102	452 467	447
0.75 ***	34550 40510	109385	12025 14555	26987	2965 3358	4145	857 905	962	377 385	395
•••	36740 48748	110078	13025 17375	27126	3137 3665	4137	880 932	938	380 392	368
0.8 ***	27800 33880	93077	10265 12572	22828	2530 2848	3440	700 733	770	290 298	305
•••	30100 41747	93413	11188 15055	22885	2672 3088	3423	715 752	744	295 302	275
0.85 ***	21815 27430	73652	8398 10333	17897	2008 2233	2615	508 527	550	. .	.
•••	23980 34348	73773	9182 12328	17900	2110 2395	2590	520 538	521	. .	.
0.9 ***	16187 20725	51145	6298 7685	12193	1370 1490	1667	. .	.	. .	.
•••	17960 26068	51156	6868 9032	12172	1427 1570	1637	. .	.	. .	.
0.95 ***	10030 12760	25588	3625 4262	5728	. .	.	. .	.	. .	.
•••	11120 15725	25563	3898 4795	5699	. .	.	. .	.	. .	.
0.98 ***	4892 5923	8807	. .	.	. .	.	. .	.	. .	.
•••	5323 6868	8779	. .	.	. .	.	. .	.	. .	.

TABLE 2: ALPHA= 0.005 POWER= 0.9 EXPECTED ACCRUAL THRU MINIMUM FOLLOW-UP= 3250

	DEL=.01			DEL=.02			DEL=.05			DEL=.10			DEL=.15		
FACT=	1.0 .75	.50 .25	.00 BIN	1.0 .75	.50 .25	.00 BIN	1.0 .75	.50 .25	.00 BIN	1.0 75	.50 .25	.00 BIN	1.0 .75	.50 .25	.00 BIN
PCONT=***			REQUIRED NUMBER OF PATIENTS												
0.01 ***	2272	2278	2294	834	839	839	274	274	279	141	141	141	100	100	100
•••	2272	2286	8779	834	839	2902	274	279	790	141	141	321	100	100	191
0.02 ***	5019	5035	5121	1611	1619	1636	436	436	441	192	192	192	124	124	127
•••	5024	5059	14493	1614	1628	4316	436	441	1009	192	192	373	124	127	213
0.05 ***	15183	15245	15996	4284	4333	4482	929	937	948	336	336	339	198	198	198
•••	15204	15359	30920	4304	4385	8378	932	945	1637	336	336	521	198	198	275
0.1 ***	33481	33632	37231	8932	9078	9845	1736	1774	1834	550	558	566	295	298	298
•••	33529	33933	55917	8984	9281	14553	1752	1801	2590	555	563	744	295	298	368
0.15 ***	49979	50248	58814	13084	13352	15229	2459	2541	2687	745	758	774	384	387	393
•••	50069	50787	77939	13176	13769	19984	2497	2603	3423	753	766	938	384	387	447
0.2 ***	63251	63670	79018	16431	16854	20237	3061	3199	3467	913	932	961	458	461	469
•••	63389	64510	96984	16578	17528	24671	3123	3309	4137	921	945	1102	458	466	513
0.25 ***	72892	73496	97010	18893	19497	24673	3532	3735	4149	1046	1078	1119	514	523	534
•••	73090	74707	113054	19099	20472	28614	3621	3897	4732	1059	1094	1235	517	531	566
0.3 ***	78921	79752	112339	20489	21314	28435	3881	4149	4718	1148	1192	1249	558	571	582
•••	79200	81423	126148	20773	22617	31813	3995	4369	5208	1167	1216	1340	566	579	606
0.35 ***	81581	82694	124738	21290	22366	31463	4117	4447	5165	1221	1273	1346	591	604	620
•••	81951	84917	136266	21664	24028	34268	4255	4718	5565	1246	1306	1414	599	612	632
0.4 ***	81231	82686	134057	21391	22756	33713	4247	4634	5490	1265	1327	1411	607	623	644
•••	81719	85583	143408	21870	24784	35979	4409	4954	5803	1294	1368	1459	615	631	645
0.45 ***	78303	80180	140208	20916	22593	35171	4284	4718	5685	1286	1351	1444	612	631	653
•••	78929	83888	147574	21512	24966	36946	4471	5081	5922	1314	1395	1474	620	639	645
0.5 ***	73269	75671	143141	19985	21981	35826	4239	4707	5750	1278	1346	1444	604	623	644
•••	74070	80329	148765	20713	24662	37169	4439	5097	5922	1311	1392	1459	615	631	632
0.55 ***	66636	69691	142835	18731	21014	35666	4122	4612	5685	1246	1314	1411	582	604	623
•••	67659	75422	146979	19584	23939	36648	4333	5016	5803	1278	1359	1414	591	612	606
0.6 ***	58944	62796	139289	17263	19779	34699	3938	4431	5495	1184	1254	1346	550	566	588
•••	60244	69618	142218	18219	22850	35384	4149	4832	5565	1216	1297	1340	558	579	566
0.65 ***	50754	55521	132521	15668	18324	32928	3694	4171	5170	1102	1167	1254	506	523	539
•••	52401	63288	134480	16694	21444	33375	3897	4553	5208	1132	1205	1235	514	531	513
0.7 ***	42651	48292	122544	14005	16686	30352	3386	3824	4723	997	1051	1124	449	461	474
•••	44661	56661	123767	15058	19730	30622	3578	4174	4732	1021	1086	1102	452	466	447
0.75 ***	35122	41346	109384	12296	14874	26986	3012	3399	4146	864	907	964	376	384	396
•••	37421	49836	110078	13319	17707	27126	3182	3694	4137	883	932	938	379	393	368
0.8 ***	28408	34699	93077	10516	12856	22829	2570	2882	3442	704	734	774	290	298	303
•••	30796	42764	93413	11454	15337	22885	2708	3109	3423	718	753	744	295	303	275
0.85 ***	22395	28164	73651	8615	10565	17897	2037	2256	2614	514	531	550	.	.	.
•••	24624	35219	73773	9403	12551	17900	2134	2408	2590	523	539	521	.	.	.
0.9 ***	16667	21301	51141	6454	7846	12193	1387	1501	1668	.	.	.	.	.	.
•••	18482	26726	51156	7031	9176	12172	1441	1574	1637	.	.	.	.	.	.
0.95 ***	10324	13095	25588	3699	4328	5726	.	.	.	.	.	.	.	.	.
•••	11434	16069	25563	3971	4848	5699	.	.	.	.	.	.	.	.	.
0.98 ***	5011	6039	8810	.	.	.	.	.	.	.	.	.	.	.	.
•••	5441	6966	8779	.	.	.	.	.	.	.	.	.	.	.	.

TABLE 2: ALPHA= 0.005 POWER= 0.9   EXPECTED ACCRUAL THRU MINIMUM FOLLOW-UP= 3500

		DEL=.01			DEL=.02			DEL=.05			DEL=.10			DEL=.15		
FACT=		1.0 .75	.50 .25	.00 BIN	1.0 .75	.50 .25	.00 BIN	1.0 .75	.50 .25	.00 BIN	1.0 75	.50 .25	.00 BIN	1.0 .75	.50 .25	.00 BIN
PCONT=***		REQUIRED NUMBER OF PATIENTS														
0.01	***	2272 2278	2295	834 837	843	277 277	277	137 137	137	99 99	99			
	•••	2278 2286	8779	837 837	2902	277 277	790	137 137	321	99 99	191			
0.02	***	5020 5037	5121	1613 1621	1633	435 440	440	190 190	190	125 125	125			
	•••	5025 5063	14493	1616 1625	4316	440 440	1009	190 190	373	125 125	213			
0.05	***	15187 15254	15998	4290 4337	4483	930 939	951	335 335	338	195 195	195			
	•••	15210 15371	30920	4308 4390	8378	933 942	1637	335 335	521	195 195	275			
0.1	***	33492 33655	37234	8945 9097	9846	1744 1779	1831	554 557	566	295 300	300			
	•••	33545 33979	55917	8998 9304	14553	1756 1800	2590	557 563	744	295 300	368			
0.15	***	50000 50289	58815	13105 13390	15228	2470 2549	2689	746 758	776	382 388	391			
	•••	50096 50870	77939	13206 13819	19984	2505 2605	3423	755 767	938	382 388	447			
0.2	***	63283 63732	79018	16465 16911	20236	3077 3214	3468	913 933	960	458 461	470			
	•••	63431 64639	96984	16619 17611	24671	3135 3319	4137	925 948	1102	458 466	513			
0.25	***	72937 73590	97008	18941 19585	24672	3555 3756	4150	1047 1079	1117	513 522	531			
	•••	73153 74894	113054	19165 20595	28614	3643 3914	4732	1065 1100	1235	519 528	566			
0.3	***	78983 79885	112338	20556 21431	28435	3914 4176	4719	1152 1196	1248	563 571	583			
	•••	79286 81675	126148	20857 22782	31813	4022 4386	5208	1175 1219	1340	566 580	606			
0.35	***	81666 82865	124737	21379 22516	31462	4153 4477	5165	1228 1280	1345	592 606	624			
	•••	82069 85257	136266	21776 24231	34268	4293 4745	5565	1254 1310	1414	598 615	632			
0.4	***	81343 82912	134056	21505 22945	33711	4290 4670	5489	1275 1333	1411	610 627	645			
	•••	81868 86024	143408	22012 25028	35979	4451 4981	5803	1301 1368	1459	618 636	645			
0.45	***	78446 80468	140207	21055 22817	35172	4334 4763	5685	1292 1359	1441	615 633	650			
	•••	79120 84440	147574	21694 25246	36946	4518 5113	5922	1324 1397	1474	624 641	645			
0.5	***	73453 76035	143138	20157 22240	35825	4293 4754	5751	1283 1353	1446	606 624	645			
	•••	74316 81013	148765	20924 24970	37169	4491 5130	5922	1318 1394	1459	615 633	632			
0.55	***	66873 70151	142832	18938 21303	35668	4176 4658	5685	1254 1318	1411	589 601	624			
	•••	67973 76233	146979	19825 24270	36648	4386 5051	5803	1283 1362	1414	592 615	606			
0.6	***	59243 63370	139288	17494 20087	34700	3993 4477	5492	1196 1263	1350	554 571	589			
	•••	60640 70545	142218	18486 23199	35384	4203 4868	5565	1228 1301	1340	563 580	566			
0.65	***	51138 56204	132521	15922 18640	32929	3748 4215	5168	1114 1170	1254	510 522	536			
	•••	52896 64292	134480	16978 21790	33375	3949 4588	5208	1140 1210	1235	513 531	513			
0.7	***	43126 49058	122546	14268 17004	30356	3436 3870	4722	1003 1056	1123	449 461	475			
	•••	45252 57709	123767	15345 20061	30622	3625 4203	4732	1030 1088	1102	452 466	447			
0.75	***	35676 42151	109386	12550 15170	26988	3056 3436	4145	869 913	965	379 388	396			
	•••	38077 50866	110078	13591 18010	27126	3223 3721	4137	890 933	938	382 391	368			
0.8	***	28991 35478	93076	10751 13119	22831	2605 2911	3441	706 738	773	291 300	303			
	•••	31462 43718	93413	11705 15604	22885	2741 3130	3423	720 755	744	295 300	275			
0.85	***	22945 28860	73651	8814 10777	17896	2062 2278	2613	513 531	548	. .	.			
	•••	25238 36038	73773	9613 12755	17900	2158 2421	2590	522 540	521	. .	.			
0.9	***	17118 21851	51141	6600 7995	12191	1403 1511	1665	. .	.	. .	.			
	•••	18976 27341	51156	7178 9307	12172	1450 1581	1637	. .	.	. .	.			
0.95	***	10608 13408	25588	3768 4390	5728	. .	.	. .	.	. .	.			
	•••	11731 16386	25563	4040 4894	5699	. .	.	. .	.	. .	.			
0.98	***	5121 6145	8808	. .	.	. .	.	. .	.	. .	.			
	•••	5553 7050	8779	. .	.	. .	.	. .	.	. .	.			

TABLE 2: ALPHA= 0.005 POWER= 0.9   EXPECTED ACCRUAL THRU MINIMUM FOLLOW-UP= 3750

		DEL=.01			DEL=.02			DEL=.05			DEL=.10			DEL=.15	
FACT=		1.0 .75	.50 .25	.00 BIN	1.0 .75	.50 .25	.00 BIN	1.0 .75	.50 .25	.00 BIN	1.0 75	.50 .25	.00 BIN	1.0 .75	.50 .25	.00 BIN
PCONT=***					REQUIRED NUMBER OF PATIENTS											
0.01	***	2272 2281	2294	837 837	841	275 275	278	138 138	138	97 97	97				
	•••	2275 2284	8779	837 841	2902	275 275	790	138 138	321	97 97	191				
0.02	***	5022 5041	5122	1615 1619	1634	438 438	438	190 190	190	125 125	125				
	•••	5028 5065	14493	1615 1628	4316	438 438	1009	190 190	373	125 125	213				
0.05	***	15190 15259	15997	4291 4343	4484	931 940	950	334 334	334	194 194	194				
	•••	15213 15387	30920	4315 4394	8378	934 944	1637	334 334	521	194 194	275				
0.1	***	33503 33678	37231	8956 9115	9847	1747 1778	1831	556 559	565	297 297	297				
	•••	33559 34025	55917	9012 9325	14553	1759 1803	2590	556 559	744	297 297	368				
0.15	***	50018 50331	58816	13128 13428	15228	2478 2556	2688	747 762	775	381 387	391				
	•••	50122 50950	77939	13231 13868	19984	2509 2613	3423	756 766	938	387 387	447				
0.2	***	63316 63800	79019	16497 16972	20234	3091 3222	3466	916 934	959	456 462	472				
	•••	63475 64765	96984	16662 17688	24671	3147 3325	4137	925 950	1102	462 466	513				
0.25	***	72981 73681	97009	18991 19672	24672	3578 3772	4147	1053 1081	1118	518 522	531				
	•••	73216 75078	113054	19225 20706	28614	3663 3925	4732	1066 1100	1235	522 528	566				
0.3	***	79047 80013	112338	20622 21547	28437	3940 4197	4718	1159 1197	1250	559 575	584				
	•••	79372 81931	126148	20941 22934	31813	4053 4403	5208	1178 1222	1340	569 578	606				
0.35	***	81753 83037	124737	21462 22666	31459	4188 4506	5163	1234 1281	1343	593 606	622				
	•••	82178 85591	136266	21888 24425	34268	4325 4765	5565	1259 1309	1414	603 612	632				
0.4	***	81456 83134	134056	21616 23125	33715	4328 4703	5487	1281 1338	1409	612 625	644				
	•••	82015 86459	143408	22156 25259	35979	4493 5009	5803	1309 1372	1459	622 634	645				
0.45	***	78593 80759	140206	21194 23031	35172	4375 4797	5684	1300 1362	1441	616 634	650				
	•••	79315 84991	147574	21865 25512	36946	4559 5140	5922	1328 1400	1474	625 640	645				
0.5	***	73638 76403	143140	20331 22488	35825	4338 4797	5750	1297 1356	1447	612 625	644				
	•••	74562 81681	148765	21128 25263	37169	4534 5163	5922	1325 1400	1459	616 634	632				
0.55	***	67109 70615	142834	19137 21578	35665	4225 4703	5688	1259 1325	1413	588 606	622				
	•••	68284 77022	146979	20059 24584	36648	4431 5084	5803	1291 1366	1414	597 612	606				
0.6	***	59543 63934	139291	17722 20378	34700	4043 4522	5491	1203 1263	1347	556 569	588				
	•••	61038 71434	142218	18743 23519	35384	4253 4900	5565	1231 1306	1340	565 578	566				
0.65	***	51518 56872	132522	16165 18940	32928	3794 4259	5172	1118 1178	1253	509 522	537				
	•••	53388 65262	134480	17247 22109	33375	3997 4615	5208	1147 1212	1235	518 531	513				
0.7	***	43597 49797	122543	14519 17294	30353	3481 3906	4722	1009 1062	1122	453 462	475				
	•••	45837 58703	123767	15616 20369	30622	3668 4231	4732	1034 1090	1102	456 466	447				
0.75	***	36213 42922	109384	12794 15447	26984	3100 3472	4147	875 916	963	378 387	397				
	•••	38716 51847	110078	13853 18293	27126	3259 3743	4137	893 940	938	381 391	368				
0.8	***	29556 36228	93078	10975 13366	22831	2641 2937	3443	709 738	772	293 297	303				
	•••	32103 44622	93413	11937 15847	22885	2772 3147	3423	725 756	744	297 303	275				
0.85	***	23472 29519	73653	9003 10975	17894	2088 2294	2613	518 531	550	. .	.				
	•••	25822 36809	73773	9809 12944	17900	2181 2431	2590	522 541	521	. .	.				
0.9	***	17547 22366	51143	6738 8131	12190	1413 1522	1666	. .	.	. .	.				
	•••	19447 27912	51156	7319 9425	12172	1465 1587	1637	. .	.	. .	.				
0.95	***	10868 13703	25587	3837 4447	5725	. .	.	. .	.	. .	.				
	•••	12012 16681	25563	4100 4938	5699	. .	.	. .	.	. .	.				
0.98	***	5225 6241	8809	. .	.	. .	.	. .	.	. .	.				
	•••	5656 7131	8779	. .	.	. .	.	. .	.	. .	.				

TABLE 2: ALPHA= 0.005 POWER= 0.9   EXPECTED ACCRUAL THRU MINIMUM FOLLOW-UP= 4000

	DEL=.01			DEL=.02			DEL=.05			DEL=.10			DEL=.15		
FACT=	1.0 .75	.50 .25	.00 BIN	1.0 .75	.50 .25	.00 BIN	1.0 .75	.50 .25	.00 BIN	1.0 75	.50 .25	.00 BIN	1.0 .75	.50 .25	.00 BIN
PCONT=•••				REQUIRED NUMBER OF PATIENTS											
0.01 ***	2273	2283	2293	837	837	843	277	277	277	137	137	137	97	97	97
•••	2277	2287	8779	837	837	2902	277	277	790	137	137	321	97	97	191
0.02 ***	5023	5043	5123	1613	1623	1637	437	437	443	193	193	193	127	127	127
•••	5033	5067	14493	1617	1627	4316	437	437	1009	193	193	373	127	127	213
0.05 ***	15197	15273	15997	4297	4347	4483	933	943	953	333	337	337	197	197	197
•••	15223	15403	30920	4317	4397	8378	937	943	1637	337	337	521	197	197	275
0.1 ***	33517	33703	37233	8967	9137	9847	1747	1783	1833	553	557	567	297	297	297
•••	33577	34067	55917	9027	9343	14553	1767	1807	2590	557	563	744	297	297	368
0.15 ***	50043	50373	58817	13147	13463	15227	2487	2563	2687	753	763	777	383	387	393
•••	50153	51033	77939	13263	13907	19984	2517	2617	3423	757	767	938	387	387	447
0.2 ***	63347	63863	79017	16533	17033	20237	3103	3237	3467	917	937	963	457	463	467
•••	63517	64893	96984	16707	17763	24671	3163	3333	4137	927	947	1102	463	467	513
0.25 ***	73027	73777	97013	19037	19753	24673	3597	3787	4147	1057	1083	1117	517	527	533
•••	73277	75263	113054	19287	20817	28614	3683	3937	4732	1067	1103	1235	523	527	566
0.3 ***	79113	80137	112337	20687	21657	28437	3963	4223	4717	1163	1203	1247	563	573	583
•••	79457	82183	126148	21027	23077	31813	4077	4417	5208	1183	1223	1340	567	577	606
0.35 ***	81837	83207	124737	21547	22807	31463	4217	4537	5167	1243	1287	1347	597	607	623
•••	82297	85927	136266	21997	24607	34268	4357	4783	5565	1263	1313	1414	603	613	632
0.4 ***	81567	83357	134057	21727	23303	33713	4367	4737	5487	1287	1343	1413	613	627	643
•••	82167	86893	143408	22293	25473	35979	4527	5033	5803	1313	1373	1459	623	637	645
0.45 ***	78737	81043	140207	21333	23243	35173	4417	4837	5683	1307	1367	1443	617	633	653
•••	79507	85527	147574	22033	25763	36946	4597	5167	5922	1337	1403	1474	627	643	645
0.5 ***	73823	76767	143137	20497	22723	35823	4387	4837	5747	1303	1363	1443	613	627	643
•••	74807	82337	148765	21327	25537	37169	4583	5193	5922	1333	1403	1459	617	637	632
0.55 ***	67343	71073	142833	19333	21837	35667	4273	4743	5687	1267	1333	1413	593	607	623
•••	68597	77793	146979	20283	24877	36648	4477	5113	5803	1297	1367	1414	597	613	606
0.6 ***	59843	64493	139293	17937	20657	34703	4093	4563	5493	1207	1273	1347	557	573	587
•••	61433	72303	142218	18987	23823	35384	4297	4927	5565	1237	1307	1340	563	577	566
0.65 ***	51897	57523	132523	16397	19227	32927	3843	4297	5173	1127	1183	1253	513	523	537
•••	53873	66193	134480	17507	22413	33375	4043	4643	5208	1153	1213	1235	517	533	513
0.7 ***	44057	50517	122543	14757	17577	30353	3523	3943	4723	1017	1063	1123	453	463	473
•••	46403	59663	123767	15873	20657	30622	3707	4257	4732	1037	1093	1102	457	467	447
0.75 ***	36743	43663	109387	13027	15707	26987	3137	3497	4143	877	917	963	377	387	397
•••	39333	52777	110078	14097	18557	27126	3297	3763	4137	897	937	938	383	393	368
0.8 ***	30097	36937	93077	11187	13597	22827	2673	2957	3443	713	743	773	293	297	303
•••	32717	45477	93413	12163	16073	22885	2797	3163	3423	727	757	744	297	303	275
0.85 ***	23977	30153	73653	9183	11163	17897	2107	2307	2617	517	533	553	.	.	.
•••	26377	37543	73773	9993	13117	17900	2203	2443	2590	527	543	521	.	.	.
0.9 ***	17963	22857	51143	6867	8257	12193	1427	1527	1667	.	.	.	.	.	.
•••	19887	28453	51156	7447	9537	12172	1473	1593	1637	.	.	.	.	.	.
0.95 ***	11123	13977	25587	3897	4497	5727	.	.	.	.	.	.	.	.	.
•••	12273	16957	25563	4157	4973	5699	.	.	.	.	.	.	.	.	.
0.98 ***	5323	6333	8807	.	.	.	.	.	.	.	.	.	.	.	.
•••	5753	7203	8779	.	.	.	.	.	.	.	.	.	.	.	.

TABLE 2: ALPHA= 0.005 POWER= 0.9   EXPECTED ACCRUAL THRU MINIMUM FOLLOW-UP= 4250

	DEL=.01			DEL=.02			DEL=.05			DEL=.10			DEL=.15		
FACT=	1.0 .75	.50 .25	.00 BIN	1.0 .75	.50 .25	.00 BIN	1.0 .75	.50 .25	.00 BIN	1.0 75	.50 .25	.00 BIN	1.0 .75	.50 .25	.00 BIN

PCONT=***   REQUIRED NUMBER OF PATIENTS

PCONT	1.0/.75	.50/.25	.00/BIN	1.0/.75	.50/.25	.00/BIN	1.0/.75	.50/.25	.00/BIN	1.0/.75	.50/.25	.00/BIN	1.0/.75	.50/.25	.00/BIN
0.01 ***	2277	2281	2292	836	836	843	273	279	279	141	141	141	99	99	99
***	2277	2288	8779	836	843	2902	273	279	790	141	141	321	99	99	191
0.02 ***	5022	5043	5124	1612	1622	1633	439	439	439	194	194	194	124	124	124
***	5033	5071	14493	1618	1629	4316	439	439	1009	194	194	373	124	124	213
0.05 ***	15201	15282	15994	4300	4353	4480	932	942	949	337	337	337	194	194	194
***	15229	15420	30920	4321	4402	8378	938	942	1637	337	337	521	194	194	275
0.1 ***	33525	33727	37233	8981	9151	9846	1750	1788	1831	556	560	566	294	301	301
***	33593	34109	55917	9045	9364	14553	1767	1809	2590	556	560	744	294	301	368
0.15 ***	50062	50412	58817	13168	13501	15229	2493	2568	2685	751	762	772	386	386	390
***	50178	51113	77939	13288	13954	19984	2525	2621	3423	758	768	938	386	390	447
0.2 ***	63381	63927	79021	16568	17088	20237	3116	3244	3467	921	938	959	460	464	471
***	63562	65022	96984	16748	17832	24671	3173	3339	4137	928	949	1102	460	464	513
0.25 ***	73075	73868	97009	19086	19833	24674	3616	3807	4147	1059	1087	1119	517	524	534
***	73341	75445	113054	19351	20917	28614	3694	3945	4732	1070	1102	1235	524	528	566
0.3 ***	79174	80268	112335	20754	21767	28436	3988	4243	4714	1165	1204	1246	566	577	588
***	79542	82436	126148	21108	23219	31813	4098	4434	5208	1183	1225	1340	570	581	606
0.35 ***	81926	83382	124741	21636	22947	31464	4247	4561	5167	1246	1289	1346	598	609	619
***	82411	86261	136266	22103	24781	34268	4385	4799	5565	1268	1314	1414	602	613	632
0.4 ***	81682	83583	134059	21838	23474	33716	4402	4767	5490	1293	1346	1410	613	630	645
***	82315	87319	143408	22426	25677	35979	4561	5054	5803	1321	1374	1459	623	634	645
0.45 ***	78883	81331	140204	21470	23442	35172	4455	4869	5681	1314	1374	1442	619	634	651
***	79701	86059	147574	22199	25996	36946	4636	5188	5922	1342	1406	1474	630	645	645
0.5 ***	74006	77134	143137	20658	22953	35824	4427	4869	5751	1310	1367	1442	613	630	645
***	75052	82978	148765	21523	25801	37169	4618	5220	5922	1335	1406	1459	619	634	632
0.55 ***	67578	71531	142835	19521	22086	35665	4317	4778	5688	1272	1335	1410	592	609	623
***	68910	78537	146979	20503	25153	36648	4519	5139	5803	1303	1374	1414	598	613	606
0.6 ***	60145	65043	139290	18151	20917	34698	4136	4597	5490	1214	1272	1346	560	570	588
***	61823	73139	142218	19224	24105	35384	4338	4954	5565	1246	1310	1340	566	581	566
0.65 ***	52272	58158	132522	16625	19493	32930	3885	4332	5171	1129	1187	1250	513	524	538
***	54350	67089	134480	17751	22692	33375	4083	4668	5208	1155	1214	1235	517	528	513
0.7 ***	44511	51216	122545	14984	17836	30352	3563	3970	4721	1023	1066	1123	453	464	475
***	46959	60576	123767	16121	20928	30622	3747	4278	4732	1044	1091	1102	460	471	447
0.75 ***	37254	44373	109388	13246	15955	26984	3169	3524	4147	885	921	963	379	390	396
***	39932	53663	110078	14336	18803	27126	3329	3779	4137	900	942	938	386	390	368
0.8 ***	30624	37620	93078	11387	13816	22830	2695	2978	3439	719	740	772	294	301	305
***	33306	46296	93413	12371	16281	22885	2823	3173	3423	730	758	744	294	301	275
0.85 ***	24466	30752	73649	9347	11340	17896	2128	2323	2617	517	534	549	.	.	.
***	26916	38232	73773	10165	13278	17900	2217	2451	2590	528	545	521	.	.	.
0.9 ***	18353	23325	51141	6988	8380	12190	1438	1533	1665	.	.	.	.	.	.
***	20318	28960	51156	7572	9633	12172	1480	1597	1637	.	.	.	.	.	.
0.95 ***	11361	14241	25588	3956	4544	5730	.	.	.	.	.	.	.	.	.
***	12523	17209	25563	4211	5008	5699	.	.	.	.	.	.	.	.	.
0.98 ***	5415	6414	8811	.	.	.	.	.	.	.	.	.	.	.	.
***	5840	7271	8779	.	.	.	.	.	.	.	.	.	.	.	.

TABLE 2: ALPHA= 0.005 POWER= 0.9   EXPECTED ACCRUAL THRU MINIMUM FOLLOW-UP= 4500

	DEL=.01			DEL=.02			DEL=.05			DEL=.10			DEL=.15		
FACT=	1.0 .75	.50 .25	.00 BIN	1.0 .75	.50 .25	.00 BIN	1.0 .75	.50 .25	.00 BIN	1.0 75	.50 .25	.00 BIN	1.0 .75	.50 .25	.00 BIN
PCONT=•••				REQUIRED NUMBER OF PATIENTS											
0.01 •••	2276	2280	2291	836	840	840	278	278	278	138	138	138	98	98	98
•••	2276	2287	8779	836	840	2902	278	278	790	138	138	321	98	98	191
0.02 •••	5025	5048	5122	1616	1623	1635	435	442	442	195	195	195	127	127	127
•••	5032	5070	14493	1616	1628	4316	435	442	1009	195	195	373	127	127	213
0.05 •••	15206	15292	15994	4305	4357	4481	937	941	948	334	334	334	195	195	195
•••	15236	15431	30920	4328	4406	8378	937	948	1637	334	334	521	195	195	275
0.1 •••	33540	33746	37230	8992	9172	9847	1758	1785	1830	555	559	566	296	296	300
•••	33607	34151	55917	9060	9379	14553	1770	1808	2590	559	566	744	296	300	368
0.15 •••	50081	50453	58818	13193	13537	15229	2501	2573	2685	750	761	773	386	386	390
•••	50205	51195	77939	13312	13991	19984	2535	2625	3423	757	768	938	386	390	447
0.2 •••	63413	63993	79016	16597	17141	20235	3131	3255	3468	926	941	960	458	465	469
•••	63604	65152	96984	16793	17902	24671	3187	3345	4137	930	948	1102	458	465	513
0.25 •••	73121	73961	97012	19133	19916	24675	3630	3817	4148	1061	1088	1117	521	525	532
•••	73403	75630	113054	19410	21018	28614	3716	3956	4732	1072	1106	1235	521	532	566
0.3 •••	79241	80396	112339	20816	21873	28436	4013	4260	4717	1173	1207	1245	566	577	581
•••	79624	82684	126148	21191	23347	31813	4121	4447	5208	1189	1223	1340	570	581	606
0.35 •••	82009	83550	124736	21720	23081	31463	4271	4582	5167	1252	1290	1346	600	611	622
•••	82522	86595	136266	22211	24945	34268	4406	4818	5565	1268	1320	1414	604	615	632
0.4 •••	81791	83805	134058	21941	23640	33713	4429	4796	5486	1297	1346	1410	615	626	645
•••	82466	87742	143408	22564	25878	35979	4593	5070	5803	1324	1376	1459	622	638	645
0.45 •••	79023	81622	140205	21603	23640	35171	4492	4897	5685	1320	1376	1443	622	638	649
•••	79890	86588	147574	22357	26220	36946	4672	5212	5922	1346	1410	1474	626	645	645
0.5 •••	74190	77498	143137	20820	23171	35823	4463	4901	5752	1313	1369	1443	615	626	645
•••	75300	83613	148765	21709	26047	37169	4654	5239	5922	1342	1403	1459	622	638	632
0.55 •••	67811	71981	142833	19706	22328	35666	4357	4811	5685	1279	1342	1414	593	611	622
•••	69225	79264	146979	20715	25417	36648	4560	5160	5803	1308	1376	1414	600	615	606
0.6 •••	60443	65584	139290	18352	21176	34703	4177	4631	5493	1223	1279	1346	559	570	588
•••	62216	73950	142218	19455	24375	35384	4380	4976	5565	1245	1313	1340	566	581	566
0.65 •••	52646	58785	132521	16838	19751	32925	3923	4361	5171	1140	1189	1252	514	525	536
•••	54825	67946	134480	17985	22958	33375	4121	4688	5208	1162	1218	1235	521	532	513
0.7 •••	44958	51888	122543	15202	18086	30356	3603	4001	4721	1027	1072	1121	453	465	476
•••	47501	61451	123767	16354	21180	30622	3776	4301	4732	1050	1095	1102	458	469	447
0.75 •••	37751	45064	109387	13458	16192	26985	3203	3551	4143	885	919	964	379	390	397
•••	40508	54514	110078	14554	19031	27126	3360	3799	4137	903	941	938	386	390	368
0.8 •••	31132	38276	93079	11580	14021	22830	2726	3000	3439	723	746	773	296	300	307
•••	33881	47073	93413	12574	16478	22885	2850	3187	3423	735	757	744	296	300	275
0.85 •••	24933	31328	73650	9510	11505	17895	2145	2336	2613	521	536	548	.	.	.
•••	27431	38888	73773	10331	13429	17900	2235	2460	2590	525	543	521	.	.	.
0.9 •••	18735	23768	51146	7106	8490	12191	1448	1538	1668	.	.	.	.	.	.
•••	20726	29445	51156	7687	9728	12172	1488	1601	1637	.	.	.	.	.	.
0.95 •••	11584	14482	25586	4008	4593	5730	.	.	.	.	.	.	.	.	.
•••	12761	17452	25563	4260	5043	5699	.	.	.	.	.	.	.	.	.
0.98 •••	5498	6495	8805	.	.	.	.	.	.	.	.	.	.	.	.
•••	5925	7331	8779	.	.	.	.	.	.	.	.	.	.	.	.

TABLE 2: ALPHA= 0.005 POWER= 0.9   EXPECTED ACCRUAL THRU MINIMUM FOLLOW-UP= 4750

	DEL=.01			DEL=.02			DEL=.05			DEL=.10			DEL=.15		
FACT=	1.0 .75	.50 .25	.00 BIN	1.0 .75	.50 .25	.00 BIN	1.0 .75	.50 .25	.00 BIN	1.0 75	.50 .25	.00 BIN	1.0 .75	.50 .25	.00 BIN
PCONT=***					REQUIRED NUMBER OF PATIENTS										
0.01 ***	2276 2276	2283 2288	2295 8779	835 839	839 839	839 2902	277 277	277 277	277 790	139 139	139 139	139 321	98 98	98 98	98 191
0.02 ***	5027 5038	5050 5074	5122 14493	1618 1618	1623 1630	1635 4316	435 435	435 443	443 1009	193 193	193 193	193 373	127 127	127 127	127 213
0.05 ***	15208 15239	15298 15445	15992 30920	4307 4330	4362 4409	4480 8378	934 934	942 946	953 1637	336 336	336 336	336 521	193 193	198 198	198 275
0.1 ***	33550 33622	33769 34196	37232 55917	9005 9069	9187 9397	9848 14553	1761 1773	1789 1808	1832 2590	554 562	562 562	566 744	293 300	300 300	300 368
0.15 ***	50104 50235	50496 51272	58813 77939	13213 13339	13565 14028	15227 19984	2509 2537	2580 2628	2687 3423	756 756	763 768	775 938	383 388	388 388	388 447
0.2 ***	63444 63646	64057 65273	79020 96984	16628 16830	17198 17963	20238 24671	3138 3190	3262 3352	3464 4137	922 934	942 953	958 1102	459 459	467 467	467 513
0.25 ***	73170 73467	74056 75813	97010 113054	19182 19471	19989 21110	24672 28614	3649 3725	3832 3962	4148 4732	1065 1077	1089 1100	1120 1235	519 526	526 530	530 566
0.3 ***	79305 79713	80528 82938	112341 126148	20880 21272	21977 23473	28437 31813	4034 4140	4278 4457	4718 5208	1172 1191	1207 1227	1250 1340	566 573	573 578	585 606
0.35 ***	82095 82634	83722 86921	124738 136266	21806 22317	23212 25100	31460 34268	4302 4433	4604 4837	5162 5565	1255 1274	1290 1322	1345 1414	597 602	609 614	621 632
0.4 ***	81905 82610	84031 88156	134060 143408	22048 22689	23801 26062	33717 35979	4461 4615	4817 5090	5490 5803	1302 1326	1350 1381	1409 1459	621 625	633 637	645 645
0.45 ***	79167 80081	81905 87099	140204 147574	21735 22518	23825 26430	35170 36946	4528 4699	4924 5233	5684 5922	1326 1350	1381 1409	1445 1474	625 633	637 645	649 645
0.5 ***	74377 75545	77860 84225	143137 148765	20975 21889	23385 26275	35823 37169	4504 4694	4936 5264	5751 5922	1322 1345	1374 1405	1445 1459	614 621	633 637	645 632
0.55 ***	68047 69532	72429 79974	142836 146979	19882 20915	22559 25658	35669 36648	4397 4592	4848 5185	5684 5803	1286 1314	1345 1374	1409 1414	597 602	609 614	621 606
0.6 ***	60744 62601	66123 74733	139290 142218	18552 19673	21414 24632	34702 35384	4219 4414	4663 5003	5494 5565	1227 1250	1279 1314	1350 1340	562 566	573 578	585 566
0.65 ***	53018 55293	59390 68776	132521 134480	17049 18212	19994 23212	32925 33375	3962 4152	4390 4710	5169 5208	1143 1167	1191 1219	1250 1235	514 519	526 530	538 513
0.7 ***	45402 48030	52543 62292	122542 123767	15410 16581	18327 21419	30356 30622	3637 3808	4029 4314	4722 4732	1029 1048	1072 1096	1124 1102	455 459	467 467	471 447
0.75 ***	38241 41072	45727 55329	109384 110078	13660 14769	16407 19253	26983 27126	3233 3388	3570 3815	4140 4137	894 906	922 942	965 938	383 383	388 395	395 368
0.8 ***	31627 34429	38910 47812	93080 93413	11764 12762	14210 16657	22832 22885	2747 2870	3020 3202	3440 3423	720 732	744 756	775 744	293 300	300 300	305 275
0.85 ***	25385 27922	31880 39512	73652 73773	9662 10489	11665 13569	17899 17900	2165 2248	2347 2466	2616 2590	526 530	538 542	550 521	. .	. .	. .
0.9 ***	19091 21117	24193 29905	51142 51156	7216 7793	8594 9817	12192 12172	1452 1500	1547 1599	1666 1637	. .	. .	. .	. .	. .	. .
0.95 ***	11800 12987	14717 17673	25587 25563	4053 4307	4627 5067	5727 5699	. .	. .	. .	. .	. .	. .	. .	. .	. .
0.98 ***	5577 6000	6563 7390	8807 8779	. .	. .	. .	. .	. .	. .	. .	. .	. .	. .	. .	. .

TABLE 2: ALPHA= 0.005 POWER= 0.9   EXPECTED ACCRUAL THRU MINIMUM FOLLOW-UP= 5000

	DEL=.01			DEL=.02			DEL=.05			DEL=.10			DEL=.15		
FACT=	1.0 .75	.50 .25	.00 BIN	1.0 .75	.50 .25	.00 BIN	1.0 .75	.50 .25	.00 BIN	1.0 75	.50 .25	.00 BIN	1.0 .75	.50 .25	.00 BIN
PCONT=***						REQUIRED NUMBER OF PATIENTS									
0.01 ***	2279	2283	2296	833	841	841	279	279	279	141	141	141	96	96	96
•••	2279	2283	8779	841	841	2902	279	279	790	141	141	321	96	96	191
0.02 ***	5029	5054	5121	1616	1621	1633	441	441	441	191	191	191	129	129	129
•••	5033	5079	14493	1621	1629	4316	441	441	1009	191	191	373	129	129	213
0.05 ***	15216	15308	15996	4316	4366	4483	933	941	954	333	333	333	196	196	196
•••	15246	15458	30920	4333	4408	8378	941	946	1637	333	333	521	196	196	275
0.1 ***	33558	33796	37233	9016	9204	9846	1758	1791	1833	558	558	566	296	296	296
•••	33641	34241	55917	9083	9408	14553	1779	1808	2590	558	566	744	296	296	368
0.15 ***	50121	50541	58816	13233	13596	15229	2508	2583	2683	754	766	779	383	391	391
•••	50258	51354	77939	13366	14058	19984	2546	2629	3423	758	771	938	383	391	447
0.2 ***	63479	64121	79021	16666	17246	20233	3146	3271	3466	929	941	958	458	466	471
•••	63691	65404	96984	16871	18021	24671	3204	3354	4137	933	954	1102	458	466	513
0.25 ***	73216	74146	97008	19229	20066	24671	3658	3841	4146	1066	1091	1121	521	529	533
•••	73529	75991	113054	19529	21196	28614	3741	3971	4732	1079	1104	1235	521	529	566
0.3 ***	79371	80654	112341	20941	22079	28433	4054	4291	4716	1179	1208	1246	566	579	583
•••	79796	83183	126148	21354	23591	31813	4158	4471	5208	1191	1229	1340	571	579	606
0.35 ***	82179	83896	124741	21883	23341	31458	4321	4621	5166	1258	1296	1346	604	608	621
•••	82754	87246	136266	22416	25246	34268	4458	4846	5565	1279	1321	1414	604	616	632
0.4 ***	82016	84254	134058	22154	23954	33716	4491	4841	5491	1308	1354	1408	621	629	641
•••	82758	88571	143408	22821	26233	35979	4646	5108	5803	1329	1379	1459	621	633	645
0.45 ***	79316	82191	140208	21866	24004	35171	4558	4954	5683	1329	1379	1441	629	641	654
•••	80279	87608	147574	22666	26633	36946	4733	5254	5922	1354	1408	1474	629	646	645
0.5 ***	74558	78221	143141	21129	23583	35821	4533	4958	5746	1321	1379	1446	616	629	646
•••	75791	84829	148765	22066	26496	37169	4721	5283	5922	1346	1408	1459	621	633	632
0.55 ***	68283	72866	142833	20058	22779	35666	4433	4871	5683	1291	1346	1408	596	608	621
•••	69841	80666	146979	21108	25896	36648	4629	5204	5803	1316	1379	1414	604	616	606
0.6 ***	61041	66646	139291	18741	21646	34704	4254	4691	5491	1233	1283	1346	566	571	591
•••	62983	75491	142218	19883	24866	35384	4446	5021	5565	1258	1316	1340	566	579	566
0.65 ***	53383	59983	132521	17246	20229	32929	3996	4421	5171	1146	1191	1254	516	529	541
•••	55746	69583	134480	18433	23446	33375	4183	4729	5208	1171	1221	1235	521	533	513
0.7 ***	45833	53183	122546	15616	18554	30354	3666	4054	4721	1033	1071	1121	454	466	471
•••	48554	63104	123767	16796	21646	30622	3841	4333	4732	1054	1096	1102	458	471	447
0.75 ***	38716	46371	109383	13854	16621	26983	3258	3596	4146	896	929	966	383	391	396
•••	41616	56108	110078	14971	19454	27126	3408	3829	4137	908	941	938	383	391	368
0.8 ***	32104	39516	93079	11941	14396	22829	2771	3033	3441	721	746	771	296	296	304
•••	34966	48521	93413	12946	16833	22885	2891	3208	3423	733	758	744	296	304	275
0.85 ***	25821	32416	73654	9808	11808	17896	2179	2358	2616	521	533	554	.	.	.
•••	28396	40108	73773	10633	13696	17900	2266	2471	2590	529	541	521	.	.	.
0.9 ***	19446	24596	51141	7316	8691	12191	1466	1554	1666	.	.	.	.	.	.
•••	21491	30333	51156	7896	9896	12172	1504	1604	1637	.	.	.	.	.	.
0.95 ***	12008	14941	25591	4104	4666	5729	.	.	.	.	.	.	.	.	.
•••	13204	17883	25563	4354	5096	5699	.	.	.	.	.	.	.	.	.
0.98 ***	5654	6633	8808	.	.	.	.	.	.	.	.	.	.	.	.
•••	6079	7441	8779	.	.	.	.	.	.	.	.	.	.	.	.

TABLE 2: ALPHA= 0.005 POWER= 0.9   EXPECTED ACCRUAL THRU MINIMUM FOLLOW-UP= 5500

PCONT=***	DEL=.01			DEL=.02			DEL=.05			DEL=.10			DEL=.15		
FACT=	1.0 .75	.50 .25	.00 BIN	1.0 .75	.50 .25	.00 BIN	1.0 .75	.50 .25	.00 BIN	1.0 75	.50 .25	.00 BIN	1.0 .75	.50 .25	.00 BIN

REQUIRED NUMBER OF PATIENTS

PCONT	D01-a	D01-b	D01-c	D02-a	D02-b	D02-c	D05-a	D05-b	D05-c	D10-a	D10-b	D10-c	D15-a	D15-b	D15-c
0.01	2278	2286	2292	834	834	843	279	279	279	141	141	141	100	100	100
	2278	2286	8779	834	843	2902	279	279	790	141	141	321	100	100	191
0.02	5028	5055	5119	1618	1626	1632	435	435	444	188	188	188	128	128	128
	5042	5078	14493	1618	1632	4316	435	435	1009	188	188	373	128	128	213
0.05	15225	15321	15995	4321	4376	4478	939	944	953	334	334	339	196	196	196
	15258	15478	30920	4340	4418	8378	939	944	1637	334	334	521	196	196	275
0.1	33587	33843	37230	9038	9230	9849	1764	1797	1833	559	559	568	298	298	298
	33669	34324	55917	9112	9436	14553	1778	1810	2590	559	559	744	298	298	368
0.15	50164	50623	58818	13273	13658	15230	2520	2589	2685	760	765	774	389	389	389
	50315	51511	77939	13415	14125	19984	2553	2635	3423	760	774	938	389	389	447
0.2	63543	64249	79017	16729	17348	20235	3166	3282	3469	930	944	958	458	463	471
	63776	65646	96984	16949	18132	24671	3221	3364	4137	939	953	1102	463	463	513
0.25	73310	74336	97010	19323	20208	24671	3689	3859	4148	1068	1090	1118	518	526	532
	73649	76349	113054	19644	21363	28614	3763	3983	4732	1082	1104	1235	526	532	566
0.3	79498	80909	112341	21069	22265	28439	4088	4321	4720	1178	1214	1247	568	573	587
	79965	83673	126148	21509	23810	31813	4189	4486	5208	1200	1228	1340	573	581	606
0.35	82353	84236	124735	22050	23585	31464	4368	4657	5165	1260	1302	1343	600	609	623
	82977	87885	136266	22614	25524	34268	4500	4871	5565	1283	1324	1414	609	614	632
0.4	82243	84695	134058	22361	24245	33710	4541	4885	5490	1315	1357	1412	623	628	642
	83059	89379	143408	23068	26560	35979	4693	5133	5803	1338	1384	1459	628	636	645
0.45	79603	82765	140204	22114	24347	35168	4615	5000	5683	1338	1384	1439	628	636	650
	80661	88600	147574	22958	27000	36946	4789	5284	5922	1357	1412	1474	636	642	645
0.5	74928	78934	143141	21426	23970	35823	4602	5014	5751	1329	1384	1448	623	628	642
	76280	85996	148765	22403	26904	37169	4780	5317	5922	1357	1412	1459	623	636	632
0.55	68754	73737	142830	20395	23192	35663	4500	4926	5688	1302	1351	1412	600	609	623
	70459	81995	146979	21487	26327	36648	4693	5243	5803	1324	1379	1414	600	614	606
0.6	61631	67668	139291	19108	22078	34700	4321	4748	5490	1241	1288	1351	568	573	587
	63749	76940	142218	20285	25309	35384	4505	5055	5565	1260	1315	1340	568	581	566
0.65	54110	61136	132518	17631	20670	32927	4060	4473	5174	1150	1200	1255	518	526	540
	56645	71105	134480	18841	23888	33375	4244	4761	5208	1173	1219	1235	526	532	513
0.7	46685	54404	122544	16000	18979	30355	3730	4101	4720	1040	1076	1123	458	463	471
	49559	64634	123767	17197	22059	30622	3895	4363	4732	1063	1095	1102	463	471	447
0.75	39631	47598	109385	14221	17018	26987	3309	3634	4143	898	930	966	380	389	394
	42670	57580	110078	15354	19823	27126	3460	3854	4137	911	944	938	389	394	368
0.8	33018	40668	93078	12269	14744	22829	2814	3062	3441	724	746	774	298	298	306
	35979	49861	93413	13286	17150	22885	2924	3227	3423	738	760	744	298	298	275
0.85	26651	33416	73649	10083	12082	17898	2209	2374	2616	526	540	554	.	.	.
	29305	41226	73773	10913	13938	17900	2286	2484	2590	532	545	521	.	.	.
0.9	20106	25364	51140	7511	8873	12192	1475	1558	1668	.	.	.	.	.	.
	22196	31139	51156	8089	10041	12172	1516	1613	1637	.	.	.	.	.	.
0.95	12398	15349	25584	4184	4734	5729	.	.	.	.	.	.	.	.	.
	13608	18269	25563	4431	5138	5699	.	.	.	.	.	.	.	.	.
0.98	5798	6755	8809	.	.	.	.	.	.	.	.	.	.	.	.
	6210	7530	8779	.	.	.	.	.	.	.	.	.	.	.	.

TABLE 2: ALPHA= 0.005 POWER= 0.9   EXPECTED ACCRUAL THRU MINIMUM FOLLOW-UP= 6000

		DEL=.01			DEL=.02			DEL=.05			DEL=.10			DEL=.15		
FACT=		1.0 .75	.50 .25	.00 BIN	1.0 .75	.50 .25	.00 BIN	1.0 .75	.50 .25	.00 BIN	1.0 75	.50 .25	.00 BIN	1.0 .75	.50 .25	.00 BIN
PCONT=•••					REQUIRED NUMBER OF PATIENTS											
0.01	•••	2275	2284	2290	835	835	844	274	274	274	139	139	139	100	100	100
	•••	2284	2290	8779	835	835	2902	274	274	790	139	139	321	100	100	191
0.02	•••	5035	5059	5119	1615	1624	1639	439	439	439	190	190	190	124	124	124
	•••	5044	5080	14493	1624	1630	4316	439	439	1009	190	190	373	124	124	213
0.05	•••	15235	15340	15994	4324	4375	4480	934	940	949	334	334	334	199	199	199
	•••	15274	15505	30920	4345	4420	8378	940	949	1637	334	334	521	199	199	275
0.1	•••	33610	33889	37234	9055	9259	9844	1774	1795	1834	559	559	565	295	295	295
	•••	33700	34399	55917	9139	9460	14553	1780	1810	2590	559	565	744	295	295	368
0.15	•••	50209	50704	58819	13315	13714	15229	2530	2599	2689	760	769	775	385	385	394
	•••	50374	51664	77939	13465	14179	19984	2560	2635	3423	760	769	938	385	385	447
0.2	•••	63604	64384	79015	16789	17440	20239	3184	3295	3469	934	940	964	460	460	469
	•••	63865	65890	96984	17029	18229	24671	3235	3370	4137	940	955	1102	460	469	513
0.25	•••	73405	74524	97009	19405	20344	24670	3715	3880	4150	1075	1099	1120	520	529	535
	•••	73774	76705	113054	19750	21505	28614	3790	3994	4732	1084	1105	1235	529	529	566
0.3	•••	79624	81169	112339	21190	22450	28435	4120	4345	4714	1189	1210	1249	565	574	580
	•••	80140	84154	126148	21655	24010	31813	4219	4504	5208	1204	1234	1340	574	580	606
0.35	•••	82525	84574	124735	22210	23815	31459	4405	4690	5164	1270	1300	1345	604	610	619
	•••	83209	88510	136266	22810	25774	34268	4534	4894	5565	1285	1324	1414	610	619	632
0.4	•••	82465	85144	134059	22564	24520	33715	4594	4924	5485	1324	1360	1414	625	634	640
	•••	83359	90169	143408	23305	26860	35979	4735	5155	5803	1339	1384	1459	625	640	645
0.45	•••	79894	83329	140209	22360	24664	35170	4669	5044	5680	1345	1390	1444	625	640	649
	•••	81040	89554	147574	23239	27340	36946	4834	5314	5922	1369	1414	1474	634	649	645
0.5	•••	75295	79639	143140	21709	24325	35824	4654	5059	5749	1339	1390	1444	619	634	640
	•••	76765	87115	148765	22720	27280	37169	4834	5344	5922	1360	1414	1459	625	640	632
0.55	•••	69220	74590	142834	20710	23575	35665	4555	4975	5689	1309	1354	1414	604	610	625
	•••	71074	83254	146979	21835	26719	36648	4744	5269	5803	1330	1384	1414	604	619	606
0.6	•••	62215	68659	139294	19450	22480	34699	4375	4789	5494	1249	1294	1345	565	574	589
	•••	64495	78310	142218	20659	25714	35384	4564	5089	5565	1270	1315	1340	574	580	566
0.65	•••	54820	62230	132520	17989	21070	32929	4120	4510	5170	1159	1204	1249	520	529	535
	•••	57520	72535	134480	19225	24280	33375	4294	4789	5208	1180	1225	1235	520	535	513
0.7	•••	47500	55564	122545	16354	19369	30355	3775	4135	4720	1045	1084	1120	460	469	475
	•••	50515	66064	123767	17575	22435	30622	3940	4390	4732	1060	1099	1102	460	469	447
0.75	•••	40510	48745	109384	14554	17374	26989	3355	3664	4144	904	934	964	385	394	394
	•••	43660	58945	110078	15709	20164	27126	3499	3874	4137	919	949	938	385	394	368
0.8	•••	33880	41749	93079	12574	15055	22825	2845	3085	3439	730	754	769	295	304	304
	•••	36940	51094	93413	13594	17425	22885	2959	3244	3423	739	760	744	295	304	275
0.85	•••	27430	34345	73654	10330	12325	17899	2230	2395	2614	529	535	550	.	.	.
	•••	30154	42250	73773	11164	14149	17900	2305	2494	2590	535	544	521	.	.	.
0.9	•••	20725	26065	51145	7684	9034	12190	1489	1570	1669	.	.	.	.	.	.
	•••	22855	31879	51156	8260	10165	12172	1525	1615	1637	.	.	.	.	.	.
0.95	•••	12760	15724	25585	4264	4795	5725	.	.	.	.	.	.	.	.	.
	•••	13975	18619	25563	4495	5179	5699	.	.	.	.	.	.	.	.	.
0.98	•••	5920	6865	8809	.	.	.	.	.	.	.	.	.	.	.	.
	•••	6334	7615	8779	.	.	.	.	.	.	.	.	.	.	.	.

TABLE 2: ALPHA= 0.005 POWER= 0.9 EXPECTED ACCRUAL THRU MINIMUM FOLLOW-UP= 6500

		DEL=.01			DEL=.02			DEL=.05			DEL=.10			DEL=.15		
FACT=		1.0 .75	.50 .25	.00 BIN	1.0 .75	.50 .25	.00 BIN	1.0 .75	.50 .25	.00 BIN	1.0 75	.50 .25	.00 BIN	1.0 .75	.50 .25	.00 BIN

PCONT=••• REQUIRED NUMBER OF PATIENTS

PCONT		DEL=.01			DEL=.02			DEL=.05			DEL=.10			DEL=.15		
0.01	***	2280	2286	2296	839	839	839	271	281	281	141	141	141	102	102	102
	•••	2280	2286	8779	839	839	2902	281	281	790	141	141	321	102	102	191
0.02	***	5032	5058	5123	1619	1630	1636	433	443	443	189	189	189	124	124	124
	•••	5048	5081	14493	1619	1630	4316	443	443	1009	189	189	373	124	124	213
0.05	***	15247	15361	15995	4333	4382	4479	937	947	947	336	336	336	200	200	200
	•••	15280	15523	30920	4360	4425	8378	937	947	1637	336	336	521	200	200	275
0.1	***	33632	33935	37233	9078	9283	9842	1776	1798	1831	557	563	563	297	297	297
	•••	33729	34477	55917	9159	9484	14553	1782	1814	2590	557	563	744	297	297	368
0.15	***	50250	50786	58813	13352	13768	15231	2540	2605	2686	758	768	774	384	384	395
	•••	50428	51816	77939	13508	14229	19984	2572	2643	3423	758	768	938	384	384	447
0.2	***	63672	64507	79018	16856	17528	20236	3196	3309	3466	931	947	963	460	466	466
	•••	63948	66132	96984	17106	18324	24671	3244	3374	4137	937	953	1102	466	466	513
0.25	***	73493	74706	97007	19494	20469	24672	3732	3894	4148	1077	1093	1116	525	531	531
	•••	73899	77046	113054	19862	21650	28614	3807	4008	4732	1083	1110	1235	525	531	566
0.3	***	79749	81423	112341	21314	22614	28432	4148	4366	4717	1191	1213	1246	573	579	579
	•••	80312	84618	126148	21802	24191	31813	4246	4522	5208	1207	1229	1340	573	579	606
0.35	***	82691	84917	124740	22365	24028	31465	4447	4717	5162	1272	1305	1343	606	612	622
	•••	83438	89120	136266	22988	26005	34268	4571	4912	5565	1288	1327	1414	606	612	632
0.4	***	82685	85583	134057	22755	24786	33713	4636	4951	5487	1327	1370	1408	622	628	644
	•••	83660	90929	143408	23524	27132	35979	4772	5178	5803	1343	1386	1459	628	638	645
0.45	***	80182	83887	140210	22592	24965	35170	4717	5081	5682	1353	1392	1441	628	638	655
	•••	81423	90474	147574	23508	27652	36946	4880	5335	5922	1370	1418	1474	638	644	645
0.5	***	75671	80328	143141	21981	24662	35826	4707	5097	5747	1343	1392	1441	622	628	644
	•••	77257	88183	148765	23031	27619	37169	4880	5373	5922	1370	1418	1459	628	638	632
0.55	***	69691	75421	142832	21016	23941	35663	4609	5016	5682	1311	1359	1408	606	612	622
	•••	71673	84456	146979	22170	27077	36648	4788	5302	5803	1337	1386	1414	606	622	606
0.6	***	62795	69620	139289	19781	22852	34698	4431	4831	5497	1256	1294	1343	563	579	590
	•••	65222	79603	142218	21006	26076	35384	4609	5113	5565	1272	1321	1340	573	579	566
0.65	***	55521	63288	132523	18324	21444	32927	4171	4555	5172	1164	1207	1256	525	531	541
	•••	58364	73877	134480	19586	24640	33375	4343	4815	5208	1181	1229	1235	525	531	513
0.7	***	48289	56658	122546	16683	19732	30349	3823	4171	4723	1051	1083	1126	460	466	476
	•••	51442	67399	123767	17928	22771	30622	3986	4408	4732	1067	1099	1102	466	466	447
0.75	***	41345	49833	109383	14873	17707	26986	3401	3693	4148	904	931	963	384	395	395
	•••	44611	60217	110078	16033	20463	27126	3537	3894	4137	921	947	938	384	395	368
0.8	***	34698	42764	93074	12858	15334	22826	2881	3108	3439	736	752	774	297	303	303
	•••	37841	52238	93413	13882	17685	22885	2984	3261	3423	742	758	744	297	303	275
0.85	***	28166	35218	73650	10567	12550	17896	2253	2410	2611	531	541	547	.	.	.
	•••	30951	43203	73773	11396	14337	17900	2328	2507	2590	531	547	521	.	.	.
0.9	***	21298	26726	51143	7843	9176	12192	1500	1571	1668	.	.	.	.	.	.
	•••	23476	32553	51156	8412	10281	12172	1538	1619	1637	.	.	.	.	.	.
0.95	***	13092	16066	25588	4327	4847	5725	.	.	.	.	.	.	.	.	.
	•••	14321	18926	25563	4561	5221	5699	.	.	.	.	.	.	.	.	.
0.98	***	6039	6966	8812	.	.	.	.	.	.	.	.	.	.	.	.
	•••	6446	7681	8779	.	.	.	.	.	.	.	.	.	.	.	.

TABLE 2: ALPHA= 0.005 POWER= 0.9   EXPECTED ACCRUAL THRU MINIMUM FOLLOW-UP= 7000

		DEL=.01			DEL=.02			DEL=.05			DEL=.10			DEL=.15		
FACT=		1.0 .75	.50 .25	.00 BIN	1.0 .75	.50 .25	.00 BIN	1.0 .75	.50 .25	.00 BIN	1.0 75	.50 .25	.00 BIN	1.0 .75	.50 .25	.00 BIN
PCONT=***					REQUIRED NUMBER OF PATIENTS											
0.01	***	2280 2286	2297	834 834	845	274 274	274	134 134	134	99 99	99					
	•••	2280 2286	8779	834 834	2902	274 274	790	134 134	321	99 99	191					
0.02	***	5034 5062	5121	1621 1621	1632	442 442	442	186 186	186	127 127	127					
	•••	5051 5086	14493	1621 1632	4316	442 442	1009	186 186	373	127 127	213					
0.05	***	15254 15370	16000	4334 4386	4485	939 939	950	337 337	337	197 197	197					
	•••	15289 15545	30920	4362 4432	8378	939 950	1637	337 337	521	197 197	275					
0.1	***	33657 33979	37234	9094 9304	9846	1779 1796	1831	554 565	565	302 302	302					
	•••	33762 34550	55917	9181 9496	14553	1790 1814	2590	565 565	744	302 302	368					
0.15	***	50289 50866	58811	13392 13819	15230	2549 2601	2689	757 764	775	390 390	390					
	•••	50481 51962	77939	13556 14274	19984	2577 2647	3423	764 775	938	390 390	447					
0.2	***	63729 64639	79017	16910 17610	20235	3214 3319	3470	932 950	956	460 466	466					
	•••	64037 66365	96984	17179 18404	24671	3260 3382	4137	939 956	1102	460 466	513					
0.25	***	73592 74894	97007	19587 20591	24669	3756 3914	4152	1079 1096	1114	519 530	530					
	•••	74019 77390	113054	19961 21764	28614	3826 4012	4732	1090 1107	1235	530 530	566					
0.3	***	79881 81677	112337	21431 22779	28431	4176 4386	4719	1195 1219	1247	571 582	582					
	•••	80476 85079	126148	21939 24365	31813	4275 4526	5208	1201 1230	1340	571 582	606					
0.35	***	82867 85254	124734	22516 24231	31459	4474 4747	5167	1282 1306	1341	606 617	624					
	•••	83661 89716	136266	23175 26209	34268	4596 4922	5565	1289 1324	1414	606 617	632					
0.4	***	82909 86024	134055	22947 25030	33710	4666 4981	5489	1335 1370	1411	624 635	641					
	•••	83952 91670	143408	23741 27381	35979	4806 5202	5803	1352 1387	1459	635 641	645					
0.45	***	80470 84442	140204	22814 25246	35169	4765 5115	5681	1359 1394	1440	635 641	652					
	•••	81806 91361	147574	23759 27941	36946	4922 5360	5922	1376 1422	1474	635 641	645					
0.5	***	76031 81012	143137	22236 24966	35827	4754 5132	5751	1352 1394	1446	624 635	641					
	•••	77740 89209	148765	23315 27935	37169	4922 5395	5922	1370 1422	1459	624 641	632					
0.55	***	70151 76235	142829	21302 24266	35670	4660 5051	5681	1317 1359	1411	600 617	624					
	•••	72280 85597	146979	22481 27410	36648	4835 5325	5803	1341 1387	1414	606 617	606					
0.6	***	63372 70547	139287	20084 23199	34696	4474 4870	5489	1265 1300	1352	571 582	589					
	•••	65945 80837	142218	21337 26412	35384	4649 5132	5565	1282 1324	1340	571 582	566					
0.65	***	56204 64289	132521	18642 21792	32929	4211 4590	5167	1166 1212	1254	519 530	536					
	•••	59190 75139	134480	19920 24966	33375	4380 4835	5208	1195 1230	1235	530 530	513					
0.7	***	49057 57709	122546	17004 20060	30356	3872 4205	4719	1055 1090	1125	460 466	477					
	•••	52330 68664	123767	18246 23076	30622	4019 4432	4732	1072 1107	1102	466 466	447					
0.75	***	42151 50866	109386	15166 18012	26990	3435 3721	4141	915 932	967	390 390	396					
	•••	45511 61412	110078	16339 20742	27126	3564 3907	4137	921 950	938	390 390	368					
0.8	***	35477 43720	93076	13119 15604	22831	2910 3126	3441	740 757	775	302 302	302					
	•••	38697 53310	93413	14151 17907	22885	3015 3266	3423	746 764	744	302 302	275					
0.85	***	28862 36037	73651	10774 12751	17896	2280 2420	2612	530 536	547	. .	.					
	•••	31697 44087	73773	11607 14501	17900	2339 2507	2590	536 547	521	. .	.					
0.9	***	21851 27340	51140	7991 9304	12191	1510 1580	1667	. .	.	. .	.					
	•••	24050 33174	51156	8562 10382	12172	1545 1621	1637	. .	.	. .	.					
0.95	***	13410 16385	25590	4386 4894	5727	. .	.	. .	.	. .	.					
	•••	14641 19220	25563	4614 5244	5699	. .	.	. .	.	. .	.					
0.98	***	6147 7046	8807	. .	.	. .	.	. .	.	. .	.					
	•••	6539 7746	8779	. .	.	. .	.	. .	.	. .	.					

TABLE 2: ALPHA= 0.005 POWER= 0.9   EXPECTED ACCRUAL THRU MINIMUM FOLLOW-UP= 7500

PCONT	FACT	DEL=.01 1.0 .75	.50 .25	.00 BIN	DEL=.02 1.0 .75	.50 .25	.00 BIN	DEL=.05 1.0 .75	.50 .25	.00 BIN	DEL=.10 1.0 75	.50 .25	.00 BIN	DEL=.15 1.0 .75	.50 .25	.00 BIN
0.01	***	2281	2281	2293	837	837	837	275	275	275	136	136	136	99	99	99
	•••	2281	2293	8779	837	837	2902	275	275	790	136	136	321	99	99	191
0.02	***	5037	5068	5124	1618	1625	1636	436	436	436	193	193	193	125	125	125
	•••	5049	5086	14493	1625	1636	4316	436	436	1009	193	193	373	125	125	213
0.05	***	15256	15387	15999	4343	4393	4486	943	943	950	331	331	331	193	193	193
	•••	15305	15556	30920	4362	4430	8378	943	950	1637	331	331	521	193	193	275
0.1	***	33680	34025	37231	9118	9324	9849	1775	1805	1831	556	556	568	293	293	293
	•••	33793	34618	55917	9200	9518	14553	1793	1812	2590	556	568	744	293	293	368
0.15	***	50330	50949	58812	13430	13868	15230	2555	2611	2686	762	762	774	387	387	387
	•••	50536	52100	77939	13599	14318	19984	2581	2649	3423	762	774	938	387	387	447
0.2	***	63800	64768	79018	16974	17686	20236	3218	3324	3462	931	950	961	462	462	474
	•••	64118	66593	96984	17243	18481	24671	3268	3387	4137	943	950	1102	462	462	513
0.25	***	73681	75080	97006	19674	20705	24668	3774	3924	4149	1081	1100	1118	518	530	530
	•••	74150	77712	113054	20068	21886	28614	3837	4025	4732	1093	1111	1235	530	530	566
0.3	***	80011	81931	112336	21549	22936	28437	4193	4400	4718	1193	1224	1250	575	575	586
	•••	80656	85524	126148	22074	24511	31813	4287	4543	5208	1205	1231	1340	575	586	606
0.35	***	83037	85587	124737	22662	24425	31456	4505	4768	5161	1280	1306	1343	605	612	624
	•••	83893	90293	136266	23337	26405	34268	4625	4936	5565	1299	1325	1414	612	612	632
0.4	***	83131	86461	134056	23124	25261	33718	4700	5011	5487	1336	1374	1411	624	631	643
	•••	84256	92386	143408	23949	27612	35979	4843	5218	5803	1355	1393	1459	631	643	645
0.45	***	80761	84987	140206	23030	25512	35168	4793	5143	5686	1362	1400	1437	631	643	650
	•••	82193	92218	147574	24005	28205	36946	4955	5375	5922	1381	1418	1474	643	643	645
0.5	***	76400	81680	143143	22486	25261	35825	4793	5161	5750	1355	1400	1449	624	631	643
	•••	78218	90193	148765	23581	28212	37169	4962	5412	5922	1374	1418	1459	631	643	632
0.55	***	70618	77018	142831	21575	24586	35668	4700	5086	5686	1325	1362	1411	605	612	624
	•••	72868	86686	146979	22775	27706	36648	4868	5337	5803	1343	1393	1414	605	612	606
0.6	***	63931	71431	139287	20375	23518	34700	4524	4899	5487	1261	1306	1343	568	575	586
	•••	66643	81999	142218	21643	26724	35384	4693	5150	5565	1280	1325	1340	575	586	566
0.65	***	56874	65262	132518	18943	22111	32930	4261	4618	5168	1175	1212	1250	518	530	537
	•••	59986	76343	134480	20225	25268	33375	4418	4861	5208	1193	1231	1235	530	530	513
0.7	***	49793	58700	122543	17293	20368	30350	3905	4231	4718	1062	1093	1118	462	462	474
	•••	53180	69849	123767	18556	23368	30622	4055	4449	4732	1074	1100	1102	462	474	447
0.75	***	42924	51849	109381	15443	18293	26986	3474	3743	4149	912	943	961	387	387	399
	•••	46374	62536	110078	16625	20993	27126	3593	3924	4137	924	950	938	387	387	368
0.8	***	36230	44618	93080	13362	15849	22831	2937	3143	3443	736	755	774	293	305	305
	•••	39518	54305	93413	14393	18118	22885	3031	3275	3423	743	762	744	293	305	275
0.85	***	29518	36811	73655	10974	12943	17893	2293	2431	2611	530	537	549	.	.	.
	•••	32412	44900	73773	11806	14656	17900	2356	2518	2590	537	549	521	.	.	.
0.9	***	22362	27912	51143	8131	9425	12193	1524	1587	1662	.	.	.	.	.	.
	•••	24593	33755	51156	8693	10468	12172	1550	1625	1637	.	.	.	.	.	.
0.95	***	13700	16681	25587	4449	4936	5724	.	.	.	.	.	.	.	.	.
	•••	14937	19475	25563	4662	5274	5699	.	.	.	.	.	.	.	.	.
0.98	***	6237	7130	8806	.	.	.	.	.	.	.	.	.	.	.	.
	•••	6631	7805	8779	.	.	.	.	.	.	.	.	.	.	.	.

PCONT=••• REQUIRED NUMBER OF PATIENTS

TABLE 2: ALPHA= 0.005 POWER= 0.9   EXPECTED ACCRUAL THRU MINIMUM FOLLOW-UP= 8000

		DEL=.01			DEL=.02			DEL=.05			DEL=.10			DEL=.15		
FACT=		1.0 .75	.50 .25	.00 BIN	1.0 .75	.50 .25	.00 BIN	1.0 .75	.50 .25	.00 BIN	1.0 75	.50 .25	.00 BIN	1.0 .75	.50 .25	.00 BIN
PCONT=***					REQUIRED NUMBER OF PATIENTS											
0.01	***	2285	2285	2293	833	833	845	273	273	273	133	133	133	93	93	93
	•••	2285	2285	8779	833	833	2902	273	273	790	133	133	321	93	93	191
0.02	***	5045	5065	5125	1625	1625	1633	433	433	445	193	193	193	125	125	125
	•••	5053	5093	14493	1625	1633	4316	433	433	1009	193	193	373	125	125	213
0.05	***	15273	15405	15993	4345	4393	4485	945	945	953	333	333	333	193	193	193
	•••	15313	15573	30920	4373	4433	8378	945	945	1637	333	333	521	193	193	275
0.1	***	33705	34065	37233	9133	9345	9845	1785	1805	1833	553	565	565	293	293	293
	•••	33825	34685	55917	9225	9533	14553	1793	1813	2590	565	565	744	293	293	368
0.15	***	50373	51033	58813	13465	13905	15225	2565	2613	2685	765	765	773	385	385	393
	•••	50593	52233	77939	13633	14365	19984	2585	2645	3423	765	773	938	385	385	447
0.2	***	63865	64893	79013	17033	17765	20233	3233	3333	3465	933	945	965	465	465	465
	•••	64205	66813	96984	17313	18553	24671	3285	3393	4137	945	953	1102	465	465	513
0.25	***	73773	75265	97013	19753	20813	24673	3785	3933	4145	1085	1105	1113	525	525	533
	•••	74273	78033	113054	20165	21993	28614	3853	4033	4732	1093	1105	1235	525	533	566
0.3	***	80133	82185	112333	21653	23073	28433	4225	4413	4713	1205	1225	1245	573	573	585
	•••	80825	85965	126148	22205	24653	31813	4313	4553	5208	1213	1233	1340	573	585	606
0.35	***	83205	85925	124733	22805	24605	31465	4533	4785	5165	1285	1313	1345	605	613	625
	•••	84125	90853	136266	23505	26585	34268	4645	4953	5565	1293	1325	1414	613	613	632
0.4	***	83353	86893	134053	23305	25473	33713	4733	5033	5485	1345	1373	1413	625	633	645
	•••	84553	93073	143408	24153	27833	35979	4873	5233	5803	1353	1393	1459	633	633	645
0.45	***	81045	85525	140205	23245	25765	35173	4833	5165	5685	1365	1405	1445	633	645	653
	•••	82573	93033	147574	24233	28445	36946	4985	5393	5922	1385	1425	1474	633	645	645
0.5	***	76765	82333	143133	22725	25533	35825	4833	5193	5765	1365	1405	1445	625	633	645
	•••	78693	91133	148765	23845	28485	37169	4993	5433	5922	1385	1425	1459	633	633	632
0.55	***	71073	77793	142833	21833	24873	35665	4745	5113	5685	1333	1365	1413	605	613	625
	•••	73453	87725	146979	23053	27993	36648	4913	5365	5803	1345	1385	1414	613	613	606
0.6	***	64493	72305	139293	20653	23825	34705	4565	4925	5493	1273	1305	1345	573	573	585
	•••	67333	83105	142218	21933	27005	35384	4725	5173	5565	1285	1325	1340	573	585	566
0.65	***	57525	66193	132525	19225	22413	32925	4293	4645	5173	1185	1213	1253	525	533	533
	•••	60753	77473	134480	20525	25545	33375	4453	4873	5208	1193	1233	1235	525	533	513
0.7	***	50513	59665	122545	17573	20653	30353	3945	4253	4725	1065	1093	1125	465	465	473
	•••	54005	70965	123767	18845	23625	30622	4085	4465	4732	1073	1105	1102	465	473	447
0.75	***	43665	52773	109385	15705	18553	26985	3493	3765	4145	913	933	965	385	393	393
	•••	47193	63585	110078	16885	21225	27126	3625	3933	4137	925	953	938	385	393	368
0.8	***	36933	45473	93073	13593	16073	22825	2953	3165	3445	745	753	773	293	305	305
	•••	40293	55245	93413	14633	18313	22885	3053	3285	3423	745	765	744	305	305	275
0.85	***	30153	37545	73653	11165	13113	17893	2305	2445	2613	533	545	553	.	.	.
	•••	33085	45673	73773	11993	14793	17900	2373	2525	2590	533	545	521	.	.	.
0.9	***	22853	28453	51145	8253	9533	12193	1525	1593	1665	.	.	.	.	.	.
	•••	25113	34285	51156	8813	10553	12172	1553	1625	1637	.	.	.	.	.	.
0.95	***	13973	16953	25585	4493	4973	5725	.	.	.	.	.	.	.	.	.
	•••	15213	19713	25563	4713	5293	5699	.	.	.	.	.	.	.	.	.
0.98	***	6333	7205	8805	.	.	.	.	.	.	.	.	.	.	.	.
	•••	6713	7853	8779	.	.	.	.	.	.	.	.	.	.	.	.

TABLE 2: ALPHA= 0.005 POWER= 0.9   EXPECTED ACCRUAL THRU MINIMUM FOLLOW-UP= 8500

	DEL=.01			DEL=.02			DEL=.05			DEL=.10			DEL=.15		
FACT=	1.0 .75	.50 .25	.00 BIN	1.0 .75	.50 .25	.00 BIN	1.0 .75	.50 .25	.00 BIN	1.0 75	.50 .25	.00 BIN	1.0 .75	.50 .25	.00 BIN
PCONT=***				REQUIRED NUMBER OF PATIENTS											
0.01 ***	2280	2288	2288	835	843	843	282	282	282	141	141	141	99	99	99
•••	2280	2288	8779	835	843	2902	282	282	790	141	141	321	99	99	191
0.02 ***	5042	5071	5127	1621	1629	1629	439	439	439	197	197	197	120	120	120
•••	5050	5093	14493	1629	1629	4316	439	439	1009	197	197	373	120	120	213
0.05 ***	15285	15420	15994	4349	4405	4476	941	941	949	333	333	333	197	197	197
•••	15327	15590	30920	4370	4434	8378	941	949	1637	333	333	521	197	197	275
0.1 ***	33730	34112	37236	9151	9364	9845	1791	1812	1833	558	558	566	303	303	303
•••	33857	34750	55917	9236	9547	14553	1799	1820	2590	558	566	744	303	303	368
0.15 ***	50411	51112	58813	13500	13954	15229	2564	2620	2683	758	771	771	388	388	388
•••	50645	52374	77939	13678	14400	19984	2585	2649	3423	771	771	938	388	388	447
0.2 ***	63926	65018	79021	17091	17835	20236	3244	3342	3470	941	949	962	460	460	473
•••	64295	67028	96984	17375	18621	24671	3286	3393	4137	941	949	1102	460	460	513
0.25 ***	73871	75443	97012	19832	20916	24677	3810	3945	4150	1090	1098	1119	524	524	537
•••	74402	78355	113054	20257	22085	28614	3873	4030	4732	1090	1111	1235	524	537	566
0.3 ***	80267	82435	112333	21766	23219	28438	4243	4434	4710	1204	1225	1246	580	580	588
•••	80998	86387	126148	22326	24791	31813	4328	4561	5208	1217	1238	1340	580	580	606
0.35 ***	83378	86260	124743	22943	24783	31464	4561	4795	5170	1289	1310	1345	609	609	622
•••	84347	91402	136266	23665	26760	34268	4668	4965	5565	1302	1331	1414	609	622	632
0.4 ***	83582	87322	134059	23474	25676	33716	4766	5050	5488	1345	1374	1408	630	630	643
•••	84844	93748	143408	24337	28035	35979	4893	5241	5803	1353	1395	1459	630	643	645
0.45 ***	81330	86055	140200	23445	25995	35175	4872	5191	5680	1374	1408	1438	630	643	651
•••	82953	93825	147574	24451	28672	36946	5008	5403	5922	1387	1416	1474	643	643	645
0.5 ***	77130	82974	143133	22956	25803	35820	4872	5220	5751	1366	1408	1438	630	630	643
•••	79170	92040	148765	24090	28723	37169	5029	5446	5922	1387	1416	1459	630	643	632
0.55 ***	71533	78533	142835	22085	25153	35663	4774	5135	5688	1331	1374	1408	609	609	622
•••	74028	88725	146979	23325	28247	36648	4944	5382	5803	1353	1387	1414	609	622	606
0.6 ***	65039	73135	139286	20916	24103	34694	4596	4957	5488	1268	1310	1345	566	580	588
•••	68006	84156	142218	22212	27270	35384	4766	5191	5565	1289	1323	1340	580	580	566
0.65 ***	58154	67092	132521	19492	22688	32930	4328	4668	5170	1183	1217	1246	524	524	537
•••	61503	78554	134480	20810	25803	33375	4490	4893	5208	1196	1238	1235	524	537	513
0.7 ***	51218	60576	122541	17835	20924	30351	3966	4277	4723	1068	1090	1119	460	473	473
•••	54796	72022	123767	19110	23870	30622	4115	4476	4732	1076	1111	1102	460	473	447
0.75 ***	44376	53662	109388	15951	18799	26980	3520	3775	4150	920	941	962	388	388	396
•••	47988	64585	110078	17141	21434	27126	3648	3945	4137	928	949	938	388	396	368
0.8 ***	37618	46296	93081	13818	16283	22828	2981	3172	3435	736	758	771	303	303	303
•••	41040	56135	93413	14846	18480	22885	3074	3300	3423	750	758	744	303	303	275
0.85 ***	30755	38235	73645	11340	13274	17898	2322	2450	2620	537	545	545	.	.	.
•••	33730	46395	73773	12161	14923	17900	2386	2521	2590	537	545	521	.	.	.
0.9 ***	23325	28956	51141	8378	9632	12190	1536	1600	1663	.	.	.	.	.	.
•••	25599	34792	51156	8931	10631	12172	1565	1629	1637	.	.	.	.	.	.
0.95 ***	14243	17205	25591	4540	5008	5730	.	.	.	.	.	.	.	.	.
•••	15476	19925	25563	4753	5318	5699	.	.	.	.	.	.	.	.	.
0.98 ***	6410	7273	8811	.	.	.	.	.	.	.	.	.	.	.	.
•••	6793	7898	8779	.	.	.	.	.	.	.	.	.	.	.	.

TABLE 2: ALPHA= 0.005 POWER= 0.9   EXPECTED ACCRUAL THRU MINIMUM FOLLOW-UP= 9000

	DEL=.01			DEL=.02			DEL=.05			DEL=.10			DEL=.15		
FACT=	1.0 .75	.50 .25	.00 BIN	1.0 .75	.50 .25	.00 BIN	1.0 .75	.50 .25	.00 BIN	1.0 75	.50 .25	.00 BIN	1.0 .75	.50 .25	.00 BIN
PCONT=•••				REQUIRED NUMBER OF PATIENTS											
0.01 •••	2279	2287	2287	839	839	839	276	276	276	141	141	141	96	96	96
•••	2287	2287	8779	839	839	2902	276	276	790	141	141	321	96	96	191
0.02 •••	5046	5069	5122	1626	1626	1635	442	442	442	195	195	195	127	127	127
•••	5055	5091	14493	1626	1635	4316	442	442	1009	195	195	373	127	127	213
0.05 •••	15292	15427	15990	4357	4402	4484	937	951	951	330	330	330	195	195	195
•••	15337	15607	30920	4380	4439	8378	937	951	1637	330	330	521	195	195	275
0.1 •••	33742	34147	37230	9172	9375	9847	1784	1806	1829	555	569	569	299	299	299
•••	33891	34814	55917	9254	9555	14553	1792	1815	2590	555	569	744	299	299	368
0.15 •••	50451	51194	58821	13537	13987	15225	2571	2625	2684	757	771	771	389	389	389
•••	50707	52499	77939	13717	14429	19984	2594	2647	3423	771	771	938	389	389	447
0.2 •••	63996	65152	79012	17137	17902	20234	3255	3345	3471	937	951	960	465	465	465
•••	64379	67245	96984	17444	18681	24671	3300	3404	4137	937	951	1102	465	465	513
0.25 •••	73964	75629	97012	19919	21021	24675	3817	3952	4146	1086	1109	1117	524	532	532
•••	74526	78652	113054	20346	22177	28614	3885	4042	4732	1095	1109	1235	524	532	566
0.3 •••	80399	82680	112335	21876	23347	28432	4259	4447	4717	1207	1221	1244	577	577	577
•••	81164	86797	126148	22447	24914	31813	4349	4574	5208	1207	1230	1340	577	577	606
0.35 •••	83549	86595	124732	23077	24945	31461	4582	4821	5167	1289	1320	1342	614	614	622
•••	84570	91927	136266	23811	26916	34268	4686	4979	5565	1297	1334	1414	614	614	632
0.4 •••	83805	87742	134061	23640	25881	33711	4799	5069	5482	1342	1379	1410	622	636	645
•••	85146	94394	143408	24517	28221	35979	4920	5257	5803	1365	1387	1459	636	636	645
0.45 •••	81622	86586	140204	23640	26219	35174	4897	5212	5685	1379	1410	1446	636	645	645
•••	83324	94596	147574	24666	28882	36946	5046	5415	5922	1387	1424	1474	636	645	645
0.5 •••	77496	83616	143137	23167	26047	35826	4897	5235	5752	1365	1401	1446	622	636	645
•••	79634	92909	148765	24329	28964	37169	5055	5460	5922	1387	1424	1459	636	636	632
0.55 •••	71984	79260	142836	22326	25417	35669	4807	5159	5685	1342	1379	1410	614	614	622
•••	74594	89669	146979	23572	28491	36648	4979	5392	5803	1356	1387	1414	614	622	606
0.6 •••	65580	73950	139290	21179	24374	34701	4627	4979	5496	1275	1311	1342	569	577	591
•••	68662	85169	142218	22484	27510	35384	4785	5204	5565	1289	1334	1340	577	577	566
0.65 •••	58785	67942	132517	19747	22956	32924	4357	4686	5167	1185	1221	1252	524	532	532
•••	62227	79575	134480	21066	26039	33375	4515	4897	5208	1199	1230	1235	524	532	513
0.7 •••	51891	61454	122541	18082	21179	30359	3997	4304	4717	1072	1095	1117	465	465	479
•••	55559	73027	123767	19365	24090	30622	4132	4484	4732	1086	1109	1102	465	465	447
0.75 •••	45060	54510	109387	16192	19027	26984	3547	3795	4146	915	937	960	389	389	397
•••	48750	65526	110078	17376	21637	27126	3660	3952	4137	929	951	938	389	397	368
0.8 •••	38279	47076	93075	14024	16476	22830	2999	3187	3435	749	757	771	299	299	307
•••	41744	56962	93413	15059	18645	22885	3089	3300	3423	749	757	744	299	299	275
0.85 •••	31326	38886	73649	11504	13425	17894	2332	2459	2616	532	546	546	.	.	.
•••	34350	47076	73773	12322	15045	17900	2391	2535	2590	532	546	521	.	.	.
0.9 •••	23766	29445	51149	8489	9726	12187	1536	1604	1671	.	.	.	.	.	.
•••	26070	35250	51156	9029	10694	12172	1567	1626	1637	.	.	.	.	.	.
0.95 •••	14482	17452	25589	4596	5046	5730	.	.	.	.	.	.	.	.	.
•••	15720	20130	25563	4799	5339	5699	.	.	.	.	.	.	.	.	.
0.98 •••	6495	7327	8804	.	.	.	.	.	.	.	.	.	.	.	.
•••	6869	7935	8779	.	.	.	.	.	.	.	.	.	.	.	.

TABLE 2: ALPHA= 0.005 POWER= 0.9   EXPECTED ACCRUAL THRU MINIMUM FOLLOW-UP= 9500

	DEL=.01			DEL=.02			DEL=.05			DEL=.10			DEL=.15		
FACT=	1.0 / .75	.50 / .25	.00 / BIN	1.0 / .75	.50 / .25	.00 / BIN	1.0 / .75	.50 / .25	.00 / BIN	1.0 / 75	.50 / .25	.00 / BIN	1.0 / .75	.50 / .25	.00 / BIN
PCONT=***					REQUIRED NUMBER OF PATIENTS										
0.01 ***	2286	2286	2295	838	838	838	277	277	277	134	134	134	101	101	101
•••	2286	2286	8779	838	838	2902	277	277	790	134	134	321	101	101	191
0.02 ***	5050	5074	5122	1621	1630	1630	434	443	443	196	196	196	125	125	125
•••	5065	5089	14493	1621	1630	4316	443	443	1009	196	196	373	125	125	213
0.05 ***	15301	15444	15990	4362	4409	4480	942	942	956	339	339	339	196	196	196
•••	15349	15619	30920	4385	4433	8378	942	942	1637	339	339	521	196	196	275
0.1 ***	33764	34192	37232	9183	9397	9848	1788	1811	1835	562	562	562	300	300	300
•••	33921	34871	55917	9269	9578	14553	1797	1820	2590	562	562	744	300	300	368
0.15 ***	50499	51268	58811	13568	14028	15230	2580	2628	2690	766	766	775	386	386	386
•••	50760	52622	77939	13758	14455	19984	2595	2652	3423	766	775	938	386	386	447
0.2 ***	64060	65271	79023	17201	17961	20241	3260	3355	3459	942	956	956	467	467	467
•••	64464	67442	96984	17495	18730	24671	3308	3403	4137	942	956	1102	467	467	513
0.25 ***	74059	75816	97010	19989	21105	24668	3830	3958	4148	1084	1099	1123	529	529	529
•••	74638	78951	113054	20431	22260	28614	3887	4044	4732	1099	1108	1235	529	529	566
0.3 ***	80528	82941	112344	21975	23471	28435	4281	4457	4718	1203	1227	1250	576	576	585
•••	81335	87202	126148	22569	25024	31813	4362	4575	5208	1218	1241	1340	576	585	606
0.35 ***	83725	86917	124741	23210	25095	31460	4599	4837	5160	1289	1322	1345	609	609	624
•••	84803	92441	136266	23955	27058	34268	4709	4979	5565	1298	1336	1414	609	624	632
0.4 ***	84034	88152	134060	23804	26060	33717	4813	5089	5493	1345	1384	1408	633	633	648
•••	85435	95015	143408	24692	28388	35979	4946	5264	5803	1360	1393	1459	633	633	645
0.45 ***	81905	87098	140203	23828	26425	35165	4923	5231	5683	1384	1408	1440	633	648	648
•••	83701	95324	147574	24873	29076	36946	5065	5430	5922	1393	1431	1474	633	648	645
0.5 ***	77859	84224	143133	23385	26274	35821	4932	5264	5754	1369	1408	1440	633	633	648
•••	80100	93733	148765	24549	29171	37169	5089	5478	5922	1393	1431	1459	633	633	632
0.55 ***	72429	79973	142839	22554	25656	35664	4851	5184	5683	1345	1369	1408	609	609	624
•••	75151	90589	146979	23813	28720	36648	5003	5407	5803	1360	1393	1414	609	624	606
0.6 ***	66126	74733	139285	21414	24635	34705	4661	5003	5493	1274	1313	1345	576	576	585
•••	69309	86124	142218	22735	27732	35384	4813	5217	5565	1298	1322	1340	576	585	566
0.65 ***	59390	68772	132517	19989	23210	32924	4385	4709	5169	1194	1218	1250	529	529	538
•••	62944	80566	134480	21319	26259	33375	4543	4908	5208	1203	1241	1235	529	538	513
0.7 ***	52541	62288	122542	18327	21414	30359	4029	4314	4718	1075	1099	1123	467	467	467
•••	56303	73988	123767	19609	24288	30622	4163	4495	4732	1084	1108	1102	467	467	447
0.75 ***	45725	55329	109384	16403	19253	26986	3569	3815	4139	918	942	965	386	395	395
•••	49478	66420	110078	17590	21818	27126	3688	3958	4137	933	956	938	386	395	368
0.8 ***	38909	47815	93083	14209	16655	22830	3023	3198	3435	743	752	775	300	300	300
•••	42433	57766	93413	15239	18793	22885	3103	3308	3423	752	766	744	300	300	275
0.85 ***	31879	39512	73655	11668	13568	17899	2343	2462	2619	538	538	553	.	.	.
•••	34928	47720	73773	12475	15159	17900	2405	2533	2590	538	553	521	.	.	.
0.9 ***	24193	29908	51140	8589	9815	12190	1550	1598	1669	.	.	.	.	.	.
•••	26511	35688	51156	9126	10750	12172	1574	1630	1637	.	.	.	.	.	.
0.95 ***	14717	17676	25585	4623	5065	5730	.	.	.	.	.	.	.	.	.
•••	15952	20322	25563	4828	5359	5699	.	.	.	.	.	.	.	.	.
0.98 ***	6561	7393	8803	.	.	.	.	.	.	.	.	.	.	.	.
•••	6927	7972	8779	.	.	.	.	.	.	.	.	.	.	.	.

TABLE 2: ALPHA= 0.005 POWER= 0.9   EXPECTED ACCRUAL THRU MINIMUM FOLLOW-UP= 10000

	DEL=.01			DEL=.02			DEL=.05			DEL=.10			DEL=.15		
FACT=	1.0 .75	.50 .25	.00 BIN	1.0 .75	.50 .25	.00 BIN	1.0 .75	.50 .25	.00 BIN	1.0 75	.50 .25	.00 BIN	1.0 .75	.50 .25	.00 BIN
PCONT=***						REQUIRED NUMBER OF PATIENTS									
0.01 ***	2282	2282	2291	841	841	841	282	282	282	141	141	141	91	91	91
***	2282	2291	8779	841	841	2902	282	282	790	141	141	321	91	91	191
0.02 ***	5057	5082	5116	1616	1632	1632	441	441	441	191	191	191	132	132	132
***	5066	5091	14493	1632	1632	4316	441	441	1009	191	191	373	132	132	213
0.05 ***	15307	15457	15991	4366	4407	4482	941	941	957	332	332	332	191	191	191
***	15366	15632	30920	4382	4441	8378	941	941	1637	332	332	521	191	191	275
0.1 ***	33791	34241	37232	9207	9407	9841	1791	1807	1832	557	566	566	291	291	291
***	33941	34932	55917	9291	9582	14553	1807	1816	2590	557	566	744	291	291	368
0.15 ***	50541	51357	58816	13591	14057	15232	2582	2632	2682	766	766	782	391	391	391
***	50816	52741	77939	13782	14491	19984	2607	2657	3423	766	766	938	391	391	447
0.2 ***	64116	65407	79016	17241	18016	20232	3266	3357	3466	941	957	957	466	466	466
***	64557	67641	96984	17557	18782	24671	3307	3407	4137	941	957	1102	466	466	513
0.25 ***	74141	75991	97007	20066	21191	24666	3841	3966	4141	1091	1107	1116	532	532	532
***	74766	79241	113054	20516	22341	28614	3907	4057	4732	1091	1107	1235	532	532	566
0.3 ***	80657	83182	112341	22082	23591	28432	4291	4466	4716	1207	1232	1241	582	582	582
***	81507	87591	126148	22666	25141	31813	4366	4582	5208	1216	1232	1340	582	582	606
0.35 ***	83891	87241	124741	23341	25241	31457	4616	4841	5166	1291	1316	1341	607	616	616
***	85032	92941	136266	24091	27191	34268	4732	4991	5565	1307	1332	1414	616	616	632
0.4 ***	84257	88566	134057	23957	26232	33716	4841	5107	5491	1357	1382	1407	632	632	641
***	85732	95632	143408	24866	28557	35979	4966	5282	5803	1366	1391	1459	632	641	645
0.45 ***	82191	87607	140207	24007	26632	35166	4957	5257	5682	1382	1407	1441	641	641	657
***	84066	96041	147574	25057	29266	36946	5091	5441	5922	1391	1432	1474	641	641	645
0.5 ***	78216	84832	143141	23582	26491	35816	4957	5282	5741	1382	1407	1441	632	632	641
***	80557	94541	148765	24766	29382	37169	5107	5491	5922	1391	1432	1459	632	641	632
0.55 ***	72866	80666	142832	22782	25891	35666	4866	5207	5682	1341	1382	1407	607	616	616
***	75691	91457	146979	24057	28932	36648	5032	5416	5803	1357	1391	1414	616	616	606
0.6 ***	66641	75491	139291	21641	24866	34707	4691	5016	5491	1282	1316	1341	566	582	591
***	69932	87041	142218	22966	27957	35384	4841	5232	5565	1291	1332	1340	582	582	566
0.65 ***	59982	69582	132516	20232	23441	32932	4416	4732	5166	1191	1216	1257	532	532	541
***	63632	81491	134480	21557	26466	33375	4566	4932	5208	1207	1232	1235	532	532	513
0.7 ***	53182	63107	122541	18557	21641	30357	4057	4332	4716	1066	1091	1116	466	466	466
***	57016	74907	123767	19841	24491	30622	4182	4507	4732	1082	1107	1102	466	466	447
0.75 ***	46366	56107	109382	16616	19457	26982	3591	3832	4141	932	941	966	391	391	391
***	50182	67282	110078	17807	21991	27126	3707	3966	4137	932	957	938	391	391	368
0.8 ***	39516	48516	93082	14391	16832	22832	3032	3207	3441	741	757	766	291	307	307
***	43082	58516	93413	15432	18932	22885	3116	3316	3423	757	766	744	307	307	275
0.85 ***	32416	40107	73657	11807	13691	17891	2357	2466	2616	532	541	557	.	.	.
***	35491	48316	73773	12616	15257	17900	2416	2541	2590	541	541	521	.	.	.
0.9 ***	24591	30332	51141	8691	9891	12191	1557	1607	1666	.	.	.	.	.	.
***	26932	36107	51156	9216	10807	12172	1582	1632	1637	.	.	.	.	.	.
0.95 ***	14941	17882	25591	4666	5091	5732	.	.	.	.	.	.	.	.	.
***	16182	20491	25563	4866	5382	5699	.	.	.	.	.	.	.	.	.
0.98 ***	6632	7441	8807	.	.	.	.	.	.	.	.	.	.	.	.
***	6991	8016	8779	.	.	.	.	.	.	.	.	.	.	.	.

TABLE 2: ALPHA= 0.005 POWER= 0.9   EXPECTED ACCRUAL THRU MINIMUM FOLLOW-UP= 11000

	DEL=.01			DEL=.02			DEL=.05			DEL=.10			DEL=.15		
FACT=	1.0 / .75	.50 / .25	.00 BIN	1.0 / .75	.50 / .25	.00 BIN	1.0 / .75	.50 / .25	.00 BIN	1.0 / 75	.50 / .25	.00 BIN	1.0 / .75	.50 / .25	.00 BIN

PCONT=•••   REQUIRED NUMBER OF PATIENTS

PCONT	DEL=.01 (1.0/.75)	DEL=.01 (.50/.25)	DEL=.01 (.00/BIN)	DEL=.02 (1.0/.75)	DEL=.02 (.50/.25)	DEL=.02 (.00/BIN)	DEL=.05 (1.0/.75)	DEL=.05 (.50/.25)	DEL=.05 (.00/BIN)	DEL=.10 (1.0/.75)	DEL=.10 (.50/.25)	DEL=.10 (.00/BIN)	DEL=.15 (1.0/.75)	DEL=.15 (.50/.25)	DEL=.15 (.00/BIN)
0.01 •••	2290	2290	2290	832	843	843	282	282	282	145	145	145	100	100	100
•••	2290	2290	8779	843	843	2902	282	282	790	145	145	321	100	100	191
0.02 •••	5050	5078	5122	1630	1630	1630	430	430	447	183	183	183	128	128	128
•••	5067	5095	14493	1630	1630	4316	430	447	1009	183	183	373	128	128	213
0.05 •••	15325	15473	15995	4380	4418	4473	942	942	953	337	337	337	200	200	200
•••	15380	15655	30920	4390	4445	8378	942	942	1637	337	337	521	200	200	275
0.1 •••	33843	34327	37225	9230	9440	9852	1795	1805	1833	557	557	568	293	293	293
•••	34008	35042	55917	9313	9605	14553	1805	1822	2590	557	568	744	293	293	368
0.15 •••	50618	51515	58813	13658	14125	15225	2592	2630	2685	760	777	777	392	392	392
•••	50920	52972	77939	13850	14538	19984	2603	2658	3423	760	777	938	392	392	447
0.2 •••	64247	65650	79015	17343	18130	20230	3280	3362	3472	942	953	953	458	458	475
•••	64725	68025	96984	17662	18872	24671	3318	3417	4137	942	953	1102	458	475	513
0.25 •••	74340	76347	97010	20203	21358	24675	3857	3978	4143	1090	1107	1118	530	530	530
•••	75010	79795	113054	20670	22485	28614	3923	4060	4732	1090	1107	1235	530	530	566
0.3 •••	80912	83673	112345	22265	23805	28442	4325	4490	4720	1217	1228	1245	568	585	585
•••	81847	88337	126148	22887	25335	31813	4390	4600	5208	1217	1245	1340	585	585	606
0.35 •••	84240	87880	124730	23585	25527	31467	4655	4875	5160	1300	1327	1338	612	612	623
•••	85477	93892	136266	24355	27435	34268	4765	4995	5565	1310	1327	1414	612	623	632
0.4 •••	84690	89382	134053	24245	26555	33705	4885	5133	5490	1355	1382	1410	623	640	640
•••	86313	96780	143408	25180	28855	35979	4995	5298	5803	1365	1393	1459	640	640	645
0.45 •••	82765	88595	140202	24345	26995	35163	4995	5287	5683	1382	1410	1437	640	640	650
•••	84800	97395	147574	25428	29597	36946	5133	5463	5922	1393	1420	1474	640	650	645
0.5 •••	78932	86000	143145	23970	26902	35823	5012	5315	5755	1382	1410	1448	623	640	640
•••	81462	96065	148765	25170	29735	37169	5150	5507	5922	1393	1420	1459	640	640	632
0.55 •••	73735	81995	142825	23190	26325	35658	4930	5243	5683	1355	1382	1410	612	612	623
•••	76760	93105	146979	24482	29305	36648	5067	5435	5803	1365	1393	1414	612	623	606
0.6 •••	67668	76935	139295	22073	25307	34695	4748	5050	5490	1283	1310	1355	568	585	585
•••	71150	88777	142218	23410	28343	35384	4885	5243	5565	1300	1327	1340	585	585	566
0.65 •••	61140	71105	132513	20670	23888	32925	4473	4765	5177	1200	1217	1255	530	530	540
•••	64945	83250	134480	22007	26847	33375	4600	4940	5208	1217	1228	1235	530	530	513
0.7 •••	54402	64632	122547	18982	22062	30350	4105	4363	4720	1080	1090	1118	458	475	475
•••	58373	76605	123767	20275	24840	30622	4225	4528	4732	1090	1107	1102	458	475	447
0.75 •••	47593	57575	109385	17013	19818	26985	3637	3857	4143	925	942	970	392	392	392
•••	51525	68867	110078	18195	22293	27126	3730	3978	4137	942	953	938	392	392	368
0.8 •••	40663	49865	93078	14747	17150	22832	3060	3225	3445	750	760	777	293	293	310
•••	44320	59902	93413	15765	19185	22885	3142	3335	3423	750	760	744	293	310	275
0.85 •••	33420	41230	73652	12080	13933	17893	2372	2482	2620	540	540	557	.	.	.
•••	36555	49435	73773	12877	15435	17900	2427	2548	2590	540	540	521	.	.	.
0.9 •••	25362	31137	51140	8873	10045	12190	1558	1613	1668	.	.	.	.	.	.
•••	27727	36857	51156	9385	10908	12172	1585	1640	1637	.	.	.	.	.	.
0.95 •••	15352	18267	25582	4737	5133	5727	.	.	.	.	.	.	.	.	.
•••	16590	20808	25563	4930	5397	5699	.	.	.	.	.	.	.	.	.
0.98 •••	6755	7525	8807	.	.	.	.	.	.	.	.	.	.	.	.
•••	7102	8075	8779	.	.	.	.	.	.	.	.	.	.	.	.

TABLE 2: ALPHA= 0.005 POWER= 0.9   EXPECTED ACCRUAL THRU MINIMUM FOLLOW-UP= 12000

	DEL=.01			DEL=.02			DEL=.05			DEL=.10			DEL=.15		
FACT=	1.0 .75	.50 .25	.00 BIN	1.0 .75	.50 .25	.00 BIN	1.0 .75	.50 .25	.00 BIN	1.0 75	.50 .25	.00 BIN	1.0 .75	.50 .25	.00 BIN
PCONT=***				REQUIRED NUMBER OF PATIENTS											
0.01 ***	2288	2288	2288	829	829	848	278	278	278	139	139	139	98	98	98
•••	2288	2288	8779	829	848	2902	278	278	790	139	139	321	98	98	191
0.02 ***	5059	5078	5119	1628	1628	1639	439	439	439	188	188	188	128	128	128
•••	5059	5089	14493	1628	1628	4316	439	439	1009	188	188	373	128	128	213
0.05 ***	15338	15499	15998	4369	4418	4478	938	949	949	338	338	338	199	199	199
•••	15398	15668	30920	4399	4448	8378	938	949	1637	338	338	521	199	199	275
0.1 ***	33889	34399	37238	9259	9458	9848	1789	1808	1838	559	559	559	289	289	289
•••	34069	35138	55917	9338	9619	14553	1808	1819	2590	559	559	744	289	289	368
0.15 ***	50708	51668	58819	13718	14179	15229	2599	2629	2689	769	769	769	379	379	398
•••	51038	53179	77939	13909	14588	19984	2618	2659	3423	769	769	938	379	379	447
0.2 ***	64388	65888	79009	17438	18229	20239	3289	3368	3469	938	949	968	458	469	469
•••	64898	68389	96984	17768	18968	24671	3338	3409	4137	949	949	1102	469	469	513
0.25 ***	74528	76699	97009	20348	21499	24668	3878	3998	4148	1099	1099	1118	529	529	529
•••	75259	80329	113054	20809	22609	28614	3938	4069	4732	1099	1118	1235	529	529	566
0.3 ***	81169	84158	112339	22448	24008	28429	4339	4508	4718	1208	1238	1249	578	578	578
•••	82178	89048	126148	23078	25508	31813	4418	4598	5208	1219	1238	1340	578	578	606
0.35 ***	84578	88508	124729	23809	25778	31459	4688	4898	5168	1298	1328	1339	608	619	619
•••	85928	94789	136266	24608	27668	34268	4778	5018	5565	1309	1328	1414	608	619	632
0.4 ***	85148	90169	134059	24518	26858	33709	4928	5149	5479	1358	1388	1418	638	638	638
•••	86888	97868	143408	25478	29108	35979	5029	5299	5803	1369	1399	1459	638	638	645
0.45 ***	83329	89558	140209	24668	27338	35168	5048	5318	5678	1388	1418	1448	638	649	649
•••	85519	98659	147574	25759	29899	36946	5168	5479	5922	1399	1429	1474	638	649	645
0.5 ***	79639	87109	143138	24319	27278	35828	5059	5348	5749	1388	1418	1448	638	638	638
•••	82339	97478	148765	25538	30068	37169	5198	5528	5922	1399	1429	1459	638	638	632
0.55 ***	74588	83258	142838	23569	26719	35659	4969	5269	5689	1358	1388	1418	608	619	619
•••	77798	94639	146979	24878	29648	36648	5108	5449	5803	1369	1399	1414	608	619	606
0.6 ***	68659	78308	139298	22478	25718	34699	4789	5089	5498	1298	1309	1339	578	578	589
•••	72308	90379	142218	23828	28688	35384	4928	5269	5565	1309	1328	1340	578	578	566
0.65 ***	62228	72529	132518	21068	24278	32929	4508	4789	5168	1208	1219	1249	529	529	529
•••	66188	84859	134480	22418	27188	33375	4639	4958	5208	1208	1238	1235	529	529	513
0.7 ***	55568	66068	122539	19369	22429	30349	4129	4388	4718	1088	1099	1118	469	469	469
•••	59659	78169	123767	20659	25148	30622	4249	4538	4732	1088	1118	1102	469	469	447
0.75 ***	48739	58939	109388	17378	20168	26989	3668	3878	4148	938	949	968	398	398	398
•••	52778	70328	110079	18559	22568	27126	3758	3998	4137	938	949	938	398	398	368
0.8 ***	41749	51098	93079	15049	17419	22819	3079	3248	3439	758	758	769	308	308	308
•••	45469	61178	93413	16069	19399	22885	3158	3338	3423	758	769	744	308	308	275
0.85 ***	34339	42248	73658	12319	14149	17899	2389	2498	2618	529	548	548	.	.	.
•••	37538	50449	73773	13118	15589	17900	2438	2558	2590	548	548	521	.	.	.
0.9 ***	26059	31879	51139	9038	10159	12188	1568	1609	1669	.	.	.	.	.	.
•••	28448	37538	51156	9529	10999	12172	1598	1639	1637	.	.	.	.	.	.
0.95 ***	15728	18619	25579	4789	5179	5719	.	.	.	.	.	.	.	.	.
•••	16958	21098	25563	4969	5419	5699	.	.	.	.	.	.	.	.	.
0.98 ***	6859	7609	8809	.	.	.	.	.	.	.	.	.	.	.	.
•••	7208	8119	8779	.	.	.	.	.	.	.	.	.	.	.	.

TABLE 2: ALPHA= 0.005 POWER= 0.9   EXPECTED ACCRUAL THRU MINIMUM FOLLOW-UP= 13000

	DEL=.01			DEL=.02			DEL=.05			DEL=.10			DEL=.15		
FACT=	1.0 .75	.50 .25	.00 BIN	1.0 .75	.50 .25	.00 BIN	1.0 .75	.50 .25	.00 BIN	1.0 75	.50 .25	.00 BIN	1.0 .75	.50 .25	.00 BIN
PCONT=***			REQUIRED NUMBER OF PATIENTS												
0.01 ***	2284	2284	2296	833	833	833	281	281	281	139	139	139	106	106	106
***	2284	2284	8779	833	833	2902	281	281	790	139	139	321	106	106	191
0.02 ***	5058	5079	5123	1634	1634	1634	443	443	443	183	183	183	118	118	118
***	5079	5091	14493	1634	1634	4316	443	443	1009	183	183	373	118	118	213
0.05 ***	15361	15523	15999	4376	4429	4473	951	951	951	334	334	334	204	204	204
***	15426	15686	30920	4396	4441	8378	951	951	1637	334	334	521	204	204	275
0.1 ***	33939	34471	37233	9283	9478	9836	1796	1808	1829	561	561	561	301	301	301
***	34134	35218	55917	9369	9629	14553	1808	1829	2590	561	561	744	301	301	368
0.15 ***	50786	51814	58813	13768	14223	15231	2609	2641	2686	768	768	768	378	378	399
***	51143	53386	77939	13963	14613	19984	2621	2653	3423	768	768	938	378	378	447
0.2 ***	64501	66126	79016	17526	18318	20236	3303	3368	3466	951	951	963	464	464	464
***	65053	68726	96984	17851	19033	24671	3336	3421	4137	951	951	1102	464	464	513
0.25 ***	74706	77046	97001	20463	21654	24676	3888	4006	4148	1093	1114	1114	529	529	529
***	75506	80816	113054	20951	22726	28614	3953	4071	4732	1093	1114	1235	529	529	566
0.3 ***	81421	84618	112341	22608	24189	28426	4364	4526	4721	1211	1223	1244	573	573	573
***	82526	89709	126148	23258	25663	31813	4441	4603	5208	1223	1244	1340	573	573	606
0.35 ***	84911	89124	124744	24026	26009	31469	4721	4916	5156	1309	1321	1341	606	606	626
***	86373	95636	136266	24839	27861	34268	4798	5026	5565	1309	1341	1414	606	626	632
0.4 ***	85581	90923	134051	24786	27126	33711	4949	5176	5481	1374	1386	1406	626	638	638
***	87466	98886	143408	25749	29356	35979	5058	5318	5803	1374	1406	1459	638	638	645
0.45 ***	83891	90468	140214	24969	27646	35174	5079	5339	5676	1386	1418	1439	638	638	659
***	86231	99828	147574	26074	30169	36946	5188	5501	5922	1406	1439	1474	638	638	645
0.5 ***	80328	88181	143139	24656	27613	35824	5091	5371	5741	1386	1418	1439	626	638	638
***	83188	98809	148765	25879	30364	37169	5221	5546	5922	1406	1439	1459	638	638	632
0.55 ***	75421	84456	142826	23941	27081	35661	5014	5306	5676	1353	1386	1406	606	626	626
***	78789	96058	146979	25241	29953	36648	5144	5469	5803	1374	1406	1414	606	626	606
0.6 ***	69624	79601	139283	22856	26074	34698	4831	5111	5501	1288	1321	1341	573	573	594
***	73406	91854	142218	24201	28999	35384	4961	5286	5565	1309	1341	1340	573	594	566
0.65 ***	63286	73881	132523	21438	24644	32931	4559	4819	5176	1211	1223	1256	529	529	541
***	67381	86341	134480	22791	27483	33375	4668	4981	5208	1211	1244	1235	529	529	513
0.7 ***	56656	67393	122546	19736	22771	30343	4169	4408	4721	1081	1093	1126	464	464	476
***	60861	79613	123767	21016	25424	30622	4278	4559	4732	1093	1114	1102	464	476	447
0.75 ***	49831	60211	109383	17701	20463	26984	3693	3888	4148	931	951	963	399	399	399
***	53959	71651	110078	18871	22803	27126	3791	4006	4137	931	951	938	399	399	368
0.8 ***	42758	52236	93068	15328	17689	22824	3108	3259	3433	756	756	768	301	301	301
***	46561	62344	93413	16336	19606	22885	3173	3336	3423	756	768	744	301	301	275
0.85 ***	35218	43201	73654	12554	14341	17896	2414	2511	2609	541	541	541	.	.	.
***	38456	51371	73773	13334	15718	17900	2458	2556	2590	541	541	521	.	.	.
0.9 ***	26724	32553	51143	9174	10279	12196	1569	1613	1666	.	.	.	.	.	.
***	29129	38143	51156	9661	11071	12172	1601	1634	1637	.	.	.	.	.	.
0.95 ***	16064	18924	25586	4851	5221	5729	.	.	.	.	.	.	.	.	.
***	17299	21341	25563	5014	5448	5699	.	.	.	.	.	.	.	.	.
0.98 ***	6964	7679	8816	.	.	.	.	.	.	.	.	.	.	.	.
***	7289	8166	8779	.	.	.	.	.	.	.	.	.	.	.	.

TABLE 2: ALPHA= 0.005 POWER= 0.9   EXPECTED ACCRUAL THRU MINIMUM FOLLOW-UP= 14000

		DEL=.01			DEL=.02			DEL=.05			DEL=.10			DEL=.15		
FACT=		1.0 .75	.50 .25	.00 BIN	1.0 .75	.50 .25	.00 BIN	1.0 .75	.50 .25	.00 BIN	1.0 75	.50 .25	.00 BIN	1.0 .75	.50 .25	.00 BIN
PCONT=***					REQUIRED NUMBER OF PATIENTS											
0.01	***	2284	2284	2297	827	827	849	267	267	267	127	127	127	92	92	92
	•••	2284	2284	8779	827	849	2902	267	267	790	127	127	321	92	92	191
0.02	***	5062	5084	5119	1619	1632	1632	442	442	442	184	184	184	127	127	127
	•••	5062	5097	14493	1632	1632	4316	442	442	1009	184	184	373	127	127	213
0.05	***	15374	15549	16004	4384	4432	4489	932	954	954	337	337	337	197	197	197
	•••	15444	15702	30920	4397	4454	8378	954	954	1637	337	337	521	197	197	275
0.1	***	33972	34554	37227	9297	9494	9844	1794	1807	1829	569	569	569	302	302	302
	•••	34182	35302	55917	9389	9647	14553	1807	1829	2590	569	569	744	302	302	368
0.15	***	50864	51962	58809	13812	14267	15234	2599	2647	2682	757	779	779	394	394	394
	•••	51249	53572	77939	14009	14652	19984	2612	2669	3423	757	779	938	394	394	447
0.2	***	64632	66369	79017	17614	18397	20239	3312	3382	3474	954	954	954	464	464	464
	•••	65227	69042	96984	17942	19097	24671	3347	3417	4137	954	954	1102	464	464	513
0.25	***	74887	77394	97007	20589	21757	24662	3907	4012	4152	1094	1107	1107	534	534	534
	•••	75749	81292	113054	21079	22829	28614	3964	4082	4732	1107	1107	1235	534	534	566
0.3	***	81677	85072	112337	22772	24369	28429	4384	4524	4712	1212	1234	1247	582	582	582
	•••	82854	90344	126148	23424	25804	31813	4454	4607	5208	1234	1234	1340	582	582	606
0.35	***	85247	89714	124727	24229	26202	31452	4747	4922	5167	1304	1317	1339	617	617	617
	•••	86809	96434	136266	25047	28044	34268	4817	5027	5565	1317	1339	1414	617	617	632
0.4	***	86017	91674	134059	25034	27379	33714	4979	5202	5482	1374	1387	1409	639	639	639
	•••	88012	99842	143408	25992	29562	35979	5084	5329	5803	1374	1387	1459	639	639	645
0.45	***	84442	91359	140197	25244	27939	35162	5119	5364	5679	1387	1422	1444	639	639	652
	•••	86927	100949	147574	26364	30402	36946	5224	5504	5922	1409	1422	1474	639	652	645
0.5	***	81012	89202	143137	24964	27939	35827	5132	5399	5749	1387	1422	1444	639	639	639
	•••	84022	100039	148765	26202	30612	37169	5259	5552	5922	1409	1422	1459	639	639	632
0.55	***	76239	85597	142822	24264	27414	35674	5049	5329	5679	1352	1387	1409	617	617	617
	•••	79739	97392	146979	25572	30227	36648	5167	5482	5803	1374	1387	1414	617	617	606
0.6	***	70547	80837	139287	23192	26412	34694	4874	5132	5482	1304	1317	1352	582	582	582
	•••	74467	93249	142218	24544	29269	35384	4992	5294	5565	1304	1339	1340	582	582	566
0.65	***	64282	75132	132519	21792	24964	32922	4594	4839	5167	1212	1234	1247	534	534	534
	•••	68504	87732	134480	23122	27742	33375	4699	4992	5208	1212	1234	1235	534	534	513
0.7	***	57702	68657	122544	20064	23074	30354	4209	4432	4712	1094	1107	1129	464	464	477
	•••	62007	80964	123767	21337	25664	30622	4314	4559	4732	1094	1107	1102	464	464	447
0.75	***	50864	61412	109384	18012	20742	26994	3719	3907	4139	932	954	967	394	394	394
	•••	55064	72892	110078	19167	23017	27126	3802	4012	4137	932	954	938	394	394	368
0.8	***	43724	53314	93074	15597	17907	22829	3124	3264	3439	757	757	779	302	302	302
	•••	47574	63407	93413	16599	19762	22885	3194	3347	3423	757	757	744	302	302	275
0.85	***	36037	44087	73649	12749	14499	17894	2424	2507	2612	534	547	547	.	.	.
	•••	39314	52229	73773	13519	15842	17900	2459	2564	2590	547	547	521	.	.	.
0.9	***	27344	33167	51144	9297	10382	12189	1584	1619	1667	.	.	.	.	.	.
	•••	29759	38719	51156	9787	11139	12172	1597	1632	1637	.	.	.	.	.	.
0.95	***	16389	19224	25594	4887	5237	5727	.	.	.	.	.	.	.	.	.
	•••	17592	21547	25563	5062	5469	5699	.	.	.	.	.	.	.	.	.
0.98	***	7044	7744	8807	.	.	.	.	.	.	.	.	.	.	.	.
	•••	7372	8212	8779	.	.	.	.	.	.	.	.	.	.	.	.

369

TABLE 2: ALPHA= 0.005 POWER= 0.9 EXPECTED ACCRUAL THRU MINIMUM FOLLOW-UP= 15000

	DEL=.01			DEL=.02			DEL=.05			DEL=.10			DEL=.15		
FACT=	1.0 .75	.50 .25	.00 BIN	1.0 .75	.50 .25	.00 BIN	1.0 .75	.50 .25	.00 BIN	1.0 75	.50 .25	.00 BIN	1.0 .75	.50 .25	.00 BIN

PCONT=*** REQUIRED NUMBER OF PATIENTS

PCONT	1.0/.75	.50/.25	.00/BIN	1.0/.75	.50/.25	.00/BIN	1.0/.75	.50/.25	.00/BIN	1.0/75	.50/.25	.00/BIN	1.0/.75	.50/.25	.00/BIN
0.01 ***	2274	2297	2297	835	835	835	272	272	272	136	136	136	99	99	99
•••	2274	2297	8779	835	835	2902	272	272	790	136	136	321	99	99	191
0.02 ***	5072	5086	5124	1622	1636	1636	436	436	436	197	197	197	122	122	122
•••	5072	5110	14493	1622	1636	4316	436	436	1009	197	197	373	122	122	213
0.05 ***	15385	15549	15999	4397	4435	4486	947	947	947	324	324	324	197	197	197
•••	15460	15722	30920	4411	4449	8378	947	947	1637	324	324	521	197	197	275
0.1 ***	34022	34622	37224	9324	9511	9849	1810	1810	1824	549	572	572	286	286	286
•••	34247	35372	55917	9399	9661	14553	1810	1824	2590	572	572	744	286	286	368
0.15 ***	50949	52097	58810	13861	14311	15235	2611	2649	2686	760	774	774	385	385	385
•••	51347	53747	77939	14049	14686	19984	2635	2672	3423	774	774	938	385	385	447
0.2 ***	64772	66586	79022	17686	18474	20236	3324	3385	3460	947	947	961	460	460	474
•••	65410	69347	96984	18024	19149	24671	3347	3422	4137	947	961	1102	460	460	513
0.25 ***	75085	77710	96999	20710	21886	24661	3924	4022	4149	1097	1111	1111	535	535	535
•••	75985	81736	113054	21197	22922	28614	3961	4074	4732	1097	1111	1235	535	535	566
0.3 ***	81924	85524	112336	22936	24511	28435	4397	4547	4711	1224	1224	1247	572	586	586
•••	83185	90947	126148	23597	25922	31813	4472	4622	5208	1224	1247	1340	572	586	606
0.35 ***	85585	90286	124735	24422	26410	31449	4772	4936	5161	1299	1322	1336	610	610	624
•••	87249	97186	136266	25247	28186	34268	4847	5049	5565	1322	1336	1414	610	624	632
0.4 ***	86461	92386	134049	25261	27610	33722	5011	5222	5485	1374	1397	1411	624	647	647
•••	88561	100749	143408	26236	29747	35979	5110	5335	5803	1374	1397	1459	624	647	645
0.45 ***	84985	92222	140199	25510	28210	35161	5147	5372	5686	1397	1411	1435	647	647	647
•••	87610	101986	147574	26635	30624	36946	5260	5522	5922	1411	1435	1474	647	647	645
0.5 ***	81685	90197	143147	25261	28210	35822	5161	5410	5747	1397	1411	1449	624	647	647
•••	84835	101199	148765	26499	30849	37169	5274	5574	5922	1411	1435	1459	624	647	632
0.55 ***	77011	86686	142824	24586	27699	35672	5086	5335	5686	1360	1397	1411	610	610	624
•••	80672	98635	146979	25899	30474	36648	5199	5499	5803	1374	1397	1414	610	624	606
0.6 ***	71424	81999	139285	23522	26724	34697	4899	5147	5485	1299	1322	1336	572	586	586
•••	75497	94524	142218	24872	29522	35384	5011	5311	5565	1322	1336	1340	572	586	566
0.65 ***	65260	76336	132511	22111	25261	32935	4622	4861	5161	1210	1224	1247	535	535	535
•••	69586	89035	134480	23447	27985	33375	4735	4997	5208	1224	1247	1235	535	535	513
0.7 ***	58697	69849	122536	20372	23372	30347	4224	4449	4711	1097	1097	1111	460	474	474
•••	63099	82210	123767	21647	25885	30622	4336	4561	4732	1097	1111	1102	474	474	447
0.75 ***	51849	62536	109374	18286	20986	26986	3736	3924	4149	947	947	961	385	385	399
•••	56110	74035	110078	19449	23222	27126	3835	4022	4137	947	961	938	385	399	368
0.8 ***	44611	54310	93085	15849	18122	22824	3136	3272	3436	760	760	774	310	310	310
•••	48511	64397	93413	16824	19922	22885	3211	3361	3423	760	760	744	310	310	275
0.85 ***	36811	44897	73660	12947	14649	17897	2424	2522	2611	535	549	549	.	.	.
•••	40111	52997	73773	13697	15961	17900	2461	2560	2590	535	549	521	.	.	.
0.9 ***	27910	33760	51136	9422	10472	12197	1585	1622	1660	.	.	.	.	.	.
•••	30324	39211	51156	9886	11199	12172	1599	1636	1637	.	.	.	.	.	.
0.95 ***	16674	19472	25585	4936	5274	5724	.	.	.	.	.	.	.	.	.
•••	17874	21760	25563	5086	5485	5699	.	.	.	.	.	.	.	.	.
0.98 ***	7135	7810	8799	.	.	.	.	.	.	.	.	.	.	.	.
•••	7435	8236	8779	.	.	.	.	.	.	.	.	.	.	.	.

TABLE 2: ALPHA= 0.005 POWER= 0.9   EXPECTED ACCRUAL THRU MINIMUM FOLLOW-UP= 17000

	DEL=.01			DEL=.02			DEL=.05			DEL=.10			DEL=.15		
FACT=	1.0 .75	.50 .25	.00 BIN	1.0 .75	.50 .25	.00 BIN	1.0 .75	.50 .25	.00 BIN	1.0 75	.50 .25	.00 BIN	1.0 .75	.50 .25	.00 BIN
PCONT=•••				REQUIRED NUMBER OF PATIENTS											
0.01 ***	2280	2280	2280	835	835	835	282	282	282	139	139	139	96	96	96
•••	2280	2280	8779	835	835	2902	282	282	790	139	139	321	96	96	191
0.02 ***	5069	5085	5127	1626	1626	1626	436	436	436	197	197	197	112	112	112
•••	5085	5111	14493	1626	1626	4316	436	436	1009	197	197	373	112	112	213
0.05 ***	15412	15582	15991	4405	4431	4474	946	946	946	325	325	325	197	197	197
•••	15481	15752	30920	4405	4447	8378	946	946	1637	325	325	521	197	197	275
0.1 ***	34112	34750	37241	9361	9547	9845	1812	1812	1839	564	564	564	309	309	309
•••	34351	35515	55917	9446	9675	14553	1812	1812	2590	564	564	744	309	309	368
0.15 ***	51112	52371	58805	13951	14392	15226	2620	2646	2689	776	776	776	394	394	394
•••	51564	54071	77939	14137	14732	19984	2620	2662	3423	776	776	938	394	394	447
0.2 ***	65010	67034	79019	17835	18626	20241	3342	3385	3470	946	946	962	452	452	479
•••	65732	69897	96984	18159	19264	24671	3369	3427	4137	946	962	1102	452	479	513
0.25 ***	75449	78355	97012	20921	22085	24677	3937	4022	4150	1090	1116	1116	521	537	537
•••	76469	82546	113054	21405	23062	28614	3980	4091	4732	1116	1116	1235	521	537	566
0.3 ***	82435	86387	112339	23216	24789	28444	4431	4559	4702	1217	1244	1244	580	580	580
•••	83837	92040	126148	23870	26149	31813	4490	4644	5208	1217	1244	1340	580	580	606
0.35 ***	86260	91402	124749	24789	26760	31461	4787	4957	5170	1302	1329	1345	606	622	622
•••	88087	98569	136266	25612	28486	34268	4872	5069	5565	1329	1329	1414	622	622	632
0.4 ***	87322	93740	134056	25681	28035	33714	5042	5239	5494	1371	1387	1414	622	649	649
•••	89644	102436	143408	26659	30075	35979	5154	5366	5803	1387	1387	1459	649	649	645
0.45 ***	86047	93825	140192	25995	28672	35175	5196	5409	5680	1414	1414	1430	649	649	649
•••	88921	103924	147574	27126	31010	36946	5297	5536	5922	1414	1430	1474	649	649	645
0.5 ***	82971	92040	143125	25809	28715	35812	5212	5451	5749	1414	1414	1430	622	649	649
•••	86371	103329	148765	27041	31265	37169	5324	5595	5922	1414	1430	1459	649	649	632
0.55 ***	78525	88725	142827	25145	28247	35669	5127	5382	5680	1371	1387	1414	606	622	622
•••	82419	100906	146979	26462	30909	36648	5255	5510	5803	1371	1387	1414	622	622	606
0.6 ***	73127	84161	139284	24109	27270	34691	4957	5196	5494	1302	1329	1345	580	580	580
•••	77404	96869	142218	25442	29947	35384	5069	5324	5565	1329	1329	1340	580	580	566
0.65 ***	67092	78551	132526	22680	25809	32922	4660	4899	5170	1217	1244	1244	521	537	537
•••	71597	91360	134480	24024	28401	33375	4771	5026	5208	1217	1244	1235	537	537	513
0.7 ***	60574	72022	122539	20921	23870	30356	4277	4474	4729	1090	1116	1116	479	479	479
•••	65121	84459	123767	22196	26276	30622	4362	4601	4732	1090	1116	1102	479	479	447
0.75 ***	53662	64585	109380	18796	21431	26972	3767	3937	4150	946	946	962	394	394	394
•••	58040	76102	110078	19944	23556	27126	3852	4049	4137	946	962	938	394	394	368
0.8 ***	46294	56127	93086	16289	18472	22834	3172	3300	3427	750	750	776	309	309	309
•••	50289	66184	93413	17240	20199	22885	3241	3369	3423	750	776	744	309	309	275
0.85 ***	38235	46395	73637	13271	14929	17904	2450	2519	2620	537	537	537	.	.	.
•••	41576	54385	73773	14010	16135	17900	2492	2561	2590	537	537	521	.	.	.
0.9 ***	28954	34792	51139	9632	10636	12182	1600	1626	1669	.	.	.	.	.	.
•••	31392	40105	51156	10084	11290	12172	1600	1642	1637	.	.	.	.	.	.
0.95 ***	17197	19917	25596	5000	5324	5722	.	.	.	.	.	.	.	.	.
•••	18387	22085	25563	5154	5510	5699	.	.	.	.	.	.	.	.	.
0.98 ***	7279	7890	8809	.	.	.	.	.	.	.	.	.	.	.	.
•••	7550	8299	8779	.	.	.	.	.	.	.	.	.	.	.	.

TABLE 2: ALPHA= 0.005 POWER= 0.9   EXPECTED ACCRUAL THRU MINIMUM FOLLOW-UP= 20000

		DEL=.01			DEL=.02			DEL=.05			DEL=.10			DEL=.15	
FACT=		1.0 .75	.50 .25	.00 BIN	1.0 .75	.50 .25	.00 BIN	1.0 .75	.50 .25	.00 BIN	1.0 75	.50 .25	.00 BIN	1.0 .75	.50 .25	.00 BIN
PCONT=•••							REQUIRED NUMBER OF PATIENTS								
0.01	•••	2282	2282	2282	832	832	832	282	282	282	132	132	132	82	82	82
	•••	2282	2282	8779	832	832	2902	282	282	790	132	132	321	82	82	191
0.02	•••	5082	5082	5113	1632	1632	1632	432	432	432	182	182	182	132	132	132
	•••	5082	5113	14493	1632	1632	4316	432	432	1009	182	182	373	132	132	213
0.05	•••	15463	15632	15982	4413	4432	4482	932	932	963	332	332	332	182	182	182
	•••	15532	15782	30920	4432	4463	8378	932	932	1637	332	332	521	182	182	275
0.1	•••	34232	34932	37232	9413	9582	9832	1813	1813	1832	563	563	563	282	282	282
	•••	34513	35682	55917	9482	9713	14553	1813	1832	2590	563	563	744	282	282	368
0.15	•••	51363	52732	58813	14063	14482	15232	2632	2663	2682	763	763	782	382	382	382
	•••	51863	54482	77939	14232	14813	19984	2632	2663	3423	763	763	938	382	382	447
0.2	•••	65413	67632	79013	18013	18782	20232	3363	3413	3463	963	963	963	463	463	463
	•••	66213	70613	96984	18363	19382	24671	3382	3432	4137	963	963	1102	463	463	513
0.25	•••	75982	79232	97013	21182	22332	24663	3963	4063	4132	1113	1113	1113	532	532	532
	•••	77163	83632	113054	21682	23263	28614	4013	4082	4732	1113	1113	1235	532	532	566
0.3	•••	83182	87582	112332	23582	25132	28432	4463	4582	4713	1232	1232	1232	582	582	582
	•••	84782	93482	126148	24232	26413	31813	4513	4632	5208	1232	1232	1340	582	582	606
0.35	•••	87232	92932	124732	25232	27182	31463	4832	4982	5163	1313	1332	1332	613	613	613
	•••	89313	100413	136266	26082	28813	34268	4913	5063	5565	1332	1332	1414	613	613	632
0.4	•••	88563	95632	134063	26232	28563	33713	5113	5282	5482	1382	1382	1413	632	632	632
	•••	91182	104613	143408	27213	30482	35979	5182	5382	5803	1382	1413	1459	632	632	645
0.45	•••	87613	96032	140213	26632	29263	35163	5263	5432	5682	1413	1432	1432	632	632	663
	•••	90763	106432	147574	27763	31463	36946	5332	5563	5922	1413	1432	1474	632	632	645
0.5	•••	84832	94532	143132	26482	29382	35813	5282	5482	5732	1413	1432	1432	632	632	632
	•••	88532	106082	148765	27732	31763	37169	5382	5613	5922	1413	1432	1459	632	632	632
0.55	•••	80663	91463	142832	25882	28932	35663	5213	5413	5682	1382	1382	1413	613	613	613
	•••	84832	103863	146979	27182	31432	36648	5313	5532	5803	1382	1413	1414	613	613	606
0.6	•••	75482	87032	139282	24863	27963	34713	5013	5232	5482	1313	1332	1332	582	582	582
	•••	80013	99913	142218	26182	30482	35384	5113	5363	5565	1313	1332	1340	582	582	566
0.65	•••	69582	81482	132513	23432	26463	32932	4732	4932	5163	1213	1232	1263	532	532	532
	•••	74313	94382	134480	24763	28913	33375	4813	5032	5208	1232	1232	1235	532	532	513
0.7	•••	63113	74913	122532	21632	24482	30363	4332	4513	4713	1082	1113	1113	463	463	463
	•••	67832	87332	123767	22882	26732	30622	4413	4613	4732	1113	1113	1102	463	463	447
0.75	•••	56113	67282	109382	19463	21982	26982	3832	3963	4132	932	963	963	382	382	382
	•••	60632	78732	110078	20563	23963	27126	3882	4063	4137	932	963	938	382	382	368
0.8	•••	48513	58513	93082	16832	18932	22832	3213	3313	3432	763	763	763	313	313	313
	•••	52613	68432	93413	17763	20513	22885	3263	3382	3423	763	763	744	313	313	275
0.85	•••	40113	48313	73663	13682	15263	17882	2463	2532	2613	532	532	563	.	.	.
	•••	43513	56113	73773	14382	16363	17900	2513	2582	2590	532	532	521	.	.	.
0.9	•••	30332	36113	51132	9882	10813	12182	1613	1632	1663	.	.	.	.	.	.
	•••	32763	41213	51156	10313	11413	12172	1613	1632	1637	.	.	.	.	.	.
0.95	•••	17882	20482	25582	5082	5382	5732	.	.	.	.	.	.	.	.	.
	•••	19032	22513	25563	5232	5532	5699	.	.	.	.	.	.	.	.	.
0.98	•••	7432	8013	8813	.	.	.	.	.	.	.	.	.	.	.	.
	•••	7713	8363	8779	.	.	.	.	.	.	.	.	.	.	.	.

TABLE 2: ALPHA= 0.005　POWER= 0.9　　EXPECTED ACCRUAL THRU MINIMUM FOLLOW-UP= 25000

		DEL=.01			DEL=.02			DEL=.05			DEL=.10			DEL=.15		
FACT=		1.0 .75	.50 .25	.00 BIN	1.0 .75	.50 .25	.00 BIN	1.0 .75	.50 .25	.00 BIN	1.0 75	.50 .25	.00 BIN	1.0 .75	.50 .25	.00 BIN
PCONT=***				REQUIRED NUMBER OF PATIENTS												
0.01	***	2290	2290	2290	829	829	829	266	266	266	141	141	141	102	102	102
	•••	2290	2290	8779	829	829	2902	266	266	790	141	141	321	102	102	191
0.02	***	5079	5102	5102	1641	1641	1641	454	454	454	204	204	204	141	141	141
	•••	5079	5102	14493	1641	1641	4316	454	454	1009	204	204	373	141	141	213
0.05	***	15516	15665	15977	4415	4454	4477	954	954	954	329	329	329	204	204	204
	•••	15579	15829	30920	4415	4454	8378	954	954	1637	329	329	521	204	204	275
0.1	***	34454	35165	37227	9477	9641	9852	1829	1829	1829	579	579	579	290	290	290
	•••	34727	35915	55917	9540	9727	14553	1829	1829	2590	579	579	744	290	290	368
0.15	***	51727	53290	58829	14204	14602	15227	2641	2665	2665	766	766	766	391	391	391
	•••	52329	55040	77939	14391	14891	19984	2641	2665	3423	766	766	938	391	391	447
0.2	***	66016	68540	79016	18266	18977	20227	3391	3415	3454	954	954	954	454	454	477
	•••	66954	71602	96984	18602	19516	24671	3391	3454	4137	954	954	1102	454	454	513
0.25	***	76891	80579	97016	21579	22665	24665	4016	4079	4141	1102	1102	1102	516	540	540
	•••	78227	85102	113054	22040	23477	28614	4040	4102	4732	1102	1102	1235	540	540	566
0.3	***	84391	89391	112329	24102	25579	28415	4516	4602	4704	1227	1227	1227	579	579	579
	•••	86227	95477	126148	24727	26727	31813	4540	4665	5208	1227	1227	1340	579	579	606
0.35	***	88829	95204	124727	25891	27766	31454	4891	5016	5165	1329	1329	1352	602	602	602
	•••	91204	102915	136266	26704	29227	34268	4954	5079	5565	1329	1329	1414	602	602	632
0.4	***	90540	98391	134040	27016	29227	33704	5165	5329	5477	1391	1391	1415	641	641	641
	•••	93516	107641	143408	27954	30977	35979	5227	5391	5803	1391	1391	1459	641	641	645
0.45	***	90016	99266	140204	27516	30040	35165	5329	5477	5665	1415	1415	1454	641	641	641
	•••	93579	109852	147574	28602	32040	36946	5391	5579	5922	1415	1415	1474	641	641	645
0.5	***	87665	98141	143141	27454	30204	35829	5352	5540	5727	1415	1415	1454	641	641	641
	•••	91727	109891	148765	28641	32391	37169	5454	5641	5922	1415	1415	1459	641	641	632
0.55	***	83852	95352	142829	26891	29790	35665	5290	5454	5665	1391	1391	1415	602	602	602
	•••	88391	107891	146979	28165	32079	36648	5352	5579	5803	1391	1391	1414	602	602	606
0.6	***	78977	91141	139290	25891	28829	34704	5102	5266	5477	1329	1329	1352	579	579	579
	•••	83790	104040	142218	27165	31141	35384	5165	5391	5565	1329	1329	1340	579	579	566
0.65	***	73204	85602	132516	24454	27329	32915	4790	4977	5165	1227	1227	1266	540	540	540
	•••	78204	98477	124480	25727	29540	33375	4891	5079	5208	1227	1227	1235	540	540	513
0.7	***	66727	78915	122540	22602	25290	30352	4391	4540	4727	1102	1102	1102	477	477	477
	•••	71665	91204	123767	23790	27329	30622	4477	4641	4732	1102	1102	1102	477	477	447
0.75	***	59602	71016	109391	20329	22704	26977	3891	4016	4141	954	954	954	391	391	391
	•••	64266	82227	110078	21352	24454	27126	3954	4079	4137	954	954	938	391	391	368
0.8	***	51665	61766	93079	17540	19516	22829	3266	3329	3454	766	766	766	290	290	290
	•••	55829	71391	93413	18415	20891	22885	3290	3391	3423	766	766	744	290	290	275
0.85	***	42727	50915	73641	14227	15665	17891	2477	2540	2602	540	540	540	.	.	.
	•••	46165	58391	73773	14891	16641	17900	2516	2579	2590	540	540	521	.	.	.
0.9	***	32227	37852	51141	10227	11040	12204	1602	1641	1665	.	.	.	.	.	.
	•••	34602	42641	51156	10602	11540	12172	1641	1641	1637	.	.	.	.	.	.
0.95	***	18766	21227	25579	5204	5454	5727	.	.	.	.	.	.	.	.	.
	•••	19852	23016	25563	5329	5579	5699	.	.	.	.	.	.	.	.	.
0.98	***	7641	8141	8790	.	.	.	.	.	.	.	.	.	.	.	.
	•••	7891	8454	8779	.	.	.	.	.	.	.	.	.	.	.	.

TABLE 3: ALPHA= 0.01 POWER= 0.8   EXPECTED ACCRUAL THRU MINIMUM FOLLOW-UP= 30

		DEL=.10			DEL=.15			DEL=.20			DEL=.25			DEL=.30	
FACT=		1.0 .75	.50 .25	.00 BIN	1.0 .75	.50 .25	.00 BIN	1.0 .75	.50 .25	.00 BIN	1.0 75	.50 .25	.00 BIN	1.0 .75	.50 .25	.00 BIN
PCONT=•••					REQUIRED	NUMBER OF	PATIENTS									
0.05	•••	211	212	228	123	124	133	86	88	94	67	68	72	55	56	60
	•••	211	214	352	123	126	186	87	89	118	68	70	83	56	57	62
0.1	•••	332	334	382	176	178	203	115	118	133	85	87	98	68	70	77
	•••	333	339	502	176	182	248	116	121	151	86	90	102	68	72	74
0.15	•••	429	432	523	215	219	264	136	140	166	98	101	118	75	79	90
	•••	430	439	633	217	226	302	138	146	179	99	106	119	77	82	84
0.2	•••	499	504	649	243	248	317	150	156	194	105	110	135	80	85	101
	•••	500	514	743	245	258	346	152	164	201	107	117	131	82	89	92
0.25	•••	544	551	755	259	267	360	158	165	217	109	116	148	83	88	109
	•••	546	565	833	262	280	382	160	176	219	112	124	141	85	94	98
0.3	•••	565	575	842	266	276	395	160	170	234	111	119	157	83	90	114
	•••	568	594	904	269	293	409	164	183	231	114	129	147	86	97	101
0.35	•••	566	579	908	264	277	420	158	170	245	109	119	163	82	89	117
	•••	570	603	954	268	299	426	163	186	239	113	130	151	85	98	102
0.4	•••	549	565	952	255	271	435	153	167	251	105	117	165	79	88	117
	•••	554	597	984	261	297	435	158	185	241	110	129	151	82	96	101
0.45	•••	517	538	974	241	261	440	145	161	251	100	113	163	75	84	115
	•••	524	577	994	248	290	435	151	181	239	105	126	147	79	94	98
0.5	•••	475	501	975	223	246	435	136	154	246	94	107	158	71	80	110
	•••	484	548	984	231	279	426	143	175	231	99	121	141	74	89	92
0.55	•••	425	457	953	203	229	421	125	144	235	87	100	150	65	75	103
	•••	436	512	954	212	264	409	132	165	219	92	114	131	69	84	84
0.6	•••	371	411	910	181	210	397	113	132	219	79	92	138	59	68	94
	•••	385	471	904	192	245	382	121	153	201	84	105	119	63	76	74
0.65	•••	318	363	845	160	189	363	101	119	197	71	83	122	52	60	82
	•••	334	426	833	171	224	346	108	139	179	75	94	102	56	68	62
0.7	•••	268	316	758	139	167	320	88	105	171	61	72	103	.	.	.
	•••	286	379	743	150	199	302	95	122	151	65	81	83	.	.	.
0.75	•••	222	269	650	117	143	267	74	88	138	.	.	.	.	.	.
	•••	240	329	633	127	171	248	80	103	118	.	.	.	.	.	.
0.8	•••	178	222	521	95	116	205	.	.	.	.	.	.	.	.	.
	•••	195	274	502	103	139	186	.	.	.	.	.	.	.	.	.
0.85	•••	136	171	372	.	.	.	.	.	.	.	.	.	.	.	.
	•••	150	211	352	.	.	.	.	.	.	.	.	.	.	.	.

TABLE 3: ALPHA= 0.01 POWER= 0.8   EXPECTED ACCRUAL THRU MINIMUM FOLLOW-UP= 40

	DEL=.10			DEL=.15			DEL=.20			DEL=.25			DEL=.30		
FACT=	1.0 .75	.50 .25	.00 BIN	1.0 .75	.50 .25	.00 BIN	1.0 .75	.50 .25	.00 BIN	1.0 75	.50 .25	.00 BIN	1.0 .75	.50 .25	.00 BIN
PCONT=***				REQUIRED NUMBER OF PATIENTS											
0.05 ***	211	212	228	123	124	133	87	88	94	68	69	73	56	57	60
•••	211	215	352	123	127	186	87	90	118	68	70	83	56	58	62
0.1 ***	333	336	382	176	180	203	116	119	133	86	89	98	68	70	77
•••	334	341	502	177	184	248	117	123	151	87	91	102	69	73	74
0.15 ***	430	434	523	217	222	264	138	142	166	99	103	118	77	80	90
•••	431	443	633	218	230	302	139	148	179	100	108	119	78	83	84
0.2 ***	501	507	649	245	252	317	152	159	194	107	113	135	82	87	101
•••	503	521	743	247	264	346	154	168	201	109	119	131	84	91	92
0.25 ***	546	556	755	262	272	360	160	169	217	112	119	148	85	91	109
•••	549	574	833	265	288	382	164	181	219	115	128	141	87	97	98
0.3 ***	568	581	842	269	282	395	164	175	234	114	123	157	86	93	114
•••	573	606	904	274	303	409	168	189	231	117	133	147	88	100	101
0.35 ***	570	587	908	268	285	420	163	176	245	113	124	163	85	93	117
•••	576	619	954	274	310	426	168	193	239	117	135	151	88	101	102
0.4 ***	554	576	952	261	281	435	158	174	251	110	122	165	82	91	117
•••	562	616	984	268	310	435	164	193	241	115	134	151	86	100	101
0.45 ***	524	552	974	248	272	440	151	169	251	105	118	163	79	88	115
•••	533	600	994	257	304	435	158	190	239	111	132	147	83	97	98
0.5 ***	484	518	975	231	259	435	143	162	246	99	113	158	74	84	110
•••	495	574	984	242	294	426	150	184	231	105	127	141	78	93	92
0.55 ***	436	477	953	212	243	421	132	152	235	92	106	150	69	78	103
•••	450	540	954	224	280	409	140	174	219	98	119	131	73	87	84
0.6 ***	385	433	910	192	224	397	121	141	219	84	98	138	63	72	94
•••	402	501	904	205	261	382	129	162	201	90	110	119	67	79	74
0.65 ***	334	387	845	171	203	363	108	128	197	75	88	122	56	63	82
•••	354	457	833	184	239	346	116	147	179	81	99	102	59	70	62
0.7 ***	286	340	758	150	180	320	95	112	171	65	76	103	.	.	.
•••	307	409	743	162	213	302	102	129	151	70	85	83	.	.	.
0.75 ***	240	293	650	127	154	267	80	94	138	.	.	.	.	.	.
•••	260	356	633	138	183	248	86	108	118	.	.	.	.	.	.
0.8 ***	195	243	521	103	125	205	.	.	.	.	.	.	.	.	.
•••	214	297	502	112	148	186	.	.	.	.	.	.	.	.	.
0.85 ***	150	188	372	.	.	.	.	.	.	.	.	.	.	.	.
•••	165	229	352	.	.	.	.	.	.	.	.	.	.	.	.

TABLE 3: ALPHA= 0.01 POWER= 0.8   EXPECTED ACCRUAL THRU MINIMUM FOLLOW-UP= 50

	DEL=.10			DEL=.15			DEL=.20			DEL=.25			DEL=.30		
FACT=	1.0 .75	.50 .25	.00 BIN	1.0 .75	.50 .25	.00 BIN	1.0 .75	.50 .25	.00 BIN	1.0 75	.50 .25	.00 BIN	1.0 .75	.50 .25	.00 BIN

PCONT=***   REQUIRED NUMBER OF PATIENTS

PCONT	1.0/.75	.50/.25	.00/BIN	1.0/.75	.50/.25	.00/BIN	1.0/.75	.50/.25	.00/BIN	1.0/75	.50/.25	.00/BIN	1.0/.75	.50/.25	.00/BIN
0.05 ***	211	213	228	123	125	133	87	89	94	68	69	73	56	57	60
***	212	216	352	124	127	186	88	90	118	68	70	83	57	58	62
0.1 ***	334	337	382	177	181	203	117	120	133	87	90	98	69	71	77
***	335	344	502	179	186	248	118	124	151	88	92	102	70	73	74
0.15 ***	431	437	523	218	224	264	139	144	166	100	104	118	78	81	90
***	433	448	633	220	233	302	141	151	179	102	109	119	79	84	84
0.2 ***	502	511	649	247	255	317	154	161	194	109	115	135	83	88	101
***	505	527	743	250	268	346	157	171	201	111	121	131	85	93	92
0.25 ***	548	560	755	264	276	360	163	173	217	114	122	148	87	93	109
***	552	583	833	269	294	382	167	185	219	117	130	141	89	98	98
0.3 ***	572	588	842	273	288	395	167	179	234	116	126	157	88	95	114
***	577	617	904	278	310	409	172	194	231	120	136	147	91	102	101
0.35 ***	574	595	908	273	292	420	167	182	245	116	127	163	87	95	117
***	581	633	954	280	319	426	172	199	239	121	138	151	91	103	102
0.4 ***	560	586	952	266	290	435	163	180	251	114	126	165	85	94	117
***	569	633	984	275	320	435	170	200	241	119	138	151	89	102	101
0.45 ***	531	565	974	255	282	440	157	176	251	109	123	163	82	91	115
***	543	620	994	265	316	435	164	197	239	115	136	147	86	100	98
0.5 ***	492	533	975	239	270	435	148	169	246	104	118	158	78	87	110
***	507	596	984	251	306	426	157	190	231	109	131	141	82	95	92
0.55 ***	447	495	953	221	254	421	138	159	235	97	111	150	72	81	103
***	464	565	954	234	292	409	147	181	219	102	123	131	76	89	84
0.6 ***	398	453	910	202	236	397	127	148	219	89	102	138	66	74	94
***	418	526	904	215	274	382	135	169	201	94	114	119	69	81	74
0.65 ***	349	408	845	181	214	363	114	134	197	79	91	122	58	66	82
***	371	483	833	194	251	346	122	153	179	84	102	102	61	72	62
0.7 ***	302	361	758	159	191	320	100	118	171	69	79	103	.	.	.
***	325	433	743	172	224	302	108	135	151	73	88	83	.	.	.
0.75 ***	255	312	650	136	164	267	84	99	138	.	.	.	.	.	.
***	278	378	633	147	192	248	91	112	118	.	.	.	.	.	.
0.8 ***	210	260	521	110	133	205	.	.	.	.	.	.	.	.	.
***	229	316	502	119	154	186	.	.	.	.	.	.	.	.	.
0.85 ***	162	201	372	.	.	.	.	.	.	.	.	.	.	.	.
***	177	242	352	.	.	.	.	.	.	.	.	.	.	.	.

TABLE 3: ALPHA= 0.01 POWER= 0.8   EXPECTED ACCRUAL THRU MINIMUM FOLLOW-UP= 60

		DEL=.10			DEL=.15			DEL=.20			DEL=.25			DEL=.30	
FACT=		1.0 .75	.50 .25	.00 BIN	1.0 .75	.50 .25	.00 BIN	1.0 .75	.50 .25	.00 BIN	1.0 75	.50 .25	.00 BIN	1.0 .75	.50 .25	.00 BIN
PCONT=•••				REQUIRED NUMBER OF PATIENTS												
0.05	***	212	214	228	124	126	133	88	89	94	68	70	73	56	57	60
	•••	212	217	352	124	128	186	88	91	118	69	71	83	57	58	62
0.1	***	334	339	382	178	182	203	118	121	133	88	90	98	70	72	77
	•••	336	346	502	179	188	248	119	125	151	89	93	102	70	74	74
0.15	***	432	439	523	219	226	264	140	146	166	101	106	118	79	82	90
	•••	434	451	633	222	235	302	142	152	179	103	110	119	80	85	84
0.2	***	504	514	649	248	258	317	155	164	194	110	117	135	85	89	101
	•••	507	533	743	252	272	346	159	173	201	113	123	131	87	94	92
0.25	***	551	565	755	267	280	360	165	176	217	116	124	148	88	94	109
	•••	556	591	833	272	299	382	169	188	219	119	132	141	91	100	98
0.3	***	575	594	842	276	293	395	170	183	234	119	129	157	90	97	114
	•••	581	627	904	282	317	409	175	198	231	123	138	147	93	103	101
0.35	***	579	603	908	277	299	420	170	186	245	119	130	163	89	98	117
	•••	587	646	954	285	326	426	176	203	239	124	141	151	93	105	102
0.4	***	565	597	952	271	297	435	167	185	251	117	129	165	88	96	117
	•••	576	648	984	281	329	435	174	205	241	122	141	151	91	104	101
0.45	***	538	577	974	261	290	440	161	181	251	113	126	163	84	94	115
	•••	552	637	994	272	325	435	169	202	239	118	139	147	88	102	98
0.5	***	501	548	975	246	279	435	154	175	246	107	121	158	80	89	110
	•••	518	616	984	259	316	426	162	196	231	113	134	141	84	97	92
0.55	***	457	512	953	229	264	421	144	165	235	100	114	150	75	84	103
	•••	477	586	954	243	302	409	152	186	219	106	126	131	78	91	84
0.6	***	411	471	910	210	245	397	132	153	219	92	105	138	68	76	94
	•••	433	548	904	224	284	382	141	174	201	98	117	119	72	83	74
0.65	***	363	426	845	189	224	363	119	139	197	83	94	122	60	68	82
	•••	387	504	833	203	260	346	128	158	179	88	104	102	63	73	62
0.7	***	316	379	758	167	199	320	105	122	171	72	82	103	.	.	.
	•••	340	454	743	180	232	302	112	138	151	76	90	83	.	.	.
0.75	***	269	329	650	143	171	267	88	103	138	.	.	.	.	.	.
	•••	293	397	633	154	199	248	94	115	118	.	.	.	.	.	.
0.8	***	222	274	521	116	139	205	.	.	.	.	.	.	.	.	.
	•••	243	331	502	125	160	186	.	.	.	.	.	.	.	.	.
0.85	***	171	211	372	.	.	.	.	.	.	.	.	.	.	.	.
	•••	188	253	352	.	.	.	.	.	.	.	.	.	.	.	.

TABLE 3: ALPHA= 0.01 POWER= 0.8   EXPECTED ACCRUAL THRU MINIMUM FOLLOW-UP= 70

		DEL=.10			DEL=.15			DEL=.20			DEL=.25			DEL=.30	
FACT=	1.0 .75	.50 .25	.00 BIN	1.0 .75	.50 .25	.00 BIN	1.0 .75	.50 .25	.00 BIN	1.0 75	.50 .25	.00 BIN	1.0 .75	.50 .25	.00 BIN	
PCONT=***				REQUIRED NUMBER OF PATIENTS												
0.05 ***	212	214	228	124	126	133	88	89	94	69	70	72	57	58	60
•••	213	218	352	125	128	186	89	91	118	69	71	83	57	58	62
0.1 ***	335	340	382	179	183	203	119	122	133	88	91	97	70	72	77
•••	337	348	502	180	189	248	120	126	151	89	93	102	71	74	74
0.15 ***	433	441	523	220	228	264	141	147	166	102	107	118	79	83	90
•••	436	455	633	223	238	302	144	153	179	104	111	119	81	86	84
0.2 ***	506	517	649	250	261	317	157	166	194	112	118	135	86	90	101
•••	510	538	743	254	275	346	160	175	201	114	124	131	88	94	92
0.25 ***	553	570	755	269	284	360	167	179	217	118	126	148	90	96	109
•••	559	598	833	275	303	382	172	190	219	121	134	141	92	101	98
0.3 ***	578	600	842	279	298	395	172	186	234	121	131	157	91	98	114
•••	585	637	904	286	322	409	178	201	231	125	140	147	94	104	101
0.35 ***	583	611	908	281	304	420	173	190	245	121	133	163	91	99	117
•••	592	657	954	290	333	426	180	207	239	126	143	151	95	106	102
0.4 ***	571	606	952	276	304	435	171	190	251	119	132	165	90	98	117
•••	583	662	984	287	336	435	179	208	241	125	143	151	93	105	101
0.45 ***	545	589	974	267	298	440	166	186	251	116	129	163	87	96	115
•••	560	653	994	279	333	435	174	206	239	121	141	147	90	103	98
0.5 ***	509	561	975	253	287	435	158	179	246	110	124	158	82	91	110
•••	528	633	984	266	325	426	167	200	231	116	136	141	86	99	92
0.55 ***	467	526	953	236	272	421	148	170	235	103	117	150	77	85	103
•••	489	604	954	250	311	409	157	191	219	109	129	131	80	92	84
0.6 ***	422	487	910	217	254	397	137	158	219	95	108	138	70	78	94
•••	446	567	904	232	292	382	146	178	201	101	119	119	73	84	74
0.65 ***	376	443	845	197	232	363	124	144	197	85	97	122	62	69	82
•••	401	523	833	211	268	346	132	162	179	90	106	102	65	74	62
0.7 ***	329	395	758	174	207	320	109	126	171	74	83	103	.	.	.
•••	355	471	743	187	239	302	116	142	151	78	91	83	.	.	.
0.75 ***	282	343	650	149	178	267	92	106	138	.	.	.	.	.	.
•••	306	412	633	161	205	248	98	118	118	.	.	.	.	.	.
0.8 ***	233	286	521	121	143	205	.	.	.	.	.	.	.	.	.
•••	254	344	502	130	164	186	.	.	.	.	.	.	.	.	.
0.85 ***	180	221	372	.	.	.	.	.	.	.	.	.	.	.	.
•••	197	262	352	.	.	.	.	.	.	.	.	.	.	.	.

TABLE 3: ALPHA= 0.01 POWER= 0.8   EXPECTED ACCRUAL THRU MINIMUM FOLLOW-UP= 80

	DEL=.10			DEL=.15			DEL=.20			DEL=.25			DEL=.30		
FACT=	1.0 .75	.50 .25	.00 BIN	1.0 .75	.50 .25	.00 BIN	1.0 .75	.50 .25	.00 BIN	1.0 75	.50 .25	.00 BIN	1.0 .75	.50 .25	.00 BIN
PCONT=•••	REQUIRED NUMBER OF PATIENTS														
0.05 ***	212	215	228	124	127	133	88	90	94	69	70	73	57	58	60
•••	213	218	352	125	129	186	89	91	118	69	71	83	57	59	62
0.1 ***	336	341	382	180	184	203	119	123	133	89	91	98	70	73	77
•••	338	350	502	181	190	248	121	127	151	90	94	102	71	74	74
0.15 ***	434	443	523	222	230	264	142	148	166	103	108	118	80	83	90
•••	437	458	633	225	239	302	145	155	179	105	112	119	82	86	84
0.2 ***	507	521	649	252	264	317	159	168	194	113	119	135	87	91	101
•••	512	543	743	256	278	346	162	176	201	116	125	131	89	95	92
0.25 ***	556	574	755	272	288	360	169	181	217	119	128	148	91	97	109
•••	562	605	833	278	307	382	174	193	219	123	135	141	93	101	98
0.3 ***	581	606	842	282	303	395	175	189	234	123	133	157	93	100	114
•••	590	646	904	290	327	409	181	204	231	127	142	147	96	105	101
0.35 ***	587	619	908	285	310	420	176	193	245	124	135	163	93	101	117
•••	598	668	954	294	338	426	183	210	239	128	145	151	96	107	102
0.4 ***	576	616	952	281	310	435	174	193	251	122	134	165	91	100	117
•••	590	675	984	292	343	435	182	212	241	127	145	151	95	107	101
0.45 ***	552	600	974	272	304	440	169	190	251	118	132	163	88	97	115
•••	569	667	994	285	340	435	178	210	239	124	143	147	92	104	98
0.5 ***	518	574	975	259	294	435	162	184	246	113	127	158	84	93	110
•••	538	649	984	273	332	426	171	204	231	119	138	141	88	100	92
0.55 ***	477	540	953	243	280	421	152	174	235	106	119	150	78	87	103
•••	501	620	954	257	318	409	162	194	219	112	131	131	82	93	84
0.6 ***	433	501	910	224	261	397	141	162	219	98	110	138	72	79	94
•••	459	584	904	239	299	382	150	181	201	103	121	119	75	85	74
0.65 ***	387	457	845	203	239	363	128	147	197	88	99	122	63	70	82
•••	414	539	833	218	275	346	136	165	179	93	108	102	66	75	62
0.7 ***	340	409	758	180	213	320	112	129	171	76	85	103	.	.	.
•••	367	487	743	194	245	302	120	144	151	80	92	83	.	.	.
0.75 ***	293	356	650	154	183	267	94	108	138	.	.	.	.	.	.
•••	318	426	633	166	210	248	100	119	118	.	.	.	.	.	.
0.8 ***	243	297	521	125	148	205	.	.	.	.	.	.	.	.	.
•••	265	355	502	135	167	186	.	.	.	.	.	.	.	.	.
0.85 ***	188	229	372	.	.	.	.	.	.	.	.	.	.	.	.
•••	204	270	352	.	.	.	.	.	.	.	.	.	.	.	.

TABLE 3: ALPHA= 0.01 POWER= 0.8   EXPECTED ACCRUAL THRU MINIMUM FOLLOW-UP= 90

		DEL=.10			DEL=.15			DEL=.20			DEL=.25			DEL=.30		
FACT=		1.0 .75	.50 .25	.00 BIN	1.0 .75	.50 .25	.00 BIN	1.0 .75	.50 .25	.00 BIN	1.0 75	.50 .25	.00 BIN	1.0 .75	.50 .25	.00 BIN
PCONT=***		REQUIRED NUMBER OF PATIENTS														
0.05	***	213	215	228	125	127	133	88	90	94	69	70	73	57	58	60
	***	214	219	352	126	129	186	89	91	118	70	71	83	57	59	62
0.1	***	336	343	382	180	185	203	120	124	133	89	92	97	71	73	77
	***	339	352	502	182	191	248	121	127	151	90	94	102	72	74	74
0.15	***	435	445	523	223	231	264	143	150	166	104	108	118	81	84	90
	***	439	461	633	226	241	302	146	155	179	106	112	119	82	87	84
0.2	***	509	524	649	254	266	317	160	169	194	114	120	135	87	92	101
	***	514	547	743	258	281	346	164	178	201	117	126	131	89	96	92
0.25	***	558	578	755	274	291	360	171	183	217	121	129	148	92	98	109
	***	565	611	833	280	310	382	176	194	219	124	136	141	94	102	98
0.3	***	584	612	842	285	307	395	177	192	234	124	134	157	94	101	114
	***	594	654	904	293	331	409	183	206	231	129	143	147	97	106	101
0.35	***	591	626	908	289	314	420	179	196	245	125	137	163	94	102	117
	***	603	678	954	299	343	426	186	213	239	130	146	151	98	108	102
0.4	***	581	624	952	285	315	435	178	197	251	124	136	165	93	101	117
	***	597	686	984	297	348	435	185	215	241	129	147	151	96	108	101
0.45	***	558	610	974	277	310	440	173	194	251	121	134	163	90	99	115
	***	577	680	994	290	346	435	181	213	239	126	145	147	94	105	98
0.5	***	525	585	975	264	301	435	166	187	246	115	129	158	86	94	110
	***	548	663	984	279	338	426	175	207	231	121	140	141	89	101	92
0.55	***	486	553	953	249	286	421	156	178	235	108	122	150	80	88	103
	***	511	635	954	264	325	409	165	197	219	114	132	131	84	94	84
0.6	***	443	514	910	230	268	397	145	166	219	100	112	138	73	81	94
	***	471	598	904	245	305	382	153	184	201	105	122	119	76	86	74
0.65	***	398	470	845	209	245	363	131	150	198	90	101	122	65	71	82
	***	426	553	833	224	281	346	139	167	179	94	109	102	67	75	62
0.7	***	351	422	758	186	219	320	115	132	171	78	87	103	.	.	.
	***	379	500	743	199	250	302	122	146	151	82	93	83	.	.	.
0.75	***	303	368	650	159	188	267	97	110	138	.	.	.	.	.	.
	***	329	438	633	171	214	248	103	121	118	.	.	.	.	.	.
0.8	***	252	307	521	129	151	205	.	.	.	.	.	.	.	.	.
	***	274	364	502	138	170	186	.	.	.	.	.	.	.	.	.
0.85	***	194	236	372	.	.	.	.	.	.	.	.	.	.	.	.
	***	211	276	352	.	.	.	.	.	.	.	.	.	.	.	.

TABLE 3: ALPHA= 0.01 POWER= 0.8   EXPECTED ACCRUAL THRU MINIMUM FOLLOW-UP= 100

		DEL=.10			DEL=.15			DEL=.20			DEL=.25			DEL=.30		
FACT=		1.0 .75	.50 .25	.00 BIN	1.0 .75	.50 .25	.00 BIN	1.0 .75	.50 .25	.00 BIN	1.0 75	.50 .25	.00 BIN	1.0 .75	.50 .25	.00 BIN
PCONT=***					REQUIRED NUMBER OF PATIENTS											
0.05	***	213	216	228	125	127	133	89	90	94	69	70	73	57	58	60
	•••	214	220	352	126	129	186	89	92	118	70	71	83	58	59	62
0.1	***	337	344	382	181	186	203	120	124	133	90	92	98	71	73	77
	•••	340	353	502	183	192	248	122	128	151	91	94	102	72	75	74
0.15	***	437	448	523	224	233	264	144	151	166	104	109	118	81	84	90
	•••	440	464	633	227	243	302	147	156	179	106	113	119	83	87	84
0.2	***	511	527	649	255	268	317	161	171	194	115	121	135	88	93	101
	•••	516	552	743	260	283	346	165	179	201	118	127	131	90	96	92
0.25	***	560	583	755	276	294	360	173	185	217	122	130	148	93	98	109
	•••	568	617	833	283	313	382	178	196	219	126	137	141	95	103	98
0.3	***	588	617	842	288	310	395	179	194	234	126	136	157	95	102	114
	•••	598	661	904	297	335	409	185	208	231	130	144	147	98	107	101
0.35	***	595	633	908	292	319	420	182	199	245	127	138	163	95	103	117
	•••	608	687	954	303	348	426	189	215	239	132	148	151	99	109	102
0.4	***	586	633	952	290	320	435	180	200	251	126	138	165	94	102	117
	•••	603	696	984	302	353	435	188	217	241	131	148	151	98	108	101
0.45	***	565	620	974	282	316	440	176	197	251	123	136	163	91	100	115
	•••	585	692	994	295	352	435	185	215	239	128	146	147	95	106	98
0.5	***	533	596	975	270	306	435	169	190	246	118	131	158	87	95	110
	•••	557	675	984	284	344	426	178	210	231	123	141	141	91	102	92
0.55	***	495	565	953	254	292	421	159	181	235	111	123	150	81	89	103
	•••	522	648	954	269	330	409	169	200	219	116	134	131	85	95	84
0.6	***	453	526	910	236	274	397	148	169	219	102	114	138	74	81	94
	•••	481	612	904	251	311	382	157	187	201	107	123	119	78	87	74
0.65	***	408	483	845	214	251	363	134	153	197	91	102	122	66	72	82
	•••	437	566	833	229	286	346	142	169	179	96	110	102	68	76	62
0.7	***	361	433	758	191	224	320	118	135	171	79	88	103	.	.	.
	•••	390	512	743	204	255	302	125	148	151	83	94	83	.	.	.
0.75	***	312	378	650	164	192	267	99	112	138	.	.	.	.	.	.
	•••	339	448	633	176	218	248	105	122	118	.	.	.	.	.	.
0.8	***	260	316	521	133	154	205	.	.	.	.	.	.	.	.	.
	•••	282	373	502	142	173	186	.	.	.	.	.	.	.	.	.
0.85	***	201	242	372	.	.	.	.	.	.	.	.	.	.	.	.
	•••	218	282	352	.	.	.	.	.	.	.	.	.	.	.	.

TABLE 3: ALPHA= 0.01 POWER= 0.8   EXPECTED ACCRUAL THRU MINIMUM FOLLOW-UP= 110

	DEL=.05			DEL=.10			DEL=.15			DEL=.20			DEL=.25		
FACT=	1.0 .75	.50 .25	.00 BIN	1.0 .75	.50 .25	.00 BIN	1.0 .75	.50 .25	.00 BIN	1.0 75	.50 .25	.00 BIN	1.0 .75	.50 .25	.00 BIN
PCONT=•••	REQUIRED NUMBER OF PATIENTS														
0.05 ***	600	603	641	213	216	228	125	128	133	89	91	94	69	71	72
•••	601	608	1104	214	220	352	126	130	186	90	92	118	70	71	83
0.1 ***	1094	1100	1237	338	345	382	182	187	203	121	125	133	90	93	98
•••	1096	1113	1747	340	354	502	184	192	248	122	128	151	91	95	102
0.15 ***	1512	1523	1813	438	449	523	225	234	264	145	151	166	105	109	118
•••	1516	1545	2309	442	467	633	229	244	302	148	157	179	107	113	119
0.2 ***	1836	1852	2339	512	530	649	257	270	317	163	172	194	116	122	135
•••	1841	1884	2791	518	556	743	262	285	346	166	180	201	119	127	131
0.25 ***	2059	2082	2798	563	587	755	278	296	361	174	186	217	123	131	148
•••	2067	2129	3192	571	622	833	285	316	382	179	197	219	127	138	141
0.3 ***	2187	2219	3182	591	622	842	291	314	395	181	196	234	127	137	157
•••	2198	2281	3513	602	668	904	300	338	409	188	209	231	132	145	147
0.35 ***	2228	2269	3484	599	639	908	295	323	420	184	201	245	129	140	163
•••	2242	2352	3754	614	695	954	306	351	426	191	217	239	133	149	151
0.4 ***	2192	2246	3702	592	641	952	293	325	435	183	202	251	128	140	165
•••	2210	2352	3915	609	706	984	306	357	435	191	219	241	133	149	151
0.45 ***	2092	2161	3834	571	629	974	286	321	440	179	199	251	125	137	163
•••	2115	2295	3995	592	703	994	300	356	435	187	218	239	130	147	147
0.5 ***	1941	2029	3879	541	606	975	274	312	435	172	193	246	119	132	158
•••	1970	2193	3995	565	687	984	289	349	426	181	212	231	125	142	141
0.55 ***	1753	1864	3836	504	575	953	259	297	421	163	184	235	112	125	150
•••	1791	2059	3915	531	660	954	275	335	409	172	202	219	118	135	131
0.6 ***	1544	1681	3705	462	538	910	241	279	397	151	171	219	104	115	138
•••	1591	1903	3754	492	624	904	256	315	382	160	189	201	109	124	119
0.65 ***	1329	1491	3488	417	494	845	219	256	363	137	156	197	93	103	122
•••	1387	1733	3513	448	578	833	234	290	346	145	171	179	98	111	102
0.7 ***	1121	1303	3185	370	444	758	195	228	320	120	137	171	80	89	103
•••	1188	1552	3192	400	523	743	209	259	302	127	150	151	84	95	83
0.75 ***	930	1119	2796	321	388	650	168	196	267	101	114	138	.	.	.
•••	1002	1361	2791	348	457	633	180	221	248	107	124	118	.	.	.
0.8 ***	756	936	2322	267	324	521	136	157	205	.	.	.	.	.	.
•••	826	1157	2309	290	380	502	145	175	186	.	.	.	.	.	.
0.85 ***	593	748	1764	206	248	372	.	.	.	.	.	.	.	.	.
•••	653	931	1747	224	287	352	.	.	.	.	.	.	.	.	.
0.9 ***	424	538	1124	.	.	.	.	.	.	.	.	.	.	.	.
•••	469	665	1104	.	.	.	.	.	.	.	.	.	.	.	.

TABLE 3: ALPHA= 0.01 POWER= 0.8   EXPECTED ACCRUAL THRU MINIMUM FOLLOW-UP= 120

		DEL=.05			DEL=.10			DEL=.15			DEL=.20			DEL=.25	
FACT=	1.0 .75	.50 .25	.00 BIN	1.0 .75	.50 .25	.00 BIN	1.0 .75	.50 .25	.00 BIN	1.0 75	.50 .25	.00 BIN	1.0 .75	.50 .25	.00 BIN
PCONT=•••				REQUIRED NUMBER OF PATIENTS											
0.05 ***	600	603	641	214	217	228	126	128	133	89	91	94	70	71	73
•••	601	609	1104	215	220	352	127	130	186	90	92	118	70	71	83
0.1 ***	1094	1101	1237	339	346	382	182	188	203	121	125	133	90	93	97
•••	1096	1115	1747	341	356	502	184	193	248	123	128	151	91	95	102
0.15 ***	1513	1525	1813	439	451	523	226	235	264	146	152	166	106	110	118
•••	1517	1549	2309	443	469	633	230	245	302	148	158	179	108	113	119
0.2 ***	1837	1855	2339	514	533	649	258	272	317	164	173	194	117	123	135
•••	1843	1890	2791	521	559	743	264	286	346	168	181	201	119	128	131
0.25 ***	2061	2087	2798	565	591	755	280	299	360	176	188	217	124	132	148
•••	2070	2137	3192	574	627	833	288	318	382	181	198	219	128	138	141
0.3 ***	2190	2224	3182	594	627	842	293	317	395	183	198	234	129	138	157
•••	2202	2292	3513	606	674	904	303	341	409	189	211	231	133	146	147
0.35 ***	2232	2277	3484	603	646	908	299	326	420	186	203	245	130	141	163
•••	2247	2366	3754	619	703	954	310	355	426	193	218	239	135	150	151
0.4 ***	2197	2256	3702	597	648	952	297	329	435	185	205	251	129	141	165
•••	2216	2370	3915	616	715	984	310	361	435	193	221	241	134	150	151
0.45 ***	2098	2174	3834	577	637	974	290	325	440	181	202	251	126	139	163
•••	2123	2317	3995	600	713	994	304	360	435	190	220	239	132	148	147
0.5 ***	1949	2045	3879	548	616	975	279	316	435	175	196	246	121	134	158
•••	1981	2220	3995	574	697	984	294	353	426	184	214	231	127	143	141
0.55 ***	1763	1884	3836	511	586	953	264	302	421	165	186	235	114	126	150
•••	1804	2090	3915	540	671	954	280	339	409	174	204	219	119	136	131
0.6 ***	1557	1704	3705	471	548	910	245	284	397	153	174	219	105	117	138
•••	1608	1937	3754	501	635	904	261	320	382	162	191	201	110	125	119
0.65 ***	1345	1517	3488	426	504	845	224	260	363	139	158	197	94	104	122
•••	1407	1768	3513	457	589	833	239	294	346	147	173	179	99	112	102
0.7 ***	1140	1331	3185	379	454	758	199	232	320	122	138	171	82	90	103
•••	1211	1588	3192	409	532	743	213	262	302	129	151	151	85	95	83
0.75 ***	951	1146	2796	329	397	650	171	199	267	103	115	138	.	.	.
•••	1025	1395	2791	356	466	633	183	223	248	108	124	118	.	.	.
0.8 ***	776	962	2322	274	331	521	139	160	205	.	.	.	.	.	.
•••	848	1187	2309	297	387	502	148	177	186	.	.	.	.	.	.
0.85 ***	610	770	1764	211	253	372	.	.	.	.	.	.	.	.	.
•••	673	956	1747	229	291	352	.	.	.	.	.	.	.	.	.
0.9 ***	438	554	1124	.	.	.	.	.	.	.	.	.	.	.	.
•••	484	681	1104	.	.	.	.	.	.	.	.	.	.	.	.

TABLE 3: ALPHA= 0.01 POWER= 0.8   EXPECTED ACCRUAL THRU MINIMUM FOLLOW-UP= 130

PCONT		DEL=.05 1.0 .75	DEL=.05 .50 .25	DEL=.05 .00 BIN	DEL=.10 1.0 .75	DEL=.10 .50 .25	DEL=.10 .00 BIN	DEL=.15 1.0 .75	DEL=.15 .50 .25	DEL=.15 .00 BIN	DEL=.20 1.0 75	DEL=.20 .50 .25	DEL=.20 .00 BIN	DEL=.25 1.0 .75	DEL=.25 .50 .25	DEL=.25 .00 BIN
								REQUIRED NUMBER OF PATIENTS								
0.05	***	600	604	641	214	217	228	126	128	133	89	91	94	70	71	73
	•••	602	610	1104	215	221	352	127	130	186	90	92	118	70	72	83
0.1	***	1095	1102	1237	339	347	382	183	188	203	122	126	133	91	93	98
	•••	1097	1117	1747	342	357	502	185	193	248	123	128	151	92	95	102
0.15	***	1514	1527	1813	440	453	523	227	236	264	147	153	166	106	110	118
	•••	1519	1552	2309	445	471	633	231	246	302	149	158	179	108	114	119
0.2	***	1839	1858	2339	516	535	649	260	274	317	165	174	194	118	124	135
	•••	1845	1896	2791	523	563	743	265	288	346	169	182	201	120	128	131
0.25	***	2064	2091	2798	567	594	755	282	301	360	177	189	217	125	133	148
	•••	2073	2145	3192	577	632	833	290	320	382	182	200	219	129	139	141
0.3	***	2193	2230	3182	597	632	842	296	320	395	185	200	234	130	139	157
	•••	2205	2303	3513	610	680	904	305	343	409	191	212	231	134	146	147
0.35	***	2235	2285	3484	607	652	908	302	330	420	188	205	245	132	142	163
	•••	2252	2381	3754	623	710	954	313	358	426	195	220	239	136	150	151
0.4	***	2202	2266	3702	601	655	952	301	333	435	188	207	251	131	142	165
	•••	2223	2389	3915	621	723	984	314	365	435	196	223	241	136	151	151
0.45	***	2104	2186	3834	583	646	974	294	330	440	184	204	251	128	140	163
	•••	2132	2339	3995	607	722	994	309	364	435	192	221	239	133	149	147
0.5	***	1957	2060	3879	555	625	975	283	321	435	177	198	246	123	135	158
	•••	1992	2246	3995	582	707	984	298	357	426	186	216	231	128	144	141
0.55	***	1773	1903	3836	519	595	953	268	307	421	168	189	235	116	128	150
	•••	1818	2119	3915	549	681	954	284	343	409	177	206	219	121	137	131
0.6	***	1570	1727	3705	479	558	910	250	288	397	156	176	219	107	118	138
	•••	1625	1969	3754	510	645	904	266	323	382	165	192	201	112	126	119
0.65	***	1361	1543	3488	435	514	845	228	265	363	141	160	197	96	105	122
	•••	1426	1802	3513	466	598	833	243	297	346	149	174	179	100	112	102
0.7	***	1159	1357	3185	387	463	758	203	236	320	124	140	171	83	90	103
	•••	1233	1621	3192	418	541	743	217	265	302	131	152	151	86	96	83
0.75	***	970	1173	2796	336	405	650	175	202	267	104	116	138	.	.	.
	•••	1047	1427	2791	364	474	633	186	226	248	110	125	118	.	.	.
0.8	***	796	986	2322	280	338	521	141	162	205	.	.	.	.	.	.
	•••	869	1215	2309	304	393	502	150	178	186	.	.	.	.	.	.
0.85	***	627	790	1764	216	258	372	.	.	.	.	.	.	.	.	.
	•••	691	978	1747	234	295	352	.	.	.	.	.	.	.	.	.
0.9	***	450	568	1124	.	.	.	.	.	.	.	.	.	.	.	.
	•••	497	696	1104	.	.	.	.	.	.	.	.	.	.	.	.

TABLE 3: ALPHA= 0.01 POWER= 0.8   EXPECTED ACCRUAL THRU MINIMUM FOLLOW-UP= 140

PCONT	FACT	DEL=.05			DEL=.10			DEL=.15			DEL=.20			DEL=.25		
		1.0 .75	.50 .25	.00 BIN	1.0 .75	.50 .25	.00 BIN	1.0 .75	.50 .25	.00 BIN	1.0 75	.50 .25	.00 BIN	1.0 .75	.50 .25	.00 BIN
						REQUIRED NUMBER OF PATIENTS										
0.05	***	601	604	641	214	218	228	126	128	133	90	91	94	70	71	72
	***	602	611	1104	216	221	352	127	130	186	90	92	118	70	72	83
0.1	***	1095	1103	1237	340	348	382	183	189	203	122	126	133	91	93	97
	***	1098	1119	1747	343	358	502	186	194	248	124	129	151	92	95	102
0.15	***	1515	1529	1813	441	455	523	228	238	264	147	153	166	107	111	118
	***	1520	1556	2309	446	473	633	232	247	302	150	159	179	109	114	119
0.2	***	1840	1861	2339	517	538	649	261	275	317	166	175	194	118	124	135
	***	1847	1902	2791	525	566	743	267	289	346	170	182	201	121	128	131
0.25	***	2066	2095	2798	570	598	755	284	303	361	179	190	217	126	134	148
	***	2075	2153	3192	580	636	833	292	322	382	184	200	219	130	139	141
0.3	***	2196	2236	3182	600	637	842	298	322	395	186	201	234	131	140	157
	***	2209	2314	3513	613	686	904	308	346	409	193	213	231	135	147	147
0.35	***	2239	2292	3484	611	657	908	304	333	420	190	207	245	133	143	163
	***	2257	2395	3754	628	717	954	316	361	426	197	221	239	137	151	151
0.4	***	2207	2275	3702	606	662	952	304	336	435	190	209	251	132	144	165
	***	2230	2406	3915	627	731	984	317	368	435	198	224	241	137	152	151
0.45	***	2111	2199	3834	589	653	974	298	333	440	186	206	251	129	141	163
	***	2140	2361	3995	613	730	994	312	368	435	195	223	239	134	150	147
0.5	***	1965	2076	3879	561	633	975	287	325	435	179	200	246	124	136	158
	***	2002	2271	3995	589	716	984	302	361	426	188	217	231	129	145	141
0.55	***	1784	1922	3836	526	604	953	272	311	421	170	191	235	117	129	150
	***	1831	2148	3915	557	691	954	288	347	409	179	208	219	122	137	131
0.6	***	1583	1748	3705	487	567	910	254	292	397	158	178	219	108	119	138
	***	1641	2000	3754	518	654	904	270	327	382	167	194	201	113	127	119
0.65	***	1376	1567	3488	442	523	845	232	268	363	144	162	197	97	106	122
	***	1446	1833	3513	475	607	833	247	301	346	151	176	179	101	113	102
0.7	***	1176	1382	3185	395	471	758	207	239	320	126	142	171	83	91	103
	***	1254	1652	3192	426	550	743	221	268	302	133	153	151	87	96	83
0.75	***	989	1197	2796	343	412	650	178	205	267	106	118	138	.	.	.
	***	1069	1456	2791	371	481	633	189	228	248	111	126	118	.	.	.
0.8	***	814	1009	2322	286	344	521	143	164	205	.	.	.	.	.	.
	***	890	1242	2309	310	398	502	152	180	186	.	.	.	.	.	.
0.85	***	643	809	1764	221	262	372	.	.	.	.	.	.	.	.	.
	***	709	999	1747	238	299	352	.	.	.	.	.	.	.	.	.
0.9	***	462	581	1124	.	.	.	.	.	.	.	.	.	.	.	.
	***	510	710	1104	.	.	.	.	.	.	.	.	.	.	.	.

TABLE 3: ALPHA= 0.01 POWER= 0.8   EXPECTED ACCRUAL THRU MINIMUM FOLLOW-UP= 150

		DEL=.05			DEL=.10			DEL=.15			DEL=.20			DEL=.25		
FACT=		1.0 .75	.50 .25	.00 BIN	1.0 .75	.50 .25	.00 BIN	1.0 .75	.50 .25	.00 BIN	1.0 75	.50 .25	.00 BIN	1.0 .75	.50 .25	.00 BIN
PCONT=***					REQUIRED NUMBER OF PATIENTS											
0.05	***	601	605	641	215	218	228	126	129	133	90	91	94	70	71	73
	•••	602	612	1104	216	221	352	127	130	186	90	92	118	70	72	83
0.1	***	1096	1105	1237	341	349	382	184	189	203	123	126	133	91	94	97
	•••	1099	1121	1747	344	359	502	186	194	248	124	129	151	92	95	102
0.15	***	1516	1531	1813	442	457	523	229	239	264	148	154	166	107	111	118
	•••	1521	1560	2309	448	475	633	233	248	302	151	159	179	109	114	119
0.2	***	1842	1864	2339	519	541	649	263	277	317	167	176	194	119	125	135
	•••	1849	1908	2791	527	568	743	268	291	346	170	183	201	121	129	131
0.25	***	2068	2099	2798	572	601	755	286	305	360	180	192	217	127	134	148
	•••	2078	2161	3192	583	640	833	294	324	382	185	201	219	130	140	141
0.3	***	2199	2242	3182	603	641	842	301	325	395	188	202	234	132	141	157
	•••	2213	2325	3513	617	691	904	310	348	409	194	214	231	136	147	147
0.35	***	2243	2300	3484	615	663	908	307	336	420	192	208	245	134	144	163
	•••	2262	2409	3754	633	723	954	319	363	426	199	222	239	138	152	151
0.4	***	2212	2285	3702	611	668	952	307	340	435	192	210	251	133	144	165
	•••	2236	2424	3915	633	738	984	320	371	435	200	226	241	138	153	151
0.45	***	2117	2211	3834	594	660	974	301	337	440	188	208	251	130	142	163
	•••	2148	2382	3995	620	738	994	316	371	435	197	224	239	136	151	147
0.5	***	1973	2091	3879	568	641	975	291	329	435	182	202	246	125	137	158
	•••	2013	2295	3995	596	725	984	306	364	426	190	219	231	131	146	141
0.55	***	1794	1940	3836	533	612	953	276	315	421	172	193	235	118	130	150
	•••	1844	2175	3915	565	699	954	292	350	409	181	209	219	123	138	131
0.6	***	1595	1770	3705	494	575	910	257	296	397	160	180	219	109	120	138
	•••	1657	2029	3754	526	663	904	274	330	382	169	195	201	114	127	119
0.65	***	1392	1590	3488	450	531	845	236	272	363	145	163	197	98	107	122
	•••	1464	1864	3513	483	615	833	251	303	346	153	177	179	102	113	102
0.7	***	1194	1407	3185	402	479	758	210	243	320	128	143	170	84	92	103
	•••	1274	1682	3192	433	557	743	224	270	302	134	154	151	88	97	83
0.75	***	1007	1221	2795	350	419	650	180	208	267	107	119	138	.	.	.
	•••	1089	1484	2791	378	487	633	192	230	248	112	127	118	.	.	.
0.8	***	831	1031	2322	292	349	521	146	166	205	.	.	.	.	.	.
	•••	909	1266	2309	316	403	502	154	181	186	.	.	.	.	.	.
0.85	***	658	827	1764	225	266	372	.	.	.	.	.	.	.	.	.
	•••	725	1019	1747	242	302	352	.	.	.	.	.	.	.	.	.
0.9	***	473	594	1124	.	.	.	.	.	.	.	.	.	.	.	.
	•••	521	723	1104	.	.	.	.	.	.	.	.	.	.	.	.

TABLE 3: ALPHA= 0.01 POWER= 0.8   EXPECTED ACCRUAL THRU MINIMUM FOLLOW-UP= 160

REQUIRED NUMBER OF PATIENTS

PCONT=***	DEL=.05 1.0/.75	.50/.25	.00/BIN	DEL=.10 1.0/.75	.50/.25	.00/BIN	DEL=.15 1.0/.75	.50/.25	.00/BIN	DEL=.20 1.0/75	.50/.25	.00/BIN	DEL=.25 1.0/.75	.50/.25	.00/BIN
0.05 ***	601	605	641	215	218	228	127	129	133	90	91	94	70	71	73
•••	603	613	1104	216	222	352	128	130	186	90	92	118	71	72	83
0.1 ***	1096	1106	1237	341	350	382	184	190	203	123	127	133	91	94	98
•••	1100	1123	1747	345	360	502	187	195	248	125	129	151	93	95	102
0.15 ***	1517	1533	1813	443	458	523	230	239	264	148	155	166	108	112	118
•••	1523	1564	2309	449	477	633	234	248	302	151	159	179	109	114	119
0.2 ***	1843	1867	2339	521	543	649	264	278	317	168	176	194	119	125	135
•••	1851	1913	2791	529	571	743	270	292	346	171	183	201	122	129	131
0.25 ***	2070	2103	2798	574	605	755	288	307	360	181	193	217	128	135	148
•••	2081	2169	3192	585	644	833	296	325	382	186	202	219	131	140	141
0.3 ***	2202	2247	3182	606	646	842	303	327	395	189	204	234	133	142	157
•••	2217	2336	3513	621	695	904	313	350	409	195	215	231	137	148	147
0.35 ***	2247	2307	3484	619	668	908	310	338	420	193	210	245	135	145	163
•••	2267	2423	3754	637	728	954	321	366	426	200	223	239	139	152	151
0.4 ***	2216	2295	3702	616	675	952	310	343	435	193	212	251	134	145	165
•••	2243	2441	3915	638	744	984	323	373	435	201	227	241	139	153	151
0.45 ***	2123	2223	3834	600	667	974	304	340	440	190	210	251	132	143	163
•••	2157	2402	3995	626	745	994	319	374	435	199	226	239	137	151	147
0.5 ***	1981	2106	3879	574	649	975	294	332	435	184	204	246	127	138	158
•••	2024	2319	3995	603	732	984	310	367	426	192	220	231	132	146	141
0.55 ***	1804	1958	3836	540	620	953	280	318	421	174	194	235	119	131	150
•••	1858	2201	3915	572	707	954	296	353	409	183	210	219	124	139	131
0.6 ***	1608	1790	3705	501	584	910	261	299	397	162	181	219	110	121	138
•••	1673	2057	3754	534	671	904	277	333	382	171	196	201	115	128	119
0.65 ***	1407	1612	3488	457	539	845	239	275	363	147	165	197	99	108	122
•••	1482	1892	3513	490	623	833	254	306	346	155	178	179	103	114	102
0.7 ***	1211	1430	3185	409	487	758	213	245	320	129	144	171	85	92	103
•••	1293	1710	3192	441	564	743	227	272	302	136	155	151	88	97	83
0.75 ***	1025	1244	2796	356	426	650	183	210	267	108	119	138	.	.	.
•••	1109	1511	2791	385	493	633	195	232	248	113	127	118	.	.	.
0.8 ***	848	1051	2322	297	355	521	148	167	205	.	.	.	.	.	.
•••	927	1290	2309	321	408	502	156	182	186	.	.	.	.	.	.
0.85 ***	673	844	1764	229	270	372	.	.	.	.	.	.	.	.	.
•••	741	1038	1747	246	305	352	.	.	.	.	.	.	.	.	.
0.9 ***	484	606	1124	.	.	.	.	.	.	.	.	.	.	.	.
•••	533	735	1104	.	.	.	.	.	.	.	.	.	.	.	.

TABLE 3: ALPHA= 0.01 POWER= 0.8   EXPECTED ACCRUAL THRU MINIMUM FOLLOW-UP= 170

		DEL=.05			DEL=.10			DEL=.15			DEL=.20			DEL=.25		
FACT=		1.0 .75	.50 .25	.00 BIN	1.0 .75	.50 .25	.00 BIN	1.0 .75	.50 .25	.00 BIN	1.0 75	.50 .25	.00 BIN	1.0 .75	.50 .25	.00 BIN
PCONT=***					REQUIRED NUMBER OF PATIENTS											
0.05	***	602	606	641	215	219	228	127	129	133	90	91	93	70	71	72
	•••	603	613	1104	217	222	352	128	131	186	91	92	118	71	72	83
0.1	***	1097	1107	1237	342	351	382	185	190	203	123	127	133	92	94	98
	•••	1100	1125	1747	345	361	502	187	195	248	125	129	151	93	96	102
0.15	***	1518	1535	1813	444	460	523	230	240	264	149	155	166	108	112	118
	•••	1524	1567	2309	450	478	633	235	249	302	152	160	179	110	114	119
0.2	***	1844	1870	2339	522	545	649	265	280	317	168	177	194	120	126	135
	•••	1853	1919	2791	531	574	743	271	293	346	172	184	201	122	129	131
0.25	***	2072	2108	2798	576	608	755	289	309	361	182	194	217	128	136	148
	•••	2084	2177	3192	588	647	833	297	327	382	187	203	219	132	141	141
0.3	***	2204	2253	3182	609	650	842	305	329	395	191	205	234	134	142	157
	•••	2221	2346	3513	624	700	904	315	352	409	197	216	231	137	148	147
0.35	***	2251	2315	3484	622	673	908	312	341	420	195	211	245	136	146	163
	•••	2272	2436	3754	642	734	954	324	368	426	202	224	239	140	153	151
0.4	***	2221	2305	3702	620	680	952	313	346	435	195	213	251	135	146	165
	•••	2249	2457	3915	643	751	984	326	376	435	203	228	241	140	154	151
0.45	***	2130	2235	3834	605	674	974	308	343	440	192	211	251	133	144	163
	•••	2165	2422	3995	632	752	994	322	376	435	200	227	239	138	152	147
0.5	***	1989	2121	3879	580	656	975	297	335	435	186	205	246	128	139	158
	•••	2034	2341	3995	610	740	984	313	369	426	194	221	231	133	147	141
0.55	***	1814	1976	3836	547	628	953	283	322	421	176	196	235	121	131	150
	•••	1871	2226	3915	579	715	954	299	356	409	185	211	219	125	139	131
0.6	***	1621	1810	3705	508	591	910	264	302	397	164	183	219	111	121	138
	•••	1689	2084	3754	541	678	904	280	335	382	172	197	201	116	128	119
0.65	***	1422	1634	3488	464	546	845	242	278	363	149	166	198	100	108	122
	•••	1500	1920	3513	497	630	833	257	308	346	156	179	179	104	114	102
0.7	***	1227	1452	3185	416	493	758	216	248	320	131	145	171	86	93	103
	•••	1312	1737	3192	447	570	743	230	274	302	137	156	151	89	97	83
0.75	***	1042	1265	2796	362	432	650	186	212	267	109	120	138	.	.	.
	•••	1128	1536	2791	391	498	633	197	233	248	114	128	118	.	.	.
0.8	***	864	1071	2322	302	360	521	150	169	205	.	.	.	.	.	.
	•••	945	1311	2309	326	412	502	158	184	186	.	.	.	.	.	.
0.85	***	687	861	1764	232	273	372	.	.	.	.	.	.	.	.	.
	•••	756	1055	1747	250	308	352	.	.	.	.	.	.	.	.	.
0.9	***	494	617	1124	.	.	.	.	.	.	.	.	.	.	.	.
	•••	543	746	1104	.	.	.	.	.	.	.	.	.	.	.	.

TABLE 3: ALPHA= 0.01 POWER= 0.8   EXPECTED ACCRUAL THRU MINIMUM FOLLOW-UP= 180

		DEL=.05			DEL=.10			DEL=.15			DEL=.20			DEL=.25		
FACT=		1.0 .75	.50 .25	.00 BIN	1.0 .75	.50 .25	.00 BIN	1.0 .75	.50 .25	.00 BIN	1.0 75	.50 .25	.00 BIN	1.0 .75	.50 .25	.00 BIN
PCONT=***		REQUIRED NUMBER OF PATIENTS														
0.05	***	602	606	641	215	219	228	127	129	133	90	91	94	70	71	73
	***	603	614	1104	217	222	352	128	131	186	91	92	118	71	72	83
0.1	***	1098	1108	1237	343	352	382	185	191	203	124	127	133	92	94	98
	***	1101	1127	1747	346	361	502	188	195	248	125	130	151	93	96	102
0.15	***	1519	1537	1813	445	461	523	231	241	264	150	155	166	108	112	118
	***	1525	1571	2309	451	480	633	235	250	302	152	160	179	110	115	119
0.2	***	1846	1873	2339	524	548	649	266	281	317	169	178	194	120	126	135
	***	1855	1924	2791	533	576	743	272	294	346	173	184	201	123	130	131
0.25	***	2074	2112	2798	578	611	755	291	310	360	183	194	217	129	136	148
	***	2087	2184	3192	591	650	833	299	328	382	188	203	219	132	141	141
0.3	***	2207	2259	3182	612	654	842	307	331	395	192	206	234	134	143	157
	***	2224	2357	3513	627	704	904	317	353	409	198	217	231	138	149	147
0.35	***	2254	2322	3484	626	678	908	314	343	420	196	213	245	137	146	163
	***	2277	2449	3754	646	739	954	326	370	426	203	225	239	141	153	151
0.4	***	2226	2314	3702	624	686	952	315	348	435	197	215	251	136	147	165
	***	2256	2473	3915	648	756	984	329	378	435	204	229	241	141	154	151
0.45	***	2136	2247	3834	610	680	974	310	346	440	194	213	251	134	145	163
	***	2174	2441	3995	637	758	994	325	379	435	202	228	239	139	153	147
0.5	***	1997	2136	3879	585	663	975	301	338	435	187	207	246	129	140	158
	***	2045	2363	3995	616	746	984	316	372	426	196	222	231	134	148	141
0.55	***	1824	1993	3836	553	635	953	286	325	421	178	197	235	122	132	150
	***	1884	2250	3915	586	722	954	302	358	409	186	212	219	126	140	131
0.6	***	1633	1830	3705	514	598	910	268	305	397	166	184	219	112	122	138
	***	1704	2109	3754	548	685	904	284	338	382	174	198	201	117	129	119
0.65	***	1436	1655	3488	470	553	845	245	281	363	150	167	198	101	109	122
	***	1517	1946	3513	504	637	833	260	311	346	158	180	179	104	115	102
0.7	***	1243	1474	3185	422	500	758	219	250	320	132	146	171	87	93	103
	***	1331	1763	3192	454	576	743	232	276	302	138	156	151	90	98	83
0.75	***	1058	1286	2796	368	438	650	188	214	267	110	121	138	.	.	.
	***	1146	1560	2791	397	504	633	199	235	248	115	128	118	.	.	.
0.8	***	880	1090	2322	307	364	521	151	170	206	.	.	.	.	.	.
	***	962	1332	2309	331	416	502	160	184	186	.	.	.	.	.	.
0.85	***	700	876	1764	236	276	372	.	.	.	.	.	.	.	.	.
	***	770	1071	1747	253	310	352	.	.	.	.	.	.	.	.	.
0.9	***	504	627	1124	.	.	.	.	.	.	.	.	.	.	.	.
	***	554	756	1104	.	.	.	.	.	.	.	.	.	.	.	.

TABLE 3: ALPHA= 0.01 POWER= 0.8   EXPECTED ACCRUAL THRU MINIMUM FOLLOW-UP= 190

		DEL=.05			DEL=.10			DEL=.15			DEL=.20			DEL=.25		
FACT=		1.0 .75	.50 .25	.00 BIN	1.0 .75	.50 .25	.00 BIN	1.0 .75	.50 .25	.00 BIN	1.0 75	.50 .25	.00 BIN	1.0 .75	.50 .25	.00 BIN
PCONT=***					REQUIRED NUMBER OF PATIENTS											
0.05	***	602	607	641	216	219	228	127	129	133	90	92	94	70	71	73
	•••	604	615	1104	217	222	352	128	131	186	91	92	118	71	72	83
0.1	***	1098	1109	1237	343	352	382	186	191	203	124	127	133	92	94	98
	•••	1102	1129	1747	347	362	502	188	196	248	125	130	151	93	96	102
0.15	***	1520	1539	1813	446	463	523	232	242	264	150	156	166	109	112	118
	•••	1526	1574	2309	453	481	633	236	250	302	153	160	179	110	115	119
0.2	***	1847	1876	2339	525	550	649	267	282	317	170	178	194	121	126	135
	•••	1857	1930	2791	534	578	743	273	295	346	174	185	201	123	130	131
0.25	***	2076	2116	2798	581	614	755	292	312	360	184	195	217	130	136	148
	•••	2089	2192	3192	593	653	833	301	329	382	189	204	219	133	141	141
0.3	***	2210	2264	3182	614	658	842	308	333	395	193	207	234	135	143	157
	•••	2228	2367	3513	631	708	904	319	355	409	199	218	231	139	149	147
0.35	***	2258	2330	3484	629	683	908	317	345	420	198	214	245	138	147	163
	•••	2282	2462	3754	650	744	954	329	372	426	205	226	239	142	154	151
0.4	***	2231	2324	3702	629	691	952	318	351	435	198	216	251	137	148	165
	•••	2262	2489	3915	653	762	984	332	380	435	206	230	241	142	155	151
0.45	***	2142	2260	3834	615	686	974	313	349	440	195	214	251	135	145	163
	•••	2182	2460	3995	643	764	994	328	381	435	203	229	239	139	153	147
0.5	***	2005	2151	3879	591	669	975	303	341	435	189	208	246	130	141	158
	•••	2055	2384	3995	622	753	984	319	374	426	197	223	231	135	148	141
0.55	***	1834	2010	3836	559	642	953	289	327	421	179	199	235	122	133	150
	•••	1896	2273	3915	592	728	954	305	361	409	188	213	219	127	140	131
0.6	***	1645	1849	3705	520	605	910	271	308	397	167	185	219	113	123	138
	•••	1719	2134	3754	555	692	904	287	340	382	175	199	201	117	129	119
0.65	***	1450	1675	3488	477	560	845	248	283	363	152	168	197	101	110	122
	•••	1534	1971	3513	511	643	833	263	313	346	159	180	179	105	115	102
0.7	***	1259	1494	3185	428	506	758	221	253	320	133	147	170	87	94	103
	•••	1348	1787	3192	460	582	743	235	278	302	140	157	151	90	98	83
0.75	***	1074	1306	2796	373	443	650	190	216	267	111	122	138	.	.	.
	•••	1164	1582	2791	402	508	633	201	236	248	116	129	118	.	.	.
0.8	***	895	1108	2322	312	369	521	153	172	206	.	.	.	.	.	.
	•••	978	1352	2309	335	420	502	161	185	186	.	.	.	.	.	.
0.85	***	713	891	1764	239	279	372	.	.	.	.	.	.	.	.	.
	•••	784	1087	1747	256	313	352	.	.	.	.	.	.	.	.	.
0.9	***	513	637	1124	.	.	.	.	.	.	.	.	.	.	.	.
	•••	563	766	1104	.	.	.	.	.	.	.	.	.	.	.	.

TABLE 3: ALPHA= 0.01 POWER= 0.8   EXPECTED ACCRUAL THRU MINIMUM FOLLOW-UP= 200

	DEL=.05			DEL=.10			DEL=.15			DEL=.20			DEL=.25		
FACT=	1.0 .75	.50 .25	.00 BIN	1.0 .75	.50 .25	.00 BIN	1.0 .75	.50 .25	.00 BIN	1.0 75	.50 .25	.00 BIN	1.0 .75	.50 .25	.00 BIN
PCONT=***				REQUIRED NUMBER OF PATIENTS											
0.05 ***	602	607	641	216	220	228	127	129	133	90	92	94	70	71	73
***	604	615	1104	217	223	352	128	131	186	91	92	118	71	72	83
0.1 ***	1099	1110	1237	344	353	382	186	192	203	124	128	133	92	94	98
***	1103	1131	1747	347	363	502	189	196	248	126	130	151	93	96	102
0.15 ***	1521	1541	1813	448	464	523	233	243	264	151	156	166	109	113	118
***	1528	1578	2309	454	483	633	237	251	302	153	161	179	111	115	119
0.2 ***	1849	1879	2339	527	552	649	268	283	317	171	179	194	121	127	135
***	1859	1935	2791	536	580	743	274	296	346	174	185	201	124	130	131
0.25 ***	2078	2120	2798	583	617	755	294	313	360	185	196	217	130	137	148
***	2092	2200	3192	596	656	833	302	331	382	190	204	219	133	142	141
0.3 ***	2213	2270	3182	617	661	842	310	335	395	194	208	234	136	144	157
***	2232	2376	3513	634	712	904	320	356	409	200	218	231	139	150	147
0.35 ***	2262	2337	3484	633	687	908	319	348	420	199	215	245	138	148	163
***	2287	2475	3754	654	748	954	331	373	426	206	227	239	143	154	151
0.4 ***	2236	2333	3702	633	696	952	320	353	435	200	217	251	138	148	165
***	2269	2504	3915	658	767	984	334	382	435	207	231	241	143	155	151
0.45 ***	2148	2271	3834	620	692	974	316	352	440	197	215	251	136	146	163
***	2190	2478	3995	648	770	994	331	383	435	205	230	239	140	153	147
0.5 ***	2013	2165	3879	596	675	975	306	344	435	190	210	246	131	141	158
***	2066	2405	3995	628	759	984	322	376	426	199	224	231	135	149	141
0.55 ***	1844	2027	3836	565	648	953	292	330	421	181	200	235	123	134	150
***	1909	2296	3915	598	735	954	308	363	409	189	214	219	128	140	131
0.6 ***	1657	1868	3705	526	612	910	274	311	397	169	187	219	114	123	138
***	1734	2157	3754	561	698	904	289	342	382	177	200	201	118	129	119
0.65 ***	1464	1695	3488	483	566	845	251	286	363	153	169	197	102	110	122
***	1551	1995	3513	517	649	833	266	314	346	160	181	179	106	115	102
0.7 ***	1274	1514	3185	433	512	758	224	255	320	135	148	171	88	94	103
***	1366	1811	3192	466	587	743	237	280	302	141	158	151	91	98	83
0.75 ***	1089	1325	2796	378	448	650	192	218	267	112	122	138	.	.	.
***	1181	1604	2791	407	513	633	203	237	248	117	129	118	.	.	.
0.8 ***	909	1125	2322	316	373	521	154	173	205	.	.	.	.	.	.
***	994	1371	2309	340	423	502	163	186	186	.	.	.	.	.	.
0.85 ***	725	905	1764	242	282	372	.	.	.	.	.	.	.	.	.
***	797	1102	1747	259	315	352	.	.	.	.	.	.	.	.	.
0.9 ***	522	647	1124	.	.	.	.	.	.	.	.	.	.	.	.
***	572	775	1104	.	.	.	.	.	.	.	.	.	.	.	.

TABLE 3: ALPHA= 0.01 POWER= 0.8   EXPECTED ACCRUAL THRU MINIMUM FOLLOW-UP= 225

		DEL=.05			DEL=.10			DEL=.15			DEL=.20			DEL=.25		
FACT=		1.0 .75	.50 .25	.00 BIN	1.0 .75	.50 .25	.00 BIN	1.0 .75	.50 .25	.00 BIN	1.0 75	.50 .25	.00 BIN	1.0 .75	.50 .25	.00 BIN
PCONT=***		REQUIRED NUMBER OF PATIENTS														
0.05	***	603	608	641	217	220	228	128	130	133	91	92	93	71	71	73
	***	605	617	1104	218	223	352	129	131	186	91	93	118	71	72	83
0.1	***	1100	1113	1237	345	355	382	187	192	203	125	128	133	93	95	97
	***	1105	1136	1747	349	364	502	189	197	248	126	130	151	94	96	102
0.15	***	1524	1546	1813	450	467	523	234	244	264	152	157	166	110	113	118
	***	1531	1586	2309	457	485	633	239	252	302	154	161	179	111	115	119
0.2	***	1853	1886	2339	530	556	649	271	285	317	172	180	194	122	127	135
	***	1864	1947	2791	541	585	743	277	297	346	176	186	201	125	131	131
0.25	***	2084	2131	2798	588	623	755	297	317	360	187	198	217	131	138	148
	***	2099	2217	3192	601	663	833	305	333	382	192	206	219	134	142	141
0.3	***	2220	2284	3182	624	670	842	314	339	395	197	210	234	137	145	157
	***	2242	2400	3513	641	720	904	325	359	409	202	220	231	141	150	147
0.35	***	2271	2356	3484	641	697	908	324	352	420	202	217	245	140	149	163
	***	2300	2505	3754	663	758	954	336	377	426	209	228	239	144	155	151
0.4	***	2248	2357	3702	642	708	952	326	358	435	203	220	251	140	150	165
	***	2285	2541	3915	668	779	984	340	386	435	210	232	241	144	156	151
0.45	***	2164	2300	3834	631	705	974	322	357	440	200	218	251	138	148	163
	***	2211	2521	3995	660	783	994	337	387	435	208	232	239	142	154	147
0.5	***	2033	2199	3879	609	690	975	313	350	435	194	212	246	133	143	158
	***	2091	2453	3995	641	773	984	329	381	426	202	226	231	137	149	141
0.55	***	1869	2067	3836	578	663	953	299	336	421	185	203	235	125	135	150
	***	1940	2349	3915	612	748	954	315	367	409	193	216	219	130	141	131
0.6	***	1687	1912	3705	540	627	910	280	317	397	172	189	219	116	124	138
	***	1770	2213	3754	576	711	904	296	347	382	180	202	201	120	130	119
0.65	***	1498	1742	3488	496	581	845	257	291	363	156	172	197	104	111	122
	***	1590	2051	3513	531	662	833	272	318	346	163	183	179	107	116	102
0.7	***	1310	1561	3185	446	525	758	230	260	320	137	150	171	89	95	103
	***	1407	1865	3192	479	599	743	243	283	302	143	159	151	92	99	83
0.75	***	1126	1370	2795	390	460	650	197	221	267	114	124	138	.	.	.
	***	1221	1654	2791	419	523	633	208	240	248	119	130	118	.	.	.
0.8	***	943	1165	2322	326	382	521	158	175	205	.	.	.	.	.	.
	***	1031	1414	2309	349	431	502	166	188	186	.	.	.	.	.	.
0.85	***	754	938	1764	249	288	372	.	.	.	.	.	.	.	.	.
	***	827	1135	1747	266	320	352	.	.	.	.	.	.	.	.	.
0.9	***	542	669	1124	.	.	.	.	.	.	.	.	.	.	.	.
	***	594	796	1104	.	.	.	.	.	.	.	.	.	.	.	.

TABLE 3: ALPHA= 0.01 POWER= 0.8   EXPECTED ACCRUAL THRU MINIMUM FOLLOW-UP= 250

PCONT	FACT	DEL=.05 1.0/.75	DEL=.05 .50/.25	DEL=.05 .00/BIN	DEL=.10 1.0/.75	DEL=.10 .50/.25	DEL=.10 .00/BIN	DEL=.15 1.0/.75	DEL=.15 .50/.25	DEL=.15 .00/BIN	DEL=.20 1.0/75	DEL=.20 .50/.25	DEL=.20 .00/BIN	DEL=.25 1.0/.75	DEL=.25 .50/.25	DEL=.25 .00/BIN
								REQUIRED NUMBER OF PATIENTS								
0.05	***	604	610	641	217	221	228	128	130	133	91	92	94	71	72	72
	•••	606	618	1104	219	224	352	129	131	186	91	93	118	71	72	83
0.1	***	1102	1116	1237	347	356	382	188	193	203	125	128	133	93	95	97
	•••	1107	1140	1747	350	366	502	190	197	248	127	130	151	94	96	102
0.15	***	1526	1550	1813	452	470	523	236	245	264	152	158	166	110	113	118
	•••	1534	1594	2309	459	488	633	240	253	302	155	162	179	112	116	119
0.2	***	1856	1893	2339	534	561	649	273	287	317	173	181	194	123	128	135
	•••	1869	1960	2791	544	589	743	279	299	346	177	187	201	125	131	131
0.25	***	2089	2141	2798	592	629	755	300	319	360	189	199	217	133	139	148
	•••	2106	2235	3192	607	669	833	308	335	382	193	207	219	135	143	141
0.3	***	2227	2298	3182	630	677	842	318	342	395	199	212	234	139	146	157
	•••	2251	2423	3513	649	727	904	328	362	409	204	221	231	142	151	147
0.35	***	2281	2374	3484	649	707	908	328	357	420	204	219	245	142	150	163
	•••	2312	2533	3754	672	767	954	340	380	426	211	230	239	145	156	151
0.4	***	2261	2380	3702	652	719	952	331	363	435	206	222	251	142	151	165
	•••	2301	2576	3915	679	789	984	345	390	435	213	234	241	146	157	151
0.45	***	2180	2328	3834	642	717	974	328	362	440	203	221	251	139	149	163
	•••	2231	2561	3995	672	794	994	342	391	435	211	233	239	144	155	147
0.5	***	2053	2233	3879	620	702	975	319	355	435	197	215	246	134	144	158
	•••	2116	2498	3995	654	784	984	334	385	426	205	228	231	139	150	141
0.55	***	1893	2105	3836	590	676	953	305	341	421	188	205	235	127	136	150
	•••	1970	2397	3915	625	761	954	320	372	409	195	218	219	131	142	131
0.6	***	1715	1953	3705	553	640	910	286	322	397	175	191	219	117	126	138
	•••	1804	2264	3754	589	723	904	301	350	382	182	203	201	121	131	119
0.65	***	1530	1785	3488	509	593	845	262	296	363	159	174	197	105	112	122
	•••	1627	2102	3513	544	673	833	277	322	346	166	184	179	108	117	102
0.7	***	1344	1604	3185	458	537	758	234	264	320	139	152	171	90	96	103
	•••	1445	1914	3192	491	609	743	247	286	302	145	160	151	93	99	83
0.75	***	1160	1411	2796	401	470	650	201	225	267	116	125	138	.	.	.
	•••	1258	1699	2791	430	531	633	211	242	248	120	131	118	.	.	.
0.8	***	974	1202	2322	334	390	521	161	178	205	.	.	.	.	.	.
	•••	1064	1453	2309	358	437	502	168	189	186	.	.	.	.	.	.
0.85	***	780	967	1764	256	294	372	.	.	.	.	.	.	.	.	.
	•••	855	1165	1747	272	324	352	.	.	.	.	.	.	.	.	.
0.9	***	561	689	1124	.	.	.	.	.	.	.	.	.	.	.	.
	•••	613	814	1104	.	.	.	.	.	.	.	.	.	.	.	.

TABLE 3: ALPHA= 0.01 POWER= 0.8   EXPECTED ACCRUAL THRU MINIMUM FOLLOW-UP= 275

		DEL=.05			DEL=.10			DEL=.15			DEL=.20			DEL=.25		
FACT=		1.0 .75	.50 .25	.00 BIN	1.0 .75	.50 .25	.00 BIN	1.0 .75	.50 .25	.00 BIN	1.0 75	.50 .25	.00 BIN	1.0 .75	.50 .25	.00 BIN
PCONT=•••					REQUIRED NUMBER OF PATIENTS											
0.05	•••	604	611	641	217	221	228	128	130	133	91	92	94	71	72	72
	•••	607	619	1104	219	224	352	129	131	186	91	93	118	71	72	83
0.1	•••	1103	1119	1237	348	358	382	189	194	202	126	129	133	93	95	98
	•••	1108	1144	1747	352	367	502	191	198	248	127	131	151	94	96	102
0.15	•••	1529	1555	1813	454	473	523	237	246	264	153	158	166	111	114	118
	•••	1537	1601	2309	462	490	633	241	254	302	156	162	179	112	116	119
0.2	•••	1860	1900	2339	537	565	649	275	289	317	175	182	194	124	128	135
	•••	1874	1971	2791	548	593	743	281	300	346	178	187	201	126	131	131
0.25	•••	2094	2151	2798	597	635	755	303	322	360	190	200	217	134	139	148
	•••	2113	2251	3192	612	674	833	311	337	382	195	207	219	136	143	141
0.3	•••	2234	2312	3182	636	684	842	321	345	395	201	213	234	140	147	157
	•••	2261	2444	3513	655	734	904	332	364	409	206	222	231	143	151	147
0.35	•••	2290	2391	3484	656	715	908	332	360	420	206	221	245	143	151	163
	•••	2325	2560	3754	679	775	954	344	383	426	213	231	239	146	156	151
0.4	•••	2273	2402	3702	660	729	952	336	367	435	208	224	251	143	152	165
	•••	2317	2608	3915	688	798	984	349	393	435	215	235	241	147	158	151
0.45	•••	2195	2356	3834	651	728	974	332	367	440	206	223	251	141	150	163
	•••	2251	2598	3995	682	804	994	347	395	435	213	235	239	145	156	147
0.5	•••	2072	2265	3879	631	714	975	324	360	435	200	217	246	136	145	158
	•••	2141	2540	3915	665	795	984	339	388	426	208	229	231	140	151	141
0.55	•••	1917	2141	3836	602	688	953	310	346	421	190	207	235	128	137	150
	•••	1999	2442	3915	637	771	954	326	375	409	198	219	219	132	143	131
0.6	•••	1743	1992	3705	565	652	910	291	326	397	177	193	219	118	126	138
	•••	1836	2310	3754	601	734	904	306	354	382	185	204	201	122	131	119
0.65	•••	1561	1826	3488	520	605	845	267	300	363	161	175	198	106	113	122
	•••	1662	2148	3513	556	683	833	282	325	346	168	185	179	109	117	102
0.7	•••	1376	1645	3185	469	547	758	239	267	320	141	153	171	91	96	103
	•••	1481	1959	3192	502	618	743	251	288	302	147	161	151	93	99	83
0.75	•••	1191	1449	2796	410	479	650	204	227	267	117	126	138	.	.	.
	•••	1293	1740	2791	439	539	633	215	244	248	121	132	118	.	.	.
0.8	•••	1003	1235	2322	342	397	521	163	179	205	.	.	.	.	.	.
	•••	1096	1488	2309	366	442	502	171	191	186	.	.	.	.	.	.
0.85	•••	805	994	1764	261	298	371	.	.	.	.	.	.	.	.	.
	•••	881	1192	1747	277	327	352	.	.	.	.	.	.	.	.	.
0.9	•••	578	706	1124	.	.	.	.	.	.	.	.	.	.	.	.
	•••	631	831	1104	.	.	.	.	.	.	.	.	.	.	.	.

TABLE 3: ALPHA= 0.01 POWER= 0.8   EXPECTED ACCRUAL THRU MINIMUM FOLLOW-UP= 300

		DEL=.05			DEL=.10			DEL=.15			DEL=.20			DEL=.25		
FACT=		1.0 .75	.50 .25	.00 BIN	1.0 .75	.50 .25	.00 BIN	1.0 .75	.50 .25	.00 BIN	1.0 75	.50 .25	.00 BIN	1.0 .75	.50 .25	.00 BIN
PCONT=•••					REQUIRED	NUMBER	OF PATIENTS									
0.05	***	605	612	641	218	221	228	129	130	133	91	92	94	71	72	73
	•••	607	620	1104	220	224	352	129	131	186	91	93	118	71	72	83
0.1	***	1105	1121	1237	349	359	382	189	194	202	126	129	133	94	95	97
	•••	1110	1147	1747	353	368	502	192	198	248	127	131	151	94	96	102
0.15	***	1531	1560	1813	457	475	523	238	248	264	154	159	166	111	114	118
	•••	1541	1608	2309	464	492	633	242	254	302	156	162	179	112	116	119
0.2	***	1864	1908	2339	541	568	649	277	291	317	176	183	194	125	129	135
	•••	1879	1982	2791	552	596	743	283	301	346	179	188	201	127	131	131
0.25	***	2099	2161	2798	601	640	755	305	324	360	192	201	217	134	140	148
	•••	2120	2266	3192	617	678	833	313	338	382	196	208	219	137	143	141
0.3	***	2242	2325	3182	641	691	842	325	348	395	202	214	234	141	147	157
	•••	2270	2465	3513	661	740	904	335	366	409	208	223	231	144	152	147
0.35	***	2300	2409	3484	663	723	908	336	363	420	208	222	245	144	152	163
	•••	2337	2586	3754	687	782	954	348	385	426	215	232	239	148	157	151
0.4	***	2285	2424	3702	668	738	952	340	371	435	210	226	251	144	153	165
	•••	2333	2638	3915	696	806	984	353	396	435	217	236	241	148	158	151
0.45	***	2211	2382	3834	660	738	974	337	371	440	208	224	251	142	151	163
	•••	2271	2633	3995	692	813	994	352	397	435	215	236	239	146	156	147
0.5	***	2091	2295	3879	641	724	975	328	364	435	202	219	246	137	146	158
	•••	2165	2579	3995	675	804	984	344	391	426	210	230	231	141	151	141
0.55	***	1940	2175	3836	612	699	953	315	350	421	193	209	235	130	138	150
	•••	2027	2484	3915	648	781	954	330	378	409	200	220	219	133	143	131
0.6	***	1770	2029	3705	575	663	910	296	330	397	180	195	219	120	127	138
	•••	1867	2353	3754	612	743	904	311	356	382	187	205	201	123	132	119
0.65	***	1590	1864	3488	531	615	845	272	303	363	163	177	197	107	113	122
	•••	1695	2191	3513	566	692	833	286	327	346	169	186	179	110	117	102
0.7	***	1407	1682	3184	479	557	758	243	270	320	143	154	170	92	97	103
	•••	1514	2000	3192	512	626	743	255	291	302	148	161	151	94	100	83
0.75	***	1221	1484	2795	419	487	650	208	230	267	118	127	138	.	.	.
	•••	1325	1778	2791	448	545	633	217	246	248	122	132	118	.	.	.
0.8	***	1031	1266	2322	349	403	521	166	181	205	.	.	.	.	.	.
	•••	1125	1520	2309	373	447	502	173	192	186	.	.	.	.	.	.
0.85	***	827	1019	1764	266	302	372	.	.	.	.	.	.	.	.	.
	•••	905	1216	1747	282	330	352	.	.	.	.	.	.	.	.	.
0.9	***	594	723	1124	.	.	.	.	.	.	.	.	.	.	.	.
	•••	647	845	1104	.	.	.	.	.	.	.	.	.	.	.	.

TABLE 3: ALPHA= 0.01 POWER= 0.8   EXPECTED ACCRUAL THRU MINIMUM FOLLOW-UP= 325

		DEL=.05			DEL=.10			DEL=.15			DEL=.20			DEL=.25		
FACT=		1.0 .75	.50 .25	.00 BIN	1.0 .75	.50 .25	.00 BIN	1.0 .75	.50 .25	.00 BIN	1.0 75	.50 .25	.00 BIN	1.0 .75	.50 .25	.00 BIN
PCONT=•••					REQUIRED NUMBER OF PATIENTS											
0.05	***	606	613	641	218	222	228	129	131	133	91	92	94	71	72	73
	•••	608	622	1104	220	224	352	130	132	186	92	93	118	71	72	83
0.1	***	1106	1124	1237	350	360	382	190	195	203	127	129	133	94	95	97
	•••	1112	1150	1747	354	368	502	192	198	248	128	131	151	95	96	102
0.15	***	1533	1564	1813	459	477	523	240	249	264	155	159	166	112	114	118
	•••	1544	1614	2309	466	494	633	244	255	302	157	162	179	113	116	119
0.2	***	1867	1915	2339	544	572	649	279	292	317	177	184	194	125	129	135
	•••	1884	1993	2791	555	599	743	285	302	346	180	188	201	127	132	131
0.25	***	2105	2171	2798	606	645	755	307	326	361	193	202	217	135	140	148
	•••	2127	2281	3192	621	682	833	315	340	382	197	209	219	138	144	141
0.3	***	2249	2338	3182	647	697	842	327	350	395	204	216	234	142	148	157
	•••	2279	2484	3513	667	745	904	337	368	409	209	223	231	145	152	147
0.35	***	2309	2426	3484	669	730	908	339	366	420	210	224	245	145	152	163
	•••	2350	2610	3754	694	788	954	351	387	426	216	233	239	148	157	151
0.4	***	2297	2445	3703	676	746	952	343	374	435	212	227	251	146	154	165
	•••	2349	2667	3915	705	813	984	357	398	435	219	237	241	149	159	151
0.45	***	2226	2407	3834	669	747	974	341	374	440	210	226	251	143	151	163
	•••	2291	2666	3995	701	821	994	355	400	435	217	237	239	147	157	147
0.5	***	2110	2324	3879	651	734	975	333	367	435	204	220	246	138	147	158
	•••	2188	2615	3995	685	813	984	348	394	426	212	232	231	142	152	141
0.55	***	1963	2207	3836	622	709	953	319	354	421	195	211	235	131	139	150
	•••	2054	2522	3915	658	790	954	334	380	409	202	221	219	134	144	131
0.6	***	1795	2064	3705	586	673	910	300	333	397	182	196	219	121	128	138
	•••	1897	2393	3754	622	752	904	315	359	382	188	206	201	124	132	119
0.65	***	1618	1899	3488	541	625	845	276	307	363	165	178	197	108	114	122
	•••	1727	2231	3513	576	700	833	289	330	346	171	187	179	111	118	102
0.7	***	1435	1717	3185	488	566	758	246	273	320	144	155	171	92	97	103
	•••	1546	2038	3192	521	633	743	258	292	302	149	162	151	95	100	83
0.75	***	1249	1517	2798	427	494	650	210	232	268	120	128	138	.	.	.
	•••	1355	1812	2791	456	551	633	220	247	248	123	132	118	.	.	.
0.8	***	1056	1295	2322	356	409	521	168	183	206	.	.	.	.	.	.
	•••	1152	1549	2309	379	452	502	174	193	186	.	.	.	.	.	.
0.85	***	849	1042	1764	271	306	372	.	.	.	.	.	.	.	.	.
	•••	927	1239	1747	286	333	352	.	.	.	.	.	.	.	.	.
0.9	***	609	737	1124	.	.	.	.	.	.	.	.	.	.	.	.
	•••	662	858	1104	.	.	.	.	.	.	.	.	.	.	.	.

TABLE 3: ALPHA= 0.01 POWER= 0.8   EXPECTED ACCRUAL THRU MINIMUM FOLLOW-UP= 350

		DEL=.05			DEL=.10			DEL=.15			DEL=.20			DEL=.25		
FACT=		1.0 .75	.50 .25	.00 BIN	1.0 .75	.50 .25	.00 BIN	1.0 .75	.50 .25	.00 BIN	1.0 75	.50 .25	.00 BIN	1.0 .75	.50 .25	.00 BIN
PCONT=•••					REQUIRED NUMBER OF PATIENTS											
0.05	•••	606	614	641	219	222	228	129	131	133	91	92	94	71	72	73
	•••	609	622	1104	220	225	352	130	132	186	92	93	118	71	72	83
0.1	•••	1107	1126	1237	351	361	382	191	195	202	127	129	133	94	96	97
	•••	1114	1154	1747	355	369	502	193	199	248	128	131	151	95	96	102
0.15	•••	1536	1569	1813	461	479	523	241	249	264	155	160	166	112	115	118
	•••	1547	1621	2309	468	496	633	244	256	302	157	163	179	113	116	119
0.2	•••	1871	1921	2339	546	575	649	280	293	317	177	184	194	126	130	135
	•••	1888	2002	2791	558	601	743	286	303	346	181	189	201	127	132	131
0.25	•••	2110	2181	2798	610	649	755	310	328	360	194	203	217	136	141	148
	•••	2134	2295	3192	626	686	833	318	341	382	198	209	219	138	144	141
0.3	•••	2256	2351	3182	652	702	842	330	353	395	205	216	234	143	149	157
	•••	2289	2503	3513	672	750	904	340	370	409	210	224	231	145	152	147
0.35	•••	2318	2442	3484	676	736	908	342	369	420	212	225	245	146	153	163
	•••	2362	2633	3754	700	794	954	354	389	426	218	234	239	149	157	151
0.4	•••	2309	2465	3702	683	754	952	347	377	435	214	228	251	146	154	165
	•••	2364	2694	3915	712	820	984	360	400	435	221	238	241	150	159	151
0.45	•••	2241	2432	3834	677	755	974	345	377	440	212	227	251	144	152	163
	•••	2310	2697	3995	709	828	994	359	402	435	219	238	239	148	157	147
0.5	•••	2129	2352	3879	659	743	975	337	371	435	206	222	246	139	147	158
	•••	2211	2649	3995	694	820	984	352	397	426	213	232	231	143	152	141
0.55	•••	1984	2238	3836	631	718	953	323	357	421	197	212	235	132	139	150
	•••	2080	2559	3915	668	797	954	338	383	409	204	222	219	135	144	131
0.6	•••	1820	2096	3705	595	682	910	304	337	397	183	198	219	122	128	138
	•••	1926	2430	3754	631	759	904	318	361	382	190	207	201	125	133	119
0.65	•••	1645	1933	3488	550	633	845	279	309	363	167	179	197	109	115	122
	•••	1757	2268	3513	585	707	833	293	332	346	172	187	179	111	118	102
0.7	•••	1463	1750	3184	497	573	758	249	275	320	146	156	171	93	97	103
	•••	1576	2073	3192	529	639	743	261	294	302	150	163	151	95	100	83
0.75	•••	1276	1548	2796	435	501	650	213	234	267	121	128	138	.	.	.
	•••	1384	1844	2791	463	556	633	222	248	248	124	133	118	.	.	.
0.8	•••	1080	1322	2322	362	414	521	169	184	206	.	.	.	.	.	.
	•••	1177	1576	2309	384	455	502	176	194	186	.	.	.	.	.	.
0.85	•••	869	1063	1764	275	309	372	.	.	.	.	.	.	.	.	.
	•••	948	1259	1747	290	335	352	.	.	.	.	.	.	.	.	.
0.9	•••	622	751	1124	.	.	.	.	.	.	.	.	.	.	.	.
	•••	676	870	1104	.	.	.	.	.	.	.	.	.	.	.	.

TABLE 3: ALPHA= 0.01 POWER= 0.8   EXPECTED ACCRUAL THRU MINIMUM FOLLOW-UP= 375

	DEL=.05			DEL=.10			DEL=.15			DEL=.20			DEL=.25		
FACT=	1.0 .75	.50 .25	.00 BIN	1.0 .75	.50 .25	.00 BIN	1.0 .75	.50 .25	.00 BIN	1.0 75	.50 .25	.00 BIN	1.0 .75	.50 .25	.00 BIN
PCONT=•••						REQUIRED NUMBER OF PATIENTS									
0.05 ***	607	614	641	219	223	228	129	131	133	92	92	94	71	72	73
•••	610	623	1104	221	225	352	130	132	186	92	93	118	71	72	83
0.1 ***	1109	1129	1237	352	362	382	191	196	203	127	130	133	94	96	97
•••	1116	1157	1747	356	370	502	193	199	248	128	131	151	95	96	102
0.15 ***	1538	1573	1813	462	481	523	242	250	264	156	160	166	112	115	118
•••	1550	1626	2309	470	497	633	245	256	302	158	163	179	113	116	119
0.2 ***	1875	1928	2339	549	578	649	282	295	317	178	185	194	126	130	135
•••	1893	2012	2791	561	604	743	287	304	346	181	189	201	128	132	131
0.25 ***	2115	2190	2798	613	653	755	312	329	360	195	204	217	136	141	148
•••	2141	2308	3192	629	689	833	319	342	382	199	209	219	139	144	141
0.3 ***	2263	2364	3182	657	707	842	332	355	395	207	217	234	143	149	157
•••	2298	2520	3513	677	754	904	342	371	409	212	225	231	146	153	147
0.35 ***	2328	2459	3484	681	742	908	345	371	420	213	226	245	147	154	163
•••	2374	2654	3754	707	799	954	356	391	426	219	234	239	150	158	151
0.4 ***	2321	2485	3703	690	760	952	350	380	435	216	230	251	148	155	165
•••	2380	2720	3915	719	825	984	363	402	435	222	239	241	151	159	151
0.45 ***	2256	2455	3834	685	763	974	348	380	440	214	229	251	145	153	163
•••	2328	2726	3995	717	835	994	362	404	435	220	238	239	149	158	147
0.5 ***	2147	2379	3879	668	751	975	340	374	435	208	223	246	140	148	158
•••	2233	2682	3995	703	827	984	355	399	426	215	233	231	144	153	141
0.55 ***	2006	2268	3836	640	727	953	327	360	421	198	213	235	133	140	150
•••	2105	2593	3915	676	804	954	341	385	409	205	223	219	136	145	131
0.6 ***	1844	2128	3705	603	690	910	307	340	397	185	199	219	122	129	137
•••	1953	2465	3754	640	766	904	322	363	382	191	208	201	125	133	119
0.65 ***	1670	1965	3488	558	641	845	283	312	363	168	180	197	109	115	122
•••	1785	2303	3513	593	713	833	296	333	346	174	188	179	112	118	102
0.7 ***	1489	1781	3185	505	581	758	252	278	320	147	157	170	94	98	103
•••	1604	2106	3192	537	645	743	264	296	302	152	163	151	95	100	83
0.75 ***	1301	1577	2795	441	507	650	215	236	267	122	129	138	.	.	.
•••	1411	1874	2791	470	560	633	225	249	248	125	133	118	.	.	.
0.8 ***	1103	1347	2322	367	419	521	171	185	205	.	.	.	.	.	.
•••	1202	1601	2309	390	459	502	178	194	186	.	.	.	.	.	.
0.85 ***	887	1083	1764	279	312	372	.	.	.	.	.	.	.	.	.
•••	967	1277	1747	294	337	352	.	.	.	.	.	.	.	.	.
0.9 ***	635	763	1124	.	.	.	.	.	.	.	.	.	.	.	.
•••	689	881	1104	.	.	.	.	.	.	.	.	.	.	.	.

TABLE 3: ALPHA= 0.01 POWER= 0.8   EXPECTED ACCRUAL THRU MINIMUM FOLLOW-UP= 400

		DEL=.05			DEL=.10			DEL=.15			DEL=.20			DEL=.25	
FACT=	1.0 .75	.50 .25	.00 BIN	1.0 .75	.50 .25	.00 BIN	1.0 .75	.50 .25	.00 BIN	1.0 75 .25	.50	.00 BIN	1.0 .75	.50 .25	.00 BIN
PCONT=***						REQUIRED NUMBER OF PATIENTS									
0.05 ***	607	615	641	220	223	228	129	131	133	92	92	94	71	72	73
•••	610	624	1104	221	225	352	130	132	186	92	93	118	72	72	83
0.1 ***	1110	1131	1237	353	363	382	192	196	203	128	130	133	94	96	98
•••	1118	1159	1747	357	371	502	194	199	248	129	131	151	95	97	102
0.15 ***	1541	1578	1813	464	483	523	243	251	264	156	161	166	113	115	118
•••	1554	1632	2309	472	498	633	246	256	302	158	163	179	114	116	119
0.2 ***	1879	1935	2339	552	580	649	283	296	317	179	185	194	127	130	135
•••	1898	2021	2791	564	606	743	289	305	346	182	189	201	128	132	131
0.25 ***	2120	2200	2798	617	656	755	313	331	360	196	204	217	137	142	148
•••	2148	2321	3192	633	692	833	321	343	382	200	210	219	139	144	141
0.3 ***	2270	2376	3182	661	712	842	335	356	395	208	218	234	144	150	157
•••	2307	2537	3513	682	758	904	344	372	409	213	225	231	147	153	147
0.35 ***	2337	2475	3484	687	748	908	348	373	420	215	227	245	148	154	163
•••	2386	2675	3754	712	804	954	359	392	426	220	235	239	151	158	151
0.4 ***	2333	2504	3702	696	767	952	353	382	435	217	231	251	148	155	165
•••	2395	2744	3915	726	831	984	366	404	435	223	240	241	152	160	151
0.45 ***	2271	2478	3834	692	770	974	352	383	440	215	230	251	146	153	163
•••	2347	2754	3995	725	840	994	365	406	435	222	239	239	150	158	147
0.5 ***	2165	2405	3879	675	759	975	344	376	435	210	224	246	141	149	158
•••	2254	2712	3995	710	833	984	358	400	426	216	234	231	145	153	141
0.55 ***	2027	2296	3836	648	735	953	330	363	421	200	214	235	134	140	150
•••	2129	2625	3915	685	811	954	345	387	409	207	224	219	137	145	131
0.6 ***	1868	2157	3705	612	698	910	311	342	397	187	200	219	123	129	138
•••	1979	2498	3754	648	772	904	325	365	382	193	208	201	126	133	119
0.65 ***	1695	1995	3488	566	649	845	286	314	363	169	181	197	110	115	122
•••	1813	2335	3513	601	719	833	299	335	346	175	188	179	113	118	102
0.7 ***	1514	1811	3185	512	587	758	255	280	320	148	158	171	94	98	103
•••	1632	2136	3192	544	650	743	266	297	302	153	164	151	96	100	83
0.75 ***	1325	1604	2796	448	513	650	218	237	267	122	129	138	.	.	.
•••	1437	1901	2791	476	565	633	226	250	248	126	133	118	.	.	.
0.8 ***	1125	1371	2322	373	423	521	173	186	205	.	.	.	.	.	.
•••	1224	1624	2309	395	462	502	179	195	186	.	.	.	.	.	.
0.85 ***	905	1102	1764	282	315	372	.	.	.	.	.	.	.	.	.
•••	986	1295	1747	297	339	352	.	.	.	.	.	.	.	.	.
0.9 ***	647	775	1124	.	.	.	.	.	.	.	.	.	.	.	.
•••	701	891	1104	.	.	.	.	.	.	.	.	.	.	.	.

TABLE 3: ALPHA= 0.01 POWER= 0.8   EXPECTED ACCRUAL THRU MINIMUM FOLLOW-UP= 425

		DEL=.05			DEL=.10			DEL=.15			DEL=.20			DEL=.25		
FACT=		1.0 .75	.50 .25	.00 BIN	1.0 .75	.50 .25	.00 BIN	1.0 .75	.50 .25	.00 BIN	1.0 75	.50 .25	.00 BIN	1.0 .75	.50 .25	.00 BIN
PCONT=•••					REQUIRED NUMBER OF PATIENTS											
0.05	***	608	616	641	220	223	228	129	131	133	92	93	94	71	72	73
	•••	611	625	1104	221	225	352	130	132	186	92	93	118	71	72	83
0.1	***	1112	1133	1237	354	363	382	192	196	203	128	130	133	95	96	98
	•••	1119	1162	1747	358	371	502	194	199	248	129	131	151	95	97	102
0.15	***	1543	1582	1813	466	484	523	243	251	264	157	161	166	113	115	118
	•••	1557	1637	2309	473	499	633	247	257	302	159	163	179	114	116	119
0.2	***	1882	1941	2339	554	583	649	284	297	317	180	186	194	127	130	135
	•••	1903	2029	2791	566	608	743	290	305	346	182	189	201	129	132	131
0.25	***	2125	2209	2798	620	660	756	315	332	360	197	205	217	137	142	148
	•••	2154	2333	3192	637	695	833	323	344	382	201	210	219	139	145	141
0.3	***	2277	2389	3182	665	716	842	337	358	395	209	219	234	144	150	157
	•••	2316	2552	3513	686	761	904	346	373	409	213	226	231	147	153	147
0.35	***	2346	2490	3485	692	753	908	350	375	419	216	228	245	148	154	163
	•••	2397	2694	3754	717	808	954	361	394	426	221	235	239	151	158	151
0.4	***	2345	2523	3702	703	773	952	356	384	435	218	232	251	149	156	165
	•••	2409	2767	3915	732	836	984	368	405	435	224	240	241	152	160	151
0.45	***	2286	2500	3834	699	777	974	355	385	440	217	231	251	147	154	163
	•••	2364	2780	3995	731	846	994	368	408	435	223	240	239	150	158	147
0.5	***	2182	2430	3879	683	766	975	347	379	435	211	225	246	142	149	158
	•••	2275	2740	3995	718	839	984	361	402	426	218	234	231	145	153	141
0.55	***	2047	2323	3836	656	742	953	333	365	421	201	215	235	134	141	150
	•••	2152	2655	3915	692	816	954	348	388	409	208	224	219	137	145	131
0.6	***	1890	2186	3705	619	705	910	314	344	397	188	201	219	124	130	138
	•••	2005	2529	3754	656	778	904	327	366	382	194	209	201	127	133	119
0.65	***	1719	2024	3488	574	655	845	289	317	363	171	182	197	111	116	122
	•••	1839	2365	3513	609	724	833	301	336	346	176	189	179	113	119	102
0.7	***	1538	1838	3185	519	594	758	257	281	320	149	158	171	94	98	103
	•••	1657	2165	3192	551	655	743	268	298	302	153	164	151	96	101	83
0.75	***	1348	1630	2796	454	518	650	220	239	267	123	130	138	.	.	.
	•••	1461	1927	2791	482	569	633	228	251	248	126	134	118	.	.	.
0.8	***	1145	1393	2322	377	427	521	174	187	205	.	.	.	.	.	.
	•••	1246	1646	2309	399	465	502	180	195	186	.	.	.	.	.	.
0.85	***	922	1119	1764	285	317	371	.	.	.	.	.	.	.	.	
	•••	1003	1311	1747	300	340	352	.	.	.	.	.	.	.	.	
0.9	***	658	786	1124	.	.	.	.	.	.	.	.	.	.	.	.
	•••	712	900	1104	.	.	.	.	.	.	.	.	.	.	.	.

TABLE 3: ALPHA= 0.01 POWER= 0.8   EXPECTED ACCRUAL THRU MINIMUM FOLLOW-UP= 450

PCONT=***		DEL=.05			DEL=.10			DEL=.15			DEL=.20			DEL=.25		
FACT=		1.0 / .75	.50 / .25	.00 / BIN	1.0 / .75	.50 / .25	.00 / BIN	1.0 / .75	.50 / .25	.00 / BIN	1.0 / 75	.50 / .25	.00 / BIN	1.0 / .75	.50 / .25	.00 / BIN
					REQUIRED NUMBER OF PATIENTS											
0.05	***	608	617	641	220	223	228	130	131	133	92	93	93	72	72	73
	***	612	625	1104	221	225	352	130	132	186	92	93	118	72	72	83
0.1	***	1113	1135	1237	355	364	382	192	197	203	128	130	133	95	96	97
	***	1121	1164	1747	359	372	502	194	199	248	129	131	151	95	97	102
0.15	***	1545	1586	1813	467	486	523	244	252	264	157	161	166	113	115	118
	***	1560	1642	2309	475	500	633	248	257	302	159	163	179	114	117	119
0.2	***	1886	1947	2339	556	585	649	285	297	317	180	186	194	127	131	135
	***	1908	2037	2791	568	609	743	291	306	346	183	190	201	129	132	131
0.25	***	2130	2217	2798	623	663	756	316	333	360	198	206	217	138	142	148
	***	2161	2344	3192	640	697	833	324	345	382	201	211	219	140	145	141
0.3	***	2284	2400	3182	670	720	842	339	360	395	210	220	234	145	150	157
	***	2325	2567	3513	691	765	904	348	374	409	214	226	231	147	153	147
0.35	***	2355	2505	3484	697	758	908	352	377	420	217	228	245	149	155	163
	***	2409	2713	3754	723	812	954	363	395	426	222	236	239	152	158	151
0.4	***	2357	2541	3702	708	779	952	358	386	435	220	232	251	150	156	165
	***	2424	2789	3915	738	840	984	371	406	435	226	241	241	153	160	151
0.45	***	2300	2521	3834	705	783	974	357	387	440	218	231	251	148	154	163
	***	2382	2804	3995	738	851	994	371	409	435	225	240	239	151	158	147
0.5	***	2199	2453	3879	690	773	975	350	381	435	212	226	246	143	149	158
	***	2295	2767	3995	725	844	984	364	404	426	219	235	231	146	154	141
0.55	***	2067	2349	3836	663	748	953	336	367	421	203	216	235	135	141	150
	***	2175	2683	3915	699	822	954	350	390	409	209	225	219	138	145	131
0.6	***	1912	2213	3705	627	711	910	316	347	397	189	202	219	124	130	138
	***	2029	2558	3754	663	783	904	330	368	382	195	210	201	127	134	119
0.65	***	1742	2051	3488	581	662	845	291	318	363	172	183	198	111	116	122
	***	1864	2394	3513	615	729	833	303	338	346	177	189	179	113	119	102
0.7	***	1561	1865	3185	525	599	758	260	283	320	150	159	171	95	99	103
	***	1682	2192	3192	557	659	743	270	299	302	154	164	151	96	101	83
0.75	***	1370	1654	2795	460	523	650	221	240	267	124	130	138	.	.	.
	***	1484	1951	2791	487	572	633	230	252	248	127	134	118	.	.	.
0.8	***	1165	1414	2322	382	431	521	175	188	205	.	.	.	.	.	.
	***	1266	1666	2309	403	467	502	181	196	186	.	.	.	.	.	.
0.85	***	938	1135	1764	288	320	372	.	.	.	.	.	.	.	.	.
	***	1019	1325	1747	302	342	352	.	.	.	.	.	.	.	.	.
0.9	***	669	796	1124	.	.	.	.	.	.	.	.	.	.	.	.
	***	723	908	1104	.	.	.	.	.	.	.	.	.	.	.	.

TABLE 3: ALPHA= 0.01 POWER= 0.8   EXPECTED ACCRUAL THRU MINIMUM FOLLOW-UP= 475

		DEL=.05			DEL=.10			DEL=.15			DEL=.20			DEL=.25		
FACT=		1.0 .75	.50 .25	.00 BIN	1.0 .75	.50 .25	.00 BIN	1.0 .75	.50 .25	.00 BIN	1.0 75	.50 .25	.00 BIN	1.0 .75	.50 .25	.00 BIN
PCONT=•••		REQUIRED NUMBER OF PATIENTS														
0.05	***	609	618	641	220	223	228	130	131	133	92	93	93	72	72	72
	•••	612	626	1104	222	225	352	130	132	186	92	93	118	72	72	83
0.1	***	1115	1138	1237	355	365	382	193	197	203	128	130	133	95	96	97
	•••	1123	1166	1747	359	372	502	195	200	248	129	131	151	95	97	102
0.15	***	1548	1590	1813	469	487	523	245	252	264	157	161	166	113	115	118
	•••	1563	1647	2309	476	501	633	248	257	302	159	164	179	114	117	119
0.2	***	1890	1954	2338	559	587	649	286	298	317	181	186	194	128	131	135
	•••	1912	2045	2791	571	611	743	292	306	346	184	190	201	129	133	131
0.25	***	2136	2226	2798	626	666	755	318	334	360	198	206	217	138	143	148
	•••	2168	2355	3192	643	699	833	325	346	382	202	211	219	140	145	141
0.3	***	2291	2412	3182	674	724	842	340	361	395	211	220	234	146	150	157
	•••	2334	2581	3513	695	767	904	349	375	409	215	226	231	148	153	147
0.35	***	2365	2519	3484	702	763	908	355	378	419	218	229	245	149	155	163
	•••	2420	2730	3754	727	815	954	365	396	426	223	236	239	152	159	151
0.4	***	2368	2559	3702	714	784	952	361	388	435	221	233	251	150	156	165
	•••	2438	2810	3915	743	845	984	373	408	435	226	241	241	153	160	151
0.45	***	2314	2541	3834	711	789	974	360	390	440	219	232	251	148	155	163
	•••	2399	2828	3995	744	856	994	373	411	435	225	241	239	151	159	147
0.5	***	2217	2476	3879	696	779	975	352	383	435	214	227	246	143	150	158
	•••	2315	2793	3995	731	849	984	366	405	426	220	235	231	146	154	141
0.55	***	2086	2373	3836	670	755	953	339	370	421	204	217	235	135	142	150
	•••	2197	2710	3915	706	827	954	353	391	409	210	225	219	138	146	131
0.6	***	1933	2239	3705	633	718	910	319	349	397	190	203	219	125	131	137
	•••	2052	2585	3754	669	788	904	333	369	382	196	210	201	128	134	119
0.65	***	1764	2077	3488	587	667	845	294	320	363	172	183	197	112	116	122
	•••	1888	2420	3513	622	734	833	305	339	346	178	190	179	114	119	102
0.7	***	1583	1890	3185	531	604	758	262	285	320	151	159	170	95	99	103
	•••	1706	2217	3192	563	663	743	272	300	302	155	165	151	97	101	83
0.75	***	1391	1677	2795	465	527	650	223	241	267	124	131	138	.	.	.
	•••	1506	1974	2791	492	575	633	231	253	248	127	134	118	.	.	.
0.8	***	1184	1434	2322	386	434	521	177	189	205	.	.	.	.	.	.
	•••	1286	1685	2309	407	469	502	182	196	186	.	.	.	.	.	.
0.85	***	953	1151	1764	291	321	371	.	.	.	.	.	.	.	.	.
	•••	1035	1339	1747	305	343	352	.	.	.	.	.	.	.	.	.
0.9	***	679	805	1124	.	.	.	.	.	.	.	.	.	.	.	.
	•••	732	916	1104	.	.	.	.	.	.	.	.	.	.	.	.

TABLE 3: ALPHA= 0.01 POWER= 0.8   EXPECTED ACCRUAL THRU MINIMUM FOLLOW-UP= 500

	DEL=.05			DEL=.10			DEL=.15			DEL=.20			DEL=.25		
FACT=	1.0 .75	.50 .25	.00 BIN	1.0 .75	.50 .25	.00 BIN	1.0 .75	.50 .25	.00 BIN	1.0 75	.50 .25	.00 BIN	1.0 .75	.50 .25	.00 BIN

PCONT=***   REQUIRED NUMBER OF PATIENTS

PCONT	DEL=.05 1.0/.75	.50/.25	.00/BIN	DEL=.10 1.0/.75	.50/.25	.00/BIN	DEL=.15 1.0/.75	.50/.25	.00/BIN	DEL=.20 1.0/.75	.50/.25	.00/BIN	DEL=.25 1.0/.75	.50/.25	.00/BIN
0.05 ***	610	618	641	221	223	228	130	131	133	92	93	93	72	72	73
•••	613	627	1104	222	226	352	131	132	186	92	93	118	72	72	83
0.1 ***	1116	1140	1237	356	366	382	193	197	203	128	130	133	95	96	98
•••	1125	1169	1747	360	373	502	195	200	248	129	132	151	96	97	102
0.15 ***	1550	1593	1813	470	488	523	245	253	264	158	162	166	113	116	118
•••	1566	1651	2309	478	502	633	249	258	302	160	164	179	114	117	119
0.2 ***	1893	1960	2339	561	589	649	287	299	317	181	187	194	128	131	135
•••	1917	2052	2791	573	612	743	293	307	346	184	190	201	129	133	131
0.25 ***	2141	2235	2798	629	669	755	319	335	360	199	207	217	139	143	148
•••	2174	2366	3192	646	702	833	327	346	382	203	211	219	141	145	141
0.3 ***	2298	2423	3182	677	728	842	342	362	395	212	221	234	146	151	157
•••	2343	2595	3513	698	770	904	351	376	409	216	227	231	148	154	147
0.35 ***	2374	2533	3484	707	767	908	357	380	420	219	230	245	150	156	163
•••	2432	2747	3754	732	819	954	367	397	426	224	237	239	153	159	151
0.4 ***	2380	2576	3703	719	789	952	363	390	435	222	234	251	151	157	165
•••	2452	2829	3915	748	848	984	375	409	435	228	242	241	154	161	151
0.45 ***	2328	2561	3834	717	794	974	363	391	440	221	233	251	149	155	163
•••	2415	2850	3995	750	860	994	375	412	435	227	241	239	152	159	147
0.5 ***	2233	2498	3879	703	784	975	355	385	435	215	228	246	144	150	158
•••	2334	2817	3995	737	854	984	368	406	426	221	236	231	147	154	141
0.55 ***	2105	2397	3836	676	761	953	341	372	421	205	218	235	136	142	150
•••	2218	2736	3915	712	831	954	355	393	409	211	226	219	139	146	131
0.6 ***	1953	2263	3705	640	723	910	322	350	397	191	203	219	126	131	138
•••	2075	2611	3754	676	793	904	335	370	382	197	210	201	128	134	119
0.65 ***	1785	2102	3488	593	673	845	296	322	363	173	184	198	112	117	122
•••	1911	2446	3513	628	738	833	308	340	346	178	190	179	114	119	102
0.7 ***	1604	1914	3185	537	609	758	263	286	320	152	160	171	96	99	103
•••	1728	2241	3192	568	667	743	274	301	302	155	165	151	97	101	83
0.75 ***	1411	1699	2796	470	531	650	225	242	268	125	131	138	.	.	.
•••	1528	1995	2791	497	578	633	233	253	248	128	134	118	.	.	.
0.8 ***	1202	1453	2322	390	437	521	178	189	205	.	.	.	.	.	.
•••	1304	1702	2309	411	472	502	183	197	186	.	.	.	.	.	.
0.85 ***	967	1165	1764	293	323	372	.	.	.	.	.	.	.	.	.
•••	1049	1352	1747	307	344	352	.	.	.	.	.	.	.	.	.
0.9 ***	689	814	1124	.	.	.	.	.	.	.	.	.	.	.	.
•••	742	923	1104	.	.	.	.	.	.	.	.	.	.	.	.

TABLE 3: ALPHA= 0.01 POWER= 0.8   EXPECTED ACCRUAL THRU MINIMUM FOLLOW-UP= 550

		DEL=.02			DEL=.05			DEL=.10			DEL=.15			DEL=.20	
FACT=	1.0 .75	.50 .25	.00 BIN	1.0 .75	.50 .25	.00 BIN	1.0 .75	.50 .25	.00 BIN	1.0 75	.50 .25	.00 BIN	1.0 .75	.50 .25	.00 BIN

PCONT=•••   REQUIRED NUMBER OF PATIENTS

PCONT	DEL=.02 1.0/.75	.50/.25	.00/BIN	DEL=.05 1.0/.75	.50/.25	.00/BIN	DEL=.10 1.0/.75	.50/.25	.00/BIN	DEL=.15 1.0/75	.50/.25	.00/BIN	DEL=.20 1.0/.75	.50/.25	.00/BIN
0.05 •••	2857	2868	3024	611	619	641	221	224	228	130	131	133	92	93	93
•••	2861	2890	5651	614	627	1104	222	226	352	131	132	186	92	93	118
0.1 •••	5944	5971	6642	1119	1144	1237	358	367	382	194	198	202	129	131	133
•••	5953	6025	9816	1128	1172	1747	362	373	502	195	200	248	129	132	151
0.15 •••	8683	8731	10273	1555	1601	1812	473	490	523	246	254	264	158	162	166
•••	8698	8826	13479	1572	1660	2309	480	504	633	250	258	302	160	164	179
0.2 •••	10863	10937	13649	1900	1971	2338	565	593	649	289	300	316	182	187	194
•••	10888	11085	16640	1926	2065	2791	577	615	743	294	308	346	184	190	201
0.25 •••	12429	12535	16643	2151	2251	2798	635	674	756	322	337	360	200	207	217
•••	12464	12746	19300	2187	2385	3192	651	705	833	329	347	382	203	212	219
0.3 •••	13388	13533	19180	2312	2444	3182	684	734	842	345	364	395	213	222	234
•••	13436	13823	21458	2360	2620	3513	705	775	904	354	378	409	217	227	231
0.35 •••	13783	13977	21221	2391	2560	3484	715	775	908	360	383	420	221	231	245
•••	13848	14363	23113	2453	2778	3754	740	825	954	371	399	426	225	237	239
0.4 •••	13678	13930	22740	2402	2608	3702	729	798	952	367	393	435	224	235	251
•••	13762	14433	24268	2479	2866	3915	758	855	984	379	411	435	229	242	241
0.45 •••	13145	13470	23723	2356	2598	3834	728	804	974	367	395	440	223	235	251
•••	13253	14112	24920	2447	2892	3995	760	868	994	379	414	435	228	242	239
0.5 •••	12265	12681	24164	2265	2540	3878	714	795	974	360	388	435	217	229	246
•••	12404	13487	25070	2370	2862	3995	749	862	984	373	409	426	223	237	231
0.55 •••	11125	11653	24058	2140	2442	3836	688	771	953	346	375	421	207	219	235
•••	11302	12641	24719	2258	2783	3915	724	839	954	359	395	409	213	226	219
0.6 •••	9815	10481	23406	1992	2310	3705	652	734	910	326	354	397	193	204	219
•••	10040	11651	23866	2117	2659	3754	687	800	904	338	372	382	198	211	201
0.65 •••	8432	9253	22210	1826	2148	3488	605	683	844	300	325	363	175	185	198
•••	8716	10577	22511	1954	2493	3513	639	745	833	311	342	346	180	191	179
0.7 •••	7071	8037	20474	1645	1959	3184	547	618	758	267	288	320	153	161	170
•••	7417	9456	20655	1771	2285	3192	578	673	743	277	302	302	157	165	151
0.75 •••	5814	6871	18202	1449	1740	2796	479	539	650	228	244	267	126	132	138
•••	6206	8301	18296	1567	2034	2791	505	583	633	235	255	248	129	135	118
0.8 •••	4695	5753	15399	1235	1488	2322	397	442	521	179	191	206	.	.	.
•••	5097	7101	15436	1339	1735	2309	417	475	502	185	198	186	.	.	.
0.85 •••	3689	4648	12071	994	1192	1764	298	327	371	.	.	.	.	.	.
•••	4060	5816	12074	1077	1376	1747	311	346	352	.	.	.	.	.	.
0.9 •••	2720	3478	8224	706	831	1124	.	.	.	.	.	.	.	.	.
•••	3017	4359	8210	759	936	1104	.	.	.	.	.	.	.	.	.
0.95 •••	1622	2048	3864	.	.	.	.	.	.	.	.	.	.	.	.
•••	1794	2499	3844	.	.	.	.	.	.	.	.	.	.	.	.

TABLE 3: ALPHA= 0.01 POWER= 0.8   EXPECTED ACCRUAL THRU MINIMUM FOLLOW-UP= 600

PCONT	FACT= 1.0 .75	.50 .25	.00 BIN	1.0 .75	.50 .25	.00 BIN	1.0 .75	.50 .25	.00 BIN	1.0 75	.50 .25	.00 BIN	1.0 .75	.50 .25	.00 BIN
	DEL=.02			DEL=.05			DEL=.10			DEL=.15			DEL=.20		
0.05	2858	2870	3023	612	620	641	221	224	228	130	131	133	92	93	94
	2862	2894	5651	615	628	1104	223	226	352	131	132	186	92	93	118
0.1	5946	5976	6642	1121	1147	1237	359	368	382	194	198	202	129	131	133
	5956	6035	9816	1131	1176	1747	363	374	502	196	200	248	130	132	151
0.15	8687	8739	10273	1560	1608	1813	475	492	523	248	254	264	159	162	166
	8704	8843	13479	1577	1667	2309	482	505	633	251	259	302	160	164	179
0.2	10870	10951	13649	1907	1982	2339	568	596	649	290	301	317	183	188	194
	10897	11111	16640	1935	2078	2791	580	617	743	295	308	346	185	191	201
0.25	12438	12554	16642	2161	2266	2798	640	678	755	324	338	360	201	208	217
	12477	12785	19300	2200	2402	3192	656	709	833	331	348	382	204	212	219
0.3	13401	13559	19180	2325	2465	3182	691	740	842	348	366	395	214	223	233
	13454	13876	21458	2376	2644	3513	712	779	904	356	379	409	218	227	231
0.35	13801	14012	21221	2409	2585	3484	722	782	908	363	385	419	222	232	245
	13871	14432	23113	2474	2807	3754	748	830	954	373	400	426	227	238	239
0.4	13701	13976	22740	2423	2638	3702	738	806	952	371	395	434	226	236	251
	13793	14522	24268	2504	2900	3915	767	862	984	382	412	435	230	243	241
0.45	13174	13529	23723	2382	2633	3834	738	813	974	371	397	440	224	236	251
	13293	14225	24920	2477	2930	3995	770	874	994	383	416	435	230	243	239
0.5	12303	12756	24163	2295	2579	3878	724	804	974	364	391	435	219	230	246
	12454	13625	25070	2405	2903	3995	759	869	984	376	410	426	224	238	231
0.55	11173	11748	24058	2174	2483	3836	699	781	953	350	378	421	209	220	235
	11366	12805	24719	2296	2826	3915	734	847	954	362	397	409	214	227	219
0.6	9877	10598	23406	2029	2353	3705	662	743	910	330	356	397	195	205	219
	10122	11837	23866	2157	2703	3754	698	807	904	342	374	382	200	212	201
0.65	8510	9390	22210	1864	2191	3488	615	692	845	303	327	363	176	186	197
	8817	10778	22511	1995	2536	3513	649	751	833	314	343	346	181	191	179
0.7	7168	8191	20474	1682	2000	3184	557	626	758	270	290	320	154	161	170
	7537	9664	20655	1810	2325	3192	587	679	743	280	304	302	157	166	151
0.75	5926	7030	18202	1484	1778	2795	487	545	650	230	245	267	127	132	138
	6337	8504	18296	1604	2069	2791	512	588	633	237	256	248	129	135	118
0.8	4811	5907	15399	1266	1520	2322	403	447	521	181	192	205	.	.	.
	5229	7288	15436	1371	1763	2309	423	478	502	186	198	186	.	.	.
0.85	3797	4784	12071	1019	1216	1764	302	330	371	.	.	.	.	.	.
	4180	5975	12074	1102	1396	1747	315	348	352	.	.	.	.	.	.
0.9	2807	3582	8224	722	845	1124	.	.	.	.	.	.	.	.	.
	3112	4475	8210	775	947	1104	.	.	.	.	.	.	.	.	.
0.95	1673	2104	3864	.	.	.	.	.	.	.	.	.	.	.	.
	1847	2554	3844	.	.	.	.	.	.	.	.	.	.	.	.

REQUIRED NUMBER OF PATIENTS

TABLE 3: ALPHA= 0.01 POWER= 0.8   EXPECTED ACCRUAL THRU MINIMUM FOLLOW-UP= 650

	DEL=.02			DEL=.05			DEL=.10			DEL=.15			DEL=.20		
FACT=	1.0 .75	.50 .25	.00 BIN	1.0 .75	.50 .25	.00 BIN	1.0 .75	.50 .25	.00 BIN	1.0 75	.50 .25	.00 BIN	1.0 .75	.50 .25	.00 BIN
PCONT=•••				REQUIRED NUMBER OF PATIENTS											
0.05 ***	2859	2872	3024	613	621	641	222	224	228	131	132	133	92	93	94
•••	2863	2897	5651	616	629	1104	223	226	352	131	132	186	93	93	118
0.1 ***	5949	5981	6642	1124	1150	1237	360	368	382	195	198	203	129	131	133
•••	5959	6044	9816	1134	1179	1747	364	374	502	197	200	248	130	132	151
0.15 ***	8691	8748	10273	1564	1614	1813	477	494	523	249	255	264	159	162	166
•••	8710	8861	13479	1583	1674	2309	484	506	633	252	259	302	161	164	179
0.2 ***	10877	10964	13649	1914	1992	2339	572	599	649	292	302	317	184	188	194
•••	10906	11138	16640	1943	2089	2791	584	619	743	297	309	346	186	191	201
0.25 ***	12448	12573	16642	2171	2281	2798	645	682	756	326	340	361	202	209	217
•••	12490	12823	19300	2212	2419	3192	661	712	833	332	349	382	205	212	219
0.3 ***	13414	13586	19180	2338	2484	3182	697	745	842	350	368	395	216	223	234
•••	13471	13928	21458	2393	2665	3513	717	783	904	359	380	409	219	228	231
0.35 ***	13818	14047	21221	2426	2610	3484	730	788	908	366	387	419	224	233	245
•••	13894	14501	23113	2495	2833	3754	755	835	954	376	402	426	228	239	239
0.4 ***	13724	14022	22740	2445	2667	3702	746	813	952	374	398	435	227	237	251
•••	13823	14611	24268	2529	2931	3915	775	867	984	385	414	435	232	244	241
0.45 ***	13204	13588	23723	2407	2666	3834	747	821	974	374	400	440	226	237	251
•••	13332	14336	24920	2507	2964	3995	779	880	994	386	418	435	231	244	239
0.5 ***	12341	12831	24164	2324	2615	3879	734	813	974	367	394	435	220	231	246
•••	12505	13760	25070	2438	2941	3995	768	875	984	380	412	426	226	238	231
0.55 ***	11222	11842	24058	2207	2522	3836	709	790	953	354	380	421	211	221	235
•••	11430	12963	24719	2331	2866	3915	744	853	954	366	398	409	216	228	219
0.6 ***	9938	10713	23406	2064	2393	3705	673	752	910	333	359	397	197	206	219
•••	10203	12015	23866	2195	2742	3754	707	814	904	345	376	382	201	212	201
0.65 ***	8588	9524	22210	1900	2231	3488	625	699	845	307	330	363	178	187	197
•••	8916	10971	22511	2033	2574	3513	658	757	833	317	345	346	182	192	179
0.7 ***	7263	8338	20474	1717	2038	3184	566	633	758	273	292	320	155	162	171
•••	7653	9861	20655	1848	2361	3192	595	684	743	282	305	302	158	166	151
0.75 ***	6034	7182	18202	1517	1812	2796	494	551	650	232	247	268	127	132	138
•••	6464	8696	18296	1638	2101	2791	519	592	633	239	256	248	130	135	118
0.8 ***	4922	6053	15399	1295	1549	2322	409	452	522	183	193	205	.	.	.
•••	5355	7464	15436	1400	1789	2309	428	481	502	188	198	186	.	.	.
0.85 ***	3899	4912	12071	1042	1238	1764	306	333	372	.	.	.	.	.	.
•••	4293	6124	12074	1125	1415	1747	318	350	352	.	.	.	.	.	.
0.9 ***	2889	3681	8224	737	858	1124	.	.	.	.	.	.	.	.	.
•••	3201	4583	8210	790	957	1104	.	.	.	.	.	.	.	.	.
0.95 ***	1720	2156	3864	.	.	.	.	.	.	.	.	.	.	.	.
•••	1897	2606	3844	.	.	.	.	.	.	.	.	.	.	.	.

TABLE 3: ALPHA= 0.01 POWER= 0.8   EXPECTED ACCRUAL THRU MINIMUM FOLLOW-UP= 700

		DEL=.02			DEL=.05			DEL=.10			DEL=.15			DEL=.20		
FACT=		1.0 .75	.50 .25	.00 BIN	1.0 .75	.50 .25	.00 BIN	1.0 .75	.50 .25	.00 BIN	1.0 75	.50 .25	.00 BIN	1.0 .75	.50 .25	.00 BIN
PCONT=***					REQUIRED NUMBER OF PATIENTS											
0.05	***	2860	2875	3023	614	622	641	222	225	228	131	131	133	92	93	93
	•••	2865	2901	5651	617	630	1104	223	226	352	131	132	186	93	93	118
0.1	***	5951	5986	6642	1126	1153	1237	361	369	382	195	198	202	129	131	133
	•••	5962	6054	9816	1137	1182	1747	365	375	502	197	200	248	130	132	151
0.15	***	8696	8756	10272	1569	1621	1813	479	495	523	249	256	264	160	163	166
	•••	8716	8878	13479	1588	1680	2309	486	507	633	252	259	302	161	164	179
0.2	***	10883	10978	13649	1921	2002	2339	575	601	649	293	303	317	184	189	194
	•••	10915	11164	16640	1951	2099	2791	586	621	743	298	309	346	186	191	201
0.25	***	12458	12592	16642	2181	2295	2798	649	686	755	327	341	360	203	209	217
	•••	12503	12861	19300	2223	2434	3192	665	714	833	334	350	382	206	212	219
0.3	***	13427	13612	19180	2351	2503	3182	702	750	842	352	369	395	216	224	234
	•••	13489	13980	21458	2408	2685	3513	723	786	904	360	381	409	220	229	231
0.35	***	13836	14082	21221	2442	2633	3484	736	794	908	369	389	420	225	234	245
	•••	13918	14570	23113	2515	2858	3754	761	839	954	378	403	426	229	239	239
0.4	***	13746	14068	22740	2465	2694	3702	754	820	952	377	400	435	228	238	251
	•••	13854	14699	24268	2553	2959	3915	782	872	984	387	415	435	233	244	241
0.45	***	13234	13647	23723	2431	2697	3834	755	828	974	377	402	440	227	238	251
	•••	13372	14446	24920	2534	2996	3995	787	885	994	389	419	435	232	244	239
0.5	***	12379	12906	24163	2352	2649	3879	743	820	975	371	397	435	222	232	246
	•••	12555	13892	25070	2469	2975	3995	777	881	984	383	414	426	227	239	231
0.55	***	11270	11935	24057	2238	2559	3836	718	797	953	357	383	421	212	222	235
	•••	11494	13117	24719	2365	2902	3915	753	859	954	369	400	409	217	228	219
0.6	***	10000	10825	23406	2096	2431	3705	681	759	910	337	361	397	198	207	219
	•••	10283	12187	23866	2230	2779	3754	716	819	904	348	377	382	202	213	201
0.65	***	8664	9655	22210	1933	2268	3488	633	707	845	310	331	363	179	187	198
	•••	9014	11153	22511	2068	2610	3513	666	762	833	320	346	346	183	192	179
0.7	***	7356	8480	20474	1750	2073	3184	573	639	758	275	294	320	156	163	170
	•••	7767	10047	20655	1881	2394	3192	603	688	743	284	306	302	159	166	151
0.75	***	6138	7328	18202	1548	1844	2795	501	556	650	234	248	268	128	133	138
	•••	6585	8877	18296	1670	2130	2791	526	595	633	240	257	248	131	135	118
0.8	***	5029	6191	15399	1322	1576	2322	414	455	521	184	194	205	.	.	.
	•••	5475	7629	15436	1427	1813	2309	433	484	502	188	199	186	.	.	.
0.85	***	3997	5033	12071	1063	1259	1764	309	334	372	.	.	.	.	.	.
	•••	4401	6263	12074	1146	1432	1747	321	351	352	.	.	.	.	.	.
0.9	***	2967	3772	8224	751	870	1124	.	.	.	.	.	.	.	.	.
	•••	3285	4684	8210	803	966	1104	.	.	.	.	.	.	.	.	.
0.95	***	1765	2204	3864	.	.	.	.	.	.	.	.	.	.	.	.
	•••	1944	2652	3844	.	.	.	.	.	.	.	.	.	.	.	.

TABLE 3: ALPHA= 0.01 POWER= 0.8   EXPECTED ACCRUAL THRU MINIMUM FOLLOW-UP= 750

		DEL=.02			DEL=.05			DEL=.10			DEL=.15			DEL=.20	
FACT=		1.0 .75	.50 .25	.00 BIN	1.0 .75	.50 .25	.00 BIN	1.0 .75	.50 .25	.00 BIN	1.0 75	.50 .25	.00 BIN	1.0 .75	.50 .25	.00 BIN
PCONT=***					REQUIRED NUMBER OF PATIENTS											
0.05	***	2861	2876	3024	614	623	641	222	225	228	130	132	133	93	93	94
	•••	2866	2904	5651	618	630	1104	223	226	352	131	132	186	93	93	118
0.1	***	5953	5991	6642	1129	1157	1237	362	370	382	196	199	203	130	131	133
	•••	5966	6063	9816	1140	1185	1747	365	375	502	197	200	248	130	132	151
0.15	***	8700	8765	10272	1573	1626	1813	481	497	523	250	256	264	160	163	166
	•••	8722	8894	13479	1594	1685	2309	488	508	633	253	259	302	161	164	179
0.2	***	10890	10991	13649	1928	2012	2339	578	604	649	295	304	317	185	189	194
	•••	10924	11191	16640	1960	2109	2791	589	622	743	299	310	346	187	191	201
0.25	***	12467	12611	16642	2190	2308	2798	653	689	755	329	342	360	204	209	217
	•••	12515	12899	19300	2235	2447	3192	669	716	833	335	350	382	206	213	219
0.3	***	13440	13639	19180	2364	2520	3182	707	754	842	355	371	395	217	225	234
	•••	13506	14031	21458	2423	2703	3513	727	790	904	362	382	409	220	229	231
0.35	***	13854	14118	21221	2459	2654	3484	742	799	908	371	391	420	226	234	245
	•••	13941	14638	23113	2533	2880	3754	767	843	954	380	404	426	230	239	239
0.4	***	13770	14114	22740	2485	2720	3702	760	825	952	379	402	435	229	239	251
	•••	13885	14785	24268	2575	2985	3915	789	876	984	390	416	435	234	244	241
0.45	***	13263	13706	23723	2455	2726	3834	763	835	974	380	404	440	229	238	251
	•••	13411	14553	24920	2560	3026	3995	794	890	994	391	420	435	233	244	239
0.5	***	12416	12980	24163	2379	2681	3879	751	827	975	374	399	435	223	233	246
	•••	12605	14021	25070	2498	3007	3995	784	886	984	385	415	426	228	239	231
0.55	***	11318	12027	24057	2268	2593	3835	727	804	953	360	385	421	213	223	235
	•••	11558	13265	24719	2397	2935	3915	760	864	954	371	401	409	218	229	219
0.6	***	10061	10936	23406	2128	2465	3705	690	766	910	340	363	397	199	208	219
	•••	10363	12352	23866	2263	2813	3754	724	824	904	350	378	382	203	213	201
0.65	***	8741	9782	22210	1965	2303	3488	641	713	844	312	334	363	180	188	198
	•••	9111	11329	22511	2102	2643	3513	673	767	833	322	347	346	184	192	179
0.7	***	7447	8617	20474	1781	2106	3184	580	645	758	278	295	320	157	163	170
	•••	7877	10225	20655	1914	2425	3192	609	692	743	286	307	302	160	167	151
0.75	***	6239	7466	18202	1577	1874	2795	507	560	650	235	249	267	129	133	139
	•••	6702	9048	18296	1699	2157	2791	531	598	633	242	258	248	131	136	118
0.8	***	5131	6322	15399	1347	1601	2322	419	459	521	185	194	205	.	.	.
	•••	5590	7785	15436	1453	1834	2309	437	486	502	190	199	186	.	.	.
0.85	***	4090	5147	12071	1083	1278	1764	312	337	371	.	.	.	.	.	.
	•••	4503	6393	12074	1165	1447	1747	324	352	352	.	.	.	.	.	.
0.9	***	3041	3859	8224	763	881	1124	.	.	.	.	.	.	.	.	.
	•••	3365	4777	8210	814	974	1104	.	.	.	.	.	.	.	.	.
0.95	***	1807	2250	3864	.	.	.	.	.	.	.	.	.	.	.	.
	•••	1987	2695	3844	.	.	.	.	.	.	.	.	.	.	.	.

TABLE 3: ALPHA= 0.01 POWER= 0.8   EXPECTED ACCRUAL THRU MINIMUM FOLLOW-UP= 800

		DEL=.02			DEL=.05			DEL=.10			DEL=.15			DEL=.20		
FACT=		1.0 .75	.50 .25	.00 BIN	1.0 .75	.50 .25	.00 BIN	1.0 .75	.50 .25	.00 BIN	1.0 75	.50 .25	.00 BIN	1.0 .75	.50 .25	.00 BIN
PCONT=•••					REQUIRED NUMBER OF PATIENTS											
0.05	•••	2862	2879	3023	615	624	641	223	225	228	131	132	133	92	93	94
	•••	2868	2907	5651	619	631	1104	224	226	352	131	132	186	93	93	118
0.1	•••	5956	5996	6642	1131	1159	1237	363	371	382	196	199	203	130	131	133
	•••	5969	6072	9816	1142	1187	1747	366	376	502	197	201	248	131	132	151
0.15	•••	8704	8774	10273	1578	1632	1813	483	498	523	251	256	264	161	163	166
	•••	8728	8911	13479	1599	1691	2309	490	509	633	253	260	302	162	165	179
0.2	•••	10897	11004	13649	1935	2021	2339	580	606	649	296	305	317	185	189	194
	•••	10933	11217	16640	1967	2117	2791	592	624	743	300	310	346	187	192	201
0.25	•••	12477	12631	16643	2200	2321	2798	656	692	755	331	343	360	204	210	217
	•••	12528	12936	19300	2246	2460	3192	672	718	833	336	351	382	207	213	219
0.3	•••	13454	13665	19180	2376	2537	3182	712	758	842	356	372	395	218	225	234
	•••	13524	14083	21458	2438	2720	3513	732	792	904	364	382	409	221	229	231
0.35	•••	13871	14153	21221	2475	2675	3484	748	804	908	373	392	420	227	235	245
	•••	13965	14705	23113	2551	2901	3754	772	846	954	382	405	426	231	240	239
0.4	•••	13793	14160	22740	2504	2744	3702	767	831	952	382	404	435	231	240	251
	•••	13915	14871	24268	2598	3010	3915	795	880	984	392	417	435	235	245	241
0.45	•••	13293	13765	23723	2478	2754	3834	770	840	974	383	406	440	230	239	251
	•••	13451	14659	24920	2586	3053	3995	801	894	994	394	421	435	234	245	239
0.5	•••	12454	13055	24163	2405	2712	3879	759	833	975	376	400	435	224	234	246
	•••	12656	14146	25070	2527	3037	3995	792	890	984	387	416	426	229	240	231
0.55	•••	11366	12119	24058	2296	2625	3836	735	811	953	363	387	421	214	224	235
	•••	11622	13409	24719	2427	2966	3915	768	868	954	374	402	409	219	229	219
0.6	•••	10122	11044	23406	2157	2498	3705	698	772	910	342	365	397	200	208	219
	•••	10442	12510	23866	2295	2844	3754	731	828	904	353	379	382	204	213	201
0.65	•••	8817	9905	22210	1995	2335	3488	649	719	845	314	335	363	181	188	197
	•••	9206	11496	22511	2133	2672	3513	680	771	833	324	348	346	185	193	179
0.7	•••	7537	8750	20474	1811	2136	3185	587	650	758	280	297	320	158	164	171
	•••	7984	10394	20655	1944	2452	3192	615	695	743	288	308	302	160	167	151
0.75	•••	6338	7600	18202	1604	1901	2796	513	565	650	237	250	267	129	133	138
	•••	6816	9210	18296	1727	2181	2791	536	601	633	243	258	248	131	136	118
0.8	•••	5229	6448	15399	1371	1624	2322	423	462	521	186	195	205	.	.	.
	•••	5700	7932	15436	1477	1854	2309	441	488	502	190	200	186	.	.	.
0.85	•••	4180	5256	12071	1102	1295	1764	315	339	372	.	.	.	.	.	.
	•••	4601	6516	12074	1184	1461	1747	326	353	352	.	.	.	.	.	.
0.9	•••	3112	3942	8224	775	891	1124	.	.	.	.	.	.	.	.	.
	•••	3441	4865	8210	826	981	1104	.	.	.	.	.	.	.	.	.
0.95	•••	1847	2292	3864	.	.	.	.	.	.	.	.	.	.	.	.
	•••	2029	2735	3844	.	.	.	.	.	.	.	.	.	.	.	.

TABLE 3: ALPHA= 0.01 POWER= 0.8   EXPECTED ACCRUAL THRU MINIMUM FOLLOW-UP= 850

	DEL=.02			DEL=.05			DEL=.10			DEL=.15			DEL=.20		
FACT=	1.0 .75	.50 .25	.00 BIN	1.0 .75	.50 .25	.00 BIN	1.0 .75	.50 .25	.00 BIN	1.0 75	.50 .25	.00 BIN	1.0 .75	.50 .25	.00 BIN

PCONT=•••   REQUIRED NUMBER OF PATIENTS

PCONT		DEL=.02			DEL=.05			DEL=.10			DEL=.15			DEL=.20		
		1.0/.75	.50/.25	.00/BIN	1.0/.75	.50/.25	.00/BIN	1.0/.75	.50/.25	.00/BIN	1.0/75	.50/.25	.00/BIN	1.0/.75	.50/.25	.00/BIN
0.05	***	2863	2881	3024	616	624	641	223	225	228	131	132	133	93	93	94
	•••	2869	2910	5651	620	631	1104	224	226	352	131	133	186	93	93	118
0.1	***	5959	6001	6642	1133	1162	1236	363	371	382	196	199	203	130	131	133
	•••	5972	6081	9816	1144	1189	1747	367	376	502	198	201	248	131	132	151
0.15	***	8708	8783	10272	1582	1637	1813	484	499	524	252	257	264	161	163	167
	•••	8733	8928	13479	1603	1695	2309	491	510	633	254	260	302	162	165	179
0.2	***	10903	11018	13649	1941	2029	2339	583	607	649	297	305	317	186	189	194
	•••	10942	11243	16640	1975	2126	2791	594	625	743	300	310	346	188	192	201
0.25	***	12487	12650	16643	2209	2333	2798	660	695	756	332	344	360	205	210	217
	•••	12541	12974	19300	2256	2472	3192	675	720	833	338	351	382	207	213	219
0.3	***	13467	13691	19180	2389	2552	3182	716	762	842	358	373	395	219	225	233
	•••	13542	14133	21458	2451	2735	3513	736	795	904	365	383	409	222	229	231
0.35	***	13889	14188	21221	2490	2694	3485	753	808	907	375	394	419	227	236	245
	•••	13989	14772	23113	2569	2920	3754	777	849	954	384	406	426	231	240	239
0.4	***	13815	14205	22740	2523	2767	3703	773	836	952	384	405	435	232	240	251
	•••	13945	14955	24268	2618	3032	3915	801	883	984	394	418	435	236	245	241
0.45	***	13322	13823	23723	2500	2780	3834	777	846	974	385	408	440	231	240	252
	•••	13490	14761	24920	2610	3079	3995	807	898	994	395	422	435	235	245	239
0.5	***	12492	13128	24163	2430	2740	3879	766	839	974	379	402	435	225	235	246
	•••	12706	14268	25070	2553	3064	3995	798	894	984	390	417	426	230	240	231
0.55	***	11414	12209	24058	2323	2655	3836	742	816	953	365	389	421	215	224	235
	•••	11685	13549	24719	2456	2994	3915	775	872	954	376	403	409	220	229	219
0.6	***	10183	11151	23406	2186	2529	3705	705	778	910	344	366	397	201	209	219
	•••	10520	12663	23866	2325	2873	3754	737	832	904	355	380	382	205	214	201
0.65	***	8892	10024	22210	2024	2365	3488	655	724	845	317	337	363	182	189	197
	•••	9299	11656	22511	2163	2700	3513	686	774	833	326	349	346	185	193	179
0.7	***	7625	8877	20474	1838	2165	3184	594	655	758	281	298	320	158	164	171
	•••	8089	10555	20655	1973	2478	3192	621	698	743	289	308	302	161	167	151
0.75	***	6432	7728	18202	1630	1927	2796	518	569	650	239	252	267	130	134	138
	•••	6925	9365	18296	1753	2204	2791	541	604	633	244	259	248	131	136	118
0.8	***	5324	6568	15399	1393	1646	2322	427	464	521	187	195	205	.	.	.
	•••	5805	8072	15435	1499	1872	2309	444	490	502	191	200	186	.	.	.
0.85	***	4265	5360	12071	1119	1311	1765	317	340	372	.	.	.	.	.	.
	•••	4694	6632	12074	1201	1473	1747	328	355	352	.	.	.	.	.	.
0.9	***	3179	4020	8224	786	900	1124	.	.	.	.	.	.	.	.	.
	•••	3513	4948	8210	836	988	1104	.	.	.	.	.	.	.	.	.
0.95	***	1885	2331	3864	.	.	.	.	.	.	.	.	.	.	.	.
	•••	2067	2772	3844	.	.	.	.	.	.	.	.	.	.	.	.

TABLE 3: ALPHA= 0.01 POWER= 0.8   EXPECTED ACCRUAL THRU MINIMUM FOLLOW-UP= 900

PCONT=***		DEL=.02			DEL=.05			DEL=.10			DEL=.15			DEL=.20		
FACT=		1.0 .75	.50 .25	.00 BIN	1.0 .75	.50 .25	.00 BIN	1.0 .75	.50 .25	.00 BIN	1.0 75	.50 .25	.00 BIN	1.0 .75	.50 .25	.00 BIN
		REQUIRED NUMBER OF PATIENTS														
0.05	***	2864	2882	3024	617	625	641	223	225	228	131	132	133	92	93	93
	•••	2871	2913	5651	621	632	1104	224	226	352	131	132	186	93	93	118
0.1	***	5961	6005	6642	1135	1164	1236	364	371	382	197	199	203	130	131	133
	•••	5976	6090	9816	1147	1191	1747	368	376	502	198	201	248	131	132	151
0.15	***	8713	8792	10272	1586	1642	1812	486	500	523	252	257	264	161	163	166
	•••	8739	8945	13479	1608	1700	2309	492	510	633	254	260	302	162	165	179
0.2	***	10910	11031	13649	1947	2037	2339	585	609	649	297	306	317	186	190	194
	•••	10950	11269	16640	1982	2133	2791	596	626	743	301	311	346	188	192	201
0.25	***	12496	12669	16642	2217	2344	2798	663	697	756	333	345	360	206	210	217
	•••	12554	13011	19300	2267	2483	3192	678	722	833	338	352	382	208	213	219
0.3	***	13480	13718	19180	2400	2567	3182	720	765	842	360	374	395	219	226	234
	•••	13559	14184	21458	2465	2750	3513	740	797	904	366	384	409	222	230	231
0.35	***	13906	14223	21221	2504	2712	3484	758	812	908	377	395	420	228	236	245
	•••	14012	14837	23113	2585	2938	3754	782	852	954	386	406	426	232	240	239
0.4	***	13838	14251	22740	2541	2789	3702	779	840	952	386	406	434	233	241	251
	•••	13976	15039	24268	2638	3054	3915	806	886	984	396	419	435	236	245	241
0.45	***	13352	13881	23723	2521	2804	3834	783	851	974	387	409	440	231	240	251
	•••	13529	14862	24920	2633	3102	3995	813	901	994	397	423	435	236	245	239
0.5	***	12530	13201	24164	2453	2767	3879	773	845	974	381	404	435	226	235	246
	•••	12756	14388	25070	2579	3090	3995	804	898	984	392	418	426	230	240	231
0.55	***	11462	12297	24058	2349	2683	3836	748	822	953	368	390	421	216	225	235
	•••	11748	13683	24719	2484	3021	3915	781	876	954	378	404	409	220	230	219
0.6	***	10243	11255	23406	2213	2558	3705	711	783	910	347	368	397	201	209	219
	•••	10598	12810	23866	2353	2899	3754	743	836	904	356	381	382	206	214	201
0.65	***	8965	10141	22210	2051	2394	3488	662	729	845	318	338	363	182	189	198
	•••	9390	11810	22511	2191	2726	3513	692	777	833	327	350	346	186	193	179
0.7	***	7710	9000	20474	1865	2192	3185	599	659	758	283	299	320	159	164	171
	•••	8191	10709	20655	2000	2502	3192	626	701	743	290	309	302	162	167	151
0.75	***	6525	7851	18201	1654	1951	2795	523	572	650	240	252	267	130	134	138
	•••	7030	9513	18296	1778	2225	2791	545	606	633	245	259	248	132	136	118
0.8	***	5415	6683	15399	1414	1666	2322	431	467	521	188	196	206	.	.	.
	•••	5907	8205	15436	1520	1889	2309	447	491	502	192	200	186	.	.	.
0.85	***	4348	5459	12071	1135	1325	1764	320	342	371	.	.	.	.	.	.
	•••	4784	6742	12074	1216	1485	1747	330	355	352	.	.	.	.	.	.
0.9	***	3243	4094	8224	796	908	1124	.	.	.	.	.	.	.	.	.
	•••	3582	5025	8210	845	994	1104	.	.	.	.	.	.	.	.	.
0.95	***	1920	2368	3864	.	.	.	.	.	.	.	.	.	.	.	.
	•••	2105	2806	3844	.	.	.	.	.	.	.	.	.	.	.	.

TABLE 3: ALPHA= 0.01 POWER= 0.8   EXPECTED ACCRUAL THRU MINIMUM FOLLOW-UP= 950

		DEL=.02			DEL=.05			DEL=.10			DEL=.15			DEL=.20		
FACT=		1.0 .75	.50 .25	.00 BIN	1.0 .75	.50 .25	.00 BIN	1.0 .75	.50 .25	.00 BIN	1.0 75 .25	.50	.00 BIN	1.0 .75	.50 .25	.00 BIN
PCONT=				REQUIRED NUMBER OF PATIENTS												
0.05	***	2865	2884	3023	618	626	641	223	225	228	131	132	133	93	93	93
	•••	2871	2915	5651	621	633	1104	224	226	352	131	133	186	93	93	118
0.1	***	5963	6010	6642	1138	1166	1237	365	372	382	197	200	203	130	131	133
	•••	5979	6099	9816	1150	1193	1747	368	376	502	198	201	248	131	132	151
0.15	***	8717	8800	10273	1590	1647	1813	487	501	523	253	257	264	161	163	166
	•••	8745	8961	13479	1612	1704	2309	494	511	633	255	260	302	162	165	179
0.2	***	10917	11045	13649	1953	2044	2338	587	610	648	298	306	317	186	190	194
	•••	10959	11294	16640	1989	2140	2791	598	627	743	302	311	346	188	192	201
0.25	***	12506	12688	16642	2226	2355	2798	666	699	755	334	346	361	206	211	217
	•••	12567	13047	19300	2276	2493	3192	680	723	833	339	352	382	209	214	219
0.3	***	13493	13744	19180	2412	2581	3182	724	767	842	361	375	395	220	226	234
	•••	13577	14233	21458	2478	2763	3513	743	799	904	368	384	409	223	230	231
0.35	***	13924	14258	21221	2519	2730	3484	762	815	908	378	396	419	229	236	245
	•••	14035	14902	23113	2602	2955	3754	786	854	954	387	407	426	233	241	239
0.4	***	13861	14296	22740	2559	2810	3702	784	844	952	388	407	435	234	241	251
	•••	14007	15120	24268	2658	3073	3915	811	889	984	397	420	435	237	246	241
0.45	***	13382	13940	23723	2541	2828	3834	789	856	974	390	410	440	232	241	251
	•••	13568	14961	24920	2655	3125	3995	818	904	994	399	424	435	236	246	239
0.5	***	12568	13273	24163	2476	2793	3878	779	849	975	383	405	435	227	235	246
	•••	12806	14504	25070	2604	3114	3995	810	901	984	393	419	426	231	241	231
0.55	***	11510	12385	24058	2373	2710	3836	755	827	953	369	391	421	217	225	235
	•••	11811	13815	24719	2510	3046	3915	787	879	954	380	405	409	221	230	219
0.6	***	10303	11357	23406	2238	2585	3705	717	788	910	349	369	397	203	210	219
	•••	10674	12952	23866	2380	2924	3754	749	838	904	358	382	382	206	215	201
0.65	***	9039	10254	22210	2077	2420	3488	667	733	844	320	338	363	183	190	197
	•••	9480	11958	22511	2218	2750	3513	697	780	833	329	350	346	186	193	179
0.7	***	7794	9120	20474	1890	2217	3184	604	663	758	285	300	320	159	165	171
	•••	8290	10857	20655	2025	2524	3192	631	704	743	292	310	302	162	167	151
0.75	***	6615	7969	18202	1677	1974	2795	527	576	650	241	253	267	131	134	139
	•••	7133	9653	18296	1801	2244	2791	549	608	633	247	260	248	133	136	118
0.8	***	5504	6794	15399	1434	1685	2322	433	469	521	189	196	205	.	.	.
	•••	6005	8332	15436	1540	1904	2309	450	492	502	192	201	186	.	.	.
0.85	***	4427	5554	12071	1151	1339	1764	321	343	371	.	.	.	.	.	.
	•••	4870	6847	12074	1231	1495	1747	331	356	352	.	.	.	.	.	.
0.9	***	3306	4165	8224	805	916	1124	.	.	.	.	.	.	.	.	.
	•••	3648	5099	8210	854	999	1104	.	.	.	.	.	.	.	.	.
0.95	***	1955	2404	3864	.	.	.	.	.	.	.	.	.	.	.	.
	•••	2139	2838	3844	.	.	.	.	.	.	.	.	.	.	.	.

TABLE 3: ALPHA= 0.01 POWER= 0.8   EXPECTED ACCRUAL THRU MINIMUM FOLLOW-UP= 1000

PCONT= ***		DEL=.02			DEL=.05			DEL=.10			DEL=.15			DEL=.20		
FACT=		1.0 .75	.50 .25	.00 BIN	1.0 .75	.50 .25	.00 BIN	1.0 .75	.50 .25	.00 BIN	1.0 75	.50 .25	.00 BIN	1.0 .75	.50 .25	.00 BIN
						REQUIRED NUMBER OF PATIENTS										
0.05	***	2866	2886	3023	618	626	641	223	226	228	131	132	133	93	93	93
	***	2873	2918	5651	622	633	1104	225	226	352	131	133	186	93	93	118
0.1	***	5966	6015	6642	1140	1169	1236	366	373	382	197	200	203	130	131	133
	***	5983	6108	9816	1151	1195	1747	369	377	502	198	201	248	131	132	151
0.15	***	8721	8809	10273	1593	1651	1813	488	502	523	253	258	264	161	164	166
	***	8751	8977	13479	1616	1708	2309	495	511	633	255	261	302	163	165	179
0.2	***	10924	11058	13650	1960	2052	2339	589	612	649	299	306	316	187	190	194
	***	10968	11320	16640	1996	2146	2791	600	628	743	303	311	346	188	192	201
0.25	***	12515	12708	16643	2235	2366	2798	669	701	755	335	346	360	206	211	216
	***	12580	13084	19300	2286	2503	3192	683	725	833	340	353	382	209	214	219
0.3	***	13506	13771	19180	2423	2595	3182	728	770	842	362	376	395	221	226	234
	***	13595	14283	21458	2490	2776	3513	746	801	904	369	385	409	223	230	231
0.35	***	13941	14293	21221	2533	2747	3485	767	819	908	380	397	420	230	237	245
	***	14059	14966	23113	2618	2971	3754	790	856	954	388	407	426	233	241	239
0.4	***	13885	14342	22740	2576	2830	3703	789	848	952	390	409	435	234	241	251
	***	14038	15200	24268	2676	3092	3915	815	891	984	399	421	435	238	246	241
0.45	***	13411	13998	23723	2561	2850	3834	794	860	974	391	412	440	233	241	251
	***	13608	15058	24920	2676	3146	3995	823	908	994	401	425	435	237	246	239
0.5	***	12606	13345	24163	2498	2817	3879	785	854	975	385	406	435	228	236	246
	***	12856	14616	25070	2627	3136	3995	815	904	984	395	420	426	232	241	231
0.55	***	11558	12472	24058	2397	2736	3836	761	831	953	371	393	421	218	226	235
	***	11873	13941	24719	2535	3070	3915	792	882	954	381	406	409	221	230	219
0.6	***	10363	11457	23406	2263	2611	3705	723	793	910	350	370	397	203	210	219
	***	10750	13090	23866	2406	2948	3754	755	842	904	360	383	382	206	215	201
0.65	***	9111	10365	22210	2102	2446	3488	673	738	845	322	340	363	184	190	198
	***	9568	12101	22511	2244	2773	3513	702	783	833	330	351	346	187	194	179
0.7	***	7877	9235	20474	1914	2241	3185	609	666	758	286	301	320	160	165	171
	***	8386	10999	20655	2050	2545	3192	635	706	743	293	310	302	162	168	151
0.75	***	6703	8084	18202	1699	1995	2796	531	578	650	242	253	268	131	135	138
	***	7231	9788	18296	1823	2262	2791	553	610	633	248	260	248	133	136	118
0.8	***	5590	6900	15399	1453	1702	2322	437	471	521	190	197	205	.	.	.
	***	6100	8453	15436	1558	1919	2309	453	494	502	193	201	186	.	.	.
0.85	***	4503	5645	12071	1165	1352	1765	323	344	371	.	.	.	.	.	.
	***	4953	6946	12074	1246	1505	1747	333	357	352	.	.	.	.	.	.
0.9	***	3365	4232	8224	815	923	1124	.	.	.	.	.	.	.	.	.
	***	3711	5169	8210	863	1005	1104	.	.	.	.	.	.	.	.	.
0.95	***	1987	2437	3864	.	.	.	.	.	.	.	.	.	.	.	.
	***	2173	2868	3844	.	.	.	.	.	.	.	.	.	.	.	.

TABLE 3: ALPHA= 0.01 POWER= 0.8   EXPECTED ACCRUAL THRU MINIMUM FOLLOW-UP= 1100

	DEL=.02			DEL=.05			DEL=.10			DEL=.15			DEL=.20		
FACT=	1.0 .75	.50 .25	.00 BIN	1.0 .75	.50 .25	.00 BIN	1.0 .75	.50 .25	.00 BIN	1.0 75	.50 .25	.00 BIN	1.0 .75	.50 .25	.00 BIN
PCONT=•••					REQUIRED NUMBER OF PATIENTS										
0.05 ***	2868	2890	3024	620	627	642	224	225	228	131	132	133	93	93	93
•••	2876	2923	5651	623	633	1104	225	227	352	131	133	186	93	93	118
0.1 ***	5971	6025	6642	1143	1172	1237	367	373	382	197	200	202	131	131	133
•••	5989	6124	9816	1156	1198	1747	370	377	502	199	201	248	131	132	151
0.15 ***	8731	8826	10273	1600	1660	1812	490	504	523	254	258	264	162	164	166
•••	8762	9008	13479	1625	1715	2309	496	512	633	256	261	302	163	165	179
0.2 ***	10938	11085	13649	1971	2065	2338	593	615	648	300	307	316	187	191	194
•••	10986	11369	16640	2009	2159	2791	603	630	743	303	312	346	189	192	201
0.25 ***	12534	12746	16642	2251	2385	2798	674	705	756	337	347	360	207	212	217
•••	12605	13155	19300	2304	2520	3192	688	727	833	342	353	382	209	214	219
0.3 ***	13533	13823	19180	2444	2620	3182	734	775	842	364	378	395	221	227	234
•••	13630	14379	21458	2514	2799	3513	752	804	904	371	386	409	224	230	231
0.35 ***	13977	14363	21221	2560	2778	3484	775	825	907	383	399	419	231	237	245
•••	14106	15091	23113	2647	3000	3754	797	860	954	390	408	426	234	241	239
0.4 ***	13930	14433	22740	2608	2866	3702	797	855	951	393	411	434	235	242	251
•••	14099	15355	24268	2711	3126	3915	824	896	984	401	422	435	239	246	241
0.45 ***	13470	14112	23723	2598	2892	3834	804	868	974	395	414	440	235	242	251
•••	13686	15246	24920	2716	3184	3995	833	912	994	404	426	435	238	246	239
0.5 ***	12681	13487	24164	2540	2862	3878	795	862	974	389	408	435	229	236	246
•••	12956	14833	25070	2671	3178	3995	825	910	984	398	421	426	233	241	231
0.55 ***	11653	12640	24058	2442	2783	3836	771	840	953	375	395	421	219	226	235
•••	11997	14184	24719	2582	3112	3915	802	888	954	384	407	409	223	230	219
0.6 ***	10481	11651	23406	2310	2659	3705	734	800	910	353	372	397	204	211	219
•••	10899	13351	23866	2454	2991	3754	764	847	904	362	384	382	208	214	201
0.65 ***	9252	10576	22210	2149	2493	3488	683	745	844	324	342	363	185	191	197
•••	9740	12371	22511	2292	2814	3513	711	788	833	333	351	346	188	194	179
0.7 ***	8037	9456	20474	1959	2285	3184	618	673	758	288	302	320	161	165	170
•••	8573	11267	20655	2095	2583	3192	643	710	743	295	311	302	163	168	151
0.75 ***	6871	8301	18202	1740	2034	2795	538	583	650	244	254	268	131	135	138
•••	7421	10042	18296	1864	2294	2791	559	613	633	249	261	248	133	136	118
0.8 ***	5753	7101	15399	1488	1735	2322	442	475	521	191	197	206	.	.	.
•••	6279	8679	15436	1593	1944	2309	458	496	502	194	202	186	.	.	.
0.85 ***	4648	5816	12071	1192	1376	1764	327	346	371	.	.	.	.	.	.
•••	5109	7132	12074	1271	1523	1747	336	358	352	.	.	.	.	.	.
0.9 ***	3477	4359	8224	830	936	1124	.	.	.	.	.	.	.	.	.
•••	3831	5298	8210	877	1013	1104	.	.	.	.	.	.	.	.	.
0.95 ***	2048	2498	3864	.	.	.	.	.	.	.	.	.	.	.	.
•••	2235	2923	3844	.	.	.	.	.	.	.	.	.	.	.	.

TABLE 3: ALPHA= 0.01   POWER= 0.8   EXPECTED ACCRUAL THRU MINIMUM FOLLOW-UP= 1200

PCONT	DEL=.02 1.0/.75	.50/.25	.00/BIN	DEL=.05 1.0/.75	.50/.25	.00/BIN	DEL=.10 1.0/.75	.50/.25	.00/BIN	DEL=.15 1.0/.75	.50/.25	.00/BIN	DEL=.20 1.0/.75	.50/.25	.00/BIN
0.05	2870	2893	3023	620	628	641	224	226	228	131	132	133	93	93	94
	2878	2928	5651	624	634	1104	225	226	352	132	133	186	93	93	118
0.1	5976	6034	6642	1147	1176	1237	367	373	382	198	200	202	130	132	133
	5995	6139	9816	1159	1200	1747	370	377	502	199	201	248	131	132	151
0.15	8739	8843	10273	1608	1667	1813	492	505	523	254	259	264	162	164	166
	8773	9039	13479	1632	1720	2309	498	513	633	256	261	302	163	165	179
0.2	10951	11111	13649	1982	2077	2338	595	617	649	301	308	316	187	190	194
	11004	11418	16640	2020	2169	2791	605	631	743	304	312	346	189	192	201
0.25	12553	12784	16642	2266	2402	2798	678	709	755	338	348	360	208	212	217
	12631	13225	19300	2320	2536	3192	692	729	833	343	354	382	210	214	219
0.3	13559	13876	19180	2464	2644	3181	739	779	842	366	379	394	223	227	233
	13665	14473	21458	2536	2821	3513	757	806	904	372	386	409	225	230	231
0.35	14011	14432	21220	2585	2806	3484	781	830	907	385	400	419	232	238	245
	14152	15212	23113	2674	3025	3754	803	863	954	392	409	426	235	241	239
0.4	13976	14522	22740	2638	2899	3702	805	862	952	395	412	434	236	243	250
	14159	15505	24268	2744	3157	3915	831	900	984	403	423	435	239	247	241
0.45	13529	14225	23723	2633	2929	3834	813	874	974	397	415	439	235	243	251
	13765	15424	24920	2754	3217	3995	840	916	994	406	427	435	239	247	239
0.5	12756	13624	24163	2579	2903	3878	804	869	974	391	410	435	230	238	246
	13054	15040	25070	2711	3214	3995	833	914	984	400	422	426	234	241	231
0.55	11748	12805	24058	2483	2826	3835	781	847	953	378	397	421	220	227	235
	12118	14413	24719	2624	3150	3915	811	892	954	386	408	409	223	231	219
0.6	10597	11837	23406	2353	2703	3705	743	807	910	356	374	397	205	211	219
	11044	13594	23866	2497	3028	3754	772	851	904	364	385	382	208	215	201
0.65	9390	10778	22210	2191	2536	3488	691	751	844	327	343	363	186	191	197
	9904	12622	22511	2335	2850	3513	718	792	833	334	352	346	188	194	179
0.7	8191	9664	20473	1999	2325	3184	625	679	758	290	304	320	161	166	170
	8749	11515	20655	2136	2616	3192	650	713	743	297	311	302	163	168	151
0.75	7030	8504	18202	1777	2069	2795	545	588	650	245	256	267	132	135	138
	7600	10276	18296	1901	2323	2791	565	616	633	250	261	248	133	136	118
0.8	5907	7288	15399	1520	1763	2322	447	478	521	192	198	205	.	.	.
	6448	8888	15436	1624	1967	2309	462	498	502	195	202	186	.	.	.
0.85	4783	5975	12071	1216	1396	1764	330	348	371	.	.	.	.	.	.
	5256	7302	12074	1294	1538	1747	338	359	352	.	.	.	.	.	.
0.9	3582	4475	8224	845	947	1124	.	.	.	.	.	.	.	.	.
	3941	5415	8210	891	1021	1104	.	.	.	.	.	.	.	.	.
0.95	2104	2554	3864	.	.	.	.	.	.	.	.	.	.	.	.
	2291	2972	3844	.	.	.	.	.	.	.	.	.	.	.	.

REQUIRED NUMBER OF PATIENTS

TABLE 3: ALPHA= 0.01 POWER= 0.8 EXPECTED ACCRUAL THRU MINIMUM FOLLOW-UP= 1300

		DEL=.02			DEL=.05			DEL=.10			DEL=.15			DEL=.20		
FACT=		1.0 .75	.50 .25	.00 BIN	1.0 .75	.50 .25	.00 BIN	1.0 .75	.50 .25	.00 BIN	1.0 75	.50 .25	.00 BIN	1.0 .75	.50 .25	.00 BIN
PCONT=***				REQUIRED NUMBER OF PATIENTS												
0.05	***	2872	2897	3023	622	629	641	224	226	228	132	133	133	93	94	94
	•••	2881	2931	5651	625	635	1104	225	227	352	132	133	186	93	94	118
0.1	***	5981	6044	6642	1150	1179	1237	368	375	382	198	200	202	131	132	133
	•••	6002	6155	9816	1163	1202	1747	371	378	502	199	201	248	132	133	151
0.15	***	8747	8860	10273	1615	1674	1813	494	506	523	255	259	264	162	164	167
	•••	8786	9068	13479	1639	1726	2309	500	514	633	257	262	302	163	165	179
0.2	***	10964	11138	13649	1992	2089	2338	599	619	648	302	309	317	188	191	194
	•••	11022	11465	16640	2032	2178	2791	608	632	743	305	313	346	189	193	201
0.25	***	12574	12823	16643	2281	2419	2798	682	712	756	340	349	361	209	212	217
	•••	12656	13292	19300	2337	2550	3192	695	731	833	344	354	382	211	214	219
0.3	***	13586	13928	19180	2484	2665	3182	745	783	842	368	380	395	224	228	234
	•••	13700	14564	21458	2557	2839	3513	762	809	904	374	387	409	226	231	231
0.35	***	14047	14501	21221	2610	2833	3484	788	835	908	388	401	419	233	239	245
	•••	14199	15329	23113	2700	3048	3754	809	866	954	394	410	426	236	241	239
0.4	***	14022	14611	22740	2668	2931	3703	813	867	952	398	414	435	237	244	251
	•••	14220	15650	24268	2775	3183	3915	838	903	984	406	423	435	240	247	241
0.45	***	13588	14337	23723	2666	2964	3834	821	880	974	400	418	440	237	244	251
	•••	13843	15596	24920	2788	3248	3995	848	921	994	408	428	435	240	247	239
0.5	***	12831	13760	24164	2616	2941	3879	812	875	974	394	412	435	232	238	246
	•••	13153	15236	25070	2749	3247	3995	841	918	984	403	422	426	235	242	231
0.55	***	11842	12964	24057	2522	2866	3836	790	853	953	380	398	421	221	227	235
	•••	12239	14629	24719	2664	3184	3915	818	896	954	388	409	409	224	232	219
0.6	***	10713	12015	23406	2393	2742	3705	752	813	910	359	375	396	206	212	219
	•••	11186	13825	23866	2538	3062	3754	780	856	904	367	386	382	209	215	201
0.65	***	9524	10970	22210	2231	2574	3488	700	757	845	330	344	363	187	192	198
	•••	10064	12858	22511	2375	2883	3513	726	796	833	336	354	346	189	194	179
0.7	***	8338	9860	20474	2038	2361	3184	633	683	758	292	305	320	162	166	171
	•••	8919	11747	20655	2174	2645	3192	656	717	743	298	312	302	164	168	151
0.75	***	7182	8696	18202	1812	2101	2796	551	591	650	247	256	267	133	135	138
	•••	7769	10494	18296	1935	2349	2791	570	618	633	251	262	248	134	136	118
0.8	***	6052	7464	15399	1550	1789	2321	452	481	522	193	198	206	.	.	.
	•••	6607	9081	15436	1653	1987	2309	466	500	502	196	201	186	.	.	.
0.85	***	4912	6124	12071	1238	1415	1764	332	349	371	.	.	.	.	.	.
	•••	5393	7459	12074	1316	1552	1747	341	360	352	.	.	.	.	.	.
0.9	***	3681	4583	8224	858	957	1124	.	.	.	.	.	.	.	.	.
	•••	4045	5523	8210	903	1028	1104	.	.	.	.	.	.	.	.	.
0.95	***	2156	2606	3864	.	.	.	.	.	.	.	.	.	.	.	.
	•••	2344	3016	3844	.	.	.	.	.	.	.	.	.	.	.	.

TABLE 3: ALPHA= 0.01 POWER= 0.8   EXPECTED ACCRUAL THRU MINIMUM FOLLOW-UP= 1400

PCONT= •••	DEL=.02			DEL=.05			DEL=.10			DEL=.15			DEL=.20		
FACT=	1.0 .75	.50 .25	.00 BIN	1.0 .75	.50 .25	.00 BIN	1.0 .75	.50 .25	.00 BIN	1.0 75	.50 .25	.00 BIN	1.0 .75	.50 .25	.00 BIN
	REQUIRED NUMBER OF PATIENTS														
0.05 •••	2874	2901	3023	622	630	641	225	226	228	131	132	133	93	93	94
•••	2884	2936	5651	626	635	1104	226	227	352	132	133	186	93	94	118
0.1 •••	5986	6054	6642	1153	1182	1236	369	374	381	199	200	202	131	132	133
•••	6009	6169	9816	1165	1204	1747	372	378	502	199	201	248	131	132	151
0.15 •••	8756	8878	10272	1620	1680	1813	495	507	523	255	259	264	163	164	166
•••	8797	9096	13479	1645	1731	2309	501	514	633	257	262	302	164	165	179
0.2 •••	10978	11164	13649	2002	2099	2339	601	621	648	303	309	317	189	192	194
•••	11040	11511	16640	2042	2187	2791	610	633	743	306	312	346	190	192	201
0.25 •••	12592	12861	16642	2295	2433	2798	686	714	755	341	350	360	209	213	217
•••	12682	13358	19300	2351	2562	3192	699	732	833	345	354	382	211	214	219
0.3 •••	13612	13980	19180	2502	2684	3181	750	787	842	369	381	395	224	228	234
•••	13736	14653	21458	2577	2856	3513	766	811	904	375	388	409	226	231	231
0.35 •••	14082	14570	21220	2633	2858	3484	794	839	907	389	402	420	234	239	245
•••	14246	15442	23113	2725	3069	3754	815	869	954	395	410	426	236	241	239
0.4 •••	14068	14699	22739	2694	2959	3702	820	871	952	400	416	435	238	244	251
•••	14282	15788	24268	2803	3208	3915	843	906	984	407	424	435	241	248	241
0.45 •••	13647	14445	23723	2697	2996	3834	828	885	974	402	419	440	238	244	251
•••	13920	15760	24920	2820	3276	3995	854	924	994	410	429	435	241	248	239
0.5 •••	12906	13892	24163	2649	2975	3879	820	881	975	396	414	435	232	239	246
•••	13249	15422	25070	2784	3277	3995	848	921	984	404	423	426	235	242	231
0.55 •••	11935	13117	24057	2558	2901	3836	797	858	953	382	400	421	222	228	235
•••	12356	14834	24719	2701	3215	3915	825	899	954	391	409	409	225	232	219
0.6 •••	10825	12187	23405	2431	2779	3706	759	819	910	361	377	397	207	213	219
•••	11323	14041	23866	2576	3093	3754	787	858	904	368	387	382	210	216	201
0.65 •••	9655	11154	22210	2268	2610	3488	707	762	844	332	346	363	187	192	198
•••	10216	13079	22511	2411	2912	3513	732	799	833	339	354	346	190	194	179
0.7 •••	8480	10047	20474	2073	2394	3184	639	688	758	294	306	320	163	166	171
•••	9080	11964	20655	2209	2672	3192	661	719	743	300	312	302	164	168	151
0.75 •••	7328	8877	18202	1844	2131	2796	556	595	650	248	257	268	133	136	138
•••	7930	10698	18296	1966	2371	2791	574	620	633	253	262	248	134	136	118
0.8 •••	6191	7629	15399	1576	1813	2321	455	484	521	193	199	206	.	.	.
•••	6758	9260	15436	1678	2005	2309	469	501	502	196	202	186	.	.	.
0.85 •••	5033	6262	12071	1259	1431	1764	334	351	372	.	.	.	.	.	.
•••	5523	7604	12074	1334	1564	1747	342	360	352	.	.	.	.	.	.
0.9 •••	3772	4684	8224	870	966	1124	.	.	.	.	.	.	.	.	.
•••	4141	5621	8210	913	1033	1104	.	.	.	.	.	.	.	.	.
0.95 •••	2204	2652	3864	.	.	.	.	.	.	.	.	.	.	.	.
•••	2392	3055	3844	.	.	.	.	.	.	.	.	.	.	.	.

TABLE 3: ALPHA= 0.01 POWER= 0.8   EXPECTED ACCRUAL THRU MINIMUM FOLLOW-UP= 1500

		DEL=.02			DEL=.05			DEL=.10			DEL=.15			DEL=.20		
FACT=		1.0 .75	.50 .25	.00 BIN	1.0 .75	.50 .25	.00 BIN	1.0 .75	.50 .25	.00 BIN	1.0 75	.50 .25	.00 BIN	1.0 .75	.50 .25	.00 BIN
PCONT=***					REQUIRED NUMBER OF PATIENTS											
0.05	***	2876	2904	3023	623	630	641	225	226	228	132	132	133	93	93	94
	•••	2887	2939	5651	626	635	1104	226	227	352	132	132	186	93	94	118
0.1	***	5990	6063	6642	1157	1185	1237	370	375	382	199	200	202	131	132	133
	•••	6015	6183	9816	1169	1207	1747	372	379	502	200	202	248	131	132	151
0.15	***	8765	8894	10272	1627	1685	1813	497	508	523	256	260	263	163	164	166
	•••	8809	9125	13479	1651	1735	2309	502	515	633	258	262	302	164	165	179
0.2	***	10991	11191	13649	2012	2108	2339	604	622	649	304	309	317	189	191	194
	•••	11058	11555	16640	2052	2195	2791	612	635	743	307	313	346	190	193	201
0.25	***	12611	12899	16642	2308	2447	2798	689	716	755	342	350	360	209	213	217
	•••	12708	13421	19300	2365	2573	3192	701	734	833	346	355	382	211	215	219
0.3	***	13639	14032	19180	2520	2703	3182	754	789	842	371	382	395	225	229	233
	•••	13771	14738	21458	2595	2870	3513	770	813	904	376	388	409	227	232	231
0.35	***	14118	14638	21221	2654	2880	3485	799	843	907	391	404	420	234	239	245
	•••	14293	15551	23113	2747	3089	3754	818	872	954	397	412	426	237	242	239
0.4	***	14114	14785	22740	2720	2985	3702	825	875	952	402	416	435	239	245	251
	•••	14342	15922	24268	2829	3230	3915	848	909	984	409	425	435	242	247	241
0.45	***	13706	14553	23723	2726	3026	3834	834	890	974	404	420	440	238	245	251
	•••	13998	15917	24920	2850	3301	3995	860	927	994	412	429	435	241	247	239
0.5	***	12980	14021	24163	2681	3007	3878	827	886	975	398	415	435	233	239	245
	•••	13345	15600	25070	2817	3304	3995	854	924	984	406	425	426	236	243	231
0.55	***	12027	13265	24057	2593	2935	3835	804	863	953	384	401	421	223	229	235
	•••	12472	15028	24719	2735	3243	3915	832	903	954	393	410	409	226	232	219
0.6	***	10936	12352	23405	2465	2812	3705	766	824	909	363	378	397	208	213	219
	•••	11457	14246	23866	2611	3121	3754	792	862	904	370	387	382	210	215	201
0.65	***	9782	11329	22210	2302	2643	3488	713	767	845	334	347	363	187	192	198
	•••	10365	13287	22511	2446	2938	3513	738	802	833	339	354	346	190	195	179
0.7	***	8617	10224	20474	2105	2424	3185	645	692	758	295	307	320	163	167	170
	•••	9235	12168	20655	2242	2696	3192	667	722	743	301	313	302	165	169	151
0.75	***	7466	9048	18202	1874	2157	2795	560	598	650	249	258	267	133	136	139
	•••	8084	10887	18296	1995	2392	2791	578	622	633	253	262	248	134	137	118
0.8	***	6322	7785	15399	1601	1835	2322	458	485	521	194	200	205	.	.	.
	•••	6900	9428	15436	1702	2021	2309	472	502	502	197	202	186	.	.	.
0.85	***	5147	6393	12071	1278	1447	1764	337	352	371	.	.	.	.	.	.
	•••	5645	7738	12074	1352	1574	1747	344	362	352	.	.	.	.	.	.
0.9	***	3860	4777	8224	881	974	1124	.	.	.	.	.	.	.	.	.
	•••	4232	5711	8210	923	1039	1104	.	.	.	.	.	.	.	.	.
0.95	***	2250	2695	3863	.	.	.	.	.	.	.	.	.	.	.	.
	•••	2437	3091	3844	.	.	.	.	.	.	.	.	.	.	.	.

TABLE 3: ALPHA= 0.01 POWER= 0.8   EXPECTED ACCRUAL THRU MINIMUM FOLLOW-UP= 1600

	DEL=.02			DEL=.05			DEL=.10			DEL=.15			DEL=.20		
FACT=	1.0 .75	.50 .25	.00 BIN	1.0 .75	.50 .25	.00 BIN	1.0 .75	.50 .25	.00 BIN	1.0 75	.50 .25	.00 BIN	1.0 .75	.50 .25	.00 BIN
PCONT=•••				REQUIRED NUMBER OF PATIENTS											
0.05 ***	2879	2907	3023	624	631	641	225	226	228	132	132	133	93	93	94
•••	2889	2943	5651	627	636	1104	226	227	352	132	133	186	93	93	118
0.1 ***	5996	6072	6642	1159	1187	1237	371	376	382	199	201	203	131	132	133
•••	6022	6196	9816	1171	1208	1747	373	379	502	200	202	248	132	132	151
0.15 ***	8774	8911	10273	1632	1691	1813	498	509	523	256	260	264	163	165	166
•••	8820	9151	13479	1657	1739	2309	503	516	633	258	262	302	164	165	179
0.2 ***	11004	11217	13649	2021	2117	2339	606	624	649	305	310	317	189	192	194
•••	11076	11598	16640	2061	2201	2791	614	635	743	307	313	346	190	193	201
0.25 ***	12631	12936	16643	2321	2460	2798	692	718	755	343	351	360	210	213	217
•••	12734	13483	19300	2379	2584	3192	704	735	833	347	356	382	211	215	219
0.3 ***	13665	14083	19180	2537	2720	3182	758	792	842	372	382	395	225	229	234
•••	13806	14821	21458	2612	2885	3513	774	815	904	377	388	409	227	231	231
0.35 ***	14153	14705	21221	2675	2901	3484	804	846	908	392	405	420	235	240	245
•••	14339	15656	23113	2768	3106	3754	823	873	954	398	412	426	237	242	239
0.4 ***	14160	14871	22740	2744	3010	3702	831	880	952	404	417	435	240	245	251
•••	14402	16050	24268	2854	3251	3915	853	912	984	410	426	435	242	248	241
0.45 ***	13765	14659	23723	2754	3053	3834	840	894	974	406	421	440	239	245	251
•••	14074	16067	24920	2878	3323	3995	865	930	994	413	430	435	242	248	239
0.5 ***	13055	14146	24163	2712	3037	3879	833	890	975	400	416	435	234	240	246
•••	13440	15769	25070	2848	3328	3995	859	928	984	408	425	426	237	243	231
0.55 ***	12119	13409	24058	2625	2966	3836	811	868	953	387	402	421	224	229	235
•••	12585	15213	24719	2768	3268	3915	837	906	954	394	411	409	226	232	219
0.6 ***	11044	12510	23406	2498	2844	3705	772	828	910	365	379	397	208	213	219
•••	11587	14440	23866	2644	3146	3754	798	864	904	372	388	382	211	216	201
0.65 ***	9905	11496	22210	2335	2672	3488	719	771	845	335	348	363	188	193	197
•••	10507	13484	22511	2478	2963	3513	743	804	833	341	355	346	191	195	179
0.7 ***	8750	10394	20474	2136	2452	3185	650	695	758	297	308	320	164	167	171
•••	9384	12359	20655	2271	2718	3192	671	724	743	302	314	302	165	169	151
0.75 ***	7600	9210	18202	1901	2181	2796	565	601	650	250	258	267	133	136	138
•••	8230	11066	18296	2022	2411	2791	582	624	633	254	263	248	135	137	118
0.8 ***	6448	7932	15399	1624	1854	2322	462	488	521	195	200	205	.	.	.
•••	7036	9585	15436	1724	2035	2309	474	503	502	197	203	186	.	.	.
0.85 ***	5256	6516	12071	1295	1461	1764	339	353	372	.	.	.	.	.	.
•••	5761	7864	12074	1368	1584	1747	346	362	352	.	.	.	.	.	.
0.9 ***	3942	4865	8224	891	981	1124	.	.	.	.	.	.	.	.	.
•••	4318	5795	8210	932	1043	1104	.	.	.	.	.	.	.	.	.
0.95 ***	2292	2735	3864	.	.	.	.	.	.	.	.	.	.	.	.
•••	2479	3124	3844	.	.	.	.	.	.	.	.	.	.	.	.

TABLE 3: ALPHA= 0.01 POWER= 0.8   EXPECTED ACCRUAL THRU MINIMUM FOLLOW-UP= 1700

		DEL=.02			DEL=.05			DEL=.10			DEL=.15			DEL=.20		
FACT=		1.0 .75	.50 .25	.00 BIN	1.0 .75	.50 .25	.00 BIN	1.0 .75	.50 .25	.00 BIN	1.0 75	.50 .25	.00 BIN	1.0 .75	.50 .25	.00 BIN
PCONT=***					REQUIRED NUMBER OF PATIENTS											
0.05	***	2880	2910	3024	624	631	641	225	226	228	131	133	133	93	93	93
	•••	2891	2945	5651	628	636	1104	226	227	352	131	133	186	93	93	118
0.1	***	6001	6081	6642	1162	1189	1236	371	376	381	199	201	203	131	131	133
	•••	6028	6208	9816	1174	1209	1747	373	379	502	199	202	248	131	133	151
0.15	***	8782	8928	10272	1637	1695	1812	499	510	524	257	260	263	163	165	167
	•••	8831	9175	13479	1661	1742	2309	505	516	633	258	262	302	163	165	179
0.2	***	11018	11243	13649	2029	2126	2339	607	624	649	305	310	316	189	192	194
	•••	11094	11638	16640	2069	2208	2791	616	636	743	308	313	346	190	193	201
0.25	***	12650	12974	16643	2333	2472	2798	695	720	755	344	352	360	210	214	216
	•••	12760	13542	19300	2390	2594	3192	707	736	833	347	356	382	211	214	219
0.3	***	13691	14133	19180	2552	2735	3182	762	794	843	373	384	395	225	229	233
	•••	13841	14900	21458	2628	2897	3513	777	816	904	378	389	409	227	231	231
0.35	***	14188	14772	21221	2694	2919	3485	807	849	907	394	406	420	236	240	245
	•••	14386	15758	23113	2788	3121	3754	826	875	954	399	412	426	238	242	239
0.4	***	14205	14955	22739	2766	3032	3703	836	883	952	405	418	435	240	245	250
	•••	14462	16173	24268	2878	3269	3915	857	913	984	411	426	435	242	248	241
0.45	***	13823	14761	23723	2779	3079	3835	845	898	974	408	422	440	240	245	252
	•••	14150	16211	24920	2905	3344	3995	870	932	994	414	430	435	242	248	239
0.5	***	13128	14268	24163	2740	3064	3879	839	894	974	403	418	435	235	240	246
	•••	13533	15931	25070	2876	3350	3995	865	930	984	409	426	426	237	243	231
0.55	***	12209	13549	24058	2655	2994	3835	816	872	953	389	403	420	224	229	235
	•••	12696	15388	24719	2798	3291	3915	843	908	954	395	411	409	226	233	219
0.6	***	11151	12663	23406	2528	2873	3705	777	832	909	367	380	397	209	214	219
	•••	11714	14624	23866	2674	3169	3754	803	867	904	373	388	382	211	216	201
0.65	***	10025	11657	22210	2365	2700	3488	724	775	845	337	348	363	189	193	197
	•••	10645	13670	22511	2507	2985	3513	747	806	833	342	356	346	191	195	179
0.7	***	8877	10555	20474	2165	2477	3184	654	698	758	299	308	320	163	167	171
	•••	9526	12541	20655	2299	2738	3192	675	726	743	303	313	302	165	169	151
0.75	***	7728	9365	18202	1927	2203	2795	568	603	650	252	259	267	134	136	138
	•••	8370	11235	18296	2046	2428	2791	585	624	633	255	263	248	135	137	118
0.8	***	6568	8072	15399	1646	1872	2322	464	490	522	195	199	205	.	.	.
	•••	7165	9732	15436	1744	2048	2309	476	505	502	197	203	186	.	.	.
0.85	***	5360	6632	12071	1311	1474	1765	340	355	372	.	.	.	.	.	.
	•••	5870	7981	12074	1382	1593	1747	347	362	352	.	.	.	.	.	.
0.9	***	4019	4948	8223	900	988	1124	.	.	.	.	.	.	.	.	.
	•••	4398	5872	8210	940	1047	1104	.	.	.	.	.	.	.	.	.
0.95	***	2331	2772	3864	.	.	.	.	.	.	.	.	.	.	.	.
	•••	2518	3153	3844	.	.	.	.	.	.	.	.	.	.	.	.

TABLE 3: ALPHA= 0.01 POWER= 0.8   EXPECTED ACCRUAL THRU MINIMUM FOLLOW-UP= 1800

	DEL=.02			DEL=.05			DEL=.10			DEL=.15			DEL=.20		
FACT=	1.0 .75	.50 .25	.00 BIN	1.0 .75	.50 .25	.00 BIN	1.0 .75	.50 .25	.00 BIN	1.0 75	.50 .25	.00 BIN	1.0 .75	.50 .25	.00 BIN
PCONT=•••				REQUIRED NUMBER OF PATIENTS											
0.05 ***	2882	2913	3024	625	632	641	225	226	228	132	132	132	93	93	93
•••	2893	2949	5651	629	636	1104	226	227	352	132	132	186	93	93	118
0.1 ***	6005	6090	6642	1164	1191	1236	371	375	382	199	201	202	132	132	132
•••	6034	6220	9816	1176	1210	1747	373	379	502	200	201	248	132	132	151
0.15 ***	8792	8945	10272	1642	1700	1812	501	510	523	258	260	264	163	165	166
•••	8844	9200	13479	1667	1746	2309	505	516	633	258	262	302	164	165	179
0.2 ***	11031	11269	13650	2037	2133	2339	609	627	649	306	310	317	190	192	195
•••	11112	11679	16640	2078	2213	2791	618	636	743	308	314	346	191	193	201
0.25 ***	12669	13011	16642	2344	2483	2798	697	722	756	345	352	360	210	213	217
•••	12784	13600	19300	2403	2602	3192	708	737	833	348	357	382	213	215	219
0.3 ***	13718	14184	19180	2567	2749	3181	765	796	843	375	384	395	226	229	234
•••	13875	14978	21458	2643	2909	3513	780	818	904	379	389	409	227	231	231
0.35 ***	14223	14838	21221	2712	2938	3484	812	852	908	395	406	420	236	240	245
•••	14433	15855	23113	2807	3136	3754	830	877	954	400	413	426	238	243	239
0.4 ***	14251	15039	22740	2789	3054	3702	840	886	951	406	420	434	240	245	251
•••	14523	16292	24268	2900	3286	3915	861	915	984	413	426	435	243	249	241
0.45 ***	13881	14862	23723	2805	3102	3834	850	901	974	409	423	440	240	245	251
•••	14226	16348	24920	2929	3363	3995	874	933	994	416	431	435	243	249	239
0.5 ***	13200	14388	24164	2767	3090	3879	845	897	974	404	418	435	235	240	246
•••	13625	16086	25070	2904	3372	3995	870	933	984	411	426	426	237	243	231
0.55 ***	12297	13683	24058	2683	3021	3836	822	876	953	390	404	420	225	229	235
•••	12804	15555	24719	2826	3313	3915	847	910	954	397	411	409	227	233	219
0.6 ***	11256	12810	23406	2558	2899	3705	783	836	910	368	381	397	209	213	219
•••	11837	14800	23866	2703	3190	3754	807	868	904	375	389	382	211	217	201
0.65 ***	10140	11810	22209	2394	2726	3489	729	777	845	337	350	363	189	193	198
•••	10779	13848	22511	2535	3005	3513	751	807	833	343	357	346	191	195	179
0.7 ***	9000	10709	20474	2191	2502	3185	659	701	758	299	309	321	164	168	171
•••	9663	12712	20655	2325	2756	3192	678	726	743	303	314	302	165	168	151
0.75 ***	7851	9513	18201	1950	2225	2796	573	606	650	252	258	267	134	136	138
•••	8505	11394	18296	2070	2443	2791	588	627	633	255	263	248	135	137	118
0.8 ***	6684	8205	15399	1666	1889	2322	467	492	521	195	200	206	.	.	.
•••	7289	9870	15436	1764	2060	2309	478	505	502	198	202	186	.	.	.
0.85 ***	5460	6742	12071	1325	1485	1764	342	355	371	.	.	.	.	.	.
•••	5976	8091	12074	1396	1601	1747	348	363	352	.	.	.	.	.	.
0.9 ***	4094	5025	8223	908	994	1124	.	.	.	.	.	.	.	.	.
•••	4475	5946	8210	947	1050	1104	.	.	.	.	.	.	.	.	.
0.95 ***	2368	2805	3864	.	.	.	.	.	.	.	.	.	.	.	.
•••	2555	3181	3844	.	.	.	.	.	.	.	.	.	.	.	.

TABLE 3: ALPHA= 0.01 POWER= 0.8   EXPECTED ACCRUAL THRU MINIMUM FOLLOW-UP= 1900

REQUIRED NUMBER OF PATIENTS

PCONT	FACT=	DEL=.02 1.0 / .75	.50 / .25	.00 / BIN	DEL=.05 1.0 / .75	.50 / .25	.00 / BIN	DEL=.10 1.0 / .75	.50 / .25	.00 / BIN	DEL=.15 1.0 / 75	.50 / .25	.00 / BIN	DEL=.20 1.0 / .75	.50 / .25	.00 / BIN
0.05	***	2884	2915	3023	626	633	641	225	227	228	132	133	133	92	94	94
	•••	2896	2951	5651	629	636	1104	225	228	352	132	133	186	94	94	118
0.1	***	6010	6099	6642	1166	1193	1237	372	376	382	199	201	203	132	132	133
	•••	6041	6231	9816	1178	1211	1747	374	379	502	201	201	248	132	133	151
0.15	***	8800	8960	10273	1647	1704	1813	501	510	524	258	260	263	163	165	166
	•••	8855	9224	13479	1672	1748	2309	505	516	633	258	262	302	163	166	179
0.2	***	11045	11294	13649	2044	2139	2338	610	627	648	306	311	317	190	192	194
	•••	11129	11717	16640	2085	2218	2791	619	638	743	308	313	346	191	193	201
0.25	***	12688	13047	16643	2355	2493	2797	699	723	755	345	353	361	211	213	217
	•••	12810	13656	19300	2413	2610	3192	711	738	833	349	356	382	212	215	219
0.3	***	13744	14233	19180	2581	2763	3182	767	799	842	375	384	395	227	230	234
	•••	13910	15052	21458	2657	2920	3513	782	818	904	380	389	409	228	231	231
0.35	***	14258	14902	21220	2730	2956	3484	816	854	908	396	407	419	236	241	246
	•••	14478	15950	23113	2825	3149	3754	833	878	954	401	413	426	239	243	239
0.4	***	14296	15120	22740	2809	3073	3702	844	889	952	407	420	434	241	246	250
	•••	14581	16406	24268	2921	3302	3915	866	918	984	413	427	435	243	248	241
0.45	***	13940	14961	23723	2828	3125	3834	856	904	973	410	424	440	241	246	251
	•••	14299	16481	24920	2953	3381	3995	878	935	994	417	431	435	243	248	239
0.5	***	13273	14504	24163	2793	3115	3878	849	901	975	405	419	436	235	241	246
	•••	13715	16234	25070	2928	3390	3995	873	934	984	412	426	426	239	243	231
0.55	***	12385	13815	24057	2709	3046	3835	826	878	953	391	405	421	225	230	235
	•••	12910	15716	24719	2852	3332	3915	851	912	954	398	412	409	228	232	219
0.6	***	11357	12952	23405	2585	2925	3705	788	838	909	369	382	396	210	215	220
	•••	11957	14967	23866	2730	3210	3754	812	870	904	375	389	382	212	217	201
0.65	***	10254	11958	22210	2420	2750	3488	733	780	844	338	350	363	190	193	197
	•••	10907	14014	22511	2561	3023	3513	755	809	833	344	356	346	191	196	179
0.7	***	9120	10857	20474	2217	2524	3184	662	704	759	300	310	320	165	167	171
	•••	9796	12874	20655	2350	2773	3192	681	729	743	305	315	302	166	168	151
0.75	***	7969	9653	18202	1973	2244	2795	576	608	650	253	260	267	134	136	139
	•••	8633	11544	18296	2091	2457	2791	590	628	633	256	263	248	135	137	118
0.8	***	6794	8331	15399	1685	1904	2321	469	493	521	196	201	205	.	.	.
	•••	7406	10002	15436	1781	2071	2309	481	505	502	198	203	186	.	.	.
0.85	***	5554	6847	12071	1339	1495	1764	343	356	372	.	.	.	.	.	.
	•••	6075	8195	12074	1409	1607	1747	349	363	352	.	.	.	.	.	.
0.9	***	4164	5099	8224	915	999	1124	.	.	.	.	.	.	.	.	.
	•••	4548	6013	8210	954	1054	1104	.	.	.	.	.	.	.	.	.
0.95	***	2403	2838	3864	.	.	.	.	.	.	.	.	.	.	.	.
	•••	2588	3206	3844	.	.	.	.	.	.	.	.	.	.	.	.

TABLE 3: ALPHA= 0.01 POWER= 0.8   EXPECTED ACCRUAL THRU MINIMUM FOLLOW-UP= 2000

	DEL=.02			DEL=.05			DEL=.10			DEL=.15			DEL=.20		
FACT=	1.0 .75	.50 .25	.00 BIN	1.0 .75	.50 .25	.00 BIN	1.0 .75	.50 .25	.00 BIN	1.0 75	.50 .25	.00 BIN	1.0 .75	.50 .25	.00 BIN

PCONT=•••   REQUIRED NUMBER OF PATIENTS

PCONT	DEL=.02 1.0/.75	.50/.25	.00/BIN	DEL=.05 1.0/.75	.50/.25	.00/BIN	DEL=.10 1.0/.75	.50/.25	.00/BIN	DEL=.15 1.0/.75	.50/.25	.00/BIN	DEL=.20 1.0/.75	.50/.25	.00/BIN
0.05 •••	2886	2919	3024	626	632	641	226	226	229	132	132	132	94	94	94
•••	2899	2954	5651	630	636	1104	226	227	352	132	132	186	94	94	118
0.1 •••	6015	6107	6642	1169	1195	1236	372	377	382	200	201	202	131	132	132
•••	6047	6241	9816	1180	1212	1747	375	379	502	200	202	248	132	132	151
0.15 •••	8809	8977	10272	1651	1707	1812	502	511	524	257	261	264	164	165	166
•••	8866	9246	13479	1676	1751	2309	506	517	633	259	262	302	165	166	179
0.2 •••	11057	11320	13650	2052	2146	2339	612	629	649	306	311	316	190	192	194
•••	11147	11755	16640	2092	2222	2791	620	637	743	309	314	346	191	194	201
0.25 •••	12707	13084	16642	2366	2502	2799	701	725	755	346	352	360	211	214	216
•••	12836	13710	19300	2424	2617	3192	712	739	833	350	356	382	212	215	219
0.3 •••	13771	14282	19180	2595	2776	3182	770	801	842	376	385	395	226	230	234
•••	13945	15125	21458	2671	2930	3513	785	820	904	380	390	409	229	232	231
0.35 •••	14292	14966	21221	2747	2971	3485	819	856	907	397	407	420	237	241	245
•••	14525	16041	23113	2841	3162	3754	836	880	954	402	414	426	239	242	239
0.4 •••	14342	15200	22740	2830	3092	3702	849	891	952	409	421	435	241	246	251
•••	14641	16516	24268	2940	3316	3915	869	919	984	415	427	435	244	249	241
0.45 •••	13997	15059	23724	2850	3146	3834	860	907	974	412	425	440	241	246	251
•••	14372	16609	24920	2975	3397	3995	882	937	994	417	432	435	244	249	239
0.5 •••	13345	14616	24164	2817	3136	3879	854	904	975	406	420	435	236	241	246
•••	13805	16376	25070	2952	3407	3995	877	936	984	412	427	426	239	244	231
0.55 •••	12472	13941	24057	2736	3070	3836	831	882	954	392	406	421	226	230	235
•••	13015	15869	24719	2877	3350	3915	855	915	954	399	412	409	227	232	219
0.6 •••	11457	13090	23406	2611	2947	3705	792	842	910	370	382	397	210	215	219
•••	12074	15126	23866	2755	3227	3754	815	872	904	376	390	382	212	217	201
0.65 •••	10365	12101	22210	2446	2772	3489	737	784	845	340	351	364	190	194	197
•••	11032	14175	22511	2586	3041	3513	759	811	833	345	356	346	192	196	179
0.7 •••	9235	10999	20474	2241	2545	3185	666	706	759	301	310	320	165	167	171
•••	9924	13030	20655	2372	2789	3192	685	730	743	305	315	302	166	169	151
0.75 •••	8084	9789	18202	1995	2262	2796	579	610	650	254	260	267	135	136	139
•••	8757	11687	18296	2111	2470	2791	594	629	633	256	264	248	135	137	118
0.8 •••	6900	8452	15399	1702	1919	2322	471	494	521	197	201	205	.	.	.
•••	7520	10126	15436	1797	2081	2309	482	507	502	199	204	186	.	.	.
0.85 •••	5645	6946	12071	1352	1505	1765	344	357	371	.	.	.	.	.	.
•••	6171	8292	12074	1421	1614	1747	350	364	352	.	.	.	.	.	.
0.9 •••	4232	5169	8224	924	1005	1124	.	.	.	.	.	.	.	.	.
•••	4617	6077	8210	960	1057	1104	.	.	.	.	.	.	.	.	.
0.95 •••	2437	2869	3864	.	.	.	.	.	.	.	.	.	.	.	.
•••	2621	3230	3844	.	.	.	.	.	.	.	.	.	.	.	.

423

TABLE 3: ALPHA= 0.01 POWER= 0.8   EXPECTED ACCRUAL THRU MINIMUM FOLLOW-UP= 2250

		DEL=.02			DEL=.05			DEL=.10			DEL=.15			DEL=.20		
FACT=		1.0 .75	.50 .25	.00 BIN	1.0 .75	.50 .25	.00 BIN	1.0 .75	.50 .25	.00 BIN	1.0 75	.50 .25	.00 BIN	1.0 .75	.50 .25	.00 BIN
PCONT=***				REQUIRED NUMBER OF PATIENTS											
0.05	***	2891	2924	3023	628	634	641	226	226	227	132	132	133	92	94	94
	•••	2904	2958	5651	629	637	1104	226	227	352	132	133	186	92	94	118
0.1	***	6027	6128	6643	1174	1198	1237	374	376	382	199	201	202	132	132	133
	•••	6063	6266	9816	1185	1214	1747	375	379	502	201	202	248	132	132	151
0.15	***	8830	9016	10272	1662	1717	1812	505	513	523	258	261	264	164	165	165
	•••	8894	9299	13479	1686	1756	2309	509	517	633	260	263	302	164	165	179
0.2	***	11091	11382	13649	2068	2161	2338	615	629	648	308	312	316	191	192	194
	•••	11190	11842	16640	2109	2233	2791	623	638	743	309	314	346	191	194	201
0.25	***	12756	13172	16643	2389	2524	2798	705	728	755	347	354	359	212	213	216
	•••	12899	13837	19300	2446	2632	3192	717	741	833	350	357	382	213	215	219
0.3	***	13836	14402	19180	2626	2805	3182	776	804	842	378	387	395	227	230	233
	•••	14031	15296	21458	2702	2952	3513	790	822	904	382	390	409	229	232	231
0.35	***	14380	15121	21222	2786	3006	3484	826	862	908	399	409	420	237	241	244
	•••	14637	16256	23113	2879	3189	3754	842	883	954	403	413	426	239	243	239
0.4	***	14455	15394	22740	2876	3134	3702	857	897	952	412	421	434	243	247	252
	•••	14785	16773	24268	2985	3349	3915	876	922	984	416	429	435	244	249	241
0.45	***	14141	15291	23723	2902	3192	3835	869	914	974	415	426	440	243	247	252
	•••	14553	16905	24920	3026	3434	3995	890	942	994	420	433	435	244	249	239
0.5	***	13521	14886	24163	2873	3188	3878	863	911	974	409	421	435	237	241	246
	•••	14022	16707	25070	3008	3447	3995	885	941	984	415	427	426	239	243	231
0.55	***	12683	14242	24058	2794	3121	3836	840	888	953	395	407	420	226	230	235
	•••	13265	16225	24719	2935	3390	3915	863	918	954	401	413	409	229	233	219
0.6	***	11698	13412	23405	2670	3000	3705	803	848	910	372	384	396	210	215	219
	•••	12352	15495	23866	2812	3268	3754	824	876	904	378	390	382	213	216	201
0.65	***	10628	12435	22210	2504	2823	3489	747	789	845	342	351	362	191	194	198
	•••	11328	14545	22511	2642	3078	3513	766	814	833	347	357	346	192	195	179
0.7	***	9509	11330	20473	2296	2592	3185	674	711	758	303	311	320	165	168	170
	•••	10225	13386	20655	2424	2822	3192	691	732	743	306	314	302	167	168	151
0.75	***	8353	10102	18202	2043	2302	2795	584	614	651	254	261	267	134	136	139
	•••	9047	12014	18296	2157	2499	2791	599	631	633	257	264	248	136	137	118
0.8	***	7149	8732	15400	1742	1950	2322	477	496	522	198	201	205	.	.	.
	•••	7784	10409	15436	1835	2103	2309	486	509	502	199	204	186	.	.	.
0.85	***	5857	7176	12071	1380	1527	1765	347	358	371	.	.	.	.	.	.
	•••	6392	8515	12074	1447	1628	1747	353	364	352	.	.	.	.	.	.
0.9	***	4389	5328	8223	939	1015	1123	.	.	.	.	.	.	.	.	.
	•••	4777	6221	8210	974	1064	1104	.	.	.	.	.	.	.	.	.
0.95	***	2513	2936	3864	.	.	.	.	.	.	.	.	.	.	.	.
	•••	2696	3282	3844	.	.	.	.	.	.	.	.	.	.	.	.

TABLE 3: ALPHA= 0.01 POWER= 0.8   EXPECTED ACCRUAL THRU MINIMUM FOLLOW-UP= 2500

PCONT=***		DEL=.02 1.0 .75	.50 .25	.00 BIN	DEL=.05 1.0 .75	.50 .25	.00 BIN	DEL=.10 1.0 .75	.50 .25	.00 BIN	DEL=.15 1.0 75	.50 .25	.00 BIN	DEL=.20 1.0 .75	.50 .25	.00 BIN
		REQUIRED NUMBER OF PATIENTS														
0.05	***	2895	2929	3023	629	634	642	226	226	227	133	133	133	93	93	93
	***	2909	2964	5651	631	637	1104	226	227	352	133	133	186	93	93	118
0.1	***	6040	6148	6642	1177	1201	1237	374	377	383	199	201	202	133	133	133
	***	6077	6287	9816	1189	1217	1747	376	379	502	201	201	248	133	133	151
0.15	***	8851	9054	10273	1670	1723	1812	506	514	523	259	261	264	164	165	167
	***	8923	9348	13479	1693	1762	2309	509	518	633	261	262	302	165	165	179
0.2	***	11124	11442	13649	2084	2173	2339	618	633	648	309	312	317	190	192	193
	***	11234	11923	16640	2123	2242	2791	624	640	743	311	314	346	192	193	201
0.25	***	12804	13259	16642	2411	2543	2798	711	729	756	348	354	361	212	214	217
	***	12961	13954	19300	2468	2646	3192	720	742	833	351	358	382	214	215	219
0.3	***	13901	14518	19181	2654	2829	3183	781	808	842	379	387	395	227	231	234
	***	14117	15451	21458	2729	2970	3513	793	823	904	383	390	409	229	233	231
0.35	***	14467	15271	21221	2820	3037	3484	833	865	908	401	409	420	239	242	245
	***	14749	16452	23113	2914	3212	3754	848	886	954	404	414	426	240	243	239
0.4	***	14567	15577	22740	2915	3170	3702	864	901	951	414	423	434	243	246	251
	***	14927	17009	24268	3024	3376	3915	883	924	984	418	429	435	245	249	241
0.45	***	14281	15512	23723	2948	3234	3834	877	918	974	417	427	440	243	246	251
	***	14727	17177	24920	3070	3464	3995	896	945	994	421	434	435	245	249	239
0.5	***	13693	15139	24164	2923	3231	3879	871	917	974	411	423	436	237	242	246
	***	14227	17008	25070	3056	3479	3995	893	943	984	417	429	426	240	243	231
0.55	***	12884	14523	24058	2846	3167	3836	849	895	952	398	409	421	227	231	236
	***	13502	16546	24719	2986	3424	3915	870	921	954	402	414	409	229	233	219
0.6	***	11927	13712	23406	2723	3045	3706	811	852	909	374	386	396	212	215	218
	***	12612	15827	23866	2864	3301	3754	831	879	904	379	390	382	214	217	201
0.65	***	10876	12742	22211	2556	2867	3489	754	793	845	343	352	364	192	195	198
	***	11604	14876	22511	2692	3111	3513	773	817	833	348	358	346	193	196	179
0.7	***	9764	11633	20473	2343	2631	3184	681	715	758	304	312	320	165	168	170
	***	10501	13704	20655	2470	2851	3192	696	736	743	308	315	302	167	170	151
0.75	***	8601	10387	18201	2086	2337	2795	590	617	649	256	261	267	136	137	139
	***	9314	12308	18296	2196	2523	2791	602	633	633	259	264	248	136	137	118
0.8	***	7377	8986	15399	1776	1977	2321	479	498	521	198	201	206	.	.	.
	***	8026	10661	15436	1867	2121	2309	489	509	502	199	202	186	.	.	.
0.85	***	6051	7383	12071	1406	1545	1764	349	359	371	.	.	.	.	.	.
	***	6593	8709	12074	1468	1640	1747	354	365	352	.	.	.	.	.	.
0.9	***	4529	5470	8224	952	1024	1124	.	.	.	.	.	.	.	.	.
	***	4921	6345	8210	986	1070	1104	.	.	.	.	.	.	.	.	.
0.95	***	2581	2995	3864	.	.	.	.	.	.	.	.	.	.	.	.
	***	2759	3326	3844	.	.	.	.	.	.	.	.	.	.	.	.

TABLE 3: ALPHA= 0.01 POWER= 0.8   EXPECTED ACCRUAL THRU MINIMUM FOLLOW-UP= 2750

	DEL=.02			DEL=.05			DEL=.10			DEL=.15			DEL=.20		
FACT=	1.0 .75	.50 .25	.00 BIN	1.0 .75	.50 .25	.00 BJN	1.0 .75	.50 .25	.00 BIN	1.0 75	.50 .25	.00 BIN	1.0 .75	.50 .25	.00 BIN
PCONT=•••				REQUIRED NUMBER OF PATIENTS											
0.05 •••	2899	2935	3023	630	635	641	226	226	228	132	132	133	92	94	94
•••	2914	2968	5651	632	637	1104	226	228	352	132	132	186	92	94	118
0.1 •••	6051	6166	6643	1182	1204	1237	374	377	381	201	201	202	132	132	133
•••	6093	6307	9816	1192	1218	1747	376	379	502	201	202	248	132	132	151
0.15 •••	8874	9090	10273	1679	1730	1813	507	515	524	259	260	264	164	164	166
•••	8949	9390	13479	1701	1766	2309	510	518	633	260	263	302	164	166	179
0.2 •••	11157	11500	13650	2096	2184	2339	620	634	649	309	312	315	190	192	194
•••	11278	11996	16640	2136	2249	2791	627	641	743	311	314	346	192	194	201
0.25 •••	12851	13342	16642	2430	2559	2797	713	731	755	350	355	360	212	214	216
•••	13023	14062	19300	2487	2656	3192	723	744	833	352	357	382	212	216	219
0.3 •••	13966	14631	19180	2679	2851	3182	785	810	841	381	388	395	228	232	233
•••	14200	15595	21458	2755	2985	3513	797	824	904	384	391	409	230	232	231
0.35 •••	14552	15413	21221	2851	3064	3485	838	869	907	401	410	419	239	242	245
•••	14860	16633	23113	2944	3230	3754	852	886	954	407	415	426	240	243	239
0.4 •••	14677	15753	22740	2952	3202	3701	871	905	951	415	424	434	243	247	250
•••	15066	17227	24268	3061	3399	3915	886	927	984	419	429	435	245	249	241
0.45 •••	14418	15719	23724	2989	3268	3834	885	923	974	419	429	439	243	247	250
•••	14896	17426	24920	3110	3490	3995	902	947	994	424	434	435	245	249	239
0.5 •••	13859	15377	24164	2968	3270	3879	879	920	974	414	424	434	239	242	245
•••	14426	17282	25070	3099	3508	3995	899	945	984	419	429	426	240	243	231
0.55 •••	13079	14784	24057	2892	3206	3835	857	899	954	400	408	421	228	232	235
•••	13727	16838	24719	3030	3454	3915	876	924	954	404	415	409	230	233	219
0.6 •••	12144	13989	23405	2770	3085	3705	817	857	910	376	386	397	212	216	219
•••	12858	16130	23866	2907	3330	3754	837	881	904	381	391	382	214	218	201
0.65 •••	11108	13024	22209	2601	2904	3488	761	797	845	345	353	364	192	194	197
•••	11861	15176	22511	2734	3138	3513	778	819	833	350	359	346	194	195	179
0.7 •••	10001	11910	20473	2387	2665	3184	687	718	758	305	312	319	166	168	170
•••	10759	13992	20655	2509	2875	3192	703	737	743	309	315	302	168	170	151
0.75 •••	8832	10647	18201	2124	2366	2796	594	620	651	257	263	267	135	137	139
•••	9561	12569	18296	2230	2543	2791	606	634	633	259	264	248	135	137	118
0.8 •••	7589	9217	15399	1807	2000	2322	483	501	522	199	202	205	.	.	.
•••	8248	10884	15436	1893	2136	2309	491	510	502	201	204	186	.	.	.
0.85 •••	6228	7569	12070	1428	1560	1765	350	360	370	.	.	.	.	.	.
•••	6778	8882	12074	1488	1649	1747	355	366	352	.	.	.	.	.	.
0.9 •••	4659	5598	8224	964	1033	1123	.	.	.	.	.	.	.	.	.
•••	5051	6454	8210	996	1074	1104	.	.	.	.	.	.	.	.	.
0.95 •••	2641	3045	3863	.	.	.	.	.	.	.	.	.	.	.	.
•••	2817	3363	3844	.	.	.	.	.	.	.	.	.	.	.	.

TABLE 3: ALPHA= 0.01 POWER= 0.8   EXPECTED ACCRUAL THRU MINIMUM FOLLOW-UP= 3000

PCONT=***   REQUIRED NUMBER OF PATIENTS

PCONT	FACT	DEL=.01			DEL=.02			DEL=.05			DEL=.10			DEL=.15		
		1.0	.50	.00	1.0	.50	.00	1.0	.50	.00	1.0	.50	.00	1.0	.50	.00
		.75	.25	BIN	.75	.25	BIN	.75	.25	BIN	75	.25	BIN	.75	.25	BIN
0.01	***	1535	1540	1547	565	565	568	185	185	185	95	95	95	65	65	65
	***	1535	1543	5922	565	565	1958	185	185	533	95	95	217	65	65	129
0.02	***	3388	3403	3455	1090	1093	1105	295	295	298	130	130	130	85	85	85
	***	3395	3422	9776	1093	1097	2911	295	298	681	130	130	252	85	85	144
0.05	***	10255	10310	10790	2905	2938	3025	628	635	640	227	227	227	133	133	133
	***	10273	10408	20855	2920	2972	5651	632	640	1104	227	227	352	133	133	186
0.1	***	22618	22760	25112	6062	6182	6643	1183	1205	1235	373	377	380	200	200	200
	***	22667	23030	37716	6107	6325	9816	1195	1220	1747	377	380	502	200	200	248
0.15	***	33778	34025	39670	8893	9125	10273	1685	1735	1813	508	515	523	260	260	265
	***	33860	34520	52569	8975	9430	13479	1708	1768	2309	512	520	635	260	260	302
0.2	***	42767	43153	53297	11192	11555	13648	2110	2195	2338	620	635	647	310	313	317
	***	42895	43925	65415	11320	12065	16640	2147	2255	2791	628	640	743	310	313	346
0.25	***	49315	49873	65432	12898	13420	16640	2447	2575	2800	715	733	755	350	355	358
	***	49502	50987	76254	13082	14162	19300	2503	2668	3192	725	745	833	350	358	382
0.3	***	53440	54208	75770	14030	14740	19180	2702	2870	3182	790	812	842	380	388	395
	***	53695	55738	85086	14282	15730	21458	2777	2998	3513	800	827	904	385	392	409
0.35	***	55303	56330	84133	14638	15550	21220	2878	3088	3485	842	872	905	403	410	418
	***	55645	58360	91910	14965	16802	23113	2972	3250	3754	857	887	954	407	415	426
0.4	***	55153	56495	90422	14785	15920	22738	2983	3230	3703	875	910	950	415	425	433
	***	55600	59120	96728	15200	17425	24268	3092	3418	3915	890	928	984	422	430	435
0.45	***	53282	55010	94570	14552	15917	23725	3025	3302	3835	890	928	973	418	430	440
	***	53860	58330	99538	15058	17653	24920	3145	3512	3995	905	947	994	425	433	435
0.5	***	50015	52220	96545	14020	15598	24163	3005	3302	3880	887	925	973	415	425	433
	***	50755	56308	100341	14615	17533	25070	3137	3530	3995	902	947	984	418	430	426
0.55	***	45707	48485	96340	13265	15028	24058	2935	3242	3835	865	902	955	400	410	422
	***	46645	53365	99136	13940	17105	24719	3070	3478	3915	883	925	954	407	415	409
0.6	***	40727	44162	93950	12350	14245	23405	2810	3122	3703	823	860	910	377	388	395
	***	41908	49765	95925	13090	16405	23866	2945	3355	3754	842	883	904	380	392	382
0.65	***	35458	39560	89383	11327	13288	22210	2642	2938	3490	767	800	845	347	355	362
	***	36913	45710	90706	12100	15448	22511	2773	3160	3513	782	820	833	350	358	346
0.7	***	30260	34907	82655	10225	12167	20473	2425	2695	3185	692	722	760	305	313	320
	***	31963	41327	83480	10997	14252	20655	2545	2897	3192	707	737	743	310	317	302
0.75	***	25393	30298	73780	9047	10888	18200	2158	2390	2795	598	620	650	257	260	268
	***	27242	36655	74247	9790	12805	18296	2263	2560	2791	610	635	633	260	265	248
0.8	***	20927	25727	62780	7783	9430	15400	1835	2020	2323	485	500	520	200	200	205
	***	22772	31640	63007	8450	11087	15436	1918	2150	2309	493	512	502	200	205	186
0.85	***	16753	21050	49678	6392	7738	12070	1445	1573	1765	350	362	370	.	.	.
	***	18430	26140	49759	6947	9035	12074	1505	1660	1747	358	365	352	.	.	.
0.9	***	12583	15970	34495	4775	5710	8222	973	1037	1123	.	.	.	.	.	.
	***	13922	19795	34504	5170	6550	8210	1003	1078	1104	.	.	.	.	.	.
0.95	***	7783	9737	17260	2695	3092	3865	.	.	.	.	.	.	.	.	.
	***	8575	11735	17242	2867	3395	3844	.	.	.	.	.	.	.	.	.
0.98	***	3695	4370	5942	.	.	.	.	.	.	.	.	.	.	.	.
	***	3985	4937	5922	.	.	.	.	.	.	.	.	.	.	.	.

TABLE 3: ALPHA= 0.01 POWER= 0.8   EXPECTED ACCRUAL THRU MINIMUM FOLLOW-UP= 3250

		DEL=.01			DEL=.02			DEL=.05			DEL=.10			DEL=.15		
FACT=		1.0 .75	.50 .25	.00 BIN	1.0 .75	.50 .25	.00 BIN	1.0 .75	.50 .25	.00 BIN	1.0 75	.50 .25	.00 BIN	1.0 .75	.50 .25	.00 BIN
PCONT=•••						REQUIRED NUMBER OF PATIENTS										
0.01	***	1533	1538	1546	563	566	566	184	184	184	95	95	95	68	68	68
	•••	1538	1541	5922	566	566	1958	184	184	533	95	95	217	68	68	129
0.02	***	3391	3407	3456	1091	1094	1102	295	295	298	127	127	127	84	84	84
	•••	3399	3423	9776	1091	1099	2911	295	298	681	127	127	252	84	84	144
0.05	***	10259	10321	10787	2906	2944	3025	631	636	639	225	225	230	133	133	133
	•••	10281	10419	20855	2922	2976	5651	631	639	1104	225	225	352	133	133	186
0.1	***	22631	22785	25112	6075	6197	6641	1189	1208	1238	376	379	379	201	201	201
	•••	22682	23072	37716	6121	6340	9816	1197	1221	1747	376	379	502	201	201	248
0.15	***	33797	34066	39672	8916	9154	10273	1693	1741	1814	509	514	523	257	263	263
	•••	33887	34599	52569	9005	9468	13479	1712	1774	2309	514	517	633	263	263	302
0.2	***	42797	43219	53297	11223	11608	13647	2118	2204	2337	623	636	647	311	314	314
	•••	42938	44051	65415	11361	12128	16640	2156	2261	2791	628	639	743	311	314	346
0.25	***	49362	49966	65433	12946	13498	16643	2464	2586	2798	718	734	753	352	355	360
	•••	49565	51169	76254	13144	14254	19300	2516	2676	3192	726	745	833	352	355	382
0.3	***	53501	54337	75771	14094	14842	19181	2724	2887	3182	794	815	842	384	387	396
	•••	53782	55987	85086	14362	15854	21458	2798	3009	3513	802	826	904	384	393	409
0.35	***	55391	56504	84137	14720	15684	21220	2906	3109	3483	848	875	907	404	412	420
	•••	55759	58689	91910	15069	16954	23113	2996	3264	3754	859	888	954	409	417	426
0.4	***	55264	56718	90421	14891	16082	22739	3017	3256	3702	880	913	953	417	425	433
	•••	55748	59537	96728	15329	17609	24268	3123	3434	3915	896	932	984	420	428	435
0.45	***	53424	55296	94569	14684	16101	23722	3061	3329	3832	896	929	972	420	428	441
	•••	54050	58844	99538	15216	17864	24920	3179	3532	3995	913	948	994	425	433	435
0.5	***	50202	52582	96547	14176	15809	24161	3044	3334	3878	891	929	972	417	425	436
	•••	50998	56921	100341	14798	17767	25070	3171	3553	3995	907	948	984	420	428	426
0.55	***	45941	48931	96341	13444	15256	24056	2971	3272	3838	867	904	953	401	412	420
	•••	46957	54074	99136	14143	17352	24719	3106	3499	3915	888	929	954	404	417	409
0.6	***	41026	44693	93952	12548	14489	23406	2849	3150	3702	826	864	907	379	387	396
	•••	42296	50548	95925	13306	16659	23866	2984	3377	3754	848	883	904	384	393	382
0.65	***	35829	40167	89386	11535	13531	22208	2679	2968	3488	769	802	842	347	355	363
	•••	37378	46543	90706	12328	15700	22511	2806	3179	3513	786	823	833	352	360	346
0.7	***	30702	35561	82653	10435	12404	20472	2459	2724	3182	696	726	758	306	314	319
	•••	32494	42166	83480	11223	14489	20655	2578	2914	3192	709	742	743	311	314	302
0.75	***	25881	30962	73781	9249	11109	18203	2188	2416	2798	604	623	647	257	263	266
	•••	27806	37467	74247	9999	13022	18296	2289	2573	2791	612	636	633	263	266	248
0.8	***	21420	26355	62779	7968	9623	15399	1858	2037	2321	490	501	523	201	201	206
	•••	23324	32384	63007	8642	11267	15436	1939	2159	2309	493	509	502	201	206	186
0.85	***	17203	21602	49679	6546	7892	12071	1465	1587	1766	352	363	371	.	.	.
	•••	18921	26766	49759	7104	9176	12074	1522	1663	1747	355	368	352	.	.	.
0.9	***	12946	16394	34496	4886	5815	8225	981	1043	1124	.	.	.	.	.	.
	•••	14311	20250	34504	5276	6636	8210	1010	1083	1104	.	.	.	.	.	.
0.95	***	8001	9967	17260	2744	3131	3862	.	.	.	.	.	.	.	.	.
	•••	8802	11963	17242	2914	3423	3844	.	.	.	.	.	.	.	.	.
0.98	***	3776	4442	5942	.	.	.	.	.	.	.	.	.	.	.	.
	•••	4065	4991	5922	.	.	.	.	.	.	.	.	.	.	.	.

TABLE 3: ALPHA= 0.01   POWER= 0.8   EXPECTED ACCRUAL THRU MINIMUM FOLLOW-UP= 3500

	DEL=.01			DEL=.02			DEL=.05			DEL=.10			DEL=.15		
FACT=	1.0 .75	.50 .25	.00 BIN	1.0 .75	.50 .25	.00 BIN	1.0 .75	.50 .25	.00 BIN	1.0 75	.50 .25	.00 BIN	1.0 .75	.50 .25	.00 BIN
PCONT=***				REQUIRED NUMBER OF PATIENTS											
0.01 ***	1534 1537	1537 1543	1546 5922	566 566	566 566	566 1958	186 186	186 186	186 533	93 93	93 93	93 217	67 67	67 67	67 129
0.02 ***	3392 3398	3410 3424	3453 9776	1091 1091	1096 1100	1105 2911	295 295	295 295	295 681	128 128	128 128	128 252	85 85	85 85	85 144
0.05 ***	10266 10287	10328 10433	10791 20855	2911 2925	2946 2978	3021 5651	633 633	636 636	641 1104	225 225	225 225	230 352	134 134	134 134	134 186
0.1 ***	22642 22695	22805 23115	25115 37716	6084 6136	6215 6355	6644 9816	1187 1201	1210 1222	1236 1747	373 379	379 379	382 502	198 198	198 204	204 248
0.15 ***	33816 33918	34110 34679	39670 52569	8936 9027	9190 9500	10270 13479	1695 1718	1744 1773	1814 2309	510 513	513 519	522 633	260 260	260 260	265 302
0.2 ***	42828 42983	43283 44176	53299 65415	11255 11404	11658 12186	13647 16640	2129 2164	2211 2263	2339 2791	624 633	636 641	650 743	309 312	312 312	318 346
0.25 ***	49408 49624	50061 51348	65435 76254	12991 13201	13574 14338	16640 19300	2479 2531	2596 2683	2797 3192	720 729	738 746	755 833	353 353	356 356	361 382
0.3 ***	53565 53868	54466 56233	75772 85086	14160 14440	14939 15966	19177 21458	2741 2815	2902 3021	3182 3513	793 808	816 828	843 904	382 388	388 391	396 409
0.35 ***	55472 55875	56671 59013	84134 91910	14802 15170	15805 17100	21221 23113	2928 3016	3130 3275	3485 3754	851 863	878 890	907 954	405 408	414 414	417 426
0.4 ***	55376 55898	56942 59949	90420 96728	14995 15455	16234 17783	22738 24268	3042 3147	3278 3450	3704 3915	886 898	913 930	951 984	417 423	426 431	435 435
0.45 ***	53570 54244	55583 59348	94567 99538	14811 15368	16281 18063	23724 24920	3091 3205	3354 3550	3835 3995	898 916	933 951	974 994	423 426	431 435	440 435
0.5 ***	50385 51246	52943 57520	96545 100341	14326 14974	16010 17978	24161 25070	3077 3200	3363 3573	3879 3995	895 913	930 951	974 984	417 423	426 431	435 426
0.55 ***	46176 47265	49373 54755	96340 99136	13618 14338	15473 17581	24056 24719	3007 3138	3301 3520	3835 3915	872 890	907 930	951 954	405 408	414 417	423 409
0.6 ***	41320 42680	45217 51295	93951 95925	12737 13516	14711 16890	23403 23866	2885 3016	3179 3398	3704 3754	834 851	869 886	907 904	379 382	388 391	396 382
0.65 ***	36196 37832	40755 47331	89384 90706	11736 12541	13758 15931	22210 22511	2715 2838	2995 3200	3488 3513	776 790	808 825	843 833	347 353	356 361	361 346
0.7 ***	31135 33008	36187 42965	82655 83480	10634 11433	12629 14711	20472 20655	2491 2605	2745 2928	3182 3192	697 711	723 741	758 743	309 312	312 318	321 302
0.75 ***	26349 28344	31593 38231	73777 74247	9438 10200	11316 13218	18203 18296	2216 2313	2435 2587	2794 2791	606 615	624 636	650 633	260 260	265 265	268 248
0.8 ***	21890 23846	26953 33078	62778 63007	8140 8817	9803 11433	15397 15436	1878 1957	2053 2173	2321 2309	487 496	505 513	522 502	198 198	204 204	204 186
0.85 ***	17628 19393	22123 27346	49676 49759	6688 7248	8035 9304	12069 12074	1476 1534	1595 1674	1765 1747	356 356	361 365	370 352	. .	. .	. .
0.9 ***	13288 14680	16794 20673	34495 34504	4985 5378	5909 6714	8222 8210	991 1018	1047 1082	1123 1104	. .	. .	. .	. .	. .	. .
0.95 ***	8201 9010	10182 12165	17258 17242	2788 2955	3165 3450	3865 3844	. .	. .	. .	. .	. .	. .	. .	. .	. .
0.98 ***	3853 4133	4509 5043	5938 5922	. .	. .	. .	. .	. .	. .	. .	. .	. .	. .	. .	. .

TABLE 3: ALPHA= 0.01 POWER= 0.8   EXPECTED ACCRUAL THRU MINIMUM FOLLOW-UP= 3750

		DEL=.01			DEL=.02			DEL=.05			DEL=.10			DEL=.15		
FACT=		1.0 .75	.50 .25	.00 BIN	1.0 .75	.50 .25	.00 BIN	1.0 .75	.50 .25	.00 BIN	1.0 75	.50 .25	.00 BIN	1.0 .75	.50 .25	.00 BIN
PCONT=***		REQUIRED NUMBER OF PATIENTS														
0.01	***	1534 1540	1540 1544	1550 5922	565 565	565 565	565 1958	184 184	184 184	184 533	97 97	97 97	97 217	68 68	68 68	68 129
0.02	***	3391 3400	3409 3425	3453 9776	1090 1094	1094 1100	1103 2911	297 297	297 297	297 681	128 128	128 128	128 252	87 87	87 87	87 144
0.05	***	10268 10291	10338 10447	10788 20855	2913 2928	2950 2978	3022 5651	631 634	634 640	640 1104	228 228	228 228	228 352	134 134	134 134	134 186
0.1	***	22656 22713	22831 23159	25113 37716	6097 6147	6228 6368	6640 9816	1193 1203	1212 1222	1234 1747	378 378	378 381	381 502	200 200	200 203	203 248
0.15	***	33837 33944	34150 34756	39672 52569	8956 9053	9218 9531	10272 13479	1703 1722	1747 1775	1812 2309	509 513	518 518	522 633	259 259	259 265	265 302
0.2	***	42865 43025	43347 44300	53300 65415	11290 11440	11706 12237	13647 16640	2140 2172	2215 2272	2337 2791	625 631	634 640	650 743	312 312	312 316	316 346
0.25	***	49456 49684	50153 51528	65434 76254	13038 13259	13643 14422	16643 19300	2491 2543	2609 2688	2797 3192	725 728	738 747	756 833	353 353	353 359	359 382
0.3	***	53631 53950	54593 56478	75772 85086	14219 14519	15034 16075	19178 21458	2759 2828	2918 3031	3181 3513	800 809	818 828	841 904	381 387	387 391	397 409
0.35	***	55559 55990	56843 59328	84134 91910	14884 15269	15925 17234	21222 23113	2950 3034	3147 3288	3484 3754	850 865	878 893	906 954	406 409	409 415	419 426
0.4	***	55488 56047	57166 60350	90419 96728	15100 15578	16375 17941	22741 24268	3068 3172	3297 3466	3700 3915	888 903	916 934	950 984	419 425	425 428	434 435
0.45	***	53716 54434	55868 59838	94568 99538	14937 15513	16447 18241	23722 24920	3119 3231	3378 3565	3831 3995	903 916	934 953	972 994	425 428	428 434	438 435
0.5	***	50572 51490	53300 58094	96547 100341	14472 15138	16197 18181	24162 25070	3109 3231	3387 3588	3878 3995	897 916	934 953	972 984	419 425	425 428	434 426
0.55	***	46413 47575	49806 55409	96340 99136	13784 14525	15678 17791	24059 24719	3040 3166	3325 3537	3837 3915	878 893	912 931	953 954	406 409	409 415	419 409
0.6	***	41618 43056	45725 52009	93950 95925	12916 13713	14928 17106	23406 23866	2918 3044	3203 3409	3706 3754	837 850	869 888	906 904	381 387	387 391	397 382
0.65	***	36556 38281	41328 48078	89384 90706	11922 12743	13972 16147	22212 22511	2744 2866	3016 3213	3490 3513	781 794	809 828	847 833	350 353	353 359	363 346
0.7	***	31553 33503	36790 43718	82656 83480	10822 11631	12837 14913	20472 20655	2519 2631	2768 2947	3184 3192	700 715	728 743	756 743	306 312	312 316	322 302
0.75	***	26806 28859	32200 38950	73778 74247	9616 10384	11506 13403	18200 18296	2238 2337	2453 2600	2797 2791	606 616	625 640	650 633	259 259	265 265	265 248
0.8	***	22338 24350	27522 33734	62781 63007	8300 8984	9968 11590	15400 15436	1900 1975	2069 2178	2322 2309	490 500	503 513	522 502	200 200	203 203	203 186
0.85	***	18040 19834	22615 27893	49675 49759	6822 7381	8168 9419	12072 12074	1493 1544	1606 1675	1765 1747	353 359	363 368	372 352	. .	. .	. .
0.9	***	13609 15025	17168 21068	34497 34504	5078 5468	5997 6784	8225 8210	997 1025	1053 1084	1122 1104	. .	. .	. .	. .	. .	. .
0.95	***	8393 9209	10384 12353	17256 17242	2828 2993	3200 3472	3865 3844	. .	. .	. .	. .	. .	. .	. .	. .	. .
0.98	***	3922 4203	4568 5088	5941 5922	. .	. .	. .	. .	. .	. .	. .	. .	. .	. .	. .	. .

TABLE 3: ALPHA= 0.01 POWER= 0.8   EXPECTED ACCRUAL THRU MINIMUM FOLLOW-UP= 4000

PCONT=***	DEL=.01 1.0 .75	DEL=.01 .50 .25	DEL=.01 .00 BIN	DEL=.02 1.0 .75	DEL=.02 .50 .25	DEL=.02 .00 BIN	DEL=.05 1.0 .75	DEL=.05 .50 .25	DEL=.05 .00 BIN	DEL=.10 1.0 75 .75	DEL=.10 .50 .25	DEL=.10 .00 BIN	DEL=.15 1.0 .75	DEL=.15 .50 .25	DEL=.15 .00 BIN
					REQUIRED NUMBER OF PATIENTS										
0.01 ***	1537	1543	1547	563	567	567	187	187	187	93	93	93	67	67	67
•••	1537	1543	5922	567	567	1958	187	187	533	93	93	217	67	67	129
0.02 ***	3397	3413	3453	1093	1097	1103	297	297	297	127	127	127	83	83	87
•••	3403	3427	9776	1093	1097	2911	297	297	681	127	127	252	83	83	144
0.05 ***	10273	10347	10787	2917	2953	3023	633	637	643	227	227	227	133	133	133
•••	10297	10453	20855	2933	2983	5651	633	637	1104	227	227	352	133	133	186
0.1 ***	22667	22853	25113	6107	6243	6643	1193	1213	1237	377	377	383	203	203	203
•••	22727	23197	37716	6157	6377	9816	1203	1223	1747	377	383	502	203	203	248
0.15 ***	33857	34193	39673	8977	9247	10273	1707	1753	1813	513	517	523	263	263	263
•••	33973	34833	52569	9077	9563	13479	1727	1777	2309	513	517	633	263	263	302
0.2 ***	42897	43413	53297	11317	11753	13647	2147	2223	2337	627	637	647	313	313	317
•••	43067	44423	65415	11477	12287	16640	2183	2273	2791	633	643	743	313	313	346
0.25 ***	49503	50247	65433	13083	13707	16643	2503	2617	2797	723	737	753	353	357	357
•••	49747	51703	76254	13313	14493	19300	2553	2693	3192	733	747	833	353	357	382
0.3 ***	53697	54723	75773	14283	15123	19177	2777	2927	3183	803	817	843	383	387	393
•••	54037	56717	85086	14593	16177	21458	2843	3037	3513	807	827	904	387	393	409
0.35 ***	55647	57013	84133	14967	16043	21223	2973	3163	3483	857	877	907	407	413	417
•••	56103	59643	91910	15367	17363	23113	3057	3297	3754	867	893	954	407	417	426
0.4 ***	55603	57387	90423	15197	16517	22737	3093	3317	3703	893	917	953	423	427	433
•••	56197	60743	96728	15697	18093	24268	3193	3477	3915	903	933	984	423	433	435
0.45 ***	53857	56153	94567	15057	16607	23723	3147	3397	3833	907	937	973	423	433	437
•••	54627	60317	99538	15653	18413	24920	3257	3577	3995	923	953	994	427	437	435
0.5 ***	50753	53647	96547	14617	16377	24163	3137	3407	3877	903	937	973	417	427	433
•••	51733	58657	100341	15297	18367	25070	3257	3603	3995	917	953	984	423	433	426
0.55 ***	46647	50233	96343	13943	15867	24057	3067	3347	3837	883	913	953	407	413	423
•••	47883	56043	99136	14697	17987	24719	3193	3553	3915	897	933	954	407	417	409
0.6 ***	41907	46217	93953	13087	15127	23407	2947	3227	3703	843	873	907	383	387	397
•••	43433	52697	95925	13897	17307	23866	3073	3427	3754	857	887	904	387	393	382
0.65 ***	36913	41873	89383	12103	14173	22207	2773	3043	3487	783	813	843	353	357	363
•••	38717	48793	90706	12933	16343	22511	2893	3227	3513	797	827	833	353	357	346
0.7 ***	31963	37367	82653	10997	13027	20473	2543	2787	3183	707	727	757	307	313	317
•••	33987	44433	83480	11823	15097	20655	2653	2957	3192	717	743	743	313	317	302
0.75 ***	27243	32777	73777	9787	11687	18203	2263	2467	2797	607	627	647	257	263	267
•••	29357	39637	74247	10563	13567	18296	2357	2613	2791	617	637	633	263	263	248
0.8 ***	22773	28063	62777	8453	10127	15397	1917	2083	2323	493	507	523	203	203	203
•••	24827	34347	63007	9143	11727	15436	1993	2187	2309	497	513	502	203	203	186
0.85 ***	18427	23083	49677	6947	8293	12073	1503	1613	1763	357	363	373	.	.	.
•••	20257	28407	49759	7507	9523	12074	1557	1683	1747	357	367	352	.	.	.
0.9 ***	13923	17517	34497	5167	6077	8223	1003	1057	1123	.	.	.	.	.	.
•••	15357	21433	34504	5557	6847	8210	1027	1087	1104	.	.	.	.	.	.
0.95 ***	8573	10573	17257	2867	3227	3863	.	.	.	.	.	.	.	.	.
•••	9393	12527	17242	3027	3493	3844	.	.	.	.	.	.	.	.	.
0.98 ***	3983	4623	5943	.	.	.	.	.	.	.	.	.	.	.	.
•••	4263	5127	5922	.	.	.	.	.	.	.	.	.	.	.	.

TABLE 3: ALPHA= 0.01 POWER= 0.8   EXPECTED ACCRUAL THRU MINIMUM FOLLOW-UP= 4250

FACT=	DEL=.01			DEL=.02			DEL=.05			DEL=.10			DEL=.15		
	1.0 .75	.50 .25	.00 BIN	1.0 .75	.60 .25	.00 BIN	1.0 .75	.50 .25	.00 BIN	1.0 75	.50 .25	.00 BIN	1.0 .75	.50 .25	.00 BIN
PCONT=•••			REQUIRED NUMBER OF PATIENTS												
0.01 ***	1537	1544	1548	566	566	566	188	188	188	92	92	92	67	67	67
•••	1537	1544	5922	566	566	1958	188	188	533	92	92	217	67	67	129
0.02 ***	3397	3414	3456	1091	1098	1102	294	294	294	131	131	131	82	82	88
•••	3403	3428	9776	1098	1102	2911	294	294	681	131	131	252	82	82	144
0.05 ***	10278	10356	10788	2918	2957	3025	634	634	641	226	226	226	131	131	131
•••	10303	10462	20855	2936	2982	5651	634	641	1104	226	226	352	131	131	186
0.1 ***	22677	22872	25114	6117	6255	6644	1197	1214	1236	375	379	379	198	198	205
•••	22745	23233	37716	6170	6389	9816	1204	1225	1747	379	379	502	198	198	248
0.15 ***	33880	34230	39670	8996	9272	10271	1714	1756	1813	513	517	524	262	262	262
•••	33997	34910	52569	9102	9587	13479	1728	1782	2309	513	517	633	262	262	302
0.2 ***	42928	43474	53298	11351	11797	13650	2153	2228	2341	630	641	651	311	315	315
•••	43109	44543	65415	11521	12332	16640	2185	2277	2791	634	641	743	311	315	346
0.25 ***	49548	50338	65432	13129	13773	16642	2515	2628	2798	726	740	758	354	358	358
•••	49813	51874	76254	13369	14563	19300	2564	2702	3192	730	747	833	354	358	382
0.3 ***	53759	54849	75770	14340	15212	19181	2791	2940	3180	804	821	843	386	390	396
•••	54120	56953	85086	14666	16270	21458	2855	3046	3513	811	832	904	386	390	409
0.35 ***	55731	57187	84136	15042	16153	21221	2989	3173	3482	857	878	906	407	411	418
•••	56213	59949	91910	15463	17481	23113	3074	3308	3754	868	896	954	411	418	426
0.4 ***	55710	57605	90422	15297	16646	22741	3116	3333	3701	896	921	953	422	428	432
•••	56348	61129	96728	15813	18229	24268	3212	3488	3915	906	938	984	422	432	435
0.45 ***	54003	56433	94570	15176	16759	23722	3169	3414	3832	910	938	974	422	432	439
•••	54822	60778	99538	15785	18569	24920	3280	3594	3995	921	953	994	428	432	435
0.5 ***	50939	53999	96546	14755	16546	24164	3163	3428	3881	906	938	974	422	428	432
•••	51981	59195	100341	15452	18537	25070	3280	3616	3995	921	953	984	422	432	426
0.55 ***	46881	50653	96340	14096	16051	24058	3095	3371	3832	885	917	953	407	411	422
•••	48188	56649	99136	14868	18172	24719	3216	3567	3915	900	932	954	411	418	409
0.6 ***	42199	46704	93949	13253	15314	23403	2972	3248	3705	843	874	910	386	390	396
•••	43799	53351	95925	14075	17492	23866	3095	3439	3754	857	889	904	386	390	382
0.65 ***	37265	42408	89385	12268	14362	22209	2798	3057	3488	783	811	843	354	358	364
•••	39146	49477	90706	13114	16525	22511	2914	3244	3513	800	825	833	354	358	346
0.7 ***	32360	37928	82655	11170	13214	20471	2568	2808	3184	708	730	758	311	315	322
•••	34453	45110	83480	11999	15271	20655	2674	2968	3192	719	747	743	311	315	302
0.75 ***	27664	33327	73777	9948	11854	18204	2281	2483	2798	613	630	651	258	262	269
•••	29838	40283	74247	10728	13724	18296	2377	2621	2791	619	641	633	262	262	248
0.8 ***	23187	28574	62780	8599	10271	15399	1937	2090	2319	496	507	524	198	205	205
•••	25284	34932	63007	9289	11861	15436	2005	2196	2309	503	513	502	198	205	186
0.85 ***	18803	23527	49675	7062	8408	12073	1516	1622	1767	358	364	368	.	.	.
•••	20662	28893	49759	7625	9619	12074	1565	1686	1747	358	368	352	.	.	.
0.9 ***	14213	17853	34496	5252	6148	8220	1013	1059	1123	.	.	.	.	.	.
•••	15668	21778	34504	5634	6903	8210	1034	1091	1104	.	.	.	.	.	.
0.95 ***	8748	10749	17258	2904	3254	3864	.	.	.	.	.	.	.	.	.
•••	9570	12689	17242	3063	3509	3844	.	.	.	.	.	.	.	.	.
0.98 ***	4045	4672	5943	.	.	.	.	.	.	.	.	.	.	.	.
•••	4317	5167	5922	.	.	.	.	.	.	.	.	.	.	.	.

TABLE 3: ALPHA= 0.01 POWER= 0.8   EXPECTED ACCRUAL THRU MINIMUM FOLLOW-UP= 4500

		DEL=.01			DEL=.02			DEL=.05			DEL=.10			DEL=.15		
FACT=		1.0 / .75	.50 / .25	.00 BIN	1.0 / .75	.50 / .25	.00 BIN	1.0 / .75	.50 / .25	.00 BIN	1.0 / 75	.50 / .25	.00 BIN	1.0 / .75	.50 / .25	.00 BIN
PCONT=***							REQUIRED NUMBER OF PATIENTS									
0.01	***	1538	1538	1549	566	566	566	188	188	188	93	93	93	64	64	64
	•••	1538	1545	5922	566	566	1958	188	188	533	93	93	217	64	64	129
0.02	***	3401	3416	3457	1095	1099	1106	296	296	296	127	131	131	86	86	86
	•••	3405	3428	9776	1095	1099	2911	296	296	681	131	131	252	86	86	144
0.05	***	10286	10365	10792	2921	2955	3023	633	638	638	228	228	228	131	131	131
	•••	10313	10477	20855	2940	2985	5651	633	638	1104	228	228	352	131	131	186
0.1	***	22688	22897	25113	6128	6263	6641	1196	1211	1234	375	379	379	199	199	199
	•••	22762	23273	37716	6184	6398	9816	1207	1223	1747	379	379	502	199	199	248
0.15	***	33900	34275	39671	9015	9300	10275	1718	1758	1815	514	514	521	262	262	262
	•••	34028	34984	52569	9127	9611	13479	1736	1781	2309	514	521	633	262	262	302
0.2	***	42960	43541	53299	11381	11843	13650	2163	2231	2336	626	638	649	311	311	318
	•••	43151	44659	65415	11557	12378	16640	2197	2280	2791	633	645	743	311	311	346
0.25	***	49593	50430	65433	13170	13834	16642	2523	2629	2798	728	739	757	352	356	356
	•••	49875	52046	76254	13418	14628	19300	2573	2703	3192	735	746	833	356	356	382
0.3	***	53823	54975	75772	14403	15296	19178	2805	2951	3180	802	825	840	386	390	397
	•••	54210	57180	85086	14741	16354	21458	2872	3052	3513	813	829	904	386	390	409
0.35	***	55815	57356	84135	15123	16253	21221	3007	3187	3484	863	881	908	408	413	420
	•••	56332	60251	91910	15551	17587	23113	3090	3315	3754	870	892	954	413	413	426
0.4	***	55826	57828	90420	15393	16770	22740	3135	3349	3705	896	919	953	420	431	435
	•••	56494	61500	96728	15922	18363	24268	3232	3502	3915	908	937	984	424	431	435
0.45	***	54150	56708	94571	15292	16905	23723	3191	3435	3833	915	941	975	424	431	442
	•••	55009	61230	99538	15915	18723	24920	3300	3603	3995	926	960	994	431	435	435
0.5	***	51123	54341	96544	14887	16710	24161	3187	3446	3878	908	941	975	420	424	435
	•••	52219	59723	100341	15600	18701	25070	3304	3630	3995	926	953	984	424	431	426
0.55	***	47111	51060	96341	14239	16224	24056	3120	3390	3833	885	919	953	408	413	420
	•••	48484	57232	99136	15026	18345	24719	3243	3581	3915	903	937	954	408	420	409
0.6	***	42488	47175	93952	13413	15495	23403	3000	3266	3705	847	874	908	386	390	397
	•••	44160	53985	95925	14246	17670	23866	3120	3450	3754	863	892	904	386	390	382
0.65	***	37605	42922	89385	12435	14543	22211	2820	3079	3491	791	813	847	352	356	363
	•••	39563	50126	90706	13290	16698	22511	2940	3255	3513	802	829	833	352	356	346
0.7	***	32752	38467	82657	11332	13384	20471	2591	2820	3187	712	735	757	311	311	318
	•••	34905	45761	83480	12169	15431	20655	2696	2978	3192	723	746	743	311	318	302
0.75	***	28076	33859	73781	10099	12011	18199	2303	2501	2793	615	633	649	262	262	266
	•••	30300	40901	74247	10886	13868	18296	2393	2629	2791	622	638	633	262	266	248
0.8	***	23588	29066	62778	8733	10410	15398	1950	2100	2321	498	510	521	199	206	206
	•••	25725	35486	63007	9431	11978	15436	2021	2201	2309	503	514	502	199	206	186
0.85	***	19162	23948	49676	7174	8513	12068	1526	1628	1763	356	363	368	.	.	.
	•••	21052	29348	49759	7736	9712	12074	1571	1691	1747	363	368	352	.	.	.
0.9	***	14498	18172	34496	5329	6218	8220	1016	1065	1121	.	.	.	.	.	.
	•••	15971	22103	34504	5711	6956	8210	1038	1095	1104	.	.	.	.	.	.
0.95	***	8906	10916	17261	2933	3281	3862	.	.	.	.	.	.	.	.	.
	•••	9735	12840	17242	3090	3525	3844	.	.	.	.	.	.	.	.	.
0.98	***	4098	4717	5943	.	.	.	.	.	.	.	.	.	.	.	.
	•••	4368	5201	5922	.	.	.	.	.	.	.	.	.	.	.	.

433

TABLE 3: ALPHA= 0.01 POWER= 0.8   EXPECTED ACCRUAL THRU MINIMUM FOLLOW-UP= 4750

	DEL=.01			DEL=.02			DEL=.05			DEL=.10			DEL=.15		
FACT=	1.0 .75	.50 .25	.00 BIN	1.0 .75	.50 .25	.00 BIN	1.0 .75	.50 .25	.00 BIN	1.0 75	.50 .25	.00 BIN	1.0 .75	.50 .25	.00 BIN

PCONT=***   REQUIRED NUMBER OF PATIENTS

PCONT															
0.01	1535	1540	1547	566	566	566	186	186	186	91	91	91	67	67	67
	1540	1547	5922	566	566	1958	186	186	533	91	91	217	67	67	129
0.02	3400	3416	3452	1096	1096	1100	293	293	300	127	127	127	87	87	87
	3404	3428	9776	1096	1100	2911	293	293	681	127	127	252	87	87	144
0.05	10287	10370	10790	2925	2960	3024	633	637	637	229	229	229	134	134	134
	10315	10482	20855	2941	2989	5651	633	637	1104	229	229	352	134	134	186
0.1	22701	22922	25112	6135	6278	6642	1195	1215	1238	376	376	383	198	198	205
	22772	23307	37716	6195	6409	9816	1207	1227	1747	376	383	502	198	198	248
0.15	33923	34315	39670	9033	9325	10275	1718	1761	1813	514	519	526	257	265	265
	34054	35051	52569	9147	9634	13479	1737	1785	2309	514	519	633	265	265	302
0.2	42991	43608	53298	11408	11883	13648	2169	2236	2335	633	637	649	312	312	317
	43197	44777	65415	11593	12417	16640	2200	2283	2791	633	645	743	312	317	346
0.25	49641	50524	65435	13213	13897	16640	2533	2640	2799	728	740	756	352	360	360
	49938	52218	76254	13474	14693	19300	2580	2711	3192	732	744	833	352	360	382
0.3	53885	55103	75770	14460	15374	19182	2818	2960	3179	804	823	839	383	388	395
	54296	57412	85086	14804	16438	21458	2882	3055	3513	815	835	904	388	395	409
0.35	55904	57526	84138	15196	16355	21217	3024	3202	3483	863	882	906	407	412	419
	56445	60547	91910	15635	17690	23113	3103	3321	3754	870	894	954	412	419	426
0.4	55935	58048	90420	15488	16894	22737	3150	3364	3701	899	922	953	424	431	435
	56647	61872	96728	16027	18485	24268	3245	3507	3915	910	934	984	424	431	435
0.45	54289	56984	94569	15405	17044	23722	3214	3447	3832	918	942	970	424	431	443
	55203	61670	99538	16039	18861	24920	3321	3613	3995	930	958	994	431	435	435
0.5	51308	54680	96547	15013	16859	24162	3210	3464	3879	910	942	977	419	431	435
	52460	60226	100341	15742	18854	25070	3321	3642	3995	930	958	984	424	431	426
0.55	47342	51462	96338	14384	16391	24055	3143	3404	3832	894	918	953	407	412	419
	48786	57799	99136	15180	18505	24719	3262	3590	3915	906	934	954	412	419	409
0.6	42777	47634	93951	13565	15667	23402	3024	3285	3701	851	875	910	383	388	395
	44515	54593	95925	14408	17832	23866	3143	3464	3754	863	894	904	388	395	382
0.65	37949	43423	89387	12591	14717	22210	2846	3095	3487	792	815	847	352	360	364
	39967	50757	90706	13450	16859	22511	2960	3262	3513	804	827	833	352	360	346
0.7	33135	38982	82653	11487	13545	20476	2609	2834	3186	716	732	756	312	317	317
	35343	46380	83480	12330	15583	20655	2715	2989	3192	720	744	743	312	317	302
0.75	28472	34370	73783	10244	12163	18200	2319	2509	2794	614	633	649	257	265	265
	30748	41487	74247	11035	14004	18296	2407	2640	2791	621	637	633	265	265	248
0.8	23972	29541	62779	8862	10537	15398	1963	2110	2319	495	507	519	198	205	205
	26152	36013	63007	9555	12092	15436	2034	2205	2309	502	514	502	205	205	186
0.85	19502	24352	49677	7283	8613	12068	1535	1635	1765	360	364	372	.	.	.
	21426	29779	49759	7841	9793	12074	1583	1694	1747	360	364	352	.	.	.
0.9	14769	18474	34493	5399	6285	8225	1017	1065	1124	.	.	.	.	.	.
	16253	22412	34504	5779	7010	8210	1041	1096	1104	.	.	.	.	.	.
0.95	9064	11071	17258	2965	3305	3863	.	.	.	.	.	.	.	.	.
	9895	12975	17242	3119	3542	3844	.	.	.	.	.	.	.	.	.
0.98	4152	4758	5941	.	.	.	.	.	.	.	.	.	.	.	.
	4421	5228	5922	.	.	.	.	.	.	.	.	.	.	.	.

TABLE 3: ALPHA= 0.01 POWER= 0.8   EXPECTED ACCRUAL THRU MINIMUM FOLLOW-UP= 5000

	DEL=.01			DEL=.02			DEL=.05			DEL=.10			DEL=.15		
FACT=	1.0 .75	.50 .25	.00 BIN	1.0 .75	.50 .25	.00 BIN	1.0 .75	.50 .25	.00 BIN	1.0 75	.50 .25	.00 BIN	1.0 .75	.50 .25	.00 BIN
PCONT=•••				REQUIRED NUMBER OF PATIENTS											
0.01 •••	1541 1541	1546	566 566	566	183 183	183	96 96	96	66 66	66				
•••	1541 1546	5922	566 566	1958	183 183	533	96 96	217	66 66	129				
0.02 •••	3404 3416	3454	1091 1096	1104	296 296	296	129 129	129	83 83	83				
•••	3408 3433	9776	1096 1104	2911	296 296	681	129 129	252	83 83	144				
0.05 •••	10291 10379	10791	2929 2966	3021	633 633	641	229 229	229	133 133	133				
•••	10321 10491	20855	2946 2991	5651	633 641	1104	229 229	352	133 133	186				
0.1 •••	22716 22941	25116	6146 6283	6641	1204 1216	1233	379 379	383	204 204	204				
•••	22791 23341	37716	6204 6416	9816	1208 1229	1747	379 379	502	204 204	248				
0.15 •••	33941 34358	39671	9054 9346	10271	1721 1758	1808	516 516	521	258 258	266				
•••	34079 35129	52569	9166 9654	13479	1741 1783	2309	516 521	633	258 266	302				
0.2 •••	43021 43671	53296	11441 11921	13646	2171 2241	2341	633 641	646	308 316	316				
•••	43241 44891	65415	11621 12458	16640	2204 2283	2791	633 641	743	316 316	346				
0.25 •••	49683 50616	65433	13258 13954	16641	2541 2646	2796	729 741	754	354 358	358				
•••	49996 52379	76254	13521 14746	19300	2591 2716	3192	733 746	833	354 358	382				
0.3 •••	53954 55233	75771	14516 15454	19179	2829 2971	3183	808 821	841	383 391	396				
•••	54379 57633	85086	14871 16516	21458	2891 3058	3513	816 833	904	391 391	409				
0.35 •••	55991 57691	84133	15271 16454	21221	3033 3208	3483	866 883	908	408 416	421				
•••	56558 60833	91910	15721 17791	23113	3116 3329	3754	871 896	954	408 416	426				
0.4 •••	56046 58266	90421	15579 17008	22741	3171 3379	3704	904 921	954	421 429	433				
•••	56791 62229	96728	16133 18604	24268	3266 3516	3915	916 933	984	429 429	435				
0.45 •••	54433 57258	94566	15508 17179	23721	3233 3466	3833	916 946	971	429 433	441				
•••	55391 62096	99538	16166 18991	24920	3333 3621	3995	929 958	994	429 433	435				
0.5 •••	51491 55016	96546	15141 17008	24166	3229 3479	3879	916 941	971	421 429	433				
•••	52704 60721	100341	15879 18996	25070	3341 3654	3995	929 958	984	429 433	426				
0.55 •••	47571 51858	96341	14521 16546	24058	3166 3421	3833	896 921	954	408 416	421				
•••	49079 58346	99136	15329 18654	24719	3283 3604	3915	908 933	954	408 416	409				
0.6 •••	43058 48079	93954	13708 15829	23404	3046 3304	3704	854 879	908	383 391	396				
•••	44871 55171	95925	14566 17983	23866	3158 3479	3754	866 891	904	383 391	382				
0.65 •••	38279 43908	89383	12741 14879	22208	2866 3108	3491	791 816	846	354 358	366				
•••	40366 51354	90706	13608 17008	22511	2979 3271	3513	804 829	833	354 358	346				
0.7 •••	33504 39483	82654	11633 13704	20471	2629 2854	3183	716 733	758	308 316	321				
•••	35771 46971	83480	12479 15729	20655	2733 2996	3192	721 746	743	316 316	302				
0.75 •••	28858 34858	73779	10383 12308	18204	2333 2521	2796	616 633	646	258 266	266				
•••	31179 42046	74247	11179 14133	18296	2421 2646	2791	621 641	633	258 266	248				
0.8 •••	24346 29991	62779	8983 10658	15396	1979 2121	2321	496 508	521	204 204	204				
•••	26558 36516	63007	9683 12196	15436	2041 2208	2309	504 516	502	204 204	186				
0.85 •••	19833 24741	49679	7383 8708	12071	1546 1641	1766	358 366	371	. .	.				
•••	21779 30196	49759	7941 9871	12074	1591 1696	1747	358 366	352	. .	.				
0.9 •••	15029 18758	34496	5471 6346	8221	1021 1071	1121	. .	.	. .	.				
•••	16529 22696	34504	5846 7054	8210	1046 1096	1104	. .	.	. .	.				
0.95 •••	9208 11216	17258	2996 3329	3866	. .	.	. .	.	. .	.				
•••	10041 13108	17242	3141 3554	3844	. .	.	. .	.	. .	.				
0.98 •••	4204 4804	5941	. .	.	. .	.	. .	.	. .	.				
•••	4466 5258	5922	. .	.	. .	.	. .	.	. .	.				

TABLE 3: ALPHA= 0.01 POWER= 0.8   EXPECTED ACCRUAL THRU MINIMUM FOLLOW-UP= 5500

PCONT	FACT=	DEL=.01 1.0 .75	.50 .25	.00 BIN	DEL=.02 1.0 .75	.50 .25	.00 BIN	DEL=.05 1.0 .75	.50 .25	.00 BIN	DEL=.10 1.0 75	.50 .25	.00 BIN	DEL=.15 1.0 .75	.50 .25	.00 BIN
						REQUIRED NUMBER OF PATIENTS										
0.01	***	1535	1544	1549	568	568	568	188	188	188	92	92	92	64	64	64
	•••	1535	1544	5922	568	568	1958	188	188	533	92	92	217	64	64	129
0.02	***	3400	3419	3455	1095	1095	1104	298	298	298	128	128	128	86	86	86
	•••	3405	3433	9776	1095	1104	2911	298	298	681	128	128	252	86	86	144
0.05	***	10303	10390	10789	2933	2965	3020	636	636	642	224	229	229	133	133	133
	•••	10335	10509	20855	2946	2993	5651	636	636	1104	224	229	352	133	133	186
0.1	***	22738	22985	25111	6164	6307	6645	1205	1219	1233	375	380	380	202	202	202
	•••	22820	23412	37716	6224	6430	9816	1214	1228	1747	380	380	502	202	202	248
0.15	***	33985	34439	39673	9093	9387	10275	1728	1764	1810	513	518	526	257	265	265
	•••	34137	35264	52569	9208	9698	13479	1742	1783	2309	518	518	633	265	265	302
0.2	***	43088	43798	53299	11499	11994	13649	2182	2250	2341	636	642	650	312	312	312
	•••	43322	45109	65415	11691	12530	16640	2218	2292	2791	636	642	743	312	312	346
0.25	***	49779	50802	65432	13341	14062	16641	2561	2658	2795	733	746	752	353	353	361
	•••	50123	52699	76254	13616	14854	19300	2603	2718	3192	738	746	833	353	361	382
0.3	***	54083	55485	75772	14634	15596	19177	2850	2988	3180	807	820	843	389	389	394
	•••	54550	58070	85086	15005	16660	21458	2910	3070	3513	815	834	904	389	394	409
0.35	***	56159	58029	84132	15409	16633	21220	3062	3227	3483	870	884	903	408	416	416
	•••	56783	61398	91910	15885	17967	23113	3139	3345	3754	875	898	954	408	416	426
0.4	***	56269	58694	90424	15753	17224	22738	3199	3400	3703	903	925	953	422	430	435
	•••	57094	62915	96728	16330	18814	24268	3290	3529	3915	917	939	984	422	430	435
0.45	***	54720	57800	94568	15720	17425	23723	3268	3488	3832	925	944	972	430	435	435
	•••	55774	62915	99538	16394	19232	24920	3373	3639	3995	930	958	994	430	435	435
0.5	***	51855	55669	96548	15376	17279	24163	3268	3510	3881	917	944	972	422	430	435
	•••	53180	61659	100341	16133	19268	25070	3378	3667	3995	930	958	984	422	430	426
0.55	***	48033	52625	96342	14785	16839	24058	3208	3455	3832	898	925	953	408	416	422
	•••	49660	59382	99136	15610	18938	24719	3318	3620	3915	911	939	954	408	416	409
0.6	***	43619	48940	93949	13988	16133	23406	3084	3331	3703	856	884	911	389	389	394
	•••	45558	56274	95925	14854	18269	23866	3194	3496	3754	870	898	904	389	394	382
0.65	***	38930	44834	89384	13025	15175	22210	2905	3139	3488	793	820	843	353	361	361
	•••	41135	52488	90706	13905	17288	22511	3007	3290	3513	807	829	833	353	361	346
0.7	***	34219	40434	82655	11911	13993	20469	2663	2878	3185	719	738	760	312	312	320
	•••	36593	48088	83480	12764	15987	20655	2759	3015	3192	724	746	743	312	320	302
0.75	***	29599	35787	73778	10646	12571	18200	2369	2539	2795	623	636	650	265	265	265
	•••	32000	43102	74247	11444	14364	18296	2443	2658	2791	628	642	633	265	265	248
0.8	***	25056	30845	62778	9216	10885	15395	1998	2135	2319	499	513	518	202	202	202
	•••	27339	37450	63007	9918	12384	15436	2066	2218	2309	504	513	502	202	202	186
0.85	***	20464	25469	49674	7566	8878	12068	1558	1645	1764	361	367	367	.	.	.
	•••	22458	30960	49759	8122	10005	12074	1604	1700	1747	361	367	352	.	.	.
0.9	***	15514	19300	34494	5600	6453	8226	1035	1076	1123	.	.	.	.	.	.
	•••	17045	23233	34504	5971	7132	8210	1049	1095	1104	.	.	.	.	.	.
0.95	***	9483	11490	17260	3043	3364	3859	.	.	.	.	.	.	.	.	.
	•••	10316	13341	17242	3185	3579	3844	.	.	.	.	.	.	.	.	.
0.98	***	4285	4871	5944	.	.	.	.	.	.	.	.	.	.	.	.
	•••	4547	5303	5922	.	.	.	.	.	.	.	.	.	.	.	.

TABLE 3: ALPHA= 0.01 POWER= 0.8   EXPECTED ACCRUAL THRU MINIMUM FOLLOW-UP= 6000

	DEL=.01			DEL=.02			DEL=.05			DEL=.10			DEL=.15		
FACT=	1.0 .75	.50 .25	.00 BIN	1.0 .75	.50 .25	.00 BIN	1.0 .75	.50 .25	.00 BIN	1.0 75	.50 .25	.00 BIN	1.0 .75	.50 .25	.00 BIN

PCONT=***   REQUIRED NUMBER OF PATIENTS

PCONT	D1a	D1b	D1c	D2a	D2b	D2c	D5a	D5b	D5c	D10a	D10b	D10c	D15a	D15b	D15c
0.01	1540	1540	1549	565	565	565	184	184	184	94	94	94	64	64	64
	1540	1540	5922	565	565	1958	184	184	533	94	94	217	64	64	129
0.02	3400	3424	3454	1090	1099	1105	295	295	295	130	130	130	85	85	85
	3409	3439	9776	1099	1099	2911	295	295	681	130	130	252	85	85	144
0.05	10309	10405	10789	2935	2974	3025	634	640	640	229	229	229	130	130	130
	10345	10525	20855	2950	2995	5651	634	640	1104	229	229	352	130	130	186
0.1	22759	23029	25114	6184	6325	6640	1204	1219	1234	379	379	379	199	199	199
	22849	23470	37716	6244	6445	9816	1210	1225	1747	379	379	502	199	199	248
0.15	34024	34519	39670	9124	9430	10270	1735	1765	1810	514	520	520	259	259	265
	34189	35395	52569	9244	9730	13479	1750	1789	2309	514	520	633	259	265	302
0.2	43150	43924	53299	11554	12064	13645	2194	2254	2335	634	640	649	310	310	310
	43414	45325	65415	11755	12589	16640	2224	2290	2791	634	640	743	310	319	346
0.25	49870	50989	65434	13420	14164	16639	2575	2665	2800	730	745	754	355	355	355
	50245	53014	76254	13705	14950	19300	2614	2725	3192	739	745	833	355	355	382
0.3	54205	55735	75769	14740	15730	19180	2869	2995	3184	814	829	844	385	394	394
	54724	58489	85086	15124	16789	21458	2929	3079	3513	820	835	904	385	394	409
0.35	56329	58360	84130	15550	16804	21220	3085	3250	3484	874	889	904	409	415	415
	57010	61930	91910	16039	18130	23113	3160	3355	3754	880	895	954	415	415	426
0.4	56494	59119	90424	15919	17425	22735	3229	3415	3700	910	925	949	424	430	430
	57385	63580	96728	16519	19009	24268	3319	3544	3915	919	940	984	424	430	435
0.45	55009	58330	94570	15919	17650	23725	3304	3514	3835	925	949	970	430	430	439
	56149	63694	99538	16609	19450	24920	3394	3655	3995	934	964	994	430	439	435
0.5	52219	56305	96544	15595	17530	24160	3304	3529	3880	925	949	970	424	430	430
	53650	62545	100341	16375	19504	25070	3409	3685	3995	934	964	984	424	430	426
0.55	48484	53365	96340	15025	17104	24055	3244	3475	3835	904	925	955	409	415	424
	50230	60349	99136	15865	19180	24719	3349	3634	3915	910	940	954	409	415	409
0.6	44164	49765	93949	14245	16405	23404	3124	3355	3700	859	880	910	385	394	394
	46219	57304	95925	15124	18520	23866	3229	3505	3754	874	895	904	385	394	382
0.65	39559	45709	89380	13285	15445	22210	2935	3160	3490	799	820	844	355	355	364
	41875	53539	90706	14170	17530	22511	3040	3304	3513	814	829	833	355	364	346
0.7	34909	41329	82654	12169	14254	20470	2695	2899	3184	724	739	760	310	319	319
	37369	49120	83480	13030	16219	20655	2785	3025	3192	730	745	743	310	319	302
0.75	30295	36655	73780	10885	12805	18199	2389	2560	2794	619	634	649	259	265	265
	32779	44065	74247	11689	14569	18296	2470	2665	2791	625	640	633	265	265	248
0.8	25729	31639	62779	9430	11089	15400	2020	2149	2320	499	514	520	199	205	205
	28060	38305	63007	10129	12559	15436	2080	2230	2309	505	514	502	205	205	186
0.85	21049	26140	49675	7735	9034	12070	1570	1660	1765	364	364	370	.	.	.
	23080	31654	49759	8290	10129	12074	1615	1705	1747	364	370	352	.	.	.
0.9	15970	19795	34495	5710	6550	8224	1039	1075	1120	.	.	.	.	.	.
	17515	23710	34504	6079	7204	8210	1054	1099	1104	.	.	.	.	.	.
0.95	9739	11734	17260	3094	3394	3865	.	.	.	.	.	.	.	.	.
	10570	13555	17242	3229	3595	3844	.	.	.	.	.	.	.	.	.
0.98	4369	4939	5944	.	.	.	.	.	.	.	.	.	.	.	.
	4624	5350	5922	.	.	.	.	.	.	.	.	.	.	.	.

TABLE 3: ALPHA= 0.01 POWER= 0.8   EXPECTED ACCRUAL THRU MINIMUM FOLLOW-UP= 6500

PCONT=***   REQUIRED NUMBER OF PATIENTS

	DEL=.01			DEL=.02			DEL=.05			DEL=.10			DEL=.15		
FACT=	1.0 .75	.50 .25	.00 BIN	1.0 .75	.50 .25	.00 BIN	1.0 .75	.50 .25	.00 BIN	1.0 75	.50 .25	.00 BIN	1.0 .75	.50 .25	.00 BIN
0.01 ***	1538	1538	1548	563	563	563	183	183	183	92	92	92	70	70	70
0.01 •••	1538	1548	5922	563	563	1958	183	183	533	92	92	217	70	70	129
0.02 ***	3407	3423	3456	1093	1099	1099	297	297	297	124	124	124	86	86	86
0.02 •••	3417	3433	9776	1099	1099	2911	297	297	681	124	124	252	86	86	144
0.05 ***	10323	10421	10784	2946	2978	3027	638	638	638	222	222	232	135	135	135
0.05 •••	10356	10535	20855	2952	2995	5651	638	638	1104	222	222	352	135	135	186
0.1 ***	22787	23069	25111	6196	6342	6641	1207	1223	1240	378	378	378	200	200	200
0.1 •••	22885	23535	37716	6261	6456	9816	1213	1229	1747	378	378	502	200	200	248
0.15 ***	34065	34601	39671	9153	9468	10275	1743	1776	1814	514	514	525	265	265	265
0.15 •••	34243	35517	52569	9283	9761	13479	1749	1792	2309	514	525	633	265	265	302
0.2 ***	43219	44048	53294	11607	12127	13644	2204	2263	2334	638	638	644	313	313	313
0.2 •••	43496	45527	65415	11808	12647	16640	2231	2296	2791	638	644	743	313	313	346
0.25 ***	49963	51166	65433	13498	14256	16645	2588	2676	2800	736	742	752	352	352	362
0.25 •••	50369	53311	76254	13791	15036	19300	2627	2735	3192	742	752	833	352	362	382
0.3 ***	54334	55986	75768	14841	15854	19180	2887	3011	3179	817	823	839	384	395	395
0.3 •••	54887	58895	85086	15237	16905	21458	2946	3082	3513	823	833	904	384	395	409
0.35 ***	56506	58689	84137	15686	16953	21217	3108	3261	3482	872	888	904	411	417	417
0.35 •••	57243	62443	91910	16190	18276	23113	3179	3358	3754	882	898	954	411	417	426
0.4 ***	56717	59534	90420	16082	17609	22738	3255	3433	3699	915	931	953	427	427	433
0.4 •••	57682	64208	96728	16693	19180	24268	3336	3553	3915	921	937	984	427	433	435
0.45 ***	55293	58846	94569	16098	17863	23719	3326	3531	3829	931	947	969	427	433	443
0.45 •••	56522	64436	99538	16807	19651	24920	3423	3667	3995	937	963	994	433	433	435
0.5 ***	52579	56918	96546	15806	17766	24158	3336	3553	3878	931	947	969	427	427	433
0.5 •••	54117	63380	100341	16602	19716	25070	3433	3699	3995	937	963	984	427	433	426
0.55 ***	48933	54074	96341	15253	17349	24055	3271	3498	3840	904	931	953	411	417	417
0.55 •••	50786	61257	99136	16108	19407	24719	3374	3651	3915	915	937	954	411	417	409
0.6 ***	44692	50548	93952	14489	16661	23405	3147	3374	3699	866	882	904	384	395	395
0.6 •••	46859	58267	95925	15377	18747	23866	3255	3521	3754	872	898	904	384	395	382
0.65 ***	40164	46545	89386	13531	15702	22208	2968	3179	3488	801	823	839	352	362	362
0.65 •••	42580	54513	90706	14424	17756	22511	3066	3320	3513	817	833	833	352	362	346
0.7 ***	35560	42163	82652	12403	14489	20469	2724	2913	3179	726	742	758	313	313	319
0.7 •••	38111	50077	83480	13271	16423	20655	2816	3033	3192	736	752	743	313	319	302
0.75 ***	30961	37467	73780	11109	13021	18205	2416	2572	2800	622	638	644	265	265	265
0.75 •••	33502	44968	74247	11906	14749	18296	2491	2676	2791	628	644	633	265	265	248
0.8 ***	26352	32381	62778	9625	11266	15399	2036	2156	2318	498	508	525	200	206	206
0.8 •••	28741	39102	63007	10313	12702	15436	2091	2237	2309	508	514	502	200	206	186
0.85 ***	21601	26768	49681	7892	9176	12068	1587	1662	1766	362	368	368	.	.	.
0.85 •••	23671	32299	49759	8444	10242	12074	1619	1711	1747	362	368	352	.	.	.
0.9 ***	16391	20252	34493	5812	6635	8227	1045	1083	1126	.	.	.	.	.	.
0.9 •••	17961	24152	34504	6169	7268	8210	1061	1099	1104	.	.	.	.	.	.
0.95 ***	9966	11965	17262	3131	3423	3862	.	.	.	.	.	.	.	.	.
0.95 •••	10801	13742	17242	3261	3618	3844	.	.	.	.	.	.	.	.	.
0.98 ***	4441	4993	5942	.	.	.	.	.	.	.	.	.	.	.	.
0.98 •••	4685	5389	5922	.	.	.	.	.	.	.	.	.	.	.	.

TABLE 3: ALPHA= 0.01 POWER= 0.8   EXPECTED ACCRUAL THRU MINIMUM FOLLOW-UP= 7000

		DEL=.01			DEL=.02			DEL=.05			DEL=.10			DEL=.15		
FACT=		1.0 .75	.50 .25	.00 BIN	1.0 .75	.50 .25	.00 BIN	1.0 .75	.50 .25	.00 BIN	1.0 75	.50 .25	.00 BIN	1.0 .75	.50 .25	.00 BIN
PCONT=•••					REQUIRED NUMBER OF PATIENTS											
0.01	***	1534	1545	1545	565	565	565	186	186	186	92	92	92	64	64	64
	•••	1545	1545	5922	565	565	1958	186	186	533	92	92	217	64	64	129
0.02	***	3406	3424	3452	1096	1096	1107	291	291	291	127	127	127	81	81	81
	•••	3417	3435	9776	1096	1096	2911	291	291	681	127	127	252	81	81	144
0.05	***	10330	10435	10791	2945	2980	3021	635	635	641	221	221	232	134	134	134
	•••	10365	10546	20855	2962	2997	5651	635	641	1104	221	221	352	134	134	186
0.1	***	22807	23111	25117	6217	6357	6644	1212	1219	1236	379	379	379	197	204	204
	•••	22912	23584	37716	6270	6469	9816	1212	1230	1747	379	379	502	204	204	248
0.15	***	34112	34679	39666	9192	9496	10266	1744	1772	1814	512	519	519	256	256	267
	•••	34305	35635	52569	9315	9787	13479	1755	1790	2309	519	519	633	256	267	302
0.2	***	43282	44175	53299	11660	12185	13644	2210	2262	2339	635	641	652	309	309	320
	•••	43586	45721	65415	11870	12692	16640	2234	2297	2791	635	641	743	309	320	346
0.25	***	50061	51350	65437	13574	14337	16636	2595	2682	2794	740	746	757	355	355	361
	•••	50492	53596	76254	13871	15107	19300	2636	2735	3192	740	746	833	355	361	382
0.3	***	54465	56232	75769	14939	15965	19174	2899	3021	3179	816	827	845	390	390	396
	•••	55060	59277	85086	15352	17004	21458	2951	3091	3513	816	834	904	390	390	409
0.35	***	56670	59015	84134	15807	17102	21221	3126	3277	3487	880	886	904	414	414	414
	•••	57464	62935	91910	16321	18404	23113	3196	3371	3754	880	897	954	414	414	426
0.4	***	56939	59949	90416	16234	17785	22737	3277	3452	3704	915	932	950	425	431	431
	•••	57971	64796	96728	16857	19342	24268	3354	3564	3915	921	939	984	425	431	435
0.45	***	55585	59347	94564	16280	18065	23724	3354	3546	3837	932	950	974	431	431	442
	•••	56897	65129	99538	16997	19832	24920	3441	3680	3995	939	967	994	431	431	435
0.5	***	52942	57516	96541	16006	17977	24161	3365	3575	3879	932	950	974	425	431	431
	•••	54570	64166	100341	16811	19909	25070	3459	3704	3995	939	956	984	425	431	426
0.55	***	49372	54751	96342	15475	17581	24056	3301	3522	3837	904	932	950	414	414	425
	•••	51332	62119	99136	16332	19611	24719	3400	3662	3915	921	939	954	414	414	409
0.6	***	45214	51297	93951	14711	16892	23402	3179	3400	3704	869	886	904	390	390	396
	•••	47482	59161	95925	15604	18957	23866	3277	3529	3754	880	897	904	390	396	382
0.65	***	40751	47331	89384	13760	15930	22212	2997	3196	3487	810	827	845	355	361	361
	•••	43254	55427	90706	14659	17949	22511	3091	3330	3513	816	834	833	355	361	346
0.7	***	36184	42967	82657	12629	14711	20469	2741	2927	3179	722	740	757	309	320	320
	•••	38809	50971	83480	13497	16612	20655	2829	3050	3192	729	746	743	320	320	302
0.75	***	31592	38231	73774	11316	13217	18205	2437	2584	2794	624	635	652	267	267	267
	•••	34200	45791	74247	12115	14915	18296	2507	2682	2791	635	641	633	267	267	248
0.8	***	26955	33080	62777	9805	11432	15394	2052	2175	2321	501	512	519	204	204	204
	•••	29387	39835	63007	10494	12839	15436	2111	2245	2309	512	519	502	204	204	186
0.85	***	22125	27346	49676	8037	9304	12069	1597	1674	1761	361	361	372	.	.	.
	•••	24225	32887	49759	8580	10336	12074	1632	1720	1747	361	372	352	.	.	.
0.9	***	16794	20672	34497	5909	6714	8219	1044	1079	1125	.	.	.	.	.	.
	•••	18369	24557	34504	6259	7320	8210	1061	1107	1104	.	.	.	.	.	.
0.95	***	10179	12167	17260	3161	3452	3861	.	.	.	.	.	.	.	.	.
	•••	11019	13906	17242	3295	3634	3844	.	.	.	.	.	.	.	.	.
0.98	***	4509	5045	5937	.	.	.	.	.	.	.	.	.	.	.	.
	•••	4747	5419	5922	.	.	.	.	.	.	.	.	.	.	.	.

TABLE 3: ALPHA= 0.01 POWER= 0.8   EXPECTED ACCRUAL THRU MINIMUM FOLLOW-UP= 7500

PCONT=***		DEL=.01 1.0 .75	DEL=.01 .50 .25	DEL=.01 .00 BIN	DEL=.02 1.0 .75	DEL=.02 .50 .25	DEL=.02 .00 BIN	DEL=.05 1.0 .75	DEL=.05 .50 .25	DEL=.05 .00 BIN	DEL=.10 1.0 75	DEL=.10 .50 .25	DEL=.10 .00 BIN	DEL=.15 1.0 .75	DEL=.15 .50 .25	DEL=.15 .00 BIN
		REQUIRED NUMBER OF PATIENTS														
0.01	***	1543	1543	1550	568	568	568	181	181	181	99	99	99	68	68	68
	***	1543	1543	5922	568	568	1958	181	181	533	99	99	217	68	68	129
0.02	***	3406	3425	3455	1093	1100	1100	293	293	293	125	125	125	87	87	87
	***	3418	3436	9776	1100	1100	2911	293	293	681	125	125	252	87	87	144
0.05	***	10336	10449	10786	2949	2975	3024	631	643	643	230	230	230	136	136	136
	***	10374	10561	20855	2968	2993	5651	631	643	1104	230	230	352	136	136	186
0.1	***	22831	23161	25111	6230	6368	6643	1212	1224	1231	380	380	380	200	200	200
	***	22943	23637	37716	6286	6481	9816	1212	1231	1747	380	380	502	200	200	248
0.15	***	34149	34756	39668	9218	9530	10268	1749	1775	1812	518	518	518	256	268	268
	***	34355	35750	52569	9350	9811	13479	1756	1793	2309	518	518	633	256	268	302
0.2	***	43343	44300	53300	11705	12237	13643	2218	2274	2337	631	643	650	312	312	312
	***	43674	45912	65415	11918	12736	16640	2243	2300	2791	643	643	743	312	312	346
0.25	***	50150	51530	65431	13643	14424	16643	2611	2686	2799	736	743	755	350	361	361
	***	50618	53862	76254	13955	15174	19300	2649	2743	3192	743	755	833	361	361	382
0.3	***	54593	56480	75774	15031	16074	19175	2918	3031	3181	818	830	837	387	387	399
	***	55231	59649	85086	15455	17105	21458	2968	3099	3513	818	837	904	387	387	409
0.35	***	56843	59330	84136	15924	17236	21218	3143	3286	3481	875	893	905	406	418	418
	***	57687	63399	91910	16449	18530	23113	3211	3380	3754	886	893	954	418	418	426
0.4	***	57162	60350	90418	16374	17937	22737	3293	3462	3699	912	931	950	425	425	436
	***	58261	65368	96728	17011	19486	24268	3380	3575	3915	924	943	984	425	436	435
0.45	***	55868	59836	94568	16449	18237	23724	3380	3568	3830	931	950	968	425	436	436
	***	57256	65799	99538	17180	19993	24920	3462	3687	3995	943	961	994	436	436	435
0.5	***	53300	58093	96549	16193	18181	24162	3387	3586	3875	931	950	968	425	425	436
	***	55018	64906	100341	17011	20086	25070	3481	3718	3995	943	961	984	425	436	426
0.55	***	49805	55411	96343	15680	17787	24061	3324	3537	3837	912	931	950	406	418	418
	***	51856	62930	99136	16543	19793	24719	3425	3668	3915	924	943	954	418	418	409
0.6	***	45725	52006	93950	14930	17105	23405	3200	3406	3706	868	886	905	387	387	399
	***	48080	60012	95925	15830	19149	23866	3305	3549	3754	875	893	904	387	399	382
0.65	***	41330	48080	89386	13974	16149	22212	3012	3211	3493	811	830	849	350	361	361
	***	43906	56281	90706	14874	18136	22511	3106	3343	3513	818	837	833	361	361	346
0.7	***	36793	43718	82655	12837	14911	20468	2768	2949	3181	725	743	755	312	312	324
	***	39481	51800	83480	13700	16775	20655	2855	3050	3192	736	743	743	312	312	302
0.75	***	32199	38949	73775	11506	13400	18200	2450	2600	2799	624	643	650	268	268	268
	***	34861	46568	74247	12305	15061	18296	2525	2686	2791	631	643	633	268	268	248
0.8	***	27518	33736	62780	9968	11593	15399	2068	2180	2318	500	511	518	200	200	200
	***	29993	40512	63007	10662	12961	15436	2124	2243	2309	511	518	502	200	200	186
0.85	***	22618	27893	49674	8168	9418	12068	1606	1674	1768	361	368	368	.	.	.
	***	24736	33436	49759	8705	10418	12074	1636	1718	1747	361	368	352	.	.	.
0.9	***	17168	21068	34493	5993	6781	8225	1055	1081	1118	.	.	.	.	.	.
	***	18762	24924	34504	6343	7362	8210	1074	1100	1104	.	.	.	.	.	.
0.95	***	10381	12350	17255	3200	3474	3868	.	.	.	.	.	.	.	.	.
	***	11218	14068	17242	3324	3650	3844	.	.	.	.	.	.	.	.	.
0.98	***	4568	5086	5937	.	.	.	.	.	.	.	.	.	.	.	.
	***	4805	5450	5922	.	.	.	.	.	.	.	.	.	.	.	.

TABLE 3: ALPHA= 0.01 POWER= 0.8   EXPECTED ACCRUAL THRU MINIMUM FOLLOW-UP= 8000

		DEL=.01			DEL=.02			DEL=.05			DEL=.10			DEL=.15		
FACT=		1.0 .75	.50 .25	.00 BIN	1.0 .75	.50 .25	.00 BIN	1.0 .75	.50 .25	.00 BIN	1.0 75	.50 .25	.00 BIN	1.0 .75	.50 .25	.00 BIN
PCONT=•••				REQUIRED NUMBER OF PATIENTS												
0.01	•••	1545 1545	1545 1545	1545 5922	565 565	565 565	565 1958	185 185	185 185	185 533	93 93	93 93	93 217	65 65	65 65	65 129
0.02	•••	3413 3413	3425 3433	3453 9776	1093 1093	1093 1105	1105 2911	293 293	293 293	293 681	125 125	125 125	125 252	85 85	85 85	85 144
0.05	•••	10345 10385	10453 10573	10785 20855	2953 2965	2985 3005	3025 5651	633 633	633 633	645 1104	225 225	225 225	225 352	133 133	133 133	133 186
0.1	•••	22853 22973	23193 23685	25113 37716	6245 6305	6373 6485	6645 9816	1213 1213	1225 1225	1233 1747	373 373	385 385	385 502	205 205	205 205	205 248
0.15	•••	34193 34413	34833 35853	39673 52569	9245 9373	9565 9833	10273 13479	1753 1765	1773 1793	1813 2309	513 513	513 525	525 633	265 265	265 265	265 302
0.2	•••	43413 43753	44425 46093	53293 65415	11753 11973	12285 12785	13645 16640	2225 2245	2273 2305	2333 2791	633 633	645 645	645 743	313 313	313 313	313 346
0.25	•••	50245 50745	51705 54133	65433 76254	13705 14025	14493 15245	16645 19300	2613 2653	2693 2745	2793 3192	733 745	745 753	753 833	353 353	353 353	353 382
0.3	•••	54725 55405	56713 59993	75773 85086	15125 15545	16173 17193	19173 21458	2925 2973	3033 3105	3185 3513	813 825	825 833	845 904	385 393	393 393	393 409
0.35	•••	57013 57913	59645 63853	84133 91910	16045 16573	17365 18633	21225 23113	3165 3225	3293 3385	3485 3754	873 885	893 893	905 954	413 413	413 413	413 426
0.4	•••	57385 58553	60745 65913	90425 96728	16513 17153	18093 19613	22733 24268	3313 3393	3473 3573	3705 3915	913 925	933 945	953 984	425 425	433 433	433 435
0.45	•••	56153 57625	60313 66425	94565 99538	16605 17345	18413 20145	23725 24920	3393 3485	3573 3693	3833 3995	933 945	953 965	973 994	433 433	433 433	433 435
0.5	•••	53645 55453	58653 65625	96545 100341	16373 17193	18365 20253	24165 25070	3405 3493	3605 3725	3873 3995	933 945	953 965	973 984	425 425	433 433	433 426
0.55	•••	50233 52373	56045 63693	96345 99136	15865 16745	17985 19965	24053 24719	3345 3445	3553 3685	3833 3915	913 925	933 945	953 954	413 413	413 413	425 409
0.6	•••	46213 48665	52693 60813	93953 95925	15125 16033	17305 19313	23405 23866	3225 3325	3425 3553	3705 3754	873 885	885 893	905 904	385 393	393 393	393 382
0.65	•••	41873 44533	48793 57093	89385 90706	14173 15073	16345 18305	22205 22511	3045 3125	3225 3345	3485 3513	813 813	825 833	845 833	353 353	353 365	365 346
0.7	•••	37365 40125	44433 52585	82653 83480	13025 13893	15093 16933	20473 20655	2785 2865	2953 3065	3185 3192	725 733	745 753	753 743	313 313	313 313	313 302
0.75	•••	32773 35485	39633 47305	73773 74247	11685 12485	13565 15193	18205 18296	2465 2533	2613 2693	2793 2791	625 633	633 645	645 633	265 265	265 265	265 248
0.8	•••	28065 30573	34345 41145	62773 63007	10125 10813	11725 13065	15393 15436	2085 2133	2185 2253	2325 2309	505 505	513 513	525 502	205 205	205 205	205 186
0.85	•••	23085 25233	28405 33933	49673 49759	8293 8825	9525 10505	12073 12074	1613 1645	1685 1725	1725 1747	365 365	365 365	373 352	. .	. .	. .
0.9	•••	17513 19125	21433 25265	34493 34504	6073 6413	6845 7413	8225 8210	1053 1073	1085 1105	1125 1104	. .	. .	. .	. .	. .	. .
0.95	•••	10573 11405	12525 14205	17253 17242	3225 3353	3493 3653	3865 3844	. .	. .	. .	. .	. .	. .	. .	. .	. .
0.98	•••	4625 4853	5125 5473	5945 5922	. .	. .	. .	. .	. .	. .	. .	. .	. .	. .	. .	. .

TABLE 3: ALPHA= 0.01   POWER= 0.8   EXPECTED ACCRUAL THRU MINIMUM FOLLOW-UP= 8500

FACT=	DEL=.01 1.0 .75	.50 .25	.00 BIN	DEL=.02 1.0 .75	.50 .25	.00 BIN	DEL=.05 1.0 .75	.50 .25	.00 BIN	DEL=.10 1.0 75	.50 .25	.00 BIN	DEL=.15 1.0 .75	.50 .25	.00 BIN
PCONT=***					REQUIRED NUMBER OF PATIENTS										
0.01 ***	1544	1544	1544	566	566	566	184	184	184	91	91	91	70	70	70
•••	1544	1544	5922	566	566	1958	184	184	533	91	91	217	70	70	129
0.02 ***	3414	3427	3456	1098	1098	1098	290	290	290	133	133	133	78	78	91
•••	3414	3435	9776	1098	1098	2911	290	290	681	133	133	252	78	91	144
0.05 ***	10355	10461	10788	2960	2981	3023	630	643	643	226	226	226	133	133	133
•••	10397	10575	20855	2968	3002	5651	630	643	1104	226	226	352	133	133	186
0.1 ***	22871	23232	25110	6253	6389	6644	1217	1225	1238	375	375	375	197	197	205
•••	22998	23729	37716	6317	6495	9816	1217	1225	1747	375	375	502	197	205	248
0.15 ***	34226	34906	39666	9271	9590	10270	1756	1778	1812	516	516	524	261	261	261
•••	34460	35961	52569	9406	9853	13479	1770	1791	2309	516	524	633	261	261	302
0.2 ***	43470	44546	53301	11800	12331	13648	2224	2280	2343	643	643	651	311	311	311
•••	43845	46267	65415	12020	12820	16640	2245	2309	2791	643	643	743	311	311	346
0.25 ***	50334	51877	65435	13776	14562	16645	2628	2705	2798	736	750	758	354	354	354
•••	50865	54385	76254	14095	15293	19300	2662	2747	3192	736	750	833	354	354	382
0.3 ***	54852	56956	75770	15208	16270	19181	2938	3045	3180	821	835	843	388	388	396
•••	55575	60335	85086	15646	17269	21458	2989	3108	3513	821	835	904	388	388	409
0.35 ***	57190	59952	84135	16156	17481	21221	3172	3308	3478	877	898	906	410	418	418
•••	58133	64274	91910	16687	18735	23113	3236	3385	3754	885	898	954	418	418	426
0.4 ***	57601	61129	90425	16645	18225	22743	3329	3491	3703	920	941	949	431	431	431
•••	58834	66433	96728	17290	19734	24268	3406	3584	3915	928	941	984	431	431	435
0.45 ***	56433	60781	94568	16759	18565	23721	3414	3597	3831	941	949	970	431	431	439
•••	57976	67028	99538	17503	20278	24920	3499	3703	3995	949	962	994	431	439	435
0.5 ***	54002	59195	96545	16546	18536	24167	3427	3618	3881	941	949	970	431	431	431
•••	55880	66293	100341	17367	20393	25070	3512	3733	3995	949	962	984	431	431	426
0.55 ***	50653	56645	96340	16050	18175	24061	3371	3563	3831	920	928	949	410	418	418
•••	52876	64423	99136	16929	20130	24719	3456	3690	3915	928	941	954	418	418	409
0.6 ***	46700	53351	93952	15314	17495	23402	3244	3435	3703	877	885	906	388	388	396
•••	49221	61567	95925	16220	19471	23866	3342	3563	3754	885	898	904	388	396	382
0.65 ***	42408	49476	89383	14358	16525	22212	3053	3244	3491	813	821	843	354	354	367
•••	45128	57848	90706	15271	18459	22511	3151	3350	3513	821	835	833	354	354	346
0.7 ***	37924	45106	82655	13210	15271	20470	2811	2968	3180	728	750	758	311	311	325
•••	40742	53322	83480	14081	17078	20655	2883	3066	3192	736	750	743	311	311	302
0.75 ***	33326	40283	73773	11850	13720	18204	2479	2620	2798	630	643	651	261	261	269
•••	36075	47988	74247	12650	15327	18296	2543	2705	2791	630	643	633	261	269	248
0.8 ***	28574	34928	62778	10270	11863	15399	2088	2195	2322	503	516	524	205	205	205
•••	31116	41741	63007	10950	13168	15436	2139	2258	2309	516	516	502	205	205	186
0.85 ***	23530	28893	49675	8408	9619	12076	1621	1685	1770	367	367	367	.	.	.
•••	25697	34410	49759	8931	10575	12074	1650	1727	1747	367	367	352	.	.	.
0.9 ***	17856	21774	34495	6147	6899	8216	1055	1090	1119	.	.	.	.	.	.
•••	19471	25578	34504	6487	7451	8210	1076	1111	1104	.	.	.	.	.	.
0.95 ***	10745	12692	17261	3257	3512	3860	.	.	.	.	.	.	.	.	.
•••	11574	14328	17242	3371	3669	3844	.	.	.	.	.	.	.	.	.
0.98 ***	4668	5170	5943	.	.	.	.	.	.	.	.	.	.	.	.
•••	4893	5496	5922	.	.	.	.	.	.	.	.	.	.	.	.

TABLE 3: ALPHA= 0.01 POWER= 0.8   EXPECTED ACCRUAL THRU MINIMUM FOLLOW-UP= 9000

	DEL=.01			DEL=.02			DEL=.05			DEL=.10			DEL=.15		
FACT=	1.0 .75	.50 .25	.00 BIN	1.0 .75	.50 .25	.00 BIN	1.0 .75	.50 .25	.00 BIN	1.0 75	.50 .25	.00 BIN	1.0 .75	.50 .25	.00 BIN
PCONT=***						REQUIRED NUMBER OF PATIENTS									
0.01 ***	1536	1545	1545	569	569	569	186	186	186	96	96	96	60	60	60
•••	1545	1545	5922	569	569	1958	186	186	533	96	96	217	60	60	129
0.02 ***	3412	3426	3457	1095	1095	1109	299	299	299	127	127	127	82	82	82
•••	3426	3435	9776	1095	1095	2911	299	299	681	127	127	252	82	82	144
0.05 ***	10365	10477	10792	2954	2985	3021	636	636	636	231	231	231	127	127	127
•••	10410	10590	20855	2976	2999	5651	636	636	1104	231	231	352	127	127	186
0.1 ***	22897	23271	25116	6261	6396	6644	1207	1221	1230	375	375	375	195	195	195
•••	23032	23775	37716	6329	6495	9816	1221	1230	1747	375	375	502	195	195	248
0.15 ***	34274	34980	39674	9299	9614	10275	1761	1784	1815	510	524	524	262	262	262
•••	34521	36051	52569	9434	9870	13479	1770	1792	2309	524	524	633	262	262	302
0.2 ***	43544	44655	53295	11841	12381	13650	2234	2279	2332	636	645	645	307	307	321
•••	43926	46432	65415	12066	12854	16640	2256	2310	2791	636	645	743	307	321	346
0.25 ***	50429	52049	65436	13830	14631	16642	2625	2706	2796	735	749	757	352	352	352
•••	50991	54622	76254	14159	15351	19300	2670	2751	3192	749	749	833	352	352	382
0.3 ***	54974	57179	75772	15292	16350	19176	2954	3052	3179	825	825	839	389	389	397
•••	55739	60666	85086	15734	17340	21458	2999	3111	3513	825	839	904	389	389	409
0.35 ***	57359	60247	84134	16251	17587	21224	3187	3314	3480	884	892	906	411	411	420
•••	58357	64694	91910	16800	18839	23113	3246	3390	3754	884	906	954	411	420	426
0.4 ***	57831	61499	90420	16769	18366	22740	3345	3502	3705	915	937	951	434	434	434
•••	59122	66930	96728	17421	19851	24268	3412	3592	3915	929	937	984	434	434	435
0.45 ***	56706	61229	94574	16904	18726	23721	3435	3606	3831	937	960	974	434	434	442
•••	58326	67605	99538	17655	20400	24920	3516	3705	3995	951	960	994	434	434	435
0.5 ***	54344	59721	96540	16710	18704	24157	3449	3629	3876	937	951	974	420	434	434
•••	56310	66930	100341	17534	20535	25070	3525	3741	3995	951	960	984	434	434	426
0.55 ***	51059	57232	96337	16229	18344	24059	3390	3584	3831	915	937	951	411	420	420
•••	53362	65107	99136	17106	20265	24719	3480	3696	3915	929	937	954	411	420	409
0.6 ***	47175	53984	93952	15495	17669	23406	3269	3449	3705	870	892	906	389	389	397
•••	49762	62286	95925	16409	19626	23866	3359	3570	3754	884	906	904	389	397	382
0.65 ***	42922	50122	89385	14541	16701	22214	3075	3255	3494	816	825	847	352	352	366
•••	45712	58574	90706	15450	18600	22511	3156	3359	3513	816	839	833	352	366	346
0.7 ***	38467	45757	82657	13380	15427	20467	2819	2976	3187	735	749	757	307	321	321
•••	41325	54015	83480	14249	17205	20655	2895	3075	3192	735	749	743	321	321	302
0.75 ***	33855	40897	73784	12007	13866	18195	2504	2625	2796	636	636	645	262	262	262
•••	36659	48629	74247	12809	15441	18296	2557	2706	2791	636	645	633	262	262	248
0.8 ***	29062	35489	62781	10410	11976	15396	2099	2197	2324	510	510	524	209	209	209
•••	31641	42306	63007	11085	13259	15436	2152	2256	2309	510	510	502	209	209	186
0.85 ***	23946	29346	49672	8511	9712	12066	1626	1694	1761	366	366	366	.	.	.
•••	26137	34859	49759	9037	10635	12074	1657	1725	1747	366	366	352	.	.	.
0.9 ***	18172	22101	34499	6216	6959	8219	1064	1095	1117	.	.	.	.	.	.
•••	19792	25867	34504	6554	7485	8210	1072	1109	1104	.	.	.	.	.	.
0.95 ***	10919	12840	17264	3277	3525	3862	.	.	.	.	.	.	.	.	.
•••	11737	14451	17242	3390	3682	3844	.	.	.	.	.	.	.	.	.
0.98 ***	4717	5204	5946	.	.	.	.	.	.	.	.	.	.	.	.
•••	4934	5519	5922	.	.	.	.	.	.	.	.	.	.	.	.

TABLE 3: ALPHA= 0.01 POWER= 0.8   EXPECTED ACCRUAL THRU MINIMUM FOLLOW-UP= 9500

		DEL=.01			DEL=.02			DEL=.05			DEL=.10			DEL=.15		
FACT=		1.0 .75	.50 .25	.00 BIN	1.0 .75	.50 .25	.00 BIN	1.0 .75	.50 .25	.00 BIN	1.0 75	.50 .25	.00 BIN	1.0 .75	.50 .25	.00 BIN
PCONT=•••				REQUIRED NUMBER OF PATIENTS												
0.01	•••	1535	1550	1550	562	562	562	182	182	182	87	87	87	63	63	63
	•••	1535	1550	5922	562	562	1958	182	182	533	87	87	217	63	63	129
0.02	•••	3412	3426	3450	1099	1099	1099	291	291	300	125	125	125	87	87	87
	•••	3426	3435	9776	1099	1099	2911	291	291	681	125	125	252	87	87	144
0.05	•••	10370	10480	10789	2960	2984	3023	633	633	633	229	229	229	134	134	134
	•••	10418	10599	20855	2975	2999	5651	633	633	1104	229	229	352	134	134	186
0.1	•••	22925	23305	25110	6276	6404	6642	1218	1227	1241	372	386	386	196	196	205
	•••	23053	23813	37716	6333	6514	9816	1218	1227	1747	372	386	502	196	205	248
0.15	•••	34310	35047	39669	9325	9634	10275	1764	1788	1811	514	514	529	268	268	268
	•••	34572	36154	52569	9459	9886	13479	1773	1797	2309	514	514	633	268	268	302
0.2	•••	43611	44775	53301	11881	12413	13648	2239	2286	2334	633	648	648	315	315	315
	•••	44015	46589	65415	12104	12888	16640	2263	2310	2791	633	648	743	315	315	346
0.25	•••	50523	52218	65438	13900	14693	16640	2643	2714	2794	743	743	752	363	363	363
	•••	51102	54854	76254	14218	15405	19300	2675	2747	3192	743	752	833	363	363	382
0.3	•••	55106	57410	75769	15373	16441	19182	2960	3055	3174	823	838	838	386	395	395
	•••	55899	60973	85086	15809	17415	21458	3008	3117	3513	823	838	904	386	395	409
0.35	•••	57529	60545	84138	16355	17685	21215	3198	3317	3483	885	894	909	410	419	419
	•••	58574	65081	91910	16902	18920	23113	3260	3403	3754	885	894	954	410	419	426
0.4	•••	58051	61875	90423	16893	18484	22735	3364	3507	3697	918	933	956	434	434	434
	•••	59405	67409	96728	17543	19942	24268	3426	3593	3915	933	942	984	434	434	435
0.45	•••	56983	61670	94564	17044	18864	23718	3450	3616	3830	942	956	965	434	434	443
	•••	58669	68154	99538	17795	20526	24920	3521	3711	3995	942	965	994	434	434	435
0.5	•••	54679	60222	96550	16854	18849	24160	3459	3640	3878	942	956	980	434	434	434
	•••	56721	67537	100341	17685	20669	25070	3545	3744	3995	942	965	984	434	434	426
0.55	•••	51458	57799	96336	16394	18508	24050	3403	3593	3830	918	933	956	410	419	419
	•••	53833	65770	99136	17273	20408	24719	3498	3697	3915	933	942	954	410	419	409
0.6	•••	47634	54593	93947	15667	17828	23400	3284	3459	3697	870	894	909	386	395	395
	•••	50294	62968	95925	16569	19752	23866	3364	3578	3754	885	894	904	386	395	382
0.65	•••	43421	50760	89387	14717	16854	22213	3094	3260	3483	814	823	847	363	363	363
	•••	46271	59248	90706	15619	18730	22511	3174	3364	3513	823	838	833	363	363	346
0.7	•••	38980	46375	82656	13544	15586	20479	2833	2984	3189	728	743	752	315	315	315
	•••	41887	54679	83480	14408	17329	20655	2904	3079	3192	743	752	743	315	315	302
0.75	•••	34373	41483	73783	12166	14004	18199	2509	2643	2794	633	633	648	268	268	268
	•••	37199	49225	74247	12950	15539	18296	2571	2714	2791	633	648	633	268	268	248
0.8	•••	29537	36011	62778	10537	12095	15396	2105	2200	2319	505	514	514	205	205	205
	•••	32140	42828	63007	11202	13339	15436	2153	2263	2309	514	514	502	205	205	186
0.85	•••	24350	29774	49677	8613	9791	12071	1630	1693	1764	363	363	372	.	.	.
	•••	26559	35260	49759	9126	10694	12074	1669	1725	1747	363	372	352	.	.	.
0.9	•••	18469	22412	34491	6285	7013	8224	1060	1099	1123	.	.	.	.	.	.
	•••	20099	26140	34504	6609	7520	8210	1075	1108	1104	.	.	.	.	.	.
0.95	•••	11074	12974	17258	3308	3545	3863	.	.	.	.	.	.	.	.	.
	•••	11890	14550	17242	3412	3688	3844	.	.	.	.	.	.	.	.	.
0.98	•••	4756	5231	5944	.	.	.	.	.	.	.	.	.	.	.	.
	•••	4970	5540	5922	.	.	.	.	.	.	.	.	.	.	.	.

TABLE 3: ALPHA= 0.01 POWER= 0.8   EXPECTED ACCRUAL THRU MINIMUM FOLLOW-UP= 10000

PCONT=***	DEL=.01 1.0/.75	DEL=.01 .50/.25	DEL=.01 .00/BIN	DEL=.02 1.0/.75	DEL=.02 .50/.25	DEL=.02 .00/BIN	DEL=.05 1.0/.75	DEL=.05 .50/.25	DEL=.05 .00/BIN	DEL=.10 1.0/75	DEL=.10 .50/.25	DEL=.10 .00/BIN	DEL=.15 1.0/.75	DEL=.15 .50/.25	DEL=.15 .00/BIN
							REQUIRED NUMBER OF PATIENTS								
0.01 ***	1541	1541	1541	566	566	566	182	182	182	91	91	91	66	66	66
•••	1541	1541	5922	566	566	1958	182	182	533	91	91	217	66	66	129
0.02 ***	3416	3432	3457	1091	1107	1107	291	291	291	132	132	132	82	82	82
•••	3416	3441	9776	1091	1107	2911	291	291	681	132	132	252	82	82	144
0.05 ***	10382	10491	10791	2966	2991	3016	632	641	641	232	232	232	132	132	132
•••	10416	10607	20855	2966	3007	5651	641	641	1104	232	232	352	132	132	186
0.1 ***	22941	23341	25116	6282	6416	6641	1216	1232	1232	382	382	382	207	207	207
•••	23091	23857	37716	6341	6516	9816	1216	1232	1747	382	382	502	207	207	248
0.15 ***	34357	35132	39666	9341	9657	10266	1757	1782	1807	516	516	516	257	266	266
•••	34632	36232	52569	9482	9907	13479	1766	1791	2309	516	516	633	257	266	302
0.2 ***	43666	44891	53291	11916	12457	13641	2241	2282	2341	641	641	641	316	316	316
•••	44091	46732	65415	12141	12916	16640	2257	2307	2791	641	641	743	316	316	346
0.25 ***	50616	52382	65432	13957	14741	16641	2641	2716	2791	741	741	757	357	357	357
•••	51232	55082	76254	14282	15441	19300	2682	2757	3192	741	757	833	357	357	382
0.3 ***	55232	57632	75766	15457	16516	19182	2966	3057	3182	816	832	841	391	391	391
•••	56066	61266	85086	15891	17482	21458	3016	3116	3513	832	832	904	391	391	409
0.35 ***	57691	60832	84132	16457	17791	21216	3207	3332	3482	882	891	907	416	416	416
•••	58791	65466	91910	17007	19007	23113	3266	3407	3754	891	907	954	416	416	426
0.4 ***	58266	62232	90416	17007	18607	22741	3382	3516	3707	916	932	957	432	432	432
•••	59682	67857	96728	17666	20041	24268	3441	3607	3915	932	941	984	432	432	435
0.45 ***	57257	62091	94566	17182	18991	23716	3466	3616	3832	941	957	966	432	432	441
•••	59016	68682	99538	17932	20632	24920	3541	3716	3995	957	966	994	432	441	435
0.5 ***	55016	60716	96541	17007	18991	24116	3482	3657	3832	941	957	966	432	432	432
•••	57116	68116	100341	17841	20782	25070	3557	3757	3995	957	966	984	432	432	426
0.55 ***	51857	58341	96341	16541	18657	24057	3416	3607	3832	916	932	957	416	416	416
•••	54307	66391	99136	17432	20532	24719	3507	3707	3915	932	941	954	416	416	409
0.6 ***	48082	55166	93957	15832	17982	23407	3307	3482	3707	882	891	907	391	391	391
•••	50807	63616	95925	16732	19882	23866	3382	3582	3754	882	907	904	391	391	382
0.65 ***	43907	51357	89382	14882	17007	22207	3107	3266	3491	816	832	841	357	357	366
•••	46807	59907	90706	15782	18857	22511	3182	3366	3513	816	832	833	357	357	346
0.7 ***	39482	46966	82657	13707	15732	20466	2857	2991	3182	732	741	757	316	316	316
•••	42441	55307	83480	14566	17441	20655	2916	3082	3192	741	757	743	316	316	302
0.75 ***	34857	42041	73782	12307	14132	18207	2516	2641	2791	632	641	641	266	266	266
•••	37732	49807	74247	13091	15641	18296	2582	2716	2791	632	641	633	266	266	248
0.8 ***	29991	36516	62782	10657	12191	15391	2116	2207	2316	507	516	516	207	207	207
•••	32616	43332	63007	11332	13416	15436	2166	2266	2309	507	516	502	207	207	186
0.85 ***	24741	30191	49682	8707	9866	12066	1641	1691	1766	366	366	366	.	.	.
•••	26966	35657	49759	9216	10757	12074	1666	1732	1747	366	366	352	.	.	.
0.9 ***	18757	22691	34491	6341	7057	8216	1066	1091	1116	.	.	.	.	.	.
•••	20391	26391	34504	6657	7541	8210	1082	1107	1104	.	.	.	.	.	.
0.95 ***	11216	13107	17257	3332	3557	3866	.	.	.	.	.	.	.	.	.
•••	12032	14657	17242	3432	3691	3844	.	.	.	.	.	.	.	.	.
0.98 ***	4807	5257	5941	.	.	.	.	.	.	.	.	.	.	.	.
•••	5007	5557	5922	.	.	.	.	.	.	.	.	.	.	.	.

TABLE 3: ALPHA= 0.01 POWER= 0.8   EXPECTED ACCRUAL THRU MINIMUM FOLLOW-UP= 11000

		DEL=.01			DEL=.02			DEL=.05			DEL=.10			DEL=.15		
FACT=		1.0 .75	.50 .25	.00 BIN	1.0 .75	.50 .25	.00 BIN	1.0 .75	.50 .25	.00 BIN	1.0 75	.50 .25	.00 BIN	1.0 .75	.50 .25	.00 BIN
PCONT=***					REQUIRED NUMBER OF PATIENTS											
0.01	***	1547	1547	1547	568	568	568	183	183	183	90	90	90	62	62	62
	•••	1547	1547	5922	568	568	1958	183	183	533	90	90	217	62	62	129
0.02	***	3417	3428	3455	1090	1107	1107	293	293	293	128	128	128	90	90	90
	•••	3428	3445	9776	1090	1107	2911	293	293	681	128	128	252	90	90	144
0.05	***	10385	10512	10787	2960	2988	3015	640	640	640	227	227	227	128	128	128
	•••	10440	10622	20855	2977	3005	5651	640	640	1104	227	227	352	128	128	186
0.1	***	22980	23410	25115	6305	6425	6645	1217	1228	1228	375	375	375	200	200	200
	•••	23145	23932	37716	6360	6525	9816	1217	1228	1747	375	375	502	200	200	248
0.15	***	34437	35262	39673	9385	9698	10275	1767	1778	1805	513	513	530	265	265	265
	•••	34723	36400	52569	9522	9935	13479	1778	1795	2309	513	513	633	265	265	302
0.2	***	43798	45107	53302	11997	12530	13647	2245	2290	2345	640	640	650	310	310	310
	•••	44255	47015	65415	12217	12970	16640	2262	2317	2791	640	640	743	310	310	346
0.25	***	50800	52697	65430	14060	14857	16645	2658	2713	2795	750	750	750	348	365	365
	•••	51470	55502	76254	14390	15528	19300	2685	2757	3192	750	750	833	365	365	382
0.3	***	55485	58070	75770	15600	16655	19175	2988	3070	3180	815	832	843	392	392	392
	•••	56393	61838	85086	16040	17590	21458	3032	3125	3513	832	832	904	392	392	409
0.35	***	58032	61398	84130	16628	17965	21220	3225	3345	3483	887	898	898	420	420	420
	•••	59225	66172	91910	17195	19147	23113	3280	3400	3754	887	898	954	420	420	426
0.4	***	58692	62910	90427	17222	18817	22733	3400	3527	3703	925	942	953	430	430	430
	•••	60215	68713	96728	17893	20220	24248	3455	3610	3915	925	942	984	430	430	435
0.45	***	57795	62910	94563	17425	19230	23723	3483	3637	3830	942	953	970	430	430	430
	•••	59682	69665	99538	18185	20825	24920	3555	3730	3995	953	970	994	430	430	435
0.5	***	55667	61662	96543	17277	19268	24163	3510	3665	3885	942	953	970	430	430	430
	•••	57905	69208	100341	18113	21000	25070	3582	3758	3995	953	970	984	430	430	426
0.55	***	52625	59380	96340	16837	18938	24053	3455	3620	3830	925	942	953	420	420	420
	•••	55200	67547	99153	17717	20753	24719	3527	3720	3915	925	942	954	420	420	409
0.6	***	48940	56272	93947	16133	18267	23410	3335	3500	3703	887	898	915	392	392	392
	•••	51773	64825	95925	17030	20110	23866	3400	3593	3754	887	898	904	392	392	382
0.65	***	44832	52488	89382	15170	17288	22210	3142	3290	3483	815	832	843	365	365	365
	•••	47830	61095	90706	16078	19065	22511	3208	3390	3513	832	832	833	365	365	346
0.7	***	40432	48088	82655	13988	15985	20467	2878	3015	3180	733	750	760	310	320	320
	•••	43468	56465	83480	14840	17635	20655	2933	3087	3192	733	750	743	320	320	302
0.75	***	35785	43100	73773	12575	14362	18195	2537	2658	2795	640	640	650	265	265	265
	•••	38710	50865	74247	13345	15820	18296	2592	2713	2791	640	640	633	265	265	248
0.8	***	30845	37445	62773	10880	12382	15390	2135	2218	2317	513	513	513	200	200	200
	•••	33513	44238	63007	11540	13565	15436	2180	2262	2309	513	513	502	200	200	186
0.85	***	25472	30955	49672	8873	10000	12063	1640	1695	1767	365	365	365	.	.	.
	•••	27710	36373	49759	9385	10842	12074	1668	1723	1747	365	365	352	.	.	.
0.9	***	19295	23228	34492	6453	7130	8230	1080	1090	1118	.	.	.	.	.	.
	•••	20935	26858	34504	6755	7597	8210	1080	1107	1104	.	.	.	.	.	.
0.95	***	11485	13345	17260	3362	3582	3857	.	.	.	.	.	.	.	.	.
	•••	12283	14830	17242	3455	3703	3844	.	.	.	.	.	.	.	.	.
0.98	***	4875	5298	5947	.	.	.	.	.	.	.	.	.	.	.	.
	•••	5067	5590	5922	.	.	.	.	.	.	.	.	.	.	.	.

TABLE 3: ALPHA= 0.01 POWER= 0.8   EXPECTED ACCRUAL THRU MINIMUM FOLLOW-UP= 12000

FACT=	DEL=.01			DEL=.02			DEL=.05			DEL=.10			DEL=.15		
	1.0 .75	.50 .25	.00 BIN	1.0 .75	.50 .25	.00 BIN	1.0 .75	.50 .25	.00 BIN	1.0 75	.50 .25	.00 BIN	1.0 .75	.50 .25	.00 BIN
PCONT=•••				REQUIRED NUMBER OF PATIENTS											
0.01 •••	1538	1538	1549	559	559	559	188	188	188	98	98	98	68	68	68
•••	1538	1549	5922	559	559	1958	188	188	533	98	98	217	68	68	129
0.02 •••	3428	3439	3458	1099	1099	1099	289	289	289	128	128	128	79	79	79
•••	3428	3439	9776	1099	1099	2911	289	289	681	128	128	252	79	79	144
0.05 •••	10399	10519	10789	2978	2989	3019	638	638	638	229	229	229	128	128	128
•••	10448	10628	20855	2978	3008	5651	638	638	1104	229	229	352	128	128	186
0.1 •••	23029	23468	25118	6319	6439	6638	1219	1219	1238	379	379	379	199	199	199
•••	23198	23989	37716	6379	6529	9816	1219	1238	1747	379	379	502	199	199	248
0.15 •••	34519	35389	39668	9428	9728	10268	1759	1789	1808	518	518	518	259	259	259
•••	34838	36548	52569	9559	9949	13479	1778	1808	2309	518	518	633	259	259	302
0.2 •••	43928	45319	53299	12068	12589	13639	2258	2288	2329	638	638	649	308	319	319
•••	44419	47288	65415	12289	13009	16640	2269	2318	2791	638	649	743	308	319	346
0.25 •••	50989	53018	65438	14168	14948	16639	2659	2719	2798	739	739	758	349	349	349
•••	51698	55879	76254	14498	15608	19300	2689	2768	3192	739	758	833	349	349	382
0.3 •••	55729	58489	75769	15728	16789	19178	2989	3079	3188	829	829	848	398	398	398
•••	56719	62359	85086	16178	17689	21458	3038	3128	3513	829	829	904	398	398	409
0.35 •••	58358	61928	84128	16808	18128	21218	3248	3349	3488	889	889	908	409	409	409
•••	59648	66829	91910	17359	19268	23113	3289	3409	3754	889	908	954	409	409	426
0.4 •••	59119	63578	90428	17419	19009	22729	3409	3548	3698	919	938	949	428	428	428
•••	60739	69499	96728	18098	20378	24268	3469	3619	3915	938	949	984	428	428	435
0.45 •••	58328	63698	94568	17648	19448	23719	3518	3649	3829	949	968	968	428	439	439
•••	60319	70568	99538	18409	21008	24920	3578	3739	3995	949	968	994	439	439	435
0.5 •••	56299	62539	96548	17528	19508	24158	3529	3679	3878	949	968	968	428	428	428
•••	58658	70208	100341	18368	21188	25070	3608	3769	3995	949	968	984	428	428	426
0.55 •••	53359	60349	96338	17108	19178	24049	3469	3638	3829	919	938	949	409	409	428
•••	56048	68618	99136	17989	20948	24719	3548	3728	3915	938	949	954	409	409	409
0.6 •••	49759	57308	93949	16399	18518	23408	3349	3499	3698	878	889	908	398	398	398
•••	52699	65918	95925	17299	20299	23866	3428	3608	3754	889	908	904	398	398	382
0.65 •••	45709	53539	89378	15439	17528	22208	3158	3308	3488	818	829	848	349	368	368
•••	48788	62198	90706	16339	19268	22511	3229	3398	3513	829	829	833	349	368	346
0.7 •••	41329	49118	82658	14258	16219	20468	2899	3019	3188	739	739	758	319	319	319
•••	44438	57499	83480	15098	17809	20655	2959	3098	3192	739	758	743	319	319	302
0.75 •••	36649	44059	73778	12799	14569	18199	2558	2659	2798	638	638	649	259	259	259
•••	39638	51829	74247	13568	15968	18296	2618	2719	2791	638	649	633	259	259	248
0.8 •••	31639	38299	62779	11089	12559	15398	2149	2228	2318	518	518	518	199	199	199
•••	34339	45068	63007	11719	13688	15436	2179	2269	2309	518	518	502	199	199	186
0.85 •••	26138	31658	49669	9038	10129	12068	1658	1699	1759	368	368	368	.	.	.
•••	28399	37009	49759	9518	10928	12074	1688	1729	1747	368	368	352	.	.	.
0.9 •••	19789	23708	34489	6548	7208	8228	1069	1099	1118	.	.	.	.	.	.
•••	21428	27259	34504	6848	7639	8210	1088	1118	1104	.	.	.	.	.	.
0.95 •••	11738	13549	17258	3398	3589	3859	.	.	.	.	.	.	.	.	.
•••	12529	14989	17242	3488	3728	3844	.	.	.	.	.	.	.	.	.
0.98 •••	4939	5348	5948	.	.	.	.	.	.	.	.	.	.	.	.
•••	5119	5618	5922	.	.	.	.	.	.	.	.	.	.	.	.

TABLE 3: ALPHA= 0.01 POWER= 0.8   EXPECTED ACCRUAL THRU MINIMUM FOLLOW-UP= 13000

	DEL=.01			DEL=.02			DEL=.05			DEL=.10			DEL=.15		
FACT=	1.0 .75	.50 .25	.00 BIN	1.0 .75	.50 .25	.00 BIN	1.0 .75	.50 .25	.00 BIN	1.0 75	.50 .25	.00 BIN	1.0 .75	.50 .25	.00 BIN
PCONT=***					REQUIRED NUMBER OF PATIENTS										
0.01 ***	1536	1548	1548	561	561	561	183	183	183	86	86	86	74	74	74
***	1536	1548	5922	561	561	1958	183	183	533	86	86	217	74	74	129
0.02 ***	3421	3433	3454	1093	1093	1093	301	301	301	118	118	118	86	86	86
***	3433	3433	9776	1093	1093	2911	301	301	681	118	118	252	86	86	144
0.05 ***	10421	10539	10778	2978	2999	3031	638	638	638	216	216	236	139	139	139
***	10474	10636	20855	2978	3011	5651	638	638	1104	216	216	352	139	139	186
0.1 ***	23063	23539	25111	6346	6456	6639	1223	1223	1244	378	378	378	204	204	204
***	23246	24038	37716	6391	6541	9816	1223	1223	1747	378	378	502	204	204	248
0.15 ***	34601	35511	39671	9466	9759	10279	1776	1796	1808	508	529	529	269	269	269
***	34926	36681	52569	9596	9966	13479	1776	1796	2309	508	529	633	269	269	302
0.2 ***	44046	45521	53288	12131	12651	13638	2263	2296	2328	638	638	638	313	313	313
***	44578	47524	65415	12359	13053	16640	2284	2316	2791	638	638	743	313	313	346
0.25 ***	51164	53309	65431	14256	15036	16624	2674	2739	2804	736	756	756	346	366	366
***	51923	56234	76254	14581	15653	19300	2706	2771	3192	736	756	833	366	366	382
0.3 ***	55986	58899	75766	15848	16909	19184	3011	3076	3173	821	833	833	399	399	399
***	57026	62843	85086	16303	17766	21458	3043	3129	3513	833	833	904	399	399	409
0.35 ***	58683	62441	84131	16953	18274	21211	3259	3356	3486	886	898	898	411	411	411
***	60048	67446	91910	17506	19379	23113	3303	3421	3754	886	898	954	411	411	426
0.4 ***	59528	64208	90424	17603	19184	22738	3433	3551	3693	931	931	951	431	431	431
***	61251	70221	96728	18274	20496	24268	3486	3628	3915	931	951	984	431	431	435
0.45 ***	58846	64436	94563	17863	19651	23713	3531	3661	3823	951	963	963	431	431	443
***	60926	71391	99538	18611	21146	24920	3596	3746	3995	951	963	994	431	431	435
0.5 ***	56916	63384	96546	17766	19716	24156	3551	3693	3876	951	963	963	431	431	431
***	59366	71131	100341	18599	21361	25070	3616	3779	3995	951	963	984	431	431	426
0.55 ***	54068	61251	96339	17343	19411	24059	3498	3649	3844	931	931	951	411	411	411
***	56851	69603	99136	18221	21134	24719	3563	3726	3915	931	951	954	411	411	409
0.6 ***	50546	58261	93946	16661	18741	23409	3368	3519	3693	886	898	898	399	399	399
***	53569	66926	95925	17559	20484	23866	3454	3616	3754	886	898	904	399	399	382
0.65 ***	46549	54511	89384	15706	17754	22206	3173	3324	3486	821	833	833	366	366	366
***	49701	63201	90706	16584	19423	22511	3238	3401	3513	821	833	833	366	366	346
0.7 ***	42161	50071	82656	14483	16421	20463	2913	3031	3173	736	756	756	313	313	313
***	45326	58476	83480	15328	17961	20655	2966	3108	3192	736	756	743	313	313	302
0.75 ***	37461	44968	73784	13021	14743	18209	2576	2674	2804	638	638	638	269	269	269
***	40483	52703	74247	13768	16096	18296	2621	2739	2791	638	638	633	269	269	248
0.8 ***	32379	39106	62778	11266	12696	15393	2154	2231	2316	508	508	529	204	204	204
***	35121	45813	63007	11904	13789	15436	2198	2284	2309	508	508	502	204	204	186
0.85 ***	26768	32293	49681	9174	10246	12066	1666	1711	1764	366	366	366	.	.	.
***	29043	37591	49759	9641	11006	12074	1678	1731	1747	366	366	352	.	.	.
0.9 ***	20256	24156	34491	6639	7268	8231	1081	1093	1126	.	.	.	.	.	.
***	21893	27634	34504	6931	7679	8210	1093	1114	1104	.	.	.	.	.	.
0.95 ***	11969	13736	17266	3421	3616	3856	.	.	.	.	.	.	.	.	.
***	12728	15121	17242	3519	3726	3844	.	.	.	.	.	.	.	.	.
0.98 ***	4993	5383	5936	.	.	.	.	.	.	.	.	.	.	.	.
***	5176	5631	5922	.	.	.	.	.	.	.	.	.	.	.	.

TABLE 3: ALPHA= 0.01 POWER= 0.8   EXPECTED ACCRUAL THRU MINIMUM FOLLOW-UP= 14000

PCONT=***		DEL=.01 1.0 .75	DEL=.01 .50 .25	DEL=.01 .00 BIN	DEL=.02 1.0 .75	DEL=.02 .50 .25	DEL=.02 .00 BIN	DEL=.05 1.0 .75	DEL=.05 .50 .25	DEL=.05 .00 BIN	DEL=.10 1.0 75	DEL=.10 .50 .25	DEL=.10 .00 BIN	DEL=.15 1.0 .75	DEL=.15 .50 .25	DEL=.15 .00 BIN
						REQUIRED NUMBER OF PATIENTS									
0.01	***	1549	1549	1549	569	569	569	184	184	184	92	92	92	57	57	57
	•••	1549	1549	5922	569	569	1958	184	184	533	92	92	217	57	57	129
0.02	***	3417	3439	3452	1094	1094	1107	289	289	289	127	127	127	79	79	79
	•••	3439	3439	9776	1094	1094	2911	289	289	681	127	127	252	79	79	144
0.05	***	10439	10544	10789	2984	2997	3019	639	639	639	219	219	232	127	127	127
	•••	10474	10649	20855	2984	2997	5651	639	639	1104	219	219	352	127	127	186
0.1	***	23109	23577	25117	6357	6462	6637	1212	1234	1234	372	372	372	197	197	197
	•••	23297	24089	37716	6414	6554	9816	1234	1234	1747	372	372	502	197	197	248
0.15	***	34672	35639	39664	9494	9787	10264	1772	1794	1807	512	512	512	254	267	267
	•••	35022	36807	52569	9634	9997	13479	1772	1794	2309	512	512	633	267	267	302
0.2	***	44179	45719	53292	12189	12692	13637	2262	2297	2332	639	639	652	302	324	324
	•••	44739	47749	65415	12399	13077	16640	2284	2319	2791	639	639	743	302	324	346
0.25	***	51354	53594	65437	14337	15107	16634	2682	2739	2787	744	744	757	359	359	359
	•••	52159	56569	76254	14674	15724	19300	2704	2774	3192	744	744	833	359	359	382
0.3	***	56232	59277	75762	15969	16997	19167	3019	3089	3172	827	827	849	394	394	394
	•••	57339	63289	85086	16402	17859	21458	3054	3137	3513	827	827	904	394	394	409
0.35	***	59019	62939	84127	17102	18397	21219	3277	3369	3487	884	897	897	407	407	407
	•••	60454	68014	91910	17662	19482	23113	3312	3417	3754	897	897	954	407	407	426
0.4	***	59942	64794	90414	17789	19342	22737	3452	3557	3697	932	932	954	429	429	429
	•••	61749	70897	96728	18454	20624	24268	3509	3627	3915	932	954	984	429	429	435
0.45	***	59347	65122	94557	18069	19832	23717	3544	3684	3837	954	967	967	429	429	442
	•••	61517	72179	99538	18817	21289	24920	3614	3754	3995	954	967	994	429	442	435
0.5	***	57514	64164	96539	17977	19902	24159	3579	3697	3872	954	954	967	429	429	429
	•••	60069	71982	100341	18804	21499	25070	3627	3789	3995	954	967	984	429	429	426
0.55	***	54749	62112	96342	17579	19609	24054	3522	3662	3837	932	932	954	407	407	429
	•••	57619	70512	99136	18454	21289	24719	3579	3732	3915	932	954	954	407	407	409
0.6	***	51297	59159	93949	16892	18957	23402	3404	3522	3697	884	897	897	394	394	394
	•••	54399	67852	95925	17767	20637	23866	3452	3614	3754	884	897	904	394	394	382
0.65	***	47329	55427	89377	15934	17942	22212	3194	3334	3487	827	827	849	359	359	359
	•••	50549	64129	90706	16809	19574	22511	3264	3404	3513	827	827	833	359	359	346
0.7	***	42967	50969	82657	14709	16612	20462	2927	3054	3172	744	744	757	324	324	324
	•••	46174	59347	83480	15527	18104	20655	2984	3102	3192	744	757	743	324	324	302
0.75	***	38229	45789	73767	13217	14919	18209	2577	2682	2787	639	639	652	267	267	267
	•••	41287	53502	74247	13952	16214	18296	2634	2739	2791	639	639	633	267	267	248
0.8	***	33084	39839	62777	11432	12832	15387	2179	2249	2319	512	512	512	197	197	197
	•••	35849	46489	63007	12049	13882	15436	2192	2284	2309	512	512	502	197	197	186
0.85	***	27344	32887	49674	9297	10334	12062	1667	1724	1759	359	372	372	.	.	.
	•••	29632	38102	49759	9774	11069	12074	1689	1737	1747	372	372	352	.	.	.
0.9	***	20672	24557	34497	6707	7324	8212	1072	1107	1129	.	.	.	.	.	.
	•••	22304	27974	34504	6987	7722	8210	1094	1107	1104	.	.	.	.	.	.
0.95	***	12167	13904	17264	3452	3627	3859	.	.	.	.	.	.	.	.	.
	•••	12924	15234	17242	3544	3732	3844	.	.	.	.	.	.	.	.	.
0.98	***	5049	5412	5937	.	.	.	.	.	.	.	.	.	.	.	.
	•••	5224	5657	5922	.	.	.	.	.	.	.	.	.	.	.	.

TABLE 3: ALPHA= 0.01 POWER= 0.8   EXPECTED ACCRUAL THRU MINIMUM FOLLOW-UP= 15000

	DEL=.01			DEL=.02			DEL=.05			DEL=.10			DEL=.15		
FACT=	1.0 .75	.50 .25	.00 BIN	1.0 .75	.50 .25	.00 BIN	1.0 .75	.50 .25	.00 BIN	1.0 75	.50 .25	.00 BIN	1.0 .75	.50 .25	.00 BIN

PCONT=***   REQUIRED NUMBER OF PATIENTS

PCONT	DEL=.01 1.0/.75	.50/.25	.00/BIN	DEL=.02 1.0/.75	.50/.25	.00/BIN	DEL=.05 1.0/.75	.50/.25	.00/BIN	DEL=.10 1.0/.75	.50/.25	.00/BIN	DEL=.15 1.0/.75	.50/.25	.00/BIN
0.01 ***	1547	1547	1547	572	572	572	174	174	174	99	99	99	61	61	61
***	1547	1547	5922	572	572	1958	174	174	533	99	99	217	61	61	129
0.02 ***	3422	3436	3460	1097	1097	1097	286	286	286	122	122	122	85	85	85
***	3436	3436	9776	1097	1097	2911	286	286	681	122	122	252	85	85	144
0.05 ***	10449	10561	10786	2972	2986	3024	647	647	647	235	235	235	136	136	136
***	10486	10660	20855	2986	3010	5651	647	647	1104	235	235	352	136	136	186
0.1 ***	23161	23635	25111	6361	6474	6647	1224	1224	1224	385	385	385	197	197	197
***	23335	24136	37716	6422	6549	9816	1224	1224	1747	385	385	502	197	197	248
0.15 ***	34749	35747	39661	9535	9811	10261	1772	1786	1810	511	511	511	272	272	272
***	35124	36924	52569	9647	9999	13479	1786	1810	2309	511	511	633	272	272	302
0.2 ***	44297	45910	53297	12235	12736	13636	2274	2297	2335	647	647	647	310	310	310
***	44897	47949	65415	12460	13111	16640	2274	2311	2791	647	647	743	310	310	346
0.25 ***	51535	53860	65424	14424	15174	16636	2686	2747	2799	736	760	760	361	361	361
***	52374	56874	76254	14747	15760	19300	2710	2761	3192	736	760	833	361	361	382
0.3 ***	56485	59649	75774	16074	17110	19172	3024	3099	3174	835	835	835	385	385	399
***	57624	63699	85086	16510	17911	21458	3061	3136	3513	835	835	904	385	385	409
0.35 ***	59335	63399	84136	17236	18535	21211	3286	3385	3474	886	886	910	422	422	422
***	60835	68536	91910	17785	19561	23113	3324	3422	3754	886	910	954	422	422	426
0.4 ***	60347	65372	90422	17935	19486	22735	3460	3572	3699	924	947	947	422	436	436
***	62222	71536	96728	18610	20724	24268	3511	3624	3915	924	947	984	422	436	435
0.45 ***	59836	65799	94561	18235	19997	23724	3572	3685	3835	947	961	961	436	436	436
***	62086	72886	99538	18985	21422	24920	3624	3760	3995	961	961	994	436	436	435
0.5 ***	58097	64899	96549	18174	20086	24160	3586	3722	3872	947	961	961	422	436	436
***	60722	72774	100341	18999	21647	25070	3647	3797	3995	961	961	984	436	436	426
0.55 ***	55411	62935	96347	17785	19786	24061	3535	3661	3835	924	947	947	422	422	422
***	58336	71349	99136	18647	21422	24719	3610	3736	3915	924	947	954	422	422	409
0.6 ***	51999	60010	93947	17110	19149	23410	3399	3549	3699	886	886	910	385	399	399
***	55172	68724	95925	17986	20785	23866	3474	3624	3754	886	910	904	385	399	382
0.65 ***	48085	56274	89386	16149	18136	22210	3211	3347	3497	835	835	849	361	361	361
***	51347	64974	90706	17011	19711	22511	3272	3399	3513	835	835	833	361	361	346
0.7 ***	43711	51797	82660	14911	16772	20461	2949	3047	3174	736	736	760	310	310	324
***	46974	60160	83480	15722	18235	20655	2986	3122	3192	736	760	743	310	324	302
0.75 ***	38949	46561	73772	13397	15061	18197	2597	2686	2799	647	647	647	272	272	272
***	42047	54235	74247	14124	16322	18296	2649	2747	2791	647	647	633	272	272	248
0.8 ***	33736	40510	62785	11597	12961	15399	2185	2236	2311	511	511	511	197	197	197
***	36511	47110	63007	12197	13960	15436	2199	2274	2309	511	511	502	197	197	186
0.85 ***	27886	33436	49674	9422	10411	12061	1674	1711	1772	361	361	361	.	.	.
***	30197	38597	49759	9872	11124	12074	1697	1735	1747	361	361	352	.	.	.
0.9 ***	21061	24924	34486	6774	7360	8222	1074	1097	1111	.	.	.	.	.	.
***	22697	28261	34504	7060	7749	8210	1097	1111	1104	.	.	.	.	.	.
0.95 ***	12347	14072	17260	3474	3647	3872	.	.	.	.	.	.	.	.	.
***	13111	15347	17242	3549	3736	3844	.	.	.	.	.	.	.	.	.
0.98 ***	5086	5447	5935	.	.	.	.	.	.	.	.	.	.	.	.
***	5260	5672	5922	.	.	.	.	.	.	.	.	.	.	.	.

TABLE 3: ALPHA= 0.01 POWER= 0.8   EXPECTED ACCRUAL THRU MINIMUM FOLLOW-UP= 17000

		DEL=.01			DEL=.02			DEL=.05			DEL=.10			DEL=.15		
FACT=		1.0 .75	.50 .25	.00 BIN	1.0 .75	.50 .25	.00 BIN	1.0 .75	.50 .25	.00 BIN	1.0 75	.50 .25	.00 BIN	1.0 .75	.50 .25	.00 BIN
PCONT=•••					REQUIRED NUMBER OF PATIENTS											
0.01	•••	1541	1541	1541	564	564	564	181	181	181	96	96	96	70	70	70
	•••	1541	1541	5922	564	564	1958	181	181	533	96	96	217	70	70	129
0.02	•••	3427	3427	3454	1090	1090	1090	282	282	282	139	139	139	70	96	96
	•••	3427	3454	9776	1090	1090	2911	282	282	681	139	139	252	70	96	144
0.05	•••	10466	10567	10780	2986	3002	3029	649	649	649	224	224	224	139	139	139
	•••	10509	10652	20855	2986	3002	5651	649	649	1104	224	224	352	139	139	186
0.1	•••	23232	23726	25102	6386	6487	6641	1217	1217	1244	367	367	367	197	197	197
	•••	23429	24236	37716	6429	6556	9816	1217	1244	1747	367	367	502	197	197	248
0.15	•••	34904	35966	39664	9590	9845	10270	1770	1796	1812	521	521	521	266	266	266
	•••	35302	37130	52569	9701	10041	13479	1796	1796	2309	521	521	633	266	266	302
0.2	•••	44551	46267	53306	12336	12820	13654	2280	2306	2349	649	649	649	309	309	309
	•••	45189	48307	65415	12549	13160	16640	2280	2322	2791	649	649	743	309	309	346
0.25	•••	51877	54385	65435	14562	15285	16645	2705	2747	2790	750	750	750	351	351	351
	•••	52812	57429	76254	14886	15864	19300	2731	2774	3192	750	750	833	351	351	382
0.3	•••	56961	60335	75762	16262	17266	19179	3045	3114	3172	835	835	835	394	394	394
	•••	58210	64457	85086	16714	18047	21458	3071	3130	3513	835	835	904	394	394	409
0.35	•••	59952	64271	84135	17479	18727	21219	3300	3385	3470	904	904	904	410	410	410
	•••	61567	69499	91910	18031	19731	23113	3342	3427	3754	904	904	954	410	410	426
0.4	•••	61126	66439	90425	18217	19731	22749	3496	3581	3709	946	946	946	436	436	436
	•••	63140	72686	96728	18881	20921	24268	3539	3640	3915	946	946	984	436	436	435
0.45	•••	60786	67034	94574	18557	20284	23726	3597	3709	3836	946	962	962	436	436	436
	•••	63182	74190	99538	19306	21617	24920	3640	3767	3995	962	962	994	436	436	435
0.5	•••	59187	66285	96545	18541	20385	24167	3624	3725	3879	946	962	962	436	436	436
	•••	61950	74216	100341	19349	21872	25070	3666	3794	3995	962	962	984	436	436	426
0.55	•••	56637	64415	96332	18175	20130	24066	3555	3682	3836	920	946	946	410	410	410
	•••	59697	72856	99136	19025	21660	24719	3624	3751	3915	946	946	954	410	410	409
0.6	•••	53349	61567	93952	17495	19476	23402	3427	3555	3709	877	904	904	394	394	394
	•••	56621	70280	95925	18345	21022	23866	3496	3624	3754	904	904	904	394	394	382
0.65	•••	49481	57854	89389	16517	18456	22212	3241	3342	3496	819	835	835	351	351	367
	•••	52839	66524	90706	17367	19944	22511	3300	3411	3513	835	835	833	351	367	346
0.7	•••	45104	53322	82647	15269	17070	20470	2960	3071	3172	750	750	750	309	309	325
	•••	48435	61610	83480	16076	18430	20655	3002	3114	3192	750	750	743	309	325	302
0.75	•••	40275	47994	73765	13712	15327	18201	2620	2705	2790	649	649	649	266	266	266
	•••	43420	55559	74247	14435	16501	18296	2662	2747	2791	649	649	633	266	266	248
0.8	•••	34920	41746	62784	11869	13160	15396	2195	2264	2322	521	521	521	197	197	197
	•••	37751	48222	63007	12437	14095	15436	2221	2280	2309	521	521	502	197	197	186
0.85	•••	28885	34410	49667	9616	10567	12081	1685	1727	1770	367	367	367	.	.	.
	•••	31206	39425	49759	10057	11231	12074	1711	1754	1747	367	367	352	.	.	.
0.9	•••	21771	25570	34495	6896	7449	8214	1090	1116	1116	.	.	.	.	.	.
	•••	23402	28784	34504	7151	7805	8210	1090	1116	1104	.	.	.	.	.	.
0.95	•••	12692	14334	17266	3512	3666	3852	.	.	.	.	.	.	.	.	.
	•••	13415	15524	17242	3581	3767	3844	.	.	.	.	.	.	.	.	.
0.98	•••	5170	5494	5935	.	.	.	.	.	.	.	.	.	.	.	.
	•••	5324	5706	5922	.	.	.	.	.	.	.	.	.	.	.	.

TABLE 3: ALPHA= 0.01 POWER= 0.8   EXPECTED ACCRUAL THRU MINIMUM FOLLOW-UP= 15000

	DEL=.01			DEL=.02			DEL=.05			DEL=.10			DEL=.15		
FACT=	1.0 .75	.50 .25	.00 BIN	1.0 .75	.50 .25	.00 BIN	1.0 .75	.50 .25	.00 BIN	1.0 75	.50 .25	.00 BIN	1.0 .75	.50 .25	.00 BIN

PCONT=•••   REQUIRED NUMBER OF PATIENTS

PCONT	1.0/.75	.50/.25	.00/BIN	1.0/.75	.50/.25	.00/BIN	1.0/.75	.50/.25	.00/BIN	1.0/75	.50/.25	.00/BIN	1.0/.75	.50/.25	.00/BIN
0.01	1547	1547	1547	572	572	572	174	174	174	99	99	99	61	61	61
	1547	1547	5922	572	572	1958	174	174	533	99	99	217	61	61	129
0.02	3422	3436	3460	1097	1097	1097	286	286	286	122	122	122	85	85	85
	3436	3436	9776	1097	1097	2911	286	286	681	122	122	252	85	85	144
0.05	10449	10561	10786	2972	2986	3024	647	647	647	235	235	235	136	136	136
	10486	10660	20855	2986	3010	5651	647	647	1104	235	235	352	136	136	186
0.1	23161	23635	25111	6361	6474	6647	1224	1224	1224	385	385	385	197	197	197
	23335	24136	37716	6422	6549	9816	1224	1224	1747	385	385	502	197	197	248
0.15	34749	35747	39661	9535	9811	10261	1772	1786	1810	511	511	511	272	272	272
	35124	36924	52569	9647	9999	13479	1786	1810	2309	511	511	633	272	272	302
0.2	44297	45910	53297	12235	12736	13636	2274	2297	2335	647	647	647	310	310	310
	44897	47949	65415	12460	13111	16640	2274	2311	2791	647	647	743	310	310	346
0.25	51535	53860	65424	14424	15174	16636	2686	2747	2799	736	760	760	361	361	361
	52374	56874	76254	14747	15760	19300	2710	2761	3192	736	760	833	361	361	382
0.3	56485	59649	75774	16074	17110	19172	3024	3099	3174	835	835	835	385	385	399
	57624	63699	85086	16510	17911	21458	3061	3136	3513	835	835	904	385	385	409
0.35	59335	63399	84136	17236	18535	21211	3286	3385	3474	886	886	910	422	422	422
	60835	68536	91910	17785	19561	23113	3324	3422	3754	886	910	954	422	422	426
0.4	60347	65372	90422	17935	19486	22735	3460	3572	3699	924	947	947	422	436	436
	62222	71536	96728	18610	20724	24268	3511	3624	3915	924	947	984	422	436	435
0.45	59836	65799	94561	18235	19997	23724	3572	3685	3835	947	961	961	436	436	436
	62086	72886	99538	18985	21422	24920	3624	3760	3995	961	961	994	436	436	435
0.5	58097	64899	96549	18174	20086	24160	3586	3722	3872	947	961	961	422	436	436
	60722	72774	100341	18999	21647	25070	3647	3797	3995	961	961	984	436	436	426
0.55	55411	62935	96347	17785	19786	24061	3535	3661	3835	924	947	947	422	422	422
	58336	71349	99136	18647	21422	24719	3610	3736	3915	924	947	954	422	422	409
0.6	51999	60010	93947	17110	19149	23410	3399	3549	3699	886	886	910	385	399	399
	55172	68724	95925	17986	20785	23866	3474	3624	3754	886	910	904	385	399	382
0.65	48085	56274	89386	16149	18136	22210	3211	3347	3497	835	835	849	361	361	361
	51347	64974	90706	17011	19711	22511	3272	3399	3513	835	835	833	361	361	346
0.7	43711	51797	82660	14911	16772	20461	2949	3047	3174	736	736	760	310	310	324
	46974	60160	83480	15722	18235	20655	2986	3122	3192	736	760	743	310	324	302
0.75	38949	46561	73772	13397	15061	18197	2597	2686	2799	647	647	647	272	272	272
	42047	54235	74247	14124	16322	18296	2649	2747	2791	647	647	633	272	272	248
0.8	33736	40510	62785	11597	12961	15399	2185	2236	2311	511	511	511	197	197	197
	36511	47110	63007	12197	13960	15436	2199	2274	2309	511	511	502	197	197	186
0.85	27886	33436	49674	9422	10411	12061	1674	1711	1772	361	361	361	.	.	.
	30197	38597	49759	9872	11124	12074	1697	1735	1747	361	361	352	.	.	.
0.9	21061	24924	34486	6774	7360	8222	1074	1097	1111	.	.	.	.	.	.
	22697	28261	34504	7060	7749	8210	1097	1111	1104	.	.	.	.	.	.
0.95	12347	14072	17260	3474	3647	3872	.	.	.	.	.	.	.	.	.
	13111	15347	17242	3549	3736	3844	.	.	.	.	.	.	.	.	.
0.98	5086	5447	5935	.	.	.	.	.	.	.	.	.	.	.	.
	5260	5672	5922	.	.	.	.	.	.	.	.	.	.	.	.

TABLE 3: ALPHA= 0.01 POWER= 0.8   EXPECTED ACCRUAL THRU MINIMUM FOLLOW-UP= 17000

	DEL=.01			DEL=.02			DEL=.05			DEL=.10			DEL=.15		
FACT=	1.0 .75	.50 .25	.00 BIN	1.0 .75	.50 .25	.00 BIN	1.0 .75	.50 .25	.00 BIN	1.0 75	.50 .25	.00 BIN	1.0 .75	.50 .25	.00 BIN
PCONT=•••				REQUIRED NUMBER OF PATIENTS											
0.01 •••	1541 1541	1541 1541	1541 5922	564 564	564 564	564 1958	181 181	181 181	181 533	96 96	96 96	96 217	70 70	70 70	70 129
0.02 •••	3427 3427	3427 3454	3454 9776	1090 1090	1090 1090	1090 2911	282 282	282 282	282 681	139 139	139 139	139 252	70 70	96 96	96 144
0.05 •••	10466 10509	10567 10652	10780 20855	2986 2986	3002 3002	3029 5651	649 649	649 649	649 1104	224 224	224 224	224 352	139 139	139 139	139 186
0.1 •••	23232 23429	23726 24236	25102 37716	6386 6429	6487 6556	6641 9816	1217 1217	1217 1244	1244 1747	367 367	367 367	367 502	197 197	197 197	197 248
0.15 •••	34904 35302	35966 37130	39664 52569	9590 9701	9845 10041	10270 13479	1770 1796	1796 1796	1812 2309	521 521	521 521	521 633	266 266	266 266	266 302
0.2 •••	44551 45189	46267 48307	53306 65415	12336 12549	12820 13160	13654 16640	2280 2280	2306 2322	2349 2791	649 649	649 649	649 743	309 309	309 309	309 346
0.25 •••	51877 52812	54385 57429	65435 76254	14562 14886	15285 15864	16645 19300	2705 2731	2747 2774	2790 3192	750 750	750 750	750 833	351 351	351 351	351 382
0.3 •••	56961 58210	60335 64457	75762 85086	16262 16714	17266 18047	19179 21458	3045 3071	3114 3130	3172 3513	835 835	835 835	835 904	394 394	394 394	394 409
0.35 •••	59952 61567	64271 69499	84135 91910	17479 18031	18727 19731	21219 23113	3300 3342	3385 3427	3470 3754	904 904	904 904	904 954	410 410	410 410	410 426
0.4 •••	61126 63140	66439 72686	90425 96728	18217 18881	19731 20921	22749 24268	3496 3539	3581 3640	3709 3915	946 946	946 946	946 984	436 436	436 436	436 435
0.45 •••	60786 63182	67034 74190	94574 99538	18557 19306	20284 21617	23726 24920	3597 3640	3709 3767	3836 3995	946 962	962 962	962 994	436 436	436 436	436 435
0.5 •••	59187 61950	66285 74216	96545 100341	18541 19349	20385 21872	24167 25070	3624 3666	3725 3794	3879 3995	946 962	962 962	962 984	436 436	436 436	436 426
0.55 •••	56637 59697	64415 72856	96332 99136	18175 19025	20130 21660	24066 24719	3555 3624	3682 3751	3836 3915	920 946	946 946	946 954	410 410	410 410	410 409
0.6 •••	53349 56621	61567 70280	93952 95925	17495 18345	19476 21022	23402 23866	3427 3496	3555 3624	3709 3754	877 904	904 904	904 904	394 394	394 394	394 382
0.65 •••	49481 52839	57854 66524	89389 90706	16517 17367	18456 19944	22212 22511	3241 3300	3342 3411	3496 3513	819 835	835 835	835 833	351 351	351 367	367 346
0.7 •••	45104 48435	53322 61610	82647 83480	15269 16076	17070 18430	20470 20655	2960 3002	3071 3114	3172 3192	750 750	750 750	750 743	309 309	309 325	325 302
0.75 •••	40275 43420	47994 55559	73765 74247	13712 14435	15327 16501	18201 18296	2620 2662	2705 2747	2790 2791	649 649	649 649	649 633	266 266	266 266	266 248
0.8 •••	34920 37751	41746 48222	62784 63007	11869 12437	13160 14095	15396 15436	2195 2221	2264 2280	2322 2309	521 521	521 521	521 502	197 197	197 197	197 186
0.85 •••	28885 31206	34410 39425	49667 49759	9616 10057	10567 11231	12081 12074	1685 1711	1727 1754	1770 1747	367 367	367 367	367 352	. .	. .	. .
0.9 •••	21771 23402	25570 28784	34495 34504	6896 7151	7449 7805	8214 8210	1090 1090	1116 1116	1116 1104	. .	. .	. .	. .	. .	. .
0.95 •••	12692 13415	14334 15524	17266 17242	3512 3581	3668 3767	3852 3844	. .	. .	. .	. .	. .	. .	. .	. .	. .
0.98 •••	5170 5324	5494 5706	5935 5922	. .	. .	. .	. .	. .	. .	. .	. .	. .	. .	. .	. .

TABLE 3: ALPHA= 0.01 POWER= 0.8   EXPECTED ACCRUAL THRU MINIMUM FOLLOW-UP= 20000

	DEL=.01			DEL=.02			DEL=.05			DEL=.10			DEL=.15		
FACT=	1.0 .75	.50 .25	.00 BIN	1.0 .75	.50 .25	.00 BIN	1.0 .75	.50 .25	.00 BIN	1.0 75	.50 .25	.00 BIN	1.0 .75	.50 .25	.00 BIN

PCONT=•••   REQUIRED NUMBER OF PATIENTS

PCONT	DEL=.01 1.0/.75	.50/.25	.00/BIN	DEL=.02 1.0/.75	.50/.25	.00/BIN	DEL=.05 1.0/.75	.50/.25	.00/BIN	DEL=.10 1.0/.75	.50/.25	.00/BIN	DEL=.15 1.0/.75	.50/.25	.00/BIN
0.01 •••	1532	1532	1532	563	563	563	182	182	182	82	82	82	63	63	63
•••	1532	1532	5922	563	563	1958	182	182	533	82	82	217	63	63	129
0.02 •••	3432	3432	3463	1113	1113	1113	282	282	282	132	132	132	82	82	82
•••	3432	3432	9776	1113	1113	2911	282	282	681	132	132	252	82	82	144
0.05 •••	10482	10613	10782	2982	3013	3013	632	632	632	232	232	232	132	132	132
•••	10532	10682	20855	2982	3013	5651	632	632	1104	232	232	352	132	132	186
0.1 •••	23332	23863	25113	6413	6513	6632	1232	1232	1232	382	382	382	213	213	213
•••	23563	24332	37716	6463	6563	9816	1232	1232	1747	382	382	502	213	213	248
0.15 •••	35132	36232	39663	9663	9913	10263	1782	1782	1813	513	513	513	263	263	263
•••	35563	37382	52569	9763	10063	13479	1782	1813	2309	513	513	633	263	263	302
0.2 •••	44882	46732	53282	12463	12913	13632	2282	2313	2332	632	632	632	313	313	313
•••	45582	48782	65415	12663	13232	16640	2282	2313	2791	632	632	743	313	313	346
0.25 •••	52382	55082	65432	14732	15432	16632	2713	2763	2782	732	763	763	363	363	363
•••	53413	58132	76254	15063	15963	19300	2732	2763	3192	732	763	833	363	363	382
0.3 •••	57632	61263	75763	16513	17482	19182	3063	3113	3182	832	832	832	382	382	382
•••	59013	65432	85086	16932	18182	21458	3082	3132	3513	832	832	904	382	382	409
0.35 •••	60832	65463	84132	17782	19013	21213	3332	3413	3482	882	913	913	413	413	413
•••	62613	70713	91910	18313	19913	23113	3363	3432	3754	882	913	954	413	413	426
0.4 •••	62232	67863	90413	18613	20032	22732	3513	3613	3713	932	932	963	432	432	432
•••	64413	74132	96728	19232	21132	24268	3563	3663	3915	932	932	984	432	432	435
0.45 •••	62082	68682	94563	18982	20632	23713	3613	3713	3832	963	963	963	432	432	432
•••	64663	75863	99538	19713	21882	24920	3663	3763	3995	963	963	994	432	432	435
0.5 •••	60713	68113	96532	18982	20782	24163	3663	3763	3882	963	963	963	432	432	432
•••	63632	76032	100341	19782	22132	25070	3713	3813	3995	963	963	984	432	432	426
0.55 •••	58332	66382	96332	18663	20532	24063	3613	3713	3832	932	932	963	413	413	413
•••	61563	74782	99136	19482	21932	24719	3663	3763	3915	932	932	954	413	413	409
0.6 •••	55163	63613	93963	17982	19882	23413	3482	3582	3713	882	913	913	382	382	382
•••	58563	72232	95925	18813	21313	23866	3532	3632	3754	882	913	904	382	382	382
0.65 •••	51363	59913	89382	17013	18863	22213	3263	3363	3482	832	832	832	363	363	363
•••	54813	68463	90706	17813	20213	22511	3313	3432	3513	832	832	833	363	363	346
0.7 •••	46963	55313	82663	15732	17432	20463	2982	3082	3182	732	763	763	313	313	313
•••	50382	63463	83480	16482	18682	20655	3032	3132	3192	732	763	743	313	313	302
0.75 •••	42032	49813	73782	14132	15632	18213	2632	2713	2782	632	632	632	263	263	263
•••	45232	57213	74247	14813	16713	18296	2682	2763	2791	632	632	633	263	263	248
0.8 •••	36513	43332	62782	12182	13413	15382	2213	2263	2313	513	513	513	213	213	213
•••	39363	49613	63007	12763	14263	15436	2232	2282	2309	513	513	502	213	213	186
0.85 •••	30182	35663	49682	9863	10763	12063	1682	1732	1763	363	363	363	.	.	.
•••	32482	40463	49759	10263	11332	12074	1713	1732	1747	363	363	352	.	.	.
0.9 •••	22682	26382	34482	7063	7532	8213	1082	1113	1113	.	.	.	.	.	.
•••	24282	29413	34504	7282	7863	8210	1113	1113	1104	.	.	.	.	.	.
0.95 •••	13113	14663	17263	3563	3682	3863	.	.	.	.	.	.	.	.	.
•••	13782	15732	17242	3613	3782	3844	.	.	.	.	.	.	.	.	.
0.98 •••	5263	5563	5932	.	.	.	.	.	.	.	.	.	.	.	.
•••	5382	5732	5922	.	.	.	.	.	.	.	.	.	.	.	.

TABLE 3: ALPHA= 0.01   POWER= 0.8   EXPECTED ACCRUAL THRU MINIMUM FOLLOW-UP= 25000

PCONT	FACT=	DEL=.01 1.0/.75	.50/.25	.00/BIN	DEL=.02 1.0/.75	.50/.25	.00/BIN	DEL=.05 1.0/.75	.50/.25	.00/BIN	DEL=.10 1.0/75	.50/.25	.00/BIN	DEL=.15 1.0/.75	.50/.25	.00/BIN
					REQUIRED NUMBER OF PATIENTS											
0.01	***	1540	1540	1540	579	579	579	165	165	165	102	102	102	79	79	79
	•••	1540	1540	5922	579	579	1958	165	165	533	102	102	217	79	79	129
0.02	***	3415	3454	3454	1102	1102	1102	290	290	290	141	141	141	79	79	79
	•••	3454	3454	9776	1102	1102	2911	290	290	681	141	141	252	79	79	144
0.05	***	10516	10641	10790	2977	3016	3016	641	641	641	227	227	227	141	141	141
	•••	10579	10704	20855	3016	3016	5651	641	641	1104	227	227	352	141	141	186
0.1	***	23516	24016	25102	6454	6540	6641	1227	1227	1227	391	391	391	204	204	204
	•••	23704	24454	37716	6477	6579	9816	1227	1227	1747	391	391	502	204	204	248
0.15	***	35454	36602	39665	9727	9954	10266	1790	1790	1790	516	516	516	266	266	266
	•••	35915	37727	52569	9852	10102	13479	1790	1790	2309	516	516	633	266	266	302
0.2	***	45415	47391	53290	12602	13040	13641	2290	2329	2329	641	641	641	329	329	329
	•••	46204	49391	65415	12790	13290	16640	2290	2329	2791	641	641	743	329	329	346
0.25	***	53165	56079	65415	14977	15641	16641	2727	2766	2790	727	766	766	352	352	352
	•••	54290	59079	76254	15266	16079	19300	2727	2766	3192	766	766	833	352	352	382
0.3	***	58704	62602	75766	16852	17727	19165	3079	3141	3165	829	829	829	391	391	391
	•••	60227	66704	85086	17227	18352	21458	3102	3141	3513	829	829	904	391	391	409
0.35	***	62204	67141	84141	18204	19329	21227	3352	3415	3477	891	891	915	415	415	415
	•••	64141	72329	91910	18704	20141	23113	3391	3454	3754	891	891	954	415	415	426
0.4	***	63891	69852	90415	19102	20454	22727	3540	3602	3704	954	954	954	415	415	415
	•••	66266	76079	96728	19704	21415	24268	3579	3665	3915	954	954	984	415	415	435
0.45	***	64079	70977	94579	19540	21079	23727	3665	3727	3829	954	954	977	415	454	454
	•••	66829	78079	99538	20227	22165	24920	3704	3790	3995	954	977	994	415	454	435
0.5	***	62954	70665	96540	19602	21266	24165	3704	3766	3891	954	954	977	415	415	415
	•••	66079	78454	100341	20352	22477	25070	3727	3829	3995	954	977	984	415	415	426
0.55	***	60829	69102	96329	19290	21040	24040	3641	3727	3829	954	954	954	415	415	415
	•••	64165	77352	99136	20079	22290	24719	3665	3790	3915	954	954	954	415	415	409
0.6	***	57790	66415	93954	18641	20391	23391	3516	3602	3704	891	891	915	391	391	391
	•••	61329	74852	95925	19415	21641	23866	3540	3641	3754	891	891	904	391	391	382
0.65	***	54040	62704	89391	17641	19352	22204	3290	3391	3477	829	829	852	352	352	352
	•••	57602	71016	90706	18415	20540	22511	3352	3454	3513	829	829	833	352	352	346
0.7	***	49602	57977	82641	16329	17891	20477	3016	3102	3165	727	766	766	329	329	329
	•••	53079	65852	83480	17016	18977	20655	3079	3141	3192	766	766	743	329	329	302
0.75	***	44516	52266	73766	14665	16040	18204	2665	2727	2790	641	641	641	266	266	266
	•••	47766	59352	74247	15290	16977	18296	2704	2766	2791	641	641	633	266	266	248
0.8	***	38704	45454	62766	12641	13727	15391	2227	2266	2329	516	516	516	204	204	204
	•••	41540	51391	63007	13141	14454	15436	2266	2290	2309	516	516	502	204	204	186
0.85	***	31977	37290	49665	10204	10977	12079	1704	1727	1766	352	352	352	.	.	.
	•••	34266	41790	49759	10540	11454	12074	1727	1727	1747	352	352	352	.	.	.
0.9	***	23954	27454	34477	7227	7665	8227	1102	1102	1102	.	.	.	.	.	.
	•••	25477	30204	34504	7454	7915	8210	1102	1102	1104	.	.	.	.	.	.
0.95	***	13641	15040	17266	3602	3727	3852	.	.	.	.	.	.	.	.	.
	•••	14290	16016	17242	3665	3790	3844	.	.	.	.	.	.	.	.	.
0.98	***	5352	5641	5954	.	.	.	.	.	.	.	.	.	.	.	.
	•••	5477	5766	5922	.	.	.	.	.	.	.	.	.	.	.	.

TABLE 4: ALPHA= 0.01 POWER= 0.9   EXPECTED ACCRUAL THRU MINIMUM FOLLOW-UP= 30

		DEL=.10			DEL=.15			DEL=.20			DEL=.25			DEL=.30		
FACT=		1.0 .75	.50 .25	.00 BIN	1.0 .75	.50 .25	.00 BIN	1.0 .75	.50 .25	.00 BIN	1.0 75	.50 .25	.00 BIN	1.0 .75	.50 .25	.00 BIN
PCONT= ***							REQUIRED NUMBER OF PATIENTS								
0.05	***	273	274	296	158	160	172	111	113	121	87	88	94	71	72	77
	•••	273	276	456	159	162	241	112	115	153	87	89	108	72	74	80
0.1	***	430	432	495	227	229	263	149	151	172	110	112	126	87	89	99
	•••	431	437	651	228	234	322	150	156	196	110	116	133	88	92	96
0.15	***	555	558	679	278	282	342	175	179	216	125	129	153	96	100	117
	•••	556	565	821	279	289	391	177	186	232	127	135	154	98	105	109
0.2	***	645	650	841	313	319	411	192	198	252	135	140	175	102	107	131
	•••	647	661	964	315	329	449	194	208	261	137	148	170	104	114	119
0.25	***	703	710	980	334	342	467	202	210	281	139	147	191	105	111	141
	•••	705	724	1081	336	356	495	204	222	284	142	157	183	107	119	126
0.3	***	730	740	1092	341	352	512	204	215	303	140	150	204	105	113	148
	•••	733	759	1172	345	371	530	208	231	300	144	162	191	108	122	131
0.35	***	730	743	1177	338	351	544	201	214	318	137	149	211	102	112	152
	•••	734	768	1237	343	375	553	206	233	310	142	163	195	106	123	132
0.4	***	707	723	1235	325	343	564	193	209	325	132	145	214	98	109	152
	•••	712	755	1276	331	372	565	199	231	313	137	161	195	103	121	131
0.45	***	665	685	1263	306	327	570	182	201	326	125	140	212	93	105	149
	•••	672	726	1289	313	361	565	189	225	310	131	157	191	98	117	126
0.5	***	608	634	1264	281	306	564	169	190	319	117	132	205	87	99	143
	•••	616	684	1276	290	345	553	177	216	300	123	150	183	92	111	119
0.55	***	541	574	1236	253	283	546	154	176	305	107	123	194	80	92	134
	•••	552	634	1237	264	324	530	163	204	284	113	141	170	85	104	109
0.6	***	469	510	1180	224	257	515	139	162	284	97	113	178	72	84	121
	•••	483	578	1172	236	300	495	148	189	261	103	130	154	77	95	96
0.65	***	397	446	1095	196	230	471	123	145	256	86	101	158	64	74	106
	•••	414	519	1081	209	273	449	132	171	232	92	117	133	68	84	80
0.7	***	329	384	983	168	202	415	106	127	221	74	87	134	.	.	.
	•••	349	459	964	181	242	391	115	150	196	79	101	108	.	.	.
0.75	***	268	324	843	141	172	347	89	107	179	.	.	.	.	.	.
	•••	289	396	821	153	208	322	96	126	153	.	.	.	.	.	.
0.8	***	213	265	676	113	140	266	.	.	.	.	.	.	.	.	.
	•••	233	329	651	124	169	241	.	.	.	.	.	.	.	.	.
0.85	***	162	204	482	.	.	.	.	.	.	.	.	.	.	.	.
	•••	178	254	456	.	.	.	.	.	.	.	.	.	.	.	.

TABLE 4: ALPHA= 0.01 POWER= 0.9   EXPECTED ACCRUAL THRU MINIMUM FOLLOW-UP= 40

	DEL=.10			DEL=.15			DEL=.20			DEL=.25			DEL=.30		
FACT=	1.0 .75	.50 .25	.00 BIN	1.0 .75	.50 .25	.00 BIN	1.0 .75	.50 .25	.00 BIN	1.0 75	.50 .25	.00 BIN	1.0 .75	.50 .25	.00 BIN
PCONT=•••	REQUIRED NUMBER OF PATIENTS														
0.05 •••	273	274	296	159	160	172	112	113	121	87	88	94	72	73	77
•••	274	277	456	159	163	241	112	116	153	88	90	108	72	74	80
0.1 •••	431	434	495	228	231	263	150	153	172	111	113	126	88	90	99
•••	432	439	651	229	236	322	151	158	196	112	117	133	89	93	96
0.15 •••	556	560	679	279	284	342	177	182	216	127	131	153	98	102	117
•••	557	570	821	281	293	391	178	189	232	128	137	154	99	107	109
0.2 •••	647	654	841	315	322	411	194	202	252	137	143	175	104	110	131
•••	649	667	964	317	336	449	197	213	261	139	152	170	106	116	119
0.25 •••	705	715	980	336	347	467	205	215	281	142	151	191	107	115	141
•••	708	734	1081	340	365	495	208	229	284	145	162	183	110	122	126
0.3 •••	733	746	1092	345	359	512	208	221	303	144	154	204	108	117	148
•••	738	771	1172	350	382	530	212	238	300	148	167	191	111	126	131
0.35 •••	734	751	1177	343	360	544	206	221	318	142	155	211	106	116	152
•••	740	784	1237	349	388	553	211	242	310	147	169	195	110	127	132
0.4 •••	712	734	1235	331	353	564	199	218	325	137	152	214	103	114	152
•••	720	776	1276	339	387	565	206	241	313	143	168	195	107	125	131
0.45 •••	672	699	1263	313	339	570	189	210	326	131	147	212	98	110	149
•••	681	751	1289	322	378	565	197	236	310	137	164	191	103	122	126
0.5 •••	617	651	1264	290	321	564	177	200	319	123	140	205	92	104	143
•••	628	713	1276	301	363	553	186	227	300	129	158	183	97	116	119
0.55 •••	552	595	1236	264	299	546	163	187	305	113	131	194	85	97	134
•••	567	667	1237	277	344	530	172	215	284	120	149	170	90	109	109
0.6 •••	483	535	1180	236	274	515	148	173	284	103	120	178	77	89	121
•••	501	614	1172	251	320	495	157	200	261	110	137	154	82	99	96
0.65 •••	414	474	1095	209	247	471	132	156	256	92	108	158	68	78	106
•••	436	556	1081	224	292	449	141	182	232	98	123	133	72	88	80
0.7 •••	349	412	983	181	218	415	115	137	221	79	93	134	.	.	.
•••	373	495	964	196	260	391	124	160	196	85	106	108	.	.	.
0.75 •••	289	352	843	153	187	347	96	115	179	.	.	.	.	.	.
•••	313	430	821	166	223	322	104	134	153	.	.	.	.	.	.
0.8 •••	233	290	676	124	151	266	.	.	.	.	.	.	.	.	.
•••	255	358	651	135	181	241	.	.	.	.	.	.	.	.	.
0.85 •••	178	224	482	.	.	.	.	.	.	.	.	.	.	.	.
•••	196	276	456	.	.	.	.	.	.	.	.	.	.	.	.

TABLE 4: ALPHA= 0.01 POWER= 0.9   EXPECTED ACCRUAL THRU MINIMUM FOLLOW-UP= 50

		DEL=.10			DEL=.15			DEL=.20			DEL=.25			DEL=.30		
FACT=		1.0 .75	.50 .25	.00 BIN	1.0 .75	.50 .25	.00 BIN	1.0 .75	.50 .25	.00 BIN	1.0 75	.50 .25	.00 BIN	1.0 .75	.50 .25	.00 BIN
PCONT=•••					REQUIRED NUMBER OF PATIENTS											
0.05	•••	273	275	296	159	161	172	112	114	121	87	89	94	72	74	77
	•••	274	278	456	160	164	241	113	116	153	88	91	108	73	75	80
0.1	•••	432	435	495	229	232	263	151	154	172	111	115	126	88	91	99
	•••	433	442	651	230	239	322	152	159	196	113	118	133	89	94	96
0.15	•••	557	563	679	281	287	342	178	184	216	128	133	153	99	103	117
	•••	559	574	821	283	297	391	180	192	232	130	139	154	101	108	109
0.2	•••	649	657	841	317	326	411	196	205	252	139	146	175	106	112	131
	•••	652	674	964	320	341	449	200	217	261	141	154	170	108	118	119
0.25	•••	708	720	980	339	352	467	207	219	281	145	154	191	109	117	141
	•••	712	743	1081	343	372	495	211	234	284	148	165	183	113	125	126
0.3	•••	737	752	1092	348	365	512	211	226	303	147	158	204	110	120	148
	•••	742	783	1172	354	391	530	217	244	300	151	171	191	114	129	131
0.35	•••	739	760	1177	347	368	544	210	228	318	145	159	211	109	120	152
	•••	746	799	1237	354	400	553	217	249	310	151	174	195	113	130	132
0.4	•••	718	745	1235	337	363	564	204	225	325	142	157	214	106	118	152
	•••	727	795	1276	346	400	565	212	249	313	148	173	195	111	129	131
0.45	•••	678	713	1263	320	351	570	195	218	326	136	152	212	102	114	149
	•••	690	773	1289	331	392	565	204	245	310	142	170	191	107	125	126
0.5	•••	625	668	1264	298	333	564	184	209	319	128	146	205	96	108	143
	•••	640	739	1276	311	379	553	193	237	300	135	163	183	101	120	119
0.55	•••	563	615	1236	274	312	546	170	196	305	119	137	194	89	101	134
	•••	581	696	1237	288	360	530	180	224	284	126	154	170	94	112	109
0.6	•••	497	558	1180	247	288	515	155	181	284	108	126	178	81	92	121
	•••	519	644	1172	263	336	495	166	209	261	116	142	154	86	102	96
0.65	•••	431	498	1095	220	261	471	139	164	256	97	113	158	71	82	106
	•••	456	588	1081	236	307	449	149	190	232	103	127	133	76	90	80
0.7	•••	367	437	983	192	231	415	121	144	221	84	98	134	.	.	.
	•••	394	525	964	208	274	391	131	167	196	90	110	108	.	.	.
0.75	•••	308	375	843	163	198	347	102	121	179	.	.	.	.	.	.
	•••	334	457	821	177	235	322	110	139	153	.	.	.	.	.	.
0.8	•••	250	311	676	132	161	266	.	.	.	.	.	.	.	.	.
	•••	274	382	651	144	190	241	.	.	.	.	.	.	.	.	.
0.85	•••	192	240	482	.	.	.	.	.	.	.	.	.	.	.	.
	•••	211	294	456	.	.	.	.	.	.	.	.	.	.	.	.

TABLE 4: ALPHA= 0.01 POWER= 0.9   EXPECTED ACCRUAL THRU MINIMUM FOLLOW-UP= 60

		DEL=.10			DEL=.15			DEL=.20			DEL=.25			DEL=.30		
FACT=		1.0 .75	.50 .25	.00 BIN	1.0 .75	.50 .25	.00 BIN	1.0 .75	.50 .25	.00 BIN	1.0 75	.50 .25	.00 BIN	1.0 .75	.50 .25	.00 BIN
PCONT=•••		REQUIRED NUMBER OF PATIENTS														
0.05	***	274	276	296	160	162	172	113	115	121	88	89	94	73	74	77
	•••	274	279	456	160	165	241	113	117	153	88	91	108	73	75	80
0.1	***	432	437	495	229	234	263	151	155	172	112	116	126	89	92	99
	•••	434	445	651	231	240	322	153	160	196	113	119	133	90	95	96
0.15	***	558	565	679	282	289	342	179	186	216	129	135	153	100	105	117
	•••	560	578	821	284	300	391	182	194	232	131	141	154	102	109	109
0.2	***	650	661	841	319	329	411	198	208	252	140	148	175	107	114	131
	•••	654	680	964	322	346	449	202	220	261	143	156	170	110	119	119
0.25	***	710	724	980	342	356	467	210	222	281	147	157	191	111	119	141
	•••	715	751	1081	347	378	495	215	238	284	151	168	183	115	127	126
0.3	***	740	759	1092	352	371	512	215	231	303	150	162	203	113	122	148
	•••	746	795	1172	359	399	530	221	250	300	154	175	191	117	131	131
0.35	***	743	768	1177	352	375	544	214	233	318	149	163	211	112	122	152
	•••	751	814	1237	360	409	553	221	255	310	155	178	195	116	132	132
0.4	***	723	755	1235	343	371	564	209	231	325	145	161	214	109	121	152
	•••	734	812	1276	353	411	565	218	256	313	152	177	195	114	131	131
0.45	***	685	726	1263	327	361	570	201	225	326	140	157	212	105	117	149
	•••	699	794	1289	339	404	565	210	252	310	147	174	191	110	128	126
0.5	***	634	684	1264	306	345	564	190	216	319	132	150	205	99	111	143
	•••	651	763	1276	321	392	553	200	244	300	140	167	183	104	122	119
0.55	***	574	634	1236	283	324	546	176	204	305	123	141	194	92	104	134
	•••	595	721	1237	299	373	530	187	232	284	131	158	170	97	115	109
0.6	***	510	578	1180	257	300	515	162	189	284	113	130	178	84	95	121
	•••	535	671	1172	274	349	495	173	216	261	120	146	154	89	105	96
0.65	***	446	519	1095	230	273	471	145	171	256	101	117	158	74	84	106
	•••	473	614	1081	247	320	449	156	196	232	108	131	133	78	92	80
0.7	***	384	459	983	202	242	415	127	150	221	88	101	134	.	.	.
	•••	412	551	964	218	285	391	137	172	196	93	112	108	.	.	.
0.75	***	324	396	843	172	208	347	107	126	179	.	.	.	.	.	.
	•••	352	481	821	187	245	322	115	144	153	.	.	.	.	.	.
0.8	***	265	329	676	140	169	266	.	.	.	.	.	.	.	.	.
	•••	290	401	651	151	197	241	.	.	.	.	.	.	.	.	.
0.85	***	204	254	482	.	.	.	.	.	.	.	.	.	.	.	.
	•••	224	308	456	.	.	.	.	.	.	.	.	.	.	.	.

TABLE 4: ALPHA= 0.01 POWER= 0.9   EXPECTED ACCRUAL THRU MINIMUM FOLLOW-UP= 70

	DEL=.10			DEL=.15			DEL=.20			DEL=.25			DEL=.30		
FACT=	1.0 .75	.50 .25	.00 BIN	1.0 .75	.50 .25	.00 BIN	1.0 .75	.50 .25	.00 BIN	1.0 75	.50 .25	.00 BIN	1.0 .75	.50 .25	.00 BIN

PCONT=•••   REQUIRED NUMBER OF PATIENTS

PCONT		1.0/.75	.50/.25	.00/BIN	1.0/.75	.50/.25	.00/BIN	1.0/.75	.50/.25	.00/BIN	1.0/.75	.50/.25	.00/BIN	1.0/.75	.50/.25	.00/BIN
0.05	***	274	277	296	160	162	172	113	115	121	88	90	94	73	74	77
	•••	275	280	456	161	165	241	114	117	153	89	91	108	73	75	80
0.1	***	433	438	495	230	235	263	152	157	172	113	116	126	90	92	99
	•••	435	447	651	232	242	322	154	161	196	114	120	133	91	95	96
0.15	***	559	567	679	283	291	342	181	188	216	130	136	153	101	106	117
	•••	562	582	821	286	303	391	183	196	232	132	142	154	103	110	109
0.2	***	652	664	841	321	333	411	200	210	252	142	150	175	109	115	131
	•••	656	686	964	325	350	449	204	222	261	145	158	170	111	121	119
0.25	***	712	729	980	344	360	467	212	226	281	149	159	191	113	121	141
	•••	718	760	1081	350	384	495	217	241	284	153	170	183	116	128	126
0.3	***	743	765	1092	355	376	512	218	235	303	152	165	204	115	124	148
	•••	750	806	1172	363	406	530	224	254	300	157	177	191	119	133	131
0.35	***	747	776	1177	356	382	544	218	238	318	152	166	211	114	125	152
	•••	757	827	1237	365	417	553	226	260	310	158	181	195	119	134	132
0.4	***	729	766	1235	348	380	564	214	237	325	149	165	214	112	123	152
	•••	741	829	1276	360	420	565	223	261	313	155	181	195	116	133	131
0.45	***	692	739	1263	333	370	570	206	231	326	143	161	212	107	120	149
	•••	708	813	1289	347	415	565	216	258	310	151	177	191	113	130	126
0.5	***	643	699	1264	314	354	564	195	222	319	136	154	205	102	114	143
	•••	663	784	1276	329	403	553	206	250	300	144	171	183	107	125	119
0.55	***	585	651	1236	291	334	546	182	210	305	127	145	194	95	107	134
	•••	609	743	1237	308	384	530	193	238	284	135	162	170	100	117	109
0.6	***	523	597	1180	266	311	515	167	195	284	117	134	178	86	97	121
	•••	550	694	1172	283	360	495	179	221	261	124	149	154	91	106	96
0.65	***	460	539	1095	239	283	471	151	177	256	105	120	158	76	86	106
	•••	490	637	1081	256	330	449	161	201	232	111	134	133	81	94	80
0.7	***	399	478	983	210	252	415	132	155	221	91	104	134	.	.	.
	•••	429	573	964	227	295	391	142	177	196	96	115	108	.	.	.
0.75	***	339	414	843	180	216	347	111	130	179	.	.	.	.	.	.
	•••	368	500	821	195	253	322	119	147	153	.	.	.	.	.	.
0.8	***	278	344	676	146	175	266	.	.	.	.	.	.	.	.	.
	•••	305	418	651	158	203	241	.	.	.	.	.	.	.	.	.
0.85	***	215	266	482	.	.	.	.	.	.	.	.	.	.	.	.
	•••	235	320	456	.	.	.	.	.	.	.	.	.	.	.	.

TABLE 4: ALPHA= 0.01 POWER= 0.9   EXPECTED ACCRUAL THRU MINIMUM FOLLOW-UP= 80

		DEL=.10			DEL=.15			DEL=.20			DEL=.25			DEL=.30	
FACT=		1.0 .75	.50 .25	.00 BIN	1.0 .75	.50 .25	.00 BIN	1.0 .75	.50 .25	.00 BIN	1.0 75	.50 .25	.00 BIN	1.0 .75	.50 .25	.00 BIN
PCONT=•••				REQUIRED	NUMBER	OF PATIENTS										
0.05	***	274	277	296	160	163	172	113	116	121	88	90	94	73	74	77
	•••	275	281	456	161	166	241	114	118	153	89	92	108	74	75	80
0.1	***	434	439	495	231	236	263	153	158	172	113	117	126	90	93	99
	•••	436	449	651	233	244	322	155	162	196	115	120	133	91	95	96
0.15	***	560	570	679	284	293	342	182	189	216	131	137	153	102	107	117
	•••	564	586	821	288	306	391	185	198	232	134	143	154	104	110	109
0.2	***	654	667	841	322	336	411	202	213	252	143	152	175	110	116	131
	•••	658	692	964	327	354	449	206	225	261	147	160	170	113	122	119
0.25	***	715	734	980	347	365	467	215	229	281	151	162	191	115	122	141
	•••	721	768	1081	353	389	495	220	244	284	155	172	183	118	129	126
0.3	***	746	771	1092	359	382	512	221	238	303	154	167	204	117	126	148
	•••	755	816	1172	367	412	530	228	257	300	160	179	191	121	134	131
0.35	***	751	784	1177	360	388	544	221	242	318	155	169	211	116	127	152
	•••	762	840	1237	371	425	553	230	264	310	161	183	195	121	136	132
0.4	***	734	776	1235	353	387	564	218	241	325	152	168	214	114	125	152
	•••	748	844	1276	366	428	565	227	266	313	159	184	195	119	135	131
0.45	***	699	751	1263	339	378	570	210	236	326	147	164	212	110	122	149
	•••	717	830	1289	354	424	565	221	263	310	154	180	191	115	132	126
0.5	***	651	713	1264	321	363	564	200	227	319	140	158	205	104	116	143
	•••	674	803	1276	337	412	553	211	255	300	147	174	183	109	126	119
0.55	***	595	667	1236	299	344	546	187	215	305	131	149	194	97	109	134
	•••	621	764	1237	316	394	530	199	243	284	138	164	170	102	118	109
0.6	***	535	614	1180	274	320	515	173	200	284	120	137	178	89	99	121
	•••	565	715	1172	292	370	495	184	226	261	127	152	154	93	108	96
0.65	***	474	556	1095	247	292	471	156	182	256	108	123	158	78	88	106
	•••	505	658	1081	265	339	449	166	205	232	114	136	133	83	95	80
0.7	***	412	495	983	218	260	415	137	160	221	93	106	134	.	.	.
	•••	445	592	964	235	303	391	146	180	196	99	116	108	.	.	.
0.75	***	352	430	843	187	223	347	115	134	179	.	.	.	.	.	.
	•••	383	518	821	202	260	322	123	150	153	.	.	.	.	.	.
0.8	***	290	358	676	151	181	266	.	.	.	.	.	.	.	.	.
	•••	317	432	651	164	208	241	.	.	.	.	.	.	.	.	.
0.85	***	224	276	482	.	.	.	.	.	.	.	.	.	.	.	.
	•••	245	330	456	.	.	.	.	.	.	.	.	.	.	.	.

TABLE 4: ALPHA= 0.01 POWER= 0.9 EXPECTED ACCRUAL THRU MINIMUM FOLLOW-UP= 90

	DEL=.10			DEL=.15			DEL=.20			DEL=.25			DEL=.30		
FACT=	1.0 .75	.50 .25	.00 BIN	1.0 .75	.50 .25	.00 BIN	1.0 .75	.50 .25	.00 BIN	1.0 75	.50 .25	.00 BIN	1.0 .75	.50 .25	.00 BIN

PCONT=•••   REQUIRED NUMBER OF PATIENTS

PCONT		DEL=.10			DEL=.15			DEL=.20			DEL=.25			DEL=.30		
0.05	•••	275	278	296	161	163	172	114	116	121	89	90	94	73	75	77
	•••	276	282	456	162	166	241	115	118	153	89	92	108	74	76	80
0.1	•••	434	441	495	232	238	263	154	158	172	114	118	126	91	93	99
	•••	437	451	651	234	245	322	155	163	196	116	121	133	92	96	96
0.15	•••	562	572	679	286	295	342	183	191	216	132	138	153	103	107	117
	•••	565	590	821	289	308	391	186	199	232	135	144	154	105	111	109
0.2	•••	655	671	841	324	338	411	204	215	252	145	153	175	111	117	131
	•••	661	697	964	329	357	449	208	226	261	148	161	170	114	122	119
0.25	•••	717	738	980	349	368	467	217	231	281	153	163	191	116	124	141
	•••	724	775	1081	356	393	495	222	247	284	157	173	183	119	130	126
0.3	•••	749	777	1092	362	386	512	223	241	303	156	169	204	118	127	148
	•••	759	825	1172	371	417	530	231	260	300	162	181	191	122	135	131
0.35	•••	755	792	1177	364	394	544	225	246	318	157	172	211	118	128	152
	•••	768	852	1237	375	431	553	233	268	310	163	185	195	123	137	132
0.4	•••	739	785	1235	356	393	564	221	246	325	155	171	214	116	127	152
	•••	755	857	1276	371	436	565	231	270	313	161	186	195	121	137	131
0.45	•••	706	762	1263	345	385	570	214	241	326	150	167	212	112	124	149
	•••	726	846	1289	361	432	565	225	267	310	157	183	191	117	133	126
0.5	•••	660	727	1264	327	371	564	204	232	319	143	161	205	106	118	143
	•••	684	820	1276	345	421	553	216	259	300	150	176	183	111	128	119
0.55	•••	605	682	1236	306	352	546	192	220	305	134	151	194	99	111	134
	•••	634	782	1237	324	402	530	204	247	284	141	167	170	104	120	109
0.6	•••	547	630	1180	281	328	515	177	205	284	123	140	178	90	101	121
	•••	578	734	1172	300	378	495	189	230	261	130	154	154	95	109	96
0.65	•••	486	573	1095	254	300	471	160	186	256	110	125	158	80	89	106
	•••	519	676	1081	273	347	449	171	209	232	117	138	133	84	96	80
0.7	•••	425	511	983	225	268	415	141	163	221	95	108	134	.	.	.
	•••	459	610	964	242	310	391	150	183	196	101	118	108	.	.	.
0.75	•••	364	444	843	193	230	347	118	137	179	.	.	.	.	.	.
	•••	396	533	821	208	265	322	126	152	153	.	.	.	.	.	.
0.8	•••	301	370	676	156	186	266	.	.	.	.	.	.	.	.	.
	•••	329	445	651	169	212	241	.	.	.	.	.	.	.	.	.
0.85	•••	233	286	482	.	.	.	.	.	.	.	.	.	.	.	.
	•••	254	339	456	.	.	.	.	.	.	.	.	.	.	.	.

TABLE 4: ALPHA= 0.01 POWER= 0.9   EXPECTED ACCRUAL THRU MINIMUM FOLLOW-UP= 100

PCONT=•••	DEL=.10 1.0/.75	.50/.25	.00 BIN	DEL=.15 1.0/.75	.50/.25	.00 BIN	DEL=.20 1.0/.75	.50/.25	.00 BIN	DEL=.25 1.0/75	.50/.25	.00 BIN	DEL=.30 1.0/.75	.50/.25	.00 BIN
							REQUIRED NUMBER OF PATIENTS								
0.05 •••	275	278	296	161	164	172	114	116	121	89	91	94	74	75	77
•••	276	283	456	162	167	241	115	118	153	90	92	108	74	76	80
0.1 •••	435	442	495	232	239	263	154	159	172	115	118	126	91	94	99
•••	438	453	651	235	246	322	156	164	196	116	121	133	92	96	96
0.15 •••	563	574	679	287	297	342	184	192	216	133	139	153	103	108	117
•••	567	593	821	291	310	391	187	200	232	136	144	154	105	111	109
0.2 •••	657	674	841	326	341	411	205	217	252	146	154	175	112	118	131
•••	663	702	964	332	360	449	210	228	261	149	162	170	115	123	119
0.25 •••	720	743	980	352	372	467	219	234	281	154	165	191	117	125	141
•••	727	782	1081	359	397	495	225	249	284	159	175	183	120	131	126
0.3 •••	752	783	1092	365	391	512	226	244	303	158	171	204	120	129	148
•••	763	834	1172	375	422	530	233	263	300	164	183	191	124	136	131
0.35 •••	760	799	1177	368	400	544	228	249	318	159	174	211	120	130	152
•••	773	863	1237	380	437	553	236	271	310	165	187	195	124	138	132
0.4 •••	745	795	1235	363	400	564	225	249	325	157	173	214	118	129	152
•••	762	870	1276	377	442	565	235	273	313	164	188	195	122	138	131
0.45 •••	713	773	1263	351	392	570	218	245	326	152	170	212	114	125	149
•••	734	860	1289	367	439	565	229	271	310	160	185	191	119	135	126
0.5 •••	668	739	1264	333	379	564	209	237	319	146	163	205	108	120	143
•••	694	836	1276	351	428	553	220	263	300	153	178	183	113	129	119
0.55 •••	615	696	1236	312	360	546	196	224	305	137	154	194	101	112	134
•••	645	799	1237	331	410	530	208	251	284	144	169	170	106	121	109
0.6 •••	558	644	1180	288	336	515	181	209	284	126	142	178	92	102	121
•••	591	751	1172	307	385	495	193	234	261	133	156	154	97	110	96
0.65 •••	498	588	1095	261	307	471	164	190	256	113	127	158	82	90	106
•••	533	693	1081	280	354	449	175	212	232	119	139	133	86	97	80
0.7 •••	437	525	983	231	274	415	144	167	221	98	110	134	.	.	.
•••	472	625	964	249	316	391	154	186	196	103	119	108	.	.	.
0.75 •••	375	457	843	198	235	347	121	139	179	.	.	.	.	.	.
•••	408	547	821	214	270	322	129	154	153	.	.	.	.	.	.
0.8 •••	311	382	676	161	190	266	.	.	.	.	.	.	.	.	.
•••	339	456	651	173	216	241	.	.	.	.	.	.	.	.	.
0.85 •••	240	294	482	.	.	.	.	.	.	.	.	.	.	.	.
•••	262	347	456	.	.	.	.	.	.	.	.	.	.	.	.

TABLE 4: ALPHA= 0.01 POWER= 0.9   EXPECTED ACCRUAL THRU MINIMUM FOLLOW-UP= 110

		DEL=.05			DEL=.10			DEL=.15			DEL=.20			DEL=.25		
FACT=		1.0 .75	.50 .25	.00 BIN	1.0 .75	.50 .25	.00 BIN	1.0 .75	.50 .25	.00 BIN	1.0 75	.50 .25	.00 BIN	1.0 .75	.50 .25	.00 BIN
PCONT=***					REQUIRED NUMBER OF PATIENTS											
0.05	***	777	780	832	276	279	296	161	164	172	114	116	121	89	91	94
	•••	778	786	1432	277	283	456	163	167	241	115	118	153	90	92	108
0.1	***	1416	1423	1604	436	443	495	233	240	263	155	160	172	115	119	126
	•••	1418	1435	2265	438	455	651	236	247	322	157	164	196	117	122	133
0.15	***	1958	1969	2351	564	576	679	288	299	342	185	193	215	134	140	153
	•••	1962	1991	2994	568	596	821	292	311	391	188	201	232	136	145	154
0.2	***	2376	2392	3033	659	677	841	328	344	411	207	218	252	147	155	175
	•••	2381	2425	3619	665	707	964	334	362	449	211	230	261	151	163	170
0.25	***	2664	2687	3629	722	747	980	354	375	467	221	236	281	156	166	191
	•••	2672	2733	4140	731	788	1081	362	400	495	227	251	284	160	176	183
0.3	***	2828	2859	4127	756	789	1092	368	395	512	228	247	303	160	173	203
	•••	2838	2922	4556	767	842	1172	378	427	530	236	265	300	166	184	191
0.35	***	2877	2919	4519	764	807	1177	372	405	544	230	253	318	161	176	211
	•••	2891	3002	4869	779	873	1237	384	442	553	239	274	310	168	189	195
0.4	***	2827	2881	4802	750	804	1235	367	405	564	228	253	325	159	176	214
	•••	2845	2988	5077	769	882	1276	382	448	565	238	276	313	166	189	195
0.45	***	2692	2762	4973	719	784	1263	356	399	570	222	249	326	155	172	212
	•••	2715	2899	5181	743	874	1289	373	445	565	233	274	310	162	186	191
0.5	***	2490	2579	5031	676	751	1264	339	385	564	212	240	319	148	165	205
	•••	2520	2751	5181	704	850	1276	358	435	553	224	266	300	155	180	183
0.55	***	2240	2353	4975	625	709	1236	318	367	546	200	228	305	139	156	194
	•••	2278	2562	5077	656	814	1237	338	417	530	212	254	284	146	170	170
0.6	***	1959	2101	4806	568	658	1180	294	343	515	185	212	284	128	144	178
	•••	2008	2347	4869	603	766	1172	314	392	495	197	237	261	135	157	154
0.65	***	1669	1843	4524	509	601	1095	267	314	471	168	193	256	115	129	158
	•••	1730	2118	4556	545	708	1081	286	360	449	178	215	232	121	140	133
0.7	***	1389	1591	4130	448	539	983	237	280	415	147	170	221	99	111	134
	•••	1462	1882	4140	484	639	964	255	321	391	157	188	196	105	120	108
0.75	***	1134	1350	3626	386	469	843	203	240	347	124	142	179	.	.	.
	•••	1215	1640	3619	419	560	821	219	275	322	131	156	153	.	.	.
0.8	***	908	1120	3011	320	392	676	165	194	266	.	.	.	.	.	.
	•••	989	1388	2994	349	466	651	177	219	241	.	.	.	.	.	.
0.85	***	703	890	2288	248	301	482	.	.	.	.	.	.	.	.	.
	•••	776	1115	2265	270	354	456	.	.	.	.	.	.	.	.	.
0.9	***	501	640	1458	.	.	.	.	.	.	.	.	.	.	.	.
	•••	556	799	1432	.	.	.	.	.	.	.	.	.	.	.	.

TABLE 4: ALPHA= 0.01 POWER= 0.9   EXPECTED ACCRUAL THRU MINIMUM FOLLOW-UP= 120

	DEL=.05			DEL=.10			DEL=.15			DEL=.20			DEL=.25		
FACT=	1.0 .75	.50 .25	.00 BIN	1.0 .75	.50 .25	.00 BIN	1.0 .75	.50 .25	.00 BIN	1.0 75	.50 .25	.00 BIN	1.0 .75	.50 .25	.00 BIN
PCONT=***							REQUIRED NUMBER OF PATIENTS								
0.05 ***	777	780	832	276	279	296	162	165	172	115	117	121	89	91	94
***	778	787	1432	277	284	456	163	167	241	115	118	153	90	92	108
0.1 ***	1417	1424	1604	437	445	495	234	240	263	155	160	172	116	119	126
***	1419	1438	2265	439	456	651	236	248	322	157	165	196	117	122	133
0.15 ***	1959	1971	2351	565	578	679	289	300	342	186	194	216	135	141	153
***	1963	1994	2994	570	599	821	293	313	391	189	202	232	137	145	154
0.2 ***	2377	2395	3033	661	680	841	329	346	411	208	220	252	148	156	175
***	2383	2431	3619	667	711	964	336	365	449	213	231	261	152	163	170
0.25 ***	2666	2691	3629	724	751	980	356	378	467	222	238	281	157	168	191
***	2674	2742	4140	734	794	1081	364	403	495	229	253	284	162	177	183
0.3 ***	2830	2865	4127	759	795	1092	371	399	512	231	250	303	162	175	203
***	2842	2933	4556	771	850	1172	382	430	530	238	268	300	167	185	191
0.35 ***	2881	2926	4519	768	814	1177	375	409	544	233	255	318	163	178	211
***	2896	3017	4869	784	882	1237	388	447	553	242	276	310	169	190	195
0.4 ***	2832	2891	4802	755	812	1234	371	411	564	231	256	325	161	177	214
***	2851	3008	5077	776	893	1276	387	453	565	241	279	313	168	191	195
0.45 ***	2698	2774	4973	726	794	1263	361	404	570	225	252	326	157	174	212
***	2724	2923	5181	751	886	1289	378	451	565	236	277	310	164	188	191
0.5 ***	2498	2595	5031	684	763	1264	345	392	564	216	244	319	150	167	205
***	2531	2780	5181	713	864	1276	363	441	553	227	269	300	158	182	183
0.55 ***	2250	2373	4975	634	721	1236	324	373	546	204	232	305	141	158	194
***	2291	2596	5077	667	828	1237	344	423	530	215	257	284	149	172	170
0.6 ***	1973	2126	4806	578	671	1180	300	349	515	189	216	284	130	146	178
***	2025	2386	4869	614	780	1172	320	398	495	200	240	261	137	158	154
0.65 ***	1686	1872	4524	519	614	1095	273	320	471	171	196	256	117	131	158
***	1751	2160	4556	556	722	1081	292	366	449	181	217	232	123	142	133
0.7 ***	1410	1622	4130	459	551	983	242	285	415	150	172	221	101	112	134
***	1487	1925	4140	495	652	964	260	326	391	160	190	196	106	121	108
0.75 ***	1157	1383	3626	396	481	843	208	245	347	126	144	179	.	.	.
***	1242	1681	3619	430	571	821	223	279	322	134	157	153	.	.	.
0.8 ***	931	1151	3011	329	401	676	169	197	266	.	.	.	.	.	.
***	1015	1425	2994	358	475	651	181	222	241	.	.	.	.	.	.
0.85 ***	725	916	2288	254	308	482	.	.	.	.	.	.	.	.	.
***	799	1146	2265	276	360	456	.	.	.	.	.	.	.	.	.
0.9 ***	517	659	1458	.	.	.	.	.	.	.	.	.	.	.	.
***	573	820	1432	.	.	.	.	.	.	.	.	.	.	.	.

TABLE 4: ALPHA= 0.01 POWER= 0.9   EXPECTED ACCRUAL THRU MINIMUM FOLLOW-UP= 130

		DEL=.05			DEL=.10			DEL=.15			DEL=.20			DEL=.25	
FACT=	1.0 .75	.50 .25	.00 BIN	1.0 .75	.50 .25	.00 BIN	1.0 .75	.50 .25	.00 BIN	1.0 75	.50 .25	.00 BIN	1.0 .75	.50 .25	.00 BIN

PCONT=•••   REQUIRED NUMBER OF PATIENTS

PCONT		DEL=.05 1.0/.75	.50/.25	.00/BIN	DEL=.10 1.0/.75	.50/.25	.00/BIN	DEL=.15 1.0/.75	.50/.25	.00/BIN	DEL=.20 1.0/75	.50/.25	.00/BIN	DEL=.25 1.0/.75	.50/.25	.00/BIN
0.05	•••	778	781	832	276	280	296	162	165	172	115	117	121	90	91	94
	•••	779	788	1432	278	285	456	163	168	241	116	119	153	90	92	108
0.1	•••	1418	1425	1604	437	446	495	235	241	263	156	161	172	116	120	126
	•••	1420	1440	2265	440	458	651	237	248	322	158	165	196	118	122	133
0.15	•••	1960	1973	2351	566	580	679	290	302	342	187	195	216	135	141	153
	•••	1964	1998	2994	571	602	821	295	314	391	190	203	232	138	146	154
0.2	•••	2379	2398	3033	662	683	841	331	348	411	209	221	252	149	157	175
	•••	2385	2437	3619	670	716	964	338	367	449	214	232	261	153	164	170
0.25	•••	2668	2696	3629	727	756	980	358	381	467	224	240	281	158	169	191
	•••	2677	2750	4140	737	800	1081	367	406	495	230	254	284	163	178	183
0.3	•••	2833	2870	4127	762	800	1092	374	402	512	233	252	303	163	176	204
	•••	2846	2944	4556	775	858	1172	385	434	530	240	269	300	169	187	191
0.35	•••	2885	2934	4519	772	821	1177	379	413	544	236	258	318	165	179	211
	•••	2901	3032	4869	789	891	1237	392	451	553	245	278	310	171	191	195
0.4	•••	2837	2900	4802	761	821	1234	376	415	564	234	259	325	163	179	214
	•••	2858	3027	5077	782	903	1276	391	458	565	244	282	313	170	192	195
0.45	•••	2705	2787	4973	732	804	1263	365	410	570	228	255	326	159	176	212
	•••	2732	2947	5181	759	898	1289	383	456	565	239	279	310	166	189	191
0.5	•••	2507	2611	5031	692	773	1264	350	397	564	219	247	319	152	169	205
	•••	2542	2809	5181	722	876	1276	369	446	553	231	272	300	160	183	183
0.55	•••	2260	2393	4975	642	732	1236	329	379	546	207	235	305	143	160	194
	•••	2305	2629	5077	677	841	1237	349	428	530	219	259	284	151	173	170
0.6	•••	1986	2150	4806	587	683	1180	305	355	515	192	219	284	132	148	178
	•••	2042	2423	4869	625	794	1172	326	403	495	203	242	261	139	160	154
0.65	•••	1703	1900	4524	529	626	1095	278	325	471	174	199	256	119	132	158
	•••	1772	2200	4556	567	735	1081	298	371	449	184	220	232	125	143	133
0.7	•••	1430	1653	4130	469	562	983	247	290	415	153	174	221	102	114	134
	•••	1511	1965	4140	506	664	964	265	330	391	162	192	196	107	122	108
0.75	•••	1179	1414	3626	405	491	843	212	249	347	128	145	179	.	.	.
	•••	1268	1720	3619	440	581	821	228	282	322	136	159	153	.	.	.
0.8	•••	954	1180	3011	337	410	676	172	200	266	.	.	.	.	.	.
	•••	1041	1460	2994	366	483	651	184	224	241	.	.	.	.	.	.
0.85	•••	745	941	2288	260	314	482	.	.	.	.	.	.	.	.	.
	•••	821	1174	2265	283	366	456	.	.	.	.	.	.	.	.	.
0.9	•••	532	677	1458	.	.	.	.	.	.	.	.	.	.	.	.
	•••	589	840	1432	.	.	.	.	.	.	.	.	.	.	.	.

TABLE 4: ALPHA= 0.01 POWER= 0.9   EXPECTED ACCRUAL THRU MINIMUM FOLLOW-UP= 140

PCONT		DEL=.05			DEL=.10			DEL=.15			DEL=.20			DEL=.25		
FACT=		1.0 / .75	.50 / .25	.00 BIN	1.0 / .75	.50 / .25	.00 BIN	1.0 / .75	.50 / .25	.00 BIN	1.0 / 75	.50 / .25	.00 BIN	1.0 / .75	.50 / .25	.00 BIN
					REQUIRED NUMBER OF PATIENTS											
0.05	•••	778	782	832	277	280	296	162	165	172	115	117	121	90	91	94
	•••	779	789	1432	278	285	456	164	168	241	116	119	153	90	92	108
0.1	•••	1418	1426	1604	438	447	495	235	242	263	157	161	172	116	120	126
	•••	1421	1442	2265	441	459	651	238	249	322	159	166	196	118	123	133
0.15	•••	1961	1975	2351	567	582	679	291	303	342	188	196	216	136	142	153
	•••	1966	2002	2994	573	604	821	296	316	391	191	203	232	138	146	154
0.2	•••	2380	2401	3033	664	686	841	333	350	411	210	222	252	150	158	175
	•••	2387	2443	3619	672	719	964	339	369	449	215	233	261	154	165	170
0.25	•••	2670	2700	3629	729	760	980	360	384	467	226	241	281	159	170	191
	•••	2680	2758	4140	740	805	1081	370	409	495	232	255	284	164	178	183
0.3	•••	2836	2876	4127	765	806	1092	376	406	512	235	254	303	165	177	204
	•••	2849	2956	4556	779	865	1172	388	437	530	243	271	300	170	187	191
0.35	•••	2888	2941	4519	776	827	1177	382	417	544	238	260	318	167	181	211
	•••	2906	3046	4869	794	899	1237	396	455	553	247	280	310	173	192	195
0.4	•••	2841	2910	4802	766	829	1235	379	420	564	237	261	325	165	181	214
	•••	2864	3045	5077	789	913	1276	396	462	565	247	284	313	172	193	195
0.45	•••	2711	2799	4973	739	813	1263	370	415	570	231	258	326	161	177	212
	•••	2741	2970	5181	766	909	1289	388	461	565	242	282	310	168	191	191
0.5	•••	2515	2627	5031	699	784	1264	354	403	564	222	250	319	154	171	205
	•••	2552	2837	5181	731	888	1276	374	451	553	234	274	300	162	184	183
0.55	•••	2271	2412	4975	651	743	1236	335	384	546	210	238	305	145	162	194
	•••	2319	2662	5077	686	853	1237	355	433	530	222	262	284	152	174	170
0.6	•••	1999	2174	4806	597	694	1180	311	360	515	195	221	284	134	149	178
	•••	2059	2459	4869	635	806	1172	331	408	495	206	244	261	141	161	154
0.65	•••	1719	1927	4524	539	637	1095	283	330	471	177	201	256	120	134	158
	•••	1793	2238	4556	578	746	1081	303	375	449	187	221	232	126	144	133
0.7	•••	1449	1682	4130	478	573	983	252	295	415	155	176	221	104	115	134
	•••	1535	2003	4140	516	675	964	270	334	391	165	194	196	109	123	108
0.75	•••	1201	1443	3626	414	501	843	216	253	347	130	147	179	.	.	.
	•••	1292	1756	3619	449	591	821	232	285	322	138	160	153	.	.	.
0.8	•••	975	1207	3011	344	418	676	175	203	266	.	.	.	.	.	.
	•••	1065	1492	2994	374	491	651	187	226	241	.	.	.	.	.	.
0.85	•••	764	964	2288	266	320	482	.	.	.	.	.	.	.	.	.
	•••	842	1201	2265	288	371	456	.	.	.	.	.	.	.	.	.
0.9	•••	546	694	1458	.	.	.	.	.	.	.	.	.	.	.	.
	•••	605	858	1432	.	.	.	.	.	.	.	.	.	.	.	.

TABLE 4: ALPHA= 0.01 POWER= 0.9   EXPECTED ACCRUAL THRU MINIMUM FOLLOW-UP= 150

		DEL=.05			DEL=.10			DEL=.15			DEL=.20			DEL=.25		
FACT=		1.0 .75	.50 .25	.00 BIN	1.0 .75	.50 .25	.00 BIN	1.0 .75	.50 .25	.00 BIN	1.0 75	.50 .25	.00 BIN	1.0 .75	.50 .25	.00 BIN
PCONT=***					REQUIRED NUMBER OF PATIENTS											
0.05	***	778	782	832	277	281	296	163	166	172	115	117	121	90	91	94
	•••	779	790	1432	278	286	456	164	168	241	116	119	153	91	92	108
0.1	***	1419	1427	1604	439	448	495	236	243	263	157	162	172	117	120	126
	•••	1422	1444	2265	442	460	651	239	250	322	159	166	196	118	123	133
0.15	***	1962	1977	2351	568	584	679	292	304	342	188	197	215	137	142	153
	•••	1967	2006	2994	574	607	821	297	317	391	192	204	232	139	147	154
0.2	***	2382	2404	3033	666	689	841	334	352	411	212	223	252	151	159	175
	•••	2389	2448	3619	674	723	964	341	370	449	217	234	261	154	165	170
0.25	***	2672	2704	3629	731	764	980	362	386	467	227	243	281	160	171	191
	•••	2683	2767	4140	743	811	1081	372	411	495	234	257	284	165	179	183
0.3	***	2839	2882	4127	768	811	1092	379	409	512	236	256	303	166	178	203
	•••	2853	2967	4556	783	871	1172	391	440	530	244	272	300	171	188	191
0.35	***	2892	2949	4519	780	834	1177	385	421	544	240	262	318	168	182	211
	•••	2911	3061	4869	799	907	1237	400	458	553	249	282	310	174	193	195
0.4	***	2846	2920	4802	771	836	1234	383	424	564	239	264	325	167	182	214
	•••	2871	3064	5077	795	922	1276	400	466	565	249	286	313	173	194	195
0.45	***	2717	2812	4973	745	822	1263	374	419	570	234	260	326	163	179	212
	•••	2749	2993	5181	773	919	1289	392	465	565	245	284	310	170	192	191
0.5	***	2523	2643	5031	706	793	1264	359	407	564	225	252	319	156	172	205
	•••	2563	2864	5181	739	899	1276	379	455	553	236	276	300	163	185	183
0.55	***	2281	2432	4975	659	754	1236	339	389	546	213	240	305	147	163	194
	•••	2332	2693	5077	695	864	1237	360	438	530	224	264	284	154	175	170
0.6	***	2012	2197	4806	605	705	1180	315	365	515	197	224	284	135	150	178
	•••	2076	2493	4869	644	817	1172	336	412	495	209	246	261	142	162	154
0.65	***	1735	1953	4524	548	648	1095	288	335	471	179	203	256	122	135	158
	•••	1813	2274	4556	587	757	1081	307	379	449	190	223	232	127	144	133
0.7	***	1468	1710	4130	487	583	983	256	299	415	158	178	221	105	116	134
	•••	1558	2039	4140	525	685	964	274	338	391	167	195	196	110	123	108
0.75	***	1222	1471	3626	422	509	843	220	256	347	132	148	179	.	.	.
	•••	1316	1791	3619	457	599	821	235	288	322	139	161	153	.	.	.
0.8	***	996	1233	3011	351	425	676	178	205	266	.	.	.	.	.	.
	•••	1087	1523	2994	382	498	651	190	228	241	.	.	.	.	.	.
0.85	***	782	986	2288	271	325	482	.	.	.	.	.	.	.	.	.
	•••	862	1226	2265	294	375	456	.	.	.	.	.	.	.	.	.
0.9	***	560	709	1458	.	.	.	.	.	.	.	.	.	.	.	.
	•••	619	874	1432	.	.	.	.	.	.	.	.	.	.	.	.

TABLE 4: ALPHA= 0.01 POWER= 0.9   EXPECTED ACCRUAL THRU MINIMUM FOLLOW-UP= 160

	DEL=.05			DEL=.10			DEL=.15			DEL=.20			DEL=.25		
FACT=	1.0 .75	.50 .25	.00 BIN	1.0 .75	.50 .25	.00 BIN	1.0 .75	.50 .25	.00 BIN	1.0 75	.50 .25	.00 BIN	1.0 .75	.50 .25	.00 BIN
PCONT=***				REQUIRED	NUMBER	OF PATIENTS									
0.05 ***	778	783	832	277	281	296	163	166	172	116	118	121	90	92	94
•••	780	790	1432	279	286	456	164	168	241	116	119	153	91	93	108
0.1 ***	1419	1429	1604	439	449	495	236	244	263	158	162	172	117	120	126
•••	1422	1447	2265	443	462	651	239	250	322	160	166	196	119	123	133
0.15 ***	1963	1979	2351	570	586	679	293	306	342	189	198	216	137	143	153
•••	1968	2010	2994	575	609	821	298	318	391	193	204	232	140	147	154
0.2 ***	2383	2407	3033	667	692	841	336	354	411	213	225	252	152	160	175
•••	2391	2454	3619	676	727	964	343	372	449	218	235	261	155	166	170
0.25 ***	2674	2708	3629	734	768	980	365	389	467	229	244	281	162	172	191
•••	2686	2775	4140	746	815	1081	374	414	495	235	258	284	166	180	183
0.3 ***	2842	2888	4127	771	816	1092	382	412	512	238	257	303	167	179	204
•••	2857	2978	4556	787	877	1172	394	443	530	246	274	300	173	189	191
0.35 ***	2896	2957	4519	784	840	1177	388	425	544	242	264	318	169	183	211
•••	2916	3076	4869	804	915	1237	403	462	553	252	284	310	175	194	195
0.4 ***	2851	2930	4802	776	844	1235	387	428	564	241	266	325	168	184	214
•••	2878	3082	5077	801	931	1276	403	470	565	252	287	313	175	195	195
0.45 ***	2724	2825	4973	751	830	1263	378	424	570	236	263	326	164	180	212
•••	2757	3015	5181	781	928	1289	397	469	565	247	286	310	171	193	191
0.5 ***	2531	2659	5031	713	803	1264	363	412	564	227	255	319	158	174	205
•••	2574	2890	5181	747	909	1276	383	460	553	239	278	300	165	186	183
0.55 ***	2291	2451	4975	667	764	1236	344	394	546	215	243	305	149	164	194
•••	2346	2723	5077	704	875	1237	364	442	530	227	266	284	155	176	170
0.6 ***	2025	2220	4806	614	715	1180	320	370	515	200	226	284	137	152	178
•••	2093	2526	4869	654	828	1172	341	416	495	211	248	261	143	162	154
0.65 ***	1751	1979	4524	556	658	1095	292	339	471	182	205	256	123	136	158
•••	1833	2308	4556	597	768	1081	312	383	449	192	225	232	129	145	133
0.7 ***	1487	1738	4130	495	592	983	260	303	415	160	180	221	106	116	134
•••	1580	2074	4140	534	694	964	278	341	391	169	196	196	111	124	108
0.75 ***	1242	1498	3626	430	518	843	223	260	347	134	150	179	.	.	.
•••	1339	1823	3619	466	608	821	239	291	322	141	162	153	.	.	.
0.8 ***	1015	1258	3011	358	432	676	181	208	266	.	.	.	.	.	.
•••	1109	1552	2994	388	504	651	192	230	241	.	.	.	.	.	.
0.85 ***	799	1008	2288	276	330	482	.	.	.	.	.	.	.	.	.
•••	881	1250	2265	299	380	456	.	.	.	.	.	.	.	.	.
0.9 ***	573	724	1458	.	.	.	.	.	.	.	.	.	.	.	.
•••	633	890	1432	.	.	.	.	.	.	.	.	.	.	.	.

TABLE 4: ALPHA= 0.01 POWER= 0.9   EXPECTED ACCRUAL THRU MINIMUM FOLLOW-UP= 170

		DEL=.05			DEL=.10			DEL=.15			DEL=.20			DEL=.25	
FACT=	1.0 .75	.50 .25	.00 BIN	1.0 .75	.50 .25	.00 BIN	1.0 .75	.50 .25	.00 BIN	1.0 75	.50 .25	.00 BIN	1.0 .75	.50 .25	.00 BIN
PCONT=***					REQUIRED NUMBER OF PATIENTS										
0.05 ***	779	783	832	277	282	296	163	166	172	116	118	121	90	92	94
•••	780	791	1432	279	286	456	164	169	241	117	119	153	91	93	108
0.1 ***	1420	1430	1604	440	450	495	237	244	263	158	163	172	117	121	126
•••	1423	1449	2265	444	463	651	240	251	322	160	167	196	119	123	133
0.15 ***	1964	1981	2351	571	588	679	294	307	342	190	198	215	138	143	153
•••	1970	2014	2994	577	611	821	299	319	391	194	205	232	140	147	154
0.2 ***	2385	2410	3033	669	695	841	337	355	411	214	225	252	152	160	175
•••	2393	2460	3619	678	730	964	344	374	449	219	235	261	156	166	170
0.25 ***	2677	2712	3629	736	771	980	366	391	467	230	246	281	162	172	191
•••	2688	2783	4140	749	820	1081	376	416	495	237	259	284	167	180	183
0.3 ***	2845	2893	4127	774	821	1092	384	415	512	240	259	303	168	180	203
•••	2861	2989	4556	791	883	1172	396	446	530	248	275	300	174	190	191
0.35 ***	2900	2964	4519	788	846	1177	391	428	544	244	266	318	171	184	211
•••	2921	3090	4869	809	922	1237	406	465	553	254	285	310	177	195	195
0.4 ***	2856	2940	4802	781	851	1235	390	432	564	244	268	325	170	185	214
•••	2884	3100	5077	807	939	1276	407	473	565	254	289	313	176	196	195
0.45 ***	2730	2837	4973	757	838	1263	382	428	570	239	265	326	166	181	212
•••	2766	3037	5181	787	937	1289	401	473	565	250	287	310	173	194	191
0.5 ***	2539	2674	5031	720	811	1264	368	416	564	230	257	319	159	175	205
•••	2585	2916	5181	755	918	1276	388	463	553	242	280	300	166	187	183
0.55 ***	2302	2470	4975	674	773	1236	348	398	546	218	245	305	150	165	194
•••	2359	2753	5077	713	885	1237	369	446	530	229	267	284	157	177	170
0.6 ***	2038	2242	4806	622	725	1180	324	374	515	202	228	284	138	153	178
•••	2109	2558	4869	662	837	1172	345	420	495	214	249	261	145	163	154
0.65 ***	1767	2004	4524	565	667	1095	296	344	471	184	207	256	124	137	158
•••	1853	2341	4556	606	777	1081	316	386	449	194	226	232	130	146	133
0.7 ***	1505	1764	4130	503	601	983	264	307	415	161	182	221	107	117	134
•••	1601	2107	4140	543	703	964	282	344	391	170	197	196	112	124	108
0.75 ***	1261	1524	3626	437	526	843	227	263	347	135	151	179	.	.	.
•••	1361	1854	3619	473	615	821	242	293	322	142	163	153	.	.	.
0.8 ***	1034	1282	3011	364	439	676	183	210	266	.	.	.	.	.	.
•••	1130	1580	2994	395	510	651	195	232	241	.	.	.	.	.	.
0.85 ***	816	1028	2288	281	335	482	.	.	.	.	.	.	.	.	.
•••	899	1272	2265	304	384	456	.	.	.	.	.	.	.	.	.
0.9 ***	585	738	1458	.	.	.	.	.	.	.	.	.	.	.	.
•••	646	905	1432	.	.	.	.	.	.	.	.	.	.	.	.

TABLE 4: ALPHA= 0.01 POWER= 0.9   EXPECTED ACCRUAL THRU MINIMUM FOLLOW-UP= 180

		DEL=.05			DEL=.10			DEL=.15			DEL=.20			DEL=.25		
FACT=		1.0 .75	.50 .25	.00 BIN	1.0 .75	.50 .25	.00 BIN	1.0 .75	.50 .25	.00 BIN	1.0 75	.50 .25	.00 BIN	1.0 .75	.50 .25	.00 BIN
PCONT=***					REQUIRED NUMBER OF PATIENTS											
0.05	***	779	784	832	278	282	296	163	166	172	116	118	121	90	92	94
	•••	780	792	1432	279	287	456	165	169	241	117	119	153	91	93	108
0.1	***	1420	1431	1604	441	451	495	238	245	263	158	163	172	118	121	126
	•••	1424	1451	2265	444	464	651	240	251	322	160	167	196	119	123	133
0.15	***	1965	1983	2351	572	590	679	295	308	342	191	199	216	138	144	153
	•••	1971	2018	2994	578	613	821	300	320	391	194	205	232	141	147	154
0.2	***	2386	2413	3033	671	697	841	338	357	411	215	226	252	153	161	175
	•••	2395	2466	3619	680	733	964	346	375	449	220	236	261	156	166	170
0.25	***	2679	2717	3629	738	775	980	368	393	467	231	247	281	163	173	191
	•••	2691	2791	4140	751	824	1081	378	417	495	238	260	284	168	181	183
0.3	***	2847	2899	4127	777	825	1092	386	417	512	242	260	303	169	181	204
	•••	2865	3000	4556	795	888	1172	399	448	530	249	276	300	175	190	191
0.35	***	2904	2972	4519	792	852	1177	394	431	544	246	268	318	172	185	211
	•••	2926	3104	4869	814	928	1237	409	467	553	255	287	310	178	196	195
0.4	***	2861	2949	4802	785	857	1235	393	436	564	246	270	325	171	186	214
	•••	2891	3118	5077	812	946	1276	411	477	565	256	291	313	177	197	195
0.45	***	2736	2850	4973	762	846	1263	386	432	570	241	267	326	167	183	212
	•••	2774	3059	5181	794	945	1289	404	476	565	252	289	310	174	194	191
0.5	***	2547	2690	5031	727	820	1264	371	420	564	232	259	319	161	176	205
	•••	2595	2941	5181	763	927	1276	392	467	553	244	282	300	168	188	183
0.55	***	2312	2489	4975	682	782	1236	352	402	546	220	247	305	151	167	194
	•••	2373	2781	5077	721	894	1237	373	449	530	232	269	284	158	178	170
0.6	***	2051	2264	4806	630	734	1180	328	378	515	205	230	284	140	154	178
	•••	2126	2589	4869	671	847	1172	349	423	495	216	251	261	146	164	154
0.65	***	1783	2028	4524	573	676	1095	300	347	471	186	209	256	125	138	158
	•••	1872	2373	4556	614	786	1081	320	389	449	196	227	232	131	146	133
0.7	***	1523	1789	4130	511	610	983	267	310	415	163	183	221	108	118	134
	•••	1622	2138	4140	551	711	964	285	347	391	172	198	196	112	125	108
0.75	***	1280	1549	3626	444	533	843	230	265	347	137	152	179	.	.	.
	•••	1383	1884	3619	481	622	821	245	295	322	144	163	153	.	.	.
0.8	***	1053	1305	3011	370	445	676	186	212	266	.	.	.	.	.	.
	•••	1151	1606	2994	401	516	651	197	233	241	.	.	.	.	.	.
0.85	***	832	1047	2288	285	339	482	.	.	.	.	.	.	.	.	.
	•••	916	1293	2265	308	387	456	.	.	.	.	.	.	.	.	.
0.9	***	597	752	1458	.	.	.	.	.	.	.	.	.	.	.	.
	•••	659	918	1432	.	.	.	.	.	.	.	.	.	.	.	.

TABLE 4: ALPHA= 0.01 POWER= 0.9   EXPECTED ACCRUAL THRU MINIMUM FOLLOW-UP= 190

		DEL=.05			DEL=.10			DEL=.15			DEL=.20			DEL=.25		
FACT=		1.0 .75	.50 .25	.00 BIN	1.0 .75	.50 .25	.00 BIN	1.0 .75	.50 .25	.00 BIN	1.0 75	.50 .25	.00 BIN	1.0 .75	.50 .25	.00 BIN
PCONT=•••					REQUIRED NUMBER OF PATIENTS											
0.05	***	779	784	832	278	282	296	164	167	172	116	118	121	90	92	94
	•••	781	793	1432	280	287	456	165	169	241	117	119	153	91	93	108
0.1	***	1421	1432	1604	441	452	495	238	245	263	159	163	172	118	121	126
	•••	1425	1453	2265	445	465	651	241	252	322	161	167	196	119	123	133
0.15	***	1966	1985	2351	573	591	679	296	309	342	191	199	215	139	144	153
	•••	1972	2021	2994	580	615	821	301	321	391	195	206	232	141	148	154
0.2	***	2388	2416	3033	672	700	841	340	358	411	216	227	252	154	161	174
	•••	2397	2472	3619	682	736	964	347	376	449	221	237	261	157	167	170
0.25	***	2681	2721	3629	740	778	980	370	395	467	233	248	281	164	174	191
	•••	2694	2799	4140	754	828	1081	380	419	495	239	260	284	168	181	183
0.3	***	2850	2905	4127	780	830	1092	389	420	512	243	262	303	170	182	203
	•••	2868	3011	4556	799	893	1172	401	450	530	251	277	300	176	191	191
0.35	***	2907	2979	4519	796	857	1177	397	434	544	248	269	318	173	186	211
	•••	2931	3118	4869	819	934	1237	412	470	553	257	288	310	179	196	195
0.4	***	2866	2959	4802	790	864	1234	397	439	564	248	272	325	172	187	214
	•••	2897	3136	5077	818	954	1276	414	479	565	258	292	313	179	198	195
0.45	***	2743	2862	4973	768	853	1263	389	435	570	243	269	326	168	184	212
	•••	2783	3080	5181	801	953	1289	408	479	565	254	290	310	175	195	191
0.5	***	2555	2705	5031	733	828	1264	375	424	564	234	261	319	162	177	205
	•••	2606	2966	5181	770	936	1276	396	470	553	246	283	300	169	189	183
0.55	***	2322	2508	4975	689	790	1236	356	406	546	222	249	305	153	168	194
	•••	2386	2808	5077	729	903	1237	377	453	530	234	270	284	159	179	170
0.6	***	2063	2286	4806	637	743	1180	332	382	515	207	232	284	141	155	178
	•••	2142	2618	4869	679	856	1172	353	426	495	218	252	261	147	165	154
0.65	***	1798	2052	4524	580	685	1095	304	351	471	188	211	256	126	138	158
	•••	1890	2404	4556	622	794	1081	324	392	449	198	229	232	132	147	133
0.7	***	1541	1813	4130	518	618	983	271	313	415	165	185	221	109	118	134
	•••	1643	2168	4140	559	719	964	289	349	391	174	199	196	113	125	108
0.75	***	1298	1573	3626	451	540	843	233	268	347	138	153	179	.	.	.
	•••	1404	1912	3619	488	629	821	248	297	322	145	164	153	.	.	.
0.8	***	1070	1327	3011	376	450	676	188	214	266	.	.	.	.	.	.
	•••	1170	1631	2994	407	521	651	199	234	241	.	.	.	.	.	.
0.85	***	847	1065	2288	290	343	482	.	.	.	.	.	.	.	.	.
	•••	933	1313	2265	312	391	456	.	.	.	.	.	.	.	.	.
0.9	***	609	764	1458	.	.	.	.	.	.	.	.	.	.	.	.
	•••	671	931	1432	.	.	.	.	.	.	.	.	.	.	.	.

TABLE 4: ALPHA= 0.01 POWER= 0.9 EXPECTED ACCRUAL THRU MINIMUM FOLLOW-UP= 200

	DEL=.05			DEL=.10			DEL=.15			DEL=.20			DEL=.25		
FACT=	1.0 .75	.50 .25	.00 BIN	1.0 .75	.50 .25	.00 BIN	1.0 .75	.50 .25	.00 BIN	1.0 75	.50 .25	.00 BIN	1.0 .75	.50 .25	.00 BIN
PCONT=•••				REQUIRED NUMBER OF PATIENTS											
0.05 •••	779	785	832	278	283	296	164	167	172	116	118	121	91	92	94
•••	781	794	1432	280	287	456	165	169	241	117	119	153	91	93	108
0.1 •••	1422	1433	1604	442	453	495	239	246	263	159	164	172	118	121	126
•••	1425	1455	2265	446	466	651	242	252	322	161	167	196	120	124	133
0.15 •••	1967	1987	2351	574	593	679	297	310	342	192	200	216	139	144	153
•••	1974	2025	2994	581	617	821	302	321	391	195	206	232	141	148	154
0.2 •••	2389	2419	3033	674	702	841	341	360	411	217	228	252	154	162	175
•••	2399	2477	3619	684	739	964	349	378	449	221	237	261	158	167	170
0.25 •••	2683	2725	3629	743	782	980	372	397	467	234	249	281	165	175	191
•••	2697	2807	4140	757	832	1081	382	421	495	240	261	284	169	182	183
0.3 •••	2853	2910	4127	783	834	1092	391	422	512	244	263	303	171	183	204
•••	2872	3022	4556	802	898	1172	404	452	530	252	278	300	176	191	191
0.35 •••	2911	2987	4519	799	863	1177	400	437	544	249	271	318	174	187	211
•••	2936	3132	4869	823	940	1237	415	472	553	259	289	310	180	197	195
0.4 •••	2871	2969	4802	795	870	1235	400	442	564	249	273	325	173	188	214
•••	2904	3153	5077	823	960	1276	417	482	565	260	293	313	180	198	195
0.45 •••	2749	2874	4973	773	860	1263	392	439	570	245	271	326	170	185	212
•••	2791	3100	5181	807	961	1289	412	482	565	256	292	310	176	196	191
0.5 •••	2563	2721	5031	739	836	1264	379	428	564	237	263	319	163	178	205
•••	2617	2990	5181	777	944	1276	399	473	553	248	284	300	170	189	183
0.55 •••	2332	2526	4975	696	799	1236	360	410	546	224	251	305	154	169	194
•••	2399	2835	5077	736	911	1237	381	456	530	236	272	284	160	179	170
0.6 •••	2076	2307	4806	644	751	1180	336	385	515	209	234	284	142	156	178
•••	2158	2647	4869	687	864	1172	357	430	495	220	253	261	148	165	154
0.65 •••	1813	2074	4524	588	693	1095	307	354	471	190	212	256	127	139	158
•••	1909	2433	4556	630	802	1081	327	395	449	200	230	232	133	147	133
0.7 •••	1558	1837	4130	525	625	983	274	316	415	167	186	221	110	119	134
•••	1663	2197	4140	566	726	964	292	352	391	175	200	196	114	125	108
0.75 •••	1316	1596	3626	457	547	843	235	270	347	139	154	179	.	.	.
•••	1424	1940	3619	494	635	821	250	299	322	146	165	153	.	.	.
0.8 •••	1087	1348	3011	382	456	676	190	216	266	.	.	.	.	.	.
•••	1189	1655	2994	412	526	651	201	236	241	.	.	.	.	.	.
0.85 •••	862	1082	2288	294	347	482	.	.	.	.	.	.	.	.	.
•••	949	1332	2265	316	394	456	.	.	.	.	.	.	.	.	.
0.9 •••	619	777	1458	.	.	.	.	.	.	.	.	.	.	.	.
•••	683	944	1432	.	.	.	.	.	.	.	.	.	.	.	.

TABLE 4: ALPHA= 0.01 POWER= 0.9   EXPECTED ACCRUAL THRU MINIMUM FOLLOW-UP= 225

	DEL=.05			DEL=.10			DEL=.15			DEL=.20			DEL=.25		
FACT=	1.0 .75	.50 .25	.00 BIN	1.0 .75	.50 .25	.00 BIN	1.0 .75	.50 .25	.00 BIN	1.0 75	.50 .25	.00 BIN	1.0 .75	.50 .25	.00 BIN
PCONT=•••				REQUIRED NUMBER OF PATIENTS											
0.05 •••	780	786	831	279	284	296	164	167	172	117	118	121	91	92	94
•••	782	795	1432	281	288	456	165	169	241	117	120	153	91	93	108
0.1 •••	1423	1436	1604	444	455	495	240	247	263	160	164	172	119	122	126
•••	1427	1460	2265	448	468	651	243	253	322	162	168	196	120	124	133
0.15 •••	1969	1992	2351	577	597	678	299	312	342	193	201	216	140	145	153
•••	1977	2034	2994	584	621	821	304	323	391	197	207	232	142	149	154
0.2 •••	2393	2426	3033	678	708	841	344	363	411	219	230	252	156	163	174
•••	2404	2491	3619	689	745	964	352	380	449	223	239	261	159	168	170
0.25 •••	2688	2735	3629	748	790	980	376	401	467	236	251	281	167	176	191
•••	2704	2827	4140	764	841	1081	386	424	495	243	263	284	171	183	183
0.3 •••	2860	2925	4127	791	844	1092	396	428	512	248	266	303	174	185	203
•••	2882	3048	4556	811	910	1172	409	457	530	255	281	300	178	192	191
0.35 •••	2921	3006	4519	809	875	1177	406	443	544	253	275	318	176	189	211
•••	2949	3165	4869	834	954	1237	421	478	553	262	291	310	182	198	195
0.4 •••	2883	2993	4802	806	885	1235	407	449	564	254	277	325	176	190	214
•••	2920	3194	5077	836	976	1276	424	488	565	264	296	313	182	200	195
0.45 •••	2765	2905	4973	787	877	1263	400	447	570	249	275	326	173	187	212
•••	2812	3150	5181	822	978	1289	419	489	565	260	294	310	179	197	191
0.5 •••	2583	2758	5031	754	854	1264	387	436	564	241	267	319	166	181	205
•••	2643	3047	5181	793	962	1276	408	480	553	252	287	300	173	191	183
0.55 •••	2358	2570	4975	712	818	1236	368	418	546	229	255	305	157	171	194
•••	2432	2898	5077	754	930	1237	389	462	530	240	275	284	163	181	170
0.6 •••	2107	2357	4806	661	770	1180	344	393	515	213	237	284	145	158	178
•••	2197	2714	4869	705	883	1172	365	436	495	224	256	261	150	166	154
0.65 •••	1850	2129	4524	605	712	1095	315	362	471	194	216	256	130	141	158
•••	1953	2502	4556	648	820	1081	335	401	449	203	232	232	135	148	133
0.7 •••	1599	1893	4130	542	643	983	282	323	415	170	189	221	111	120	134
•••	1710	2265	4140	583	743	964	299	357	391	178	202	196	116	126	108
0.75 •••	1359	1650	3626	472	563	843	242	276	347	142	156	179	.	.	.
•••	1471	2003	3619	509	649	821	257	303	322	149	166	153	.	.	.
0.8 •••	1128	1397	3011	394	468	676	194	219	266	.	.	.	.	.	.
•••	1233	1710	2994	425	536	651	205	238	241	.	.	.	.	.	.
0.85 •••	897	1123	2288	303	356	482	.	.	.	.	.	.	.	.	.
•••	986	1376	2265	325	401	456	.	.	.	.	.	.	.	.	.
0.9 •••	645	805	1458	.	.	.	.	.	.	.	.	.	.	.	.
•••	709	972	1432	.	.	.	.	.	.	.	.	.	.	.	.

TABLE 4: ALPHA= 0.01 POWER= 0.9   EXPECTED ACCRUAL THRU MINIMUM FOLLOW-UP= 250

		DEL=.05			DEL=.10			DEL=.15			DEL=.20			DEL=.25	
FACT=	1.0 .75	.50 .25	.00 BIN	1.0 .75	.50 .25	.00 BIN	1.0 .75	.50 .25	.00 BIN	1.0 75	.50 .25	.00 BIN	1.0 .75	.50 .25	.00 BIN
PCONT=•••				REQUIRED NUMBER OF PATIENTS											
0.05 •••	781	787	832	280	284	296	165	168	172	117	119	121	91	92	94
•••	783	797	1432	282	289	456	166	170	241	118	120	153	92	93	108
0.1 •••	1424	1439	1604	445	457	495	241	248	263	161	165	172	119	122	126
•••	1429	1465	2265	450	470	651	244	254	322	163	168	196	121	124	133
0.15 •••	1972	1997	2351	579	600	679	301	314	342	195	202	216	141	146	153
•••	1980	2043	2994	587	624	821	306	325	391	198	208	232	143	149	154
0.2 •••	2397	2434	3033	682	714	841	347	366	411	220	231	252	157	164	175
•••	2409	2504	3619	694	751	964	355	382	449	225	240	261	160	168	170
0.25 •••	2693	2746	3629	754	797	980	380	405	467	239	253	281	168	177	191
•••	2711	2846	4140	770	849	1081	390	428	495	245	264	284	172	183	183
0.3 •••	2867	2939	4127	798	854	1092	401	432	512	251	269	303	175	186	204
•••	2891	3073	4556	819	919	1172	414	461	530	258	282	300	180	193	191
0.35 •••	2930	3024	4519	817	887	1177	411	449	544	257	277	318	179	191	211
•••	2962	3198	4869	844	966	1237	427	482	553	266	293	310	184	199	195
0.4 •••	2896	3017	4802	817	898	1234	413	456	564	257	280	325	178	192	214
•••	2937	3234	5077	848	990	1276	431	494	565	267	298	313	184	201	195
0.45 •••	2781	2935	4973	799	892	1264	407	454	570	254	278	326	175	189	212
•••	2833	3197	5181	835	993	1289	427	494	565	264	297	310	181	199	191
0.5 •••	2603	2794	5031	768	870	1264	394	443	564	245	271	319	168	182	205
•••	2669	3100	5181	809	979	1276	415	486	553	256	290	300	175	192	183
0.55 •••	2383	2613	4975	727	835	1236	376	426	546	233	258	305	159	172	194
•••	2464	2957	5077	770	947	1237	397	468	530	244	277	284	165	182	170
0.6 •••	2138	2405	4806	677	787	1180	352	400	515	217	241	284	147	159	178
•••	2235	2776	4869	722	899	1172	373	442	495	228	258	261	152	167	154
0.65 •••	1886	2180	4524	620	728	1095	323	368	471	197	219	256	132	142	158
•••	1996	2565	4556	664	836	1081	342	406	449	207	234	232	136	149	133
0.7 •••	1638	1945	4130	557	658	983	288	328	415	173	191	221	113	122	134
•••	1755	2326	4140	599	757	964	305	361	391	181	204	196	117	127	108
0.75 •••	1399	1701	3626	486	576	843	247	280	347	144	158	179	.	.	.
•••	1516	2060	3619	523	661	821	262	306	322	151	167	153	.	.	.
0.8 •••	1165	1443	3011	406	479	676	199	223	266	.	.	.	.	.	.
•••	1274	1760	2994	437	545	651	209	241	241	.	.	.	.	.	.
0.85 •••	929	1160	2288	311	363	482	.	.	.	.	.	.	.	.	.
•••	1021	1415	2265	333	406	456	.	.	.	.	.	.	.	.	.
0.9 •••	668	830	1458	.	.	.	.	.	.	.	.	.	.	.	.
•••	734	997	1432	.	.	.	.	.	.	.	.	.	.	.	.

TABLE 4: ALPHA= 0.01 POWER= 0.9   EXPECTED ACCRUAL THRU MINIMUM FOLLOW-UP= 275

		DEL=.05			DEL=.10			DEL=.15			DEL=.20			DEL=.25		
FACT=		1.0 .75	.50 .25	.00 BIN	1.0 .75	.50 .25	.00 BIN	1.0 .75	.50 .25	.00 BIN	1.0 75	.50 .25	.00 BIN	1.0 .75	.50 .25	.00 BIN
PCONT=•••				REQUIRED	NUMBER OF	PATIENTS										
0.05	***	781	788	832	280	285	296	165	168	172	117	119	121	91	92	94
	•••	784	799	1432	282	289	456	166	170	241	118	120	153	92	93	108
0.1	***	1426	1442	1604	447	459	495	242	249	263	161	166	172	120	122	126
	•••	1431	1470	2265	451	471	651	245	255	322	163	168	196	121	124	133
0.15	***	1974	2001	2351	582	604	679	303	315	342	196	203	216	142	146	153
	•••	1983	2051	2994	590	628	821	308	326	391	199	208	232	144	149	154
0.2	***	2400	2441	3033	685	718	841	349	368	411	222	233	252	158	165	175
	•••	2414	2517	3619	698	756	964	357	384	449	227	241	261	161	169	170
0.25	***	2699	2756	3629	759	804	980	383	408	467	241	255	281	169	178	191
	•••	2718	2864	4140	776	855	1081	394	430	495	247	266	284	173	184	183
0.3	***	2875	2953	4127	804	863	1092	405	436	512	253	271	303	177	187	204
	•••	2901	3097	4556	827	928	1172	418	464	530	261	284	300	182	194	191
0.35	***	2939	3043	4519	826	897	1177	417	454	544	260	280	318	180	192	211
	•••	2974	3228	4869	854	977	1237	432	486	553	268	295	310	186	200	195
0.4	***	2908	3041	4802	827	911	1234	419	461	564	261	283	325	180	193	213
	•••	2953	3271	5077	860	1002	1276	437	498	565	271	300	313	186	202	195
0.45	***	2796	2964	4973	811	906	1264	414	460	570	257	281	326	177	190	212
	•••	2854	3241	5181	848	1007	1289	433	499	565	268	299	310	183	199	191
0.5	***	2623	2830	5030	781	885	1264	401	450	564	249	274	319	171	184	205
	•••	2695	3150	5181	823	993	1276	422	491	553	260	292	300	177	193	183
0.55	***	2408	2654	4975	741	850	1236	383	432	546	237	261	305	161	174	194
	•••	2495	3011	5077	785	961	1237	404	473	530	248	279	284	167	183	170
0.6	***	2168	2450	4806	692	803	1180	359	407	515	221	244	284	149	160	178
	•••	2271	2833	4869	737	914	1172	379	446	495	231	260	261	154	168	154
0.65	***	1920	2228	4524	635	744	1095	329	374	471	200	221	256	133	143	158
	•••	2036	2623	4556	679	850	1081	348	410	449	210	236	232	138	150	133
0.7	***	1675	1994	4130	570	672	983	294	333	415	176	193	221	114	122	134
	•••	1797	2383	4140	612	769	964	311	365	391	184	205	196	118	127	108
0.75	***	1436	1747	3626	498	588	843	252	285	347	147	160	179	.	.	.
	•••	1557	2112	3619	536	671	821	266	309	322	153	168	153	.	.	.
0.8	***	1200	1484	3011	416	489	676	202	226	266	.	.	.	.	.	.
	•••	1312	1806	2994	447	553	651	213	243	241	.	.	.	.	.	.
0.85	***	959	1195	2288	319	370	482	.	.	.	.	.	.	.	.	.
	•••	1053	1451	2265	341	411	456	.	.	.	.	.	.	.	.	.
0.9	***	689	853	1458	.	.	.	.	.	.	.	.	.	.	.	.
	•••	756	1019	1432	.	.	.	.	.	.	.	.	.	.	.	.

TABLE 4: ALPHA= 0.01 POWER= 0.9   EXPECTED ACCRUAL THRU MINIMUM FOLLOW-UP= 300

		DEL=.05			DEL=.10			DEL=.15			DEL=.20			DEL=.25		
FACT=		1.0 .75	.50 .25	.00 BIN	1.0 .75	.50 .25	.00 BIN	1.0 .75	.50 .25	.00 BIN	1.0 75	.50 .25	.00 BIN	1.0 .75	.50 .25	.00 BIN
PCONT=***					REQUIRED NUMBER OF PATIENTS											
0.05	***	782	790	832	281	286	295	166	168	172	117	119	121	91	92	94
	•••	785	800	1432	283	289	456	167	170	241	118	120	153	92	93	108
0.1	***	1427	1444	1604	448	460	495	243	250	262	162	166	172	120	123	126
	•••	1433	1474	2265	453	473	651	246	255	322	164	169	196	121	124	133
0.15	***	1977	2006	2351	584	607	679	304	317	342	197	204	215	142	147	153
	•••	1987	2059	2994	593	630	821	310	327	391	200	209	232	144	149	154
0.2	***	2404	2448	3033	689	723	841	352	370	411	223	234	252	159	165	175
	•••	2419	2530	3619	702	760	964	360	386	449	228	241	261	162	169	170
0.25	***	2704	2767	3629	764	811	980	386	411	467	243	257	281	171	179	191
	•••	2725	2882	4140	782	862	1081	397	433	495	249	267	284	175	184	183
0.3	***	2882	2967	4127	811	871	1092	409	440	512	256	272	303	178	188	203
	•••	2910	3121	4556	834	936	1172	422	467	530	263	285	300	183	195	191
0.35	***	2949	3061	4519	834	907	1177	421	458	544	262	282	318	182	193	211
	•••	2987	3258	4869	862	986	1237	437	490	553	271	297	310	187	201	195
0.4	***	2920	3064	4802	836	922	1234	424	466	564	264	286	325	182	194	214
	•••	2969	3307	5077	870	1013	1276	442	502	565	273	302	313	188	203	195
0.45	***	2812	2993	4973	822	919	1263	419	465	570	260	284	326	179	192	212
	•••	2874	3283	5181	860	1019	1289	439	504	565	271	301	310	185	200	191
0.5	***	2643	2864	5031	793	899	1264	407	455	564	252	276	319	172	185	205
	•••	2721	3197	5181	836	1006	1276	428	496	553	263	294	300	178	194	183
0.55	***	2432	2693	4975	754	864	1236	389	438	546	240	264	305	163	175	194
	•••	2526	3062	5077	799	975	1237	410	478	530	251	281	284	169	183	170
0.6	***	2197	2493	4806	705	817	1180	365	412	515	224	246	284	150	162	178
	•••	2306	2887	4869	751	927	1172	385	451	495	234	262	261	155	169	154
0.65	***	1953	2274	4524	648	757	1095	335	379	471	203	223	256	135	144	158
	•••	2074	2677	4556	693	862	1081	354	414	449	212	237	232	139	151	133
0.7	***	1710	2039	4130	583	685	983	299	338	415	178	195	221	115	123	133
	•••	1837	2435	4140	625	780	964	316	368	391	186	206	196	119	128	108
0.75	***	1471	1791	3626	509	599	843	256	288	347	148	161	179	.	.	.
	•••	1596	2160	3619	547	681	821	270	312	322	154	169	153	.	.	.
0.8	***	1233	1523	3011	425	498	676	205	228	266	.	.	.	.	.	.
	•••	1348	1847	2994	456	560	651	216	244	241	.	.	.	.	.	.
0.85	***	986	1226	2288	325	375	482	.	.	.	.	.	.	.	.	.
	•••	1082	1483	2265	347	416	456	.	.	.	.	.	.	.	.	.
0.9	***	709	874	1458	.	.	.	.	.	.	.	.	.	.	.	.
	•••	777	1039	1432	.	.	.	.	.	.	.	.	.	.	.	.

TABLE 4: ALPHA= 0.01 POWER= 0.9   EXPECTED ACCRUAL THRU MINIMUM FOLLOW-UP= 325

		DEL=.05			DEL=.10			DEL=.15			DEL=.20			DEL=.25		
FACT=		1.0 .75	.50 .25	.00 BIN	1.0 .75	.50 .25	.00 BIN	1.0 .75	.50 .25	.00 BIN	1.0 75	.50 .25	.00 BIN	1.0 .75	.50 .25	.00 BIN
PCONT=***					REQUIRED NUMBER OF PATIENTS											
0.05	***	783	791	832	281	286	296	166	168	172	118	119	121	92	93	94
	•••	785	802	1432	283	290	456	167	170	241	118	120	153	92	93	108
0.1	***	1429	1447	1604	449	462	495	244	250	263	162	166	172	121	123	126
	•••	1435	1478	2265	454	474	651	247	255	322	164	169	196	122	125	133
0.15	***	1979	2011	2351	587	610	679	306	318	342	198	205	216	143	147	153
	•••	1990	2067	2994	596	633	821	311	328	391	201	209	232	145	150	154
0.2	***	2408	2456	3033	693	727	841	354	372	411	225	235	252	160	166	175
	•••	2424	2542	3619	706	764	964	362	388	449	229	242	261	162	170	170
0.25	***	2709	2777	3629	769	816	980	389	414	467	245	258	281	172	180	191
	•••	2732	2899	4140	787	867	1081	400	435	495	251	268	284	175	185	183
0.3	***	2889	2981	4127	817	879	1092	413	444	512	258	274	303	180	189	203
	•••	2920	3143	4556	841	944	1172	426	470	530	265	286	300	184	196	191
0.35	***	2959	3079	4519	842	916	1177	426	462	544	265	284	318	184	194	211
	•••	2999	3286	4869	871	995	1237	441	493	553	273	298	310	188	202	195
0.4	***	2932	3087	4802	845	933	1235	429	471	564	266	288	325	184	196	214
	•••	2985	3342	5077	880	1023	1276	447	506	565	276	303	313	189	203	195
0.45	***	2828	3021	4973	832	930	1263	425	470	570	263	286	326	181	193	212
	•••	2895	3322	5181	872	1030	1289	444	508	565	273	302	310	186	201	191
0.5	***	2663	2897	5031	805	911	1264	413	460	564	255	279	319	174	186	205
	•••	2746	3242	5181	848	1018	1276	433	499	553	266	295	300	180	195	183
0.55	***	2456	2731	4975	766	877	1236	395	443	546	243	266	305	164	177	194
	•••	2556	3110	5077	811	987	1237	416	482	530	253	282	284	170	184	170
0.6	***	2226	2534	4806	718	830	1180	371	417	515	227	248	284	152	163	178
	•••	2340	2937	4869	764	938	1172	391	454	495	236	263	261	157	170	154
0.65	***	1985	2317	4524	660	770	1095	340	384	471	206	225	256	136	145	158
	•••	2111	2727	4556	706	873	1081	359	418	449	215	238	232	140	151	133
0.7	***	1744	2082	4130	595	697	983	304	342	415	181	196	221	117	124	134
	•••	1875	2484	4140	637	790	964	320	371	391	188	207	196	120	128	108
0.75	***	1505	1831	3626	520	609	843	260	291	347	150	162	179	.	.	.
	•••	1633	2205	3619	558	689	821	274	314	322	156	170	153	.	.	.
0.8	***	1264	1559	3011	434	506	676	208	230	266	.	.	.	.	.	.
	•••	1381	1886	2994	464	567	651	218	246	241	.	.	.	.	.	.
0.85	***	1013	1255	2288	331	381	482	.	.	.	.	.	.	.	.	.
	•••	1110	1512	2265	353	420	456	.	.	.	.	.	.	.	.	.
0.9	***	728	894	1458	.	.	.	.	.	.	.	.	.	.	.	.
	•••	796	1057	1432	.	.	.	.	.	.	.	.	.	.	.	.

TABLE 4: ALPHA= 0.01 POWER= 0.9   EXPECTED ACCRUAL THRU MINIMUM FOLLOW-UP= 350

		DEL=.05			DEL=.10			DEL=.15			DEL=.20			DEL=.25		
FACT=		1.0 .75	.50 .25	.00 BIN	1.0 .75	.50 .25	.00 BIN	1.0 .75	.50 .25	.00 BIN	1.0 75	.50 .25	.00 BIN	1.0 .75	.50 .25	.00 BIN
PCONT=•••					REQUIRED	NUMBER	OF PATIENTS									
0.05	•••	783	792	832	282	286	296	166	169	172	118	119	121	92	93	94
	•••	786	803	1432	284	290	456	167	170	241	118	120	153	92	93	108
0.1	•••	1430	1450	1604	451	463	495	244	251	263	163	167	172	121	123	126
	•••	1437	1482	2265	456	475	651	247	256	322	165	169	196	122	125	133
0.15	•••	1982	2016	2351	589	612	678	307	319	342	199	205	216	143	147	153
	•••	1993	2074	2994	598	635	821	313	329	391	202	210	232	145	150	154
0.2	•••	2411	2463	3033	696	731	841	356	374	411	226	236	252	160	166	174
	•••	2429	2554	3619	710	768	964	364	389	449	230	243	261	163	170	170
0.25	•••	2714	2787	3629	773	822	980	392	416	467	246	259	281	173	180	191
	•••	2739	2915	4140	792	872	1081	402	436	495	252	269	284	176	185	183
0.3	•••	2896	2995	4127	823	886	1092	416	447	512	260	276	303	181	190	204
	•••	2929	3165	4556	848	950	1172	429	472	530	267	287	300	185	196	191
0.35	•••	2968	3097	4519	849	925	1177	430	466	544	267	286	318	185	195	211
	•••	3012	3314	4869	879	1003	1237	445	496	553	275	299	310	190	202	195
0.4	•••	2945	3109	4802	854	943	1235	434	475	564	269	290	325	185	197	214
	•••	3001	3374	5077	890	1032	1276	451	509	565	278	305	313	190	204	195
0.45	•••	2843	3048	4973	842	941	1263	430	474	570	266	288	326	182	194	212
	•••	2915	3360	5181	882	1040	1289	449	511	565	276	304	310	188	202	191
0.5	•••	2682	2929	5031	816	923	1264	418	465	565	258	281	319	176	188	205
	•••	2770	3284	5181	859	1028	1276	439	503	553	268	297	300	181	195	183
0.55	•••	2480	2767	4975	778	890	1236	400	447	546	246	268	305	166	177	194
	•••	2585	3156	5077	824	998	1237	421	485	530	256	284	284	171	185	170
0.6	•••	2253	2574	4806	729	842	1180	376	422	515	229	250	284	153	164	178
	•••	2373	2984	4869	776	949	1172	396	458	495	239	265	261	158	170	154
0.65	•••	2016	2357	4524	672	782	1095	345	388	471	208	227	256	137	146	158
	•••	2147	2774	4556	717	883	1081	364	421	449	217	239	232	141	152	133
0.7	•••	1776	2123	4130	606	707	983	308	345	415	183	198	221	118	124	134
	•••	1911	2529	4140	648	799	964	325	373	391	190	208	196	121	129	108
0.75	•••	1537	1869	3626	530	619	843	264	294	347	152	163	179	.	.	.
	•••	1668	2246	3619	567	696	821	277	316	322	157	170	153	.	.	.
0.8	•••	1294	1593	3011	442	513	676	211	232	266	.	.	.	.	.	.
	•••	1413	1921	2994	472	572	651	221	247	241	.	.	.	.	.	.
0.85	•••	1037	1282	2288	337	385	482	.	.	.	.	.	.	.	.	.
	•••	1136	1539	2265	358	423	456	.	.	.	.	.	.	.	.	.
0.9	•••	745	911	1458	.	.	.	.	.	.	.	.	.	.	.	.
	•••	813	1073	1432	.	.	.	.	.	.	.	.	.	.	.	.

TABLE 4: ALPHA= 0.01 POWER= 0.9   EXPECTED ACCRUAL THRU MINIMUM FOLLOW-UP= 375

		DEL=.05			DEL=.10			DEL=.15			DEL=.20			DEL=.25		
FACT=		1.0 .75	.50 .25	.00 BIN	1.0 .75	.50 .25	.00 BIN	1.0 .75	.50 .25	.00 BIN	1.0 75	.50 .25	.00 BIN	1.0 .75	.50 .25	.00 BIN
PCONT=•••					REQUIRED NUMBER OF PATIENTS											
0.05	•••	784	793	831	282	287	296	167	169	172	118	119	121	92	93	94
	•••	787	804	1432	284	290	456	167	170	241	118	120	153	92	93	108
0.1	•••	1432	1453	1604	452	465	495	245	252	263	163	167	172	121	123	126
	•••	1439	1486	2265	457	476	651	248	256	322	165	169	196	122	125	133
0.15	•••	1984	2020	2351	591	615	679	309	320	342	199	206	215	144	148	153
	•••	1997	2081	2994	600	637	821	314	329	391	202	210	232	146	150	154
0.2	•••	2415	2470	3033	699	735	841	358	376	410	227	237	252	161	167	175
	•••	2434	2565	3619	714	771	964	366	390	449	231	243	261	164	170	170
0.25	•••	2720	2797	3629	778	827	980	394	419	467	248	260	281	174	181	191
	•••	2746	2931	4140	797	877	1081	405	438	495	253	269	284	177	186	183
0.3	•••	2903	3008	4127	829	892	1092	419	450	512	261	277	303	182	191	204
	•••	2939	3185	4556	854	956	1172	432	474	530	268	288	300	186	197	191
0.35	•••	2977	3115	4519	856	933	1177	433	469	544	269	287	318	186	196	211
	•••	3024	3340	4869	887	1010	1237	449	498	553	277	300	310	191	203	195
0.4	•••	2957	3131	4802	862	952	1235	438	479	564	271	291	325	186	197	214
	•••	3017	3405	5077	898	1041	1276	455	512	565	280	306	313	192	205	195
0.45	•••	2859	3074	4973	851	951	1263	434	478	570	268	290	326	183	195	212
	•••	2935	3396	5181	892	1049	1289	453	514	565	278	305	310	189	202	191
0.5	•••	2702	2960	5030	826	934	1264	423	469	565	261	283	319	177	189	205
	•••	2794	3323	5181	870	1038	1276	443	506	553	271	298	300	182	196	183
0.55	•••	2503	2801	4975	788	901	1236	405	452	546	248	270	305	167	178	194
	•••	2613	3198	5077	835	1007	1237	425	488	530	258	285	284	172	185	170
0.6	•••	2280	2611	4806	740	853	1180	381	426	515	232	252	284	154	164	178
	•••	2405	3028	4869	787	959	1172	400	461	495	241	266	261	159	171	154
0.65	•••	2046	2396	4524	683	792	1095	350	392	471	210	228	256	138	147	158
	•••	2180	2818	4556	728	892	1081	368	423	449	219	240	232	142	152	133
0.7	•••	1807	2161	4130	616	717	983	312	349	415	184	199	221	118	125	133
	•••	1945	2571	4140	658	808	964	328	376	391	191	209	196	122	129	108
0.75	•••	1567	1905	3626	539	627	843	267	297	347	153	164	179	.	.	.
	•••	1701	2285	3619	576	703	821	280	317	322	158	171	153	.	.	.
0.8	•••	1322	1625	3011	449	520	676	213	234	266	.	.	.	.	.	.
	•••	1443	1954	2994	479	577	651	223	248	241	.	.	.	.	.	.
0.85	•••	1060	1308	2288	342	390	482	.	.	.	.	.	.	.	.	.
	•••	1160	1564	2265	363	426	456	.	.	.	.	.	.	.	.	.
0.9	•••	761	928	1458	.	.	.	.	.	.	.	.	.	.	.	.
	•••	830	1088	1432	.	.	.	.	.	.	.	.	.	.	.	.

TABLE 4: ALPHA= 0.01 POWER= 0.9   EXPECTED ACCRUAL THRU MINIMUM FOLLOW-UP= 400

		DEL=.05			DEL=.10			DEL=.15			DEL=.20			DEL=.25		
FACT=		1.0 .75	.50 .25	.00 BIN	1.0 .75	.50 .25	.00 BIN	1.0 .75	.50 .25	.00 BIN	1.0 75	.50 .25	.00 BIN	1.0 .75	.50 .25	.00 BIN
PCONT=•••		REQUIRED NUMBER OF PATIENTS														
0.05	***	785	794	832	283	287	296	167	169	172	118	119	121	92	93	94
	•••	788	805	1432	285	291	456	168	171	241	119	120	153	92	93	108
0.1	***	1433	1455	1604	453	466	495	246	252	263	164	167	172	121	124	126
	•••	1441	1489	2265	458	477	651	249	257	322	165	170	196	122	125	133
0.15	***	1987	2025	2351	593	617	679	310	321	342	200	206	216	144	148	153
	•••	2000	2088	2994	603	639	821	315	330	391	203	210	232	146	150	154
0.2	***	2419	2477	3033	702	739	841	360	378	411	228	237	252	162	167	175
	•••	2439	2575	3619	717	774	964	367	391	449	232	244	261	164	171	170
0.25	***	2725	2807	3629	782	832	980	397	421	467	249	261	281	175	182	191
	•••	2753	2946	4140	802	881	1081	407	440	495	255	270	284	178	186	183
0.3	***	2910	3022	4127	834	898	1092	422	452	512	263	278	303	183	191	204
	•••	2948	3206	4556	860	962	1172	435	476	530	270	289	300	187	197	191
0.35	***	2987	3132	4519	863	940	1177	437	472	544	271	289	318	187	197	211
	•••	3037	3364	4869	894	1017	1237	452	501	553	279	301	310	192	203	195
0.4	***	2969	3153	4802	870	960	1235	442	482	564	273	293	325	188	198	214
	•••	3033	3435	5077	907	1048	1276	459	514	565	282	307	313	193	205	195
0.45	***	2874	3100	4973	860	961	1263	439	482	570	271	292	326	185	196	212
	•••	2954	3430	5181	901	1058	1289	458	517	565	280	306	310	190	203	191
0.5	***	2721	2990	5031	836	944	1264	428	473	564	263	284	319	178	189	205
	•••	2818	3361	5181	880	1047	1276	448	509	553	273	299	300	183	196	183
0.55	***	2526	2835	4975	799	911	1236	410	456	546	251	272	305	169	179	194
	•••	2640	3239	5077	845	1017	1237	430	491	530	260	286	284	173	186	170
0.6	***	2307	2647	4806	751	864	1180	385	430	515	234	253	284	156	165	178
	•••	2435	3070	5077	798	968	1172	405	463	495	243	267	261	160	171	154
0.65	***	2074	2433	4524	693	802	1095	354	395	471	212	230	256	139	147	158
	•••	2213	2860	4556	739	901	1081	372	426	449	220	241	232	143	152	133
0.7	***	1837	2197	4130	625	726	983	316	352	415	186	200	221	119	125	134
	•••	1978	2611	4140	668	815	964	332	378	391	193	210	196	122	129	108
0.75	***	1596	1940	3626	547	635	843	270	299	347	154	165	179	.	.	.
	•••	1732	2321	3619	584	709	821	283	319	322	159	171	153	.	.	.
0.8	***	1348	1655	3011	456	526	676	216	236	266	.	.	.	.	.	.
	•••	1471	1985	2994	486	582	651	225	249	241	.	.	.	.	.	.
0.85	***	1082	1332	2288	347	394	482	.	.	.	.	.	.	.	.	.
	•••	1184	1588	2265	368	429	456	.	.	.	.	.	.	.	.	.
0.9	***	777	944	1458	.	.	.	.	.	.	.	.	.	.	.	.
	•••	846	1102	1432	.	.	.	.	.	.	.	.	.	.	.	.

TABLE 4: ALPHA= 0.01 POWER= 0.9   EXPECTED ACCRUAL THRU MINIMUM FOLLOW-UP= 425

	DEL=.05			DEL=.10			DEL=.15			DEL=.20			DEL=.25		
FACT=	1.0 .75	.50 .25	.00 BIN	1.0 .75	.50 .25	.00 BIN	1.0 .75	.50 .25	.00 BIN	1.0 75	.50 .25	.00 BIN	1.0 .75	.50 .25	.00 BIN
PCONT=***						REQUIRED NUMBER OF PATIENTS									
0.05 ***	785	795	832	283	288	295	167	169	172	118	120	121	92	93	94
•••	789	806	1432	285	291	456	168	171	241	119	120	153	93	93	108
0.1 ***	1434	1458	1604	454	467	495	247	253	263	164	167	172	121	124	126
•••	1443	1493	2265	459	478	651	249	257	322	166	170	196	122	125	133
0.15 ***	1989	2029	2351	595	619	679	311	322	342	201	207	215	145	148	153
•••	2003	2095	2994	605	641	821	316	331	391	203	211	232	146	150	154
0.2 ***	2423	2484	3033	705	742	841	361	379	410	229	238	252	162	167	175
•••	2444	2586	3619	720	777	964	369	392	449	233	244	261	165	171	170
0.25 ***	2730	2817	3629	786	836	980	399	423	467	250	262	281	175	182	191
•••	2760	2960	4140	806	885	1081	409	441	495	256	271	284	178	186	183
0.3 ***	2917	3035	4127	839	904	1092	425	455	512	265	280	303	184	192	204
•••	2958	3224	4556	866	967	1172	438	478	530	271	290	300	188	197	191
0.35 ***	2996	3149	4519	869	947	1177	440	475	544	273	290	318	188	198	211
•••	3049	3388	4869	901	1023	1237	455	503	553	281	302	310	193	204	195
0.4 ***	2981	3174	4802	878	969	1234	446	485	563	275	295	325	189	199	214
•••	3049	3463	5077	915	1055	1276	463	516	565	284	308	313	193	206	195
0.45 ***	2890	3125	4973	869	970	1263	443	486	570	273	293	326	186	197	212
•••	2974	3462	5181	910	1065	1289	461	519	565	282	307	310	191	203	191
0.5 ***	2739	3019	5031	845	953	1264	432	477	564	265	286	319	179	190	205
•••	2841	3397	5181	890	1055	1276	452	511	553	274	300	300	184	197	183
0.55 ***	2548	2867	4975	808	921	1236	414	459	546	253	273	305	170	180	194
•••	2667	3277	5077	855	1025	1237	434	494	530	262	287	284	174	186	170
0.6 ***	2332	2681	4806	761	873	1180	390	433	515	236	255	284	156	166	178
•••	2465	3109	4869	808	976	1172	409	466	495	244	268	261	161	171	154
0.65 ***	2102	2468	4524	703	811	1096	358	398	471	214	231	256	140	148	158
•••	2244	2899	4556	748	908	1081	376	428	449	222	242	232	144	153	133
0.7 ***	1866	2232	4130	634	734	983	319	354	415	187	201	221	120	126	133
•••	2009	2648	4140	677	822	964	335	379	391	194	210	196	122	129	108
0.75 ***	1624	1972	3626	555	642	843	273	301	347	155	165	179	.	.	.
•••	1762	2355	3619	592	715	821	286	320	322	160	172	153	.	.	.
0.8 ***	1373	1684	3011	462	531	676	218	237	266	.	.	.	.	.	.
•••	1498	2013	2994	492	586	651	226	250	241	.	.	.	.	.	.
0.85 ***	1104	1354	2288	351	397	482	.	.	.	.	.	.	.	.	.
•••	1206	1609	2265	372	431	456	.	.	.	.	.	.	.	.	.
0.9 ***	791	958	1458	.	.	.	.	.	.	.	.	.	.	.	.
•••	860	1115	1432	.	.	.	.	.	.	.	.	.	.	.	.

TABLE 4: ALPHA= 0.01   POWER= 0.9   EXPECTED ACCRUAL THRU MINIMUM FOLLOW-UP= 450

		DEL=.05			DEL=.10			DEL=.15			DEL=.20			DEL=.25	
FACT=	1.0 .75	.50 .25	.00 BIN	1.0 .75	.50 .25	.00 BIN	1.0 .75	.50 .25	.00 BIN	1.0 75	.50 .25	.00 BIN	1.0 .75	.50 .25	.00 BIN
PCONT=***		REQUIRED NUMBER OF PATIENTS														
0.05 ***	786	795	831	284	288	296	167	169	172	118	120	121	92	93	94
•••	789	807	1432	285	291	456	168	171	241	119	120	153	93	93	108
0.1 ***	1436	1460	1604	455	468	495	247	253	262	164	168	172	122	124	126
•••	1445	1496	2265	460	479	651	250	257	322	166	170	196	123	125	133
0.15 ***	1992	2034	2351	597	621	678	312	323	342	201	207	216	145	149	153
•••	2006	2101	2994	607	642	821	317	331	391	204	211	232	147	150	154
0.2 ***	2426	2491	3033	708	745	841	363	380	411	230	239	252	163	168	174
•••	2448	2595	3619	723	779	964	370	393	449	234	244	261	165	171	170
0.25 ***	2736	2827	3629	790	841	980	401	424	468	251	263	281	176	183	191
•••	2767	2975	4140	811	889	1081	411	442	495	257	271	284	179	186	183
0.3 ***	2925	3048	4127	844	910	1092	428	457	512	266	280	303	185	192	203
•••	2967	3243	4556	871	972	1172	440	479	530	273	290	300	188	198	191
0.35 ***	3006	3165	4519	875	954	1177	443	478	544	275	291	318	189	198	211
•••	3061	3411	4869	907	1029	1237	458	504	553	282	303	310	193	204	195
0.4 ***	2993	3194	4802	885	976	1234	449	488	564	277	296	325	190	200	213
•••	3064	3490	5077	922	1062	1276	466	518	565	285	309	313	194	206	195
0.45 ***	2905	3150	4973	877	978	1263	447	489	570	275	294	326	187	197	212
•••	2993	3493	5181	919	1073	1289	465	522	565	284	308	310	192	204	191
0.5 ***	2758	3047	5031	854	962	1264	436	480	564	267	287	319	181	191	205
•••	2864	3431	5181	899	1063	1276	455	514	553	276	301	300	185	197	183
0.55 ***	2570	2898	4975	818	930	1236	418	462	546	255	275	305	171	181	194
•••	2693	3313	5077	864	1033	1237	438	496	530	264	288	284	175	186	170
0.6 ***	2357	2714	4806	770	883	1180	393	436	515	237	256	284	158	166	178
•••	2493	3146	4869	817	984	1172	412	468	495	246	268	261	162	172	154
0.65 ***	2129	2502	4524	712	820	1095	362	401	471	216	232	256	141	148	158
•••	2274	2936	4556	757	915	1081	379	430	449	223	243	232	144	153	133
0.7 ***	1893	2265	4130	643	743	983	323	357	415	189	202	221	120	126	134
•••	2039	2683	4140	685	828	964	338	381	391	195	211	196	123	130	108
0.75 ***	1650	2003	3626	563	649	843	276	303	347	156	166	179	.	.	.
•••	1791	2387	3619	599	720	821	288	322	322	161	172	153	.	.	.
0.8 ***	1397	1710	3011	468	536	676	219	238	266	.	.	.	.	.	.
•••	1523	2040	2994	498	590	651	228	251	241	.	.	.	.	.	.
0.85 ***	1123	1376	2288	356	401	482	.	.	.	.	.	.	.	.	.
•••	1226	1630	2265	375	434	456	.	.	.	.	.	.	.	.	.
0.9 ***	805	972	1458	.	.	.	.	.	.	.	.	.	.	.	.
•••	874	1127	1432	.	.	.	.	.	.	.	.	.	.	.	.

481

TABLE 4: ALPHA= 0.01 POWER= 0.9   EXPECTED ACCRUAL THRU MINIMUM FOLLOW-UP= 475

	DEL=.05			DEL=.10			DEL=.15			DEL=.20			DEL=.25		
FACT=	1.0 .75	.50 .25	.00 BIN	1.0 .75	.50 .25	.00 BIN	1.0 .75	.50 .25	.00 BIN	1.0 75	.50 .25	.00 BIN	1.0 .75	.50 .25	.00 BIN

PCONT=••• REQUIRED NUMBER OF PATIENTS

PCONT															
0.05	786	796	832	284	288	296	167	169	172	118	120	121	92	93	94
	790	808	1432	286	291	456	168	171	241	119	120	153	93	93	108
0.1	1438	1463	1604	456	469	495	248	254	262	165	168	172	122	124	126
	1446	1499	2265	461	479	651	250	257	322	166	170	196	123	125	133
0.15	1994	2038	2351	599	623	679	313	324	342	202	207	216	145	149	153
	2009	2107	2994	609	644	821	318	332	391	204	211	232	147	150	154
0.2	2430	2498	3033	711	748	841	364	381	411	231	239	252	163	168	175
	2453	2605	3619	726	782	964	372	394	449	235	245	261	166	171	170
0.25	2740	2836	3629	794	845	980	403	426	467	252	264	281	176	183	191
	2773	2988	4140	814	892	1081	413	443	495	257	271	284	179	187	183
0.3	2932	3060	4127	849	915	1092	430	459	512	267	281	303	185	193	203
	2976	3261	4556	876	976	1172	443	481	530	274	291	300	189	198	191
0.35	3015	3182	4519	881	960	1177	446	480	544	276	292	318	190	199	211
	3073	3433	4869	913	1034	1237	461	506	553	283	303	310	194	204	195
0.4	3005	3214	4802	892	983	1234	452	491	564	279	297	325	191	200	213
	3079	3516	5077	929	1068	1276	469	520	565	287	309	313	195	206	195
0.45	2920	3173	4973	885	986	1264	450	492	570	276	296	326	188	198	212
	3011	3523	5181	927	1079	1289	469	523	565	285	309	310	192	204	191
0.5	2776	3074	5031	862	971	1264	440	483	564	269	289	319	181	191	205
	2886	3463	5181	907	1070	1276	459	516	553	278	302	300	186	198	183
0.55	2592	2928	4975	826	939	1236	422	466	546	257	276	305	172	181	194
	2718	3348	5077	873	1040	1237	441	498	530	265	289	284	176	187	170
0.6	2381	2746	4806	779	891	1180	397	439	514	239	257	284	158	167	178
	2521	3182	4869	826	990	1172	416	470	495	248	269	261	162	172	154
0.65	2155	2534	4524	720	828	1096	365	403	471	217	233	256	141	149	158
	2303	2971	4556	766	922	1081	382	431	449	224	243	232	145	153	133
0.7	1920	2296	4130	651	750	983	326	359	415	190	203	221	121	127	134
	2068	2716	4140	693	834	964	340	383	391	196	211	196	124	130	108
0.75	1676	2032	3626	569	655	843	278	305	347	157	167	179	.	.	.
	1818	2417	3619	606	725	821	290	323	322	162	172	153	.	.	.
0.8	1421	1736	3011	474	541	676	221	240	266	.	.	.	.	.	.
	1548	2065	2994	503	593	651	229	251	241	.	.	.	.	.	.
0.85	1142	1396	2288	359	403	482	.	.	.	.	.	.	.	.	.
	1246	1649	2265	379	436	456	.	.	.	.	.	.	.	.	.
0.9	818	984	1458	.	.	.	.	.	.	.	.	.	.	.	.
	887	1138	1432	.	.	.	.	.	.	.	.	.	.	.	.

TABLE 4: ALPHA= 0.01 POWER= 0.9   EXPECTED ACCRUAL THRU MINIMUM FOLLOW-UP= 500

PCONT		DEL=.05			DEL=.10			DEL=.15			DEL=.20			DEL=.25		
FACT=		1.0 .75	.50 .25	.00 BIN	1.0 .75	.50 .25	.00 BIN	1.0 .75	.50 .25	.00 BIN	1.0 75	.50 .25	.00 BIN	1.0 .75	.50 .25	.00 BIN
PCONT=***					REQUIRED NUMBER OF PATIENTS											
0.05	***	787	797	832	284	288	296	168	170	172	118	120	121	92	93	94
	***	791	808	1432	286	292	456	168	171	241	119	120	153	93	93	108
0.1	***	1439	1465	1604	457	470	495	248	254	263	165	168	172	122	124	126
	***	1448	1502	2265	463	480	651	251	258	322	167	170	196	123	125	133
0.15	***	1997	2043	2351	600	624	678	314	325	342	202	208	216	146	149	153
	***	2013	2113	2994	610	645	821	319	332	391	205	211	232	147	151	154
0.2	***	2434	2504	3033	713	751	841	366	383	411	231	240	252	164	168	175
	***	2458	2614	3619	729	784	964	373	394	449	235	245	261	166	171	170
0.25	***	2746	2846	3629	797	848	980	405	428	468	253	264	281	177	183	191
	***	2780	3001	4140	818	895	1081	415	444	495	258	272	284	180	187	183
0.3	***	2939	3073	4127	854	919	1092	432	461	512	268	282	303	186	193	203
	***	2985	3278	4556	881	980	1172	445	482	530	275	292	300	189	198	191
0.35	***	3024	3198	4519	887	966	1177	449	483	544	277	293	318	191	199	211
	***	3085	3454	4869	919	1039	1237	463	508	553	285	304	310	195	205	195
0.4	***	3017	3234	4802	898	990	1234	456	493	563	280	298	325	192	201	213
	***	3094	3541	5077	936	1073	1276	472	522	565	288	310	313	196	207	195
0.45	***	2935	3197	4973	892	993	1263	453	494	570	278	297	326	189	198	212
	***	3030	3551	5181	934	1086	1289	472	525	565	287	310	310	193	204	191
0.5	***	2794	3100	5031	870	978	1264	443	486	564	271	290	319	182	192	205
	***	2908	3494	5181	915	1076	1276	462	518	553	279	303	300	187	198	183
0.55	***	2613	2957	4975	835	947	1236	426	468	546	258	277	305	173	182	194
	***	2743	3380	5077	882	1047	1237	444	500	530	267	289	284	177	187	170
0.6	***	2405	2776	4806	787	899	1180	400	442	515	241	258	284	159	168	178
	***	2548	3215	4869	834	997	1172	419	472	495	249	270	261	163	173	154
0.65	***	2180	2565	4524	728	836	1095	368	406	471	218	234	256	142	149	158
	***	2330	3004	4556	774	928	1081	385	433	449	226	244	232	145	153	133
0.7	***	1945	2326	4130	658	757	983	328	361	415	191	204	221	122	127	133
	***	2096	2748	4140	700	839	964	343	384	391	197	212	196	124	130	108
0.75	***	1701	2060	3626	576	661	843	280	306	347	158	167	179	.	.	.
	***	1844	2445	3619	613	729	821	292	324	322	162	173	153	.	.	.
0.8	***	1443	1760	3011	479	545	676	223	241	266	.	.	.	.	.	.
	***	1571	2089	2994	508	597	651	231	252	241	.	.	.	.	.	.
0.85	***	1160	1415	2288	363	406	482	.	.	.	.	.	.	.	.	.
	***	1265	1666	2265	382	438	456	.	.	.	.	.	.	.	.	.
0.9	***	830	997	1458	.	.	.	.	.	.	.	.	.	.	.	.
	***	900	1148	1432	.	.	.	.	.	.	.	.	.	.	.	.

483

TABLE 4: ALPHA= 0.01 POWER= 0.9   EXPECTED ACCRUAL THRU MINIMUM FOLLOW-UP= 550

	DEL=.02			DEL=.05			DEL=.10			DEL=.15			DEL=.20		
FACT=	1.0 .75	.50 .25	.00 BIN	1.0 .75	.50 .25	.00 BIN	1.0 .75	.50 .25	.00 BIN	1.0 75	.50 .25	.00 BIN	1.0 .75	.50 .25	.00 BIN

PCONT=***   REQUIRED NUMBER OF PATIENTS

PCONT	1.0	.50	.00	1.0	.50	.00	1.0	.50	.00	1.0	.50	.00	1.0	.50	.00
0.05	3702	3713	3921	788	799	831	285	289	296	168	170	172	119	120	121
	3706	3736	7329	792	810	1432	287	292	456	169	171	241	119	121	153
0.1	7701	7728	8615	1442	1470	1604	459	471	495	249	255	263	166	168	172
	7710	7782	12731	1452	1507	2265	464	481	651	252	258	322	167	170	196
0.15	11247	11295	13324	2001	2051	2351	604	628	679	316	326	342	203	208	215
	11263	11390	17482	2019	2123	2994	614	648	821	320	333	391	206	212	232
0.2	14068	14142	17703	2441	2517	3033	718	756	841	368	384	411	233	241	252
	14093	14289	21583	2468	2631	3619	734	788	964	375	396	449	236	246	261
0.25	16089	16195	21585	2756	2864	3629	804	855	980	408	430	467	255	266	281
	16124	16406	25032	2794	3025	4140	825	901	1081	418	446	495	260	272	284
0.3	17321	17466	24877	2953	3097	4127	863	928	1092	437	464	512	270	284	303
	17369	17757	27831	3004	3309	4556	890	987	1172	448	484	530	277	292	300
0.35	17820	18013	27524	3043	3228	4519	897	976	1177	454	486	544	280	295	318
	17884	18400	29979	3109	3493	4869	930	1048	1237	468	510	553	287	305	310
0.4	17665	17918	29494	3041	3271	4802	911	1002	1234	461	498	564	283	300	325
	17749	18422	31475	3124	3588	5077	949	1084	1276	477	525	565	291	311	313
0.45	16952	17278	30769	2964	3241	4973	906	1007	1264	459	499	570	281	299	326
	17061	17926	32322	3066	3604	5181	948	1097	1289	477	529	565	289	311	310
0.5	15785	16200	31340	2830	3150	5030	885	993	1264	450	491	564	274	292	319
	15923	17023	32517	2950	3553	5181	930	1088	1276	468	521	553	282	304	300
0.55	14272	14802	31203	2654	3011	4975	850	961	1236	432	473	546	261	279	305
	14449	15831	32061	2790	3442	5077	897	1058	1237	450	503	530	269	291	284
0.6	12530	13204	30357	2450	2833	4806	803	914	1180	407	446	514	244	260	284
	12755	14457	30955	2599	3278	4869	850	1009	1172	424	475	495	251	271	261
0.65	10679	11529	28807	2228	2623	4524	743	850	1095	374	410	471	221	236	256
	10967	12996	29197	2384	3066	4556	789	939	1081	390	436	449	228	245	232
0.7	8847	9885	26555	1994	2383	4130	672	769	983	333	365	415	193	205	221
	9210	11507	26789	2148	2807	4140	714	849	964	347	386	391	199	212	196
0.75	7155	8339	23608	1747	2112	3626	588	671	843	285	309	347	159	168	179
	7586	10017	23730	1894	2498	3619	624	737	821	296	326	322	164	173	153
0.8	5676	6901	19973	1484	1806	3011	489	553	676	225	243	266	.	.	.
	6137	8513	20021	1614	2133	2994	517	602	651	233	253	241	.	.	.
0.85	4392	5528	15656	1194	1451	2288	369	411	482	.	.	.	.	.	.
	4828	6947	15660	1300	1699	2265	388	441	456	.	.	.	.	.	.
0.9	3207	4120	10667	853	1019	1458	.	.	.	.	.	.	.	.	.
	3563	5209	10648	923	1167	1432	.	.	.	.	.	.	.	.	.
0.95	1914	2443	5011	.	.	.	.	.	.	.	.	.	.	.	.
	2125	3022	4986	.	.	.	.	.	.	.	.	.	.	.	.

TABLE 4: ALPHA= 0.01 POWER= 0.9   EXPECTED ACCRUAL THRU MINIMUM FOLLOW-UP= 600

		DEL=.02			DEL=.05			DEL=.10			DEL=.15			DEL=.20		
FACT=		1.0 .75	.50 .25	.00 BIN	1.0 .75	.50 .25	.00 BIN	1.0 .75	.50 .25	.00 BIN	1.0 75	.50 .25	.00 BIN	1.0 .75	.50 .25	.00 BIN
PCONT=•••					REQUIRED NUMBER OF PATIENTS											
0.05	•••	3703	3715	3921	790	800	832	286	289	295	168	170	172	119	120	121
	•••	3707	3740	7329	794	811	1432	287	292	456	169	171	241	119	121	153
0.1	•••	7703	7733	8615	1444	1474	1604	460	473	495	250	255	262	166	169	172
	•••	7713	7792	12731	1455	1511	2265	466	482	651	252	258	322	167	170	196
0.15	•••	11251	11303	13324	2006	2059	2351	607	630	679	317	327	342	204	209	215
	•••	11269	11408	17482	2025	2133	2994	617	649	821	321	334	391	206	212	232
0.2	•••	14075	14155	17703	2448	2530	3033	723	760	841	370	386	410	234	241	252
	•••	14101	14316	21583	2477	2647	3619	739	791	964	377	397	449	237	246	261
0.25	•••	16098	16214	21586	2767	2882	3629	811	862	980	411	433	467	257	267	281
	•••	16137	16445	25032	2807	3047	4140	832	905	1081	421	447	495	261	273	284
0.3	•••	17334	17492	24877	2967	3121	4127	871	936	1092	440	467	512	272	285	303
	•••	17387	17809	27831	3022	3339	4556	898	993	1172	452	486	530	278	293	300
0.35	•••	17837	18048	27524	3061	3258	4519	907	986	1177	458	490	544	282	297	318
	•••	17908	18470	29979	3132	3530	4869	940	1055	1237	472	513	553	289	306	310
0.4	•••	17688	17963	29494	3064	3307	4802	922	1013	1234	466	502	563	286	302	325
	•••	17780	18514	31475	3152	3631	5077	960	1093	1276	482	528	565	293	312	313
0.45	•••	16982	17337	30769	2993	3283	4973	919	1019	1263	465	503	570	284	301	326
	•••	17100	18043	32322	3100	3653	5181	961	1106	1289	482	532	565	292	312	310
0.5	•••	15823	16276	31340	2864	3197	5030	899	1006	1264	455	496	564	276	294	319
	•••	15974	17169	32517	2990	3606	5181	944	1098	1276	473	524	553	284	305	300
0.55	•••	14320	14898	31203	2693	3062	4975	864	975	1236	437	478	546	263	281	305
	•••	14513	16008	32061	2835	3498	5077	911	1069	1237	455	506	530	271	292	284
0.6	•••	12591	13324	30358	2493	2887	4806	817	926	1180	412	451	515	246	262	284
	•••	12837	14665	30955	2647	3335	4869	864	1019	1172	430	478	495	253	272	261
0.65	•••	10758	11676	28807	2273	2677	4524	757	862	1095	379	414	471	223	237	256
	•••	11071	13228	29197	2433	3122	4556	802	948	1081	395	439	449	230	245	232
0.7	•••	8947	10056	26555	2039	2435	4130	685	780	983	338	368	415	195	206	221
	•••	9338	11752	26789	2197	2860	4140	726	857	964	352	388	391	200	213	196
0.75	•••	7277	8522	23608	1790	2160	3626	599	680	843	288	311	347	161	169	179
	•••	7733	10261	23730	1939	2545	3619	635	744	821	299	327	322	164	174	153
0.8	•••	5808	7082	19973	1523	1847	3011	498	560	676	228	244	266	.	.	.
	•••	6289	8741	20021	1655	2173	2994	526	607	651	235	254	241	.	.	.
0.85	•••	4519	5691	15656	1226	1483	2288	375	416	482	.	.	.	.	.	.
	•••	4970	7143	15660	1332	1728	2265	394	444	456	.	.	.	.	.	.
0.9	•••	3311	4248	10666	874	1039	1457	.	.	.	.	.	.	.	.	.
	•••	3677	5356	10648	944	1183	1432	.	.	.	.	.	.	.	.	.
0.95	•••	1976	2513	5011	.	.	.	.	.	.	.	.	.	.	.	.
	•••	2191	3095	4986	.	.	.	.	.	.	.	.	.	.	.	.

TABLE 4: ALPHA= 0.01 POWER= 0.9   EXPECTED ACCRUAL THRU MINIMUM FOLLOW-UP= 650

REQUIRED NUMBER OF PATIENTS

PCONT= ***	DEL=.02			DEL=.05			DEL=.10			DEL=.15			DEL=.20		
FACT=	1.0 .75	.50 .25	.00 BIN	1.0 .75	.50 .25	.00 BIN	1.0 .75	.50 .25	.00 BIN	1.0 75	.50 .25	.00 BIN	1.0 .75	.50 .25	.00 BIN
0.05	3704	3717	3921	790	802	831	286	290	296	168	170	172	119	120	121
	3708	3744	7329	795	812	1432	288	292	456	169	171	241	120	120	153
0.1	7706	7738	8615	1447	1478	1604	462	474	495	250	255	263	166	169	172
	7717	7802	12731	1459	1516	2265	467	483	651	253	259	322	168	171	196
0.15	11256	11312	13324	2011	2067	2351	610	633	679	318	328	342	205	210	216
	11274	11425	17482	2031	2142	2994	619	652	821	322	334	391	207	212	232
0.2	14081	14169	17703	2456	2542	3033	727	764	841	372	387	411	235	242	252
	14111	14343	21583	2486	2662	3619	743	794	964	379	398	449	238	246	261
0.25	16108	16233	21586	2777	2899	3629	816	867	980	414	435	467	258	268	281
	16149	16483	25032	2820	3069	4140	838	910	1081	423	449	495	262	274	284
0.3	17347	17519	24877	2981	3143	4127	879	944	1092	444	470	512	274	286	303
	17404	17862	27831	3039	3366	4556	906	999	1172	455	488	530	280	294	300
0.35	17855	18083	27524	3079	3286	4519	916	995	1177	462	493	544	284	298	318
	17931	18540	29979	3154	3564	4869	950	1062	1237	476	515	553	291	307	310
0.4	17711	18010	29494	3087	3342	4802	933	1023	1234	471	506	564	288	303	325
	17811	18605	31475	3180	3671	5077	971	1100	1276	487	530	565	295	314	313
0.45	17012	17396	30769	3021	3322	4973	930	1030	1263	470	508	570	286	302	326
	17140	18158	32322	3134	3698	5181	973	1115	1289	487	534	565	294	313	310
0.5	15860	16352	31340	2897	3242	5030	912	1017	1264	461	500	565	279	295	319
	16024	17313	32517	3028	3655	5181	957	1108	1276	478	527	553	286	306	300
0.55	14368	14994	31203	2731	3110	4975	877	987	1236	443	482	546	266	282	305
	14577	16181	32061	2877	3550	5077	924	1078	1237	460	509	530	274	293	284
0.6	12653	13444	30358	2534	2937	4806	830	938	1180	417	454	515	248	263	284
	12919	14866	30955	2692	3388	4869	877	1028	1172	434	480	495	255	273	261
0.65	10837	11821	28807	2317	2727	4524	770	873	1095	384	418	471	225	238	256
	11175	13450	29197	2480	3174	4556	814	957	1081	399	441	449	231	246	232
0.7	9047	10221	26555	2082	2484	4130	697	790	983	342	371	415	197	207	221
	9464	11985	26789	2243	2908	4140	737	864	964	355	390	391	201	214	196
0.75	7395	8699	23608	1831	2205	3626	609	689	843	291	314	347	162	170	179
	7875	10491	23730	1982	2589	3619	644	750	821	302	328	322	166	174	153
0.8	5935	7255	19973	1559	1886	3011	506	567	676	230	246	266	.	.	.
	6435	8956	20021	1693	2209	2994	533	612	651	237	255	241	.	.	.
0.85	4639	5845	15656	1255	1512	2288	380	419	482	.	.	.	.	.	.
	5104	7327	15660	1362	1754	2265	398	446	456	.	.	.	.	.	.
0.9	3410	4368	10666	894	1057	1458	.	.	.	.	.	.	.	.	.
	3785	5493	10648	963	1198	1432	.	.	.	.	.	.	.	.	.
0.95	2034	2579	5011	.	.	.	.	.	.	.	.	.	.	.	.
	2253	3162	4986	.	.	.	.	.	.	.	.	.	.	.	.

TABLE 4: ALPHA= 0.01 POWER= 0.9   EXPECTED ACCRUAL THRU MINIMUM FOLLOW-UP= 700

	DEL=.02			DEL=.05			DEL=.10			DEL=.15			DEL=.20		
FACT=	1.0 .75	.50 .25	.00 BIN	1.0 .75	.50 .25	.00 BIN	1.0 .75	.50 .25	.00 BIN	1.0 75	.50 .25	.00 BIN	1.0 .75	.50 .25	.00 BIN
PCONT=***	REQUIRED NUMBER OF PATIENTS														
0.05 ***	3705 3710	3719 3747	3921 7329	792 796	803 814	831 1432	286 288	290 292	296 456	169 170	170 171	172 241	119 120	120 121	121 153
0.1 ***	7708 7720	7743 7812	8615 12731	1450 1462	1482 1520	1604 2265	464 468	475 484	495 651	251 254	256 259	263 322	166 168	169 170	172 196
0.15 ***	11260 11280	11321 11443	13323 17482	2016 2037	2074 2150	2351 2994	612 622	635 653	678 821	319 324	329 335	342 391	205 207	210 212	215 232
0.2 ***	14088 14120	14182 14370	17703 21583	2463 2495	2553 2675	3033 3619	731 747	768 797	841 964	374 381	389 398	411 449	236 239	243 247	252 261
0.25 ***	16118 16162	16252 16522	21586 25032	2787 2833	2915 3088	3629 4140	822 843	872 914	980 1081	416 425	436 450	467 495	259 264	268 274	281 284
0.3 ***	17360 17422	17545 17915	24877 27831	2994 3056	3165 3392	4127 4556	886 913	950 1004	1092 1172	447 458	472 489	512 530	276 281	287 294	303 300
0.35 ***	17872 17954	18118 18610	27524 29979	3097 3176	3313 3595	4519 4869	925 958	1003 1068	1177 1237	466 479	496 516	544 553	286 292	299 308	318 310
0.4 ***	17734 17841	18055 18695	29494 31475	3109 3208	3374 3708	4802 5077	943 981	1032 1108	1234 1276	475 490	509 532	564 565	290 297	305 314	325 313
0.45 ***	17041 17179	17455 18273	30769 32322	3048 3166	3360 3740	4973 5181	941 983	1040 1123	1263 1289	474 491	511 536	570 565	288 296	304 314	326 310
0.5 ***	15898 16074	16427 17455	31341 32517	2929 3065	3284 3701	5031 5181	923 968	1028 1116	1264 1276	465 482	503 529	565 553	281 288	297 307	319 300
0.55 ***	14417 14642	15089 16350	31203 32061	2767 2918	3155 3597	4975 5077	890 936	998 1087	1236 1237	447 464	485 511	546 530	268 275	284 293	305 284
0.6 ***	12714 13001	13563 15061	30357 30955	2574 2735	2984 3436	4806 4869	842 888	949 1036	1180 1172	422 438	458 482	515 495	250 257	264 274	284 261
0.65 ***	10915 11277	11963 13663	28807 29197	2357 2523	2774 3222	4524 4556	782 825	883 964	1095 1081	388 403	421 443	471 449	227 233	240 247	256 232
0.7 ***	9145 9588	10382 12206	26555 26789	2123 2286	2529 2953	4130 4140	707 747	799 871	983 964	345 358	373 392	415 391	198 203	208 214	221 196
0.75 ***	7511 8013	8867 10710	23608 23730	1869 2022	2246 2628	3626 3619	618 653	696 755	843 821	294 304	316 330	347 322	163 166	170 174	179 153
0.8 ***	6057 6575	7419 9158	19973 20021	1593 1727	1921 2242	3011 2994	513 539	572 615	676 651	232 239	247 256	266 241	. .	. .	. .
0.85 ***	4755 5232	5991 7499	15656 15660	1283 1389	1539 1778	2288 2265	385 402	423 449	482 456	. .	. .	. .	. .	. .	. .
0.9 ***	3503 3887	4481 5620	10666 10648	912 980	1073 1211	1458 1432	. .	. .	. .	. .	. .	. .	. .	. .	. .
0.95 ***	2089 2311	2641 3225	5011 4986	. .	. .	. .	. .	. .	. .	. .	. .	. .	. .	. .	. .

TABLE 4: ALPHA= 0.01 POWER= 0.9   EXPECTED ACCRUAL THRU MINIMUM FOLLOW-UP= 750

		DEL=.02			DEL=.05			DEL=.10			DEL=.15			DEL=.20		
FACT=		1.0 .75	.50 .25	.00 BIN	1.0 .75	.50 .25	.00 BIN	1.0 .75	.50 .25	.00 BIN	1.0 75	.50 .25	.00 BIN	1.0 .75	.50 .25	.00 BIN
PCONT=***							REQUIRED NUMBER OF PATIENTS								
0.05	***	3706	3722	3921	793	804	831	287	290	295	169	170	172	119	120	121
	•••	3711	3751	7329	797	814	1432	289	293	456	169	171	241	120	121	153
0.1	***	7711	7748	8615	1452	1486	1604	465	476	495	251	256	263	167	169	172
	•••	7723	7822	12731	1465	1523	2265	470	484	651	254	259	322	168	171	196
0.15	***	11265	11330	13324	2020	2081	2351	615	637	679	320	329	342	205	210	215
	•••	11286	11460	17482	2043	2158	2994	625	654	821	325	335	391	208	213	232
0.2	***	14095	14195	17704	2470	2565	3033	735	771	841	376	390	410	236	243	252
	•••	14128	14396	21583	2504	2688	3619	751	799	964	382	399	449	240	247	261
0.25	***	16127	16271	21586	2797	2931	3629	827	877	979	419	438	468	260	269	281
	•••	16175	16560	25032	2845	3106	4140	849	917	1081	428	451	495	265	274	284
0.3	***	17374	17572	24877	3008	3185	4127	892	956	1092	450	474	512	277	288	303
	•••	17440	17967	27831	3073	3416	4556	919	1009	1172	460	490	530	282	295	300
0.35	***	17890	18154	27524	3115	3340	4519	933	1010	1177	469	499	544	288	300	318
	•••	17978	18679	29979	3198	3625	4869	966	1074	1237	483	518	553	293	308	310
0.4	***	17757	18101	29494	3131	3405	4802	952	1041	1234	479	512	564	291	306	325
	•••	17872	18785	31475	3234	3743	5077	990	1114	1276	493	534	565	298	315	313
0.45	***	17071	17514	30769	3074	3396	4973	951	1049	1264	478	514	570	290	305	325
	•••	17219	18386	32322	3197	3779	5181	994	1130	1289	494	538	565	297	314	310
0.5	***	15936	16503	31340	2960	3323	5030	934	1038	1264	469	506	565	282	298	319
	•••	16125	17594	32517	3100	3743	5181	979	1123	1276	486	531	553	290	308	300
0.55	***	14465	15184	31203	2801	3198	4975	901	1008	1236	452	488	546	270	285	305
	•••	14706	16516	32061	2957	3642	5077	947	1094	1237	469	513	530	277	294	284
0.6	***	12776	13680	30358	2611	3028	4806	853	959	1180	426	460	514	252	265	284
	•••	13082	15249	30955	2776	3482	4869	899	1043	1172	442	484	495	259	274	261
0.65	***	10993	12102	28807	2396	2818	4524	792	892	1095	392	423	471	228	240	256
	•••	11379	13868	29197	2565	3265	4556	835	971	1081	406	444	449	234	248	232
0.7	***	9243	10538	26555	2161	2571	4130	717	807	983	349	376	415	199	209	221
	•••	9708	12419	26789	2326	2994	4140	757	877	964	361	393	391	204	214	196
0.75	***	7624	9030	23608	1905	2285	3625	627	703	843	296	318	347	164	171	179
	•••	8146	10917	23730	2060	2665	3619	661	760	821	306	331	322	167	175	153
0.8	***	6175	7576	19973	1625	1954	3011	520	577	676	234	248	266	.	.	.
	•••	6709	9349	20021	1760	2271	2994	545	619	651	241	257	241	.	.	.
0.85	***	4864	6130	15656	1308	1564	2288	390	426	482	.	.	.	.	.	.
	•••	5354	7662	15660	1415	1799	2265	406	450	456	.	.	.	.	.	.
0.9	***	3592	4588	10667	928	1088	1458	.	.	.	.	.	.	.	.	.
	•••	3984	5740	10648	997	1222	1432	.	.	.	.	.	.	.	.	.
0.95	***	2141	2698	5011	.	.	.	.	.	.	.	.	.	.	.	.
	•••	2366	3282	4986	.	.	.	.	.	.	.	.	.	.	.	.

TABLE 4: ALPHA= 0.01 POWER= 0.9   EXPECTED ACCRUAL THRU MINIMUM FOLLOW-UP= 800

		DEL=.02			DEL=.05			DEL=.10			DEL=.15			DEL=.20		
FACT=		1.0 .75	.50 .25	.00 BIN	1.0 .75	.50 .25	.00 BIN	1.0 .75	.50 .25	.00 BIN	1.0 75	.50 .25	.00 BIN	1.0 .75	.50 .25	.00 BIN
PCONT=•••		REQUIRED NUMBER OF PATIENTS														
0.05	•••	3707	3724	3921	794	805	832	287	291	296	169	171	172	119	120	121
	•••	3713	3755	7329	798	815	1432	289	293	456	170	171	241	120	121	153
0.1	•••	7713	7753	8615	1455	1489	1604	466	477	495	252	257	263	167	170	172
	•••	7726	7831	12731	1468	1527	2265	471	485	651	254	259	322	168	171	196
0.15	•••	11269	11338	13324	2025	2088	2351	617	639	679	321	330	342	206	210	216
	•••	11292	11477	17482	2048	2165	2994	627	656	821	325	336	391	208	213	232
0.2	•••	14102	14209	17703	2477	2575	3033	739	774	841	378	391	411	237	244	252
	•••	14137	14423	21583	2513	2700	3619	754	801	964	384	400	449	240	247	261
0.25	•••	16137	16291	21586	2807	2946	3629	832	881	980	421	440	467	261	270	281
	•••	16188	16598	25032	2858	3123	4140	853	920	1081	429	452	495	265	275	284
0.3	•••	17387	17598	24877	3022	3206	4127	898	962	1092	452	476	512	278	289	303
	•••	17457	18020	27831	3089	3438	4556	926	1013	1172	463	492	530	283	295	300
0.35	•••	17908	18189	27524	3132	3364	4519	940	1017	1177	472	501	544	289	301	318
	•••	18001	18749	29979	3218	3652	4869	973	1079	1237	485	520	553	295	309	310
0.4	•••	17780	18147	29494	3153	3435	4802	960	1048	1235	482	514	564	293	307	325
	•••	17903	18874	31475	3259	3775	5077	998	1120	1276	497	536	565	300	315	313
0.45	•••	17100	17573	30769	3100	3430	4973	961	1058	1263	482	517	570	292	306	326
	•••	17258	18499	32322	3226	3815	5181	1003	1136	1289	498	540	565	298	315	310
0.5	•••	15974	16578	31340	2990	3361	5031	944	1047	1264	473	509	564	284	299	319
	•••	16175	17731	32517	3134	3782	5181	988	1130	1276	489	533	553	291	308	300
0.55	•••	14513	15279	31203	2835	3239	4975	911	1017	1236	456	491	546	272	286	305
	•••	14770	16677	32061	2993	3683	5077	957	1101	1237	472	515	530	278	295	284
0.6	•••	12837	13796	30358	2647	3070	4806	864	968	1180	430	463	515	253	267	284
	•••	13163	15432	30955	2815	3523	4869	909	1050	1172	445	486	495	260	275	261
0.65	•••	11071	12238	28807	2433	2860	4524	802	901	1095	395	426	471	230	241	256
	•••	11479	14065	29197	2604	3306	4556	845	977	1081	409	446	449	235	248	232
0.7	•••	9338	10688	26555	2197	2611	4130	726	815	983	352	378	415	200	210	221
	•••	9827	12621	26789	2365	3032	4140	765	882	964	364	394	391	205	215	196
0.75	•••	7733	9186	23608	1940	2321	3626	635	709	843	299	319	347	165	171	179
	•••	8276	11115	23730	2095	2699	3619	668	764	821	308	332	322	168	175	153
0.8	•••	6289	7726	19973	1655	1985	3011	526	582	676	236	249	266	.	.	.
	•••	6838	9531	20021	1791	2299	2994	551	622	651	242	257	241	.	.	.
0.85	•••	4970	6262	15656	1332	1588	2288	394	429	482	.	.	.	.	.	.
	•••	5471	7816	15660	1439	1819	2265	410	452	456	.	.	.	.	.	.
0.9	•••	3677	4690	10667	944	1102	1458	.	.	.	.	.	.	.	.	.
	•••	4076	5852	10648	1012	1233	1432	.	.	.	.	.	.	.	.	.
0.95	•••	2191	2753	5011	.	.	.	.	.	.	.	.	.	.	.	.
	•••	2418	3336	4986	.	.	.	.	.	.	.	.	.	.	.	.

TABLE 4: ALPHA= 0.01 POWER= 0.9   EXPECTED ACCRUAL THRU MINIMUM FOLLOW-UP= 850

		DEL=.02			DEL=.05			DEL=.10			DEL=.15			DEL=.20		
FACT=		1.0 .75	.50 .25	.00 BIN	1.0 .75	.50 .25	.00 BIN	1.0 .75	.50 .25	.00 BIN	1.0 75	.50 .25	.00 BIN	1.0 .75	.50 .25	.00 BIN
PCONT=***					REQUIRED NUMBER OF PATIENTS											
0.05	***	3708	3726	3921	794	806	832	288	291	295	169	171	172	120	120	121
	•••	3714	3758	7329	799	816	1432	289	293	456	170	171	241	120	121	153
0.1	***	7716	7757	8615	1458	1493	1604	467	478	495	253	257	263	168	170	172
	•••	7730	7841	12731	1471	1530	2265	472	485	651	255	259	322	169	171	196
0.15	***	11273	11347	13323	2029	2095	2351	619	641	679	322	331	342	207	210	215
	•••	11298	11494	17482	2054	2171	2994	629	657	821	326	336	391	208	213	232
0.2	***	14108	14222	17703	2484	2586	3033	742	777	841	379	392	410	238	244	252
	•••	14146	14450	21583	2521	2711	3619	757	803	964	385	400	449	241	248	261
0.25	***	16147	16310	21585	2817	2960	3629	836	885	980	423	441	467	262	271	281
	•••	16201	16637	25032	2870	3139	4140	858	923	1081	431	453	495	266	275	284
0.3	***	17400	17624	24877	3035	3224	4127	904	967	1092	454	478	512	280	290	303
	•••	17475	18072	27831	3105	3459	4556	931	1017	1172	465	493	530	284	295	300
0.35	***	17925	18224	27524	3149	3388	4519	947	1023	1177	475	503	544	290	302	318
	•••	18025	18818	29979	3239	3678	4869	980	1083	1237	488	521	553	295	309	310
0.4	***	17803	18193	29494	3174	3463	4802	969	1055	1234	485	516	563	294	308	325
	•••	17933	18963	31475	3284	3805	5077	1006	1125	1276	499	537	565	301	316	313
0.45	***	17130	17632	30769	3126	3462	4973	970	1065	1263	486	519	570	293	307	326
	•••	17298	18609	32322	3255	3849	5181	1011	1142	1289	501	542	565	300	316	310
0.5	***	16012	16653	31340	3019	3397	5031	953	1055	1264	477	511	564	286	300	319
	•••	16226	17866	32517	3166	3819	5181	997	1136	1276	493	535	553	292	309	300
0.55	***	14561	15372	31203	2867	3277	4975	921	1025	1236	459	494	546	273	287	305
	•••	14834	16835	32061	3029	3722	5077	966	1107	1237	475	516	530	280	295	284
0.6	***	12899	13910	30357	2681	3109	4806	873	976	1179	433	465	514	255	267	284
	•••	13244	15608	30955	2852	3562	4869	918	1056	1172	448	487	495	261	275	261
0.65	***	11149	12371	28807	2468	2899	4524	811	908	1096	398	428	471	231	242	256
	•••	11578	14255	29197	2641	3344	4556	854	983	1081	412	447	449	236	248	232
0.7	***	9433	10835	26555	2232	2648	4130	734	821	983	354	379	415	201	210	221
	•••	9943	12816	26789	2401	3067	4140	773	887	964	366	395	391	205	215	196
0.75	***	7840	9336	23608	1972	2355	3625	642	715	843	301	321	346	165	172	179
	•••	8401	11305	23730	2129	2730	3619	674	768	821	310	332	322	169	175	153
0.8	***	6399	7870	19973	1684	2013	3011	531	586	676	237	250	266	.	.	.
	•••	6962	9705	20021	1820	2324	2994	556	624	651	243	257	241	.	.	.
0.85	***	5071	6388	15656	1354	1609	2288	397	431	481	.	.	.	.	.	.
	•••	5583	7962	15660	1462	1837	2265	413	454	456	.	.	.	.	.	.
0.9	***	3758	4786	10667	958	1115	1458	.	.	.	.	.	.	.	.	.
	•••	4164	5959	10648	1025	1243	1432	.	.	.	.	.	.	.	.	.
0.95	***	2238	2804	5012	.	.	.	.	.	.	.	.	.	.	.	.
	•••	2467	3386	4986	.	.	.	.	.	.	.	.	.	.	.	.

TABLE 4: ALPHA= 0.01 POWER= 0.9   EXPECTED ACCRUAL THRU MINIMUM FOLLOW-UP= 900

	DEL=.02			DEL=.05			DEL=.10			DEL=.15			DEL=.20		
FACT=	1.0 .75	.50 .25	.00 BIN	1.0 .75	.50 .25	.00 BIN	1.0 .75	.50 .25	.00 BIN	1.0 75	.50 .25	.00 BIN	1.0 .75	.50 .25	.00 BIN
PCONT=•••					REQUIRED NUMBER OF PATIENTS										
0.05 •••	3709	3728	3921	795	807	831	288	291	296	170	171	172	119	120	121
•••	3716	3762	7329	800	816	1432	289	293	456	170	171	241	120	120	153
0.1 •••	7718	7763	8615	1460	1496	1604	468	479	495	253	257	262	168	170	172
•••	7733	7850	12731	1474	1532	2265	473	486	651	255	260	322	169	171	196
0.15 •••	11278	11356	13323	2034	2101	2351	621	642	678	323	332	342	207	211	216
•••	11304	11511	17482	2060	2178	2994	630	658	821	327	336	391	209	213	232
0.2 •••	14115	14236	17703	2491	2595	3033	745	779	841	380	393	411	239	244	252
•••	14156	14476	21583	2530	2721	3619	760	805	964	386	401	449	242	248	261
0.25 •••	16156	16329	21585	2827	2975	3629	840	889	980	424	442	468	263	271	281
•••	16214	16675	25032	2882	3154	4140	861	926	1081	433	453	495	267	276	284
0.3 •••	17413	17651	24877	3048	3243	4127	910	972	1092	457	479	512	280	290	303
•••	17492	18124	27831	3121	3479	4556	936	1019	1172	467	494	530	285	296	300
0.35 •••	17943	18260	27524	3165	3411	4519	954	1028	1178	478	504	544	291	303	318
•••	18048	18887	29979	3258	3702	4869	986	1088	1237	490	522	553	297	309	310
0.4 •••	17826	18239	29494	3194	3491	4802	976	1062	1234	488	518	564	296	308	325
•••	17964	19051	31475	3307	3834	5077	1013	1130	1276	502	539	565	302	316	313
0.45 •••	17159	17691	30770	3150	3493	4973	978	1073	1263	489	522	570	294	308	326
•••	17337	18719	32322	3282	3881	5181	1019	1147	1289	504	543	565	301	316	310
0.5 •••	16049	16727	31340	3047	3431	5031	962	1063	1264	480	514	564	287	301	319
•••	16276	17997	32517	3197	3854	5181	1006	1142	1276	496	536	553	294	309	300
0.55 •••	14609	15465	31203	2898	3313	4975	930	1033	1236	462	496	546	275	288	305
•••	14898	16989	32061	3062	3758	5077	975	1113	1237	478	518	530	281	296	284
0.6 •••	12960	14023	30358	2714	3146	4806	883	983	1180	436	468	515	256	269	284
•••	13325	15780	30955	2887	3599	4869	927	1061	1172	451	488	495	262	276	261
0.65 •••	11226	12502	28807	2502	2936	4524	820	915	1095	401	430	471	232	243	256
•••	11676	14438	29197	2677	3379	4556	862	987	1081	414	448	449	237	249	232
0.7 •••	9526	10977	26555	2265	2683	4130	743	828	983	357	381	415	202	210	221
•••	10056	13002	26789	2435	3100	4140	780	891	964	368	396	391	206	216	196
0.75 •••	7945	9482	23608	2003	2387	3626	649	720	843	303	322	347	166	172	179
•••	8523	11485	23730	2160	2759	3619	681	771	821	312	333	322	169	176	153
0.8 •••	6506	8008	19973	1710	2040	3011	536	590	676	238	251	266	.	.	.
•••	7082	9870	20021	1847	2348	2994	560	627	651	244	258	241	.	.	.
0.85 •••	5169	6509	15656	1376	1630	2288	401	434	482	.	.	.	.	.	.
•••	5691	8101	15660	1483	1854	2265	416	455	456	.	.	.	.	.	.
0.9 •••	3836	4879	10666	972	1127	1458	.	.	.	.	.	.	.	.	.
•••	4248	6059	10648	1038	1252	1432	.	.	.	.	.	.	.	.	.
0.95 •••	2282	2852	5012	.	.	.	.	.	.	.	.	.	.	.	.
•••	2513	3432	4986	.	.	.	.	.	.	.	.	.	.	.	.

TABLE 4: ALPHA= 0.01 POWER= 0.9   EXPECTED ACCRUAL THRU MINIMUM FOLLOW-UP= 950

		DEL=.02			DEL=.05			DEL=.10			DEL=.15			DEL=.20		
FACT=		1.0 .75	.50 .25	.00 BIN	1.0 .75	.50 .25	.00 BIN	1.0 .75	.50 .25	.00 BIN	1.0 75	.50 .25	.00 BIN	1.0 .75	.50 .25	.00 BIN
PCONT=•••						REQUIRED NUMBER OF PATIENTS									
0.05	•••	3710	3730	3921	796	808	831	288	291	296	169	171	172	120	120	121
	•••	3717	3765	7329	801	817	1432	290	293	456	170	172	241	120	121	153
0.1	•••	7721	7767	8615	1463	1499	1604	469	479	495	254	257	262	168	170	172
	•••	7736	7860	12731	1477	1535	2265	473	486	651	255	260	322	169	171	196
0.15	•••	11282	11364	13323	2039	2107	2351	622	644	679	324	331	342	207	211	216
	•••	11309	11528	17482	2065	2183	2994	632	659	821	327	336	391	209	213	232
0.2	•••	14122	14249	17703	2497	2605	3033	748	781	841	381	394	410	239	245	252
	•••	14164	14503	21583	2538	2731	3619	762	806	964	387	401	449	242	248	261
0.25	•••	16165	16348	21585	2837	2988	3629	844	892	980	426	443	467	264	272	281
	•••	16227	16712	25032	2893	3168	4140	865	928	1081	434	454	495	267	276	284
0.3	•••	17426	17677	24877	3060	3260	4127	914	976	1092	458	481	512	281	291	303
	•••	17510	18175	27831	3136	3497	4556	941	1022	1172	469	495	530	286	296	300
0.35	•••	17961	18294	27524	3181	3433	4519	960	1034	1177	480	506	544	292	304	318
	•••	18071	18954	29979	3277	3725	4869	992	1091	1237	492	523	553	298	310	310
0.4	•••	17849	18285	29494	3214	3516	4802	983	1068	1235	491	520	564	297	310	325
	•••	17994	19138	31475	3331	3860	5077	1020	1134	1276	504	539	565	303	317	313
0.45	•••	17189	17750	30769	3173	3523	4973	986	1079	1264	492	523	570	296	309	325
	•••	17376	18827	32322	3309	3911	5181	1027	1152	1289	506	544	565	302	317	310
0.5	•••	16087	16802	31340	3073	3464	5030	971	1070	1264	483	516	564	289	302	319
	•••	16327	18127	32517	3227	3886	5181	1014	1147	1276	498	538	553	295	310	300
0.55	•••	14657	15558	31203	2928	3348	4974	939	1040	1236	466	498	546	276	289	305
	•••	14962	17138	32061	3095	3792	5077	983	1117	1237	481	519	530	281	296	284
0.6	•••	13021	14134	30357	2746	3181	4806	891	990	1180	439	470	514	257	269	284
	•••	13404	15947	30955	2921	3633	4869	935	1066	1172	453	490	495	262	276	261
0.65	•••	11303	12630	28806	2534	2971	4524	828	922	1096	403	431	471	233	243	256
	•••	11773	14615	29197	2711	3412	4556	869	992	1081	416	449	449	238	249	232
0.7	•••	9618	11115	26555	2296	2717	4130	749	834	983	359	382	415	203	211	221
	•••	10167	13182	26789	2468	3131	4140	787	895	964	370	397	391	207	216	196
0.75	•••	8047	9622	23608	2032	2417	3625	655	725	843	305	323	347	166	172	179
	•••	8640	11658	23730	2191	2786	3619	686	774	821	313	334	322	169	176	153
0.8	•••	6609	8141	19973	1736	2065	3011	541	593	676	239	251	266	.	.	.
	•••	7198	10027	20021	1873	2370	2994	565	629	651	245	258	241	.	.	.
0.85	•••	5263	6625	15656	1396	1648	2288	403	436	482	.	.	.	.	.	.
	•••	5795	8234	15660	1502	1870	2265	418	456	456	.	.	.	.	.	.
0.9	•••	3911	4967	10666	984	1138	1458	.	.	.	.	.	.	.	.	.
	•••	4328	6154	10648	1051	1260	1432	.	.	.	.	.	.	.	.	.
0.95	•••	2325	2898	5011	.	.	.	.	.	.	.	.	.	.	.	.
	•••	2558	3477	4986	.	.	.	.	.	.	.	.	.	.	.	.

TABLE 4: ALPHA= 0.01 POWER= 0.9   EXPECTED ACCRUAL THRU MINIMUM FOLLOW-UP= 1000

		DEL=.02			DEL=.05			DEL=.10			DEL=.15			DEL=.20		
FACT=		1.0 .75	.50 .25	.00 BIN	1.0 .75	.50 .25	.00 BIN	1.0 .75	.50 .25	.00 BIN	1.0 75	.50 .25	.00 BIN	1.0 .75	.50 .25	.00 BIN
PCONT=***				REQUIRED NUMBER OF PATIENTS											
0.05	***	3711	3732	3921	797	808	831	288	291	296	170	171	172	120	120	121
	•••	3718	3768	7329	802	818	1432	290	293	456	170	171	241	120	121	153
0.1	***	7723	7773	8615	1465	1501	1604	470	480	495	254	258	263	168	170	172
	•••	7740	7869	12731	1480	1538	2265	475	487	651	256	260	322	169	171	196
0.15	***	11286	11373	13323	2043	2113	2351	625	645	678	325	332	342	208	211	216
	•••	11315	11545	17482	2070	2189	2994	634	660	821	328	337	391	210	213	232
0.2	***	14128	14263	17703	2505	2614	3033	751	784	841	383	395	411	240	245	252
	•••	14173	14529	21583	2546	2740	3619	765	808	964	388	402	449	242	248	261
0.25	***	16175	16368	21586	2846	3001	3629	848	895	980	428	444	468	265	272	281
	•••	16240	16750	25032	2905	3181	4140	869	930	1081	435	455	495	268	276	284
0.3	***	17440	17704	24877	3073	3278	4127	920	980	1092	461	482	512	282	291	303
	•••	17528	18227	27831	3151	3515	4556	946	1026	1172	470	495	530	286	297	300
0.35	***	17978	18330	27524	3198	3454	4519	966	1039	1177	483	508	544	293	304	318
	•••	18095	19022	29979	3296	3747	4869	998	1095	1237	494	524	553	298	311	310
0.4	***	17872	18331	29494	3234	3541	4802	990	1073	1235	493	522	563	298	310	325
	•••	18025	19224	31475	3353	3886	5077	1026	1138	1276	506	541	565	304	317	313
0.45	***	17219	17809	30770	3196	3551	4973	993	1086	1263	495	525	570	297	310	326
	•••	17416	18933	32322	3335	3940	5181	1034	1156	1289	509	545	565	303	317	310
0.5	***	16125	16876	31340	3100	3495	5031	978	1076	1264	486	518	565	290	303	319
	•••	16377	18254	32517	3256	3917	5181	1021	1151	1276	501	539	553	296	310	300
0.55	***	14706	15650	31203	2956	3380	4975	947	1046	1236	468	500	546	277	290	305
	•••	15026	17285	32061	3126	3824	5077	991	1123	1237	483	521	530	283	296	284
0.6	***	13082	14243	30358	2776	3215	4806	899	997	1180	441	472	515	258	270	284
	•••	13484	16108	30955	2953	3665	4869	942	1071	1172	456	491	495	264	276	261
0.65	***	11378	12755	28807	2565	3004	4524	836	928	1095	406	433	471	234	244	256
	•••	11869	14786	29197	2743	3444	4556	876	996	1081	418	450	449	239	250	232
0.7	***	9708	11250	26555	2326	2748	4130	756	839	983	361	384	415	204	211	221
	•••	10276	13355	26789	2499	3160	4140	793	899	964	372	398	391	208	216	196
0.75	***	8146	9758	23608	2060	2445	3626	661	730	843	306	324	346	167	173	179
	•••	8756	11825	23730	2219	2811	3619	691	778	821	315	335	322	170	176	153
0.8	***	6709	8270	19973	1760	2090	3011	545	596	676	241	252	266	.	.	.
	•••	7310	10178	20021	1898	2390	2994	569	631	651	246	259	241	.	.	.
0.85	***	5355	6736	15656	1415	1666	2288	406	438	482	.	.	.	.	.	.
	•••	5895	8360	15660	1521	1884	2265	421	458	456	.	.	.	.	.	.
0.9	***	3983	5051	10666	996	1148	1458	.	.	.	.	.	.	.	.	.
	•••	4406	6245	10648	1063	1268	1432	.	.	.	.	.	.	.	.	.
0.95	***	2366	2941	5011	.	.	.	.	.	.	.	.	.	.	.	.
	•••	2600	3518	4986	.	.	.	.	.	.	.	.	.	.	.	.

TABLE 4: ALPHA= 0.01 POWER= 0.9   EXPECTED ACCRUAL THRU MINIMUM FOLLOW-UP= 1100

		DEL=.02			DEL=.05			DEL=.10			DEL=.15			DEL=.20		
FACT=		1.0 .75	.50 .25	.00 BIN	1.0 .75	.50 .25	.00 BIN	1.0 .75	.50 .25	.00 BIN	1.0 75	.50 .25	.00 BIN	1.0 .75	.50 .25	.00 BIN
PCONT=•••				REQUIRED NUMBER OF PATIENTS												
0.05	•••	3713	3736	3921	799	810	831	289	291	296	170	170	172	120	120	121
	•••	3721	3774	7329	804	819	1432	290	294	456	170	171	241	120	121	153
0.1	•••	7728	7782	8615	1470	1506	1604	471	481	495	254	258	263	169	170	172
	•••	7746	7887	12731	1484	1542	2265	476	488	651	257	260	322	169	171	196
0.15	•••	11295	11390	13324	2051	2123	2351	628	648	679	326	333	342	208	212	215
	•••	11327	11579	17482	2079	2199	2994	637	661	821	329	337	391	210	213	232
0.2	•••	14142	14289	17703	2517	2631	3033	756	788	841	384	395	411	241	246	252
	•••	14191	14581	21583	2561	2756	3619	770	811	964	389	402	449	243	248	261
0.25	•••	16195	16406	21586	2864	3025	3629	855	901	980	430	445	467	265	272	280
	•••	16265	16825	25032	2925	3205	4140	875	934	1081	437	456	495	269	276	284
0.3	•••	17466	17757	24877	3097	3309	4127	928	987	1092	464	484	512	284	292	302
	•••	17563	18329	27831	3178	3547	4556	954	1031	1172	473	496	530	288	297	300
0.35	•••	18013	18400	27524	3228	3493	4519	976	1048	1177	486	510	544	295	305	318
	•••	18142	19156	29979	3331	3786	4869	1008	1101	1237	497	525	553	300	311	310
0.4	•••	17918	18422	29494	3271	3587	4802	1002	1083	1234	498	525	564	300	312	325
	•••	18086	19393	31475	3395	3932	5077	1038	1145	1276	510	543	565	305	318	313
0.45	•••	17278	17926	30769	3241	3604	4973	1006	1097	1264	499	529	570	299	311	326
	•••	17494	19141	32322	3384	3992	5181	1046	1164	1289	513	547	565	305	318	310
0.5	•••	16200	17023	31340	3150	3552	5031	993	1088	1264	491	521	565	291	304	319
	•••	16477	18501	32517	3310	3973	5181	1035	1159	1276	505	540	553	298	311	300
0.55	•••	14802	15830	31203	3011	3442	4975	961	1058	1236	473	503	546	279	291	305
	•••	15152	17566	32061	3184	3882	5077	1005	1131	1237	488	522	530	285	297	284
0.6	•••	13204	14458	30357	2833	3278	4806	914	1009	1180	446	475	514	260	271	284
	•••	13641	16417	30955	3013	3724	4869	956	1079	1172	459	493	495	265	277	261
0.65	•••	11529	12996	28807	2623	3066	4524	850	939	1095	411	436	471	236	245	256
	•••	12056	15111	29197	2804	3501	4556	889	1004	1081	422	452	449	240	250	232
0.7	•••	9885	11507	26555	2383	2806	4130	769	848	983	364	386	415	205	213	221
	•••	10487	13682	26789	2557	3213	4140	804	906	964	375	400	391	208	217	196
0.75	•••	8339	10017	23608	2112	2498	3626	671	737	843	309	326	346	168	173	179
	•••	8976	12139	23730	2272	2857	3619	701	782	821	317	335	322	170	176	153
0.8	•••	6901	8513	19973	1805	2133	3011	554	602	676	243	253	266	.	.	.
	•••	7524	10462	20021	1943	2428	2994	576	635	651	247	259	241	.	.	.
0.85	•••	5527	6946	15657	1451	1699	2288	411	441	482	.	.	.	.	.	.
	•••	6084	8596	15660	1556	1910	2265	425	459	456	.	.	.	.	.	.
0.9	•••	4120	5209	10667	1019	1167	1457	.	.	.	.	.	.	.	.	.
	•••	4553	6413	10648	1083	1281	1432	.	.	.	.	.	.	.	.	.
0.95	•••	2443	3022	5011	.	.	.	.	.	.	.	.	.	.	.	.
	•••	2679	3593	4986	.	.	.	.	.	.	.	.	.	.	.	.

TABLE 4: ALPHA= 0.01   POWER= 0.9   EXPECTED ACCRUAL THRU MINIMUM FOLLOW-UP= 1200

PCONT=***	DEL=.02 1.0/.75	DEL=.02 .50/.25	DEL=.02 .00/BIN	DEL=.05 1.0/.75	DEL=.05 .50/.25	DEL=.05 .00/BIN	DEL=.10 1.0/.75	DEL=.10 .50/.25	DEL=.10 .00/BIN	DEL=.15 1.0/.75	DEL=.15 .50/.25	DEL=.15 .00/BIN	DEL=.20 1.0/.75	DEL=.20 .50/.25	DEL=.20 .00/BIN
					REQUIRED	NUMBER OF	PATIENTS								
0.05 ***	3715	3739	3921	800	811	832	289	292	295	169	171	172	120	121	121
•••	3724	3780	7329	805	820	1432	291	294	456	170	172	241	120	121	153
0.1 ***	7733	7792	8615	1474	1511	1603	472	482	495	255	258	262	169	170	172
•••	7753	7905	12731	1489	1546	2265	477	488	651	256	260	322	169	171	196
0.15 ***	11303	11408	13324	2059	2133	2350	630	649	679	327	334	342	208	211	215
•••	11338	11611	17482	2088	2207	2994	639	662	821	330	337	391	210	214	232
0.2 ***	14155	14316	17703	2530	2647	3033	760	791	841	386	397	410	241	246	252
•••	14209	14632	21583	2575	2771	3619	774	813	964	391	403	449	244	249	261
0.25 ***	16213	16444	21586	2881	3047	3629	862	905	979	433	447	467	267	273	280
•••	16291	16899	25032	2946	3227	4140	881	937	1081	439	457	495	270	277	284
0.3 ***	17492	17809	24877	3121	3339	4126	936	993	1092	467	486	512	285	293	303
•••	17598	18429	27831	3205	3575	4556	961	1035	1172	475	498	530	289	298	300
0.35 ***	18048	18470	27523	3258	3529	4519	986	1055	1177	490	513	544	297	306	318
•••	18189	19287	29979	3364	3822	4869	1017	1106	1237	500	527	553	301	311	310
0.4 ***	17963	18514	29494	3307	3631	4802	1013	1093	1234	502	528	563	301	312	325
•••	18147	19558	31475	3435	3973	5077	1048	1151	1276	514	544	565	307	319	313
0.45 ***	17337	18043	30769	3283	3652	4972	1019	1106	1263	503	532	570	301	312	325
•••	17573	19342	32322	3430	4039	5181	1057	1171	1289	517	549	565	306	319	310
0.5 ***	16276	17169	31340	3197	3606	5030	1006	1098	1264	496	524	564	294	305	319
•••	16578	18737	32517	3361	4024	5181	1047	1166	1276	508	542	553	299	311	300
0.55 ***	14898	16008	31203	3062	3498	4975	975	1069	1236	478	506	546	280	292	305
•••	15278	17834	32061	3238	3935	5077	1016	1138	1237	491	524	530	286	298	284
0.6 ***	13324	14665	30358	2887	3335	4806	926	1018	1180	451	478	514	262	271	284
•••	13795	16708	30955	3070	3777	4869	967	1086	1172	463	494	495	266	277	261
0.65 ***	11676	13228	28807	2677	3122	4524	862	948	1095	414	439	471	237	245	256
•••	12238	15416	29197	2860	3552	4556	901	1010	1081	425	453	449	241	250	232
0.7 ***	10056	11752	26554	2435	2860	4130	780	857	983	368	388	415	206	213	221
•••	10688	13987	26789	2611	3259	4140	814	910	964	377	400	391	209	217	196
0.75 ***	8522	10261	23608	2160	2545	3625	680	744	843	311	327	346	169	174	179
•••	9186	12431	23730	2321	2898	3619	709	787	821	319	336	322	171	176	153
0.8 ***	7082	8741	19972	1847	2173	3011	560	607	676	244	254	266	.	.	.
•••	7726	10726	20021	1984	2460	2994	582	637	651	249	260	241	.	.	.
0.85 ***	5691	7143	15656	1483	1728	2288	415	444	481	.	.	.	.	.	.
•••	6262	8814	15660	1588	1933	2265	429	461	456	.	.	.	.	.	.
0.9 ***	4248	5356	10666	1039	1183	1457	.	.	.	.	.	.	.	.	.
•••	4690	6567	10648	1102	1292	1432	.	.	.	.	.	.	.	.	.
0.95 ***	2513	3095	5011	.	.	.	.	.	.	.	.	.	.	.	.
•••	2752	3661	4986	.	.	.	.	.	.	.	.	.	.	.	.

TABLE 4: ALPHA= 0.01 POWER= 0.9 EXPECTED ACCRUAL THRU MINIMUM FOLLOW-UP= 1300

		DEL=.02			DEL=.05			DEL=.10			DEL=.15			DEL=.20		
FACT=		1.0	.50	.00	1.0	.50	.00	1.0	.50	.00	1.0	.50	.00	1.0	.50	.00
		.75	.25	BIN	.75	.25	BIN	.75	.25	BIN	75	.25	BIN	.75	.25	BIN
PCONT=***		REQUIRED NUMBER OF PATIENTS														
0.05	***	3717	3744	3921	802	812	831	290	292	296	170	172	172	120	120	121
	***	3726	3786	7329	806	821	1432	291	294	456	171	172	241	120	121	153
0.1	***	7738	7803	8615	1478	1515	1604	474	483	495	255	258	263	169	171	172
	***	7759	7922	12731	1493	1550	2265	478	488	651	257	261	322	170	172	196
0.15	***	11313	11425	13323	2067	2142	2351	633	652	679	328	334	342	210	212	215
	***	11350	11644	17482	2097	2215	2994	641	664	821	331	338	391	211	214	232
0.2	***	14168	14343	17704	2542	2662	3033	764	794	841	388	397	410	242	246	252
	***	14227	14683	21583	2589	2785	3619	778	815	964	393	404	449	245	249	261
0.25	***	16233	16483	21586	2899	3069	3630	867	910	980	435	448	467	267	274	281
	***	16317	16971	25032	2965	3247	4140	887	940	1081	441	458	495	271	277	284
0.3	***	17519	17862	24877	3144	3366	4127	943	999	1092	470	487	512	286	294	303
	***	17633	18528	27831	3230	3602	4556	968	1038	1172	478	499	530	290	298	300
0.35	***	18083	18540	27524	3287	3564	4519	994	1062	1177	493	514	544	298	307	318
	***	18236	19415	29979	3396	3854	4869	1025	1111	1237	503	528	553	302	312	310
0.4	***	18010	18605	29494	3342	3671	4802	1023	1100	1234	505	530	564	303	314	325
	***	18208	19719	31475	3473	4011	5077	1058	1156	1276	517	545	565	308	319	313
0.45	***	17396	18159	30769	3322	3698	4972	1030	1116	1264	508	534	570	302	313	326
	***	17652	19537	32322	3473	4081	5181	1068	1177	1289	520	551	565	307	319	310
0.5	***	16352	17313	31341	3242	3655	5030	1017	1108	1264	500	526	565	295	306	318
	***	16677	18965	32517	3409	4070	5181	1058	1173	1276	512	544	553	301	312	300
0.55	***	14994	16181	31203	3110	3550	4975	986	1078	1236	482	509	546	282	292	305
	***	15403	18089	32061	3289	3983	5077	1028	1144	1237	494	526	530	287	298	284
0.6	***	13445	14865	30358	2937	3388	4806	939	1028	1180	454	480	514	263	273	284
	***	13947	16985	30955	3122	3825	4869	978	1092	1172	466	496	495	268	278	261
0.65	***	11821	13449	28806	2728	3174	4524	873	957	1095	418	440	471	238	246	256
	***	12415	15703	29197	2911	3598	4556	911	1016	1081	428	455	449	242	251	232
0.7	***	10221	11985	26555	2484	2908	4130	791	864	983	370	390	415	207	214	221
	***	10883	14274	26789	2660	3302	4140	824	916	964	380	401	391	211	217	196
0.75	***	8699	10491	23608	2205	2589	3625	689	750	843	314	328	347	170	174	179
	***	9385	12704	23730	2366	2935	3619	717	791	821	321	337	322	172	176	153
0.8	***	7255	8955	19973	1886	2208	3011	566	612	676	245	255	266	.	.	.
	***	7916	10970	20021	2022	2489	2994	588	640	651	250	260	241	.	.	.
0.85	***	5845	7327	15656	1512	1754	2288	419	446	482	.	.	.	.	.	.
	***	6428	9015	15660	1616	1953	2265	432	462	456	.	.	.	.	.	.
0.9	***	4368	5492	10666	1057	1198	1458	.	.	.	.	.	.	.	.	.
	***	4817	6708	10648	1119	1303	1432	.	.	.	.	.	.	.	.	.
0.95	***	2579	3162	5011	.	.	.	.	.	.	.	.	.	.	.	.
	***	2820	3722	4986	.	.	.	.	.	.	.	.	.	.	.	.

TABLE 4: ALPHA= 0.01 POWER= 0.9   EXPECTED ACCRUAL THRU MINIMUM FOLLOW-UP= 1400

	DEL=.02			DEL=.05			DEL=.10			DEL=.15			DEL=.20		
FACT=	1.0 .75	.50 .25	.00 BIN	1.0 .75	.50 .25	.00 BIN	1.0 .75	.50 .25	.00 BIN	1.0 75	.50 .25	.00 BIN	1.0 .75	.50 .25	.00 BIN
PCONT=•••	REQUIRED NUMBER OF PATIENTS														
0.05 ***	3720 3748 3921			802 814 831			290 292 296			171 171 172			120 121 121		
•••	3729 3790 7329			808 821 1432			291 294 456			171 171 241			121 121 153		
0.1 ***	7743 7812 8615			1482 1520 1604			475 484 495			256 259 262			169 171 172		
•••	7766 7939 12731			1498 1552 2265			479 489 651			257 261 322			170 171 196		
0.15 ***	11321 11442 13324			2075 2150 2351			635 653 678			329 335 342			210 213 215		
•••	11362 11675 17482			2105 2222 2994			643 664 821			332 339 391			211 213 232		
0.2 ***	14182 14370 17703			2553 2675 3033			767 797 841			388 398 410			242 247 252		
•••	14245 14733 21583			2602 2796 3619			780 816 964			394 404 449			245 249 261		
0.25 ***	16252 16522 21585			2915 3088 3629			872 913 980			437 450 467			269 274 281		
•••	16342 17043 25032			2984 3265 4140			891 942 1081			443 458 495			271 277 284		
0.3 ***	17545 17915 24877			3165 3392 4126			950 1004 1092			472 489 512			287 294 303		
•••	17669 18624 27831			3255 3625 4556			975 1042 1172			480 500 530			290 298 300		
0.35 ***	18119 18609 27524			3314 3595 4519			1003 1068 1177			496 516 544			299 308 318		
•••	18282 19540 29979			3426 3883 4869			1032 1115 1237			506 529 553			304 312 310		
0.4 ***	18056 18695 29494			3374 3708 4802			1032 1108 1235			508 532 563			304 314 325		
•••	18270 19876 31475			3508 4046 5077			1066 1161 1276			520 547 565			309 319 313		
0.45 ***	17455 18273 30769			3360 3740 4973			1040 1123 1263			511 536 570			304 314 325		
•••	17730 19725 32322			3513 4119 5181			1077 1182 1289			522 552 565			309 319 310		
0.5 ***	16427 17454 31341			3284 3700 5030			1028 1116 1263			503 529 564			297 307 318		
•••	16777 19183 32517			3453 4111 5181			1067 1178 1276			515 545 553			302 312 300		
0.55 ***	15089 16350 31203			3155 3597 4974			997 1087 1236			485 511 546			283 293 305		
•••	15528 18334 32061			3336 4026 5077			1038 1150 1237			497 527 530			289 299 284		
0.6 ***	13562 15060 30357			2984 3436 4805			949 1036 1179			458 482 514			264 274 284		
•••	14097 17248 30955			3170 3867 4869			988 1097 1172			469 497 495			269 278 261		
0.65 ***	11963 13663 28807			2775 3222 4524			883 964 1095			421 443 471			240 247 256		
•••	12587 15975 29197			2959 3639 4556			920 1020 1081			430 456 449			243 251 232		
0.7 ***	10382 12206 26554			2529 2953 4130			799 871 983			374 392 415			208 214 221		
•••	11070 14543 26789			2705 3340 4140			832 920 964			382 402 391			211 217 196		
0.75 ***	8867 10710 23608			2246 2628 3626			696 755 843			316 330 346			171 174 179		
•••	9576 12959 23730			2407 2968 3619			724 794 821			322 338 322			172 177 153		
0.8 ***	7419 9158 19973			1921 2242 3011			572 615 676			247 255 266			. . .		
•••	8097 11198 20021			2057 2516 2994			592 643 651			251 261 241			. . .		
0.85 ***	5991 7500 15656			1539 1778 2288			423 449 482			. . .			. . .		
•••	6587 9202 15660			1642 1970 2265			435 464 456			. . .			. . .		
0.9 ***	4481 5620 10666			1073 1211 1458			. . .			. . .			. . .		
•••	4938 6839 10648			1134 1312 1432			. . .			. . .			. . .		
0.95 ***	2641 3225 5011			. . .			. . .			. . .			. . .		
•••	2882 3777 4986			. . .			. . .			. . .			. . .		

497

TABLE 4: ALPHA= 0.01 POWER= 0.9   EXPECTED ACCRUAL THRU MINIMUM FOLLOW-UP= 1500

		DEL=.02			DEL=.05			DEL=.10			DEL=.15			DEL=.20		
FACT=		1.0 .75	.50 .25	.00 BIN	1.0 .75	.50 .25	.00 BIN	1.0 .75	.50 .25	.00 BIN	1.0 75	.50 .25	.00 BIN	1.0 .75	.50 .25	.00 BIN
PCONT=•••				REQUIRED NUMBER OF PATIENTS												
0.05	•••	3722	3751	3922	804	815	832	290	292	295	170	172	172	120	121	121
	•••	3732	3795	7329	808	822	1432	292	294	456	170	172	241	120	121	153
0.1	•••	7747	7822	8615	1486	1523	1604	476	485	495	256	260	262	170	170	172
	•••	7773	7955	12731	1502	1555	2265	480	489	651	258	260	322	170	172	196
0.15	•••	11330	11460	13324	2081	2158	2351	637	654	679	329	335	342	210	213	215
	•••	11373	11707	17482	2112	2228	2994	645	665	821	332	338	391	211	214	232
0.2	•••	14195	14396	17704	2565	2688	3033	770	799	841	390	399	410	243	247	252
	•••	14262	14782	21583	2614	2808	3619	784	817	964	395	405	449	245	249	261
0.25	•••	16271	16660	21586	2930	3106	3629	877	917	980	438	451	468	269	275	281
	•••	16368	17113	25032	3001	3280	4140	895	945	1081	444	458	495	272	277	284
0.3	•••	17572	17967	24877	3185	3416	4127	956	1009	1092	474	490	512	288	295	303
	•••	17704	18719	27831	3277	3647	4556	980	1044	1172	482	500	530	292	299	300
0.35	•••	18154	18680	27524	3339	3624	4519	1010	1073	1177	499	518	544	300	308	318
	•••	18330	19662	29979	3454	3910	4869	1039	1118	1237	507	530	553	304	313	310
0.4	•••	18101	18785	29494	3405	3742	4802	1040	1114	1235	512	534	563	305	315	325
	•••	18331	20027	31475	3541	4077	5077	1073	1165	1276	522	547	565	310	320	313
0.45	•••	17514	18386	30770	3395	3779	4972	1049	1130	1264	514	538	570	305	314	325
	•••	17809	19907	32322	3551	4155	5181	1085	1187	1289	525	553	565	309	320	310
0.5	•••	16503	17594	31340	3323	3742	5030	1038	1123	1264	506	532	564	298	307	319
	•••	16876	19392	32517	3494	4148	5181	1076	1183	1276	517	547	553	303	313	300
0.55	•••	15184	16516	31203	3198	3642	4974	1008	1094	1235	488	513	545	285	294	305
	•••	15650	18566	32061	3380	4065	5077	1046	1155	1237	500	529	530	290	299	284
0.6	•••	13680	15249	30358	3028	3482	4805	959	1043	1179	460	484	515	265	274	284
	•••	14243	17497	30955	3215	3907	4869	997	1102	1172	472	499	495	270	278	261
0.65	•••	12102	13867	28807	2818	3265	4524	892	971	1095	423	444	470	240	247	256
	•••	12755	16232	29197	3004	3677	4556	928	1025	1081	433	457	449	244	251	232
0.7	•••	10537	12419	26555	2570	2993	4130	807	877	983	376	393	415	209	215	221
	•••	11250	14797	26789	2748	3375	4140	839	923	964	384	403	391	212	217	196
0.75	•••	9030	10917	23608	2285	2665	3625	703	760	843	318	331	347	170	174	179
	•••	9757	13199	23730	2445	2997	3619	729	797	821	323	338	322	172	177	153
0.8	•••	7576	9350	19972	1954	2272	3011	577	619	676	248	257	266	.	.	.
	•••	8270	11413	20021	2090	2539	2994	596	645	651	252	262	241	.	.	.
0.85	•••	6129	7662	15656	1565	1799	2288	427	450	482	.	.	.	.	.	.
	•••	6736	9378	15660	1666	1987	2265	438	465	456	.	.	.	.	.	.
0.9	•••	4588	5740	10667	1088	1222	1458	.	.	.	.	.	.	.	.	.
	•••	5051	6959	10648	1148	1320	1432	.	.	.	.	.	.	.	.	.
0.95	•••	2698	3282	5011	.	.	.	.	.	.	.	.	.	.	.	.
	•••	2941	3828	4986	.	.	.	.	.	.	.	.	.	.	.	.

TABLE 4: ALPHA= 0.01 POWER= 0.9   EXPECTED ACCRUAL THRU MINIMUM FOLLOW-UP= 1600

	DEL=.02			DEL=.05			DEL=.10			DEL=.15			DEL=.20		
FACT=	1.0 .75	.50 .25	.00 BIN	1.0 .75	.50 .25	.00 BIN	1.0 .75	.50 .25	.00 BIN	1.0 75	.50 .25	.00 BIN	1.0 .75	.50 .25	.00 BIN
PCONT=***				REQUIRED NUMBER OF PATIENTS											
0.05 ***	3724	3755	3921	805	815	832	291	293	296	171	171	172	120	121	121
***	3735	3799	7329	810	822	1432	292	294	456	171	172	241	120	121	153
0.1 ***	7753	7831	8615	1489	1527	1604	477	485	495	257	259	263	170	171	172
***	7779	7970	12731	1505	1557	2265	481	490	651	258	261	322	170	172	196
0.15 ***	11338	11477	13324	2088	2165	2351	639	656	679	330	336	342	210	213	216
***	11385	11737	17482	2120	2234	2994	647	666	821	333	339	391	212	214	232
0.2 ***	14209	14423	17703	2575	2700	3033	774	801	841	391	400	411	244	247	252
***	14280	14830	21583	2626	2817	3619	787	819	964	395	405	449	246	250	261
0.25 ***	16291	16598	21586	2946	3123	3629	881	920	980	440	452	467	270	275	281
***	16394	17182	25032	3017	3295	4140	899	947	1081	445	459	495	272	278	284
0.3 ***	17598	18020	24877	3206	3438	4127	962	1013	1092	476	492	512	289	295	303
***	17739	18812	27831	3299	3667	4556	985	1047	1172	483	501	530	292	299	300
0.35 ***	18189	18749	27524	3364	3652	4519	1017	1079	1177	501	520	544	301	309	318
***	18376	19782	29979	3480	3935	4869	1045	1121	1237	510	531	553	305	313	310
0.4 ***	18147	18874	29494	3435	3775	4802	1048	1120	1235	514	536	564	307	315	325
***	18392	20175	31475	3573	4106	5077	1080	1169	1276	524	549	565	311	320	313
0.45 ***	17573	18499	30769	3430	3815	4973	1058	1136	1263	517	540	570	306	315	326
***	17887	20082	32322	3587	4187	5181	1093	1191	1289	528	554	565	311	320	310
0.5 ***	16578	17731	31340	3361	3782	5031	1047	1130	1264	509	533	564	299	308	319
***	16975	19594	32517	3534	4184	5181	1084	1187	1276	520	548	553	304	313	300
0.55 ***	15279	16677	31203	3239	3683	4975	1017	1101	1236	491	515	546	286	295	305
***	15771	18790	32061	3422	4101	5077	1055	1159	1237	502	529	530	290	300	284
0.6 ***	13796	15432	30358	3070	3523	4806	968	1050	1180	463	486	515	267	275	284
***	14387	17735	30955	3258	3943	4869	1005	1106	1172	474	499	495	271	279	261
0.65 ***	12238	14065	28807	2860	3306	4524	901	977	1095	426	446	471	241	248	256
***	12917	16476	29197	3046	3712	4556	935	1029	1081	435	458	449	245	252	232
0.7 ***	10688	12621	26555	2611	3032	4130	815	882	983	378	394	415	210	215	221
***	11423	15039	26789	2788	3406	4140	846	927	964	386	404	391	212	218	196
0.75 ***	9186	11115	23608	2321	2699	3626	709	764	843	319	332	347	171	175	179
***	9932	13426	23730	2481	3025	3619	735	800	821	325	339	322	173	177	153
0.8 ***	7726	9531	19973	1985	2299	3011	582	622	676	249	257	266	.	.	.
***	8434	11614	20021	2119	2560	2994	601	647	651	253	262	241	.	.	.
0.85 ***	6262	7816	15656	1588	1819	2288	429	452	482	.	.	.	.	.	.
***	6878	9541	15660	1689	2001	2265	440	466	456	.	.	.	.	.	.
0.9 ***	4690	5852	10667	1102	1233	1458	.	.	.	.	.	.	.	.	.
***	5158	7072	10648	1161	1327	1432	.	.	.	.	.	.	.	.	.
0.95 ***	2753	3336	5011	.	.	.	.	.	.	.	.	.	.	.	.
***	2996	3874	4986	.	.	.	.	.	.	.	.	.	.	.	.

TABLE 4: ALPHA= 0.01 POWER= 0.9 EXPECTED ACCRUAL THRU MINIMUM FOLLOW-UP= 1700

		DEL=.02			DEL=.05			DEL=.10			DEL=.15			DEL=.20		
FACT=		1.0 .75	.50 .25	.00 BIN	1.0 .75	.50 .25	.00 BIN	1.0 .75	.50 .25	.00 BIN	1.0 75	.50 .25	.00 BIN	1.0 .75	.50 .25	.00 BIN
PCONT=•••					REQUIRED NUMBER OF PATIENTS											
0.05	***	3726	3758	3922	806	816	832	291	293	295	171	171	172	120	121	121
	•••	3738	3803	7329	811	822	1432	292	294	456	171	172	241	120	121	153
0.1	***	7757	7841	8614	1493	1530	1604	478	486	495	257	259	262	170	171	172
	•••	7786	7985	12731	1508	1559	2265	481	490	651	258	261	322	170	171	196
0.15	***	11347	11494	13323	2095	2171	2351	641	656	679	330	335	342	210	212	216
	•••	11396	11766	17482	2126	2239	2994	648	667	821	333	339	391	211	214	232
0.2	***	14222	14450	17703	2586	2710	3033	777	803	841	392	401	410	244	248	252
	•••	14298	14877	21583	2637	2826	3619	789	820	964	396	405	449	245	250	261
0.25	***	16310	16636	21585	2960	3138	3629	885	923	979	441	452	467	271	275	280
	•••	16419	17250	25032	3032	3308	4140	902	947	1081	446	460	495	273	278	284
0.3	***	17624	18072	24877	3225	3459	4126	967	1017	1092	478	493	512	290	295	303
	•••	17775	18903	27831	3319	3684	4556	989	1049	1172	484	501	530	292	299	300
0.35	***	18224	18818	27524	3388	3678	4519	1023	1083	1177	503	520	544	301	309	318
	•••	18423	19897	29979	3506	3958	4869	1051	1124	1237	511	531	553	306	313	310
0.4	***	18193	18963	29494	3463	3805	4802	1055	1125	1234	516	537	563	308	316	325
	•••	18452	20317	31475	3603	4133	5077	1087	1172	1276	526	549	565	311	321	313
0.45	***	17632	18610	30769	3463	3849	4972	1066	1142	1263	520	542	571	307	316	326
	•••	17964	20251	32322	3621	4217	5181	1100	1194	1289	530	554	565	311	321	310
0.5	***	16653	17866	31341	3397	3820	5031	1055	1136	1264	511	534	564	301	309	318
	•••	17072	19787	32517	3571	4216	5181	1091	1191	1276	522	548	553	305	313	300
0.55	***	15372	16835	31203	3276	3722	4974	1025	1107	1236	494	516	546	287	295	305
	•••	15890	19003	32061	3461	4134	5077	1062	1163	1237	505	530	530	291	299	284
0.6	***	13910	15608	30357	3109	3563	4806	976	1056	1179	465	488	514	267	275	284
	•••	14528	17962	30955	3298	3976	4869	1013	1110	1172	476	500	495	271	279	261
0.65	***	12371	14254	28806	2898	3344	4524	908	983	1095	428	447	471	242	248	256
	•••	13074	16708	29197	3085	3743	4556	942	1032	1081	437	458	449	245	252	232
0.7	***	10835	12816	26555	2647	3067	4130	821	887	983	379	395	415	210	216	221
	•••	11591	15267	26789	2825	3435	4140	852	930	964	386	405	391	212	218	196
0.75	***	9336	11305	23608	2354	2730	3625	715	768	843	321	333	346	172	175	180
	•••	10100	13640	23730	2514	3049	3619	739	802	821	326	339	322	173	177	153
0.8	***	7870	9705	19973	2013	2324	3011	586	624	675	250	257	267	.	.	.
	•••	8590	11804	20021	2147	2579	2994	605	648	651	254	261	241	.	.	.
0.85	***	6387	7962	15656	1610	1837	2288	431	454	481	.	.	.	.	.	.
	•••	7014	9695	15660	1710	2014	2265	442	466	456	.	.	.	.	.	.
0.9	***	4787	5958	10666	1115	1243	1457	.	.	.	.	.	.	.	.	.
	•••	5259	7177	10648	1172	1333	1432	.	.	.	.	.	.	.	.	.
0.95	***	2804	3386	5012	.	.	.	.	.	.	.	.	.	.	.	.
	•••	3047	3917	4986	.	.	.	.	.	.	.	.	.	.	.	.

TABLE 4: ALPHA= 0.01 POWER= 0.9   EXPECTED ACCRUAL THRU MINIMUM FOLLOW-UP= 1800

	DEL=.02			DEL=.05			DEL=.10			DEL=.15			DEL=.20		
FACT=	1.0 .75	.50 .25	.00 BIN	1.0 .75	.50 .25	.00 BIN	1.0 .75	.50 .25	.00 BIN	1.0 75	.50 .25	.00 BIN	1.0 .75	.50 .25	.00 BIN
PCONT=•••				REQUIRED NUMBER OF PATIENTS											
0.05 •••	3728	3762	3921	807	816	831	291	294	296	171	171	172	120	120	121
•••	3739	3807	7329	811	823	1432	292	294	456	171	172	241	120	121	153
0.1 •••	7762	7850	8615	1496	1532	1604	479	486	495	258	260	262	170	171	172
•••	7793	7998	12731	1512	1561	2265	483	490	651	258	261	322	171	172	196
0.15 •••	11355	11511	13323	2101	2178	2351	642	658	678	332	336	342	211	213	216
•••	11409	11794	17482	2133	2244	2994	649	667	821	334	339	391	211	215	232
0.2 •••	14235	14476	17703	2595	2721	3033	780	805	841	393	400	411	244	247	252
•••	14316	14922	21583	2647	2835	3619	791	821	964	397	406	449	246	249	261
0.25 •••	16329	16674	21585	2974	3154	3629	888	926	980	442	453	468	271	276	281
•••	16445	17315	25032	3048	3321	4140	906	949	1081	447	460	495	273	278	284
0.3 •••	17651	18123	24877	3243	3478	4126	972	1019	1092	479	494	512	290	296	303
•••	17810	18991	27831	3339	3701	4556	993	1052	1172	486	503	530	294	299	300
0.35 •••	18260	18886	27524	3411	3702	4519	1028	1088	1178	504	522	544	303	309	318
•••	18470	20011	29979	3530	3978	4869	1055	1127	1237	513	532	553	306	314	310
0.4 •••	18240	19050	29494	3491	3834	4803	1062	1129	1234	519	539	564	308	316	325
•••	18514	20456	31475	3631	4157	5077	1092	1176	1276	528	550	565	312	321	313
0.45 •••	17691	18719	30770	3493	3881	4972	1073	1147	1263	522	543	570	308	316	326
•••	18042	20415	32322	3653	4245	5181	1107	1198	1289	532	555	565	312	321	310
0.5 •••	16728	17997	31340	3431	3854	5031	1063	1142	1264	514	537	564	301	309	319
•••	17169	19974	32517	3606	4245	5181	1098	1194	1276	524	549	553	305	314	300
0.55 •••	15465	16989	31203	3313	3757	4974	1032	1113	1236	496	519	546	288	296	305
•••	16008	19208	32061	3498	4164	5077	1068	1167	1237	506	531	530	291	300	284
0.6 •••	14023	15780	30358	3147	3599	4806	983	1061	1180	468	488	515	269	276	285
•••	14665	18179	30955	3336	4007	4869	1019	1113	1172	478	501	495	272	280	261
0.65 •••	12502	14438	28806	2936	3379	4524	915	987	1095	429	447	471	243	249	256
•••	13227	16929	29197	3122	3772	4556	948	1035	1081	438	459	449	245	252	232
0.7 •••	10977	13002	26554	2683	3100	4130	828	891	983	381	396	415	210	216	222
•••	11751	15483	26789	2859	3462	4140	857	933	964	388	405	391	213	218	196
0.75 •••	9481	11485	23608	2387	2758	3626	720	771	843	321	333	346	172	175	179
•••	10261	13843	23730	2546	3072	3619	744	804	821	327	339	322	174	177	153
0.8 •••	8007	9870	19973	2040	2348	3012	589	627	676	251	258	267	.	.	.
•••	8741	11983	20021	2173	2598	2994	607	649	651	254	262	241	.	.	.
0.85 •••	6509	8101	15657	1630	1854	2288	434	456	481	.	.	.	.	.	.
•••	7143	9840	15660	1728	2026	2265	444	468	456	.	.	.	.	.	.
0.9 •••	4879	6059	10666	1127	1252	1458	.	.	.	.	.	.	.	.	.
•••	5356	7275	10648	1183	1340	1432	.	.	.	.	.	.	.	.	.
0.95 •••	2852	3432	5012	.	.	.	.	.	.	.	.	.	.	.	.
•••	3095	3957	4986	.	.	.	.	.	.	.	.	.	.	.	.

TABLE 4: ALPHA= 0.01 POWER= 0.9   EXPECTED ACCRUAL THRU MINIMUM FOLLOW-UP= 1900

		DEL=.02			DEL=.05			DEL=.10			DEL=.15			DEL=.20		
FACT=		1.0 .75	.50 .25	.00 BIN	1.0 .75	.50 .25	.00 BIN	1.0 .75	.50 .25	.00 BIN	1.0 75	.50 .25	.00 BIN	1.0 .75	.50 .25	.00 BIN
PCONT=•••				REQUIRED NUMBER OF PATIENTS												
0.05	***	3730	3765	3921	807	817	831	291	293	296	171	172	172	120	121	121
	•••	3743	3811	7329	812	824	1432	292	294	456	171	172	241	121	121	153
0.1	***	7767	7860	8615	1498	1535	1604	479	486	495	258	260	262	170	171	172
	•••	7799	8013	12731	1514	1564	2265	483	490	651	258	261	322	171	172	196
0.15	***	11364	11528	13323	2106	2184	2351	643	659	679	331	336	342	211	213	216
	•••	11420	11821	17482	2139	2249	2994	650	668	821	334	339	391	212	215	232
0.2	***	14249	14503	17703	2605	2731	3032	781	806	840	394	401	410	244	248	251
	•••	14334	14967	21583	2656	2842	3619	793	823	964	398	406	449	247	250	261
0.25	***	16348	16712	21585	2987	3168	3629	892	928	980	443	453	467	272	275	281
	•••	16470	17379	25032	3061	3333	4140	908	951	1081	448	460	495	273	277	284
0.3	***	17677	18175	24877	3260	3497	4126	976	1022	1092	481	495	512	291	296	303
	•••	17844	19077	27831	3357	3716	4556	997	1053	1172	486	503	530	293	299	300
0.35	***	18294	18954	27523	3433	3725	4519	1034	1091	1177	505	524	543	304	310	318
	•••	18517	20121	29979	3553	3997	4869	1060	1129	1237	514	533	553	306	313	310
0.4	***	18285	19137	29494	3516	3860	4802	1067	1134	1235	520	539	564	310	317	325
	•••	18574	20591	31475	3657	4180	5077	1098	1178	1276	529	551	565	313	320	313
0.45	***	17750	18826	30769	3523	3911	4973	1079	1151	1263	524	543	570	308	317	325
	•••	18120	20573	32322	3683	4270	5181	1113	1200	1289	533	557	565	312	320	310
0.5	***	16802	18127	31340	3464	3887	5030	1070	1147	1263	516	538	564	301	310	319
	•••	17265	20153	32517	3640	4272	5181	1104	1198	1276	526	550	553	306	315	300
0.55	***	15558	17138	31202	3347	3792	4974	1040	1117	1236	498	519	546	288	296	305
	•••	16123	19405	32061	3533	4193	5077	1075	1170	1237	508	532	530	292	300	284
0.6	***	14133	15947	30357	3181	3632	4806	990	1066	1180	470	490	514	269	277	284
	•••	14800	18386	30955	3371	4035	4869	1025	1116	1172	479	501	495	273	280	261
0.65	***	12630	14614	28806	2971	3412	4524	921	992	1096	431	448	471	243	249	256
	•••	13377	17140	29197	3157	3800	4556	954	1037	1081	440	459	449	246	253	232
0.7	***	11115	13182	26554	2717	3131	4130	833	895	983	382	398	415	211	216	220
	•••	11908	15690	26789	2892	3486	4140	862	934	964	389	406	391	213	218	196
0.75	***	9622	11658	23609	2416	2785	3625	725	774	843	323	334	346	172	175	179
	•••	10415	14036	23730	2574	3093	3619	748	806	821	327	339	322	174	178	153
0.8	***	8141	10027	19972	2065	2370	3011	593	629	676	251	258	266	.	.	.
	•••	8886	12154	20021	2196	2614	2994	610	650	651	255	262	241	.	.	.
0.85	***	6625	8234	15656	1648	1870	2288	436	456	482	.	.	.	.	.	.
	•••	7267	9977	15660	1745	2037	2265	445	469	456	.	.	.	.	.	.
0.9	***	4967	6155	10666	1137	1260	1458	.	.	.	.	.	.	.	.	.
	•••	5448	7367	10648	1193	1344	1432	.	.	.	.	.	.	.	.	.
0.95	***	2897	3477	5011	.	.	.	.	.	.	.	.	.	.	.	.
	•••	3141	3993	4986	.	.	.	.	.	.	.	.	.	.	.	.

TABLE 4: ALPHA= 0.01 POWER= 0.9 EXPECTED ACCRUAL THRU MINIMUM FOLLOW-UP= 2000

	DEL=.02			DEL=.05			DEL=.10			DEL=.15			DEL=.20		
FACT=	1.0 .75	.50 .25	.00 BIN	1.0 .75	.50 .25	.00 BIN	1.0 .75	.50 .25	.00 BIN	1.0 75	.50 .25	.00 BIN	1.0 .75	.50 .25	.00 BIN
PCONT=•••				REQUIRED NUMBER OF PATIENTS											
0.05 •••	3732 3769 3921			809 817 831			291 294 296			171 171 172			120 121 121		
•••	3745 3814 7329			812 824 1432			292 295 456			171 172 241			121 121 153		
0.1 •••	7772 7869 8615			1501 1537 1604			480 487 495			257 260 262			170 171 172		
•••	7805 8026 12731			1517 1565 2265			484 491 651			259 261 322			171 171 196		
0.15 •••	11374 11545 13324			2112 2189 2351			645 660 679			332 337 342			211 214 216		
•••	11431 11849 17482			2145 2252 2994			652 669 821			335 340 391			212 215 232		
0.2 •••	14262 14529 17704			2614 2740 3034			784 809 841			395 402 411			245 249 252		
•••	14352 15010 21583			2666 2850 3619			795 824 964			397 406 449			246 250 261		
0.25 •••	16367 16750 21586			3001 3181 3629			895 930 980			444 455 467			272 276 281		
•••	16496 17441 25032			3075 3344 4140			911 952 1081			449 461 495			274 279 284		
0.3 •••	17704 18227 24877			3277 3515 4127			980 1026 1092			482 495 512			291 297 302		
•••	17880 19161 27831			3375 3731 4556			1001 1055 1172			489 504 530			294 300 300		
0.35 •••	18330 19022 27524			3454 3747 4519			1039 1095 1177			507 524 544			304 311 317		
•••	18564 20227 29979			3575 4016 4869			1065 1131 1237			515 534 553			307 314 310		
0.4 •••	18331 19224 29494			3541 3886 4802			1074 1137 1235			522 541 564			310 317 325		
•••	18635 20721 31475			3684 4201 5077			1102 1181 1276			531 551 565			314 321 313		
0.45 •••	17809 18934 30770			3551 3940 4972			1086 1156 1264			525 545 570			310 317 326		
•••	18196 20727 32322			3712 4294 5181			1119 1204 1289			535 557 565			314 321 310		
0.5 •••	16876 18254 31340			3495 3917 5031			1076 1151 1264			517 539 565			302 310 319		
•••	17360 20326 32517			3671 4299 5181			1110 1201 1276			527 551 553			306 315 300		
0.55 •••	15650 17285 31204			3380 3824 4975			1046 1122 1236			500 521 546			290 296 305		
•••	16239 19594 32061			3566 4220 5077			1081 1172 1237			510 532 530			292 301 284		
0.6 •••	14244 16107 30357			3215 3665 4806			997 1071 1180			472 491 515			270 276 284		
•••	14931 18585 30955			3405 4061 4869			1031 1119 1172			481 502 495			274 280 261		
0.65 •••	12755 14786 28807			3004 3444 4524			929 996 1095			434 450 471			244 250 256		
•••	13521 17342 29197			3190 3825 4556			960 1041 1081			441 460 449			246 252 232		
0.7 •••	11250 13355 26555			2749 3160 4130			839 899 984			384 399 415			211 216 221		
•••	12060 15887 26789			2924 3510 4140			867 937 964			391 406 391			214 219 196		
0.75 •••	9757 11825 23609			2445 2811 3626			730 777 844			324 335 346			172 176 179		
•••	10565 14220 23730			2602 3112 3619			752 807 821			329 340 322			175 177 153		
0.8 •••	8270 10179 19972			2090 2390 3011			596 631 676			252 259 266			. . .		
•••	9024 12315 20021			2220 2629 2994			612 652 651			255 262 241			. . .		
0.85 •••	6736 8360 15656			1666 1884 2289			437 457 482			. . .			. . .		
•••	7386 10106 15660			1762 2047 2265			447 469 456			. . .			. . .		
0.9 •••	5051 6245 10666			1149 1267 1457			. . .			. . .			. . .		
•••	5536 7454 10648			1202 1350 1432			. . .			. . .			. . .		
0.95 •••	2941 3517 5011			. . .			. . .			. . .			. . .		
•••	3184 4027 4986			. . .			. . .			. . .			. . .		

TABLE 4: ALPHA= 0.01 POWER= 0.9   EXPECTED ACCRUAL THRU MINIMUM FOLLOW-UP= 2250

		DEL=.02			DEL=.05			DEL=.10			DEL=.15			DEL=.20		
FACT=		1.0 .75	.50 .25	.00 BIN	1.0 .75	.50 .25	.00 BIN	1.0 .75	.50 .25	.00 BIN	1.0 75	.50 .25	.00 BIN	1.0 .75	.50 .25	.00 BIN
PCONT=***				REQUIRED	NUMBER	OF	PATIENTS									
0.05	***	3737	3776	3922	809	820	831	292	294	295	171	171	173	120	120	120
	•••	3751	3822	7329	814	825	1432	292	295	456	171	171	241	120	120	153
0.1	***	7784	7891	8614	1507	1542	1604	480	488	494	258	260	263	170	171	173
	•••	7821	8057	12731	1523	1569	2265	485	491	651	260	261	322	171	171	196
0.15	***	11395	11587	13324	2126	2201	2351	648	662	679	333	337	342	212	213	215
	•••	11459	11912	17482	2158	2261	2994	654	669	821	336	340	391	212	215	232
0.2	***	14295	14594	17703	2635	2760	3033	789	811	840	396	403	410	246	249	252
	•••	14397	15114	21583	2688	2865	3619	798	825	964	399	406	449	247	250	261
0.25	***	16416	16844	21585	3030	3211	3629	902	935	980	447	455	468	272	277	281
	•••	16559	17592	25032	3106	3368	4140	916	955	1081	451	461	495	274	278	284
0.3	***	17769	18354	24878	3317	3554	4127	988	1032	1092	485	497	511	292	298	303
	•••	17967	19362	27831	3416	3763	4556	1008	1059	1172	491	505	530	295	300	300
0.35	***	18417	19189	27524	3503	3795	4519	1050	1102	1177	511	525	544	305	312	317
	•••	18679	20483	29979	3625	4055	4869	1074	1136	1237	519	534	553	308	314	310
0.4	***	18446	19436	29494	3598	3943	4802	1085	1147	1234	525	542	564	312	317	326
	•••	18784	21030	31475	3743	4248	5077	1113	1186	1276	534	552	565	314	322	313
0.45	***	17954	19192	30770	3616	4003	4972	1099	1166	1264	530	548	570	311	317	326
	•••	18387	21088	32322	3778	4347	5181	1130	1209	1289	538	558	565	314	322	310
0.5	***	17060	18561	31341	3566	3987	5030	1091	1161	1264	522	541	565	305	311	319
	•••	17593	20733	32517	3743	4355	5181	1123	1208	1276	531	552	553	308	314	300
0.55	***	15875	17634	31203	3456	3897	4975	1062	1133	1236	505	523	545	291	298	305
	•••	16516	20036	32061	3642	4279	5077	1094	1178	1237	513	534	530	294	300	284
0.6	***	14509	16491	30358	3293	3737	4806	1011	1081	1180	475	493	514	271	277	284
	•••	15249	19049	30955	3481	4120	4869	1043	1124	1172	483	503	495	274	281	261
0.65	***	13055	15189	28807	3081	3514	4524	942	1005	1095	437	452	471	244	250	255
	•••	13868	17811	29197	3265	3880	4556	972	1046	1081	444	461	449	247	253	232
0.7	***	11570	13760	26555	2820	3225	4130	851	907	983	387	401	415	212	216	221
	•••	12418	16343	26789	2994	3560	4140	877	942	964	393	407	391	215	219	196
0.75	***	10079	12214	23608	2510	2868	3625	739	784	843	326	336	347	173	176	179
	•••	10918	14644	23730	2665	3155	3619	761	811	821	330	342	322	174	178	153
0.8	***	8571	10529	19973	2144	2435	3011	603	635	676	253	260	267	.	.	.
	•••	9350	12687	20021	2271	2662	2994	618	655	651	257	263	241	.	.	.
0.85	***	6997	8652	15657	1707	1917	2288	441	460	482	.	.	.	.	.	.
	•••	7662	10402	15660	1799	2069	2265	449	471	456	.	.	.	.	.	.
0.9	***	5248	6453	10666	1171	1284	1458	.	.	.	.	.	.	.	.	.
	•••	5740	7651	10648	1222	1360	1432	.	.	.	.	.	.	.	.	.
0.95	***	3040	3611	5011	.	.	.	.	.	.	.	.	.	.	.	.
	•••	3282	4103	4986	.	.	.	.	.	.	.	.	.	.	.	.

TABLE 4: ALPHA= 0.01 POWER= 0.9   EXPECTED ACCRUAL THRU MINIMUM FOLLOW-UP= 2500

		DEL=.02			DEL=.05			DEL=.10			DEL=.15			DEL=.20		
FACT=		1.0 .75	.50 .25	.00 BIN	1.0 .75	.50 .25	.00 BIN	1.0 .75	.50 .25	.00 BIN	1.0 75	.50 .25	.00 BIN	1.0 .75	.50 .25	.00 BIN
PCONT=***					REQUIRED NUMBER OF PATIENTS											
0.05	***	3742	3783	3921	812	820	831	292	293	295	171	171	171	120	121	121
	...	3758	3827	7329	815	824	1432	293	295	456	171	171	241	120	121	153
0.1	***	7796	7914	8615	1514	1548	1604	483	489	495	259	261	262	170	171	171
	...	7837	8086	12731	1529	1571	2265	486	492	651	259	262	322	171	171	196
0.15	***	11417	11627	13323	2137	2212	2351	649	662	677	334	337	342	212	214	215
	...	11489	11971	17482	2170	2268	2994	656	670	821	336	340	391	212	215	232
0.2	***	14329	14658	17702	2654	2777	3033	793	814	842	396	402	411	246	249	251
	...	14440	15211	21583	2708	2879	3619	802	826	964	399	408	449	248	249	261
0.25	***	16464	16936	21586	3059	3237	3629	908	939	979	448	458	467	273	276	281
	...	16624	17733	25032	3134	3387	4140	921	958	1081	452	462	495	274	279	284
0.3	***	17836	18479	24877	3352	3589	4126	996	1037	1092	487	498	512	293	298	302
	...	18054	19549	27831	3452	3790	4556	1015	1062	1172	492	504	530	295	299	300
0.35	***	18504	19351	27523	3546	3839	4518	1059	1109	1177	514	527	543	306	312	318
	...	18795	20720	29979	3670	4089	4869	1083	1140	1237	520	536	553	309	315	310
0.4	***	18559	19639	29493	3651	3993	4801	1096	1154	1234	529	545	564	312	318	324
	...	18934	21318	31475	3795	4289	5077	1123	1190	1276	537	554	565	315	321	313
0.45	***	18101	19440	30770	3676	4061	4973	1111	1174	1264	533	549	570	312	318	326
	...	18573	21421	32322	3839	4392	5181	1140	1214	1289	540	559	565	315	321	310
0.5	***	17242	18851	31340	3631	4046	5031	1102	1170	1264	526	543	564	306	312	318
	...	17821	21108	32517	3808	4402	5181	1134	1212	1276	534	552	553	309	315	300
0.55	***	16095	17964	31202	3524	3959	4974	1073	1142	1236	508	524	546	292	298	304
	...	16783	20439	32061	3709	4327	5077	1104	1184	1237	515	536	530	295	301	284
0.6	***	14767	16849	30358	3362	3801	4806	1023	1089	1179	479	495	515	271	277	284
	...	15549	19470	30955	3549	4170	4869	1054	1129	1172	487	504	495	274	281	261
0.65	***	13340	15562	28808	3148	3574	4524	952	1014	1095	440	454	471	246	251	256
	...	14192	18236	29197	3331	3927	4556	981	1051	1081	446	462	449	248	252	232
0.7	***	11870	14133	26554	2884	3281	4131	861	914	983	389	401	415	214	217	221
	...	12751	16754	26789	3056	3602	4140	886	945	964	395	408	391	215	218	196
0.75	***	10377	12570	23608	2567	2917	3626	746	789	843	327	337	346	174	176	179
	...	11242	15024	23730	2720	3192	3619	767	814	821	333	342	322	174	177	153
0.8	***	8849	10849	19973	2192	2474	3011	609	639	676	254	261	267	.	.	.
	...	9648	13018	20021	2317	2690	2994	623	656	651	258	264	241	.	.	.
0.85	***	7237	8917	15656	1742	1943	2289	445	462	483	.	.	.	.	.	.
	...	7914	10664	15660	1831	2087	2265	452	471	456	.	.	.	.	.	.
0.9	***	5426	6639	10667	1190	1298	1458	.	.	.	.	.	.	.	.	.
	...	5923	7823	10648	1240	1368	1432	.	.	.	.	.	.	.	.	.
0.95	***	3129	3692	5011	.	.	.	.	.	.	.	.	.	.	.	.
	...	3370	4167	4986	.	.	.	.	.	.	.	.	.	.	.	.

TABLE 4: ALPHA= 0.01 POWER= 0.9   EXPECTED ACCRUAL THRU MINIMUM FOLLOW-UP= 2750

		DEL=.02			DEL=.05			DEL=.10			DEL=.15			DEL=.20		
FACT=		1.0 .75	.50 .25	.00 BIN	1.0 .75	.50 .25	.00 BIN	1.0 .75	.50 .25	.00 BIN	1.0 75	.50 .25	.00 BIN	1.0 .75	.50 .25	.00 BIN
PCONT=***					REQUIRED	NUMBER OF	PATIENTS									
0.05	***	3746	3789	3921	813	821	831	291	294	295	171	171	171	120	122	122
	•••	3762	3834	7329	816	826	1432	294	295	456	171	171	241	120	122	153
0.1	***	7809	7935	8616	1519	1552	1603	484	489	494	259	260	263	170	171	171
	•••	7854	8110	12731	1532	1574	2265	486	491	651	259	260	322	171	171	196
0.15	***	11438	11668	13323	2148	2220	2350	652	665	679	335	338	342	212	214	216
	•••	11517	12027	17482	2179	2275	2994	658	672	821	336	340	391	212	214	232
0.2	***	14363	14721	17702	2672	2794	3033	795	816	841	398	404	410	247	249	252
	•••	14485	15302	21583	2724	2890	3619	806	828	964	401	407	449	249	250	261
0.25	***	16512	17025	21585	3083	3260	3629	912	941	979	450	459	467	274	278	281
	•••	16687	17864	25032	3158	3404	4140	926	958	1081	453	462	495	276	280	284
0.3	***	17902	18600	24876	3385	3619	4126	1003	1041	1092	489	500	511	294	298	302
	•••	18141	19726	27831	3485	3814	4556	1020	1065	1172	494	505	530	297	300	300
0.35	***	18593	19509	27523	3588	3877	4520	1067	1113	1177	517	529	544	307	312	318
	•••	18909	20943	29979	3710	4117	4869	1089	1143	1237	522	535	553	311	315	310
0.4	***	18672	19837	29493	3698	4037	4801	1106	1160	1233	532	546	563	314	319	326
	•••	19079	21585	31475	3842	4322	5077	1130	1194	1276	539	555	565	315	322	313
0.45	***	18244	19678	30768	3729	4110	4972	1122	1180	1263	535	551	570	314	319	326
	•••	18754	21731	32322	3890	4430	5181	1149	1218	1289	542	560	565	315	322	310
0.5	***	17419	19129	31341	3690	4100	5030	1113	1177	1264	529	544	565	307	312	319
	•••	18041	21451	32517	3865	4446	5181	1143	1216	1276	535	555	553	309	315	300
0.55	***	16309	18273	31204	3587	4014	4975	1084	1147	1235	510	527	546	294	298	305
	•••	17039	20810	32061	3769	4372	5077	1115	1189	1237	518	535	530	295	302	284
0.6	***	15013	17183	30358	3425	3856	4805	1034	1096	1180	483	496	515	273	278	283
	•••	15836	19856	30955	3611	4212	4869	1064	1134	1172	489	505	495	276	281	261
0.65	***	13610	15908	28808	3210	3629	4523	962	1019	1096	441	455	470	247	250	256
	•••	14497	18622	29197	3391	3968	4556	989	1055	1081	448	463	449	249	254	232
0.7	***	12153	14478	26554	2942	3330	4130	869	919	982	391	401	415	214	218	221
	•••	13062	17128	26789	3110	3640	4140	893	948	964	397	408	391	216	219	196
0.75	***	10655	12897	23608	2619	2959	3626	754	793	844	329	338	346	175	177	178
	•••	11544	15370	23730	2769	3222	3619	773	816	821	333	342	322	175	178	153
0.8	***	9109	11142	19973	2234	2509	3010	615	642	676	256	260	266	.	.	.
	•••	9923	13316	20021	2356	2714	2994	627	658	651	257	264	241	.	.	.
0.85	***	7457	9157	15656	1772	1965	2289	448	463	483	.	.	.	.	.	.
	•••	8146	10898	15660	1859	2103	2265	455	472	456	.	.	.	.	.	.
0.9	***	5589	6808	10666	1208	1309	1457	.	.	.	.	.	.	.	.	.
	•••	6090	7974	10648	1254	1376	1432	.	.	.	.	.	.	.	.	.
0.95	***	3210	3763	5011	.	.	.	.	.	.	.	.	.	.	.	.
	•••	3447	4223	4986	.	.	.	.	.	.	.	.	.	.	.	.

TABLE 4: ALPHA= 0.01 POWER= 0.9   EXPECTED ACCRUAL THRU MINIMUM FOLLOW-UP= 3000

	DEL=.01			DEL=.02			DEL=.05			DEL=.10			DEL=.15		
FACT=	1.0 .75	.50 .25	.00 BIN	1.0 .75	.50 .25	.00 BIN	1.0 .75	.50 .25	.00 BIN	1.0 75	.50 .25	.00 BIN	1.0 .75	.50 .25	.00 BIN
PCONT=•••				REQUIRED NUMBER OF PATIENTS											
0.01 •••	1990	1993	2008	730	733	737	242	242	242	122	122	122	85	85	85
•••	1990	2000	7680	733	733	2539	242	242	691	122	122	281	85	85	167
0.02 •••	4390	4408	4480	1412	1420	1430	385	385	385	167	167	167	110	110	110
•••	4397	4430	12679	1412	1423	3775	385	385	883	167	167	326	110	110	186
0.05 •••	13285	13340	13993	3752	3793	3920	815	823	830	290	295	295	170	170	170
•••	13303	13445	27050	3767	3838	7329	815	827	1432	295	295	456	170	170	241
0.1 •••	29297	29435	32570	7820	7955	8615	1525	1555	1603	485	490	493	260	260	260
•••	29342	29713	48918	7870	8135	12731	1535	1577	2265	485	493	651	260	260	322
0.15 •••	43735	43985	51455	11458	11705	13322	2158	2230	2350	655	665	677	335	340	343
•••	43817	44480	68183	11545	12077	17482	2188	2278	2994	658	670	821	335	340	391
0.2 •••	55352	55742	69130	14395	14780	17702	2687	2807	3032	797	815	842	400	403	410
•••	55483	56515	84845	14530	15385	21583	2740	2900	3619	808	827	964	403	407	449
0.25 •••	63797	64355	84868	16558	17113	21587	3107	3280	3628	917	943	980	452	460	467
•••	63980	65473	98903	16750	17987	25032	3182	3418	4140	928	962	1081	455	463	495
0.3 •••	69080	69853	98278	17968	18718	24875	3415	3647	4127	1007	1045	1093	490	500	512
•••	69340	71390	110358	18227	19892	27831	3515	3835	4556	1025	1067	1172	493	505	530
0.35 •••	71425	72452	109123	18680	19663	27523	3625	3910	4517	1075	1120	1175	520	530	542
•••	71765	74500	119210	19022	21152	29979	3748	4142	4869	1093	1145	1237	523	538	553
0.4 •••	71135	72478	117275	18785	20027	29492	3740	4078	4802	1112	1165	1235	535	545	565
•••	71582	75148	125458	19225	21835	31475	3887	4352	5077	1138	1198	1276	542	553	565
0.45 •••	68593	70325	122657	18385	19907	30770	3778	4153	4970	1130	1187	1262	538	553	568
•••	69170	73735	129102	18932	22018	32322	3940	4465	5181	1157	1220	1289	545	560	565
0.5 •••	64213	66430	125222	17593	19393	31340	3740	4150	5030	1123	1183	1262	530	545	565
•••	64952	70705	130144	18253	21770	32517	3917	4480	5181	1150	1220	1276	538	553	553
0.55 •••	58445	61258	124955	16517	18565	31202	3643	4063	4975	1093	1153	1235	512	527	545
•••	59387	66493	128582	17285	21152	32061	3823	4408	5077	1123	1190	1237	520	535	530
0.6 •••	51755	55295	121855	15250	17495	30358	3482	3905	4805	1045	1100	1180	482	497	515
•••	52952	61487	124416	16108	20210	30955	3665	4250	4869	1070	1138	1172	490	505	495
0.65 •••	44642	49000	115933	13865	16232	28805	3265	3677	4525	970	1025	1093	445	455	470
•••	46153	56000	117648	14785	18977	29197	3445	4003	4556	995	1055	1081	448	463	449
0.7 •••	37610	42730	107203	12418	14795	26555	2995	3373	4130	875	925	985	392	403	415
•••	39445	50230	108275	13355	17470	26789	3160	3670	4140	898	950	964	400	407	391
0.75 •••	31075	36677	95695	10918	13198	23608	2665	2998	3625	760	797	842	332	340	347
•••	33152	44252	96300	11825	15685	23730	2810	3250	3619	778	820	821	335	343	322
0.8 •••	25220	30850	81425	9350	11413	19970	2270	2537	3010	617	643	677	257	260	265
•••	27362	38012	81721	10180	13585	20021	2390	2732	2994	632	658	651	257	265	241
0.85 •••	19940	25078	64430	7663	9377	15655	1798	1985	2290	448	463	482	.	.	.
•••	21928	31330	64538	8360	11110	15660	1885	2117	2265	455	475	456	.	.	.
0.9 •••	14867	18985	44740	5740	6958	10667	1220	1318	1457	.	.	.	.	.	.
•••	16480	23770	44753	6245	8110	10648	1265	1382	1432	.	.	.	.	.	.
0.95 •••	9212	11653	22385	3283	3827	5012	.	.	.	.	.	.	.	.	.
•••	10190	14260	22364	3515	4270	4986	.	.	.	.	.	.	.	.	.
0.98 •••	4453	5353	7705	.	.	.	.	.	.	.	.	.	.	.	.
•••	4832	6148	7680	.	.	.	.	.	.	.	.	.	.	.	.

TABLE 4: ALPHA= 0.01 POWER= 0.9 EXPECTED ACCRUAL THRU MINIMUM FOLLOW-UP= 3250

REQUIRED NUMBER OF PATIENTS

PCONT	DEL=.01 1.0/.75	.50/.25	.00/BIN	DEL=.02 1.0/.75	.50/.25	.00/BIN	DEL=.05 1.0/.75	.50/.25	.00/BIN	DEL=.10 1.0/75	.50/.25	.00/BIN	DEL=.15 1.0/.75	.50/.25	.00/BIN
0.01 ***	1988	1993	2009	729	734	737	241	241	241	119	119	119	87	87	87
•••	1993	2001	7680	734	734	2539	241	241	691	119	119	281	87	87	167
0.02 ***	4393	4409	4479	1411	1419	1433	384	384	384	168	168	168	111	111	111
•••	4398	4434	12679	1416	1424	3775	384	384	883	168	168	326	111	111	186
0.05 ***	13290	13352	13994	3756	3800	3922	815	823	831	290	295	295	173	173	173
•••	13311	13461	27050	3773	3846	7329	818	826	1432	295	295	456	173	173	241
0.1 ***	29309	29459	32571	7835	7973	8615	1525	1557	1603	485	490	493	257	263	263
•••	29358	29759	48918	7884	8155	12731	1541	1579	2265	485	493	651	257	263	322
0.15 ***	43756	44027	51453	11483	11743	13322	2167	2237	2351	656	664	677	336	339	344
•••	43845	44563	68183	11573	12125	17482	2196	2286	2994	661	672	821	336	339	391
0.2 ***	55386	55805	69130	14427	14842	17702	2703	2817	3033	802	818	839	401	404	409
•••	55524	56645	84845	14571	15464	21583	2752	2906	3619	810	831	964	401	409	449
0.25 ***	63841	64450	84868	16610	17198	21586	3126	3296	3629	921	945	981	452	458	466
•••	64044	65656	98903	16813	18100	25032	3199	3431	4140	932	961	1081	458	461	495
0.3 ***	69146	69983	98277	18032	18836	24876	3442	3670	4125	1013	1046	1091	490	501	509
•••	69426	71644	110358	18311	20047	27831	3540	3849	4556	1029	1067	1172	498	506	530
0.35 ***	71511	72624	109124	18766	19811	27522	3659	3938	4520	1078	1124	1176	517	531	542
•••	71879	74837	119210	19132	21347	29979	3781	4166	4869	1099	1148	1237	526	539	553
0.4 ***	71246	72700	117279	18896	20209	29491	3781	4114	4799	1119	1167	1232	534	547	563
•••	71733	75584	125458	19364	22070	31475	3922	4377	5077	1143	1200	1276	542	555	565
0.45 ***	68735	70612	122658	18528	20123	30767	3824	4195	4970	1135	1192	1262	539	555	571
•••	69361	74284	129102	19107	22284	32322	3984	4496	5181	1164	1224	1289	547	563	565
0.5 ***	64396	66793	125220	17764	19644	31341	3792	4190	5032	1132	1189	1265	534	547	563
•••	65198	71376	130144	18457	22065	32517	3962	4512	5181	1156	1221	1276	539	555	553
0.55 ***	58681	61720	124957	16716	18844	31203	3691	4109	4975	1102	1159	1238	514	531	547
•••	59700	67286	128582	17520	21469	32061	3873	4442	5077	1127	1197	1237	523	539	530
0.6 ***	52054	55865	121856	15476	17791	30358	3532	3951	4804	1051	1108	1181	485	498	514
•••	53351	62386	124416	16366	20537	30955	3716	4284	4869	1078	1140	1172	493	506	495
0.65 ***	45023	49671	115933	14111	16537	28806	3318	3719	4523	978	1029	1094	444	458	469
•••	46643	56978	117648	15058	19302	29197	3491	4036	4556	1002	1059	1081	452	466	449
0.7 ***	38079	43474	107204	12669	15094	26555	3041	3415	4130	883	929	981	396	404	417
•••	40026	51239	108275	13628	17783	26789	3204	3699	4140	904	953	964	401	409	391
0.75 ***	31617	37454	95694	11161	13482	23609	2708	3033	3626	766	799	842	331	339	347
•••	33794	45237	96300	12087	15971	23730	2849	3272	3619	783	818	821	336	344	322
0.8 ***	25786	31601	81426	9574	11662	19974	2305	2565	3012	623	647	677	257	263	266
•••	28004	38929	81721	10419	13834	20021	2419	2752	2994	631	661	651	257	263	241
0.85 ***	20472	25746	64434	7854	9582	15654	1823	2004	2289	452	466	482	.	.	.
•••	22517	32113	64538	8558	11299	15660	1907	2126	2265	458	474	456	.	.	.
0.9 ***	15302	19506	44742	5880	7099	10666	1238	1327	1457	.	.	.	.	.	.
•••	16954	24353	44753	6384	8233	10648	1278	1387	1432	.	.	.	.	.	.
0.95 ***	9476	11954	22387	3350	3886	5011	.	.	.	.	.	.	.	.	.
•••	10476	14557	22364	3581	4312	4986	.	.	.	.	.	.	.	.	.
0.98 ***	4561	5449	7705	.	.	.	.	.	.	.	.	.	.	.	.
•••	4937	6229	7680	.	.	.	.	.	.	.	.	.	.	.	.

TABLE 4: ALPHA= 0.01 POWER= 0.9   EXPECTED ACCRUAL THRU MINIMUM FOLLOW-UP= 3500

		DEL=.01			DEL=.02			DEL=.05			DEL=.10			DEL=.15		
FACT=		1.0 .75	.50 .25	.00 BIN	1.0 .75	.50 .25	.00 BIN	1.0 .75	.50 .25	.00 BIN	1.0 75	.50 .25	.00 BIN	1.0 .75	.50 .25	.00 BIN
PCONT=***					REQUIRED NUMBER OF PATIENTS											
0.01	***	1989 1992	1992 2001	2006 7680	732 732	732 732	738 2539	242 242	242 242	242 691	120 120	120 120	120 281	85 85	85 85	85 167
0.02	***	4395 4398	4413 4433	4483 12679	1411 1415	1420 1423	1432 3775	382 382	382 382	388 883	169 169	169 169	169 326	111 111	111 111	111 186
0.05	***	13294 13315	13358 13478	13994 27050	3760 3777	3803 3847	3923 7329	816 820	825 825	828 1432	291 295	295 295	295 456	172 172	172 172	172 241
0.1	***	29318 29376	29481 29805	32570 48918	7846 7898	7991 8175	8616 12731	1528 1543	1560 1581	1604 2265	484 487	487 493	496 651	260 260	260 260	260 322
0.15	***	43779 43875	44068 44645	51453 68183	11503 11600	11780 12168	13323 17482	2173 2202	2243 2290	2351 2994	659 662	668 671	676 821	335 338	338 338	344 391
0.2	***	55416 55568	55871 56773	69128 84845	14461 14615	14898 15543	17704 21583	2715 2768	2829 2916	3033 3619	802 811	820 828	843 964	400 400	405 408	408 449
0.25	***	63890 64105	64543 65841	84869 98903	16654 16873	17284 18211	21583 25032	3147 3217	3313 3441	3628 4140	925 933	948 965	977 1081	452 458	458 461	466 495
0.3	***	69210 69513	70108 71896	98277 110358	18098 18395	18946 20192	24879 27831	3468 3567	3695 3865	4127 4556	1018 1035	1053 1070	1091 1172	493 496	501 505	510 530
0.35	***	71595 71992	72794 75174	109124 119210	18850 19244	19953 21528	27525 29979	3690 3809	3966 4185	4518 4869	1082 1105	1126 1149	1175 1237	522 528	531 536	545 553
0.4	***	71359 71878	72925 76014	117279 125458	19008 19501	20385 22289	29493 31475	3818 3961	4145 4404	4801 5077	1126 1149	1175 1201	1236 1276	536 545	548 557	563 565
0.45	***	68883 69556	70898 74824	122655 129102	18666 19273	20332 22537	30768 32322	3865 4022	4229 4521	4973 5181	1143 1170	1196 1228	1263 1289	540 548	554 563	571 565
0.5	***	64581 65444	67162 72033	125224 130144	17931 18658	19883 22345	31340 32517	3835 4005	4229 4544	5028 5181	1140 1166	1193 1228	1263 1276	536 540	548 557	563 553
0.55	***	58916 60013	62180 68055	124956 128582	16911 17748	19107 21764	31205 32061	3739 3917	4150 4474	4973 5077	1108 1135	1166 1196	1236 1237	519 522	531 536	545 530
0.6	***	52354 53745	56423 63256	121855 124416	15695 16613	18071 20840	30356 30955	3581 3760	3993 4311	4806 4869	1056 1082	1108 1143	1178 1172	487 493	501 505	513 495
0.65	***	45401 47130	50324 57910	115935 117648	14347 15315	16820 19606	28808 29197	3363 3532	3756 4063	4521 4556	986 1009	1035 1061	1096 1081	449 452	458 466	470 449
0.7	***	38541 40597	44199 52196	107202 108275	12909 13889	15376 18075	26553 26789	3083 3243	3450 3721	4127 4140	890 907	930 956	983 964	396 400	405 408	414 391
0.75	***	32145 34411	38200 46171	95693 96300	11395 12335	13743 16234	23610 23730	2745 2885	3060 3293	3625 3619	767 785	802 820	843 821	330 335	338 344	347 322
0.8	***	26331 28624	32316 39789	81425 81721	9788 10637	11894 14058	19973 20021	2333 2447	2587 2768	3013 2994	624 636	650 662	676 651	256 260	260 265	265 241
0.85	***	20980 23076	26378 32845	64432 64538	8030 8744	9768 11477	15656 15660	1843 1922	2018 2138	2286 2265	452 461	466 475	484 456	. .	. .	. .
0.9	***	15712 17401	19996 24896	44741 44753	6008 6516	7225 8341	10663 10648	1248 1289	1336 1394	1458 1432	. .	. .	. .	. .	. .	. .
0.95	***	9727 10739	12230 14834	22385 22364	3410 3637	3935 4351	5011 4986	. .	. .	. .	. .	. .	. .	. .	. .	. .
0.98	***	4658 5034	5541 6303	7706 7680	. .	. .	. .	. .	. .	. .	. .	. .	. .	. .	. .	. .

TABLE 4: ALPHA= 0.01 POWER= 0.9   EXPECTED ACCRUAL THRU MINIMUM FOLLOW-UP= 3750

	DEL=.01			DEL=.02			DEL=.05			DEL=.10			DEL=.15		
FACT=	1.0 .75	.50 .25	.00 BIN	1.0 .75	.50 .25	.00 BIN	1.0 .75	.50 .25	.00 BIN	1.0 75	.50 .25	.00 BIN	1.0 .75	.50 .25	.00 BIN
PCONT=***					REQUIRED NUMBER OF PATIENTS										
0.01 ***	1990 1994	1994 2000	2009 7680	734 734	734 734	734 2539	241 241	241 241	241 691	119 119	119 119	119 281	87 87	87 87	87 167
0.02 ***	4394 4413	4403 4437	4478 12679	1413 1418	1418 1422	1431 3775	381 381	381 387	387 883	166 166	166 166	166 326	109 109	109 109	109 186
0.05 ***	13300 13366	13325 13488	13994 27050	3762 3809	3781 3850	3922 7329	818 822	818 828	831 1432	293 293	293 293	293 456	172 172	172 172	172 241
0.1 ***	29331 29506	29388 29847	32572 48918	7859 8009	7915 8191	8613 12731	1534 1563	1550 1581	1606 2265	484 490	490 494	494 651	259 259	259 259	259 322
0.15 ***	43797 44106	43900 44725	51453 68183	11525 11815	11628 12209	13325 17482	2181 2247	2209 2290	2350 2994	659 668	663 672	678 821	334 340	334 340	340 391
0.2 ***	55450 55934	55609 56900	69128 84845	14497 14956	14656 15612	17703 21583	2725 2838	2778 2922	3031 3619	803 822	813 831	841 964	400 406	400 406	409 449
0.25 ***	63934 64634	64169 66025	84869 98903	16703 17365	16934 18312	21584 25032	3162 3331	3237 3453	3631 4140	925 950	940 963	978 1081	453 462	456 462	466 495
0.3 ***	69275 70240	69597 72153	98275 110358	18162 19056	18481 20328	24878 27831	3494 3715	3588 3878	4128 4556	1019 1053	1038 1072	1090 1172	494 503	500 509	513 530
0.35 ***	71678 72963	72109 75509	109122 119210	18934 20093	19353 21700	27522 29979	3719 3991	3837 4203	4516 4869	1090 1128	1109 1150	1178 1237	522 531	528 537	541 553
0.4 ***	71472 73150	72031 76441	117278 125458	19118 20556	19638 22493	29491 31475	3856 4175	3991 4422	4803 5077	1131 1178	1156 1203	1234 1276	537 550	547 556	565 565
0.45 ***	69025 71187	69747 75359	122656 129102	18800 20534	19441 22769	30772 32322	3903 4263	4062 4544	4972 5181	1150 1197	1175 1231	1263 1289	541 556	550 565	569 565
0.5 ***	64769 67525	65693 72678	125225 130144	18097 20106	18850 22600	31338 32517	3878 4263	4047 4568	5031 5181	1147 1197	1169 1225	1263 1276	537 550	541 556	565 553
0.55 ***	59150 62631	60325 68800	124956 128582	17103 19356	17965 22038	31203 32061	3784 4188	3959 4497	4975 5077	1118 1169	1141 1197	1234 1237	518 531	522 537	547 530
0.6 ***	52653 56975	54138 64090	121853 124416	15906 18334	16850 21125	30359 30955	3625 4028	3800 4338	4806 4869	1066 1113	1090 1147	1178 1172	490 500	494 509	513 495
0.65 ***	45781 50959	47609 58803	115934 117648	14572 17088	15559 19887	28806 29197	3406 3794	3575 4084	4525 4556	991 1038	1015 1066	1094 1081	447 456	453 466	472 449
0.7 ***	38997 44890	41150 53106	107206 108275	13137 15640	14131 18344	26553 26789	3125 3481	3278 3743	4131 4140	893 934	912 959	981 964	397 406	400 409	415 391
0.75 ***	32656 38909	35003 47059	95693 96300	11613 13990	12569 16478	23609 23730	2778 3087	2918 3312	3625 3619	775 803	790 822	841 821	334 340	334 344	344 322
0.8 ***	26856 32997	29215 40600	81428 81721	9987 12109	10850 14266	19972 20021	2365 2609	2472 2781	3012 2994	631 650	640 663	678 651	259 259	259 265	265 241
0.85 ***	21462 26978	23613 33531	64431 64538	8200 9944	8918 11637	15653 15660	1863 2031	1943 2144	2290 2265	456 466	462 475	481 456	. .	. .	. .
0.9 ***	16103 20463	17825 25400	44740 44753	6128 7343	6640 8444	10666 10648	1259 1343	1297 1394	1456 1432	. .	. .	. .	. .	. .	. .
0.95 ***	9968 12490	10984 15087	22384 22364	3466 3981	3691 4384	5009 4986	. .	. .	. .	. .	. .	. .	. .	. .	. .
0.98 ***	4747 5622	5122 6368	7703 7680	. .	. .	. .	. .	. .	. .	. .	. .	. .	. .	. .	. .

TABLE 4: ALPHA= 0.01 POWER= 0.9   EXPECTED ACCRUAL THRU MINIMUM FOLLOW-UP= 4000

		DEL=.01			DEL=.02			DEL=.05			DEL=.10			DEL=.15		
FACT=		1.0 .75	.50 .25	.00 BIN	1.0 .75	.50 .25	.00 BIN	1.0 .75	.50 .25	.00 BIN	1.0 75	.50 .25	.00 BIN	1.0 .75	.50 .25	.00 BIN
PCONT=•••					REQUIRED NUMBER OF PATIENTS											
0.01	•••	1993	1997	2007	733	733	737	243	243	243	123	123	123	87	87	87
	•••	1993	2003	7680	733	733	2539	243	243	691	123	123	281	87	87	167
0.02	•••	4397	4417	4483	1413	1423	1433	383	383	387	167	167	167	113	113	113
	•••	4403	4437	12679	1417	1427	3775	383	383	883	167	167	326	113	113	186
0.05	•••	13303	13377	13993	3767	3813	3923	817	823	833	293	293	297	173	173	173
	•••	13327	13503	27050	3787	3853	7329	823	827	1432	293	293	456	173	173	241
0.1	•••	29343	29527	32573	7867	8027	8613	1537	1563	1603	487	493	493	257	263	263
	•••	29407	29893	48918	7927	8207	12731	1547	1583	2265	487	493	651	263	263	322
0.15	•••	43817	44153	51453	11543	11847	13323	2187	2253	2353	657	667	677	337	337	343
	•••	43927	44807	68183	11653	12247	17482	2217	2297	2994	663	673	821	337	343	391
0.2	•••	55483	55997	69127	14527	15007	17703	2737	2847	3033	807	823	843	403	407	413
	•••	55653	57027	84845	14697	15677	21583	2787	2927	3619	813	833	964	403	407	449
0.25	•••	63983	64727	84867	16747	17443	21587	3183	3343	3627	927	953	977	453	463	467
	•••	64233	66207	98903	16997	18407	25032	3253	3463	4140	943	963	1081	457	463	495
0.3	•••	69337	70367	98277	18227	19163	24877	3513	3733	4127	1027	1053	1093	493	503	513
	•••	69683	72403	110358	18563	20457	27831	3607	3893	4556	1037	1073	1172	497	507	530
0.35	•••	71767	73137	109123	19023	20227	27523	3747	4017	4517	1093	1133	1177	523	533	543
	•••	72223	75837	119210	19457	21863	29979	3863	4217	4869	1113	1153	1237	527	537	553
0.4	•••	71583	73373	117277	19223	20723	29493	3887	4203	4803	1137	1183	1233	543	553	563
	•••	72177	76863	125458	19773	22683	31475	4023	4443	5077	1157	1207	1276	547	557	565
0.45	•••	69167	71477	122657	18933	20727	30767	3937	4293	4973	1157	1203	1263	543	557	567
	•••	69937	75883	129102	19597	22993	32322	4093	4563	5181	1177	1233	1289	553	563	565
0.5	•••	64953	67887	125223	18253	20327	31337	3917	4297	5033	1153	1203	1263	537	553	563
	•••	65937	73303	130144	19037	22843	32517	4083	4593	5181	1173	1227	1276	543	557	553
0.55	•••	59387	63083	124957	17283	19593	31203	3823	4217	4973	1123	1173	1237	523	533	547
	•••	60637	69527	128582	18173	22297	32061	3997	4523	5077	1147	1203	1237	527	537	530
0.6	•••	52953	57513	121857	16107	18583	30357	3663	4063	4807	1073	1117	1177	493	503	513
	•••	54527	64893	124416	17073	21387	30955	3837	4363	4869	1093	1147	1172	497	507	495
0.65	•••	46153	51577	115933	14787	17343	28807	3443	3823	4523	997	1043	1093	447	457	473
	•••	48077	59657	117648	15797	20153	29197	3613	4107	4556	1017	1067	1081	453	463	449
0.7	•••	39443	45563	107203	13353	15887	26553	3157	3507	4127	897	937	983	397	407	413
	•••	41687	53977	108275	14367	18593	26789	3313	3763	4140	917	957	964	403	407	391
0.75	•••	33153	39593	95693	11823	14217	23607	2813	3113	3627	777	807	843	333	337	347
	•••	35583	47903	96300	12787	16707	23730	2947	3327	3619	793	823	821	337	343	322
0.8	•••	27363	33647	81427	10177	12313	19973	2387	2627	3013	633	653	677	257	263	267
	•••	29783	41373	81721	11047	14463	20021	2497	2793	2994	643	663	651	263	263	241
0.85	•••	21927	27547	64433	8357	10107	15657	1883	2047	2287	457	467	483	.	.	.
	•••	24123	34183	64538	9077	11783	15660	1957	2153	2265	463	473	456	.	.	.
0.9	•••	16483	20903	44743	6243	7453	10667	1267	1347	1457	.	.	.	.	.	.
	•••	18227	25877	44753	6753	8533	10648	1307	1397	1432	.	.	.	.	.	.
0.95	•••	10193	12733	22387	3517	4027	5013	.	.	.	.	.	.	.	.	.
	•••	11223	15323	22364	3743	4417	4986	.	.	.	.	.	.	.	.	.
0.98	•••	4833	5703	7707	.	.	.	.	.	.	.	.	.	.	.	.
	•••	5203	6427	7680	.	.	.	.	.	.	.	.	.	.	.	.

TABLE 4: ALPHA= 0.01 POWER= 0.9   EXPECTED ACCRUAL THRU MINIMUM FOLLOW-UP= 4250

	DEL=.01			DEL=.02			DEL=.05			DEL=.10			DEL=.15		
FACT=	1.0 .75	.50 .25	.00 BIN	1.0 .75	.50 .25	.00 BIN	1.0 .75	.50 .25	.00 BIN	1.0 75	.50 .25	.00 BIN	1.0 .75	.50 .25	.00 BIN

PCONT=•••   REQUIRED NUMBER OF PATIENTS

PCONT	DEL=.01 1.0/.75	.50/.25	.00/BIN	DEL=.02 1.0/.75	.50/.25	.00/BIN	DEL=.05 1.0/.75	.50/.25	.00/BIN	DEL=.10 1.0/75	.50/.25	.00/BIN	DEL=.15 1.0/.75	.50/.25	.00/BIN
0.01 •••	1990	1994	2005	730	736	736	241	241	241	120	120	120	88	88	88
•••	1994	2001	7680	730	736	2539	241	241	691	120	120	281	88	88	167
0.02 •••	4395	4417	4480	1416	1420	1431	386	386	386	167	167	167	109	109	109
•••	4406	4438	12679	1416	1427	3775	386	386	883	167	167	326	109	109	186
0.05 •••	13310	13384	13996	3768	3818	3924	815	825	832	294	294	294	173	173	173
•••	13338	13518	27050	3790	3860	7329	821	825	1432	294	294	456	173	173	241
0.1 •••	29353	29551	32573	7880	8040	8613	1537	1565	1601	485	492	496	258	262	262
•••	29424	29934	48918	7940	8227	12731	1554	1586	2265	485	492	651	258	262	322
0.15 •••	43842	44193	51453	11567	11882	13320	2196	2256	2351	662	666	677	337	337	343
•••	43959	44887	68183	11680	12286	17482	2224	2298	2994	666	673	821	337	343	391
0.2 •••	55512	56065	69129	14559	15063	17704	2748	2855	3031	811	825	843	400	407	411
•••	55699	57155	84845	14740	15739	21583	2798	2936	3619	815	832	964	400	407	449
0.25 •••	64029	64820	84869	16795	17517	21587	3195	3354	3630	932	953	981	453	460	464
•••	64295	66392	98903	17056	18501	25032	3265	3471	4140	942	963	1081	460	464	495
0.3 •••	69406	70493	98278	18289	19260	24876	3535	3747	4126	1027	1055	1091	496	503	513
•••	69767	72650	110358	18639	20577	27831	3630	3907	4556	1044	1076	1172	503	507	530
0.35 •••	71853	73309	109126	19107	20354	27522	3768	4034	4519	1098	1133	1176	524	534	545
•••	72338	76163	119210	19557	22012	29979	3885	4232	4869	1112	1155	1237	528	538	553
0.4 •••	71694	73596	117275	19330	20881	29492	3913	4225	4799	1140	1183	1236	538	549	560
•••	72328	77279	125458	19897	22868	31475	4051	4459	5077	1161	1208	1276	545	556	565
0.45 •••	69314	71764	122658	19064	20913	30767	3970	4321	4969	1161	1208	1261	545	556	570
•••	70132	76390	129102	19755	23198	32322	4126	4583	5181	1183	1236	1289	549	566	565
0.5 •••	65138	68252	125223	18410	20535	31340	3949	4328	5029	1155	1204	1261	538	549	566
•••	66180	73915	130144	19217	23074	32517	4115	4608	5181	1176	1229	1276	545	556	553
0.55 •••	59620	63530	124957	17460	19819	31202	3860	4247	4976	1129	1176	1236	524	534	545
•••	60948	70228	128582	18374	22539	32061	4030	4544	5077	1151	1204	1237	528	538	530
0.6 •••	53249	58041	121855	16302	18820	30359	3701	4094	4806	1076	1123	1176	492	503	513
•••	54913	65670	124416	17290	21640	30955	3875	4381	4869	1098	1151	1172	496	507	495
0.65 •••	46519	52183	115933	14988	17581	28808	3482	3853	4523	1002	1044	1098	449	460	471
•••	48542	60474	117648	16019	20397	29197	3648	4130	4556	1023	1066	1081	453	464	449
0.7 •••	39883	46211	107203	13561	16121	26555	3195	3535	4130	900	938	981	400	407	418
•••	42216	54807	108275	14585	18824	26789	3343	3779	4140	921	959	964	400	411	391
0.75 •••	33635	40251	95692	12024	14436	23605	2840	3138	3626	779	811	843	333	343	347
•••	36139	48702	96300	12998	16918	23730	2972	3343	3619	793	825	821	337	343	322
0.8 •••	27851	34269	81427	10356	12509	19972	2415	2649	3010	634	651	677	258	262	269
•••	30327	42099	81721	11234	14638	20021	2521	2802	2994	641	662	651	262	262	241
0.85 •••	22373	28096	64433	8507	10256	15654	1898	2058	2288	460	471	481	.	.	.
•••	24611	34793	64538	9230	11918	15660	1973	2160	2265	464	475	456	.	.	.
0.9 •••	16837	21321	44738	6350	7558	10664	1278	1353	1459	.	.	.	.	.	.
•••	18612	26325	44753	6860	8620	10648	1314	1399	1432	.	.	.	.	.	.
0.95 •••	10405	12959	22383	3567	4066	5012	.	.	.	.	.	.	.	.	.
•••	11446	15541	22364	3786	4444	4986	.	.	.	.	.	.	.	.	.
0.98 •••	4912	5773	7706	.	.	.	.	.	.	.	.	.	.	.	.
•••	5277	6484	7680	.	.	.	.	.	.	.	.	.	.	.	.

TABLE 4: ALPHA= 0.01 POWER= 0.9   EXPECTED ACCRUAL THRU MINIMUM FOLLOW-UP= 4500

	DEL=.01			DEL=.02			DEL=.05			DEL=.10			DEL=.15		
FACT=	1.0 .75	.50 .25	.00 BIN	1.0 .75	.50 .25	.00 BIN	1.0 .75	.50 .25	.00 BIN	1.0 75	.50 .25	.00 BIN	1.0 .75	.50 .25	.00 BIN
PCONT=***			REQUIRED NUMBER OF PATIENTS												
0.01 ***	1995	1999	2006	735	735	735	240	240	244	120	120	120	86	86	86
***	1995	1999	7680	735	735	2539	240	240	691	120	120	281	86	86	167
0.02 ***	4402	4418	4481	1414	1421	1432	386	386	386	165	165	165	109	109	109
***	4406	4440	12679	1421	1425	3775	386	386	883	165	165	326	109	109	186
0.05 ***	13312	13395	13991	3776	3821	3923	818	825	829	296	296	296	172	172	172
***	13339	13530	27050	3795	3862	7329	825	829	1432	296	296	456	172	172	241
0.1 ***	29366	29573	32572	7890	8058	8614	1545	1567	1605	487	491	491	262	262	262
***	29438	29973	48918	7957	8238	12731	1556	1583	2265	487	491	651	262	262	322
0.15 ***	43860	44231	51454	11584	11910	13323	2201	2258	2348	660	667	678	334	341	341
***	43984	44970	68183	11708	12315	17482	2231	2303	2994	667	671	821	341	341	391
0.2 ***	55549	56130	69128	14595	15112	17704	2760	2865	3034	813	825	840	401	408	408
***	55740	57277	84845	14779	15798	21583	2809	2940	3619	818	829	964	401	408	449
0.25 ***	64076	64916	84866	16845	17591	21585	3210	3367	3630	937	953	982	453	458	469
***	64353	66570	98903	17115	18588	25032	3281	3480	4140	941	964	1081	458	465	495
0.3 ***	69465	70624	98276	18352	19365	24877	3551	3761	4125	1031	1061	1095	498	503	510
***	69855	72896	110358	18716	20692	27831	3648	3911	4556	1043	1072	1172	498	510	530
0.35 ***	71936	73477	109121	19189	20483	27525	3795	4053	4519	1099	1133	1178	525	532	543
***	72453	76485	119210	19661	22159	29979	3911	4245	4869	1117	1155	1237	532	536	553
0.4 ***	71805	73819	117278	19436	21030	29494	3941	4249	4800	1144	1185	1234	543	555	566
***	72480	77689	125458	20028	23036	31475	4076	4474	5077	1166	1207	1276	548	559	565
0.45 ***	69461	72048	122655	19189	21086	30772	4001	4346	4969	1166	1207	1263	548	559	570
***	70327	76897	129102	19905	23396	32322	4155	4605	5181	1185	1234	1289	555	566	565
0.5 ***	65321	68610	125220	18559	20730	31339	3986	4357	5032	1162	1207	1263	543	555	566
***	66428	74512	130144	19391	23291	32517	4148	4631	5181	1185	1234	1276	548	559	553
0.55 ***	59858	63971	124957	17632	20033	31204	3896	4278	4976	1133	1178	1234	521	532	543
***	61260	70912	128582	18566	22766	32061	4065	4564	5077	1155	1207	1237	525	536	530
0.6 ***	53546	58564	121856	16489	19050	30356	3738	4121	4807	1083	1121	1178	491	503	514
***	55297	66416	124416	17497	21873	30955	3907	4402	4869	1099	1151	1172	498	510	495
0.65 ***	46886	52770	115935	15191	17812	28808	3513	3878	4526	1005	1043	1095	453	458	469
***	49001	61264	117648	16230	20629	29197	3675	4148	4556	1027	1072	1081	458	465	449
0.7 ***	40312	46841	107205	13762	16343	26553	3225	3558	4132	908	941	982	401	408	413
***	42731	55601	108275	14797	19043	26789	3371	3799	4140	926	960	964	401	408	391
0.75 ***	34102	40886	95696	12214	14644	23606	2865	3153	3626	784	813	840	334	341	345
***	36678	49470	96300	13200	17115	23730	2996	3356	3619	795	825	821	341	345	322
0.8 ***	28320	34867	81424	10526	12686	19972	2433	2663	3011	633	656	678	262	262	266
***	30851	42798	81721	11411	14808	20021	2539	2816	2994	645	667	651	262	262	241
0.85 ***	22800	28616	64432	8655	10403	15656	1916	2066	2287	458	469	480	.	.	.
***	25080	35378	64538	9379	12045	15660	1988	2168	2265	465	476	456	.	.	.
0.9 ***	17182	21720	44738	6454	7653	10668	1286	1358	1459	.	.	.	.	.	.
***	18982	26749	44753	6960	8693	10648	1320	1403	1432	.	.	.	.	.	.
0.95 ***	10605	13177	22384	3608	4103	5010	.	.	.	.	.	.	.	.	.
***	11651	15746	22364	3828	4470	4986	.	.	.	.	.	.	.	.	.
0.98 ***	4987	5835	7703	.	.	.	.	.	.	.	.	.	.	.	.
***	5351	6533	7680	.	.	.	.	.	.	.	.	.	.	.	.

TABLE 4: ALPHA= 0.01 POWER= 0.9   EXPECTED ACCRUAL THRU MINIMUM FOLLOW-UP= 4750

	DEL=.01			DEL=.02			DEL=.05			DEL=.10			DEL=.15		
FACT=	1.0 / .75	.50 / .25	.00 / BIN	1.0 / .75	.50 / .25	.00 / BIN	1.0 / .75	.50 / .25	.00 / BIN	1.0 / 75	.50 / .25	.00 / BIN	1.0 / .75	.50 / .25	.00 / BIN
PCONT=***						REQUIRED NUMBER OF PATIENTS									
0.01 ***	1991	1998	2010	732	732	732	241	241	241	122	122	122	87	87	87
***	1991	2003	7680	732	732	2539	241	241	691	122	122	281	87	87	167
0.02 ***	4402	4421	4480	1417	1421	1433	383	383	383	170	170	170	110	110	110
***	4409	4445	12679	1417	1428	3775	383	383	883	170	170	326	110	110	186
0.05 ***	13315	13403	13992	3780	3827	3922	815	823	827	293	293	293	170	170	170
***	13344	13541	27050	3796	3863	7329	823	827	1432	293	293	456	170	170	241
0.1 ***	29375	29600	32569	7900	8071	8613	1547	1571	1607	490	490	495	257	257	265
***	29453	30016	48918	7964	8249	12731	1559	1587	2265	490	490	651	257	265	322
0.15 ***	43882	44273	51450	11605	11942	13320	2205	2264	2347	661	668	680	336	340	340
***	44012	45045	68183	11729	12346	17482	2236	2300	2994	668	673	821	336	340	391
0.2 ***	55578	56196	69128	14626	15160	17702	2770	2870	3032	811	823	839	400	407	412
***	55785	57400	84845	14823	15849	21583	2818	2941	3619	815	835	964	407	407	449
0.25 ***	64121	65007	84867	16890	17662	21585	3222	3376	3630	934	953	977	455	459	467
***	64418	66753	98903	17167	18671	25032	3293	3483	4140	946	965	1081	459	467	495
0.3 ***	69532	70747	98274	18414	19455	24874	3570	3780	4124	1037	1060	1089	495	502	514
***	69940	73142	110358	18794	20801	27831	3661	3922	4556	1048	1077	1172	502	507	530
0.35 ***	72025	73652	109123	19269	20607	27522	3815	4069	4520	1108	1136	1179	526	538	542
***	72564	76804	119210	19763	22293	29979	3927	4259	4869	1120	1155	1237	530	538	553
0.4 ***	71918	74044	117274	19538	21177	29494	3970	4267	4801	1148	1191	1231	542	554	562
***	72631	78093	125458	20148	23200	31475	4100	4492	5077	1167	1207	1276	550	554	565
0.45 ***	69603	72334	122653	19317	21260	30772	4034	4366	4972	1172	1215	1262	550	562	566
***	70517	77393	129102	20053	23580	32322	4183	4615	5181	1191	1238	1289	554	566	565
0.5 ***	65506	68966	125225	18707	20927	31342	4017	4378	5031	1167	1207	1262	542	554	562
***	66674	75094	130144	19562	23492	32517	4176	4647	5181	1184	1231	1276	550	554	553
0.55 ***	60091	64402	124957	17804	20243	31204	3927	4302	4972	1136	1179	1238	526	530	542
***	61563	71574	128582	18754	22982	32061	4093	4580	5077	1160	1207	1237	530	538	530
0.6 ***	53845	59074	121857	16676	19265	30356	3768	4148	4805	1084	1124	1179	495	502	514
***	55673	67137	124416	17697	22091	30955	3939	4421	4869	1108	1155	1172	495	507	495
0.65 ***	47247	53346	115932	15382	18030	28805	3547	3903	4520	1005	1048	1096	455	459	471
***	49451	62027	117648	16438	20844	29197	3708	4164	4556	1029	1072	1081	455	467	449
0.7 ***	40735	47449	107204	13949	16557	26556	3257	3582	4129	910	942	982	400	407	412
***	43228	56362	108275	15002	19253	26789	3400	3815	4140	922	965	964	400	412	391
0.75 ***	34560	41495	95692	12394	14840	23611	2894	3174	3625	787	811	839	336	340	348
***	37200	50199	96300	13387	17298	23730	3020	3369	3619	799	827	821	336	340	322
0.8 ***	28777	35443	81423	10695	12857	19970	2454	2675	3012	637	657	673	257	265	265
***	31353	43459	81721	11582	14966	20021	2557	2822	2994	645	661	651	257	265	241
0.85 ***	23212	29114	64430	8784	10537	15655	1927	2074	2288	459	471	483	.	.	.
***	25527	35937	64538	9515	12163	15660	1998	2169	2265	467	478	456	.	.	.
0.9 ***	17507	22095	44741	6547	7739	10667	1290	1362	1457	.	.	.	.	.	.
***	19336	27150	44753	7050	8767	10648	1326	1409	1432	.	.	.	.	.	.
0.95 ***	10802	13379	22388	3654	4136	5007	.	.	.	.	.	.	.	.	.
***	11855	15940	22364	3867	4492	4986	.	.	.	.	.	.	.	.	.
0.98 ***	5055	5898	7703	.	.	.	.	.	.	.	.	.	.	.	.
***	5418	6575	7680	.	.	.	.	.	.	.	.	.	.	.	.

TABLE 4: ALPHA= 0.01 POWER= 0.9   EXPECTED ACCRUAL THRU MINIMUM FOLLOW-UP= 5000

	DEL=.01			DEL=.02			DEL=.05			DEL=.10			DEL=.15		
FACT=	1.0 .75	.50 .25	.00 BIN	1.0 .75	.50 .25	.00 BIN	1.0 .75	.50 .25	.00 BIN	1.0 75	.50 .25	.00 BIN	1.0 .75	.50 .25	.00 BIN

PCONT=***   REQUIRED NUMBER OF PATIENTS

PCONT	DEL=.01 1.0/.75	.50/.25	.00 BIN	DEL=.02 1.0/.75	.50/.25	.00 BIN	DEL=.05 1.0/.75	.50/.25	.00 BIN	DEL=.10 1.0/.75	.50/.25	.00 BIN	DEL=.15 1.0/.75	.50/.25	.00 BIN
0.01	1991	1996	2008	733	733	733	241	241	241	121	121	121	83	83	83
	1996	2004	7680	733	733	2539	241	241	691	121	121	281	83	83	167
0.02	4404	4421	4479	1416	1421	1429	383	383	383	166	166	166	108	108	108
	4408	4446	12679	1421	1429	3775	383	383	883	166	166	326	108	108	186
0.05	13321	13408	13991	3783	3829	3921	821	821	829	291	296	296	171	171	171
	13354	13554	27050	3804	3866	7329	821	829	1432	296	296	456	171	171	241
0.1	29391	29621	32571	7916	8083	8616	1546	1571	1604	491	491	496	258	258	258
	29466	30054	48918	7979	8266	12731	1558	1583	2265	491	491	651	258	258	322
0.15	43904	44316	51454	11629	11971	13321	2208	2266	2354	658	671	679	333	341	341
	44041	45121	68183	11754	12379	17482	2233	2304	2994	666	671	821	341	341	391
0.2	55608	56258	69129	14658	15208	17704	2779	2879	3033	816	829	841	404	408	408
	55829	57521	84845	14858	15904	21583	2821	2946	3619	821	833	964	404	408	449
0.25	64166	65104	84866	16933	17733	21583	3233	3383	3629	941	958	979	458	458	466
	64479	66929	98903	17229	18746	25032	3304	3491	4140	946	966	1081	458	466	495
0.3	69596	70879	98279	18479	19546	24879	3591	3791	4129	1033	1058	1091	496	504	508
	70021	73379	110358	18871	20904	27831	3679	3933	4556	1046	1079	1172	504	508	530
0.35	72108	73821	109121	19354	20721	27521	3841	4091	4516	1108	1141	1179	529	533	541
	72679	77121	119210	19858	22421	29979	3946	4271	4869	1121	1158	1237	529	541	553
0.4	72029	74266	117279	19641	21316	29491	3991	4291	4804	1154	1191	1233	546	554	566
	72779	78491	125458	20271	23354	31475	4121	4504	5077	1171	1208	1276	546	558	565
0.45	69746	72616	122658	19441	21421	30771	4058	4391	4971	1171	1216	1266	546	558	571
	70708	77871	129102	20196	23754	32322	4208	4633	5181	1191	1233	1289	554	566	565
0.5	65691	69321	125221	18854	21108	31341	4046	4404	5029	1171	1208	1266	541	554	566
	66916	75658	130144	19721	23683	32517	4204	4666	5181	1191	1233	1276	546	558	553
0.55	60321	64833	124954	17966	20441	31204	3958	4329	4971	1141	1183	1233	521	533	546
	61871	72216	128582	18933	23183	32061	4121	4596	5077	1158	1208	1237	529	541	530
0.6	54133	59579	121854	16846	19471	30358	3804	4171	4804	1091	1129	1179	496	504	516
	56054	67841	124416	17883	22296	30955	3966	4433	4869	1108	1154	1172	496	508	495
0.65	47608	53904	115933	15558	18233	28808	3571	3929	4521	1016	1054	1096	454	458	471
	49891	62754	117648	16633	21046	29197	3733	4179	4556	1029	1071	1081	458	466	449
0.7	41146	48041	107204	14133	16754	26554	3279	3604	4129	916	946	983	404	408	416
	43721	57096	108275	15191	19446	26789	3429	3829	4140	929	966	964	404	408	391
0.75	35004	42079	95696	12571	15021	23608	2916	3191	3629	791	816	841	333	341	346
	37708	50896	96300	13571	17479	23730	3041	3379	3619	804	829	821	341	341	322
0.8	29216	35991	81429	10846	13016	19971	2471	2691	3008	641	654	679	258	266	266
	31846	44091	81721	11741	15108	20021	2571	2829	2994	646	666	651	258	266	241
0.85	23608	29591	64433	8916	10666	15654	1941	2083	2291	458	471	483	.	.	.
	25958	36466	64538	9646	12271	15660	2008	2179	2265	466	479	456	.	.	.
0.9	17829	22458	44741	6641	7821	10666	1296	1366	1458	.	.	.	.	.	.
	19671	27529	44753	7141	8833	10648	1329	1408	1432	.	.	.	.	.	.
0.95	10983	13571	22383	3691	4166	5008	.	.	.	.	.	.	.	.	.
	12046	16121	22364	3904	4516	4986	.	.	.	.	.	.	.	.	.
0.98	5121	5954	7704	.	.	.	.	.	.	.	.	.	.	.	.
	5483	6621	7680	.	.	.	.	.	.	.	.	.	.	.	.

TABLE 4: ALPHA= 0.01 POWER= 0.9   EXPECTED ACCRUAL THRU MINIMUM FOLLOW-UP= 5500

	DEL=.01			DEL=.02			DEL=.05			DEL=.10			DEL=.15		
FACT=	1.0 / .75	.50 / .25	.00 BIN	1.0 / .75	.50 / .25	.00 BIN	1.0 / .75	.50 / .25	.00 BIN	1.0 / 75	.50 / .25	.00 BIN	1.0 / .75	.50 / .25	.00 BIN
PCONT=•••				REQUIRED NUMBER OF PATIENTS											
0.01 •••	1989	1998	2003	733	733	733	243	243	243	119	119	119	86	86	86
•••	1998	2003	7680	733	733	2539	243	243	691	119	119	281	86	86	167
0.02 •••	4404	4423	4478	1420	1420	1434	380	380	389	169	169	169	114	114	114
•••	4418	4445	12679	1420	1425	3775	380	380	883	169	169	326	114	114	186
0.05 •••	13333	13429	13993	3790	3832	3923	820	829	829	293	293	293	169	169	169
•••	13369	13575	27050	3804	3868	7329	820	829	1432	293	293	456	169	169	241
0.1 •••	29415	29668	32569	7938	8108	8617	1549	1571	1604	490	490	490	257	257	265
•••	29498	30135	48918	8006	8287	12731	1563	1585	2265	490	490	651	257	265	322
0.15 •••	43940	44403	51456	11669	12027	13319	2218	2273	2347	664	669	678	339	339	339
•••	44092	45274	68183	11801	12434	17482	2245	2305	2994	669	678	821	339	339	391
0.2 •••	55678	56384	69130	14722	15299	17700	2795	2891	3034	815	829	843	403	408	408
•••	55911	57768	84845	14936	15995	21583	2836	2952	3619	820	834	964	403	408	449
0.25 •••	64263	65289	84869	17026	17865	21583	3263	3405	3625	939	958	980	458	463	463
•••	64601	67274	98903	17334	18888	25032	3323	3502	4140	953	966	1081	458	463	495
0.3 •••	69722	71133	98275	18599	19727	24878	3620	3813	4129	1040	1063	1090	499	504	513
•••	70198	73855	110358	19020	21096	27831	3708	3950	4556	1049	1076	1172	504	504	530
0.35 •••	72279	74163	109124	19507	20945	27523	3873	4115	4519	1109	1145	1178	526	532	545
•••	72906	77738	119210	20043	22664	29979	3983	4285	4869	1123	1159	1237	532	540	553
0.4 •••	72252	74708	117278	19837	21583	29489	4038	4321	4803	1159	1192	1233	545	554	559
•••	73077	79264	125458	20497	23640	31475	4162	4528	5077	1178	1214	1276	545	559	565
0.45 •••	70038	73181	122654	19680	21729	30768	4107	4431	4973	1178	1219	1260	554	559	568
•••	71091	78810	129102	20469	24080	32322	4253	4657	5181	1200	1241	1289	554	568	565
0.5 •••	66059	70019	125225	19130	21454	31340	4101	4445	5028	1178	1214	1260	545	554	568
•••	67406	76756	130144	20038	24039	32517	4253	4693	5181	1192	1241	1276	545	559	553
0.55 •••	60793	65674	124955	18269	20808	31203	4010	4368	4973	1145	1186	1233	526	532	545
•••	62484	73443	128582	19273	23558	32061	4175	4624	5077	1164	1214	1237	532	540	530
0.6 •••	54720	60550	121856	17183	19859	30355	3854	4211	4803	1095	1131	1178	499	504	513
•••	56791	69166	124416	18250	22683	30955	4019	4464	4869	1118	1159	1172	499	513	495
0.65 •••	48313	54976	115935	15904	18621	28810	3625	3969	4519	1021	1054	1095	458	463	471
•••	50747	64139	117648	16999	21426	29197	3785	4203	4556	1035	1076	1081	458	463	449
0.7 •••	41955	49165	107204	14474	17128	26555	3331	3639	4129	917	944	980	403	408	416
•••	44664	58474	108275	15555	19795	26789	3469	3854	4140	930	966	964	403	408	391
0.75 •••	35864	43198	95695	12893	15368	23604	2960	3221	3625	793	815	843	339	339	348
•••	38674	52218	96300	13910	17796	23730	3075	3400	3619	801	829	821	339	348	322
0.8 •••	30053	37038	81423	11141	13314	19974	2506	2713	3007	642	655	678	257	265	265
•••	32775	45283	81721	12040	15376	20021	2603	2842	2994	650	664	651	265	265	241
0.85 •••	24369	30493	64428	9153	10899	15657	1962	2099	2286	463	471	485	.	.	.
•••	26780	37450	64538	9882	12475	15660	2030	2190	2265	463	477	456	.	.	.
0.9 •••	18420	23145	44738	6810	7970	10665	1310	1379	1453	.	.	.	.	.	.
•••	20313	28233	44753	7305	8955	10648	1338	1412	1432	.	.	.	.	.	.
0.95 •••	11334	13933	22389	3763	4225	5009	.	.	.	.	.	.	.	.	.
•••	12406	16449	22364	3969	4555	4986	.	.	.	.	.	.	.	.	.
0.98 •••	5243	6059	7704	.	.	.	.	.	.	.	.	.	.	.	.
•••	5600	6700	7680	.	.	.	.	.	.	.	.	.	.	.	.

TABLE 4: ALPHA= 0.01 POWER= 0.9 EXPECTED ACCRUAL THRU MINIMUM FOLLOW-UP= 6000														
	DEL=.01			DEL=.02			DEL=.05			DEL=.10			DEL=.15		
FACT=	1.0 .75	.50 .25	.00 BIN	1.0 .75	.50 .25	.00 BIN	1.0 .75	.50 .25	.00 BIN	1.0 75	.50 .25	.00 BIN	1.0 .75	.50 .25	.00 BIN
PCONT=***	REQUIRED NUMBER OF PATIENTS														
0.01 ***	1990	1999	2005	730	730	739	244	244	244	124	124	124	85	85	85
•••	1999	1999	7680	730	730	2539	244	244	691	124	124	281	85	85	167
0.02 ***	4405	4429	4480	1420	1420	1429	385	385	385	169	169	169	109	109	109
•••	4414	4450	12679	1420	1429	3775	385	385	883	169	169	326	109	109	186
0.05 ***	13339	13444	13990	3790	3835	3919	820	829	829	295	295	295	169	169	169
•••	13375	13594	27050	3814	3874	7329	820	829	1432	295	295	456	169	169	241
0.1 ***	29434	29710	32569	7954	8134	8614	1555	1579	1600	490	490	490	259	259	259
•••	29530	30205	48918	8029	8305	12731	1564	1585	2265	490	490	651	259	259	322
0.15 ***	43984	44479	51454	11704	12079	13324	2230	2275	2350	664	670	679	340	340	340
•••	44149	45424	68183	11845	12484	17482	2254	2314	2994	670	670	821	340	340	391
0.2 ***	55744	56515	69130	14779	15385	17704	2809	2899	3034	814	829	844	400	409	409
•••	55999	58000	84845	15010	16084	21583	2845	2959	3619	820	835	964	409	409	449
0.25 ***	64354	65470	84865	17110	17989	21589	3280	3415	3625	940	964	979	460	460	469
•••	64729	67615	98903	17440	19015	25032	3340	3514	4140	949	970	1081	460	460	495
0.3 ***	69850	71389	98275	18715	19894	24874	3649	3835	4129	1045	1069	1090	499	505	514
•••	70369	74320	110358	19159	21274	27831	3730	3964	4556	1054	1075	1172	505	505	530
0.35 ***	72454	74500	109120	19660	21154	27520	3910	4144	4519	1120	1144	1174	529	535	544
•••	73135	78340	119210	20224	22879	29979	4015	4309	4869	1129	1159	1237	535	535	553
0.4 ***	72475	75145	117274	20029	21835	29494	4075	4354	4804	1165	1195	1234	544	550	565
•••	73375	80014	125458	20719	23899	31475	4204	4540	5077	1180	1210	1276	550	559	565
0.45 ***	70324	73735	122659	19909	22015	30769	4150	4465	4969	1189	1219	1264	550	559	565
•••	71479	79705	129102	20725	24370	32322	4294	4684	5181	1204	1240	1289	559	565	565
0.5 ***	66430	70705	125224	19390	21769	31339	4150	4480	5029	1180	1219	1264	544	550	565
•••	67885	77794	130144	20329	24355	32517	4300	4714	5181	1204	1240	1276	550	559	553
0.55 ***	61255	66490	124954	18565	21154	31204	4060	4405	4975	1150	1189	1234	529	535	544
•••	63085	74605	128582	19594	23899	32061	4219	4645	5077	1174	1210	1237	529	544	530
0.6 ***	55294	61489	121855	17494	20209	30355	3904	4249	4804	1099	1135	1180	499	505	514
•••	57514	70420	124416	18580	23029	30955	4060	4489	4869	1120	1159	1172	499	505	495
0.65 ***	49000	55999	115930	16234	18979	28804	3679	4000	4525	1024	1054	1090	454	460	469
•••	51580	65434	117648	17344	21760	29197	3820	4225	4556	1039	1075	1081	460	469	449
0.7 ***	42730	50230	107200	14794	17470	26554	3370	3670	4129	925	949	985	400	409	415
•••	45565	59764	108275	15889	20119	26789	3505	3874	4140	934	964	964	409	409	391
0.75 ***	36679	44254	95695	13195	15685	23605	2995	3250	3625	799	820	844	340	340	349
•••	39595	53434	96300	14215	18079	23730	3109	3415	3619	805	829	821	340	340	322
0.8 ***	30850	38014	81424	11410	13585	19969	2539	2734	3010	640	655	679	259	265	265
•••	33649	46375	81721	12310	15610	20021	2629	2860	2994	649	664	651	259	265	241
0.85 ***	25075	31330	64429	9379	11110	15655	1984	2119	2290	460	475	484	.	.	.
•••	27550	38350	64538	10105	12649	15660	2044	2194	2265	469	475	456	.	.	.
0.9 ***	18985	23770	44740	6955	8110	10669	1315	1384	1459	.	.	.	.	.	.
•••	20905	28870	44753	7450	9055	10648	1345	1414	1432	.	.	.	.	.	.
0.95 ***	11650	14260	22384	3829	4270	5014	.	.	.	.	.	.	.	.	.
•••	12730	16744	22364	4024	4585	4986	.	.	.	.	.	.	.	.	.
0.98 ***	5350	6145	7705	.	.	.	.	.	.	.	.	.	.	.	.
•••	5704	6760	7680	.	.	.	.	.	.	.	.	.	.	.	.

TABLE 4: ALPHA= 0.01 POWER= 0.9   EXPECTED ACCRUAL THRU MINIMUM FOLLOW-UP= 6500

	DEL=.01			DEL=.02			DEL=.05			DEL=.10			DEL=.15		
FACT=	1.0 / .75	.50 / .25	.00 BIN	1.0 / .75	.50 / .25	.00 BIN	1.0 / .75	.50 / .25	.00 BIN	1.0 / 75	.50 / .25	.00 BIN	1.0 / .75	.50 / .25	.00 BIN
PCONT=***			REQUIRED NUMBER OF PATIENTS												
0.01	1993	2003	2009	736	736	736	238	238	238	118	118	118	86	86	86
	1993	2003	7680	736	736	2539	238	238	691	118	118	281	86	86	167
0.02	4408	4431	4479	1418	1424	1435	384	384	384	167	167	167	108	108	108
	4414	4447	12679	1418	1424	3775	384	384	883	167	167	326	108	108	186
0.05	13352	13460	13996	3797	3846	3921	823	823	833	297	297	297	173	173	173
	13384	13612	27050	3813	3878	7329	823	823	1432	297	297	456	173	173	241
0.1	29456	29758	32570	7973	8152	8617	1554	1581	1603	492	492	492	265	265	265
	29563	30278	48918	8048	8325	12731	1565	1587	2265	492	492	651	265	265	322
0.15	44026	44562	51452	11743	12127	13319	2237	2286	2351	661	671	677	336	336	346
	44205	45559	68183	11889	12523	17482	2253	2312	2994	671	677	821	336	336	391
0.2	55807	56642	69132	14841	15464	17701	2816	2903	3033	817	833	839	401	411	411
	56083	58228	84845	15074	16157	21583	2865	2962	3619	823	833	964	401	411	449
0.25	64452	65655	84868	17197	18097	21585	3293	3433	3628	947	963	980	460	460	466
	64848	67946	98903	17544	19131	25032	3358	3521	4140	953	969	1081	460	466	495
0.3	69983	71641	98274	18838	20047	24873	3667	3846	4122	1045	1067	1093	498	508	508
	70536	74771	110358	19293	21428	27831	3748	3970	4556	1061	1077	1172	498	508	530
0.35	72626	74836	109123	19813	21347	27522	3937	4165	4522	1126	1148	1175	531	541	541
	73363	78921	119210	20398	23080	29979	4041	4317	4869	1132	1158	1237	531	541	553
0.4	72697	75583	117281	20209	22072	29488	4116	4376	4798	1164	1197	1229	547	557	563
	73672	80735	125458	20924	24136	31475	4230	4561	5077	1181	1213	1276	547	557	565
0.45	70611	74283	122660	20122	22283	30766	4197	4496	4967	1191	1223	1262	557	563	573
	71862	80572	129102	20967	24640	32322	4327	4701	5181	1207	1240	1289	557	563	565
0.5	66792	71375	125217	19641	22062	31341	4187	4512	5032	1191	1223	1262	547	557	563
	68368	78785	130144	20599	24656	32517	4333	4733	5181	1207	1240	1276	547	557	553
0.55	61722	67286	124957	18844	21471	31205	4106	4441	4977	1158	1197	1240	531	541	547
	63672	75703	128582	19895	24207	32061	4262	4668	5077	1175	1213	1237	531	541	530
0.6	55862	62388	121853	17788	20534	30360	3953	4284	4804	1110	1142	1181	498	508	514
	58218	71592	124416	18893	23340	30955	4100	4506	4869	1126	1158	1172	498	508	495
0.65	49671	56977	115932	16537	19299	28806	3716	4035	4522	1028	1061	1093	460	466	466
	52378	66646	117648	17658	22072	29197	3862	4246	4556	1045	1077	1081	460	466	449
0.7	43473	51241	107206	15091	17782	26557	3417	3699	4132	931	953	980	401	411	417
	46421	60958	108275	16196	20404	26789	3547	3888	4140	937	969	964	411	411	391
0.75	37451	45234	95691	13482	15968	23606	3033	3271	3628	801	817	839	336	346	346
	40467	54572	96300	14506	18335	23730	3141	3433	3619	807	833	821	336	346	322
0.8	31601	38929	81423	11662	13833	19976	2562	2751	3011	644	661	677	265	265	265
	34471	47390	81721	12566	15822	20021	2653	2871	2994	655	671	651	265	265	241
0.85	25745	32115	64436	9582	11298	15653	2003	2123	2286	466	476	482	.	.	.
	28269	39183	64538	10307	12810	15660	2058	2198	2265	466	476	456	.	.	.
0.9	19505	24353	44741	7096	8233	10665	1327	1386	1457	.	.	.	.	.	.
	21455	29456	44753	7583	9153	10648	1353	1418	1432	.	.	.	.	.	.
0.95	11954	14554	22387	3888	4311	5010	.	.	.	.	.	.	.	.	.
	13037	17008	22364	4073	4609	4986	.	.	.	.	.	.	.	.	.
0.98	5448	6228	7707	.	.	.	.	.	.	.	.	.	.	.	.
	5796	6819	7680	.	.	.	.	.	.	.	.	.	.	.	.

TABLE 4: ALPHA= 0.01 POWER= 0.9   EXPECTED ACCRUAL THRU MINIMUM FOLLOW-UP= 7000

	DEL=.01			DEL=.02			DEL=.05			DEL=.10			DEL=.15		
FACT=	1.0 .75	.50 .25	.00 BIN	1.0 .75	.50 .25	.00 BIN	1.0 .75	.50 .25	.00 BIN	1.0 75	.50 .25	.00 BIN	1.0 .75	.50 .25	.00 BIN
PCONT=•••					REQUIRED NUMBER OF PATIENTS										
0.01 •••	1989 2000	2006	729 729	740	239 239	239	116 116	116	81 81	81					
•••	2000 2000	7680	729 729	2539	239 239	691	116 116	281	81 81	167					
0.02 •••	4415 4432	4485	1422 1422	1429	379 379	390	169 169	169	110 110	110					
•••	4421 4450	12679	1422 1429	3775	379 379	883	169 169	326	110 110	186					
0.05 •••	13357 13480	13994	3802 3844	3925	827 827	827	291 291	291	169 169	169					
•••	13399 13626	27050	3820 3879	7329	827 827	1432	291 291	456	169 169	241					
0.1 •••	29481 29807	32572	7991 8177	8615	1562 1580	1604	484 495	495	256 256	256					
•••	29586 30350	48918	8061 8341	12731	1569 1586	2265	495 495	651	256 256	322					
0.15 •••	44070 44647	51455	11782 12167	13322	2245 2286	2350	670 670	676	337 337	344					
•••	44262 45697	68183	11929 12559	17482	2262 2315	2994	670 676	821	337 337	391					
0.2 •••	55871 56775	69130	14897 15545	17704	2829 2916	3032	816 827	845	407 407	407					
•••	56169 58455	84845	15149 16227	21583	2864 2969	3619	827 834	964	407 407	449					
0.25 •••	64545 65840	84869	17284 18211	21582	3312 3441	3627	950 967	974	460 460	466					
•••	64971 68272	98903	17634 19237	25032	3371 3529	4140	956 967	1081	460 466	495					
0.3 •••	70110 71895	98274	18946 20189	24879	3697 3861	4124	1055 1072	1090	501 501	512					
•••	70705 75202	110358	19430 21571	27831	3774 3984	4556	1061 1079	1172	501 512	530					
0.35 •••	72794 75174	109124	19955 21530	27521	3966 4187	4520	1125 1149	1177	530 536	547					
•••	73592 79479	119210	20567 23262	29979	4065 4334	4869	1131 1160	1237	536 536	553					
0.4 •••	72927 76014	117279	20381 22289	29492	4141 4404	4520	1177 1201	1236	547 554	565					
•••	73966 81421	125458	21127 24347	31475	4264 4579	5077	1184 1219	1276	554 554	565					
0.45 •••	70897 74824	122651	20329 22534	30770	4229 4520	4975	1195 1230	1265	554 565	571					
•••	72234 81397	129102	21204 24890	32322	4362 4719	5181	1212 1247	1289	554 565	565					
0.5 •••	67159 72035	125224	19885 22341	31336	4229 4544	5027	1195 1230	1265	547 554	565					
•••	68850 79724	130144	20865 24914	32517	4369 4754	5181	1212 1247	1276	554 554	553					
0.55 •••	62182 68051	124955	19104 21764	31207	4152 4474	4975	1166 1195	1236	530 536	547					
•••	64254 76749	128582	20171 24487	32061	4292 4684	5077	1177 1212	1237	536 536	530					
0.6 •••	56425 63256	121857	18071 20836	30356	3995 4310	4806	1107 1142	1177	501 501	512					
•••	58910 72706	124416	19191 23619	30955	4135 4526	4869	1125 1160	1172	501 512	495					
0.65 •••	50324 57912	115931	16822 19605	28810	3756 4065	4520	1037 1061	1096	460 466	466					
•••	53152 67782	117648	17960 22352	29197	3896 4264	4556	1044 1079	1081	460 466	449					
0.7 •••	44199 52196	107199	15376 18071	26552	3452 3721	4124	932 956	985	407 407	414					
•••	47244 62084	108275	16490 20672	26789	3575 3907	4140	939 967	964	407 414	391					
0.75 •••	38196 46170	95695	13742 16234	23612	3056 3295	3627	799 816	845	337 344	344					
•••	41294 55626	96300	14775 18572	23730	3172 3441	3619	810 834	821	337 344	322					
0.8 •••	32316 39789	81421	11894 14057	19972	2584 2770	3015	652 659	676	256 267	267					
•••	35250 48329	81721	12797 16017	20021	2671 2875	2994	652 670	651	267 267	241					
0.85 •••	26377 32841	64429	9770 11474	15656	2017 2140	2286	466 477	484	. .	.					
•••	28950 39957	64538	10494 12944	15660	2076 2210	2265	466 477	456	. .	.					
0.9 •••	19996 24896	44741	7221 8341	10662	1335 1394	1457	. .	.	. .	.					
•••	21974 30000	44753	7711 9227	10648	1359 1422	1432	. .	.	. .	.					
0.95 •••	12226 14834	22387	3931 4351	5010	. .	.	. .	.	. .	.					
•••	13311 17242	22364	4124 4631	4986	. .	.	. .	.	. .	.					
0.98 •••	5541 6305	7705	. .	.	. .	.	. .	.	. .	.					
•••	5874 6871	7680	. .	.	. .	.	. .	.	. .	.					

TABLE 4: ALPHA= 0.01 POWER= 0.9   EXPECTED ACCRUAL THRU MINIMUM FOLLOW-UP= 7500

	DEL=.01			DEL=.02			DEL=.05			DEL=.10			DEL=.15		
FACT=	1.0 .75	.50 .25	.00 BIN	1.0 .75	.50 .25	.00 BIN	1.0 .75	.50 .25	.00 BIN	1.0 75	.50 .25	.00 BIN	1.0 .75	.50 .25	.00 BIN
PCONT=***				REQUIRED NUMBER OF PATIENTS											
0.01 ***	1993	2000	2011	736	736	736	237	237	237	118	118	118	87	87	87
•••	2000	2000	7680	736	736	2539	237	237	691	118	118	281	87	87	167
0.02 ***	4411	4437	4475	1418	1418	1430	380	387	387	162	162	162	106	106	106
•••	4418	4456	12679	1418	1430	3775	380	387	883	162	162	326	106	106	186
0.05 ***	13362	13486	13993	3811	3849	3924	818	830	830	293	293	293	174	174	174
•••	13411	13643	27050	3830	3886	7329	818	830	1432	293	293	456	174	174	241
0.1 ***	29506	29843	32574	8011	8187	8611	1561	1580	1606	493	493	493	256	256	256
•••	29618	30406	48918	8086	8356	12731	1568	1587	2265	493	493	651	256	256	322
0.15 ***	44105	44724	51455	11818	12211	13325	2243	2293	2349	668	668	680	343	343	343
•••	44318	45830	68183	11968	12593	17482	2262	2318	2994	668	680	821	343	343	391
0.2 ***	55936	56900	69125	14956	15612	17705	2836	2918	3031	818	830	837	406	406	406
•••	56255	58662	84845	15211	16299	21583	2881	2975	3619	830	837	964	406	406	449
0.25 ***	64636	66024	84868	17368	18312	21586	3331	3455	3631	950	961	980	462	462	462
•••	65105	68581	98903	17731	19336	25032	3387	3530	4140	961	968	1081	462	462	495
0.3 ***	70243	72155	98274	19055	20330	24875	3718	3875	4130	1055	1074	1093	500	511	511
•••	70880	75624	110358	19550	21706	27831	3793	3987	4556	1062	1081	1172	500	511	530
0.35 ***	72961	75511	109118	20093	21699	27518	3987	4205	4512	1130	1149	1175	530	537	537
•••	73824	80030	119210	20724	23431	29979	4093	4343	4869	1137	1168	1237	537	537	553
0.4 ***	73149	76437	117275	20555	22493	29487	4175	4418	4805	1175	1205	1231	549	556	568
•••	74262	82093	125458	21312	24549	31475	4287	4587	5077	1186	1212	1276	556	556	565
0.45 ***	71187	75361	122656	20536	22768	30774	4261	4543	4974	1193	1231	1261	556	568	568
•••	72612	82186	129102	21418	25111	32322	4393	4730	5181	1212	1243	1289	556	568	565
0.5 ***	67524	72680	125225	20105	22599	31336	4261	4568	5030	1193	1224	1261	549	556	568
•••	69324	80630	130144	21106	25168	32517	4400	4768	5181	1212	1243	1276	549	556	553
0.55 ***	62630	68799	124955	19355	22036	31205	4186	4493	4974	1168	1193	1231	530	537	549
•••	64831	77743	128582	20443	24743	32061	4325	4711	5077	1186	1212	1237	537	537	530
0.6 ***	56975	64093	121850	18331	21125	30361	4025	4336	4805	1111	1149	1175	500	511	511
•••	59574	73756	124416	19468	23881	30955	4168	4543	4869	1130	1156	1172	500	511	495
0.65 ***	50956	58805	115936	17086	19887	28805	3793	4081	4524	1036	1062	1093	455	462	474
•••	53900	68855	117648	18237	22606	29197	3924	4280	4556	1055	1081	1081	462	462	449
0.7 ***	44893	53105	107206	15643	18343	26555	3481	3743	4130	931	961	980	406	406	418
•••	48043	63143	108275	16756	20911	26789	3605	3912	4140	943	968	964	406	406	391
0.75 ***	38911	47056	95693	13993	16475	23611	3087	3312	3624	800	818	837	343	343	343
•••	42080	56618	96300	15024	18781	23730	3193	3455	3619	811	830	821	343	343	322
0.8 ***	32993	40599	81425	12106	14262	19974	2611	2780	3012	650	661	680	256	268	268
•••	35993	49212	81721	13018	16193	20021	2686	2881	2994	650	668	651	268	268	241
0.85 ***	26975	33530	64430	9943	11637	15650	2030	2143	2293	462	474	481	.	.	.
•••	29593	40674	64538	10662	13074	15660	2086	2206	2265	474	474	456	.	.	.
0.9 ***	20461	25400	44743	7343	8443	10662	1343	1393	1456	.	.	.	.	.	.
•••	22456	30493	44753	7824	9305	10648	1362	1418	1432	.	.	.	.	.	.
0.95 ***	12493	15087	22381	3980	4381	5011	.	.	.	.	.	.	.	.	.
•••	13568	17461	22364	4168	4655	4986	.	.	.	.	.	.	.	.	.
0.98 ***	5618	6368	7700	.	.	.	.	.	.	.	.	.	.	.	.
•••	5956	6912	7680	.	.	.	.	.	.	.	.	.	.	.	.

TABLE 4: ALPHA= 0.01 POWER= 0.9 EXPECTED ACCRUAL THRU MINIMUM FOLLOW-UP= 8000

		DEL=.01			DEL=.02			DEL=.05			DEL=.10			DEL=.15		
FACT=		1.0 .75	.50 .25	.00 BIN	1.0 .75	.50 .25	.00 BIN	1.0 .75	.50 .25	.00 BIN	1.0 75	.50 .25	.00 BIN	1.0 .75	.50 .25	.00 BIN
PCONT=***					REQUIRED NUMBER OF PATIENTS											
0.01	***	1993	2005	2005	733	733	733	245	245	245	125	125	125	85	85	85
	•••	1993	2005	7680	733	733	2539	245	245	691	125	125	281	85	85	167
0.02	***	4413	4433	4485	1425	1425	1433	385	385	385	165	165	165	113	113	113
	•••	4425	4453	12679	1425	1425	3775	385	385	883	165	165	326	113	113	186
0.05	***	13373	13505	13993	3813	3853	3925	825	825	833	293	293	293	173	173	173
	•••	13425	13653	27050	3833	3885	7329	825	825	1432	293	293	456	173	173	241
0.1	***	29525	29893	32573	8025	8205	8613	1565	1585	1605	493	493	493	265	265	265
	•••	29653	30473	48918	8105	8365	12731	1573	1593	2265	493	493	651	265	265	322
0.15	***	44153	44805	51453	11845	12245	13325	2253	2293	2353	665	673	673	333	345	345
	•••	44373	45965	68183	12005	12633	17482	2273	2325	2994	673	673	821	333	345	391
0.2	***	55993	57025	69125	15005	15673	17705	2845	2925	3033	825	833	845	405	405	413
	•••	56345	58873	84845	15273	16353	21583	2885	2973	3619	825	833	964	405	405	449
0.25	***	64725	66205	84865	17445	18405	21585	3345	3465	3625	953	965	973	465	465	465
	•••	65225	68885	98903	17825	19425	25032	3393	3533	4140	953	973	1081	465	465	495
0.3	***	70365	72405	98273	19165	20453	24873	3733	3893	4125	1053	1073	1093	505	505	513
	•••	71045	76033	110358	19665	21825	27831	3805	3993	4556	1065	1085	1172	505	505	530
0.35	***	73133	75833	109125	20225	21865	27525	4013	4213	4513	1133	1153	1173	533	533	545
	•••	74045	80553	119210	20865	23585	29979	4105	4353	4869	1145	1165	1237	533	545	553
0.4	***	73373	76865	117273	20725	22685	29493	4205	4445	4805	1185	1205	1233	553	553	565
	•••	74553	82733	125458	21493	24733	31475	4313	4605	5077	1193	1213	1276	553	553	565
0.45	***	71473	75885	122653	20725	22993	30765	4293	4565	4973	1205	1233	1265	553	565	565
	•••	72993	82945	129102	21633	25313	32322	4413	4745	5181	1213	1245	1289	553	565	565
0.5	***	67885	73305	125225	20325	22845	31333	4293	4593	5033	1205	1225	1265	553	553	565
	•••	69785	81493	130144	21333	25393	32517	4433	4785	5181	1213	1245	1276	553	565	553
0.55	***	63085	69525	124953	19593	22293	31205	4213	4525	4973	1173	1205	1233	533	533	545
	•••	65393	78685	128582	20685	24985	32061	4353	4725	5077	1185	1213	1237	533	545	530
0.6	***	57513	64893	121853	18585	21385	30353	4065	4365	4805	1113	1145	1173	505	505	513
	•••	60233	74753	124416	19733	24125	30955	4193	4553	4869	1133	1165	1172	505	513	495
0.65	***	51573	59653	115933	17345	20153	28805	3825	4105	4525	1045	1065	1093	453	465	473
	•••	54625	69873	117648	18493	22833	29197	3953	4293	4556	1053	1073	1081	465	465	449
0.7	***	45565	53973	107205	15885	18593	26553	3505	3765	4125	933	953	985	405	405	413
	•••	48793	64133	108275	17005	21125	26789	3625	3925	4140	945	973	964	405	413	391
0.75	***	39593	47905	95693	14213	16705	23605	3113	3325	3625	805	825	845	333	345	345
	•••	42833	57553	96300	15253	18973	23730	3213	3465	3619	813	833	821	345	345	322
0.8	***	33645	41373	81425	12313	14465	19973	2625	2793	3013	653	665	673	265	265	265
	•••	36693	50045	81721	13213	16353	20021	2705	2893	2994	653	665	651	265	265	241
0.85	***	27545	34185	64433	10105	11785	15653	2045	2153	2285	465	473	485	.	.	.
	•••	30193	41345	64538	10825	13193	15660	2093	2213	2265	473	473	456	.	.	.
0.9	***	20905	25873	44745	7453	8533	10665	1345	1393	1453	.	.	.	.	.	.
	•••	22925	30953	44753	7925	9373	10648	1373	1425	1432	.	.	.	.	.	.
0.95	***	12733	15325	22385	4025	4413	5013	.	.	.	.	.	.	.	.	.
	•••	13813	17665	22364	4205	4673	4986	.	.	.	.	.	.	.	.	.
0.98	***	5705	6425	7705	.	.	.	.	.	.	.	.	.	.	.	.
	•••	6025	6953	7680	.	.	.	.	.	.	.	.	.	.	.	.

TABLE 4: ALPHA= 0.01 POWER= 0.9   EXPECTED ACCRUAL THRU MINIMUM FOLLOW-UP= 8500

		DEL=.01			DEL=.02			DEL=.05			DEL=.10			DEL=.15	
FACT=	1.0 .75	.50 .25	.00 BIN	1.0 .75	.50 .25	.00 BIN	1.0 .75	.50 .25	.00 BIN	1.0 75	.50 .25	.00 BIN	1.0 .75	.50 .25	.00 BIN

PCONT=•••   REQUIRED NUMBER OF PATIENTS

PCONT		1.0	.50	.00	1.0	.50	.00	1.0	.50	.00	1.0	.50	.00	1.0	.50	.00
0.01	•••	1990 2003	2003	2003	736 736	736	736	240 240	240	240	120 120	120	120	91 91	91	91
	•••	2003 2003		7680	736 736		2539	240 240		691	120 120		281	91 91		167
0.02	•••	4413 4434	4476		1416 1430	1430		388 388	388		163 163	163		112 112	112	
	•••	4426 4455	12679		1416 1430	3775		388 388	883		163 163	326		112 112	186	
0.05	•••	13380 13521	13996		3818 3860	3924		821 821	835		290 290	290		176 176	176	
	•••	13436 13670	27050		3831 3881	7329		821 835	1432		290 290	456		176 176	241	
0.1	•••	29551 29934	32569		8038 8230	8612		1565 1586	1600		495 495	495		261 261	261	
	•••	29679 30529	48918		8123 8378	12731		1578 1586	2265		495 495	651		261 261	322	
0.15	•••	44193 44886	51452		11885 12288	13316		2258 2301	2351		665 673	673		333 346	346	
	•••	44426 46084	68183		12041 12658	17482		2280 2322	2994		673 673	821		333 346	391	
0.2	•••	56063 57155	69132		15059 15739	17707		2853 2938	3031		821 835	843		410 410	410	
	•••	56433 59081	84845		15327 16411	21583		2896 2981	3619		821 835	964		410 410	449	
0.25	•••	64818 66391	84865		17516 18501	21583		3350 3470	3626		949 962	983		460 460	460	
	•••	65350 69183	98903		17906 19513	25032		3406 3541	4140		962 970	1081		460 460	495	
0.3	•••	70492 72646	98274		19258 20576	24876		3746 3903	4128		1055 1076	1090		503 503	516	
	•••	71223 76429	110358		19776 21936	27831		3818 4001	4556		1068 1076	1172		503 503	530	
0.35	•••	73305 76166	109125		20350 22008	27525		4030 4235	4519		1132 1153	1175		537 537	545	
	•••	74275 81061	119210		21009 23721	29979		4128 4362	4869		1140 1161	1237		537 537	553	
0.4	•••	73595 77279	117271		20881 22871	29488		4221 4455	4795		1183 1204	1238		545 558	558	
	•••	74856 83356	125458		21668 24898	31475		4328 4617	5077		1196 1217	1276		558 558	565	
0.45	•••	71767 76386	122661		20916 23198	30763		4320 4583	4965		1204 1238	1260		558 566	566	
	•••	73361 83667	129102		21830 25514	32322		4447 4753	5181		1217 1246	1289		558 566	565	
0.5	•••	68248 73913	125219		20533 23070	31336		4328 4604	5029		1204 1225	1260		545 558	566	
	•••	70245 82315	130144		21561 25599	32517		4455 4795	5181		1217 1246	1276		558 558	553	
0.55	•••	63530 70224	124956		19819 22539	31201		4243 4540	4978		1175 1204	1238		537 537	545	
	•••	65945 79574	128582		20924 25208	32061		4383 4731	5077		1183 1217	1237		537 545	530	
0.6	•••	58040 65668	121853		18820 21638	30359		4094 4383	4808		1119 1153	1175		503 503	516	
	•••	60866 75698	124416		19981 24345	30955		4221 4575	4869		1132 1161	1172		503 516	495	
0.65	•••	52183 60470	115933		17580 20393	28808		3852 4128	4519		1047 1068	1098		460 460	473	
	•••	55320 70832	117648		18748 23062	29197		3980 4306	4556		1055 1076	1081		460 473	449	
0.7	•••	46211 54810	107199		16121 18820	26555		3533 3775	4128		941 962	983		410 410	418	
	•••	49526 65081	108275		17248 21328	26789		3648 3937	4140		949 970	964		410 410	391	
0.75	•••	40253 48698	95695		14435 16921	23601		3138 3342	3626		813 821	843		346 346	346	
	•••	43555 58443	96300		15476 19152	23730		3236 3470	3619		813 835	821		346 346	322	
0.8	•••	34269 42102	81423		12509 14634	19968		2649 2798	3010		651 665	673		261 261	269	
	•••	37371 50823	81721		13401 16496	20021		2726 2896	2994		651 673	651		261 261	241	
0.85	•••	28098 34792	64436		10256 11914	15654		2054 2160	2288		473 473	481		.	.	.
	•••	30776 41975	64538		10971 13295	15660		2110 2216	2265		473 481	456		.	.	.
0.9	•••	21320 26321	44737		7558 8620	10660		1353 1395	1459		.	.	.	.	.	.
	•••	23360 31379	44753		8017 9428	10648		1374 1430	1432		.	.	.	.	.	.
0.95	•••	12955 15540	22382		4065 4447	5008		.	.	.	.	.	.	.	.	.
	•••	14039 17843	22364		4235 4689	4986		.	.	.	.	.	.	.	.	.
0.98	•••	5773 6487	7706		.	.	.	.	.	.	.	.	.	.	.	.
	•••	6091 6997	7680		.	.	.	.	.	.	.	.	.	.	.	.

TABLE 4: ALPHA= 0.01 POWER= 0.9   EXPECTED ACCRUAL THRU MINIMUM FOLLOW-UP= 9000

	DEL=.01			DEL=.02			DEL=.05			DEL=.10			DEL=.15		
FACT=	1.0 .75	.50 .25	.00 BIN	1.0 .75	.50 .25	.00 BIN	1.0 .75	.50 .25	.00 BIN	1.0 75	.50 .25	.00 BIN	1.0 .75	.50 .25	.00 BIN
PCONT=***			REQUIRED NUMBER OF PATIENTS												
0.01 ***	1995	1995	2009	735	735	735	240	240	240	119	119	119	82	82	82
***	1995	2009	7680	735	735	2539	240	240	691	119	119	281	82	82	167
0.02 ***	4416	4439	4484	1424	1424	1432	389	389	389	164	164	164	105	105	105
***	4425	4461	12679	1424	1424	3775	389	389	883	164	164	326	105	105	186
0.05 ***	13394	13529	13987	3817	3862	3921	825	825	825	299	299	299	172	172	172
***	13447	13686	27050	3840	3885	7329	825	825	1432	299	299	456	172	172	241
0.1 ***	29571	29976	32572	8061	8241	8610	1567	1581	1604	487	487	487	262	262	262
***	29715	30584	48918	8137	8385	12731	1581	1590	2265	487	487	651	262	262	322
0.15 ***	44227	44970	51450	11909	12314	13326	2256	2301	2346	667	667	681	344	344	344
***	44475	46199	68183	12075	12682	17482	2279	2324	2994	667	681	821	344	344	391
0.2 ***	56130	57277	69126	15112	15801	17700	2864	2940	3030	825	825	839	411	411	411
***	56512	59280	84845	15382	16454	21583	2895	2985	3619	825	839	964	411	411	449
0.25 ***	64919	66570	84862	17587	18591	21584	3367	3480	3629	951	960	982	456	465	465
***	65467	69472	98903	17984	19581	25032	3412	3547	4140	960	974	1081	465	465	495
0.3 ***	70620	72892	98272	19365	20692	24877	3764	3907	4124	1064	1072	1095	501	510	510
***	71385	76821	110358	19896	22042	27831	3831	4011	4556	1064	1086	1172	501	510	530
0.35 ***	73477	76484	109117	20481	22155	27524	4056	4245	4515	1131	1154	1176	532	532	546
***	74504	81546	119210	21156	23856	29979	4146	4371	4869	1140	1162	1237	532	546	553
0.4 ***	73815	77685	117276	21030	23032	29490	4245	4470	4799	1185	1207	1230	555	555	569
***	75142	83954	125458	21831	25057	31475	4349	4619	5077	1199	1221	1276	555	555	565
0.45 ***	72051	76897	122654	21089	23392	30772	4349	4605	4965	1207	1230	1266	555	569	569
***	73739	84367	129102	22020	25687	32322	4461	4762	5181	1221	1244	1289	555	569	565
0.5 ***	68609	74512	125219	20729	23294	31335	4357	4627	5032	1207	1230	1266	555	555	569
***	70701	83099	130144	21772	25791	32517	4484	4857	5181	1221	1244	1276	555	555	553
0.55 ***	63974	70912	124957	20031	22762	31200	4281	4560	4979	1176	1207	1230	532	532	546
***	66494	80444	128582	21156	25409	32061	4402	4740	5077	1185	1221	1237	532	546	530
0.6 ***	58560	66412	121852	19050	21876	30359	4124	4402	4807	1117	1154	1176	501	510	510
***	61485	76596	124416	20211	24554	30955	4245	4582	4869	1140	1162	1172	501	510	495
0.65 ***	52769	61260	115935	17812	20625	28806	3876	4146	4529	1041	1072	1095	456	465	465
***	55995	71745	117648	18974	23257	29197	3997	4312	4556	1050	1086	1081	465	465	449
0.7 ***	46837	55604	107205	16341	19041	26556	3561	3795	4132	937	960	982	411	411	411
***	50226	65976	108275	17466	21516	26789	3674	3952	4140	951	974	964	411	411	391
0.75 ***	40889	49470	95699	14640	17115	23609	3156	3359	3629	816	825	839	344	344	344
***	44250	59271	96300	15689	19320	23730	3246	3480	3619	816	839	821	344	344	322
0.8 ***	34867	42801	81420	12682	14811	19972	2661	2819	3007	659	667	681	262	262	262
***	38009	51554	81721	13582	16634	20021	2729	2909	2994	659	667	651	262	262	241
0.85 ***	28612	35376	64432	10401	12044	15652	2062	2166	2287	465	479	479	.	.	.
***	31326	42562	64538	11107	13394	15660	2121	2220	2265	479	479	456	.	.	.
0.9 ***	21719	26745	44736	7656	8691	10671	1356	1401	1455	.	.	.	.	.	.
***	23766	31785	44753	8106	9487	10648	1379	1432	1432	.	.	.	.	.	.
0.95 ***	13177	15742	22380	4101	4470	5010	.	.	.	.	.	.	.	.	.
***	14257	18015	22364	4267	4709	4986	.	.	.	.	.	.	.	.	.
0.98 ***	5834	6531	7701	.	.	.	.	.	.	.	.	.	.	.	.
***	6149	7026	7680	.	.	.	.	.	.	.	.	.	.	.	.

TABLE 4: ALPHA= 0.01 POWER= 0.9   EXPECTED ACCRUAL THRU MINIMUM FOLLOW-UP= 9500

		DEL=.01			DEL=.02			DEL=.05			DEL=.10			DEL=.15	
FACT=	1.0 .75	.50 .25	.00 BIN	1.0 .75	.50 .25	.00 BIN	1.0 .75	.50 .25	.00 BIN	1.0 75	.50 .25	.00 BIN	1.0 .75	.50 .25	.00 BIN
PCONT=•••				REQUIRED NUMBER OF PATIENTS												
0.01 ***	2001	2001	2010	728	728	728	244	244	244	125	125	125	87	87	87
•••	2001	2001	7680	728	728	2539	244	244	691	125	125	281	87	87	167
0.02 ***	4424	4448	4480	1417	1431	1431	386	386	386	173	173	173	110	110	110
•••	4433	4457	12679	1417	1431	3775	386	386	883	173	173	326	110	110	186
0.05 ***	13401	13544	13995	3830	3863	3925	823	823	823	291	291	291	173	173	173
•••	13458	13695	27050	3839	3887	7329	823	823	1432	291	291	456	173	173	241
0.1 ***	29599	30012	32568	8067	8248	8613	1574	1583	1607	490	490	490	253	268	268
•••	29741	30629	48918	8153	8399	12731	1574	1598	2265	490	490	651	268	268	322
0.15 ***	44276	45045	51449	11938	12342	13315	2263	2295	2343	671	671	680	339	339	339
•••	44538	46319	68183	12104	12713	17482	2286	2319	2994	671	671	821	339	339	391
0.2 ***	56199	57395	69128	15159	15848	17700	2865	2937	3032	823	838	838	410	410	410
•••	56603	59462	84845	15444	16498	21583	2904	2984	3619	823	838	964	410	410	449
0.25 ***	65010	66753	84865	17662	18674	21580	3379	3483	3625	956	965	980	458	467	467
•••	65589	69745	98903	18065	19657	25032	3426	3554	4140	956	965	1081	458	467	495
0.3 ***	70743	73142	98269	19458	20797	24873	3783	3925	4124	1060	1075	1084	505	505	514
•••	71565	77194	110358	19989	22141	27831	3839	4020	4556	1060	1084	1172	505	505	530
0.35 ***	73655	76799	109123	20607	22293	27518	4068	4258	4519	1132	1155	1179	538	538	538
•••	74724	82024	119210	21286	23979	29979	4163	4376	4869	1146	1170	1237	538	538	553
0.4 ***	74044	78096	117269	21177	23195	29489	4267	4495	4804	1194	1203	1227	553	553	562
•••	75436	84533	125458	21984	25205	31475	4376	4623	5077	1194	1218	1276	553	562	565
0.45 ***	72334	77393	122651	21263	23575	30772	4362	4614	4970	1218	1241	1265	562	562	562
•••	74106	85040	129102	22198	25855	32322	4480	4780	5181	1227	1250	1289	562	562	565
0.5 ***	68962	75089	125225	20930	23495	31342	4376	4647	5027	1203	1227	1265	553	553	562
•••	71147	83853	130144	21975	25974	32517	4504	4813	5181	1218	1250	1276	553	562	553
0.55 ***	64402	71574	124955	20241	22982	31199	4305	4575	4970	1179	1203	1241	529	538	538
•••	67029	81255	128582	21367	25594	32061	4433	4756	5077	1194	1218	1237	538	538	530
0.6 ***	59073	67133	121853	19268	22094	30359	4148	4424	4804	1123	1155	1179	505	505	514
•••	62089	77464	124416	20431	24754	30955	4267	4590	4869	1132	1170	1172	505	514	495
0.65 ***	53349	62027	115930	18033	20844	28800	3901	4163	4519	1051	1075	1099	458	467	467
•••	56659	72619	117648	19196	23448	29197	4020	4329	4556	1060	1084	1081	467	467	449
0.7 ***	47444	56365	107199	16560	19253	26559	3578	3815	4124	942	965	980	410	410	410
•••	50903	66815	108275	17676	21690	26789	3688	3958	4140	956	965	964	410	410	391
0.75 ***	41498	50199	95695	14835	17296	23614	3174	3364	3625	814	823	838	339	339	348
•••	44918	60055	96300	15880	19467	23730	3260	3483	3619	823	838	821	339	348	322
0.8 ***	35441	43454	81421	12855	14969	19965	2675	2818	3008	657	657	671	268	268	268
•••	38633	52242	81721	13758	16759	20021	2747	2913	2994	657	671	651	268	268	241
0.85 ***	29109	35940	64425	10537	12166	15658	2073	2168	2286	467	481	481	.	.	.
•••	31855	43122	64538	11240	13482	15660	2120	2224	2265	467	481	456	.	.	.
0.9 ***	22094	27153	44737	7734	8770	10670	1360	1408	1455	.	.	.	.	.	.
•••	24160	32164	44753	8185	9530	10648	1384	1431	1432	.	.	.	.	.	.
0.95 ***	13378	15943	22388	4139	4495	5003	.	.	.	.	.	.	.	.	.
•••	14455	18175	22364	4305	4718	4986	.	.	.	.	.	.	.	.	.
0.98 ***	5896	6570	7701	.	.	.	.	.	.	.	.	.	.	.	.
•••	6205	7060	7680	.	.	.	.	.	.	.	.	.	.	.	.

TABLE 4: ALPHA= 0.01 POWER= 0.9   EXPECTED ACCRUAL THRU MINIMUM FOLLOW-UP= 10000

	DEL=.01			DEL=.02			DEL=.05			DEL=.10			DEL=.15		
FACT=	1.0 .75	.50 .25	.00 BIN	1.0 .75	.50 .25	.00 BIN	1.0 .75	.50 .25	.00 BIN	1.0 75	.50 .25	.00 BIN	1.0 .75	.50 .25	.00 BIN
PCONT=•••				REQUIRED NUMBER OF PATIENTS											
0.01 •••	1991	2007	2007	732	732	732	241	241	241	116	116	116	82	82	82
•••	1991	2007	7680	732	732	2539	241	241	691	116	116	281	82	82	167
0.02 •••	4416	4441	4482	1416	1432	1432	382	382	382	166	166	166	107	107	107
•••	4432	4457	12679	1416	1432	3775	382	382	883	166	166	326	107	107	186
0.05 •••	13407	13557	13991	3832	3866	3916	816	832	832	291	291	291	166	166	166
•••	13466	13707	27050	3841	3891	7329	832	832	1432	291	291	456	166	166	241
0.1 •••	29616	30057	32566	8082	8266	8616	1566	1582	1607	491	491	491	257	257	257
•••	29766	30682	48918	8157	8407	12731	1582	1591	2265	491	491	651	257	257	322
0.15 •••	44316	45116	51457	11966	12382	13316	2266	2307	2357	666	666	682	341	341	341
•••	44591	46432	68183	12141	12732	17482	2282	2332	2994	666	682	821	341	341	391
0.2 •••	56257	57516	69132	15207	15907	17707	2882	2941	3032	832	832	841	407	407	407
•••	56682	59641	84845	15491	16541	21583	2907	2982	3619	832	832	964	407	407	449
0.25 •••	65107	66932	84866	17732	18741	21582	3382	3491	3632	957	966	982	457	466	466
•••	65716	70016	98903	18141	19716	25032	3432	3557	4140	957	966	1081	466	466	495
0.3 •••	70882	73382	98282	19541	20907	24882	3791	3932	4132	1057	1082	1091	507	507	507
•••	71732	77557	110358	20091	22232	27831	3857	4016	4556	1066	1082	1172	507	507	530
0.35 •••	73816	77116	109116	20716	22416	27516	4091	4266	4516	1141	1157	1182	532	541	541
•••	74957	82482	119210	21407	24091	29979	4166	4382	4869	1141	1166	1237	532	541	553
0.4 •••	74266	78491	117282	21316	23357	29491	4291	4507	4807	1191	1207	1232	557	557	566
•••	75732	85091	125458	22141	25341	31475	4382	4641	5077	1207	1216	1276	557	557	565
0.45 •••	72616	77866	122657	21416	23757	30766	4391	4632	4966	1216	1232	1266	557	566	566
•••	74466	85682	129102	22366	26007	32322	4507	4782	5181	1232	1241	1289	557	566	565
0.5 •••	69316	75657	125216	21107	23682	31341	4407	4666	5032	1207	1232	1266	557	557	566
•••	71591	84582	130144	22157	26141	32517	4516	4832	5181	1216	1241	1276	557	557	553
0.55 •••	64832	72216	124957	20441	23182	31207	4332	4591	4966	1182	1207	1232	532	541	541
•••	67541	82041	128582	21566	25766	32061	4457	4766	5077	1191	1216	1237	532	541	530
0.6 •••	59582	67841	121857	19466	22291	30357	4166	4432	4807	1132	1157	1182	507	507	516
•••	62682	78282	124416	20641	24932	30955	4291	4607	4869	1141	1166	1172	507	507	495
0.65 •••	53907	62757	115932	18232	21041	28807	3932	4182	4516	1057	1066	1091	457	466	466
•••	57291	73441	117648	19407	23616	29197	4041	4382	4556	1057	1082	1081	466	466	449
0.7 •••	48041	57091	107207	16757	19441	26557	3607	3832	4132	941	966	982	407	407	416
•••	51566	67632	108275	17882	21857	26789	3707	3966	4140	957	966	964	407	416	391
0.75 •••	42082	50891	95691	15016	17482	23607	3191	3382	3632	816	832	841	341	341	341
•••	45557	60807	96300	16057	19607	23730	3282	3491	3619	816	832	821	341	341	322
0.8 •••	35991	44091	81432	13016	15107	19966	2691	2832	3007	657	666	682	266	266	266
•••	39216	52907	81721	13907	16866	20021	2757	2916	2994	657	666	651	266	266	241
0.85 •••	29591	36466	64432	10666	12266	15657	2082	2182	2291	466	482	482	.	.	.
•••	32357	43641	64538	11357	13566	15660	2132	2232	2265	466	482	456	.	.	.
0.9 •••	22457	27532	44741	7816	8832	10666	1366	1407	1457	.	.	.	.	.	.
•••	24532	32516	44753	8266	9582	10648	1391	1432	1432	.	.	.	.	.	.
0.95 •••	13566	16116	22382	4166	4516	5007	.	.	.	.	.	.	.	.	.
•••	14657	18316	22364	4332	4732	4986	.	.	.	.	.	.	.	.	.
0.98 •••	5957	6616	7707	.	.	.	.	.	.	.	.	.	.	.	.
•••	6257	7082	7680	.	.	.	.	.	.	.	.	.	.	.	.

TABLE 4: ALPHA= 0.01 POWER= 0.9   EXPECTED ACCRUAL THRU MINIMUM FOLLOW-UP= 11000

		DEL=.01			DEL=.02			DEL=.05			DEL=.10			DEL=.15		
FACT=		1.0 .75	.50 .25	.00 BIN	1.0 .75	.50 .25	.00 BIN	1.0 .75	.50 .25	.00 BIN	1.0 75	.50 .25	.00 BIN	1.0 .75	.50 .25	.00 BIN
PCONT=•••					REQUIRED NUMBER OF PATIENTS											
0.01	•••	1998	1998	1998	733	733	733	238	238	238	117	117	117	90	90	90
	•••	1998	1998	7680	733	733	2539	238	238	691	117	117	281	90	90	167
0.02	•••	4418	4445	4473	1420	1420	1437	375	375	392	172	172	172	117	117	117
	•••	4435	4462	12679	1420	1420	3775	375	375	883	172	172	326	117	117	186
0.05	•••	13427	13575	13988	3830	3868	3923	832	832	832	293	293	293	172	172	172
	•••	13482	13713	27050	3857	3895	7329	832	832	1432	293	293	456	172	172	241
0.1	•••	29663	30130	32567	8103	8285	8615	1575	1585	1602	485	485	485	255	265	265
	•••	29828	30780	48918	8185	8422	12731	1575	1602	2265	485	485	651	265	265	322
0.15	•••	44403	45272	51460	12025	12437	13317	2273	2300	2345	667	678	678	337	337	337
	•••	44695	46630	68183	12200	12778	17482	2290	2328	2994	667	678	821	337	337	391
0.2	•••	56382	57768	69125	15297	15995	17700	2895	2950	3032	832	832	843	403	403	403
	•••	56860	59995	84845	15583	16628	21583	2922	2988	3619	832	832	964	403	403	449
0.25	•••	65292	67272	84872	17865	18883	21578	3400	3500	3620	953	970	980	458	458	458
	•••	65963	70528	98903	18278	19835	25032	3445	3555	4140	970	970	1081	458	458	495
0.3	•••	71133	73855	98275	19725	21100	24878	3813	3950	4132	1063	1080	1090	502	502	513
	•••	72068	78245	110358	20285	22392	27831	3868	4033	4556	1063	1080	1172	502	513	530
0.35	•••	74158	77733	109127	20945	22667	27518	4115	4280	4517	1145	1162	1173	530	540	540
	•••	75395	83360	119210	21650	24300	29979	4198	4390	4869	1145	1162	1237	540	540	553
0.4	•••	74708	79262	117278	21578	23640	29487	4325	4528	4803	1190	1217	1228	557	557	557
	•••	76303	86137	125458	22420	25593	31475	4418	4655	5077	1200	1217	1276	557	557	565
0.45	•••	73185	78805	122657	21732	24080	30763	4435	4655	4968	1217	1245	1255	557	568	568
	•••	75175	86907	129102	22695	26297	32322	4528	4803	5181	1228	1245	1289	557	568	565
0.5	•••	70022	76760	125225	21457	24042	31340	4445	4693	5023	1217	1245	1255	557	557	568
	•••	72470	85945	130144	22513	26445	32517	4555	4847	5181	1228	1245	1276	557	557	553
0.55	•••	65677	73443	124950	20808	23558	31203	4363	4627	4968	1190	1217	1228	530	540	540
	•••	68548	83525	128582	21952	26088	32061	4490	4775	5077	1200	1217	1237	540	540	530
0.6	•••	60545	69170	121860	19862	22678	30350	4215	4462	4803	1135	1162	1173	502	513	513
	•••	63818	79823	124416	21028	25252	30955	4325	4610	4869	1145	1162	1172	502	513	495
0.65	•••	54980	64137	115930	18625	21430	28810	3967	4198	4517	1052	1080	1090	458	458	475
	•••	58510	75000	117648	19790	23932	29197	4077	4352	4556	1063	1080	1081	458	475	449
0.7	•••	49160	58472	107202	17123	19790	26555	3637	3857	4132	942	970	980	403	403	420
	•••	52807	69125	108275	18250	22145	26789	3730	3978	4140	953	970	964	403	420	391
0.75	•••	43193	52213	95690	15363	17800	23602	3225	3400	3620	815	832	843	337	348	348
	•••	46768	62195	96300	16397	19862	23730	3307	3500	3619	815	832	821	337	348	322
0.8	•••	37033	45283	81418	13317	15380	19972	2713	2840	3005	650	667	678	265	265	265
	•••	40333	54110	81721	14197	17068	20021	2768	2922	2994	667	667	651	265	265	241
0.85	•••	30488	37445	64423	10897	12475	15655	2097	2190	2290	475	475	485	.	.	.
	•••	33310	44612	64538	11585	13713	15660	2135	2235	2265	475	475	456	.	.	.
0.9	•••	23145	28233	44733	7965	8955	10660	1382	1410	1448	.	.	.	.	.	.
	•••	25235	33155	44753	8405	9660	10648	1393	1437	1432	.	.	.	.	.	.
0.95	•••	13933	16452	22392	4225	4555	5012	.	.	.	.	.	.	.	.	.
	•••	15005	18580	22364	4380	4765	4986	.	.	.	.	.	.	.	.	.
0.98	•••	6057	6700	7707	.	.	.	.	.	.	.	.	.	.	.	.
	•••	6343	7130	7680	.	.	.	.	.	.	.	.	.	.	.	.

TABLE 4: ALPHA= 0.01 POWER= 0.9 EXPECTED ACCRUAL THRU MINIMUM FOLLOW-UP= 12000

	DEL=.01			DEL=.02			DEL=.05			DEL=.10			DEL=.15		
FACT=	1.0 .75	.50 .25	.00 BIN	1.0 .75	.50 .25	.00 BIN	1.0 .75	.50 .25	.00 BIN	1.0 75	.50 .25	.00 BIN	1.0 .75	.50 .25	.00 BIN
PCONT=•••				REQUIRED NUMBER OF PATIENTS											
0.01 ***	1999 1999	1999	728 728	739	248 248	248	128 128	128	79 79	79					
•••	1999 1999	7680	728 739	2539	248 248	691	128 128	281	79 79	167					
0.02 ***	4429 4448	4478	1418 1429	1429	379 379	379	169 169	169	109 109	109					
•••	4429 4459	12679	1429 1429	3775	379 379	883	169 169	326	109 109	186					
0.05 ***	13448 13598	13988	3829 3878	3919	829 829	829	289 289	289	169 169	169					
•••	13508 13729	27050	3848 3889	7329	829 829	1432	289 289	456	169 169	241					
0.1 ***	29708 30199	32569	8138 8299	8618	1579 1579	1598	488 488	488	259 259	259					
•••	29888 30859	48918	8209 8438	12731	1579 1598	2265	488 488	651	259 259	322					
0.15 ***	44479 45428	51458	12079 12488	13328	2269 2318	2348	668 668	679	338 338	338					
•••	44809 46819	68183	12248 12818	17482	2299 2329	2994	668 679	821	338 338	391					
0.2 ***	56509 57998	69128	15379 16088	17708	2899 2959	3038	829 829	848	409 409	409					
•••	57019 60319	84845	15679 16688	21583	2929 2989	3619	829 829	964	409 409	449					
0.25 ***	65468 67609	84859	17989 19009	21589	3409 3518	3619	968 968	979	458 458	469					
•••	66199 70999	98903	18409 19939	25032	3458 3559	4140	968 968	1081	458 469	495					
0.3 ***	71389 74318	98269	19898 21278	24878	3829 3968	4129	1069 1069	1088	499 499	518					
•••	72398 78878	110358	20449 22538	27831	3889 4039	4556	1069 1088	1172	499 499	530					
0.35 ***	74498 78338	109118	21158 22879	27518	4148 4309	4519	1148 1159	1178	529 529	548					
•••	75829 84169	119210	21859 24488	29979	4219 4399	4869	1148 1159	1237	529 548	553					
0.4 ***	75139 80018	117278	21829 23899	29498	4358 4538	4808	1189 1208	1238	548 559	559					
•••	76868 87128	125458	22688 25808	31475	4448 4658	5077	1208 1219	1276	559 559	565					
0.45 ***	73729 79699	122659	22009 24368	30769	4459 4688	4969	1219 1238	1268	559 559	559					
•••	75878 88039	129102	22988 26539	32322	4568 4808	5181	1238 1249	1289	559 559	565					
0.5 ***	70699 77798	125228	21769 24349	31339	4478 4718	5029	1219 1238	1268	548 559	559					
•••	73298 87218	130144	22838 26719	32517	4598 4849	5181	1219 1249	1276	559 559	553					
0.55 ***	66488 74599	124958	21158 23899	31208	4399 4639	4969	1189 1208	1238	529 548	548					
•••	69529 84878	128582	22298 26378	32061	4519 4789	5077	1208 1219	1237	529 548	530					
0.6 ***	61489 70418	121849	20209 23029	30349	4249 4489	4808	1129 1159	1178	499 499	518					
•••	64898 81229	124416	21379 25538	30955	4358 4628	4869	1148 1159	1172	499 518	495					
0.65 ***	55999 65438	115928	18979 21758	28808	3998 4219	4519	1058 1069	1088	458 469	469					
•••	59659 76418	117648	20149 24218	29197	4099 4358	4556	1069 1088	1081	458 469	449					
0.7 ***	50228 59768	107198	17468 20119	26558	3668 3878	4129	949 968	979	409 409	409					
•••	53978 70508	108275	18589 22399	26789	3758 3998	4140	949 968	964	409 409	391					
0.75 ***	44258 53438	95689	15679 18079	23599	3248 3409	3619	818 829	848	338 338	349					
•••	47899 63469	96300	16699 20089	23730	3319 3518	3619	818 829	821	338 349	322					
0.8 ***	38018 46369	81428	13579 15608	19969	2738 2858	3008	649 668	679	259 259	259					
•••	41378 55219	81721	14468 17258	20021	2798 2929	2994	668 668	651	259 259	241					
0.85 ***	31328 38348	64429	11108 12649	15649	2119 2198	2288	469 469	488	.	.	.			
•••	34178 45488	64538	11779 13838	15660	2149 2239	2265	469 469	456	.	.	.			
0.9 ***	23768 28868	44738	8108 9049	10669	1388 1418	1459	.	.	.	.	.	.		
•••	25879 33728	44753	8528 9728	10648	1399 1429	1432	.	.	.	.	.	.		
0.95 ***	14258 16748	22388	4268 4579	5018	.	.	.	.	.	.	.	.	.	.	
•••	15319 18799	22364	4418 4778	4986	.	.	.	.	.	.	.	.	.	.	
0.98 ***	6139 6758	7699	.	.	.	.	.	.	.	.	.	.	.	.	.
•••	6428 7178	7680	.	.	.	.	.	.	.	.	.	.	.	.	.

TABLE 4: ALPHA= 0.01 POWER= 0.9   EXPECTED ACCRUAL THRU MINIMUM FOLLOW-UP= 13000

REQUIRED NUMBER OF PATIENTS

PCONT		DEL=.01			DEL=.02			DEL=.05			DEL=.10			DEL=.15		
FACT=		1.0 .75	.50 .25	.00 BIN	1.0 .75	.50 .25	.00 BIN	1.0 .75	.50 .25	.00 BIN	1.0 75	.50 .25	.00 BIN	1.0 .75	.50 .25	.00 BIN
0.01	***	2003 2003	2003 2003	2003 7680	736 736	736 736	736 2539	236 236	236 236	236 691	118 118	118 118	118 281	86 86	86 86	86 167
0.02	***	4429 4441	4441 4461	4473 12679	1418 1418	1418 1418	1439 3775	378 378	378 378	378 883	171 171	171 171	171 326	106 106	106 106	106 186
0.05	***	13464 13529	13606 13756	13996 27050	3844 3856	3876 3888	3921 7329	821 821	821 833	833 1432	301 301	301 301	301 456	171 171	171 171	171 241
0.1	***	29758 29941	30278 30949	32574 48918	8146 8231	8329 8438	8621 12731	1581 1581	1581 1601	1601 2265	496 496	496 496	496 651	269 269	269 269	269 322
0.15	***	44566 44903	45553 47004	51456 68183	12131 12294	12521 12846	13313 17482	2284 2296	2316 2328	2349 2994	671 671	671 671	671 821	334 334	334 334	346 391
0.2	***	56636 57188	58228 60621	69136 84845	15458 15751	16161 16746	17701 21583	2901 2934	2966 2999	3031 3619	833 833	833 833	833 964	411 411	411 411	411 449
0.25	***	65659 66451	67946 71444	84866 98903	18091 18534	19131 20041	21589 25032	3433 3466	3519 3563	3628 4140	963 963	963 963	984 1081	464 464	464 464	464 495
0.3	***	71639 72723	74771 79483	98268 110358	20041 20614	21426 22673	24871 27831	3844 3909	3974 4039	4116 4556	1061 1081	1081 1081	1093 1172	508 508	508 508	508 530
0.35	***	74836 76266	78919 84931	109123 119210	21341 22056	23084 24644	27516 29979	4169 4234	4311 4408	4526 4869	1146 1158	1158 1158	1179 1237	541 541	541 541	541 553
0.4	***	75583 77424	80739 88031	117281 125458	22076 22921	24136 26009	29486 31475	4376 4461	4559 4668	4798 5077	1191 1211	1211 1223	1223 1276	561 561	561 561	561 565
0.45	***	74283 76558	80576 89103	122664 129102	22283 23258	24644 26768	30766 32322	4494 4591	4701 4819	4961 5181	1223 1223	1244 1256	1256 1289	561 561	561 561	573 565
0.5	***	71379 74109	78789 88388	125211 130144	22056 23149	24656 26963	31339 32517	4506 4624	4733 4863	5026 5181	1223 1223	1244 1256	1256 1276	561 561	561 561	561 553
0.55	***	67284 70448	75701 86146	124951 128582	21471 22608	24201 26638	31209 32061	4441 4538	4668 4819	4981 5077	1191 1211	1211 1223	1244 1237	541 541	541 541	541 530
0.6	***	62388 65919	71586 82538	121851 124416	20528 21719	23344 25793	30364 30955	4278 4396	4506 4636	4798 4869	1146 1146	1158 1158	1179 1172	508 508	508 508	508 495
0.65	***	56981 60751	66646 77728	115936 117648	19293 20484	22076 24461	28804 29197	4039 4136	4246 4376	4526 4556	1061 1061	1081 1081	1093 1081	464 464	464 464	464 449
0.7	***	51241 55076	60958 71769	107206 108275	17786 18903	20398 22629	26561 26789	3693 3791	3888 4006	4136 4140	951 963	963 963	984 964	411 411	411 411	411 391
0.75	***	45228 48954	54576 64631	95689 96300	15966 16986	18339 20289	23604 23730	3271 3336	3433 3519	3628 3619	821 821	833 833	833 821	346 346	346 346	346 322
0.8	***	38923 42336	47394 56213	81421 81721	13833 14699	15816 17408	19976 20021	2751 2804	2869 2934	3011 2994	659 659	671 671	671 651	269 269	269 269	269 241
0.85	***	32119 34991	39183 46268	64436 64538	11298 11969	12814 13951	15653 15660	2121 2166	2198 2251	2284 2265	476 476	476 476	476 456	. .	. .	. .
0.9	***	24351 26464	29454 34243	44741 44753	8231 8633	9153 9791	10669 10648	1386 1406	1418 1439	1451 1432	. .	. .	. .	. .	. .	. .
0.95	***	14548 15609	17006 19001	22381 22364	4311 4461	4603 4798	5014 4986	. .	. .	. .	. .	. .	. .	. .	. .	. .
0.98	***	6228 6488	6813 7203	7711 7680	. .	. .	. .	. .	. .	. .	. .	. .	. .	. .	. .	. .

TABLE 4: ALPHA= 0.01 POWER= 0.9   EXPECTED ACCRUAL THRU MINIMUM FOLLOW-UP= 14000

	DEL=.01			DEL=.02			DEL=.05			DEL=.10			DEL=.15		
FACT=	1.0 .75	.50 .25	.00 BIN	1.0 .75	.50 .25	.00 BIN	1.0 .75	.50 .25	.00 BIN	1.0 75	.50 .25	.00 BIN	1.0 .75	.50 .25	.00 BIN
PCONT=***				REQUIRED NUMBER OF PATIENTS											
0.01 ***	2004	2004	2004	722	722	744	232	232	232	114	114	114	79	79	79
***	2004	2004	7680	722	744	2539	232	232	691	114	114	281	79	79	167
0.02 ***	4432	4454	4489	1422	1422	1422	372	372	394	162	162	162	114	114	114
***	4432	4467	12679	1422	1422	3775	372	394	883	162	162	326	114	114	186
0.05 ***	13484	13624	13987	3837	3872	3929	827	827	827	289	289	289	162	162	162
***	13532	13764	27050	3859	3894	7329	827	827	1432	289	289	456	162	162	241
0.1 ***	29807	30354	32572	8177	8339	8619	1584	1584	1597	499	499	499	254	254	254
***	30004	31019	48918	8247	8457	12731	1584	1597	2265	499	499	651	254	254	322
0.15 ***	44647	45697	51459	12167	12552	13322	2284	2319	2354	674	674	674	337	337	337
***	45019	47167	68183	12342	12867	17482	2297	2332	2994	674	674	821	337	337	391
0.2 ***	56779	58459	69134	15549	16227	17697	2914	2962	3032	827	827	849	407	407	407
***	57352	60909	84845	15829	16809	21583	2949	2997	3619	827	827	964	407	407	449
0.25 ***	65844	68272	84862	18209	19237	21582	3439	3522	3627	967	967	967	464	464	464
***	66684	71864	98903	18642	20112	25032	3487	3579	4140	967	967	1081	464	464	495
0.3 ***	71899	75202	98267	20182	21569	24872	3859	3977	4117	1072	1072	1094	499	512	512
***	73054	80054	110358	20764	22794	27831	3929	4047	4556	1072	1094	1172	512	512	530
0.35 ***	75167	79472	109117	21534	23262	27519	4187	4327	4524	1142	1164	1177	534	534	547
***	76694	85654	119210	22247	24789	29979	4257	4419	4869	1164	1164	1237	534	534	553
0.4 ***	76007	81419	117272	22282	24347	29492	4397	4572	4804	1199	1212	1234	547	547	569
***	77954	88909	125458	23144	26189	31475	4489	4677	5077	1212	1234	1276	547	569	565
0.45 ***	74817	81397	122649	22527	24894	30774	4524	4712	4979	1234	1247	1269	569	569	569
***	77219	90099	129102	23507	26959	32322	4607	4839	5181	1234	1247	1289	569	569	565
0.5 ***	72039	79717	125217	22339	24907	31334	4537	4747	5027	1234	1247	1269	547	547	569
***	74909	89482	130144	23424	27182	32517	4604	4874	5181	1234	1247	1276	547	569	553
0.55 ***	68049	76742	124959	21757	24487	31207	4467	4677	4979	1199	1212	1234	534	534	547
***	71352	87312	128582	22912	26854	32061	4572	4817	5077	1199	1234	1237	534	547	530
0.6 ***	63254	72704	121857	20834	23612	30354	4314	4524	4804	1142	1164	1177	499	512	512
***	66894	83764	124416	22024	26027	30955	4419	4664	4869	1142	1164	1172	512	512	495
0.65 ***	57912	67782	115929	19609	22352	28814	4069	4257	4524	1059	1072	1094	464	464	464
***	61762	78947	117648	20777	24684	29197	4152	4384	4556	1072	1094	1081	464	464	449
0.7 ***	52194	62077	107192	18069	20672	26552	3719	3907	4117	954	967	989	407	407	407
***	56114	72927	108275	19189	22829	26789	3802	4012	4140	954	967	964	407	407	391
0.75 ***	46174	55624	95699	16227	18572	23612	3299	3439	3627	814	827	849	337	337	337
***	49954	65704	96300	17242	20462	23730	3369	3522	3619	827	827	821	337	337	322
0.8 ***	39782	48322	81419	14057	16017	19972	2774	2879	3019	652	674	674	267	267	267
***	43234	57142	81721	14919	17557	20021	2822	2949	2994	652	674	651	267	267	241
0.85 ***	32839	39957	64422	11467	12937	15654	2144	2214	2284	477	477	477	.	.	.
***	35757	46992	64538	12119	14044	15660	2179	2249	2265	477	477	456	.	.	.
0.9 ***	24894	30004	44739	8339	9227	10662	1387	1422	1457	.	.	.	.	.	.
***	27007	34707	44753	8737	9844	10648	1409	1444	1432	.	.	.	.	.	.
0.95 ***	14827	17242	22387	4349	4629	5014	.	.	.	.	.	.	.	.	.
***	15877	19189	22364	4489	4804	4986	.	.	.	.	.	.	.	.	.
0.98 ***	6309	6869	7709	.	.	.	.	.	.	.	.	.	.	.	.
***	6567	7232	7680	.	.	.	.	.	.	.	.	.	.	.	.

529

TABLE 4: ALPHA= 0.01 POWER= 0.9 EXPECTED ACCRUAL THRU MINIMUM FOLLOW-UP= 15000

PCONT	FACT	DEL=.01 1.0/.75	.50/.25	.00/BIN	DEL=.02 1.0/.75	.50/.25	.00/BIN	DEL=.05 1.0/.75	.50/.25	.00/BIN	DEL=.10 1.0/75	.50/.25	.00/BIN	DEL=.15 1.0/.75	.50/.25	.00/BIN
0.01	•••	1997	1997	2011	736	736	736	235	235	235	122	122	122	85	85	85
	•••	1997	1997	7680	736	736	2539	235	235	691	122	122	281	85	85	167
0.02	•••	4435	4449	4472	1411	1435	1435	385	385	385	160	160	160	99	99	99
	•••	4449	4472	12679	1435	1435	3775	385	385	883	160	160	326	99	99	186
0.05	•••	13486	13636	13997	3849	3886	3924	835	835	835	286	286	286	174	174	174
	•••	13547	13772	27050	3872	3886	7329	835	835	1432	286	286	456	174	174	241
0.1	•••	29836	30399	32574	8185	8349	8611	1585	1585	1599	497	497	497	249	249	249
	•••	30047	31074	48918	8260	8461	12731	1585	1599	2265	497	497	651	249	249	322
0.15	•••	44724	45835	51460	12211	12586	13322	2297	2311	2349	661	685	685	347	347	347
	•••	45122	47311	68183	12385	12886	17482	2297	2335	2994	661	685	821	347	347	391
0.2	•••	56897	58660	69122	15610	16299	17710	2911	2972	3024	835	835	835	399	399	399
	•••	57511	61172	84845	15910	16847	21583	2949	3010	3619	835	835	964	399	399	449
0.25	•••	66024	68574	84872	18310	19336	21586	3460	3535	3624	961	961	985	460	460	460
	•••	66924	72272	98903	18736	20199	25032	3497	3572	4140	961	985	1081	460	460	495
0.3	•••	72160	75624	98274	20335	21699	24872	3872	3985	4135	1074	1074	1097	511	511	511
	•••	73374	80597	110358	20897	22885	27831	3924	4060	4556	1074	1074	1172	511	511	530
0.35	•••	75511	80035	109111	21699	23424	27511	4210	4336	4510	1149	1172	1172	535	535	535
	•••	77124	86311	119210	22411	24924	29979	4261	4435	4869	1149	1172	1237	535	535	553
0.4	•••	76435	82097	117272	22486	24549	29485	4411	4585	4810	1210	1210	1224	549	549	572
	•••	78497	89710	125458	23349	26335	31475	4510	4697	5077	1210	1224	1276	549	549	565
0.45	•••	75361	82186	122649	22772	25111	30774	4547	4735	4974	1224	1247	1261	572	572	572
	•••	77874	91022	129102	23747	27136	32322	4636	4847	5181	1224	1247	1289	572	572	565
0.5	•••	72685	80635	125222	22599	25172	31336	4561	4772	5035	1224	1247	1261	549	549	572
	•••	75661	90511	130144	23686	27385	32517	4660	4885	5181	1224	1247	1276	549	549	553
0.55	•••	68799	77747	124960	22036	24736	31210	4486	4711	4974	1186	1210	1224	535	535	549
	•••	72211	88411	128582	23185	27061	32061	4599	4824	5077	1210	1224	1237	535	535	530
0.6	•••	64097	73749	121847	21122	23874	30361	4336	4547	4810	1149	1149	1172	511	511	511
	•••	67847	84886	124416	22299	26222	30955	4435	4660	4869	1149	1172	1172	511	511	495
0.65	•••	58810	68860	115936	19885	22599	28810	4074	4285	4524	1060	1074	1097	460	460	474
	•••	62747	80072	117648	21047	24872	29197	4172	4397	4556	1074	1074	1081	460	474	449
0.7	•••	53110	63136	107199	18347	20911	26560	3736	3910	4135	961	961	985	399	399	422
	•••	57099	74011	108275	19449	23011	26789	3835	4022	4140	961	985	964	399	422	391
0.75	•••	47049	56611	95686	16472	18774	23611	3310	3460	3624	811	835	835	347	347	347
	•••	50897	66699	96300	17485	20611	23730	3385	3535	3619	835	835	821	347	347	322
0.8	•••	40599	49210	81422	14260	16186	19974	2785	2874	3010	661	661	685	272	272	272
	•••	44086	57999	81721	15099	17672	20021	2822	2949	2994	661	661	651	272	272	241
0.85	•••	33535	40674	64435	11635	13074	15647	2147	2199	2297	474	474	474	.	.	.
	•••	36460	47649	64538	12272	14124	15660	2185	2236	2265	474	474	456	.	.	.
0.9	•••	25397	30497	44747	8447	9310	10660	1397	1411	1449	.	.	.	.	.	.
	•••	27535	35147	44753	8836	9886	10648	1411	1435	1432	.	.	.	.	.	.
0.95	•••	15085	17461	22374	4374	4660	5011	.	.	.	.	.	.	.	.	.
	•••	16111	19360	22364	4510	4824	4986	.	.	.	.	.	.	.	.	.
0.98	•••	6361	6910	7697	.	.	.	.	.	.	.	.	.	.	.	.
	•••	6624	7261	7680	.	.	.	.	.	.	.	.	.	.	.	.

PCONT=••• REQUIRED NUMBER OF PATIENTS

TABLE 4: ALPHA= 0.01 POWER= 0.9   EXPECTED ACCRUAL THRU MINIMUM FOLLOW-UP= 17000

PCONT=•••	DEL=.01 1.0/.75	DEL=.01 .50/.25	DEL=.01 .00/BIN	DEL=.02 1.0/.75	DEL=.02 .50/.25	DEL=.02 .00/BIN	DEL=.05 1.0/.75	DEL=.05 .50/.25	DEL=.05 .00/BIN	DEL=.10 1.0/75	DEL=.10 .50/.25	DEL=.10 .00/BIN	DEL=.15 1.0/.75	DEL=.15 .50/.25	DEL=.15 .00/BIN
	REQUIRED NUMBER OF PATIENTS														
0.01 •••	2009	2009	2009	734	734	734	240	240	240	112	112	112	96	96	96
•••	2009	2009	7680	734	734	2539	240	240	691	112	112	281	96	96	167
0.02 •••	4431	4447	4474	1430	1430	1430	394	394	394	155	155	155	112	112	112
•••	4447	4474	12679	1430	1430	3775	394	394	883	155	155	326	112	112	186
0.05 •••	13526	13670	13994	3852	3879	3921	819	835	835	282	282	282	181	181	181
•••	13585	13797	27050	3879	3895	7329	819	835	1432	282	282	456	181	181	241
0.1 •••	29931	30526	32566	8230	8384	8612	1584	1584	1600	495	495	495	266	266	266
•••	30160	31206	48918	8299	8485	12731	1584	1600	2265	495	495	651	266	266	322
0.15 •••	44891	46081	51452	12294	12650	13314	2306	2322	2349	665	665	665	351	351	351
•••	45316	47611	68183	12437	12931	17482	2306	2322	2994	665	665	821	351	351	391
0.2 •••	57147	59086	69132	15736	16416	17707	2944	2986	3029	835	835	835	410	410	410
•••	57854	61652	84845	16034	16926	21583	2960	3002	3619	835	835	964	410	410	449
0.25 •••	66396	69175	84857	18499	19519	21575	3470	3539	3624	962	962	989	452	452	452
•••	67390	72984	98903	18924	20326	25032	3496	3581	4140	962	962	1081	452	452	495
0.3 •••	72644	76426	98271	20581	21941	24874	3895	4006	4134	1074	1074	1090	495	495	521
•••	74004	81569	110358	21150	23062	27831	3964	4065	4556	1074	1090	1172	495	521	530
0.35 •••	76171	81059	109125	22000	23726	27525	4235	4362	4516	1159	1159	1175	537	537	537
•••	77930	87561	119210	22749	25145	29979	4304	4431	4869	1159	1175	1237	537	537	553
0.4 •••	77276	83354	117269	22876	24890	29480	4447	4617	4787	1201	1217	1244	564	564	564
•••	79502	91190	125458	23726	26616	31475	4532	4702	5077	1217	1217	1276	564	564	565
0.45 •••	76384	83667	122666	23190	25511	30755	4575	4745	4957	1244	1244	1260	564	564	564
•••	79104	92720	129102	24167	27466	32322	4660	4856	5181	1244	1260	1289	564	564	565
0.5 •••	73919	82307	125216	23062	25596	31334	4601	4787	5026	1217	1244	1260	564	564	564
•••	77106	92380	130144	24151	27721	32517	4702	4899	5181	1244	1260	1276	564	564	553
0.55 •••	70221	79571	124961	22536	25214	31206	4532	4729	4984	1201	1217	1244	537	537	537
•••	73834	90409	128582	23684	27424	32061	4644	4856	5077	1201	1217	1237	537	537	530
0.6 •••	65674	75704	121859	21644	24337	30356	4389	4575	4814	1159	1159	1175	495	521	521
•••	69600	86940	124416	22791	26590	30955	4474	4686	4869	1159	1175	1172	495	521	495
0.65 •••	60462	70832	115925	20385	23062	28800	4134	4304	4516	1074	1074	1090	452	479	479
•••	64585	82121	117648	21532	25214	29197	4219	4405	4556	1074	1090	1081	452	479	449
0.7 •••	54810	65079	107196	18812	21320	26547	3767	3937	4134	962	962	989	410	410	410
•••	58916	75959	108275	19901	23317	26789	3852	4022	4140	962	962	964	410	410	391
0.75 •••	48690	58449	95695	16926	19152	23599	3342	3470	3624	819	835	835	351	351	351
•••	52626	68479	96300	17904	20895	23730	3411	3539	3619	835	835	821	351	351	322
0.8 •••	42102	50815	81415	14631	16501	19960	2790	2901	3002	665	665	665	266	266	266
•••	45656	59511	81721	15455	17904	20021	2859	2944	2994	665	665	651	266	266	241
0.85 •••	34792	41975	64441	11911	13287	15651	2152	2221	2280	479	479	479	.	.	.
•••	37751	48817	64538	12522	14291	15660	2195	2264	2265	479	479	456	.	.	.
0.9 •••	26319	31376	44737	8612	9420	10652	1387	1430	1456	.	.	.	.	.	.
•••	28444	35897	44753	8995	9972	10648	1414	1430	1432	.	.	.	.	.	.
0.95 •••	15540	17835	22382	4447	4686	5000	.	.	.	.	.	.	.	.	.
•••	16544	19620	22364	4559	4830	4986	.	.	.	.	.	.	.	.	.
0.98 •••	6487	6997	7704	.	.	.	.	.	.	.	.	.	.	.	.
•••	6726	7321	7680	.	.	.	.	.	.	.	.	.	.	.	.

TABLE 4: ALPHA= 0.01 POWER= 0.9   EXPECTED ACCRUAL THRU MINIMUM FOLLOW-UP= 20000

	DEL=.01			DEL=.02			DEL=.05			DEL=.10			DEL=.15		
FACT=	1.0 .75	.50 .25	.00 BIN	1.0 .75	.50 .25	.00 BIN	1.0 .75	.50 .25	.00 BIN	1.0 75	.50 .25	.00 BIN	1.0 .75	.50 .25	.00 BIN

PCONT=***   REQUIRED NUMBER OF PATIENTS

PCONT	DEL=.01 1.0/.75	.50/.25	.00/BIN	DEL=.02 1.0/.75	.50/.25	.00/BIN	DEL=.05 1.0/.75	.50/.25	.00/BIN	DEL=.10 1.0/.75	.50/.25	.00/BIN	DEL=.15 1.0/.75	.50/.25	.00/BIN
0.01 ***	2013	2013	2013	732	732	732	232	232	232	113	113	113	82	82	82
•••	2013	2013	7680	732	732	2539	232	232	691	113	113	281	82	82	167
0.02 ***	4432	4463	4482	1432	1432	1432	382	382	382	163	163	163	113	113	113
•••	4463	4463	12679	1432	1432	3775	382	382	883	163	163	326	113	113	186
0.05 ***	13563	13713	13982	3863	3882	3913	832	832	832	282	282	282	163	163	163
•••	13613	13813	27050	3882	3913	7329	832	832	1432	282	282	456	163	163	241
0.1 ***	30063	30682	32563	8263	8413	8613	1582	1582	1613	482	482	482	263	263	263
•••	30313	31332	48918	8332	8513	12731	1582	1582	2265	482	482	651	263	263	322
0.15 ***	45113	46432	51463	12382	12732	13313	2313	2332	2363	663	682	682	332	332	332
•••	45613	47963	68183	12532	12982	17482	2313	2332	2994	663	682	821	332	332	391
0.2 ***	57513	59632	69132	15913	16532	17713	2932	2982	3032	832	832	832	413	413	413
•••	58313	62282	84845	16182	17032	21583	2963	3013	3619	832	832	964	413	413	449
0.25 ***	66932	70013	84863	18732	19713	21582	3482	3563	3632	963	963	982	463	463	463
•••	68063	73932	98903	19163	20482	25032	3513	3582	4140	963	982	1081	463	463	495
0.3 ***	73382	77563	98282	20913	22232	24882	3932	4013	4132	1082	1082	1082	513	513	513
•••	74913	82832	110358	21482	23282	27831	3982	4063	4556	1082	1082	1172	513	513	530
0.35 ***	77113	82482	109113	22413	24082	27513	4263	4382	4513	1163	1163	1182	532	532	532
•••	79113	89163	119210	23132	25432	29979	4313	4432	4869	1163	1163	1237	532	532	553
0.4 ***	78482	85082	117282	23363	25332	29482	4513	4632	4813	1213	1213	1232	563	563	563
•••	80963	93113	125458	24213	26932	31475	4563	4713	5077	1213	1232	1276	563	563	565
0.45 ***	77863	85682	122663	23763	26013	30763	4632	4782	4963	1232	1232	1263	563	563	563
•••	80832	94913	129102	24732	27832	32322	4713	4863	5181	1232	1263	1289	563	563	565
0.5 ***	75663	84582	125213	23682	26132	31332	4663	4832	5032	1232	1232	1263	563	563	563
•••	79113	94813	130144	24732	28132	32517	4732	4913	5181	1232	1263	1276	563	563	553
0.55 ***	72213	82032	124963	23182	25763	31213	4582	4763	4963	1213	1213	1232	532	532	532
•••	76063	92982	128582	24313	27863	32061	4682	4863	5077	1213	1232	1237	532	532	530
0.6 ***	67832	78282	121863	22282	24932	30363	4432	4613	4813	1163	1163	1182	513	513	513
•••	71963	89582	124416	23432	27013	30955	4513	4682	4869	1163	1163	1172	513	513	495
0.65 ***	62763	73432	115932	21032	23613	28813	4182	4332	4513	1063	1082	1082	463	463	463
•••	67032	84732	117648	22163	25632	29197	4263	4413	4556	1082	1082	1081	463	463	449
0.7 ***	57082	67632	107213	19432	21863	26563	3832	3963	4132	963	963	982	413	413	413
•••	61332	78463	108275	20482	23713	26789	3882	4032	4140	963	982	964	413	413	391
0.75 ***	50882	60813	95682	17482	19613	23613	3382	3482	3632	832	832	832	332	332	332
•••	54932	70732	96300	18413	21213	23730	3432	3563	3619	832	832	821	332	332	322
0.8 ***	44082	52913	81432	15113	16863	19963	2832	2913	3013	663	663	682	263	263	263
•••	47713	61432	81721	15882	18163	20021	2863	2963	2994	663	663	651	263	263	241
0.85 ***	36463	43632	64432	12263	13563	15663	2182	2232	2282	482	482	482	.	.	.
•••	39463	50282	64538	12863	14463	15660	2213	2263	2265	482	482	456	.	.	.
0.9 ***	27532	32513	44732	8832	9582	10663	1413	1432	1463	.	.	.	.	.	.
•••	29632	36813	44753	9182	10063	10648	1413	1432	1432	.	.	.	.	.	.
0.95 ***	16113	18313	22382	4513	4732	5013	.	.	.	.	.	.	.	.	.
•••	17082	19963	22364	4613	4863	4986	.	.	.	.	.	.	.	.	.
0.98 ***	6613	7082	7713	.	.	.	.	.	.	.	.	.	.	.	.
•••	6832	7363	7680	.	.	.	.	.	.	.	.	.	.	.	.

TABLE 4: ALPHA= 0.01 POWER= 0.9   EXPECTED ACCRUAL THRU MINIMUM FOLLOW-UP= 25000

	DEL=.01			DEL=.02			DEL=.05			DEL=.10			DEL=.15		
FACT=	1.0 / .75	.50 / .25	.00 / BIN	1.0 / .75	.50 / .25	.00 / BIN	1.0 / .75	.50 / .25	.00 / BIN	1.0 / 75	.50 / .25	.00 / BIN	1.0 / .75	.50 / .25	.00 / BIN
PCONT=***				REQUIRED NUMBER OF PATIENTS											
0.01 ***	2016	2016	2016	727	727	727	227	227	227	102	102	102	79	79	79
...	2016	2016	7680	727	727	2539	227	227	691	102	102	281	79	79	167
0.02 ***	4454	4454	4477	1415	1415	1415	391	391	391	165	165	165	102	102	102
...	4454	4477	12679	1415	1415	3775	391	391	883	165	165	326	102	102	186
0.05 ***	13602	13727	13977	3852	3891	3915	829	829	829	290	290	290	165	165	165
...	13665	13852	27050	3891	3915	7329	829	829	1432	290	290	456	165	165	241
0.1 ***	30227	30891	32579	8329	8454	8602	1579	1602	1602	477	477	477	266	266	266
...	30516	31516	48918	8391	8516	12731	1579	1602	2265	477	477	651	266	266	322
0.15 ***	45477	46915	51454	12516	12829	13329	2329	2329	2352	665	665	665	329	329	329
...	46040	48415	68183	12641	13040	17482	2329	2329	2994	665	665	821	329	329	391
0.2 ***	58102	60477	69141	16102	16727	17704	2954	2977	3040	829	829	829	415	415	415
...	59016	63141	84845	16391	17141	21583	2977	3016	3619	829	829	964	415	415	449
0.25 ***	67790	71227	84852	19079	19977	21579	3516	3579	3641	977	977	977	454	454	454
...	69079	75204	98903	19477	20665	25032	3540	3602	4140	977	977	1081	454	454	495
0.3 ***	74540	79204	98266	21352	22602	24891	3954	4040	4141	1079	1079	1079	516	516	516
...	76290	84579	110358	21891	23540	27831	4016	4079	4556	1079	1079	1172	516	516	530
0.35 ***	78641	84540	109102	22977	24579	27516	4329	4415	4516	1165	1165	1165	540	540	540
...	80891	91352	119210	23665	25766	29979	4352	4454	4869	1165	1165	1237	540	540	553
0.4 ***	80391	87579	117266	24016	25915	29477	4540	4665	4977	1204	1227	1227	540	540	579
...	83141	95727	125458	24852	27352	31475	4602	4727	5077	1227	1227	1276	540	540	565
0.45 ***	80141	88579	122665	24516	26665	30766	4704	4829	4977	1227	1266	1266	579	579	579
...	83415	97891	129102	25454	28290	32322	4766	4891	5181	1227	1266	1289	579	579	565
0.5 ***	78290	87829	125227	24516	26852	31329	4727	4852	5016	1227	1266	1266	540	540	579
...	82040	98079	130144	25540	28641	32517	4790	4954	5181	1227	1266	1276	540	579	553
0.55 ***	75165	85516	124954	24040	26516	31204	4665	4790	4977	1204	1227	1227	540	540	540
...	79290	96454	128582	25141	28391	32061	4727	4891	5077	1227	1227	1237	540	540	530
0.6 ***	71016	81891	121852	23165	25665	30352	4477	4641	4790	1165	1165	1165	516	516	516
...	75391	93141	124416	24266	27540	30955	4579	4727	4869	1165	1165	1172	516	516	495
0.65 ***	66040	77079	115915	21915	24329	28790	4227	4352	4516	1079	1079	1102	454	454	477
...	70516	88227	117648	22977	26141	29197	4290	4454	4556	1079	1079	1081	454	477	449
0.7 ***	60352	71141	107204	20266	22516	26540	3891	3977	4141	954	977	977	415	415	415
...	64766	81766	108275	21266	24165	26789	3915	4040	4140	977	977	964	415	415	391
0.75 ***	54016	64079	95704	18204	20204	23602	3415	3516	3641	829	829	829	352	352	352
...	58141	73704	96300	19102	21602	23730	3454	3579	3619	829	829	821	352	352	322
0.8 ***	46891	55727	81415	15727	17329	19977	2852	2915	3016	665	665	665	266	266	266
...	50579	63954	81721	16454	18454	20021	2891	2977	2994	665	665	651	266	266	241
0.85 ***	38766	45891	64415	12727	13891	15641	2204	2227	2290	477	477	477	.	.	.
...	41766	52204	64538	13266	14665	15660	2227	2266	2265	477	477	456	.	.	.
0.9 ***	29165	33977	44727	9102	9766	10665	1415	1415	1454	.	.	.	.	.	.
...	31227	37977	44753	9415	10165	10648	1415	1454	1432	.	.	.	.	.	.
0.95 ***	16891	18915	22391	4602	4790	5016	.	.	.	.	.	.	.	.	.
...	17790	20352	22364	4704	4891	4986	.	.	.	.	.	.	.	.	.
0.98 ***	6790	7204	7704	.	.	.	.	.	.	.	.	.	.	.	.
...	6977	7415	7680	.	.	.	.	.	.	.	.	.	.	.	.

TABLE 5: ALPHA= 0.025 POWER= 0.8   EXPECTED ACCRUAL THRU MINIMUM FOLLOW-UP= 30

		DEL=.10			DEL=.15			DEL=.20			DEL=.25			DEL=.30		
FACT=		1.0 .75	.50 .25	.00 BIN	1.0 .75	.50 .25	.00 BIN	1.0 .75	.50 .25	.00 BIN	1.0 75	.50 .25	.00 BIN	1.0 .75	.50 .25	.00 BIN
PCONT=•••					REQUIRED NUMBER OF PATIENTS											
0.05	•••	165 165	166 168	178 275	96 97	97 99	104 145	68 68	69 70	73 93	53 53	54 55	57 65	44 44	45 45	47 48
0.1	•••	260 261	263 267	299 393	138 139	140 144	159 194	91 92	93 96	104 118	67 68	69 71	76 80	53 54	55 57	60 58
0.15	•••	336 337	339 346	409 495	169 171	173 179	206 236	107 109	111 116	130 140	77 78	80 84	92 93	60 61	63 65	71 66
0.2	•••	391 393	396 406	507 581	191 193	197 206	248 271	119 120	124 131	152 157	84 85	88 93	105 103	64 65	68 71	79 72
0.25	•••	427 429	434 448	591 652	205 207	212 224	282 299	125 128	132 141	170 171	87 90	93 100	116 110	66 68	71 75	85 76
0.3	•••	444 447	454 472	659 707	210 214	220 236	309 320	128 131	136 147	183 181	89 91	96 103	123 115	67 69	72 78	89 79
0.35	•••	445 450	458 482	710 746	209 214	222 241	328 334	127 131	137 150	192 187	88 91	96 105	127 118	66 69	72 79	92 80
0.4	•••	433 438	449 479	745 770	203 209	219 241	340 341	123 128	136 150	196 189	85 89	95 105	129 118	64 67	71 78	92 79
0.45	•••	409 416	430 466	762 778	193 200	211 237	344 341	118 123	132 148	197 187	82 86	92 102	128 115	61 64	69 76	90 76
0.5	•••	377 386	403 446	762 770	180 188	201 228	341 334	111 117	126 143	193 181	77 82	88 99	124 110	58 61	65 72	86 72
0.55	•••	340 351	371 419	745 746	165 174	188 217	329 320	103 109	118 135	184 171	72 76	82 93	117 103	54 57	61 68	81 66
0.6	•••	300 313	336 388	712 707	149 159	174 202	311 299	94 100	109 126	171 157	66 70	76 86	108 93	49 52	56 62	73 58
0.65	•••	260 275	300 354	661 652	133 142	157 185	284 271	84 90	99 114	155 140	59 63	68 77	96 80	43 46	49 55	64 48
0.7	•••	221 237	263 317	593 581	116 125	139 165	250 236	73 79	87 101	133 118	51 54	59 66	81 65	. .	. .	. .
0.75	•••	185 201	226 276	509 495	99 107	120 142	209 194	62 66	73 84	108 93	. .	. .	. .	. .	. .	. .
0.8	•••	151 165	187 230	408 393	80 87	97 115	161 145	. .	. .	. .	. .	. .	. .	. .	. .	. .
0.85	•••	116 127	145 177	291 275	. .	. .	. .	. .	. .	. .	. .	. .	. .	. .	. .	. .

TABLE 5: ALPHA= 0.025 POWER= 0.8   EXPECTED ACCRUAL THRU MINIMUM FOLLOW-UP= 40

	DEL=.10			DEL=.15			DEL=.20			DEL=.25			DEL=.30		
FACT=	1.0 .75	.50 .25	.00 BIN	1.0 .75	.50 .25	.00 BIN	1.0 .75	.50 .25	.00 BIN	1.0 75	.50 .25	.00 BIN	1.0 .75	.50 .25	.00 BIN
PCONT=•••	REQUIRED NUMBER OF PATIENTS														
0.05 •••	165	167	178	97	98	104	68	70	73	53	54	57	44	45	47
•••	166	169	275	97	100	145	69	71	93	54	55	65	44	46	48
0.1 •••	261	264	299	139	142	159	92	94	104	68	70	76	54	56	60
•••	262	269	393	140	146	194	93	97	118	69	72	80	55	57	58
0.15 •••	337	342	409	171	175	207	109	113	130	78	82	92	61	64	71
•••	339	350	495	172	182	236	110	118	140	80	85	93	62	66	66
0.2 •••	393	400	507	193	200	248	121	127	152	85	90	105	65	69	79
•••	395	413	581	196	210	271	123	134	157	87	95	103	67	73	72
0.25 •••	429	439	591	207	217	282	128	136	170	90	96	116	68	73	85
•••	432	457	652	210	230	299	131	145	171	92	102	110	70	77	76
0.3 •••	447	460	659	214	226	309	131	141	183	91	99	123	69	75	89
•••	452	484	707	218	243	320	135	152	181	94	107	115	71	80	79
0.35 •••	450	466	710	214	229	328	131	143	192	91	100	127	69	75	92
•••	455	496	746	219	250	334	135	156	187	95	109	118	71	81	80
0.4 •••	438	460	745	209	227	340	128	142	196	89	99	129	67	74	92
•••	446	496	770	216	251	341	133	157	189	93	109	118	70	80	79
0.45 •••	416	443	762	200	221	344	123	138	197	86	96	128	64	72	90
•••	425	486	778	208	248	341	129	154	187	90	106	115	68	78	76
0.5 •••	386	418	762	188	212	341	117	133	193	82	92	124	61	68	86
•••	397	468	770	197	241	334	123	150	181	86	103	110	64	75	72
0.55 •••	351	389	746	174	200	329	109	125	184	76	87	117	57	64	81
•••	364	444	746	184	230	320	116	142	171	81	97	103	60	70	66
0.6 •••	313	356	712	159	185	311	100	116	171	70	80	108	52	58	73
•••	329	414	707	169	215	299	107	133	157	74	89	93	55	64	58
0.65 •••	275	321	661	142	169	284	90	105	155	63	72	96	46	52	64
•••	292	380	652	153	197	271	96	120	140	66	80	80	48	56	48
0.7 •••	237	284	593	125	150	250	79	93	133	54	62	81	.	.	.
•••	255	341	581	135	176	236	85	106	118	58	69	65	.	.	.
0.75 •••	201	246	509	107	129	209	67	78	108	.	.	.	.	.	.
•••	219	298	495	116	151	194	71	88	93	.	.	.	.	.	.
0.8 •••	165	205	408	87	104	161	.	.	.	.	.	.	.	.	.
•••	181	249	393	94	121	145	.	.	.	.	.	.	.	.	.
0.85 •••	127	158	291	.	.	.	.	.	.	.	.	.	.	.	.
•••	140	191	275	.	.	.	.	.	.	.	.	.	.	.	.

TABLE 5: ALPHA= 0.025 POWER= 0.8   EXPECTED ACCRUAL THRU MINIMUM FOLLOW-UP= 50

		DEL=.10			DEL=.15			DEL=.20			DEL=.25			DEL=.30		
FACT=		1.0 .75	.50 .25	.00 BIN	1.0 .75	.50 .25	.00 BIN	1.0 .75	.50 .25	.00 BIN	1.0 75	.50 .25	.00 BIN	1.0 .75	.50 .25	.00 BIN
PCONT=•••					REQUIRED NUMBER OF PATIENTS											
0.05	•••	166	167	178	97	99	104	69	70	73	54	55	57	44	45	47
	•••	166	170	275	98	100	145	69	71	93	54	55	65	45	46	48
0.1	•••	262	265	299	140	143	159	93	95	104	69	71	76	55	56	60
	•••	263	271	393	141	147	194	94	98	118	70	73	80	55	58	58
0.15	•••	338	344	409	172	177	207	110	115	130	80	83	92	62	65	71
	•••	340	354	495	174	185	236	112	120	140	81	86	93	63	67	66
0.2	•••	395	403	507	195	203	248	122	129	152	87	92	105	67	70	79
	•••	398	418	581	198	214	271	125	136	157	89	97	103	68	74	72
0.25	•••	432	443	591	210	221	282	130	139	170	92	98	116	70	74	85
	•••	436	464	652	214	235	299	133	148	171	94	104	110	72	78	76
0.3	•••	451	466	659	217	231	309	134	144	183	94	101	123	71	76	89
	•••	456	494	707	222	250	320	138	156	181	97	109	115	73	81	79
0.35	•••	454	474	710	218	235	328	134	147	192	94	103	127	71	77	92
	•••	461	509	746	225	257	334	139	160	187	98	111	118	73	82	80
0.4	•••	444	470	745	214	235	340	132	147	196	92	102	129	69	76	92
	•••	453	511	770	222	260	341	138	161	189	96	111	118	72	82	79
0.45	•••	423	455	762	206	230	344	128	144	197	89	100	128	67	74	90
	•••	434	504	778	215	257	341	134	159	187	94	109	115	70	80	76
0.5	•••	395	433	762	195	221	341	122	138	193	85	96	124	63	71	86
	•••	408	487	770	205	250	334	128	155	181	90	105	110	67	77	72
0.55	•••	361	405	746	182	209	329	114	131	184	80	90	117	59	66	81
	•••	377	464	746	192	239	320	121	147	171	84	100	103	62	72	66
0.6	•••	325	373	712	167	195	311	105	122	171	73	83	108	54	60	73
	•••	343	435	707	178	225	299	112	137	157	77	92	93	57	65	58
0.65	•••	288	339	661	150	178	284	95	110	155	66	75	96	48	53	64
	•••	307	400	652	161	206	271	101	125	140	70	82	80	50	58	48
0.7	•••	251	302	593	133	158	250	83	97	133	57	65	81	.	.	.
	•••	271	361	581	143	184	236	89	109	118	60	71	65	.	.	.
0.75	•••	215	262	509	114	136	209	70	81	108	.	.	.	.	.	.
	•••	233	315	495	123	158	194	75	91	93	.	.	.	.	.	.
0.8	•••	177	218	408	92	110	161	.	.	.	.	.	.	.	.	.
	•••	194	263	393	100	126	145	.	.	.	.	.	.	.	.	.
0.85	•••	137	169	291	.	.	.	.	.	.	.	.	.	.	.	.
	•••	150	201	275	.	.	.	.	.	.	.	.	.	.	.	.

TABLE 5: ALPHA= 0.025 POWER= 0.8   EXPECTED ACCRUAL THRU MINIMUM FOLLOW-UP= 60

		DEL=.10			DEL=.15			DEL=.20			DEL=.25			DEL=.30		
FACT=		1.0 .75	.50 .25	.00 BIN	1.0 .75	.50 .25	.00 BIN	1.0 .75	.50 .25	.00 BIN	1.0 75	.50 .25	.00 BIN	1.0 .75	.50 .25	.00 BIN
PCONT=•••		REQUIRED NUMBER OF PATIENTS														
0.05	***	166	168	178	97	99	104	69	70	73	54	55	57	44	45	47
	•••	167	171	275	98	101	145	70	71	93	54	56	65	45	46	48
0.1	***	263	267	299	140	144	159	93	96	104	69	71	76	55	57	60
	•••	264	273	393	142	148	194	94	99	118	70	73	80	56	58	58
0.15	***	340	346	409	173	179	206	111	116	130	80	84	92	63	65	71
	•••	342	358	495	175	187	236	113	121	140	82	87	93	64	67	66
0.2	***	396	406	507	197	206	248	124	131	152	88	93	105	68	71	79
	•••	400	423	581	200	217	271	127	138	157	90	98	103	69	74	72
0.25	***	434	448	591	212	224	282	132	141	170	93	100	116	71	75	85
	•••	439	471	652	217	239	299	136	150	171	96	105	110	73	79	76
0.3	***	454	472	659	220	236	309	136	148	183	96	103	123	72	78	89
	•••	460	503	707	226	255	320	141	159	181	99	110	115	75	82	79
0.35	***	458	482	710	222	241	328	137	151	192	96	105	127	72	79	92
	•••	466	520	746	229	263	334	143	164	187	100	113	118	75	84	80
0.4	***	449	479	745	219	241	340	136	150	196	95	105	129	71	78	92
	•••	460	525	770	227	267	341	142	165	189	99	113	118	74	83	79
0.45	***	430	466	762	211	237	344	132	148	197	92	102	128	69	76	90
	•••	443	519	778	221	265	341	138	163	187	96	112	115	72	81	76
0.5	***	403	446	762	201	228	341	126	143	193	88	99	124	65	72	86
	•••	418	504	770	212	258	334	133	159	181	92	108	110	68	78	72
0.55	***	371	419	745	188	217	329	118	135	184	82	93	117	61	68	81
	•••	389	481	746	200	247	320	125	151	171	87	102	103	64	73	66
0.6	***	336	388	712	174	202	311	109	126	171	76	86	108	56	62	73
	•••	356	452	707	185	232	299	116	141	157	80	94	93	58	66	58
0.65	***	300	354	661	157	185	284	99	114	155	68	77	96	49	55	64
	•••	321	418	652	169	214	271	105	128	140	72	84	80	52	58	48
0.7	***	263	317	593	139	165	250	87	101	133	59	66	81	.	.	.
	•••	284	377	581	150	191	236	93	112	118	62	72	65	.	.	.
0.75	***	226	276	509	120	142	209	73	84	108	.	.	.	.	.	.
	•••	246	330	495	129	163	194	78	93	93	.	.	.	.	.	.
0.8	***	187	230	408	97	115	161	.	.	.	.	.	.	.	.	.
	•••	205	275	393	104	130	145	.	.	.	.	.	.	.	.	.
0.85	***	145	177	291	.	.	.	.	.	.	.	.	.	.	.	.
	•••	158	209	275	.	.	.	.	.	.	.	.	.	.	.	.

TABLE 5: ALPHA= 0.025 POWER= 0.8   EXPECTED ACCRUAL THRU MINIMUM FOLLOW-UP= 70

		DEL=.10			DEL=.15			DEL=.20			DEL=.25			DEL=.30		
FACT=		1.0 .75	.50 .25	.00 BIN	1.0 .75	.50 .25	.00 BIN	1.0 .75	.50 .25	.00 BIN	1.0 75	.50 .25	.00 BIN	1.0 .75	.50 .25	.00 BIN
PCONT=•••		REQUIRED NUMBER OF PATIENTS															
0.05	***	166	169	178	98	99	104	69	71	73	54	55	57	45	45	47
	•••	167	171	275	98	101	145	70	72	93	54	56	65	45	46	48
0.1	***	263	268	299	141	145	159	94	97	104	70	72	76	55	57	60
	•••	265	275	393	143	149	194	95	99	118	71	74	80	56	58	58
0.15	***	341	348	409	174	181	206	112	117	130	81	85	92	63	66	71
	•••	343	361	495	177	189	236	114	122	140	83	88	93	64	68	66
0.2	***	398	410	507	198	208	248	125	132	152	89	94	105	68	72	79
	•••	402	428	581	202	220	271	128	139	157	91	99	103	70	75	72
0.25	***	436	452	591	214	227	282	134	143	170	95	101	116	72	76	85
	•••	442	478	652	219	243	299	138	152	171	97	106	110	74	80	76
0.3	***	457	478	659	223	240	309	139	150	183	97	105	123	74	79	89
	•••	464	511	707	229	259	320	143	161	181	101	112	115	76	83	79
0.35	***	462	489	710	226	246	328	140	153	192	98	107	127	74	80	92
	•••	472	530	746	233	268	334	146	166	187	102	115	118	76	84	80
0.4	***	454	488	745	223	247	340	139	154	196	97	107	129	73	79	92
	•••	466	536	770	232	272	341	145	168	189	101	115	118	75	84	79
0.45	***	436	477	762	217	243	344	135	151	197	94	105	128	70	77	90
	•••	451	532	778	227	271	341	142	166	187	99	113	115	73	82	76
0.5	***	411	457	762	207	235	341	130	146	193	90	101	124	67	74	86
	•••	428	518	770	218	264	334	137	162	181	95	109	110	70	79	72
0.55	***	380	432	745	194	224	329	122	139	184	85	95	117	63	69	81
	•••	400	496	746	206	254	320	129	154	171	89	103	103	65	74	66
0.6	***	346	402	712	180	209	311	113	130	171	78	88	108	57	63	73
	•••	368	468	707	192	239	299	120	144	157	82	95	93	60	67	58
0.65	***	311	368	661	163	192	284	102	118	155	70	79	96	51	56	64
	•••	333	432	652	175	219	271	109	131	140	74	85	80	53	59	48
0.7	***	274	330	593	145	171	250	90	103	133	61	68	81	.	.	.
	•••	296	391	581	156	196	236	96	114	118	64	73	65	.	.	.
0.75	***	237	287	509	125	147	209	76	86	108	.	.	.	.	.	.
	•••	257	342	495	134	167	194	80	95	93	.	.	.	.	.	.
0.8	***	197	240	408	101	118	161	.	.	.	.	.	.	.	.	.
	•••	214	285	393	108	133	145	.	.	.	.	.	.	.	.	.
0.85	***	152	184	291	.	.	.	.	.	.	.	.	.	.	.	.
	•••	165	216	275	.	.	.	.	.	.	.	.	.	.	.	.

TABLE 5: ALPHA= 0.025 POWER= 0.8   EXPECTED ACCRUAL THRU MINIMUM FOLLOW-UP= 80

	DEL=.10			DEL=.15			DEL=.20			DEL=.25			DEL=.30		
FACT=	1.0 .75	.50 .25	.00 BIN	1.0 .75	.50 .25	.00 BIN	1.0 .75	.50 .25	.00 BIN	1.0 75	.50 .25	.00 BIN	1.0 .75	.50 .25	.00 BIN
PCONT=***			REQUIRED	NUMBER	OF	PATIENTS									
0.05 ***	167	169	178	98	100	104	70	71	73	54	55	57	45	46	47
***	168	172	275	99	101	145	70	72	93	55	56	65	45	46	48
0.1 ***	264	269	299	142	146	159	94	97	104	70	72	76	56	57	60
***	266	276	393	143	150	194	96	100	118	71	74	80	57	59	58
0.15 ***	342	350	409	175	182	207	113	118	130	82	85	92	64	66	71
***	345	364	495	178	190	236	115	122	140	83	88	93	65	68	66
0.2 ***	400	413	507	200	210	248	127	134	152	90	95	105	69	73	79
***	404	432	581	204	222	271	129	140	157	92	99	103	71	75	72
0.25 ***	439	457	591	217	230	282	136	145	170	96	102	116	73	77	85
***	445	484	652	222	246	299	139	154	171	99	107	110	75	80	76
0.3 ***	460	484	659	226	243	309	141	152	183	99	107	123	75	80	89
***	468	518	707	233	262	320	146	163	181	102	113	115	77	84	79
0.35 ***	466	496	710	229	250	328	143	156	192	100	109	127	75	81	92
***	477	539	746	237	273	334	148	168	187	104	116	118	78	85	80
0.4 ***	460	496	745	227	251	340	142	157	196	99	109	129	74	80	92
***	473	547	770	237	277	341	148	170	189	103	116	118	77	85	79
0.45 ***	443	486	762	221	248	344	138	154	197	96	106	128	72	78	90
***	459	543	778	232	276	341	145	169	187	101	115	115	75	83	76
0.5 ***	418	468	762	212	241	341	133	150	193	92	103	124	68	75	86
***	437	530	770	223	270	334	140	164	181	97	111	110	71	80	72
0.55 ***	389	444	746	200	230	329	125	142	184	87	97	117	64	70	81
***	410	509	746	212	259	320	133	157	171	91	105	103	67	75	66
0.6 ***	356	414	712	185	215	311	116	133	171	80	89	108	58	64	73
***	378	481	707	197	244	299	123	146	157	84	97	93	61	68	58
0.65 ***	321	380	661	169	197	284	105	120	155	72	80	96	52	56	64
***	344	445	652	180	224	271	112	133	140	76	86	80	54	60	48
0.7 ***	284	341	593	150	176	250	93	106	133	62	69	81	.	.	.
***	307	402	581	161	200	236	98	116	118	65	74	65	.	.	.
0.75 ***	246	298	509	129	151	209	78	88	108	.	.	.	.	.	.
***	267	352	495	138	171	194	82	96	93	.	.	.	.	.	.
0.8 ***	205	249	408	104	121	161	.	.	.	.	.	.	.	.	.
***	222	293	393	112	136	145	.	.	.	.	.	.	.	.	.
0.85 ***	158	191	291	.	.	.	.	.	.	.	.	.	.	.	.
***	172	222	275	.	.	.	.	.	.	.	.	.	.	.	.

TABLE 5: ALPHA= 0.025 POWER= 0.8   EXPECTED ACCRUAL THRU MINIMUM FOLLOW-UP= 90

REQUIRED NUMBER OF PATIENTS

PCONT	DEL=.10			DEL=.15			DEL=.20			DEL=.25			DEL=.30		
FACT=	1.0 / .75	.50 / .25	.00 / BIN	1.0 / .75	.50 / .25	.00 / BIN	1.0 / .75	.50 / .25	.00 / BIN	1.0 / 75	.50 / .25	.00 / BIN	1.0 / .75	.50 / .25	.00 / BIN
0.05	167	170	178	98	100	104	70	71	73	54	55	57	45	46	47
	168	172	275	99	102	145	70	72	93	55	56	65	45	46	48
0.1	265	270	299	142	147	159	95	98	104	71	73	76	56	58	60
	267	278	393	144	151	194	96	100	118	72	74	80	57	59	58
0.15	343	352	409	176	184	207	114	119	130	83	86	92	64	67	71
	346	366	495	179	191	236	116	123	140	84	89	93	65	68	66
0.2	401	416	507	202	212	248	128	135	152	91	96	105	70	73	79
	406	436	581	206	224	271	131	141	157	93	100	103	71	76	72
0.25	441	461	591	219	233	282	137	147	170	97	103	116	74	78	85
	448	489	652	224	248	299	141	155	171	100	108	110	75	81	76
0.3	463	489	659	229	247	309	143	154	183	100	108	123	76	81	89
	472	525	707	236	266	320	147	164	181	103	114	115	78	84	79
0.35	470	503	710	232	254	328	145	158	192	101	110	127	76	82	92
	482	547	746	241	276	334	150	170	187	105	117	118	79	86	80
0.4	465	504	745	231	256	340	144	159	196	101	110	129	75	81	92
	479	556	770	241	281	341	150	172	189	105	117	118	78	86	79
0.45	449	495	762	226	253	344	141	157	197	98	108	128	73	79	90
	466	554	778	237	280	341	148	171	187	102	116	115	76	84	76
0.5	426	478	762	216	246	341	136	152	192	94	104	124	70	76	86
	446	542	770	228	275	334	143	167	181	99	112	110	72	80	72
0.55	397	454	745	205	235	329	128	145	184	89	98	117	65	71	81
	419	521	746	217	264	320	135	159	171	93	106	103	68	75	66
0.6	365	425	712	190	220	310	119	135	171	82	91	108	59	65	73
	388	492	707	202	249	299	126	148	157	86	98	93	62	68	58
0.65	330	390	661	173	202	284	108	123	155	73	81	96	53	57	64
	354	456	652	185	229	271	114	135	140	77	87	80	55	60	48
0.7	293	351	593	154	180	250	95	108	133	63	70	81	.	.	.
	317	413	581	165	204	236	101	118	118	66	74	65	.	.	.
0.75	254	307	509	133	155	209	80	90	108	.	.	.	.	.	.
	276	361	495	142	174	194	84	97	93	.	.	.	.	.	.
0.8	212	256	408	107	124	161	.	.	.	.	.	.	.	.	.
	230	300	393	115	138	145	.	.	.	.	.	.	.	.	.
0.85	164	196	291	.	.	.	.	.	.	.	.	.	.	.	.
	177	226	275	.	.	.	.	.	.	.	.	.	.	.	.

TABLE 5: ALPHA= 0.025 POWER= 0.8   EXPECTED ACCRUAL THRU MINIMUM FOLLOW-UP= 100

	DEL=.10			DEL=.15			DEL=.20			DEL=.25			DEL=.30		
FACT=	1.0 .75	.50 .25	.00 BIN	1.0 .75	.50 .25	.00 BIN	1.0 .75	.50 .25	.00 BIN	1.0 75	.50 .25	.00 BIN	1.0 .75	.50 .25	.00 BIN
PCONT=•••	REQUIRED NUMBER OF PATIENTS														
0.05 ***	167 168	170 173	178 275	99 99	100 102	104 145	70 70	71 72	73 93	55 55	55 56	57 65	45 45	46 46	47 48
0.1 ***	265 268	271 279	299 393	143 145	147 151	159 194	95 97	98 101	104 118	71 72	73 74	76 80	56 57	58 59	60 58
0.15 ***	344 348	354 368	409 495	177 180	185 192	207 236	115 117	120 124	130 140	83 85	86 89	92 93	65 66	67 69	71 66
0.2 ***	403 409	418 440	507 581	203 207	214 225	248 271	129 132	136 142	152 157	92 94	97 100	105 103	70 72	74 76	79 72
0.25 ***	443 451	464 493	591 652	221 226	235 250	282 299	139 143	148 156	170 171	98 101	104 109	116 110	74 76	78 81	85 76
0.3 ***	466 476	494 531	659 707	231 238	250 268	309 320	144 149	156 166	183 181	101 105	109 114	123 115	76 79	81 85	89 79
0.35 ***	474 487	509 554	710 746	235 244	257 280	328 334	147 153	160 172	192 187	103 106	111 118	127 118	77 79	82 86	92 80
0.4 ***	470 485	511 564	745 770	235 245	260 285	340 341	147 153	161 174	196 189	102 106	111 118	129 118	76 79	82 86	92 79
0.45 ***	455 473	504 563	762 778	230 241	257 284	344 341	144 150	159 173	197 187	100 104	109 117	128 115	74 77	80 84	90 76
0.5 ***	433 454	487 552	762 770	221 233	250 279	341 334	138 145	155 169	193 181	96 100	105 113	124 110	71 73	77 81	86 72
0.55 ***	405 428	464 531	746 746	209 222	239 268	329 320	131 138	147 161	184 171	90 94	100 107	117 103	66 69	72 76	81 66
0.6 ***	373 397	435 503	712 707	195 207	225 253	311 299	122 128	137 150	171 157	83 87	92 98	108 93	60 63	65 69	73 58
0.65 ***	339 363	400 466	661 652	178 190	206 232	284 271	110 117	125 136	155 140	75 78	82 88	96 80	53 55	58 60	64 48
0.7 ***	302 325	361 422	593 581	158 169	184 207	250 236	97 103	109 119	133 118	65 67	71 75	81 65	. .	. .	. .
0.75 ***	262 284	315 369	509 495	136 145	158 176	209 194	81 86	91 98	108 93	. .	. .	. .	. .	. .	. .
0.8 ***	218 237	263 306	408 393	110 117	126 139	161 145	. .	. .	. .	. .	. .	. .	. .	. .	. .
0.85 ***	169 182	201 231	291 275	. .	. .	. .	. .	. .	. .	. .	. .	. .	. .	. .	. .

TABLE 5: ALPHA= 0.025 POWER= 0.8   EXPECTED ACCRUAL THRU MINIMUM FOLLOW-UP= 110

PCONT=***		DEL=.05			DEL=.10			DEL=.15			DEL=.20			DEL=.25		
FACT=		1.0 / .75	.50 / .25	.00 / BIN	1.0 / .75	.50 / .25	.00 / BIN	1.0 / .75	.50 / .25	.00 / BIN	1.0 / 75	.50 / .25	.00 / BIN	1.0 / .75	.50 / .25	.00 / BIN
		REQUIRED NUMBER OF PATIENTS														
0.05	***	470	473	502	168	170	178	99	100	104	70	71	73	55	56	57
	•••	471	478	864	169	173	275	100	102	145	71	72	93	55	56	65
0.1	***	857	863	967	266	272	299	144	148	159	96	99	104	71	73	76
	•••	859	875	1366	268	280	393	145	152	194	97	101	118	72	75	80
0.15	***	1185	1196	1418	345	356	409	178	186	207	115	120	130	84	87	92
	•••	1189	1217	1806	349	370	495	181	193	236	117	124	140	85	89	93
0.2	***	1439	1456	1829	405	421	507	204	216	248	130	137	152	93	97	105
	•••	1445	1488	2182	411	443	581	209	226	271	133	143	157	95	101	103
0.25	***	1616	1639	2188	446	468	591	223	237	282	140	149	170	99	105	116
	•••	1623	1684	2496	454	498	652	229	252	299	144	157	171	101	109	110
0.3	***	1718	1749	2489	469	498	659	234	252	309	146	157	183	102	110	123
	•••	1728	1811	2748	480	537	707	241	271	320	151	167	181	106	115	115
0.35	***	1752	1793	2725	478	515	710	238	261	328	149	162	192	104	112	127
	•••	1765	1874	2936	492	561	746	247	282	334	154	173	187	108	118	118
0.4	***	1726	1780	2896	474	518	745	238	263	340	149	163	196	104	112	129
	•••	1744	1883	3062	491	572	770	248	288	341	155	176	189	107	119	118
0.45	***	1651	1720	2999	461	511	762	233	261	344	146	161	197	101	111	128
	•••	1674	1848	3124	480	571	778	245	288	341	152	175	187	105	118	115
0.5	***	1537	1624	3034	439	496	762	225	254	341	141	157	193	97	107	124
	•••	1567	1777	3124	461	561	770	237	282	334	148	170	181	101	114	110
0.55	***	1396	1504	3000	412	473	746	213	243	329	133	149	184	92	101	117
	•••	1433	1681	3062	436	541	746	226	272	320	140	163	171	96	108	103
0.6	***	1238	1369	2898	381	444	712	199	229	311	124	139	171	85	93	108
	•••	1284	1565	2936	406	512	707	211	256	299	131	152	157	88	99	93
0.65	***	1077	1227	2728	347	409	661	182	210	284	113	127	155	76	83	96
	•••	1132	1436	2748	372	475	652	194	235	271	119	137	140	79	88	80
0.7	***	921	1083	2491	309	369	593	162	188	250	99	111	133	65	71	81
	•••	982	1294	2496	333	430	581	173	210	236	104	120	118	68	75	65
0.75	***	775	938	2186	269	323	509	139	161	209	83	92	108	.	.	.
	•••	837	1140	2182	291	376	495	148	178	194	87	99	93	.	.	.
0.8	***	638	790	1816	224	269	408	112	128	161	.	.	.	.	.	.
	•••	697	972	1806	243	312	393	119	141	145	.	.	.	.	.	.
0.85	***	504	634	1380	173	205	291	.	.	.	.	.	.	.	.	.
	•••	555	783	1366	187	234	275	.	.	.	.	.	.	.	.	.
0.9	***	362	455	879	.	.	.	.	.	.	.	.	.	.	.	.
	•••	399	556	864	.	.	.	.	.	.	.	.	.	.	.	.

TABLE 5: ALPHA= 0.025 POWER= 0.8   EXPECTED ACCRUAL THRU MINIMUM FOLLOW-UP= 120

		DEL=.05			DEL=.10			DEL=.15			DEL=.20			DEL=.25		
FACT=		1.0 .75	.50 .25	.00 BIN	1.0 .75	.50 .25	.00 BIN	1.0 .75	.50 .25	.00 BIN	1.0 75	.50 .25	.00 BIN	1.0 .75	.50 .25	.00 BIN
PCONT=•••						REQUIRED NUMBER OF PATIENTS											
0.05	•••	470	473	502	168	171	178	99	101	104	70	71	73	55	56	57
	•••	471	479	864	169	173	275	100	102	145	71	72	93	55	56	65
0.1	•••	857	864	967	267	273	299	144	148	159	96	99	104	71	73	76
	•••	860	878	1366	269	281	393	146	152	194	97	101	118	72	75	80
0.15	•••	1186	1198	1418	346	358	409	179	187	206	116	121	130	84	87	92
	•••	1190	1221	1806	350	372	495	182	194	236	118	124	140	85	89	93
0.2	•••	1441	1459	1829	406	423	507	206	217	248	131	138	152	93	98	105
	•••	1447	1493	2182	413	445	581	210	228	271	134	143	157	95	101	103
0.25	•••	1618	1643	2188	448	471	591	224	239	282	141	150	170	100	105	116
	•••	1626	1692	2496	457	502	652	230	254	299	145	158	171	102	110	110
0.3	•••	1720	1755	2489	472	503	659	236	255	309	148	159	183	103	110	123
	•••	1732	1821	2748	484	542	707	243	273	320	152	168	181	106	116	115
0.35	•••	1755	1801	2725	482	520	710	241	263	328	151	164	192	105	113	127
	•••	1770	1888	2936	496	567	746	250	285	334	156	174	187	109	119	118
0.4	•••	1731	1790	2896	479	525	745	241	267	340	150	165	196	105	113	129
	•••	1751	1900	3062	496	579	770	251	291	341	157	177	189	109	120	118
0.45	•••	1657	1733	2999	466	519	762	237	265	344	148	163	197	102	112	128
	•••	1683	1868	3124	486	579	778	248	291	341	154	176	187	106	118	115
0.5	•••	1545	1640	3034	446	504	762	228	258	340	143	159	193	99	108	124
	•••	1577	1801	3124	468	569	770	241	285	334	150	172	181	103	114	110
0.55	•••	1406	1522	3000	419	481	745	217	247	329	135	151	184	93	102	117
	•••	1446	1708	3062	444	549	746	229	275	320	142	164	171	97	108	103
0.6	•••	1251	1390	2898	388	452	712	202	232	310	126	141	171	86	94	108
	•••	1300	1594	2936	414	521	707	215	259	299	133	153	157	89	100	93
0.65	•••	1093	1250	2728	354	418	661	185	214	284	114	128	155	77	84	96
	•••	1150	1465	2748	380	483	652	197	238	271	120	139	140	80	89	80
0.7	•••	938	1107	2491	317	377	593	165	191	250	101	112	133	66	72	81
	•••	1002	1323	2496	341	438	581	176	212	236	106	121	118	69	76	65
0.75	•••	793	961	2186	276	330	509	142	163	209	84	93	108	.	.	.
	•••	857	1168	2182	298	383	495	151	180	194	88	99	93	.	.	.
0.8	•••	655	812	1816	230	275	408	115	130	161	.	.	.	.	.	.
	•••	716	997	1806	249	317	393	121	142	145	.	.	.	.	.	.
0.85	•••	519	652	1380	177	209	291	.	.	.	.	.	.	.	.	.
	•••	571	802	1366	191	237	275	.	.	.	.	.	.	.	.	.
0.9	•••	373	468	879	.	.	.	.	.	.	.	.	.	.	.	.
	•••	411	568	864	.	.	.	.	.	.	.	.	.	.	.	.

TABLE 5: ALPHA= 0.025 POWER= 0.8   EXPECTED ACCRUAL THRU MINIMUM FOLLOW-UP= 130

		DEL=.05			DEL=.10			DEL=.15			DEL=.20			DEL=.25		
FACT=		1.0 .75	.50 .25	.00 BIN	1.0 .75	.50 .25	.00 BIN	1.0 .75	.50 .25	.00 BIN	1.0 75	.50 .25	.00 BIN	1.0 .75	.50 .25	.00 BIN
PCONT=•••				REQUIRED	NUMBER	OF	PATIENTS									
0.05	•••	470	474	502	168	171	178	99	101	104	70	72	73	55	56	57
	•••	472	479	864	169	174	275	100	102	145	71	72	93	55	56	65
0.1	•••	858	865	967	267	274	299	145	149	159	96	99	104	72	74	76
	•••	860	880	1366	270	282	393	146	153	194	98	101	118	73	75	80
0.15	•••	1187	1200	1418	347	359	409	180	188	206	116	121	130	84	87	92
	•••	1192	1225	1806	352	374	495	183	195	236	119	125	140	86	90	93
0.2	•••	1442	1461	1829	408	426	507	207	218	248	132	138	152	94	98	105
	•••	1449	1499	2182	415	448	581	212	229	271	135	144	157	96	101	103
0.25	•••	1620	1647	2188	450	475	591	226	241	282	142	151	170	100	106	116
	•••	1629	1700	2496	459	505	652	232	255	299	146	158	171	103	110	110
0.3	•••	1723	1760	2489	475	507	659	238	257	309	149	160	183	104	111	123
	•••	1736	1832	2748	487	546	707	246	275	320	154	169	181	107	116	115
0.35	•••	1759	1808	2725	486	525	710	243	266	328	152	165	192	106	114	127
	•••	1776	1901	2936	501	572	746	253	287	334	158	175	187	109	119	118
0.4	•••	1736	1800	2896	484	531	745	244	270	340	152	167	196	106	114	129
	•••	1757	1917	3062	502	585	770	254	293	341	158	178	189	109	120	118
0.45	•••	1664	1745	2999	472	525	762	240	268	344	150	165	197	104	112	128
	•••	1691	1888	3124	492	586	778	251	294	341	156	177	187	108	119	115
0.5	•••	1553	1655	3034	452	511	762	232	261	341	145	160	193	100	109	124
	•••	1588	1825	3124	475	577	770	244	288	334	151	173	181	104	115	110
0.55	•••	1416	1540	3000	426	489	746	220	251	329	137	153	184	94	103	117
	•••	1459	1734	3062	451	557	746	233	278	320	144	165	171	98	109	103
0.6	•••	1264	1410	2898	395	460	712	206	236	310	128	143	171	87	95	108
	•••	1316	1622	2936	421	528	707	218	262	299	134	154	157	90	100	93
0.65	•••	1107	1272	2728	361	425	661	189	217	284	116	129	155	78	85	96
	•••	1168	1493	2748	387	491	652	200	241	271	122	140	140	81	89	80
0.7	•••	955	1129	2491	323	384	593	168	193	250	102	113	133	67	73	81
	•••	1021	1351	2496	348	444	581	179	214	236	107	122	118	70	76	65
0.75	•••	810	983	2186	282	336	509	145	165	209	85	94	108	.	.	.
	•••	877	1194	2182	304	388	495	154	182	194	89	100	93	.	.	.
0.8	•••	671	832	1816	235	280	408	116	132	161	.	.	.	.	.	.
	•••	734	1019	1806	254	321	393	123	143	145	.	.	.	.	.	.
0.85	•••	533	669	1380	181	213	291	.	.	.	.	.	.	.	.	.
	•••	587	820	1366	194	240	275	.	.	.	.	.	.	.	.	.
0.9	•••	383	479	879	.	.	.	.	.	.	.	.	.	.	.	.
	•••	422	580	864	.	.	.	.	.	.	.	.	.	.	.	.

TABLE 5: ALPHA= 0.025 POWER= 0.8   EXPECTED ACCRUAL THRU MINIMUM FOLLOW-UP= 140

		DEL=.05			DEL=.10			DEL=.15			DEL=.20			DEL=.25		
FACT=		1.0 .75	.50 .25	.00 BIN	1.0 .75	.50 .25	.00 BIN	1.0 .75	.50 .25	.00 BIN	1.0 75	.50 .25	.00 BIN	1.0 .75	.50 .25	.00 BIN
PCONT=•••					REQUIRED NUMBER OF PATIENTS											
0.05	•••	471	474	502	169	171	178	99	101	104	71	72	73	55	56	57
	•••	472	480	864	170	174	275	100	102	145	71	72	93	55	56	65
0.1	•••	858	867	967	268	275	299	145	149	159	97	99	104	72	74	76
	•••	861	882	1366	271	283	393	147	153	194	98	101	118	73	75	80
0.15	•••	1188	1202	1418	348	361	409	181	189	207	117	122	130	85	88	92
	•••	1193	1228	1806	353	375	495	184	195	236	119	125	140	86	90	93
0.2	•••	1444	1464	1829	410	428	507	208	220	248	132	139	152	94	99	105
	•••	1451	1504	2182	417	450	581	213	230	271	135	144	157	96	102	103
0.25	•••	1622	1651	2188	452	478	591	228	243	282	143	152	170	101	106	116
	•••	1632	1708	2496	462	509	652	234	257	299	147	159	171	103	110	110
0.3	•••	1726	1766	2489	478	511	659	240	259	309	150	161	183	105	112	123
	•••	1740	1842	2748	491	550	707	248	276	320	155	170	181	108	116	115
0.35	•••	1763	1816	2725	489	530	710	246	268	328	153	166	192	107	115	127
	•••	1781	1915	2936	505	578	746	255	289	334	159	176	187	110	120	118
0.4	•••	1741	1809	2896	488	536	745	247	272	340	154	168	196	107	115	129
	•••	1764	1933	3062	507	591	770	257	296	341	160	179	189	110	121	118
0.45	•••	1670	1757	2999	477	532	762	243	271	344	151	166	197	105	113	128
	•••	1699	1908	3124	498	593	778	254	296	341	158	178	187	109	119	115
0.5	•••	1561	1670	3033	457	518	762	235	264	341	146	162	193	101	109	124
	•••	1598	1847	3124	481	583	770	247	291	334	153	174	181	105	116	110
0.55	•••	1426	1558	3000	432	496	746	224	254	329	139	154	184	95	103	117
	•••	1472	1758	3062	458	564	746	236	280	320	146	166	171	99	109	103
0.6	•••	1276	1430	2898	402	468	712	209	239	311	130	144	171	88	95	108
	•••	1332	1648	2936	428	535	707	222	264	299	136	155	157	91	101	93
0.65	•••	1122	1293	2728	368	432	661	192	219	284	118	131	155	79	85	96
	•••	1185	1520	2748	394	498	652	204	243	271	123	140	140	82	90	80
0.7	•••	971	1151	2491	330	391	593	171	196	250	103	114	133	68	73	81
	•••	1039	1377	2496	355	450	581	182	216	236	108	122	118	70	76	65
0.75	•••	827	1004	2186	287	342	509	147	167	209	86	95	108	.	.	.
	•••	895	1218	2182	310	393	495	156	183	194	90	101	93	.	.	.
0.8	•••	687	851	1816	240	285	408	118	133	161	.	.	.	.	.	.
	•••	751	1040	1806	259	325	393	125	144	145	.	.	.	.	.	.
0.85	•••	547	684	1380	184	216	291	.	.	.	.	.	.	.	.	.
	•••	601	837	1366	198	243	275	.	.	.	.	.	.	.	.	.
0.9	•••	393	490	879	.	.	.	.	.	.	.	.	.	.	.	.
	•••	432	591	864	.	.	.	.	.	.	.	.	.	.	.	.

TABLE 5: ALPHA= 0.025 POWER= 0.8   EXPECTED ACCRUAL THRU MINIMUM FOLLOW-UP= 150

	DEL=.05			DEL=.10			DEL=.15			DEL=.20			DEL=.25		
FACT=	1.0 .75	.50 .25	.00 BIN	1.0 .75	.50 .25	.00 BIN	1.0 .75	.50 .25	.00 BIN	1.0 75	.50 .25	.00 BIN	1.0 .75	.50 .25	.00 BIN
PCONT=•••						REQUIRED NUMBER OF PATIENTS									
0.05 ***	471	475	502	169	172	178	100	101	104	71	72	73	55	56	57
•••	472	481	864	170	174	275	100	102	145	71	72	93	55	56	65
0.1 ***	859	868	967	269	276	299	145	150	158	97	100	104	72	74	76
•••	862	883	1366	271	283	393	147	153	194	98	102	118	73	75	80
0.15 ***	1189	1204	1418	349	362	409	182	189	206	118	122	130	85	88	92
•••	1194	1232	1806	354	377	495	185	196	236	119	125	140	86	90	93
0.2 ***	1445	1467	1829	411	430	507	209	221	248	133	140	152	95	99	105
•••	1453	1510	2182	418	453	581	214	231	271	136	145	157	97	102	103
0.25 ***	1624	1656	2188	454	481	591	229	244	282	144	153	170	102	107	116
•••	1635	1715	2496	464	512	652	235	258	299	148	160	171	104	111	110
0.3 ***	1729	1772	2489	481	515	659	242	261	309	151	162	183	106	112	123
•••	1743	1852	2748	494	554	707	250	278	320	156	170	181	109	117	115
0.35 ***	1767	1823	2725	493	535	710	248	271	328	155	167	192	108	115	127
•••	1786	1927	2936	509	582	746	257	291	334	160	177	187	111	120	118
0.4 ***	1746	1819	2896	492	542	745	249	275	340	155	169	196	108	116	129
•••	1770	1949	3062	511	597	770	260	298	341	161	180	189	111	121	118
0.45 ***	1676	1769	2999	482	538	762	245	273	344	153	168	197	106	114	128
•••	1708	1926	3124	504	599	778	257	298	341	159	179	187	109	120	115
0.5 ***	1569	1684	3034	463	524	762	238	267	341	148	163	193	102	110	124
•••	1609	1868	3124	487	590	770	250	293	334	155	175	181	105	116	110
0.55 ***	1436	1574	3000	438	503	745	227	257	329	141	156	184	96	104	117
•••	1485	1781	3062	464	571	746	239	282	320	147	167	171	100	110	103
0.6 ***	1288	1449	2898	408	474	712	212	241	310	131	145	171	89	96	108
•••	1347	1672	2936	435	542	707	225	266	299	137	156	157	92	101	93
0.65 ***	1136	1313	2728	374	439	661	195	222	284	119	132	155	79	86	95
•••	1202	1545	2748	400	504	652	206	245	271	125	141	140	82	90	80
0.7 ***	987	1172	2491	335	397	593	174	198	250	105	115	133	68	73	81
•••	1057	1401	2496	361	456	581	184	218	236	109	123	118	71	77	65
0.75 ***	842	1024	2186	293	347	509	149	169	209	87	95	108	.	.	.
•••	913	1241	2182	315	398	495	158	185	194	91	101	93	.	.	.
0.8 ***	702	869	1816	244	289	408	120	134	161	.	.	.	.	.	.
•••	767	1060	1806	263	329	393	126	145	145	.	.	.	.	.	.
0.85 ***	559	699	1380	188	219	291	.	.	.	.	.	.	.	.	.
•••	615	852	1366	201	245	275	.	.	.	.	.	.	.	.	.
0.9 ***	402	500	879	.	.	.	.	.	.	.	.	.	.	.	.
•••	442	600	864	.	.	.	.	.	.	.	.	.	.	.	.

TABLE 5: ALPHA= 0.025 POWER= 0.8   EXPECTED ACCRUAL THRU MINIMUM FOLLOW-UP= 160

PCONT	FACT	DEL=.05 1.0/.75	.50/.25	.00/BIN	DEL=.10 1.0/.75	.50/.25	.00/BIN	DEL=.15 1.0/.75	.50/.25	.00/BIN	DEL=.20 1.0/.75	.50/.25	.00/BIN	DEL=.25 1.0/.75	.50/.25	.00/BIN
					REQUIRED NUMBER OF PATIENTS											
0.05	***	471	475	502	169	172	178	100	101	104	71	72	73	55	56	57
	•••	473	482	864	170	174	275	100	103	145	71	72	93	56	56	65
0.1	***	860	869	967	269	276	299	146	150	159	97	100	104	72	74	76
	•••	863	885	1366	272	284	393	148	154	194	98	102	118	73	75	80
0.15	***	1190	1206	1418	350	364	409	182	190	207	118	122	130	85	88	92
	•••	1195	1235	1806	355	378	495	186	196	236	120	126	140	87	90	93
0.2	***	1447	1470	1829	413	432	507	210	222	248	134	140	152	95	99	105
	•••	1455	1515	2182	420	455	581	215	232	271	137	145	157	97	102	103
0.25	***	1626	1660	2188	457	484	591	230	246	282	145	154	170	102	107	116
	•••	1637	1723	2496	467	514	652	237	259	299	149	160	171	105	111	110
0.3	***	1732	1777	2489	484	518	659	243	262	309	152	163	183	107	113	123
	•••	1747	1862	2748	497	558	707	251	279	320	157	171	181	109	117	115
0.35	***	1770	1831	2725	496	539	710	250	273	328	156	168	192	109	116	127
	•••	1791	1940	2936	513	587	746	260	293	334	161	178	187	112	121	118
0.4	***	1751	1828	2896	496	547	745	251	277	340	157	170	196	109	116	129
	•••	1777	1964	3062	516	602	770	262	300	341	163	181	189	112	122	118
0.45	***	1683	1781	2999	486	543	762	248	276	344	154	169	197	106	115	128
	•••	1716	1944	3124	509	604	778	260	300	341	161	180	187	110	120	115
0.5	***	1577	1698	3034	468	530	762	241	270	341	150	164	193	103	111	124
	•••	1619	1888	3124	493	596	770	253	295	334	156	176	181	106	116	110
0.55	***	1446	1591	3000	444	509	746	230	259	329	142	157	184	97	105	117
	•••	1498	1803	3062	470	577	746	242	285	320	149	168	171	100	110	103
0.6	***	1300	1467	2898	414	481	712	215	244	311	133	146	171	89	97	108
	•••	1361	1696	2936	441	548	707	227	268	299	139	157	157	93	101	93
0.65	***	1150	1333	2728	380	445	661	197	224	284	120	133	155	80	86	96
	•••	1219	1569	2748	406	509	652	209	247	271	126	142	140	83	90	80
0.7	***	1002	1191	2491	341	402	593	176	200	250	106	116	133	69	74	81
	•••	1074	1424	2496	366	461	581	186	219	236	110	124	118	71	77	65
0.75	***	858	1043	2186	298	352	509	151	171	209	88	96	108	.	.	.
	•••	930	1262	2182	320	403	495	160	186	194	92	101	93	.	.	.
0.8	***	716	886	1816	249	293	408	121	136	161	.	.	.	.	.	.
	•••	783	1079	1806	267	332	393	128	146	145	.	.	.	.	.	.
0.85	***	571	713	1380	191	222	291	.	.	.	.	.	.	.	.	.
	•••	628	867	1366	204	247	275	.	.	.	.	.	.	.	.	.
0.9	***	411	509	879	.	.	.	.	.	.	.	.	.	.	.	.
	•••	451	610	864	.	.	.	.	.	.	.	.	.	.	.	.

TABLE 5: ALPHA= 0.025 POWER= 0.8   EXPECTED ACCRUAL THRU MINIMUM FOLLOW-UP= 170

		DEL=.05			DEL=.10			DEL=.15			DEL=.20			DEL=.25		
FACT=		1.0 .75	.50 .25	.00 BIN	1.0 .75	.50 .25	.00 BIN	1.0 .75	.50 .25	.00 BIN	1.0 75	.50 .25	.00 BIN	1.0 .75	.50 .25	.00 BIN
PCONT=•••					REQUIRED NUMBER OF PATIENTS											
0.05	***	471	476	502	169	172	178	100	101	104	71	72	73	55	56	57
	•••	473	482	864	170	175	275	101	103	145	71	72	93	56	56	65
0.1	***	860	870	967	270	277	299	146	150	159	98	100	104	72	74	76
	•••	864	887	1366	273	285	393	148	154	194	99	102	118	73	75	80
0.15	***	1191	1208	1418	351	365	409	183	191	206	118	123	130	86	88	92
	•••	1197	1238	1806	357	379	495	186	197	236	120	126	140	87	90	93
0.2	***	1448	1473	1829	414	434	507	211	223	248	134	141	152	96	99	105
	•••	1457	1520	2182	422	457	581	216	232	271	137	145	157	97	102	103
0.25	***	1628	1664	2188	459	486	591	232	247	282	146	154	170	103	108	116
	•••	1640	1730	2496	469	517	652	238	260	299	150	161	171	105	111	110
0.3	***	1735	1783	2489	486	522	659	245	264	309	153	164	183	107	113	123
	•••	1751	1872	2748	500	561	707	253	280	320	158	172	181	110	117	115
0.35	***	1774	1838	2725	500	543	710	252	275	328	157	169	192	109	116	127
	•••	1796	1952	2936	516	591	746	262	294	334	163	178	187	112	121	118
0.4	***	1756	1838	2896	500	551	745	254	279	340	158	171	196	109	117	129
	•••	1783	1979	3062	520	606	770	264	301	341	164	181	189	113	122	118
0.45	***	1689	1792	2999	491	549	762	251	278	344	156	170	197	107	115	128
	•••	1724	1962	3124	514	609	778	262	302	341	162	181	187	111	121	115
0.5	***	1585	1712	3033	473	536	762	243	272	341	151	166	193	103	111	124
	•••	1630	1908	3124	498	601	770	256	297	334	157	177	181	107	117	110
0.55	***	1456	1607	3000	449	515	745	232	262	329	144	158	184	98	105	117
	•••	1510	1825	3062	476	582	746	245	286	320	150	169	171	101	110	103
0.6	***	1312	1485	2898	419	487	712	218	246	310	134	147	171	90	97	108
	•••	1376	1718	2936	447	553	707	230	270	299	140	157	157	93	102	93
0.65	***	1164	1351	2728	385	451	661	200	227	284	122	134	155	81	87	96
	•••	1234	1591	2748	412	515	652	211	248	271	127	143	140	84	91	80
0.7	***	1016	1210	2491	346	408	593	178	202	250	107	117	133	69	74	81
	•••	1091	1446	2496	372	466	581	189	221	236	111	124	118	72	77	65
0.75	***	872	1061	2186	303	357	509	153	172	209	89	97	108	.	.	.
	•••	946	1282	2182	325	407	495	161	187	194	92	102	93	.	.	.
0.8	***	730	902	1816	252	297	408	123	137	161	.	.	.	.	.	.
	•••	798	1096	1806	271	335	393	129	147	145	.	.	.	.	.	.
0.85	***	583	726	1380	194	224	291	.	.	.	.	.	.	.	.	.
	•••	640	880	1366	207	249	275	.	.	.	.	.	.	.	.	.
0.9	***	419	518	879	.	.	.	.	.	.	.	.	.	.	.	.
	•••	460	618	864	.	.	.	.	.	.	.	.	.	.	.	.

TABLE 5: ALPHA= 0.025 POWER= 0.8   EXPECTED ACCRUAL THRU MINIMUM FOLLOW-UP= 180

		DEL=.05			DEL=.10			DEL=.15			DEL=.20			DEL=.25		
FACT=		1.0 .75	.50 .25	.00 BIN	1.0 .75	.50 .25	.00 BIN	1.0 .75	.50 .25	.00 BIN	1.0 75	.50 .25	.00 BIN	1.0 .75	.50 .25	.00 BIN
PCONT=***					REQUIRED NUMBER OF PATIENTS											
0.05	***	472	476	502	170	172	178	100	102	104	71	72	73	55	56	57
	•••	473	483	864	171	175	275	101	103	145	71	73	93	56	56	65
0.1	***	861	871	967	270	278	299	147	151	159	98	100	104	73	74	76
	•••	864	889	1366	273	285	393	148	154	194	99	102	118	73	75	80
0.15	***	1192	1210	1418	352	366	409	184	191	207	119	123	130	86	89	92
	•••	1198	1242	1806	358	380	495	187	197	236	121	126	140	87	90	93
0.2	***	1450	1476	1829	416	436	507	212	224	248	135	141	152	96	100	105
	•••	1459	1525	2182	423	458	581	217	233	271	138	146	157	98	102	103
0.25	***	1630	1668	2188	461	489	591	233	248	282	147	155	170	103	108	116
	•••	1643	1737	2496	471	519	652	239	261	299	150	161	171	105	111	110
0.3	***	1738	1789	2489	489	525	659	247	266	309	154	164	183	108	114	123
	•••	1755	1881	2748	503	564	707	255	282	320	159	172	181	110	118	115
0.35	***	1778	1845	2725	503	547	710	254	276	328	158	170	192	110	117	127
	•••	1801	1964	2936	520	595	746	263	296	334	164	179	187	113	121	118
0.4	***	1761	1847	2896	504	556	745	256	281	340	159	172	196	110	117	129
	•••	1790	1993	3062	525	611	770	267	303	341	165	182	189	113	122	118
0.45	***	1695	1804	2999	495	554	762	253	280	344	157	171	197	108	116	128
	•••	1733	1978	3124	519	614	778	265	304	341	163	182	187	112	121	115
0.5	***	1593	1726	3034	478	542	762	246	275	341	152	167	192	104	112	124
	•••	1640	1926	3124	504	606	770	258	299	334	159	177	181	108	117	110
0.55	***	1466	1623	3000	454	521	746	235	264	329	145	159	184	98	106	117
	•••	1522	1845	3062	481	588	746	247	288	320	151	169	171	102	111	103
0.6	***	1324	1502	2898	425	492	712	220	249	310	135	148	171	91	98	108
	•••	1390	1739	2936	452	559	707	232	272	299	141	158	157	94	102	93
0.65	***	1177	1370	2728	390	456	661	202	229	284	123	135	155	81	87	96
	•••	1250	1613	2748	418	519	652	213	250	271	128	143	140	84	91	80
0.7	***	1030	1228	2491	351	413	593	180	204	250	108	118	134	70	74	81
	•••	1107	1467	2496	377	470	581	191	222	236	112	125	118	72	77	65
0.75	***	886	1078	2186	307	361	509	155	174	209	90	97	108	.	.	.
	•••	961	1301	2182	330	410	495	163	188	194	93	102	93	.	.	.
0.8	***	743	917	1816	256	300	408	124	138	161	.	.	.	.	.	.
	•••	812	1113	1806	275	338	393	130	147	145	.	.	.	.	.	.
0.85	***	594	738	1380	196	226	291	.	.	.	.	.	.	.	.	.
	•••	652	893	1366	209	251	275	.	.	.	.	.	.	.	.	.
0.9	***	427	527	879	.	.	.	.	.	.	.	.	.	.	.	.
	•••	468	626	864	.	.	.	.	.	.	.	.	.	.	.	.

TABLE 5: ALPHA= 0.025 POWER= 0.8   EXPECTED ACCRUAL THRU MINIMUM FOLLOW-UP= 190

		DEL=.05			DEL=.10			DEL=.15			DEL=.20			DEL=.25		
FACT=		1.0 .75	.50 .25	.00 BIN	1.0 .75	.50 .25	.00 BIN	1.0 .75	.50 .25	.00 BIN	1.0 75	.50 .25	.00 BIN	1.0 .75	.50 .25	.00 BIN
PCONT=***					REQUIRED NUMBER OF PATIENTS											
0.05	***	472	477	502	170	173	178	100	102	104	71	72	73	55	56	57
	•••	474	483	864	171	175	275	101	103	145	71	73	93	56	56	65
0.1	***	861	872	967	271	278	299	147	151	158	98	100	104	73	74	76
	•••	865	891	1366	274	286	393	149	154	194	99	102	118	73	75	80
0.15	***	1193	1212	1418	353	367	409	184	192	206	119	123	130	86	89	92
	•••	1199	1245	1806	359	381	495	187	198	236	121	126	140	87	90	93
0.2	***	1451	1479	1829	417	438	507	213	224	248	135	142	152	96	100	105
	•••	1461	1530	2182	425	460	581	218	234	271	138	146	157	98	102	103
0.25	***	1633	1672	2188	462	491	591	234	249	282	147	155	170	104	108	116
	•••	1646	1744	2496	474	522	652	241	262	299	151	161	171	106	112	110
0.3	***	1740	1794	2489	491	528	659	248	267	309	155	165	183	108	114	123
	•••	1759	1890	2748	506	567	707	256	283	320	160	173	181	111	118	115
0.35	***	1782	1853	2725	506	551	710	256	278	328	159	171	192	111	117	127
	•••	1806	1975	2936	523	598	746	265	297	334	165	180	187	114	122	118
0.4	***	1765	1856	2896	508	560	744	258	283	340	160	173	196	111	118	129
	•••	1796	2007	3062	529	615	770	269	304	341	166	183	189	114	123	118
0.45	***	1702	1815	2999	500	558	762	255	282	344	158	172	197	109	116	128
	•••	1741	1994	3124	523	619	778	267	305	341	164	182	187	112	121	115
0.5	***	1601	1739	3034	483	547	762	248	277	341	154	168	193	105	112	124
	•••	1650	1944	3124	509	611	770	260	301	334	160	178	181	108	117	110
0.55	***	1475	1638	3000	459	526	745	237	266	329	146	160	184	99	106	117
	•••	1534	1864	3062	486	592	746	249	290	320	152	170	171	102	111	103
0.6	***	1335	1519	2898	430	498	712	222	251	310	136	149	171	91	98	108
	•••	1403	1759	2936	458	563	707	235	273	299	142	159	157	94	102	93
0.65	***	1190	1387	2728	396	462	661	204	230	284	124	135	155	82	88	96
	•••	1265	1633	2748	423	524	652	216	251	271	129	144	140	84	91	80
0.7	***	1044	1246	2491	356	418	593	182	205	250	109	118	133	70	75	81
	•••	1122	1486	2496	382	474	581	192	223	236	113	125	118	72	78	65
0.75	***	900	1095	2186	311	365	509	156	175	209	90	98	108	.	.	.
	•••	976	1319	2182	334	414	495	165	189	194	94	102	93	.	.	.
0.8	***	755	932	1816	260	303	408	125	139	161	.	.	.	.	.	.
	•••	825	1128	1806	278	341	393	131	148	145	.	.	.	.	.	.
0.85	***	605	750	1380	199	229	291	.	.	.	.	.	.	.	.	.
	•••	663	905	1366	212	252	275	.	.	.	.	.	.	.	.	.
0.9	***	435	535	879	.	.	.	.	.	.	.	.	.	.	.	.
	•••	476	633	864	.	.	.	.	.	.	.	.	.	.	.	.

| TABLE 5: ALPHA= 0.025 POWER= 0.8   EXPECTED ACCRUAL THRU MINIMUM FOLLOW-UP= 200 |

		DEL=.05			DEL=.10			DEL=.15			DEL=.20			DEL=.25		
FACT=		1.0 .75	.50 .25	.00 BIN	1.0 .75	.50 .25	.00 BIN	1.0 .75	.50 .25	.00 BIN	1.0 75	.50 .25	.00 BIN	1.0 .75	.50 .25	.00 BIN
PCONT=•••							REQUIRED NUMBER OF PATIENTS								
0.05	•••	472	477	502	170	173	178	100	102	104	71	72	73	55	56	57
	•••	474	484	864	171	175	275	101	103	145	72	73	93	56	56	65
0.1	•••	862	873	967	271	279	299	147	151	159	98	101	104	73	74	76
	•••	866	892	1366	274	286	393	149	154	194	99	102	118	74	75	80
0.15	•••	1194	1214	1418	354	368	409	185	192	207	120	124	130	86	89	92
	•••	1201	1248	1806	360	382	495	188	198	236	121	126	140	88	90	93
0.2	•••	1453	1482	1829	418	440	507	214	225	248	136	142	152	97	100	105
	•••	1463	1535	2182	427	462	581	219	234	271	139	146	157	98	103	103
0.25	•••	1635	1676	2188	464	493	591	235	250	282	148	156	170	104	109	116
	•••	1649	1751	2496	476	524	652	242	263	299	152	162	171	106	112	110
0.3	•••	1743	1800	2489	494	531	659	250	268	309	156	166	183	109	114	123
	•••	1762	1899	2748	508	570	707	258	284	320	160	173	181	111	118	115
0.35	•••	1786	1860	2725	509	554	710	257	280	328	160	172	192	111	118	127
	•••	1811	1986	2936	527	601	746	267	298	334	165	180	187	114	122	118
0.4	•••	1770	1865	2896	511	564	745	260	285	340	161	174	196	111	118	129
	•••	1803	2020	3062	532	619	770	270	306	341	167	183	189	114	123	118
0.45	•••	1708	1826	2999	504	563	762	257	284	344	159	173	197	109	117	128
	•••	1749	2010	3124	527	623	778	269	307	341	165	183	187	113	122	115
0.5	•••	1609	1752	3034	487	552	762	250	279	341	155	169	193	105	113	124
	•••	1660	1962	3124	513	616	770	262	302	334	161	179	181	109	118	110
0.55	•••	1485	1653	3000	464	531	746	239	268	329	147	161	184	100	107	117
	•••	1546	1883	3062	491	597	746	252	291	320	153	171	171	103	111	103
0.6	•••	1347	1535	2898	435	503	712	225	253	311	137	150	171	92	98	108
	•••	1417	1779	2936	463	568	707	237	275	299	143	159	157	95	103	93
0.65	•••	1202	1404	2728	400	466	661	206	232	284	125	136	155	82	88	96
	•••	1279	1652	2748	428	528	652	218	252	271	130	144	140	85	91	80
0.7	•••	1057	1262	2491	361	422	593	184	207	250	109	119	133	71	75	81
	•••	1137	1505	2496	386	478	581	194	224	236	114	125	118	73	78	65
0.75	•••	913	1111	2186	315	369	509	158	176	209	91	98	108	.	.	.
	•••	991	1337	2182	338	417	495	166	190	194	94	103	93	.	.	.
0.8	•••	767	946	1816	263	306	408	126	139	161	.	.	.	.	.	.
	•••	838	1143	1806	282	343	393	132	149	145	.	.	.	.	.	.
0.85	•••	615	762	1380	201	231	291	.	.	.	.	.	.	.	.	.
	•••	674	917	1366	214	254	275	.	.	.	.	.	.	.	.	.
0.9	•••	442	542	879	.	.	.	.	.	.	.	.	.	.	.	.
	•••	483	640	864	.	.	.	.	.	.	.	.	.	.	.	.

TABLE 5: ALPHA= 0.025 POWER= 0.8   EXPECTED ACCRUAL THRU MINIMUM FOLLOW-UP= 225

	DEL=.05			DEL=.10			DEL=.15			DEL=.20			DEL=.25		
FACT=	1.0 .75	.50 .25	.00 BIN	1.0 .75	.50 .25	.00 BIN	1.0 .75	.50 .25	.00 BIN	1.0 75	.50 .25	.00 BIN	1.0 .75	.50 .25	.00 BIN

PCONT=***   REQUIRED NUMBER OF PATIENTS

PCONT	1.0/.75	.50/.25	.00/BIN	1.0/.75	.50/.25	.00/BIN	1.0/.75	.50/.25	.00/BIN	1.0/.75	.50/.25	.00/BIN	1.0/.75	.50/.25	.00/BIN
0.05 ***	473	478	502	170	173	178	101	102	104	71	72	73	56	56	57
•••	475	485	864	172	175	275	101	103	145	72	73	93	56	56	65
0.1 ***	863	876	967	273	280	299	148	152	158	99	101	104	73	75	76
•••	868	896	1366	276	287	393	150	155	194	100	102	118	74	75	80
0.15 ***	1197	1218	1418	356	371	409	186	193	206	120	124	130	87	89	92
•••	1204	1255	1806	362	384	495	189	199	236	122	127	140	88	91	93
0.2 ***	1456	1489	1829	422	443	507	216	227	248	137	143	152	97	101	105
•••	1467	1546	2182	430	465	581	221	235	271	140	147	157	99	103	103
0.25 ***	1640	1686	2188	469	499	591	238	253	282	149	157	170	105	109	116
•••	1656	1767	2496	481	529	652	244	264	299	153	163	171	107	112	110
0.3 ***	1751	1813	2489	500	538	659	253	271	309	158	167	183	110	115	123
•••	1772	1920	2748	515	577	707	261	286	320	162	174	181	112	119	115
0.35 ***	1795	1877	2725	516	562	710	261	283	328	162	173	192	112	118	127
•••	1823	2013	2936	534	609	746	271	300	334	167	181	187	115	122	118
0.4 ***	1783	1887	2896	520	574	745	264	289	340	164	176	196	113	119	129
•••	1819	2052	3062	542	627	770	275	309	341	169	185	189	116	124	118
0.45 ***	1723	1853	2999	513	573	762	262	289	344	162	175	197	111	118	128
•••	1769	2046	3124	538	633	778	273	310	341	168	184	187	114	122	115
0.5 ***	1628	1784	3034	498	563	762	255	283	341	157	171	192	107	114	124
•••	1684	2002	3124	524	626	770	267	305	334	163	180	181	110	118	110
0.55 ***	1508	1688	3000	475	543	745	244	272	329	150	163	184	101	108	117
•••	1574	1927	3062	503	607	746	257	294	320	156	172	171	104	112	103
0.6 ***	1374	1573	2898	446	514	712	230	257	311	140	152	171	93	99	108
•••	1449	1824	2936	474	578	707	241	278	299	145	160	157	96	103	93
0.65 ***	1232	1443	2728	411	477	661	211	236	284	127	138	155	84	89	95
•••	1313	1698	2748	439	538	652	222	255	271	132	145	140	86	92	80
0.7 ***	1089	1301	2491	371	432	593	188	210	250	111	120	133	72	75	81
•••	1172	1549	2496	397	487	581	198	227	236	115	126	118	73	78	65
0.75 ***	944	1148	2186	325	378	509	161	179	209	92	99	108	.	.	.
•••	1024	1376	2182	347	424	495	169	192	194	95	103	93	.	.	.
0.8 ***	796	979	1816	271	313	408	129	141	161	.	.	.	.	.	.
•••	869	1177	1806	289	348	393	134	150	145	.	.	.	.	.	.
0.85 ***	639	788	1380	206	235	291	.	.	.	.	.	.	.	.	.
•••	699	942	1366	219	257	275	.	.	.	.	.	.	.	.	.
0.9 ***	459	559	879	.	.	.	.	.	.	.	.	.	.	.	.
•••	500	656	864	.	.	.	.	.	.	.	.	.	.	.	.

TABLE 5: ALPHA= 0.025 POWER= 0.8   EXPECTED ACCRUAL THRU MINIMUM FOLLOW-UP= 250

		DEL=.05			DEL=.10			DEL=.15			DEL=.20			DEL=.25		
FACT=		1.0 .75	.50 .25	.00 BIN	1.0 .75	.50 .25	.00 BIN	1.0 .75	.50 .25	.00 BIN	1.0 75	.50 .25	.00 BIN	1.0 .75	.50 .25	.00 BIN
PCONT=***					REQUIRED NUMBER OF PATIENTS											
0.05	***	474	479	502	171	174	178	101	102	104	72	72	73	56	56	57
	•••	476	486	864	172	176	275	101	103	145	72	73	93	56	57	65
0.1	***	865	879	967	274	282	299	149	152	159	99	101	104	74	75	76
	•••	870	899	1366	277	288	393	150	155	194	100	102	118	74	76	80
0.15	***	1199	1223	1418	359	373	409	187	194	207	121	125	130	87	89	92
	•••	1207	1262	1806	364	386	495	190	199	236	123	127	140	88	91	93
0.2	***	1460	1496	1829	425	447	507	218	228	248	138	144	152	98	101	105
	•••	1472	1557	2182	434	468	581	222	236	271	141	147	157	99	103	103
0.25	***	1645	1696	2189	473	503	591	240	255	282	151	158	170	106	110	116
	•••	1662	1782	2496	485	533	652	247	266	299	154	163	171	108	112	110
0.3	***	1758	1827	2489	505	544	659	256	274	309	159	168	183	111	116	123
	•••	1781	1940	2748	521	582	707	264	288	320	163	175	181	113	119	115
0.35	***	1804	1895	2725	523	570	710	265	286	328	164	175	192	113	119	127
	•••	1836	2037	2936	542	616	746	274	303	334	169	182	187	116	123	118
0.4	***	1795	1909	2896	528	582	745	268	292	340	166	177	196	114	120	129
	•••	1834	2081	3062	550	635	770	279	311	341	171	186	189	117	124	118
0.45	***	1739	1878	2999	522	583	762	266	292	344	164	177	197	112	119	128
	•••	1788	2080	3124	547	641	778	278	313	341	170	185	187	115	123	115
0.5	***	1647	1813	3034	507	573	762	260	287	341	159	172	193	108	115	124
	•••	1708	2040	3124	534	634	770	272	308	334	165	181	181	111	119	110
0.55	***	1531	1721	3000	485	553	746	249	276	329	152	164	184	102	109	117
	•••	1602	1967	3062	513	616	746	261	297	320	158	173	171	105	112	103
0.6	***	1400	1608	2898	456	524	712	234	260	311	142	154	172	94	100	108
	•••	1479	1865	2936	485	587	707	246	280	299	147	161	157	97	104	93
0.65	***	1261	1480	2728	422	487	661	215	239	284	129	139	155	84	89	96
	•••	1345	1739	2748	449	546	652	226	258	271	134	146	140	87	92	80
0.7	***	1118	1337	2491	380	441	593	192	213	250	113	121	134	72	76	81
	•••	1204	1588	2496	406	494	581	201	229	236	117	127	118	74	78	65
0.75	***	972	1181	2186	333	386	509	164	181	209	94	100	108	.	.	.
	•••	1055	1412	2182	355	430	495	172	193	194	97	104	93	.	.	.
0.8	***	822	1008	1816	277	319	408	131	143	161	.	.	.	.	.	.
	•••	897	1207	1806	295	353	393	136	151	145	.	.	.	.	.	.
0.85	***	660	811	1380	211	239	291	.	.	.	.	.	.	.	.	.
	•••	722	965	1366	223	260	275	.	.	.	.	.	.	.	.	.
0.9	***	474	574	879	.	.	.	.	.	.	.	.	.	.	.	.
	•••	515	669	864	.	.	.	.	.	.	.	.	.	.	.	.

TABLE 5: ALPHA= 0.025 POWER= 0.8   EXPECTED ACCRUAL THRU MINIMUM FOLLOW-UP= 275

		DEL=.05			DEL=.10			DEL=.15			DEL=.20			DEL=.25		
FACT=		1.0 .75	.50 .25	.00 BIN	1.0 .75	.50 .25	.00 BIN	1.0 .75	.50 .25	.00 BIN	1.0 75	.50 .25	.00 BIN	1.0 .75	.50 .25	.00 BIN
PCONT=***		REQUIRED NUMBER OF PATIENTS														
0.05	***	474 480 502			171 174 178			101 102 104			72 72 73			56 56 57		
	•••	476 487 864			172 176 275			102 103 145			72 73 93			56 56 65		
0.1	***	866 881 967			275 282 299			149 153 158			99 101 104			74 75 76		
	•••	871 903 1366			278 289 393			151 155 194			100 103 118			74 76 80		
0.15	***	1201 1227 1418			360 375 409			188 195 206			122 125 130			88 90 92		
	•••	1210 1268 1806			366 388 495			191 200 236			123 127 140			89 91 93		
0.2	***	1464 1503 1829			428 450 507			219 230 248			139 144 152			98 101 105		
	•••	1477 1567 2182			437 470 581			224 237 271			141 148 157			100 103 103		
0.25	***	1650 1706 2189			477 508 591			242 256 282			152 159 169			106 110 116		
	•••	1669 1796 2496			490 537 652			249 267 299			155 164 171			108 113 110		
0.3	***	1765 1840 2489			510 549 659			258 276 309			161 169 183			112 116 123		
	•••	1790 1958 2748			526 587 707			266 289 320			165 175 181			114 119 115		
0.35	***	1814 1911 2725			529 576 710			268 289 328			166 176 192			114 120 127		
	•••	1848 2060 2936			548 621 746			277 305 334			171 183 187			117 123 118		
0.4	***	1807 1929 2896			535 590 745			272 295 340			168 179 196			115 121 129		
	•••	1850 2109 3062			557 641 770			282 313 341			173 186 189			118 124 118		
0.45	***	1754 1903 2999			530 591 762			270 295 344			166 178 197			113 119 128		
	•••	1807 2111 3124			555 648 778			281 315 341			171 186 187			116 123 115		
0.5	***	1666 1841 3033			516 582 762			264 290 341			161 174 193			109 115 124		
	•••	1730 2074 3124			543 642 770			275 310 334			167 182 181			112 119 110		
0.55	***	1553 1752 3000			495 562 745			253 279 329			154 166 184			103 109 117		
	•••	1628 2003 3062			523 624 746			265 300 320			159 174 171			106 113 103		
0.6	***	1425 1641 2898			466 534 712			238 264 310			144 155 171			95 101 108		
	•••	1508 1903 2936			494 594 707			249 283 299			149 162 157			98 104 93		
0.65	***	1288 1513 2728			431 496 661			219 242 284			131 140 155			85 90 96		
	•••	1375 1776 2748			458 553 652			229 260 271			135 147 140			87 92 80		
0.7	***	1146 1370 2491			389 449 593			195 216 250			114 122 133			73 76 81		
	•••	1234 1623 2496			414 500 581			204 230 236			118 127 118			74 78 65		
0.75	***	999 1212 2186			340 392 509			167 183 209			95 100 108			. . .		
	•••	1084 1444 2182			363 435 495			174 194 194			97 104 93			. . .		
0.8	***	846 1035 1816			283 324 408			133 144 161			. . .			. . .		
	•••	922 1234 1806			301 356 393			138 151 145			. . .			. . .		
0.85	***	680 833 1380			215 242 291			. . .			. . .			. . .		
	•••	742 986 1366			227 262 275			. . .			. . .			. . .		
0.9	***	487 588 879			. . .			. . .			. . .			. . .		
	•••	529 681 864			. . .			. . .			. . .			. . .		

TABLE 5: ALPHA= 0.025 POWER= 0.8   EXPECTED ACCRUAL THRU MINIMUM FOLLOW-UP= 300

		DEL=.05			DEL=.10			DEL=.15			DEL=.20			DEL=.25	
FACT=	1.0 .75	.50 .25	.00 BIN	1.0 .75	.50 .25	.00 BIN	1.0 .75	.50 .25	.00 BIN	1.0 75	.50 .25	.00 BIN	1.0 .75	.50 .25	.00 BIN
PCONT=***				REQUIRED NUMBER OF PATIENTS											
0.05 ***	475	481	502	172	174	178	101	102	104	72	72	73	56	56	57
•••	477	488	864	173	176	275	102	103	145	72	73	93	56	56	65
0.1 ***	868	883	967	276	283	299	150	153	158	100	102	104	74	75	76
•••	873	905	1366	279	289	393	151	156	194	100	103	118	74	76	80
0.15 ***	1204	1232	1418	362	377	409	189	196	206	122	125	130	88	90	92
•••	1213	1274	1806	368	389	495	192	200	236	124	127	140	89	91	93
0.2 ***	1467	1510	1829	430	453	507	221	231	248	140	145	152	99	102	105
•••	1482	1576	2182	439	473	581	225	238	271	142	148	157	100	103	103
0.25 ***	1656	1715	2188	481	511	591	244	258	282	153	160	169	107	111	115
•••	1676	1809	2496	493	540	652	250	268	299	156	164	171	109	113	110
0.3 ***	1772	1852	2488	515	554	659	261	278	309	162	170	183	112	117	123
•••	1800	1975	2748	531	591	707	268	291	320	166	176	181	114	120	115
0.35 ***	1823	1927	2725	535	582	710	271	291	328	167	177	192	115	120	127
•••	1860	2081	2936	554	626	746	280	306	334	172	184	187	118	124	118
0.4 ***	1819	1949	2896	541	597	745	275	298	340	169	180	196	116	121	129
•••	1865	2134	3062	564	647	770	285	315	341	174	187	189	118	125	118
0.45 ***	1769	1926	2999	538	599	762	273	298	344	168	179	197	114	120	128
•••	1826	2140	3124	563	654	778	284	317	341	173	187	187	117	124	115
0.5 ***	1684	1868	3034	524	590	762	267	293	340	163	175	193	110	116	124
•••	1752	2105	3124	552	649	770	279	312	334	169	183	181	113	120	110
0.55 ***	1574	1781	3000	503	571	745	256	282	329	156	167	184	104	110	117
•••	1653	2037	3062	531	631	746	268	301	320	161	175	171	107	113	103
0.6 ***	1449	1672	2898	474	542	712	241	266	310	145	156	171	96	101	108
•••	1535	1937	2936	503	601	707	253	285	299	150	163	157	98	104	93
0.65 ***	1313	1545	2728	439	504	661	222	245	284	132	141	154	86	90	95
•••	1404	1810	2748	466	559	652	232	261	271	136	147	140	88	93	80
0.7 ***	1171	1401	2491	397	456	593	198	218	250	115	123	133	73	77	81
•••	1262	1655	2496	422	506	581	207	232	236	119	128	118	75	79	65
0.75 ***	1024	1241	2186	347	398	509	169	185	209	95	101	108	.	.	.
•••	1111	1473	2182	369	440	495	176	195	194	98	104	93	.	.	.
0.8 ***	869	1060	1816	289	329	408	134	145	161	.	.	.	.	.	.
•••	946	1259	1806	306	360	393	139	152	145	.	.	.	.	.	.
0.85 ***	699	852	1380	219	245	291	.	.	.	.	.	.	.	.	.
•••	762	1004	1366	230	264	275	.	.	.	.	.	.	.	.	.
0.9 ***	500	600	879	.	.	.	.	.	.	.	.	.	.	.	.
•••	542	692	864	.	.	.	.	.	.	.	.	.	.	.	.

TABLE 5: ALPHA= 0.025 POWER= 0.8   EXPECTED ACCRUAL THRU MINIMUM FOLLOW-UP= 325

	DEL=.05			DEL=.10			DEL=.15			DEL=.20			DEL=.25		
FACT=	1.0 .75	.50 .25	.00 BIN	1.0 .75	.50 .25	.00 BIN	1.0 .75	.50 .25	.00 BIN	1.0 75	.50 .25	.00 BIN	1.0 .75	.50 .25	.00 BIN
PCONT=•••				REQUIRED NUMBER OF PATIENTS											
0.05 ***	476	482	502	172	174	178	101	103	104	72	73	73	56	56	57
•••	478	489	864	173	176	275	102	103	145	72	73	93	56	56	65
0.1 ***	869	886	967	277	284	299	150	154	159	100	102	104	74	75	76
•••	875	908	1366	280	290	393	152	156	194	101	103	118	75	76	80
0.15 ***	1206	1236	1418	364	378	409	190	197	207	122	126	130	88	90	92
•••	1217	1279	1806	370	390	495	193	201	236	124	128	140	89	91	93
0.2 ***	1471	1516	1829	433	455	507	222	232	248	140	145	152	99	102	105
•••	1487	1585	2182	442	475	581	226	239	271	143	148	157	101	104	103
0.25 ***	1661	1725	2188	484	515	591	246	259	282	154	160	170	107	111	116
•••	1683	1821	2496	497	543	652	252	269	299	157	164	171	109	113	110
0.3 ***	1779	1865	2489	519	559	659	263	280	309	163	171	183	113	117	123
•••	1809	1992	2748	536	595	707	270	292	320	167	176	181	115	120	115
0.35 ***	1832	1943	2725	540	588	710	273	293	328	169	178	192	116	121	127
•••	1872	2101	2936	560	631	746	282	307	334	173	184	187	118	124	118
0.4 ***	1831	1968	2896	548	603	745	278	300	340	171	181	196	116	122	129
•••	1880	2158	3062	571	652	770	287	316	341	175	188	189	119	125	118
0.45 ***	1784	1949	2999	545	606	762	277	301	344	169	180	197	115	120	128
•••	1844	2166	3124	570	660	778	287	318	341	174	187	187	117	124	115
0.5 ***	1702	1893	3033	532	597	762	271	296	341	165	176	193	111	117	124
•••	1773	2135	3124	559	655	770	282	314	334	170	183	181	114	120	110
0.55 ***	1595	1809	3000	511	578	745	260	285	329	157	168	184	105	110	117
•••	1676	2068	3062	539	637	746	271	303	320	162	175	171	107	113	103
0.6 ***	1472	1701	2898	482	549	712	245	269	311	147	157	171	97	102	108
•••	1561	1969	2936	511	607	707	255	286	299	151	164	157	99	104	93
0.65 ***	1338	1574	2728	447	511	661	225	247	284	133	142	155	86	91	95
•••	1430	1841	2748	474	565	652	235	263	271	137	148	140	88	93	80
0.7 ***	1196	1430	2491	404	463	593	201	220	250	116	124	133	74	77	81
•••	1289	1685	2496	429	511	581	209	233	236	120	128	118	75	79	65
0.75 ***	1048	1267	2186	353	404	509	171	186	209	96	102	108	.	.	.
•••	1136	1500	2182	375	444	495	178	196	194	99	105	93	.	.	.
0.8 ***	890	1083	1816	294	333	408	136	146	161	.	.	.	.	.	.
•••	968	1281	1806	311	363	393	141	153	145	.	.	.	.	.	.
0.85 ***	716	870	1380	222	248	291	.	.	.	.	.	.	.	.	.
•••	779	1020	1366	234	266	275	.	.	.	.	.	.	.	.	.
0.9 ***	512	612	879	.	.	.	.	.	.	.	.	.	.	.	.
•••	554	701	864	.	.	.	.	.	.	.	.	.	.	.	.

TABLE 5: ALPHA= 0.025 POWER= 0.8   EXPECTED ACCRUAL THRU MINIMUM FOLLOW-UP= 350

		DEL=.05			DEL=.10			DEL=.15			DEL=.20			DEL=.25	
FACT=	1.0 .75	.50 .25	.00 BIN	1.0 .75	.50 .25	.00 BIN	1.0 .75	.50 .25	.00 BIN	1.0 75	.50 .25	.00 BIN	1.0 .75	.50 .25	.00 BIN
PCONT=•••						REQUIRED NUMBER OF PATIENTS									
0.05 ***	476	482	502	172	175	178	101	103	104	72	73	73	56	56	57
•••	479	489	864	173	176	275	102	103	145	72	73	93	56	57	65
0.1 ***	871	888	967	278	285	299	151	154	159	100	102	104	74	75	76
•••	877	911	1366	281	291	393	152	156	194	101	103	118	75	76	80
0.15 ***	1209	1240	1418	365	380	409	191	197	206	123	126	130	89	90	92
•••	1220	1284	1806	371	391	495	194	201	236	124	128	140	89	91	93
0.2 ***	1475	1523	1829	435	458	507	223	232	248	141	146	152	100	102	105
•••	1492	1593	2182	444	476	581	227	239	271	143	148	157	101	104	103
0.25 ***	1666	1734	2189	488	518	591	248	260	282	155	161	170	108	111	116
•••	1690	1833	2496	500	545	652	253	270	299	157	165	171	110	113	110
0.3 ***	1786	1876	2489	523	563	659	265	281	309	164	172	183	113	118	123
•••	1818	2007	2748	540	598	707	272	293	320	168	177	181	115	120	115
0.35 ***	1842	1958	2725	545	593	710	276	295	328	170	179	192	117	121	127
•••	1883	2120	2936	565	635	746	284	309	334	174	185	187	119	124	118
0.4 ***	1842	1986	2896	554	608	745	280	302	340	172	182	196	117	122	129
•••	1895	2180	3062	577	657	770	290	318	341	176	188	189	120	125	118
0.45 ***	1798	1970	2999	551	612	762	279	303	344	171	181	197	115	121	128
•••	1861	2192	3124	577	665	778	290	320	341	176	188	187	118	124	115
0.5 ***	1719	1917	3034	539	604	762	273	298	341	166	177	192	112	117	124
•••	1793	2162	3124	566	660	770	284	316	334	171	184	181	114	120	110
0.55 ***	1615	1835	3000	518	585	745	263	287	329	159	169	184	106	111	117
•••	1699	2096	3062	546	642	746	274	305	320	163	176	171	108	114	103
0.6 ***	1494	1729	2898	489	556	712	248	271	311	148	158	171	97	102	108
•••	1585	1998	2936	518	612	707	258	288	299	152	164	157	99	105	93
0.65 ***	1361	1602	2728	454	517	661	228	249	284	134	143	155	87	91	96
•••	1456	1870	2748	481	570	652	237	264	271	138	148	140	89	93	80
0.7 ***	1219	1457	2491	410	468	593	203	221	250	117	124	133	74	77	81
•••	1314	1712	2496	435	515	581	211	234	236	120	129	118	76	79	65
0.75 ***	1070	1292	2186	359	408	509	173	188	209	97	102	108	.	.	.
•••	1159	1524	2182	381	447	495	180	197	194	99	105	93	.	.	.
0.8 ***	910	1104	1816	298	337	408	137	147	161	.	.	.	.	.	.
•••	989	1301	1806	315	365	393	142	153	145	.	.	.	.	.	.
0.85 ***	732	887	1380	225	250	291	.	.	.	.	.	.	.	.	.
•••	796	1035	1366	236	267	275	.	.	.	.	.	.	.	.	.
0.9 ***	523	622	879	.	.	.	.	.	.	.	.	.	.	.	.
•••	564	710	864	.	.	.	.	.	.	.	.	.	.	.	.

TABLE 5: ALPHA= 0.025 POWER= 0.8   EXPECTED ACCRUAL THRU MINIMUM FOLLOW-UP= 375

		DEL=.05			DEL=.10			DEL=.15			DEL=.20			DEL=.25		
FACT=		1.0 .75	.50 .25	.00 BIN	1.0 .75	.50 .25	.00 BIN	1.0 .75	.50 .25	.00 BIN	1.0 75	.50 .25	.00 BIN	1.0 .75	.50 .25	.00 BIN
PCONT=***					REQUIRED	NUMBER OF	PATIENTS									
0.05	***	477	483	502	172	175	178	102	103	104	72	73	73	56	56	57
	•••	479	490	864	174	176	275	102	103	145	72	73	93	56	57	65
0.1	***	872	890	967	278	286	299	151	154	159	100	102	104	74	75	76
	•••	879	913	1366	281	291	393	152	156	194	101	103	118	75	76	80
0.15	***	1211	1244	1418	367	381	409	192	197	206	123	126	130	89	90	92
	•••	1223	1289	1806	373	392	495	194	201	236	125	128	140	89	91	93
0.2	***	1478	1529	1829	437	460	507	224	233	248	141	146	152	100	102	105
	•••	1496	1600	2182	447	478	581	228	240	271	144	149	157	101	104	103
0.25	***	1671	1742	2188	490	521	591	249	261	282	155	161	170	108	111	116
	•••	1696	1843	2496	503	547	652	255	270	299	158	165	171	110	113	110
0.3	***	1793	1888	2488	527	567	659	267	283	309	165	172	183	114	118	123
	•••	1827	2021	2748	544	601	707	274	294	320	168	177	181	116	120	115
0.35	***	1851	1972	2725	550	597	710	278	296	328	171	179	192	117	122	127
	•••	1895	2138	2936	570	638	746	286	310	334	175	185	187	119	124	118
0.4	***	1854	2004	2896	559	614	745	283	304	340	173	183	196	118	123	129
	•••	1909	2200	3062	582	661	770	292	319	341	178	189	189	120	125	118
0.45	***	1812	1990	2999	557	618	762	282	305	344	172	182	197	116	121	128
	•••	1878	2215	3124	583	670	778	292	321	341	177	189	187	118	124	115
0.5	***	1736	1940	3033	545	610	762	276	300	340	167	178	193	112	117	124
	•••	1813	2188	3124	573	665	770	287	317	334	172	184	181	115	120	110
0.55	***	1634	1860	3000	525	591	745	265	289	329	160	170	184	106	111	117
	•••	1721	2123	3062	553	647	746	276	306	320	164	176	171	108	114	103
0.6	***	1515	1754	2898	496	562	711	250	273	310	149	159	171	98	102	108
	•••	1608	2026	2936	524	617	707	260	289	299	153	164	157	100	105	93
0.65	***	1383	1628	2728	460	523	661	230	251	284	135	144	155	88	91	95
	•••	1480	1897	2748	487	574	652	239	265	271	139	148	140	89	93	80
0.7	***	1241	1482	2491	416	473	593	205	223	250	118	125	133	75	77	81
	•••	1337	1738	2496	441	519	581	213	235	236	121	129	118	76	79	65
0.75	***	1091	1315	2186	364	413	509	175	189	209	97	102	108	.	.	.
	•••	1181	1547	2182	385	450	495	181	198	194	100	105	93	.	.	.
0.8	***	928	1124	1816	302	340	408	138	148	161	.	.	.	.	.	.
	•••	1008	1320	1806	319	368	393	143	154	145	.	.	.	.	.	.
0.85	***	747	902	1380	228	252	291	.	.	.	.	.	.	.	.	.
	•••	811	1049	1366	239	268	275	.	.	.	.	.	.	.	.	.
0.9	***	533	631	879	.	.	.	.	.	.	.	.	.	.	.	.
	•••	574	718	864	.	.	.	.	.	.	.	.	.	.	.	.

TABLE 5: ALPHA= 0.025 POWER= 0.8   EXPECTED ACCRUAL THRU MINIMUM FOLLOW-UP= 400

	DEL=.05			DEL=.10			DEL=.15			DEL=.20			DEL=.25		
FACT=	1.0 .75	.50 .25	.00 BIN	1.0 .75	.50 .25	.00 BIN	1.0 .75	.50 .25	.00 BIN	1.0 75	.50 .25	.00 BIN	1.0 .75	.50 .25	.00 BIN
PCONT=•••				REQUIRED NUMBER OF PATIENTS											
0.05 •••	477	484	502	173	175	178	102	103	104	72	73	73	56	56	57
•••	480	490	864	174	177	275	102	103	145	72	73	93	56	57	65
0.1 •••	873	892	967	279	286	299	151	154	159	101	102	104	74	75	76
•••	880	915	1366	282	292	393	153	156	194	101	103	118	75	76	80
0.15 •••	1214	1248	1418	368	382	409	192	198	207	124	126	130	89	90	92
•••	1226	1293	1806	374	393	495	195	202	236	125	128	140	90	91	93
0.2 •••	1482	1535	1829	440	462	507	225	234	248	142	146	152	100	103	105
•••	1501	1607	2182	449	479	581	229	240	271	144	149	157	101	104	103
0.25 •••	1676	1751	2188	493	524	591	250	263	282	156	162	170	109	112	116
•••	1703	1854	2496	506	550	652	256	271	299	159	165	171	110	114	110
0.3 •••	1800	1899	2489	531	570	659	268	284	309	166	173	183	114	118	123
•••	1835	2034	2748	548	603	707	275	295	320	169	177	181	116	120	115
0.35 •••	1860	1986	2725	554	601	710	280	298	328	172	180	192	118	122	127
•••	1906	2154	2936	574	642	746	288	311	334	176	185	187	120	124	118
0.4 •••	1865	2020	2896	564	619	745	285	306	340	174	183	196	118	123	129
•••	1922	2220	3062	587	665	770	294	320	341	178	189	189	121	126	118
0.45 •••	1826	2010	2999	563	623	762	284	307	344	173	183	197	117	122	128
•••	1895	2237	3124	588	674	778	294	323	341	178	189	187	119	125	115
0.5 •••	1752	1962	3034	552	616	762	279	302	341	169	179	193	113	118	124
•••	1832	2212	3124	579	669	770	289	318	334	173	185	181	115	121	110
0.55 •••	1653	1883	3000	531	597	746	268	291	329	161	171	184	107	111	117
•••	1742	2148	3062	559	652	746	278	307	320	165	177	171	109	114	103
0.6 •••	1535	1779	2898	503	568	712	253	275	311	150	159	171	98	103	108
•••	1631	2051	2936	531	621	707	263	290	299	154	165	157	100	105	93
0.65 •••	1404	1652	2728	466	528	661	232	252	284	136	144	155	88	91	96
•••	1503	1922	2748	493	578	652	241	266	271	140	149	140	90	93	80
0.7 •••	1262	1505	2491	422	478	593	207	224	250	119	125	133	75	78	81
•••	1360	1761	2496	446	523	581	215	236	236	122	129	118	76	79	65
0.75 •••	1111	1337	2186	369	417	509	176	190	209	98	103	108	.	.	.
•••	1202	1568	2182	390	453	495	183	199	194	100	105	93	.	.	.
0.8 •••	946	1143	1816	306	343	408	139	149	161	.	.	.	.	.	.
•••	1027	1337	1806	323	370	393	144	154	145	.	.	.	.	.	.
0.85 •••	762	917	1380	231	254	291	.	.	.	.	.	.	.	.	.
•••	826	1062	1366	241	270	275	.	.	.	.	.	.	.	.	.
0.9 •••	542	640	879	.	.	.	.	.	.	.	.	.	.	.	.
•••	584	725	864	.	.	.	.	.	.	.	.	.	.	.	.

TABLE 5: ALPHA= 0.025 POWER= 0.8 — EXPECTED ACCRUAL THRU MINIMUM FOLLOW-UP= 425

		DEL=.05			DEL=.10			DEL=.15			DEL=.20			DEL=.25		
FACT=		1.0 .75	.50 .25	.00 BIN	1.0 .75	.50 .25	.00 BIN	1.0 .75	.50 .25	.00 BIN	1.0 75	.50 .25	.00 BIN	1.0 .75	.50 .25	.00 BIN
PCONT=***								REQUIRED NUMBER OF PATIENTS								
0.05	***	478	484	502	173	175	178	102	103	104	72	73	73	56	56	57
	•••	480	491	864	174	177	275	102	103	145	72	73	93	56	57	65
0.1	***	875	894	967	280	287	299	152	155	159	101	102	104	74	75	76
	•••	882	917	1366	283	292	393	153	156	194	102	103	118	75	76	80
0.15	***	1216	1251	1418	370	383	409	193	198	206	124	127	130	89	91	93
	•••	1229	1297	1806	375	394	495	195	202	236	125	128	140	90	91	93
0.2	***	1486	1541	1829	442	463	507	226	235	248	142	147	152	101	103	105
	•••	1505	1614	2182	451	481	581	230	240	271	144	149	157	102	104	103
0.25	***	1681	1759	2189	496	527	591	251	263	282	156	162	170	109	112	116
	•••	1709	1863	2496	509	552	652	257	272	299	159	166	171	110	114	110
0.3	***	1807	1910	2488	535	574	659	270	285	309	167	173	183	115	118	123
	•••	1844	2047	2748	551	606	707	277	295	320	170	178	181	116	120	115
0.35	***	1869	2000	2725	558	605	710	281	299	328	173	181	192	118	122	127
	•••	1917	2170	2936	578	645	746	289	312	334	176	186	187	120	125	118
0.4	***	1876	2037	2896	569	623	745	287	307	340	175	184	196	119	123	129
	•••	1936	2238	3062	592	668	770	296	321	341	179	190	189	121	126	118
0.45	***	1839	2028	2999	568	628	762	286	308	344	174	183	197	117	122	128
	•••	1911	2258	3124	594	678	778	297	324	341	178	189	187	119	125	115
0.5	***	1768	1982	3034	557	621	762	281	304	341	170	179	192	113	118	124
	•••	1850	2234	3124	584	673	770	291	319	334	174	185	181	116	121	110
0.55	***	1671	1905	3000	537	602	745	270	293	329	162	171	184	107	112	117
	•••	1762	2172	3062	565	656	746	281	308	320	166	177	171	109	114	103
0.6	***	1554	1802	2898	509	573	712	255	276	311	151	160	171	99	103	108
	•••	1652	2075	2936	536	625	707	265	291	299	155	165	157	101	105	93
0.65	***	1424	1676	2728	472	533	661	234	254	284	137	145	155	88	91	95
	•••	1524	1945	2748	499	582	652	243	267	271	141	149	140	90	93	80
0.7	***	1282	1527	2491	427	483	593	209	226	250	120	126	133	75	78	81
	•••	1381	1783	2496	451	526	581	216	237	236	122	129	118	77	79	65
0.75	***	1130	1357	2186	374	421	509	178	191	209	99	103	108	.	.	.
	•••	1222	1587	2182	394	456	495	184	199	194	101	105	93	.	.	.
0.8	***	963	1160	1816	310	346	408	140	149	161	.	.	.	.	.	.
	•••	1044	1353	1806	326	371	393	145	154	145	.	.	.	.	.	.
0.85	***	775	930	1380	233	255	291	.	.	.	.	.	.	.	.	.
	•••	839	1073	1366	243	271	275	.	.	.	.	.	.	.	.	.
0.9	***	551	648	879	.	.	.	.	.	.	.	.	.	.	.	.
	•••	592	731	864	.	.	.	.	.	.	.	.	.	.	.	.

TABLE 5: ALPHA= 0.025 POWER= 0.8   EXPECTED ACCRUAL THRU MINIMUM FOLLOW-UP= 450

		DEL=.05			DEL=.10			DEL=.15			DEL=.20			DEL=.25		
FACT=		1.0 .75	.50 .25	.00 BIN	1.0 .75	.50 .25	.00 BIN	1.0 .75	.50 .25	.00 BIN	1.0 75	.50 .25	.00 BIN	1.0 .75	.50 .25	.00 BIN
PCONT=•••					REQUIRED NUMBER OF PATIENTS											
0.05	•••	478	485	501	173	175	178	102	103	104	72	73	73	56	56	57
	•••	481	491	864	174	177	275	102	104	145	72	73	93	56	57	65
0.1	•••	876	896	967	280	287	299	152	155	158	101	102	104	75	75	76
	•••	883	919	1366	283	292	393	153	156	194	102	103	118	75	76	80
0.15	•••	1218	1255	1418	371	384	409	193	199	207	124	127	130	89	91	92
	•••	1232	1301	1806	377	395	495	196	202	236	126	128	140	90	91	93
0.2	•••	1489	1546	1829	443	465	507	227	235	248	143	147	152	101	103	105
	•••	1510	1620	2182	453	482	581	231	241	271	145	149	157	102	104	103
0.25	•••	1686	1767	2188	499	529	591	253	264	282	157	163	170	109	112	116
	•••	1715	1872	2496	512	553	652	258	272	299	159	166	171	111	114	110
0.3	•••	1813	1920	2489	538	577	659	271	286	309	167	174	183	115	119	123
	•••	1852	2059	2748	554	608	707	278	296	320	170	178	181	117	121	115
0.35	•••	1877	2013	2725	562	609	710	283	300	328	173	181	192	118	122	127
	•••	1927	2184	2936	582	648	746	291	312	334	177	186	187	120	125	118
0.4	•••	1887	2052	2896	574	627	744	289	309	340	176	185	196	119	123	129
	•••	1949	2255	3062	597	672	770	298	322	341	180	190	189	121	126	118
0.45	•••	1853	2046	2999	573	633	762	289	310	344	175	184	197	118	122	128
	•••	1926	2277	3124	599	681	778	298	325	341	179	190	187	120	125	115
0.5	•••	1783	2002	3034	563	626	762	283	305	341	171	180	192	114	118	124
	•••	1868	2255	3124	590	677	770	293	320	334	175	186	181	116	121	110
0.55	•••	1688	1927	3000	543	607	745	272	294	329	163	172	184	108	112	117
	•••	1781	2194	3062	571	660	746	282	309	320	167	177	171	110	114	103
0.6	•••	1573	1824	2898	514	578	712	257	278	311	152	160	171	99	103	108
	•••	1672	2098	2936	542	629	707	266	292	299	156	165	157	101	105	93
0.65	•••	1443	1698	2728	477	538	661	236	255	284	138	145	154	89	92	95
	•••	1545	1967	2748	504	585	652	245	268	271	141	149	140	90	93	80
0.7	•••	1301	1549	2491	432	487	593	210	227	250	120	126	133	75	78	81
	•••	1401	1804	2496	456	529	581	218	237	236	123	129	118	77	79	65
0.75	•••	1148	1376	2187	378	424	509	179	192	209	99	103	108	.	.	.
	•••	1241	1605	2182	398	458	495	185	199	194	101	105	93	.	.	.
0.8	•••	979	1177	1816	313	348	408	141	150	161	.	.	.	.	.	.
	•••	1060	1369	1806	329	373	393	145	155	145	.	.	.	.	.	.
0.85	•••	788	942	1380	235	257	291	.	.	.	.	.	.	.	.	.
	•••	852	1084	1366	245	272	275	.	.	.	.	.	.	.	.	.
0.9	•••	559	656	879	.	.	.	.	.	.	.	.	.	.	.	.
	•••	600	737	864	.	.	.	.	.	.	.	.	.	.	.	.

TABLE 5: ALPHA= 0.025 POWER= 0.8   EXPECTED ACCRUAL THRU MINIMUM FOLLOW-UP= 475

	DEL=.05			DEL=.10			DEL=.15			DEL=.20			DEL=.25		
FACT=	1.0 .75	.50 .25	.00 BIN	1.0 .75	.50 .25	.00 BIN	1.0 .75	.50 .25	.00 BIN	1.0 75	.50 .25	.00 BIN	1.0 .75	.50 .25	.00 BIN
PCONT=***				REQUIRED NUMBER OF PATIENTS											
0.05 ***	479	485	502	173	175	178	102	103	104	72	73	73	56	56	57
•••	481	492	864	174	177	275	102	104	145	72	73	93	56	57	65
0.1 ***	877	898	967	281	288	299	152	155	159	101	102	104	74	75	76
•••	885	920	1366	284	292	393	153	156	194	102	103	118	75	76	80
0.15 ***	1220	1258	1418	372	385	409	194	199	206	124	127	130	89	91	92
•••	1234	1305	1806	378	395	495	196	203	236	126	129	140	90	91	93
0.2 ***	1493	1552	1829	445	466	507	227	236	248	143	147	152	101	103	105
•••	1514	1626	2182	454	483	581	231	241	271	145	149	157	102	104	103
0.25 ***	1691	1774	2189	501	531	591	254	265	282	158	163	169	110	112	115
•••	1722	1881	2496	514	555	652	259	273	299	160	166	171	111	114	110
0.3 ***	1820	1930	2488	541	579	659	273	287	309	168	174	183	115	119	123
•••	1860	2070	2748	557	610	707	279	297	320	171	178	181	117	121	115
0.35 ***	1886	2025	2725	566	612	710	285	302	328	174	182	192	119	123	127
•••	1938	2198	2936	586	650	746	292	313	334	178	186	187	121	125	118
0.4 ***	1898	2067	2896	578	631	745	290	310	340	177	185	196	120	124	129
•••	1962	2272	3062	601	675	770	299	323	341	181	190	189	122	126	118
0.45 ***	1866	2063	2999	578	637	762	290	311	344	176	185	197	118	122	128
•••	1941	2295	3124	603	685	778	300	325	341	180	190	187	120	125	115
0.5 ***	1799	2021	3034	568	630	762	285	307	340	171	181	192	114	118	124
•••	1885	2275	3124	595	680	770	295	321	334	175	186	181	116	121	110
0.55 ***	1705	1947	3000	548	612	745	274	296	329	164	172	184	108	112	117
•••	1800	2215	3062	576	663	746	284	311	320	168	178	171	110	115	103
0.6 ***	1591	1845	2898	520	583	712	259	279	311	153	161	171	100	103	108
•••	1692	2119	2936	547	632	707	268	293	299	156	166	157	101	105	93
0.65 ***	1462	1719	2728	482	542	661	238	257	284	138	146	154	89	92	96
•••	1565	1988	2748	509	588	652	246	269	271	142	150	140	90	93	80
0.7 ***	1320	1569	2491	437	490	593	212	228	250	121	127	133	76	78	81
•••	1421	1823	2496	460	531	581	219	238	236	124	130	118	77	79	65
0.75 ***	1165	1394	2186	382	427	509	180	192	209	99	103	108	.	.	.
•••	1258	1622	2182	402	460	495	186	200	194	101	106	93	.	.	.
0.8 ***	994	1192	1816	316	350	408	142	150	161	.	.	.	.	.	.
•••	1076	1383	1806	332	375	393	146	155	145	.	.	.	.	.	.
0.85 ***	800	954	1380	237	258	291	.	.	.	.	.	.	.	.	.
•••	865	1094	1366	247	273	275	.	.	.	.	.	.	.	.	.
0.9 ***	567	663	879	.	.	.	.	.	.	.	.	.	.	.	.
•••	608	742	864	.	.	.	.	.	.	.	.	.	.	.	.

TABLE 5: ALPHA= 0.025 POWER= 0.8   EXPECTED ACCRUAL THRU MINIMUM FOLLOW-UP= 500

		DEL=.05			DEL=.10			DEL=.15			DEL=.20			DEL=.25	
FACT=		1.0 .75	.50 .25	.00 BIN	1.0 .75	.50 .25	.00 BIN	1.0 .75	.50 .25	.00 BIN	1.0 75	.50 .25	.00 BIN	1.0 .75	.50 .25	.00 BIN
PCONT=•••		REQUIRED NUMBER OF PATIENTS														
0.05	***	479	486	502	173	176	178	102	103	104	72	73	73	56	57	57
	•••	482	492	864	174	177	275	103	103	145	73	73	93	56	57	65
0.1	***	878	899	967	282	288	299	153	155	158	101	103	104	75	76	76
	•••	887	922	1366	284	293	393	154	157	194	102	103	118	75	76	80
0.15	***	1223	1262	1418	373	386	409	194	199	207	125	127	130	89	91	93
	•••	1237	1308	1806	379	396	495	197	203	236	126	128	140	90	92	93
0.2	***	1496	1557	1829	447	468	508	228	236	248	143	147	152	101	103	105
	•••	1518	1632	2182	456	484	581	232	242	271	145	149	157	102	104	103
0.25	***	1696	1782	2188	503	533	591	255	266	282	158	163	170	110	113	116
	•••	1728	1889	2496	516	556	652	260	273	299	160	166	171	111	114	110
0.3	***	1827	1940	2488	544	582	659	274	288	309	168	175	183	116	119	123
	•••	1868	2081	2748	560	612	707	280	297	320	171	178	181	117	121	115
0.35	***	1895	2037	2725	570	616	710	286	303	328	175	182	192	119	123	127
	•••	1948	2212	2936	589	652	746	294	314	334	178	187	187	121	125	118
0.4	***	1909	2081	2896	582	635	745	292	311	340	178	186	196	120	124	129
	•••	1974	2287	3062	605	677	770	301	324	341	181	191	189	122	126	118
0.45	***	1878	2080	2999	583	641	762	292	313	344	177	185	197	118	123	128
	•••	1956	2313	3124	608	688	778	302	326	341	181	190	187	121	125	115
0.5	***	1813	2040	3033	573	634	762	287	308	341	172	181	193	115	119	124
	•••	1901	2294	3124	599	684	770	297	322	334	176	186	181	117	121	110
0.55	***	1721	1967	3000	553	616	746	276	297	329	164	173	184	108	113	117
	•••	1818	2235	3062	580	666	746	286	311	320	168	178	171	110	115	103
0.6	***	1608	1865	2898	524	587	712	260	280	311	153	161	172	100	103	108
	•••	1711	2138	2936	552	635	707	270	294	299	157	166	157	102	106	93
0.65	***	1480	1738	2728	487	546	661	239	258	284	139	146	155	89	92	96
	•••	1584	2007	2748	513	591	652	248	269	271	142	150	140	91	94	80
0.7	***	1338	1588	2491	441	494	593	213	228	250	121	127	133	76	78	81
	•••	1439	1841	2496	464	534	581	220	238	236	124	130	118	77	79	65
0.75	***	1181	1412	2186	386	430	509	181	193	209	100	104	108	.	.	.
	•••	1276	1638	2182	405	463	495	187	200	194	102	106	93	.	.	.
0.8	***	1008	1207	1816	319	353	408	143	151	161	.	.	.	.	.	.
	•••	1090	1396	1806	334	376	393	147	155	145	.	.	.	.	.	.
0.85	***	811	965	1380	239	260	291	.	.	.	.	.	.	.	.	.
	•••	876	1104	1366	248	273	275	.	.	.	.	.	.	.	.	.
0.9	***	574	669	879	.	.	.	.	.	.	.	.	.	.	.	.
	•••	615	747	864	.	.	.	.	.	.	.	.	.	.	.	.

TABLE 5: ALPHA= 0.025 POWER= 0.8   EXPECTED ACCRUAL THRU MINIMUM FOLLOW-UP= 550

		DEL=.02			DEL=.05			DEL=.10			DEL=.15			DEL=.20		
FACT=		1.0 .75	.50 .25	.00 BIN	1.0 .75	.50 .25	.00 BIN	1.0 .75	.50 .25	.00 BIN	1.0 75	.50 .25	.00 BIN	1.0 .75	.50 .25	.00 BIN
PCONT=•••				REQUIRED	NUMBER	OF PATIENTS										
0.05	***	2237 2248 2365			480 487 501			174 176 178			102 103 104			72 73 73		
	•••	2241 2269 4419			483 492 864			175 177 275			103 103 145			72 73 93		
0.1	***	4654 4681 5195			881 903 967			283 289 299			153 155 158			101 103 104		
	•••	4664 4735 7677			890 925 1366			285 293 393			154 157 194			102 103 118		
0.15	***	6801 6849 8034			1227 1268 1418			375 388 409			195 200 206			125 127 130		
	•••	6817 6944 10542			1243 1314 1806			380 397 495			197 203 236			126 129 140		
0.2	***	8512 8586 10675			1503 1567 1829			450 470 507			230 237 248			144 148 152		
	•••	8537 8733 13014			1527 1642 2182			459 486 581			233 242 271			146 150 157		
0.25	***	9743 9849 13016			1706 1796 2189			508 536 591			256 267 282			159 164 169		
	•••	9779 10060 15094			1739 1904 2496			520 558 652			261 274 299			161 166 171		
0.3	***	10502 10647 15001			1840 1958 2489			549 587 659			276 289 309			169 175 183		
	•••	10550 10936 16781			1884 2101 2748			565 616 707			282 298 320			172 179 181		
0.35	***	10822 11015 16596			1911 2060 2725			576 621 710			289 305 328			176 183 192		
	•••	10886 11398 18076			1968 2236 2936			596 657 746			296 315 334			179 187 187		
0.4	***	10752 11005 17784			1929 2109 2896			590 641 745			295 313 340			179 186 196		
	•••	10836 11500 18979			1998 2316 3062			612 682 770			303 325 341			182 191 189		
0.45	***	10351 10676 18553			1903 2111 2999			591 648 762			295 315 344			178 186 197		
	•••	10460 11303 19489			1983 2345 3124			616 693 778			304 328 341			182 191 187		
0.5	***	9683 10097 18898			1841 2074 3033			582 642 762			290 310 341			173 182 192		
	•••	9822 10871 19607			1933 2329 3124			608 689 770			299 324 334			177 187 181		
0.55	***	8816 9339 18815			1752 2003 3000			562 624 745			279 300 329			166 174 184		
	•••	8993 10266 19332			1851 2271 3062			589 672 746			289 313 320			170 179 171		
0.6	***	7824 8472 18305			1641 1903 2898			533 594 712			264 283 310			155 162 171		
	•••	8046 9540 18665			1746 2175 2936			560 641 707			272 295 299			158 166 157		
0.65	***	6780 7557 17370			1513 1776 2728			496 553 661			242 259 284			140 147 155		
	•••	7055 8732 17606			1619 2043 2748			521 596 652			250 270 271			143 150 140		
0.7	***	5758 6639 16012			1370 1623 2491			449 500 593			215 230 250			122 127 133		
	•••	6080 7867 16153			1473 1874 2496			472 538 581			222 239 236			125 130 118		
0.75	***	4806 5738 14235			1212 1444 2186			392 435 509			183 194 209			100 104 108		
	•••	5156 6951 14309			1307 1668 2182			411 466 495			188 201 194			102 106 93		
0.8	***	3938 4848 12043			1035 1234 1816			324 356 408			144 151 161			. . .		
	•••	4288 5975 12072			1118 1419 1806			339 378 393			147 156 145			. . .		
0.85	***	3131 3942 9441			833 985 1380			242 262 291			. . .			. . .		
	•••	3447 4905 9443			897 1121 1366			251 275 275			. . .			. . .		
0.9	***	2325 2955 6432			588 681 879			. . .			. . .			. . .		
	•••	2574 3668 6421			628 756 864			. . .			. . .			. . .		
0.95	***	1383 1727 3022			. . .			. . .			. . .			. . .		
	•••	1522 2076 3007			. . .			. . .			. . .			. . .		

TABLE 5: ALPHA= 0.025 POWER= 0.8   EXPECTED ACCRUAL THRU MINIMUM FOLLOW-UP= 600

	DEL=.02			DEL=.05			DEL=.10			DEL=.15			DEL=.20		
FACT=	1.0 .75	.50 .25	.00 BIN	1.0 .75	.50 .25	.00 BIN	1.0 .75	.50 .25	.00 BIN	1.0 75	.50 .25	.00 BIN	1.0 .75	.50 .25	.00 BIN

PCONT=***   REQUIRED NUMBER OF PATIENTS

PCONT	DEL=.02 1.0/.75	.50/.25	.00/BIN	DEL=.05 1.0/.75	.50/.25	.00/BIN	DEL=.10 1.0/.75	.50/.25	.00/BIN	DEL=.15 1.0/.75	.50/.25	.00/BIN	DEL=.20 1.0/.75	.50/.25	.00/BIN
0.05 ***	2238	2250	2365	481	488	502	174	176	178	102	103	104	72	73	73
***	2242	2272	4419	484	493	864	175	177	275	103	104	145	73	73	93
0.1 ***	4657	4687	5195	883	905	967	283	289	299	153	155	158	101	103	104
***	4667	4744	7677	892	927	1366	286	293	393	154	157	194	102	103	118
0.15 ***	6805	6857	8034	1232	1274	1418	377	389	409	196	200	206	125	127	130
***	6823	6961	10542	1248	1320	1806	382	398	495	198	203	236	126	129	140
0.2 ***	8519	8599	10675	1510	1576	1829	452	473	507	230	238	248	145	148	152
***	8546	8759	13014	1535	1652	2182	461	487	581	234	242	271	146	150	157
0.25 ***	9753	9868	13016	1715	1809	2188	511	540	591	258	268	282	160	164	169
***	9791	10098	15094	1751	1918	2496	524	561	652	262	274	299	162	167	171
0.3 ***	10515	10673	15000	1852	1975	2488	554	591	659	278	290	309	170	176	183
***	10568	10987	16781	1899	2118	2748	570	618	707	284	299	320	173	179	181
0.35 ***	10839	11050	16596	1927	2081	2725	582	626	710	291	306	328	177	184	192
***	10910	11466	18076	1986	2258	2936	601	660	746	298	316	334	180	187	187
0.4 ***	10775	11050	17784	1949	2134	2896	596	647	745	298	315	340	180	187	196
***	10867	11587	18979	2020	2342	3062	619	686	770	305	326	341	183	191	189
0.45 ***	10381	10735	18553	1926	2140	2999	599	654	762	298	317	344	179	187	197
***	10499	11410	19489	2009	2374	3124	623	697	778	307	329	341	183	191	187
0.5 ***	9721	10172	18898	1868	2105	3034	590	649	762	293	312	340	175	182	193
***	9872	10999	19607	1962	2360	3124	616	694	770	302	325	334	178	187	181
0.55 ***	8864	9431	18815	1781	2036	3000	571	631	745	282	301	329	167	175	184
***	9056	10414	19332	1883	2304	3062	597	677	746	291	314	320	170	179	171
0.6 ***	7885	8582	18305	1672	1937	2898	542	601	712	266	284	310	156	163	171
***	8126	9703	18665	1779	2208	2936	568	646	707	275	296	299	159	167	157
0.65 ***	6857	7683	17370	1545	1810	2728	503	559	661	245	261	284	141	147	154
***	7151	8905	17606	1652	2075	2748	528	601	652	252	272	271	144	151	140
0.7 ***	5849	6775	16012	1401	1655	2491	456	506	593	218	232	250	123	128	133
***	6190	8042	16153	1505	1904	2496	478	542	581	224	240	236	125	130	118
0.75 ***	4906	5876	14235	1241	1473	2186	398	440	509	185	195	209	101	104	108
***	5273	7121	14309	1336	1694	2182	417	469	495	190	202	194	103	106	93
0.8 ***	4040	4979	12043	1060	1259	1816	329	360	408	145	152	161	.	.	.
***	4402	6129	12072	1143	1440	1806	343	380	393	148	156	145	.	.	.
0.85 ***	3223	4055	9440	852	1004	1380	245	264	290	.	.	.	.	.	.
***	3548	5033	9443	916	1135	1366	254	276	275	.	.	.	.	.	.
0.9 ***	2398	3041	6432	600	692	879	.	.	.	.	.	.	.	.	.
***	2653	3760	6421	640	764	864	.	.	.	.	.	.	.	.	.
0.95 ***	1424	1771	3022	.	.	.	.	.	.	.	.	.	.	.	.
***	1565	2119	3007	.	.	.	.	.	.	.	.	.	.	.	.

TABLE 5: ALPHA= 0.025 POWER= 0.8   EXPECTED ACCRUAL THRU MINIMUM FOLLOW-UP= 650

	DEL=.02			DEL=.05			DEL=.10			DEL=.15			DEL=.20		
FACT=	1.0 .75	.50 .25	.00 BIN	1.0 .75	.50 .25	.00 BIN	1.0 .75	.50 .25	.00 BIN	1.0 75	.50 .25	.00 BIN	1.0 .75	.50 .25	.00 BIN
PCONT=•••			REQUIRED NUMBER OF PATIENTS												
0.05 •••	2239	2252	2365	482	489	502	175	176	178	103	103	104	73	73	73
•••	2244	2275	4419	484	494	864	175	177	275	103	103	145	73	73	93
0.1 •••	4660	4692	5195	886	908	967	284	290	299	153	156	159	102	103	104
•••	4670	4753	7677	895	929	1366	287	294	393	155	157	194	102	103	118
0.15 •••	6809	6866	8034	1236	1279	1418	378	390	409	197	201	207	126	128	130
•••	6828	6978	10542	1253	1325	1806	384	399	495	198	203	236	127	129	140
0.2 •••	8525	8613	10675	1516	1585	1829	455	475	507	232	239	248	145	148	152
•••	8555	8785	13014	1542	1660	2182	464	489	581	235	243	271	146	150	157
0.25 •••	9762	9888	13016	1725	1821	2188	515	543	591	259	269	282	160	164	170
•••	9804	10135	15094	1761	1930	2496	527	563	652	263	275	299	162	167	171
0.3 •••	10528	10700	15000	1865	1992	2489	559	595	659	280	292	309	171	176	183
•••	10585	11039	16781	1913	2135	2748	575	621	707	285	300	320	174	179	181
0.35 •••	10857	11086	16596	1943	2101	2725	588	631	710	293	307	328	178	184	192
•••	10933	11533	18076	2004	2278	2936	606	663	746	300	317	334	181	188	187
0.4 •••	10798	11096	17784	1968	2157	2896	603	652	744	300	316	340	181	188	197
•••	10897	11671	18979	2042	2365	3062	625	690	770	307	327	341	184	192	189
0.45 •••	10410	10793	18553	1949	2166	2999	606	660	762	301	318	344	180	188	197
•••	10538	11515	19489	2034	2400	3124	630	701	778	309	330	341	184	192	187
0.5 •••	9759	10246	18898	1893	2135	3033	597	655	762	296	314	341	176	184	192
•••	9922	11123	19607	1989	2389	3124	622	698	770	304	326	334	179	188	181
0.55 •••	8913	9522	18815	1809	2068	3000	578	637	745	285	303	329	168	175	184
•••	9120	10555	19332	1913	2334	3062	604	681	746	294	315	320	172	179	171
0.6 •••	7946	8690	18305	1701	1969	2898	549	607	712	269	286	311	157	164	171
•••	8205	9859	18665	1810	2238	2936	575	650	707	277	297	299	160	167	157
0.65 •••	6932	7805	17370	1575	1841	2728	510	565	661	247	263	284	142	148	155
•••	7245	9069	17606	1683	2104	2748	535	604	652	254	272	271	145	151	140
0.7 •••	5937	6905	16012	1430	1685	2491	463	511	593	220	233	250	124	128	133
•••	6295	8208	16153	1535	1931	2496	484	545	581	226	241	236	126	131	118
0.75 •••	5003	6007	14235	1267	1500	2186	404	444	509	186	197	209	101	105	108
•••	5384	7279	14309	1364	1717	2182	422	471	495	191	202	194	103	106	93
0.8 •••	4136	5102	12043	1083	1281	1816	333	363	408	146	153	161	.	.	.
•••	4509	6272	12072	1166	1459	1806	346	383	393	149	157	145	.	.	.
0.85 •••	3311	4162	9441	870	1020	1380	248	266	291	.	.	.	.	.	.
•••	3644	5153	9443	934	1149	1366	256	277	275	.	.	.	.	.	.
0.9 •••	2467	3121	6432	612	701	879	.	.	.	.	.	.	.	.	.
•••	2727	3846	6421	651	771	864	.	.	.	.	.	.	.	.	.
0.95 •••	1463	1812	3022	.	.	.	.	.	.	.	.	.	.	.	.
•••	1606	2157	3007	.	.	.	.	.	.	.	.	.	.	.	.

TABLE 5: ALPHA= 0.025 POWER= 0.8   EXPECTED ACCRUAL THRU MINIMUM FOLLOW-UP= 700

	DEL=.02			DEL=.05			DEL=.10			DEL=.15			DEL=.20		
FACT=	1.0 .75	.50 .25	.00 BIN	1.0 .75	.50 .25	.00 BIN	1.0 .75	.50 .25	.00 BIN	1.0 75	.50 .25	.00 BIN	1.0 .75	.50 .25	.00 BIN
PCONT=***	REQUIRED NUMBER OF PATIENTS														
0.05 ***	2240 2254	2364	482 489	502	175 176	178	103 103	104	72 73	73
•••	2245 2278	4419	485 494	864	175 177	275	103 103	145	73 73	93

TABLE 5: ALPHA= 0.025 POWER= 0.8   EXPECTED ACCRUAL THRU MINIMUM FOLLOW-UP= 750

		DEL=.02			DEL=.05			DEL=.10			DEL=.15			DEL=.20		
FACT=		1.0 .75	.50 .25	.00 BIN	1.0 .75	.50 .25	.00 BIN	1.0 .75	.50 .25	.00 BIN	1.0 75	.50 .25	.00 BIN	1.0 .75	.50 .25	.00 BIN
PCONT=***					REQUIRED NUMBER OF PATIENTS											
0.05	***	2241	2256	2365	483	490	502	175	176	178	103	103	104	72	73	73
	•••	2246	2281	4419	486	495	864	175	177	275	103	104	145	73	73	93
0.1	***	4664	4701	5195	890	913	967	286	291	299	154	156	159	102	103	104
	•••	4676	4771	7677	899	933	1366	288	295	393	155	157	194	102	103	118
0.15	***	6818	6883	8034	1244	1289	1418	381	392	409	198	201	206	126	128	130
	•••	6840	7010	10542	1262	1333	1806	386	400	495	199	204	236	127	129	140
0.2	***	8539	8640	10675	1529	1600	1829	460	478	507	233	240	248	146	149	152
	•••	8573	8837	13014	1557	1675	2182	468	491	581	236	244	271	147	150	157
0.25	***	9782	9926	13015	1742	1843	2188	521	547	591	261	270	282	161	165	169
	•••	9830	10209	15094	1781	1951	2496	533	566	652	265	276	299	163	167	171
0.3	***	10555	10753	15000	1888	2021	2488	567	601	659	282	294	309	172	177	183
	•••	10621	11139	16781	1940	2163	2748	582	625	707	288	301	320	175	180	181
0.35	***	10892	11155	16596	1972	2138	2725	597	638	710	296	310	328	179	185	192
	•••	10980	11664	18076	2037	2314	2936	616	668	746	303	319	334	182	188	187
0.4	***	10844	11187	17784	2004	2200	2896	614	661	745	304	319	340	183	189	196
	•••	10959	11836	18979	2081	2407	3062	635	696	770	311	329	341	185	192	189
0.45	***	10470	10910	18553	1990	2215	2999	618	670	762	305	321	344	182	189	197
	•••	10617	11715	19489	2080	2447	3124	641	708	778	312	332	341	185	192	187
0.5	***	9834	10391	18897	1940	2188	3034	610	665	762	300	317	340	178	184	192
	•••	10023	11359	19607	2040	2439	3124	634	705	770	308	328	334	181	188	181
0.55	***	9009	9699	18815	1860	2123	3000	591	647	745	289	306	329	170	176	184
	•••	9246	10822	19332	1967	2386	3062	616	688	746	297	317	320	173	180	171
0.6	***	8066	8896	18305	1754	2026	2898	562	617	711	273	289	310	159	164	171
	•••	8359	10150	18665	1865	2290	2936	587	656	707	280	299	299	161	168	157
0.65	***	7079	8035	17370	1628	1897	2728	523	574	661	251	265	284	144	148	154
	•••	7427	9373	17606	1738	2154	2748	546	611	652	258	274	271	146	151	140
0.7	***	6108	7149	16012	1481	1737	2491	473	519	593	223	235	250	125	129	133
	•••	6497	8511	16153	1588	1977	2496	494	550	581	229	242	236	127	131	118
0.75	***	5186	6249	14235	1315	1547	2186	413	450	509	189	198	209	102	105	108
	•••	5592	7569	14309	1412	1758	2182	430	476	495	193	203	194	104	107	93
0.8	***	4317	5329	12043	1124	1320	1816	340	368	408	148	154	160	.	.	.
	•••	4710	6534	12072	1207	1491	1806	353	385	393	151	157	145	.	.	.
0.85	***	3473	4357	9440	902	1049	1380	252	268	291	.	.	.	.	.	.
	•••	3820	5369	9443	965	1171	1366	259	279	275	.	.	.	.	.	.
0.9	***	2594	3267	6432	631	717	879	.	.	.	.	.	.	.	.	.
	•••	2862	3998	6421	669	783	864	.	.	.	.	.	.	.	.	.
0.95	***	1534	1885	3022	.	.	.	.	.	.	.	.	.	.	.	.
	•••	1678	2224	3007	.	.	.	.	.	.	.	.	.	.	.	.

TABLE 5: ALPHA= 0.025 POWER= 0.8   EXPECTED ACCRUAL THRU MINIMUM FOLLOW-UP= 800

		DEL=.02			DEL=.05			DEL=.10			DEL=.15			DEL=.20		
FACT=		1.0 .75	.50 .25	.00 BIN	1.0 .75	.50 .25	.00 BIN	1.0 .75	.50 .25	.00 BIN	1.0 75	.50 .25	.00 BIN	1.0 .75	.50 .25	.00 BIN
PCONT=•••					REQUIRED NUMBER OF PATIENTS											
0.05	•••	2242	2258	2365	484	490	502	175	177	178	103	103	104	73	73	73
	•••	2248	2283	4419	487	495	864	176	177	275	103	104	145	73	73	93
0.1	•••	4667	4706	5195	892	915	967	286	292	299	154	156	159	102	103	104
	•••	4680	4780	7677	901	935	1366	289	295	393	155	157	194	103	104	118
0.15	•••	6823	6892	8034	1248	1293	1418	382	393	409	198	202	207	126	128	130
	•••	6846	7026	10542	1266	1337	1806	387	400	495	200	204	236	127	129	140
0.2	•••	8546	8653	10675	1535	1607	1829	462	479	507	234	240	248	146	149	152
	•••	8582	8862	13014	1563	1681	2182	470	492	581	237	244	271	148	150	157
0.25	•••	9791	9945	13016	1751	1854	2188	524	550	591	263	271	282	162	165	170
	•••	9843	10245	15094	1791	1961	2496	535	567	652	267	276	299	163	167	171
0.3	•••	10568	10779	15000	1899	2034	2489	570	603	659	284	295	309	173	177	183
	•••	10638	11188	16781	1952	2176	2748	585	627	707	289	301	320	175	180	181
0.35	•••	10910	11191	16596	1986	2154	2725	601	642	710	298	311	328	180	185	192
	•••	11003	11727	18076	2053	2329	2936	619	671	746	304	319	334	183	188	187
0.4	•••	10867	11233	17784	2020	2220	2896	619	665	745	306	320	340	183	189	196
	•••	10989	11916	18979	2100	2425	3062	639	698	770	312	329	341	186	193	189
0.45	•••	10499	10968	18553	2010	2237	2999	623	674	762	307	323	344	183	189	197
	•••	10657	11811	19489	2101	2467	3124	646	711	778	314	332	341	186	193	187
0.5	•••	9872	10462	18898	1962	2212	3034	616	669	762	302	318	341	179	185	193
	•••	10073	11471	19607	2063	2461	3124	640	708	770	310	329	334	182	189	181
0.55	•••	9056	9785	18815	1883	2148	3000	597	652	746	291	307	329	171	177	184
	•••	9308	10948	19332	1991	2409	3062	621	691	746	299	318	320	174	180	171
0.6	•••	8126	8996	18305	1779	2051	2898	568	621	712	275	290	311	159	165	171
	•••	8435	10285	18665	1891	2313	2936	592	659	707	282	300	299	162	168	157
0.65	•••	7151	8145	17370	1652	1922	2728	528	578	661	252	266	284	144	149	155
	•••	7514	9514	17606	1764	2176	2748	551	613	652	259	275	271	146	152	140
0.7	•••	6190	7263	16012	1505	1761	2491	478	523	593	224	236	250	125	129	133
	•••	6593	8652	16153	1612	1998	2496	498	553	581	230	243	236	127	131	118
0.75	•••	5273	6362	14235	1337	1568	2186	417	453	509	190	199	209	103	105	108
	•••	5691	7702	14309	1434	1775	2182	434	478	495	194	204	194	104	107	93
0.8	•••	4402	5433	12043	1143	1337	1816	343	370	408	149	154	161	.	.	.
	•••	4803	6653	12072	1225	1506	1806	355	387	393	151	157	145	.	.	.
0.85	•••	3549	4447	9441	917	1062	1380	254	270	291	.	.	.	.	.	.
	•••	3903	5467	9443	979	1181	1366	261	279	275	.	.	.	.	.	.
0.9	•••	2653	3333	6432	640	725	879	.	.	.	.	.	.	.	.	.
	•••	2925	4067	6421	677	787	864	.	.	.	.	.	.	.	.	.
0.95	•••	1566	1918	3022	.	.	.	.	.	.	.	.	.	.	.	.
	•••	1711	2254	3007	.	.	.	.	.	.	.	.	.	.	.	.

TABLE 5: ALPHA= 0.025 POWER= 0.8   EXPECTED ACCRUAL THRU MINIMUM FOLLOW-UP= 850

		DEL=.02			DEL=.05			DEL=.10			DEL=.15			DEL=.20		
FACT=		1.0 .75	.50 .25	.00 BIN	1.0 .75	.50 .25	.00 BIN	1.0 .75	.50 .25	.00 BIN	1.0 75	.50 .25	.00 BIN	1.0 .75	.50 .25	.00 BIN

PCONT=•••   REQUIRED NUMBER OF PATIENTS

PCONT		.02 A	.02 B	.02 C	.05 A	.05 B	.05 C	.10 A	.10 B	.10 C	.15 A	.15 B	.15 C	.20 A	.20 B	.20 C
0.05	***	2243	2260	2365	484	491	502	175	176	178	103	103	104	72	73	73
	•••	2249	2286	4419	487	495	864	176	178	275	103	104	145	73	73	93
0.1	***	4669	4711	5195	894	917	967	287	292	299	155	156	158	102	103	104
	•••	4683	4788	7677	903	936	1366	289	295	393	155	157	194	103	103	118
0.15	***	6827	6901	8034	1251	1297	1417	383	394	409	198	202	206	127	128	130
	•••	6852	7042	10542	1270	1341	1806	388	401	495	200	204	236	128	129	140
0.2	***	8552	8667	10675	1541	1614	1829	463	481	508	235	240	248	147	149	152
	•••	8590	8887	13014	1570	1687	2182	471	493	581	238	244	271	148	151	157
0.25	***	9801	9965	13016	1759	1863	2188	527	552	591	263	272	282	162	165	170
	•••	9855	10281	15094	1800	1970	2496	538	569	652	267	276	299	164	168	171
0.3	***	10581	10805	15000	1909	2047	2488	573	606	658	285	295	309	173	178	183
	•••	10656	11236	16781	1964	2187	2748	588	629	707	290	301	320	175	180	181
0.35	***	10928	11225	16596	2000	2170	2725	605	645	710	299	311	328	181	186	192
	•••	11027	11790	18076	2067	2344	2936	623	672	746	305	320	334	183	189	187
0.4	***	10890	11278	17784	2037	2238	2895	623	668	745	307	321	340	184	190	196
	•••	11020	11994	18979	2117	2442	3062	644	700	770	314	329	341	187	193	189
0.45	***	10528	11025	18553	2028	2258	2998	628	678	762	308	324	344	184	189	197
	•••	10696	11904	19489	2121	2486	3124	650	713	778	316	333	341	186	193	187
0.5	***	9910	10533	18898	1982	2234	3034	621	673	763	304	320	341	179	185	192
	•••	10122	11579	19607	2085	2481	3124	644	711	770	311	329	334	182	189	181
0.55	***	9104	9869	18815	1905	2172	3000	602	656	746	293	308	329	171	177	184
	•••	9370	11069	19332	2015	2430	3062	627	694	746	300	318	320	174	180	171
0.6	***	8185	9093	18305	1802	2075	2898	573	626	712	276	291	310	159	165	171
	•••	8509	10415	18665	1914	2334	2936	597	662	707	283	300	299	162	168	157
0.65	***	7222	8250	17370	1676	1945	2728	533	582	661	254	267	284	145	149	155
	•••	7600	9648	17606	1787	2197	2748	555	616	652	260	275	271	147	152	140
0.7	***	6270	7373	16012	1528	1783	2491	482	526	593	225	237	250	125	129	134
	•••	6685	8785	16153	1634	2016	2496	502	555	581	231	243	236	127	131	118
0.75	***	5357	6470	14235	1357	1587	2186	420	456	509	191	199	209	103	105	108
	•••	5785	7829	14309	1454	1791	2182	437	479	495	195	204	194	104	107	93
0.8	***	4483	5534	12043	1160	1353	1816	345	372	408	149	154	161	.	.	.
	•••	4893	6766	12072	1243	1518	1806	358	388	393	152	157	145	.	.	.
0.85	***	3621	4532	9440	930	1073	1380	255	271	291	.	.	.	.	.	.
	•••	3981	5560	9443	992	1189	1366	263	280	275	.	.	.	.	.	.
0.9	***	2708	3397	6432	648	731	879	.	.	.	.	.	.	.	.	.
	•••	2984	4131	6421	685	792	864	.	.	.	.	.	.	.	.	.
0.95	***	1596	1948	3022	.	.	.	.	.	.	.	.	.	.	.	.
	•••	1742	2281	3007	.	.	.	.	.	.	.	.	.	.	.	.

TABLE 5: ALPHA= 0.025 POWER= 0.8   EXPECTED ACCRUAL THRU MINIMUM FOLLOW-UP= 900

	DEL=.02			DEL=.05			DEL=.10			DEL=.15			DEL=.20		
FACT=	1.0 .75	.50 .25	.00 BIN	1.0 .75	.50 .25	.00 BIN	1.0 .75	.50 .25	.00 BIN	1.0 75	.50 .25	.00 BIN	1.0 .75	.50 .25	.00 BIN

PCONT=***   REQUIRED NUMBER OF PATIENTS

PCONT	DEL=.02 1.0/.75	.50/.25	.00/BIN	DEL=.05 1.0/.75	.50/.25	.00/BIN	DEL=.10 1.0/.75	.50/.25	.00/BIN	DEL=.15 1.0/.75	.50/.25	.00/BIN	DEL=.20 1.0/.75	.50/.25	.00/BIN
0.05 ***	2244	2262	2364	485	491	501	175	177	179	103	104	104	73	73	73
***	2250	2288	4419	488	496	864	176	177	275	103	104	145	73	73	93
0.1 ***	4672	4716	5195	896	919	967	287	292	299	155	156	158	102	103	104
***	4686	4796	7677	905	938	1366	289	295	393	155	158	194	102	104	118
0.15 ***	6831	6909	8034	1255	1301	1418	384	395	409	199	202	207	127	128	130
***	6858	7057	10542	1274	1343	1806	389	401	495	200	204	236	127	129	140
0.2 ***	8559	8679	10675	1546	1620	1829	465	482	507	235	241	248	147	149	152
***	8599	8911	13014	1576	1693	2182	473	494	581	238	244	271	148	150	157
0.25 ***	9811	9984	13016	1767	1872	2188	529	553	591	264	272	282	163	165	170
***	9869	10316	15094	1809	1977	2496	540	569	652	268	277	299	164	168	171
0.3 ***	10594	10832	15000	1920	2059	2489	577	608	659	286	296	309	174	178	183
***	10674	11283	16781	1976	2198	2748	591	630	707	290	302	320	176	180	181
0.35 ***	10945	11260	16596	2013	2184	2725	609	648	710	300	312	328	181	186	192
***	11051	11851	18076	2081	2357	2936	626	674	746	306	320	334	183	189	187
0.4 ***	10913	11323	17784	2052	2255	2896	627	672	744	308	322	340	185	190	197
***	11051	12069	18979	2134	2457	3062	647	703	770	315	330	341	187	193	189
0.45 ***	10558	11081	18553	2046	2277	2999	632	681	762	310	325	344	184	190	197
***	10735	11996	19489	2140	2504	3124	654	716	778	317	334	341	187	193	187
0.5 ***	9947	10603	18897	2002	2255	3034	626	677	762	305	320	341	180	186	192
***	10172	11684	19607	2106	2501	3124	649	713	770	312	330	334	182	189	181
0.55 ***	9152	9951	18815	1927	2194	3000	607	659	746	294	309	329	172	177	184
***	9431	11186	19332	2036	2449	3062	631	696	746	302	318	320	174	181	171
0.6 ***	8244	9187	18305	1824	2098	2898	578	629	712	278	292	311	161	165	171
***	8582	10540	18665	1937	2354	2936	601	665	707	285	300	299	163	168	157
0.65 ***	7291	8353	17370	1698	1967	2728	538	585	660	255	268	284	145	149	154
***	7683	9776	17606	1810	2216	2748	559	618	652	261	275	271	147	152	140
0.7 ***	6347	7479	16012	1549	1803	2491	487	529	593	227	237	251	126	129	134
***	6776	8912	16153	1656	2034	2496	506	557	581	231	243	236	128	131	118
0.75 ***	5438	6574	14235	1376	1605	2187	424	458	509	191	199	209	103	105	108
***	5876	7948	14309	1473	1806	2182	440	480	495	195	204	194	104	107	93
0.8 ***	4562	5629	12043	1177	1369	1816	348	373	408	150	155	161	.	.	.
***	4979	6872	12072	1259	1530	1806	360	389	393	152	158	145	.	.	.
0.85 ***	3690	4613	9441	942	1084	1380	257	272	290	.	.	.	.	.	.
***	4055	5647	9443	1004	1197	1366	264	280	275	.	.	.	.	.	.
0.9 ***	2762	3456	6432	656	737	879	.	.	.	.	.	.	.	.	.
***	3041	4191	6421	692	795	864	.	.	.	.	.	.	.	.	.
0.95 ***	1625	1977	3022	.	.	.	.	.	.	.	.	.	.	.	.
***	1771	2306	3007	.	.	.	.	.	.	.	.	.	.	.	.

TABLE 5: ALPHA= 0.025 POWER= 0.8   EXPECTED ACCRUAL THRU MINIMUM FOLLOW-UP= 950

		DEL=.02			DEL=.05			DEL=.10			DEL=.15			DEL=.20		
FACT=		1.0 .75	.50 .25	.00 BIN	1.0 .75	.50 .25	.00 BIN	1.0 .75	.50 .25	.00 BIN	1.0 75	.50 .25	.00 BIN	1.0 .75	.50 .25	.00 BIN
PCONT=•••						REQUIRED NUMBER OF PATIENTS											
0.05	***	2245	2263	2364	485	492	502	175	177	178	103	103	104	72	73	73
	•••	2252	2290	4419	488	496	864	176	178	275	103	103	145	72	73	93
0.1	***	4674	4721	5195	898	920	967	287	292	299	155	156	159	102	103	104
	•••	4689	4803	7677	907	939	1366	290	295	393	156	158	194	103	103	118
0.15	***	6835	6918	8034	1258	1305	1417	386	395	409	199	203	206	127	128	130
	•••	6863	7072	10542	1277	1347	1806	390	401	495	201	204	236	128	129	140
0.2	***	8565	8693	10674	1552	1626	1829	466	483	507	236	241	248	147	149	152
	•••	8608	8935	13014	1582	1698	2182	474	494	581	238	244	271	148	150	157
0.25	***	9820	10003	13016	1774	1881	2189	531	555	591	265	273	282	163	166	169
	•••	9881	10350	15094	1817	1985	2496	542	571	652	268	277	299	164	167	171
0.3	***	10608	10858	15001	1930	2070	2489	579	610	659	287	296	309	174	178	183
	•••	10691	11330	16781	1986	2208	2748	593	631	707	292	302	320	176	181	181
0.35	***	10963	11295	16596	2025	2198	2725	612	650	710	302	313	329	182	186	192
	•••	11073	11911	18076	2095	2369	2936	629	676	746	307	320	334	184	189	187
0.4	***	10936	11368	17784	2067	2272	2896	631	675	745	310	323	340	185	190	196
	•••	11081	12143	18979	2149	2472	3062	651	704	770	316	331	341	188	193	189
0.45	***	10588	11138	18553	2063	2295	2998	637	685	762	311	325	344	185	190	197
	•••	10774	12084	19489	2158	2520	3124	658	717	778	318	334	341	187	193	187
0.5	***	9985	10672	18897	2021	2275	3034	630	680	762	306	321	340	181	186	192
	•••	10221	11786	19607	2125	2518	3124	653	716	770	314	330	334	183	189	181
0.55	***	9199	10033	18815	1947	2215	3000	612	663	745	296	311	329	172	178	184
	•••	9492	11297	19332	2058	2468	3062	635	698	746	303	319	320	175	181	171
0.6	***	8301	9279	18305	1845	2119	2898	583	633	711	279	293	311	161	166	171
	•••	8654	10660	18665	1958	2373	2936	605	666	707	286	301	299	163	168	157
0.65	***	7360	8452	17370	1718	1988	2728	542	589	660	257	268	284	146	150	154
	•••	7765	9900	17606	1831	2233	2748	563	620	652	262	276	271	147	152	140
0.7	***	6423	7581	16012	1569	1823	2491	490	532	593	228	238	250	127	129	133
	•••	6863	9033	16153	1675	2050	2496	509	558	581	232	243	236	128	131	118
0.75	***	5516	6674	14235	1394	1622	2186	427	460	508	192	200	209	103	106	108
	•••	5964	8063	14309	1491	1820	2182	443	482	495	196	204	194	105	107	93
0.8	***	4637	5721	12043	1192	1382	1816	350	375	408	150	155	160	.	.	.
	•••	5062	6974	12072	1274	1541	1806	362	390	393	153	158	145	.	.	.
0.85	***	3757	4691	9441	954	1094	1380	258	273	291	.	.	.	.	.	.
	•••	4127	5730	9443	1015	1205	1366	265	281	275	.	.	.	.	.	.
0.9	***	2813	3513	6432	663	742	879	.	.	.	.	.	.	.	.	.
	•••	3095	4248	6421	698	799	864	.	.	.	.	.	.	.	.	.
0.95	***	1652	2004	3022	.	.	.	.	.	.	.	.	.	.	.	.
	•••	1799	2329	3007	.	.	.	.	.	.	.	.	.	.	.	.

TABLE 5: ALPHA= 0.025 POWER= 0.8   EXPECTED ACCRUAL THRU MINIMUM FOLLOW-UP= 1000

	DEL=.02			DEL=.05			DEL=.10			DEL=.15			DEL=.20		
FACT=	1.0 .75	.50 .25	.00 BIN	1.0 .75	.50 .25	.00 BIN	1.0 .75	.50 .25	.00 BIN	1.0 75	.50 .25	.00 BIN	1.0 .75	.50 .25	.00 BIN
PCONT=***					REQUIRED NUMBER OF PATIENTS										
0.05 ***	2246	2266	2365	486	492	501	176	177	178	103	103	104	73	73	73
•••	2253	2293	4419	489	496	864	176	178	275	103	104	145	73	73	93
0.1 ***	4676	4726	5195	900	922	967	288	293	299	155	156	158	103	103	104
•••	4693	4811	7677	909	940	1366	290	296	393	156	158	194	103	104	118
0.15 ***	6840	6926	8034	1261	1308	1418	386	396	410	200	203	206	127	128	130
•••	6869	7088	10542	1281	1349	1806	391	402	495	201	205	236	128	130	140
0.2 ***	8573	8706	10675	1557	1632	1829	468	484	508	236	241	248	147	150	152
•••	8617	8959	13014	1587	1702	2182	475	495	581	239	245	271	148	151	157
0.25 ***	9830	10022	13016	1781	1889	2188	533	556	591	266	273	282	163	166	170
•••	9894	10385	15094	1825	1992	2496	543	571	652	269	277	299	165	168	171
0.3 ***	10621	10884	15000	1940	2081	2488	582	612	659	288	297	309	175	178	183
•••	10709	11376	16781	1996	2218	2748	596	633	707	292	303	320	176	181	181
0.35 ***	10980	11330	16596	2037	2211	2725	616	652	710	303	314	328	182	186	192
•••	11097	11970	18076	2108	2381	2936	632	678	746	308	321	334	185	189	187
0.4 ***	10959	11412	17785	2081	2287	2896	635	677	745	311	324	340	186	191	196
•••	11111	12216	18979	2165	2486	3062	654	706	770	317	331	341	188	193	189
0.45 ***	10617	11194	18553	2080	2313	2999	641	688	762	313	326	344	185	190	196
•••	10813	12170	19489	2175	2536	3124	662	720	778	319	335	341	188	193	187
0.5 ***	10023	10739	18898	2040	2294	3033	635	684	762	308	322	341	181	186	193
•••	10270	11883	19607	2144	2535	3124	656	718	770	315	331	334	183	190	181
0.55 ***	9246	10112	18815	1966	2235	3000	616	666	746	297	311	330	173	178	184
•••	9551	11406	19332	2078	2485	3062	639	700	746	304	320	320	175	181	171
0.6 ***	8359	9368	18305	1865	2138	2898	586	635	711	280	294	311	161	166	171
•••	8725	10775	18665	1979	2390	2936	609	668	707	286	301	299	163	169	157
0.65 ***	7427	8548	17370	1738	2007	2728	546	591	661	258	270	284	146	150	155
•••	7845	10018	17606	1851	2250	2748	566	621	652	263	276	271	148	152	140
0.7 ***	6497	7680	16012	1588	1841	2491	494	534	593	228	238	250	126	130	133
•••	6948	9150	16153	1695	2065	2496	512	560	581	233	244	236	128	131	118
0.75 ***	5593	6770	14235	1412	1638	2186	430	463	509	193	200	210	104	106	108
•••	6048	8172	14309	1508	1833	2182	445	483	495	196	205	194	105	107	93
0.8 ***	4710	5809	12043	1207	1396	1816	353	376	408	151	155	161	.	.	.
•••	5141	7071	12072	1288	1551	1806	363	391	393	153	158	145	.	.	.
0.85 ***	3821	4765	9441	965	1104	1380	260	273	291	.	.	.	.	.	.
•••	4196	5808	9443	1026	1211	1366	266	281	275	.	.	.	.	.	.
0.9 ***	2863	3567	6432	670	747	879	.	.	.	.	.	.	.	.	.
•••	3146	4302	6421	704	803	864	.	.	.	.	.	.	.	.	.
0.95 ***	1678	2030	3022	.	.	.	.	.	.	.	.	.	.	.	.
•••	1825	2351	3007	.	.	.	.	.	.	.	.	.	.	.	.

TABLE 5: ALPHA= 0.025 POWER= 0.8   EXPECTED ACCRUAL THRU MINIMUM FOLLOW-UP= 1100

PCONT=***	DEL=.02 1.0 / .75	.50 / .25	.00 / BIN	DEL=.05 1.0 / .75	.50 / .25	.00 / BIN	DEL=.10 1.0 / .75	.50 / .25	.00 / BIN	DEL=.15 1.0 / 75	.50 / .25	.00 / BIN	DEL=.20 1.0 / .75	.50 / .25	.00 / BIN
					REQUIRED NUMBER OF PATIENTS										
0.05	2248	2269	2364	487	492	501	176	177	178	103	103	104	73	73	73
	2256	2296	4419	489	496	864	176	177	275	103	104	145	73	73	93
0.1	4681	4735	5195	903	925	967	289	293	298	155	157	158	103	103	104
	4700	4826	7677	912	942	1366	291	296	393	156	158	194	103	104	118
0.15	6849	6944	8034	1268	1314	1418	388	397	409	200	203	206	127	129	130
	6880	7116	10542	1287	1354	1806	392	402	495	202	204	236	128	129	140
0.2	8585	8733	10675	1567	1643	1829	470	485	507	237	242	247	148	150	152
	8635	9004	13014	1598	1710	2182	477	496	581	239	245	271	148	151	157
0.25	9849	10060	13016	1796	1904	2188	536	558	591	267	274	282	164	166	169
	9920	10450	15094	1840	2005	2496	547	573	652	270	278	299	165	168	171
0.3	10647	10936	15001	1958	2100	2489	587	615	659	290	298	309	175	179	183
	10744	11464	16781	2017	2234	2748	599	635	707	294	303	320	177	181	181
0.35	11015	11398	16596	2060	2236	2725	621	657	710	305	315	328	183	187	192
	11144	12082	18076	2132	2402	2936	637	680	746	309	321	334	185	189	187
0.4	11005	11500	17784	2109	2315	2896	642	682	745	313	324	340	186	191	196
	11172	12354	18979	2193	2510	3062	660	709	770	318	332	341	188	193	189
0.45	10676	11303	18553	2111	2345	2999	648	692	762	315	328	344	186	191	197
	10891	12334	19489	2208	2563	3124	668	723	778	321	335	341	188	194	187
0.5	10097	10872	18897	2073	2329	3033	642	689	763	310	324	340	181	187	192
	10367	12071	19607	2179	2564	3124	664	721	770	316	331	334	184	190	181
0.55	9339	10266	18815	2003	2271	3000	624	672	745	300	313	329	174	179	184
	9670	11611	19332	2115	2516	3062	646	704	746	306	320	320	176	181	171
0.6	8472	9540	18305	1902	2175	2898	594	641	712	283	295	310	162	166	171
	8863	10993	18665	2017	2420	2936	615	672	707	288	302	299	164	169	157
0.65	7557	8732	17370	1776	2043	2728	553	596	661	259	270	284	147	150	155
	7998	10240	17606	1888	2279	2748	573	625	652	265	277	271	148	152	140
0.7	6639	7867	16012	1623	1874	2491	500	538	593	230	239	250	127	130	133
	7110	9367	16153	1729	2091	2496	518	562	581	235	245	236	129	132	118
0.75	5738	6951	14235	1444	1668	2186	435	466	509	194	201	209	104	106	108
	6210	8376	14309	1539	1856	2182	450	485	495	197	205	194	105	107	93
0.8	4848	5974	12043	1234	1419	1816	356	378	408	151	155	161	.	.	.
	5292	7251	12072	1314	1569	1806	367	392	393	153	158	145	.	.	.
0.85	3942	4905	9441	985	1121	1380	262	275	291	.	.	.	.	.	.
	4326	5954	9443	1045	1224	1366	268	282	275	.	.	.	.	.	.
0.9	2955	3668	6432	681	756	879	.	.	.	.	.	.	.	.	.
	3244	4401	6421	715	808	864	.	.	.	.	.	.	.	.	.
0.95	1726	2076	3022	.	.	.	.	.	.	.	.	.	.	.	.
	1874	2391	3007	.	.	.	.	.	.	.	.	.	.	.	.

TABLE 5: ALPHA= 0.025 POWER= 0.8   EXPECTED ACCRUAL THRU MINIMUM FOLLOW-UP= 1200

	DEL=.02			DEL=.05			DEL=.10			DEL=.15			DEL=.20		
FACT=	1.0 .75	.50 .25	.00 BIN	1.0 .75	.50 .25	.00 BIN	1.0 .75	.50 .25	.00 BIN	1.0 75	.50 .25	.00 BIN	1.0 .75	.50 .25	.00 BIN
PCONT=•••	REQUIRED NUMBER OF PATIENTS														
0.05 •••	2250 2272	2365	487 493	502	176 177	178	103 103	104	73 73	73				
•••	2258 2299	4419	490 497	864	176 178	275	103 103	145	73 73	93				
0.1 •••	4687 4744	5194	905 927	967	289 293	298	155 157	158	103 103	104				
•••	4706 4839	7677	915 943	1366	292 296	393	156 157	194	103 103	118				
0.15 •••	6857 6961	8034	1273 1320	1417	389 397	409	200 203	206	127 129	130				
•••	6892 7143	10542	1293 1358	1806	393 403	495	202 205	236	128 129	140				
0.2 •••	8599 8759	10675	1576 1651	1829	472 487	507	238 242	247	148 150	152				
•••	8653 9048	13014	1607 1718	2182	479 496	581	240 245	271	148 151	157				
0.25 •••	9868 10098	13015	1809 1918	2188	540 561	591	268 274	282	164 166	169				
•••	9945 10513	15094	1853 2015	2496	550 574	652	271 277	299	165 168	171				
0.3 •••	10673 10987	15000	1975 2118	2488	591 618	658	290 298	309	175 179	183				
•••	10779 11548	16781	2034 2249	2748	603 636	707	295 304	320	177 181	181				
0.35 •••	11050 11466	16596	2081 2258	2725	626 660	710	306 316	328	184 187	192				
•••	11191 12190	18076	2154 2420	2936	641 682	746	310 322	334	185 190	187				
0.4 •••	11050 11587	17784	2134 2341	2896	647 686	745	315 326	340	187 191	196				
•••	11233 12487	18979	2220 2532	3062	664 712	770	320 332	341	189 193	189				
0.45 •••	10735 11410	18553	2140 2374	2998	654 697	762	316 329	344	187 191	196				
•••	10968 12489	19489	2236 2587	3124	674 726	778	322 336	341	189 193	187				
0.5 •••	10171 10999	18898	2105 2360	3034	649 694	762	312 325	340	182 187	193				
•••	10462 12247	19607	2212 2590	3124	669 724	770	318 332	334	184 190	181				
0.55 •••	9431 10414	18814	2036 2304	3000	631 676	745	301 313	329	175 179	184				
•••	9784 11803	19332	2148 2543	3062	652 707	746	307 321	320	177 181	171				
0.6 •••	8582 9703	18305	1937 2208	2898	601 646	712	284 296	310	163 166	171				
•••	8995 11194	18665	2051 2447	2936	621 675	707	290 303	299	165 169	157				
0.65 •••	7683 8905	17370	1810 2074	2728	559 601	661	261 271	284	147 151	154				
•••	8145 10445	17606	1921 2305	2748	578 628	652	266 277	271	148 152	140				
0.7 •••	6775 8042	16012	1655 1903	2491	505 541	593	232 240	250	127 130	133				
•••	7263 9568	16153	1761 2115	2496	523 565	581	235 245	236	129 132	118				
0.75 •••	5875 7120	14235	1473 1693	2186	439 469	508	195 202	209	104 106	108				
•••	6362 8564	14309	1567 1876	2182	453 487	495	199 205	194	105 107	93				
0.8 •••	4978 6129	12043	1258 1440	1816	360 380	408	152 156	160	. .	.				
•••	5433 7417	12072	1337 1585	1806	370 393	393	154 158	145	. .	.				
0.85 •••	4055 5033	9440	1003 1135	1380	264 276	290	. .	.	. .	.				
•••	4447 6086	9443	1062 1234	1366	269 283	275	. .	.	. .	.				
0.9 •••	3040 3760	6432	691 764	879	. .	.	. .	.	. .	.				
•••	3333 4489	6421	724 814	864	. .	.	. .	.	. .	.				
0.95 •••	1771 2119	3022	. .	.	. .	.	. .	.	. .	.				
•••	1918 2426	3007	. .	.	. .	.	. .	.	. .	.				

TABLE 5: ALPHA= 0.025 POWER= 0.8   EXPECTED ACCRUAL THRU MINIMUM FOLLOW-UP= 1300

PCONT	DEL=.02 1.0/.75	.50/.25	.00/BIN	DEL=.05 1.0/.75	.50/.25	.00/BIN	DEL=.10 1.0/.75	.50/.25	.00/BIN	DEL=.15 1.0/75	.50/.25	.00/BIN	DEL=.20 1.0/.75	.50/.25	.00/BIN
						REQUIRED	NUMBER	OF	PATIENTS					
0.05	2252	2275	2364	488	494	501	176	177	178	103	103	104	73	73	73
	2260	2303	4419	491	497	864	176	178	275	103	104	145	73	73	93
0.1	4691	4753	5195	908	929	967	290	294	299	156	157	159	102	103	104
	4712	4852	7677	917	945	1366	292	297	393	156	158	194	103	104	118
0.15	6867	6978	8034	1279	1324	1418	390	399	409	201	203	206	128	129	130
	6904	7169	10542	1298	1362	1806	394	404	495	202	205	236	128	129	140
0.2	8612	8786	10675	1584	1660	1829	474	488	507	239	243	248	148	150	152
	8671	9090	13014	1616	1725	2182	481	497	581	240	245	271	149	151	157
0.25	9887	10135	13016	1821	1930	2188	543	563	591	269	275	282	164	167	170
	9971	10574	15094	1866	2026	2496	552	575	652	271	278	299	166	168	171
0.3	10700	11039	15000	1992	2135	2489	595	621	659	292	300	309	176	180	183
	10814	11630	16781	2051	2262	2748	607	638	707	296	304	320	178	181	181
0.35	11086	11533	16596	2101	2278	2725	630	663	710	307	317	328	185	188	192
	11237	12294	18076	2175	2437	2936	646	684	746	312	323	334	186	189	187
0.4	11096	11672	17784	2157	2365	2896	653	690	744	316	327	340	188	192	197
	11293	12613	18979	2244	2551	3062	669	714	770	322	333	341	189	194	189
0.45	10793	11515	18554	2166	2400	2999	660	701	762	318	330	344	188	192	197
	11043	12636	19489	2265	2610	3124	679	728	778	324	336	341	189	194	187
0.5	10246	11123	18898	2135	2389	3033	655	698	762	315	326	341	184	188	193
	10556	12413	19607	2241	2614	3124	674	726	770	319	333	334	185	190	181
0.55	9522	10555	18815	2068	2333	3000	637	681	745	303	315	329	175	180	185
	9896	11984	19332	2179	2567	3062	657	709	746	309	322	320	177	182	171
0.6	8690	9859	18305	1969	2239	2898	607	650	712	286	297	310	163	167	172
	9124	11384	18665	2083	2472	2936	627	678	707	292	303	299	165	169	157
0.65	7805	9069	17370	1841	2104	2728	565	604	661	263	272	284	148	151	154
	8285	10637	17606	1953	2328	2748	583	630	652	267	278	271	149	153	140
0.7	6906	8208	16012	1685	1930	2490	511	545	593	233	240	250	128	131	133
	7409	9755	16153	1790	2136	2496	526	567	581	237	245	236	129	132	118
0.75	6007	7279	14235	1500	1717	2187	444	471	509	197	202	209	105	107	108
	6506	8738	14309	1593	1894	2182	457	488	495	199	206	194	106	107	93
0.8	5102	6272	12043	1280	1459	1816	362	383	408	153	157	161	.	.	.
	5566	7569	12072	1358	1598	1806	372	394	393	154	159	145	.	.	.
0.85	4162	5153	9441	1020	1149	1380	266	277	291	.	.	.	.	.	.
	4560	6207	9443	1077	1243	1366	271	284	275	.	.	.	.	.	.
0.9	3121	3846	6432	701	771	879	.	.	.	.	.	.	.	.	.
	3417	4570	6421	733	818	864	.	.	.	.	.	.	.	.	.
0.95	1812	2157	3022	.	.	.	.	.	.	.	.	.	.	.	.
	1958	2458	3007	.	.	.	.	.	.	.	.	.	.	.	.

TABLE 5: ALPHA= 0.025 POWER= 0.8   EXPECTED ACCRUAL THRU MINIMUM FOLLOW-UP= 1400

		DEL=.02			DEL=.05			DEL=.10			DEL=.15			DEL=.20		
FACT=		1.0 .75	.50 .25	.00 BIN	1.0 .75	.50 .25	.00 BIN	1.0 .75	.50 .25	.00 BIN	1.0 75	.50 .25	.00 BIN	1.0 .75	.50 .25	.00 BIN
PCONT=				REQUIRED	NUMBER	OF	PATIENTS									
0.05	***	2254	2278	2364	489	494	501	176	178	178	103	103	104	73	73	73
	•••	2263	2306	4419	492	498	864	177	178	275	103	104	145	73	73	93
0.1	***	4696	4763	5195	911	931	967	290	294	298	156	157	158	103	103	104
	•••	4719	4863	7677	920	947	1366	292	297	393	157	157	194	103	103	118
0.15	***	6875	6994	8034	1284	1329	1417	391	399	409	201	204	206	128	129	130
	•••	6915	7193	10542	1304	1365	1806	395	404	495	202	205	236	129	129	140
0.2	***	8626	8811	10675	1592	1668	1829	476	490	507	239	243	248	149	150	152
	•••	8689	9131	13014	1624	1731	2182	483	498	581	241	245	271	150	151	157
0.25	***	9907	10173	13016	1832	1941	2188	545	564	591	269	276	282	164	167	170
	•••	9997	10632	15094	1878	2034	2496	554	577	652	272	278	299	166	168	171
0.3	***	10727	11089	15000	2006	2150	2488	598	623	659	293	300	309	177	179	183
	•••	10849	11708	16781	2067	2274	2748	610	640	707	297	304	320	178	181	181
0.35	***	11120	11599	16596	2120	2297	2725	634	666	710	309	318	328	185	188	192
	•••	11283	12393	18076	2194	2452	2936	649	686	746	313	323	334	186	190	187
0.4	***	11142	11755	17784	2180	2387	2895	657	693	745	318	328	340	188	192	196
	•••	11353	12732	18979	2266	2569	3062	674	716	770	323	333	341	190	194	189
0.45	***	10852	11616	18553	2192	2425	2999	665	704	762	320	331	344	188	192	197
	•••	11119	12775	19489	2289	2629	3124	683	731	778	325	337	341	190	194	187
0.5	***	10319	11243	18897	2162	2415	3034	660	702	762	316	327	340	184	188	192
	•••	10649	12568	19607	2269	2635	3124	679	729	770	321	333	334	185	190	181
0.55	***	9611	10692	18815	2096	2361	2999	642	684	745	304	316	329	176	179	184
	•••	10006	12153	19332	2208	2589	3062	661	712	746	310	322	320	178	182	171
0.6	***	8795	10007	18305	1998	2265	2898	612	654	711	288	298	311	164	167	171
	•••	9248	11561	18665	2111	2494	2936	631	680	707	292	304	299	165	170	157
0.65	***	7922	9225	17370	1870	2130	2728	570	608	661	264	273	284	148	151	155
	•••	8419	10816	17606	1981	2348	2748	587	632	652	269	278	271	150	153	140
0.7	***	7030	8364	16012	1712	1955	2490	515	548	593	234	241	250	129	131	133
	•••	7548	9929	16153	1816	2154	2496	530	569	581	238	246	236	129	132	118
0.75	***	6130	7429	14235	1524	1739	2187	447	473	508	197	203	209	105	107	108
	•••	6641	8899	14309	1617	1909	2182	459	490	495	199	206	194	106	108	93
0.8	***	5218	6407	12043	1301	1476	1816	365	384	408	153	157	161	.	.	.
	•••	5691	7710	12072	1378	1611	1806	374	395	393	155	158	145	.	.	.
0.85	***	4262	5265	9440	1035	1160	1380	267	278	290	.	.	.	.	.	.
	•••	4665	6319	9443	1091	1251	1366	272	284	275	.	.	.	.	.	.
0.9	***	3196	3924	6432	710	777	879	.	.	.	.	.	.	.	.	.
	•••	3494	4644	6421	740	822	864	.	.	.	.	.	.	.	.	.
0.95	***	1850	2192	3022	.	.	.	.	.	.	.	.	.	.	.	.
	•••	1995	2486	3007	.	.	.	.	.	.	.	.	.	.	.	.

TABLE 5: ALPHA= 0.025 POWER= 0.8   EXPECTED ACCRUAL THRU MINIMUM FOLLOW-UP= 1500

		DEL=.02			DEL=.05			DEL=.10			DEL=.15			DEL=.20	
FACT=	1.0 .75	.50 .25	.00 BIN	1.0 .75	.50 .25	.00 BIN	1.0 .75	.50 .25	.00 BIN	1.0 75	.50 .25	.00 BIN	1.0 .75	.50 .25	.00 BIN
PCONT=***							REQUIRED NUMBER OF PATIENTS								
0.05 ***	2257	2281	2364	489	495	502	176	177	178	103	104	104	73	73	73
···	2266	2308	4419	492	498	864	177	178	275	103	104	145	73	73	93
0.1 ***	4702	4771	5195	913	933	967	290	294	299	157	157	158	103	103	104
···	4726	4875	7677	922	948	1366	292	296	393	157	157	194	103	104	118
0.15 ***	6883	7010	8034	1289	1333	1417	392	400	410	202	203	206	128	129	130
···	6926	7217	10542	1308	1368	1806	395	404	495	202	205	236	128	129	140
0.2 ***	8640	8837	10674	1600	1674	1829	478	491	507	240	244	247	149	150	152
···	8707	9170	13014	1632	1735	2182	484	499	581	242	245	271	149	151	157
0.25 ***	9926	10209	13015	1843	1951	2188	547	566	590	270	275	282	165	167	170
···	10022	10688	15094	1889	2042	2496	556	577	652	273	278	299	166	169	171
0.3 ***	10753	11138	15000	2021	2163	2488	601	625	659	293	301	308	177	180	183
···	10884	11782	16781	2081	2285	2748	612	640	707	297	305	320	178	181	181
0.35 ***	11155	11663	16597	2137	2314	2725	638	668	710	310	319	328	185	188	192
···	11330	12487	18076	2212	2465	2936	652	687	746	314	323	334	187	190	187
0.4 ***	11187	11836	17784	2200	2407	2896	661	695	744	319	329	340	188	192	196
···	11412	12847	18979	2287	2585	3062	677	718	770	323	334	341	190	194	189
0.45 ***	10910	11715	18553	2215	2447	2999	669	708	762	322	332	344	188	192	197
···	11194	12907	19489	2313	2647	3124	687	732	778	326	337	341	190	194	187
0.5 ***	10391	11359	18897	2188	2438	3034	665	705	762	317	328	340	185	188	192
···	10739	12716	19607	2294	2654	3124	683	731	770	322	334	334	187	190	181
0.55 ***	9698	10822	18815	2123	2386	3000	647	688	745	307	317	329	176	180	184
···	10112	12313	19332	2235	2609	3062	667	714	746	311	322	320	178	182	171
0.6 ***	8896	10149	18305	2026	2290	2898	617	656	712	289	299	310	164	168	172
···	9367	11728	18665	2138	2512	2936	635	682	707	293	305	299	166	170	157
0.65 ***	8035	9373	17370	1897	2154	2728	575	611	661	265	274	284	148	151	155
···	8548	10984	17606	2007	2367	2748	592	634	652	269	278	271	150	153	140
0.7 ***	7148	8512	16012	1737	1977	2491	519	550	593	235	242	250	128	131	133
···	7680	10091	16153	1841	2171	2496	533	570	581	238	245	236	130	132	118
0.75 ***	6249	7569	14235	1547	1758	2186	450	476	509	198	203	209	105	107	108
···	6770	9050	14309	1639	1924	2182	462	491	495	200	206	194	106	108	93
0.8 ***	5329	6533	12043	1320	1492	1816	367	385	408	154	157	160	.	.	.
···	5809	7840	12072	1396	1621	1806	376	395	393	155	158	145	.	.	.
0.85 ***	4357	5369	9440	1049	1171	1380	268	278	290	.	.	.	.	.	.
···	4765	6423	9443	1103	1258	1366	274	284	275	.	.	.	.	.	.
0.9 ***	3267	3998	6432	717	783	879	.	.	.	.	.	.	.	.	.
···	3567	4712	6421	747	825	864	.	.	.	.	.	.	.	.	.
0.95 ***	1884	2224	3022	.	.	.	.	.	.	.	.	.	.	.	.
···	2030	2510	3007	.	.	.	.	.	.	.	.	.	.	.	.

TABLE 5: ALPHA= 0.025 POWER= 0.8   EXPECTED ACCRUAL THRU MINIMUM FOLLOW-UP= 1600

		DEL=.02			DEL=.05			DEL=.10			DEL=.15			DEL=.20		
FACT=		1.0 .75	.50 .25	.00 BIN	1.0 .75	.50 .25	.00 BIN	1.0 .75	.50 .25	.00 BIN	1.0 75	.50 .25	.00 BIN	1.0 .75	.50 .25	.00 BIN
PCONT=***								REQUIRED NUMBER OF PATIENTS								
0.05	***	2258	2283	2365	490	495	502	177	177	178	103	104	104	73	73	73
	•••	2268	2311	4419	493	498	864	177	178	275	104	104	145	73	73	93
0.1	***	4706	4780	5195	915	935	967	292	295	299	156	157	159	103	104	104
	•••	4732	4885	7677	924	949	1366	293	297	393	157	158	194	103	104	118
0.15	***	6892	7026	8034	1293	1337	1418	393	400	409	202	204	207	128	129	130
	•••	6938	7240	10542	1312	1371	1806	397	405	495	203	205	236	129	130	140
0.2	***	8653	8862	10675	1607	1681	1829	479	492	507	240	244	248	149	150	152
	•••	8724	9206	13014	1639	1740	2182	485	499	581	242	246	271	150	151	157
0.25	***	9945	10245	13016	1854	1961	2188	550	567	591	271	276	282	165	167	170
	•••	10048	10741	15094	1899	2049	2496	558	578	652	273	279	299	166	168	171
0.3	***	10779	11188	15000	2034	2176	2489	603	627	659	295	301	309	177	180	183
	•••	10919	11854	16781	2094	2294	2748	614	642	707	298	305	320	179	181	181
0.35	***	11191	11727	16596	2154	2329	2725	642	671	710	311	319	328	185	188	192
	•••	11376	12578	18076	2228	2477	2936	655	689	746	315	323	334	187	190	187
0.4	***	11233	11916	17784	2220	2425	2896	665	698	745	320	329	340	189	193	196
	•••	11471	12955	18979	2307	2599	3062	680	719	770	325	334	341	191	194	189
0.45	***	10968	11811	18553	2237	2467	2999	674	711	762	323	332	344	189	193	197
	•••	11267	13033	19489	2335	2663	3124	691	734	778	327	338	341	191	195	187
0.5	***	10462	11471	18898	2212	2461	3034	669	708	762	318	329	341	185	189	193
	•••	10828	12857	19607	2318	2671	3124	687	733	770	323	334	334	187	190	181
0.55	***	9785	10948	18815	2148	2409	3000	652	691	746	307	318	329	177	180	184
	•••	10215	12464	19332	2260	2627	3062	670	716	746	312	323	320	178	182	171
0.6	***	8996	10285	18305	2051	2313	2898	621	659	712	290	300	311	165	168	171
	•••	9483	11885	18665	2164	2531	2936	639	683	707	295	305	299	166	170	157
0.65	***	8145	9514	17370	1922	2176	2728	578	613	661	266	275	284	149	152	155
	•••	8672	11142	17606	2031	2384	2748	595	635	652	270	279	271	150	153	140
0.7	***	7263	8652	16012	1761	1998	2491	523	553	593	236	243	250	129	131	133
	•••	7806	10244	16153	1864	2186	2496	537	571	581	239	246	236	130	132	118
0.75	***	6362	7702	14235	1568	1775	2186	453	478	509	199	204	209	105	107	108
	•••	6892	9190	14309	1658	1937	2182	465	492	495	201	206	194	106	107	93
0.8	***	5433	6653	12043	1337	1506	1816	370	387	408	154	157	161	.	.	.
	•••	5921	7962	12072	1412	1631	1806	378	397	393	156	159	145	.	.	.
0.85	***	4447	5467	9441	1062	1181	1380	270	279	291	.	.	.	.	.	.
	•••	4859	6519	9443	1115	1265	1366	274	285	275	.	.	.	.	.	.
0.9	***	3333	4067	6432	725	787	879	.	.	.	.	.	.	.	.	.
	•••	3635	4775	6421	753	828	864	.	.	.	.	.	.	.	.	.
0.95	***	1918	2254	3022	.	.	.	.	.	.	.	.	.	.	.	.
	•••	2061	2534	3007	.	.	.	.	.	.	.	.	.	.	.	.

TABLE 5: ALPHA= 0.025 POWER= 0.8   EXPECTED ACCRUAL THRU MINIMUM FOLLOW-UP= 1700

	DEL=.02			DEL=.05			DEL=.10			DEL=.15			DEL=.20		
FACT=	1.0 .75	.50 .25	.00 BIN	1.0 .75	.50 .25	.00 BIN	1.0 .75	.50 .25	.00 BIN	1.0 75	.50 .25	.00 BIN	1.0 .75	.50 .25	.00 BIN
PCONT=•••						REQUIRED	NUMBER	OF PATIENTS						
0.05 •••	2260	2285	2365	491	495	501	176	177	178	103	104	104	73	73	73
•••	2269	2313	4419	493	498	864	177	178	275	104	104	145	73	73	93
0.1 •••	4711	4787	5195	917	936	967	292	295	299	156	157	158	103	103	104
•••	4738	4895	7677	925	950	1366	293	296	393	157	158	194	103	104	118
0.15 •••	6901	7042	8034	1297	1341	1417	394	401	409	202	204	206	129	129	131
•••	6950	7261	10542	1316	1372	1806	397	405	495	203	205	236	129	129	140
0.2 •••	8667	8886	10675	1614	1687	1829	481	493	508	240	244	248	148	151	152
•••	8742	9241	13014	1646	1744	2182	486	499	581	242	245	271	150	151	157
0.25 •••	9965	10280	13015	1863	1970	2188	551	568	590	272	276	282	165	168	170
•••	10072	10793	15094	1908	2056	2496	560	579	652	274	279	299	167	169	171
0.3 •••	10805	11236	15000	2047	2188	2488	605	629	658	295	301	309	177	180	182
•••	10953	11922	16781	2107	2303	2748	616	643	707	299	305	320	178	182	181
0.35 •••	11225	11789	16596	2169	2344	2725	645	673	709	311	320	328	186	189	192
•••	11421	12664	18076	2244	2488	2936	658	690	746	316	324	334	187	190	187
0.4 •••	11278	11993	17784	2237	2441	2895	668	700	745	321	329	340	190	193	197
•••	11529	13059	18979	2324	2611	3062	683	720	770	325	335	341	191	194	189
0.45 •••	11024	11904	18553	2258	2486	2998	678	713	762	324	333	344	189	193	197
•••	11339	13153	19489	2356	2677	3124	695	735	778	328	338	341	191	194	187
0.5 •••	10533	11579	18897	2234	2481	3033	673	711	763	320	329	341	185	189	192
•••	10915	12990	19607	2339	2687	3124	690	734	770	324	335	334	187	190	181
0.55 •••	9868	11069	18814	2171	2430	2999	656	694	746	308	318	329	177	180	184
•••	10316	12607	19332	2282	2642	3062	673	717	746	313	324	320	178	182	171
0.6 •••	9093	10414	18304	2075	2334	2898	626	662	712	291	299	310	165	168	171
•••	9595	12034	18665	2186	2547	2936	643	684	707	295	305	299	167	170	157
0.65 •••	8250	9649	17369	1945	2197	2728	582	616	661	267	275	284	148	152	155
•••	8791	11290	17606	2054	2399	2748	598	636	652	271	279	271	151	153	140
0.7 •••	7373	8784	16012	1783	2016	2490	526	554	593	237	243	250	129	131	134
•••	7926	10387	16153	1885	2200	2496	539	573	581	240	246	236	131	133	118
0.75 •••	6470	7828	14235	1587	1791	2186	456	479	509	199	204	209	105	107	108
•••	7009	9322	14309	1676	1948	2182	467	493	495	202	206	194	106	107	93
0.8 •••	5533	6766	12043	1353	1518	1816	372	388	408	154	157	160	.	.	.
•••	6028	8076	12072	1427	1640	1806	379	397	393	156	159	145	.	.	.
0.85 •••	4532	5560	9440	1073	1189	1380	271	280	291	.	.	.	.	.	.
•••	4949	6608	9443	1126	1270	1366	275	284	275	.	.	.	.	.	.
0.9 •••	3397	4131	6432	731	792	879	.	.	.	.	.	.	.	.	.
•••	3699	4832	6421	758	831	864	.	.	.	.	.	.	.	.	.
0.95 •••	1948	2281	3021	.	.	.	.	.	.	.	.	.	.	.	.
•••	2091	2555	3007	.	.	.	.	.	.	.	.	.	.	.	.

TABLE 5: ALPHA= 0.025 POWER= 0.8   EXPECTED ACCRUAL THRU MINIMUM FOLLOW-UP= 1800

		DEL=.02			DEL=.05			DEL=.10			DEL=.15			DEL=.20		
FACT=		1.0 .75	.50 .25	.00 BIN	1.0 .75	.50 .25	.00 BIN	1.0 .75	.50 .25	.00 BIN	1.0 75	.50 .25	.00 BIN	1.0 .75	.50 .25	.00 BIN
PCONT=***					REQUIRED NUMBER OF PATIENTS											
0.05	***	2262	2288	2364	492	496	501	177	177	179	103	103	105	73	73	73
	•••	2272	2315	4419	494	498	864	177	177	275	103	103	145	73	73	93
0.1	***	4716	4796	5195	919	938	967	292	294	299	156	157	159	103	103	105
	•••	4744	4904	7677	927	951	1366	294	297	393	157	159	194	103	103	118
0.15	***	6909	7057	8034	1302	1343	1417	395	402	409	202	204	207	128	129	130
	•••	6960	7281	10542	1320	1374	1806	398	405	495	204	206	236	129	129	140
0.2	***	8679	8911	10675	1620	1693	1829	481	494	507	240	244	247	150	150	152
	•••	8759	9276	13014	1651	1748	2182	487	499	581	243	246	271	150	152	157
0.25	***	9984	10316	13016	1872	1977	2188	553	569	591	272	276	282	165	168	170
	•••	10098	10842	15094	1918	2061	2496	561	579	652	274	279	299	166	168	171
0.3	***	10831	11283	15000	2058	2198	2488	609	630	659	296	301	309	177	180	183
	•••	10988	11988	16781	2118	2310	2748	618	643	707	299	305	320	179	181	181
0.35	***	11260	11850	16596	2184	2357	2724	648	674	710	312	319	328	186	189	192
	•••	11466	12746	18076	2258	2499	2936	660	690	746	316	324	334	188	190	187
0.4	***	11323	12069	17784	2256	2457	2895	672	703	744	321	330	339	190	193	197
	•••	11586	13158	18979	2342	2623	3062	686	722	770	326	335	341	191	195	189
0.45	***	11081	11996	18553	2277	2504	2999	681	715	762	325	334	344	190	193	197
	•••	11409	13267	19489	2373	2691	3124	697	737	778	330	339	341	191	195	187
0.5	***	10603	11684	18897	2256	2501	3034	677	713	762	321	330	341	186	189	192
	•••	10999	13116	19607	2360	2701	3124	694	735	770	325	335	334	188	191	181
0.55	***	9951	11186	18814	2193	2449	3000	659	696	746	309	318	330	177	181	184
	•••	10414	12743	19332	2304	2657	3062	677	719	746	314	324	320	179	182	171
0.6	***	9186	10540	18305	2098	2355	2898	629	665	712	292	300	310	165	168	171
	•••	9703	12174	18665	2208	2562	2936	645	686	707	296	306	299	166	170	157
0.65	***	8353	9776	17370	1968	2216	2728	585	618	660	267	276	285	150	152	154
	•••	8905	11431	17606	2074	2413	2748	600	638	652	272	280	271	150	153	140
0.7	***	7479	8912	16012	1803	2034	2490	528	557	593	237	243	251	129	132	134
	•••	8043	10522	16153	1903	2213	2496	542	573	581	240	246	236	130	132	118
0.75	***	6574	7948	14235	1605	1806	2187	458	480	508	199	204	209	105	107	108
	•••	7121	9447	14309	1694	1959	2182	469	494	495	201	207	194	105	108	93
0.8	***	5629	6873	12043	1369	1530	1815	373	389	408	155	157	161	.	.	.
	•••	6129	8183	12072	1440	1648	1806	380	398	393	156	159	145	.	.	.
0.85	***	4614	5647	9441	1084	1197	1380	272	280	290	.	.	.	.	.	.
	•••	5033	6691	9443	1135	1275	1366	276	285	275	.	.	.	.	.	.
0.9	***	3456	4191	6432	737	795	879	.	.	.	.	.	.	.	.	.
	•••	3761	4886	6421	764	834	864	.	.	.	.	.	.	.	.	.
0.95	***	1977	2306	3021	.	.	.	.	.	.	.	.	.	.	.	.
	•••	2118	2574	3007	.	.	.	.	.	.	.	.	.	.	.	.

581

TABLE 5: ALPHA= 0.025 POWER= 0.8   EXPECTED ACCRUAL THRU MINIMUM FOLLOW-UP= 1900

		DEL=.02			DEL=.05			DEL=.10			DEL=.15			DEL=.20		
FACT=		1.0 .75	.50 .25	.00 BIN	1.0 .75	.50 .25	.00 BIN	1.0 .75	.50 .25	.00 BIN	1.0 75	.50 .25	.00 BIN	1.0 .75	.50 .25	.00 BIN
PCONT=***					REQUIRED NUMBER OF PATIENTS											
0.05	***	2263	2291	2364	491	496	502	177	178	178	103	103	104	73	73	73
	•••	2274	2317	4419	494	498	864	177	178	275	103	104	145	73	73	93
0.1	***	4721	4803	5195	920	939	968	292	296	299	156	158	159	103	103	104
	•••	4751	4913	7677	928	951	1366	294	296	393	156	158	194	103	104	118
0.15	***	6918	7072	8034	1305	1346	1417	395	401	410	203	204	206	128	129	130
	•••	6972	7300	10542	1322	1377	1806	399	405	495	203	205	236	129	129	140
0.2	***	8694	8934	10674	1626	1698	1828	483	494	507	241	244	248	149	151	152
	•••	8777	9307	13014	1657	1751	2182	488	500	581	242	246	271	149	152	157
0.25	***	10003	10350	13016	1881	1985	2188	554	571	591	273	277	282	166	167	170
	•••	10123	10888	15094	1926	2066	2496	562	581	652	274	279	299	167	168	171
0.3	***	10858	11330	15000	2070	2208	2489	610	631	659	296	303	308	178	180	182
	•••	11022	12050	16781	2129	2319	2748	619	643	707	299	305	320	179	182	181
0.35	***	11295	11910	16596	2198	2369	2725	649	676	710	313	320	329	186	189	192
	•••	11510	12825	18076	2272	2507	2936	662	692	746	317	324	334	187	190	187
0.4	***	11368	12143	17784	2272	2472	2896	674	704	744	323	331	339	190	193	196
	•••	11643	13252	18979	2357	2635	3062	688	723	770	326	334	341	192	194	189
0.45	***	11137	12084	18553	2295	2519	2998	685	717	762	325	334	344	190	193	197
	•••	11480	13376	19489	2391	2702	3124	700	738	778	330	339	341	191	194	187
0.5	***	10672	11786	18897	2275	2517	3034	680	716	762	322	330	341	186	189	192
	•••	11083	13237	19607	2379	2714	3124	697	737	770	325	334	334	187	191	181
0.55	***	10033	11297	18814	2215	2467	2999	662	698	745	311	319	329	178	180	184
	•••	10509	12871	19332	2324	2671	3062	679	719	746	315	324	320	179	182	171
0.6	***	9279	10660	18305	2118	2372	2897	633	666	711	293	301	311	166	168	171
	•••	9807	12307	18665	2229	2574	2936	648	687	707	296	305	299	167	170	157
0.65	***	8452	9900	17369	1987	2234	2728	589	619	660	268	275	284	149	152	154
	•••	9015	11563	17606	2094	2426	2748	603	638	652	272	280	271	151	153	140
0.7	***	7581	9033	16012	1823	2049	2491	532	558	592	237	243	250	129	132	133
	•••	8153	10651	16153	1922	2224	2496	543	574	581	241	247	236	130	133	118
0.75	***	6673	8063	14235	1622	1820	2186	460	482	508	201	204	209	106	106	108
	•••	7227	9563	14309	1710	1968	2182	470	495	495	201	206	194	106	108	93
0.8	***	5721	6974	12043	1382	1541	1816	375	389	408	155	158	160	.	.	.
	•••	6226	8284	12072	1453	1655	1806	382	399	393	156	159	145	.	.	.
0.85	***	4690	5730	9440	1094	1205	1379	273	281	291	.	.	.	.	.	.
	•••	5114	6770	9443	1144	1280	1366	277	286	275	.	.	.	.	.	.
0.9	***	3512	4248	6431	742	799	878	.	.	.	.	.	.	.	.	.
	•••	3818	4936	6421	768	836	864	.	.	.	.	.	.	.	.	.
0.95	***	2004	2329	3022	.	.	.	.	.	.	.	.	.	.	.	.
	•••	2144	2592	3007	.	.	.	.	.	.	.	.	.	.	.	.

TABLE 5: ALPHA= 0.025 POWER= 0.8　　EXPECTED ACCRUAL THRU MINIMUM FOLLOW-UP= 2000

	DEL=.02			DEL=.05			DEL=.10			DEL=.15			DEL=.20		
FACT=	1.0 .75	.50 .25	.00 BIN	1.0 .75	.50 .25	.00 BIN	1.0 .75	.50 .25	.00 BIN	1.0 75	.50 .25	.00 BIN	1.0 .75	.50 .25	.00 BIN
PCONT=•••				REQUIRED NUMBER OF PATIENTS											
0.05 •••	2266	2292	2365	492	496	501	177	177	179	104	104	104	72	74	74
•••	2276	2319	4419	494	499	864	177	179	275	104	104	145	74	74	93
0.1 •••	4726	4811	5195	922	940	967	292	296	299	156	157	159	104	104	104
•••	4756	4921	7677	930	952	1366	294	297	393	157	159	194	104	104	118
0.15 •••	6926	7087	8034	1309	1349	1417	396	402	410	202	205	206	129	130	130
•••	6984	7319	10542	1326	1379	1806	399	406	495	204	205	236	129	130	140
0.2 •••	8706	8959	10675	1632	1702	1829	484	495	507	241	245	247	150	151	152
•••	8794	9339	13014	1662	1755	2182	489	501	581	242	246	271	150	151	157
0.25 •••	10022	10385	13016	1889	1992	2189	556	571	591	272	277	282	166	167	170
•••	10147	10934	15094	1934	2071	2496	564	581	652	275	280	299	167	169	171
0.3 •••	10884	11376	15000	2081	2217	2489	612	632	659	297	302	309	179	181	182
•••	11056	12111	16781	2140	2325	2748	622	645	707	300	306	320	180	181	181
0.35 •••	11330	11970	16596	2211	2381	2725	652	677	710	314	321	329	186	189	192
•••	11555	12901	18076	2285	2516	2936	664	692	746	317	325	334	187	190	187
0.4 •••	11412	12216	17785	2287	2486	2896	677	706	745	324	331	340	191	194	196
•••	11700	13342	18979	2372	2645	3062	691	724	770	327	336	341	192	195	189
0.45 •••	11194	12170	18554	2312	2536	2999	687	720	762	326	335	344	190	194	196
•••	11549	13480	19489	2409	2715	3124	702	739	778	330	339	341	192	195	187
0.5 •••	10739	11884	18897	2294	2535	3034	684	717	762	322	331	341	186	190	192
•••	11164	13352	19607	2397	2726	3124	700	739	770	326	335	334	187	191	181
0.55 •••	10112	11406	18815	2235	2485	3000	666	700	746	311	320	330	179	181	184
•••	10601	12995	19332	2344	2684	3062	682	721	746	315	325	320	180	182	171
0.6 •••	9367	10775	18305	2139	2390	2897	635	669	711	294	301	311	166	169	171
•••	9910	12435	18665	2247	2587	2936	651	689	707	297	306	299	167	170	157
0.65 •••	8549	10017	17370	2007	2250	2729	591	621	661	270	276	284	150	152	155
•••	9122	11691	17606	2112	2437	2748	606	640	652	272	280	271	151	154	140
0.7 •••	7680	9150	16012	1841	2065	2491	534	560	594	239	244	250	130	131	134
•••	8261	10771	16153	1939	2235	2496	546	575	581	241	247	236	131	132	118
0.75 •••	6770	8172	14235	1639	1832	2186	462	484	509	200	205	210	106	107	109
•••	7330	9674	14309	1725	1977	2182	472	495	495	202	207	194	106	107	93
0.8 •••	5809	7071	12044	1396	1551	1816	376	391	407	155	159	161	.	.	.
•••	6319	8379	12072	1465	1662	1806	384	399	393	156	160	145	.	.	.
0.85 •••	4765	5809	9441	1104	1211	1380	274	281	291	.	.	.	.	.	.
•••	5191	6844	9443	1152	1285	1366	277	286	275	.	.	.	.	.	.
0.9 •••	3567	4302	6432	747	802	879	.	.	.	.	.	.	.	.	.
•••	3872	4982	6421	774	837	864	.	.	.	.	.	.	.	.	.
0.95 •••	2030	2351	3022	.	.	.	.	.	.	.	.	.	.	.	.
•••	2169	2609	3007	.	.	.	.	.	.	.	.	.	.	.	.

TABLE 5: ALPHA= 0.025 POWER= 0.8   EXPECTED ACCRUAL THRU MINIMUM FOLLOW-UP= 2250

	DEL=.02			DEL=.05			DEL=.10			DEL=.15			DEL=.20		
FACT=	1.0 .75	.50 .25	.00 BIN	1.0 .75	.50 .25	.00 BIN	1.0 .75	.50 .25	.00 BIN	1.0 75	.50 .25	.00 BIN	1.0 .75	.50 .25	.00 BIN
PCONT=***				REQUIRED	NUMBER	OF PATIENTS									
0.05 ***	2269	2297	2365	493	496	502	177	178	178	104	104	104	73	73	73
***	2280	2323	4419	494	499	864	177	178	275	104	104	145	73	73	93
0.1 ***	4737	4829	5194	925	942	967	294	297	299	157	157	159	104	104	104
***	4771	4940	7677	933	953	1366	295	298	393	157	159	194	104	104	118
0.15 ***	6948	7122	8034	1316	1355	1417	398	403	409	204	205	207	129	129	131
***	7010	7362	10542	1333	1382	1806	401	406	495	204	205	236	129	129	140
0.2 ***	8740	9015	10675	1645	1712	1829	486	496	507	241	244	247	150	150	151
***	8836	9410	13014	1675	1762	2182	491	502	581	243	246	271	150	151	157
0.25 ***	10070	10467	13016	1908	2008	2188	559	573	590	274	278	282	167	168	170
***	10209	11039	15094	1951	2082	2496	567	582	652	275	280	299	167	168	171
0.3 ***	10949	11486	15000	2105	2238	2489	615	635	659	298	303	309	178	181	182
***	11139	12251	16781	2162	2339	2748	626	646	707	300	306	320	179	181	181
0.35 ***	11415	12110	16596	2241	2407	2725	658	680	710	314	322	329	187	190	192
***	11663	13077	18076	2313	2534	2936	668	694	746	319	325	334	188	191	187
0.4 ***	11522	12389	17785	2323	2516	2895	683	710	745	325	331	340	191	194	196
***	11836	13553	18979	2407	2666	3062	696	727	770	329	336	341	192	195	189
0.45 ***	11330	12373	18554	2352	2569	2999	694	724	762	327	336	344	191	194	196
***	11715	13721	19489	2446	2739	3124	708	741	778	331	340	341	192	195	187
0.5 ***	10904	12116	18897	2337	2570	3033	690	721	762	325	331	340	187	190	192
***	11359	13618	19607	2438	2753	3124	705	741	770	327	336	334	188	191	181
0.55 ***	10303	11660	18815	2279	2522	2999	673	704	745	313	320	329	178	181	184
***	10822	13277	19332	2386	2711	3062	687	724	746	316	325	320	179	182	171
0.6 ***	9582	11044	18305	2184	2427	2898	642	673	711	295	302	311	167	168	171
***	10150	12726	18665	2291	2614	2936	657	690	707	299	306	299	167	170	157
0.65 ***	8776	10292	17369	2052	2286	2728	597	626	660	271	277	284	150	153	154
***	9372	11981	17606	2154	2463	2748	612	642	652	274	281	271	151	153	140
0.7 ***	7911	9419	16013	1881	2098	2490	539	564	593	240	244	250	131	132	133
***	8512	11049	16153	1977	2258	2496	551	578	581	241	247	236	131	132	118
0.75 ***	6994	8425	14235	1675	1861	2187	466	486	509	201	205	209	106	106	108
***	7570	9927	14309	1757	1996	2182	477	496	495	204	207	194	106	108	93
0.8 ***	6014	7294	12043	1424	1573	1815	379	392	407	156	159	160	.	.	.
***	6533	8594	12072	1492	1676	1806	385	399	393	157	160	145	.	.	.
0.85 ***	4938	5987	9440	1124	1226	1379	275	282	291	.	.	.	.	.	.
***	5369	7010	9443	1171	1293	1366	278	286	275	.	.	.	.	.	.
0.9 ***	3691	4424	6432	758	809	879	.	.	.	.	.	.	.	.	.
***	3998	5086	6421	783	842	864	.	.	.	.	.	.	.	.	.
0.95 ***	2088	2400	3022	.	.	.	.	.	.	.	.	.	.	.	.
***	2224	2643	3007	.	.	.	.	.	.	.	.	.	.	.	.

TABLE 5: ALPHA= 0.025 POWER= 0.8   EXPECTED ACCRUAL THRU MINIMUM FOLLOW-UP= 2500

		DEL=.02			DEL=.05			DEL=.10			DEL=.15			DEL=.20		
FACT=		1.0 .75	.50 .25	.00 BIN	1.0 .75	.50 .25	.00 BIN	1.0 .75	.50 .25	.00 BIN	1.0 75	.50 .25	.00 BIN	1.0 .75	.50 .25	.00 BIN
PCONT=***					REQUIRED NUMBER OF PATIENTS											
0.05	***	2273	2301	2365	493	496	501	177	177	177	104	104	104	73	73	73
	•••	2286	2326	4419	495	499	864	177	177	275	104	104	145	73	73	93
0.1	***	4749	4845	5195	927	945	967	293	296	298	158	158	159	102	104	104
	•••	4786	4956	7677	936	954	1366	295	298	393	158	158	194	102	104	118
0.15	***	6968	7156	8034	1321	1361	1418	398	402	409	202	204	206	129	129	129
	•••	7037	7398	10542	1339	1386	1806	401	406	495	204	206	236	129	129	140
0.2	***	8773	9070	10674	1656	1721	1829	489	496	508	242	245	248	149	151	151
	•••	8877	9474	13014	1686	1768	2182	492	501	581	243	246	271	149	151	157
0.25	***	10117	10543	13015	1923	2020	2189	562	574	590	274	277	283	167	168	170
	•••	10268	11134	15094	1967	2092	2496	568	583	652	276	279	299	167	168	171
0.3	***	11014	11590	14999	2126	2256	2489	620	637	659	299	304	309	179	181	183
	•••	11220	12379	16781	2184	2352	2748	627	646	707	301	306	320	179	183	181
0.35	***	11499	12243	16596	2268	2429	2724	662	683	711	317	321	327	187	190	192
	•••	11768	13237	18076	2339	2549	2936	671	696	746	318	324	334	189	190	187
0.4	***	11629	12551	17784	2354	2542	2895	689	714	745	326	333	340	192	193	196
	•••	11968	13743	18979	2436	2686	3062	699	727	770	329	336	341	193	195	189
0.45	***	11462	12564	18552	2387	2599	2998	699	727	762	329	336	343	192	193	196
	•••	11873	13937	19489	2481	2759	3124	712	743	778	333	340	341	193	195	187
0.5	***	11062	12331	18898	2374	2602	3034	696	724	762	324	333	340	187	190	192
	•••	11543	13858	19607	2474	2776	3124	711	743	770	329	337	334	189	192	181
0.55	***	10486	11895	18815	2320	2556	2999	679	709	745	314	321	329	179	181	184
	•••	11029	13531	19332	2423	2734	3062	693	726	746	318	324	320	181	183	171
0.6	***	9783	11290	18304	2223	2461	2898	648	676	712	296	302	311	167	168	171
	•••	10371	12987	18665	2327	2637	2936	661	693	707	299	308	299	168	170	157
0.65	***	8989	10543	17370	2090	2317	2727	602	629	661	271	277	284	151	152	154
	•••	9604	12239	17606	2190	2484	2748	615	643	652	274	281	271	151	152	140
0.7	***	8126	9664	16012	1918	2126	2490	543	565	593	240	245	249	131	133	134
	•••	8742	11295	16153	2011	2276	2496	554	579	581	243	248	236	131	133	118
0.75	***	7201	8652	14236	1706	1886	2186	470	487	509	202	206	209	106	108	108
	•••	7787	10149	14309	1786	2012	2182	479	498	495	204	208	194	108	108	93
0.8	***	6201	7493	12043	1449	1592	1815	381	393	408	156	159	161	.	.	.
	•••	6729	8784	12072	1514	1689	1806	387	401	393	158	159	145	.	.	.
0.85	***	5095	6148	9440	1142	1239	1379	276	283	290	.	.	.	.	.	.
	•••	5529	7154	9443	1187	1301	1366	279	287	275	.	.	.	.	.	.
0.9	***	3804	4531	6433	767	815	879	.	.	.	.	.	.	.	.	.
	•••	4111	5176	6421	790	845	864	.	.	.	.	.	.	.	.	.
0.95	***	2139	2442	3021	.	.	.	.	.	.	.	.	.	.	.	.
	•••	2271	2673	3007	.	.	.	.	.	.	.	.	.	.	.	.

TABLE 5: ALPHA= 0.025 POWER= 0.8   EXPECTED ACCRUAL THRU MINIMUM FOLLOW-UP= 2750

	DEL=.02			DEL=.05			DEL=.10			DEL=.15			DEL=.20		
FACT=	1.0 / .75	.50 / .25	.00 BIN	1.0 / .75	.50 / .25	.00 BIN	1.0 / .75	.50 / .25	.00 BIN	1.0 / 75	.50 / .25	.00 BIN	1.0 / .75	.50 / .25	.00 BIN

PCONT=•••   REQUIRED NUMBER OF PATIENTS

PCONT	DEL.02 (a)	DEL.02 (b)	DEL.02 (c)	DEL.05 (a)	DEL.05 (b)	DEL.05 (c)	DEL.10 (a)	DEL.10 (b)	DEL.10 (c)	DEL.15 (a)	DEL.15 (b)	DEL.15 (c)	DEL.20 (a)	DEL.20 (b)	DEL.20 (c)
0.05 •••	2277	2304	2364	494	498	501	177	178	178	104	104	104	74	74	74
•••	2289	2329	4419	496	500	864	178	178	275	104	104	145	74	74	93
0.1 •••	4760	4860	5195	931	947	967	294	297	298	157	157	157	102	104	104
•••	4798	4970	7677	938	955	1366	295	297	393	157	157	194	102	104	118
0.15 •••	6990	7187	8035	1328	1364	1418	398	404	408	204	205	205	129	130	130
•••	7062	7431	10542	1345	1388	1806	401	407	495	204	205	236	129	130	140
0.2 •••	8805	9121	10675	1666	1728	1828	489	498	507	243	245	247	150	150	153
•••	8920	9534	13014	1694	1773	2182	493	503	581	243	247	271	150	150	157
0.25 •••	10163	10618	13016	1938	2033	2189	563	577	590	274	278	281	166	168	170
•••	10328	11219	15094	1979	2098	2496	570	584	652	276	280	299	168	168	171
0.3 •••	11077	11689	15000	2146	2271	2488	621	639	658	300	304	309	180	181	184
•••	11298	12495	16781	2201	2363	2748	630	648	707	302	307	320	180	181	181
0.35 •••	11582	12369	16595	2292	2449	2725	665	685	709	318	322	328	188	190	192
•••	11871	13381	18076	2361	2562	2936	675	697	746	319	326	334	188	190	187
0.4 •••	11734	12703	17784	2381	2564	2896	692	716	744	328	333	340	192	194	195
•••	12094	13916	18979	2463	2701	3062	703	730	770	331	336	341	194	195	189
0.45 •••	11590	12741	18554	2419	2624	2999	704	730	762	331	336	343	192	194	197
•••	12026	14134	19489	2509	2777	3124	716	745	778	333	340	341	194	195	187
0.5 •••	11212	12531	18897	2409	2629	3033	700	728	762	326	333	340	188	190	192
•••	11718	14072	19607	2505	2794	3124	714	744	770	329	336	334	188	192	181
0.55 •••	10658	12112	18815	2354	2584	3000	683	711	745	315	322	329	180	181	184
•••	11223	13760	19332	2456	2755	3062	697	727	746	319	326	320	180	184	171
0.6 •••	9971	11518	18304	2260	2488	2897	652	679	711	298	304	311	168	170	171
•••	10580	13222	18665	2361	2656	2936	665	694	707	300	307	299	168	170	157
0.65 •••	9186	10772	17369	2124	2344	2727	606	632	661	273	278	285	150	153	154
•••	9819	12470	17606	2222	2504	2748	618	645	652	276	281	271	153	154	140
0.7 •••	8325	9885	16011	1948	2150	2490	548	569	593	242	245	250	130	132	133
•••	8953	11515	16153	2040	2292	2496	558	580	581	243	249	236	132	132	118
0.75 •••	7392	8860	14236	1734	1906	2185	474	489	508	202	205	209	106	108	108
•••	7987	10348	14309	1811	2026	2182	480	498	495	204	208	194	106	108	93
0.8 •••	6374	7675	12043	1473	1608	1816	383	395	407	156	159	161	.	.	.
•••	6907	8953	12072	1534	1697	1806	390	401	393	157	159	145	.	.	.
0.85 •••	5238	6292	9441	1158	1249	1380	278	283	290	.	.	.	.	.	.
•••	5675	7282	9443	1199	1308	1366	281	287	275	.	.	.	.	.	.
0.9 •••	3906	4626	6431	775	821	879	.	.	.	.	.	.	.	.	.
•••	4210	5254	6421	797	848	864	.	.	.	.	.	.	.	.	.
0.95 •••	2184	2480	3021	.	.	.	.	.	.	.	.	.	.	.	.
•••	2315	2700	3007	.	.	.	.	.	.	.	.	.	.	.	.

TABLE 5: ALPHA= 0.025 POWER= 0.8   EXPECTED ACCRUAL THRU MINIMUM FOLLOW-UP= 3000

	DEL=.01			DEL=.02			DEL=.05			DEL=.10			DEL=.15		
FACT=	1.0 .75	.50 .25	.00 BIN	1.0 .75	.50 .25	.00 BIN	1.0 .75	.50 .25	.00 BIN	1.0 75	.50 .25	.00 BIN	1.0 .75	.50 .25	.00 BIN
PCONT=•••				REQUIRED NUMBER OF PATIENTS											
0.01 ***	1202	1205	1210	440	440	445	145	145	145	73	73	73	50	50	50
•••	1202	1205	4631	440	445	1531	145	145	417	73	73	170	50	50	101
0.02 ***	2653	2668	2702	853	857	865	230	230	230	100	100	103	65	65	65
•••	2660	2680	7645	857	860	2277	230	230	532	100	103	197	65	65	113
0.05 ***	8035	8087	8440	2282	2308	2365	493	497	500	178	178	178	103	103	103
•••	8050	8170	16310	2293	2330	4419	497	500	864	178	178	275	103	103	145
0.1 ***	17720	17860	19640	4772	4873	5195	932	947	965	295	295	298	155	155	160
•••	17765	18122	29497	4810	4982	7677	940	958	1366	295	298	393	155	160	194
0.15 ***	26470	26720	31025	7010	7217	8035	1333	1367	1415	400	403	410	205	205	205
•••	26552	27205	41113	7085	7460	10542	1348	1390	1806	403	407	495	205	205	236
0.2 ***	33530	33917	41683	8837	9170	10675	1675	1735	1828	490	497	508	242	245	245
•••	33658	34678	51159	8957	9587	13014	1700	1775	2182	493	500	581	245	245	271
0.25 ***	38690	39250	51175	10210	10690	13015	1952	2042	2188	565	575	590	275	280	283
•••	38875	40345	59636	10385	11297	15094	1993	2105	2496	572	583	652	275	280	299
0.3 ***	41960	42730	59260	11140	11780	14998	2162	2285	2488	625	640	658	302	305	310
•••	42215	44233	66543	11375	12598	16781	2218	2372	2748	632	647	707	302	305	320
0.35 ***	43475	44503	65800	11665	12485	16595	2312	2465	2725	670	688	710	317	325	328
•••	43817	46483	71881	11968	13513	18076	2380	2575	2936	677	700	746	320	325	334
0.4 ***	43427	44765	70715	11837	12845	17785	2405	2585	2897	695	718	745	328	332	340
•••	43873	47305	75648	12215	14072	18979	2485	2713	3062	707	730	770	332	335	341
0.45 ***	42047	43768	73960	11713	12905	18553	2447	2645	2998	707	733	763	332	335	343
•••	42625	46925	77846	12170	14315	19489	2537	2792	3124	718	745	778	335	340	341
0.5 ***	39598	41777	75505	11357	12715	18898	2440	2653	3032	703	730	763	328	332	340
•••	40333	45587	78474	11882	14270	19607	2533	2810	3124	718	745	770	332	335	334
0.55 ***	36362	39065	75343	10820	12313	18815	2387	2608	2998	688	715	745	317	320	328
•••	37288	43505	77532	11405	13967	19332	2485	2770	3062	700	730	746	320	325	320
0.6 ***	32627	35893	73475	10150	11728	18305	2290	2510	2897	655	680	710	298	305	310
•••	33775	40862	75020	10775	13435	18665	2390	2672	2936	670	695	707	302	305	299
0.65 ***	28685	32470	69905	9373	10982	17368	2155	2368	2728	610	632	662	272	280	283
•••	30058	37798	70939	10018	12680	17606	2248	2518	2748	620	647	652	275	280	271
0.7 ***	24790	28930	64640	8510	10090	16010	1978	2170	2492	550	568	595	242	245	250
•••	26335	34385	65287	9148	11713	16153	2065	2308	2496	560	580	581	242	250	236
0.75 ***	21080	25340	57700	7570	9050	14233	1757	1922	2185	475	490	508	205	205	208
•••	22705	30650	58067	8173	10528	14309	1832	2038	2182	482	500	495	205	208	194
0.8 ***	17590	21670	49097	6535	7840	12043	1490	1622	1817	385	395	407	155	160	160
•••	19172	26552	49276	7070	9103	12072	1550	1708	1806	392	400	393	160	160	145
0.85 ***	14215	17815	38852	5368	6422	9440	1172	1258	1378	280	283	290	.	.	.
•••	15628	21955	38915	5807	7393	9443	1210	1315	1366	280	287	275	.	.	.
0.9 ***	10730	13520	26980	4000	4712	6433	782	823	880	.	.	.	.	.	.
•••	11845	16577	26985	4303	5323	6421	800	850	864	.	.	.	.	.	.
0.95 ***	6613	8173	13498	2222	2510	3020	.	.	.	.	.	.	.	.	.
•••	7250	9710	13485	2350	2720	3007	.	.	.	.	.	.	.	.	.
0.98 ***	3085	3587	4648	.	.	.	.	.	.	.	.	.	.	.	.
•••	3302	3988	4631	.	.	.	.	.	.	.	.	.	.	.	.

TABLE 5: ALPHA= 0.025 POWER= 0.8 EXPECTED ACCRUAL THRU MINIMUM FOLLOW-UP= 3250

		DEL=.01			DEL=.02			DEL=.05			DEL=.10			DEL=.15		
FACT=		1.0 .75	.50 .25	.00 BIN	1.0 .75	.50 .25	.00 BIN	1.0 .75	.50 .25	.00 BIN	1.0 75	.50 .25	.00 BIN	1.0 .75	.50 .25	.00 BIN
PCONT=***					REQUIRED NUMBER OF PATIENTS											
0.01	***	1200	1205	1208	441	441	444	144	144	144	76	76	76	51	51	51
	***	1205	1208	4631	441	444	1531	144	144	417	76	76	170	51	51	101
0.02	***	2654	2668	2700	856	859	864	230	233	233	100	100	100	68	68	68
	***	2662	2684	7645	856	859	2277	230	233	532	100	100	197	68	68	113
0.05	***	8038	8095	8436	2286	2310	2362	493	498	501	176	176	176	103	103	103
	***	8057	8179	16310	2294	2334	4419	498	498	864	176	176	275	103	103	145
0.1	***	17731	17881	19641	4780	4886	5194	937	948	964	295	295	298	157	157	157
	***	17783	18162	29497	4824	4994	7677	940	956	1366	295	298	393	157	157	194
0.15	***	26490	26761	31027	7031	7245	8033	1338	1371	1416	401	404	409	206	206	206
	***	26579	27278	41113	7112	7489	10542	1351	1392	1806	401	409	495	206	206	236
0.2	***	33562	33981	41684	8867	9216	10674	1684	1741	1831	490	498	506	241	246	246
	***	33705	34797	51159	8997	9634	13014	1709	1782	2182	493	501	581	246	246	271
0.25	***	38734	39344	51174	10256	10755	13014	1964	2050	2188	566	579	591	274	279	282
	***	38938	40522	59636	10438	11369	15094	2001	2110	2496	571	582	652	279	279	299
0.3	***	42025	42857	59258	11199	11873	15001	2180	2297	2489	628	639	661	303	303	306
	***	42304	44471	66543	11451	12697	16781	2232	2378	2748	636	647	707	303	306	320
0.35	***	43561	44674	65799	11743	12599	16597	2334	2481	2724	672	688	709	319	322	328
	***	43929	46794	71881	12063	13636	18076	2399	2581	2936	680	696	746	319	328	334
0.4	***	43536	44986	70714	11933	12981	17783	2427	2603	2895	696	721	745	328	336	339
	***	44024	47691	75648	12331	14216	18979	2505	2724	3062	709	729	770	331	336	341
0.45	***	42191	44051	73959	11836	13062	18552	2473	2668	3001	712	734	761	331	339	344
	***	42816	47399	77846	12307	14476	19489	2557	2806	3124	721	745	778	336	339	341
0.5	***	39783	42126	75503	11499	12892	18896	2464	2676	3033	709	734	761	328	336	339
	***	40579	46139	78474	12039	14449	19607	2557	2825	3124	721	745	770	331	336	334
0.55	***	36592	39490	75346	10979	12499	18815	2416	2630	3001	693	718	745	319	322	328
	***	37592	44124	77532	11576	14159	19332	2508	2784	3062	704	729	746	319	328	320
0.6	***	32917	36381	73477	10316	11922	18303	2318	2532	2898	661	685	712	298	303	311
	***	34147	41532	75020	10958	13628	18665	2416	2687	2936	672	696	707	303	306	299
0.65	***	29041	33009	69905	9549	11177	17369	2180	2386	2727	615	636	661	274	279	282
	***	30488	38494	70939	10202	12873	17606	2272	2532	2748	623	647	652	279	282	271
0.7	***	25193	29499	64640	8683	10281	16012	2001	2188	2489	555	571	591	241	246	249
	***	26807	35081	65287	9333	11892	16153	2086	2318	2496	563	582	581	246	249	236
0.75	***	21512	25905	57701	7732	9224	14232	1777	1939	2188	477	493	509	206	206	209
	***	23194	31316	58067	8342	10690	14309	1850	2045	2182	485	501	495	206	209	194
0.8	***	18016	22200	49097	6681	7989	12044	1509	1631	1814	387	396	409	157	160	160
	***	19641	27151	49276	7221	9236	12072	1566	1712	1806	393	401	393	157	160	145
0.85	***	14595	18271	38851	5490	6543	9439	1184	1265	1379	279	287	290	.	.	.
	***	16044	22455	38915	5929	7497	9443	1221	1319	1366	282	287	275	.	.	.
0.9	***	11031	13867	26978	4081	4788	6432	786	826	880	.	.	.	.	.	.
	***	12166	16935	26985	4385	5384	6421	807	851	864	.	.	.	.	.	.
0.95	***	6790	8355	13498	2261	2538	3020	.	.	.	.	.	.	.	.	.
	***	7432	9877	13485	2383	2741	3007	.	.	.	.	.	.	.	.	.
0.98	***	3147	3637	4645	.	.	.	.	.	.	.	.	.	.	.	.
	***	3361	4027	4631	.	.	.	.	.	.	.	.	.	.	.	.

TABLE 5: ALPHA= 0.025 POWER= 0.8   EXPECTED ACCRUAL THRU MINIMUM FOLLOW-UP= 3500

		DEL=.01			DEL=.02			DEL=.05			DEL=.10			DEL=.15	
FACT=	1.0 .75	.50 .25	.00 BIN	1.0 .75	.50 .25	.00 BIN	1.0 .75	.50 .25	.00 BIN	1.0 75	.50 .25	.00 BIN	1.0 .75	.50 .25	.00 BIN
PCONT=•••				REQUIRED NUMBER OF PATIENTS												
0.01 ***	1201	1205	1210	440	443	443	146	146	146	73	73	73	50	50	50
•••	1205	1205	4631	443	443	1531	146	146	417	73	73	170	50	50	101
0.02 ***	2657	2671	2701	855	860	863	230	230	233	102	102	102	67	67	67
•••	2663	2683	7645	855	860	2277	230	233	532	102	102	197	67	67	113
0.05 ***	8044	8105	8438	2286	2313	2365	496	496	501	178	178	178	102	102	102
•••	8065	8193	16310	2298	2333	4419	496	501	864	178	178	275	102	102	145
0.1 ***	17742	17905	19641	4792	4897	5195	939	951	965	295	295	300	155	155	160
•••	17800	18197	29497	4833	5002	7677	942	956	1366	295	295	393	155	160	194
0.15 ***	26510	26804	31025	7050	7268	8035	1341	1371	1415	400	405	408	204	204	207
•••	26606	27355	41113	7134	7513	10542	1359	1394	1806	405	405	495	204	204	236
0.2 ***	33594	34049	41682	8896	9260	10672	1691	1747	1826	493	501	505	242	248	248
•••	33746	34918	51159	9033	9680	13014	1718	1782	2182	496	501	581	242	248	271
0.25 ***	38783	39433	51173	10296	10818	13014	1975	2059	2190	566	580	589	277	277	283
•••	39001	40690	59636	10494	11439	15094	2010	2115	2496	575	583	652	277	283	299
0.3 ***	42090	42986	59258	11258	11955	15000	2193	2307	2488	627	641	659	300	303	309
•••	42388	44701	66543	11521	12786	16781	2243	2386	2748	636	650	707	303	309	320
0.35 ***	43648	44841	65797	11818	12702	16596	2351	2491	2724	671	688	711	318	321	326
•••	44045	47095	71881	12156	13749	18076	2412	2593	2936	680	697	746	321	326	334
0.4 ***	43651	45208	70715	12028	13110	17783	2447	2619	2893	703	720	746	330	335	338
•••	44173	48070	75648	12445	14353	18979	2526	2736	3062	711	732	770	330	338	341
0.45 ***	42335	44330	73958	11950	13210	18553	2496	2683	2998	715	738	764	335	338	344
•••	43009	47851	77846	12436	14627	19489	2578	2815	3124	723	746	778	335	338	341
0.5 ***	39967	42475	75506	11631	13052	18897	2491	2692	3033	711	732	764	330	335	338
•••	40821	46666	78474	12186	14615	19607	2584	2838	3124	723	746	770	330	338	334
0.55 ***	36826	39903	75343	11127	12676	18815	2438	2648	2998	694	720	746	318	321	330
•••	37893	44715	77532	11740	14335	19332	2535	2797	3062	706	732	746	321	326	320
0.6 ***	33209	36858	73476	10476	12104	18302	2342	2552	2899	662	685	711	300	303	309
•••	34507	42172	75020	11127	13805	18665	2438	2701	2936	671	697	707	303	309	299
0.65 ***	29385	33527	69906	9715	11360	17371	2208	2403	2727	615	636	659	274	277	283
•••	30908	39153	70939	10380	13043	17606	2295	2543	2748	627	650	652	277	283	271
0.7 ***	25582	30041	64642	8849	10453	16010	2024	2208	2491	554	571	592	242	248	251
•••	27262	35738	65287	9505	12060	16153	2106	2330	2496	563	583	581	242	248	236
0.75 ***	21925	26440	57700	7890	9386	14233	1800	1954	2185	478	493	510	204	207	207
•••	23663	31940	58067	8502	10835	14309	1870	2053	2182	487	501	495	204	207	194
0.8 ***	18416	22695	49099	6819	8131	12043	1525	1642	1814	388	396	408	155	160	160
•••	20087	27714	49276	7361	9360	12072	1578	1721	1806	391	400	393	160	160	145
0.85 ***	14956	18696	38853	5603	6653	9438	1193	1271	1380	277	286	291	.	.	.
•••	16435	22919	38915	6043	7589	9443	1231	1324	1366	283	286	275	.	.	.
0.9 ***	11316	14186	26979	4162	4859	6434	793	834	878	.	.	.	.	.	.
•••	12466	17261	26985	4460	5436	6421	811	855	864	.	.	.	.	.	.
0.95 ***	6953	8525	13498	2295	2566	3021	.	.	.	.	.	.	.	.	.
•••	7601	10030	13485	2412	2759	3007	.	.	.	.	.	.	.	.	.
0.98 ***	3200	3686	4649	.	.	.	.	.	.	.	.	.	.	.	.
•••	3415	4063	4631	.	.	.	.	.	.	.	.	.	.	.	.

TABLE 5: ALPHA= 0.025 POWER= 0.8   EXPECTED ACCRUAL THRU MINIMUM FOLLOW-UP= 3750

		DEL=.01			DEL=.02			DEL=.05			DEL=.10			DEL=.15		
FACT=		1.0 .75	.50 .25	.00 BIN	1.0 .75	.50 .25	.00 BIN	1.0 .75	.50 .25	.00 BIN	1.0 75	.50 .25	.00 BIN	1.0 .75	.50 .25	.00 BIN

PCONT=•••   REQUIRED NUMBER OF PATIENTS

PCONT		DEL=.01 1.0/.75	.50/.25	.00/BIN	DEL=.02 1.0/.75	.50/.25	.00/BIN	DEL=.05 1.0/.75	.50/.25	.00/BIN	DEL=.10 1.0/.75	.50/.25	.00/BIN	DEL=.15 1.0/.75	.50/.25	.00/BIN
0.01	•••	1203	1206	1212	443	443	443	147	147	147	72	72	72	53	53	53
	•••	1203	1206	4631	443	443	1531	147	147	417	72	72	170	53	53	101
0.02	•••	2659	2669	2703	856	859	865	231	231	231	100	100	100	68	68	68
	•••	2665	2684	7645	856	859	2277	231	231	532	100	100	197	68	68	113
0.05	•••	8047	8112	8440	2290	2318	2365	494	500	500	175	175	175	106	106	106
	•••	8069	8200	16310	2300	2337	4419	494	500	864	175	175	275	106	106	145
0.1	•••	17753	17928	19638	4803	4909	5197	940	950	968	293	297	297	156	156	156
	•••	17815	18231	29497	4844	5013	7677	944	959	1366	297	297	393	156	156	194
0.15	•••	26534	26843	31025	7066	7297	8031	1347	1375	1418	400	406	409	203	203	203
	•••	26637	27425	41113	7156	7540	10542	1362	1394	1806	400	406	495	203	203	236
0.2	•••	33625	34113	41684	8928	9297	10675	1694	1750	1831	494	500	509	247	247	247
	•••	33790	35031	51159	9068	9719	13014	1722	1784	2182	494	503	581	247	247	271
0.25	•••	38828	39528	51172	10343	10878	13015	1981	2065	2187	569	578	588	275	278	284
	•••	39063	40859	59636	10544	11497	15094	2018	2122	2496	575	584	652	278	278	299
0.3	•••	42153	43113	59256	11318	12034	14997	2206	2318	2487	631	644	659	303	306	306
	•••	42475	44931	66543	11590	12869	16781	2256	2393	2748	634	650	707	303	306	320
0.35	•••	43731	45012	65800	11894	12803	16597	2365	2506	2725	678	691	709	322	325	325
	•••	44159	47393	71881	12241	13850	18076	2431	2600	2936	681	700	746	322	325	334
0.4	•••	43759	45428	70713	12125	13231	17781	2468	2631	2894	706	725	743	331	334	340
	•••	44322	48438	75648	12550	14472	18979	2543	2744	3062	715	734	770	331	334	341
0.45	•••	42481	44609	73956	12063	13347	18550	2515	2697	2997	715	738	762	334	340	344
	•••	43197	48293	77846	12565	14768	19489	2600	2828	3124	728	747	778	334	340	341
0.5	•••	40150	42813	75503	11759	13206	18897	2515	2712	3034	715	738	762	331	334	340
	•••	41065	47172	78474	12331	14768	19607	2603	2867	3124	725	747	770	331	334	334
0.55	•••	37056	40306	75344	11272	12841	18813	2463	2669	2997	697	719	743	316	325	331
	•••	38197	45278	77532	11894	14491	19332	2556	2809	3062	709	734	746	322	325	320
0.6	•••	33494	37315	73475	10628	12275	18303	2369	2572	2900	663	687	709	303	306	312
	•••	34868	42775	75020	11290	13972	18665	2459	2712	2936	678	697	707	303	306	299
0.65	•••	29725	34028	69906	9869	11528	17369	2228	2422	2725	622	640	659	275	278	284
	•••	31315	39781	70939	10544	13206	17606	2318	2553	2748	631	650	652	278	284	271
0.7	•••	25966	30556	64643	9003	10619	16009	2047	2219	2491	556	575	593	241	247	250
	•••	27706	36359	65287	9663	12209	16153	2125	2337	2496	565	584	581	247	247	236
0.75	•••	22325	26947	57700	8031	9531	14234	1816	1966	2187	481	494	509	203	209	209
	•••	24106	32528	58067	8650	10972	14309	1887	2065	2182	484	500	495	203	209	194
0.8	•••	18803	23168	49100	6950	8256	12044	1540	1653	1816	391	397	406	156	156	162
	•••	20509	28234	49276	7493	9472	12072	1591	1728	1806	391	400	393	156	162	145
0.85	•••	15303	19100	38853	5706	6753	9438	1203	1278	1381	278	284	288	.	.	.
	•••	16806	23350	38915	6147	7672	9443	1240	1325	1366	284	288	275	.	.	.
0.9	•••	11584	14491	26978	4234	4925	6434	800	837	878	.	.	.	.	.	.
	•••	12753	17566	26985	4531	5487	6421	813	856	864	.	.	.	.	.	.
0.95	•••	7109	8678	13497	2322	2584	3022	.	.	.	.	.	.	.	.	.
	•••	7759	10165	13485	2440	2772	3007	.	.	.	.	.	.	.	.	.
0.98	•••	3256	3728	4647	.	.	.	.	.	.	.	.	.	.	.	.
	•••	3462	4094	4631	.	.	.	.	.	.	.	.	.	.	.	.

TABLE 5: ALPHA= 0.025 POWER= 0.8   EXPECTED ACCRUAL THRU MINIMUM FOLLOW-UP= 4000

	DEL=.01			DEL=.02			DEL=.05			DEL=.10			DEL=.15		
FACT=	1.0 .75	.50 .25	.00 BIN	1.0 .75	.50 .25	.00 BIN	1.0 .75	.50 .25	.00 BIN	1.0 75	.50 .25	.00 BIN	1.0 .75	.50 .25	.00 BIN
PCONT=***				REQUIRED NUMBER OF PATIENTS											
0.01 ***	1203 1203	1207 1207	1213 4631	443 443	443 443	443 1531	147 147	147 147	147 417	73 73	73 73	73 170	53 53	53 53	53 101
0.02 ***	2657 2667	2673 2683	2703 7645	857 857	857 863	863 2277	233 233	233 233	233 532	103 103	103 103	103 197	67 67	67 67	67 113
0.05 ***	8053 8077	8117 8207	8437 16310	2293 2303	2317 2337	2363 4419	497 497	497 497	503 864	177 177	177 177	177 275	103 103	103 103	103 145
0.1 ***	17767 17827	17953 18267	19637 29497	4813 4857	4923 5023	5193 7677	937 947	953 957	967 1366	297 297	297 297	297 393	157 157	157 157	157 194
0.15 ***	26553 26663	26883 27497	31027 41113	7087 7177	7317 7557	8033 10542	1347 1363	1377 1397	1417 1806	403 403	407 407	407 495	203 203	203 207	207 236
0.2 ***	33657 33833	34177 35147	41683 51159	8957 9103	9337 9757	10673 13014	1703 1727	1753 1787	1827 2182	493 497	503 503	507 581	243 243	247 247	247 271
0.25 ***	38877 39123	39617 41023	51173 59636	10383 10593	10933 11553	13017 15094	1993 2027	2073 2123	2187 2496	573 577	583 587	593 652	277 277	277 283	283 299
0.3 ***	42217 42557	43243 45153	59257 66543	11377 11657	12113 12943	14997 16781	2217 2267	2323 2397	2487 2748	633 637	643 653	657 707	303 303	307 307	307 320
0.35 ***	43817 44273	45183 47677	65797 71881	11967 12327	12903 13947	16597 18076	2383 2443	2517 2607	2723 2936	677 683	693 703	707 746	323 323	323 327	327 334
0.4 ***	43873 44467	45643 48793	70717 75648	12217 12653	13343 14587	17783 18979	2487 2557	2643 2753	2897 3062	707 713	723 733	743 770	333 333	337 337	337 341
0.45 ***	42623 43387	44877 48717	73957 77846	12167 12683	13477 14897	18553 19489	2537 2617	2713 2837	2997 3124	717 727	737 747	763 778	333 337	337 343	343 341
0.5 ***	40333 41303	43147 47663	75507 78474	11883 12463	13353 14907	18897 19607	2533 2623	2727 2857	3033 3124	717 727	737 747	763 770	333 333	333 337	343 334
0.55 ***	37287 38487	40697 45817	75343 77532	11407 12043	12993 14643	18813 19332	2483 2573	2683 2817	2997 3062	697 707	723 733	747 746	317 323	323 327	327 320
0.6 ***	33777 35217	37757 43353	73477 75020	10773 11443	12433 14123	18303 18665	2387 2477	2587 2723	2897 2936	667 677	687 697	713 707	303 303	307 307	313 299
0.65 ***	30057 31713	34507 40373	69903 70939	10017 10697	11693 13353	17367 17606	2247 2333	2437 2563	2727 2748	623 627	637 647	663 652	277 277	277 283	283 271
0.7 ***	26333 28127	31053 36943	64643 65287	9147 9813	10773 12347	16013 16153	2063 2143	2233 2347	2493 2496	557 567	573 583	593 581	243 247	247 247	247 236
0.75 ***	22707 24533	27433 33083	57703 58067	8173 8793	9673 11097	14233 14309	1833 1897	1977 2067	2187 2182	483 487	493 503	507 495	203 207	207 207	207 194
0.8 ***	19173 20913	23613 28733	49097 49276	7073 7617	8377 9577	12043 12072	1553 1603	1663 1733	1817 1806	393 393	397 403	407 393	157 157	157 157	163 145
0.85 ***	15627 17157	19483 23757	38853 38915	5807 6247	6843 7747	9443 9443	1213 1247	1283 1327	1377 1366	283 283	287 287	293 275	. .	. .	. .
0.9 ***	11843 13023	14773 17853	26977 26985	4303 4597	4983 5533	6433 6421	803 817	837 857	877 864	. .	. .	. .	. .	. .	. .
0.95 ***	7253 7903	8823 10293	13497 13485	2353 2467	2607 2783	3023 3007	. .	. .	. .	. .	. .	. .	. .	. .	. .
0.98 ***	3303 3507	3767 4123	4647 4631	. .	. .	. .	. .	. .	. .	. .	. .	. .	. .	. .	. .

TABLE 5: ALPHA= 0.025 POWER= 0.8   EXPECTED ACCRUAL THRU MINIMUM FOLLOW-UP= 4250

REQUIRED NUMBER OF PATIENTS

PCONT=***	DEL=.01			DEL=.02			DEL=.05			DEL=.10			DEL=.15		
FACT=	1.0 .75	.50 .25	.00 BIN	1.0 .75	.50 .25	.00 BIN	1.0 .75	.50 .25	.00 BIN	1.0 75	.50 .25	.00 BIN	1.0 .75	.50 .25	.00 BIN
0.01 ***	1204	1204	1208	443	443	443	145	145	145	71	71	71	50	50	50
•••	1204	1208	4631	443	443	1531	145	145	417	71	71	170	50	50	101
0.02 ***	2659	2674	2702	857	857	864	230	230	230	103	103	103	67	67	67
•••	2663	2685	7645	857	864	2277	230	230	532	103	103	197	67	67	113
0.05 ***	8057	8125	8439	2292	2319	2366	496	496	503	177	177	177	103	103	103
•••	8082	8216	16310	2309	2341	4419	496	503	864	177	177	275	103	103	145
0.1 ***	17779	17974	19638	4820	4933	5192	942	953	963	294	294	301	156	156	156
•••	17843	18303	29497	4863	5029	7677	949	959	1366	294	294	393	156	156	194
0.15 ***	26576	26927	31028	7105	7338	8036	1353	1378	1416	400	407	407	205	205	205
•••	26693	27564	41113	7196	7579	10542	1363	1399	1806	400	407	495	205	205	236
0.2 ***	33695	34241	41685	8985	9374	10675	1707	1756	1831	496	503	507	241	248	248
•••	33876	35257	51159	9134	9789	13014	1728	1788	2182	496	503	581	248	248	271
0.25 ***	38923	39713	51173	10426	10989	13012	2001	2075	2185	570	581	592	279	279	279
•••	39188	41186	59636	10643	11606	15094	2037	2128	2496	577	588	652	279	279	299
0.3 ***	42280	43368	59259	11429	12183	14999	2228	2334	2489	634	645	655	301	305	311
•••	42645	45372	66543	11723	13019	16781	2277	2404	2748	641	651	707	305	305	320
0.35 ***	43906	45351	65797	12041	12991	16593	2394	2525	2723	677	694	708	322	322	326
•••	44388	47958	71881	12407	14039	18076	2451	2610	2936	687	698	746	322	326	334
0.4 ***	43984	45861	70717	12300	13448	17783	2500	2653	2893	708	726	747	333	337	337
•••	44618	49137	75648	12753	14698	18979	2568	2759	3062	715	736	770	333	337	341
0.45 ***	42769	45149	73957	12275	13603	18554	2553	2727	2999	719	740	762	333	337	343
•••	43580	49127	77846	12796	15020	19489	2632	2844	3124	730	751	778	337	343	341
0.5 ***	40516	43474	75504	12003	13486	18898	2553	2738	3031	719	740	762	333	337	343
•••	41540	48128	78474	12594	15042	19607	2638	2865	3124	730	751	770	333	337	334
0.55 ***	37520	41079	75345	11535	13140	18813	2504	2695	2999	704	719	747	322	326	326
•••	38778	46339	77532	12179	14783	19332	2589	2829	3062	708	730	746	322	326	320
0.6 ***	34056	38189	73475	10915	12583	18303	2408	2600	2897	673	687	708	301	305	311
•••	35558	43899	75020	11588	14262	18665	2493	2727	2936	677	698	707	305	305	299
0.65 ***	30380	34970	69905	10161	11839	17368	2266	2451	2727	623	641	662	279	279	283
•••	32095	40935	70939	10845	13490	17606	2351	2574	2748	630	651	652	279	283	271
0.7 ***	26693	31528	64639	9289	10915	16008	2079	2245	2489	560	577	592	241	248	252
•••	28535	37499	65287	9959	12477	16153	2160	2355	2496	566	581	581	248	248	236
0.75 ***	23074	27894	57701	8301	9803	14234	1845	1983	2185	485	496	507	205	205	209
•••	24944	33603	58067	8922	11213	14309	1909	2075	2182	492	503	495	205	209	194
0.8 ***	19525	24041	49101	7186	8493	12041	1565	1671	1813	390	400	407	156	156	163
•••	21300	29194	49276	7732	9672	12072	1612	1735	1806	396	400	393	156	163	145
0.85 ***	15941	19844	38852	5900	6931	9438	1218	1289	1378	279	283	290	.	.	.
•••	17492	24137	38915	6336	7817	9443	1250	1331	1366	283	290	275	.	.	.
0.9 ***	12088	15042	26980	4363	5033	6431	804	836	878	.	.	.	.	.	.
•••	13278	18119	26985	4657	5571	6421	821	857	864	.	.	.	.	.	.
0.95 ***	7388	8960	13497	2377	2628	3021	.	.	.	.	.	.	.	.	.
•••	8040	10409	13485	2489	2798	3007	.	.	.	.	.	.	.	.	.
0.98 ***	3343	3807	4646	.	.	.	.	.	.	.	.	.	.	.	.
•••	3545	4147	4631	.	.	.	.	.	.	.	.	.	.	.	.

TABLE 5: ALPHA= 0.025 POWER= 0.8   EXPECTED ACCRUAL THRU MINIMUM FOLLOW-UP= 4500

	DEL=.01			DEL=.02			DEL=.05			DEL=.10			DEL=.15		
FACT=	1.0 .75	.50 .25	.00 BIN	1.0 .75	.50 .25	.00 BIN	1.0 .75	.50 .25	.00 BIN	1.0 75	.50 .25	.00 BIN	1.0 .75	.50 .25	.00 BIN
PCONT=***				REQUIRED NUMBER OF PATIENTS											
0.01 ***	1200	1207	1211	442	442	442	143	143	143	75	75	75	53	53	53
•••	1207	1207	4631	442	442	1531	143	143	417	75	75	170	53	53	101
0.02 ***	2663	2674	2703	858	858	863	233	233	233	98	105	105	64	64	64
•••	2670	2685	7645	858	863	2277	233	233	532	98	105	197	64	64	113
0.05 ***	8058	8130	8441	2298	2321	2366	498	498	503	176	176	176	105	105	105
•••	8085	8227	16310	2310	2343	4419	498	498	864	176	176	275	105	105	145
0.1 ***	17790	17996	19639	4830	4942	5194	941	953	964	296	296	300	154	161	161
•••	17861	18334	29497	4875	5036	7677	948	960	1366	296	296	393	154	161	194
0.15 ***	26591	26963	31024	7125	7361	8036	1353	1380	1414	401	408	408	206	206	206
•••	26722	27633	41113	7215	7597	10542	1369	1398	1806	401	408	495	206	206	236
0.2 ***	33724	34305	41666	9015	9408	10673	1713	1763	1830	498	503	510	244	244	244
•••	33915	35366	51159	9172	9825	13014	1736	1792	2182	498	503	581	244	244	271
0.25 ***	38966	39806	51173	10466	11040	13013	2006	2085	2186	570	581	588	278	278	285
•••	39248	41340	59636	10691	11651	15094	2040	2130	2496	577	588	652	278	278	299
0.3 ***	42348	43496	59257	11483	12248	15000	2235	2336	2490	633	645	660	300	307	307
•••	42731	45581	66543	11782	13080	16781	2287	2404	2748	638	649	707	307	307	320
0.35 ***	43991	45514	65798	12108	13076	16597	2404	2535	2726	678	694	712	323	323	330
•••	44501	48232	71881	12484	14122	18076	2467	2618	2936	690	701	746	323	323	334
0.4 ***	44096	46072	70714	12390	13553	17783	2516	2663	2895	712	728	746	330	334	341
•••	44767	49474	75648	12844	14790	18979	2584	2764	3062	716	735	770	334	334	341
0.45 ***	42911	45413	73961	12371	13721	18555	2568	2737	3000	723	739	761	334	341	345
•••	43770	49519	77846	12907	15135	19489	2647	2854	3124	735	750	778	334	341	341
0.5 ***	40699	43793	75506	12113	13616	18896	2568	2753	3034	723	739	761	330	334	341
•••	41779	48581	78474	12716	15161	19607	2651	2876	3124	728	750	770	334	341	334
0.55 ***	37747	41453	75345	11658	13278	18813	2523	2708	3000	705	723	746	318	323	330
•••	39068	46830	77532	12311	14910	19332	2606	2838	3062	712	735	746	323	330	320
0.6 ***	34327	38606	73477	11044	12727	18307	2426	2613	2899	671	690	712	300	307	311
•••	35895	44430	75020	11726	14392	18665	2512	2737	2936	683	701	707	307	307	299
0.65 ***	30698	35418	69904	10290	11978	17366	2287	2460	2726	626	645	660	278	278	285
•••	32471	41471	70939	10983	13620	17606	2366	2580	2748	633	649	652	278	285	271
0.7 ***	27037	31980	64639	9420	11051	16012	2096	2258	2490	566	577	593	244	244	251
•••	28931	38028	65287	10088	12596	16153	2168	2359	2496	570	581	581	244	251	236
0.75 ***	23430	28335	57698	8423	9926	14235	1860	1995	2186	487	498	510	206	206	210
•••	25338	34102	58067	9048	11314	14309	1920	2085	2182	491	503	495	206	206	194
0.8 ***	19864	24443	49098	7293	8591	12041	1571	1673	1815	390	397	408	161	161	161
•••	21671	29636	49276	7838	9757	12072	1623	1740	1806	397	401	393	161	161	145
0.85 ***	16241	20190	38850	5988	7012	9442	1223	1290	1380	285	285	289	.	.	.
•••	17816	24495	38915	6420	7878	9443	1256	1335	1366	285	289	275	.	.	.
0.9 ***	12315	15296	26981	4425	5088	6431	806	840	881	.	.	.	.	.	.
•••	13519	18368	26985	4710	5606	6421	825	858	864	.	.	.	.	.	.
0.95 ***	7518	9086	13496	2400	2640	3023	.	.	.	.	.	.	.	.	.
•••	8175	10522	13485	2512	2809	3007	.	.	.	.	.	.	.	.	.
0.98 ***	3390	3840	4650	.	.	.	.	.	.	.	.	.	.	.	.
•••	3585	4170	4631	.	.	.	.	.	.	.	.	.	.	.	.

TABLE 5: ALPHA= 0.025 POWER= 0.8   EXPECTED ACCRUAL THRU MINIMUM FOLLOW-UP= 4750

		DEL=.01			DEL=.02			DEL=.05			DEL=.10			DEL=.15		
FACT=		1.0 .75	.50 .25	.00 BIN	1.0 .75	.50 .25	.00 BIN	1.0 .75	.50 .25	.00 BIN	1.0 75	.50 .25	.00 BIN	1.0 .75	.50 .25	.00 BIN
PCONT=•••					REQUIRED NUMBER OF PATIENTS											
0.01	***	1203	1207	1207	443	443	443	146	146	146	75	75	75	51	51	51
	•••	1203	1207	4631	443	443	1531	146	146	417	75	75	170	51	51	101
0.02	***	2663	2675	2699	858	858	863	234	234	234	103	103	103	67	67	67
	•••	2668	2687	7645	858	863	2277	234	234	532	103	103	197	67	67	113
0.05	***	8067	8138	8439	2300	2324	2367	495	495	502	174	174	174	103	103	103
	•••	8090	8233	16310	2312	2343	4419	495	502	864	174	174	275	103	103	145
0.1	***	17804	18018	19637	4837	4948	5193	942	953	965	293	293	300	158	158	158
	•••	17875	18362	29497	4884	5043	7677	946	958	1366	293	300	393	158	158	194
0.15	***	26615	27007	31025	7140	7378	8035	1357	1385	1417	400	407	407	205	205	205
	•••	26746	27696	41113	7235	7615	10540	1369	1397	1806	407	407	495	205	205	236
0.2	***	33757	34370	41685	9040	9444	10672	1718	1765	1825	495	502	507	245	245	245
	•••	33959	35474	51159	9199	9852	13014	1737	1797	2182	495	502	581	245	245	271
0.25	***	39017	39896	51173	10505	11087	13018	2015	2086	2188	573	578	590	277	281	281
	•••	39310	41495	59636	10731	11700	15094	2046	2134	2496	578	585	652	277	281	299
0.3	***	42409	43620	59260	11539	12318	15002	2248	2347	2490	637	645	657	305	305	305
	•••	42817	45786	66543	11843	13142	16781	2295	2407	2748	637	649	707	305	305	320
0.35	***	44076	45679	65795	12175	13161	16597	2419	2545	2723	680	692	709	324	324	329
	•••	44618	48494	71881	12560	14199	18076	2473	2620	2936	685	704	746	324	324	334
0.4	***	44207	46285	70712	12472	13648	17785	2525	2675	2894	709	728	744	329	336	340
	•••	44915	49795	75648	12935	14887	18979	2597	2775	3062	720	732	770	336	336	341
0.45	***	43055	45675	73961	12472	13830	18552	2585	2751	2996	728	740	763	336	340	340
	•••	43957	49902	77846	13011	15239	19489	2656	2858	3124	732	752	778	336	340	341
0.5	***	40882	44107	75505	12223	13743	18897	2585	2763	3032	720	740	763	329	336	340
	•••	42010	49012	78474	12833	15279	19607	2668	2882	3124	732	752	770	336	336	334
0.55	***	37972	41815	75343	11776	13410	18813	2537	2723	3000	704	720	744	324	324	329
	•••	39350	47306	77532	12437	15030	19332	2620	2846	3062	716	732	746	324	329	320
0.6	***	34600	39013	73474	11170	12857	18303	2442	2628	2894	673	692	709	300	305	312
	•••	36222	44931	75020	11859	14515	18665	2525	2747	2936	680	704	707	305	305	299
0.65	***	31009	35847	69904	10418	12116	17369	2300	2473	2727	625	645	661	277	281	281
	•••	32830	41986	70939	11118	13735	17606	2378	2585	2748	633	649	652	277	281	271
0.7	***	27375	32422	64644	9544	11178	16011	2110	2264	2490	566	578	590	245	245	253
	•••	29311	38530	65287	10220	12710	16153	2181	2367	2496	573	585	581	245	245	236
0.75	***	23777	28757	57704	8542	10042	14234	1872	2003	2188	483	495	507	205	205	210
	•••	25717	34572	58067	9164	11415	14309	1932	2086	2182	490	502	495	205	205	194
0.8	***	20195	24834	49099	7394	8689	12045	1583	1682	1813	395	400	407	158	158	158
	•••	22024	30052	49276	7940	9836	12072	1630	1742	1806	395	400	393	158	158	145
0.85	***	16526	20516	38851	6072	7085	9437	1231	1298	1381	281	288	288	.	.	.
	•••	18125	24834	38915	6504	7936	9443	1262	1333	1366	281	288	275	.	.	.
0.9	***	12543	15536	26976	4480	5133	6432	811	839	875	.	.	.	.	.	.
	•••	13755	18600	26985	4765	5644	6421	827	858	864	.	.	.	.	.	.
0.95	***	7639	9207	13498	2419	2656	3020	.	.	.	.	.	.	.	.	.
	•••	8297	10624	13485	2533	2818	3007	.	.	.	.	.	.	.	.	.
0.98	***	3428	3867	4647	.	.	.	.	.	.	.	.	.	.	.	.
	•••	3625	4188	4631	.	.	.	.	.	.	.	.	.	.	.	.

TABLE 5: ALPHA= 0.025 POWER= 0.8   EXPECTED ACCRUAL THRU MINIMUM FOLLOW-UP= 5000

		DEL=.01			DEL=.02			DEL=.05			DEL=.10			DEL=.15		
FACT=		1.0 .75	.50 .25	.00 BIN	1.0 .75	.50 .25	.00 BIN	1.0 .75	.50 .25	.00 BIN	1.0 75	.50 .25	.00 BIN	1.0 .75	.50 .25	.00 BIN
PCONT=***				REQUIRED NUMBER OF PATIENTS												
0.01	***	1204	1208	1208	441	441	441	146	146	146	71	71	71	54	54	54
	•••	1204	1208	4631	441	441	1531	146	146	417	71	71	170	54	54	101
0.02	***	2666	2679	2704	858	858	866	233	233	233	104	104	104	66	66	66
	•••	2671	2683	7645	858	858	2277	233	233	532	104	104	197	66	66	113
0.05	***	8071	8146	8441	2304	2329	2366	496	496	504	179	179	179	104	104	104
	•••	8096	8241	16310	2308	2341	4419	496	496	864	179	179	275	104	104	145
0.1	***	17816	18041	19641	4846	4954	5196	946	954	966	296	296	296	158	158	158
	•••	17891	18396	29497	4891	5046	7677	946	958	1366	296	296	393	158	158	194
0.15	***	26633	27046	31029	7154	7396	8033	1358	1383	1416	404	404	408	204	204	204
	•••	26771	27758	41113	7254	7629	10542	1371	1404	1806	404	408	495	204	204	236
0.2	***	33791	34429	41683	9071	9471	10671	1721	1766	1829	496	504	508	246	246	246
	•••	34004	35571	51159	9229	9879	13014	1741	1796	2182	496	504	581	246	246	271
0.25	***	39058	39983	51171	10541	11133	13016	2021	2091	2191	571	583	591	279	279	283
	•••	39371	41641	59636	10779	11741	15094	2054	2133	2496	579	583	652	279	279	299
0.3	***	42471	43741	59258	11591	12379	14996	2254	2354	2491	633	646	658	304	304	308
	•••	42904	45991	66543	11896	13204	16781	2296	2416	2748	641	654	707	304	308	320
0.35	***	44158	45841	65796	12241	13233	16596	2429	2546	2721	683	696	708	321	321	329
	•••	44729	48754	71881	12633	14271	18076	2483	2629	2936	691	704	746	321	329	334
0.4	***	44321	46496	70716	12554	13741	17783	2541	2683	2896	716	729	746	333	333	341
	•••	45058	50108	75648	13021	14971	18979	2608	2779	3062	721	733	770	333	341	341
0.45	***	43196	45933	73958	12566	13933	18554	2596	2758	2996	729	741	758	333	341	341
	•••	44146	50271	77846	13116	15333	19489	2671	2866	3124	733	754	778	341	341	341
0.5	***	41066	44416	75504	12329	13858	18896	2604	2779	3033	721	741	758	333	333	341
	•••	42241	49429	78474	12946	15383	19607	2683	2891	3124	733	754	770	333	341	334
0.55	***	38196	42171	75346	11896	13529	18816	2554	2733	2996	708	729	746	321	321	329
	•••	39629	47766	77532	12558	15141	19332	2633	2854	3062	716	733	746	321	329	320
0.6	***	34866	39404	73479	11291	12983	18304	2458	2633	2896	679	691	708	304	308	308
	•••	36541	45408	75020	11983	14629	18665	2541	2754	2936	683	704	707	304	308	299
0.65	***	31316	36266	69904	10541	12241	17371	2316	2483	2729	629	641	658	279	279	283
	•••	33183	42471	70939	11241	13846	17606	2391	2591	2748	633	654	652	279	283	271
0.7	***	27704	32841	64641	9666	11296	16008	2129	2279	2491	566	579	591	246	246	246
	•••	29683	39008	65287	10341	12808	16153	2196	2371	2496	571	583	581	246	246	236
0.75	***	24108	29166	57704	8654	10146	14233	1883	2008	2183	483	496	508	204	208	208
	•••	26083	35021	58067	9279	11504	14309	1946	2091	2182	491	504	495	204	208	194
0.8	***	20508	25204	49096	7491	8783	12041	1591	1691	1816	391	404	408	158	158	158
	•••	22366	30454	49276	8041	9908	12072	1633	1746	1806	396	404	393	158	158	145
0.85	***	16804	20829	38854	6146	7154	9441	1241	1304	1379	283	283	291	.	.	.
	•••	18416	25154	38915	6579	7991	9443	1266	1341	1366	283	291	275	.	.	.
0.9	***	12754	15766	26979	4529	5179	6433	816	846	879	.	.	.	.	.	.
	•••	13979	18816	26985	4816	5671	6421	829	858	864	.	.	.	.	.	.
0.95	***	7758	9316	13496	2441	2671	3021	.	.	.	.	.	.	.	.	.
	•••	8416	10716	13485	2546	2829	3007	.	.	.	.	.	.	.	.	.
0.98	***	3458	3896	4646	.	.	.	.	.	.	.	.	.	.	.	.
	•••	3654	4208	4631	.	.	.	.	.	.	.	.	.	.	.	.

TABLE 5: ALPHA= 0.025 POWER= 0.8   EXPECTED ACCRUAL THRU MINIMUM FOLLOW-UP= 5500

	DEL=.01			DEL=.02			DEL=.05			DEL=.10			DEL=.15		
FACT=	1.0 .75	.50 .25	.00 BIN	1.0 .75	.50 .25	.00 BIN	1.0 .75	.50 .25	.00 BIN	1.0 75	.50 .25	.00 BIN	1.0 .75	.50 .25	.00 BIN

PCONT=***   REQUIRED NUMBER OF PATIENTS

PCONT															
0.01 ***	1205	1205	1214	444	444	444	147	147	147	73	73	73	50	50	50
•••	1205	1205	4631	444	444	1531	147	147	417	73	73	170	50	50	101
0.02 ***	2663	2677	2699	856	862	862	229	229	229	100	100	100	64	64	64
•••	2671	2690	7645	856	862	2277	229	229	532	100	100	197	64	64	113
0.05 ***	8080	8158	8438	2305	2328	2360	499	499	499	174	174	174	105	105	105
•••	8108	8254	16310	2314	2341	4419	499	499	864	174	174	275	105	105	145
0.1 ***	17838	18077	19639	4858	4968	5193	944	953	966	298	298	298	155	155	155
•••	17920	18448	29497	4904	5064	7677	953	958	1366	298	298	393	155	155	194
0.15 ***	26679	27124	31024	7187	7429	8034	1365	1384	1420	403	408	408	202	202	202
•••	26830	27875	41113	7283	7654	10542	1379	1398	1806	403	408	495	202	202	236
0.2 ***	33856	34558	41680	9120	9533	10674	1728	1769	1824	499	504	504	243	243	243
•••	34090	35768	51159	9285	9931	13014	1750	1797	2182	499	504	581	243	243	271
0.25 ***	39155	40168	51173	10619	11215	13017	2030	2099	2190	573	581	587	279	279	279
•••	39494	41928	59636	10858	11820	15094	2066	2140	2496	581	587	652	279	279	299
0.3 ***	42601	43990	59258	11691	12494	14997	2273	2360	2484	636	650	655	306	306	306
•••	43069	46374	66543	12008	13305	16781	2314	2415	2748	642	650	707	306	306	320
0.35 ***	44334	46168	65798	12370	13383	16592	2451	2561	2726	683	697	710	320	325	325
•••	44953	49243	71881	12769	14405	18076	2498	2635	2936	691	705	746	325	325	334
0.4 ***	44545	46905	70712	12700	13919	17783	2561	2699	2897	719	733	746	334	334	339
•••	45357	50705	75648	13190	15129	18979	2630	2787	3062	719	738	770	334	339	341
0.45 ***	43487	46438	73957	12742	14130	18553	2622	2773	3001	733	746	760	334	339	339
•••	44513	50975	77846	13305	15514	19489	2690	2878	3124	738	752	778	339	339	341
0.5 ***	41424	45013	75505	12530	14070	18896	2630	2795	3034	724	746	760	334	334	339
•••	42698	50219	78474	13154	15583	19607	2704	2897	3124	738	752	770	334	339	334
0.55 ***	38633	42854	75345	12109	13759	18814	2580	2754	3001	710	724	746	320	325	325
•••	40173	48624	77532	12783	15349	19332	2663	2864	3062	719	738	746	325	325	320
0.6 ***	35388	40154	73475	11518	13223	18305	2484	2658	2897	678	691	710	306	306	312
•••	37162	46314	75020	12219	14832	18665	2567	2768	2936	683	705	707	306	306	299
0.65 ***	31904	37052	69900	10770	12467	17370	2341	2506	2726	628	642	664	279	279	284
•••	33862	43390	70939	11477	14048	17606	2415	2603	2748	636	650	652	279	284	271
0.7 ***	28334	33642	64643	9882	11513	16009	2149	2292	2493	568	581	595	243	251	251
•••	30391	39906	65287	10564	12998	16153	2218	2383	2496	573	587	581	243	251	236
0.75 ***	24740	29938	57699	8859	10349	14235	1907	2025	2182	490	499	504	202	210	210
•••	26780	35855	58067	9483	11669	14309	1962	2099	2182	490	504	495	202	210	194
0.8 ***	21110	25900	49097	7676	8955	12040	1604	1695	1819	394	403	408	160	160	160
•••	23013	31189	49276	8218	10047	12072	1654	1750	1806	394	403	393	160	160	145
0.85 ***	17329	21418	38853	6293	7283	9442	1247	1310	1379	284	284	293	.	.	.
•••	18970	25749	38915	6719	8089	9443	1274	1343	1366	284	284	275	.	.	.
0.9 ***	13154	16188	26981	4624	5256	6430	820	848	875	.	.	.	.	.	.
•••	14392	19218	26985	4904	5729	6421	834	862	864	.	.	.	.	.	.
0.95 ***	7970	9524	13498	2479	2699	3020	.	.	.	.	.	.	.	.	.
•••	8630	10885	13485	2580	2842	3007	.	.	.	.	.	.	.	.	.
0.98 ***	3529	3942	4643	.	.	.	.	.	.	.	.	.	.	.	.
•••	3716	4239	4631	.	.	.	.	.	.	.	.	.	.	.	.

TABLE 5: ALPHA= 0.025 POWER= 0.8   EXPECTED ACCRUAL THRU MINIMUM FOLLOW-UP= 6000

	DEL=.01			DEL=.02			DEL=.05			DEL=.10			DEL=.15		
FACT=	1.0 .75	.50 .25	.00 BIN	1.0 .75	.50 .25	.00 BIN	1.0 .75	.50 .25	.00 BIN	1.0 75	.50 .25	.00 BIN	1.0 .75	.50 .25	.00 BIN
PCONT=***			REQUIRED	NUMBER	OF PATIENTS										
0.01 ***	1204	1204	1210	439	445	445	145	145	145	70	70	70	49	49	49
***	1204	1210	4631	445	445	1531	145	145	417	70	70	170	49	49	101
0.02 ***	2665	2680	2704	859	859	865	229	229	229	100	100	100	64	64	64
***	2674	2689	7645	859	859	2277	229	229	532	100	100	197	64	64	113
0.05 ***	8089	8170	8440	2305	2329	2365	499	499	499	175	175	175	100	100	100
***	8119	8260	16310	2320	2344	4419	499	499	864	175	175	275	100	100	145
0.1 ***	17860	18124	19639	4870	4984	5194	949	955	964	295	295	295	154	160	160
***	17950	18499	29497	4924	5065	7677	949	964	1366	295	295	393	160	160	194
0.15 ***	26719	27205	31024	7219	7459	8035	1369	1390	1414	400	409	409	205	205	205
***	26884	27985	41113	7315	7675	10542	1375	1405	1806	409	409	495	205	205	236
0.2 ***	33919	34675	41680	9169	9589	10675	1735	1774	1825	499	499	505	244	244	244
***	34174	35959	51159	9340	9979	13014	1750	1804	2182	499	505	581	244	244	271
0.25 ***	39250	40345	51175	10690	11299	13015	2044	2104	2185	574	580	589	280	280	280
***	39619	42199	59636	10930	11890	15094	2074	2140	2496	580	589	652	280	280	299
0.3 ***	42730	44230	59260	11779	12595	14995	2284	2374	2485	640	649	655	304	304	310
***	43240	46744	66543	12109	13399	16781	2320	2425	2748	640	655	707	304	304	320
0.35 ***	44500	46480	65800	12484	13510	16594	2464	2575	2725	685	700	709	325	325	325
***	45184	49705	71881	12904	14524	18076	2515	2644	2936	694	700	746	325	325	334
0.4 ***	44764	47305	70714	12844	14074	17785	2584	2710	2899	715	730	745	334	334	340
***	45640	51274	75648	13339	15274	18979	2644	2794	3062	724	739	770	334	340	341
0.45 ***	43765	46924	73960	12904	14314	18550	2644	2794	2995	730	745	760	334	340	340
***	44875	51625	77846	13480	15679	19489	2710	2884	3124	739	754	778	340	340	341
0.5 ***	41779	45589	75505	12715	14269	18895	2650	2809	3034	730	745	760	334	334	340
***	43144	50959	78474	13354	15754	19607	2725	2914	3124	739	754	770	334	340	334
0.55 ***	39064	43504	75340	12310	13969	18814	2605	2770	2995	715	730	745	319	325	325
***	40699	49420	77532	12994	15529	19332	2680	2875	3062	724	739	746	325	325	320
0.6 ***	35890	40864	73474	11725	13435	18304	2509	2674	2899	679	694	709	304	304	310
***	37759	47155	75020	12430	15019	18665	2584	2770	2936	685	700	707	304	310	299
0.65 ***	32470	37795	69904	10984	12679	17365	2365	2515	2725	634	649	664	280	280	280
***	34510	44239	70939	11689	14230	17606	2434	2614	2748	640	655	652	280	280	271
0.7 ***	28930	34384	64639	10090	11710	16009	2170	2305	2494	565	580	595	244	250	250
***	31054	40729	65287	10774	13165	16153	2230	2389	2496	574	589	581	244	250	236
0.75 ***	25339	30649	57700	9049	10525	14230	1924	2035	2185	490	499	505	205	205	205
***	27430	36625	58067	9670	11815	14309	1975	2104	2182	490	505	495	205	205	194
0.8 ***	21670	26554	49099	7840	9100	12040	1624	1705	1819	394	400	409	160	160	160
***	23614	31855	49276	8380	10165	12072	1660	1759	1806	400	400	393	160	160	145
0.85 ***	17815	21955	38854	6424	7390	9439	1255	1315	1375	280	289	289	.	.	.
***	19480	26284	38915	6844	8170	9443	1285	1345	1366	289	289	275	.	.	.
0.9 ***	13519	16579	26980	4714	5320	6430	820	850	880	.	.	.	.	.	.
***	14770	19585	26985	4984	5770	6421	835	865	864	.	.	.	.	.	.
0.95 ***	8170	9709	13495	2509	2719	3019	.	.	.	.	.	.	.	.	.
***	8824	11035	13485	2605	2854	3007	.	.	.	.	.	.	.	.	.
0.98 ***	3589	3985	4645	.	.	.	.	.	.	.	.	.	.	.	.
***	3769	4270	4631	.	.	.	.	.	.	.	.	.	.	.	.

TABLE 5: ALPHA= 0.025 POWER= 0.8 — EXPECTED ACCRUAL THRU MINIMUM FOLLOW-UP= 6500

	DEL=.01			DEL=.02			DEL=.05			DEL=.10			DEL=.15		
FACT=	1.0 .75	.50 .25	.00 BIN	1.0 .75	.50 .25	.00 BIN	1.0 .75	.50 .25	.00 BIN	1.0 75	.50 .25	.00 BIN	1.0 .75	.50 .25	.00 BIN

PCONT=*** REQUIRED NUMBER OF PATIENTS

PCONT	1.0/.75	.50/.25	.00/BIN	1.0/.75	.50/.25	.00/BIN	1.0/.75	.50/.25	.00/BIN	1.0/75	.50/.25	.00/BIN	1.0/.75	.50/.25	.00/BIN
0.01 ***	1207	1207	1207	443	443	443	141	141	141	76	76	76	53	53	53
***	1207	1207	4631	443	443	1531	141	141	417	76	76	170	53	53	101
0.02 ***	2670	2686	2702	856	856	866	232	232	232	102	102	102	70	70	70
***	2676	2692	7645	856	866	2277	232	232	532	102	102	197	70	70	113
0.05 ***	8097	8178	8438	2312	2334	2381	498	498	498	173	173	173	102	102	102
***	8130	8276	16310	2318	2345	4419	498	498	864	173	173	275	102	102	145
0.1 ***	17880	18162	19641	4886	4993	5194	947	953	963	297	297	297	157	157	157
***	17977	18546	29497	4934	5075	7677	953	963	1366	297	297	393	157	157	194
0.15 ***	26758	27278	31026	7242	7486	8032	1370	1392	1418	401	411	411	206	206	206
***	26937	28091	41113	7350	7697	10542	1376	1402	1806	401	411	495	206	206	236
0.2 ***	33983	34796	41686	9218	9631	10671	1743	1782	1831	498	498	508	248	248	248
***	34260	36134	51159	9387	10015	13014	1760	1798	2182	498	508	581	248	248	271
0.25 ***	39346	40522	51176	10752	11369	13011	2052	2107	2188	579	579	590	281	281	281
***	39742	42456	59636	11006	11948	15094	2074	2150	2496	579	590	652	281	281	299
0.3 ***	42856	44471	59258	11873	12696	15003	2296	2377	2491	638	644	661	303	303	303
***	43408	47087	66543	12208	13482	16781	2334	2426	2748	644	655	707	303	303	320
0.35 ***	44676	46794	65801	12598	13638	16596	2481	2578	2724	687	693	709	319	330	330
***	45407	50142	71881	13021	14630	18076	2529	2643	2936	693	703	746	319	330	334
0.4 ***	44985	47688	70714	12978	14213	17782	2605	2724	2897	720	726	742	336	336	336
***	45933	51799	75648	13482	15399	18979	2659	2800	3062	726	736	770	336	336	341
0.45 ***	44048	47396	73958	13059	14473	18552	2670	2806	3001	736	742	758	336	336	346
***	45234	52248	77846	13644	15816	19489	2735	2897	3124	742	752	778	336	346	341
0.5 ***	42125	46138	75502	12891	14451	18893	2676	2822	3033	736	742	758	336	336	336
***	43577	51647	78474	13531	15913	19607	2741	2919	3124	742	752	770	336	336	334
0.55 ***	39492	44123	75346	12501	14158	18812	2627	2783	3001	720	726	742	319	330	330
***	41204	50168	77532	13183	15692	19332	2702	2881	3062	720	736	746	319	330	320
0.6 ***	36378	41529	73477	11922	13628	18302	2529	2686	2897	687	693	709	303	303	313
***	38328	47932	75020	12631	15182	18665	2605	2783	2936	687	703	707	303	303	299
0.65 ***	33008	38491	69902	11174	12875	17366	2383	2529	2724	638	644	661	281	281	281
***	35121	45023	70939	11889	14392	17606	2458	2621	2748	638	655	652	281	281	271
0.7 ***	29498	35078	64637	10281	11889	16011	2188	2318	2491	573	579	590	248	248	248
***	31682	41491	65287	10957	13313	16153	2247	2399	2496	573	590	581	248	248	236
0.75 ***	25907	31318	57698	9224	10687	14229	1938	2042	2188	492	498	508	206	206	206
***	28042	37331	58067	9842	11948	14309	1987	2107	2182	498	508	495	206	206	194
0.8 ***	22202	27148	49096	7989	9235	12046	1630	1711	1814	395	401	411	157	157	157
***	24174	32472	49276	8526	10275	12072	1668	1760	1806	401	401	393	157	157	145
0.85 ***	18270	22452	38848	6543	7496	9436	1262	1321	1376	287	287	287	.	.	.
***	19960	26774	38915	6960	8249	9443	1288	1343	1366	287	287	275	.	.	.
0.9 ***	13866	16937	26980	4788	5383	6429	823	850	882	.	.	.	.	.	.
***	15123	19911	26985	5048	5812	6421	839	866	864	.	.	.	.	.	.
0.95 ***	8357	9874	13498	2540	2741	3017	.	.	.	.	.	.	.	.	.
***	9007	11168	13485	2627	2865	3007	.	.	.	.	.	.	.	.	.
0.98 ***	3634	4024	4642	.	.	.	.	.	.	.	.	.	.	.	.
***	3813	4295	4631	.	.	.	.	.	.	.	.	.	.	.	.

598   *CRC Handbook of Sample Size Guidelines for Clinical Trials*

TABLE 5: ALPHA= 0.025 POWER= 0.8   EXPECTED ACCRUAL THRU MINIMUM FOLLOW-UP= 7000

	DEL=.01			DEL=.02			DEL=.05			DEL=.10			DEL=.15		
FACT=	1.0 .75	.50 .25	.00 BIN	1.0 .75	.50 .25	.00 BIN	1.0 .75	.50 .25	.00 BIN	1.0 75	.50 .25	.00 BIN	1.0 .75	.50 .25	.00 BIN
PCONT=•••				REQUIRED NUMBER OF PATIENTS											
0.01 •••	1201 1201	1201 1212	1212 4631	442 442	442 442	442 1531	145 145	145 145	145 417	75 75	75 75	75 170	46 46	46 46	46 101
0.02 •••	2671 2671	2682 2689	2700 7645	862 862	862 862	862 2277	232 232	232 232	232 532	99 99	99 99	99 197	64 64	64 64	64 113
0.05 •••	8107 8142	8195 8282	8440 16310	2315 2321	2332 2350	2367 4419	495 495	501 501	501 864	180 180	180 180	180 275	99 99	99 99	99 145
0.1 •••	17907 18012	18194 18590	19640 29497	4894 4946	4999 5086	5191 7677	950 950	956 956	967 1366	291 291	291 302	302 393	151 162	162 162	162 194
0.15 •••	26804 26990	27357 28186	31021 41113	7267 7372	7512 7722	8037 10542	1370 1387	1394 1405	1411 1806	407 407	407 407	407 495	204 204	204 204	204 236
0.2 •••	34049 34346	34917 36300	41679 51159	9262 9426	9682 10050	10669 13014	1744 1761	1779 1807	1825 2182	501 501	501 501	501 581	250 250	250 250	250 271
0.25 •••	39432 39870	40692 42705	51175 59636	10820 11071	11439 11999	13014 15094	2059 2087	2111 2146	2192 2496	582 582	582 589	589 652	274 274	285 285	285 299
0.3 •••	42985 43580	44700 47419	59260 66543	11957 12296	12786 13556	15002 16781	2304 2339	2385 2437	2490 2748	641 641	652 652	659 707	302 302	309 309	309 320
0.35 •••	44840 45627	47097 50562	65794 71881	12699 13130	13749 14722	16595 18076	2490 2542	2595 2654	2724 2936	687 694	694 705	711 746	320 326	326 326	326 334
0.4 •••	45207 46211	48066 52301	70711 75648	13112 13620	14355 15510	17785 18979	2619 2671	2735 2811	2892 3062	722 729	729 740	746 770	337 337	337 337	337 341
0.45 •••	44332 45592	47850 52826	73960 77846	13206 13795	14624 15947	18555 19489	2682 2741	2811 2899	2997 3124	740 740	746 757	764 778	337 337	337 344	344 341
0.5 •••	42477 44000	46666 52295	75506 78474	13049 13696	14617 16052	18894 19607	2689 2759	2840 2927	3032 3124	729 740	746 757	764 770	337 337	337 337	337 334
0.55 •••	39905 41696	44717 50866	75342 77532	12675 13364	14337 15842	18817 19332	2647 2717	2794 2892	2997 3062	722 722	729 740	746 746	320 326	326 326	326 320
0.6 •••	36860 38879	42169 48655	73476 75020	12104 12815	13801 15335	18299 18665	2549 2619	2700 2787	2899 2936	687 687	694 705	711 707	302 302	309 309	309 299
0.65 •••	33524 35705	39152 45750	69906 70939	11362 12069	13042 14536	17371 17606	2402 2472	2542 2630	2724 2748	635 641	652 652	659 652	274 285	285 285	285 271
0.7 •••	30041 32275	35740 42197	64639 65287	10452 11135	12062 13445	16006 16153	2210 2262	2332 2402	2490 2496	571 571	582 589	589 581	250 250	250 250	250 236
0.75 •••	26436 28624	31942 37980	57702 58067	9385 10004	10837 12062	14232 14309	1954 2000	2052 2111	2181 2182	495 495	501 501	512 495	204 204	204 204	204 194
0.8 •••	22691 24704	27714 33045	49099 49276	8131 8656	9356 10365	12045 12072	1639 1685	1720 1761	1814 1806	396 396	396 407	407 393	162 162	162 162	162 145
0.85 •••	18695 20410	22919 27224	38855 38915	6655 7057	7589 8317	9437 9443	1271 1300	1324 1352	1376 1366	285 285	285 291	291 275	. .	. .	. .
0.9 •••	14186 15457	17260 20206	26979 26985	4859 5115	5436 5850	6434 6421	834 845	851 862	880 864	. .	. .	. .	. .	. .	. .
0.95 •••	8527 9164	10032 11292	13497 13485	2566 2654	2759 2875	3021 3007	. .	. .	. .	. .	. .	. .	. .	. .	. .
0.98 •••	3686 3855	4065 4316	4649 4631	. .	. .	. .	. .	. .	. .	. .	. .	. .	. .	. .	. .

TABLE 5: ALPHA= 0.025 POWER= 0.8   EXPECTED ACCRUAL THRU MINIMUM FOLLOW-UP= 7500

	DEL=.01			DEL=.02			DEL=.05			DEL=.10			DEL=.15		
FACT=	1.0 .75	.50 .25	.00 BIN	1.0 .75	.50 .25	.00 BIN	1.0 .75	.50 .25	.00 BIN	1.0 75	.50 .25	.00 BIN	1.0 .75	.50 .25	.00 BIN

PCONT=***   REQUIRED NUMBER OF PATIENTS

PCONT	DEL=.01 1.0/.75	.50/.25	.00/BIN	DEL=.02 1.0/.75	.50/.25	.00/BIN	DEL=.05 1.0/.75	.50/.25	.00/BIN	DEL=.10 1.0/.75	.50/.25	.00/BIN	DEL=.15 1.0/.75	.50/.25	.00/BIN
0.01	1205	1205	1212	443	443	443	143	143	143	68	68	68	50	50	50
	1205	1205	4631	443	443	1531	143	143	417	68	68	170	50	50	101
0.02	2668	2686	2705	856	856	868	230	230	230	99	99	99	68	68	68
	2675	2693	7645	856	856	2277	230	230	532	99	99	197	68	68	113
0.05	8112	8199	8443	2318	2337	2368	500	500	500	174	174	174	106	106	106
	8143	8293	16310	2330	2349	4419	500	500	864	174	174	275	106	106	145
0.1	17930	18230	19636	4906	5011	5199	950	961	968	293	293	293	155	155	155
	18043	18631	29497	4955	5086	7677	950	961	1366	293	293	393	155	155	194
0.15	26843	27425	31025	7299	7543	8030	1374	1393	1418	406	406	406	200	200	200
	27043	28280	41113	7400	7737	10542	1381	1400	1806	406	406	495	200	200	236
0.2	34111	35030	41686	9293	9718	10674	1749	1786	1831	500	500	511	249	249	249
	34430	36455	51159	9474	10081	13014	1768	1805	2182	500	500	581	249	249	271
0.25	39530	40861	51174	10880	11499	13018	2068	2124	2187	575	586	586	275	275	286
	39987	42931	59636	11131	12050	15094	2093	2150	2496	586	586	652	275	275	299
0.3	43111	44930	59255	12031	12868	14993	2318	2393	2487	643	650	661	305	305	305
	43737	47724	66543	12380	13625	16781	2349	2431	2748	643	650	707	305	305	320
0.35	45012	47393	65799	12800	13850	16599	2506	2600	2724	687	699	706	324	324	324
	45837	50956	71881	13231	14806	18076	2543	2656	2936	699	706	746	324	324	334
0.4	45425	48436	70711	13231	14468	17780	2630	2743	2893	725	736	743	331	331	343
	46493	52775	75648	13737	15612	18979	2686	2818	3062	725	736	770	331	343	341
0.45	44611	48293	73955	13343	14768	18549	2693	2825	2993	736	743	762	343	343	343
	45931	53375	77846	13936	16062	19489	2761	2900	3124	743	755	778	343	343	341
0.5	42811	47168	75500	13205	14768	18893	2712	2843	3031	736	743	762	331	331	343
	44412	52906	78474	13861	16175	19607	2780	2930	3124	743	755	770	331	343	334
0.55	40306	45275	75343	12837	14487	18811	2668	2806	2993	718	736	743	324	324	331
	42174	51518	77532	13531	15980	19332	2731	2893	3062	725	736	746	324	324	320
0.6	37318	42774	73475	12275	13974	18305	2574	2712	2900	687	699	706	305	305	312
	39406	49336	75020	12987	15474	18665	2637	2799	2936	687	706	707	305	305	299
0.65	34025	39781	69905	11525	13205	17368	2424	2555	2724	643	650	661	275	286	286
	36268	46430	70939	12237	14668	17606	2480	2637	2748	643	650	652	275	286	271
0.7	30556	36361	64643	10618	12211	16006	2218	2337	2487	575	586	593	249	249	249
	32843	42849	65287	11293	13568	16153	2274	2405	2496	575	586	581	249	249	236
0.75	26949	32525	57699	9530	10974	14236	1962	2068	2187	493	500	511	211	211	211
	29168	38581	58067	10149	12162	14309	2011	2124	2182	500	500	495	211	211	194
0.8	23168	28231	49100	8255	9474	12043	1655	1730	1812	399	399	406	155	162	162
	25205	33568	49276	8780	10449	12072	1693	1768	1806	399	406	393	155	162	145
0.85	19100	23349	38855	6755	7674	9436	1280	1325	1381	286	286	286	.	.	.
	20825	27643	38915	7156	8375	9443	1299	1355	1366	286	286	275	.	.	.
0.9	14487	17562	26975	4925	5487	6436	837	856	875	.	.	.	.	.	.
	15762	20480	26985	5180	5881	6421	849	868	864	.	.	.	.	.	.
0.95	8675	10168	13493	2581	2768	3024	.	.	.	.	.	.	.	.	.
	9312	11393	13485	2675	2881	3007	.	.	.	.	.	.	.	.	.
0.98	3725	4093	4643	.	.	.	.	.	.	.	.	.	.	.	.
	3893	4336	4631	.	.	.	.	.	.	.	.	.	.	.	.

TABLE 5: ALPHA= 0.025 POWER= 0.8   EXPECTED ACCRUAL THRU MINIMUM FOLLOW-UP= 8000

PCONT=•••   REQUIRED NUMBER OF PATIENTS

PCONT	DEL=.01 1.0/.75	.50/.25	.00/BIN	DEL=.02 1.0/.75	.50/.25	.00/BIN	DEL=.05 1.0/.75	.50/.25	.00/BIN	DEL=.10 1.0/.75	.50/.25	.00/BIN	DEL=.15 1.0/.75	.50/.25	.00/BIN
0.01	1205	1205	1213	445	445	445	145	145	145	73	73	73	53	53	53
	1205	1205	4631	445	445	1531	145	145	417	73	73	170	53	53	101
0.02	2673	2685	2705	853	865	865	233	233	233	105	105	105	65	65	65
	2673	2693	7645	853	865	2277	233	233	532	105	105	197	65	65	113
0.05	8113	8205	8433	2313	2333	2365	493	493	505	173	173	173	105	105	105
	8153	8293	16310	2325	2345	4419	493	505	864	173	173	275	105	105	145
0.1	17953	18265	19633	4925	5025	5193	953	953	965	293	293	293	153	153	153
	18065	18665	29497	4965	5093	7677	953	965	1366	293	293	393	153	153	194
0.15	26885	27493	31025	7313	7553	8033	1373	1393	1413	405	405	405	205	205	205
	27093	28365	41113	7425	7753	10542	1385	1405	1806	405	405	495	205	205	236
0.2	34173	35145	41685	9333	9753	10673	1753	1785	1825	505	505	505	245	245	245
	34513	36605	51159	9513	10105	13014	1773	1805	2182	505	505	581	245	245	271
0.25	39613	41025	51173	10933	11553	13013	2073	2125	2185	585	585	593	273	285	285
	40105	43153	59636	11193	12093	15094	2093	2153	2496	585	585	652	273	285	299
0.3	43245	45153	59253	12113	12945	14993	2325	2393	2485	645	653	653	305	305	305
	43905	48025	66543	12453	13685	16781	2353	2433	2748	645	653	707	305	305	320
0.35	45185	47673	65793	12905	13945	16593	2513	2605	2725	693	705	705	325	325	325
	46053	51325	71881	13333	14885	18076	2553	2665	2936	693	705	746	325	325	334
0.4	45645	48793	70713	13345	14585	17785	2645	2753	2893	725	733	745	333	333	333
	46765	53225	75648	13853	15705	18979	2693	2813	3062	725	733	770	333	333	341
0.45	44873	48713	73953	13473	14893	18553	2713	2833	2993	733	745	765	333	345	345
	46273	53893	77846	14073	16173	19489	2773	2913	3124	745	753	778	333	345	341
0.5	43145	47665	75505	13353	14905	18893	2725	2853	3033	733	745	765	333	333	345
	44813	53473	78474	14005	16293	19607	2785	2933	3124	745	753	770	333	333	334
0.55	40693	45813	75345	12993	14645	18813	2685	2813	2993	725	733	745	325	325	325
	42633	52133	77532	13685	16093	19332	2745	2905	3062	725	733	746	325	325	320
0.6	37753	43353	73473	12433	14125	18305	2585	2725	2893	685	693	713	305	305	313
	39905	49973	75020	13145	15593	18665	2653	2805	2936	693	705	707	305	305	299
0.65	34505	40373	69905	11693	13353	17365	2433	2565	2725	633	645	665	273	285	285
	36793	47073	70939	12393	14785	17606	2493	2633	2748	645	653	652	285	285	271
0.7	31053	36945	64645	10773	12345	16013	2233	2345	2493	573	585	593	245	245	245
	33373	43465	65287	11445	13673	16153	2285	2413	2496	573	585	581	245	245	236
0.75	27433	33085	57705	9673	11093	14233	1973	2065	2185	493	505	505	205	205	205
	29685	39145	58067	10285	12265	14309	2025	2125	2182	493	505	495	205	205	194
0.8	23613	28733	49093	8373	9573	12045	1665	1733	1813	393	405	405	153	153	165
	25673	34053	49276	8893	10525	12072	1693	1773	1806	405	405	393	153	153	145
0.85	19485	23753	38853	6845	7745	9445	1285	1325	1373	285	285	293	.	.	.
	21225	28025	38915	7245	8425	9443	1305	1353	1366	285	285	275	.	.	.
0.9	14773	17853	26973	4985	5533	6433	833	853	873	.	.	.	.	.	.
	16053	20725	26985	5225	5913	6421	845	865	864	.	.	.	.	.	.
0.95	8825	10293	13493	2605	2785	3025	.	.	.	.	.	.	.	.	.
	9453	11493	13485	2693	2893	3007	.	.	.	.	.	.	.	.	.
0.98	3765	4125	4645	.	.	.	.	.	.	.	.	.	.	.	.
	3933	4353	4631	.	.	.	.	.	.	.	.	.	.	.	.

TABLE 5: ALPHA= 0.025 POWER= 0.8   EXPECTED ACCRUAL THRU MINIMUM FOLLOW-UP= 8500

		DEL=.01			DEL=.02			DEL=.05			DEL=.10			DEL=.15		
FACT=		1.0 .75	.50 .25	.00 BIN	1.0 .75	.50 .25	.00 BIN	1.0 .75	.50 .25	.00 BIN	1.0 75	.50 .25	.00 BIN	1.0 .75	.50 .25	.00 BIN
PCONT=•••						REQUIRED NUMBER OF PATIENTS										
0.01	•••	1204	1204	1204	439	439	439	141	141	141	70	70	70	48	48	48
	•••	1204	1204	4631	439	439	1531	141	141	417	70	70	170	48	48	101
0.02	•••	2670	2683	2705	856	864	864	226	226	226	99	99	99	70	70	70
	•••	2683	2691	7645	856	864	2277	226	226	532	99	99	197	70	70	113
0.05	•••	8123	8216	8442	2322	2343	2365	495	503	503	176	176	176	99	99	99
	•••	8166	8301	16310	2330	2351	4419	495	503	864	176	176	275	99	99	145
0.1	•••	17970	18302	19641	4936	5029	5191	949	962	962	290	290	303	155	155	155
	•••	18090	18706	29497	4978	5106	7677	949	962	1366	290	303	393	155	155	194
0.15	•••	26930	27567	31031	7337	7579	8038	1374	1395	1416	410	410	410	205	205	205
	•••	27150	28446	41113	7443	7762	10542	1387	1408	1806	410	410	495	205	205	236
0.2	•••	34240	35260	41685	9377	9789	10673	1756	1791	1833	503	503	503	248	248	248
	•••	34601	36747	51159	9547	10129	13014	1770	1812	2182	503	503	581	248	248	271
0.25	•••	39709	41188	51176	10992	11608	13011	2075	2131	2181	580	588	588	282	282	282
	•••	40232	43364	59636	11247	12140	15094	2096	2152	2496	580	588	652	282	282	299
0.3	•••	43364	45375	59259	12182	13019	14995	2330	2407	2492	643	651	651	303	303	311
	•••	44065	48307	66543	12530	13741	16781	2365	2436	2748	651	651	707	303	303	320
0.35	•••	45353	47954	65796	12990	14039	16589	2521	2606	2726	694	694	707	325	325	325
	•••	46275	51686	71881	13423	14966	18076	2564	2662	2936	694	707	746	325	325	334
0.4	•••	45863	49136	70713	13444	14698	17779	2649	2755	2896	728	736	750	333	333	333
	•••	47040	53649	75648	13967	15795	18979	2705	2819	3062	728	736	770	333	333	341
0.45	•••	45149	49123	73956	13606	15016	18557	2726	2840	3002	736	750	758	333	346	346
	•••	46607	54385	77846	14201	16270	19489	2776	2917	3124	750	758	778	346	346	341
0.5	•••	43470	48124	75507	13486	15038	18897	2734	2861	3031	736	750	758	333	333	346
	•••	45205	54023	78474	14137	16398	19607	2798	2946	3124	750	758	770	333	333	334
0.55	•••	41082	46339	75345	13138	14783	18812	2691	2832	3002	715	728	750	325	325	325
	•••	43080	52714	77532	13826	16206	19332	2755	2904	3062	728	736	746	325	325	320
0.6	•••	38192	43895	73475	12586	14265	18302	2598	2726	2896	686	694	707	303	303	311
	•••	40389	50568	75020	13295	15710	18665	2662	2811	2936	694	707	707	303	311	299
0.65	•••	34970	40933	69905	11842	13486	17367	2450	2577	2726	643	651	665	282	282	282
	•••	37308	47670	70939	12543	14889	17606	2513	2641	2748	643	651	652	282	282	271
0.7	•••	31528	37499	64635	10915	12480	16007	2245	2351	2492	580	580	588	248	248	248
	•••	33886	44044	65287	11587	13776	16153	2301	2415	2496	580	588	581	248	248	236
0.75	•••	27894	33602	57700	9802	11213	14230	1982	2075	2181	495	503	503	205	205	205
	•••	30181	39680	58067	10405	12352	14309	2025	2131	2182	495	503	495	205	205	194
0.8	•••	24040	29190	49101	8493	9675	12041	1671	1735	1812	396	396	410	155	163	163
	•••	26122	34503	49276	9003	10588	12072	1706	1770	1806	396	410	393	155	163	145
0.85	•••	19840	24133	38851	6933	7813	9441	1289	1331	1374	282	290	290	.	.	.
	•••	21596	28375	38915	7316	8471	9443	1310	1353	1366	282	290	275	.	.	.
0.9	•••	15038	18119	26980	5029	5573	6431	835	856	877	.	.	.	.	.	.
	•••	16326	20958	26985	5276	5943	6421	843	864	864	.	.	.	.	.	.
0.95	•••	8960	10405	13500	2628	2798	3023	.	.	.	.	.	.	.	.	.
	•••	9590	11574	13485	2705	2896	3007	.	.	.	.	.	.	.	.	.
0.98	•••	3810	4150	4646	.	.	.	.	.	.	.	.	.	.	.	.
	•••	3958	4370	4631	.	.	.	.	.	.	.	.	.	.	.	.

TABLE 5: ALPHA= 0.025 POWER= 0.8   EXPECTED ACCRUAL THRU MINIMUM FOLLOW-UP= 9000

		DEL=.01			DEL=.02			DEL=.05			DEL=.10			DEL=.15		
FACT=		1.0 .75	.50 .25	.00 BIN	1.0 .75	.50 .25	.00 BIN	1.0 .75	.50 .25	.00 BIN	1.0 75	.50 .25	.00 BIN	1.0 .75	.50 .25	.00 BIN
PCONT=***						REQUIRED NUMBER OF PATIENTS										
0.01	***	1207	1207	1207	442	442	442	141	141	141	74	74	74	51	51	51
	•••	1207	1207	4631	442	442	1531	141	141	417	74	74	170	51	51	101
0.02	***	2670	2684	2706	861	861	861	231	231	231	105	105	105	60	60	60
	•••	2684	2692	7645	861	861	2277	231	231	532	105	105	197	60	60	113
0.05	***	8129	8227	8444	2324	2346	2369	501	501	501	172	172	172	105	105	105
	•••	8174	8309	16310	2332	2355	4419	501	501	864	172	172	275	105	105	145
0.1	***	17992	18330	19635	4942	5032	5190	951	960	960	299	299	299	164	164	164
	•••	18119	18735	29497	4979	5100	7677	960	960	1366	299	299	393	164	164	194
0.15	***	26961	27636	31020	7364	7597	8039	1379	1401	1410	411	411	411	209	209	209
	•••	27209	28522	41113	7462	7777	10542	1387	1410	1806	411	411	495	209	209	236
0.2	***	34305	35362	41685	9411	9825	10671	1761	1792	1829	501	501	510	240	240	240
	•••	34679	36884	51159	9591	10162	13014	1770	1806	2182	501	501	581	240	240	271
0.25	***	39809	41339	51171	11040	11647	13011	2085	2130	2189	577	591	591	276	276	285
	•••	40349	43552	59636	11301	12179	15094	2107	2152	2496	577	591	652	276	276	299
0.3	***	43499	45577	59257	12246	13079	15000	2332	2400	2490	645	645	659	307	307	307
	•••	44227	48570	66543	12592	13799	16781	2369	2445	2748	645	659	707	307	307	320
0.35	***	45510	48232	65796	13079	14122	16597	2535	2616	2729	690	704	712	321	321	330
	•••	46477	52026	71881	13515	15022	18076	2571	2670	2936	704	704	746	321	330	334
0.4	***	46072	49470	70710	13551	14789	17781	2661	2760	2895	726	735	749	330	330	344
	•••	47301	54060	75648	14069	15877	18979	2715	2827	3062	726	735	770	330	344	341
0.45	***	45411	49515	73964	13717	15135	18555	2737	2850	2999	735	749	757	344	344	344
	•••	46927	54839	77846	14316	16364	19489	2796	2917	3124	749	757	778	344	344	341
0.5	***	43791	48584	75502	13619	15157	18892	2751	2872	3030	735	749	757	330	344	344
	•••	45591	54532	78474	14271	16499	19607	2805	2954	3124	749	757	770	330	344	334
0.55	***	41451	46829	75345	13281	14910	18816	2706	2841	2999	726	735	749	321	330	330
	•••	43507	53264	77532	13965	16305	19332	2774	2909	3062	726	735	746	321	330	320
0.6	***	38602	44430	73477	12727	14392	18307	2616	2737	2895	690	704	712	307	307	307
	•••	40866	51135	75020	13439	15810	18665	2670	2819	2936	690	704	707	307	307	299
0.65	***	35421	41474	69900	11976	13619	17362	2459	2580	2729	645	645	659	276	285	285
	•••	37792	48232	70939	12682	14991	17606	2512	2647	2748	645	659	652	276	285	271
0.7	***	31979	38031	64635	11054	12592	16012	2256	2355	2490	577	577	591	240	254	254
	•••	34386	44579	65287	11715	13866	16153	2310	2422	2496	577	591	581	254	254	236
0.75	***	28334	34102	57696	9929	11310	14235	1995	2085	2189	501	501	510	209	209	209
	•••	30651	40169	58067	10522	12426	14309	2040	2130	2182	501	501	495	209	209	194
0.8	***	24441	29639	49101	8587	9757	12044	1671	1739	1815	397	397	411	164	164	164
	•••	26556	34935	49276	9105	10657	12072	1702	1770	1806	397	411	393	164	164	145
0.85	***	20189	24495	38850	7012	7881	9442	1289	1334	1379	285	285	285	.	.	.
	•••	21952	28702	38915	7395	8520	9443	1311	1356	1366	285	285	275	.	.	.
0.9	***	15292	18366	26984	5091	5609	6427	839	861	884	.	.	.	.	.	.
	•••	16575	21179	26985	5325	5969	6421	847	870	864	.	.	.	.	.	.
0.95	***	9082	10522	13492	2639	2805	3021	.	.	.	.	.	.	.	.	.
	•••	9712	11661	13485	2715	2909	3007	.	.	.	.	.	.	.	.	.
0.98	***	3840	4169	4650	.	.	.	.	.	.	.	.	.	.	.	.
	•••	3989	4380	4631	.	.	.	.	.	.	.	.	.	.	.	.

TABLE 5: ALPHA= 0.025 POWER= 0.8   EXPECTED ACCRUAL THRU MINIMUM FOLLOW-UP= 9500

	DEL=.01			DEL=.02			DEL=.05			DEL=.10			DEL=.15		
FACT=	1.0 / .75	.50 / .25	.00 BIN	1.0 / .75	.50 / .25	.00 BIN	1.0 / .75	.50 / .25	.00 BIN	1.0 / 75	.50 / .25	.00 BIN	1.0 / .75	.50 / .25	.00 BIN

PCONT=•••   REQUIRED NUMBER OF PATIENTS

PCONT	1.0/.75	.50/.25	.00 BIN	1.0/.75	.50/.25	.00 BIN	1.0/.75	.50/.25	.00 BIN	1.0/.75	.50/.25	.00 BIN	1.0/.75	.50/.25	.00 BIN
0.01 •••	1203	1203	1203	443	443	443	149	149	149	78	78	78	54	54	54
•••	1203	1203	4631	443	443	1531	149	149	417	78	78	170	54	54	101
0.02 •••	2675	2690	2699	861	861	861	229	229	229	101	101	101	63	63	63
•••	2675	2690	7645	861	861	2277	229	229	532	101	101	197	63	63	113
0.05 •••	8138	8233	8438	2319	2343	2367	490	505	505	173	173	173	101	101	101
•••	8176	8319	16310	2334	2358	4419	505	505	864	173	173	275	101	101	145
0.1 •••	18018	18365	19633	4946	5041	5193	956	956	965	291	300	300	158	158	158
•••	18151	18769	29497	4994	5113	7677	956	965	1366	300	300	393	158	158	194
0.15 •••	27010	27699	31024	7378	7615	8034	1384	1393	1417	410	410	410	205	205	205
•••	27257	28601	41113	7473	7782	10542	1393	1408	1806	410	410	495	205	205	236
0.2 •••	34373	35474	41688	9444	9848	10670	1764	1797	1820	505	505	505	244	244	244
•••	34762	37009	51159	9610	10180	13014	1773	1811	2182	505	505	581	244	244	271
0.25 •••	39892	41498	51173	11083	11700	13021	2082	2129	2191	576	585	585	277	277	277
•••	40462	43739	59636	11344	12214	15094	2105	2153	2496	585	585	652	277	277	299
0.3 •••	43620	45782	59263	12318	13140	15002	2343	2405	2485	648	648	657	300	300	300
•••	44395	48822	66543	12665	13838	16781	2381	2453	2748	648	657	707	300	300	320
0.35 •••	45678	48489	65794	13164	14194	16593	2548	2619	2723	695	704	704	324	324	324
•••	46684	52337	71881	13591	15088	18076	2580	2666	2936	695	704	746	324	324	334
0.4 •••	46280	49795	70710	13648	14883	17780	2675	2770	2889	728	728	743	339	339	339
•••	47563	54441	75648	14170	15943	18979	2723	2833	3062	728	743	770	339	339	341
0.45 •••	45678	49905	73964	13829	15239	18555	2747	2856	2999	743	752	766	339	339	339
•••	47245	55282	77846	14423	16441	19489	2794	2928	3124	743	752	778	339	339	341
0.5 •••	44110	49012	75508	13743	15278	18897	2761	2880	3032	743	752	766	339	339	339
•••	45963	55020	78474	14384	16593	19607	2818	2951	3124	743	752	770	339	339	334
0.55 •••	41815	47302	75341	13410	15025	18816	2723	2842	2999	719	728	743	324	324	324
•••	43920	53785	77532	14099	16403	19332	2785	2913	3062	728	743	746	324	324	320
0.6 •••	39013	44927	73474	12855	14518	18303	2628	2747	2889	695	704	704	300	300	315
•••	41317	51672	75020	13568	15904	18665	2675	2818	2936	695	704	707	300	315	299
0.65 •••	35845	41982	69903	12119	13734	17368	2476	2580	2723	648	648	657	277	277	277
•••	38268	48765	70939	12808	15088	17606	2524	2652	2748	648	657	652	277	277	271
0.7 •••	32425	38529	64639	11178	12713	16014	2263	2367	2485	576	585	585	244	244	253
•••	34857	45093	65287	11834	13948	16153	2310	2429	2496	576	585	581	244	244	236
0.75 •••	28753	34572	57704	10038	11415	14233	2001	2082	2191	490	505	505	205	205	205
•••	31095	40643	58067	10632	12499	14309	2049	2129	2182	505	505	495	205	205	194
0.8 •••	24834	30050	49098	8684	9839	12048	1678	1740	1811	395	395	410	158	158	158
•••	26948	35332	49276	9198	10718	12072	1716	1773	1806	395	410	393	158	158	145
0.85 •••	20512	24834	38847	7084	7939	9435	1298	1336	1384	291	291	291	.	.	.
•••	22293	29014	38915	7464	8556	9443	1313	1360	1366	291	291	275	.	.	.
0.9 •••	15539	18603	26972	5136	5644	6428	838	861	870	.	.	.	.	.	.
•••	16821	21367	26985	5359	5977	6421	847	870	864	.	.	.	.	.	.
0.95 •••	9207	10623	13496	2652	2818	3023	.	.	.	.	.	.	.	.	.
•••	9824	11739	13485	2738	2913	3007	.	.	.	.	.	.	.	.	.
0.98 •••	3863	4186	4647	.	.	.	.	.	.	.	.	.	.	.	.
•••	4020	4400	4631	.	.	.	.	.	.	.	.	.	.	.	.

TABLE 5: ALPHA= 0.025 POWER= 0.8   EXPECTED ACCRUAL THRU MINIMUM FOLLOW-UP= 10000

	DEL=.01			DEL=.02			DEL=.05			DEL=.10			DEL=.15		
FACT=	1.0 .75	.50 .25	.00 BIN	1.0 .75	.50 .25	.00 BIN	1.0 .75	.50 .25	.00 BIN	1.0 75	.50 .25	.00 BIN	1.0 .75	.50 .25	.00 BIN
PCONT=***				REQUIRED NUMBER OF PATIENTS											
0.01 ***	1207	1207	1207	441	441	441	141	141	141	66	66	66	57	57	57
***	1207	1207	4631	441	441	1531	141	141	417	66	66	170	57	57	101
0.02 ***	2682	2682	2707	857	857	866	232	232	232	107	107	107	66	66	66
***	2682	2691	7645	857	857	2277	232	232	532	107	107	197	66	66	113
0.05 ***	8141	8241	8441	2332	2341	2366	491	491	507	182	182	182	107	107	107
***	8182	8316	16310	2332	2357	4419	491	507	864	182	182	275	107	107	145
0.1 ***	18041	18391	19641	4957	5041	5191	957	957	966	291	291	291	157	157	157
***	18166	18791	29497	4991	5116	7677	957	966	1366	291	291	393	157	157	194
0.15 ***	27041	27757	31032	7391	7632	8032	1382	1407	1416	407	407	407	207	207	207
***	27307	28666	41113	7491	7791	10542	1391	1407	1806	407	407	495	207	207	236
0.2 ***	34432	35566	41682	9466	9882	10666	1766	1791	1832	507	507	507	241	241	241
***	34841	37132	51159	9641	10207	13014	1782	1807	2182	507	507	581	241	241	271
0.25 ***	39982	41641	51166	11132	11741	13016	2091	2132	2191	582	582	591	282	282	282
***	40582	43932	59636	11391	12241	15094	2107	2157	2496	582	591	652	282	282	299
0.3 ***	43741	45991	59257	12382	13207	14991	2357	2416	2491	641	657	657	307	307	307
***	44541	49066	66543	12732	13891	16781	2382	2441	2748	641	657	707	307	307	320
0.35 ***	45841	48757	65791	13232	14266	16591	2541	2632	2716	691	707	707	316	332	332
***	46891	52641	71881	13666	15141	18076	2582	2666	2936	691	707	746	332	332	334
0.4 ***	46491	50107	70716	13741	14966	17782	2682	2782	2891	732	732	741	332	341	341
***	47816	54807	75648	14266	16016	18979	2732	2832	3062	732	741	770	332	341	341
0.45 ***	45932	50266	73957	13932	15332	18557	2757	2866	2991	741	757	757	341	341	341
***	47557	55707	77846	14532	16516	19489	2807	2932	3124	741	757	778	341	341	341
0.5 ***	44416	49432	75507	13857	15382	18891	2782	2891	3032	741	757	757	332	341	341
***	46316	55482	78474	14507	16682	19607	2832	2957	3124	741	757	770	332	341	334
0.55 ***	42166	47766	75341	13532	15141	18816	2732	2857	2991	732	732	741	316	332	332
***	44332	54266	77532	14216	16491	19332	2791	2916	3062	732	741	746	332	332	320
0.6 ***	39407	45407	73482	12982	14632	18307	2632	2757	2891	691	707	707	307	307	307
***	41757	52182	75020	13691	15991	18665	2691	2816	2936	691	707	707	307	307	299
0.65 ***	36266	42466	69907	12241	13841	17366	2482	2591	2732	641	657	657	282	282	282
***	38716	49266	70939	12932	15166	17606	2532	2657	2748	641	657	652	282	282	271
0.7 ***	32841	39007	64641	11291	12807	16007	2282	2366	2491	582	582	591	241	241	241
***	35307	45566	65287	11957	14032	16153	2316	2432	2496	582	591	581	241	241	236
0.75 ***	29166	35016	57707	10141	11507	14232	2007	2091	2182	491	507	507	207	207	207
***	31532	41082	58067	10741	12566	14309	2041	2132	2182	491	507	495	207	207	194
0.8 ***	25207	30457	49091	8782	9907	12041	1691	1741	1816	407	407	407	157	157	157
***	27341	35707	49276	9282	10766	12072	1716	1782	1806	407	407	393	157	157	145
0.85 ***	20832	25157	38857	7157	7991	9441	1307	1341	1382	282	291	291	.	.	.
***	22616	29307	38915	7532	8591	9443	1316	1357	1366	282	291	275	.	.	.
0.9 ***	15766	18816	26982	5182	5666	6432	841	857	882	.	.	.	.	.	.
***	17041	21557	26985	5407	6007	6421	857	866	864	.	.	.	.	.	.
0.95 ***	9316	10716	13491	2666	2832	3016	.	.	.	.	.	.	.	.	.
***	9932	11807	13485	2741	2916	3007	.	.	.	.	.	.	.	.	.
0.98 ***	3891	4207	4641	.	.	.	.	.	.	.	.	.	.	.	.
***	4041	4407	4631	.	.	.	.	.	.	.	.	.	.	.	.

TABLE 5: ALPHA= 0.025 POWER= 0.8   EXPECTED ACCRUAL THRU MINIMUM FOLLOW-UP= 11000

FACT=	DEL=.01			DEL=.02			DEL=.05			DEL=.10			DEL=.15		
	1.0 .75	.50 .25	.00 BIN	1.0 .75	.50 .25	.00 BIN	1.0 .75	.50 .25	.00 BIN	1.0 75	.50 .25	.00 BIN	1.0 .75	.50 .25	.00 BIN

PCONT=•••   REQUIRED NUMBER OF PATIENTS

PCONT		1.0/.75	.50/.25	.00/BIN	1.0/.75	.50/.25	.00/BIN	1.0/.75	.50/.25	.00/BIN	1.0/.75	.50/.25	.00/BIN	1.0/.75	.50/.25	.00/BIN
0.01	***	1200	1200	1217	447	447	447	145	145	145	73	73	73	45	45	45
	•••	1200	1200	4631	447	447	1531	145	145	417	73	73	170	45	45	101
0.02	***	2675	2685	2702	860	860	860	227	227	227	100	100	100	62	62	62
	•••	2685	2685	7645	860	860	2277	227	227	532	100	100	197	62	62	113
0.05	***	8158	8257	8433	2328	2345	2355	502	502	502	172	172	172	100	100	100
	•••	8202	8323	16310	2328	2355	4419	502	502	864	172	172	275	100	100	145
0.1	***	18075	18443	19642	4968	5067	5188	953	953	970	293	293	293	155	155	155
	•••	18223	18845	29497	5012	5122	7677	953	970	1366	293	293	393	155	155	194
0.15	***	27122	27875	31027	7432	7652	8037	1382	1393	1420	403	403	403	200	200	200
	•••	27397	28800	41113	7525	7817	10542	1393	1410	1806	403	403	495	200	200	236
0.2	***	34558	35768	41680	9533	9935	10677	1767	1795	1822	502	502	502	238	238	238
	•••	34998	37352	51159	9698	10237	13014	1778	1805	2182	502	502	581	238	238	271
0.25	***	40168	41928	51168	11210	11815	13015	2097	2135	2190	585	585	585	282	282	282
	•••	40800	44255	59636	11475	12300	15094	2125	2163	2496	585	585	652	282	282	299
0.3	***	43990	46372	59253	12492	13300	14995	2355	2410	2482	650	650	650	310	310	310
	•••	44860	49518	66543	12833	13960	16781	2383	2455	2748	650	650	707	310	310	320
0.35	***	46163	49243	65798	13383	14400	16590	2565	2630	2730	695	705	705	320	320	320
	•••	47290	53220	71881	13812	15242	18076	2592	2675	2936	695	705	746	320	320	334
0.4	***	46905	50700	70710	13922	15132	17783	2702	2785	2895	733	733	750	337	337	337
	•••	48308	55485	75648	14428	16133	18979	2740	2840	3062	733	733	770	337	337	341
0.45	***	46438	50975	73955	14125	15517	18553	2768	2878	3005	750	750	760	337	337	337
	•••	48143	56475	77846	14720	16655	19489	2823	2933	3124	750	760	778	337	337	341
0.5	***	45008	50222	75505	14070	15583	18900	2795	2895	3032	750	750	760	337	337	337
	•••	47005	56338	78474	14720	16820	19607	2840	2960	3124	750	760	770	337	337	334
0.55	***	42852	48627	75340	13757	15352	18817	2757	2867	3005	722	733	750	320	320	320
	•••	45090	55183	77532	14445	16655	19332	2812	2922	3062	733	733	746	320	320	320
0.6	***	40157	46317	73470	13218	14830	18305	2658	2768	2895	695	705	705	310	310	310
	•••	42577	53120	75020	13922	16150	18665	2702	2823	2936	695	705	707	310	310	299
0.65	***	37050	43385	69895	12465	14043	17370	2510	2603	2730	640	650	667	282	282	282
	•••	39580	50195	70939	13152	15325	17606	2548	2658	2748	650	650	652	282	282	271
0.7	***	33640	39910	64643	11513	12998	16012	2290	2383	2493	585	585	595	255	255	255
	•••	36153	46455	65287	12162	14170	16153	2328	2427	2496	585	585	581	255	255	236
0.75	***	29938	35850	57702	10347	11667	14235	2025	2097	2180	502	502	502	210	210	210
	•••	32330	41873	58067	10925	12685	14309	2053	2135	2182	502	502	495	210	210	194
0.8	***	25895	31192	49095	8955	10045	12035	1695	1750	1822	403	403	403	155	155	155
	•••	28068	36390	49276	9440	10853	12072	1723	1778	1806	403	403	393	155	155	145
0.85	***	21413	25747	38848	7278	8092	9440	1310	1338	1382	282	282	293	.	.	.
	•••	23217	29828	38915	7635	8653	9443	1327	1355	1366	282	293	275	.	.	.
0.9	***	16188	19213	26985	5260	5727	6425	843	860	870	.	.	.	.	.	.
	•••	17470	21880	26985	5463	6040	6421	860	870	864	.	.	.	.	.	.
0.95	***	9522	10880	13493	2702	2840	3015	.	.	.	.	.	.	.	.	.
	•••	10127	11925	13485	2768	2922	3007	.	.	.	.	.	.	.	.	.
0.98	***	3940	4242	4638	.	.	.	.	.	.	.	.	.	.	.	.
	•••	4077	4418	4631	.	.	.	.	.	.	.	.	.	.	.	.

TABLE 5: ALPHA= 0.025 POWER= 0.8   EXPECTED ACCRUAL THRU MINIMUM FOLLOW-UP= 12000

	DEL=.01			DEL=.02			DEL=.05			DEL=.10			DEL=.15		
FACT=	1.0 .75	.50 .25	.00 BIN	1.0 .75	.50 .25	.00 BIN	1.0 .75	.50 .25	.00 BIN	1.0 75	.50 .25	.00 BIN	1.0 .75	.50 .25	.00 BIN
PCONT=•••				REQUIRED NUMBER OF PATIENTS											
0.01 •••	1208 1208	1208 1208	1208 4631	439 439	439 439	439 1531	139 139	139 139	139 417	68 68	68 68	68 170	49 49	49 49	49 101
0.02 •••	2678 2678	2689 2689	2708 7645	859 859	859 859	859 2277	229 229	229 229	229 532	98 98	98 98	98 197	68 68	68 68	68 113
0.05 •••	8168 8209	8258 8329	8438 16310	2329 2329	2348 2348	2359 4419	499 499	499 499	499 864	169 169	169 169	169 275	98 98	98 98	98 145
0.1 •••	18128 18259	18499 18889	19639 29497	4988 5018	5059 5119	5198 7677	949 949	968 968	968 1366	289 289	289 289	289 393	158 158	158 158	158 194
0.15 •••	27199 27499	27979 28909	31028 41113	7459 7549	7669 7838	8029 10542	1388 1399	1399 1399	1418 1806	409 409	409 409	409 495	199 199	199 199	199 236
0.2 •••	34669 35149	35959 37549	41678 51159	9589 9758	9979 10268	10669 13014	1778 1789	1808 1808	1819 2182	499 499	499 499	499 581	248 248	248 248	248 271
0.25 •••	40339 41018	42199 44558	51169 59636	11299 11558	11888 12338	13009 15094	2108 2119	2138 2168	2179 2496	578 589	589 589	589 652	278 278	278 278	278 299
0.3 •••	44228 45158	46748 49928	59258 66543	12589 12938	13399 14029	14989 16781	2378 2389	2419 2449	2479 2748	649 649	649 649	649 707	308 308	308 308	308 320
0.35 •••	46478 47678	49699 53738	65798 71881	13508 13939	14528 15338	16598 18076	2569 2599	2648 2678	2719 2936	698 698	698 709	709 746	319 319	319 319	319 334
0.4 •••	47299 48788	51278 56108	70718 75648	14078 14588	15278 16238	17779 18979	2708 2749	2798 2839	2899 3062	728 728	739 739	739 770	338 338	338 338	338 341
0.45 •••	46928 48709	51619 57188	73958 77846	14318 14899	15679 16778	18548 19489	2798 2839	2888 2929	2989 3124	739 739	758 758	758 778	338 338	338 338	338 341
0.5 •••	45589 47659	50959 57128	75499 78474	14269 14899	15758 16958	18889 19607	2809 2858	2918 2959	3038 3124	739 739	758 758	758 770	338 338	338 338	338 334
0.55 •••	43508 45818	49418 56018	75338 77532	13969 14648	15529 16789	18818 19332	2768 2809	2869 2929	2989 3062	728 728	739 739	739 746	319 319	319 319	319 320
0.6 •••	40868 43358	47149 53959	73478 75020	13429 14119	15019 16279	18308 18665	2678 2719	2768 2828	2899 2936	698 698	698 709	709 707	308 308	308 308	308 299
0.65 •••	37789 40369	44239 51038	69908 70939	12679 13358	14228 15458	17359 17606	2509 2558	2618 2659	2719 2748	649 649	649 649	668 652	278 278	278 278	278 271
0.7 •••	34388 36938	40729 47258	64639 65287	11708 12349	13159 14288	16009 16153	2299 2348	2389 2438	2498 2496	578 578	589 589	589 581	248 248	248 248	248 236
0.75 •••	30649 33079	36619 42608	57698 58067	10519 11089	11809 12788	14228 14309	2029 2059	2108 2138	2179 2182	499 499	499 499	499 495	199 199	199 199	199 194
0.8 •••	26558 28729	31849 37009	49099 49276	9098 9578	10159 10939	12038 12072	1699 1729	1759 1789	1819 1806	398 398	398 409	409 393	158 158	158 158	158 145
0.85 •••	21949 23749	26288 30308	38858 38915	7388 7748	8168 8708	9439 9443	1309 1328	1339 1358	1369 1366	289 289	289 289	289 275	. .	. .	. .
0.9 •••	16579 17858	19579 22178	26978 26985	5318 5528	5768 6068	6428 6421	848 859	859 878	878 864	. .	. .	. .	. .	. .	. .
0.95 •••	9709 10298	11029 12038	13489 13485	2719 2779	2858 2929	3019 3007	. .	. .	. .	. .	. .	. .	. .	. .	. .
0.98 •••	3979 4118	4268 4448	4639 4631	. .	. .	. .	. .	. .	. .	. .	. .	. .	. .	. .	. .

TABLE 5: ALPHA= 0.025 POWER= 0.8   EXPECTED ACCRUAL THRU MINIMUM FOLLOW-UP= 13000

REQUIRED NUMBER OF PATIENTS

PCONT / FACT	DEL=.01 1.0 .75	.50 .25	.00 BIN	DEL=.02 1.0 .75	.50 .25	.00 BIN	DEL=.05 1.0 .75	.50 .25	.00 BIN	DEL=.10 1.0 .75	.50 .25	.00 BIN	DEL=.15 1.0 .75	.50 .25	.00 BIN
0.01 ***	1211	1211	1211	443	443	443	139	139	139	74	74	74	53	53	53
•••	1211	1211	4631	443	443	1531	139	139	417	74	74	170	53	53	101
0.02 ***	2686	2686	2706	854	866	866	236	236	236	106	106	106	74	74	74
•••	2686	2686	7645	854	866	2277	236	236	532	106	106	197	74	74	113
0.05 ***	8178	8276	8438	2328	2349	2361	496	496	496	171	171	171	106	106	106
•••	8211	8341	16310	2328	2349	4419	496	496	864	171	171	275	106	106	145
0.1 ***	18156	18546	19639	4993	5079	5188	951	963	963	301	301	301	151	151	151
•••	18306	18936	29497	5026	5123	7677	963	963	1366	301	301	393	151	151	194
0.15 ***	27276	28089	31026	7484	7691	8036	1386	1406	1418	411	411	411	204	204	204
•••	27581	29011	41113	7581	7841	10542	1406	1406	1806	411	411	495	204	204	236
0.2 ***	34796	36128	41686	9629	10019	10669	1776	1796	1829	496	508	508	248	248	248
•••	35283	37741	51159	9803	10291	13014	1796	1808	2182	496	508	581	248	248	271
0.25 ***	40516	42454	51176	11363	11948	13009	2101	2154	2186	573	594	594	281	281	281
•••	41231	44838	59636	11623	12391	15094	2121	2166	2496	594	594	652	281	281	299
0.3 ***	44469	47081	59256	12696	13476	15003	2381	2426	2491	638	659	659	301	301	301
•••	45444	50298	66543	13041	14093	16781	2393	2458	2748	659	659	707	301	301	320
0.35 ***	46788	50136	65801	13638	14634	16596	2576	2641	2718	691	703	703	334	334	334
•••	48044	54219	71881	14061	15414	18076	2609	2686	2936	703	703	746	334	334	334
0.4 ***	47686	51793	70708	14211	15393	17786	2718	2804	2901	724	736	736	334	334	334
•••	49246	56668	75648	14731	16336	18979	2771	2848	3062	736	736	770	334	334	341
0.45 ***	47394	52248	73958	14471	15816	18546	2804	2901	2999	736	756	756	334	346	346
•••	49258	57838	77846	15056	16888	19489	2848	2946	3124	756	756	778	346	346	341
0.5 ***	46138	51651	75506	14451	15913	18891	2816	2913	3031	736	756	756	334	334	334
•••	48283	57838	78474	15089	17071	19607	2869	2966	3124	756	756	770	334	334	334
0.55 ***	44123	50168	75344	14158	15686	18806	2783	2881	2999	724	736	736	334	334	334
•••	46496	56766	77532	14829	16909	19332	2836	2934	3062	736	736	746	334	334	320
0.6 ***	41523	47926	73471	13626	15186	18306	2686	2783	2901	691	703	703	301	301	313
•••	44079	54739	75020	14309	16401	18665	2739	2836	2936	703	703	707	301	313	299
0.65 ***	38489	45021	69896	12879	14386	17364	2523	2621	2718	638	659	659	281	281	281
•••	41121	51814	70939	13529	15556	17606	2576	2674	2748	659	659	652	281	281	271
0.7 ***	35076	41491	64631	11883	13313	16011	2316	2393	2491	573	594	594	248	248	248
•••	37676	47991	65287	12521	14386	16153	2361	2446	2496	573	594	581	248	248	236
0.75 ***	31318	37331	57696	10681	11948	14223	2036	2101	2186	496	508	508	204	204	204
•••	33776	43266	58067	11254	12879	14309	2068	2154	2182	496	508	495	204	204	194
0.8 ***	27146	32476	49096	9239	10279	12046	1711	1764	1808	399	399	411	151	151	151
•••	29336	37558	49276	9694	11006	12072	1731	1796	1806	399	399	393	151	151	145
0.85 ***	22446	26768	38846	7496	8243	9434	1321	1341	1374	281	281	281	.	.	.
•••	24254	30721	38915	7841	8763	9443	1321	1353	1366	281	281	275	.	.	.
0.9 ***	16941	19911	26984	5383	5806	6423	854	866	886	.	.	.	.	.	.
•••	18209	22434	26985	5578	6086	6421	854	866	864	.	.	.	.	.	.
0.95 ***	9868	11168	13496	2739	2869	3011	.	.	.	.	.	.	.	.	.
•••	10453	12111	13485	2804	2934	3007	.	.	.	.	.	.	.	.	.
0.98 ***	4018	4299	4636	.	.	.	.	.	.	.	.	.	.	.	.
•••	4148	4461	4631	.	.	.	.	.	.	.	.	.	.	.	.

TABLE 5: ALPHA= 0.025 POWER= 0.8　　EXPECTED ACCRUAL THRU MINIMUM FOLLOW-UP= 14000

		DEL=.01			DEL=.02			DEL=.05			DEL=.10			DEL=.15		
FACT=		1.0 .75	.50 .25	.00 BIN	1.0 .75	.50 .25	.00 BIN	1.0 .75	.50 .25	.00 BIN	1.0 75	.50 .25	.00 BIN	1.0 .75	.50 .25	.00 BIN
PCONT=***					REQUIRED NUMBER OF PATIENTS											
0.01	***	1199	1212	1212	442	442	442	149	149	149	79	79	79	44	44	44
	•••	1212	1212	4631	442	442	1531	149	149	417	79	79	170	44	44	101
0.02	***	2682	2682	2704	862	862	862	232	232	232	92	92	92	57	57	57
	•••	2682	2704	7645	862	862	2277	232	232	532	92	92	197	57	57	113
0.05	***	8199	8282	8444	2332	2354	2367	499	499	499	184	184	184	92	92	92
	•••	8234	8339	16310	2332	2354	4419	499	499	864	184	184	275	92	92	145
0.1	***	18187	18594	19644	4992	5084	5189	954	954	967	289	302	302	162	162	162
	•••	18349	18979	29497	5049	5132	7677	954	967	1366	302	302	393	162	162	194
0.15	***	27357	28184	31019	7512	7722	8037	1387	1409	1409	407	407	407	197	197	197
	•••	27672	29107	41113	7604	7849	10542	1387	1409	1806	407	407	495	197	197	236
0.2	***	34917	36304	41672	9682	10054	10662	1772	1807	1829	499	499	499	254	254	254
	•••	35429	37914	51159	9844	10312	13014	1794	1807	2182	499	499	581	254	254	271
0.25	***	40692	42709	51179	11432	11992	13007	2109	2144	2192	582	582	582	289	289	289
	•••	41449	45102	59636	11677	12412	15094	2122	2157	2496	582	582	652	289	289	299
0.3	***	44704	47412	59264	12784	13554	15002	2389	2437	2494	652	652	652	302	302	302
	•••	45719	50654	66543	13112	14149	16781	2402	2459	2748	652	652	707	302	302	320
0.35	***	47097	50562	65787	13742	14722	16599	2599	2647	2717	687	709	709	324	324	324
	•••	48414	54644	71881	14162	15479	18076	2612	2682	2936	709	709	746	324	324	334
0.4	***	48064	52299	70709	14359	15514	17789	2739	2809	2892	722	744	744	337	337	337
	•••	49687	57199	75648	14849	16424	18979	2774	2844	3062	722	744	770	337	337	341
0.45	***	47854	52824	73964	14617	15947	18559	2809	2892	2997	744	757	757	337	337	337
	•••	49779	58437	77846	15199	16984	19489	2857	2949	3124	744	757	778	337	337	341
0.5	***	46664	52299	75504	14617	16052	18887	2844	2927	3032	744	757	757	337	337	337
	•••	48869	58494	78474	15234	17172	19607	2879	2984	3124	744	757	770	337	337	334
0.55	***	44717	50864	75342	14337	15842	18817	2787	2892	2997	722	744	744	324	324	324
	•••	47154	57457	77532	14989	17019	19332	2844	2949	3062	722	744	746	324	324	320
0.6	***	42162	48659	73474	13799	15339	18292	2704	2787	2892	687	709	709	302	302	302
	•••	44752	55449	75020	14477	16507	18665	2739	2844	2936	709	709	707	302	302	299
0.65	***	39152	45754	69904	13042	14534	17369	2542	2634	2717	652	652	652	289	289	289
	•••	41812	52509	70939	13694	15667	17606	2577	2669	2748	652	652	652	289	289	271
0.7	***	35744	42197	64632	12062	13449	16004	2332	2402	2494	582	582	582	254	254	254
	•••	38369	48637	65287	12679	14477	16153	2367	2437	2496	582	582	581	254	254	236
0.75	***	31942	37984	57702	10837	12062	14232	2052	2109	2179	499	499	512	197	197	197
	•••	34414	43864	58067	11384	12959	14309	2087	2144	2182	499	499	495	197	197	194
0.8	***	27707	33049	49092	9354	10369	12049	1724	1759	1807	394	407	407	162	162	162
	•••	29912	38067	49276	9809	11069	12072	1737	1794	1806	407	407	393	162	162	145
0.85	***	22912	27217	38859	7582	8317	9437	1317	1352	1374	289	289	289	.	.	.
	•••	24719	31102	38915	7919	8807	9443	1339	1352	1366	289	289	275	.	.	.
0.9	***	17264	20204	26972	5434	5854	6427	849	862	884	.	.	.	.	.	.
	•••	18524	22667	26985	5622	6112	6421	862	862	864	.	.	.	.	.	.
0.95	***	10032	11292	13497	2752	2879	3019	.	.	.	.	.	.	.	.	.
	•••	10592	12202	13485	2809	2949	3007	.	.	.	.	.	.	.	.	.
0.98	***	4069	4314	4642	.	.	.	.	.	.	.	.	.	.	.	.
	•••	4174	4467	4631	.	.	.	.	.	.	.	.	.	.	.	.

TABLE 5: ALPHA= 0.025 POWER= 0.8   EXPECTED ACCRUAL THRU MINIMUM FOLLOW-UP= 15000

		DEL=.01			DEL=.02			DEL=.05			DEL=.10			DEL=.15	
FACT=		1.0 .75	.50 .25	.00 BIN	1.0 .75	.50 .25	.00 BIN	1.0 .75	.50 .25	.00 BIN	1.0 75	.50 .25	.00 BIN	1.0 .75	.50 .25	.00 BIN
PCONT=***				REQUIRED NUMBER OF PATIENTS												
0.01	***	1210	1210	1210	436	436	436	136	136	136	61	61	61	47	47	47
	•••	1210	1210	4631	436	436	1531	136	136	417	61	61	170	47	47	101
0.02	***	2686	2686	2710	849	849	872	235	235	235	99	99	99	61	61	61
	•••	2686	2686	7645	849	849	2277	235	235	532	99	99	197	61	61	113
0.05	***	8199	8297	8447	2335	2349	2372	497	497	497	174	174	174	99	99	99
	•••	8236	8349	16310	2335	2349	4419	497	497	864	174	174	275	99	99	145
0.1	***	18235	18624	19636	5011	5086	5199	961	961	961	286	286	286	160	160	160
	•••	18399	18999	29497	5049	5147	7677	961	961	1366	286	286	393	160	160	194
0.15	***	27422	28285	31022	7547	7735	8035	1397	1397	1411	399	399	399	197	197	197
	•••	27760	29185	41113	7622	7861	10542	1397	1411	1806	399	399	495	197	197	236
0.2	***	35035	36460	41686	9722	10074	10674	1786	1810	1824	497	497	511	249	249	249
	•••	35574	38049	51159	9872	10336	13014	1786	1810	2182	497	497	581	249	249	271
0.25	***	40861	42924	51174	11499	12047	13022	2124	2147	2185	586	586	586	272	272	286
	•••	41635	45324	59636	11747	12460	15094	2124	2161	2496	586	586	652	272	272	299
0.3	***	44935	47724	59260	12872	13622	14986	2386	2424	2485	647	647	661	310	310	310
	•••	45985	50972	66543	13210	14185	16781	2410	2461	2748	647	647	707	310	310	320
0.35	***	47386	50949	65799	13847	14799	16599	2597	2649	2724	699	699	699	324	324	324
	•••	48760	55060	71881	14274	15535	18076	2635	2686	2936	699	699	746	324	324	334
0.4	***	48436	52772	70711	14461	15610	17785	2747	2822	2897	736	736	736	324	347	347
	•••	50110	57685	75648	14972	16486	18979	2785	2860	3062	736	736	770	347	347	341
0.45	***	48286	53372	73960	14761	16060	18549	2822	2897	2986	736	760	760	347	347	347
	•••	50274	58997	77846	15324	17049	19489	2860	2949	3124	760	760	778	347	347	341
0.5	***	47161	52899	75497	14761	16172	18886	2836	2935	3024	736	760	760	324	347	347
	•••	49435	59086	78474	15385	17260	19607	2897	2972	3124	760	760	770	347	347	334
0.55	***	45272	51511	75347	14485	15985	18811	2799	2897	2986	736	736	736	324	324	324
	•••	47761	58111	77532	15136	17110	19332	2860	2949	3062	736	736	746	324	324	320
0.6	***	42774	49336	73472	13974	15474	18310	2710	2799	2897	699	699	699	310	310	310
	•••	45399	56110	75020	14635	16599	18665	2747	2836	2936	699	699	707	310	310	299
0.65	***	39774	46435	69910	13210	14672	17372	2560	2635	2724	647	647	661	286	286	286
	•••	42474	53147	70939	13847	15760	17606	2597	2672	2748	647	661	652	286	286	271
0.7	***	36361	42849	64636	12211	13561	15999	2335	2410	2485	586	586	586	249	249	249
	•••	39010	49261	65287	12811	14574	16153	2372	2447	2496	586	586	581	249	249	236
0.75	***	32522	38574	57699	10974	12160	14236	2072	2124	2185	497	497	511	211	211	211
	•••	35011	44410	58067	11499	13022	14309	2086	2147	2182	497	511	495	211	211	194
0.8	***	28224	33572	49097	9474	10449	12047	1735	1772	1810	399	399	399	160	160	160
	•••	30460	38522	49276	9910	11124	12072	1749	1786	1806	399	399	393	160	160	145
0.85	***	23349	27647	38860	7674	8372	9456	1322	1360	1374	286	286	286	.	.	.
	•••	25149	31449	38915	7997	8836	9443	1336	1360	1366	286	286	275	.	.	.
0.9	***	17560	20485	26972	5485	5874	6436	849	872	872	.	.	.	.	.	.
	•••	18811	22885	26985	5672	6136	6421	849	872	864	.	.	.	.	.	.
0.95	***	10172	11386	13486	2761	2874	3024	.	.	.	.	.	.	.	.	.
	•••	10711	12272	13485	2822	2949	3007	.	.	.	.	.	.	.	.	.
0.98	***	4097	4336	4636	.	.	.	.	.	.	.	.	.	.	.	.
	•••	4210	4472	4631	.	.	.	.	.	.	.	.	.	.	.	.

TABLE 5: ALPHA= 0.025 POWER= 0.8   EXPECTED ACCRUAL THRU MINIMUM FOLLOW-UP= 17000

		DEL=.01			DEL=.02			DEL=.05			DEL=.10			DEL=.15		
FACT=		1.0 .75	.50 .25	.00 BIN	1.0 .75	.50 .25	.00 BIN	1.0 .75	.50 .25	.00 BIN	1.0 75	.50 .25	.00 BIN	1.0 .75	.50 .25	.00 BIN
PCONT=***					REQUIRED NUMBER OF PATIENTS											
0.01	***	1201	1201	1201	436	436	436	139	139	139	70	70	70	54	54	54
	•••	1201	1201	4631	436	436	1531	139	139	417	70	70	170	54	54	101
0.02	***	2689	2689	2705	861	861	861	224	224	224	96	96	96	70	70	70
	•••	2689	2689	7645	861	861	2277	224	224	532	96	96	197	70	70	113
0.05	***	8214	8299	8442	2349	2349	2365	495	495	495	181	181	181	96	96	96
	•••	8256	8357	16310	2349	2349	4419	495	495	864	181	181	275	96	96	145
0.1	***	18302	18711	19646	5026	5111	5196	962	962	962	282	309	309	155	155	155
	•••	18472	19067	29497	5069	5154	7677	962	962	1366	309	309	393	155	155	194
0.15	***	27567	28444	31036	7576	7762	8044	1387	1414	1414	410	410	410	197	197	197
	•••	27907	29336	41113	7661	7890	10542	1387	1414	1806	410	410	495	197	197	236
0.2	***	35260	36747	41677	9786	10126	10679	1796	1812	1839	495	495	495	240	240	240
	•••	35839	38346	51159	9956	10381	13014	1796	1812	2182	495	495	581	240	240	271
0.25	***	41194	43361	51181	11614	12140	13016	2136	2152	2179	580	580	580	282	282	282
	•••	42017	45757	59636	11842	12506	15094	2136	2179	2496	580	580	652	282	282	299
0.3	***	45375	48307	59256	13016	13739	14987	2407	2434	2492	649	649	649	309	309	309
	•••	46506	51564	66543	13330	14265	16781	2407	2450	2748	649	649	707	309	309	320
0.35	***	47951	51691	65801	14036	14971	16586	2604	2662	2731	691	707	707	325	325	325
	•••	49396	55787	71881	14435	15625	18076	2646	2689	2936	707	707	746	325	325	334
0.4	***	49141	53646	70705	14690	15795	17776	2747	2816	2901	734	734	750	325	325	325
	•••	50900	58550	75648	15184	16602	18979	2790	2859	3062	734	734	770	325	325	341
0.45	***	49115	54385	73961	15014	16262	18557	2832	2917	3002	750	750	750	351	351	351
	•••	51197	59979	77846	15566	17197	19489	2875	2960	3124	750	750	778	351	351	341
0.5	***	48121	54029	75507	15030	16390	18897	2859	2944	3029	750	750	750	325	325	351
	•••	50475	60191	78474	15651	17410	19607	2901	2986	3124	750	750	770	325	325	334
0.55	***	46336	52711	75337	14775	16204	18812	2832	2901	3002	734	734	750	325	325	325
	•••	48902	59256	77532	15412	17266	19332	2859	2944	3062	734	734	746	325	325	320
0.6	***	43887	50560	73467	14265	15710	18302	2731	2816	2901	691	707	707	309	309	309
	•••	46607	57275	75020	14902	16772	18665	2774	2859	2936	707	707	707	309	309	299
0.65	***	40939	47670	69897	13484	14886	17367	2577	2646	2731	649	649	665	282	282	282
	•••	43675	54300	70939	14121	15906	17606	2604	2689	2748	649	649	652	282	282	271
0.7	***	37496	44041	64627	12480	13781	16007	2349	2407	2492	580	580	580	240	240	240
	•••	40190	50331	65287	13059	14716	16153	2391	2450	2496	580	580	581	240	240	236
0.75	***	33602	39680	57700	11205	12352	14222	2067	2136	2179	495	495	495	197	197	197
	•••	36110	45375	58067	11715	13144	14309	2094	2152	2182	495	495	495	197	197	194
0.8	***	29182	34495	49099	9675	10594	12039	1727	1770	1812	394	410	410	155	155	155
	•••	31419	39324	49276	10084	11231	12072	1754	1796	1806	410	410	393	155	155	145
0.85	***	24125	28375	38856	7805	8469	9446	1329	1345	1371	282	282	282	.	.	.
	•••	25936	32056	38915	8129	8910	9443	1345	1371	1366	282	282	275	.	.	.
0.9	***	18116	20964	26972	5579	5935	6429	861	861	877	.	.	.	.	.	.
	•••	19349	23232	26985	5749	6174	6421	861	877	864	.	.	.	.	.	.
0.95	***	10397	11571	13500	2790	2901	3029	.	.	.	.	.	.	.	.	.
	•••	10934	12395	13485	2832	2960	3007	.	.	.	.	.	.	.	.	.
0.98	***	4150	4362	4644	.	.	.	.	.	.	.	.	.	.	.	.
	•••	4261	4490	4631	.	.	.	.	.	.	.	.	.	.	.	.

| TABLE 5: ALPHA= 0.025 POWER= 0.8 | | | EXPECTED ACCRUAL THRU MINIMUM FOLLOW-UP= 20000 | | | | | | | | | | | |

		DEL=.01			DEL=.02			DEL=.05			DEL=.10			DEL=.15		
FACT=		1.0 .75	.50 .25	.00 BIN	1.0 .75	.50 .25	.00 BIN	1.0 .75	.50 .25	.00 BIN	1.0 75	.50 .25	.00 BIN	1.0 .75	.50 .25	.00 BIN
PCONT=***					REQUIRED NUMBER OF PATIENTS											
0.01	***	1213	1213	1213	432	432	432	132	132	132	63	63	63	63	63	63
	•••	1213	1213	4631	432	432	1531	132	132	417	63	63	170	63	63	101
0.02	***	2682	2682	2713	863	863	863	232	232	232	113	113	113	63	63	63
	•••	2682	2682	7645	863	863	2277	232	232	532	113	113	197	63	63	113
0.05	***	8232	8313	8432	2332	2363	2363	482	513	513	182	182	182	113	113	113
	•••	8263	8363	16310	2332	2363	4419	513	513	864	182	182	275	113	113	145
0.1	***	18382	18782	19632	5032	5113	5182	963	963	963	282	282	282	163	163	163
	•••	18563	19132	29497	5082	5163	7677	963	963	1366	282	282	393	163	163	194
0.15	***	27763	28663	31032	7632	7782	8032	1413	1413	1413	413	413	413	213	213	213
	•••	28132	29532	41113	7713	7913	10542	1413	1413	1806	413	413	495	213	213	236
0.2	***	35563	37132	41682	9882	10213	10663	1782	1813	1832	513	513	513	232	232	232
	•••	36182	38682	51159	10032	10413	13014	1813	1813	2182	513	513	581	232	232	271
0.25	***	41632	43932	51163	11732	12232	13013	2132	2163	2182	582	582	582	282	282	282
	•••	42532	46282	59636	11963	12582	15094	2132	2163	2496	582	582	652	282	282	299
0.3	***	45982	49063	59263	13213	13882	14982	2413	2432	2482	663	663	663	313	313	313
	•••	47182	52282	66543	13513	14363	16781	2432	2463	2748	663	663	707	313	313	320
0.35	***	48763	52632	65782	14263	15132	16582	2632	2663	2713	713	713	713	332	332	332
	•••	50282	56713	71881	14663	15763	18076	2663	2682	2936	713	713	746	332	332	334
0.4	***	50113	54813	70713	14963	16013	17782	2782	2832	2882	732	732	732	332	332	332
	•••	51982	59663	75648	15432	16763	18979	2813	2863	3062	732	732	770	332	332	341
0.45	***	50263	55713	73963	15332	16513	18563	2863	2932	2982	763	763	763	332	332	332
	•••	52432	61232	77846	15863	17363	19489	2882	2963	3124	763	763	778	332	332	341
0.5	***	49432	55482	75513	15382	16682	18882	2882	2963	3032	763	763	763	332	332	332
	•••	51863	61563	78474	15963	17613	19607	2913	2982	3124	763	763	770	332	332	334
0.55	***	47763	54263	75332	15132	16482	18813	2863	2913	2982	732	732	732	332	332	332
	•••	50413	60682	77532	15732	17463	19332	2882	2963	3062	732	732	746	332	332	320
0.6	***	45413	52182	73482	14632	15982	18313	2763	2813	2882	713	713	713	313	313	313
	•••	48182	58732	75020	15232	16963	18665	2782	2863	2936	713	713	707	313	313	299
0.65	***	42463	49263	69913	13832	15163	17363	2582	2663	2732	663	663	663	282	282	282
	•••	45263	55732	70939	14432	16082	17606	2613	2682	2748	663	663	652	282	282	271
0.7	***	39013	45563	64632	12813	14032	16013	2363	2432	2482	582	582	582	232	232	232
	•••	41732	51682	65287	13363	14882	16153	2382	2463	2496	582	582	581	232	232	236
0.75	***	35013	41082	57713	11513	12563	14232	2082	2132	2182	513	513	513	213	213	213
	•••	37563	46582	58067	11982	13282	14309	2113	2163	2182	513	513	495	213	213	194
0.8	***	30463	35713	49082	9913	10763	12032	1732	1782	1813	413	413	413	163	163	163
	•••	32663	40313	49276	10313	11332	12072	1763	1782	1806	413	413	393	163	163	145
0.85	***	25163	29313	38863	7982	8582	9432	1332	1363	1382	282	282	282	.	.	.
	•••	26932	32782	38915	8263	8982	9443	1332	1363	1366	282	282	275	.	.	.
0.9	***	18813	21563	26982	5663	6013	6432	863	863	882	.	.	.	.	.	.
	•••	20013	23682	26985	5832	6213	6421	863	863	864	.	.	.	.	.	.
0.95	***	10713	11813	13482	2832	2913	3013	.	.	.	.	.	.	.	.	.
	•••	11213	12532	13485	2863	2963	3007	.	.	.	.	.	.	.	.	.
0.98	***	4213	4413	4632	.	.	.	.	.	.	.	.	.	.	.	.
	•••	4313	4513	4631	.	.	.	.	.	.	.	.	.	.	.	.

TABLE 5: ALPHA= 0.025 POWER= 0.8   EXPECTED ACCRUAL THRU MINIMUM FOLLOW-UP= 25000

		DEL=.01			DEL=.02			DEL=.05			DEL=.10			DEL=.15		
FACT=		1.0 .75	.50 .25	.00 BIN	1.0 .75	.50 .25	.00 BIN	1.0 .75	.50 .25	.00 BIN	1.0 75	.50 .25	.00 BIN	1.0 .75	.50 .25	.00 BIN
PCONT=***				REQUIRED NUMBER OF PATIENTS												
0.01	***	1204	1204	1204	454	454	454	141	141	141	79	79	79	40	40	40
	***	1204	1204	4631	454	454	1531	141	141	417	79	79	170	40	40	101
0.02	***	2704	2704	2704	852	852	852	227	227	227	102	102	102	79	79	79
	***	2704	2704	7645	852	852	2277	227	227	532	102	102	197	79	79	113
0.05	***	8266	8329	8454	2352	2352	2352	516	516	516	165	165	165	102	102	102
	***	8290	8391	16310	2352	2352	4419	516	516	864	165	165	275	102	102	145
0.1	***	18516	18915	19641	5079	5141	5204	954	954	954	290	290	290	165	165	165
	***	18704	19204	29497	5102	5165	7677	954	954	1366	290	290	393	165	165	194
0.15	***	28040	28954	31016	7704	7829	8040	1391	1415	1415	415	415	415	204	204	204
	***	28415	29766	41113	7766	7915	10542	1415	1415	1806	415	415	495	204	204	236
0.2	***	36040	37641	41665	9977	10266	10665	1790	1829	1829	516	516	516	227	227	227
	***	36704	39102	51159	10141	10454	13014	1790	1829	2182	516	516	581	227	227	271
0.25	***	42329	44704	51165	11915	12352	13016	2141	2165	2204	579	579	579	266	266	290
	***	43290	46954	59636	12102	12665	15094	2141	2165	2496	579	579	652	266	290	299
0.3	***	46915	50102	59266	13454	14079	14977	2415	2454	2477	641	641	665	290	290	290
	***	48204	53204	66543	13727	14477	16781	2454	2477	2748	641	665	707	290	290	320
0.35	***	49915	53977	65790	14579	15352	16602	2641	2665	2727	704	704	704	329	329	329
	***	51579	57891	71881	14915	15915	18076	2665	2704	2936	704	704	746	329	329	334
0.4	***	51540	56391	70704	15329	16290	17790	2790	2852	2891	727	727	727	329	329	329
	***	53516	61079	75648	15766	16954	18979	2829	2852	3062	727	727	770	329	329	341
0.45	***	51954	57516	73954	15727	16829	18540	2891	2954	2977	766	766	766	329	329	329
	***	54227	62852	77846	16227	17579	19489	2915	2977	3124	766	766	778	329	329	341
0.5	***	51290	57477	75516	15829	17016	18891	2915	2977	3040	766	766	766	329	329	329
	***	53852	63329	78474	16352	17829	19607	2954	3016	3124	766	766	770	329	329	334
0.55	***	49790	56391	75352	15602	16852	18829	2891	2915	2977	727	727	727	329	329	329
	***	52516	62579	77532	16165	17704	19332	2915	2954	3062	727	727	746	329	329	320
0.6	***	47540	54352	73477	15102	16352	18290	2766	2829	2891	704	704	704	290	290	290
	***	50391	60641	75020	15665	17204	18665	2790	2852	2936	704	704	707	290	290	299
0.65	***	44641	51415	69891	14329	15516	17352	2602	2665	2727	641	665	665	290	290	290
	***	47477	57602	70939	14852	16329	17606	2641	2704	2748	641	665	652	290	290	271
0.7	***	41102	47641	64641	13227	14329	16016	2391	2454	2477	579	579	579	227	227	227
	***	43852	53415	65287	13727	15079	16153	2415	2454	2496	579	579	581	227	227	236
0.75	***	36977	42954	57704	11891	12829	14227	2102	2141	2165	516	516	516	204	204	204
	***	39516	48102	58067	12329	13454	14309	2141	2165	2182	516	516	495	204	204	194
0.8	***	32165	37290	49102	10227	10977	12040	1766	1790	1829	391	391	415	165	165	165
	***	34352	41579	49276	10579	11454	12072	1766	1790	1806	391	415	393	165	165	145
0.85	***	26540	30516	38852	8204	8727	9454	1352	1352	1391	290	290	290	.	.	.
	***	28266	33704	38915	8454	9040	9443	1352	1352	1366	290	290	275	.	.	.
0.9	***	19766	22329	26977	5790	6079	6415	852	852	891	.	.	.	.	.	.
	***	20891	24204	26985	5915	6227	6421	852	891	864	.	.	.	.	.	.
0.95	***	11102	12079	13477	2852	2915	3016	.	.	.	.	.	.	.	.	.
	***	11540	12704	13485	2891	2977	3007	.	.	.	.	.	.	.	.	.
0.98	***	4290	4454	4641	.	.	.	.	.	.	.	.	.	.	.	.
	***	4352	4540	4631	.	.	.	.	.	.	.	.	.	.	.	.

TABLE 6: ALPHA= 0.025 POWER= 0.9   EXPECTED ACCRUAL THRU MINIMUM FOLLOW-UP= 30

	DEL=.10			DEL=.15			DEL=.20			DEL=.25			DEL=.30		
FACT=	1.0 .75	.50 .25	.00 BIN	1.0 .75	.50 .25	.00 BIN	1.0 .75	.50 .25	.00 BIN	1.0 75	.50 .25	.00 BIN	1.0 .75	.50 .25	.00 BIN
PCONT=***	REQUIRED NUMBER OF PATIENTS														
0.05 ***	220	222	239	128	129	139	90	92	98	70	71	76	58	59	62
•••	221	224	368	129	131	194	91	93	124	71	73	87	58	60	65
0.1 ***	348	350	400	184	186	212	121	123	139	89	91	102	71	73	80
•••	348	354	526	185	190	260	122	127	158	90	94	107	71	75	78
0.15 ***	449	452	548	225	229	276	142	146	174	102	106	124	79	82	94
•••	450	459	662	226	236	316	144	152	187	103	110	124	80	86	88
0.2 ***	522	527	679	254	260	332	157	162	203	110	115	141	84	88	106
•••	524	537	778	256	270	362	159	171	211	112	122	138	85	93	96
0.25 ***	569	576	791	271	279	377	165	172	227	114	121	155	86	92	114
•••	571	590	873	274	293	400	167	183	229	117	130	148	88	98	102
0.3 ***	591	601	882	278	288	413	167	177	245	115	124	164	86	93	120
•••	594	620	946	281	306	428	171	191	242	118	134	154	89	101	106
0.35 ***	592	604	950	276	289	439	165	177	257	114	124	170	85	93	123
•••	596	629	999	280	311	446	169	194	250	117	136	158	88	102	107
0.4 ***	574	590	997	266	283	455	159	174	263	110	121	173	82	91	123
•••	579	622	1030	272	309	456	165	193	253	114	135	158	86	100	106
0.45 ***	541	561	1020	251	271	461	151	168	263	104	117	171	78	88	121
•••	548	601	1041	258	302	456	157	189	250	109	131	154	82	97	102
0.5 ***	496	522	1020	232	256	456	141	159	258	98	111	166	73	83	116
•••	505	569	1030	241	289	446	148	181	242	103	126	148	77	93	96
0.55 ***	443	476	998	211	238	441	130	149	246	90	104	157	68	77	108
•••	454	531	999	221	273	428	137	171	229	96	119	138	72	87	88
0.6 ***	387	426	952	188	218	416	117	137	229	82	96	144	61	71	98
•••	401	488	946	199	254	400	125	159	211	87	109	124	65	79	78
0.65 ***	331	376	884	166	196	380	104	124	207	73	86	128	54	63	86
•••	347	441	873	177	232	362	112	144	187	78	98	107	58	70	65
0.7 ***	278	327	794	143	173	335	91	109	179	63	74	108	.	.	.
•••	296	392	778	155	206	316	98	127	158	68	85	87	.	.	.
0.75 ***	229	278	681	121	148	280	76	91	145	.	.	.	.	.	.
•••	248	340	662	132	177	260	82	107	124	.	.	.	.	.	.
0.8 ***	184	229	546	98	120	215	.	.	.	.	.	.	.	.	.
•••	202	283	526	107	143	194	.	.	.	.	.	.	.	.	.
0.85 ***	140	177	389	.	.	.	.	.	.	.	.	.	.	.	.
•••	155	219	368	.	.	.	.	.	.	.	.	.	.	.	.

TABLE 6: ALPHA= 0.025 POWER= 0.9   EXPECTED ACCRUAL THRU MINIMUM FOLLOW-UP= 40

	DEL=.10			DEL=.15			DEL=.20			DEL=.25			DEL=.30		
FACT=	1.0 .75	.50 .25	.00 BIN	1.0 .75	.50 .25	.00 BIN	1.0 .75	.50 .25	.00 BIN	1.0 75	.50 .25	.00 BIN	1.0 .75	.50 .25	.00 BIN
PCONT=***				REQUIRED NUMBER OF PATIENTS											
0.05 ***	221	222	239	129	130	139	91	92	98	71	72	76	58	59	62
•••	221	225	368	129	132	194	91	94	124	71	73	87	59	60	65
0.1 ***	349	351	400	185	188	212	122	125	139	90	93	102	71	74	80
•••	349	357	526	186	193	260	123	128	158	91	96	107	72	76	78
0.15 ***	450	454	548	226	232	276	144	149	174	103	107	124	80	84	94
•••	451	463	662	228	240	316	145	155	187	105	112	124	81	87	88
0.2 ***	524	530	679	256	263	332	159	166	203	112	118	141	86	90	106
•••	526	544	778	258	275	362	161	175	211	114	125	138	87	95	96
0.25 ***	571	581	791	274	284	377	167	177	227	117	124	155	88	95	114
•••	574	599	873	277	300	400	171	189	229	120	133	148	91	101	102
0.3 ***	595	607	882	281	294	413	171	182	245	118	128	164	89	97	120
•••	599	632	946	286	315	428	175	197	242	122	138	154	92	104	106
0.35 ***	596	613	950	280	297	439	169	184	257	117	128	170	88	97	123
•••	602	645	999	286	322	446	175	201	250	122	140	158	92	105	107
0.4 ***	579	601	997	272	293	455	165	181	263	114	127	173	86	95	123
•••	587	641	1030	279	322	456	171	201	253	119	140	158	89	104	106
0.45 ***	548	575	1020	258	283	461	157	176	263	109	123	171	82	92	121
•••	557	624	1041	267	316	456	165	198	250	115	137	154	86	101	102
0.5 ***	505	539	1020	241	269	456	148	168	258	103	117	166	77	87	116
•••	516	596	1030	251	305	446	156	191	242	109	132	148	81	97	96
0.55 ***	454	496	998	221	252	441	137	158	246	96	110	157	72	82	108
•••	469	561	999	232	290	428	145	181	229	102	124	138	76	90	88
0.6 ***	401	449	952	199	232	416	125	146	229	87	101	144	65	74	98
•••	418	519	946	212	271	400	133	168	211	93	115	124	69	83	78
0.65 ***	347	401	884	177	210	380	112	132	207	78	91	128	58	66	86
•••	367	473	873	190	248	362	120	153	187	83	103	107	61	73	65
0.7 ***	296	352	794	155	186	335	98	116	179	68	79	108	.	.	.
•••	317	423	778	167	221	316	105	134	158	72	89	87	.	.	.
0.75 ***	248	302	681	132	160	280	82	98	145	.	.	.	.	.	.
•••	269	368	662	143	190	260	89	112	124	.	.	.	.	.	.
0.8 ***	202	251	546	107	130	215	.	.	.	.	.	.	.	.	.
•••	221	307	526	116	153	194	.	.	.	.	.	.	.	.	.
0.85 ***	155	194	389	.	.	.	.	.	.	.	.	.	.	.	.
•••	170	237	368	.	.	.	.	.	.	.	.	.	.	.	.

TABLE 6: ALPHA= 0.025 POWER= 0.9   EXPECTED ACCRUAL THRU MINIMUM FOLLOW-UP= 50

	DEL=.10			DEL=.15			DEL=.20			DEL=.25			DEL=.30		
FACT=	1.0	.50	.00	1.0	.50	.00	1.0	.50	.00	1.0	.50	.00	1.0	.50	.00
	.75	.25	BIN	.75	.25	BIN	.75	.25	BIN	75	.25	BIN	.75	.25	BIN

PCONT=***   REQUIRED NUMBER OF PATIENTS

PCONT	1.0/.75	.50/.25	.00/BIN	1.0/.75	.50/.25	.00/BIN	1.0/.75	.50/.25	.00/BIN	1.0/.75	.50/.25	.00/BIN	1.0/.75	.50/.25	.00/BIN
0.05 ***	221	223	239	129	131	139	91	93	98	71	72	76	59	60	62
•••	222	226	368	130	133	194	92	94	124	72	74	87	59	61	65
0.1 ***	349	353	400	185	189	212	122	126	139	91	94	102	72	74	80
•••	350	359	526	187	195	260	124	130	158	92	96	107	73	76	78
0.15 ***	451	457	548	228	234	276	145	151	174	104	109	124	81	85	94
•••	453	468	662	230	243	316	147	157	187	106	114	124	83	88	88
0.2 ***	525	534	679	258	267	332	161	168	203	114	120	141	87	92	106
•••	528	550	778	261	280	362	163	178	211	116	127	138	89	97	96
0.25 ***	574	585	791	276	288	377	170	180	227	119	127	155	90	97	114
•••	578	608	873	280	306	400	174	193	229	122	136	148	93	103	102
0.3 ***	598	614	882	285	300	413	174	187	245	121	131	164	91	99	120
•••	603	643	946	290	323	428	179	202	242	125	142	154	95	106	106
0.35 ***	600	621	950	285	304	439	173	189	257	121	132	170	91	99	123
•••	607	659	999	292	332	446	179	207	250	125	144	158	94	107	107
0.4 ***	585	611	997	277	301	455	170	188	263	118	131	173	89	98	123
•••	594	658	1030	286	333	456	177	208	253	123	144	158	92	107	106
0.45 ***	554	588	1020	265	293	461	163	183	263	114	128	171	85	95	121
•••	566	644	1041	275	328	456	171	205	250	119	141	154	89	104	102
0.5 ***	513	555	1020	249	280	456	154	175	258	108	122	166	80	90	116
•••	528	619	1030	260	318	446	163	198	242	114	136	148	85	99	96
0.55 ***	465	514	998	230	263	441	143	165	246	100	115	157	75	85	108
•••	483	586	999	243	303	428	152	188	229	106	128	138	79	93	88
0.6 ***	414	470	952	209	244	416	132	153	229	92	106	144	68	77	98
•••	434	546	946	223	284	400	140	175	211	98	118	124	72	85	78
0.65 ***	362	422	884	187	222	380	118	139	207	82	95	128	60	68	86
•••	385	500	873	201	260	362	127	159	187	88	106	107	64	75	65
0.7 ***	312	374	794	164	197	335	104	122	179	71	82	108	.	.	.
•••	336	448	778	178	232	316	111	140	158	76	91	87	.	.	.
0.75 ***	264	323	681	140	169	280	87	103	145	.	.	.	.	.	.
•••	287	391	662	152	199	260	94	117	124	.	.	.	.	.	.
0.8 ***	216	268	546	114	137	215	.	.	.	.	.	.	.	.	.
•••	237	327	526	123	160	194	.	.	.	.	.	.	.	.	.
0.85 ***	167	207	389	.	.	.	.	.	.	.	.	.	.	.	.
•••	183	251	368	.	.	.	.	.	.	.	.	.	.	.	.

TABLE 6: ALPHA= 0.025　POWER= 0.9　　EXPECTED ACCRUAL THRU MINIMUM FOLLOW-UP= 60

		DEL=.10			DEL=.15			DEL=.20			DEL=.25			DEL=.30		
FACT=		1.0 .75	.50 .25	.00 BIN	1.0 .75	.50 .25	.00 BIN	1.0 .75	.50 .25	.00 BIN	1.0 75	.50 .25	.00 BIN	1.0 .75	.50 .25	.00 BIN
PCONT=***		REQUIRED NUMBER OF PATIENTS														
0.05	***	221	224	239	129	131	139	92	93	98	71	73	76	59	60	62
	•••	222	227	368	130	134	194	92	95	124	72	74	87	59	61	65
0.1	***	350	354	400	186	190	212	123	127	139	91	94	102	73	75	80
	•••	351	362	526	188	196	260	125	131	158	93	97	107	74	77	78
0.15	***	452	459	548	229	236	276	146	152	174	106	110	124	82	86	94
	•••	454	472	662	232	246	316	149	159	187	107	115	124	84	89	88
0.2	***	527	537	679	260	270	332	162	171	203	115	122	141	88	93	106
	•••	530	556	778	263	284	362	166	180	211	118	128	138	90	98	96
0.25	***	576	590	791	279	293	377	172	183	227	121	130	155	92	98	114
	•••	581	616	873	284	312	400	176	196	229	124	138	148	95	104	102
0.3	***	601	620	882	288	306	413	177	191	245	124	134	164	93	101	120
	•••	607	654	946	294	330	428	182	206	242	128	144	154	97	108	106
0.35	***	604	629	950	289	311	439	177	194	257	124	136	170	93	102	123
	•••	613	672	999	297	340	446	184	212	250	128	147	158	97	109	107
0.4	***	590	622	997	283	309	455	174	193	263	121	135	173	91	100	123
	•••	601	674	1030	293	342	456	181	213	253	127	147	158	95	108	106
0.45	***	561	601	1020	271	302	461	168	188	263	117	131	171	88	97	121
	•••	575	662	1041	283	338	456	176	210	250	123	144	154	92	106	102
0.5	***	522	569	1020	256	289	456	159	181	258	111	126	166	83	93	116
	•••	539	639	1030	269	329	446	168	204	242	117	139	148	87	101	96
0.55	***	476	531	998	238	273	441	149	171	246	104	119	157	77	87	108
	•••	496	607	999	252	314	428	158	194	229	110	131	138	82	95	88
0.6	***	426	488	952	218	254	416	137	159	229	96	109	144	71	79	98
	•••	449	568	946	232	294	400	146	181	211	101	121	124	74	86	78
0.65	***	376	441	884	196	232	380	124	144	207	86	98	128	63	70	86
	•••	401	522	873	210	270	362	132	164	187	91	109	107	66	76	65
0.7	***	327	392	794	173	206	335	109	127	179	74	85	108	.	.	.
	•••	352	470	778	186	241	316	116	144	158	79	93	87	.	.	.
0.75	***	278	340	681	148	177	280	91	107	145	.	.	.	.	.	.
	•••	302	410	662	160	207	260	98	120	124	.	.	.	.	.	.
0.8	***	229	283	546	120	143	215	.	.	.	.	.	.	.	.	.
	•••	251	343	526	130	166	194	.	.	.	.	.	.	.	.	.
0.85	***	177	219	389	.	.	.	.	.	.	.	.	.	.	.	.
	•••	194	262	368	.	.	.	.	.	.	.	.	.	.	.	.

TABLE 6: ALPHA= 0.025 POWER= 0.9   EXPECTED ACCRUAL THRU MINIMUM FOLLOW-UP= 70

		DEL=.10			DEL=.15			DEL=.20			DEL=.25			DEL=.30		
FACT=		1.0 .75	.50 .25	.00 BIN	1.0 .75	.50 .25	.00 BIN	1.0 .75	.50 .25	.00 BIN	1.0 75	.50 .25	.00 BIN	1.0 .75	.50 .25	.00 BIN
PCONT=•••		REQUIRED NUMBER OF PATIENTS														
0.05	***	222	224	239	130	132	139	92	94	98	72	73	76	59	60	62
	•••	223	228	368	131	134	194	93	95	124	72	74	87	60	61	65
0.1	***	351	355	400	187	192	212	124	128	139	92	95	102	73	75	80
	•••	352	364	526	189	197	260	125	132	158	93	98	107	74	77	78
0.15	***	453	461	548	230	238	276	147	154	174	107	111	124	83	86	94
	•••	456	475	662	233	248	316	150	160	187	109	116	124	84	90	88
0.2	***	529	541	679	261	273	332	164	173	203	117	123	141	89	94	106
	•••	533	562	778	265	287	362	167	182	211	119	130	138	92	99	96
0.25	***	578	595	791	281	296	377	174	186	227	123	131	155	93	100	114
	•••	584	624	873	287	316	400	179	199	229	126	139	148	96	105	102
0.3	***	604	626	882	291	311	413	180	194	245	126	136	164	95	103	120
	•••	612	664	946	298	336	428	185	209	242	130	146	154	98	109	106
0.35	***	609	637	950	293	317	439	181	198	257	126	138	170	95	103	123
	•••	618	685	999	302	346	446	187	215	250	131	149	158	99	111	107
0.4	***	596	632	997	288	316	455	178	197	263	124	137	172	93	102	123
	•••	608	689	1030	298	350	456	186	217	253	130	149	158	97	110	106
0.45	***	568	612	1020	277	309	460	172	193	263	120	134	171	90	99	121
	•••	584	679	1041	290	347	456	181	215	250	126	147	154	94	107	102
0.5	***	531	583	1020	263	298	456	164	186	258	115	129	166	85	95	116
	•••	549	657	1030	276	337	446	173	208	242	121	142	148	89	103	96
0.55	***	486	546	998	245	282	441	154	177	246	107	122	157	80	89	108
	•••	508	627	999	260	323	428	163	198	229	113	134	138	84	96	88
0.6	***	438	504	952	225	263	416	142	164	229	99	112	144	73	81	98
	•••	463	588	946	240	303	400	151	185	211	104	124	124	76	88	78
0.65	***	389	458	884	203	240	380	128	149	207	88	101	128	64	72	86
	•••	416	541	873	218	278	362	137	168	187	94	111	107	68	77	65
0.7	***	340	408	794	180	214	335	113	131	179	77	87	108	.	.	.
	•••	367	488	778	194	249	316	120	147	158	81	95	87	.	.	.
0.75	***	291	355	681	154	184	280	95	110	145	.	.	.	.	.	.
	•••	316	427	662	166	213	260	101	122	124	.	.	.	.	.	.
0.8	***	240	296	546	125	149	215	.	.	.	.	.	.	.	.	.
	•••	263	356	526	135	170	194	.	.	.	.	.	.	.	.	.
0.85	***	186	228	389	.	.	.	.	.	.	.	.	.	.	.	.
	•••	203	272	368	.	.	.	.	.	.	.	.	.	.	.	.

TABLE 6: ALPHA= 0.025 POWER= 0.9   EXPECTED ACCRUAL THRU MINIMUM FOLLOW-UP= 80

PCONT=***		DEL=.10			DEL=.15			DEL=.20			DEL=.25			DEL=.30		
FACT=		1.0 .75	.50 .25	.00 BIN	1.0 .75	.50 .25	.00 BIN	1.0 .75	.50 .25	.00 BIN	1.0 75	.50 .25	.00 BIN	1.0 .75	.50 .25	.00 BIN
						REQUIRED NUMBER OF PATIENTS										
0.05	***	222	225	239	130	132	139	92	94	98	72	73	76	59	60	62
	•••	223	228	368	131	135	194	93	95	124	72	74	87	60	61	65
0.1	***	351	357	400	188	193	212	125	128	139	93	96	102	74	76	80
	•••	353	366	526	190	198	260	126	132	158	94	98	107	75	78	78
0.15	***	454	463	548	232	240	276	149	155	174	107	112	124	84	87	94
	•••	457	479	662	235	250	316	151	161	187	109	116	124	85	90	88
0.2	***	530	544	679	263	275	332	166	175	203	118	125	141	90	95	106
	•••	535	567	778	268	290	362	169	184	211	121	131	138	92	99	96
0.25	***	581	599	791	284	300	377	177	189	227	124	133	155	95	101	114
	•••	587	631	873	290	320	400	181	201	229	128	141	148	97	106	102
0.3	***	607	632	882	294	315	413	182	197	245	128	138	164	97	104	120
	•••	616	673	946	302	341	428	188	212	242	132	148	154	100	110	106
0.35	***	613	645	950	297	322	439	184	201	257	128	140	170	97	105	123
	•••	624	696	999	307	352	446	191	219	250	133	151	158	100	112	107
0.4	***	601	641	997	293	322	455	181	201	263	127	140	173	95	104	123
	•••	615	702	1030	304	357	456	189	221	253	132	152	158	99	111	106
0.45	***	575	624	1020	283	316	461	176	198	263	123	137	171	92	101	121
	•••	592	694	1041	296	354	456	185	218	250	129	149	154	96	109	102
0.5	***	539	596	1020	269	305	456	168	191	258	117	132	166	87	97	116
	•••	560	674	1030	283	345	446	177	212	242	123	144	148	91	104	96
0.55	***	496	561	998	252	290	441	158	181	246	110	124	157	82	90	108
	•••	520	644	999	267	331	428	168	202	229	116	136	138	85	97	88
0.6	***	449	519	952	232	271	416	146	168	229	101	115	144	74	83	98
	•••	476	605	946	248	311	400	155	189	211	107	126	124	78	89	78
0.65	***	401	473	884	210	248	380	132	153	207	91	103	128	66	73	86
	•••	429	558	873	225	285	362	141	171	187	96	112	107	69	78	65
0.7	***	352	423	794	186	221	335	116	134	179	79	89	108	.	.	.
	•••	380	504	778	200	255	316	124	150	158	83	96	87	.	.	.
0.75	***	302	368	681	160	190	280	98	112	145	.	.	.	.	.	.
	•••	329	441	662	172	218	260	104	124	124	.	.	.	.	.	.
0.8	***	251	307	546	130	153	215	.	.	.	.	.	.	.	.	.
	•••	273	367	526	139	174	194	.	.	.	.	.	.	.	.	.
0.85	***	194	237	389	.	.	.	.	.	.	.	.	.	.	.	.
	•••	211	280	368	.	.	.	.	.	.	.	.	.	.	.	.

TABLE 6: ALPHA= 0.025 POWER= 0.9   EXPECTED ACCRUAL THRU MINIMUM FOLLOW-UP= 90

	DEL=.10			DEL=.15			DEL=.20			DEL=.25			DEL=.30		
FACT=	1.0 .75	.50 .25	.00 BIN	1.0 .75	.50 .25	.00 BIN	1.0 .75	.50 .25	.00 BIN	1.0 75	.50 .25	.00 BIN	1.0 .75	.50 .25	.00 BIN

PCONT=•••   REQUIRED NUMBER OF PATIENTS

PCONT	1.0/.75	.50/.25	.00/BIN	1.0/.75	.50/.25	.00/BIN	1.0/.75	.50/.25	.00/BIN	1.0/.75	.50/.25	.00/BIN	1.0/.75	.50/.25	.00/BIN
0.05 •••	223	225	239	130	133	139	92	94	98	72	73	76	60	61	62
•••	224	229	368	131	135	194	93	96	124	73	74	87	60	61	65
0.1 •••	352	358	400	188	194	212	125	129	139	93	96	102	74	76	80
•••	354	367	526	190	199	260	127	133	158	94	98	107	75	78	78
0.15 •••	455	465	548	233	241	276	150	156	174	108	113	124	84	88	94
•••	459	482	662	236	252	316	152	162	187	110	117	124	86	90	88
0.2 •••	532	547	679	265	278	332	167	176	203	119	126	141	91	96	106
•••	537	571	778	270	293	362	171	186	211	122	131	138	93	100	96
0.25 •••	583	604	791	286	303	377	178	191	227	126	135	155	96	102	114
•••	590	637	873	292	324	400	183	203	229	130	142	148	98	107	102
0.3 •••	610	638	882	297	319	413	185	200	245	130	140	164	98	105	120
•••	620	681	946	306	345	428	191	215	242	134	149	154	101	111	106
0.35 •••	617	652	950	301	327	439	186	204	257	130	142	170	98	106	123
•••	629	706	999	311	357	446	194	221	250	136	153	158	102	113	107
0.4 •••	606	650	997	297	328	455	185	205	263	129	142	173	97	105	123
•••	622	713	1030	309	362	456	193	224	253	135	153	158	100	112	106
0.45 •••	582	634	1020	288	323	461	180	201	263	125	139	171	93	103	120
•••	601	707	1041	302	360	456	189	222	250	131	151	154	97	110	102
0.5 •••	547	608	1020	275	312	456	172	195	258	120	134	166	89	98	116
•••	569	688	1030	289	352	446	181	215	242	126	146	148	93	105	96
0.55 •••	505	574	998	258	297	441	162	185	246	113	126	157	83	92	108
•••	531	659	999	273	337	428	171	205	229	119	138	138	87	98	88
0.6 •••	460	533	952	238	278	416	150	172	229	104	117	144	76	84	98
•••	488	620	946	254	317	400	159	192	211	109	127	124	79	90	78
0.65 •••	412	487	884	216	254	380	136	156	207	93	105	128	67	74	85
•••	441	573	873	232	292	362	144	174	187	98	114	107	70	79	65
0.7 •••	363	436	794	192	227	335	119	137	179	80	90	108	.	.	.
•••	392	518	778	206	260	316	127	152	158	85	97	87	.	.	.
0.75 •••	313	380	681	165	195	280	100	115	145	.	.	.	.	.	.
•••	340	453	662	177	222	260	107	126	124	.	.	.	.	.	.
0.8 •••	260	317	546	134	157	215	.	.	.	.	.	.	.	.	.
•••	283	377	526	143	177	194	.	.	.	.	.	.	.	.	.
0.85 •••	201	244	389	.	.	.	.	.	.	.	.	.	.	.	.
•••	219	287	368	.	.	.	.	.	.	.	.	.	.	.	.

TABLE 6: ALPHA= 0.025 POWER= 0.9   EXPECTED ACCRUAL THRU MINIMUM FOLLOW-UP= 100

PCONT=	FACT=	DEL=.10 1.0/.75	.50/.25	.00/BIN	DEL=.15 1.0/.75	.50/.25	.00/BIN	DEL=.20 1.0/.75	.50/.25	.00/BIN	DEL=.25 1.0/.75	.50/.25	.00/BIN	DEL=.30 1.0/.75	.50/.25	.00/BIN
					REQUIRED NUMBER OF PATIENTS											
0.05	***	223	226	239	131	133	139	93	94	98	72	74	76	60	61	62
	•••	224	230	368	132	135	194	93	96	124	73	75	87	60	61	65
0.1	***	353	359	400	189	195	212	126	130	139	94	96	102	74	76	80
	•••	355	369	526	191	200	260	127	133	158	95	99	107	75	78	78
0.15	***	457	468	548	234	243	276	151	157	174	109	114	124	85	88	94
	•••	460	485	662	237	253	316	153	163	187	111	118	124	86	91	88
0.2	***	534	550	679	267	280	332	168	178	203	120	127	141	92	97	106
	•••	539	576	778	272	295	362	172	187	211	123	132	138	94	100	96
0.25	***	585	608	791	288	306	377	180	193	227	127	136	155	97	103	114
	•••	593	643	873	295	327	400	185	204	229	131	143	148	99	107	102
0.3	***	614	643	882	300	323	413	187	202	245	131	142	164	99	106	120
	•••	624	689	946	309	349	428	193	217	242	136	150	154	102	112	106
0.35	***	621	659	950	304	332	439	189	207	257	132	144	170	99	107	123
	•••	635	715	999	315	362	446	196	224	250	137	154	158	103	113	107
0.4	***	611	658	997	301	333	455	188	208	263	131	144	173	98	107	123
	•••	628	724	1030	314	367	456	196	226	253	137	155	158	102	113	106
0.45	***	588	644	1020	293	328	461	183	205	263	128	141	171	95	104	121
	•••	609	719	1041	307	366	456	192	224	250	133	152	154	99	111	102
0.5	***	555	619	1020	280	318	456	175	198	258	122	136	166	90	99	116
	•••	579	701	1030	295	357	446	185	218	242	128	147	148	94	106	96
0.55	***	514	586	998	263	303	441	165	188	246	115	128	157	85	93	108
	•••	541	673	999	279	343	428	175	208	229	121	139	138	88	99	88
0.6	***	470	546	952	244	284	416	153	175	229	106	118	144	77	85	98
	•••	499	634	946	260	323	400	163	194	211	111	128	124	81	90	78
0.65	***	422	500	884	222	260	380	139	159	207	95	106	128	68	75	86
	•••	453	587	873	238	297	362	148	176	187	100	115	107	71	79	65
0.7	***	374	448	794	197	232	335	122	140	179	82	91	108	.	.	.
	•••	403	530	778	212	265	316	130	154	158	86	98	87	.	.	.
0.75	***	323	391	681	169	199	280	103	117	145	.	.	.	.	.	.
	•••	350	464	662	182	226	260	109	128	124	.	.	.	.	.	.
0.8	***	268	327	546	137	160	215	.	.	.	.	.	.	.	.	.
	•••	292	386	526	147	180	194	.	.	.	.	.	.	.	.	.
0.85	***	207	251	389	.	.	.	.	.	.	.	.	.	.	.	.
	•••	225	293	368	.	.	.	.	.	.	.	.	.	.	.	.

TABLE 6: ALPHA= 0.025 POWER= 0.9   EXPECTED ACCRUAL THRU MINIMUM FOLLOW-UP= 110

	DEL=.05			DEL=.10			DEL=.15			DEL=.20			DEL=.25		
FACT=	1.0 .75	.50 .25	.00 BIN	1.0 .75	.50 .25	.00 BIN	1.0 .75	.50 .25	.00 BIN	1.0 75	.50 .25	.00 BIN	1.0 .75	.50 .25	.00 BIN
PCONT=***			REQUIRED NUMBER OF PATIENTS												
0.05 ***	628	631	671	223	226	239	131	133	139	93	95	98	72	74	76
•••	629	636	1156	224	230	368	132	136	194	94	96	124	73	75	87
0.1 ***	1145	1151	1295	353	360	400	190	195	212	126	130	139	94	97	102
•••	1147	1164	1829	356	370	526	192	201	260	128	134	158	95	99	107
0.15 ***	1583	1594	1898	458	470	548	235	244	276	151	158	174	110	114	124
•••	1586	1615	2417	462	487	662	239	255	316	154	164	187	112	118	124
0.2 ***	1921	1937	2448	536	553	679	268	282	332	170	179	203	121	127	141
•••	1927	1970	2922	542	580	778	273	297	362	174	188	211	124	133	138
0.25 ***	2155	2178	2930	588	612	791	290	309	377	182	194	227	128	137	155
•••	2163	2224	3342	596	649	873	298	329	400	187	206	229	132	144	148
0.3 ***	2289	2320	3331	617	649	882	303	327	413	189	204	245	133	143	164
•••	2299	2383	3678	628	696	946	312	352	428	195	218	242	137	151	154
0.35 ***	2331	2372	3648	625	666	950	308	336	439	192	209	257	134	146	170
•••	2344	2455	3930	640	724	999	319	366	446	199	226	250	139	155	158
0.4 ***	2292	2346	3876	617	667	997	305	338	455	190	210	263	133	146	173
•••	2310	2453	4098	635	734	1030	318	372	456	199	229	253	138	156	158
0.45 ***	2187	2256	4014	594	654	1020	297	333	461	186	207	263	129	143	171
•••	2210	2390	4182	616	730	1041	312	371	456	195	227	250	135	154	154
0.5 ***	2027	2116	4061	562	630	1020	285	324	456	178	201	258	124	138	166
•••	2057	2281	4182	588	713	1030	300	363	446	188	221	242	130	148	148
0.55 ***	1830	1941	4016	523	597	998	269	309	441	169	191	246	117	130	157
•••	1868	2139	4098	551	685	999	285	348	428	178	211	229	122	140	138
0.6 ***	1610	1748	3879	479	557	952	249	289	416	156	178	229	108	120	144
•••	1657	1974	3930	509	647	946	266	328	400	166	197	211	113	130	124
0.65 ***	1383	1547	3652	432	511	884	227	265	380	142	162	207	97	108	128
•••	1441	1795	3678	463	599	873	243	302	362	150	178	187	101	116	107
0.7 ***	1164	1349	3334	383	460	794	202	237	335	125	142	179	83	92	108
•••	1232	1606	3342	414	542	778	217	269	316	132	156	158	87	99	87
0.75 ***	963	1156	2927	331	401	681	174	203	280	105	118	145	.	.	.
•••	1036	1407	2922	360	474	662	186	229	260	111	129	124	.	.	.
0.8 ***	781	966	2431	276	335	546	141	163	215	.	.	.	.	.	.
•••	852	1195	2417	300	394	526	150	182	194	.	.	.	.	.	.
0.85 ***	611	772	1847	213	257	389	.	.	.	.	.	.	.	.	.
•••	674	962	1829	231	298	368	.	.	.	.	.	.	.	.	.
0.9 ***	437	555	1177	.	.	.	.	.	.	.	.	.	.	.	.
•••	484	687	1156	.	.	.	.	.	.	.	.	.	.	.	.

TABLE 6: ALPHA= 0.025 POWER= 0.9   EXPECTED ACCRUAL THRU MINIMUM FOLLOW-UP= 120

	DEL=.05			DEL=.10			DEL=.15			DEL=.20			DEL=.25		
FACT=	1.0 .75	.50 .25	.00 BIN	1.0 .75	.50 .25	.00 BIN	1.0 .75	.50 .25	.00 BIN	1.0 75	.50 .25	.00 BIN	1.0 .75	.50 .25	.00 BIN
PCONT=•••				REQUIRED NUMBER OF PATIENTS											
0.05 ***	628 631		671	223 227		239	131 134		139	93 95		98	73 74		76
•••	629 637		1156	225 231		368	132 136		194	94 96		124	73 75		87
0.1 ***	1145 1152		1295	354 362		400	190 196		212	127 131		139	94 97		102
•••	1147 1166		1829	357 372		526	193 202		260	128 134		158	96 99		107
0.15 ***	1584 1596		1898	459 472		548	236 246		276	152 159		174	110 115		124
•••	1588 1619		2417	463 490		662	240 256		316	155 165		187	112 118		124
0.2 ***	1923 1940		2448	537 556		679	270 284		332	171 180		203	122 128		141
•••	1928 1976		2922	544 583		778	275 299		362	175 189		211	125 133		138
0.25 ***	2157 2182		2930	590 616		791	292 312		377	183 196		227	130 138		155
•••	2165 2232		3342	599 654		873	300 332		400	189 207		229	133 144		148
0.3 ***	2291 2326		3331	620 654		882	306 330		413	191 206		245	134 144		164
•••	2303 2394		3678	632 702		946	315 355		428	197 220		242	138 152		154
0.35 ***	2334 2380		3648	629 672		950	311 340		439	194 212		257	136 147		170
•••	2349 2469		3930	645 732		999	322 370		446	201 228		250	140 156		158
0.4 ***	2297 2356		3876	622 674		997	309 342		455	193 213		263	135 147		172
•••	2317 2471		4098	641 744		1030	322 376		456	201 230		253	140 157		158
0.45 ***	2193 2269		4014	601 662		1020	302 338		460	188 210		263	131 144		171
•••	2218 2413		4182	624 741		1041	316 375		456	198 229		250	137 155		154
0.5 ***	2036 2132		4061	569 639		1020	289 329		456	181 204		258	126 139		166
•••	2068 2309		4182	596 724		1030	305 367		446	191 223		242	132 150		148
0.55 ***	1840 1961		4016	531 607		998	273 314		441	171 194		246	118 131		157
•••	1881 2171		4098	561 697		999	290 353		428	181 213		229	124 141		138
0.6 ***	1623 1771		3879	488 568		952	254 294		415	159 181		229	109 121		144
•••	1674 2009		3930	519 658		946	271 332		400	168 198		211	115 130		124
0.65 ***	1399 1574		3652	441 522		884	232 270		380	144 164		207	98 109		128
•••	1461 1832		3678	473 610		873	248 306		362	153 180		187	103 116		107
0.7 ***	1183 1378		3334	392 470		794	206 241		335	127 144		178	85 93		108
•••	1255 1642		3342	423 552		778	221 272		316	134 157		158	88 99		87
0.75 ***	984 1185		2927	340 410		681	177 207		280	106 120		145	.	.	.
•••	1060 1442		2922	368 483		662	190 232		260	112 130		124	.	.	.
0.8 ***	802 993		2431	283 343		546	143 166		215	.	.	.	.	.	.
•••	875 1226		2417	307 401		526	153 184		194	.	.	.	.	.	.
0.85 ***	629 794		1847	219 262		389	.	.	.	.	.	.	.	.	.
•••	694 987		1829	237 303		368	.	.	.	.	.	.	.	.	.
0.9 ***	451 571		1177	.	.	.	.	.	.	.	.	.	.	.	.
•••	499 704		1156	.	.	.	.	.	.	.	.	.	.	.	.

TABLE 6: ALPHA= 0.025 POWER= 0.9   EXPECTED ACCRUAL THRU MINIMUM FOLLOW-UP= 130

	DEL=.05			DEL=.10			DEL=.15			DEL=.20			DEL=.25		
FACT=	1.0 .75	.50 .25	.00 BIN	1.0 .75	.50 .25	.00 BIN	1.0 .75	.50 .25	.00 BIN	1.0 75	.50 .25	.00 BIN	1.0 .75	.50 .25	.00 BIN

PCONT=•••   REQUIRED NUMBER OF PATIENTS

PCONT	1.0/.75	.50/.25	.00/BIN	1.0/.75	.50/.25	.00/BIN	1.0/.75	.50/.25	.00/BIN	1.0/.75	.50/.25	.00/BIN	1.0/.75	.50/.25	.00/BIN
0.05 •••	628	632	671	224	227	239	132	134	139	93	95	98	73	74	76
•••	630	638	1156	225	231	368	133	136	194	94	96	124	73	75	87
0.1 •••	1146	1153	1295	355	363	400	191	197	212	127	131	139	95	97	102
•••	1148	1168	1829	358	373	526	193	202	260	129	134	158	96	99	107
0.15 •••	1585	1598	1898	460	473	548	237	247	276	153	160	174	111	115	124
•••	1589	1623	2417	465	492	662	241	257	316	156	165	187	113	119	124
0.2 •••	1924	1943	2448	539	559	679	271	286	332	172	181	203	123	129	141
•••	1930	1982	2922	546	587	778	277	301	362	176	190	211	125	134	138
0.25 •••	2159	2186	2930	593	620	791	295	314	377	185	197	227	131	139	155
•••	2168	2241	3342	602	659	873	302	334	400	190	208	229	134	145	148
0.3 •••	2294	2331	3331	623	659	882	308	333	413	193	208	245	135	145	164
•••	2307	2405	3678	636	709	946	318	358	428	199	221	242	139	153	154
0.35 •••	2338	2387	3648	633	679	950	314	343	439	196	214	257	137	148	170
•••	2355	2484	3930	650	739	999	326	373	446	203	229	250	142	157	158
0.4 •••	2302	2366	3876	627	682	997	313	346	455	195	215	263	136	148	173
•••	2323	2490	4098	647	752	1030	326	380	456	204	232	253	141	158	158
0.45 •••	2199	2281	4014	607	671	1020	306	343	460	191	212	263	133	146	171
•••	2227	2436	4182	631	750	1041	321	379	456	200	231	250	139	156	154
0.5 •••	2044	2148	4061	576	649	1020	294	333	456	184	206	258	128	141	166
•••	2079	2335	4182	604	735	1030	310	371	446	193	225	242	133	151	148
0.55 •••	1851	1980	4016	539	617	998	278	318	441	174	196	246	120	133	157
•••	1895	2201	4098	569	707	999	295	357	428	184	215	229	126	142	138
0.6 •••	1636	1794	3879	496	578	952	259	299	416	162	183	229	111	123	144
•••	1691	2042	3930	529	669	946	275	336	400	171	200	211	116	131	124
0.65 •••	1415	1600	3652	450	532	884	236	274	380	147	166	207	99	110	128
•••	1481	1866	3678	483	620	873	252	309	362	155	182	187	104	117	107
0.7 •••	1202	1405	3334	401	479	794	210	245	335	129	146	179	86	94	108
•••	1277	1677	3342	432	561	778	225	276	316	136	159	158	89	100	87
0.75 •••	1004	1212	2927	348	419	681	181	210	280	108	121	145	.	.	.
•••	1083	1475	2922	376	491	662	193	235	260	114	131	124	.	.	.
0.8 •••	821	1018	2431	290	349	546	146	168	215	.	.	.	.	.	.
•••	897	1256	2417	314	408	526	156	186	194	.	.	.	.	.	.
0.85 •••	647	815	1847	224	267	389	.	.	.	.	.	.	.	.	.
•••	713	1011	1829	242	307	368	.	.	.	.	.	.	.	.	.
0.9 •••	464	586	1177	.	.	.	.	.	.	.	.	.	.	.	.
•••	512	720	1156	.	.	.	.	.	.	.	.	.	.	.	.

TABLE 6: ALPHA= 0.025 POWER= 0.9   EXPECTED ACCRUAL THRU MINIMUM FOLLOW-UP= 140

		DEL=.05			DEL=.10			DEL=.15			DEL=.20			DEL=.25	
FACT=	1.0 .75	.50 .25	.00 BIN	1.0 .75	.50 .25	.00 BIN	1.0 .75	.50 .25	.00 BIN	1.0 75	.50 .25	.00 BIN	1.0 .75	.50 .25	.00 BIN
PCONT=***				REQUIRED NUMBER OF PATIENTS											
0.05 ***	629	632	671	224	228	239	132	134	139	94	95	98	73	74	76
•••	630	639	1156	225	231	368	133	136	194	94	96	124	74	75	87
0.1 ***	1146	1154	1295	355	364	400	192	197	212	128	132	139	95	98	102
•••	1149	1170	1829	358	374	526	194	203	260	129	135	158	96	99	107
0.15 ***	1586	1600	1898	461	475	548	238	248	276	154	160	174	111	116	124
•••	1590	1627	2417	466	494	662	242	258	316	156	166	187	113	119	124
0.2 ***	1926	1946	2448	541	561	679	273	287	332	173	182	203	123	130	141
•••	1932	1987	2922	548	590	778	278	302	362	177	190	211	126	134	138
0.25 ***	2161	2191	2930	595	624	791	296	316	377	186	199	227	132	139	155
•••	2171	2249	3342	605	663	873	304	336	400	191	209	229	135	146	148
0.3 ***	2297	2337	3331	626	664	882	311	336	413	194	209	245	136	146	164
•••	2310	2416	3678	640	714	946	321	361	428	201	223	242	141	153	154
0.35 ***	2342	2395	3648	637	685	950	317	346	439	198	216	257	138	149	170
•••	2360	2498	3930	655	746	999	329	376	446	205	231	250	143	158	158
0.4 ***	2307	2376	3876	632	689	997	316	350	455	197	217	263	137	149	172
•••	2330	2508	4098	653	760	1030	330	383	456	206	234	253	143	159	158
0.45 ***	2206	2294	4014	613	679	1020	309	347	461	193	215	263	134	147	171
•••	2235	2458	4182	638	759	1041	325	383	456	202	232	250	140	157	154
0.5 ***	2052	2163	4061	583	657	1020	298	337	456	186	208	258	129	142	166
•••	2089	2361	4182	612	744	1030	314	375	446	196	227	242	135	151	148
0.55 ***	1861	1999	4016	546	627	998	282	323	441	177	198	246	122	134	157
•••	1908	2230	4098	578	717	999	299	361	428	186	216	229	127	143	138
0.6 ***	1649	1816	3879	504	588	952	263	303	416	164	185	229	112	124	144
•••	1707	2074	3930	537	679	946	280	340	400	173	202	211	117	132	124
0.65 ***	1431	1624	3652	458	541	884	240	278	380	149	168	207	101	111	128
•••	1501	1899	3678	491	630	873	256	313	362	157	183	187	105	118	107
0.7 ***	1220	1431	3334	408	488	794	214	249	335	131	147	179	87	95	108
•••	1299	1709	3342	440	570	778	229	279	316	138	160	158	90	100	87
0.75 ***	1023	1237	2927	355	427	681	184	213	280	110	122	145	.	.	.
•••	1105	1505	2922	384	499	662	196	237	260	115	132	124	.	.	.
0.8 ***	840	1042	2431	296	356	546	149	170	215	.	.	.	.	.	.
•••	918	1283	2417	321	413	526	158	187	194	.	.	.	.	.	.
0.85 ***	663	835	1847	228	272	389	.	.	.	.	.	.	.	.	.
•••	731	1033	1829	246	311	368	.	.	.	.	.	.	.	.	.
0.9 ***	476	600	1177	.	.	.	.	.	.	.	.	.	.	.	.
•••	525	734	1156	.	.	.	.	.	.	.	.	.	.	.	.

TABLE 6: ALPHA= 0.025 POWER= 0.9   EXPECTED ACCRUAL THRU MINIMUM FOLLOW-UP= 150

		DEL=.05			DEL=.10			DEL=.15			DEL=.20			DEL=.25		
FACT=		1.0 .75	.50 .25	.00 BIN	1.0 .75	.50 .25	.00 BIN	1.0 .75	.50 .25	.00 BIN	1.0 75	.50 .25	.00 BIN	1.0 .75	.50 .25	.00 BIN
PCONT=***				REQUIRED	NUMBER OF	PATIENTS										
0.05	***	629	633	671	224	228	239	132	134	139	94	95	98	73	74	76
	•••	630	640	1156	226	232	368	133	136	194	94	96	124	74	75	87
0.1	***	1147	1156	1295	356	365	400	192	198	212	128	132	139	95	98	102
	•••	1150	1172	1829	359	375	526	194	203	260	130	135	158	96	100	107
0.15	***	1587	1601	1898	462	477	548	239	249	276	154	161	174	112	116	124
	•••	1592	1630	2417	467	496	662	243	259	316	157	166	187	114	119	124
0.2	***	1927	1949	2448	542	564	679	274	289	332	174	183	203	124	130	141
	•••	1934	1993	2922	550	593	778	280	303	362	178	191	211	127	135	138
0.25	***	2163	2195	2930	597	627	791	298	318	377	187	200	227	132	140	155
	•••	2174	2257	3342	608	667	873	306	338	400	193	210	229	136	146	148
0.3	***	2300	2343	3331	629	668	882	313	338	413	196	211	245	137	147	164
	•••	2314	2427	3678	643	720	946	323	363	428	202	224	242	142	154	154
0.35	***	2346	2402	3648	641	690	950	320	349	439	200	217	257	139	150	170
	•••	2365	2512	3930	659	752	999	332	379	446	207	232	250	144	158	158
0.4	***	2312	2386	3876	636	695	997	319	353	455	199	219	263	139	151	172
	•••	2336	2525	4098	658	767	1030	333	386	456	208	235	253	144	159	158
0.45	***	2212	2306	4014	618	686	1020	313	350	460	196	217	263	136	148	171
	•••	2243	2479	4182	644	767	1041	328	386	456	205	234	250	141	157	154
0.5	***	2060	2179	4061	590	666	1020	302	341	456	189	210	258	130	143	166
	•••	2100	2386	4182	619	753	1030	318	379	446	198	228	242	136	152	148
0.55	***	1871	2018	4016	554	635	998	286	327	441	179	200	246	123	135	157
	•••	1922	2258	4098	586	726	999	303	364	428	188	218	229	128	144	138
0.6	***	1661	1838	3879	512	596	952	267	307	416	166	187	229	113	125	144
	•••	1724	2104	3930	545	688	946	284	343	400	175	203	211	118	133	124
0.65	***	1446	1648	3652	466	550	884	244	282	380	151	170	207	102	112	128
	•••	1520	1930	3678	500	638	873	260	316	362	159	184	187	106	118	107
0.7	***	1238	1456	3334	416	496	794	218	252	335	133	149	179	88	96	108
	•••	1319	1740	3342	448	578	778	232	281	316	140	161	158	91	101	87
0.75	***	1042	1262	2927	362	434	681	187	215	280	111	123	145	.	.	.
	•••	1126	1534	2922	391	505	662	199	239	260	117	132	124	.	.	.
0.8	***	858	1064	2431	302	362	546	151	172	215	.	.	.	.	.	.
	•••	938	1309	2417	327	419	526	160	189	194	.	.	.	.	.	.
0.85	***	679	854	1847	233	276	389	.	.	.	.	.	.	.	.	.
	•••	748	1053	1829	251	314	368	.	.	.	.	.	.	.	.	.
0.9	***	488	613	1177	.	.	.	.	.	.	.	.	.	.	.	.
	•••	538	748	1156	.	.	.	.	.	.	.	.	.	.	.	.

TABLE 6: ALPHA= 0.025 POWER= 0.9   EXPECTED ACCRUAL THRU MINIMUM FOLLOW-UP= 160

		DEL=.05			DEL=.10			DEL=.15			DEL=.20			DEL=.25		
FACT=		1.0 .75	.50 .25	.00 BIN	1.0 .75	.50 .25	.00 BIN	1.0 .75	.50 .25	.00 BIN	1.0 75	.50 .25	.00 BIN	1.0 .75	.50 .25	.00 BIN
PCONT=***					REQUIRED	NUMBER OF	PATIENTS									
0.05	***	629	633	671	225	228	239	132	135	139	94	95	98	73	74	76
	•••	631	641	1156	226	232	368	133	137	194	95	97	124	74	75	87
0.1	***	1148	1157	1295	357	366	400	193	198	212	128	132	139	96	98	102
	•••	1151	1175	1829	360	376	526	195	204	260	130	135	158	97	100	107
0.15	***	1588	1603	1898	463	479	548	240	250	276	155	161	174	112	116	124
	•••	1593	1634	2417	469	498	662	244	259	316	158	167	187	114	120	124
0.2	***	1929	1952	2448	544	567	679	275	290	332	175	184	203	125	131	141
	•••	1936	1999	2922	552	596	778	281	305	362	179	192	211	127	135	138
0.25	***	2165	2199	2930	599	631	791	300	320	377	189	201	227	133	141	155
	•••	2177	2265	3342	611	671	873	308	340	400	194	211	229	137	147	148
0.3	***	2303	2349	3331	632	673	882	315	341	413	197	212	245	138	148	164
	•••	2318	2438	3678	647	725	946	326	365	428	204	225	242	142	155	154
0.35	***	2349	2410	3648	645	696	950	322	352	439	201	219	257	140	151	170
	•••	2370	2526	3930	664	758	999	335	381	446	209	233	250	145	159	158
0.4	***	2317	2395	3876	641	702	997	322	357	455	201	221	263	140	152	173
	•••	2343	2543	4098	664	774	1030	336	389	456	209	237	253	145	160	158
0.45	***	2218	2318	4014	624	694	1020	316	354	461	198	218	263	137	149	171
	•••	2252	2500	4182	651	775	1041	332	389	456	206	235	250	142	158	154
0.5	***	2068	2194	4061	596	674	1020	305	345	456	191	212	258	132	144	166
	•••	2111	2410	4182	626	761	1030	322	382	446	200	229	242	137	153	148
0.55	***	1881	2036	4016	561	644	998	290	331	441	181	202	246	124	136	157
	•••	1935	2285	4098	593	735	999	307	367	428	190	219	229	129	145	138
0.6	***	1674	1859	3879	519	605	952	271	311	416	168	189	229	115	126	144
	•••	1740	2133	3930	553	696	946	287	346	400	177	204	211	119	133	124
0.65	***	1461	1671	3652	473	558	884	248	285	380	153	171	207	103	112	128
	•••	1538	1960	3678	508	647	873	264	319	362	161	185	187	107	119	107
0.7	***	1255	1480	3334	423	504	794	221	255	335	134	150	179	89	96	108
	•••	1339	1770	3342	456	585	778	235	284	316	141	162	158	92	101	87
0.75	***	1060	1285	2927	368	441	681	190	218	280	112	124	145	.	.	.
	•••	1147	1562	2922	398	512	662	202	241	260	118	133	124	.	.	.
0.8	***	875	1085	2431	307	367	546	153	174	215	.	.	.	.	.	.
	•••	957	1333	2417	332	424	526	162	190	194	.	.	.	.	.	.
0.85	***	694	871	1847	237	280	389	.	.	.	.	.	.	.	.	.
	•••	764	1073	1829	255	317	368	.	.	.	.	.	.	.	.	.
0.9	***	499	625	1177	.	.	.	.	.	.	.	.	.	.	.	.
	•••	549	760	1156	.	.	.	.	.	.	.	.	.	.	.	.

TABLE 6: ALPHA= 0.025 POWER= 0.9   EXPECTED ACCRUAL THRU MINIMUM FOLLOW-UP= 170

		DEL=.05			DEL=.10			DEL=.15			DEL=.20			DEL=.25		
FACT=		1.0 .75	.50 .25	.00 BIN	1.0 .75	.50 .25	.00 BIN	1.0 .75	.50 .25	.00 BIN	1.0 75	.50 .25	.00 BIN	1.0 .75	.50 .25	.00 BIN
PCONT=•••					REQUIRED NUMBER OF PATIENTS											
0.05	***	629	634	671	225	229	239	133	135	139	94	96	98	73	74	76
	•••	631	641	1156	226	232	368	134	137	194	95	97	124	74	75	87
0.1	***	1148	1158	1295	357	366	400	193	199	212	129	133	139	96	98	102
	•••	1151	1177	1829	361	377	526	195	204	260	130	135	158	97	100	107
0.15	***	1589	1605	1898	464	480	548	241	251	276	156	162	174	113	117	124
	•••	1594	1638	2417	470	500	662	245	260	316	158	167	187	115	120	124
0.2	***	1930	1955	2448	545	569	679	276	292	331	176	185	203	125	131	141
	•••	1938	2004	2922	554	599	778	283	306	362	180	192	211	128	135	138
0.25	***	2167	2203	2930	602	634	791	302	322	377	190	202	227	134	141	155
	•••	2179	2273	3342	613	675	873	310	341	400	195	212	229	137	147	148
0.3	***	2306	2354	3331	635	677	882	317	343	413	199	213	245	139	148	164
	•••	2322	2448	3678	651	729	946	328	367	428	205	226	242	143	155	154
0.35	***	2353	2418	3648	649	701	950	325	355	439	203	220	257	141	152	170
	•••	2375	2540	3930	668	764	999	337	383	446	210	234	250	146	160	158
0.4	***	2322	2405	3876	645	708	997	325	359	455	203	222	263	141	152	173
	•••	2350	2559	4098	669	781	1030	339	392	456	211	238	253	146	161	158
0.45	***	2224	2331	4014	629	700	1020	320	357	460	199	220	263	138	150	171
	•••	2260	2520	4182	657	782	1041	335	392	456	208	237	250	143	159	154
0.5	***	2076	2209	4061	602	681	1020	309	348	456	193	214	258	133	145	166
	•••	2121	2433	4182	633	769	1030	325	385	446	202	231	242	138	154	148
0.55	***	1891	2054	4016	567	651	998	293	334	441	183	204	246	125	137	157
	•••	1948	2311	4098	601	742	999	310	370	428	192	220	229	130	145	138
0.6	***	1687	1879	3879	526	613	952	274	314	416	170	190	229	116	126	144
	•••	1756	2160	3930	561	704	946	291	349	400	179	206	211	120	134	124
0.65	***	1476	1694	3652	480	566	884	251	289	380	155	173	207	104	113	128
	•••	1556	1988	3678	515	654	873	267	321	362	163	186	187	108	119	107
0.7	***	1272	1503	3334	430	511	794	224	258	335	136	151	178	89	97	108
	•••	1359	1797	3342	463	592	778	238	286	316	143	162	158	93	102	87
0.75	***	1077	1308	2927	375	447	681	192	220	280	114	125	145	.	.	.
	•••	1166	1588	2922	404	517	662	204	243	260	119	134	124	.	.	.
0.8	***	892	1106	2431	313	373	546	155	176	215	.	.	.	.	.	.
	•••	976	1356	2417	338	428	526	164	191	194	.	.	.	.	.	.
0.85	***	708	888	1847	241	283	389	.	.	.	.	.	.	.	.	.
	•••	779	1091	1829	259	320	368	.	.	.	.	.	.	.	.	.
0.9	***	509	637	1177	.	.	.	.	.	.	.	.	.	.	.	.
	•••	560	772	1156	.	.	.	.	.	.	.	.	.	.	.	.

TABLE 6: ALPHA= 0.025 POWER= 0.9   EXPECTED ACCRUAL THRU MINIMUM FOLLOW-UP= 180

	DEL=.05			DEL=.10			DEL=.15			DEL=.20			DEL=.25		
FACT=	1.0 .75	.50 .25	.00 BIN	1.0 .75	.50 .25	.00 BIN	1.0 .75	.50 .25	.00 BIN	1.0 75	.50 .25	.00 BIN	1.0 .75	.50 .25	.00 BIN
PCONT=*			REQUIRED NUMBER OF PATIENTS												
0.05 ***	630	634	671	225	229	239	133	135	139	94	96	98	73	74	76
•••	631	642	1156	227	233	368	134	137	194	95	97	124	74	75	87
0.1 ***	1149	1159	1295	358	367	400	194	199	212	129	133	139	96	98	102
•••	1152	1179	1829	362	378	526	196	204	260	131	135	158	97	100	107
0.15 ***	1590	1607	1898	465	482	548	242	252	276	156	162	174	113	117	123
•••	1596	1641	2417	471	501	662	246	261	316	159	167	187	115	120	124
0.2 ***	1932	1958	2448	547	571	679	278	293	332	176	186	203	126	132	141
•••	1940	2010	2922	556	601	778	284	307	362	180	193	211	128	136	138
0.25 ***	2170	2207	2930	604	637	791	303	324	377	191	203	227	135	142	155
•••	2182	2281	3342	616	678	873	312	343	400	196	212	229	138	147	148
0.3 ***	2309	2360	3331	638	681	882	319	345	413	200	215	245	140	149	164
•••	2326	2459	3678	654	734	946	330	369	428	206	226	242	144	155	154
0.35 ***	2357	2425	3648	652	706	950	327	357	439	204	221	257	142	153	170
•••	2380	2553	3930	672	769	999	340	386	446	212	235	250	147	160	158
0.4 ***	2327	2415	3876	650	713	997	328	362	455	205	224	263	142	153	173
•••	2356	2576	4098	674	787	1030	342	394	456	213	239	253	147	161	158
0.45 ***	2231	2343	4014	634	707	1020	323	360	461	201	222	263	139	151	171
•••	2269	2539	4182	663	789	1041	338	394	456	210	238	250	144	159	154
0.5 ***	2084	2224	4061	608	688	1020	312	352	456	195	215	258	134	146	166
•••	2132	2456	4182	639	776	1030	329	387	446	204	232	242	139	154	148
0.55 ***	1901	2072	4016	574	659	998	297	337	441	185	205	246	126	138	157
•••	1961	2335	4098	607	750	999	314	373	428	194	222	229	132	146	138
0.6 ***	1699	1899	3879	533	620	952	278	317	416	172	192	229	117	127	144
•••	1771	2187	3930	568	711	946	294	352	400	181	207	211	121	134	124
0.65 ***	1491	1715	3652	487	573	884	254	292	380	156	174	207	105	114	128
•••	1574	2015	3678	522	661	873	270	323	362	164	187	187	109	120	107
0.7 ***	1288	1525	3334	436	518	794	227	260	335	137	152	179	90	97	108
•••	1378	1824	3342	470	598	778	241	288	316	144	163	158	93	102	87
0.75 ***	1094	1329	2927	380	453	681	195	222	280	115	126	145	.	.	.
•••	1185	1613	2922	410	523	662	207	244	260	120	134	124	.	.	.
0.8 ***	908	1125	2431	317	377	546	157	177	215	.	.	.	.	.	.
•••	993	1377	2417	343	432	526	166	192	194	.	.	.	.	.	.
0.85 ***	722	904	1847	244	287	389	.	.	.	.	.	.	.	.	.
•••	794	1108	1829	262	323	368	.	.	.	.	.	.	.	.	.
0.9 ***	519	648	1177	.	.	.	.	.	.	.	.	.	.	.	.
•••	571	783	1156	.	.	.	.	.	.	.	.	.	.	.	.

TABLE 6: ALPHA= 0.025 POWER= 0.9   EXPECTED ACCRUAL THRU MINIMUM FOLLOW-UP= 190

	DEL=.05			DEL=.10			DEL=.15			DEL=.20			DEL=.25		
FACT=	1.0 .75	.50 .25	.00 BIN	1.0 .75	.50 .25	.00 BIN	1.0 .75	.50 .25	.00 BIN	1.0 75	.50 .25	.00 BIN	1.0 .75	.50 .25	.00 BIN
PCONT=***					REQUIRED NUMBER OF PATIENTS										
0.05 ***	630	635	671	226	229	239	133	135	139	94	96	98	73	75	76
***	632	643	1156	227	233	368	134	137	194	95	97	124	74	75	87
0.1 ***	1149	1160	1295	359	368	400	194	200	212	130	133	139	96	98	102
***	1153	1181	1829	362	378	526	196	205	260	131	136	158	97	100	107
0.15 ***	1591	1609	1898	467	483	548	242	253	276	157	163	174	113	117	124
***	1597	1645	2417	473	502	662	246	261	316	159	168	187	115	120	124
0.2 ***	1933	1961	2448	549	573	679	279	294	331	177	186	203	126	132	141
***	1942	2015	2922	558	603	778	285	308	362	181	193	211	129	136	138
0.25 ***	2172	2212	2929	606	640	791	305	325	377	192	204	227	135	142	155
***	2185	2288	3342	619	681	873	313	344	400	197	213	229	138	148	148
0.3 ***	2311	2366	3331	641	685	882	321	347	413	201	216	245	141	150	164
***	2329	2469	3678	657	738	946	332	370	428	207	227	242	145	156	154
0.35 ***	2361	2432	3648	656	710	950	329	360	439	206	223	257	143	153	170
***	2385	2566	3930	677	774	999	342	387	446	213	236	250	148	160	158
0.4 ***	2332	2424	3876	654	719	997	331	365	455	206	225	263	143	154	173
***	2363	2592	4098	679	793	1030	345	396	456	214	240	253	148	162	158
0.45 ***	2237	2355	4014	639	713	1020	325	363	460	203	223	263	140	152	171
***	2277	2559	4182	668	795	1041	341	397	456	212	239	250	145	160	154
0.5 ***	2092	2239	4061	614	695	1020	315	355	456	196	217	258	135	146	166
***	2142	2478	4182	646	782	1030	332	390	446	205	233	242	140	155	148
0.55 ***	1912	2089	4016	580	666	998	300	340	441	187	207	246	127	139	157
***	1974	2359	4098	614	757	999	317	375	428	195	222	229	132	146	138
0.6 ***	1712	1919	3879	539	628	952	281	320	416	174	193	229	117	128	144
***	1786	2212	3930	575	718	946	297	354	400	182	208	211	122	135	124
0.65 ***	1506	1736	3652	493	580	884	257	294	380	158	175	207	105	114	128
***	1591	2041	3678	529	667	873	273	325	362	165	188	187	109	120	107
0.7 ***	1304	1546	3334	443	524	794	230	263	335	139	153	179	91	98	108
***	1396	1849	3342	476	604	778	244	290	316	145	164	158	94	102	87
0.75 ***	1110	1350	2927	386	459	681	197	224	280	116	127	145	.	.	.
***	1203	1636	2922	416	528	662	209	246	260	121	135	124	.	.	.
0.8 ***	923	1144	2431	322	382	546	158	178	215	.	.	.	.	.	.
***	1010	1398	2417	347	436	526	167	193	194	.	.	.	.	.	.
0.85 ***	735	920	1847	248	290	389	.	.	.	.	.	.	.	.	.
***	808	1124	1829	266	325	368	.	.	.	.	.	.	.	.	.
0.9 ***	529	658	1177	.	.	.	.	.	.	.	.	.	.	.	.
***	581	793	1156	.	.	.	.	.	.	.	.	.	.	.	.

TABLE 6: ALPHA= 0.025 POWER= 0.9   EXPECTED ACCRUAL THRU MINIMUM FOLLOW-UP= 200

	DEL=.05			DEL=.10			DEL=.15			DEL=.20			DEL=.25		
FACT=	1.0 .75	.50 .25	.00 BIN	1.0 .75	.50 .25	.00 BIN	1.0 .75	.50 .25	.00 BIN	1.0 75	.50 .25	.00 BIN	1.0 .75	.50 .25	.00 BIN
PCONT=•••						REQUIRED NUMBER OF PATIENTS									
0.05 •••	630	635	671	226	230	239	133	135	139	94	96	98	74	75	76
•••	632	644	1156	227	233	368	134	137	194	95	97	124	74	75	87
0.1 •••	1150	1161	1295	359	369	400	195	200	212	130	133	139	96	99	102
•••	1154	1182	1829	363	379	526	197	205	260	131	136	158	97	100	107
0.15 •••	1592	1611	1898	468	485	548	243	253	276	157	163	174	114	118	124
•••	1598	1648	2417	474	504	662	247	262	316	160	168	187	115	120	124
0.2 •••	1934	1964	2448	550	576	679	280	295	332	178	187	203	127	132	141
•••	1944	2021	2922	560	606	778	286	309	362	182	194	211	129	136	138
0.25 •••	2174	2216	2930	608	643	791	306	327	377	193	204	227	136	143	155
•••	2188	2296	3342	621	684	873	315	345	400	198	214	229	139	148	148
0.3 •••	2314	2371	3331	643	689	882	323	349	413	202	217	245	142	150	164
•••	2333	2479	3678	661	742	946	334	372	428	208	228	242	145	156	154
0.35 •••	2365	2440	3648	659	715	950	332	362	439	207	224	257	144	154	170
•••	2390	2579	3930	681	779	999	344	389	446	214	237	250	149	161	158
0.4 •••	2337	2434	3876	658	724	997	333	367	455	208	226	263	144	155	173
•••	2369	2608	4098	684	798	1030	347	398	456	216	241	253	149	162	158
0.45 •••	2243	2367	4014	644	719	1020	328	366	461	205	224	263	141	152	171
•••	2285	2577	4182	674	801	1041	344	399	456	213	240	250	146	160	154
0.5 •••	2100	2253	4061	619	701	1020	318	357	456	198	218	258	136	147	166
•••	2153	2499	4182	652	789	1030	335	392	446	207	234	242	141	155	148
0.55 •••	1922	2106	4016	586	673	998	303	343	441	188	208	246	128	139	157
•••	1987	2383	4098	621	763	999	320	378	428	197	223	229	133	147	138
0.6 •••	1724	1938	3879	546	634	952	284	323	416	175	194	229	118	128	144
•••	1801	2237	3930	581	725	946	300	356	400	184	208	211	123	135	124
0.65 •••	1520	1756	3652	500	587	884	260	297	380	159	176	207	106	115	128
•••	1608	2066	3678	535	674	873	276	327	362	167	189	187	110	120	107
0.7 •••	1319	1567	3334	448	530	794	232	265	335	140	154	179	91	98	108
•••	1414	1874	3342	482	610	778	246	291	316	146	164	158	94	102	87
0.75 •••	1126	1369	2927	391	464	681	199	226	280	117	128	145	.	.	.
•••	1221	1659	2922	422	533	662	211	247	260	122	135	124	.	.	.
0.8 •••	938	1161	2431	327	386	546	160	180	215	.	.	.	.	.	.
•••	1026	1418	2417	352	440	526	169	194	194	.	.	.	.	.	.
0.85 •••	748	934	1847	251	293	389	.	.	.	.	.	.	.	.	.
•••	822	1140	1829	269	328	368	.	.	.	.	.	.	.	.	.
0.9 •••	538	668	1177	.	.	.	.	.	.	.	.	.	.	.	.
•••	591	803	1156	.	.	.	.	.	.	.	.	.	.	.	.

TABLE 6: ALPHA= 0.025 POWER= 0.9   EXPECTED ACCRUAL THRU MINIMUM FOLLOW-UP= 225

		DEL=.05			DEL=.10			DEL=.15			DEL=.20			DEL=.25		
FACT=		1.0 .75	.50 .25	.00 BIN	1.0 .75	.50 .25	.00 BIN	1.0 .75	.50 .25	.00 BIN	1.0 75	.50 .25	.00 BIN	1.0 .75	.50 .25	.00 BIN
PCONT=•••					REQUIRED	NUMBER	OF PATIENTS									
0.05	•••	631	637	671	226	230	239	133	136	139	95	96	98	74	75	76
	•••	633	645	1156	228	233	368	134	137	194	95	97	124	74	75	87
0.1	•••	1151	1164	1295	361	371	400	195	201	212	130	134	139	97	99	102
	•••	1156	1187	1829	365	381	526	198	206	260	132	136	158	98	100	107
0.15	•••	1594	1616	1898	470	488	548	245	255	276	158	164	174	114	118	124
	•••	1602	1657	2417	477	507	662	249	263	316	161	168	187	116	120	124
0.2	•••	1938	1971	2448	554	581	679	282	298	332	179	188	203	128	133	141
	•••	1949	2034	2922	564	611	778	289	311	362	183	194	211	130	137	138
0.25	•••	2179	2226	2930	613	650	791	310	330	377	195	206	227	137	144	155
	•••	2195	2314	3342	627	691	873	318	348	400	200	215	229	140	149	148
0.3	•••	2321	2385	3331	650	698	882	327	353	413	205	219	245	143	151	164
	•••	2343	2503	3678	668	750	946	338	375	428	211	229	242	147	157	154
0.35	•••	2374	2458	3648	668	726	950	337	367	439	210	226	257	146	155	170
	•••	2402	2610	3930	690	790	999	349	393	446	217	239	250	150	162	158
0.4	•••	2349	2457	3876	669	737	997	339	373	455	211	229	263	146	156	173
	•••	2386	2645	4098	695	810	1030	353	403	456	219	243	253	151	163	158
0.45	•••	2259	2396	4014	656	733	1020	335	372	461	208	227	263	143	154	171
	•••	2306	2621	4182	686	815	1041	350	404	456	217	242	250	148	161	154
0.5	•••	2120	2288	4061	632	716	1020	325	364	456	202	221	258	138	149	166
	•••	2179	2549	4182	666	803	1030	341	397	446	210	236	242	143	156	148
0.55	•••	1946	2147	4016	600	688	998	310	349	441	192	211	246	130	141	157
	•••	2018	2437	4098	635	778	999	327	383	428	200	226	229	135	148	138
0.6	•••	1754	1983	3879	560	650	952	291	329	416	179	197	229	120	130	144
	•••	1838	2294	3930	597	739	946	307	361	400	187	210	211	125	136	124
0.65	•••	1554	1805	3652	514	602	884	267	303	380	162	179	207	108	116	128
	•••	1648	2124	3678	550	687	873	282	332	362	170	191	187	111	121	107
0.7	•••	1356	1615	3334	462	545	794	238	270	335	142	156	179	93	99	108
	•••	1456	1930	3342	496	622	778	252	295	316	149	166	158	96	103	87
0.75	•••	1164	1416	2927	404	476	681	204	230	280	119	129	145	.	.	.
	•••	1262	1711	2922	434	543	662	216	250	260	123	136	124	.	.	.
0.8	•••	973	1203	2431	337	396	546	164	183	215	.	.	.	.	.	.
	•••	1064	1463	2417	362	448	526	172	196	194	.	.	.	.	.	.
0.85	•••	777	968	1847	258	299	389	.	.	.	.	.	.	.	.	.
	•••	854	1175	1829	276	333	368	.	.	.	.	.	.	.	.	.
0.9	•••	559	691	1177	.	.	.	.	.	.	.	.	.	.	.	.
	•••	613	825	1156	.	.	.	.	.	.	.	.	.	.	.	.

TABLE 6: ALPHA= 0.025 POWER= 0.9   EXPECTED ACCRUAL THRU MINIMUM FOLLOW-UP= 250

PCONT=***		DEL=.05			DEL=.10			DEL=.15			DEL=.20			DEL=.25		
FACT=		1.0 .75	.50 .25	.00 BIN	1.0 .75	.50 .25	.00 BIN	1.0 .75	.50 .25	.00 BIN	1.0 75	.50 .25	.00 BIN	1.0 .75	.50 .25	.00 BIN
					REQUIRED NUMBER OF PATIENTS											
0.05	***	632	638	671	227	231	239	134	136	139	95	96	98	74	75	76
	•••	634	647	1156	229	234	368	135	137	194	95	97	124	74	75	87
0.1	***	1153	1167	1295	362	372	400	196	202	212	131	134	139	97	99	102
	•••	1158	1191	1829	366	382	526	199	206	260	132	136	158	98	101	107
0.15	***	1597	1621	1898	472	491	548	246	256	276	159	165	174	115	119	124
	•••	1605	1665	2417	480	510	662	251	264	316	162	169	187	117	121	124
0.2	***	1942	1979	2449	557	585	679	285	300	332	181	189	203	129	134	141
	•••	1954	2046	2922	568	615	778	291	312	362	185	195	211	131	137	138
0.25	***	2184	2237	2930	618	656	791	313	333	377	197	208	227	138	145	155
	•••	2202	2332	3342	633	697	873	321	350	400	202	216	229	141	149	148
0.3	***	2329	2399	3331	657	706	882	331	357	413	207	221	245	144	152	164
	•••	2352	2526	3678	676	758	946	342	378	428	213	231	242	148	158	154
0.35	***	2384	2477	3648	676	735	950	342	371	439	213	228	257	148	157	170
	•••	2415	2639	3930	699	799	999	354	397	446	220	240	250	152	163	158
0.4	***	2361	2481	3876	678	748	997	344	378	455	214	231	263	148	157	172
	•••	2402	2681	4098	706	821	1030	359	407	456	222	244	253	152	164	158
0.45	***	2275	2425	4014	667	745	1020	340	377	461	211	230	263	145	155	171
	•••	2327	2663	4182	698	826	1041	356	408	456	219	244	250	150	162	154
0.5	***	2140	2322	4061	644	730	1020	331	369	456	205	224	258	140	150	166
	•••	2204	2595	4182	679	816	1030	347	401	446	213	238	242	144	157	148
0.55	***	1971	2186	4016	612	702	998	316	355	441	195	214	246	132	142	157
	•••	2049	2487	4098	649	791	999	333	387	428	203	227	229	137	148	138
0.6	***	1783	2026	3879	573	664	952	297	334	416	182	199	229	122	131	144
	•••	1873	2347	3930	610	752	946	313	365	400	190	212	211	126	137	124
0.65	***	1587	1849	3652	527	615	884	272	308	380	165	181	207	109	117	128
	•••	1686	2177	3678	564	699	873	288	336	362	172	192	187	113	122	107
0.7	***	1392	1660	3334	474	557	794	243	274	335	145	158	179	94	100	108
	•••	1495	1981	3342	509	633	778	257	298	316	151	167	158	97	104	87
0.75	***	1199	1458	2927	415	487	681	208	234	280	120	130	145	.	.	.
	•••	1300	1758	2922	445	552	662	219	252	260	125	137	124	.	.	.
0.8	***	1006	1241	2431	346	404	546	167	185	215	.	.	.	.	.	.
	•••	1099	1504	2417	371	455	526	175	198	194	.	.	.	.	.	.
0.85	***	805	999	1847	265	305	389	.	.	.	.	.	.	.	.	.
	•••	883	1207	1829	282	337	368	.	.	.	.	.	.	.	.	.
0.9	***	579	712	1177	.	.	.	.	.	.	.	.	.	.	.	.
	•••	633	844	1156	.	.	.	.	.	.	.	.	.	.	.	.

TABLE 6: ALPHA= 0.025 POWER= 0.9   EXPECTED ACCRUAL THRU MINIMUM FOLLOW-UP= 275

		DEL=.05			DEL=.10			DEL=.15			DEL=.20			DEL=.25		
FACT=		1.0 .75	.50 .25	.00 BIN	1.0 .75	.50 .25	.00 BIN	1.0 .75	.50 .25	.00 BIN	1.0 75	.50 .25	.00 BIN	1.0 .75	.50 .25	.00 BIN
PCONT=•••					REQUIRED NUMBER OF PATIENTS											
0.05	***	632	639	671	227	231	239	134	136	139	95	96	98	74	75	76
	•••	635	648	1156	229	234	368	135	138	194	96	97	124	74	75	87
0.1	***	1154	1170	1295	363	374	400	197	203	212	132	135	139	98	99	102
	•••	1159	1195	1829	367	383	526	200	207	260	133	136	158	98	101	107
0.15	***	1599	1626	1898	475	494	548	248	257	276	160	166	174	116	119	123
	•••	1608	1672	2417	482	512	662	252	265	316	162	169	187	117	121	124
0.2	***	1946	1986	2448	561	589	679	287	302	332	182	190	203	129	134	141
	•••	1959	2058	2922	572	619	778	293	314	362	186	196	211	132	137	138
0.25	***	2190	2247	2929	623	662	791	316	336	377	198	209	227	139	145	155
	•••	2209	2348	3342	638	703	873	324	352	400	203	217	229	142	150	148
0.3	***	2336	2413	3331	663	713	882	335	360	413	209	222	245	146	153	164
	•••	2362	2548	3678	682	765	946	345	380	428	215	232	242	149	158	154
0.35	***	2393	2495	3648	683	744	950	346	375	439	215	230	257	149	157	170
	•••	2427	2667	3930	707	807	999	358	399	446	222	241	250	153	163	158
0.4	***	2373	2503	3876	687	758	997	349	382	455	217	233	263	149	158	172
	•••	2418	2714	4098	715	831	1030	363	410	456	224	246	253	153	165	158
0.45	***	2290	2452	4014	677	757	1020	346	382	461	214	232	263	147	156	171
	•••	2347	2701	4182	709	837	1041	361	411	456	222	245	250	151	163	154
0.5	***	2159	2355	4061	655	742	1020	336	374	456	208	226	257	142	151	166
	•••	2229	2638	4182	690	827	1030	353	405	446	216	239	242	146	158	148
0.55	***	1995	2223	4016	624	715	998	322	360	441	198	216	246	134	143	157
	•••	2078	2534	4098	661	802	999	338	391	428	206	229	229	138	149	138
0.6	***	1811	2066	3879	585	676	952	302	339	415	184	201	229	123	132	144
	•••	1906	2395	3930	623	763	946	318	369	400	192	213	211	127	137	124
0.65	***	1618	1891	3652	539	628	884	277	312	380	167	183	207	110	118	128
	•••	1722	2226	3678	576	710	873	292	339	362	174	193	187	114	122	107
0.7	***	1424	1701	3334	486	568	794	248	278	335	147	159	178	95	100	108
	•••	1532	2028	3342	520	643	778	261	301	316	153	168	158	97	104	87
0.75	***	1231	1498	2927	425	497	681	212	237	280	122	131	145	.	.	.
	•••	1336	1801	2922	455	560	662	223	254	260	126	137	124	.	.	.
0.8	***	1036	1276	2431	354	412	546	170	187	215	.	.	.	.	.	.
	•••	1131	1540	2417	379	461	526	177	199	194	.	.	.	.	.	.
0.85	***	830	1027	1847	271	310	389	.	.	.	.	.	.	.	.	.
	•••	909	1235	1829	288	341	368	.	.	.	.	.	.	.	.	.
0.9	***	596	731	1177	.	.	.	.	.	.	.	.	.	.	.	.
	•••	651	862	1156	.	.	.	.	.	.	.	.	.	.	.	.

TABLE 6: ALPHA= 0.025 POWER= 0.9   EXPECTED ACCRUAL THRU MINIMUM FOLLOW-UP= 300

		DEL=.05			DEL=.10			DEL=.15			DEL=.20			DEL=.25	
FACT=	1.0 .75	.50 .25	.00 BIN	1.0 .75	.50 .25	.00 BIN	1.0 .75	.50 .25	.00 BIN	1.0 75	.50 .25	.00 BIN	1.0 .75	.50 .25	.00 BIN
PCONT=***				REQUIRED NUMBER OF PATIENTS											
0.05 ***	633	640	671	228	232	239	134	136	139	95	96	98	74	75	76
•••	635	649	1156	229	235	368	135	138	194	96	97	124	75	75	87
0.1 ***	1156	1172	1295	365	375	400	198	203	212	132	135	139	98	100	102
•••	1161	1199	1829	369	384	526	200	207	260	133	137	158	99	101	107
0.15 ***	1601	1630	1898	477	496	548	249	259	276	161	166	174	116	119	124
•••	1611	1680	2417	484	514	662	253	266	316	163	170	187	118	121	124
0.2 ***	1949	1993	2448	564	593	679	289	303	331	183	191	203	130	135	141
•••	1964	2069	2922	576	622	778	295	315	362	187	196	211	132	137	138
0.25 ***	2195	2257	2929	627	667	791	318	338	377	200	210	227	140	146	154
•••	2216	2364	3342	643	707	873	327	353	400	204	217	229	143	150	148
0.3 ***	2343	2427	3331	668	720	882	338	363	413	211	224	245	147	154	164
•••	2371	2569	3678	689	771	946	349	382	428	217	232	242	150	159	154
0.35 ***	2402	2512	3648	690	752	950	349	379	439	217	232	256	150	158	170
•••	2440	2693	3930	715	815	999	362	402	446	224	242	250	154	164	158
0.4 ***	2386	2525	3876	695	767	997	353	386	455	219	235	263	151	159	172
•••	2434	2746	4098	724	839	1030	367	413	456	226	247	253	155	165	158
0.45 ***	2306	2479	4014	686	767	1020	350	386	460	217	234	263	148	157	171
•••	2367	2737	4182	719	846	1041	366	415	456	224	246	250	152	163	154
0.5 ***	2179	2386	4061	666	753	1020	341	379	456	210	228	258	143	152	166
•••	2253	2679	4182	701	837	1030	357	408	446	218	241	242	147	158	148
0.55 ***	2018	2258	4016	635	726	998	327	364	441	200	218	246	135	144	157
•••	2106	2577	4098	673	812	999	343	394	428	208	230	229	139	150	138
0.6 ***	1838	2104	3879	596	688	952	307	343	415	187	203	229	125	133	144
•••	1938	2440	3930	634	773	946	323	372	400	194	214	211	128	138	124
0.65 ***	1648	1930	3652	550	638	884	282	316	380	170	184	207	112	118	128
•••	1756	2270	3678	587	719	873	297	341	362	176	194	187	115	123	107
0.7 ***	1456	1740	3334	496	578	794	252	281	335	149	161	178	96	101	108
•••	1567	2071	3342	530	651	778	265	303	316	154	169	158	98	104	87
0.75 ***	1262	1534	2927	434	505	681	215	239	280	123	132	145	.	.	.
•••	1369	1840	2922	464	567	662	226	256	260	127	138	124	.	.	.
0.8 ***	1064	1309	2431	362	419	546	172	189	215	.	.	.	.	.	.
•••	1161	1574	2417	386	466	526	180	200	194	.	.	.	.	.	.
0.85 ***	854	1053	1847	276	314	389	.	.	.	.	.	.	.	.	.
•••	934	1260	1829	293	344	368	.	.	.	.	.	.	.	.	.
0.9 ***	613	748	1177	.	.	.	.	.	.	.	.	.	.	.	.
•••	668	877	1156	.	.	.	.	.	.	.	.	.	.	.	.

TABLE 6: ALPHA= 0.025 POWER= 0.9   EXPECTED ACCRUAL THRU MINIMUM FOLLOW-UP= 325

		DEL=.05			DEL=.10			DEL=.15			DEL=.20			DEL=.25		
FACT=		1.0 .75	.50 .25	.00 BIN	1.0 .75	.50 .25	.00 BIN	1.0 .75	.50 .25	.00 BIN	1.0 75	.50 .25	.00 BIN	1.0 .75	.50 .25	.00 BIN
PCONT=***					REQUIRED NUMBER OF PATIENTS											
0.05	***	634	641	671	229	232	239	135	136	139	95	96	98	74	75	76
	•••	636	650	1156	230	235	368	135	138	194	96	97	124	75	75	87
0.1	***	1157	1175	1295	366	376	400	199	204	212	132	135	139	98	100	102
	•••	1163	1202	1829	370	385	526	201	207	260	134	137	158	99	101	107
0.15	***	1604	1635	1898	479	498	548	250	260	276	161	167	174	117	120	123
	•••	1615	1686	2417	487	516	662	254	267	316	164	170	187	118	121	124
0.2	***	1953	2000	2448	567	597	679	291	305	331	184	192	203	131	135	141
	•••	1969	2080	2922	579	625	778	297	316	362	188	197	211	133	138	138
0.25	***	2200	2267	2930	632	672	791	321	340	377	201	211	227	141	147	155
	•••	2223	2379	3342	648	712	873	329	355	400	206	218	229	144	150	148
0.3	***	2350	2440	3331	674	726	882	341	365	413	213	225	245	148	155	164
	•••	2381	2589	3678	695	777	946	352	385	428	218	233	242	151	159	154
0.35	***	2412	2530	3648	697	760	950	353	382	439	219	233	257	151	159	170
	•••	2452	2718	3930	722	821	999	365	404	446	225	243	250	155	164	158
0.4	***	2398	2547	3876	703	776	996	357	390	455	221	237	263	152	160	172
	•••	2450	2775	4098	733	847	1030	371	415	456	228	248	253	156	166	158
0.45	***	2322	2505	4014	695	777	1020	355	390	460	219	236	263	149	158	171
	•••	2387	2772	4182	728	855	1041	370	417	456	226	247	250	153	164	154
0.5	***	2198	2416	4061	676	763	1020	346	382	456	213	230	258	144	153	166
	•••	2277	2716	4182	711	846	1030	362	411	446	220	242	242	148	159	148
0.55	***	2041	2291	4016	646	736	998	331	368	441	203	220	246	136	145	157
	•••	2134	2618	4098	683	821	999	348	397	428	210	231	229	140	150	138
0.6	***	1864	2140	3879	607	698	952	312	347	416	189	205	229	126	133	144
	•••	1968	2481	3930	645	782	946	327	374	400	196	216	211	129	138	124
0.65	***	1677	1967	3652	560	648	884	286	319	380	172	186	207	112	119	128
	•••	1789	2312	3678	597	728	873	301	344	362	178	195	187	116	123	107
0.7	***	1485	1777	3334	506	587	794	255	284	335	150	162	179	96	101	108
	•••	1599	2111	3342	540	658	778	268	305	316	156	169	158	99	104	87
0.75	***	1291	1568	2927	442	513	681	219	242	280	125	133	145	.	.	.
	•••	1401	1877	2922	472	573	662	229	258	260	129	138	124	.	.	.
0.8	***	1090	1339	2431	369	425	546	174	190	215	.	.	.	.	.	.
	•••	1190	1604	2417	393	470	526	182	201	194	.	.	.	.	.	.
0.85	***	876	1077	1847	281	318	389	.	.	.	.	.	.	.	.	.
	•••	957	1284	1829	297	346	368	.	.	.	.	.	.	.	.	.
0.9	***	628	763	1177	.	.	.	.	.	.	.	.	.	.	.	.
	•••	684	891	1156	.	.	.	.	.	.	.	.	.	.	.	.

TABLE 6: ALPHA= 0.025 POWER= 0.9   EXPECTED ACCRUAL THRU MINIMUM FOLLOW-UP= 350

		DEL=.05			DEL=.10			DEL=.15			DEL=.20			DEL=.25		
FACT=		1.0 .75	.50 .25	.00 BIN	1.0 .75	.50 .25	.00 BIN	1.0 .75	.50 .25	.00 BIN	1.0 75	.50 .25	.00 BIN	1.0 .75	.50 .25	.00 BIN
PCONT=•••					REQUIRED NUMBER OF PATIENTS											
0.05	***	634	642	671	229	232	239	135	137	139	96	97	98	74	75	76
	•••	637	651	1156	230	235	368	136	138	194	96	97	124	75	75	87
0.1	***	1158	1177	1295	367	377	400	199	204	212	133	135	139	98	100	102
	•••	1165	1206	1829	371	386	526	201	208	260	134	137	158	99	101	107
0.15	***	1606	1640	1898	481	500	548	251	260	276	162	167	174	117	120	124
	•••	1618	1693	2417	489	518	662	255	267	316	164	170	187	118	122	124
0.2	***	1957	2007	2448	570	600	679	292	306	332	185	192	203	131	135	141
	•••	1974	2090	2922	582	628	778	298	317	362	188	197	211	133	138	138
0.25	***	2205	2277	2930	636	677	791	323	342	377	202	212	227	142	147	155
	•••	2230	2394	3342	652	716	873	331	356	400	207	218	229	144	150	148
0.3	***	2357	2453	3331	679	731	882	344	368	413	214	226	245	149	155	164
	•••	2390	2608	3678	700	782	946	354	386	428	219	234	242	152	159	154
0.35	***	2421	2546	3648	703	767	950	356	384	439	221	235	257	152	160	170
	•••	2465	2741	3930	729	827	999	369	406	446	227	244	250	156	165	158
0.4	***	2410	2568	3876	711	784	997	361	393	455	223	238	263	153	161	173
	•••	2465	2803	4098	741	854	1030	375	418	456	230	249	253	157	166	158
0.45	***	2337	2530	4014	704	785	1020	359	393	460	221	237	263	150	159	171
	•••	2406	2804	4182	737	862	1041	374	420	456	228	248	250	154	164	154
0.5	***	2217	2445	4061	685	772	1020	350	386	456	215	231	258	145	154	166
	•••	2300	2752	4182	721	854	1030	366	414	446	222	243	242	149	159	148
0.55	***	2063	2323	4016	655	746	998	336	372	441	205	221	246	137	145	157
	•••	2160	2655	4098	693	830	999	351	399	428	212	232	229	141	151	138
0.6	***	1889	2173	3879	617	708	952	316	350	416	191	206	229	127	134	144
	•••	1998	2520	3930	655	790	946	331	377	400	198	216	211	130	139	124
0.65	***	1704	2002	3652	570	657	884	290	322	380	173	187	207	113	120	128
	•••	1820	2351	3678	607	735	873	304	346	362	180	196	187	116	123	107
0.7	***	1514	1811	3334	514	595	794	259	287	335	152	163	178	97	102	108
	•••	1630	2148	3342	549	665	778	271	307	316	157	170	158	99	105	87
0.75	***	1318	1601	2927	450	520	681	221	244	280	126	134	145	.	.	.
	•••	1430	1910	2922	480	579	662	231	259	260	129	139	124	.	.	.
0.8	***	1115	1367	2431	375	430	546	176	192	215	.	.	.	.	.	.
	•••	1216	1633	2417	399	474	526	183	202	194	.	.	.	.	.	.
0.85	***	896	1099	1847	285	322	389	.	.	.	.	.	.	.	.	.
	•••	979	1305	1829	301	349	368	.	.	.	.	.	.	.	.	.
0.9	***	643	777	1177	.	.	.	.	.	.	.	.	.	.	.	.
	•••	699	903	1156	.	.	.	.	.	.	.	.	.	.	.	.

TABLE 6: ALPHA= 0.025 POWER= 0.9   EXPECTED ACCRUAL THRU MINIMUM FOLLOW-UP= 375

		DEL=.05			DEL=.10			DEL=.15			DEL=.20			DEL=.25		
FACT=		1.0 .75	.50 .25	.00 BIN	1.0 .75	.50 .25	.00 BIN	1.0 .75	.50 .25	.00 BIN	1.0 75	.50 .25	.00 BIN	1.0 .75	.50 .25	.00 BIN
PCONT=•••					REQUIRED NUMBER OF PATIENTS											
0.05	•••	635	643	671	229	233	239	135	137	139	96	97	98	74	75	76
	•••	638	652	1156	231	235	368	136	138	194	96	97	124	75	75	87
0.1	•••	1160	1180	1295	368	378	400	200	205	212	133	136	139	99	100	102
	•••	1167	1209	1829	372	387	526	202	208	260	134	137	158	99	101	107
0.15	•••	1609	1644	1898	483	502	548	252	261	276	163	167	174	117	120	124
	•••	1621	1699	2417	491	519	662	256	268	316	165	170	187	118	122	124
0.2	•••	1960	2014	2448	573	603	679	294	307	332	186	193	203	132	136	141
	•••	1979	2100	2922	585	630	778	300	317	362	189	198	211	133	138	138
0.25	•••	2210	2286	2930	640	680	791	325	343	377	203	213	227	142	148	155
	•••	2237	2407	3342	656	719	873	333	358	400	208	219	229	145	151	148
0.3	•••	2364	2466	3331	684	737	882	346	370	413	215	227	245	149	156	164
	•••	2399	2626	3678	706	786	946	357	388	428	221	235	242	152	160	154
0.35	•••	2431	2563	3648	709	773	950	359	387	439	222	236	257	153	160	170
	•••	2477	2764	3930	735	833	999	371	408	446	228	245	250	156	165	158
0.4	•••	2422	2588	3876	718	791	997	364	395	455	225	239	263	154	162	172
	•••	2480	2830	4098	748	860	1030	378	419	456	231	250	253	157	167	158
0.45	•••	2352	2554	4014	712	793	1020	362	396	460	223	238	263	152	160	171
	•••	2425	2834	4182	745	869	1041	377	422	456	230	249	250	155	165	154
0.5	•••	2235	2472	4061	693	781	1020	354	389	456	216	233	257	146	154	166
	•••	2322	2785	4182	730	861	1030	369	416	446	224	243	242	150	160	148
0.55	•••	2085	2353	4016	664	755	998	340	375	441	207	222	246	138	146	157
	•••	2186	2691	4098	702	837	999	355	401	428	214	233	229	142	151	138
0.6	•••	1914	2206	3879	626	717	952	319	354	415	193	207	229	128	135	144
	•••	2026	2557	3930	664	797	946	334	379	400	199	217	211	131	139	124
0.65	•••	1731	2035	3652	579	666	884	294	325	380	175	188	207	114	120	128
	•••	1849	2387	3678	615	742	873	308	348	362	181	196	187	117	124	107
0.7	•••	1541	1843	3334	523	603	794	262	289	335	153	164	178	98	102	108
	•••	1660	2182	3342	557	671	778	274	309	316	158	170	158	100	105	87
0.75	•••	1345	1631	2927	457	527	681	224	245	280	127	134	145	.	.	.
	•••	1458	1941	2922	487	584	662	234	260	260	130	139	124	.	.	.
0.8	•••	1139	1393	2431	381	435	546	178	193	215	.	.	.	.	.	.
	•••	1241	1659	2417	404	478	526	185	203	194	.	.	.	.	.	.
0.85	•••	916	1120	1847	289	325	389	.	.	.	.	.	.	.	.	.
	•••	999	1325	1829	305	351	368	.	.	.	.	.	.	.	.	.
0.9	•••	656	791	1177	.	.	.	.	.	.	.	.	.	.	.	.
	•••	712	915	1156	.	.	.	.	.	.	.	.	.	.	.	.

TABLE 6: ALPHA= 0.025 POWER= 0.9   EXPECTED ACCRUAL THRU MINIMUM FOLLOW-UP= 400

		DEL=.05			DEL=.10			DEL=.15			DEL=.20			DEL=.25		
FACT=		1.0 .75	.50 .25	.00 BIN	1.0 .75	.50 .25	.00 BIN	1.0 .75	.50 .25	.00 BIN	1.0 75	.50 .25	.00 BIN	1.0 .75	.50 .25	.00 BIN
PCONT=•••					REQUIRED NUMBER OF PATIENTS											
0.05	•••	635	644	671	230	233	239	135	137	139	96	97	98	75	75	76
	•••	638	653	1156	231	235	368	136	138	194	96	97	124	75	76	87
0.1	•••	1161	1182	1295	369	379	400	200	205	212	133	136	139	99	100	102
	•••	1169	1212	1829	373	387	526	202	208	260	134	137	158	99	101	107
0.15	•••	1611	1648	1898	485	504	548	253	262	276	163	168	174	118	120	124
	•••	1624	1705	2417	493	521	662	257	268	316	165	171	187	119	122	124
0.2	•••	1964	2021	2448	576	606	679	295	309	332	187	194	203	132	136	141
	•••	1984	2109	2922	588	633	778	301	318	362	190	198	211	134	138	138
0.25	•••	2216	2296	2930	643	684	791	327	345	377	204	214	227	143	148	155
	•••	2243	2421	3342	660	722	873	335	359	400	209	219	229	145	151	148
0.3	•••	2371	2479	3331	689	742	882	349	372	413	217	228	245	150	156	164
	•••	2409	2643	3678	711	790	946	359	389	428	222	235	242	153	160	154
0.35	•••	2440	2579	3648	715	779	950	362	389	439	224	237	257	154	161	170
	•••	2489	2785	3930	741	838	999	374	410	446	230	246	250	157	165	158
0.4	•••	2434	2608	3876	724	798	997	367	398	455	226	241	263	155	162	173
	•••	2496	2855	4098	755	866	1030	381	421	456	233	250	253	158	167	158
0.45	•••	2367	2577	4014	719	801	1020	366	399	461	224	240	263	152	160	171
	•••	2443	2863	4182	753	876	1041	380	424	456	231	250	250	156	165	154
0.5	•••	2253	2499	4061	701	789	1020	357	392	456	218	234	258	147	155	166
	•••	2344	2817	4182	738	868	1030	373	418	446	225	244	242	151	160	148
0.55	•••	2106	2383	4016	673	763	998	343	378	441	208	223	246	139	147	157
	•••	2211	2724	4098	711	844	999	358	403	428	215	234	229	143	151	138
0.6	•••	1938	2237	3879	634	725	952	323	356	416	194	208	229	128	135	144
	•••	2053	2591	3930	672	804	946	338	381	400	201	218	211	132	139	124
0.65	•••	1756	2066	3652	587	674	884	297	327	380	176	189	207	115	120	128
	•••	1877	2420	3678	624	748	873	311	350	362	182	197	187	117	124	107
0.7	•••	1567	1874	3334	530	610	794	265	291	335	154	164	179	98	102	108
	•••	1688	2214	3342	564	677	778	277	310	316	159	171	158	100	105	87
0.75	•••	1369	1659	2927	464	533	681	226	247	280	128	135	145	.	.	.
	•••	1485	1970	2922	494	588	662	236	261	260	131	140	124	.	.	.
0.8	•••	1161	1418	2431	386	440	546	180	194	215	.	.	.	.	.	.
	•••	1265	1683	2417	410	481	526	186	204	194	.	.	.	.	.	.
0.85	•••	934	1140	1847	293	328	389	.	.	.	.	.	.	.	.	.
	•••	1018	1343	1829	308	353	368	.	.	.	.	.	.	.	.	.
0.9	•••	668	803	1177	.	.	.	.	.	.	.	.	.	.	.	.
	•••	725	925	1156	.	.	.	.	.	.	.	.	.	.	.	.

TABLE 6: ALPHA= 0.025 POWER= 0.9   EXPECTED ACCRUAL THRU MINIMUM FOLLOW-UP= 425

		DEL=.05			DEL=.10			DEL=.15			DEL=.20			DEL=.25		
FACT=		1.0 .75	.50 .25	.00 BIN	1.0 .75	.50 .25	.00 BIN	1.0 .75	.50 .25	.00 BIN	1.0 75	.50 .25	.00 BIN	1.0 .75	.50 .25	.00 BIN
PCONT=•••				REQUIRED	NUMBER	OF	PATIENTS									
0.05	***	636	644	671	230	233	239	136	137	139	96	97	98	75	75	76
	•••	639	654	1156	231	236	368	136	138	194	96	97	124	75	76	87
0.1	***	1163	1185	1295	370	380	400	201	205	212	133	136	139	99	100	102
	•••	1171	1215	1829	374	388	526	203	208	260	135	137	158	99	101	107
0.15	***	1614	1652	1898	486	506	548	254	263	276	164	168	174	118	120	124
	•••	1627	1710	2417	494	522	662	258	269	316	166	171	187	119	122	124
0.2	***	1968	2027	2448	578	608	679	297	309	332	187	194	203	133	136	141
	•••	1988	2118	2922	591	635	778	302	319	362	190	198	211	134	138	138
0.25	***	2221	2305	2930	647	688	791	328	346	377	205	214	227	144	148	155
	•••	2250	2433	3342	664	725	873	336	359	400	209	220	229	146	151	148
0.3	***	2379	2491	3331	693	746	882	351	374	413	218	229	245	151	156	164
	•••	2418	2660	3678	715	794	946	361	390	428	223	236	242	154	160	154
0.35	***	2449	2594	3648	720	784	951	365	391	439	225	238	257	155	161	170
	•••	2501	2805	3930	747	842	999	376	411	446	231	246	250	158	166	158
0.4	***	2446	2627	3876	731	804	997	370	400	455	228	242	263	155	163	172
	•••	2511	2879	4098	761	871	1030	384	423	456	234	251	253	159	167	158
0.45	***	2382	2600	4014	726	808	1020	369	401	460	226	241	263	153	161	171
	•••	2461	2890	4182	760	881	1041	383	426	456	233	251	250	157	166	154
0.5	***	2271	2524	4061	709	796	1020	361	394	456	220	235	257	148	155	166
	•••	2365	2846	4182	745	874	1030	376	420	446	227	245	242	152	160	148
0.55	***	2127	2410	4016	681	771	998	346	380	441	210	224	246	140	147	156
	•••	2235	2756	4098	719	850	999	361	405	428	217	234	229	143	152	138
0.6	***	1961	2266	3879	642	732	952	326	359	416	196	209	229	129	136	144
	•••	2079	2623	3930	680	810	946	341	382	400	202	218	211	132	139	124
0.65	***	1781	2096	3652	595	681	884	300	330	380	178	190	207	115	121	128
	•••	1904	2452	3678	631	754	873	313	351	362	183	197	187	118	124	107
0.7	***	1591	1903	3334	538	616	794	267	293	335	155	165	179	99	103	108
	•••	1715	2244	3342	571	682	778	279	311	316	160	171	158	101	105	87
0.75	***	1393	1686	2927	470	538	681	228	249	280	128	136	145	.	.	.
	•••	1510	1997	2922	500	592	662	238	262	260	132	140	124	.	.	.
0.8	***	1183	1441	2431	391	444	546	181	195	215	.	.	.	.	.	.
	•••	1287	1706	2417	414	484	526	188	204	194	.	.	.	.	.	.
0.85	***	952	1158	1847	296	330	389	.	.	.	.	.	.	.	.	.
	•••	1036	1360	1829	311	355	368	.	.	.	.	.	.	.	.	.
0.9	***	680	814	1177	.	.	.	.	.	.	.	.	.	.	.	.
	•••	736	935	1156	.	.	.	.	.	.	.	.	.	.	.	.

TABLE 6: ALPHA= 0.025 POWER= 0.9   EXPECTED ACCRUAL THRU MINIMUM FOLLOW-UP= 450

	DEL=.05			DEL=.10			DEL=.15			DEL=.20			DEL=.25		
FACT=	1.0 .75	.50 .25	.00 BIN	1.0 .75	.50 .25	.00 BIN	1.0 .75	.50 .25	.00 BIN	1.0 75	.50 .25	.00 BIN	1.0 .75	.50 .25	.00 BIN
PCONT=•••					REQUIRED NUMBER OF PATIENTS										
0.05 •••	636	645	671	230	234	239	136	137	139	96	97	98	75	75	76
•••	640	654	1156	231	236	368	136	138	194	96	97	124	75	75	87
0.1 •••	1164	1187	1295	371	381	400	201	206	212	134	136	139	99	100	102
•••	1172	1217	1829	375	388	526	203	208	260	135	138	158	100	101	107
0.15 •••	1616	1657	1898	488	507	548	255	263	276	164	168	174	118	120	123
•••	1630	1715	2417	496	523	662	259	269	316	166	171	187	119	122	124
0.2 •••	1971	2034	2448	581	611	679	298	311	332	188	194	203	133	136	141
•••	1993	2126	2922	593	636	778	303	320	362	191	199	211	135	139	138
0.25 •••	2226	2314	2930	650	691	791	330	348	378	206	215	227	144	149	154
•••	2257	2445	3342	667	728	873	338	360	400	210	220	229	146	151	148
0.3 •••	2385	2503	3331	698	750	882	353	375	414	219	229	245	151	157	164
•••	2427	2675	3678	720	798	946	363	391	428	224	236	242	154	160	154
0.35 •••	2458	2610	3648	726	789	950	367	393	439	226	239	257	155	162	170
•••	2512	2824	3930	752	847	999	379	412	446	232	247	250	158	166	158
0.4 •••	2457	2645	3876	737	810	996	373	402	455	229	243	263	156	163	172
•••	2525	2901	4098	768	876	1030	386	424	456	235	252	253	159	167	158
0.45 •••	2396	2621	4014	733	815	1020	372	404	460	227	242	263	154	161	171
•••	2479	2916	4182	767	887	1041	386	427	456	234	251	250	157	166	154
0.5 •••	2288	2549	4061	716	803	1020	364	397	456	221	236	258	149	156	166
•••	2386	2874	4182	753	880	1030	379	421	446	228	246	242	152	160	148
0.55 •••	2147	2437	4016	688	778	998	349	383	441	211	226	246	141	148	157
•••	2258	2785	4098	726	856	999	364	407	428	218	235	229	144	152	138
0.6 •••	1983	2294	3879	650	739	952	329	361	415	197	210	229	130	136	144
•••	2104	2653	3930	688	816	946	343	384	400	203	219	211	133	140	124
0.65 •••	1805	2124	3652	602	687	884	303	332	380	179	190	207	116	121	128
•••	1930	2482	3678	639	759	873	316	352	362	184	198	187	118	124	107
0.7 •••	1615	1930	3334	545	622	794	270	295	335	156	166	179	99	103	108
•••	1740	2272	3342	578	686	778	281	312	260	161	172	158	101	105	87
0.75 •••	1416	1711	2927	477	543	681	230	250	280	129	136	145	.	.	.
•••	1534	2022	2922	505	596	662	239	263	260	132	140	124	.	.	.
0.8 •••	1203	1463	2431	396	448	546	183	196	215	.	.	.	.	.	.
•••	1308	1727	2417	419	487	526	189	204	194	.	.	.	.	.	.
0.85 •••	968	1175	1847	299	333	389	.	.	.	.	.	.	.	.	.
•••	1053	1375	1829	314	356	368	.	.	.	.	.	.	.	.	.
0.9 •••	691	825	1177	.	.	.	.	.	.	.	.	.	.	.	.
•••	748	944	1156	.	.	.	.	.	.	.	.	.	.	.	.

TABLE 6: ALPHA= 0.025 POWER= 0.9   EXPECTED ACCRUAL THRU MINIMUM FOLLOW-UP= 475

		DEL=.05			DEL=.10			DEL=.15			DEL=.20			DEL=.25		
FACT=		1.0 .75	.50 .25	.00 BIN	1.0 .75	.50 .25	.00 BIN	1.0 .75	.50 .25	.00 BIN	1.0 75	.50 .25	.00 BIN	1.0 .75	.50 .25	.00 BIN
PCONT=•••					REQUIRED NUMBER OF PATIENTS											
0.05	***	637	646	671	230	234	239	136	137	139	96	97	98	75	75	76
	•••	641	655	1156	232	236	368	137	138	194	96	97	124	75	75	87
0.1	***	1166	1189	1295	371	381	400	202	206	212	134	136	139	99	100	102
	•••	1174	1219	1829	376	389	526	204	209	260	135	137	158	100	101	107
0.15	***	1619	1661	1898	489	509	548	256	264	276	165	169	174	118	121	124
	•••	1633	1720	2417	498	524	662	259	269	316	167	171	187	119	122	124
0.2	***	1975	2040	2448	583	613	679	299	311	332	188	195	203	133	137	141
	•••	1998	2134	2922	596	638	778	304	320	362	191	199	211	135	139	138
0.25	***	2231	2323	2930	653	694	791	332	349	377	207	215	227	144	149	155
	•••	2264	2456	3342	670	730	873	339	361	400	211	221	229	146	151	148
0.3	***	2392	2515	3331	701	754	882	355	376	413	220	230	245	152	157	165
	•••	2436	2690	3678	724	801	946	365	392	428	224	237	242	154	161	154
0.35	***	2468	2625	3648	731	794	950	369	395	439	227	239	257	156	162	170
	•••	2524	2843	3930	757	851	999	381	414	446	233	247	250	159	166	158
0.4	***	2469	2663	3876	742	816	997	376	405	455	230	243	263	157	164	172
	•••	2540	2923	4098	773	881	1030	388	426	456	236	252	253	160	168	158
0.45	***	2410	2642	4014	739	821	1020	374	406	460	229	243	263	155	162	171
	•••	2496	2940	4182	773	891	1041	389	429	456	235	252	250	158	166	154
0.5	***	2306	2572	4061	723	810	1020	367	399	456	223	237	257	149	156	166
	•••	2406	2901	4182	760	885	1030	381	423	456	229	246	242	153	161	148
0.55	***	2167	2463	4016	695	785	998	352	385	441	213	226	246	141	148	156
	•••	2280	2813	4098	733	861	999	367	408	428	219	235	229	145	152	138
0.6	***	2005	2321	3879	657	746	952	332	363	416	198	211	229	130	136	144
	•••	2128	2682	3930	695	821	946	346	385	400	204	219	211	133	140	124
0.65	***	1827	2151	3652	609	694	884	305	334	380	180	191	207	116	121	128
	•••	1955	2510	3678	645	764	873	318	354	362	185	198	187	119	124	107
0.7	***	1638	1956	3334	551	628	794	272	297	335	157	166	178	99	103	108
	•••	1765	2299	3342	584	691	778	283	313	316	162	172	158	101	105	87
0.75	***	1438	1735	2927	482	548	681	232	251	280	130	136	145	.	.	.
	•••	1557	2046	2922	511	599	662	241	264	260	133	140	124	.	.	.
0.8	***	1223	1484	2431	400	451	546	184	197	215	.	.	.	.	.	.
	•••	1329	1747	2417	423	490	526	190	205	194	.	.	.	.	.	.
0.85	***	984	1191	1847	302	335	389	.	.	.	.	.	.	.	.	.
	•••	1069	1390	1829	317	358	368	.	.	.	.	.	.	.	.	.
0.9	***	702	835	1177	.	.	.	.	.	.	.	.	.	.	.	.
	•••	758	952	1156	.	.	.	.	.	.	.	.	.	.	.	.

TABLE 6: ALPHA= 0.025 POWER= 0.9   EXPECTED ACCRUAL THRU MINIMUM FOLLOW-UP= 500

PCONT=***   REQUIRED NUMBER OF PATIENTS

PCONT	FACT	DEL=.05 1.0/.75	DEL=.05 .50/.25	DEL=.05 .00/BIN	DEL=.10 1.0/.75	DEL=.10 .50/.25	DEL=.10 .00/BIN	DEL=.15 1.0/.75	DEL=.15 .50/.25	DEL=.15 .00/BIN	DEL=.20 1.0/75	DEL=.20 .50/.25	DEL=.20 .00/BIN	DEL=.25 1.0/.75	DEL=.25 .50/.25	DEL=.25 .00/BIN
0.05	***	638	647	671	231	234	239	136	137	139	96	97	98	75	75	76
	•••	641	655	1156	232	236	368	137	138	194	97	98	124	75	76	87
0.1	***	1167	1191	1295	372	382	400	202	206	212	134	136	139	99	101	102
	•••	1176	1222	1829	377	389	526	204	209	260	135	138	158	100	101	107
0.15	***	1621	1665	1898	491	510	548	256	264	276	165	169	174	118	121	123
	•••	1637	1725	2417	499	525	662	260	270	316	167	171	187	120	122	124
0.2	***	1979	2046	2448	585	615	679	300	312	332	189	195	203	133	137	141
	•••	2003	2142	2922	598	640	778	305	321	362	192	199	211	135	139	138
0.25	***	2237	2332	2930	656	698	791	333	350	378	208	216	227	145	149	155
	•••	2270	2467	3342	673	733	873	341	362	400	212	221	229	147	152	148
0.3	***	2399	2526	3331	706	758	882	357	378	413	221	231	245	152	158	164
	•••	2445	2704	3678	728	804	946	366	393	428	225	237	242	155	161	154
0.35	***	2477	2639	3648	735	799	950	371	397	439	228	240	257	157	163	170
	•••	2535	2860	3930	762	854	999	383	415	446	234	248	250	159	166	158
0.4	***	2481	2681	3876	748	821	997	378	407	455	231	244	263	158	164	173
	•••	2554	2944	4098	779	885	1030	391	427	456	237	253	253	161	168	158
0.45	***	2425	2663	4014	745	826	1020	377	408	461	230	243	263	155	162	171
	•••	2513	2963	4182	779	896	1041	391	430	456	236	252	250	158	166	154
0.5	***	2322	2595	4061	730	816	1020	369	401	456	224	238	258	150	157	166
	•••	2426	2927	4182	766	890	1030	384	424	446	230	247	242	153	161	148
0.55	***	2186	2488	4016	702	791	998	355	387	441	214	227	246	142	148	157
	•••	2302	2840	4098	740	866	999	369	410	428	220	236	229	145	152	138
0.6	***	2026	2347	3879	664	752	953	334	365	416	199	212	229	131	137	144
	•••	2151	2709	3930	702	826	946	348	387	400	205	220	211	134	140	124
0.65	***	1849	2177	3652	615	699	884	308	336	380	181	192	207	117	122	128
	•••	1979	2537	3678	652	768	873	320	355	362	186	199	187	119	125	107
0.7	***	1660	1981	3334	557	633	794	274	298	335	158	167	178	100	103	108
	•••	1788	2324	3342	590	694	778	285	314	316	162	172	158	102	106	87
0.75	***	1458	1758	2927	487	552	681	234	253	280	130	137	145	.	.	.
	•••	1579	2069	2922	516	603	662	242	265	260	133	141	124	.	.	.
0.8	***	1241	1503	2431	404	455	546	185	198	215	.	.	.	.	.	.
	•••	1348	1766	2417	427	492	526	191	206	194	.	.	.	.	.	.
0.85	***	999	1207	1847	305	337	389	.	.	.	.	.	.	.	.	.
	•••	1085	1403	1829	319	359	368	.	.	.	.	.	.	.	.	.
0.9	***	712	844	1177	.	.	.	.	.	.	.	.	.	.	.	.
	•••	768	960	1156	.	.	.	.	.	.	.	.	.	.	.	.

TABLE 6: ALPHA= 0.025 POWER= 0.9   EXPECTED ACCRUAL THRU MINIMUM FOLLOW-UP= 550

		DEL=.02			DEL=.05			DEL=.10			DEL=.15			DEL=.20		
FACT=		1.0 .75	.50 .25	.00 BIN	1.0 .75	.50 .25	.00 BIN	1.0 .75	.50 .25	.00 BIN	1.0 75	.50 .25	.00 BIN	1.0 .75	.50 .25	.00 BIN

PCONT=***   REQUIRED NUMBER OF PATIENTS

PCONT		DEL=.02 1.0/.75	.50/.25	.00/BIN	DEL=.05 1.0/.75	.50/.25	.00/BIN	DEL=.10 1.0/.75	.50/.25	.00/BIN	DEL=.15 1.0/.75	.50/.25	.00/BIN	DEL=.20 1.0/.75	.50/.25	.00/BIN
0.05	***	2991	3002	3165	639	648	671	231	234	239	136	137	139	96	97	98
	•••	2994	3024	5916	642	657	1156	233	236	368	137	138	194	96	98	124
0.1	***	6221	6249	6954	1170	1195	1295	374	383	400	203	206	212	135	136	139
	•••	6231	6303	10277	1179	1226	1829	378	390	526	204	209	260	135	138	158
0.15	***	9088	9136	10755	1626	1672	1898	494	512	548	257	265	276	166	169	174
	•••	9104	9231	14112	1643	1733	2417	501	527	662	261	270	316	167	171	187
0.2	***	11370	11444	14290	1986	2058	2448	589	619	679	302	313	332	190	196	203
	•••	11395	11591	17422	2012	2156	2922	602	642	778	307	321	362	193	199	211
0.25	***	13007	13113	17424	2247	2348	2929	662	703	791	335	352	377	209	217	227
	•••	13043	13325	20206	2283	2487	3342	679	737	873	343	363	400	212	221	229
0.3	***	14009	14155	20081	2413	2549	3331	713	765	882	360	380	413	222	232	245
	•••	14058	14445	22465	2462	2731	3678	735	809	946	369	395	428	226	237	242
0.35	***	14422	14615	22217	2494	2667	3648	744	807	950	375	399	439	230	241	257
	•••	14486	15001	24199	2557	2893	3930	771	861	999	386	417	446	235	248	250
0.4	***	14308	14561	23808	2503	2714	3876	758	831	996	382	410	455	233	246	263
	•••	14392	15064	25407	2581	2982	4098	789	892	1030	395	429	456	239	253	253
0.45	***	13747	14072	24838	2452	2701	4014	756	837	1020	382	411	461	232	245	263
	•••	13855	14716	26090	2546	3007	4182	791	904	1041	395	432	456	238	253	250
0.5	***	12822	13238	25298	2355	2638	4061	742	827	1020	374	405	456	226	239	257
	•••	12961	14047	26248	2463	2974	4182	778	899	1030	388	426	446	232	247	242
0.55	***	11623	12152	25187	2223	2534	4016	715	802	998	360	391	441	216	228	246
	•••	11800	13148	25880	2344	2890	4098	752	875	999	374	412	428	222	237	229
0.6	***	10245	10912	24505	2066	2395	3880	676	763	952	339	368	415	201	213	229
	•••	10470	12099	24987	2195	2760	3930	714	834	946	353	389	400	207	221	211
0.65	***	8787	9614	23253	1891	2226	3652	628	710	884	312	338	380	182	193	206
	•••	9072	10966	23569	2024	2586	3678	663	776	873	324	357	362	188	199	187
0.7	***	7353	8333	21436	1702	2028	3334	568	643	794	278	301	335	159	168	178
	•••	7702	9787	21625	1832	2370	3342	600	701	778	288	316	316	163	173	158
0.75	***	6029	7108	19057	1498	1801	2927	497	560	681	236	254	280	131	137	145
	•••	6427	8580	19156	1621	2110	2922	525	608	662	245	266	260	134	141	124
0.8	***	4853	5940	16122	1276	1540	2431	412	461	546	187	199	215	.	.	.
	•••	5266	7332	16161	1384	1800	2417	433	496	526	192	206	194	.	.	.
0.85	***	3804	4793	12638	1027	1235	1847	310	341	389	.	.	.	.	.	.
	•••	4186	6003	12641	1113	1428	1829	324	362	368	.	.	.	.	.	.
0.9	***	2800	3584	8610	731	862	1177	.	.	.	.	.	.	.	.	.
	•••	3107	4499	8596	786	973	1156	.	.	.	.	.	.	.	.	.
0.95	***	1671	2114	4045	.	.	.	.	.	.	.	.	.	.	.	.
	•••	1848	2585	4025	.	.	.	.	.	.	.	.	.	.	.	.

TABLE 6: ALPHA= 0.025 POWER= 0.9   EXPECTED ACCRUAL THRU MINIMUM FOLLOW-UP= 600

	DEL=.02			DEL=.05			DEL=.10			DEL=.15			DEL=.20		
FACT=	1.0 .75	.50 .25	.00 BIN	1.0 .75	.50 .25	.00 BIN	1.0 .75	.50 .25	.00 BIN	1.0 75	.50 .25	.00 BIN	1.0 .75	.50 .25	.00 BIN
PCONT=•••				REQUIRED NUMBER OF PATIENTS											
0.05 ***	2992	3004	3165	640	649	671	232	235	239	136	137	139	96	97	98
•••	2996	3028	5916	643	657	1156	233	236	368	137	138	194	97	97	124
0.1 ***	6224	6254	6954	1172	1199	1295	375	384	400	203	207	212	135	137	139
•••	6234	6313	10277	1182	1229	1829	379	391	526	205	209	260	136	138	158
0.15 ***	9092	9145	10755	1630	1679	1898	496	514	548	259	266	276	166	170	174
•••	9110	9249	14112	1648	1741	2417	504	528	662	262	271	316	168	172	187
0.2 ***	11377	11457	14290	1993	2069	2448	593	622	679	303	315	331	191	196	203
•••	11404	11618	17422	2021	2168	2922	605	645	778	308	322	362	193	200	211
0.25 ***	13017	13133	17424	2257	2364	2929	667	707	791	338	353	377	210	217	227
•••	13055	13363	20206	2296	2506	3342	684	740	873	345	364	400	214	221	229
0.3 ***	14023	14181	20081	2427	2569	3331	719	771	881	363	382	413	224	232	245
•••	14075	14498	22465	2479	2755	3678	742	814	946	371	396	428	228	238	242
0.35 ***	14439	14650	22217	2512	2693	3648	752	815	950	379	402	439	232	242	256
•••	14509	15071	24199	2579	2923	3930	779	866	999	389	418	446	237	249	250
0.4 ***	14331	14607	23808	2525	2746	3876	767	839	997	386	413	455	235	247	263
•••	14423	15154	25407	2608	3017	4098	798	899	1030	398	431	456	241	254	253
0.45 ***	13777	14131	24838	2479	2737	4014	767	846	1020	386	415	460	234	246	263
•••	13895	14830	26090	2577	3046	4182	801	911	1041	399	434	456	239	254	250
0.5 ***	12860	13313	25298	2386	2678	4061	753	837	1020	379	408	455	228	241	257
•••	13011	14187	26248	2498	3017	4182	789	906	1030	392	429	446	234	248	242
0.55 ***	11671	12247	25187	2258	2577	4016	726	812	998	364	394	440	218	230	246
•••	11864	13314	25880	2383	2935	4098	763	883	999	377	414	428	223	237	229
0.6 ***	10306	11030	24505	2104	2440	3879	688	773	952	343	371	415	203	214	229
•••	10551	12289	24987	2236	2806	3930	725	841	946	356	391	400	208	221	211
0.65 ***	8866	9754	23253	1930	2270	3652	638	719	884	316	341	380	184	194	206
•••	9174	11173	23569	2066	2631	3678	674	783	873	327	358	362	189	200	187
0.7 ***	7450	8489	21436	1740	2071	3334	578	651	794	281	303	335	161	169	178
•••	7824	10002	21625	1874	2413	3342	610	707	778	291	317	316	164	173	158
0.75 ***	6142	7271	19057	1534	1840	2927	505	567	681	239	256	280	132	138	145
•••	6562	8791	19156	1659	2147	2922	532	613	662	247	267	260	135	141	124
0.8 ***	4972	6098	16122	1309	1574	2431	419	466	546	189	200	215	.	.	.
•••	5401	7527	16161	1417	1830	2417	440	499	526	194	207	194	.	.	.
0.85 ***	3915	4933	12638	1053	1260	1847	314	344	389	.	.	.	.	.	.
•••	4309	6168	12641	1139	1450	1829	328	364	368	.	.	.	.	.	.
0.9 ***	2890	3692	8610	748	877	1177	.	.	.	.	.	.	.	.	.
•••	3205	4621	8596	803	986	1156	.	.	.	.	.	.	.	.	.
0.95 ***	1723	2172	4045	.	.	.	.	.	.	.	.	.	.	.	.
•••	1904	2644	4025	.	.	.	.	.	.	.	.	.	.	.	.

TABLE 6: ALPHA= 0.025 POWER= 0.9   EXPECTED ACCRUAL THRU MINIMUM FOLLOW-UP= 650

	DEL=.02			DEL=.05			DEL=.10			DEL=.15			DEL=.20		
FACT=	1.0	.50	.00	1.0	.50	.00	1.0	.50	.00	1.0	.50	.00	1.0	.50	.00
	.75	.25	BIN	.75	.25	BIN	.75	.25	BIN	75	.25	BIN	.75	.25	BIN
PCONT=***				REQUIRED NUMBER OF PATIENTS											
0.05 ***	2993	3006	3165	641	650	671	232	235	239	136	138	139	97	97	98
•••	2997	3031	5916	645	658	1156	233	237	368	137	138	194	97	97	124
0.1 ***	6226	6259	6954	1175	1202	1295	376	385	400	204	207	212	135	137	139
•••	6237	6322	10277	1185	1233	1829	380	392	526	205	210	260	136	138	158
0.15 ***	9097	9154	10755	1635	1686	1898	498	516	548	259	267	276	166	170	174
•••	9116	9266	14112	1654	1748	2417	506	529	662	263	271	316	168	172	187
0.2 ***	11384	11471	14290	2000	2080	2448	597	625	679	305	316	331	192	197	203
•••	11413	11645	17422	2030	2180	2922	609	647	778	310	323	362	194	200	211
0.25 ***	13027	13152	17424	2267	2379	2930	672	712	791	340	355	377	211	218	227
•••	13068	13402	20206	2308	2523	3342	689	743	873	347	365	400	214	222	229
0.3 ***	14036	14208	20081	2440	2589	3331	726	777	881	366	385	413	225	233	244
•••	14093	14550	22465	2495	2777	3678	748	818	946	374	397	428	229	239	242
0.35 ***	14457	14685	22218	2530	2718	3648	760	821	951	382	405	439	233	243	257
•••	14533	15140	24199	2600	2950	3930	786	871	999	392	419	446	238	249	250
0.4 ***	14354	14653	23808	2547	2775	3876	776	847	996	389	415	455	237	248	263
•••	14454	15243	25407	2633	3050	4098	806	904	1030	401	432	456	242	255	253
0.45 ***	13806	14191	24838	2505	2772	4014	777	855	1020	390	418	461	236	247	263
•••	13934	14942	26090	2607	3083	4182	810	918	1041	402	436	456	241	255	250
0.5 ***	12898	13388	25298	2416	2716	4061	763	846	1020	383	411	456	230	242	257
•••	13061	14324	26248	2532	3057	4182	799	913	1030	396	431	446	236	249	242
0.55 ***	11719	12341	25188	2291	2618	4016	736	821	998	368	397	441	220	231	246
•••	11928	13476	25880	2420	2976	4098	773	889	999	381	416	428	225	238	229
0.6 ***	10368	11146	24505	2140	2481	3880	698	782	952	347	374	415	205	216	229
•••	10633	12472	24987	2275	2848	3930	735	848	946	359	392	400	210	222	211
0.65 ***	8943	9890	23253	1967	2312	3652	648	728	884	319	344	380	185	195	207
•••	9275	11370	23569	2105	2672	3678	683	789	873	331	360	362	190	201	187
0.7 ***	7546	8640	21436	1777	2111	3334	587	658	794	284	305	335	162	169	179
•••	7942	10205	21625	1912	2451	3342	619	712	778	294	318	316	165	174	158
0.75 ***	6252	7428	19057	1568	1877	2927	513	573	681	242	258	280	133	138	145
•••	6691	8989	19156	1694	2181	2922	540	617	662	249	268	260	136	141	124
0.8 ***	5086	6249	16122	1338	1605	2431	425	470	546	190	201	215	.	.	.
•••	5531	7709	16161	1448	1858	2417	445	502	526	195	207	194	.	.	.
0.85 ***	4020	5065	12638	1077	1284	1847	318	346	389	.	.	.	.	.	.
•••	4426	6322	12641	1164	1470	1829	331	365	368	.	.	.	.	.	.
0.9 ***	2975	3794	8610	763	891	1177	.	.	.	.	.	.	.	.	.
•••	3298	4734	8596	818	996	1156	.	.	.	.	.	.	.	.	.
0.95 ***	1773	2226	4045	.	.	.	.	.	.	.	.	.	.	.	.
•••	1956	2697	4025	.	.	.	.	.	.	.	.	.	.	.	.

TABLE 6: ALPHA= 0.025 POWER= 0.9   EXPECTED ACCRUAL THRU MINIMUM FOLLOW-UP= 700

		DEL=.02			DEL=.05			DEL=.10			DEL=.15			DEL=.20		
FACT=		1.0 .75	.50 .25	.00 BIN	1.0 .75	.50 .25	.00 BIN	1.0 .75	.50 .25	.00 BIN	1.0 75	.50 .25	.00 BIN	1.0 .75	.50 .25	.00 BIN
PCONT=***					REQUIRED NUMBER OF PATIENTS											
0.05	***	2994	3008	3166	642	651	671	233	235	239	137	138	139	96	97	98
	***	2998	3035	5916	646	659	1156	233	236	368	137	138	194	97	97	124
0.1	***	6229	6263	6954	1178	1206	1295	377	386	400	204	208	212	135	137	139
	***	6240	6332	10277	1188	1236	1829	381	392	526	206	210	260	136	138	158
0.15	***	9101	9162	10755	1640	1693	1898	500	518	548	261	267	276	167	170	174
	***	9122	9283	14112	1660	1755	2417	508	530	662	264	271	316	169	172	187
0.2	***	11390	11484	14290	2007	2090	2448	600	628	679	306	317	331	192	197	203
	***	11422	11671	17422	2038	2191	2922	612	649	778	311	324	362	195	200	211
0.25	***	13036	13171	17424	2277	2394	2930	677	716	791	342	356	377	212	219	227
	***	13081	13440	20206	2320	2538	3342	693	746	873	348	366	400	215	222	229
0.3	***	14049	14234	20081	2453	2608	3331	731	782	882	368	386	413	226	234	245
	***	14111	14602	22465	2511	2798	3678	753	821	946	376	398	428	230	239	242
0.35	***	14474	14720	22218	2546	2741	3648	767	828	950	384	406	439	235	244	257
	***	14556	15209	24199	2620	2976	3930	793	876	999	394	421	446	239	250	250
0.4	***	14377	14698	23808	2567	2803	3876	784	854	996	393	418	455	238	249	263
	***	14485	15332	25407	2657	3080	4098	814	909	1030	404	434	456	243	255	253
0.45	***	13836	14250	24838	2530	2804	4014	785	862	1020	393	420	460	237	248	263
	***	13974	15052	26090	2635	3117	4182	818	923	1041	405	438	456	242	255	250
0.5	***	12935	13463	25298	2445	2752	4061	772	854	1020	386	414	456	231	243	257
	***	13112	14458	26248	2564	3093	4182	807	919	1030	398	432	446	236	250	242
0.55	***	11768	12435	25188	2323	2655	4016	746	830	998	372	399	441	221	232	246
	***	11992	13632	25880	2454	3015	4098	782	895	999	384	418	428	226	239	229
0.6	***	10429	11260	24505	2173	2520	3880	708	790	952	350	376	415	206	216	229
	***	10713	12647	24987	2312	2886	3930	744	854	946	362	394	400	211	222	211
0.65	***	9021	10023	23253	2002	2350	3652	657	735	884	322	346	380	187	196	207
	***	9374	11558	23569	2142	2709	3678	691	795	873	333	361	362	191	201	187
0.7	***	7640	8786	21436	1811	2147	3334	595	665	794	287	307	335	163	170	178
	***	8057	10397	21625	1948	2485	3342	626	717	778	296	320	316	166	174	158
0.75	***	6358	7577	19057	1601	1910	2927	520	579	681	243	259	280	134	139	145
	***	6815	9176	19156	1727	2212	2922	546	621	662	251	269	260	136	142	124
0.8	***	5196	6392	16122	1367	1633	2431	430	474	546	192	202	215	.	.	.
	***	5654	7880	16161	1477	1883	2417	450	505	526	197	208	194	.	.	.
0.85	***	4121	5190	12638	1099	1305	1847	322	349	389	.	.	.	.	.	.
	***	4537	6466	12641	1186	1488	1829	334	367	368	.	.	.	.	.	.
0.9	***	3056	3890	8610	777	903	1177	.	.	.	.	.	.	.	.	.
	***	3385	4838	8596	831	1006	1156	.	.	.	.	.	.	.	.	.
0.95	***	1819	2277	4045	.	.	.	.	.	.	.	.	.	.	.	.
	***	2004	2746	4025	.	.	.	.	.	.	.	.	.	.	.	.

TABLE 6: ALPHA= 0.025 POWER= 0.9   EXPECTED ACCRUAL THRU MINIMUM FOLLOW-UP= 750

		DEL=.02			DEL=.05			DEL=.10			DEL=.15			DEL=.20		
FACT=		1.0 .75	.50 .25	.00 BIN	1.0 .75	.50 .25	.00 BIN	1.0 .75	.50 .25	.00 BIN	1.0 75	.50 .25	.00 BIN	1.0 .75	.50 .25	.00 BIN
PCONT=●●●		REQUIRED NUMBER OF PATIENTS														
0.05	●●●	2995	3010	3165	643	652	671	233	235	239	137	138	139	97	97	98
	●●●	3000	3038	5916	647	660	1156	234	237	368	137	139	194	97	98	124
0.1	●●●	6231	6268	6954	1180	1209	1295	378	387	400	205	208	212	136	137	139
	●●●	6244	6341	10277	1191	1238	1829	382	393	526	206	210	260	136	138	158
0.15	●●●	9106	9171	10755	1644	1699	1898	502	519	548	261	268	276	168	170	174
	●●●	9127	9300	14112	1665	1761	2417	510	531	662	264	272	316	169	172	187
0.2	●●●	11397	11498	14290	2014	2100	2449	603	630	679	307	318	332	193	198	203
	●●●	11431	11698	17422	2046	2201	2922	615	650	778	312	324	362	195	200	211
0.25	●●●	13046	13190	17424	2286	2407	2929	680	719	791	343	357	378	213	219	227
	●●●	13094	13478	20206	2332	2553	3342	697	748	873	350	366	400	216	223	229
0.3	●●●	14062	14260	20081	2466	2626	3331	737	786	881	370	387	413	227	235	244
	●●●	14128	14654	22465	2526	2817	3678	758	825	946	378	399	428	230	239	242
0.35	●●●	14492	14755	22218	2563	2764	3648	773	833	950	387	408	439	235	245	257
	●●●	14580	15278	24199	2639	3000	3930	799	880	999	397	422	446	240	250	250
0.4	●●●	14400	14744	23808	2588	2830	3876	791	860	997	395	419	455	239	250	263
	●●●	14515	15419	25407	2680	3108	4098	821	914	1030	407	435	456	244	256	253
0.45	●●●	13865	14308	24837	2554	2834	4014	793	869	1020	396	422	460	238	249	263
	●●●	14013	15161	26090	2663	3147	4182	826	928	1041	408	439	456	244	256	250
0.5	●●●	12973	13538	25298	2472	2785	4061	781	861	1020	389	416	456	233	244	258
	●●●	13162	14589	26248	2595	3126	4182	816	924	1030	401	433	446	238	250	242
0.55	●●●	11815	12528	25188	2353	2691	4016	755	837	998	375	401	441	222	233	246
	●●●	12056	13784	25880	2487	3049	4098	791	901	999	387	419	428	227	239	229
0.6	●●●	10490	11372	24505	2206	2557	3879	716	797	952	354	379	415	207	217	229
	●●●	10793	12817	24987	2346	2921	3930	752	859	946	365	395	400	212	223	211
0.65	●●●	9098	10152	23253	2035	2387	3652	666	742	884	325	348	380	188	196	206
	●●●	9471	11738	23569	2177	2744	3678	700	799	873	335	363	362	192	201	187
0.7	●●●	7733	8926	21436	1843	2182	3334	603	671	794	289	309	335	164	170	178
	●●●	8170	10581	21625	1981	2517	3342	633	722	778	298	320	316	167	174	158
0.75	●●●	6461	7720	19057	1630	1941	2927	527	584	681	245	260	280	134	139	145
	●●●	6935	9353	19156	1759	2240	2922	552	625	662	252	269	260	137	142	124
0.8	●●●	5300	6527	16122	1393	1659	2431	435	478	546	193	203	215	.	.	.
	●●●	5772	8042	16161	1504	1906	2417	454	507	526	198	208	194	.	.	.
0.85	●●●	4218	5309	12638	1120	1324	1847	325	351	389	.	.	.	.	.	.
	●●●	4643	6602	12641	1206	1504	1829	337	368	368	.	.	.	.	.	.
0.9	●●●	3132	3980	8610	790	915	1177	.	.	.	.	.	.	.	.	.
	●●●	3468	4936	8596	844	1015	1156	.	.	.	.	.	.	.	.	.
0.95	●●●	1863	2324	4045	.	.	.	.	.	.	.	.	.	.	.	.
	●●●	2050	2792	4025	.	.	.	.	.	.	.	.	.	.	.	.

TABLE 6: ALPHA= 0.025 POWER= 0.9   EXPECTED ACCRUAL THRU MINIMUM FOLLOW-UP= 800

		DEL=.02			DEL=.05			DEL=.10			DEL=.15			DEL=.20		
FACT=		1.0 .75	.50 .25	.00 BIN	1.0 .75	.50 .25	.00 BIN	1.0 .75	.50 .25	.00 BIN	1.0 75	.50 .25	.00 BIN	1.0 .75	.50 .25	.00 BIN
PCONT=•••					REQUIRED NUMBER OF PATIENTS											
0.05	***	2996	3012	3165	644	653	671	233	235	239	137	138	139	97	97	98
	•••	3001	3041	5916	647	660	1156	234	237	368	137	139	194	97	98	124
0.1	***	6234	6274	6954	1182	1212	1295	379	387	400	205	208	212	136	137	139
	•••	6247	6351	10277	1194	1241	1829	383	393	526	206	210	260	137	138	158
0.15	***	9110	9179	10755	1648	1705	1898	504	521	548	262	268	276	168	171	174
	•••	9133	9317	14112	1670	1766	2417	511	532	662	265	272	316	169	172	187
0.2	***	11404	11511	14290	2021	2109	2448	606	633	679	309	318	332	194	198	203
	•••	11440	11724	17422	2054	2210	2922	618	652	778	313	324	362	196	201	211
0.25	***	13056	13210	17424	2296	2421	2930	684	722	791	345	359	377	214	219	227
	•••	13107	13516	20206	2343	2566	3342	701	751	873	351	367	400	216	223	229
0.3	***	14076	14287	20081	2479	2643	3331	742	790	882	372	389	413	228	235	245
	•••	14146	14706	22465	2541	2835	3678	763	828	946	380	400	428	231	240	242
0.35	***	14509	14791	22218	2579	2785	3648	779	838	950	389	410	439	237	246	257
	•••	14603	15345	24199	2658	3022	3930	805	883	999	399	423	446	241	251	250
0.4	***	14423	14790	23808	2608	2855	3876	798	866	997	398	421	455	241	250	263
	•••	14546	15506	25407	2703	3133	4098	828	918	1030	409	436	456	245	256	253
0.45	***	13895	14367	24838	2577	2863	4014	801	876	1020	399	424	461	240	250	263
	•••	14053	15268	26090	2689	3176	4182	833	933	1041	410	440	456	245	256	250
0.5	***	13011	13612	25298	2499	2817	4061	789	868	1020	392	418	456	234	244	258
	•••	13213	14717	26248	2624	3158	4182	823	929	1030	404	435	446	239	251	242
0.55	***	11864	12620	25188	2383	2724	4016	763	844	998	378	403	441	223	234	246
	•••	12120	13932	25880	2519	3082	4098	799	905	999	390	420	428	228	240	229
0.6	***	10551	11482	24505	2237	2591	3879	725	804	952	356	381	416	208	218	229
	•••	10873	12980	24987	2379	2954	3930	760	864	946	368	396	400	213	223	211
0.65	***	9174	10278	23253	2066	2420	3652	674	748	884	327	350	380	189	197	207
	•••	9567	11911	23569	2210	2776	3678	707	804	873	338	363	362	193	202	187
0.7	***	7824	9062	21436	1874	2214	3334	610	677	794	291	310	335	164	171	179
	•••	8279	10756	21625	2013	2547	3342	640	725	778	300	321	316	168	175	158
0.75	***	6562	7857	19057	1659	1970	2927	533	588	681	247	261	280	135	140	145
	•••	7051	9522	19156	1787	2266	2922	558	627	662	254	270	260	137	142	124
0.8	***	5401	6657	16122	1418	1683	2431	440	481	546	194	204	215	.	.	.
	•••	5885	8195	16161	1528	1927	2417	459	509	526	199	209	194	.	.	.
0.85	***	4309	5422	12638	1140	1343	1847	328	353	389	.	.	.	.	.	.
	•••	4744	6730	12641	1226	1519	1829	339	369	368	.	.	.	.	.	.
0.9	***	3205	4065	8610	803	925	1177	.	.	.	.	.	.	.	.	.
	•••	3546	5028	8596	856	1022	1156	.	.	.	.	.	.	.	.	.
0.95	***	1904	2368	4045	.	.	.	.	.	.	.	.	.	.	.	.
	•••	2093	2834	4025	.	.	.	.	.	.	.	.	.	.	.	.

TABLE 6: ALPHA= 0.025 POWER= 0.9   EXPECTED ACCRUAL THRU MINIMUM FOLLOW-UP= 850

		DEL=.02			DEL=.05			DEL=.10			DEL=.15			DEL=.20		
FACT=		1.0 .75	.50 .25	.00 BIN	1.0 .75	.50 .25	.00 BIN	1.0 .75	.50 .25	.00 BIN	1.0 75	.50 .25	.00 BIN	1.0 .75	.50 .25	.00 BIN
PCONT=•••					REQUIRED NUMBER OF PATIENTS											
0.05	***	2997	3014	3165	644	654	671	233	236	239	137	138	139	97	97	98
	•••	3003	3044	5916	648	661	1156	235	237	368	137	138	194	97	97	124
0.1	***	6237	6278	6954	1185	1215	1295	380	388	400	205	208	212	136	137	139
	•••	6250	6360	10277	1196	1243	1829	384	393	526	207	210	260	137	138	158
0.15	***	9114	9188	10755	1652	1710	1898	505	522	548	263	269	276	168	171	174
	•••	9139	9334	14112	1675	1771	2417	513	533	662	265	272	316	169	172	187
0.2	***	11410	11525	14290	2027	2118	2448	608	634	679	309	319	332	194	198	203
	•••	11449	11750	17422	2062	2219	2922	620	654	778	314	325	362	196	201	211
0.25	***	13065	13229	17424	2305	2433	2929	688	725	791	346	359	377	214	220	227
	•••	13120	13553	20206	2354	2578	3342	705	752	873	352	368	400	217	223	229
0.3	***	14089	14313	20081	2491	2660	3331	746	794	882	374	390	413	229	236	244
	•••	14164	14757	22465	2555	2851	3678	767	830	946	381	401	428	232	240	242
0.35	***	14527	14826	22218	2594	2805	3648	784	842	951	391	411	440	238	246	257
	•••	14626	15412	24199	2675	3042	3930	810	886	999	400	424	446	242	251	250
0.4	***	14446	14836	23808	2627	2879	3876	804	871	997	400	423	455	242	251	263
	•••	14576	15591	25407	2725	3157	4098	834	922	1030	411	437	456	246	256	253
0.45	***	13924	14426	24838	2600	2890	4014	808	882	1020	401	426	460	241	250	263
	•••	14092	15373	26090	2713	3203	4182	840	937	1041	412	441	456	246	256	250
0.5	***	13049	13686	25298	2524	2846	4061	797	874	1020	394	420	456	235	245	257
	•••	13263	14841	26248	2652	3187	4182	831	933	1030	406	436	446	240	251	242
0.55	***	11912	12710	25188	2411	2756	4016	771	850	998	380	405	441	224	234	246
	•••	12183	14074	25880	2549	3112	4098	805	910	999	392	421	428	229	240	229
0.6	***	10612	11590	24505	2266	2623	3879	732	810	952	359	382	416	209	219	229
	•••	10952	13136	24987	2411	2985	3930	767	868	946	369	397	400	214	223	211
0.65	***	9249	10400	23253	2096	2452	3652	681	754	884	329	351	380	190	197	207
	•••	9661	12076	23569	2241	2805	3678	713	808	873	340	364	362	193	202	187
0.7	***	7913	9192	21436	1903	2244	3334	616	682	794	293	311	335	165	171	179
	•••	8385	10923	21625	2043	2574	3342	646	729	778	301	322	316	168	174	158
0.75	***	6659	7989	19057	1686	1997	2927	538	593	681	249	262	280	136	140	145
	•••	7163	9682	19156	1814	2290	2922	562	630	662	255	271	260	138	142	124
0.8	***	5499	6781	16122	1441	1706	2431	444	484	546	195	204	215	.	.	.
	•••	5994	8340	16161	1552	1946	2417	462	511	526	199	209	194	.	.	.
0.85	***	4397	5529	12638	1158	1360	1847	330	355	389	.	.	.	.	.	.
	•••	4840	6851	12641	1243	1532	1829	342	371	368	.	.	.	.	.	.
0.9	***	3275	4147	8610	814	935	1177	.	.	.	.	.	.	.	.	.
	•••	3621	5115	8596	867	1029	1156	.	.	.	.	.	.	.	.	.
0.95	***	1943	2409	4045	.	.	.	.	.	.	.	.	.	.	.	.
	•••	2134	2873	4025	.	.	.	.	.	.	.	.	.	.	.	.

TABLE 6: ALPHA= 0.025 POWER= 0.9   EXPECTED ACCRUAL THRU MINIMUM FOLLOW-UP= 900

		DEL=.02			DEL=.05			DEL=.10			DEL=.15			DEL=.20	
FACT=	1.0 .75	.50 .25	.00 BIN	1.0 .75	.50 .25	.00 BIN	1.0 .75	.50 .25	.00 BIN	1.0 75	.50 .25	.00 BIN	1.0 .75	.50 .25	.00 BIN
PCONT=***					REQUIRED NUMBER OF PATIENTS											
0.05 ***	2998	3016	3165	645	654	671	234	236	239	137	138	139	97	98	98
•••	3004	3047	5916	649	661	1156	235	237	368	137	138	194	97	98	124
0.1 ***	6239	6283	6954	1187	1217	1295	381	388	399	206	208	212	136	137	139
•••	6254	6369	10277	1199	1245	1829	384	393	526	207	210	260	137	138	158
0.15 ***	9118	9197	10755	1657	1715	1898	507	523	548	263	269	276	168	171	174
•••	9145	9351	14112	1680	1776	2417	514	534	662	266	272	316	170	172	187
0.2 ***	11417	11538	14291	2034	2126	2448	611	636	679	311	320	332	194	199	203
•••	11457	11777	17422	2069	2227	2922	622	655	778	315	325	362	197	201	211
0.25 ***	13075	13248	17424	2314	2445	2930	692	728	791	348	360	378	215	220	227
•••	13133	13591	20206	2364	2591	3342	707	755	873	353	368	400	217	224	229
0.3 ***	14102	14340	20081	2503	2675	3331	750	798	882	375	391	414	229	236	245
•••	14181	14807	22465	2569	2867	3678	771	833	946	383	401	428	233	240	242
0.35 ***	14545	14861	22218	2610	2825	3648	789	847	950	393	413	440	239	246	257
•••	14650	15479	24199	2693	3061	3930	815	890	999	402	424	446	243	251	250
0.4 ***	14469	14882	23808	2645	2901	3876	810	876	996	402	424	455	243	252	263
•••	14607	15675	25407	2746	3179	4098	839	925	1030	413	438	456	247	257	253
0.45 ***	13954	14484	24837	2621	2916	4014	815	887	1020	404	427	460	242	251	263
•••	14131	15476	26090	2737	3228	4182	846	941	1041	415	442	456	246	257	250
0.5 ***	13087	13760	25299	2549	2874	4061	803	880	1020	397	422	456	236	245	258
•••	13313	14963	26248	2679	3214	4182	837	937	1030	408	437	446	240	251	242
0.55 ***	11960	12800	25188	2437	2785	4016	778	856	998	383	407	441	226	235	246
•••	12246	14213	25880	2577	3141	4098	812	914	999	394	422	428	230	240	229
0.6 ***	10673	11696	24505	2294	2654	3879	739	816	953	361	384	415	210	219	229
•••	11030	13288	24987	2440	3013	3930	773	872	946	371	398	400	215	224	211
0.65 ***	9324	10519	23253	2124	2482	3652	687	759	884	332	352	380	190	198	207
•••	9755	12236	23569	2270	2832	3678	720	811	873	342	365	362	194	202	187
0.7 ***	8000	9319	21435	1930	2272	3334	622	686	794	295	312	335	165	172	179
•••	8489	11082	21625	2071	2600	3342	651	731	778	303	323	316	168	175	158
0.75 ***	6753	8116	19056	1711	2022	2927	543	596	681	250	263	280	136	140	145
•••	7272	9835	19156	1841	2312	2922	567	632	662	256	271	260	138	143	124
0.8 ***	5593	6900	16122	1463	1727	2431	448	487	546	196	204	215	.	.	.
•••	6099	8479	16161	1574	1964	2417	466	513	526	200	209	194	.	.	.
0.85 ***	4482	5632	12638	1175	1375	1847	333	356	389	.	.	.	.	.	.
•••	4933	6966	12641	1260	1545	1829	344	371	368	.	.	.	.	.	.
0.9 ***	3342	4223	8610	825	944	1177	.	.	.	.	.	.	.	.	.
•••	3692	5196	8596	877	1036	1156	.	.	.	.	.	.	.	.	.
0.95 ***	1981	2448	4045	.	.	.	.	.	.	.	.	.	.	.	.
•••	2172	2909	4025	.	.	.	.	.	.	.	.	.	.	.	.

TABLE 6: ALPHA= 0.025 POWER= 0.9   EXPECTED ACCRUAL THRU MINIMUM FOLLOW-UP= 950

		DEL=.02			DEL=.05			DEL=.10			DEL=.15			DEL=.20		
FACT=		1.0 .75	.50 .25	.00 BIN	1.0 .75	.50 .25	.00 BIN	1.0 .75	.50 .25	.00 BIN	1.0 75	.50 .25	.00 BIN	1.0 .75	.50 .25	.00 BIN
PCONT=•••				REQUIRED NUMBER OF PATIENTS												
0.05	***	2998	3018	3165	646	654	671	234	236	239	137	138	139	97	97	98
	•••	3006	3050	5916	650	661	1156	235	237	368	138	139	194	97	97	124
0.1	***	6241	6289	6954	1189	1219	1294	381	389	400	205	209	212	136	137	139
	•••	6257	6378	10277	1201	1247	1829	385	394	526	207	210	260	137	139	158
0.15	***	9123	9206	10755	1661	1720	1898	508	524	547	264	269	276	169	171	174
	•••	9150	9367	14112	1684	1780	2417	515	534	662	266	273	316	170	172	187
0.2	***	11424	11552	14290	2040	2134	2448	613	638	679	311	320	331	195	198	203
	•••	11466	11802	17422	2077	2234	2922	624	656	778	315	325	362	197	201	211
0.25	***	13085	13268	17424	2323	2456	2930	694	730	791	349	361	377	215	220	227
	•••	13146	13627	20206	2375	2601	3342	710	756	873	355	368	400	217	223	229
0.3	***	14115	14366	20081	2515	2690	3331	754	801	882	376	392	413	230	236	245
	•••	14199	14858	22465	2583	2881	3678	775	834	946	384	401	428	233	241	242
0.35	***	14562	14896	22218	2624	2843	3648	794	850	950	395	413	439	239	247	257
	•••	14674	15544	24199	2709	3079	3930	819	892	999	403	425	446	243	251	250
0.4	***	14492	14928	23808	2663	2923	3876	816	881	996	405	426	455	243	252	262
	•••	14637	15757	25407	2765	3200	4098	844	928	1030	414	439	456	248	257	253
0.45	***	13984	14543	24837	2642	2940	4014	821	891	1020	406	429	460	242	252	263
	•••	14170	15576	26090	2761	3251	4182	852	944	1041	416	443	456	247	257	250
0.5	***	13124	13832	25299	2572	2901	4061	809	885	1020	399	423	456	237	246	257
	•••	13363	15081	26248	2704	3239	4182	843	941	1030	410	438	446	241	251	242
0.55	***	12008	12889	25188	2463	2813	4016	785	861	998	385	408	441	226	235	247
	•••	12310	14347	25880	2604	3167	4098	818	918	999	395	423	428	230	241	229
0.6	***	10733	11800	24505	2320	2682	3879	746	821	952	363	386	416	211	219	229
	•••	11107	13434	24987	2468	3040	3930	779	875	946	374	399	400	215	224	211
0.65	***	9398	10635	23253	2151	2510	3652	694	764	884	334	353	380	191	198	207
	•••	9846	12389	23569	2299	2858	3678	725	814	873	343	366	362	194	202	187
0.7	***	8086	9442	21436	1957	2299	3334	628	691	793	296	313	335	166	172	178
	•••	8590	11236	21625	2098	2623	3342	656	734	778	304	323	316	169	175	158
0.75	***	6846	8238	19057	1735	2046	2927	547	599	680	251	264	280	136	140	145
	•••	7376	9982	19156	1865	2332	2922	571	635	662	257	272	260	139	143	124
0.8	***	5684	7015	16122	1484	1747	2431	451	489	546	197	205	215	.	.	.
	•••	6200	8610	16161	1594	1980	2417	469	514	526	201	210	194	.	.	.
0.85	***	4564	5730	12638	1191	1389	1847	335	357	389	.	.	.	.	.	.
	•••	5022	7075	12641	1276	1556	1829	346	372	368	.	.	.	.	.	.
0.9	***	3406	4297	8610	835	952	1177	.	.	.	.	.	.	.	.	.
	•••	3761	5273	8596	887	1041	1156	.	.	.	.	.	.	.	.	.
0.95	***	2016	2485	4045	.	.	.	.	.	.	.	.	.	.	.	.
	•••	2209	2943	4025	.	.	.	.	.	.	.	.	.	.	.	.

TABLE 6: ALPHA= 0.025 POWER= 0.9   EXPECTED ACCRUAL THRU MINIMUM FOLLOW-UP= 1000

		DEL=.02			DEL=.05			DEL=.10			DEL=.15			DEL=.20		
FACT=		1.0 .75	.50 .25	.00 BIN	1.0 .75	.50 .25	.00 BIN	1.0 .75	.50 .25	.00 BIN	1.0 75	.50 .25	.00 BIN	1.0 .75	.50 .25	.00 BIN
PCONT=***					REQUIRED	NUMBER	OF PATIENTS									
0.05	***	3000	3020	3165	646	655	671	234	236	239	137	138	139	97	98	98
	•••	3007	3053	5916	650	662	1156	235	237	368	138	139	194	97	98	124
0.1	***	6244	6293	6955	1191	1221	1295	382	390	400	206	209	212	136	138	139
	•••	6260	6386	10277	1204	1249	1829	386	394	526	208	210	260	137	138	158
0.15	***	9127	9214	10755	1665	1725	1898	510	525	548	265	270	276	169	171	174
	•••	9156	9383	14112	1688	1785	2417	516	535	662	267	273	316	170	173	187
0.2	***	11431	11565	14290	2046	2141	2448	615	640	679	312	321	331	195	199	203
	•••	11475	11828	17422	2083	2241	2922	626	657	778	316	326	362	197	201	211
0.25	***	13094	13286	17424	2331	2467	2930	698	733	791	350	361	378	216	221	227
	•••	13158	13664	20206	2385	2611	3342	713	758	873	355	369	400	218	224	229
0.3	***	14128	14393	20081	2526	2704	3331	758	804	881	378	393	413	231	237	245
	•••	14216	14908	22465	2596	2895	3678	778	836	946	385	402	428	234	241	242
0.35	***	14580	14931	22218	2639	2860	3648	799	854	950	396	415	440	240	248	256
	•••	14697	15609	24199	2726	3096	3930	823	895	999	405	426	446	244	252	250
0.4	***	14515	14973	23808	2681	2944	3876	821	885	996	406	427	455	245	253	263
	•••	14668	15839	25407	2785	3220	4098	850	931	1030	416	440	456	248	258	253
0.45	***	14013	14601	24838	2663	2963	4014	826	896	1020	408	430	461	243	252	263
	•••	14210	15675	26090	2783	3274	4182	857	948	1041	418	444	456	248	258	250
0.5	***	13162	13905	25298	2595	2926	4061	816	890	1020	401	424	456	238	246	258
	•••	13413	15197	26248	2728	3263	4182	849	944	1030	412	438	446	242	252	242
0.55	***	12056	12976	25188	2488	2840	4016	791	866	998	387	410	441	227	236	246
	•••	12373	14478	25880	2630	3192	4098	825	921	999	398	424	428	231	241	229
0.6	***	10793	11901	24505	2346	2710	3880	752	826	953	365	386	416	212	220	230
	•••	11184	13576	24987	2495	3065	3930	785	878	946	375	400	400	216	225	211
0.65	***	9471	10748	23253	2177	2537	3652	700	768	885	336	355	380	192	199	206
	•••	9935	12536	23569	2325	2882	3678	730	817	873	345	366	362	195	203	187
0.7	***	8170	9561	21436	1981	2324	3334	633	695	794	298	315	335	167	172	178
	•••	8690	11383	21625	2123	2645	3342	661	737	778	306	324	316	170	175	158
0.75	***	6935	8356	19057	1758	2069	2927	552	603	681	253	265	280	136	141	145
	•••	7478	10121	19156	1888	2351	2922	575	636	662	258	272	260	138	143	124
0.8	***	5772	7125	16122	1503	1766	2431	455	491	546	198	206	215	.	.	.
	•••	6297	8736	16161	1614	1995	2417	471	516	526	201	210	194	.	.	.
0.85	***	4643	5825	12638	1206	1403	1847	337	359	389	.	.	.	.	.	.
	•••	5108	7180	12641	1291	1566	1829	348	373	368	.	.	.	.	.	.
0.9	***	3468	4368	8610	845	960	1177	.	.	.	.	.	.	.	.	.
	•••	3826	5346	8596	895	1047	1156	.	.	.	.	.	.	.	.	.
0.95	***	2050	2520	4045	.	.	.	.	.	.	.	.	.	.	.	.
	•••	2243	2975	4025	.	.	.	.	.	.	.	.	.	.	.	.

653

TABLE 6: ALPHA= 0.025 POWER= 0.9   EXPECTED ACCRUAL THRU MINIMUM FOLLOW-UP= 1100

	DEL=.02			DEL=.05			DEL=.10			DEL=.15			DEL=.20		
FACT=	1.0 .75	.50 .25	.00 BIN	1.0 .75	.50 .25	.00 BIN	1.0 .75	.50 .25	.00 BIN	1.0 75	.50 .25	.00 BIN	1.0 .75	.50 .25	.00 BIN

PCONT=*** REQUIRED NUMBER OF PATIENTS

PCONT	.02 a	.02 b	.02 c	.05 a	.05 b	.05 c	.10 a	.10 b	.10 c	.15 a	.15 b	.15 c	.20 a	.20 b	.20 c
0.05 ***	3002	3024	3165	648	657	671	235	236	239	137	138	139	97	98	98
•••	3009	3058	5916	652	663	1156	235	237	368	137	139	194	97	98	124
0.1 ***	6249	6303	6954	1195	1226	1295	383	390	400	206	209	212	136	137	139
•••	6267	6403	10277	1208	1252	1829	386	395	526	208	210	260	137	138	158
0.15 ***	9135	9231	10755	1672	1733	1897	512	527	548	265	270	276	169	171	174
•••	9168	9415	14112	1697	1792	2417	518	536	662	268	273	316	170	173	187
0.2 ***	11444	11591	14290	2058	2155	2448	619	642	679	313	321	331	196	199	203
•••	11493	11878	17422	2096	2254	2922	629	659	778	317	326	362	197	201	211
0.25 ***	13113	13325	17424	2348	2487	2930	703	736	791	352	363	378	217	221	227
•••	13184	13736	20206	2403	2630	3342	718	760	873	357	369	400	219	224	229
0.3 ***	14155	14445	20081	2549	2731	3331	765	809	881	380	395	413	232	237	245
•••	14252	15005	22465	2620	2919	3678	785	840	946	387	403	428	235	241	242
0.35 ***	14615	15001	22217	2667	2893	3648	807	861	950	400	417	439	241	248	257
•••	14744	15735	24199	2756	3126	3930	831	899	999	408	427	446	245	252	250
0.4 ***	14561	15064	23808	2714	2982	3876	830	892	996	410	429	455	246	253	263
•••	14729	15997	25407	2821	3256	4098	858	936	1030	419	441	456	250	258	253
0.45 ***	14073	14716	24837	2701	3007	4014	837	904	1020	411	433	461	245	253	263
•••	14289	15866	26090	2824	3314	4182	867	953	1041	422	445	456	249	258	250
0.5 ***	13238	14047	25298	2638	2974	4060	827	899	1020	405	426	456	239	247	257
•••	13513	15419	26248	2774	3306	4182	859	950	1030	415	440	446	243	252	242
0.55 ***	12152	13148	25187	2534	2890	4016	802	874	998	390	412	441	228	236	246
•••	12497	14727	25880	2679	3237	4098	835	927	999	401	425	428	232	241	229
0.6 ***	10912	12099	24505	2395	2760	3880	763	834	952	368	389	415	213	221	229
•••	11335	13845	24987	2545	3110	3930	795	884	946	378	401	400	217	225	211
0.65 ***	9614	10966	23253	2226	2586	3652	710	776	884	338	357	380	193	199	206
•••	10109	12815	23569	2375	2925	3678	740	822	873	347	368	362	196	203	187
0.7 ***	8332	9787	21436	2028	2370	3334	643	701	793	301	316	335	168	173	178
•••	8880	11660	21625	2171	2685	3342	669	741	778	308	324	316	170	175	158
0.75 ***	7107	8580	19057	1801	2110	2927	560	609	681	254	266	280	137	141	144
•••	7673	10385	19156	1930	2386	2922	582	640	662	260	272	260	139	142	124
0.8 ***	5940	7332	16122	1540	1800	2431	461	496	546	199	206	215	.	.	.
•••	6483	8973	16161	1650	2023	2417	477	518	526	202	210	194	.	.	.
0.85 ***	4793	6003	12638	1235	1428	1847	340	362	389	.	.	.	.	.	.
•••	5270	7374	12641	1318	1585	1829	351	374	368	.	.	.	.	.	.
0.9 ***	3584	4499	8610	862	973	1176	.	.	.	.	.	.	.	.	.
•••	3950	5482	8596	911	1056	1156	.	.	.	.	.	.	.	.	.
0.95 ***	2113	2585	4045	.	.	.	.	.	.	.	.	.	.	.	.
•••	2309	3033	4025	.	.	.	.	.	.	.	.	.	.	.	.

TABLE 6: ALPHA= 0.025 POWER= 0.9   EXPECTED ACCRUAL THRU MINIMUM FOLLOW-UP= 1200

		DEL=.02			DEL=.05			DEL=.10			DEL=.15			DEL=.20	
FACT=	1.0 .75	.50 .25	.00 BIN	1.0 .75	.50 .25	.00 BIN	1.0 .75	.50 .25	.00 BIN	1.0 75	.50 .25	.00 BIN	1.0 .75	.50 .25	.00 BIN
PCONT=•••						REQUIRED NUMBER OF PATIENTS									
0.05 ***	3004	3028	3165	649	657	671	235	236	238	137	138	139	97	97	97
•••	3012	3062	5916	652	663	1156	235	238	368	138	139	194	97	97	124
0.1 ***	6253	6313	6954	1199	1229	1294	384	391	400	207	209	212	136	138	139
•••	6274	6419	10277	1211	1255	1829	387	395	526	208	211	260	137	138	158
0.15 ***	9145	9249	10755	1679	1741	1897	514	528	547	265	271	276	169	172	174
•••	9179	9446	14112	1705	1798	2417	520	537	662	268	273	316	170	172	187
0.2 ***	11457	11618	14290	2069	2168	2448	622	645	679	315	322	331	196	199	203
•••	11511	11927	17422	2109	2265	2922	632	660	778	318	326	362	198	201	211
0.25 ***	13132	13363	17424	2364	2506	2929	707	740	790	353	364	377	217	221	226
•••	13210	13807	20206	2420	2647	3342	722	762	873	358	370	400	219	224	229
0.3 ***	14181	14497	20081	2569	2755	3331	771	814	881	382	396	413	232	238	244
•••	14287	15100	22465	2643	2941	3678	790	843	946	388	404	428	235	241	242
0.35 ***	14650	15070	22217	2693	2923	3648	814	866	950	402	418	439	242	249	256
•••	14791	15859	24199	2785	3154	3930	838	902	999	409	427	446	245	253	250
0.4 ***	14607	15154	23808	2746	3017	3876	839	898	997	412	430	455	247	254	262
•••	14790	16150	25407	2854	3288	4098	865	940	1030	421	442	456	250	258	253
0.45 ***	14131	14830	24838	2737	3046	4014	846	911	1020	415	434	460	246	253	263
•••	14367	16049	26090	2863	3350	4182	875	958	1041	424	446	456	250	258	250
0.5 ***	13313	14187	25298	2678	3016	4060	837	906	1020	408	429	455	241	248	257
•••	13612	15631	26248	2816	3345	4182	868	955	1030	418	441	446	244	253	242
0.55 ***	12247	13314	25187	2577	2935	4015	812	883	997	394	414	440	229	237	246
•••	12619	14962	25880	2724	3277	4098	844	931	999	403	426	428	233	241	229
0.6 ***	11030	12289	24505	2440	2806	3879	773	841	952	371	391	415	214	221	229
•••	11482	14098	24987	2590	3150	3930	804	889	946	380	402	400	217	225	211
0.65 ***	9754	11173	23253	2270	2631	3652	719	783	884	341	358	380	194	199	206
•••	10277	13075	23569	2420	2964	3678	748	826	873	349	368	362	196	203	187
0.7 ***	8489	10002	21436	2071	2413	3334	651	707	793	303	317	335	169	173	178
•••	9061	11918	21625	2214	2720	3342	676	745	778	310	325	316	171	175	158
0.75 ***	7271	8791	19057	1840	2147	2926	567	613	681	256	267	280	138	141	145
•••	7857	10629	19156	1969	2416	2922	588	643	662	261	273	260	139	142	124
0.8 ***	6098	7527	16122	1573	1830	2431	466	499	546	200	207	215	.	.	.
•••	6657	9190	16161	1683	2047	2417	481	520	526	203	211	194	.	.	.
0.85 ***	4933	6168	12638	1260	1450	1847	343	364	388	.	.	.	.	.	.
•••	5422	7551	12641	1342	1602	1829	353	375	368	.	.	.	.	.	.
0.9 ***	3692	4621	8610	877	985	1177	.	.	.	.	.	.	.	.	.
•••	4065	5605	8596	925	1064	1156	.	.	.	.	.	.	.	.	.
0.95 ***	2172	2644	4045	.	.	.	.	.	.	.	.	.	.	.	.
•••	2368	3085	4025	.	.	.	.	.	.	.	.	.	.	.	.

TABLE 6: ALPHA= 0.025 POWER= 0.9   EXPECTED ACCRUAL THRU MINIMUM FOLLOW-UP= 1300

		DEL=.02			DEL=.05			DEL=.10			DEL=.15			DEL=.20		
FACT=		1.0 .75	.50 .25	.00 BIN	1.0 .75	.50 .25	.00 BIN	1.0 .75	.50 .25	.00 BIN	1.0 75	.50 .25	.00 BIN	1.0 .75	.50 .25	.00 BIN
PCONT=***					REQUIRED NUMBER OF PATIENTS											
0.05	***	3006	3032	3165	650	658	671	235	237	239	137	138	139	97	97	97
	•••	3015	3067	5916	653	664	1156	236	237	368	138	139	194	97	97	124
0.1	***	6259	6322	6954	1202	1233	1294	385	392	400	207	210	212	136	138	139
	•••	6280	6435	10277	1215	1258	1829	388	396	526	208	211	260	137	138	158
0.15	***	9154	9266	10755	1686	1748	1898	516	529	548	266	271	276	170	172	174
	•••	9191	9476	14112	1712	1804	2417	523	538	662	269	274	316	171	173	187
0.2	***	11471	11645	14290	2080	2180	2448	625	647	679	316	323	331	197	200	203
	•••	11529	11975	17422	2121	2275	2922	635	661	778	319	327	362	198	201	211
0.25	***	13152	13401	17424	2379	2523	2930	712	744	791	355	365	377	218	222	227
	•••	13236	13875	20206	2437	2661	3342	726	765	873	360	370	400	220	224	229
0.3	***	14207	14550	20081	2590	2777	3331	777	817	882	384	397	414	233	239	245
	•••	14322	15193	22465	2665	2962	3678	796	846	946	390	405	428	236	241	242
0.35	***	14685	15140	22218	2718	2950	3648	822	871	951	405	419	440	243	250	257
	•••	14838	15978	24199	2811	3178	3930	844	905	999	411	429	446	246	253	250
0.4	***	14653	15243	23808	2775	3050	3877	847	904	996	415	432	455	248	254	263
	•••	14851	16297	25407	2887	3317	4098	873	944	1030	423	443	456	251	258	253
0.45	***	14190	14942	24837	2772	3083	4014	855	918	1020	418	436	461	247	254	263
	•••	14446	16225	26090	2898	3383	4182	883	962	1041	426	448	456	251	258	250
0.5	***	13388	14324	25298	2716	3057	4061	846	913	1020	411	431	456	241	249	258
	•••	13710	15833	26248	2856	3380	4182	876	959	1030	420	442	446	245	253	242
0.55	***	12341	13476	25187	2618	2976	4015	822	889	998	396	416	440	231	238	246
	•••	12741	15186	25880	2766	3313	4098	852	936	999	406	427	428	234	242	229
0.6	***	11146	12472	24505	2481	2848	3880	782	848	952	375	393	415	215	222	229
	•••	11625	14336	24987	2633	3186	3930	812	893	946	383	403	400	219	225	211
0.65	***	9891	11370	23253	2312	2672	3652	728	789	884	344	360	380	195	201	206
	•••	10440	13320	23569	2462	2999	3678	756	830	873	352	370	362	198	203	187
0.7	***	8640	10205	21435	2111	2450	3334	658	713	794	305	318	335	169	174	179
	•••	9235	12159	21625	2254	2752	3342	683	748	778	311	327	316	172	176	158
0.75	***	7428	8989	19056	1877	2181	2927	573	617	681	258	268	279	138	141	145
	•••	8032	10856	19156	2005	2444	2922	593	646	662	263	274	260	140	143	124
0.8	***	6249	7709	16123	1605	1858	2431	471	502	546	201	207	215	.	.	.
	•••	6821	9392	16161	1713	2069	2417	485	523	526	204	211	194	.	.	.
0.85	***	5065	6322	12638	1284	1470	1847	346	365	389	.	.	.	.	.	.
	•••	5564	7715	12641	1365	1617	1829	355	376	368	.	.	.	.	.	.
0.9	***	3794	4734	8610	890	996	1176	.	.	.	.	.	.	.	.	.
	•••	4172	5718	8596	938	1072	1156	.	.	.	.	.	.	.	.	.
0.95	***	2226	2697	4046	.	.	.	.	.	.	.	.	.	.	.	.
	•••	2423	3131	4025	.	.	.	.	.	.	.	.	.	.	.	.

TABLE 6: ALPHA= 0.025 POWER= 0.9   EXPECTED ACCRUAL THRU MINIMUM FOLLOW-UP= 1400

	DEL=.02			DEL=.05			DEL=.10			DEL=.15			DEL=.20		
FACT=	1.0 .75	.50 .25	.00 BIN	1.0 .75	.50 .25	.00 BIN	1.0 .75	.50 .25	.00 BIN	1.0 75	.50 .25	.00 BIN	1.0 .75	.50 .25	.00 BIN
PCONT=•••				REQUIRED NUMBER OF PATIENTS											
0.05 ***	3008 3034 3166			651 659 671			235 236 239			137 138 139			97 97 98		
•••	3018 3071 5916			654 664 1156			235 238 368			138 139 194			97 98 124		
0.1 ***	6263 6331 6954			1206 1235 1295			386 392 400			207 210 212			137 138 139		
•••	6287 6450 10277			1219 1260 1829			388 395 526			208 211 260			137 138 158		
0.15 ***	9162 9283 10755			1693 1754 1898			518 530 548			267 271 276			170 172 174		
•••	9202 9505 14112			1718 1809 2417			523 538 662			269 274 316			171 173 187		
0.2 ***	11484 11672 14290			2090 2191 2448			627 649 679			317 324 332			197 200 203		
•••	11546 12022 17422			2131 2284 2922			638 662 778			320 327 362			199 201 211		
0.25 ***	13171 13440 17424			2394 2538 2929			716 746 791			356 366 377			219 222 227		
•••	13261 13942 20206			2453 2675 3342			730 766 873			360 371 400			220 225 229		
0.3 ***	14234 14602 20081			2608 2798 3331			781 822 882			386 398 413			234 239 245		
•••	14357 15283 22465			2685 2979 3678			800 848 946			392 405 428			236 241 242		
0.35 ***	14720 15209 22218			2741 2976 3648			828 876 950			406 421 439			244 250 256		
•••	14885 16094 24199			2837 3201 3930			850 908 999			413 430 446			247 253 250		
0.4 ***	14698 15332 23808			2803 3080 3876			854 909 997			417 434 455			248 255 262		
•••	14913 16439 25407			2916 3343 4098			879 948 1030			425 444 456			252 259 253		
0.45 ***	14249 15053 24838			2803 3117 4014			862 923 1020			420 437 460			248 255 263		
•••	14523 16393 26090			2932 3412 4182			890 965 1041			428 448 456			251 259 250		
0.5 ***	13463 14458 25298			2752 3093 4061			854 919 1020			414 432 456			242 249 257		
•••	13808 16025 26248			2893 3412 4182			883 962 1030			423 443 446			246 254 242		
0.55 ***	12435 13632 25188			2656 3014 4015			829 895 997			399 417 441			232 239 246		
•••	12860 15397 25880			2804 3346 4098			859 940 999			408 428 428			235 242 229		
0.6 ***	11259 12647 24505			2520 2886 3880			790 854 952			376 394 416			216 222 229		
•••	11765 14560 24987			2673 3218 3930			819 897 946			385 404 400			220 226 211		
0.65 ***	10023 11558 23253			2350 2709 3652			735 794 885			346 361 381			196 200 206		
•••	10597 13549 23569			2501 3030 3678			762 834 873			353 370 362			198 204 187		
0.7 ***	8786 10398 21436			2147 2485 3334			665 717 794			307 319 335			170 174 178		
•••	9401 12385 21625			2290 2781 3342			689 752 778			313 327 316			171 176 158		
0.75 ***	7577 9176 19057			1910 2212 2927			578 621 681			259 269 280			139 142 144		
•••	8198 11068 19156			2039 2468 2922			598 648 662			263 274 260			140 143 124		
0.8 ***	6392 7880 16122			1633 1883 2431			474 505 546			202 208 215			. . .		
•••	6977 9579 16161			1740 2088 2417			488 523 526			205 212 194			. . .		
0.85 ***	5190 6466 12638			1305 1488 1847			349 367 388			. . .			. . .		
•••	5698 7867 12641			1385 1629 1829			357 377 368			. . .			. . .		
0.9 ***	3889 4838 8610			903 1006 1177			. . .			. . .			. . .		
•••	4273 5821 8596			949 1078 1156			. . .			. . .			. . .		
0.95 ***	2277 2747 4045			. . .			. . .			. . .			. . .		
•••	2474 3173 4025			. . .			. . .			. . .			. . .		

TABLE 6: ALPHA= 0.025 POWER= 0.9   EXPECTED ACCRUAL THRU MINIMUM FOLLOW-UP= 1500

		DEL=.02			DEL=.05			DEL=.10			DEL=.15			DEL=.20		
FACT=		1.0	.50	.00	1.0	.50	.00	1.0	.50	.00	1.0	.50	.00	1.0	.50	.00
		.75	.25	BIN	.75	.25	BIN	.75	.25	BIN	75	.25	BIN	.75	.25	BIN
PCONT=***						REQUIRED	NUMBER	OF	PATIENTS							
0.05	***	3010	3038	3165	652	660	671	235	237	239	138	139	139	97	97	97
	•••	3020	3075	5916	655	665	1156	236	238	368	138	139	194	97	97	124
0.1	***	6268	6341	6954	1208	1238	1295	387	393	399	208	210	212	137	138	139
	•••	6293	6464	10277	1222	1262	1829	389	395	526	209	211	260	138	139	158
0.15	***	9170	9300	10755	1699	1760	1897	519	532	547	268	272	277	170	172	174
	•••	9214	9533	14112	1725	1814	2417	525	539	662	270	274	316	172	173	187
0.2	***	11497	11698	14290	2100	2201	2449	630	650	679	318	324	332	198	200	203
	•••	11565	12067	17422	2141	2292	2922	639	664	778	320	327	362	199	202	211
0.25	***	13190	13477	17424	2407	2553	2930	719	748	791	357	367	378	219	223	227
	•••	13286	14007	20206	2467	2687	3342	732	768	873	362	371	400	220	225	229
0.3	***	14260	14654	20081	2626	2817	3331	787	825	881	387	399	413	234	239	245
	•••	14392	15370	22465	2704	2995	3678	803	850	946	393	406	428	237	242	242
0.35	***	14755	15277	22218	2764	3000	3648	832	879	950	408	422	440	245	250	257
	•••	14932	16205	24199	2860	3220	3930	854	910	999	414	430	446	247	253	250
0.4	***	14744	15419	23808	2829	3108	3877	860	914	997	419	435	455	249	256	262
	•••	14973	16576	25407	2944	3367	4098	885	950	1030	427	444	456	252	259	253
0.45	***	14308	15161	24837	2834	3147	4014	869	928	1020	422	439	460	249	256	263
	•••	14600	16554	26090	2963	3439	4182	896	968	1041	430	449	456	252	259	250
0.5	***	13537	14589	25298	2785	3127	4061	862	923	1020	416	433	455	244	250	258
	•••	13905	16208	26248	2927	3440	4182	890	965	1030	424	443	446	247	254	242
0.55	***	12528	13784	25188	2690	3050	4016	837	901	997	401	419	440	232	239	247
	•••	12977	15598	25880	2840	3376	4098	866	943	999	410	429	428	236	243	229
0.6	***	11372	12817	24505	2557	2921	3879	797	859	952	379	395	415	217	223	230
	•••	11902	14772	24987	2709	3247	3930	826	900	946	386	405	400	220	226	211
0.65	***	10152	11738	23253	2387	2744	3652	742	800	884	348	363	380	196	202	206
	•••	10748	13766	23569	2537	3058	3678	769	837	873	354	371	362	199	203	187
0.7	***	8926	10580	21436	2182	2517	3334	671	722	794	308	320	335	170	174	178
	•••	9560	12597	21625	2324	2807	3342	695	754	778	314	327	316	172	176	158
0.75	***	7720	9353	19057	1940	2240	2927	584	624	680	260	269	280	139	142	144
	•••	8356	11266	19156	2069	2490	2922	603	650	662	264	275	260	140	143	124
0.8	***	6527	8042	16122	1659	1906	2431	478	507	545	202	208	215	.	.	.
	•••	7124	9754	16161	1765	2105	2417	491	525	526	205	212	194	.	.	.
0.85	***	5309	6602	12638	1325	1504	1847	352	368	389	.	.	.	.	.	.
	•••	5825	8008	12641	1403	1640	1829	359	378	368	.	.	.	.	.	.
0.9	***	3980	4936	8610	915	1014	1177	.	.	.	.	.	.	.	.	.
	•••	4368	5917	8596	960	1084	1156	.	.	.	.	.	.	.	.	.
0.95	***	2324	2792	4045	.	.	.	.	.	.	.	.	.	.	.	.
	•••	2520	3211	4025	.	.	.	.	.	.	.	.	.	.	.	.

TABLE 6: ALPHA= 0.025 POWER= 0.9   EXPECTED ACCRUAL THRU MINIMUM FOLLOW-UP= 1600

	DEL=.02			DEL=.05			DEL=.10			DEL=.15			DEL=.20		
FACT=	1.0 .75	.50 .25	.00 BIN	1.0 .75	.50 .25	.00 BIN	1.0 .75	.50 .25	.00 BIN	1.0 75	.50 .25	.00 BIN	1.0 .75	.50 .25	.00 BIN
PCONT=•••						REQUIRED NUMBER OF PATIENTS									
0.05 •••	3012	3041	3165	653	660	671	235	237	239	138	139	139	97	98	98
•••	3023	3078	5916	656	665	1156	236	238	368	138	139	194	97	98	124
0.1 •••	6274	6351	6954	1212	1241	1295	387	393	400	208	210	212	137	138	139
•••	6300	6477	10277	1224	1263	1829	390	396	526	209	211	260	138	139	158
0.15 •••	9179	9317	10755	1705	1766	1898	521	532	548	268	272	276	171	172	174
•••	9226	9561	14112	1731	1818	2417	526	540	662	270	274	316	171	173	187
0.2 •••	11511	11724	14290	2109	2210	2448	633	652	679	318	324	332	198	201	203
•••	11583	12110	17422	2151	2299	2922	642	664	778	321	328	362	199	202	211
0.25 •••	13210	13516	17424	2421	2566	2930	722	751	791	359	367	377	219	223	227
•••	13312	14070	20206	2481	2698	3342	735	769	873	363	372	400	221	225	229
0.3 •••	14287	14706	20081	2643	2835	3331	790	828	882	389	400	413	235	240	245
•••	14428	15455	22465	2722	3010	3678	807	852	946	394	406	428	237	242	242
0.35 •••	14791	15345	22218	2785	3022	3648	838	883	950	410	423	439	246	251	257
•••	14978	16313	24199	2882	3239	3930	859	913	999	416	431	446	248	254	250
0.4 •••	14790	15506	23808	2855	3133	3876	866	918	997	421	436	455	250	256	263
•••	15034	16708	25407	2970	3388	4098	890	953	1030	428	445	456	253	259	253
0.45 •••	14367	15268	24838	2863	3176	4014	876	933	1020	424	440	461	250	256	263
•••	14678	16709	26090	2993	3463	4182	902	971	1041	432	450	456	253	259	250
0.5 •••	13612	14717	25298	2817	3158	4061	868	929	1020	418	435	456	244	251	258
•••	14000	16383	26248	2959	3466	4182	896	969	1030	426	445	446	247	254	242
0.55 •••	12620	13932	25188	2724	3082	4016	844	905	998	403	420	441	234	240	246
•••	13091	15789	25880	2874	3403	4098	872	946	999	411	430	428	236	243	229
0.6 •••	11482	12980	24505	2591	2954	3879	804	864	952	381	396	416	218	223	229
•••	12034	14974	24987	2744	3275	3930	831	903	946	388	405	400	220	226	211
0.65 •••	10278	11911	23253	2420	2776	3652	748	804	884	350	363	380	197	202	207
•••	10894	13971	23569	2571	3084	3678	774	840	873	356	371	362	199	204	187
0.7 •••	9062	10756	21436	2214	2547	3334	677	725	794	310	321	335	171	175	179
•••	9714	12797	21625	2356	2830	3342	699	756	778	315	328	316	173	176	158
0.75 •••	7857	9522	19057	1970	2266	2927	588	627	681	261	270	280	140	142	145
•••	8507	11453	19156	2097	2510	2922	606	652	662	266	275	260	141	143	124
0.8 •••	6657	8195	16122	1683	1927	2431	481	509	546	204	209	215	.	.	.
•••	7265	9918	16161	1789	2120	2417	495	526	526	206	212	194	.	.	.
0.85 •••	5422	6730	12638	1343	1519	1847	353	369	389	.	.	.	.	.	.
•••	5945	8140	12641	1420	1651	1829	361	379	368	.	.	.	.	.	.
0.9 •••	4065	5028	8610	925	1022	1177	.	.	.	.	.	.	.	.	.
•••	4457	6005	8596	969	1089	1156	.	.	.	.	.	.	.	.	.
0.95 •••	2368	2834	4045	.	.	.	.	.	.	.	.	.	.	.	.
•••	2564	3246	4025	.	.	.	.	.	.	.	.	.	.	.	.

TABLE 6: ALPHA= 0.025 POWER= 0.9   EXPECTED ACCRUAL THRU MINIMUM FOLLOW-UP= 1700

		DEL=.02			DEL=.05			DEL=.10			DEL=.15			DEL=.20		
FACT=		1.0 .75	.50 .25	.00 BIN	1.0 .75	.50 .25	.00 BIN	1.0 .75	.50 .25	.00 BIN	1.0 75	.50 .25	.00 BIN	1.0 .75	.50 .25	.00 BIN
PCONT=***								REQUIRED NUMBER OF PATIENTS								
0.05	***	3014	3044	3165	653	661	671	236	237	239	138	138	139	97	97	97
	•••	3025	3081	5916	656	665	1156	236	238	368	138	139	194	97	97	124
0.1	***	6278	6360	6954	1214	1243	1295	388	393	399	208	210	212	137	138	139
	•••	6306	6489	10277	1227	1264	1829	391	396	526	209	211	260	138	139	158
0.15	***	9188	9334	10754	1710	1771	1897	522	533	548	269	272	276	171	172	174
	•••	9237	9587	14112	1736	1821	2417	527	539	662	270	274	316	172	173	187
0.2	***	11525	11750	14290	2118	2220	2448	634	653	679	318	325	331	199	201	203
	•••	11600	12153	17422	2160	2305	2922	644	665	778	322	328	362	199	202	211
0.25	***	13229	13553	17424	2433	2579	2929	726	752	790	359	367	377	220	223	227
	•••	13338	14131	20206	2494	2707	3342	738	770	873	363	372	400	221	225	229
0.3	***	14313	14757	20081	2659	2851	3331	794	830	882	390	401	413	236	240	244
	•••	14462	15537	22465	2739	3024	3678	811	853	946	395	407	428	238	242	242
0.35	***	14826	15412	22218	2805	3042	3648	843	886	951	411	424	440	246	250	257
	•••	15025	16417	24199	2904	3255	3930	862	915	999	418	431	446	248	254	250
0.4	***	14836	15591	23808	2879	3157	3876	871	922	996	423	437	454	250	256	262
	•••	15094	16835	25407	2994	3408	4098	894	955	1030	429	445	456	254	259	253
0.45	***	14426	15373	24838	2890	3203	4014	882	937	1020	426	441	460	250	256	263
	•••	14754	16857	26090	3021	3485	4182	907	974	1041	433	450	456	254	259	250
0.5	***	13686	14841	25298	2846	3186	4061	874	933	1020	420	435	456	245	250	257
	•••	14094	16551	26248	2989	3490	4182	901	972	1030	427	445	446	248	254	242
0.55	***	12711	14074	25187	2756	3112	4016	850	909	998	405	420	441	233	240	246
	•••	13204	15971	25880	2906	3427	4098	877	949	999	413	430	428	237	243	229
0.6	***	11589	13136	24505	2623	2985	3879	809	868	952	382	397	415	219	223	229
	•••	12163	15165	24987	2775	3299	3930	837	905	946	390	406	400	221	226	211
0.65	***	10399	12076	23253	2452	2805	3652	754	807	884	350	364	380	197	202	207
	•••	11036	14164	23569	2602	3108	3678	779	843	873	357	372	362	199	204	187
0.7	***	9192	10922	21436	2244	2574	3334	682	729	794	311	322	335	171	174	178
	•••	9861	12986	21625	2385	2851	3342	703	758	778	316	328	316	173	176	158
0.75	***	7989	9683	19057	1997	2290	2927	593	630	681	262	271	280	140	142	144
	•••	8652	11629	19156	2123	2528	2922	610	653	662	267	275	260	141	143	124
0.8	***	6781	8340	16122	1706	1946	2431	484	511	546	204	209	214	.	.	.
	•••	7399	10072	16161	1810	2135	2417	497	527	526	207	212	194	.	.	.
0.85	***	5529	6851	12638	1360	1532	1848	355	371	389	.	.	.	.	.	.
	•••	6059	8263	12641	1436	1661	1829	362	379	368	.	.	.	.	.	.
0.9	***	4147	5115	8610	935	1030	1177	.	.	.	.	.	.	.	.	.
	•••	4541	6087	8596	977	1093	1156	.	.	.	.	.	.	.	.	.
0.95	***	2409	2873	4045	.	.	.	.	.	.	.	.	.	.	.	.
	•••	2605	3278	4025	.	.	.	.	.	.	.	.	.	.	.	.

TABLE 6: ALPHA= 0.025 POWER= 0.9   EXPECTED ACCRUAL THRU MINIMUM FOLLOW-UP= 1800

		DEL=.02			DEL=.05			DEL=.10			DEL=.15			DEL=.20		
FACT=		1.0 .75	.50 .25	.00 BIN	1.0 .75	.50 .25	.00 BIN	1.0 .75	.50 .25	.00 BIN	1.0 75	.50 .25	.00 BIN	1.0 .75	.50 .25	.00 BIN
PCONT=***						REQUIRED	NUMBER OF	PATIENTS								
0.05	***	3016 3048 3165	654 661 672	236 237 238	138 138 139	98 98 98
	•••	3027 3084 5916	657 666 1156	236 238 368	138 139 194	98 98 124
0.1	***	6283 6369 6954	1217 1245 1295	388 393 399	208 210 213	137 138 139
	•••	6312 6502 10277	1230 1266 1829	391 397 526	209 211 260	138 138 158
0.15	***	9197 9351 10755	1716 1776 1898	523 534 548	269 272 276	171 172 174
	•••	9249 9612 14112	1741 1824 2417	528 540 662	271 274 316	172 173 187
0.2	***	11538 11776 14291	2126 2227 2448	636 654 679	319 325 332	199 201 204
	•••	11618 12194 17422	2169 2312 2922	645 666 778	323 328 362	200 202 211
0.25	***	13248 13591 17424	2445 2591 2929	728 755 791	360 368 378	220 224 227
	•••	13364 14190 20206	2505 2717 3342	740 771 873	364 372 400	222 225 229
0.3	***	14340 14807 20081	2675 2866 3331	798 832 882	391 402 414	236 240 245
	•••	14498 15616 22465	2755 3036 3678	813 855 946	396 407 428	238 242 242
0.35	***	14861 15479 22218	2825 3061 3648	847 890 951	413 424 440	246 251 256
	•••	15070 16518 24199	2922 3271 3930	866 917 999	418 432 446	249 254 250
0.4	***	14881 15675 23808	2901 3179 3876	876 924 996	424 438 456	252 256 263
	•••	15153 16957 25407	3017 3426 4098	899 957 1030	431 447 456	254 260 253
0.45	***	14484 15475 24837	2916 3228 4014	886 940 1020	427 442 460	251 256 263
	•••	14829 17000 26090	3046 3505 4182	911 976 1041	434 451 456	254 260 250
0.5	***	13760 14964 25299	2874 3214 4061	879 937 1020	422 436 456	245 251 258
	•••	14187 16710 26248	3017 3512 4182	906 974 1030	429 445 456	249 254 242
0.55	***	12800 14213 25188	2785 3141 4016	856 913 998	407 422 441	235 240 246
	•••	13314 16146 25880	2935 3450 4098	883 951 999	414 431 428	237 243 229
0.6	***	11695 13288 24504	2654 3012 3879	816 872 953	384 398 415	219 224 229
	•••	12289 15347 24987	2805 3322 3930	841 908 946	390 406 400	222 226 211
0.65	***	10518 12235 23253	2481 2832 3651	759 811 884	352 366 380	198 202 207
	•••	11172 14348 23569	2631 3129 3678	783 843 873	359 372 362	200 204 187
0.7	***	9319 11082 21435	2272 2600 3334	686 731 794	312 323 335	172 175 179
	•••	10002 13164 21625	2413 2871 3342	708 759 778	317 328 316	173 177 158
0.75	***	8115 9834 19056	2022 2312 2927	596 632 681	263 271 280	141 143 145
	•••	8790 11796 19156	2148 2544 2922	613 654 662	267 276 260	141 144 124
0.8	***	6900 8479 16122	1727 1964 2431	487 513 546	204 209 215	. . .
	•••	7527 10218 16161	1830 2148 2417	499 528 526	207 213 194	. . .
0.85	***	5631 6966 12638	1374 1545 1847	357 371 389	. . .	. . .
	•••	6168 8379 12641	1450 1669 1829	363 380 368	. . .	. . .
0.9	***	4223 5196 8610	944 1036 1176	. . .	. . .	. . .
	•••	4621 6164 8596	985 1097 1156	. . .	. . .	. . .
0.95	***	2448 2909 4045	. . .	. . .	. . .	. . .
	•••	2643 3307 4025	. . .	. . .	. . .	. . .

TABLE 6: ALPHA= 0.025 POWER= 0.9   EXPECTED ACCRUAL THRU MINIMUM FOLLOW-UP= 1900

	DEL=.02			DEL=.05			DEL=.10			DEL=.15			DEL=.20		
FACT=	1.0 .75	.50 .25	.00 BIN	1.0 .75	.50 .25	.00 BIN	1.0 .75	.50 .25	.00 BIN	1.0 75	.50 .25	.00 BIN	1.0 .75	.50 .25	.00 BIN
PCONT=***						REQUIRED NUMBER OF PATIENTS									
0.05 ***	3018	3051	3165	654	661	671	236	237	239	137	139	139	97	97	98
***	3030	3087	5916	657	666	1156	236	237	368	139	139	194	97	97	124
0.1 ***	6289	6378	6954	1219	1246	1294	389	394	400	209	210	212	137	139	139
***	6319	6513	10277	1231	1267	1829	391	396	526	209	211	260	137	139	158
0.15 ***	9205	9367	10755	1721	1780	1897	524	534	547	269	273	277	171	172	174
***	9260	9636	14112	1745	1827	2417	529	540	662	270	274	316	172	173	187
0.2 ***	11552	11802	14290	2134	2234	2448	638	655	679	320	325	331	198	201	203
***	11636	12233	17422	2177	2318	2922	646	666	778	323	329	362	199	201	211
0.25 ***	13268	13627	17424	2455	2602	2929	730	756	790	361	368	377	220	223	227
***	13389	14247	20206	2517	2725	3342	742	771	873	364	372	400	222	225	229
0.3 ***	14366	14858	20082	2690	2880	3331	801	835	882	391	401	413	236	241	244
***	14532	15694	22465	2770	3047	3678	817	856	946	396	407	428	239	242	242
0.35 ***	14896	15544	22218	2842	3079	3648	850	892	950	413	425	439	247	251	256
***	15116	16615	24199	2941	3286	3930	869	918	999	419	432	446	249	254	250
0.4 ***	14928	15757	23808	2923	3200	3876	881	928	996	426	439	455	251	258	262
***	15213	17075	25407	3040	3443	4098	902	959	1030	432	446	456	254	260	253
0.45 ***	14543	15576	24837	2940	3251	4013	892	944	1020	429	443	460	251	258	263
***	14904	17137	26090	3070	3524	4182	915	978	1041	436	451	456	254	260	250
0.5 ***	13832	15081	25298	2901	3239	4061	885	940	1020	422	438	456	246	251	258
***	14278	16864	26248	3043	3533	4182	911	976	1030	429	446	446	249	254	242
0.55 ***	12889	14347	25188	2813	3167	4016	861	918	997	408	422	440	235	241	247
***	13422	16311	25880	2963	3471	4098	887	953	999	415	431	428	237	243	229
0.6 ***	11800	13434	24505	2682	3040	3879	820	875	952	386	399	415	220	224	229
***	12411	15522	24987	2834	3343	3930	847	909	946	391	407	400	222	227	211
0.65 ***	10635	12389	23253	2510	2858	3651	763	814	885	353	365	380	198	201	206
***	11305	14524	23569	2659	3149	3678	787	847	873	360	372	362	201	204	187
0.7 ***	9442	11236	21435	2299	2623	3334	691	733	793	313	323	334	172	175	178
***	10138	13334	21625	2438	2889	3342	711	761	778	318	329	316	173	177	158
0.75 ***	8238	9982	19057	2046	2332	2927	600	635	680	263	272	280	140	142	144
***	8924	11953	19156	2170	2560	2922	616	655	662	267	275	260	141	144	124
0.8 ***	7014	8610	16123	1747	1980	2431	489	514	546	205	210	215	.	.	.
***	7649	10355	16161	1849	2158	2417	501	529	526	208	212	194	.	.	.
0.85 ***	5730	7075	12638	1389	1557	1847	357	372	389	.	.	.	.	.	.
***	6272	8488	12641	1464	1676	1829	364	380	368	.	.	.	.	.	.
0.9 ***	4297	5274	8610	952	1041	1177	.	.	.	.	.	.	.	.	.
***	4696	6235	8596	992	1101	1156	.	.	.	.	.	.	.	.	.
0.95 ***	2485	2942	4046	.	.	.	.	.	.	.	.	.	.	.	.
***	2680	3334	4025	.	.	.	.	.	.	.	.	.	.	.	.

TABLE 6: ALPHA= 0.025 POWER= 0.9   EXPECTED ACCRUAL THRU MINIMUM FOLLOW-UP= 2000

		DEL=.02			DEL=.05			DEL=.10			DEL=.15			DEL=.20		
FACT=		1.0 .75	.50 .25	.00 BIN	1.0 .75	.50 .25	.00 BIN	1.0 .75	.50 .25	.00 BIN	1.0 75	.50 .25	.00 BIN	1.0 .75	.50 .25	.00 BIN
PCONT=•••					REQUIRED NUMBER OF PATIENTS											
0.05	***	3020	3052	3165	655	662	671	236	237	239	139	139	139	97	97	97
	•••	3032	3090	5916	659	666	1156	236	237	368	139	139	194	97	97	124
0.1	***	6294	6386	6955	1221	1249	1295	390	394	400	209	210	212	137	139	139
	•••	6326	6525	10277	1234	1269	1829	392	397	526	210	211	260	139	139	158
0.15	***	9214	9384	10755	1725	1785	1897	525	535	547	270	272	276	171	172	174
	•••	9272	9660	14112	1751	1831	2417	530	541	662	271	275	316	172	174	187
0.2	***	11565	11827	14290	2141	2241	2449	640	657	679	321	326	331	199	201	204
	•••	11654	12272	17422	2184	2322	2922	647	667	778	324	329	362	200	202	211
0.25	***	13286	13664	17424	2467	2611	2930	732	757	791	361	369	377	221	224	227
	•••	13415	14304	20206	2527	2732	3342	744	772	873	365	374	400	222	225	229
0.3	***	14392	14907	20081	2704	2895	3331	804	836	881	394	402	414	237	241	245
	•••	14567	15767	22465	2785	3057	3678	819	857	946	397	407	428	239	242	242
0.35	***	14931	15609	22217	2860	3096	3649	854	895	950	415	426	440	247	252	256
	•••	15164	16710	24199	2960	3299	3930	872	920	999	420	432	446	250	254	250
0.4	***	14974	15839	23809	2944	3220	3876	885	931	996	427	440	455	252	257	262
	•••	15272	17189	25407	3060	3459	4098	906	961	1030	434	447	456	255	260	253
0.45	***	14601	15675	24837	2964	3274	4014	896	947	1020	430	444	461	252	257	264
	•••	14979	17269	26090	3095	3541	4182	920	980	1041	436	452	456	255	260	250
0.5	***	13905	15197	25299	2926	3262	4061	890	944	1020	424	439	456	246	252	257
	•••	14370	17012	26248	3069	3551	4182	915	979	1030	431	446	446	249	255	242
0.55	***	12976	14477	25187	2840	3192	4016	866	921	997	410	424	441	236	241	246
	•••	13529	16471	25880	2990	3491	4098	891	956	999	416	432	428	239	244	229
0.6	***	11901	13576	24505	2710	3065	3880	826	879	952	386	400	416	220	225	230
	•••	12531	15687	24987	2861	3362	3930	850	912	946	392	407	400	222	227	211
0.65	***	10749	12536	23254	2537	2882	3652	769	817	885	355	366	380	199	202	206
	•••	11434	14691	23569	2685	3167	3678	791	847	873	360	374	362	201	205	187
0.7	***	9561	11382	21436	2324	2645	3334	695	737	794	315	324	335	172	175	179
	•••	10270	13496	21625	2462	2905	3342	715	762	778	319	330	316	174	177	158
0.75	***	8356	10121	19057	2069	2351	2927	602	636	681	265	272	280	141	142	145
	•••	9052	12102	19156	2191	2574	2922	619	657	662	269	276	260	141	144	124
0.8	***	7125	8736	16122	1766	1995	2431	491	516	546	206	210	215	.	.	.
	•••	7767	10485	16161	1866	2170	2417	504	530	526	207	212	194	.	.	.
0.85	***	5825	7180	12639	1404	1566	1847	359	372	389	.	.	.	.	.	.
	•••	6371	8590	12641	1476	1684	1829	366	381	368	.	.	.	.	.	.
0.9	***	4367	5346	8610	960	1047	1177	.	.	.	.	.	.	.	.	.
	•••	4770	6302	8596	1000	1104	1156	.	.	.	.	.	.	.	.	.
0.95	***	2520	2975	4045	.	.	.	.	.	.	.	.	.	.	.	.
	•••	2714	3360	4025	.	.	.	.	.	.	.	.	.	.	.	.

TABLE 6: ALPHA= 0.025 POWER= 0.9 EXPECTED ACCRUAL THRU MINIMUM FOLLOW-UP= 2250

		DEL=.02			DEL=.05			DEL=.10			DEL=.15			DEL=.20		
FACT=		1.0 .75	.50 .25	.00 BIN	1.0 .75	.50 .25	.00 BIN	1.0 .75	.50 .25	.00 BIN	1.0 75	.50 .25	.00 BIN	1.0 .75	.50 .25	.00 BIN
PCONT=***					REQUIRED NUMBER OF PATIENTS											
0.05	***	3025	3059	3165	657	663	672	236	237	239	139	139	139	97	98	98
	***	3039	3096	5916	659	666	1156	237	237	368	139	139	194	98	98	124
0.1	***	6305	6408	6954	1226	1253	1295	390	395	399	209	210	212	137	139	139
	***	6342	6550	10277	1239	1271	1829	392	398	526	209	210	260	137	139	158
0.15	***	9236	9423	10754	1735	1794	1898	527	537	548	269	272	277	171	173	174
	***	9300	9714	14112	1760	1838	2417	531	541	662	271	274	316	173	173	187
0.2	***	11598	11891	14290	2158	2257	2448	644	659	679	322	326	331	199	201	204
	***	11698	12362	17422	2201	2334	2922	651	668	778	325	329	362	201	202	211
0.25	***	13335	13754	17425	2491	2634	2929	738	761	792	362	370	378	222	223	226
	***	13477	14435	20206	2552	2750	3342	748	775	873	367	374	400	223	224	229
0.3	***	14459	15029	20081	2736	2924	3331	809	840	882	395	403	413	237	241	244
	***	14654	15944	22465	2817	3081	3678	825	859	946	399	407	428	239	243	242
0.35	***	15018	15766	22217	2901	3133	3647	862	899	950	417	427	440	249	253	257
	***	15277	16931	24199	2999	3328	3930	880	922	999	421	433	446	250	254	250
0.4	***	15086	16035	23807	2991	3264	3877	894	938	997	430	441	455	253	258	263
	***	15419	17456	25407	3107	3493	4098	914	964	1030	435	448	456	255	260	253
0.45	***	14744	15913	24837	3017	3324	4015	907	955	1019	433	446	461	253	258	263
	***	15161	17578	26090	3147	3580	4182	928	984	1041	438	452	456	255	260	250
0.5	***	14082	15474	25298	2985	3317	4061	899	950	1020	427	440	455	247	253	257
	***	14590	17355	26248	3126	3593	4182	924	983	1030	433	447	446	250	254	242
0.55	***	13190	14787	25187	2902	3248	4016	877	928	998	413	426	441	237	241	246
	***	13784	16841	25880	3050	3534	4098	901	960	999	419	433	428	239	244	229
0.6	***	12147	13909	24505	2772	3120	3880	837	885	952	389	402	416	221	224	229
	***	12816	16072	24987	2922	3405	3930	859	916	946	395	407	400	223	227	211
0.65	***	11018	12882	23253	2598	2936	3652	779	824	884	357	368	381	199	202	207
	***	11738	15077	23569	2743	3207	3678	800	852	873	362	374	362	201	205	187
0.7	***	9842	11727	21435	2382	2694	3334	703	742	794	316	325	334	173	176	178
	***	10580	13870	21625	2517	2941	3342	721	766	778	320	330	316	174	177	158
0.75	***	8634	10448	19057	2120	2395	2926	610	641	680	266	272	280	140	143	145
	***	9353	12447	19156	2240	2606	2922	624	659	662	269	277	260	142	143	124
0.8	***	7382	9028	16122	1808	2029	2431	496	519	545	207	210	215	.	.	.
	***	8042	10783	16161	1905	2193	2417	507	531	526	208	212	194	.	.	.
0.85	***	6045	7419	12638	1434	1590	1847	362	374	389	.	.	.	.	.	.
	***	6602	8824	12641	1504	1698	1829	368	381	368	.	.	.	.	.	.
0.9	***	4530	5514	8610	977	1059	1177	.	.	.	.	.	.	.	.	.
	***	4935	6454	8596	1015	1110	1156	.	.	.	.	.	.	.	.	.
0.95	***	2600	3047	4046	.	.	.	.	.	.	.	.	.	.	.	.
	***	2791	3416	4025	.	.	.	.	.	.	.	.	.	.	.	.

TABLE 6: ALPHA= 0.025 POWER= 0.9   EXPECTED ACCRUAL THRU MINIMUM FOLLOW-UP= 2500

		DEL=.02			DEL=.05			DEL=.10			DEL=.15			DEL=.20		
FACT=		1.0 .75	.50 .25	.00 BIN	1.0 .75	.50 .25	.00 BIN	1.0 .75	.50 .25	.00 BIN	1.0 75	.50 .25	.00 BIN	1.0 .75	.50 .25	.00 BIN
PCONT=***					REQUIRED NUMBER OF PATIENTS											
0.05	***	3029	3065	3165	658	664	671	237	237	239	139	139	139	98	98	98
	•••	3043	3101	5916	661	667	1156	237	239	368	139	139	194	98	98	124
0.1	***	6317	6427	6954	1231	1256	1295	392	395	399	209	211	212	137	139	139
	•••	6358	6573	10277	1242	1273	1829	393	398	526	211	211	260	139	139	158
0.15	***	9258	9462	10754	1745	1801	1898	529	537	548	271	273	276	171	173	174
	•••	9327	9765	14112	1770	1842	2417	533	542	662	271	274	316	171	173	187
0.2	***	11631	11951	14290	2174	2270	2448	646	661	679	323	326	331	199	201	202
	•••	11742	12446	17422	2217	2343	2922	652	670	778	324	329	362	201	202	211
0.25	***	13383	13842	17424	2514	2654	2929	742	764	790	364	370	377	221	224	226
	•••	13540	14556	20206	2574	2764	3342	751	776	873	367	373	400	223	226	229
0.3	***	14524	15146	20081	2767	2951	3331	815	845	881	396	404	414	239	242	245
	•••	14740	16106	22465	2846	3101	3678	829	862	946	399	409	428	240	243	242
0.35	***	15106	15918	22218	2937	3167	3648	868	904	949	418	427	439	249	252	256
	•••	15390	17136	24199	3036	3352	3930	886	924	999	423	434	446	251	254	250
0.4	***	15198	16224	23808	3034	3302	3876	901	942	996	433	442	454	254	259	262
	•••	15562	17699	25407	3149	3521	4098	920	967	1030	437	448	456	256	261	253
0.45	***	14886	16139	24837	3065	3367	4014	915	959	1020	436	446	461	254	259	264
	•••	15339	17859	26090	3195	3614	4182	936	987	1041	440	452	456	256	261	250
0.5	***	14256	15734	25298	3037	3364	4061	909	958	1020	429	442	456	248	252	258
	•••	14799	17668	26248	3177	3627	4182	931	986	1030	436	448	446	251	256	242
0.55	***	13395	15076	25187	2956	3296	4015	886	934	998	415	426	440	237	242	246
	•••	14027	17176	25880	3102	3570	4098	909	964	999	421	434	428	240	243	229
0.6	***	12381	14218	24504	2827	3168	3879	845	892	952	392	402	415	221	224	229
	•••	13084	16418	24987	2974	3440	3930	867	920	946	396	409	400	223	227	211
0.65	***	11271	13199	23252	2652	2983	3651	786	829	884	359	368	381	199	202	206
	•••	12021	15424	23569	2795	3242	3678	806	854	873	364	374	362	201	204	187
0.7	***	10104	12040	21436	2433	2737	3334	711	746	793	318	326	336	173	176	177
	•••	10868	14202	21625	2565	2973	3342	727	768	778	321	331	316	174	177	158
0.75	***	8892	10745	19058	2165	2431	2926	615	645	681	267	273	279	142	143	145
	•••	9629	12752	19156	2283	2631	2922	629	662	662	270	276	260	142	143	124
0.8	***	7620	9293	16121	1845	2058	2431	501	521	545	208	211	215	.	.	.
	•••	8292	11046	16161	1940	2212	2417	511	533	526	209	214	194	.	.	.
0.85	***	6246	7636	12639	1461	1609	1846	364	376	389	.	.	.	.	.	.
	•••	6812	9029	12641	1527	1712	1829	370	383	368	.	.	.	.	.	.
0.9	***	4677	5662	8611	992	1068	1176	.	.	.	.	.	.	.	.	.
	•••	5087	6586	8596	1026	1117	1156	.	.	.	.	.	.	.	.	.
0.95	***	2671	3109	4045	.	.	.	.	.	.	.	.	.	.	.	.
	•••	2861	3462	4025	.	.	.	.	.	.	.	.	.	.	.	.

TABLE 6: ALPHA= 0.025 POWER= 0.9   EXPECTED ACCRUAL THRU MINIMUM FOLLOW-UP= 2750

		DEL=.02			DEL=.05			DEL=.10			DEL=.15			DEL=.20		
FACT=		1.0 .75	.50 .25	.00 BIN	1.0 .75	.50 .25	.00 BIN	1.0 .75	.50 .25	.00 BIN	1.0 75	.50 .25	.00 BIN	1.0 .75	.50 .25	.00 BIN
PCONT=***				REQUIRED	NUMBER	OF	PATIENTS									
0.05	***	3033	3069	3165	658	665	672	236	236	239	139	139	139	98	98	98
	•••	3048	3105	5916	661	668	1156	236	239	368	139	139	194	98	98	124
0.1	***	6330	6447	6953	1235	1259	1295	391	395	400	209	211	212	139	139	139
	•••	6371	6594	10277	1246	1275	1829	393	398	526	211	211	260	139	139	158
0.15	***	9279	9499	10755	1752	1807	1897	531	538	548	271	273	276	171	173	173
	•••	9356	9810	14112	1776	1847	2417	534	542	662	273	274	316	173	173	187
0.2	***	11665	12010	14291	2189	2282	2449	649	663	679	322	328	331	201	202	204
	•••	11785	12522	17422	2229	2350	2922	654	670	778	324	329	362	201	202	211
0.25	***	13430	13927	17424	2535	2670	2930	745	766	790	366	370	377	223	225	226
	•••	13604	14669	20206	2593	2777	3342	754	778	873	367	374	400	223	226	229
0.3	***	14590	15260	20081	2793	2975	3330	821	847	881	398	405	414	239	242	245
	•••	14824	16255	22465	2872	3117	3678	833	864	946	401	408	428	240	243	242
0.35	***	15192	16065	22218	2969	3195	3649	874	907	950	421	429	439	250	252	257
	•••	15501	17324	24199	3068	3374	3930	890	927	999	424	434	446	250	254	250
0.4	***	15310	16405	23807	3072	3337	3877	909	947	996	434	443	455	256	259	263
	•••	15702	17926	25407	3186	3547	4098	926	969	1030	438	450	456	257	260	253
0.45	***	15025	16351	24837	3109	3404	4014	923	964	1020	438	448	460	256	259	263
	•••	15509	18118	26090	3236	3642	4182	941	989	1041	443	453	456	257	260	250
0.5	***	14425	15977	25298	3085	3404	4061	917	962	1020	431	443	455	249	254	257
	•••	15003	17955	26248	3222	3659	4182	938	989	1030	438	448	446	250	256	242
0.55	***	13593	15344	25188	3006	3337	4016	893	938	998	417	428	441	239	242	247
	•••	14258	17481	25880	3150	3602	4098	916	965	999	422	434	428	240	243	229
0.6	***	12605	14505	24505	2876	3210	3879	852	896	951	393	404	415	223	226	230
	•••	13337	16735	24987	3021	3471	3930	872	923	946	398	408	400	225	228	211
0.65	***	11511	13494	23253	2700	3023	3652	793	833	885	360	370	379	201	204	205
	•••	12287	15738	23569	2841	3270	3678	813	857	873	366	374	362	202	205	187
0.7	***	10350	12330	21435	2476	2773	3334	716	751	793	319	326	335	173	177	178
	•••	11133	14504	21625	2607	2999	3342	731	771	778	322	331	316	175	177	158
0.75	***	9129	11016	19057	2205	2463	2927	620	648	680	267	274	280	142	144	144
	•••	9884	13027	19156	2318	2653	2922	634	663	662	271	276	260	142	144	124
0.8	***	7839	9534	16121	1876	2082	2432	505	524	546	208	211	214	.	.	.
	•••	8523	11283	16161	1969	2229	2417	514	534	526	209	212	194	.	.	.
0.85	***	6431	7830	12638	1483	1625	1847	366	377	390	.	.	.	.	.	.
	•••	7004	9212	12641	1548	1721	1829	370	383	368	.	.	.	.	.	.
0.9	***	4812	5795	8610	1003	1077	1177	.	.	.	.	.	.	.	.	.
	•••	5223	6702	8596	1037	1122	1156	.	.	.	.	.	.	.	.	.
0.95	***	2734	3164	4045	.	.	.	.	.	.	.	.	.	.	.	.
	•••	2921	3502	4025	.	.	.	.	.	.	.	.	.	.	.	.

TABLE 6: ALPHA= 0.025 POWER= 0.9   EXPECTED ACCRUAL THRU MINIMUM FOLLOW-UP= 3000

PCONT=***	DEL=.01 1.0/.75	DEL=.01 .50/.25	DEL=.01 .00 BIN	DEL=.02 1.0/.75	DEL=.02 .50/.25	DEL=.02 .00 BIN	DEL=.05 1.0/.75	DEL=.05 .50/.25	DEL=.05 .00 BIN	DEL=.10 1.0/75	DEL=.10 .50/.25	DEL=.10 .00 BIN	DEL=.15 1.0/.75	DEL=.15 .50/.25	DEL=.15 .00 BIN
						REQUIRED NUMBER OF PATIENTS									
0.01 ***	1607	1610	1622	590	590	595	193	197	197	100	100	100	70	70	70
•••	1607	1615	6200	590	595	2049	193	197	558	100	100	227	70	70	135
0.02 ***	3550	3565	3617	1142	1145	1157	310	310	310	137	137	137	88	88	88
•••	3553	3580	10235	1142	1150	3048	310	310	712	137	137	264	88	88	151
0.05 ***	10735	10790	11297	3040	3073	3163	658	665	670	238	238	238	137	137	137
•••	10753	10888	21835	3050	3107	5916	662	665	1156	238	238	368	137	137	194
0.1 ***	23675	23815	26293	6340	6463	6955	1240	1262	1295	392	395	400	208	212	212
•••	23720	24088	39487	6385	6613	10277	1247	1277	1829	392	395	526	208	212	260
0.15 ***	35353	35600	41533	9298	9535	10753	1760	1813	1895	530	538	545	272	272	275
•••	35435	36095	55038	9385	9853	14112	1783	1850	2417	535	542	662	272	275	316
0.2 ***	44758	45145	55802	11698	12065	14290	2200	2293	2447	650	662	677	325	328	332
•••	44885	45917	68488	11825	12595	17422	2240	2357	2922	658	670	778	325	328	362
0.25 ***	51605	52165	68507	13475	14008	17425	2552	2687	2930	748	767	790	365	370	377
•••	51790	53278	79836	13663	14773	20206	2612	2788	3342	755	778	873	370	373	400
0.3 ***	55910	56683	79330	14653	15370	20080	2818	2995	3332	823	850	880	400	407	415
•••	56170	58213	89082	14908	16393	22465	2893	3130	3678	835	865	946	403	410	428
0.35 ***	57853	58880	88085	15275	16205	22217	2998	3220	3647	880	910	950	422	430	440
•••	58195	60913	96227	15610	17500	24199	3095	3392	3930	895	928	999	425	433	446
0.4 ***	57680	59023	94667	15418	16577	23807	3107	3365	3875	913	950	995	433	445	455
•••	58127	61660	101271	15838	18133	25407	3220	3568	4098	932	973	1030	440	448	456
0.45 ***	55700	57433	99010	15160	16555	24838	3148	3437	4015	928	970	1018	437	448	460
•••	56278	60770	104213	15673	18358	26090	3272	3665	4182	947	992	1041	445	455	456
0.5 ***	52262	54470	101080	14590	16210	25300	3125	3440	4060	925	965	1018	433	445	455
•••	53000	58595	105054	15197	18220	26248	3260	3685	4182	943	992	1030	437	448	446
0.55 ***	47720	50507	100865	13783	15598	25187	3050	3377	4015	902	943	995	418	430	440
•••	48662	55457	103793	14477	17762	25880	3193	3628	4098	920	970	999	422	433	428
0.6 ***	42470	45928	98365	12815	14773	24505	2920	3245	3880	857	898	950	395	403	415
•••	43655	51643	100430	13577	17023	24987	3065	3497	3930	880	925	946	400	410	400
0.65 ***	36913	41068	93583	11740	13765	23252	2743	3058	3650	800	838	883	362	370	380
•••	38380	47372	94967	12535	16025	23569	2882	3295	3678	815	857	873	365	373	362
0.7 ***	31430	36163	86537	10580	12598	21437	2518	2807	3332	722	752	793	320	328	335
•••	33155	42770	87401	11383	14777	21625	2645	3020	3342	737	770	778	325	332	316
0.75 ***	26305	31333	77245	9355	11267	19055	2240	2488	2927	625	650	680	268	275	280
•••	28195	37892	77734	10120	13277	19156	2350	2672	2922	635	665	662	272	275	260
0.8 ***	21625	26563	65728	8042	9752	16123	1907	2105	2432	508	523	545	208	212	215
•••	23518	32687	65966	8735	11495	16161	1993	2245	2417	515	535	526	208	212	194
0.85 ***	17275	21715	52010	6602	8008	12640	1502	1640	1847	370	377	388	.	.	.
•••	19003	26995	52096	7180	9377	12641	1565	1730	1829	373	385	368	.	.	.
0.9 ***	12958	16468	36115	4937	5915	8608	1015	1082	1175	.	.	.	.	.	.
•••	14345	20450	36125	5345	6805	8596	1048	1127	1156	.	.	.	.	.	.
0.95 ***	8020	10052	18070	2792	3212	4045	.	.	.	.	.	.	.	.	.
•••	8840	12152	18052	2975	3538	4025	.	.	.	.	.	.	.	.	.
0.98 ***	3823	4532	6220	.	.	.	.	.	.	.	.	.	.	.	.
•••	4123	5135	6200	.	.	.	.	.	.	.	.	.	.	.	.

TABLE 6: ALPHA= 0.025 POWER= 0.9   EXPECTED ACCRUAL THRU MINIMUM FOLLOW-UP= 3250

		DEL=.01			DEL=.02			DEL=.05			DEL=.10			DEL=.15		
FACT=		1.0 .75	.50 .25	.00 BIN	1.0 .75	.50 .25	.00 BIN	1.0 .75	.50 .25	.00 BIN	1.0 75	.50 .25	.00 BIN	1.0 .75	.50 .25	.00 BIN
PCONT=***				REQUIRED NUMBER OF PATIENTS												
0.01	***	1606	1611	1619	591	591	596	192	198	198	100	100	100	71	71	71
	•••	1606	1614	6200	591	591	2049	192	198	558	100	100	227	71	71	135
0.02	***	3548	3564	3618	1140	1148	1156	311	311	311	136	136	136	87	87	87
	•••	3556	3581	10235	1143	1151	3048	311	311	712	136	136	264	87	87	151
0.05	***	10739	10801	11296	3041	3077	3166	661	664	672	238	238	238	136	141	141
	•••	10760	10901	21835	3058	3109	5916	664	669	1156	238	238	368	141	141	194
0.1	***	23687	23836	26290	6351	6481	6952	1241	1262	1294	393	396	401	209	209	214
	•••	23739	24134	39487	6400	6627	10277	1254	1278	1829	396	396	526	209	209	260
0.15	***	35374	35642	41532	9322	9566	10755	1766	1817	1899	531	539	547	271	274	274
	•••	35463	36175	55038	9411	9891	14112	1790	1855	2417	534	542	662	274	274	316
0.2	***	44788	45210	55800	11730	12120	14289	2213	2302	2448	653	664	680	322	328	331
	•••	44929	46042	68488	11868	12656	17422	2253	2362	2922	656	672	778	328	331	362
0.25	***	51653	52257	68507	13526	14086	17423	2570	2700	2931	750	769	791	368	371	376
	•••	51851	53460	79836	13726	14866	20206	2627	2798	3342	758	777	873	368	376	400
0.3	***	55976	56813	79330	14717	15476	20079	2838	3012	3329	826	851	880	401	404	412
	•••	56255	58465	89082	14988	16524	22465	2914	3142	3678	839	867	946	404	409	428
0.35	***	57937	59050	88086	15362	16337	22216	3025	3244	3646	883	913	948	425	428	441
	•••	58311	61244	96227	15716	17661	24199	3123	3407	3930	896	932	999	425	433	446
0.4	***	57791	59245	94667	15524	16740	23809	3139	3394	3809	921	953	997	436	444	452
	•••	58278	62081	101271	15971	18327	25407	3247	3589	4098	937	972	1030	441	449	456
0.45	***	55846	57723	99009	15294	16748	24836	3182	3467	4011	932	972	1018	441	449	461
	•••	56471	61298	104213	15833	18576	26090	3309	3686	4182	953	994	1041	444	452	456
0.5	***	52444	54833	101081	14749	16426	25299	3163	3472	4060	929	969	1021	433	444	458
	•••	53246	59221	105054	15383	18463	26248	3301	3708	4182	948	994	1030	441	449	446
0.55	***	47956	50957	100866	13969	15833	25185	3090	3410	4016	907	945	997	420	428	441
	•••	48972	56182	103793	14684	18019	25880	3228	3651	4098	924	969	999	425	433	428
0.6	***	42767	46469	98364	13019	15021	24502	2960	3280	3878	864	904	953	396	404	417
	•••	44048	52449	100430	13802	17287	24987	3101	3521	3930	883	924	946	401	409	400
0.65	***	37288	41687	93581	11954	14018	23251	2781	3090	3651	802	839	883	363	371	379
	•••	38851	48227	94967	12770	16288	23569	2919	3318	3678	823	859	873	368	376	362
0.7	***	31877	36833	86537	10796	12843	21436	2554	2833	3334	726	758	794	322	328	336
	•••	33697	43642	87401	11616	15029	21625	2679	3041	3342	742	774	778	322	331	316
0.75	***	26802	32018	77247	9561	11499	19056	2272	2513	2928	628	653	680	271	274	279
	•••	28768	38734	77734	10341	13506	19156	2378	2687	2922	639	664	662	271	279	260
0.8	***	22127	27216	65729	8233	9956	16123	1931	2123	2432	509	526	547	209	214	214
	•••	24085	33456	65966	8935	11689	16161	2018	2256	2417	517	534	526	209	214	194
0.85	***	17739	22284	52011	6763	8171	12637	1522	1652	1847	368	379	387	.	.	.
	•••	19514	27644	52096	7342	9525	12641	1582	1741	1829	371	384	368	.	.	.
0.9	***	13336	16911	36113	5051	6026	8610	1021	1091	1176	.	.	.	.	.	.
	•••	14749	20927	36125	5457	6896	8596	1054	1132	1156	.	.	.	.	.	.
0.95	***	8244	10297	18067	2841	3253	4044	.	.	.	.	.	.	.	.	.
	•••	9078	12388	18052	3025	3569	4025	.	.	.	.	.	.	.	.	.
0.98	***	3906	4609	6221	.	.	.	.	.	.	.	.	.	.	.	.
	•••	4206	5194	6200	.	.	.	.	.	.	.	.	.	.	.	.

TABLE 6: ALPHA= 0.025 POWER= 0.9   EXPECTED ACCRUAL THRU MINIMUM FOLLOW-UP= 3500

		DEL=.01			DEL=.02			DEL=.05			DEL=.10			DEL=.15		
FACT=		1.0 .75	.50 .25	.00 BIN	1.0 .75	.50 .25	.00 BIN	1.0 .75	.50 .25	.00 BIN	1.0 75	.50 .25	.00 BIN	1.0 .75	.50 .25	.00 BIN
PCONT=	***				REQUIRED NUMBER OF PATIENTS											
0.01	***	1607	1613	1621	592	592	592	195	195	195	99	99	99	67	67	67
	•••	1607	1616	6200	592	592	2049	195	195	558	99	99	227	67	67	135
0.02	***	3550	3567	3616	1140	1149	1158	309	309	312	134	134	137	90	90	90
	•••	3558	3585	10235	1143	1152	3048	309	309	712	134	134	264	90	90	151
0.05	***	10742	10809	11293	3048	3083	3165	659	668	671	239	239	239	137	137	137
	•••	10765	10914	21835	3060	3112	5916	662	668	1156	239	239	368	137	137	194
0.1	***	23698	23864	26291	6364	6495	6953	1245	1266	1292	391	396	400	207	213	213
	•••	23753	24173	39487	6411	6644	10277	1254	1280	1829	396	396	526	213	213	260
0.15	***	35391	35685	41533	9342	9601	10756	1773	1823	1896	531	540	548	274	274	277
	•••	35487	36257	55038	9435	9925	14112	1796	1858	2417	536	545	662	274	274	316
0.2	***	44820	45275	55801	11763	12174	14291	2225	2307	2447	653	668	680	326	326	330
	•••	44972	46168	68488	11911	12720	17422	2260	2368	2922	659	671	778	326	330	362
0.25	***	51698	52348	68506	13574	14160	17424	2584	2710	2928	755	773	790	365	373	379
	•••	51916	53640	79836	13784	14960	20206	2640	2803	3342	758	781	873	370	373	400
0.3	***	56041	56939	79330	14781	15578	20078	2858	3030	3331	828	855	881	400	405	414
	•••	56338	58710	89082	15070	16645	22465	2934	3156	3678	843	869	946	405	408	428
0.35	***	58024	59223	88089	15446	16470	22219	3051	3261	3646	886	916	951	423	431	440
	•••	58421	61571	96227	15817	17812	24199	3144	3418	3930	898	930	999	426	435	446
0.4	***	57905	59471	94669	15633	16893	23806	3170	3418	3873	921	956	995	435	443	452
	•••	58426	62495	101271	16097	18509	25407	3278	3602	4098	939	974	1030	440	449	456
0.45	***	55988	58006	99009	15423	16928	24838	3214	3494	4013	939	974	1021	440	449	461
	•••	56662	61807	104213	15989	18783	26090	3340	3704	4182	956	995	1041	443	452	456
0.5	***	52628	55192	101083	14904	16631	25299	3200	3503	4063	933	974	1021	435	443	458
	•••	53491	59830	105054	15560	18684	26248	3331	3725	4182	951	995	1030	440	449	446
0.55	***	48192	51403	100864	14143	16059	25188	3126	3436	4013	913	951	995	423	431	440
	•••	49283	56881	103793	14886	18258	25880	3266	3672	4098	930	974	999	426	435	428
0.6	***	43065	46999	98365	13210	15257	24506	2998	3310	3879	869	907	951	396	405	414
	•••	44435	53220	100430	14015	17532	24987	3135	3541	3930	886	930	946	400	408	400
0.65	***	37654	42286	93584	12156	14256	23251	2820	3118	3651	808	843	886	365	370	379
	•••	39311	49038	94967	12991	16531	23569	2951	3336	3678	825	863	873	370	373	362
0.7	***	32316	37479	86535	11001	13075	21435	2587	2858	3331	729	758	793	321	326	335
	•••	34218	44465	87401	11833	15257	21625	2710	3056	3342	741	776	778	326	330	316
0.75	***	27280	32666	77248	9759	11713	19055	2298	2535	2925	633	653	680	268	274	277
	•••	29324	39530	77734	10550	13714	19156	2409	2701	2922	641	668	662	274	277	260
0.8	***	22613	27831	65727	8411	10147	16120	1954	2141	2430	510	528	545	207	213	216
	•••	24625	34180	65966	9120	11868	16161	2041	2263	2417	519	536	526	213	213	194
0.85	***	18180	22826	52013	6910	8318	12638	1537	1665	1849	370	379	388	.	.	.
	•••	19996	28251	52096	7493	9657	12641	1595	1747	1829	373	382	368	.	.	.
0.9	***	13688	17322	36114	5156	6128	8607	1030	1096	1175	.	.	.	.	.	.
	•••	15131	21370	36125	5562	6980	8596	1061	1131	1156	.	.	.	.	.	.
0.95	***	8455	10523	18071	2890	3293	4045	.	.	.	.	.	.	.	.	.
	•••	9298	12603	18052	3068	3593	4025	.	.	.	.	.	.	.	.	.
0.98	***	3984	4675	6218	.	.	.	.	.	.	.	.	.	.	.	.
	•••	4281	5247	6200	.	.	.	.	.	.	.	.	.	.	.	.

TABLE 6: ALPHA= 0.025 POWER= 0.9   EXPECTED ACCRUAL THRU MINIMUM FOLLOW-UP= 3750

		DEL=.01			DEL=.02			DEL=.05			DEL=.10			DEL=.15		
FACT=		1.0 .75	.50 .25	.00 BIN	1.0 .75	.50 .25	.00 BIN	1.0 .75	.50 .25	.00 BIN	1.0 75	.50 .25	.00 BIN	1.0 .75	.50 .25	.00 BIN
PCONT=***						REQUIRED NUMBER OF PATIENTS										
0.01	***	1606 1609	1609 1615	1619 6200	593 593	593 593	593 2049	194 194	194 194	194 558	97 97	97 97	97 227	68 68	68 68	68 135
0.02	***	3550 3559	3569 3588	3616 10235	1141 1147	1147 1150	1156 3048	312 312	312 312	312 712	134 134	134 134	134 264	87 87	87 87	87 151
0.05	***	10747 10769	10816 10925	11294 21835	3050 3063	3087 3119	3166 5916	663 663	663 668	672 1156	237 237	237 237	237 368	138 138	138 138	138 194
0.1	***	23712 23768	23884 24213	26290 39487	6378 6428	6509 6659	6953 10277	1244 1253	1268 1278	1297 1829	391 397	397 397	400 526	209 209	209 213	213 260
0.15	***	35412 35519	35725 36334	41534 55038	9363 9462	9631 9959	10756 14112	1778 1803	1825 1859	1897 2417	531 537	541 541	547 662	275 275	275 275	275 316
0.2	***	44853 45016	45340 46291	55803 68488	11797 11950	12222 12775	14290 17422	2234 2272	2318 2375	2450 2922	653 659	668 672	678 778	325 325	325 331	331 362
0.25	***	51743 51978	52441 53819	68506 79836	13619 13840	14234 15040	17425 20206	2600 2656	2722 2809	2928 3342	756 762	772 781	790 873	368 368	372 372	378 400
0.3	***	56106 56425	57068 58956	79328 89082	14843 15147	15672 16756	20078 22465	2875 2950	3044 3166	3331 3678	831 847	856 869	878 946	400 406	406 409	415 428
0.35	***	58109 58534	59393 61891	88084 96227	15528 15916	16591 17950	22216 24199	3072 3166	3284 3434	3650 3930	893 903	916 931	950 999	425 428	434 434	438 446
0.4	***	58015 58578	59693 62903	94666 101271	15738 16225	17047 18672	23806 25407	3194 3303	3438 3622	3875 4098	925 940	959 978	997 1030	438 443	447 453	453 456
0.45	***	56134 56856	58291 62303	99012 104213	15550 16137	17103 18972	24837 26090	3247 3368	3518 3725	4015 4182	944 959	978 997	1019 1041	443 447	453 456	462 456
0.5	***	52816 53734	55550 60419	101078 105054	15053 15734	16825 18893	25297 26248	3231 3363	3528 3747	4062 4182	940 959	978 997	1019 1030	438 443	447 453	456 446
0.55	***	48425 49591	51841 57550	100863 103793	14313 15078	16272 18481	25188 25880	3162 3297	3466 3691	4015 4098	916 934	953 972	997 999	425 425	428 434	438 428
0.6	***	43362 44815	47515 53956	98365 100430	13400 14219	15481 17759	24503 24987	3031 3166	3334 3559	3878 3930	875 893	906 931	953 946	397 400	406 409	415 400
0.65	***	38018 39766	42869 49812	93584 94967	12350 13197	14481 16756	23253 23569	2853 2984	3143 3353	3650 3678	813 828	847 865	884 873	363 368	372 378	381 362
0.7	***	32740 34728	38097 45241	86538 87401	11197 12040	13291 15472	21434 21625	2618 2734	2884 3072	3334 3342	734 747	762 775	794 778	322 325	331 331	334 316
0.75	***	27743 29853	33288 40278	77243 77734	9944 10747	11913 13906	19056 19156	2328 2431	2556 2716	2928 2922	634 644	653 668	681 662	269 275	275 278	278 260
0.8	***	23075 25141	28418 34859	65731 65966	8575 9293	10319 12025	16122 16161	1975 2056	2153 2275	2431 2417	513 522	528 537	547 526	209 209	213 213	213 194
0.85	***	18603 20459	23331 28822	52009 52096	7047 7634	8459 9781	12640 12641	1553 1609	1675 1750	1844 1829	372 378	378 381	387 368	. .	. .	. .
0.9	***	14022 15490	17712 21781	36115 36125	5253 5659	6218 7053	8609 8596	1038 1066	1100 1137	1175 1156	. .	. .	. .	. .	. .	. .
0.95	***	8656 9503	10731 12803	18068 18052	2931 3109	3325 3622	4043 4025	. .	. .	. .	. .	. .	. .	. .	. .	. .
0.98	***	4056 4353	4741 5294	6218 6200	. .	. .	. .	. .	. .	. .	. .	. .	. .	. .	. .	. .

TABLE 6: ALPHA= 0.025 POWER= 0.9   EXPECTED ACCRUAL THRU MINIMUM FOLLOW-UP= 4000

		DEL=.01			DEL=.02			DEL=.05			DEL=.10			DEL=.15		
FACT=		1.0 .75	.50 .25	.00 BIN	1.0 .75	.50 .25	.00 BIN	1.0 .75	.50 .25	.00 BIN	1.0 75	.50 .25	.00 BIN	1.0 .75	.50 .25	.00 BIN
PCONT=•••					REQUIRED NUMBER OF PATIENTS											
0.01	***	1607	1613	1623	593	593	593	193	197	197	97	97	97	67	67	67
	•••	1607	1617	6200	593	593	2049	197	197	558	97	97	227	67	67	135
0.02	***	3553	3573	3617	1143	1147	1157	307	313	313	137	137	137	87	87	87
	•••	3563	3587	10235	1147	1153	3048	307	313	712	137	137	264	87	87	151
0.05	***	10753	10827	11297	3053	3087	3163	663	667	673	237	237	237	137	137	137
	•••	10777	10937	21835	3067	3117	5916	663	667	1156	237	237	368	137	137	194
0.1	***	23723	23907	26293	6387	6523	6953	1247	1267	1293	393	397	397	207	213	213
	•••	23783	24257	39487	6437	6673	10277	1257	1283	1829	397	397	526	213	213	260
0.15	***	35433	35767	41533	9383	9657	10753	1783	1833	1897	533	543	547	273	273	277
	•••	35547	36413	55038	9487	9987	14112	1807	1863	2417	537	543	662	273	273	316
0.2	***	44887	45403	55803	11827	12273	14287	2243	2323	2447	657	667	677	327	327	333
	•••	45057	46417	68488	11993	12827	17422	2277	2377	2922	663	673	778	327	327	362
0.25	***	51793	52537	68507	13663	14303	17423	2613	2733	2927	757	773	793	367	373	377
	•••	52037	53997	79836	13897	15117	20206	2667	2817	3342	763	783	873	373	373	400
0.3	***	56167	57197	79327	14907	15767	20083	2893	3057	3333	837	857	883	403	407	413
	•••	56513	59197	89082	15223	16863	22465	2967	3173	3678	847	867	946	403	407	428
0.35	***	58193	59563	88087	15607	16707	22217	3097	3297	3647	893	917	947	427	433	437
	•••	58653	62207	96227	16017	18083	24199	3187	3443	3930	907	933	999	427	437	446
0.4	***	58127	59917	94667	15837	17187	23807	3217	3457	3877	933	963	997	437	447	453
	•••	58723	63303	101271	16343	18833	25407	3327	3633	4098	943	977	1030	443	453	456
0.45	***	56277	58577	99007	15673	17267	24837	3273	3543	4013	947	977	1017	443	453	463
	•••	57047	62793	104213	16283	19153	26090	3393	3737	4182	963	997	1041	447	457	456
0.5	***	52997	55907	101083	15197	17013	25297	3263	3553	4063	943	977	1017	437	447	457
	•••	53983	60993	105054	15897	19093	26248	3393	3763	4182	957	997	1030	443	453	446
0.55	***	48663	52273	100867	14477	16473	25187	3193	3493	4017	923	957	997	423	433	443
	•••	49897	58203	103793	15257	18687	25880	3323	3707	4098	937	977	999	427	437	428
0.6	***	43657	48017	98363	13577	15687	24503	3063	3363	3877	877	913	953	397	407	417
	•••	45193	54663	100430	14413	17973	24987	3197	3577	3930	893	933	946	403	413	400
0.65	***	38377	43433	93583	12537	14693	23253	2883	3167	3653	817	847	883	367	373	377
	•••	40207	50553	94967	13397	16963	23569	3007	3373	3678	833	863	873	367	377	362
0.7	***	33157	38693	86537	11383	13497	21437	2643	2903	3333	737	763	793	323	327	333
	•••	35223	45983	87401	12237	15667	21625	2763	3087	3342	747	777	778	327	333	316
0.75	***	28193	33883	77247	10123	12103	19057	2353	2573	2927	637	657	683	273	277	277
	•••	30363	40987	77734	10927	14083	19156	2453	2727	2922	647	667	662	273	277	260
0.8	***	23517	28977	65727	8737	10483	16123	1993	2167	2433	517	527	547	207	213	213
	•••	25637	35503	65966	9457	12177	16161	2073	2283	2417	523	537	526	213	213	194
0.85	***	19003	23817	52013	7177	8587	12637	1567	1683	1847	373	383	387	.	.	.
	•••	20893	29357	52096	7767	9893	12641	1623	1757	1829	377	383	368	.	.	.
0.9	***	14343	18077	36117	5347	6303	8607	1047	1103	1177	.	.	.	.	.	.
	•••	15833	22167	36125	5753	7123	8596	1073	1137	1156	.	.	.	.	.	.
0.95	***	8843	10927	18067	2973	3357	4043	.	.	.	.	.	.	.	.	.
	•••	9697	12987	18052	3147	3643	4025	.	.	.	.	.	.	.	.	.
0.98	***	4123	4797	6223	.	.	.	.	.	.	.	.	.	.	.	.
	•••	4417	5337	6200	.	.	.	.	.	.	.	.	.	.	.	.

TABLE 6: ALPHA= 0.025 POWER= 0.9   EXPECTED ACCRUAL THRU MINIMUM FOLLOW-UP= 4250

		DEL=.01			DEL=.02			DEL=.05			DEL=.10			DEL=.15		
FACT=		1.0 .75	.50 .25	.00 BIN	1.0 .75	.50 .25	.00 BIN	1.0 .75	.50 .25	.00 BIN	1.0 75	.50 .25	.00 BIN	1.0 .75	.50 .25	.00 BIN
PCONT=***				REQUIRED NUMBER OF PATIENTS													
0.01	***	1608 1612	1618	592	592	592	194	194	194	99	99	99	71	71	71
	•••	1612 1618	6200	592	592	2049	194	194	558	99	99	227	71	71	135
0.02	***	3556 3573	3616	1144	1151	1155	311	311	311	135	135	135	88	88	88
	•••	3563 3588	10235	1144	1151	3048	311	311	712	135	135	264	88	88	151
0.05	***	10756 10834	11298	3057	3095	3163	662	666	673	237	237	237	141	141	141
	•••	10781 10947	21835	3074	3120	5916	662	666	1156	237	237	368	141	141	194
0.1	***	23733 23931	26293	6399	6538	6952	1250	1268	1293	396	396	400	209	209	209
	•••	23797 24296	39487	6453	6680	10277	1261	1282	1829	396	396	526	209	209	260
0.15	***	35452 35809	41536	9400	9687	10756	1788	1835	1898	534	538	549	273	273	273
	•••	35576 36489	55038	9513	10016	14112	1809	1863	2417	538	545	662	273	273	316
0.2	***	44919 45468	55799	11861	12318	14287	2249	2330	2447	655	666	677	326	326	333
	•••	45100 46534	68488	12031	12881	17422	2288	2383	2922	662	673	778	326	333	362
0.25	***	51836 52629	68507	13709	14368	17422	2621	2744	2929	758	772	789	368	375	375
	•••	52102 54173	79836	13954	15190	20206	2674	2823	3342	768	783	873	368	375	400
0.3	***	56235 57325	79329	14967	15856	20078	2908	3067	3329	836	857	878	400	407	411
	•••	56596 59439	89082	15297	16961	22465	2982	3180	3678	847	868	946	407	411	428
0.35	***	58281 59737	88084	15690	16823	22220	3116	3312	3648	896	921	949	428	432	439
	•••	58763 62521	96227	16111	18208	24199	3205	3456	3930	906	932	999	428	432	446
0.4	***	58239 60134	94665	15941	17326	23807	3244	3478	3875	932	963	995	439	449	453
	•••	58876 63693	101271	16461	18979	25407	3343	3641	4098	949	981	1030	443	449	456
0.45	***	56426 58859	99011	15796	17428	24838	3301	3563	4013	949	981	1017	443	453	460
	•••	57240 63264	104213	16419	19319	26090	3414	3754	4182	963	1002	1041	449	453	456
0.5	***	53185 56256	101079	15335	17188	25295	3290	3573	4062	949	981	1017	439	449	453
	•••	54227 61547	105054	16058	19270	26248	3418	3775	4182	963	1002	1030	443	449	446
0.55	***	48893 52697	100866	14634	16663	25188	3223	3513	4013	921	959	995	422	432	439
	•••	50206 58827	103793	15431	18877	25880	3350	3722	4098	938	974	999	428	439	428
0.6	***	43948 48510	98363	13745	15881	24504	3095	3386	3881	878	910	953	400	407	418
	•••	45563 55338	100430	14595	18165	24987	3223	3594	3930	896	932	946	400	411	400
0.65	***	38731 43973	93582	12711	14889	23251	2908	3191	3652	821	847	885	368	375	379
	•••	40644 51258	94967	13586	17156	23569	3035	3386	3678	836	864	873	368	375	362
0.7	***	33561 39267	86537	11557	13688	21434	2670	2925	3333	740	762	793	322	326	333
	•••	35697 46689	87401	12417	15849	21625	2787	3099	3342	751	779	778	326	333	316
0.75	***	28627 34453	77247	10288	12279	19058	2373	2589	2925	641	655	683	273	273	279
	•••	30858 41664	77734	11100	14245	19156	2472	2738	2922	645	666	662	273	279	260
0.8	***	23945 29509	65729	8886	10639	16121	2011	2181	2430	517	528	545	209	209	216
	•••	26109 36111	65966	9608	12311	16161	2090	2292	2417	524	538	526	209	216	194
0.85	***	19387 24281	52013	7303	8709	12636	1580	1693	1845	375	379	390	.	.	.
	•••	21310 29863	52096	7891	9995	12641	1633	1760	1829	375	386	368	.	.	.
0.9	***	14648 18427	36118	5433	6382	8609	1055	1108	1176	.	.	.	.	.	.
	•••	16157 22528	36125	5836	7186	8596	1080	1140	1156	.	.	.	.	.	.
0.95	***	9017 11117	18070	3010	3386	4045	.	.	.	.	.	.	.	.	.
	•••	9878 13157	18052	3180	3662	4025	.	.	.	.	.	.	.	.	.
0.98	***	4189 4852	6219	.	.	.	.	.	.	.	.	.	.	.	.
	•••	4476 5379	6200	.	.	.	.	.	.	.	.	.	.	.	.

TABLE 6: ALPHA= 0.025 POWER= 0.9   EXPECTED ACCRUAL THRU MINIMUM FOLLOW-UP= 4500

		DEL=.01			DEL=.02			DEL=.05			DEL=.10			DEL=.15		
FACT=		1.0 .75	.50 .25	.00 BIN	1.0 .75	.50 .25	.00 BIN	1.0 .75	.50 .25	.00 BIN	1.0 75	.50 .25	.00 BIN	1.0 .75	.50 .25	.00 BIN
PCONT=***						REQUIRED NUMBER OF PATIENTS										
0.01	***	1612	1612	1623	593	593	593	195	195	195	98	98	98	71	71	71
	***	1612	1616	6200	593	593	2049	195	195	558	98	98	227	71	71	135
0.02	***	3558	3574	3615	1144	1151	1155	311	311	311	138	138	138	86	86	86
	***	3563	3592	10235	1144	1151	3048	311	311	712	138	138	264	86	86	151
0.05	***	10763	10841	11298	3056	3097	3165	660	667	671	240	240	240	138	138	138
	***	10792	10961	21835	3075	3124	5916	667	667	1156	240	240	368	138	138	194
0.1	***	23745	23955	26295	6409	6551	6956	1252	1268	1297	397	397	397	210	210	210
	***	23813	24330	39487	6465	6690	10277	1263	1279	1829	397	397	526	210	210	260
0.15	***	35475	35850	41531	9424	9712	10751	1792	1837	1898	536	543	548	273	273	278
	***	35603	36566	55038	9532	10043	14112	1815	1864	2417	536	543	662	273	273	316
0.2	***	44951	45532	55803	11888	12360	14291	2258	2332	2449	660	667	678	323	330	330
	***	45143	46657	68488	12068	12923	17422	2291	2381	2922	660	671	778	330	330	362
0.25	***	51881	52721	68505	13755	14437	17423	2636	2748	2928	761	773	791	368	375	379
	***	52163	54345	79836	14010	15263	20206	2685	2827	3342	768	784	873	368	375	400
0.3	***	56298	57450	79331	15026	15945	20078	2921	3079	3333	840	858	881	401	408	413
	***	56681	59673	89082	15371	17051	22465	2996	3187	3678	851	870	946	408	408	428
0.35	***	58368	59903	88084	15765	16928	22215	3131	3326	3648	896	919	948	424	431	442
	***	58879	62828	96227	16203	18323	24199	3221	3461	3930	908	937	999	431	435	446
0.4	***	58350	60360	94665	16035	17456	23808	3266	3491	3878	937	964	998	442	446	453
	***	59021	64076	101271	16575	19117	25407	3367	3653	4098	948	982	1030	442	453	456
0.45	***	56568	59138	99008	15911	17576	24836	3322	3581	4013	953	982	1020	446	453	458
	***	57435	63728	104213	16552	19477	26090	3439	3761	4182	971	998	1041	446	458	456
0.5	***	53366	56602	101078	15472	17355	25298	3315	3592	4058	948	982	1020	442	446	453
	***	54469	62085	105054	16208	19443	26248	3439	3788	4182	964	998	1030	442	453	446
0.55	***	49125	53115	100864	14786	16838	25185	3248	3536	4013	926	960	998	424	431	442
	***	50509	59430	103793	15596	19061	25880	3378	3738	4098	941	975	999	431	435	428
0.6	***	44238	48990	98362	13908	16073	24506	3120	3405	3878	885	915	953	401	408	413
	***	45930	55988	100430	14775	18352	24987	3248	3603	3930	896	930	946	401	413	400
0.65	***	39086	44501	93581	12885	15078	23250	2933	3210	3653	825	851	885	368	375	379
	***	41066	51933	94967	13766	17340	23569	3056	3394	3678	836	870	873	368	375	362
0.7	***	33960	39817	86538	11726	13868	21435	2692	2940	3333	739	768	795	323	330	334
	***	36161	47366	87401	12596	16023	21625	2805	3108	3342	750	780	778	330	334	316
0.75	***	29051	35002	77246	10448	12446	19054	2393	2606	2928	638	660	678	273	278	278
	***	31335	42303	77734	11265	14403	19156	2490	2748	2922	649	667	662	273	278	260
0.8	***	24360	30018	65730	9026	10781	16125	2028	2190	2433	521	532	543	210	210	217
	***	26565	36690	65966	9750	12439	16161	2107	2298	2417	525	536	526	210	210	194
0.85	***	19758	24720	52012	7417	8823	12637	1590	1695	1848	375	379	390	.	.	.
	***	21716	30345	52096	8006	10088	12641	1639	1770	1829	379	386	368	.	.	.
0.9	***	14943	18757	36116	5516	6454	8610	1061	1110	1178	.	.	.	.	.	.
	***	16466	22868	36125	5914	7241	8596	1083	1140	1156	.	.	.	.	.	.
0.95	***	9188	11287	18071	3045	3416	4046	.	.	.	.	.	.	.	.	.
	***	10054	13312	18052	3210	3675	4025	.	.	.	.	.	.	.	.	.
0.98	***	4245	4901	6218	.	.	.	.	.	.	.	.	.	.	.	.
	***	4530	5415	6200	.	.	.	.	.	.	.	.	.	.	.	.

673

TABLE 6: ALPHA= 0.025 POWER= 0.9   EXPECTED ACCRUAL THRU MINIMUM FOLLOW-UP= 4750

	DEL=.01			DEL=.02			DEL=.05			DEL=.10			DEL=.15		
FACT=	1.0 .75	.50 .25	.00 BIN	1.0 .75	.50 .25	.00 BIN	1.0 .75	.50 .25	.00 BIN	1.0 75	.50 .25	.00 BIN	1.0 .75	.50 .25	.00 BIN
PCONT=***						REQUIRED NUMBER OF PATIENTS									
0.01 ***	1611 1611	1611 1618	1618 6200	590 590	590 590	590 2049	193 193	193 193	193 558	98 98	98 98	98 227	67 67	67 67	67 135
0.02 ***	3559 3566	3578 3590	3618 10235	1143 1148	1148 1148	1155 3048	312 312	312 312	312 712	134 134	134 134	134 264	87 87	87 87	87 151
0.05 ***	10767 10798	10850 10969	11297 21835	3060 3079	3095 3127	3167 5916	661 661	668 668	668 1156	234 234	241 241	241 368	139 139	139 139	139 194
0.1 ***	23758 23829	23979 24371	26295 39487	6416 6475	6563 6701	6955 10277	1255 1262	1274 1279	1290 1829	395 395	395 400	400 526	210 210	210 210	210 260
0.15 ***	35498 35628	35890 36638	41535 55038	9444 9555	9741 10066	10755 14112	1797 1813	1837 1868	1896 2417	538 538	542 542	550 662	269 277	277 277	277 316
0.2 ***	44986 45188	45596 46772	55804 68488	11919 12104	12405 12964	14289 17422	2264 2295	2335 2390	2450 2922	661 661	668 673	680 778	324 329	329 329	329 362
0.25 ***	51933 52222	52816 54514	68503 79836	13795 14056	14495 15327	17424 20206	2644 2692	2758 2834	2929 3342	763 768	775 780	792 873	372 372	372 376	376 400
0.3 ***	56362 56766	57578 59905	79328 89082	15089 15441	16027 17139	20084 22465	2937 3008	3091 3198	3328 3678	839 851	858 870	882 946	400 407	407 412	412 428
0.35 ***	58452 58991	60072 63124	88085 96227	15845 16296	17037 18433	22214 24199	3150 3233	3340 3471	3649 3930	899 910	922 934	946 999	424 431	431 435	435 446
0.4 ***	58464 59169	60578 64449	94664 101271	16130 16688	17578 19245	23805 25407	3285 3380	3507 3665	3875 4098	942 953	965 982	994 1030	443 443	447 447	455 456
0.45 ***	56711 57625	59419 64176	99010 104213	16027 16680	17721 19621	24839 26090	3345 3459	3594 3772	4010 4182	958 970	982 1001	1017 1041	447 447	455 455	459 456
0.5 ***	53552 54712	56949 62608	101083 105054	15607 16355	17519 19602	25297 26248	3340 3459	3613 3803	4057 4182	953 970	982 1001	1017 1030	443 443	447 455	455 446
0.55 ***	49360 50809	53524 60012	100862 103793	14935 15754	17013 19229	25190 25880	3274 3400	3554 3749	4017 4098	930 946	958 977	994 999	424 431	431 435	443 428
0.6 ***	44527 46292	49463 56616	98364 100430	14068 14942	16248 18521	24506 24987	3143 3269	3423 3618	3879 3930	887 899	918 934	953 946	400 407	407 412	412 400
0.65 ***	39428 41483	45014 52579	93583 94967	13042 13937	15255 17507	23255 23569	2960 3079	3226 3412	3649 3678	827 839	851 870	882 873	364 372	372 376	376 362
0.7 ***	34350 36614	40355 48007	86537 87401	11883 12762	14040 16182	21438 21625	2715 2822	2960 3119	3333 3342	744 756	768 780	792 778	324 329	329 329	336 316
0.75 ***	29458 31793	35526 42912	77243 77734	10600 11420	12603 14543	19055 19156	2414 2509	2616 2751	2925 2922	645 649	661 668	680 662	269 277	277 277	281 260
0.8 ***	24755 27000	30510 37236	65732 65966	9164 9888	10917 12560	16122 16161	2046 2117	2205 2307	2430 2417	519 526	530 538	542 526	210 210	210 210	217 194
0.85 ***	20112 22095	25135 30795	52009 52096	7532 8119	8926 10180	12638 12641	1599 1647	1706 1773	1849 1829	376 376	383 383	388 368	. .	. .	. .
0.9 ***	15220 16764	19075 23188	36115 36125	5589 5988	6523 7295	8613 8596	1065 1089	1112 1143	1179 1156	. .	. .	. .	. .	. .	. .
0.95 ***	9349 10216	11451 13462	18070 18052	3079 3238	3440 3697	4045 4025	. .	. .	. .	. .	. .	. .	. .	. .	. .
0.98 ***	4302 4580	4948 5447	6219 6200	. .	. .	. .	. .	. .	. .	. .	. .	. .	. .	. .	. .

TABLE 6: ALPHA= 0.025 POWER= 0.9   EXPECTED ACCRUAL THRU MINIMUM FOLLOW-UP= 5000

		DEL=.01			DEL=.02			DEL=.05			DEL=.10			DEL=.15		
FACT=		1.0 .75	.50 .25	.00 BIN	1.0 .75	.50 .25	.00 BIN	1.0 .75	.50 .25	.00 BIN	1.0 75	.50 .25	.00 BIN	1.0 .75	.50 .25	.00 BIN
PCONT=***						REQUIRED NUMBER OF PATIENTS										
0.01	***	1608	1616	1621	591	591	591	196	196	196	96	96	96	71	71	71
	•••	1608	1616	6200	591	591	2049	196	196	558	96	96	227	71	71	135
0.02	***	3558	3579	3616	1146	1146	1154	308	308	308	133	133	133	91	91	91
	•••	3566	3591	10235	1146	1154	3048	308	308	712	133	133	264	91	91	151
0.05	***	10771	10858	11296	3066	3104	3166	666	666	671	233	241	241	141	141	141
	•••	10804	10979	21835	3079	3129	5916	666	666	1156	233	241	368	141	141	194
0.1	***	23771	23996	26291	6429	6571	6954	1254	1271	1296	396	396	396	208	208	208
	•••	23846	24408	39487	6483	6708	10277	1266	1283	1829	396	396	526	208	208	260
0.15	***	35516	35933	41533	9458	9766	10754	1804	1841	1896	533	541	546	271	271	279
	•••	35654	36708	55038	9579	10091	14112	1821	1866	2417	541	546	662	271	279	316
0.2	***	45016	45658	55804	11954	12446	14291	2271	2341	2446	658	671	679	329	329	329
	•••	45229	46891	68488	12141	13008	17422	2304	2391	2922	666	671	778	329	329	362
0.25	***	51979	52908	68508	13841	14554	17421	2654	2766	2929	766	779	791	371	371	379
	•••	52291	54683	79836	14108	15383	20206	2704	2833	3342	771	783	873	371	379	400
0.3	***	56429	57708	79329	15146	16104	20079	2954	3104	3329	846	858	879	404	408	416
	•••	56854	60133	89082	15508	17221	22465	3016	3204	3678	854	871	946	404	408	428
0.35	***	58533	60246	88083	15916	17133	22216	3166	3354	3646	904	921	946	429	433	441
	•••	59108	63421	96227	16383	18533	24199	3246	3479	3930	916	933	999	429	433	446
0.4	***	58579	60796	94666	16221	17696	23808	3304	3521	3879	941	966	996	441	446	454
	•••	59321	64816	101271	16791	19366	25407	3404	3671	4098	954	979	1030	446	454	456
0.45	***	56858	59691	99008	16141	17858	24833	3366	3616	4016	958	983	1021	446	454	458
	•••	57816	64616	104213	16808	19758	26090	3479	3783	4182	971	1004	1041	446	458	456
0.5	***	53733	57283	101079	15733	17666	25296	3366	3629	4058	958	983	1021	441	446	454
	•••	54954	63108	105054	16496	19758	26248	3483	3816	4182	971	1004	1030	446	454	446
0.55	***	49591	53921	100866	15079	17179	25183	3296	3571	4016	933	966	996	429	433	441
	•••	51108	60571	103793	15908	19391	25880	3421	3758	4098	946	979	999	429	433	428
0.6	***	44816	49916	98366	14216	16416	24504	3166	3441	3879	891	921	954	404	408	416
	•••	46646	57216	100430	15104	18691	24987	3291	3629	3930	904	933	946	404	408	400
0.65	***	39766	45516	93583	13196	15421	23254	2983	3241	3654	829	854	883	366	371	379
	•••	41891	53204	94967	14104	17666	23569	3096	3421	3678	841	866	873	371	379	362
0.7	***	34729	40866	86533	12041	14204	21433	2733	2971	3333	746	766	791	329	329	333
	•••	37054	48621	87401	12921	16333	21625	2841	3129	3342	758	779	778	329	333	316
0.75	***	29854	36033	77246	10746	12754	19058	2429	2629	2929	646	658	679	271	279	279
	•••	32233	43496	77734	11571	14679	19156	2521	2758	2922	654	671	662	271	279	260
0.8	***	25141	30979	65729	9291	11046	16121	2058	2208	2429	521	533	546	208	216	216
	•••	27421	37758	65966	10021	12671	16161	2129	2308	2417	529	541	526	208	216	194
0.85	***	20458	25541	52008	7633	9029	12641	1608	1708	1846	379	383	391	.	.	.
	•••	22466	31229	52096	8221	10258	12641	1658	1771	1829	379	383	368	.	.	.
0.9	***	15491	19371	36116	5658	6583	8608	1066	1116	1179	.	.	.	.	.	.
	•••	17046	23496	36125	6058	7341	8596	1091	1146	1156	.	.	.	.	.	.
0.95	***	9504	11608	18071	3108	3458	4046	.	.	.	.	.	.	.	.	.
	•••	10371	13604	18052	3266	3708	4025	.	.	.	.	.	.	.	.	.
0.98	***	4354	4991	6221	.	.	.	.	.	.	.	.	.	.	.	.
	•••	4633	5479	6200	.	.	.	.	.	.	.	.	.	.	.	.

675

TABLE 6: ALPHA= 0.025 POWER= 0.9   EXPECTED ACCRUAL THRU MINIMUM FOLLOW-UP= 5500

		DEL=.01			DEL=.02			DEL=.05			DEL=.10			DEL=.15		
FACT=		1.0 .75	.50 .25	.00 BIN	1.0 .75	.50 .25	.00 BIN	1.0 .75	.50 .25	.00 BIN	1.0 75	.50 .25	.00 BIN	1.0 .75	.50 .25	.00 BIN
PCONT=***					REQUIRED NUMBER OF PATIENTS											
0.01	***	1613	1613	1618	595	595	595	196	196	196	100	100	100	73	73	73
	•••	1613	1618	6200	595	595	2049	196	196	558	100	100	227	73	73	135
0.02	***	3557	3579	3620	1145	1150	1159	312	312	312	133	133	133	86	86	86
	•••	3570	3593	10235	1145	1150	3048	312	312	712	133	133	264	86	86	151
0.05	***	10784	10872	11293	3070	3103	3166	664	669	669	238	238	238	141	141	141
	•••	10811	10995	21835	3084	3130	5916	664	669	1156	238	238	368	141	141	194
0.1	***	23791	24044	26294	6444	6595	6953	1260	1274	1296	394	394	403	210	210	210
	•••	23879	24479	39487	6508	6728	10277	1269	1283	1829	394	394	526	210	210	260
0.15	***	35561	36015	41534	9497	9808	10756	1805	1846	1893	540	540	545	270	270	279
	•••	35713	36845	55038	9620	10129	14112	1824	1865	2417	540	545	662	270	279	316
0.2	***	45082	45791	55801	12008	12522	14290	2278	2347	2451	664	669	678	325	325	334
	•••	45315	47117	68488	12205	13080	17422	2314	2396	2922	664	678	778	325	325	362
0.25	***	52067	53093	68506	13924	14667	17425	2671	2773	2933	765	779	788	367	375	375
	•••	52410	55009	79836	14208	15500	20206	2718	2842	3342	774	779	873	375	375	400
0.3	***	56558	57960	79328	15258	16256	20079	2974	3117	3331	848	862	884	403	408	416
	•••	57025	60578	89082	15643	17370	22465	3043	3213	3678	856	870	946	408	408	428
0.35	***	58708	60578	88086	16064	17320	22215	3194	3373	3648	903	925	953	430	435	435
	•••	59335	63988	96227	16550	18723	24199	3276	3496	3930	917	939	999	430	435	446
0.4	***	58799	61233	94664	16408	17925	23805	3337	3543	3873	944	966	994	444	449	458
	•••	59615	65523	101271	16999	19589	25407	3433	3689	4098	958	980	1030	444	449	456
0.45	***	57149	60234	99009	16353	18118	24836	3405	3639	4010	966	985	1021	449	449	458
	•••	58199	65454	104213	17045	20015	26090	3510	3804	4182	980	1008	1041	449	458	456
0.5	***	54102	57952	101080	15973	17953	25295	3405	3661	4060	958	985	1021	444	449	458
	•••	55430	64079	105054	16765	20038	26248	3515	3832	4182	972	999	1030	444	449	446
0.55	***	50054	54707	100865	15340	17480	25185	3337	3598	4019	939	966	999	430	435	444
	•••	51695	61645	103793	16201	19685	25880	3455	3777	4098	953	980	999	430	435	428
0.6	***	45379	50802	98363	14502	16738	24506	3208	3469	3881	898	925	953	403	408	416
	•••	47345	58359	100430	15404	18984	24987	3331	3648	3930	911	939	946	408	408	400
0.65	***	40429	46470	93586	13493	15739	23255	3020	3268	3653	834	856	884	367	375	380
	•••	42675	54377	94967	14405	17961	23569	3139	3433	3678	843	870	873	375	375	362
0.7	***	35465	41850	86538	12329	14502	21432	2773	3001	3331	752	774	793	325	334	334
	•••	37890	49784	87401	13223	16605	21625	2878	3144	3342	760	779	778	325	334	316
0.75	***	30611	36997	77243	11018	13025	19053	2465	2649	2924	650	664	678	270	279	279
	•••	33086	44595	77734	11848	14923	19156	2548	2773	2922	655	669	662	270	279	260
0.8	***	25873	31863	65729	9533	11284	16119	2080	2231	2429	526	532	545	210	210	215
	•••	28224	38738	65966	10261	12874	16161	2149	2319	2417	526	540	526	210	215	194
0.85	***	21110	26294	52012	7828	9208	12640	1626	1723	1846	375	380	389	.	.	.
	•••	23164	32028	52096	8419	10413	12641	1673	1778	1829	380	389	368	.	.	.
0.9	***	15995	19933	36117	5793	6700	8611	1076	1123	1178	.	.	.	.	.	.
	•••	17582	24058	36125	6191	7429	8596	1095	1145	1156	.	.	.	.	.	.
0.95	***	9785	11889	18071	3166	3502	4046	.	.	.	.	.	.	.	.	.
	•••	10665	13850	18052	3318	3735	4025	.	.	.	.	.	.	.	.	.
0.98	***	4445	5064	6219	.	.	.	.	.	.	.	.	.	.	.	.
	•••	4720	5531	6200	.	.	.	.	.	.	.	.	.	.	.	.

TABLE 6: ALPHA= 0.025 POWER= 0.9   EXPECTED ACCRUAL THRU MINIMUM FOLLOW-UP= 6000

	DEL=.01			DEL=.02			DEL=.05			DEL=.10			DEL=.15		
FACT=	1.0 .75	.50 .25	.00 BIN	1.0 .75	.50 .25	.00 BIN	1.0 .75	.50 .25	.00 BIN	1.0 75	.50 .25	.00 BIN	1.0 .75	.50 .25	.00 BIN
PCONT=***			REQUIRED NUMBER OF PATIENTS												
0.01 ***	1609	1615	1624	589	595	595	199	199	199	100	100	100	70	70	70
***	1609	1615	6200	589	595	2049	199	199	558	100	100	227	70	70	135
0.02 ***	3565	3580	3619	1144	1150	1159	310	310	310	139	139	139	85	85	85
***	3574	3595	10235	1150	1150	3048	310	310	712	139	139	264	85	85	151
0.05 ***	10789	10885	11299	3070	3109	3160	664	664	670	235	235	235	139	139	139
***	10825	11014	21835	3085	3130	5916	664	670	1156	235	235	368	139	139	194
0.1 ***	23815	24085	26290	6460	6610	6955	1264	1279	1294	394	394	400	214	214	214
***	23905	24544	39487	6520	6745	10277	1270	1285	1829	394	400	526	214	214	260
0.15 ***	35599	36094	41530	9535	9850	10750	1810	1849	1894	535	544	544	274	274	274
***	35764	36985	55038	9655	10165	14112	1834	1870	2417	544	544	662	274	274	316
0.2 ***	45145	45919	55804	12064	12595	14290	2290	2359	2449	664	670	679	325	325	334
***	45400	47335	68488	12274	13144	17422	2320	2395	2922	664	670	778	325	334	362
0.25 ***	52165	53275	68509	14005	14770	17425	2689	2785	2929	769	775	790	370	370	379
***	52534	55330	79836	14305	15595	20206	2734	2854	3342	769	784	873	370	379	400
0.3 ***	56680	58210	79330	15370	16390	20080	2995	3130	3334	850	865	880	409	409	415
***	57199	61000	89082	15769	17509	22465	3055	3220	3678	859	874	946	409	409	428
0.35 ***	58879	60910	88084	16204	17500	22219	3220	3394	3649	910	925	949	430	430	439
***	59560	64540	96227	16705	18895	24199	3295	3505	3930	919	940	999	430	439	446
0.4 ***	59020	61660	94669	16579	18130	23809	3364	3565	3874	949	970	994	445	445	454
***	59914	66199	101271	17185	19795	25407	3460	3700	4098	964	985	1030	445	454	456
0.45 ***	57430	60769	99010	16555	18355	24835	3439	3664	4015	970	994	1015	445	454	460
***	58579	66259	104213	17269	20245	26090	3544	3820	4182	979	1009	1041	454	454	456
0.5 ***	54469	58594	101080	16210	18220	25300	3439	3685	4060	964	994	1015	445	445	454
***	55909	64990	105054	17014	20284	26248	3550	3850	4182	979	1000	1030	445	454	446
0.55 ***	50509	55459	100864	15595	17764	25189	3379	3625	4015	940	970	994	430	430	439
***	52270	62644	103793	16474	19945	25880	3490	3799	4098	955	979	999	430	439	428
0.6 ***	45925	51640	98365	14770	17020	24505	3244	3499	3880	895	925	949	400	409	415
***	48019	59425	100430	15685	19255	24987	3364	3664	3930	910	940	946	409	409	400
0.65 ***	41065	47374	93580	13765	16024	23254	3055	3295	3649	835	859	880	370	370	379
***	43429	55465	94967	14689	18220	23569	3169	3454	3678	844	874	873	370	379	362
0.7 ***	36160	42769	86539	12595	14779	21439	2809	3019	3334	754	769	790	325	334	334
***	38689	50854	87401	13495	16849	21625	2905	3160	3342	760	784	778	325	334	316
0.75 ***	31330	37894	77245	11269	13279	19054	2485	2674	2929	649	664	679	274	274	280
***	33880	45604	77734	12100	15139	19156	2575	2785	2922	655	670	662	274	280	260
0.8 ***	26560	32689	65725	9754	11494	16120	2104	2245	2434	520	535	544	214	214	214
***	28975	39634	65966	10480	13054	16161	2170	2329	2417	529	535	526	214	214	194
0.85 ***	21715	26995	52009	8005	9379	12640	1639	1729	1849	379	385	385	.	.	.
***	23815	32755	52096	8590	10540	12641	1684	1789	1829	379	385	368	.	.	.
0.9 ***	16465	20449	36115	5914	6805	8605	1084	1129	1174	.	.	.	.	.	.
***	18079	24565	36125	6304	7504	8596	1105	1150	1156	.	.	.	.	.	.
0.95 ***	10054	12154	18070	3214	3535	4045	.	.	.	.	.	.	.	.	.
***	10930	14074	18052	3355	3760	4025	.	.	.	.	.	.	.	.	.
0.98 ***	4534	5134	6220	.	.	.	.	.	.	.	.	.	.	.	.
***	4795	5575	6200	.	.	.	.	.	.	.	.	.	.	.	.

TABLE 6: ALPHA= 0.025 POWER= 0.9   EXPECTED ACCRUAL THRU MINIMUM FOLLOW-UP= 6500

PCONT	FACT=	DEL=.01 1.0 / .75	DEL=.01 .50 / .25	DEL=.01 .00 / BIN	DEL=.02 1.0 / .75	DEL=.02 .50 / .25	DEL=.02 .00 / BIN	DEL=.05 1.0 / .75	DEL=.05 .50 / .25	DEL=.05 .00 / BIN	DEL=.10 1.0 / 75	DEL=.10 .50 / .25	DEL=.10 .00 / BIN	DEL=.15 1.0 / .75	DEL=.15 .50 / .25	DEL=.15 .00 / BIN
PCONT=•••				REQUIRED NUMBER OF PATIENTS												
0.01	•••	1613	1613	1619	590	590	596	200	200	200	102	102	102	70	70	70
	•••	1613	1619	6200	590	596	2049	200	200	558	102	102	227	70	70	135
0.02	•••	3563	3580	3618	1148	1148	1158	313	313	313	135	135	135	86	86	86
	•••	3569	3596	10235	1148	1148	3048	313	313	712	135	135	264	86	86	151
0.05	•••	10801	10898	11298	3076	3108	3163	661	671	671	238	238	238	141	141	141
	•••	10833	11022	21835	3092	3131	5916	661	671	1156	238	238	368	141	141	194
0.1	•••	23833	24136	26287	6478	6624	6949	1262	1278	1294	395	395	401	206	206	216
	•••	23941	24597	39487	6543	6754	10277	1272	1288	1829	395	401	526	206	216	260
0.15	•••	35641	36177	41529	9566	9891	10752	1814	1857	1896	541	541	547	271	271	271
	•••	35820	37109	55038	9696	10199	14112	1831	1873	2417	541	547	662	271	271	316
0.2	•••	45212	46041	55797	12117	12653	14288	2302	2361	2448	661	671	677	330	330	330
	•••	45488	47542	68488	12328	13206	17422	2328	2399	2922	671	677	778	330	330	362
0.25	•••	52254	53457	68504	14083	14863	17425	2702	2800	2930	768	774	791	368	378	378
	•••	52661	55634	79836	14392	15686	20206	2741	2854	3342	774	785	873	368	378	400
0.3	•••	56815	58462	79327	15475	16521	20079	3011	3141	3326	850	866	882	401	411	411
	•••	57367	61413	89082	15887	17626	22465	3076	3228	3678	856	872	946	411	411	428
0.35	•••	59047	61241	88086	16336	17658	22218	3244	3407	3645	915	931	947	427	433	443
	•••	59788	65070	96227	16856	19050	24199	3320	3515	3930	921	937	999	433	433	446
0.4	•••	59242	62080	94667	16742	18324	23811	3391	3586	3878	953	969	996	443	449	449
	•••	60211	66841	101271	17366	19976	25407	3482	3716	4098	963	986	1030	449	449	456
0.45	•••	57725	61300	99006	16748	18578	24835	3466	3683	4008	969	996	1018	449	449	460
	•••	58949	67009	104213	17479	20453	26090	3569	3829	4182	980	1002	1041	449	460	456
0.5	•••	54832	59220	101080	16423	18465	25296	3472	3710	4057	969	996	1018	443	449	460
	•••	56376	65850	105054	17246	20512	26248	3580	3862	4182	980	1002	1030	449	449	446
0.55	•••	50954	56181	100868	15832	18016	25182	3407	3651	4018	947	969	996	427	433	443
	•••	52833	63591	103793	16716	20187	25880	3521	3813	4098	953	986	999	433	433	428
0.6	•••	46469	52449	98366	15020	17284	24499	3277	3521	3878	904	921	953	401	411	417
	•••	48673	60422	100430	15946	19494	24987	3391	3677	3930	915	937	946	411	411	400
0.65	•••	41686	48224	93578	14018	16287	23248	3092	3320	3651	839	856	882	368	378	378
	•••	44156	56490	94967	14955	18454	23569	3196	3466	3678	850	872	873	368	378	362
0.7	•••	36833	43642	86536	12842	15026	21438	2832	3043	3336	758	774	791	330	330	336
	•••	39449	51858	87401	13752	17073	21625	2930	3173	3342	768	785	778	330	330	316
0.75	•••	32017	38734	77247	11499	13508	19056	2513	2686	2930	655	661	677	271	281	281
	•••	34639	46534	77734	12338	15334	19156	2594	2800	2922	661	671	662	271	281	260
0.8	•••	27213	33453	65726	9956	11688	16125	2123	2253	2432	525	531	547	216	216	216
	•••	29683	40457	65966	10687	13216	16161	2182	2334	2417	531	541	526	216	216	194
0.85	•••	22283	27646	52011	8168	9527	12637	1652	1743	1847	378	384	384	.	.	.
	•••	24428	33431	52096	8747	10654	12641	1695	1792	1829	378	384	368	.	.	.
0.9	•••	16911	20924	36112	6023	6895	8607	1093	1132	1175	.	.	.	.	.	.
	•••	18536	25019	36125	6407	7567	8596	1110	1148	1156	.	.	.	.	.	.
0.95	•••	10297	12387	18064	3255	3569	4041	.	.	.	.	.	.	.	.	.
	•••	11174	14272	18052	3401	3775	4025	.	.	.	.	.	.	.	.	.
0.98	•••	4609	5194	6218	.	.	.	.	.	.	.	.	.	.	.	.
	•••	4869	5617	6200	.	.	.	.	.	.	.	.	.	.	.	.

TABLE 6: ALPHA= 0.025 POWER= 0.9   EXPECTED ACCRUAL THRU MINIMUM FOLLOW-UP= 7000

	DEL=.01			DEL=.02			DEL=.05			DEL=.10			DEL=.15		
FACT=	1.0 .75	.50 .25	.00 BIN	1.0 .75	.50 .25	.00 BIN	1.0 .75	.50 .25	.00 BIN	1.0 75	.50 .25	.00 BIN	1.0 .75	.50 .25	.00 BIN
PCONT=***						REQUIRED NUMBER OF PATIENTS									
0.01 ***	1615	1615	1621	589	589	589	197	197	197	99	99	99	64	64	64
•••	1615	1615	6200	589	589	2049	197	197	558	99	99	227	64	64	135
0.02 ***	3564	3581	3616	1149	1149	1160	309	309	309	134	134	134	92	92	92
•••	3575	3599	10235	1149	1149	3048	309	309	712	134	134	264	92	92	151
0.05 ***	10809	10914	11292	3085	3109	3161	670	670	670	239	239	239	134	134	134
•••	10844	11036	21835	3091	3137	5916	670	670	1156	239	239	368	134	134	194
0.1 ***	23864	24172	26290	6497	6644	6952	1265	1282	1289	396	396	396	215	215	215
•••	23969	24662	39487	6556	6766	10277	1271	1282	1829	396	396	526	215	215	260
0.15 ***	35687	36254	41532	9601	9927	10756	1825	1860	1895	536	547	547	274	274	274
•••	35880	37234	55038	9735	10225	14112	1842	1877	2417	536	547	662	274	274	316
0.2 ***	45277	46170	55801	12174	12716	14291	2304	2367	2444	670	670	676	326	326	326
•••	45575	47745	68488	12395	13259	17422	2339	2402	2922	670	676	778	326	326	362
0.25 ***	52347	53642	68506	14162	14956	17424	2706	2805	2927	775	781	792	372	372	379
•••	52785	55924	79836	14477	15772	20206	2752	2857	3342	775	781	873	372	372	400
0.3 ***	56939	58706	79332	15580	16647	20077	3032	3155	3330	851	869	880	407	407	414
•••	57534	61804	89082	16000	17739	22465	3085	3231	3678	862	869	946	407	407	428
0.35 ***	59225	61570	88089	16472	17809	22219	3260	3417	3645	915	932	950	431	431	442
•••	60019	65566	96227	16997	19185	24199	3336	3522	3930	921	939	999	431	431	446
0.4 ***	59470	62497	94669	16892	18509	23805	3417	3599	3872	956	974	991	442	449	449
•••	60502	67456	101271	17540	20147	25407	3505	3721	4098	967	985	1030	449	449	456
0.45 ***	58006	61804	99009	16927	18782	24837	3494	3704	4012	974	991	1020	449	449	460
•••	59319	67736	104213	17669	20644	26090	3592	3844	4182	985	1009	1041	449	460	456
0.5 ***	55189	59826	101085	16630	18684	25299	3505	3721	4065	974	991	1020	442	449	460
•••	56834	66669	105054	17459	20714	26248	3599	3872	4182	985	1009	1030	449	449	446
0.55 ***	51402	56880	100864	16059	18257	25187	3435	3669	4012	950	974	991	431	431	442
•••	53386	64481	103793	16951	20399	25880	3546	3826	4098	956	985	999	431	442	428
0.6 ***	46999	53222	98361	15254	17529	24505	3312	3540	3879	904	932	950	407	407	414
•••	49302	61360	100430	16192	19716	24987	3417	3697	3930	915	939	946	407	414	400
0.65 ***	42285	49040	93584	14256	16531	23251	3120	3336	3651	845	862	886	372	372	379
•••	44846	57440	94967	15195	18666	23569	3214	3476	3678	851	869	873	372	379	362
0.7 ***	37479	44461	86531	13077	15254	21431	2857	3056	3330	757	775	792	326	326	337
•••	40174	52785	87401	13987	17266	21625	2951	3179	3342	764	781	778	326	337	316
0.75 ***	32666	39526	77250	11712	13714	19051	2531	2700	2927	652	670	676	274	274	274
•••	35355	47412	77734	12552	15510	19156	2612	2805	2922	659	670	662	274	274	260
0.8 ***	27830	34182	65724	10144	11870	16122	2140	2262	2426	530	536	547	215	215	215
•••	30350	41224	65966	10872	13357	16161	2199	2339	2417	530	536	526	215	215	194
0.85 ***	22825	28250	52015	8317	9654	12640	1667	1744	1849	379	379	390	.	.	.
•••	25001	34049	52096	8895	10756	12641	1702	1790	1829	379	390	368	.	.	.
0.9 ***	17319	21372	36114	6130	6976	8604	1096	1131	1177	.	.	.	.	.	.
•••	18964	25450	36125	6497	7624	8596	1114	1149	1156	.	.	.	.	.	.
0.95 ***	10522	12605	18071	3295	3592	4047	.	.	.	.	.	.	.	.	.
•••	11397	14449	18052	3435	3791	4025	.	.	.	.	.	.	.	.	.
0.98 ***	4677	5244	6217	.	.	.	.	.	.	.	.	.	.	.	.
•••	4929	5657	6200	.	.	.	.	.	.	.	.	.	.	.	.

TABLE 6: ALPHA= 0.025 POWER= 0.9   EXPECTED ACCRUAL THRU MINIMUM FOLLOW-UP= 7500

	DEL=.01			DEL=.02			DEL=.05			DEL=.10			DEL=.15		
FACT=	1.0 .75	.50 .25	.00 BIN	1.0 .75	.50 .25	.00 BIN	1.0 .75	.50 .25	.00 BIN	1.0 75	.50 .25	.00 BIN	1.0 .75	.50 .25	.00 BIN

PCONT=***   REQUIRED NUMBER OF PATIENTS

PCONT	DEL=.01 1.0/.75	.50/.25	.00/BIN	DEL=.02 1.0/.75	.50/.25	.00/BIN	DEL=.05 1.0/.75	.50/.25	.00/BIN	DEL=.10 1.0/.75	.50/.25	.00/BIN	DEL=.15 1.0/.75	.50/.25	.00/BIN
0.01 ***	1606	1618	1618	593	593	593	193	193	193	99	99	99	68	68	68
***	1618	1618	6200	593	593	2049	193	193	558	99	99	227	68	68	135
0.02 ***	3568	3586	3612	1149	1149	1156	312	312	312	136	136	136	87	87	87
***	3575	3593	10235	1149	1149	3048	312	312	712	136	136	264	87	87	151
0.05 ***	10812	10925	11293	3087	3118	3162	661	668	668	237	237	237	136	136	136
***	10861	11049	21835	3099	3136	5916	668	668	1156	237	237	368	136	136	194
0.1 ***	23881	24211	26293	6511	6661	6950	1268	1280	1299	399	399	399	211	211	211
***	23993	24718	39487	6575	6774	10277	1268	1287	1829	399	399	526	211	211	260
0.15 ***	35724	36331	41536	9631	9961	10756	1824	1861	1899	537	537	549	275	275	275
***	35930	37343	55038	9762	10250	14112	1843	1880	2417	537	549	662	275	275	316
0.2 ***	45343	46287	55805	12218	12774	14293	2318	2375	2450	668	668	680	324	331	331
***	45661	47937	68488	12443	13306	17422	2337	2405	2922	668	680	778	331	331	362
0.25 ***	52437	53818	68506	14236	15043	17424	2724	2806	2930	774	781	793	368	368	380
***	52906	56206	79836	14555	15837	20206	2761	2862	3342	774	781	873	368	380	400
0.3 ***	57068	58955	79325	15668	16756	20075	3043	3162	3331	856	868	875	406	406	418
***	57706	62187	89082	16100	17843	22465	3099	3237	3678	856	875	946	406	406	428
0.35 ***	59393	61887	88081	16587	17949	22212	3286	3436	3650	912	931	950	436	436	436
***	60249	66050	96227	17131	19318	24199	3350	3530	3930	924	943	999	436	436	446
0.4 ***	59693	62900	94662	17049	18668	23806	3436	3624	3875	961	980	999	443	455	455
***	60793	68037	101271	17693	20300	25407	3518	3736	4098	968	987	1030	443	455	456
0.45 ***	58287	62300	99012	17105	18968	24837	3518	3725	4018	980	999	1018	455	455	462
***	59693	68424	104213	17855	20818	26090	3612	3849	4182	987	1006	1041	455	455	456
0.5 ***	55550	60418	101075	16824	18893	25299	3530	3743	4062	980	999	1018	443	455	455
***	57286	67437	105054	17668	20911	26248	3624	3886	4182	987	1006	1030	443	455	446
0.55 ***	51837	57549	100861	16268	18481	25186	3462	3687	4018	950	968	999	425	436	436
***	53918	65318	103793	17180	20600	25880	3568	3837	4098	961	987	999	436	436	428
0.6 ***	47518	53956	98368	15481	17761	24500	3331	3556	3875	905	931	950	406	406	418
***	49918	62243	100430	16418	19918	24987	3436	3706	3930	924	943	946	406	418	400
0.65 ***	42868	49812	93586	14480	16756	23255	3143	3350	3650	849	868	886	368	380	380
***	45511	58325	94967	15425	18868	23569	3237	3493	3678	856	875	873	368	380	362
0.7 ***	38093	45237	86536	13287	15474	21436	2881	3068	3331	762	774	793	331	331	331
***	40868	53656	87401	14199	17450	21625	2975	3193	3342	762	781	778	331	331	316
0.75 ***	33286	40280	77243	11911	13906	19055	2555	2712	2930	650	668	680	275	275	275
***	36031	48218	77734	12755	15668	19156	2630	2806	2922	661	668	662	275	275	260
0.8 ***	28418	34861	65731	10318	12024	16118	2150	2274	2431	530	537	549	211	211	211
***	30980	41937	65966	11049	13486	16161	2206	2349	2417	530	537	526	211	211	194
0.85 ***	23330	28824	52006	8461	9781	12643	1674	1749	1843	380	380	387	.	.	.
***	25543	34625	52096	9031	10850	12641	1711	1793	1829	380	387	368	.	.	.
0.9 ***	17712	21781	36118	6218	7055	8611	1100	1137	1175	.	.	.	.	.	.
***	19374	25831	36125	6586	7681	8596	1118	1156	1156	.	.	.	.	.	.
0.95 ***	10730	12800	18068	3324	3624	4043	.	.	.	.	.	.	.	.	.
***	11611	14618	18052	3462	3811	4025	.	.	.	.	.	.	.	.	.
0.98 ***	4737	5293	6218	.	.	.	.	.	.	.	.	.	.	.	.
***	4993	5686	6200	.	.	.	.	.	.	.	.	.	.	.	.

TABLE 6: ALPHA= 0.025 POWER= 0.9   EXPECTED ACCRUAL THRU MINIMUM FOLLOW-UP= 8000

		DEL=.01			DEL=.02			DEL=.05			DEL=.10			DEL=.15		
FACT=		1.0 .75	.50 .25	.00 BIN	1.0 .75	.50 .25	.00 BIN	1.0 .75	.50 .25	.00 BIN	1.0 75	.50 .25	.00 BIN	1.0 .75	.50 .25	.00 BIN
PCONT=•••					REQUIRED NUMBER OF PATIENTS											
0.01	•••	1613	1613	1625	593	593	593	193	193	193	93	93	93	65	65	65
	•••	1613	1613	6200	593	593	2049	193	193	558	93	93	227	65	65	135
0.02	•••	3573	3585	3613	1145	1153	1153	313	313	313	133	133	133	85	85	85
	•••	3573	3605	10235	1145	1153	3048	313	313	712	133	133	264	85	85	151
0.05	•••	10825	10933	11293	3085	3113	3165	665	665	673	233	233	233	133	133	133
	•••	10865	11053	21835	3105	3145	5916	665	665	1156	233	233	368	133	133	194
0.1	•••	23905	24253	26293	6525	6673	6953	1265	1285	1293	393	393	393	213	213	213
	•••	24025	24765	39487	6585	6785	10277	1273	1285	1829	393	393	526	213	213	260
0.15	•••	35765	36413	41533	9653	9985	10753	1833	1865	1893	545	545	545	273	273	273
	•••	35985	37465	55038	9793	10273	14112	1845	1873	2417	545	545	662	273	273	316
0.2	•••	45405	46413	55805	12273	12825	14285	2325	2373	2445	665	673	673	325	325	333
	•••	45745	48125	68488	12493	13353	17422	2345	2413	2922	665	673	778	325	333	362
0.25	•••	52533	53993	68505	14305	15113	17425	2733	2813	2925	773	785	793	373	373	373
	•••	53033	56473	79836	14633	15913	20206	2773	2865	3342	773	785	873	373	373	400
0.3	•••	57193	59193	79325	15765	16865	20085	3053	3173	3333	853	865	885	405	405	413
	•••	57873	62553	89082	16205	17933	22465	3113	3245	3678	865	873	946	405	413	428
0.35	•••	59565	62205	88085	16705	18085	22213	3293	3445	3645	913	933	945	433	433	433
	•••	60465	66513	96227	17265	19433	24199	3365	3533	3930	925	945	999	433	433	446
0.4	•••	59913	63305	94665	17185	18833	23805	3453	3633	3873	965	973	993	445	453	453
	•••	61085	68605	101271	17853	20433	25407	3533	3745	4098	965	985	1030	445	453	456
0.45	•••	58573	62793	99005	17265	19153	24833	3545	3733	4013	973	993	1013	453	453	465
	•••	60053	69073	104213	18033	20973	26090	3633	3865	4182	985	1005	1041	453	453	456
0.5	•••	55905	60993	101085	17013	19093	25293	3553	3765	4065	973	993	1013	445	453	453
	•••	57733	68173	105054	17865	21085	26248	3645	3893	4182	985	1005	1030	445	453	446
0.55	•••	52273	58205	100865	16473	18685	25185	3493	3705	4013	953	973	993	433	433	445
	•••	54445	66113	103793	17385	20785	25880	3593	3845	4098	965	985	999	433	433	428
0.6	•••	48013	54665	98365	15685	17973	24505	3365	3573	3873	913	933	953	405	413	413
	•••	50513	63073	100430	16633	20093	24987	3465	3713	3930	925	945	946	405	413	400
0.65	•••	43433	50553	93585	14693	16965	23253	3165	3373	3653	845	865	885	373	373	373
	•••	46153	59173	94967	15633	19045	23569	3265	3493	3678	853	873	873	373	373	362
0.7	•••	38693	45985	86533	13493	15665	21433	2905	3085	3333	765	773	793	325	333	333
	•••	41525	54473	87401	14405	17613	21625	2985	3193	3342	765	785	778	333	333	316
0.75	•••	33885	40985	77245	12105	14085	19053	2573	2725	2925	653	665	685	273	273	273
	•••	36685	48985	77734	12933	15813	19156	2645	2813	2922	665	673	662	273	273	260
0.8	•••	28973	35505	65725	10485	12173	16125	2165	2285	2433	525	533	545	213	213	213
	•••	31573	42605	65966	11205	13605	16161	2225	2353	2417	533	545	526	213	213	194
0.85	•••	23813	29353	52013	8585	9893	12633	1685	1753	1845	385	385	385	.	.	.
	•••	26053	35153	52096	9153	10933	12641	1713	1793	1829	385	385	368	.	.	.
0.9	•••	18073	22165	36113	6305	7125	8605	1105	1133	1173	.	.	.	.	.	.
	•••	19753	26193	36125	6665	7725	8596	1125	1153	1156	.	.	.	.	.	.
0.95	•••	10925	12985	18065	3353	3645	4045	.	.	.	.	.	.	.	.	.
	•••	11805	14765	18052	3485	3825	4025	.	.	.	.	.	.	.	.	.
0.98	•••	4793	5333	6225	.	.	.	.	.	.	.	.	.	.	.	.
	•••	5045	5713	6200	.	.	.	.	.	.	.	.	.	.	.	.

TABLE 6: ALPHA= 0.025 POWER= 0.9   EXPECTED ACCRUAL THRU MINIMUM FOLLOW-UP= 8500

	DEL=.01			DEL=.02			DEL=.05			DEL=.10			DEL=.15		
FACT=	1.0 .75	.50 .25	.00 BIN	1.0 .75	.50 .25	.00 BIN	1.0 .75	.50 .25	.00 BIN	1.0 75	.50 .25	.00 BIN	1.0 .75	.50 .25	.00 BIN
PCONT=***			REQUIRED NUMBER OF PATIENTS												
0.01 ***	1608	1621	1621	588	588	588	197	197	197	99	99	99	70	70	70
•••	1608	1621	6200	588	588	2049	197	197	558	99	99	227	70	70	135
0.02 ***	3576	3584	3618	1153	1153	1153	311	311	311	133	133	133	91	91	91
•••	3576	3597	10235	1153	1153	3048	311	311	712	133	133	264	91	91	151
0.05 ***	10830	10950	11298	3095	3116	3159	665	665	673	240	240	240	141	141	141
•••	10873	11064	21835	3108	3138	5916	665	673	1156	240	240	368	141	141	194
0.1 ***	23933	24295	26292	6538	6678	6955	1268	1281	1289	396	396	396	205	205	205
•••	24061	24813	39487	6601	6793	10277	1281	1289	1829	396	396	526	205	205	260
0.15 ***	35812	36492	41536	9683	10015	10758	1833	1863	1897	537	545	545	269	269	269
•••	36046	37563	55038	9823	10299	14112	1841	1876	2417	545	545	662	269	269	316
0.2 ***	45468	46530	55795	12318	12883	14286	2330	2386	2450	665	673	673	325	333	333
•••	45829	48294	68488	12543	13393	17422	2351	2415	2922	673	673	778	325	333	362
0.25 ***	52629	54172	68503	14371	15186	17418	2747	2819	2925	771	779	792	375	375	375
•••	53152	56730	79836	14698	15973	20206	2776	2875	3342	779	779	873	375	375	400
0.3 ***	57325	59442	79332	15858	16963	20074	3066	3180	3329	856	864	877	410	410	410
•••	58048	62906	89082	16305	18013	22465	3116	3244	3678	864	877	946	410	410	428
0.35 ***	59740	62523	88087	16823	18204	22220	3308	3456	3648	920	928	949	431	431	439
•••	60696	64965	96227	17388	19543	24199	3385	3541	3930	928	941	999	431	439	446
0.4 ***	60130	63692	94661	17325	18982	23806	3478	3640	3873	962	983	991	452	452	452
•••	61376	69140	101271	17991	20563	25407	3555	3746	4098	970	983	1030	452	452	456
0.45 ***	58855	63267	99010	17431	19322	24834	3563	3754	4009	983	1005	1013	452	452	460
•••	60420	69693	104213	18196	21115	26090	3648	3873	4182	991	1005	1041	452	460	456
0.5 ***	56255	61546	101079	17184	19266	25293	3576	3775	4065	983	1005	1013	452	452	452
•••	58167	68877	105054	18047	21243	26248	3669	3903	4182	991	1005	1030	452	452	446
0.55 ***	52693	58826	100866	16666	18876	25187	3512	3725	4009	962	970	991	431	439	439
•••	54958	66866	103793	17580	20945	25880	3605	3852	4098	962	983	999	431	439	428
0.6 ***	48506	55341	98359	15880	18161	24507	3385	3597	3881	906	928	949	410	410	418
•••	51091	63862	100430	16836	20265	24987	3478	3725	3930	920	941	946	410	410	400
0.65 ***	43972	51261	93578	14889	17155	23253	3193	3385	3648	843	864	885	375	375	375
•••	46777	59960	94967	15837	19203	23569	3278	3499	3678	856	877	873	375	375	362
0.7 ***	39263	46692	86536	13691	15845	21434	2925	3095	3329	758	779	792	325	333	333
•••	42166	55243	87401	14591	17771	21625	3002	3201	3342	771	779	778	333	333	316
0.75 ***	34452	41664	77250	12275	14243	19054	2585	2734	2925	651	665	686	269	282	282
•••	37300	49696	77734	13117	15943	19156	2662	2819	2922	665	673	662	282	282	260
0.8 ***	29509	36110	65732	10639	12310	16121	2181	2288	2428	524	537	545	205	218	218
•••	32144	43236	65966	11353	13712	16161	2237	2351	2417	537	537	526	218	218	194
0.85 ***	24281	29862	52013	8705	9993	12636	1693	1756	1841	375	388	388	.	.	.
•••	26534	35650	52096	9271	11013	12641	1727	1799	1829	375	388	368	.	.	.
0.9 ***	18430	22531	36118	6381	7188	8612	1111	1140	1175	.	.	.	.	.	.
•••	20108	26526	36125	6742	7770	8596	1119	1153	1156	.	.	.	.	.	.
0.95 ***	11120	13160	18068	3385	3661	4043	.	.	.	.	.	.	.	.	.
•••	11978	14902	18052	3512	3831	4025	.	.	.	.	.	.	.	.	.
0.98 ***	4851	5382	6219	.	.	.	.	.	.	.	.	.	.	.	.
•••	5093	5743	6200	.	.	.	.	.	.	.	.	.	.	.	.

TABLE 6: ALPHA= 0.025 POWER= 0.9 EXPECTED ACCRUAL THRU MINIMUM FOLLOW-UP= 9000

	DEL=.01			DEL=.02			DEL=.05			DEL=.10			DEL=.15		
FACT=	1.0 .75	.50 .25	.00 BIN	1.0 .75	.50 .25	.00 BIN	1.0 .75	.50 .25	.00 BIN	1.0 75	.50 .25	.00 BIN	1.0 .75	.50 .25	.00 BIN
PCONT=***				REQUIRED NUMBER OF PATIENTS											
0.01 ***	1612 1612	1612 1612	1626 6200	591 591	591 591	591 2049	195 195	195 195	195 558	96 96	96 96	96 227	74 74	74 74	74 135
0.02 ***	3570 3584	3592 3606	3615 10235	1154 1154	1154 1154	1154 3048	307 307	307 307	307 712	141 141	141 141	141 264	82 82	82 82	82 151
0.05 ***	10837 10882	10964 11076	11301 21835	3097 3111	3120 3142	3165 5916	667 667	667 667	667 1156	240 240	240 240	240 368	141 141	141 141	141 194
0.1 ***	23955 24090	24329 24855	26295 39487	6554 6607	6689 6801	6959 10277	1266 1275	1275 1289	1297 1829	397 397	397 397	397 526	209 209	209 209	209 260
0.15 ***	35849 36096	36569 37671	41527 55038	9712 9847	10041 10320	10747 14112	1837 1851	1860 1882	1896 2417	546 546	546 546	546 662	276 276	276 276	276 316
0.2 ***	45532 45915	46657 48471	55806 68488	12359 12592	12921 13425	14294 17422	2332 2355	2377 2414	2445 2922	667 667	667 681	681 778	330 330	330 330	330 362
0.25 ***	52724 53272	54344 56985	68505 79836	14437 14775	15261 16026	17421 20206	2751 2782	2827 2872	2931 3342	771 780	780 780	794 873	375 375	375 375	375 400
0.3 ***	57449 58214	59676 63240	79327 89082	15945 16395	17047 18096	20076 22465	3075 3134	3187 3255	3336 3678	861 861	870 870	884 946	411 411	411 411	411 428
0.35 ***	59901 60914	62826 67380	88080 96227	16926 17497	18321 19635	22214 24199	3322 3390	3457 3547	3651 3930	915 929	937 937	951 999	434 434	434 434	442 446
0.4 ***	60360 61656	64072 69652	94664 101271	17452 18127	19117 20684	23811 25407	3494 3570	3651 3750	3876 4098	960 974	982 982	996 1030	442 442	456 456	456 456
0.45 ***	59136 60765	63726 70296	99006 104213	17579 18352	19477 21255	24832 26090	3584 3660	3764 3876	4011 4182	982 996	996 1005	1019 1041	456 456	456 456	456 456
0.5 ***	56602 58596	62084 69540	101076 105054	17354 18217	19446 21390	25296 26248	3592 3682	3786 3907	4056 4182	982 996	996 1005	1019 1030	442 442	456 456	456 446
0.55 ***	53115 55455	59429 67582	100860 103793	16836 17759	19064 21097	25184 25880	3539 3629	3741 3862	4011 4098	960 974	974 982	996 999	434 434	434 434	442 428
0.6 ***	48989 51644	55986 64612	98362 100430	16071 17025	18352 20422	24509 24987	3404 3494	3606 3727	3876 3930	915 929	929 937	951 946	411 411	411 411	411 400
0.65 ***	44497 47369	51936 60720	93584 94967	15081 16026	17340 19356	23249 23569	3210 3291	3390 3516	3651 3678	847 861	870 870	884 873	375 375	375 375	375 362
0.7 ***	39817 42765	47369 55972	86541 87401	13866 14775	16026 17902	21435 21625	2940 3021	3111 3210	3336 3342	771 771	780 780	794 778	330 330	330 330	330 316
0.75 ***	35002 37896	42306 50370	77249 77734	12449 13281	14406 16071	19050 19156	2602 2670	2751 2827	2931 2922	659 667	667 667	681 662	276 276	276 276	276 260
0.8 ***	30021 32685	36690 43822	65729 65966	10784 11490	12435 13807	16125 16161	2189 2242	2301 2355	2436 2417	532 532	532 546	546 526	209 209	209 217	217 194
0.85 ***	24720 26992	30345 36119	52012 52096	8826 9375	10086 11085	12637 12641	1694 1725	1770 1806	1851 1829	375 389	389 389	389 368	. .	. .	. .
0.9 ***	18757 20445	22866 26835	36119 36125	6450 6801	7237 7814	8610 8596	1109 1131	1140 1154	1176 1156	. .	. .	. .	. .	. .	. .
0.95 ***	11287 12156	13312 15022	18074 18052	3412 3539	3674 3840	4042 4025	. .	. .	. .	. .	. .	. .	. .	. .	. .
0.98 ***	4897 5136	5415 5766	6216 6200	. .	. .	. .	. .	. .	. .	. .	. .	. .	. .	. .	. .

TABLE 6: ALPHA= 0.025 POWER= 0.9   EXPECTED ACCRUAL THRU MINIMUM FOLLOW-UP= 9500

PCONT=***		DEL=.01			DEL=.02			DEL=.05			DEL=.10			DEL=.15		
FACT=		1.0 .75	.50 .25	.00 BIN	1.0 .75	.50 .25	.00 BIN	1.0 .75	.50 .25	.00 BIN	1.0 75	.50 .25	.00 BIN	1.0 .75	.50 .25	.00 BIN
						REQUIRED	NUMBER	OF	PATIENTS							
0.01	***	1607	1621	1621	585	585	585	196	196	196	101	101	101	63	63	63
	•••	1607	1621	6200	585	585	2049	196	196	558	101	101	227	63	63	135
0.02	***	3578	3593	3616	1146	1146	1155	315	315	315	134	134	134	87	87	87
	•••	3578	3602	10235	1146	1155	3048	315	315	712	134	134	264	87	87	151
0.05	***	10845	10964	11297	3094	3127	3165	671	671	671	244	244	244	134	134	134
	•••	10893	11083	21835	3103	3141	5916	671	671	1156	244	244	368	134	134	194
0.1	***	23979	24374	26298	6561	6704	6950	1274	1274	1289	395	395	395	205	205	205
	•••	24113	24896	39487	6618	6808	10277	1274	1289	1829	395	395	526	205	205	260
0.15	***	35893	36638	41530	9744	10062	10750	1835	1868	1892	538	538	553	277	277	277
	•••	36154	37769	55038	9872	10338	14112	1844	1883	2417	538	538	662	277	277	316
0.2	***	45592	46770	55804	12404	12959	14289	2334	2390	2453	671	671	680	324	324	324
	•••	45995	48632	68488	12641	13458	17422	2358	2414	2922	671	671	778	324	324	362
0.25	***	52812	54513	68501	14494	15325	17424	2761	2833	2928	775	775	790	372	372	372
	•••	53396	57229	79836	14835	16085	20206	2794	2880	3342	775	790	873	372	372	400
0.3	***	57576	59904	79331	16023	17139	20084	3094	3198	3331	861	870	885	410	410	410
	•••	58384	63561	89082	16474	18175	22465	3141	3260	3678	861	870	946	410	410	428
0.35	***	60070	63119	88080	17035	18436	22213	3340	3474	3649	918	933	942	434	434	434
	•••	61139	67789	96227	17605	19728	24199	3403	3554	3930	933	942	999	434	434	446
0.4	***	60578	64449	94659	17581	19244	23804	3507	3664	3878	965	980	989	443	443	458
	•••	61946	70149	101271	18270	20797	25407	3578	3759	4098	965	989	1030	443	458	456
0.45	***	59414	64179	99005	17724	19624	24834	3593	3768	4005	980	1004	1013	458	458	458
	•••	61124	70862	104213	18508	21381	26090	3673	3887	4182	989	1013	1041	458	458	456
0.5	***	56944	62611	101086	17519	19600	25300	3616	3806	4053	980	1004	1013	443	458	458
	•••	59010	70173	105054	18374	21524	26248	3697	3925	4142	989	1013	1030	443	458	446
0.55	***	53524	60008	100858	17011	19229	25190	3554	3744	4020	956	980	989	434	434	443
	•••	55947	68273	103793	17938	21248	25880	3640	3863	4098	965	989	999	434	434	428
0.6	***	49463	56612	98364	16251	18517	24502	3426	3616	3878	918	933	956	410	410	410
	•••	52185	65319	100430	17201	20559	24987	3507	3735	3930	918	942	946	410	410	400
0.65	***	45013	52574	93581	15254	17510	23258	3222	3412	3649	847	870	885	372	372	372
	•••	47943	61433	94967	16204	19490	23569	3308	3521	3678	861	870	873	372	372	362
0.7	***	40358	48005	86537	14043	16180	21438	2960	3118	3331	766	775	790	324	324	339
	•••	43359	56659	87401	14945	18033	21625	3032	3222	3342	775	790	778	324	339	316
0.75	***	35522	42908	77241	12603	14541	19054	2619	2747	2928	657	671	680	277	277	277
	•••	38458	51007	77734	13434	16180	19156	2675	2833	2922	657	671	662	277	277	260
0.8	***	30510	37232	65732	10917	12555	16118	2200	2310	2429	529	538	538	205	205	220
	•••	33209	44371	65966	11629	13900	16161	2248	2367	2417	538	538	526	205	220	194
0.85	***	25134	30795	52004	8922	10180	12641	1702	1773	1844	386	386	386	.	.	.
	•••	27438	36558	52096	9468	11145	12641	1740	1811	1829	386	386	368	.	.	.
0.9	***	19078	23186	36115	6523	7298	8613	1108	1146	1179	.	.	.	.	.	.
	•••	20773	27129	36125	6870	7844	8596	1132	1155	1156	.	.	.	.	.	.
0.95	***	11454	13458	18065	3435	3697	4044	.	.	.	.	.	.	.	.	.
	•••	12309	15144	18052	3554	3854	4025	.	.	.	.	.	.	.	.	.
0.98	***	4946	5445	6214	.	.	.	.	.	.	.	.	.	.	.	.
	•••	5169	5787	6200	.	.	.	.	.	.	.	.	.	.	.	.

TABLE 6: ALPHA= 0.025 POWER= 0.9   EXPECTED ACCRUAL THRU MINIMUM FOLLOW-UP= 10000

PCONT	FACT	DEL=.01 1.0 .75	.50 .25	.00 BIN	DEL=.02 1.0 .75	.50 .25	.00 BIN	DEL=.05 1.0 .75	.50 .25	.00 BIN	DEL=.10 1.0 75	.50 .25	.00 BIN	DEL=.15 1.0 .75	.50 .25	.00 BIN
							REQUIRED NUMBER OF PATIENTS								
0.01	•••	1616	1616	1616	591	591	591	191	191	191	91	91	91	66	66	66
	•••	1616	1616	6200	591	591	2049	191	191	558	91	91	227	66	66	135
0.02	•••	3582	3591	3616	1141	1157	1157	307	307	307	132	132	132	91	91	91
	•••	3582	3607	10235	1157	1157	3048	307	307	712	132	132	264	91	91	151
0.05	•••	10857	10982	11291	3107	3132	3166	666	666	666	241	241	241	141	141	141
	•••	10907	11091	21835	3116	3141	5916	666	666	1156	241	241	368	141	141	194
0.1	•••	23991	24407	26291	6566	6707	6957	1266	1282	1291	391	391	391	207	207	207
	•••	24141	24941	39487	6632	6816	10277	1282	1291	1829	391	391	526	207	207	260
0.15	•••	35932	36707	41532	9766	10091	10757	1841	1866	1891	541	541	541	266	282	282
	•••	36207	37857	55038	9907	10357	14112	1857	1882	2417	541	541	662	266	282	316
0.2	•••	45657	46891	55807	12441	13007	14291	2341	2391	2441	666	666	682	332	332	332
	•••	46082	48791	68488	12682	13491	17422	2366	2416	2922	666	682	778	332	332	362
0.25	•••	52907	54682	68507	14557	15382	17416	2766	2832	2932	782	782	791	366	382	382
	•••	53516	57457	79836	14891	16132	20206	2791	2882	3342	782	782	873	366	382	400
0.3	•••	57707	60132	79332	16107	17216	20082	3107	3207	3332	857	866	882	407	407	416
	•••	58541	63866	89082	16566	18241	22465	3141	3257	3678	866	882	946	407	407	428
0.35	•••	60241	63416	88082	17132	18532	22216	3357	3482	3641	916	932	941	432	432	441
	•••	61357	68182	96227	17707	19816	24199	3407	3557	3930	932	941	999	432	432	446
0.4	•••	60791	64816	94666	17691	19366	23807	3516	3666	3882	966	982	991	441	457	457
	•••	62216	70616	101271	18391	20891	25407	3591	3766	4098	966	991	1030	441	457	456
0.45	•••	59691	64616	99007	17857	19757	24832	3616	3782	4016	982	1007	1016	457	457	457
	•••	61466	71407	104213	18641	21491	26090	3691	3891	4182	991	1007	1041	457	457	456
0.5	•••	57282	63107	101082	17666	19757	25291	3632	3816	4057	982	1007	1016	441	457	457
	•••	59432	70782	105054	18532	21657	26248	3716	3932	4182	991	1007	1030	441	457	446
0.55	•••	53916	60566	100866	17182	19391	25182	3566	3757	4016	966	982	991	432	432	441
	•••	56416	68916	103793	18107	21382	25880	3657	3882	4098	966	991	999	432	441	428
0.6	•••	49916	57216	98366	16416	18691	24507	3441	3632	3882	916	932	957	407	407	416
	•••	52707	65991	100430	17366	20691	24987	3532	3741	3930	932	941	946	407	416	400
0.65	•••	45516	53207	93582	15416	17666	23257	3241	3416	3657	857	866	882	366	382	382
	•••	48507	62107	94967	16366	19616	23569	3316	3532	3678	857	882	873	382	382	362
0.7	•••	40866	48616	86532	14207	16332	21432	2966	3132	3332	766	782	791	332	332	332
	•••	43916	57316	87401	15107	18157	21625	3041	3216	3342	766	782	778	332	332	316
0.75	•••	36032	43491	77241	12757	14682	19057	2632	2757	2932	657	666	682	282	282	282
	•••	39007	51607	77734	13582	16282	19156	2691	2841	2922	666	682	662	282	282	260
0.8	•••	30982	37757	65732	11041	12666	16116	2207	2307	2432	532	541	541	216	216	216
	•••	33707	44891	65966	11741	13982	16161	2257	2366	2417	532	541	526	216	216	194
0.85	•••	25541	31232	52007	9032	10257	12641	1707	1766	1841	382	382	391	.	.	
	•••	27857	36966	52096	9566	11207	12641	1741	1807	1829	382	382	368	.	.	
0.9	•••	19366	23491	36116	6582	7341	8607	1116	1141	1182	.	.	.	.	.	.
	•••	21082	27391	36125	6916	7882	8596	1132	1157	1156	.	.	.	.	.	.
0.95	•••	11607	13607	18066	3457	3707	4041	.	.	.	.	.	.	.	.	.
	•••	12457	15241	18052	3582	3866	4025	.	.	.	.	.	.	.	.	.
0.98	•••	4991	5482	6216	.	.	.	.	.	.	.	.	.	.	.	.
	•••	5207	5807	6200	.	.	.	.	.	.	.	.	.	.	.	.

TABLE 6: ALPHA= 0.025 POWER= 0.9   EXPECTED ACCRUAL THRU MINIMUM FOLLOW-UP= 11000

	DEL=.01			DEL=.02			DEL=.05			DEL=.10			DEL=.15		
FACT=	1.0 .75	.50 .25	.00 BIN	1.0 .75	.50 .25	.00 BIN	1.0 .75	.50 .25	.00 BIN	1.0 75	.50 .25	.00 BIN	1.0 .75	.50 .25	.00 BIN

PCONT=•••   REQUIRED NUMBER OF PATIENTS

PCONT															
0.01 •••	1613	1613	1613	595	595	595	200	200	200	100	100	100	73	73	73
•••	1613	1613	6200	595	595	2049	200	200	558	100	100	227	73	73	135
0.02 •••	3582	3593	3620	1145	1145	1162	310	310	310	128	128	128	90	90	90
•••	3582	3610	10235	1145	1145	3048	310	310	712	128	128	264	90	90	151
0.05 •••	10870	10990	11293	3098	3125	3170	667	667	667	238	238	238	145	145	145
•••	10925	11100	21835	3115	3142	5916	667	667	1156	238	238	368	145	145	194
0.1 •••	24042	24482	26297	6590	6728	6948	1272	1283	1300	392	392	403	210	210	210
•••	24207	25015	39487	6645	6827	10277	1283	1283	1829	392	392	526	210	210	260
0.15 •••	36015	36840	41532	9808	10127	10760	1850	1860	1888	540	540	540	265	282	282
•••	36307	38023	55038	9945	10385	14112	1860	1877	2417	540	540	662	265	282	316
0.2 •••	45795	47115	55805	12520	13080	14290	2345	2400	2455	667	678	678	320	320	337
•••	46245	49078	68488	12750	13548	17422	2372	2427	2922	667	678	778	320	337	362
0.25 •••	53093	55007	68510	14665	15500	17425	2768	2840	2933	777	777	788	375	375	375
•••	53753	57895	79836	15005	16215	20206	2812	2878	3342	777	788	873	375	375	400
0.3 •••	57960	60573	79328	16260	17370	20082	3115	3208	3335	860	870	887	403	403	420
•••	58868	64450	89082	16710	18360	22465	3153	3263	3678	870	870	946	403	403	428
0.35 •••	60573	63983	88090	17315	18718	22210	3373	3500	3648	925	942	953	430	430	430
•••	61783	68922	96227	17910	19972	24199	3428	3565	3930	925	942	999	430	430	446
0.4 •••	61233	65523	94662	17920	19587	23805	3538	3692	3868	970	980	997	447	447	458
•••	62762	71507	101271	18625	21083	25407	3610	3775	4098	970	980	1030	447	447	456
0.45 •••	60232	65457	99007	18113	20010	24840	3637	3802	4005	980	1008	1025	447	458	458
•••	62140	72425	104213	18910	21705	26090	3720	3895	4182	997	1008	1041	458	458	456
0.5 •••	57950	64082	101080	17948	20038	25290	3665	3830	4060	980	997	1025	447	447	458
•••	60232	71907	105054	18828	21880	26248	3730	3940	4182	997	1008	1030	447	447	446
0.55 •••	54705	61645	100860	17480	19680	25180	3593	3775	4022	970	980	997	430	430	447
•••	57328	70132	103793	18405	21622	25880	3692	3885	4098	970	980	999	430	430	428
0.6 •••	50800	58362	98358	16738	18982	24510	3472	3648	3885	925	942	953	403	403	420
•••	53715	67255	100430	17690	20935	24987	3555	3758	3930	925	942	946	403	420	400
0.65 •••	46465	54375	93590	15737	17965	23255	3263	3428	3648	860	870	887	375	375	375
•••	49562	63367	94967	16683	19862	23569	3345	3538	3678	860	870	873	375	375	362
0.7 •••	41845	49782	86533	14500	16600	21430	3005	3142	3335	777	777	788	337	337	337
•••	44980	58527	87401	15407	18360	21625	3070	3235	3342	777	788	778	337	337	316
0.75 •••	36995	44595	77238	13025	14923	19048	2647	2768	2922	667	667	678	282	282	282
•••	40030	52725	77734	13840	16463	19156	2713	2840	2922	667	678	662	282	282	260
0.8 •••	31863	38738	65732	11282	12877	16122	2235	2317	2427	530	540	540	210	210	210
•••	34640	45860	65966	11970	14125	16161	2273	2372	2417	540	540	526	210	210	194
0.85 •••	26297	32028	52010	9203	10413	12640	1723	1778	1850	375	392	392	.	.	.
•••	28635	37720	52096	9742	11310	12641	1750	1805	1829	375	392	368	.	.	.
0.9 •••	19928	24053	36115	6700	7432	8615	1118	1145	1173	.	.	.	.	.	.
•••	21650	27892	36125	7030	7927	8596	1135	1162	1156	.	.	.	.	.	.
0.95 •••	11887	13850	18075	3500	3730	4050	.	.	.	.	.	.	.	.	.
•••	12740	15435	18052	3610	3885	4025	.	.	.	.	.	.	.	.	.
0.98 •••	5067	5535	6222	.	.	.	.	.	.	.	.	.	.	.	.
•••	5270	5837	6200	.	.	.	.	.	.	.	.	.	.	.	.

TABLE 6: ALPHA= 0.025 POWER= 0.9   EXPECTED ACCRUAL THRU MINIMUM FOLLOW-UP= 12000

		DEL=.01			DEL=.02			DEL=.05			DEL=.10			DEL=.15	
FACT=	1.0 .75	.50 .25	.00 BIN	1.0 .75	.50 .25	.00 BIN	1.0 .75	.50 .25	.00 BIN	1.0 .75	.50 .25	.00 BIN	1.0 .75	.50 .25	.00 BIN
PCONT=***				REQUIRED NUMBER OF PATIENTS											
0.01 ***	1609	1609	1628	589	589	589	199	199	199	98	98	98	68	68	68
•••	1609	1609	6200	589	589	2049	199	199	558	98	98	227	68	68	135
0.02 ***	3578	3589	3619	1148	1148	1159	308	308	308	139	139	139	79	79	79
•••	3589	3608	10235	1148	1148	3048	308	308	712	139	139	264	79	79	151
0.05 ***	10879	11018	11299	3109	3128	3158	668	668	668	229	229	229	139	139	139
•••	10939	11119	21835	3109	3139	5916	668	668	1156	229	229	368	139	139	194
0.1 ***	24079	24548	26288	6608	6739	6949	1279	1279	1298	398	398	398	218	218	218
•••	24259	25088	39487	6668	6829	10277	1279	1279	1829	398	398	526	218	218	260
0.15 ***	36098	36979	41528	9848	10159	10748	1849	1868	1898	548	548	548	278	278	278
•••	36409	38179	55038	9979	10399	14112	1868	1879	2417	548	548	662	278	278	316
0.2 ***	45919	47329	55808	12589	13148	14288	2359	2389	2449	668	668	679	319	338	338
•••	46418	49358	68488	12829	13598	17422	2378	2419	2922	668	679	778	319	338	362
0.25 ***	53269	55328	68509	14768	15589	17419	2779	2858	2929	769	788	788	368	379	379
•••	53989	58298	79836	15109	16298	20206	2809	2888	3342	788	788	873	368	379	400
0.3 ***	58208	60998	79328	16388	17509	20078	3128	3218	3338	859	878	878	409	409	409
•••	59198	64999	89082	16868	18469	22465	3169	3278	3678	859	878	946	409	409	428
0.35 ***	60908	64538	88088	17498	18889	22219	3398	3499	3649	919	938	949	428	439	439
•••	62209	69608	96227	18079	20108	24199	3439	3578	3930	938	938	999	439	439	446
0.4 ***	61658	66199	94669	18128	19789	23809	3559	3698	3878	968	979	998	439	458	458
•••	63308	72319	101271	18829	21248	25407	3638	3788	4098	979	979	1030	458	458	456
0.45 ***	60769	66259	99008	18349	20239	24829	3668	3818	4009	998	1009	1009	458	458	458
•••	62798	73369	104213	19148	21889	26090	3739	3908	4182	998	1009	1041	458	458	456
0.5 ***	58598	64988	101078	18218	20288	25298	3679	3848	4058	998	998	1009	439	458	458
•••	60998	72949	105054	19088	22088	26248	3758	3949	4182	998	1009	1030	458	458	446
0.55 ***	55459	62648	100868	17768	19939	25189	3619	3799	4009	968	979	998	428	439	439
•••	58208	71258	103793	18679	21829	25880	3709	3889	4098	979	979	999	439	439	428
0.6 ***	51638	59419	98359	17018	19249	24499	3499	3668	3878	919	938	949	409	409	409
•••	54668	68408	100430	17978	21158	24987	3578	3758	3930	938	938	946	409	409	400
0.65 ***	47378	55459	93578	16028	18218	23258	3289	3458	3649	859	878	878	368	379	379
•••	50558	64519	94967	16958	20059	23569	3368	3548	3678	859	878	873	379	379	362
0.7 ***	42769	50858	86539	14779	16849	21439	3019	3158	3338	769	788	788	338	338	338
•••	45979	59629	87401	15668	18559	21625	3079	3248	3342	769	788	778	338	338	316
0.75 ***	37898	45608	77239	13279	15139	19058	2678	2779	2929	668	668	679	278	278	278
•••	40988	53738	77734	14078	16628	19156	2719	2858	2922	668	679	662	278	278	260
0.8 ***	32689	39638	65719	11498	13058	16118	2239	2329	2438	529	529	548	218	218	218
•••	35498	46729	65966	12169	14258	16161	2288	2378	2417	529	548	526	218	218	194
0.85 ***	26989	32749	52009	9379	10538	12638	1729	1789	1849	379	379	379	.	.	.
•••	29359	38389	52096	9889	11389	12641	1759	1808	1829	379	379	368	.	.	.
0.9 ***	20449	24559	36109	6799	7508	8599	1129	1148	1178	.	.	.	.	.	.
•••	22159	28328	36125	7118	7988	8596	1129	1159	1156	.	.	.	.	.	.
0.95 ***	12158	14078	18068	3529	3758	4039	.	.	.	.	.	.	.	.	.
•••	12979	15608	18052	3638	3889	4025	.	.	.	.	.	.	.	.	.
0.98 ***	5138	5569	6218	.	.	.	.	.	.	.	.	.	.	.	.
•••	5329	5869	6200	.	.	.	.	.	.	.	.	.	.	.	.

TABLE 6: ALPHA= 0.025 POWER= 0.9   EXPECTED ACCRUAL THRU MINIMUM FOLLOW-UP= 13000

	DEL=.01			DEL=.02			DEL=.05			DEL=.10			DEL=.15		
FACT=	1.0 .75	.50 .25	.00 BIN	1.0 .75	.50 .25	.00 BIN	1.0 .75	.50 .25	.00 BIN	1.0 75	.50 .25	.00 BIN	1.0 .75	.50 .25	.00 BIN
PCONT=•••				REQUIRED NUMBER OF PATIENTS											
0.01 •••	1613	1613	1613	594	594	594	204	204	204	106	106	106	74	74	74
•••	1613	1613	6200	594	594	2049	204	204	558	106	106	227	74	74	135
0.02 •••	3584	3596	3616	1146	1146	1158	313	313	313	139	139	139	86	86	86
•••	3584	3596	10235	1146	1158	3048	313	313	712	139	139	264	86	86	151
0.05 •••	10896	11026	11298	3108	3129	3161	671	671	671	236	236	236	139	139	139
•••	10941	11124	21835	3129	3141	5916	671	671	1156	236	236	368	139	139	194
0.1 •••	24136	24591	26281	6618	6748	6943	1276	1288	1288	399	399	399	204	216	216
•••	24298	25143	39487	6683	6846	10277	1276	1288	1829	399	399	526	204	216	260
0.15 •••	36181	37103	41523	9889	10193	10746	1861	1873	1894	541	541	541	269	269	269
•••	36506	38326	55038	10019	10421	14112	1861	1894	2417	541	541	662	269	269	316
0.2 •••	46041	47536	55791	12651	13204	14288	2361	2393	2446	671	671	671	334	334	334
•••	46581	49604	68488	12891	13638	17422	2381	2426	2922	671	671	778	334	334	362
0.25 •••	53451	55628	68498	14861	15686	17429	2804	2848	2934	768	789	789	378	378	378
•••	54231	58671	79836	15219	16356	20206	2816	2881	3342	789	789	873	378	378	400
0.3 •••	58456	61413	79321	16519	17624	20073	3141	3226	3324	866	866	886	411	411	411
•••	59516	65496	89082	16986	18566	22465	3173	3271	3678	866	866	946	411	411	428
0.35 •••	61239	65074	88084	17656	19054	22218	3401	3519	3649	931	931	951	431	431	443
•••	62616	70241	96227	18241	20224	24199	3454	3584	3930	931	951	999	431	431	446
0.4 •••	62084	66841	94661	18318	19976	23811	3584	3714	3876	963	984	996	443	443	443
•••	63818	73081	101271	19021	21394	25407	3649	3791	4098	984	984	1030	443	443	456
0.45 •••	61304	67003	99004	18578	20451	24839	3681	3823	4006	996	996	1016	443	464	464
•••	63416	74251	104213	19379	22056	26090	3758	3921	4182	996	1016	1041	464	464	456
0.5 •••	59224	65854	101084	18469	20516	25294	3714	3856	4051	996	996	1016	443	443	464
•••	61726	73926	105054	19326	22251	26248	3779	3953	4182	996	1016	1030	443	443	446
0.55 •••	56181	63591	100868	18014	20191	25176	3649	3811	4018	963	984	996	431	431	443
•••	59029	72289	103793	18936	22011	25880	3726	3909	4098	984	984	999	431	443	428
0.6 •••	52443	60426	98366	17278	19488	24493	3519	3681	3876	919	931	951	411	411	411
•••	55551	69461	100430	18221	21341	24987	3596	3779	3930	931	951	946	411	411	400
0.65 •••	48218	56494	93576	16291	18448	23246	3324	3466	3649	854	866	886	378	378	378
•••	51489	65573	94967	17213	20236	23569	3389	3551	3678	866	886	873	378	378	362
0.7 •••	43636	51858	86536	15024	17071	21438	3043	3173	3336	768	789	789	334	334	334
•••	46918	60633	87401	15913	18729	21625	3096	3238	3342	768	789	778	334	334	316
0.75 •••	38728	46528	77241	13508	15328	19054	2686	2804	2934	659	671	671	281	281	281
•••	41881	54653	77734	14288	16779	19156	2739	2848	2922	671	671	662	281	281	260
0.8 •••	33451	40451	65724	11688	13216	16129	2251	2328	2426	529	541	541	216	216	216
•••	36311	47524	65966	12359	14374	16161	2296	2381	2417	541	541	526	216	216	194
0.85 •••	27646	33431	52009	9531	10648	12631	1743	1796	1841	378	378	378	.	.	.
•••	30018	39009	52096	10031	11481	12641	1764	1808	1829	378	378	368	.	.	.
0.9 •••	20918	25013	36116	6899	7561	8601	1126	1146	1179	.	.	.	.	.	.
•••	22641	28718	36125	7203	8016	8596	1146	1158	1156	.	.	.	.	.	.
0.95 •••	12391	14276	18058	3563	3779	4039	.	.	.	.	.	.	.	.	.
•••	13204	15751	18052	3661	3909	4025	.	.	.	.	.	.	.	.	.
0.98 •••	5188	5611	6216	.	.	.	.	.	.	.	.	.	.	.	.
•••	5383	5891	6200	.	.	.	.	.	.	.	.	.	.	.	.

TABLE 6: ALPHA= 0.025 POWER= 0.9   EXPECTED ACCRUAL THRU MINIMUM FOLLOW-UP= 14000

		DEL=.01			DEL=.02			DEL=.05			DEL=.10			DEL=.15		
FACT=		1.0 .75	.50 .25	.00 BIN	1.0 .75	.50 .25	.00 BIN	1.0 .75	.50 .25	.00 BIN	1.0 75	.50 .25	.00 BIN	1.0 .75	.50 .25	.00 BIN
PCONT=•••					REQUIRED NUMBER OF PATIENTS											
0.01	•••	1619	1619	1619	582	582	582	197	197	197	92	92	92	57	57	57
	•••	1619	1619	6200	582	582	2049	197	197	558	92	92	227	57	57	135
0.02	•••	3579	3592	3614	1142	1142	1164	302	302	302	127	127	127	92	92	92
	•••	3592	3614	10235	1142	1142	3048	302	302	712	127	127	264	92	92	151
0.05	•••	10907	11034	11292	3102	3137	3159	674	674	674	232	232	232	127	127	127
	•••	10964	11139	21835	3124	3159	5916	674	674	1156	232	232	368	127	127	194
0.1	•••	24172	24662	26294	6637	6764	6952	1282	1282	1282	394	394	394	219	219	219
	•••	24347	25187	39487	6694	6847	10277	1282	1282	1829	394	394	526	219	219	260
0.15	•••	36247	37227	41532	9927	10229	10754	1864	1877	1899	547	547	547	267	267	267
	•••	36619	38452	55038	10054	10452	14112	1864	1877	2417	547	547	662	267	267	316
0.2	•••	46174	47749	55799	12714	13252	14289	2367	2402	2437	674	674	674	324	324	324
	•••	46734	49849	68488	12959	13672	17422	2389	2424	2922	674	674	778	324	324	362
0.25	•••	53642	55917	68504	14954	15772	17417	2809	2857	2927	779	779	792	372	372	372
	•••	54469	59019	79836	15304	16424	20206	2822	2892	3342	779	792	873	372	372	400
0.3	•••	58704	61797	79332	16647	17732	20077	3159	3229	3334	862	862	884	407	407	407
	•••	59824	65962	89082	17102	18642	22465	3194	3277	3678	862	884	946	407	407	428
0.35	•••	61574	65564	88082	17802	19189	22212	3417	3522	3649	932	932	954	429	429	442
	•••	63022	70827	96227	18397	20344	24199	3474	3579	3930	932	954	999	429	442	446
0.4	•••	62497	67454	94662	18502	20147	23809	3592	3719	3872	967	989	989	442	442	442
	•••	64317	73789	101271	19202	21512	25407	3662	3789	4098	967	989	1030	442	442	456
0.45	•••	61797	67734	99002	18782	20637	24837	3697	3837	4012	989	1002	1024	442	464	464
	•••	64024	75062	104213	19574	22199	26090	3767	3929	4182	1002	1002	1041	464	464	456
0.5	•••	59824	66662	101089	18677	20707	25292	3719	3872	4069	989	1002	1024	442	442	464
	•••	62427	74817	105054	19552	22422	26248	3802	3964	4182	1002	1002	1030	442	442	446
0.55	•••	56884	64479	100857	18257	20392	25187	3662	3824	4012	967	989	989	429	442	442
	•••	59824	73229	103793	19167	22177	25880	3754	3907	4098	967	989	999	429	442	428
0.6	•••	53222	61364	98359	17522	19714	24509	3544	3697	3872	932	932	954	407	407	407
	•••	56407	70442	100430	18467	21499	24987	3614	3789	3930	932	954	946	407	407	400
0.65	•••	49044	57444	93577	16529	18664	23249	3334	3474	3649	862	862	884	372	372	372
	•••	52369	66544	94967	17452	20392	23569	3404	3557	3678	862	884	873	372	372	362
0.7	•••	44459	52789	86529	15247	17264	21429	3054	3172	3334	779	779	792	324	337	337
	•••	47797	61574	87401	16122	18874	21625	3124	3242	3342	779	792	778	324	337	316
0.75	•••	39524	47412	77254	13707	15514	19049	2704	2809	2927	674	674	674	267	267	267
	•••	42709	55497	77734	14499	16892	19156	2752	2857	2922	674	674	662	267	267	260
0.8	•••	34182	41217	65717	11874	13357	16122	2262	2332	2424	534	534	547	219	219	219
	•••	37052	48239	65966	12517	14477	16161	2297	2389	2417	534	547	526	219	219	194
0.85	•••	28254	34042	52019	9647	10754	12644	1737	1794	1842	372	394	394	.	.	.
	•••	30647	39559	52096	10159	11537	12641	1772	1807	1829	372	394	368	.	.	.
0.9	•••	21372	25454	36107	6974	7617	8597	1129	1142	1177	.	.	.	.	.	.
	•••	23087	29059	36125	7267	8059	8596	1142	1164	1156	.	.	.	.	.	.
0.95	•••	12609	14442	18069	3592	3789	4047	.	.	.	.	.	.	.	.	.
	•••	13414	15877	18052	3684	3907	4025	.	.	.	.	.	.	.	.	.
0.98	•••	5237	5657	6217	.	.	.	.	.	.	.	.	.	.	.	.
	•••	5434	5902	6200	.	.	.	.	.	.	.	.	.	.	.	.

TABLE 6: ALPHA= 0.025 POWER= 0.9   EXPECTED ACCRUAL THRU MINIMUM FOLLOW-UP= 15000

PCONT=***		DEL=.01			DEL=.02			DEL=.05			DEL=.10			DEL=.15		
FACT=		1.0 .75	.50 .25	.00 BIN	1.0 .75	.50 .25	.00 BIN	1.0 .75	.50 .25	.00 BIN	1.0 75	.50 .25	.00 BIN	1.0 .75	.50 .25	.00 BIN
							REQUIRED NUMBER OF PATIENTS									
0.01	***	1622	1622	1622	586	586	586	197	197	197	99	99	99	61	61	61
	•••	1622	1622	6200	586	586	2049	197	197	558	99	99	227	61	61	135
0.02	***	3586	3586	3610	1149	1149	1149	310	310	310	136	136	136	85	85	85
	•••	3586	3610	10235	1149	1149	3048	310	310	712	136	136	264	85	85	151
0.05	***	10922	11049	11297	3122	3136	3160	661	661	661	235	235	235	136	136	136
	•••	10974	11147	21835	3122	3160	5916	661	661	1156	235	235	368	136	136	194
0.1	***	24211	24722	26297	6661	6774	6947	1285	1285	1299	399	399	399	211	211	211
	•••	24399	25247	39487	6699	6849	10277	1285	1285	1829	399	399	526	211	211	260
0.15	***	36324	37336	41536	9961	10247	10749	1861	1885	1899	535	549	549	272	272	272
	•••	36699	38574	55038	10097	10472	14112	1861	1885	2417	549	549	662	272	272	316
0.2	***	46285	47935	55810	12774	13299	14297	2372	2410	2447	661	685	685	324	324	324
	•••	46885	50049	68488	12999	13711	17422	2386	2424	2922	661	685	778	324	324	362
0.25	***	53822	56199	68499	15047	15835	17424	2799	2860	2935	774	774	797	361	385	385
	•••	54685	59335	79836	15385	16472	20206	2836	2897	3342	774	797	873	385	385	400
0.3	***	58960	62185	79322	16749	17836	20072	3160	3235	3324	872	872	872	399	399	422
	•••	60136	66399	89082	17222	18722	22465	3197	3286	3678	872	872	946	399	399	428
0.35	***	61885	66047	88074	17949	19322	22210	3436	3535	3647	924	947	947	436	436	436
	•••	63422	71386	96227	18535	20424	24199	3474	3586	3930	924	947	999	436	436	446
0.4	***	62897	68035	94660	18661	20297	23799	3624	3736	3872	985	985	999	460	460	460
	•••	64810	74461	101271	19360	21624	25407	3661	3797	4098	985	985	1030	460	460	456
0.45	***	62297	68424	99010	18961	20822	24835	3722	3849	4022	999	999	1022	460	460	460
	•••	64622	75811	104213	19749	22336	26090	3774	3924	4182	999	1022	1041	460	460	456
0.5	***	60422	67435	101072	18886	20911	25299	3736	3886	4060	999	999	1022	460	460	460
	•••	63099	75647	105054	19749	22561	26248	3811	3961	4182	999	1022	1030	460	460	446
0.55	***	57549	65311	100861	18474	20597	25186	3685	3835	4022	961	985	999	436	436	436
	•••	60572	74110	103793	19397	22322	25880	3760	3924	4098	985	985	999	436	436	428
0.6	***	53949	62236	98372	17761	19922	24497	3549	3699	3872	924	947	947	399	422	422
	•••	57211	71349	100430	18685	21661	24987	3624	3774	3930	924	947	946	399	422	400
0.65	***	49810	58322	93586	16749	18872	23260	3347	3497	3647	872	872	886	385	385	385
	•••	53199	67435	94967	17672	20536	23569	3422	3572	3678	872	872	873	385	385	362
0.7	***	45235	53649	86536	15474	17447	21436	3061	3197	3324	774	774	797	324	324	324
	•••	48624	62424	87401	16336	18999	21625	3122	3249	3342	774	797	778	324	324	316
0.75	***	40285	48211	77236	13899	15661	19060	2710	2799	2935	661	661	685	272	272	272
	•••	43486	56274	77734	14672	17011	19156	2761	2860	2922	661	685	662	272	272	260
0.8	***	34861	41935	65724	12024	13486	16111	2274	2349	2424	535	535	549	211	211	211
	•••	37749	48886	65966	12661	14560	16161	2311	2386	2417	535	535	526	211	211	194
0.85	***	28824	34622	51999	9774	10847	12647	1749	1786	1847	385	385	385	.	.	.
	•••	31224	40074	52096	10261	11611	12641	1772	1824	1829	385	385	368	.	.	.
0.9	***	21774	25824	36122	7060	7674	8611	1135	1149	1172	.	.	.	.	.	.
	•••	23499	29386	36125	7336	8086	8596	1149	1172	1156	.	.	.	.	.	.
0.95	***	12797	14611	18061	3624	3811	4036	.	.	.	.	.	.	.	.	.
	•••	13599	15999	18052	3699	3924	4025	.	.	.	.	.	.	.	.	.
0.98	***	5297	5686	6211	.	.	.	.	.	.	.	.	.	.	.	.
	•••	5485	5935	6200	.	.	.	.	.	.	.	.	.	.	.	.

TABLE 6: ALPHA= 0.025 POWER= 0.9   EXPECTED ACCRUAL THRU MINIMUM FOLLOW-UP= 17000

		DEL=.01			DEL=.02			DEL=.05			DEL=.10			DEL=.15		
FACT=		1.0 .75	.50 .25	.00 BIN	1.0 .75	.50 .25	.00 BIN	1.0 .75	.50 .25	.00 BIN	1.0 75	.50 .25	.00 BIN	1.0 .75	.50 .25	.00 BIN
PCONT=***							REQUIRED NUMBER OF PATIENTS								
0.01	***	1626	1626	1626	580	580	580	197	197	197	96	96	96	70	70	70
	***	1626	1626	6200	580	580	2049	197	197	558	96	96	227	70	70	135
0.02	***	3581	3597	3624	1159	1159	1159	309	309	309	139	139	139	96	96	96
	***	3597	3597	10235	1159	1159	3048	309	309	712	139	139	264	96	96	151
0.05	***	10950	11061	11290	3114	3130	3156	665	665	665	240	240	240	139	139	139
	***	10992	11162	21835	3130	3156	5916	665	665	1156	240	240	368	139	139	194
0.1	***	24295	24805	26292	6684	6785	6955	1286	1286	1286	394	394	394	197	197	197
	***	24491	25341	39487	6726	6870	10277	1286	1286	1829	394	394	526	197	197	260
0.15	***	36492	37555	41534	10015	10296	10764	1855	1881	1897	537	537	537	266	266	266
	***	36901	38787	55038	10142	10482	14112	1881	1881	2417	537	537	662	266	266	316
0.2	***	46522	48291	55787	12889	13399	14291	2391	2407	2450	665	665	665	325	325	325
	***	47186	50432	68488	13101	13755	17422	2391	2434	2922	665	665	778	325	325	362
0.25	***	54172	56722	68495	15184	15965	17410	2816	2875	2917	776	776	792	367	367	367
	***	55107	59910	79836	15524	16560	20206	2832	2901	3342	776	792	873	367	367	400
0.3	***	59442	62911	79332	16969	18005	20071	3172	3241	3326	861	877	877	410	410	410
	***	60717	67204	89082	17410	18839	22465	3215	3284	3678	877	877	946	410	410	428
0.35	***	62529	66965	88087	18201	19535	22212	3454	3539	3640	920	946	946	436	436	436
	***	64186	72389	96227	18770	20597	24199	3496	3597	3930	946	946	999	436	436	446
0.4	***	63692	69132	94659	18982	20555	23811	3640	3751	3879	989	989	989	452	452	452
	***	65759	75661	101271	19662	21830	25407	3682	3810	4098	989	989	1030	452	452	456
0.45	***	63267	69685	99010	19322	21107	24831	3751	3879	4006	1005	1005	1005	452	452	452
	***	65732	77165	104213	20087	22552	26090	3810	3937	4182	1005	1005	1041	452	452	456
0.5	***	61551	68877	101076	19264	21235	25299	3767	3895	4065	1005	1005	1005	452	452	452
	***	64399	77149	105054	20114	22807	26248	3836	3980	4182	1005	1005	1030	452	452	446
0.55	***	58831	66864	100864	18881	20937	25187	3725	3852	4006	962	989	989	436	436	436
	***	61976	75704	103793	19774	22595	25880	3794	3937	4098	989	989	999	436	436	428
0.6	***	55346	63862	98356	18159	20257	24507	3597	3725	3879	920	946	946	410	410	410
	***	58720	72984	100430	19067	21915	24987	3640	3794	3930	946	946	946	410	410	400
0.65	***	51266	59952	93570	17155	19195	23259	3385	3496	3640	861	877	877	367	367	367
	***	54751	69047	94967	18047	20794	23569	3454	3581	3678	861	877	873	367	367	362
0.7	***	46692	55235	86541	15837	17776	21431	3087	3199	3326	776	776	792	325	325	325
	***	50161	63947	87401	16687	19221	21625	3156	3257	3342	776	792	778	325	325	316
0.75	***	41661	49694	77250	14249	15949	19051	2731	2816	2917	665	665	691	282	282	282
	***	44934	57657	77734	14987	17197	19156	2774	2875	2922	665	665	662	282	282	260
0.8	***	36110	43234	65732	12310	13712	16119	2280	2349	2434	537	537	537	224	224	224
	***	39042	50076	65966	12931	14716	16161	2322	2391	2417	537	537	526	224	224	194
0.85	***	29862	35642	52005	9999	11019	12634	1754	1796	1839	394	394	394	.	.	.
	***	32285	40955	52096	10466	11715	12641	1770	1812	1829	394	394	368	.	.	.
0.9	***	22536	26531	36110	7194	7762	8612	1132	1159	1175	.	.	.	.	.	.
	***	24236	29931	36125	7449	8145	8596	1159	1159	1156	.	.	.	.	.	.
0.95	***	13160	14902	18074	3666	3836	4049	.	.	.	.	.	.	.	.	.
	***	13925	16204	18052	3751	3937	4025	.	.	.	.	.	.	.	.	.
0.98	***	5382	5749	6216	.	.	.	.	.	.	.	.	.	.	.	.
	***	5552	5961	6200	.	.	.	.	.	.	.	.	.	.	.	.

TABLE 6: ALPHA= 0.025 POWER= 0.9   EXPECTED ACCRUAL THRU MINIMUM FOLLOW-UP= 20000

	DEL=.01			DEL=.02			DEL=.05			DEL=.10			DEL=.15		
FACT=	1.0 .75	.50 .25	.00 BIN	1.0 .75	.50 .25	.00 BIN	1.0 .75	.50 .25	.00 BIN	1.0 75	.50 .25	.00 BIN	1.0 .75	.50 .25	.00 BIN

PCONT=•••   REQUIRED NUMBER OF PATIENTS

PCONT		DEL=.01			DEL=.02			DEL=.05			DEL=.10			DEL=.15		
0.01	***	1613	1613	1613	582	582	582	182	182	182	82	82	82	63	63	63
	•••	1613	1613	6200	582	582	2049	182	182	558	82	82	227	63	63	135
0.02	***	3582	3613	3613	1163	1163	1163	313	313	313	132	132	132	82	82	82
	•••	3582	3613	10235	1163	1163	3048	313	313	712	132	132	264	82	82	151
0.05	***	10982	11082	11282	3132	3132	3163	663	663	663	232	232	232	132	132	132
	•••	11032	11182	21835	3132	3163	5916	663	663	1156	232	232	368	132	132	194
0.1	***	24413	24932	26282	6713	6813	6963	1282	1282	1282	382	382	382	213	213	213
	•••	24613	25432	39487	6763	6882	10277	1282	1282	1829	382	382	526	213	213	260
0.15	***	36713	37863	41532	10082	10363	10763	1863	1882	1882	532	532	532	282	282	282
	•••	37163	39063	55038	10213	10532	14112	1863	1882	2417	532	532	662	282	282	316
0.2	***	46882	48782	55813	13013	13482	14282	2382	2413	2432	663	682	682	332	332	332
	•••	47613	50932	68488	13213	13832	17422	2413	2432	2922	663	682	778	332	332	362
0.25	***	54682	57463	68513	15382	16132	17413	2832	2882	2932	782	782	782	382	382	382
	•••	55732	60663	79836	15713	16663	20206	2863	2913	3342	782	782	873	382	382	400
0.3	***	60132	63863	79332	17213	18232	20082	3213	3263	3332	863	882	882	413	413	413
	•••	61532	68213	89082	17663	18982	22465	3232	3282	3678	863	882	946	413	413	428
0.35	***	63413	68182	88082	18532	19813	22213	3482	3563	3632	932	932	932	432	432	432
	•••	65232	73682	96227	19082	20782	24199	3513	3613	3930	932	932	999	432	432	446
0.4	***	64813	70613	94663	19363	20882	23813	3663	3763	3882	982	982	982	463	463	463
	•••	67032	77182	101271	20032	22063	25407	3713	3813	4098	982	982	1030	463	463	456
0.45	***	64613	71413	99013	19763	21482	24832	3782	3882	4013	1013	1013	1013	463	463	463
	•••	67263	78932	104213	20513	22832	26090	3832	3932	4182	1013	1013	1041	463	463	456
0.5	***	63113	70782	101082	19763	21663	25282	3813	3932	4063	1013	1013	1013	463	463	463
	•••	66132	79063	105054	20582	23113	26248	3863	3982	4182	1013	1013	1030	463	463	446
0.55	***	60563	68913	100863	19382	21382	25182	3763	3882	4013	982	982	982	432	432	432
	•••	63882	77732	103793	20263	22882	25880	3813	3932	4098	982	982	999	432	432	428
0.6	***	57213	65982	98363	18682	20682	24513	3632	3732	3882	932	932	963	413	413	413
	•••	60732	75063	100430	19563	22213	24987	3682	3813	3930	932	932	946	413	413	400
0.65	***	53213	62113	93582	17663	19613	23263	3413	3532	3663	863	882	882	382	382	382
	•••	56813	71113	94967	18532	21082	23569	3463	3582	3678	863	882	873	382	382	362
0.7	***	48613	57313	86532	16332	18163	21432	3132	3213	3332	782	782	782	332	332	332
	•••	52163	65882	87401	17132	19482	21625	3182	3282	3342	782	782	778	332	332	316
0.75	***	43482	51613	77232	14682	16282	19063	2763	2832	2932	663	682	682	282	282	282
	•••	46832	59413	77734	15382	17432	19156	2782	2882	2922	663	682	662	282	282	260
0.8	***	37763	44882	65732	12663	13982	16113	2313	2363	2432	532	532	532	213	213	213
	•••	40713	51532	65966	13263	14882	16161	2332	2382	2417	532	532	526	213	213	194
0.85	***	31232	36963	52013	10263	11213	12632	1763	1813	1832	382	382	382	.	.	.
	•••	33632	42063	52096	10682	11832	12641	1782	1832	1829	382	382	368	.	.	.
0.9	***	23482	27382	36113	7332	7882	8613	1132	1163	1182	.	.	.	.	.	.
	•••	25163	30613	36125	7582	8213	8596	1163	1163	1156	.	.	.	.	.	.
0.95	***	13613	15232	18063	3713	3863	4032	.	.	.	.	.	.	.	.	.
	•••	14332	16432	18052	3782	3932	4025	.	.	.	.	.	.	.	.	.
0.98	***	5482	5813	6213	.	.	.	.	.	.	.	.	.	.	.	.
	•••	5632	5982	6200	.	.	.	.	.	.	.	.	.	.	.	.

TABLE 6: ALPHA= 0.025 POWER= 0.9   EXPECTED ACCRUAL THRU MINIMUM FOLLOW-UP= 25000

PCONT	DEL=.01 1.0 .75	DEL=.01 .50 .25	DEL=.01 .00 BIN	DEL=.02 1.0 .75	DEL=.02 .50 .25	DEL=.02 .00 BIN	DEL=.05 1.0 .75	DEL=.05 .50 .25	DEL=.05 .00 BIN	DEL=.10 1.0 75	DEL=.10 .50 .25	DEL=.10 .00 BIN	DEL=.15 1.0 .75	DEL=.15 .50 .25	DEL=.15 .00 BIN
						REQUIRED NUMBER OF PATIENTS									
0.01	1602	1602	1602	579	579	579	204	204	204	102	102	102	79	79	79
	1602	1602	6200	579	579	2049	204	204	558	102	102	227	79	79	135
0.02	3602	3602	3602	1141	1141	1141	290	290	290	141	141	141	79	79	79
	3602	3602	10235	1141	1141	3048	290	290	712	141	141	264	79	79	151
0.05	11016	11141	11290	3141	3141	3165	665	665	665	227	227	227	141	141	141
	11079	11204	21835	3141	3165	5916	665	665	1156	227	227	368	141	141	194
0.1	24579	25102	26290	6727	6829	6954	1290	1290	1290	391	391	391	204	204	204
	24790	25579	39487	6790	6891	10277	1290	1290	1829	391	391	526	204	204	260
0.15	37040	38266	41540	10165	10415	10766	1852	1891	1891	540	540	540	266	266	266
	37540	39415	55038	10290	10579	14112	1891	1891	2417	540	540	662	266	266	316
0.2	47454	49477	55790	13165	13602	14290	2391	2415	2454	665	665	665	329	329	329
	48227	51579	68488	13391	13915	17422	2415	2415	2922	665	665	778	329	329	362
0.25	55477	58477	68516	15641	16329	17415	2852	2891	2915	790	790	790	391	391	391
	56641	61665	79836	15954	16790	20206	2852	2915	3342	790	790	873	391	391	400
0.3	61204	65266	79329	17579	18516	20079	3227	3266	3329	852	891	891	415	415	415
	62790	69579	89082	17977	19165	22465	3227	3290	3678	852	891	946	415	415	428
0.35	64790	69915	88079	18977	20165	22204	3516	3579	3641	954	954	954	415	454	454
	66829	75391	96227	19516	21040	24199	3540	3602	3930	954	954	999	415	454	446
0.4	66516	72704	94665	19891	21329	23790	3704	3790	3891	977	977	977	454	454	454
	68954	79227	101271	20516	22352	25407	3727	3829	4098	977	977	1030	454	454	456
0.45	66641	73829	99016	20352	21977	24829	3829	3915	4016	1016	1016	1016	454	454	454
	69477	81266	104213	21079	23165	26090	3852	3954	4182	1016	1016	1041	454	454	456
0.5	65415	73454	101079	20391	22165	25290	3852	3954	4040	1016	1016	1016	454	454	454
	68641	81641	105054	21204	23454	26248	3891	4016	4182	1016	1016	1030	454	454	446
0.55	63141	71766	100852	20079	21915	25165	3790	3891	4016	977	977	977	415	454	454
	66602	80454	103793	20891	23266	25880	3852	3954	4098	977	977	999	454	454	428
0.6	59915	68954	98352	19391	21227	24516	3665	3766	3891	954	954	954	415	415	415
	63602	77829	100430	20204	22602	24987	3727	3829	3930	954	954	946	415	415	400
0.65	55977	65040	93579	18329	20141	23266	3454	3540	3641	852	891	891	391	391	391
	59704	73829	94967	19141	21454	23569	3516	3602	3678	852	891	873	391	391	362
0.7	51352	60141	86540	16954	18641	21415	3165	3227	3329	790	790	790	329	329	329
	54977	68454	87401	17704	19829	21625	3204	3290	3342	790	790	778	329	329	316
0.75	46079	54204	77227	15227	16704	19040	2790	2852	2915	665	665	665	266	266	266
	49454	61665	77734	15891	17704	19156	2829	2891	2922	665	665	662	266	266	260
0.8	40040	47141	65727	13141	14329	16102	2329	2391	2415	540	540	540	204	204	204
	43016	53415	65966	13665	15102	16161	2352	2391	2417	540	540	526	204	204	194
0.85	33102	38704	52016	10602	11454	12641	1790	1829	1852	391	391	391	.	.	.
	35477	43477	52096	10977	11977	12641	1790	1829	1829	391	391	368	.	.	.
0.9	24790	28516	36102	7540	8016	8602	1141	1165	1165	.	.	.	.	.	.
	26415	31454	36125	7766	8290	8596	1165	1165	1156	.	.	.	.	.	.
0.95	14165	15665	18079	3766	3891	4040	.	.	.	.	.	.	.	.	.
	14852	16704	18052	3829	3954	4025	.	.	.	.	.	.	.	.	.
0.98	5602	5891	6227	.	.	.	.	.	.	.	.	.	.	.	.
	5727	6040	6200	.	.	.	.	.	.	.	.	.	.	.	.

TABLE 7: ALPHA= 0.05 POWER= 0.8   EXPECTED ACCRUAL THRU MINIMUM FOLLOW-UP= 30

		DEL=.10			DEL=.15			DEL=.20			DEL=.25			DEL=.30		
FACT=		1.0 .75	.50 .25	.00 BIN	1.0 .75	.50 .25	.00 BIN	1.0 .75	.50 .25	.00 BIN	1.0 75	.50 .25	.00 BIN	1.0 .75	.50 .25	.00 BIN
PCONT=***					REQUIRED NUMBER OF PATIENTS											
0.05	***	130	131	141	76	77	82	54	55	58	42	43	45	35	35	37
	***	131	133	217	77	79	115	54	56	73	42	44	51	35	36	38
0.1	***	206	208	235	109	112	125	72	74	82	54	55	60	43	44	47
	***	206	212	310	110	115	153	73	77	93	54	57	63	43	45	46
0.15	***	266	269	323	134	138	163	86	89	103	62	64	73	48	50	56
	***	267	276	390	136	143	186	87	93	110	63	67	73	49	52	52
0.2	***	309	315	400	152	157	195	95	99	120	67	71	83	51	54	62
	***	311	324	458	154	165	213	96	105	124	69	75	81	53	57	57
0.25	***	338	345	466	163	170	222	100	106	134	70	75	91	53	57	67
	***	340	358	514	165	181	235	103	114	135	72	80	87	55	61	60
0.3	***	352	362	519	168	177	243	103	110	144	72	78	97	54	59	71
	***	355	379	557	171	191	252	106	119	143	74	84	91	56	63	62
0.35	***	354	366	559	168	180	259	102	112	151	71	78	100	54	59	72
	***	358	389	588	172	196	263	106	122	147	74	85	93	56	63	63
0.4	***	345	361	587	164	178	268	100	111	155	70	78	102	52	58	72
	***	350	389	606	169	197	268	104	123	149	73	85	93	55	63	62
0.45	***	327	347	600	156	173	271	96	108	155	67	75	101	50	56	71
	***	334	380	613	163	194	268	101	121	147	71	83	91	53	61	60
0.5	***	303	327	601	147	165	268	91	104	152	64	72	98	48	54	68
	***	311	366	606	154	188	263	96	117	143	67	80	87	50	59	57
0.55	***	275	304	587	136	156	260	85	98	145	59	68	92	44	50	64
	***	285	346	588	144	179	252	90	111	135	63	76	81	47	55	52
0.6	***	245	278	561	124	144	245	78	91	135	54	63	85	40	46	58
	***	257	323	557	132	168	235	83	104	124	58	70	73	43	50	46
0.65	***	214	250	521	111	131	224	70	82	122	49	56	75	36	41	51
	***	228	296	514	119	154	213	75	94	110	52	63	63	38	44	38
0.7	***	185	221	467	97	117	197	62	72	105	42	49	64	.	.	.
	***	199	265	458	105	137	186	66	83	93	45	54	51	.	.	.
0.75	***	156	191	401	83	100	165	52	61	85	.	.	.	.	.	.
	***	170	232	390	90	118	153	56	69	73	.	.	.	.	.	.
0.8	***	128	159	321	68	81	127	.	.	.	.	.	.	.	.	.
	***	140	193	310	73	95	115	.	.	.	.	.	.	.	.	.
0.85	***	99	123	229	.	.	.	.	.	.	.	.	.	.	.	.
	***	108	148	217	.	.	.	.	.	.	.	.	.	.	.	.

TABLE 7: ALPHA= 0.05 POWER= 0.8   EXPECTED ACCRUAL THRU MINIMUM FOLLOW-UP= 40

		DEL=.10			DEL=.15			DEL=.20			DEL=.25			DEL=.30		
FACT=		1.0 .75	.50 .25	.00 BIN	1.0 .75	.50 .25	.00 BIN	1.0 .75	.50 .25	.00 BIN	1.0 75	.50 .25	.00 BIN	1.0 .75	.50 .25	.00 BIN
PCONT=•••					REQUIRED NUMBER OF PATIENTS											
0.05	***	131	132	141	77	78	82	54	55	58	42	43	45	35	36	37
	•••	131	134	217	77	79	115	55	56	73	43	44	51	35	36	38
0.1	***	206	209	235	110	113	125	73	75	82	54	56	60	43	45	47
	•••	207	214	310	111	116	153	74	78	93	55	58	63	44	46	46
0.15	***	267	271	323	136	140	163	87	90	103	63	66	73	49	51	56
	•••	268	279	390	137	146	186	88	94	110	64	68	73	50	53	52
0.2	***	311	318	400	154	160	195	97	102	120	69	73	83	53	56	62
	•••	313	330	458	156	169	213	99	107	124	70	76	81	54	58	57
0.25	***	340	350	466	165	174	222	103	109	134	72	77	91	55	59	67
	•••	343	366	514	169	186	235	105	117	135	74	82	87	57	62	60
0.3	***	355	368	519	171	182	243	106	114	144	74	80	97	56	60	71
	•••	360	390	557	175	197	252	109	123	143	77	86	91	58	64	62
0.35	***	358	374	559	172	186	259	106	116	151	74	81	100	56	61	72
	•••	363	402	588	177	203	263	110	127	147	77	88	93	58	65	63
0.4	***	350	371	587	169	185	268	104	116	155	73	81	102	55	60	72
	•••	357	404	606	175	205	268	109	127	149	76	88	93	57	65	62
0.45	***	334	359	600	163	181	271	101	113	155	71	79	101	53	59	71
	•••	343	398	613	170	203	268	106	126	147	74	86	91	55	63	60
0.5	***	311	342	601	154	175	268	96	109	152	67	76	98	50	56	68
	•••	322	385	606	162	198	263	102	122	143	71	83	87	53	61	57
0.55	***	285	320	587	144	165	260	90	104	145	63	71	92	47	52	64
	•••	298	367	588	152	189	252	96	116	135	67	79	81	49	57	52
0.6	***	257	295	561	132	154	245	83	96	135	58	66	85	43	48	58
	•••	271	344	557	141	178	235	89	109	124	61	73	73	45	52	46
0.65	***	228	268	521	119	141	224	75	87	122	52	59	75	38	42	51
	•••	243	317	514	128	163	213	80	99	110	55	65	63	40	46	38
0.7	***	199	239	467	105	125	197	66	77	105	45	51	64	.	.	.
	•••	214	285	458	113	146	186	71	86	93	48	56	51	.	.	.
0.75	***	170	207	401	90	108	165	56	64	85	.	.	.	.	.	.
	•••	185	249	390	97	125	153	59	72	73	.	.	.	.	.	.
0.8	***	140	173	321	73	87	127	.	.	.	.	.	.	.	.	.
	•••	153	208	310	79	100	115	.	.	.	.	.	.	.	.	.
0.85	***	108	133	229	.	.	.	.	.	.	.	.	.	.	.	.
	•••	119	159	217	.	.	.	.	.	.	.	.	.	.	.	.

TABLE 7: ALPHA= 0.05 POWER= 0.8 EXPECTED ACCRUAL THRU MINIMUM FOLLOW-UP= 50

		DEL=.10			DEL=.15			DEL=.20			DEL=.25			DEL=.30		
FACT=		1.0	.50	.00	1.0	.50	.00	1.0	.50	.00	1.0	.50	.00	1.0	.50	.00
		.75	.25	BIN	.75	.25	BIN	.75	.25	BIN	75	.25	BIN	.75	.25	BIN
PCONT=***		REQUIRED NUMBER OF PATIENTS														
0.05	***	131	133	141	77	78	82	55	56	58	43	43	45	35	36	37
	***	132	135	217	77	80	115	55	56	73	43	44	51	35	36	38
0.1	***	207	211	235	111	114	125	74	76	82	55	57	60	44	45	47
	***	208	216	310	112	117	153	75	78	93	56	58	63	44	46	46
0.15	***	268	273	323	137	142	163	88	92	103	64	67	73	50	52	56
	***	270	283	390	139	148	186	89	95	110	65	69	73	50	53	52
0.2	***	313	321	400	156	163	195	98	104	120	70	74	83	54	57	62
	***	316	335	458	158	172	213	100	109	124	72	77	81	55	59	57
0.25	***	343	354	466	168	178	222	105	112	134	74	79	91	56	60	67
	***	347	373	514	172	190	235	108	119	135	76	83	87	58	63	60
0.3	***	358	374	519	174	187	243	108	117	144	76	82	97	57	62	71
	***	364	399	557	179	202	252	112	126	143	79	87	91	59	65	62
0.35	***	362	382	559	176	191	259	109	120	151	76	83	100	57	62	72
	***	369	413	588	182	209	263	113	130	147	79	90	93	60	66	63
0.4	***	355	380	587	174	192	268	108	120	155	75	83	102	57	62	72
	***	364	417	606	181	212	268	113	131	149	79	90	93	59	66	62
0.45	***	340	370	600	168	188	271	105	118	155	73	81	101	55	60	71
	***	351	412	613	176	210	268	110	130	147	77	88	91	57	64	60
0.5	***	320	354	601	160	182	268	100	114	152	70	78	98	52	58	68
	***	332	401	606	169	205	263	106	126	143	74	85	87	54	62	57
0.55	***	295	334	587	150	173	260	94	108	145	66	74	92	49	54	64
	***	309	384	588	159	197	252	100	120	135	69	81	81	51	58	52
0.6	***	268	310	561	139	162	245	87	100	135	61	68	85	44	49	58
	***	284	361	557	148	185	235	93	112	124	64	75	73	47	53	46
0.65	***	239	283	521	126	148	224	79	91	122	54	61	75	39	43	51
	***	256	333	514	135	170	213	84	102	110	57	67	63	41	46	38
0.7	***	211	253	467	112	132	197	70	80	105	47	53	64	.	.	.
	***	227	301	458	120	152	186	74	89	93	50	57	51	.	.	.
0.75	***	181	221	401	96	113	165	59	67	85	.	.	.	.	.	.
	***	197	263	390	103	130	153	62	74	73	.	.	.	.	.	.
0.8	***	150	184	321	78	91	127	.	.	.	.	.	.	.	.	.
	***	164	219	310	83	103	115	.	.	.	.	.	.	.	.	.
0.85	***	116	142	229	.	.	.	.	.	.	.	.	.	.	.	.
	***	127	167	217	.	.	.	.	.	.	.	.	.	.	.	.

TABLE 7: ALPHA= 0.05 POWER= 0.8   EXPECTED ACCRUAL THRU MINIMUM FOLLOW-UP= 60

PCONT=***	DEL=.10 1.0 .75	.50 .25	.00 BIN	DEL=.15 1.0 .75	.50 .25	.00 BIN	DEL=.20 1.0 .75	.50 .25	.00 BIN	DEL=.25 1.0 75	.50 .25	.00 BIN	DEL=.30 1.0 .75	.50 .25	.00 BIN
	REQUIRED NUMBER OF PATIENTS														
0.05 ***	131	133	141	77	79	82	55	56	58	43	44	45	35	36	37
•••	132	135	217	78	80	115	55	57	73	43	44	51	36	36	38
0.1 ***	208	212	235	112	115	125	74	77	82	55	57	60	44	45	47
•••	209	217	310	113	118	153	75	79	93	56	58	63	45	46	46
0.15 ***	269	276	323	138	143	163	89	93	103	64	67	73	50	52	56
•••	271	286	390	140	149	186	90	96	110	66	69	73	51	54	52
0.2 ***	314	324	400	157	165	195	99	105	120	71	75	83	54	57	62
•••	318	339	458	160	174	213	102	110	124	73	78	81	56	59	57
0.25 ***	345	358	466	170	181	222	106	114	134	75	80	91	57	61	67
•••	350	379	514	174	193	235	109	121	135	77	84	87	59	63	60
0.3 ***	362	379	519	177	191	243	110	119	144	78	84	97	59	63	71
•••	368	406	557	182	206	252	114	128	143	80	89	91	60	66	62
0.35 ***	366	389	559	180	196	259	112	122	151	78	85	100	59	63	72
•••	374	422	588	186	214	263	116	132	147	81	91	93	61	67	63
0.4 ***	361	389	587	178	197	268	111	123	155	78	85	102	58	63	72
•••	371	428	606	185	217	268	116	134	149	81	91	93	60	67	62
0.45 ***	347	380	600	173	194	271	108	121	155	75	83	101	56	61	71
•••	359	425	613	181	216	268	113	132	147	79	90	91	59	65	60
0.5 ***	327	366	601	165	188	268	104	117	152	72	80	98	54	59	68
•••	342	414	606	175	211	263	109	129	143	76	87	87	56	63	57
0.55 ***	304	346	587	156	179	260	98	111	145	68	76	92	50	55	64
•••	320	398	588	165	203	252	104	123	135	71	82	81	52	59	52
0.6 ***	278	323	561	144	168	245	91	104	135	63	70	85	46	50	58
•••	295	375	557	154	191	235	96	115	124	66	76	73	48	53	46
0.65 ***	250	296	521	131	154	224	82	94	122	56	63	75	41	44	51
•••	268	347	514	141	175	213	87	104	110	59	68	63	42	47	38
0.7 ***	221	265	467	117	137	197	72	83	105	49	54	64	.	.	.
•••	239	314	458	125	157	186	77	91	93	51	58	51	.	.	.
0.75 ***	191	232	401	100	118	165	61	69	85	.	.	.	.	.	.
•••	207	275	390	108	134	153	64	75	73	.	.	.	.	.	.
0.8 ***	159	193	321	81	95	127	.	.	.	.	.	.	.	.	.
•••	173	228	310	87	106	115	.	.	.	.	.	.	.	.	.
0.85 ***	123	148	229	.	.	.	.	.	.	.	.	.	.	.	.
•••	133	173	217	.	.	.	.	.	.	.	.	.	.	.	.

TABLE 7: ALPHA= 0.05 POWER= 0.8   EXPECTED ACCRUAL THRU MINIMUM FOLLOW-UP= 70

	DEL=.10			DEL=.15			DEL=.20			DEL=.25			DEL=.30		
FACT=	1.0 .75	.50 .25	.00 BIN	1.0 .75	.50 .25	.00 BIN	1.0 .75	.50 .25	.00 BIN	1.0 75	.50 .25	.00 BIN	1.0 .75	.50 .25	.00 BIN
PCONT=***			REQUIRED NUMBER OF PATIENTS												
0.05 ***	132	134	141	77	79	82	55	56	58	43	44	45	36	36	37
•••	132	136	217	78	80	115	55	57	73	43	44	51	36	36	38
0.1 ***	208	213	235	112	116	125	75	77	82	56	57	60	44	45	47
•••	210	219	310	114	119	153	76	79	93	56	59	63	45	46	46
0.15 ***	270	278	323	139	145	163	90	94	103	65	68	73	51	52	56
•••	273	288	390	141	151	186	91	97	110	66	70	73	51	54	52
0.2 ***	316	327	400	159	167	195	101	106	120	72	76	83	55	58	62
•••	320	343	458	162	176	213	103	111	124	73	79	81	56	60	57
0.25 ***	347	362	466	172	183	222	108	115	134	76	81	91	58	61	67
•••	353	385	514	177	195	235	111	122	135	78	85	87	59	64	60
0.3 ***	365	385	519	180	194	243	112	121	144	79	85	97	60	64	71
•••	372	413	557	186	209	252	116	130	143	81	90	91	61	66	62
0.35 ***	370	395	559	183	200	259	114	125	151	80	87	100	60	64	72
•••	379	430	588	190	218	263	118	134	147	83	92	93	62	68	63
0.4 ***	366	396	587	182	201	268	113	125	155	79	87	102	59	64	72
•••	377	437	606	190	221	268	118	136	149	82	92	93	61	67	62
0.45 ***	353	389	600	177	199	271	111	124	155	77	85	101	57	62	71
•••	367	435	613	186	221	268	116	135	147	81	91	91	60	66	60
0.5 ***	335	376	601	170	193	268	107	120	152	74	82	98	55	60	68
•••	350	426	606	180	216	263	112	131	143	78	88	87	57	63	57
0.55 ***	312	357	587	161	185	260	101	114	145	70	77	92	51	56	64
•••	329	409	588	171	208	252	106	125	135	73	83	81	53	59	52
0.6 ***	287	334	561	149	173	245	94	106	135	64	72	85	47	51	58
•••	305	387	557	159	195	235	99	117	124	68	77	73	49	54	46
0.65 ***	259	307	521	136	159	224	85	97	122	58	64	75	41	45	51
•••	278	359	514	146	180	213	90	106	110	61	69	63	43	47	38
0.7 ***	230	276	467	121	142	197	75	85	105	50	55	64	.	.	.
•••	249	324	458	130	160	186	79	93	93	52	59	51	.	.	.
0.75 ***	200	241	401	104	122	165	63	71	85	.	.	.	.	.	.
•••	216	284	390	112	137	153	66	76	73	.	.	.	.	.	.
0.8 ***	166	201	321	84	98	127	.	.	.	.	.	.	.	.	.
•••	181	236	310	90	108	115	.	.	.	.	.	.	.	.	.
0.85 ***	128	154	229	.	.	.	.	.	.	.	.	.	.	.	.
•••	139	178	217	.	.	.	.	.	.	.	.	.	.	.	.

TABLE 7: ALPHA= 0.05 POWER= 0.8   EXPECTED ACCRUAL THRU MINIMUM FOLLOW-UP= 80

	DEL=.10			DEL=.15			DEL=.20			DEL=.25			DEL=.30		
FACT=	1.0 .75	.50 .25	.00 BIN	1.0 .75	.50 .25	.00 BIN	1.0 .75	.50 .25	.00 BIN	1.0 75	.50 .25	.00 BIN	1.0 .75	.50 .25	.00 BIN
PCONT=***				REQUIRED NUMBER OF PATIENTS											
0.05 ***	132	134	141	78	79	82	55	56	58	43	44	45	36	36	37
•••	133	136	217	78	80	115	56	57	73	43	44	51	36	36	38
0.1 ***	209	214	235	113	116	125	75	78	82	56	58	60	45	46	47
•••	211	220	310	114	119	153	76	79	93	57	59	63	45	46	46
0.15 ***	271	279	323	140	146	163	90	94	103	66	68	73	51	53	56
•••	274	290	390	142	152	186	92	98	110	67	70	73	52	54	52
0.2 ***	318	330	400	160	169	195	102	107	120	73	76	83	56	58	62
•••	322	347	458	164	178	213	104	112	124	74	79	81	57	60	57
0.25 ***	350	366	466	174	186	222	109	117	134	77	82	91	59	62	67
•••	356	389	514	179	198	235	113	123	135	79	86	87	60	64	60
0.3 ***	368	390	519	182	197	243	114	123	144	80	86	97	60	64	71
•••	376	419	557	188	212	252	118	131	143	83	90	91	62	67	62
0.35 ***	374	402	559	186	203	259	116	127	151	81	88	100	61	65	72
•••	384	437	588	193	221	263	121	136	147	84	93	93	63	68	63
0.4 ***	371	404	587	185	205	268	116	127	155	81	88	102	60	65	72
•••	383	446	606	193	225	268	121	137	149	84	93	93	62	68	62
0.45 ***	359	398	600	181	203	271	113	126	155	79	86	101	59	63	71
•••	374	445	613	190	225	268	119	137	147	82	92	91	61	67	60
0.5 ***	342	385	601	175	198	268	109	122	152	76	83	98	56	61	68
•••	358	436	606	184	220	263	115	133	143	79	89	87	58	64	57
0.55 ***	320	367	587	165	189	260	104	116	145	71	79	92	52	57	64
•••	338	420	588	175	212	252	109	127	135	75	84	81	54	60	52
0.6 ***	295	344	561	154	178	245	96	109	135	66	73	85	48	52	58
•••	314	397	557	164	199	235	102	119	124	69	78	73	50	54	46
0.65 ***	268	317	521	141	163	224	87	99	122	59	65	75	42	46	51
•••	287	369	514	150	183	213	92	108	110	62	69	63	44	48	38
0.7 ***	239	285	467	125	146	197	77	86	105	51	56	64	.	.	.
•••	257	334	458	134	163	186	81	94	93	53	59	51	.	.	.
0.75 ***	207	249	401	108	125	165	64	72	85	.	.	.	.	.	.
•••	224	292	390	115	139	153	68	77	73	.	.	.	.	.	.
0.8 ***	173	208	321	87	100	127	.	.	.	.	.	.	.	.	.
•••	187	242	310	93	110	115	.	.	.	.	.	.	.	.	.
0.85 ***	133	159	229	.	.	.	.	.	.	.	.	.	.	.	.
•••	144	182	217	.	.	.	.	.	.	.	.	.	.	.	.

TABLE 7: ALPHA= 0.05 POWER= 0.8 EXPECTED ACCRUAL THRU MINIMUM FOLLOW-UP= 90

		DEL=.10			DEL=.15			DEL=.20			DEL=.25			DEL=.30		
FACT=		1.0	.50	.00	1.0	.50	.00	1.0	.50	.00	1.0	.50	.00	1.0	.50	.00
		.75	.25	BIN	.75	.25	BIN	.75	.25	BIN	75	.25	BIN	.75	.25	BIN
PCONT=***		REQUIRED NUMBER OF PATIENTS														
0.05	***	132	134	141	78	79	82	55	56	58	43	44	45	36	36	37
	***	133	137	217	79	80	115	56	57	73	44	44	51	36	37	38
0.1	***	210	215	235	113	117	125	76	78	82	56	58	60	45	46	47
	***	212	221	310	115	120	153	77	80	93	57	59	63	45	47	46
0.15	***	272	281	323	141	147	163	91	95	103	66	69	73	51	53	56
	***	276	292	390	143	153	186	93	98	110	67	70	73	52	54	52
0.2	***	319	333	400	162	170	195	103	108	120	73	77	83	56	58	62
	***	324	350	458	165	179	213	105	113	124	75	79	81	57	60	57
0.25	***	352	370	466	176	188	222	111	118	134	78	83	91	59	62	67
	***	358	393	514	181	199	235	114	124	135	80	86	87	61	64	60
0.3	***	371	394	519	185	200	243	116	125	144	81	87	97	61	65	71
	***	379	424	557	191	214	252	119	132	143	84	91	91	63	67	62
0.35	***	378	407	559	189	206	259	118	128	151	82	89	100	62	66	72
	***	389	444	588	196	223	263	122	137	147	85	93	93	63	69	63
0.4	***	375	410	587	189	209	268	118	129	155	82	89	102	61	65	72
	***	389	453	606	197	228	268	123	139	149	85	94	93	63	68	62
0.45	***	365	405	600	185	207	271	116	128	155	80	87	101	59	64	71
	***	380	453	613	194	228	268	121	138	147	83	93	91	61	67	60
0.5	***	348	393	601	178	202	268	112	124	152	77	84	98	57	61	68
	***	366	445	606	188	223	263	117	135	143	80	90	87	59	64	57
0.55	***	327	376	587	169	193	259	106	118	145	73	80	92	53	57	64
	***	346	429	588	179	215	252	111	129	135	76	85	81	55	60	52
0.6	***	303	353	561	158	182	245	98	110	135	67	74	85	48	52	58
	***	323	406	557	168	203	235	104	120	124	70	78	73	50	55	46
0.65	***	276	325	521	145	167	224	89	100	122	60	66	75	43	46	51
	***	296	377	514	154	186	213	94	109	110	63	70	63	44	48	38
0.7	***	246	294	467	129	149	197	79	88	105	52	57	64	.	.	.
	***	265	342	458	137	166	186	83	95	93	54	60	51	.	.	.
0.75	***	214	257	401	111	127	165	66	73	85	.	.	.	.	.	.
	***	232	299	390	118	141	153	69	78	73	.	.	.	.	.	.
0.8	***	179	214	321	89	102	127	.	.	.	.	.	.	.	.	.
	***	193	247	310	95	112	115	.	.	.	.	.	.	.	.	.
0.85	***	138	163	229	.	.	.	.	.	.	.	.	.	.	.	.
	***	148	186	217	.	.	.	.	.	.	.	.	.	.	.	.

TABLE 7: ALPHA= 0.05 POWER= 0.8   EXPECTED ACCRUAL THRU MINIMUM FOLLOW-UP= 100

		DEL=.10			DEL=.15			DEL=.20			DEL=.25			DEL=.30		
FACT=		1.0 .75	.50 .25	.00 BIN	1.0 .75	.50 .25	.00 BIN	1.0 .75	.50 .25	.00 BIN	1.0 75	.50 .25	.00 BIN	1.0 .75	.50 .25	.00 BIN
PCONT=***					REQUIRED NUMBER OF PATIENTS											
0.05	***	133	135	141	78	80	82	56	56	58	43	44	45	36	36	37
	•••	133	137	217	79	81	115	56	57	73	44	44	51	36	37	38
0.1	***	211	216	235	114	117	125	76	78	82	57	58	60	45	46	47
	•••	213	222	310	115	120	153	77	80	93	57	59	63	45	47	46
0.15	***	273	283	323	142	148	163	92	95	103	67	69	73	52	53	56
	•••	277	294	390	144	153	186	93	98	110	68	71	73	52	54	52
0.2	***	321	335	400	163	172	195	104	109	120	74	77	83	57	59	62
	•••	326	352	458	167	180	213	106	113	124	75	80	81	58	60	57
0.25	***	354	373	466	178	190	222	112	119	134	79	83	91	60	63	67
	•••	361	397	514	183	201	235	115	125	135	81	87	87	61	65	60
0.3	***	374	399	519	187	202	243	117	126	144	82	87	97	62	65	71
	•••	383	429	557	193	216	252	121	133	143	84	91	91	63	68	62
0.35	***	382	413	559	191	209	259	120	130	151	83	90	100	62	66	72
	•••	393	450	588	199	226	263	124	138	147	86	94	93	64	69	63
0.4	***	380	417	587	192	212	268	120	131	155	83	90	102	62	66	72
	•••	394	460	606	200	231	268	124	140	149	86	95	93	64	69	62
0.45	***	370	412	600	188	210	271	118	130	155	81	88	101	60	64	71
	•••	387	460	613	197	231	268	123	139	147	85	94	91	62	67	60
0.5	***	354	401	601	182	205	268	114	126	152	78	85	98	58	62	68
	•••	373	452	606	192	227	263	119	136	143	82	91	87	59	65	57
0.55	***	334	384	587	173	197	260	108	120	145	74	81	92	54	58	64
	•••	354	437	588	183	218	252	113	130	135	77	86	81	56	60	52
0.6	***	310	361	561	162	185	245	100	112	135	68	75	85	49	53	58
	•••	330	414	557	171	206	235	105	121	124	71	79	73	51	55	46
0.65	***	283	333	521	148	170	224	91	102	122	61	67	75	43	46	51
	•••	303	385	514	157	189	213	96	110	110	64	70	63	45	48	38
0.7	***	253	301	467	132	152	197	80	89	105	53	57	64	.	.	.
	•••	273	349	458	140	168	186	84	96	93	55	60	51	.	.	.
0.75	***	221	263	401	113	130	165	67	74	85	.	.	.	.	.	.
	•••	238	305	390	120	143	153	70	79	73	.	.	.	.	.	.
0.8	***	184	219	321	91	103	127	.	.	.	.	.	.	.	.	.
	•••	199	252	310	97	113	115	.	.	.	.	.	.	.	.	.
0.85	***	142	167	229	.	.	.	.	.	.	.	.	.	.	.	.
	•••	152	189	217	.	.	.	.	.	.	.	;	.	.	.	.

TABLE 7: ALPHA= 0.05 POWER= 0.8   EXPECTED ACCRUAL THRU MINIMUM FOLLOW-UP= 110

		DEL=.05			DEL=.10			DEL=.15			DEL=.20			DEL=.25		
FACT=		1.0 .75	.50 .25	.00 BIN	1.0 .75	.50 .25	.00 BIN	1.0 .75	.50 .25	.00 BIN	1.0 75	.50 .25	.00 BIN	1.0 .75	.50 .25	.00 BIN
PCONT=***					REQUIRED NUMBER OF PATIENTS											
0.05	***	371	374	395	133	135	141	78	80	82	56	56	58	43	44	45
	•••	372	378	681	134	137	217	79	81	115	56	57	73	44	44	51
0.1	***	676	683	762	211	217	235	114	118	125	76	78	82	57	58	60
	•••	678	694	1076	213	223	310	116	121	153	77	80	93	57	59	63
0.15	***	936	947	1117	274	284	323	143	149	163	92	96	103	67	69	73
	•••	940	967	1422	278	296	390	145	154	186	94	99	110	68	71	73
0.2	***	1137	1154	1441	323	337	400	164	173	195	104	110	120	74	78	83
	•••	1143	1185	1719	328	355	458	168	181	213	107	114	124	76	80	81
0.25	***	1278	1301	1724	356	376	466	179	191	222	113	120	134	80	84	91
	•••	1285	1345	1967	364	401	514	184	202	235	116	125	135	82	87	87
0.3	***	1360	1391	1960	377	403	519	189	204	243	118	127	144	83	88	97
	•••	1370	1451	2164	386	434	557	195	218	252	122	134	143	85	92	91
0.35	***	1389	1430	2147	385	417	559	194	212	259	121	131	151	84	90	100
	•••	1402	1508	2313	398	455	588	201	228	263	125	139	147	87	95	93
0.4	***	1371	1425	2281	384	422	587	194	214	268	121	132	155	84	91	102
	•••	1389	1522	2412	399	466	606	203	233	268	126	141	149	87	95	93
0.45	***	1315	1384	2362	375	419	600	191	213	271	119	131	155	83	89	101
	•••	1338	1502	2461	392	467	613	200	233	268	124	141	147	86	94	91
0.5	***	1230	1315	2390	360	408	601	185	208	268	115	128	152	79	86	98
	•••	1259	1454	2461	379	459	606	195	229	263	121	137	143	82	91	87
0.55	***	1123	1226	2363	340	391	587	176	200	259	110	122	145	75	82	92
	•••	1159	1384	2412	360	444	588	186	221	252	115	131	135	78	86	81
0.6	***	1005	1126	2283	316	368	561	165	188	245	102	113	135	69	75	85
	•••	1049	1297	2313	337	422	557	175	208	235	107	122	124	72	79	73
0.65	***	884	1018	2149	289	340	521	151	173	224	93	103	122	62	67	75
	•••	933	1197	2164	310	392	514	160	191	213	97	111	110	64	71	63
0.7	***	765	906	1962	259	308	467	135	154	197	82	90	105	53	58	64
	•••	818	1084	1967	279	355	458	143	170	186	85	97	93	55	60	51
0.75	***	651	791	1722	226	269	401	116	132	165	68	75	85	.	.	.
	•••	705	959	1719	244	310	390	123	145	153	71	79	73	.	.	.
0.8	***	541	670	1431	189	224	321	93	105	127	.	.	.	.	.	.
	•••	591	819	1422	204	256	310	98	114	115	.	.	.	.	.	.
0.85	***	430	539	1087	145	170	229	.	.	.	.	.	.	.	.	.
	•••	473	659	1076	156	191	217	.	.	.	.	.	.	.	.	.
0.9	***	310	386	693	.	.	.	.	.	.	.	.	.	.	.	.
	•••	340	465	681	.	.	.	.	.	.	.	.	.	.	.	.

TABLE 7: ALPHA= 0.05   POWER= 0.8   EXPECTED ACCRUAL THRU MINIMUM FOLLOW-UP= 120

		DEL=.05			DEL=.10			DEL=.15			DEL=.20			DEL=.25	
FACT=	1.0 .75	.50 .25	.00 BIN	1.0 .75	.50 .25	.00 BIN	1.0 .75	.50 .25	.00 BIN	1.0 75	.50 .25	.00 BIN	1.0 .75	.50 .25	.00 BIN
PCONT=•••					REQUIRED NUMBER OF PATIENTS										
0.05 •••	371	374	395	133	135	141	79	80	82	56	57	58	43	44	45
•••	372	379	681	134	137	217	79	81	115	56	57	73	44	44	51
0.1 •••	677	684	762	212	217	235	115	118	125	77	79	82	57	58	60
•••	679	696	1076	214	223	310	116	121	153	78	80	93	58	59	63
0.15 •••	937	949	1117	276	286	323	143	149	163	93	96	103	67	69	73
•••	941	971	1422	279	297	390	146	154	186	94	99	110	68	71	73
0.2 •••	1139	1157	1441	324	339	400	165	174	195	105	110	120	75	78	83
•••	1145	1190	1719	330	357	458	169	182	213	107	114	124	76	80	81
0.25 •••	1280	1305	1724	358	379	466	181	193	222	114	121	134	80	84	91
•••	1288	1353	1967	366	404	514	186	203	235	117	126	135	82	87	87
0.3 •••	1363	1397	1960	379	406	519	191	206	243	119	128	144	84	89	97
•••	1374	1461	2164	390	437	557	197	219	252	123	134	143	86	92	91
0.35 •••	1392	1438	2147	389	422	559	196	214	259	122	132	151	85	91	100
•••	1408	1521	2313	402	460	588	203	230	263	127	140	147	88	95	93
0.4 •••	1376	1435	2281	389	428	587	197	217	268	123	134	155	85	91	102
•••	1396	1538	2412	404	471	606	205	235	268	127	142	149	88	96	93
0.45 •••	1322	1396	2362	380	425	600	194	216	271	121	132	155	83	90	101
•••	1347	1521	2461	398	473	613	203	235	268	126	141	147	86	95	91
0.5 •••	1238	1329	2390	366	414	601	188	211	268	117	129	152	80	87	98
•••	1269	1475	2461	385	466	606	198	231	263	122	138	143	83	91	87
0.55 •••	1133	1243	2363	346	397	587	179	203	259	111	123	145	76	82	92
•••	1172	1407	2412	367	451	588	189	223	252	116	132	135	79	87	81
0.6 •••	1017	1145	2283	323	375	561	168	191	245	104	115	135	70	76	85
•••	1064	1322	2313	344	428	557	178	210	235	109	123	124	73	80	73
0.65 •••	898	1038	2149	296	347	521	154	175	224	94	104	122	63	68	75
•••	950	1222	2164	317	398	514	163	193	213	99	111	110	65	71	63
0.7 •••	780	927	1962	265	314	467	137	157	197	83	91	105	54	58	64
•••	836	1108	1967	285	360	458	146	172	186	86	97	93	56	61	51
0.75 •••	666	810	1722	232	275	401	118	134	165	69	75	85	.	.	.
•••	722	981	1719	249	315	390	125	146	153	72	80	73	.	.	.
0.8 •••	556	688	1430	193	228	321	95	106	127	.	.	.	.	.	.
•••	607	839	1422	208	260	310	100	115	115	.	.	.	.	.	.
0.85 •••	443	553	1087	148	173	229	.	.	.	.	.	.	.	.	.
•••	487	674	1076	159	193	217	.	.	.	.	.	.	.	.	.
0.9 •••	319	396	693	.	.	.	.	.	.	.	.	.	.	.	.
•••	350	475	681	.	.	.	.	.	.	.	.	.	.	.	.

TABLE 7: ALPHA= 0.05 POWER= 0.8   EXPECTED ACCRUAL THRU MINIMUM FOLLOW-UP= 130

	DEL=.05			DEL=.10			DEL=.15			DEL=.20			DEL=.25		
FACT=	1.0 .75	.50 .25	.00 BIN	1.0 .75	.50 .25	.00 BIN	1.0 .75	.50 .25	.00 BIN	1.0 75	.50 .25	.00 BIN	1.0 .75	.50 .25	.00 BIN
PCONT=***				REQUIRED NUMBER OF PATIENTS											
0.05 ***	371	375	395	133	136	141	79	80	82	56	57	58	44	44	45
...	373	380	681	134	138	217	79	81	115	56	57	73	44	44	51
0.1 ***	678	685	762	212	218	235	115	119	125	77	79	82	57	58	60
...	680	698	1076	215	224	310	117	121	153	78	80	93	58	59	63
0.15 ***	938	951	1117	277	287	323	144	150	163	93	97	103	67	70	73
...	942	974	1422	281	298	390	147	155	186	95	99	110	68	71	73
0.2 ***	1140	1159	1441	326	341	400	166	175	195	106	111	120	75	78	83
...	1147	1196	1719	332	359	458	170	183	213	108	115	124	77	80	81
0.25 ***	1282	1309	1724	360	382	466	182	194	222	115	121	134	81	85	91
...	1291	1360	1967	369	406	514	187	205	235	118	126	135	83	88	87
0.3 ***	1365	1402	1960	382	410	519	192	207	243	120	129	144	84	89	97
...	1378	1471	2164	393	441	557	199	221	252	124	135	143	87	93	91
0.35 ***	1396	1445	2147	392	426	559	198	216	259	124	133	151	86	91	100
...	1413	1533	2313	405	464	588	205	231	263	128	140	147	88	95	93
0.4 ***	1381	1444	2281	393	433	587	199	219	268	124	135	155	86	92	102
...	1402	1553	2412	408	476	606	208	237	268	129	143	149	89	96	93
0.45 ***	1328	1407	2362	385	430	600	197	218	271	122	134	155	84	91	101
...	1355	1538	2461	403	478	613	206	237	268	127	142	147	87	95	91
0.5 ***	1246	1343	2390	371	420	601	191	214	268	119	130	152	81	88	98
...	1280	1495	2461	391	472	606	200	233	263	124	139	143	84	92	87
0.55 ***	1143	1260	2363	352	404	587	182	205	259	113	124	145	77	83	92
...	1185	1429	2412	373	457	588	192	225	252	118	133	135	80	87	81
0.6 ***	1029	1163	2283	328	381	561	171	193	245	105	116	135	71	76	85
...	1078	1345	2313	350	434	557	180	212	235	110	124	124	73	80	73
0.65 ***	911	1058	2149	301	353	521	156	178	224	95	105	122	63	68	75
...	966	1245	2164	323	403	514	166	195	213	100	112	110	66	71	63
0.7 ***	795	946	1962	271	319	467	140	158	197	84	92	105	55	58	64
...	853	1131	1967	291	365	458	148	173	186	87	98	93	56	61	51
0.75 ***	681	829	1722	237	279	401	120	135	165	70	76	85	.	.	.
...	739	1002	1719	254	319	390	127	147	153	73	80	73	.	.	.
0.8 ***	570	704	1431	197	232	321	96	107	127	.	.	.	.	.	.
...	623	857	1422	212	263	310	101	115	115	.	.	.	.	.	.
0.85 ***	455	567	1087	151	176	229	.	.	.	.	.	.	.	.	.
...	500	688	1076	162	196	217	.	.	.	.	.	.	.	.	.
0.9 ***	327	405	693	.	.	.	.	.	.	.	.	.	.	.	.
...	359	484	681	.	.	.	.	.	.	.	.	.	.	.	.

TABLE 7: ALPHA= 0.05 POWER= 0.8   EXPECTED ACCRUAL THRU MINIMUM FOLLOW-UP= 140

	DEL=.05			DEL=.10			DEL=.15			DEL=.20			DEL=.25		
FACT=	1.0	.50	.00	1.0	.50	.00	1.0	.50	.00	1.0	.50	.00	1.0	.50	.00
	.75	.25	BIN	.75	.25	BIN	.75	.25	BIN	75	.25	BIN	.75	.25	BIN
PCONT=***				REQUIRED NUMBER OF PATIENTS											
0.05 ***	372	375	395	134	136	141	79	80	82	56	57	58	44	44	45
•••	373	380	681	134	138	217	79	81	115	56	57	73	44	44	51
0.1 ***	678	686	762	213	219	235	116	119	125	77	79	82	57	59	60
•••	681	700	1076	215	225	310	117	121	153	78	80	93	58	59	63
0.15 ***	939	953	1117	278	288	323	145	151	163	94	97	103	68	70	73
•••	944	978	1422	282	299	390	147	155	186	95	99	110	69	71	73
0.2 ***	1142	1162	1441	327	343	400	167	176	195	106	111	120	76	79	83
•••	1149	1201	1719	333	361	458	171	183	213	109	115	124	77	81	81
0.25 ***	1284	1313	1724	363	385	466	183	195	222	115	122	134	81	85	91
•••	1294	1367	1967	371	409	514	188	205	235	118	127	135	83	88	87
0.3 ***	1368	1408	1960	385	413	519	194	209	243	121	130	144	85	90	97
•••	1382	1480	2164	396	444	557	200	222	252	125	136	143	87	93	91
0.35 ***	1400	1452	2147	396	430	559	200	218	259	125	134	151	87	92	100
•••	1418	1545	2313	409	468	588	207	233	263	129	141	147	89	96	93
0.4 ***	1386	1453	2281	396	437	587	201	221	268	125	136	155	87	92	102
•••	1409	1568	2412	413	480	606	210	238	268	130	144	149	89	97	93
0.45 ***	1334	1419	2362	389	435	600	199	221	271	124	135	155	85	91	101
•••	1363	1555	2461	408	483	613	208	239	268	129	143	147	88	95	91
0.5 ***	1254	1357	2390	376	426	601	193	216	268	120	131	152	82	88	98
•••	1290	1514	2461	396	477	606	203	235	263	125	140	143	85	92	87
0.55 ***	1153	1275	2363	357	409	587	185	208	260	114	125	145	77	83	92
•••	1197	1450	2412	378	462	588	194	227	252	119	133	135	80	87	81
0.6 ***	1041	1180	2283	334	387	561	173	195	245	106	117	135	71	77	85
•••	1092	1366	2313	356	439	557	183	214	235	111	125	124	74	80	73
0.65 ***	925	1076	2149	307	359	521	159	180	224	97	106	122	64	69	75
•••	982	1267	2164	328	408	514	168	197	213	101	113	110	66	72	63
0.7 ***	809	964	1962	276	324	467	142	160	197	85	93	105	55	59	64
•••	869	1152	1967	296	370	458	150	175	186	88	98	93	57	61	51
0.75 ***	696	846	1722	241	284	401	122	137	165	71	76	85	.	.	.
•••	755	1022	1719	259	323	390	128	148	153	73	80	73	.	.	.
0.8 ***	583	720	1430	201	236	321	98	108	127	.	.	.	.	.	.
•••	637	874	1422	216	266	310	102	116	115	.	.	.	.	.	.
0.85 ***	466	580	1087	154	178	229	.	.	.	.	.	.	.	.	.
•••	512	701	1076	165	197	217	.	.	.	.	.	.	.	.	.
0.9 ***	335	413	693	.	.	.	.	.	.	.	.	.	.	.	.
•••	367	492	681	.	.	.	.	.	.	.	.	.	.	.	.

TABLE 7: ALPHA= 0.05 POWER= 0.8   EXPECTED ACCRUAL THRU MINIMUM FOLLOW-UP= 150

	DEL=.05			DEL=.10			DEL=.15			DEL=.20			DEL=.25		
FACT=	1.0 .75	.50 .25	.00 BIN	1.0 .75	.50 .25	.00 BIN	1.0 .75	.50 .25	.00 BIN	1.0 75	.50 .25	.00 BIN	1.0 .75	.50 .25	.00 BIN

PCONT=•••   REQUIRED NUMBER OF PATIENTS

PCONT	1.0/.75	.50/.25	.00/BIN	1.0/.75	.50/.25	.00/BIN	1.0/.75	.50/.25	.00/BIN	1.0/75	.50/.25	.00/BIN	1.0/.75	.50/.25	.00/BIN
0.05 ***	372	376	395	134	136	141	79	80	82	56	57	58	44	44	45
•••	373	381	681	135	138	217	80	81	115	56	57	73	44	44	51
0.1 ***	679	687	762	213	219	235	116	119	125	77	79	82	57	59	60
•••	682	702	1076	216	225	310	117	122	153	78	80	93	58	59	63
0.15 ***	940	955	1117	278	289	323	145	151	163	94	97	103	68	70	73
•••	945	981	1422	283	300	390	148	156	186	95	100	110	69	71	73
0.2 ***	1143	1165	1441	329	345	400	168	177	195	107	112	120	76	79	83
•••	1151	1205	1719	335	362	458	172	184	213	109	115	124	77	81	81
0.25 ***	1286	1317	1724	364	387	466	185	196	222	116	123	134	82	85	91
•••	1297	1374	1967	373	411	514	190	206	235	119	127	135	83	88	87
0.3 ***	1371	1414	1960	387	416	519	196	210	243	122	130	144	85	90	97
•••	1385	1489	2164	399	447	557	202	223	252	126	136	143	87	93	91
0.35 ***	1404	1460	2147	399	434	559	202	219	259	126	135	151	87	92	100
•••	1423	1556	2313	413	471	588	209	234	263	130	142	147	90	96	93
0.4 ***	1391	1462	2281	400	442	587	203	223	268	126	137	155	87	93	102
•••	1415	1582	2412	417	485	606	212	240	268	131	144	149	90	97	93
0.45 ***	1341	1430	2362	394	440	600	201	223	271	125	136	155	86	92	101
•••	1372	1571	2461	412	488	613	210	241	268	130	144	147	88	96	91
0.5 ***	1261	1370	2390	381	431	601	196	218	268	121	132	152	83	89	98
•••	1300	1532	2461	401	482	606	205	237	263	126	140	143	85	93	87
0.55 ***	1163	1291	2363	362	415	587	187	210	259	115	126	145	78	84	92
•••	1209	1469	2412	383	467	588	197	228	252	120	134	135	81	88	81
0.6 ***	1052	1197	2283	339	392	561	175	198	245	107	118	135	72	77	85
•••	1106	1387	2313	361	444	557	185	215	235	112	125	124	74	81	73
0.65 ***	938	1093	2149	312	364	521	161	182	224	98	107	122	65	69	75
•••	997	1287	2164	333	413	514	170	198	213	102	113	110	67	72	63
0.7 ***	823	982	1962	281	329	467	144	162	197	86	93	105	55	59	64
•••	884	1172	1967	301	374	458	152	176	186	89	98	93	57	61	51
0.75 ***	709	863	1722	245	288	401	123	138	165	71	77	85	.	.	.
•••	770	1040	1719	263	326	390	130	149	153	74	81	73	.	.	.
0.8 ***	595	735	1430	205	239	321	99	109	127	.	.	.	.	.	.
•••	651	889	1422	219	268	310	103	117	115	.	.	.	.	.	.
0.85 ***	477	592	1087	157	180	229	.	.	.	.	.	.	.	.	.
•••	523	713	1076	167	199	217	.	.	.	.	.	.	.	.	.
0.9 ***	343	421	693	.	.	.	.	.	.	.	.	.	.	.	.
•••	375	499	681	.	.	.	.	.	.	.	.	.	.	.	.

TABLE 7: ALPHA= 0.05 POWER= 0.8   EXPECTED ACCRUAL THRU MINIMUM FOLLOW-UP= 160

	DEL=.05			DEL=.10			DEL=.15			DEL=.20			DEL=.25		
FACT=	1.0 .75	.50 .25	.00 BIN	1.0 .75	.50 .25	.00 BIN	1.0 .75	.50 .25	.00 BIN	1.0 75	.50 .25	.00 BIN	1.0 .75	.50 .25	.00 BIN
PCONT=•••				REQUIRED NUMBER OF PATIENTS											
0.05 •••	372	376	395	134	136	141	79	80	82	56	57	58	44	44	45
•••	374	381	681	135	138	217	80	81	115	56	57	73	44	45	51
0.1 •••	679	688	762	214	220	235	116	119	125	78	79	82	58	59	60
•••	682	703	1076	216	226	310	118	122	153	78	81	93	58	59	63
0.15 •••	941	956	1117	279	290	323	146	152	163	94	98	103	68	70	73
•••	946	984	1422	284	301	390	148	156	186	96	100	110	69	71	73
0.2 •••	1145	1168	1441	330	347	400	169	178	195	107	112	120	76	79	83
•••	1153	1210	1719	337	364	458	173	185	213	109	115	124	78	81	81
0.25 •••	1288	1322	1724	366	389	466	186	198	222	117	123	134	82	86	91
•••	1299	1381	1967	375	413	514	191	207	235	120	128	135	84	88	87
0.3 •••	1374	1419	1960	390	419	519	197	212	243	123	131	144	86	90	97
•••	1389	1498	2164	401	450	557	203	224	252	127	136	143	88	93	91
0.35 •••	1408	1467	2147	402	437	559	203	221	259	127	136	151	88	93	100
•••	1428	1567	2313	416	475	588	211	235	263	131	142	147	90	96	93
0.4 •••	1396	1472	2281	404	446	587	205	225	268	127	137	155	88	93	102
•••	1422	1595	2412	420	488	606	214	241	268	132	145	149	90	97	93
0.45 •••	1347	1441	2362	398	445	600	203	225	271	126	137	155	86	92	101
•••	1380	1587	2461	417	492	613	212	242	268	131	144	147	89	96	91
0.5 •••	1269	1383	2390	385	436	601	198	220	268	122	133	152	83	89	98
•••	1310	1550	2461	406	486	606	207	238	263	127	141	143	86	93	87
0.55 •••	1172	1305	2363	367	420	587	189	212	260	116	127	145	79	84	92
•••	1221	1488	2412	388	471	588	199	230	252	121	135	135	81	88	81
0.6 •••	1064	1213	2283	344	397	561	178	199	245	109	119	135	73	78	85
•••	1119	1406	2313	366	448	557	187	217	235	113	126	124	75	81	73
0.65 •••	950	1110	2149	317	369	521	163	183	224	99	108	122	65	69	75
•••	1011	1306	2164	338	417	514	172	199	213	103	114	110	67	72	63
0.7 •••	836	998	1962	285	334	467	146	163	197	86	94	105	56	59	64
•••	899	1190	1967	305	378	458	153	177	186	90	99	93	57	61	51
0.75 •••	722	879	1722	249	292	401	125	139	165	72	77	85	.	.	.
•••	784	1057	1719	267	329	390	131	150	153	74	81	73	.	.	.
0.8 •••	608	749	1431	208	242	321	100	110	127	.	.	.	.	.	.
•••	664	904	1422	223	271	310	104	117	115	.	.	.	.	.	.
0.85 •••	487	603	1087	159	182	229	.	.	.	.	.	.	.	.	.
•••	533	725	1076	169	200	217	.	.	.	.	.	.	.	.	.
0.9 •••	350	429	693	.	.	.	.	.	.	.	.	.	.	.	.
•••	382	506	681	.	.	.	.	.	.	.	.	.	.	.	.

TABLE 7: ALPHA= 0.05 POWER= 0.8   EXPECTED ACCRUAL THRU MINIMUM FOLLOW-UP= 170

		DEL=.05			DEL=.10			DEL=.15			DEL=.20			DEL=.25		
FACT=		1.0 .75	.50 .25	.00 BIN	1.0 .75	.50 .25	.00 BIN	1.0 .75	.50 .25	.00 BIN	1.0 75	.50 .25	.00 BIN	1.0 .75	.50 .25	.00 BIN
PCONT=***					REQUIRED NUMBER OF PATIENTS											
0.05	***	372	376	395	134	136	141	79	80	82	56	57	58	44	44	45
	•••	374	382	681	135	138	217	80	81	115	57	57	73	44	45	51
0.1	***	680	689	762	214	221	235	116	120	125	78	79	82	58	59	60
	•••	683	705	1076	217	226	310	118	122	153	79	81	93	58	59	63
0.15	***	942	958	1117	280	291	323	146	152	163	95	98	103	68	70	73
	•••	947	987	1422	285	302	390	149	157	186	96	100	110	69	72	73
0.2	***	1146	1171	1441	331	348	400	170	178	195	108	112	120	77	79	83
	•••	1155	1215	1719	338	365	458	173	185	213	110	116	124	78	81	81
0.25	***	1290	1326	1724	368	391	466	187	198	222	117	124	134	82	86	91
	•••	1302	1387	1967	377	415	514	192	208	235	120	128	135	84	88	87
0.3	***	1377	1425	1960	392	422	519	198	213	243	124	131	144	86	91	97
	•••	1393	1507	2164	404	453	557	205	225	252	127	137	143	88	93	91
0.35	***	1411	1474	2147	405	441	559	205	222	259	127	136	151	88	93	100
	•••	1433	1578	2313	419	478	588	212	236	263	131	143	147	91	96	93
0.4	***	1401	1480	2281	407	449	587	207	226	268	128	138	155	88	94	102
	•••	1428	1608	2412	424	492	606	215	242	268	133	145	149	91	97	93
0.45	***	1353	1452	2362	402	449	600	205	226	271	127	137	155	87	93	101
	•••	1388	1602	2461	421	496	613	214	243	268	132	145	147	90	96	91
0.5	***	1277	1396	2390	389	440	601	200	222	268	123	134	152	84	89	98
	•••	1320	1566	2461	410	490	606	209	240	263	128	141	143	86	93	87
0.55	***	1182	1320	2363	371	424	587	191	213	259	117	128	145	79	85	92
	•••	1232	1506	2412	393	476	588	201	231	252	122	135	135	82	88	81
0.6	***	1075	1228	2283	348	402	561	180	201	245	110	119	135	73	78	85
	•••	1132	1424	2313	371	453	557	189	218	235	114	126	124	75	81	73
0.65	***	962	1126	2149	321	373	521	165	185	224	99	108	122	66	70	75
	•••	1025	1325	2164	343	421	514	174	200	213	103	114	110	67	72	63
0.7	***	849	1014	1962	289	338	467	147	165	197	87	94	105	56	59	64
	•••	913	1208	1967	310	381	458	155	178	186	91	99	93	58	61	51
0.75	***	735	894	1722	253	295	401	126	140	165	72	78	85	.	.	.
	•••	797	1073	1719	271	332	390	133	151	153	75	81	73	.	.	.
0.8	***	619	762	1430	211	245	321	101	111	127	.	.	.	.	.	.
	•••	676	918	1422	226	273	310	105	118	115	.	.	.	.	.	.
0.85	***	496	613	1087	161	184	229	.	.	.	.	.	.	.	.	.
	•••	544	735	1076	171	202	217	.	.	.	.	.	.	.	.	.
0.9	***	357	436	693	.	.	.	.	.	.	.	.	.	.	.	.
	•••	389	512	681	.	.	.	.	.	.	.	.	.	.	.	.

TABLE 7: ALPHA= 0.05 POWER= 0.8   EXPECTED ACCRUAL THRU MINIMUM FOLLOW-UP= 180

	DEL=.05			DEL=.10			DEL=.15			DEL=.20			DEL=.25		
FACT=	1.0 .75	.50 .25	.00 BIN	1.0 .75	.50 .25	.00 BIN	1.0 .75	.50 .25	.00 BIN	1.0 75	.50 .25	.00 BIN	1.0 .75	.50 .25	.00 BIN
PCONT=•••				REQUIRED NUMBER OF PATIENTS											
0.05 •••	373	377	395	134	137	141	79	80	82	56	57	58	44	44	45
•••	374	382	681	135	138	217	80	81	115	57	57	73	44	45	51
0.1 •••	680	690	762	215	221	235	117	120	125	78	80	82	58	59	60
•••	684	706	1076	217	226	310	118	122	153	79	81	93	58	60	63
0.15 •••	943	960	1117	281	292	323	147	153	163	95	98	103	69	70	73
•••	949	989	1422	286	303	390	149	157	186	96	100	110	69	72	73
0.2 •••	1148	1174	1441	333	350	400	170	179	195	108	113	120	77	80	83
•••	1157	1219	1719	339	367	458	174	186	213	110	116	124	78	81	81
0.25 •••	1292	1330	1724	370	393	465	188	199	222	118	124	134	83	86	91
•••	1305	1394	1967	379	417	514	193	208	235	121	128	135	84	89	87
0.3 •••	1380	1430	1960	394	424	519	200	214	243	125	132	144	87	91	97
•••	1397	1515	2164	406	455	557	206	226	252	128	137	143	89	94	91
0.35 •••	1415	1481	2147	407	444	559	206	223	259	128	137	151	89	93	100
•••	1438	1588	2313	422	481	588	214	237	263	132	143	147	91	97	93
0.4 •••	1406	1489	2281	410	453	587	209	228	268	129	139	155	89	94	102
•••	1434	1620	2412	428	495	606	217	243	268	134	146	149	91	98	93
0.45 •••	1359	1462	2362	405	453	600	207	228	271	128	138	155	87	93	101
•••	1396	1616	2461	425	499	613	216	245	268	132	145	147	90	96	91
0.5 •••	1285	1408	2390	393	445	601	202	224	268	124	135	152	84	90	98
•••	1329	1582	2461	414	494	606	211	241	263	129	142	143	87	93	87
0.55 •••	1191	1333	2363	375	429	587	193	215	260	118	129	145	80	85	92
•••	1243	1522	2412	398	479	588	203	232	252	123	136	135	82	88	81
0.6 •••	1085	1243	2283	353	406	561	182	203	245	110	120	135	74	78	85
•••	1145	1442	2313	375	456	557	191	219	235	115	127	124	76	81	73
0.65 •••	974	1141	2149	325	377	521	167	186	224	100	109	122	66	70	75
•••	1038	1342	2164	347	425	514	175	201	213	104	114	110	68	72	63
0.7 •••	861	1029	1962	294	342	467	149	166	197	88	95	105	57	60	64
•••	927	1225	1967	314	384	458	156	179	186	91	100	93	58	62	51
0.75 •••	747	908	1722	257	299	401	127	141	165	73	78	85	.	.	.
•••	810	1088	1719	275	335	390	134	151	153	75	81	73	.	.	.
0.8 •••	630	774	1431	214	247	321	102	112	127	.	.	.	.	.	.
•••	688	931	1422	228	275	310	106	118	115	.	.	.	.	.	.
0.85 •••	506	623	1087	163	186	229	.	.	.	.	.	.	.	.	.
•••	553	745	1076	173	203	217	.	.	.	.	.	.	.	.	.
0.9 •••	363	442	693	.	.	.	.	.	.	.	.	.	.	.	.
•••	396	518	681	.	.	.	.	.	.	.	.	.	.	.	.

TABLE 7: ALPHA= 0.05 POWER= 0.8   EXPECTED ACCRUAL THRU MINIMUM FOLLOW-UP= 190

		DEL=.05			DEL=.10			DEL=.15			DEL=.20			DEL=.25		
FACT=		1.0 .75	.50 .25	.00 BIN	1.0 .75	.50 .25	.00 BIN	1.0 .75	.50 .25	.00 BIN	1.0 75	.50 .25	.00 BIN	1.0 .75	.50 .25	.00 BIN
PCONT=***					REQUIRED NUMBER OF PATIENTS											
0.05	***	373 377 395			135 137 141			79 80 82			56 57 58			44 44 45		
	•••	374 383 681			135 138 217			80 81 115			57 57 73			44 45 51		
0.1	***	681 691 762			215 222 235			117 120 125			78 80 82			58 59 60		
	•••	685 708 1076			218 227 310			118 122 153			79 81 93			58 60 63		
0.15	***	944 962 1117			282 293 323			147 153 163			95 98 103			69 70 73		
	•••	950 992 1422			287 304 390			150 157 186			97 100 110			70 72 73		
0.2	***	1149 1177 1441			334 351 400			171 179 195			109 113 120			77 80 83		
	•••	1158 1224 1719			341 368 458			175 186 213			111 116 124			78 81 81		
0.25	***	1294 1334 1724			372 395 466			189 200 222			118 124 134			83 87 91		
	•••	1308 1400 1967			381 419 514			194 209 235			121 128 135			85 89 87		
0.3	***	1383 1435 1960			396 427 519			201 215 243			125 132 144			87 91 97		
	•••	1401 1523 2164			409 457 557			207 226 252			128 137 143			89 94 91		
0.35	***	1419 1488 2147			410 447 559			208 225 259			129 137 151			89 94 100		
	•••	1443 1598 2313			425 483 588			215 238 263			133 143 147			91 97 93		
0.4	***	1410 1498 2281			414 456 587			210 229 268			130 139 155			89 94 102		
	•••	1441 1632 2412			431 498 606			218 244 268			134 146 149			92 98 93		
0.45	***	1365 1473 2362			409 457 600			209 229 271			129 139 155			88 93 101		
	•••	1403 1629 2461			428 503 613			218 246 268			133 146 147			90 97 91		
0.5	***	1292 1420 2390			397 449 601			203 225 268			125 135 152			85 90 98		
	•••	1339 1597 2461			418 497 606			213 242 263			130 142 143			87 94 87		
0.55	***	1200 1347 2363			380 433 587			195 217 260			119 129 145			80 85 92		
	•••	1254 1538 2412			402 483 588			204 234 252			124 136 135			83 89 81		
0.6	***	1096 1257 2283			357 411 561			183 204 245			111 121 135			74 79 85		
	•••	1157 1458 2313			379 460 557			192 220 235			116 127 124			76 82 73		
0.65	***	986 1156 2149			329 381 521			168 188 224			101 109 122			66 70 75		
	•••	1051 1359 2164			351 428 514			177 202 213			105 115 110			68 73 63		
0.7	***	873 1044 1962			297 345 467			150 167 197			89 95 105			57 60 64		
	•••	940 1240 1967			317 387 458			158 180 186			92 100 93			58 62 51		
0.75	***	758 921 1722			260 302 401			129 142 165			73 78 85			. . .		
	•••	823 1103 1719			278 337 390			135 152 153			76 82 73			. . .		
0.8	***	640 786 1431			217 250 321			103 112 127			. . .			. . .		
	•••	699 943 1422			231 277 310			107 119 115			. . .			. . .		
0.85	***	514 633 1087			165 187 229			. . .			. . .			. . .		
	•••	562 754 1076			175 204 217			. . .			. . .			. . .		
0.9	***	369 448 693			. . .			. . .			. . .			. . .		
	•••	402 524 681			. . .			. . .			. . .			. . .		

TABLE 7: ALPHA= 0.05 POWER= 0.8   EXPECTED ACCRUAL THRU MINIMUM FOLLOW-UP= 200

		DEL=.05			DEL=.10			DEL=.15			DEL=.20			DEL=.25		
FACT=		1.0 .75	.50 .25	.00 BIN	1.0 .75	.50 .25	.00 BIN	1.0 .75	.50 .25	.00 BIN	1.0 75	.50 .25	.00 BIN	1.0 .75	.50 .25	.00 BIN
PCONT=•••					REQUIRED NUMBER OF PATIENTS											
0.05	***	373	378	395	135	137	141	80	81	82	56	57	58	44	44	45
	•••	375	383	681	136	138	217	80	81	115	57	57	73	44	45	51
0.1	***	682	692	762	216	222	235	117	120	125	78	80	82	58	59	60
	•••	685	709	1076	218	227	310	119	122	153	79	81	93	58	60	63
0.15	***	945	964	1117	283	294	323	148	153	163	95	98	103	69	71	73
	•••	951	995	1422	287	304	390	150	157	186	97	100	110	70	72	73
0.2	***	1151	1180	1441	335	352	400	172	180	195	109	113	120	77	80	83
	•••	1160	1228	1719	342	369	458	175	186	213	111	116	124	78	81	81
0.25	***	1297	1338	1724	373	397	466	190	201	222	119	125	134	83	87	91
	•••	1311	1405	1967	383	420	514	195	210	235	122	129	135	85	89	87
0.3	***	1385	1441	1960	399	429	519	202	216	243	126	133	144	87	91	97
	•••	1404	1530	2164	411	459	557	208	227	252	129	138	143	89	94	91
0.35	***	1423	1495	2147	413	450	559	209	226	259	130	138	151	90	94	100
	•••	1447	1608	2313	428	486	588	216	239	263	133	144	147	92	97	93
0.4	***	1415	1506	2281	417	460	587	212	231	268	131	140	155	90	95	102
	•••	1447	1643	2412	434	501	606	220	245	268	135	146	149	92	98	93
0.45	***	1372	1483	2362	412	460	600	210	231	271	130	139	155	88	94	101
	•••	1411	1642	2461	432	506	613	219	247	268	134	146	147	91	97	91
0.5	***	1300	1432	2390	401	452	601	205	227	268	126	136	152	85	91	98
	•••	1348	1611	2461	422	501	606	214	243	263	131	143	143	88	94	87
0.55	***	1209	1360	2363	384	437	587	197	218	260	120	130	145	81	86	92
	•••	1265	1554	2412	406	486	588	206	235	252	125	136	135	83	89	81
0.6	***	1106	1271	2283	361	414	561	185	206	245	112	121	135	75	79	85
	•••	1169	1474	2313	383	463	557	194	221	235	116	127	124	77	82	73
0.65	***	997	1170	2149	333	385	521	170	189	224	102	110	122	67	70	75
	•••	1064	1374	2164	355	431	514	178	203	213	105	115	110	68	73	63
0.7	***	884	1058	1962	301	349	467	152	168	197	89	96	105	57	60	64
	•••	952	1255	1967	321	390	458	159	180	186	92	100	93	58	62	51
0.75	***	770	935	1722	263	305	401	130	143	165	74	79	85	.	.	.
	•••	835	1116	1719	281	339	390	136	152	153	76	82	73	.	.	.
0.8	***	651	798	1431	219	252	321	103	113	127	.	.	.	.	.	.
	•••	710	954	1422	234	278	310	108	119	115	.	.	.	.	.	.
0.85	***	523	642	1087	167	189	229	.	.	.	.	.	.	.	.	.
	•••	571	763	1076	176	205	217	.	.	.	.	.	.	.	.	.
0.9	***	375	454	693	.	.	.	.	.	.	.	.	.	.	.	.
	•••	408	529	681	.	.	.	.	.	.	.	.	.	.	.	.

TABLE 7: ALPHA= 0.05 POWER= 0.8   EXPECTED ACCRUAL THRU MINIMUM FOLLOW-UP= 225

	DEL=.05			DEL=.10			DEL=.15			DEL=.20			DEL=.25		
FACT=	1.0 .75	.50 .25	.00 BIN	1.0 .75	.50 .25	.00 BIN	1.0 .75	.50 .25	.00 BIN	1.0 75	.50 .25	.00 BIN	1.0 .75	.50 .25	.00 BIN

PCONT=***   REQUIRED NUMBER OF PATIENTS

PCONT	D.05 1.0/.75	.50/.25	.00/BIN	D.10 1.0/.75	.50/.25	.00/BIN	D.15 1.0/.75	.50/.25	.00/BIN	D.20 1.0/.75	.50/.25	.00/BIN	D.25 1.0/.75	.50/.25	.00/BIN
0.05 ***	374	379	395	135	137	141	80	81	82	57	57	58	44	44	45
•••	376	384	681	136	139	217	80	81	115	57	57	73	44	45	51
0.1 ***	683	695	762	217	223	235	118	121	125	78	80	82	58	59	60
•••	687	712	1076	219	228	310	119	123	153	79	81	93	59	60	63
0.15 ***	947	968	1117	285	296	323	149	154	163	96	99	103	69	71	73
•••	955	1001	1422	289	306	390	151	158	186	97	101	110	70	72	73
0.2 ***	1154	1186	1441	338	355	400	173	181	195	110	114	120	78	80	83
•••	1165	1237	1719	345	372	458	177	187	213	112	117	124	79	82	81
0.25 ***	1302	1347	1724	377	401	466	192	203	222	120	126	134	84	87	91
•••	1317	1419	1967	387	424	514	196	211	235	123	129	135	86	89	87
0.3 ***	1393	1454	1960	404	435	519	204	218	243	127	134	144	88	92	97
•••	1414	1548	2164	414	464	557	210	228	252	130	138	143	90	94	91
0.35 ***	1432	1511	2147	419	456	559	212	228	259	131	139	151	90	95	100
•••	1460	1630	2313	434	491	588	219	241	263	135	144	147	92	97	93
0.4 ***	1427	1526	2281	424	467	587	215	233	268	133	141	155	91	95	102
•••	1463	1670	2412	442	507	606	223	247	268	137	147	149	93	98	93
0.45 ***	1387	1507	2362	420	468	600	214	234	271	131	141	155	89	94	101
•••	1430	1673	2461	440	513	613	223	249	268	136	147	147	92	97	91
0.5 ***	1319	1460	2390	410	461	601	209	230	268	128	137	152	86	91	98
•••	1370	1644	2461	431	508	606	218	245	263	132	144	143	89	94	87
0.55 ***	1231	1390	2363	392	446	587	201	221	259	122	131	145	82	86	92
•••	1291	1589	2412	415	494	588	210	237	252	126	137	135	84	89	81
0.6 ***	1131	1304	2283	370	423	561	189	209	245	114	122	135	75	79	85
•••	1197	1511	2313	392	471	557	198	223	235	118	128	124	77	82	73
0.65 ***	1023	1203	2149	342	393	521	174	192	224	103	111	122	68	71	75
•••	1093	1410	2164	364	438	514	182	205	213	107	116	110	69	73	63
0.7 ***	911	1090	1962	309	356	467	155	171	197	90	97	105	58	60	64
•••	982	1290	1967	329	396	458	162	182	186	93	101	93	59	62	51
0.75 ***	796	965	1722	271	311	401	132	145	165	75	79	85	.	.	.
•••	863	1148	1719	288	345	390	138	154	153	77	82	73	.	.	.
0.8 ***	674	824	1430	225	257	321	105	114	127	.	.	.	.	.	.
•••	735	981	1422	239	282	310	109	120	115	.	.	.	.	.	.
0.85 ***	542	663	1087	171	192	229	.	.	.	.	.	.	.	.	.
•••	591	783	1076	180	207	217	.	.	.	.	.	.	.	.	.
0.9 ***	388	468	693	.	.	.	.	.	.	.	.	.	.	.	.
•••	421	540	681	.	.	.	.	.	.	.	.	.	.	.	.

TABLE 7: ALPHA= 0.05 POWER= 0.8   EXPECTED ACCRUAL THRU MINIMUM FOLLOW-UP= 250

	DEL=.05			DEL=.10			DEL=.15			DEL=.20			DEL=.25		
FACT=	1.0 .75	.50 .25	.00 BIN	1.0 .75	.50 .25	.00 BIN	1.0 .75	.50 .25	.00 BIN	1.0 75	.50 .25	.00 BIN	1.0 .75	.50 .25	.00 BIN
PCONT=***	REQUIRED NUMBER OF PATIENTS														
0.05 ***	374	379	395	135	137	141	80	81	82	57	57	58	44	44	45
•••	376	385	681	136	139	217	80	82	115	57	57	73	44	45	51
0.1 ***	684	697	762	218	224	235	118	121	125	79	80	82	58	59	60
•••	689	715	1076	220	229	310	119	123	153	79	81	93	59	60	63
0.15 ***	950	973	1117	286	298	323	150	155	163	97	99	103	70	71	73
•••	958	1006	1422	291	307	390	152	158	186	98	101	110	70	72	73
0.2 ***	1158	1193	1441	340	358	400	175	182	195	111	114	120	78	80	83
•••	1170	1246	1719	348	374	458	178	188	213	112	117	124	79	82	81
0.25 ***	1307	1357	1724	381	405	466	194	204	222	121	126	134	85	87	91
•••	1324	1432	1967	391	427	514	198	212	235	123	130	135	86	89	87
0.3 ***	1400	1466	1960	408	439	519	207	220	243	128	135	144	89	92	97
•••	1423	1565	2164	421	467	557	213	230	252	131	139	143	91	94	91
0.35 ***	1441	1527	2147	424	462	559	215	230	259	133	140	151	91	95	100
•••	1472	1651	2313	440	496	588	222	242	263	136	145	147	93	98	93
0.4 ***	1439	1546	2281	430	474	587	218	236	268	134	142	155	92	96	102
•••	1477	1694	2412	448	513	606	226	249	268	138	148	149	94	99	93
0.45 ***	1402	1530	2362	427	476	600	217	236	271	133	142	155	90	95	101
•••	1448	1700	2461	448	519	613	226	251	268	137	148	147	92	98	91
0.5 ***	1336	1485	2390	417	469	601	212	232	268	129	139	152	87	92	98
•••	1392	1675	2461	439	514	606	221	247	263	134	144	143	89	94	87
0.55 ***	1252	1418	2363	401	454	587	204	224	259	124	132	145	83	87	92
•••	1315	1622	2412	423	500	588	213	239	252	128	138	135	84	89	81
0.6 ***	1154	1333	2283	378	431	561	192	211	245	115	123	135	76	80	85
•••	1223	1544	2313	400	477	557	201	225	235	119	129	124	78	82	73
0.65 ***	1048	1233	2149	350	401	521	177	194	224	105	112	122	68	71	75
•••	1121	1443	2164	372	444	514	184	207	213	108	116	110	70	73	63
0.7 ***	937	1120	1962	317	363	467	158	173	197	92	97	105	58	61	64
•••	1009	1321	1967	336	401	458	164	183	186	94	101	93	59	62	51
0.75 ***	820	992	1722	277	317	401	134	147	165	76	80	85	.	.	.
•••	889	1175	1719	294	349	390	140	155	153	78	82	73	.	.	.
0.8 ***	696	848	1431	230	261	321	107	115	127	.	.	.	.	.	.
•••	758	1004	1422	244	285	310	111	120	115	.	.	.	.	.	.
0.85 ***	560	681	1087	174	194	229	.	.	.	.	.	.	.	.	.
•••	610	800	1076	183	209	217	.	.	.	.	.	.	.	.	.
0.9 ***	400	479	693	.	.	.	.	.	.	.	.	.	.	.	.
•••	434	550	681	.	.	.	.	.	.	.	.	.	.	.	.

TABLE 7: ALPHA= 0.05 POWER= 0.8	EXPECTED ACCRUAL THRU MINIMUM FOLLOW-UP= 275

		DEL=.05			DEL=.10			DEL=.15			DEL=.20			DEL=.25		
FACT=		1.0 .75	.50 .25	.00 BIN	1.0 .75	.50 .25	.00 BIN	1.0 .75	.50 .25	.00 BIN	1.0 75	.50 .25	.00 BIN	1.0 .75	.50 .25	.00 BIN
PCONT=•••				REQUIRED NUMBER OF PATIENTS												
0.05	•••	375	380	395	136	138	141	80	81	82	57	57	58	44	44	45
	•••	377	385	681	136	139	217	80	81	115	57	57	73	44	45	51
0.1	•••	686	700	762	219	224	235	119	121	125	79	80	82	58	59	60
	•••	691	717	1076	221	229	310	120	123	153	80	81	93	59	60	63
0.15	•••	952	977	1117	288	299	322	150	155	163	97	99	103	70	71	73
	•••	961	1011	1422	293	308	390	153	158	186	98	101	110	70	72	73
0.2	•••	1162	1199	1441	343	360	400	176	183	195	111	115	120	79	81	83
	•••	1175	1254	1719	350	375	458	179	188	213	113	117	124	79	82	81
0.25	•••	1312	1365	1724	384	408	465	195	205	222	122	127	134	85	88	91
	•••	1331	1443	1967	394	430	514	200	212	235	124	130	135	86	89	87
0.3	•••	1407	1478	1960	412	443	519	209	221	243	129	135	144	89	93	97
	•••	1432	1580	2164	425	471	557	214	231	252	132	139	143	91	95	91
0.35	•••	1451	1542	2147	429	467	559	217	232	259	134	141	151	92	96	100
	•••	1483	1670	2313	445	500	588	224	243	263	137	145	147	94	98	93
0.4	•••	1451	1564	2281	436	479	587	221	238	268	135	143	155	92	96	102
	•••	1492	1716	2412	454	518	606	228	250	268	139	148	149	94	99	93
0.45	•••	1416	1551	2362	434	482	600	220	239	271	134	143	155	91	95	101
	•••	1466	1726	2461	454	524	613	228	252	268	138	148	147	93	98	91
0.5	•••	1354	1510	2390	424	475	601	215	235	268	131	139	152	88	92	98
	•••	1412	1702	2461	446	520	606	224	249	263	135	145	143	90	95	87
0.55	•••	1272	1445	2363	408	461	587	207	226	260	125	133	145	83	87	92
	•••	1338	1651	2412	430	506	588	216	240	252	129	139	135	85	90	81
0.6	•••	1176	1361	2283	386	438	561	195	213	245	117	124	135	77	80	85
	•••	1248	1573	2164	408	482	557	203	227	235	120	129	124	78	83	73
0.65	•••	1071	1261	2149	357	407	520	179	196	224	106	113	122	69	72	75
	•••	1146	1472	2164	379	449	514	187	208	213	109	117	110	70	73	63
0.7	•••	960	1147	1962	323	369	467	160	175	197	92	98	105	59	61	64
	•••	1034	1348	1967	343	406	458	166	184	186	95	101	93	60	62	51
0.75	•••	842	1017	1722	283	322	401	136	148	165	76	80	85	.	.	.
	•••	913	1200	1719	300	352	390	142	155	153	78	83	73	.	.	.
0.8	•••	716	870	1431	235	265	321	108	116	127	.	.	.	.	.	.
	•••	778	1025	1422	248	288	310	112	121	115	.	.	.	.	.	.
0.85	•••	576	698	1087	177	197	229	.	.	.	.	.	.	.	.	.
	•••	627	815	1076	186	211	217	.	.	.	.	.	.	.	.	.
0.9	•••	411	490	693	.	.	.	.	.	.	.	.	.	.	.	.
	•••	444	559	681	.	.	.	.	.	.	.	.	.	.	.	.

TABLE 7: ALPHA= 0.05 POWER= 0.8   EXPECTED ACCRUAL THRU MINIMUM FOLLOW-UP= 300

	DEL=.05			DEL=.10			DEL=.15			DEL=.20			DEL=.25		
FACT=	1.0 .75	.50 .25	.00 BIN	1.0 .75	.50 .25	.00 BIN	1.0 .75	.50 .25	.00 BIN	1.0 75	.50 .25	.00 BIN	1.0 .75	.50 .25	.00 BIN

PCONT=***   REQUIRED NUMBER OF PATIENTS

PCONT	.05(1)	.05(2)	.05(3)	.10(1)	.10(2)	.10(3)	.15(1)	.15(2)	.15(3)	.20(1)	.20(2)	.20(3)	.25(1)	.25(2)	.25(3)
0.05 ***	376	381	395	136	138	141	80	81	82	57	57	58	44	44	45
•••	378	386	681	137	139	217	81	82	115	57	58	73	44	45	51
0.1 ***	687	702	762	219	225	235	119	121	125	79	80	82	59	59	60
•••	692	719	1076	222	229	310	120	123	153	80	81	93	59	60	63
0.15 ***	955	981	1117	289	300	322	151	156	163	97	100	103	70	71	73
•••	964	1016	1422	294	309	390	153	159	186	98	101	110	71	72	73
0.2 ***	1165	1205	1441	345	362	400	177	184	195	112	115	120	79	81	83
•••	1180	1262	1719	352	377	458	180	189	213	113	117	124	80	82	81
0.25 ***	1317	1374	1724	387	411	466	196	206	222	123	127	134	85	88	91
•••	1337	1454	1967	397	432	514	201	213	235	125	130	135	87	89	87
0.3 ***	1414	1489	1960	416	447	519	210	223	243	130	136	144	90	93	97
•••	1441	1594	2164	429	474	557	216	232	252	133	140	143	91	95	91
0.35 ***	1459	1556	2147	434	471	559	219	234	259	135	142	151	92	96	100
•••	1495	1687	2313	450	504	588	226	244	263	138	146	147	94	98	93
0.4 ***	1462	1582	2281	442	484	586	223	240	268	136	144	155	93	97	102
•••	1506	1737	2412	460	522	606	231	252	268	140	149	149	95	99	93
0.45 ***	1430	1571	2362	440	488	600	223	241	271	136	144	155	92	96	101
•••	1483	1749	2461	460	529	613	231	253	268	139	149	147	94	98	91
0.5 ***	1370	1532	2389	431	481	601	218	237	268	132	140	152	89	93	98
•••	1432	1728	2461	452	525	606	226	250	263	136	145	143	91	95	87
0.55 ***	1291	1469	2363	415	467	587	210	228	259	126	134	145	84	88	92
•••	1360	1677	2412	437	511	588	218	241	252	130	139	135	86	90	81
0.6 ***	1197	1387	2283	392	444	561	198	215	245	118	125	135	77	81	85
•••	1271	1600	2313	414	487	557	206	228	235	121	130	124	79	83	73
0.65 ***	1093	1287	2149	364	413	520	182	198	224	107	113	122	69	72	75
•••	1170	1499	2164	385	453	514	189	209	213	110	117	110	70	73	63
0.7 ***	982	1172	1962	329	374	467	162	176	197	93	98	105	59	61	64
•••	1058	1373	1967	349	410	458	168	185	186	96	102	93	60	62	51
0.75 ***	863	1040	1722	288	326	401	138	149	165	77	81	85	.	.	.
•••	934	1222	1719	305	355	390	143	156	153	79	83	73	.	.	.
0.8 ***	735	889	1430	239	268	321	109	117	127	.	.	.	.	.	.
•••	798	1043	1422	252	290	310	113	121	115	.	.	.	.	.	.
0.85 ***	592	713	1087	180	199	229	.	.	.	.	.	.	.	.	.
•••	642	829	1076	189	212	217	.	.	.	.	.	.	.	.	.
0.9 ***	421	499	693	.	.	.	.	.	.	.	.	.	.	.	.
•••	454	567	681	.	.	.	.	.	.	.	.	.	.	.	.

TABLE 7: ALPHA= 0.05 POWER= 0.8   EXPECTED ACCRUAL THRU MINIMUM FOLLOW-UP= 325

		DEL=.05			DEL=.10			DEL=.15			DEL=.20			DEL=.25		
FACT=		1.0 .75	.50 .25	.00 BIN	1.0 .75	.50 .25	.00 BIN	1.0 .75	.50 .25	.00 BIN	1.0 75	.50 .25	.00 BIN	1.0 .75	.50 .25	.00 BIN
PCONT=•••		REQUIRED NUMBER OF PATIENTS														
0.05	•••	376	381	395	136	138	141	80	81	82	57	57	58	44	44	45
	•••	378	387	681	137	139	217	81	82	115	57	57	73	44	45	51
0.1	•••	689	704	762	220	226	235	119	122	125	79	81	82	59	60	60
	•••	694	721	1076	223	230	310	120	123	153	80	81	93	59	60	63
0.15	•••	957	985	1117	291	302	323	152	156	163	98	100	103	70	71	73
	•••	967	1020	1422	295	310	390	154	159	186	99	101	110	71	72	73
0.2	•••	1169	1211	1441	347	364	400	178	185	195	112	116	120	79	81	83
	•••	1184	1269	1719	354	378	458	181	189	213	114	117	124	80	82	81
0.25	•••	1323	1382	1724	390	414	466	198	207	222	123	128	134	86	88	91
	•••	1344	1464	1967	400	434	514	202	214	235	125	130	135	87	90	87
0.3	•••	1420	1500	1960	420	451	519	212	224	244	131	136	144	90	93	97
	•••	1449	1607	2164	433	476	557	218	232	252	134	140	143	92	95	91
0.35	•••	1469	1570	2147	438	476	559	221	235	259	136	142	151	93	96	100
	•••	1506	1703	2313	454	507	588	227	245	263	139	146	147	94	98	93
0.4	•••	1474	1598	2281	447	489	587	225	242	268	138	145	155	93	97	102
	•••	1520	1756	2412	465	525	606	233	253	268	141	149	149	95	99	93
0.45	•••	1444	1591	2362	446	493	600	225	242	271	137	144	155	92	96	101
	•••	1499	1770	2461	466	533	613	233	255	268	140	149	147	94	98	91
0.5	•••	1387	1554	2390	437	487	601	221	239	268	133	141	152	89	93	98
	•••	1451	1751	2461	458	529	606	229	251	263	137	146	143	91	95	87
0.55	•••	1309	1492	2363	421	472	587	212	230	260	127	135	145	84	88	92
	•••	1380	1702	2412	443	515	588	220	243	252	131	140	135	86	90	81
0.6	•••	1217	1411	2283	398	450	561	200	217	245	119	126	135	78	81	85
	•••	1293	1625	2313	420	491	557	208	229	235	122	130	124	79	83	73
0.65	•••	1114	1311	2149	370	418	521	184	200	224	108	114	122	69	72	75
	•••	1192	1523	2164	391	457	514	191	210	213	110	118	110	71	74	63
0.7	•••	1002	1195	1962	335	379	467	164	177	197	94	99	105	59	61	64
	•••	1080	1396	1967	354	413	458	170	186	186	96	102	93	60	63	51
0.75	•••	883	1061	1722	293	330	401	140	150	165	78	81	85	.	.	.
	•••	955	1243	1719	309	358	390	144	157	153	79	83	73	.	.	.
0.8	•••	752	908	1431	243	271	321	110	117	127	.	.	.	.	.	.
	•••	816	1060	1422	255	292	310	114	122	115	.	.	.	.	.	.
0.85	•••	606	727	1087	183	201	229	.	.	.	.	.	.	.	.	.
	•••	656	841	1076	191	213	217	.	.	.	.	.	.	.	.	.
0.9	•••	431	508	693	.	.	.	.	.	.	.	.	.	.	.	.
	•••	463	573	681	.	.	.	.	.	.	.	.	.	.	.	.

TABLE 7: ALPHA= 0.05 POWER= 0.8   EXPECTED ACCRUAL THRU MINIMUM FOLLOW-UP= 350

	DEL=.05			DEL=.10			DEL=.15			DEL=.20			DEL=.25		
FACT=	1.0 .75	.50 .25	.00 BIN	1.0 .75	.50 .25	.00 BIN	1.0 .75	.50 .25	.00 BIN	1.0 75	.50 .25	.00 BIN	1.0 .75	.50 .25	.00 BIN
PCONT=•••				REQUIRED NUMBER OF PATIENTS											
0.05 •••	377	382	395	136	138	141	80	81	82	57	57	58	44	45	45
•••	379	387	681	137	139	217	81	82	115	57	57	73	44	45	51
0.1 •••	690	705	762	221	226	236	120	122	125	80	81	82	59	59	60
•••	696	723	1076	223	230	310	121	123	153	80	81	93	59	60	63
0.15 •••	959	988	1117	292	303	323	152	157	163	98	100	103	70	71	73
•••	970	1024	1422	297	311	390	154	159	186	99	101	110	71	72	73
0.2 •••	1172	1217	1441	349	366	400	179	185	195	113	116	120	79	81	83
•••	1189	1275	1719	356	379	458	182	190	213	114	118	124	80	82	81
0.25 •••	1328	1390	1724	392	416	465	199	208	222	124	128	134	86	88	91
•••	1350	1473	1967	403	435	514	203	214	235	126	131	135	87	90	87
0.3 •••	1427	1511	1960	423	454	519	213	225	243	132	137	144	91	94	97
•••	1458	1620	2164	436	479	557	219	233	252	134	140	143	92	95	91
0.35 •••	1477	1583	2147	442	479	559	223	237	259	137	143	151	93	96	100
•••	1516	1718	2313	458	510	588	229	246	263	139	147	147	95	98	93
0.4 •••	1485	1614	2281	451	493	587	227	243	268	139	145	155	94	97	102
•••	1533	1774	2412	469	529	606	234	254	268	142	150	149	96	99	93
0.45 •••	1457	1609	2362	451	498	600	227	244	271	138	145	155	93	96	101
•••	1515	1790	2461	471	536	613	235	256	268	141	150	147	94	98	91
0.5 •••	1402	1574	2390	442	492	601	223	240	268	134	142	152	90	93	98
•••	1468	1773	2461	464	533	606	231	252	263	138	146	143	91	95	87
0.55 •••	1327	1514	2363	427	478	587	214	232	260	128	135	145	85	88	92
•••	1400	1725	2412	449	519	588	222	244	252	132	140	135	87	90	81
0.6 •••	1236	1433	2283	404	454	561	202	219	245	120	126	135	78	81	85
•••	1314	1648	2313	426	495	557	209	230	235	123	130	124	80	83	73
0.65 •••	1133	1333	2149	375	423	521	186	201	224	108	114	122	70	72	75
•••	1213	1546	2164	396	461	514	193	211	213	111	118	110	71	74	63
0.7 •••	1022	1216	1962	340	383	467	166	178	197	95	99	105	59	61	64
•••	1100	1417	1967	358	416	458	171	187	186	97	102	93	61	62	51
0.75 •••	901	1081	1722	297	334	401	141	151	165	78	81	85	.	.	.
•••	974	1261	1719	313	361	390	145	157	153	80	83	73	.	.	.
0.8 •••	768	924	1431	246	274	321	111	118	127	.	.	.	.	.	.
•••	832	1076	1422	258	294	310	114	122	115	.	.	.	.	.	.
0.85 •••	618	740	1087	185	202	229	.	.	.	.	.	.	.	.	.
•••	669	852	1076	193	214	217	.	.	.	.	.	.	.	.	.
0.9 •••	439	515	693	.	.	.	.	.	.	.	.	.	.	.	.
•••	472	579	681	.	.	.	.	.	.	.	.	.	.	.	.

TABLE 7: ALPHA= 0.05 POWER= 0.8   EXPECTED ACCRUAL THRU MINIMUM FOLLOW-UP= 375

	DEL=.05			DEL=.10			DEL=.15			DEL=.20			DEL=.25		
FACT=	1.0 .75	.50 .25	.00 BIN	1.0 .75	.50 .25	.00 BIN	1.0 .75	.50 .25	.00 BIN	1.0 75	.50 .25	.00 BIN	1.0 .75	.50 .25	.00 BIN

PCONT=•••   REQUIRED NUMBER OF PATIENTS

PCONT	DEL=.05			DEL=.10			DEL=.15			DEL=.20			DEL=.25		
	1.0	.50	.00	1.0	.50	.00	1.0	.50	.00	1.0	.50	.00	1.0	.50	.00
	.75	.25	BIN	.75	.25	BIN	.75	.25	BIN	75	.25	BIN	.75	.25	BIN
0.05 •••	377	383	395	137	138	141	80	81	82	57	57	58	44	45	45
•••	379	387	681	137	139	217	81	82	115	57	58	73	44	45	51
0.1 •••	691	707	762	221	227	235	120	122	125	80	81	82	59	59	60
•••	697	725	1076	224	230	310	121	123	153	80	81	93	59	60	63
0.15 •••	962	992	1117	293	304	323	153	157	163	98	100	103	70	72	73
•••	973	1028	1422	298	312	390	155	160	186	99	101	110	71	72	73
0.2 •••	1176	1222	1441	351	368	400	179	186	195	113	116	120	80	81	83
•••	1193	1281	1719	358	380	458	182	190	213	114	118	124	80	82	81
0.25 •••	1333	1398	1724	395	418	466	200	209	222	124	128	134	86	88	91
•••	1356	1482	1967	405	437	514	204	215	235	126	131	135	88	90	87
0.3 •••	1434	1521	1960	426	456	519	215	226	243	132	137	144	91	94	97
•••	1466	1631	2164	439	481	557	220	234	252	135	140	143	92	95	91
0.35 •••	1486	1596	2147	446	482	559	224	238	259	137	143	151	94	97	100
•••	1527	1732	2313	462	512	588	230	247	263	140	147	147	95	99	93
0.4 •••	1495	1629	2281	455	497	587	229	244	268	139	146	155	94	98	102
•••	1546	1790	2412	474	531	606	236	254	268	142	150	149	96	100	93
0.45 •••	1470	1626	2362	456	502	600	229	245	271	139	146	155	93	97	101
•••	1530	1809	2461	475	539	613	236	257	268	142	150	147	95	99	91
0.5 •••	1417	1593	2390	448	497	601	225	242	268	135	142	152	90	93	98
•••	1485	1793	2461	469	536	606	232	253	263	138	147	143	92	95	87
0.55 •••	1343	1535	2363	432	482	587	216	233	260	129	136	145	85	88	92
•••	1418	1745	2412	454	523	588	224	245	252	132	140	135	87	90	81
0.6 •••	1254	1454	2283	410	459	560	204	220	245	120	127	135	79	81	85
•••	1333	1670	2313	431	498	557	211	231	235	123	131	124	80	83	73
0.65 •••	1152	1355	2149	380	427	520	187	202	224	109	115	122	70	73	75
•••	1233	1566	2164	401	464	514	194	212	213	112	118	110	71	74	63
0.7 •••	1040	1236	1962	344	387	467	167	179	197	95	100	105	60	62	64
•••	1120	1437	1967	363	419	458	173	187	186	97	102	93	61	63	51
0.75 •••	918	1099	1722	301	337	401	142	152	165	78	81	85	.	.	.
•••	992	1279	1719	317	363	390	146	158	153	80	83	73	.	.	.
0.8 •••	783	940	1430	249	276	321	112	118	127	.	.	.	.	.	.
•••	848	1090	1422	261	295	310	115	122	115	.	.	.	.	.	.
0.85 •••	631	752	1087	187	204	229	.	.	.	.	.	.	.	.	.
•••	681	862	1076	194	215	217	.	.	.	.	.	.	.	.	.
0.9 •••	447	522	693	.	.	.	.	.	.	.	.	.	.	.	.
•••	479	585	681	.	.	.	.	.	.	.	.	.	.	.	.

TABLE 7: ALPHA= 0.05 POWER= 0.8   EXPECTED ACCRUAL THRU MINIMUM FOLLOW-UP= 400

		DEL=.05			DEL=.10			DEL=.15			DEL=.20			DEL=.25		
FACT=		1.0 .75	.50 .25	.00 BIN	1.0 .75	.50 .25	.00 BIN	1.0 .75	.50 .25	.00 BIN	1.0 75	.50 .25	.00 BIN	1.0 .75	.50 .25	.00 BIN
PCONT=•••					REQUIRED	NUMBER OF	PATIENTS									
0.05	•••	378	383	395	137	138	141	81	81	82	57	57	58	44	45	45
	•••	380	388	681	138	139	217	81	82	115	57	58	73	44	45	51
0.1	•••	692	709	762	222	227	235	120	122	125	80	81	82	59	60	60
	•••	699	727	1076	224	231	310	121	124	153	80	81	93	59	60	63
0.15	•••	964	995	1117	294	304	323	153	157	163	98	100	103	71	72	73
	•••	975	1031	1422	299	312	390	155	160	186	99	101	110	71	72	73
0.2	•••	1180	1228	1441	352	369	400	180	186	195	113	116	120	80	81	83
	•••	1197	1287	1719	360	382	458	183	190	213	115	118	124	81	82	81
0.25	•••	1338	1405	1724	397	420	466	201	210	222	125	129	134	87	89	91
	•••	1363	1490	1967	407	439	514	205	215	235	127	131	135	88	90	87
0.3	•••	1441	1530	1960	429	459	519	216	227	243	133	138	144	91	94	97
	•••	1474	1642	2164	442	483	557	221	234	252	135	141	143	93	95	91
0.35	•••	1495	1608	2147	450	486	559	226	239	259	138	144	151	94	97	100
	•••	1537	1745	2313	465	514	588	232	248	263	141	147	147	95	99	93
0.4	•••	1506	1643	2281	460	501	587	231	245	268	140	146	155	95	98	102
	•••	1558	1805	2412	477	534	606	237	255	268	143	150	149	96	100	93
0.45	•••	1483	1642	2362	460	506	600	231	247	271	139	146	155	94	97	101
	•••	1544	1826	2461	480	542	613	238	257	268	143	150	147	95	99	91
0.5	•••	1432	1611	2390	452	501	601	227	243	268	136	143	152	91	94	98
	•••	1502	1812	2461	473	539	606	234	254	263	139	147	143	92	96	87
0.55	•••	1360	1554	2363	437	486	587	218	235	260	130	136	145	86	89	92
	•••	1436	1765	2412	458	526	588	226	246	252	133	140	135	87	90	81
0.6	•••	1271	1474	2283	414	463	561	206	221	245	121	127	135	79	82	85
	•••	1352	1689	2313	436	501	557	213	232	235	124	131	124	80	83	73
0.65	•••	1170	1374	2149	385	431	521	189	203	224	110	115	122	70	73	75
	•••	1252	1586	2164	405	467	514	196	213	213	112	118	110	72	74	63
0.7	•••	1058	1255	1962	349	390	467	168	180	197	96	100	105	60	62	64
	•••	1138	1455	1967	367	421	458	174	188	186	98	102	93	61	63	51
0.75	•••	935	1116	1722	305	339	401	143	152	165	79	82	85	.	.	.
	•••	1009	1295	1719	320	365	390	147	158	153	80	84	73	.	.	.
0.8	•••	798	954	1431	252	278	321	113	119	127	.	.	.	.	.	.
	•••	863	1102	1422	264	297	310	116	123	115	.	.	.	.	.	.
0.85	•••	642	763	1087	189	205	229	.	.	.	.	.	.	.	.	.
	•••	693	872	1076	196	216	217	.	.	.	.	.	.	.	.	.
0.9	•••	454	529	693	.	.	.	.	.	.	.	.	.	.	.	.
	•••	486	590	681	.	.	.	.	.	.	.	.	.	.	.	.

TABLE 7: ALPHA= 0.05 POWER= 0.8   EXPECTED ACCRUAL THRU MINIMUM FOLLOW-UP= 425

	DEL=.05			DEL=.10			DEL=.15			DEL=.20			DEL=.25		
FACT=	1.0 .75	.50 .25	.00 BIN	1.0 .75	.50 .25	.00 BIN	1.0 .75	.50 .25	.00 BIN	1.0 75	.50 .25	.00 BIN	1.0 .75	.50 .25	.00 BIN
PCONT=•••				REQUIRED NUMBER OF PATIENTS											
0.05 •••	378	383	395	137	138	141	81	81	82	57	57	58	44	45	45
•••	380	388	681	138	139	217	81	82	115	57	58	73	44	45	51
0.1 •••	694	711	762	222	227	235	120	122	125	80	81	82	59	60	60
•••	700	728	1076	225	231	310	121	124	153	80	81	93	59	60	63
0.15 •••	966	998	1117	295	305	323	154	158	163	99	101	103	71	72	73
•••	978	1034	1422	299	313	390	155	160	186	99	102	110	71	72	73
0.2 •••	1183	1233	1441	354	370	400	181	187	195	113	116	120	80	81	83
•••	1201	1292	1719	361	383	458	184	190	213	115	118	124	81	82	81
0.25 •••	1342	1412	1724	399	422	465	202	210	222	125	129	134	87	89	91
•••	1368	1498	1967	409	440	514	206	215	235	127	131	135	88	90	87
0.3 •••	1447	1540	1960	432	461	519	217	228	243	133	138	144	92	94	97
•••	1482	1651	2164	445	484	557	222	235	252	136	141	143	93	95	91
0.35 •••	1503	1619	2147	453	489	560	227	240	259	138	144	151	94	97	101
•••	1547	1757	2313	468	516	588	233	248	263	141	147	147	96	99	93
0.4 •••	1516	1657	2281	463	504	587	232	246	268	141	147	155	95	98	102
•••	1570	1820	2412	481	537	606	239	256	268	144	150	149	96	100	93
0.45 •••	1495	1658	2362	464	510	600	232	248	271	140	146	155	94	97	101
•••	1558	1842	2461	484	545	613	239	258	268	143	150	147	95	99	91
0.5 •••	1446	1628	2390	457	504	601	228	244	268	137	143	152	91	94	98
•••	1517	1829	2461	478	542	606	235	255	263	140	147	143	92	96	87
0.55 •••	1375	1572	2363	442	490	587	220	236	260	130	137	145	86	89	92
•••	1453	1784	2412	463	528	588	227	246	252	134	141	135	87	91	81
0.6 •••	1288	1493	2283	419	467	561	207	222	245	122	128	135	79	82	85
•••	1370	1708	2313	440	504	557	214	232	235	125	131	124	81	83	73
0.65 •••	1187	1393	2149	389	435	520	190	204	224	110	115	122	71	73	75
•••	1270	1604	2164	409	469	514	197	213	213	113	119	110	72	74	63
0.7 •••	1074	1273	1962	353	393	467	170	181	197	96	100	105	60	62	64
•••	1155	1471	1967	371	424	458	175	188	186	98	103	93	61	63	51
0.75 •••	950	1132	1722	308	342	401	144	153	165	79	82	85	.	.	.
•••	1025	1309	1719	323	367	390	148	158	153	81	84	73	.	.	.
0.8 •••	811	968	1430	255	280	321	113	119	127	.	.	.	.	.	.
•••	876	1114	1422	266	298	310	116	123	115	.	.	.	.	.	.
0.85 •••	653	773	1087	190	206	229	.	.	.	.	.	.	.	.	.
•••	703	880	1076	198	216	217	.	.	.	.	.	.	.	.	.
0.9 •••	461	535	693	.	.	.	.	.	.	.	.	.	.	.	.
•••	493	594	681	.	.	.	.	.	.	.	.	.	.	.	.

TABLE 7: ALPHA= 0.05 POWER= 0.8   EXPECTED ACCRUAL THRU MINIMUM FOLLOW-UP= 450

		DEL=.05			DEL=.10			DEL=.15			DEL=.20			DEL=.25		
FACT=		1.0 .75	.50 .25	.00 BIN	1.0 .75	.50 .25	.00 BIN	1.0 .75	.50 .25	.00 BIN	1.0 75	.50 .25	.00 BIN	1.0 .75	.50 .25	.00 BIN
PCONT=•••					REQUIRED NUMBER OF PATIENTS											
0.05	•••	379	384	395	137	139	141	81	81	82	57	57	58	45	45	45
	•••	381	388	681	138	140	217	81	82	115	57	58	73	45	45	51
0.1	•••	695	712	762	223	228	235	121	123	125	80	81	82	59	60	60
	•••	702	729	1076	225	231	310	122	124	153	81	81	93	59	60	63
0.15	•••	968	1001	1117	296	306	323	154	158	163	99	100	103	71	72	73
	•••	981	1037	1422	300	313	390	156	160	186	100	102	110	71	72	73
0.2	•••	1186	1237	1441	355	372	400	181	187	195	114	117	120	80	82	83
	•••	1206	1297	1719	363	383	458	184	191	213	115	118	124	81	82	81
0.25	•••	1347	1419	1724	401	424	465	203	211	222	126	129	134	87	89	91
	•••	1374	1504	1967	411	441	514	206	216	235	127	131	135	88	90	87
0.3	•••	1454	1548	1960	435	464	519	218	228	243	134	138	144	92	94	97
	•••	1489	1661	2164	447	486	557	223	235	252	136	141	143	93	95	91
0.35	•••	1511	1630	2147	456	491	559	228	240	259	139	144	151	95	97	100
	•••	1556	1769	2313	471	518	588	234	249	263	141	147	147	96	99	93
0.4	•••	1526	1670	2281	467	507	586	234	247	268	141	147	155	95	98	102
	•••	1581	1833	2412	485	539	606	240	257	268	144	151	149	97	100	93
0.45	•••	1507	1673	2362	468	513	600	234	249	271	141	147	155	94	97	101
	•••	1571	1857	2461	488	547	613	240	259	268	144	151	147	96	99	91
0.5	•••	1459	1644	2390	461	508	600	230	245	268	137	144	152	91	94	98
	•••	1532	1845	2461	482	545	606	237	255	263	140	147	143	93	96	87
0.55	•••	1390	1589	2363	446	494	587	221	237	259	131	137	145	86	89	92
	•••	1469	1800	2412	467	531	588	228	247	252	134	141	135	87	91	81
0.6	•••	1304	1511	2283	423	471	561	209	223	245	122	128	135	79	82	85
	•••	1387	1725	2313	444	507	557	216	233	235	125	131	124	81	83	73
0.65	•••	1203	1410	2149	393	438	521	192	205	224	111	116	122	71	73	75
	•••	1287	1620	2164	413	471	514	198	213	213	113	119	110	72	74	63
0.7	•••	1090	1290	1962	356	396	467	171	182	197	96	100	105	60	62	64
	•••	1172	1487	1967	374	425	458	176	189	186	99	103	93	61	63	51
0.75	•••	965	1148	1722	311	345	401	145	154	165	79	82	85	.	.	.
	•••	1040	1323	1719	326	368	390	149	159	153	81	84	73	.	.	.
0.8	•••	824	981	1431	257	282	321	114	120	127	.	.	.	.	.	.
	•••	889	1125	1422	268	299	310	117	123	115	.	.	.	.	.	.
0.85	•••	663	783	1087	192	207	229	.	.	.	.	.	.	.	.	.
	•••	714	888	1076	199	217	217	.	.	.	.	.	.	.	.	.
0.9	•••	468	540	693	.	.	.	.	.	.	.	.	.	.	.	.
	•••	499	599	681	.	.	.	.	.	.	.	.	.	.	.	.

TABLE 7: ALPHA= 0.05 POWER= 0.8   EXPECTED ACCRUAL THRU MINIMUM FOLLOW-UP= 475

		DEL=.05			DEL=.10			DEL=.15			DEL=.20			DEL=.25		
FACT=		1.0 .75	.50 .25	.00 BIN	1.0 .75	.50 .25	.00 BIN	1.0 .75	.50 .25	.00 BIN	1.0 75	.50 .25	.00 BIN	1.0 .75	.50 .25	.00 BIN
PCONT=***				REQUIRED NUMBER OF PATIENTS												
0.05	***	379	384	395	137	139	141	81	81	82	57	57	58	45	45	45
	•••	381	389	681	138	140	217	81	82	115	57	58	73	45	45	51
0.1	***	696	713	762	223	228	235	121	123	125	80	81	82	59	60	60
	•••	703	731	1076	226	231	310	122	124	153	80	81	93	59	60	63
0.15	***	970	1004	1117	297	307	322	154	158	163	99	101	103	71	72	73
	•••	983	1040	1422	301	314	390	156	160	186	100	102	110	71	72	73
0.2	***	1190	1242	1441	357	373	400	182	188	195	114	117	120	80	82	83
	•••	1210	1302	1719	364	384	458	184	191	213	115	118	124	81	82	81
0.25	***	1352	1426	1724	403	425	466	203	211	222	126	129	134	87	89	91
	•••	1380	1511	1967	413	442	514	207	216	235	127	131	135	88	90	87
0.3	***	1460	1557	1960	437	466	519	219	229	243	134	139	144	92	94	97
	•••	1497	1670	2164	450	487	557	224	235	252	136	141	143	93	96	91
0.35	***	1519	1641	2147	459	494	559	229	241	259	140	145	151	95	97	100
	•••	1565	1780	2313	474	520	588	235	249	263	142	148	147	96	99	93
0.4	***	1536	1682	2281	470	510	587	235	248	268	142	148	155	96	98	102
	•••	1592	1846	2412	488	541	606	241	257	268	145	151	149	97	100	93
0.45	***	1518	1687	2362	472	516	600	235	250	271	141	147	155	94	97	101
	•••	1584	1872	2461	491	549	613	242	259	268	144	151	147	96	99	91
0.5	***	1473	1660	2390	465	511	601	231	246	268	138	144	152	91	94	98
	•••	1547	1860	2461	485	547	606	238	256	263	141	148	143	93	96	87
0.55	***	1404	1606	2363	450	497	587	223	238	260	132	138	145	86	89	92
	•••	1485	1816	2412	471	533	588	229	247	252	134	141	135	88	91	81
0.6	***	1319	1527	2283	427	474	561	210	224	245	123	128	135	80	82	85
	•••	1403	1741	2313	448	509	557	217	233	235	126	131	124	81	83	73
0.65	***	1218	1427	2149	397	441	520	193	206	224	111	116	122	71	73	75
	•••	1303	1636	2164	416	473	514	199	214	213	114	119	110	72	74	63
0.7	***	1105	1305	1962	359	399	467	172	183	197	97	101	105	61	62	64
	•••	1187	1501	1967	377	427	458	177	189	186	99	103	93	61	63	51
0.75	***	979	1162	1722	314	347	401	146	154	165	80	82	85	.	.	.
	•••	1054	1335	1719	329	370	390	150	159	153	81	84	73	.	.	.
0.8	***	836	993	1430	259	283	321	115	120	127	.	.	.	.	.	.
	•••	902	1136	1422	270	300	310	117	123	115	.	.	.	.	.	.
0.85	***	672	792	1087	193	208	229	.	.	.	.	.	.	.	.	.
	•••	723	895	1076	200	218	217	.	.	.	.	.	.	.	.	.
0.9	***	473	545	693	.	.	.	.	.	.	.	.	.	.	.	.
	•••	505	602	681	.	.	.	.	.	.	.	.	.	.	.	.

TABLE 7: ALPHA= 0.05 POWER= 0.8   EXPECTED ACCRUAL THRU MINIMUM FOLLOW-UP= 500

		DEL=.05			DEL=.10			DEL=.15			DEL=.20			DEL=.25		
FACT=		1.0 .75	.50 .25	.00 BIN	1.0 .75	.50 .25	.00 BIN	1.0 .75	.50 .25	.00 BIN	1.0 75	.50 .25	.00 BIN	1.0 .75	.50 .25	.00 BIN
PCONT=•••					REQUIRED NUMBER OF PATIENTS											
0.05	•••	379	385	395	138	139	141	81	82	82	57	58	58	44	45	45
	•••	382	389	681	138	140	217	81	82	115	57	58	73	44	45	51
0.1	•••	697	715	762	224	228	235	121	123	125	80	81	82	59	60	60
	•••	704	732	1076	226	232	310	122	124	153	81	82	93	59	60	63
0.15	•••	973	1006	1117	298	307	323	155	158	163	99	101	103	71	72	73
	•••	986	1043	1422	302	314	390	156	160	186	100	102	110	72	73	73
0.2	•••	1193	1246	1441	358	373	400	183	188	195	114	117	120	80	82	83
	•••	1213	1306	1719	365	385	458	185	191	213	116	118	124	81	83	81
0.25	•••	1357	1432	1724	405	427	466	204	212	222	126	130	134	88	89	91
	•••	1385	1518	1967	415	443	514	208	217	235	128	132	135	88	90	87
0.3	•••	1466	1565	1960	439	468	519	220	230	243	135	139	144	92	94	97
	•••	1504	1678	2164	452	489	557	224	236	252	137	141	143	93	96	91
0.35	•••	1527	1651	2147	462	496	559	230	242	259	140	145	151	95	98	100
	•••	1574	1790	2313	477	522	588	236	250	263	143	148	147	96	99	93
0.4	•••	1546	1694	2281	473	513	587	236	249	268	143	148	155	96	98	102
	•••	1603	1858	2412	491	543	606	242	258	268	145	151	149	97	100	93
0.45	•••	1530	1700	2362	476	519	600	236	251	271	142	148	155	95	98	101
	•••	1597	1885	2461	494	552	613	243	260	268	145	151	147	96	99	91
0.5	•••	1485	1675	2390	469	514	601	233	247	268	138	144	152	92	94	98
	•••	1561	1875	2461	489	549	606	239	257	263	141	148	143	93	96	87
0.55	•••	1418	1622	2363	454	500	587	224	239	259	132	138	145	87	89	93
	•••	1500	1832	2412	474	536	588	231	248	252	135	141	135	88	91	81
0.6	•••	1333	1543	2283	431	477	561	211	225	245	123	129	135	80	82	85
	•••	1418	1756	2313	451	511	557	218	234	235	126	132	124	81	83	73
0.65	•••	1233	1443	2149	401	444	521	194	207	224	112	116	122	71	73	75
	•••	1319	1651	2164	420	475	514	200	215	213	114	119	110	72	74	63
0.7	•••	1120	1321	1962	363	401	467	173	183	197	97	101	105	61	62	64
	•••	1202	1515	1967	380	429	458	178	190	186	99	103	93	61	63	51
0.75	•••	992	1175	1722	317	349	401	147	155	165	80	83	85	.	.	.
	•••	1068	1348	1719	331	371	390	150	159	153	81	84	73	.	.	.
0.8	•••	848	1004	1431	261	285	321	115	120	127	.	.	.	.	.	.
	•••	913	1145	1422	272	301	310	118	123	115	.	.	.	.	.	.
0.85	•••	681	800	1087	194	209	229	.	.	.	.	.	.	.	.	.
	•••	732	902	1076	201	218	217	.	.	.	.	.	.	.	.	.
0.9	•••	479	550	693	.	.	.	.	.	.	.	.	.	.	.	.
	•••	510	606	681	.	.	.	.	.	.	.	.	.	.	.	.

TABLE 7: ALPHA= 0.05 POWER= 0.8 EXPECTED ACCRUAL THRU MINIMUM FOLLOW-UP= 550

		DEL=.02			DEL=.05			DEL=.10			DEL=.15			DEL=.20		
FACT=		1.0 .75	.50 .25	.00 BIN	1.0 .75	.50 .25	.00 BIN	1.0 .75	.50 .25	.00 BIN	1.0 75	.50 .25	.00 BIN	1.0 .75	.50 .25	.00 BIN
PCONT=***					REQUIRED NUMBER OF PATIENTS											
0.05	***	1764	1776	1863	380	385	395	138	139	140	81	81	82	57	57	58
	•••	1768	1794	3481	382	389	681	138	140	217	81	82	115	57	58	73
0.1	***	3672	3699	4092	699	717	762	224	229	235	121	123	125	80	81	82
	•••	3681	3751	6047	707	734	1076	226	232	310	122	124	153	81	81	93
0.15	***	5367	5415	6328	976	1011	1117	299	308	322	155	158	163	99	101	103
	•••	5383	5509	8304	990	1047	1422	303	314	390	157	160	186	100	102	110
0.2	***	6721	6795	8409	1199	1254	1441	360	375	400	183	188	195	115	117	120
	•••	6745	6940	10251	1221	1314	1719	367	386	458	186	192	213	116	118	124
0.25	***	7697	7803	10253	1365	1443	1724	408	430	465	205	212	222	127	130	134
	•••	7733	8012	11890	1396	1529	1967	418	445	514	209	217	235	128	132	135
0.3	***	8303	8449	11816	1478	1581	1960	443	471	519	221	231	243	135	139	144
	•••	8352	8734	13219	1518	1693	2164	455	491	557	226	236	252	137	142	143
0.35	***	8566	8759	13073	1542	1669	2147	467	500	559	232	243	259	141	145	151
	•••	8630	9135	14239	1592	1809	2313	481	525	588	237	250	263	143	148	147
0.4	***	8523	8775	14009	1564	1716	2281	479	518	587	238	250	268	143	148	155
	•••	8607	9257	14950	1624	1880	2412	496	546	606	244	258	268	146	151	149
0.45	***	8223	8547	14614	1551	1726	2362	482	524	600	239	252	271	143	148	155
	•••	8331	9147	15352	1620	1909	2461	500	555	613	245	261	268	145	151	147
0.5	***	7716	8126	14886	1510	1702	2390	475	520	600	235	249	268	140	145	152
	•••	7854	8853	15445	1587	1902	2461	495	553	606	241	257	263	142	148	143
0.55	***	7057	7568	14820	1445	1651	2363	461	506	587	226	240	259	133	138	145
	•••	7232	8418	15228	1528	1859	2412	481	539	588	233	249	252	136	142	135
0.6	***	6305	6924	14419	1361	1573	2283	438	482	561	213	226	245	124	129	135
	•••	6522	7879	14703	1447	1784	2313	457	514	557	220	235	235	127	132	124
0.65	***	5517	6237	13682	1261	1472	2149	407	449	520	196	208	224	113	117	122
	•••	5777	7262	13868	1348	1677	2164	426	479	514	202	215	213	114	119	110
0.7	***	4743	5534	12613	1147	1348	1962	369	406	467	175	184	197	98	101	105
	•••	5037	6584	12724	1230	1540	1967	385	432	458	179	190	186	100	103	93
0.75	***	4012	4826	11213	1017	1200	1722	322	352	401	148	155	165	80	83	85
	•••	4322	5847	11271	1093	1369	1719	335	373	390	151	160	153	81	84	73
0.8	***	3328	4107	9486	870	1025	1431	265	288	321	116	121	127	.	.	.
	•••	3630	5043	9509	935	1163	1422	276	302	310	118	124	115	.	.	.
0.85	***	2671	3355	7436	698	815	1087	197	211	229	.	.	.	.	.	.
	•••	2939	4144	7438	748	914	1076	203	219	217	.	.	.	.	.	.
0.9	***	1993	2516	5066	490	559	693	.	.	.	.	.	.	.	.	.
	•••	2201	3090	5058	520	612	681	.	.	.	.	.	.	.	.	.
0.95	***	1180	1456	2380	.	.	.	.	.	.	.	.	.	.	.	.
	•••	1293	1726	2368	.	.	.	.	.	.	.	.	.	.	.	.

TABLE 7: ALPHA= 0.05 POWER= 0.8   EXPECTED ACCRUAL THRU MINIMUM FOLLOW-UP= 600

	DEL=.02			DEL=.05			DEL=.10			DEL=.15			DEL=.20		
FACT=	1.0 .75	.50 .25	.00 BIN	1.0 .75	.50 .25	.00 BIN	1.0 .75	.50 .25	.00 BIN	1.0 75	.50 .25	.00 BIN	1.0 .75	.50 .25	.00 BIN
PCONT=•••	REQUIRED NUMBER OF PATIENTS														
0.05 ***	1765	1778	1863	381	386	395	138	139	140	81	82	82	57	58	58
•••	1770	1797	3481	383	390	681	138	140	217	81	82	115	57	58	73
0.1 ***	3674	3704	4092	701	719	762	225	229	235	121	123	125	80	81	82
•••	3685	3760	6047	709	736	1076	227	232	310	122	124	153	81	82	93
0.15 ***	5372	5424	6328	980	1016	1117	300	309	322	156	159	163	100	101	103
•••	5389	5525	8304	995	1051	1422	304	315	390	157	161	186	100	102	110
0.2 ***	6727	6808	8408	1205	1262	1441	362	377	400	184	189	195	115	117	120
•••	6754	6965	10251	1228	1321	1719	369	387	458	186	192	213	116	118	124
0.25 ***	7707	7822	10252	1374	1454	1724	411	432	466	206	213	222	127	130	134
•••	7745	8048	11890	1405	1539	1967	420	446	514	209	217	235	128	132	135
0.3 ***	8316	8475	11816	1489	1594	1960	447	474	519	223	232	243	136	140	144
•••	8369	8783	13219	1530	1706	2164	459	493	557	227	237	252	138	142	143
0.35 ***	8583	8794	13073	1556	1687	2147	471	503	559	234	244	259	142	146	151
•••	8653	9199	14239	1607	1825	2313	485	527	588	239	251	263	143	148	147
0.4 ***	8546	8821	14009	1582	1737	2281	484	521	586	240	251	268	144	149	155
•••	8638	9338	14950	1643	1899	2412	501	548	606	245	259	268	146	152	149
0.45 ***	8252	8605	14614	1571	1749	2362	488	529	600	241	253	271	143	149	155
•••	8371	9246	15352	1642	1931	2461	506	558	613	247	262	268	146	152	147
0.5 ***	7753	8198	14886	1532	1728	2389	481	525	601	237	250	268	140	145	152
•••	7904	8968	15445	1611	1925	2461	501	556	606	243	258	263	143	148	143
0.55 ***	7105	7655	14821	1469	1677	2363	467	511	587	228	241	259	134	139	145
•••	7295	8548	15228	1554	1884	2412	486	542	588	235	250	252	136	142	135
0.6 ***	6365	7026	14419	1387	1600	2283	444	487	560	215	228	245	125	130	135
•••	6598	8020	14703	1474	1808	2313	463	518	557	221	236	235	127	132	124
0.65 ***	5590	6350	13682	1287	1499	2149	413	453	520	198	209	224	113	117	122
•••	5866	7409	13868	1374	1701	2164	431	482	514	203	216	213	115	119	110
0.7 ***	4826	5653	12613	1172	1373	1962	374	410	467	176	185	197	98	101	105
•••	5135	6731	12724	1255	1561	1967	390	434	458	180	191	186	100	103	93
0.75 ***	4101	4943	11213	1040	1222	1722	326	355	401	149	156	165	80	83	85
•••	4424	5987	11271	1116	1388	1719	339	375	390	152	160	153	82	84	73
0.8 ***	3416	4217	9487	889	1043	1430	268	290	321	116	121	127	.	.	.
•••	3727	5169	9509	954	1178	1422	278	304	310	119	124	115	.	.	.
0.85 ***	2750	3449	7436	713	829	1087	199	212	229	.	.	.	.	.	.
•••	3025	4248	7438	763	924	1076	205	220	217	.	.	.	.	.	.
0.9 ***	2054	2586	5066	499	566	692	.	.	.	.	.	.	.	.	.
•••	2267	3162	5058	529	617	681	.	.	.	.	.	.	.	.	.
0.95 ***	1214	1491	2380	.	.	.	.	.	.	.	.	.	.	.	.
•••	1328	1757	2368	.	.	.	.	.	.	.	.	.	.	.	.

TABLE 7: ALPHA= 0.05 POWER= 0.8   EXPECTED ACCRUAL THRU MINIMUM FOLLOW-UP= 650

		DEL=.02			DEL=.05			DEL=.10			DEL=.15			DEL=.20		
FACT=		1.0 .75	.50 .25	.00 BIN	1.0 .75	.50 .25	.00 BIN	1.0 .75	.50 .25	.00 BIN	1.0 75	.50 .25	.00 BIN	1.0 .75	.50 .25	.00 BIN
PCONT=***		REQUIRED NUMBER OF PATIENTS														
0.05	***	1767	1780	1863	381	387	395	138	139	141	81	81	82	57	58	58
	•••	1771	1800	3481	384	390	681	138	140	217	81	82	115	58	58	73
0.1	***	3677	3709	4092	704	721	762	226	230	236	122	123	125	81	81	82
	•••	3688	3768	6047	711	737	1076	228	232	310	123	124	153	81	81	93
0.15	***	5376	5433	6328	985	1020	1117	302	310	322	156	159	163	100	101	103
	•••	5395	5541	8304	999	1055	1422	305	315	390	158	161	186	101	102	110
0.2	***	6734	6821	8409	1211	1269	1441	364	379	400	185	189	195	116	117	120
	•••	6763	6991	10251	1234	1327	1719	371	388	458	187	192	213	116	119	124
0.25	***	7717	7842	10252	1382	1464	1724	414	434	465	207	214	222	128	130	133
	•••	7758	8085	11890	1414	1548	1967	423	448	514	210	218	235	129	132	135
0.3	***	8330	8501	11816	1500	1607	1960	451	476	519	224	232	244	136	140	144
	•••	8387	8832	13219	1543	1719	2164	462	495	557	228	237	252	138	142	143
0.35	***	8601	8829	13073	1570	1703	2147	476	507	559	236	245	259	142	146	151
	•••	8677	9263	14239	1623	1841	2313	489	529	588	240	251	263	144	149	147
0.4	***	8569	8866	14009	1598	1756	2281	489	525	587	242	253	268	145	149	155
	•••	8668	9417	14950	1661	1917	2412	505	551	606	247	260	268	147	152	149
0.45	***	8282	8662	14615	1591	1770	2362	493	532	600	242	255	271	145	149	155
	•••	8410	9341	15352	1663	1952	2461	510	561	613	248	262	268	146	152	147
0.5	***	7791	8270	14886	1554	1751	2389	487	529	601	239	251	268	141	146	152
	•••	7954	9079	15445	1634	1947	2461	506	559	606	244	259	263	143	149	143
0.55	***	7153	7741	14821	1492	1702	2363	472	515	587	230	243	259	135	140	145
	•••	7357	8672	15228	1578	1906	2412	491	545	588	236	250	252	137	142	135
0.6	***	6424	7125	14419	1411	1625	2283	450	491	561	217	229	245	126	130	135
	•••	6674	8154	14703	1499	1831	2313	468	521	557	223	236	235	128	132	124
0.65	***	5661	6458	13682	1311	1523	2149	418	457	521	200	210	224	114	118	122
	•••	5953	7548	13868	1399	1723	2164	436	484	514	205	216	213	116	120	110
0.7	***	4907	5765	12613	1195	1396	1962	379	413	467	177	186	197	99	102	105
	•••	5230	6868	12724	1279	1581	1967	394	437	458	181	191	186	100	103	93
0.75	***	4187	5055	11213	1061	1243	1722	330	358	401	150	157	165	81	83	85
	•••	4520	6118	11271	1137	1405	1719	343	377	390	153	161	153	82	84	73
0.8	***	3500	4320	9487	907	1060	1431	271	292	321	117	122	127	.	.	.
	•••	3819	5286	9509	972	1191	1422	281	305	310	119	124	115	.	.	.
0.85	***	2824	3537	7436	727	841	1087	201	213	229	.	.	.	.	.	.
	•••	3106	4344	7438	777	934	1076	207	220	217	.	.	.	.	.	.
0.9	***	2112	2651	5067	508	574	692	.	.	.	.	.	.	.	.	.
	•••	2328	3230	5058	537	622	681	.	.	.	.	.	.	.	.	.
0.95	***	1245	1523	2381	.	.	.	.	.	.	.	.	.	.	.	.
	•••	1360	1786	2368	.	.	.	.	.	.	.	.	.	.	.	.

TABLE 7: ALPHA= 0.05 POWER= 0.8   EXPECTED ACCRUAL THRU MINIMUM FOLLOW-UP= 700

	DEL=.02			DEL=.05			DEL=.10			DEL=.15			DEL=.20		
FACT=	1.0 .75	.50 .25	.00 BIN	1.0 .75	.50 .25	.00 BIN	1.0 .75	.50 .25	.00 BIN	1.0 75	.50 .25	.00 BIN	1.0 .75	.50 .25	.00 BIN
PCONT=***				REQUIRED NUMBER OF PATIENTS											
0.05 ***	1768	1781	1863	382	387	395	138	139	141	81	82	82	57	58	58
***	1773	1802	3481	384	390	681	138	140	217	81	82	115	58	58	73
0.1 ***	3680	3714	4092	705	723	762	226	230	236	122	123	125	81	81	82
***	3691	3776	6047	713	738	1076	228	233	310	123	124	153	81	82	93
0.15 ***	5380	5441	6328	988	1024	1117	303	311	323	156	159	163	100	101	103
***	5401	5556	8304	1003	1058	1422	306	316	390	158	161	186	100	102	110
0.2 ***	6741	6835	8409	1217	1276	1441	366	380	400	185	190	195	116	117	120
***	6772	7015	10251	1241	1332	1719	372	388	458	187	192	213	117	119	124
0.25 ***	7726	7861	10253	1390	1473	1724	416	436	465	208	214	222	128	131	134
***	7771	8120	11890	1423	1556	1967	425	449	514	211	218	235	129	132	135
0.3 ***	8343	8528	11816	1511	1620	1960	453	479	519	225	233	243	137	140	144
***	8405	8880	13219	1554	1730	2164	465	496	557	229	238	252	138	142	143
0.35 ***	8618	8864	13073	1583	1718	2147	479	509	559	236	246	259	143	147	151
***	8700	9324	14239	1637	1854	2313	493	531	588	241	252	263	145	149	147
0.4 ***	8592	8911	14009	1614	1773	2281	493	529	586	243	254	268	145	150	155
***	8699	9494	14950	1678	1933	2412	509	553	606	248	260	268	147	152	149
0.45 ***	8311	8719	14614	1609	1790	2362	498	536	600	244	256	271	145	149	155
***	8449	9433	15352	1682	1969	2461	515	563	613	250	263	268	147	152	147
0.5 ***	7829	8339	14886	1574	1773	2389	492	533	600	240	252	268	142	146	152
***	8003	9185	15445	1655	1966	2461	510	562	606	246	260	263	144	149	143
0.55 ***	7200	7824	14821	1514	1724	2363	478	519	587	232	244	260	135	140	145
***	7418	8790	15228	1601	1926	2412	496	548	588	237	251	252	138	142	135
0.6 ***	6483	7220	14419	1433	1648	2283	454	495	561	219	230	245	126	130	135
***	6747	8280	14703	1522	1851	2313	473	523	557	224	237	235	128	133	124
0.65 ***	5731	6562	13682	1333	1546	2149	423	460	520	201	211	224	114	118	122
***	6037	7678	13868	1422	1742	2164	440	486	514	206	217	213	116	120	110
0.7 ***	4986	5873	12613	1216	1417	1962	383	416	467	178	187	197	99	102	105
***	5321	6998	12724	1300	1599	1967	398	438	458	182	192	186	101	103	93
0.75 ***	4269	5160	11213	1081	1262	1722	334	361	401	151	157	165	81	83	86
***	4612	6240	11271	1157	1420	1719	346	378	390	154	161	153	82	84	73
0.8 ***	3579	4418	9487	924	1076	1430	274	294	321	118	122	127	.	.	.
***	3907	5395	9509	989	1203	1422	283	306	310	120	124	115	.	.	.
0.85 ***	2895	3620	7436	740	852	1087	202	214	229	.	.	.	.	.	.
***	3181	4433	7438	789	942	1076	208	221	217	.	.	.	.	.	.
0.9 ***	2166	2712	5066	515	579	693	.	.	.	.	.	.	.	.	.
***	2385	3291	5058	544	626	681	.	.	.	.	.	.	.	.	.
0.95 ***	1275	1552	2380	.	.	.	.	.	.	.	.	.	.	.	.
***	1390	1812	2368	.	.	.	.	.	.	.	.	.	.	.	.

TABLE 7: ALPHA= 0.05 POWER= 0.8   EXPECTED ACCRUAL THRU MINIMUM FOLLOW-UP= 750

		DEL=.02			DEL=.05			DEL=.10			DEL=.15			DEL=.20		
FACT=		1.0 .75	.50 .25	.00 BIN	1.0 .75	.50 .25	.00 BIN	1.0 .75	.50 .25	.00 BIN	1.0 75	.50 .25	.00 BIN	1.0 .75	.50 .25	.00 BIN
PCONT=***		REQUIRED NUMBER OF PATIENTS														
0.05	***	1769	1783	1863	383	387	395	139	139	141	81	82	82	57	57	58
	•••	1774	1804	3481	385	391	681	139	140	217	81	82	115	57	58	73
0.1	***	3682	3719	4092	707	725	762	227	230	235	122	124	125	81	81	82
	•••	3694	3784	6047	715	740	1076	229	233	310	123	124	153	81	82	93
0.15	***	5385	5450	6328	992	1028	1117	304	311	323	157	160	163	100	101	102
	•••	5406	5572	8304	1006	1061	1422	307	317	390	158	161	186	100	102	110
0.2	***	6747	6848	8409	1222	1281	1441	368	380	400	186	190	195	116	118	120
	•••	6781	7039	10251	1246	1338	1719	373	389	458	188	192	213	117	119	124
0.25	***	7736	7880	10253	1398	1482	1724	418	437	466	209	215	222	128	131	134
	•••	7784	8154	11890	1432	1564	1967	427	450	514	212	218	235	130	132	135
0.3	***	8356	8554	11816	1521	1631	1960	456	481	519	226	234	244	138	140	144
	•••	8422	8926	13219	1565	1740	2164	468	498	557	229	238	252	139	142	143
0.35	***	8636	8899	13073	1596	1732	2147	483	512	559	238	247	259	143	147	151
	•••	8724	9384	14239	1651	1867	2313	496	533	588	242	252	263	145	149	147
0.4	***	8615	8956	14009	1629	1790	2281	498	531	587	244	254	268	146	150	155
	•••	8730	9568	14950	1694	1948	2412	513	555	606	249	261	268	148	152	149
0.45	***	8341	8775	14614	1626	1809	2362	502	539	600	245	257	271	145	150	155
	•••	8488	9521	15352	1700	1986	2461	519	565	613	250	263	268	147	152	147
0.5	***	7867	8408	14886	1593	1793	2389	497	536	601	242	253	268	142	146	152
	•••	8053	9287	15445	1675	1984	2461	514	564	606	247	260	263	144	149	143
0.55	***	7248	7906	14821	1534	1745	2363	482	522	587	233	244	259	136	140	145
	•••	7479	8903	15228	1622	1945	2412	500	550	588	239	251	252	138	143	135
0.6	***	6541	7312	14419	1454	1669	2283	459	499	560	220	231	244	127	130	135
	•••	6819	8401	14703	1543	1870	2313	477	525	557	225	237	235	129	133	124
0.65	***	5800	6661	13683	1354	1566	2149	427	464	520	202	212	224	115	118	122
	•••	6119	7802	13868	1443	1760	2164	444	489	514	207	217	213	116	120	110
0.7	***	5062	5975	12613	1236	1436	1962	386	419	467	179	187	197	100	102	105
	•••	5409	7120	12724	1321	1615	1967	401	440	458	183	192	186	101	104	93
0.75	***	4348	5260	11213	1099	1279	1722	337	363	400	152	158	165	81	83	85
	•••	4700	6355	11271	1175	1434	1719	349	380	390	154	161	153	82	84	73
0.8	***	3655	4510	9486	940	1090	1430	276	295	321	118	122	127	.	.	.
	•••	3990	5497	9509	1004	1214	1422	285	307	310	120	124	115	.	.	.
0.85	***	2961	3698	7436	752	862	1087	204	215	229	.	.	.	.	.	.
	•••	3253	4516	7438	800	949	1076	209	221	217	.	.	.	.	.	.
0.9	***	2218	2769	5066	522	585	693	.	.	.	.	.	.	.	.	.
	•••	2440	3348	5058	550	630	681	.	.	.	.	.	.	.	.	.
0.95	***	1302	1579	2380	.	.	.	.	.	.	.	.	.	.	.	.
	•••	1418	1836	2368	.	.	.	.	.	.	.	.	.	.	.	.

TABLE 7: ALPHA= 0.05 POWER= 0.8   EXPECTED ACCRUAL THRU MINIMUM FOLLOW-UP= 800

		DEL=.02			DEL=.05			DEL=.10			DEL=.15			DEL=.20		
FACT=		1.0 .75	.50 .25	.00 BIN	1.0 .75	.50 .25	.00 BIN	1.0 .75	.50 .25	.00 BIN	1.0 75	.50 .25	.00 BIN	1.0 .75	.50 .25	.00 BIN
PCONT=***					REQUIRED NUMBER OF PATIENTS											
0.05	***	1770	1785	1863	383	388	395	138	139	141	81	82	82	57	58	58
	•••	1775	1806	3481	385	391	681	139	140	217	81	82	115	58	58	73
0.1	***	3685	3724	4092	709	727	762	227	231	235	122	124	125	81	81	82
	•••	3698	3792	6047	716	741	1076	229	233	310	123	124	153	81	82	93
0.15	***	5389	5458	6328	995	1031	1117	304	312	323	157	160	163	100	101	103
	•••	5412	5586	8304	1010	1064	1422	308	317	390	159	161	186	101	102	110
0.2	***	6754	6861	8409	1228	1287	1441	369	382	400	186	190	195	116	118	120
	•••	6790	7063	10251	1252	1342	1719	375	390	458	188	193	213	117	119	124
0.25	***	7746	7899	10253	1405	1490	1724	420	439	466	210	215	222	129	131	134
	•••	7797	8188	11890	1440	1571	1967	429	451	514	212	219	235	130	132	135
0.3	***	8369	8580	11816	1530	1642	1960	459	483	519	227	234	243	138	141	144
	•••	8440	8972	13219	1575	1749	2164	470	499	557	230	239	252	139	142	143
0.35	***	8653	8933	13073	1608	1745	2147	486	514	559	239	248	259	144	147	151
	•••	8747	9442	14239	1664	1878	2313	499	534	588	243	253	263	145	149	147
0.4	***	8638	9000	14009	1643	1805	2281	501	534	587	245	255	268	146	150	155
	•••	8760	9640	14950	1709	1961	2412	516	557	606	250	261	268	148	152	149
0.45	***	8371	8831	14615	1642	1826	2362	506	542	600	247	257	271	146	150	155
	•••	8527	9607	15352	1718	2001	2461	522	567	613	252	264	268	148	152	147
0.5	***	7904	8476	14886	1611	1812	2390	501	539	601	243	254	268	143	147	152
	•••	8102	9384	15445	1694	2000	2461	518	566	606	248	261	263	145	149	143
0.55	***	7295	7985	14821	1554	1765	2363	486	526	587	235	246	260	136	140	145
	•••	7539	9010	15228	1641	1962	2412	504	552	588	240	252	252	138	143	135
0.6	***	6599	7401	14419	1474	1689	2283	463	501	561	221	232	245	127	131	135
	•••	6890	8515	14703	1564	1886	2313	481	527	557	226	238	235	129	133	124
0.65	***	5867	6757	13682	1374	1586	2149	431	467	521	203	213	224	115	118	122
	•••	6198	7919	13868	1463	1776	2164	447	490	514	208	218	213	117	120	110
0.7	***	5136	6073	12613	1255	1455	1962	390	421	467	180	188	197	100	102	105
	•••	5493	7235	12724	1339	1630	1967	404	442	458	184	192	186	101	104	93
0.75	***	4424	5356	11213	1116	1295	1722	339	365	401	152	158	165	82	84	85
	•••	4785	6463	11271	1192	1447	1719	351	381	390	155	161	153	83	84	73
0.8	***	3728	4597	9487	954	1102	1431	278	297	321	119	123	127	.	.	.
	•••	4069	5593	9509	1018	1224	1422	287	308	310	121	125	115	.	.	.
0.85	***	3025	3771	7436	763	872	1087	205	216	229	.	.	.	.	.	.
	•••	3322	4594	7438	811	956	1076	210	222	217	.	.	.	.	.	.
0.9	***	2267	2823	5066	529	590	693	.	.	.	.	.	.	.	.	.
	•••	2491	3401	5058	556	633	681	.	.	.	.	.	.	.	.	.
0.95	***	1328	1605	2380	.	.	.	.	.	.	.	.	.	.	.	.
	•••	1444	1858	2368	.	.	.	.	.	.	.	.	.	.	.	.

TABLE 7: ALPHA= 0.05 POWER= 0.8   EXPECTED ACCRUAL THRU MINIMUM FOLLOW-UP= 850

		DEL=.02			DEL=.05			DEL=.10			DEL=.15			DEL=.20		
FACT=		1.0 .75	.50 .25	.00 BIN	1.0 .75	.50 .25	.00 BIN	1.0 .75	.50 .25	.00 BIN	1.0 75	.50 .25	.00 BIN	1.0 .75	.50 .25	.00 BIN
PCONT=•••					REQUIRED	NUMBER	OF PATIENTS									
0.05	•••	1771	1787	1863	383	388	395	138	139	140	82	82	82	57	57	57
	•••	1777	1808	3481	385	391	681	139	140	217	82	82	115	57	57	73
0.1	•••	3687	3728	4092	711	728	762	227	231	236	122	123	125	81	82	82
	•••	3701	3799	6047	718	742	1076	229	233	310	123	124	153	81	82	93
0.15	•••	5393	5467	6328	998	1035	1117	305	312	323	157	160	163	101	102	103
	•••	5418	5601	8304	1013	1066	1422	309	317	390	158	161	186	101	102	110
0.2	•••	6761	6874	8409	1233	1292	1441	371	383	400	187	190	195	117	118	120
	•••	6799	7086	10251	1257	1346	1719	376	390	458	189	193	213	117	119	124
0.25	•••	7755	7919	10253	1412	1498	1724	422	440	465	210	215	222	129	131	134
	•••	7809	8221	11890	1447	1577	1967	430	451	514	213	219	235	130	133	135
0.3	•••	8382	8606	11816	1540	1651	1960	461	484	519	227	235	243	138	141	144
	•••	8457	9016	13219	1585	1757	2164	472	499	557	231	239	252	139	142	143
0.35	•••	8671	8967	13073	1619	1757	2146	488	516	560	240	248	259	144	147	151
	•••	8771	9499	14239	1676	1889	2313	501	535	588	243	253	263	146	149	147
0.4	•••	8661	9044	14009	1657	1820	2281	504	537	587	246	256	268	147	151	155
	•••	8790	9709	14950	1723	1974	2412	519	559	606	251	261	268	148	153	149
0.45	•••	8400	8886	14614	1658	1842	2362	510	545	600	248	258	271	146	151	155
	•••	8566	9689	15352	1734	2015	2461	526	569	613	253	264	268	148	153	147
0.5	•••	7941	8542	14886	1628	1829	2390	504	542	600	244	255	269	143	147	152
	•••	8150	9478	15445	1711	2015	2461	521	567	606	249	261	263	145	150	143
0.55	•••	7341	8062	14821	1572	1784	2363	491	528	587	236	246	259	137	141	145
	•••	7598	9113	15228	1660	1977	2412	508	554	588	241	253	252	139	142	135
0.6	•••	6655	7486	14419	1493	1708	2283	467	504	561	222	232	244	128	131	135
	•••	6959	8624	14703	1583	1902	2313	484	529	557	227	238	235	129	133	124
0.65	•••	5932	6849	13683	1393	1603	2149	435	469	520	204	213	224	116	119	122
	•••	6275	8030	13868	1481	1790	2164	450	492	514	208	218	213	117	120	110
0.7	•••	5207	6167	12613	1273	1471	1962	393	424	467	181	188	197	100	103	105
	•••	5574	7344	12724	1357	1643	1967	407	443	458	185	192	186	102	104	93
0.75	•••	4497	5446	11213	1132	1309	1722	342	367	401	153	158	165	82	84	85
	•••	4866	6565	11271	1208	1459	1719	354	382	390	156	162	153	83	85	73
0.8	•••	3797	4680	9487	968	1114	1430	280	298	321	119	123	127	.	.	.
	•••	4145	5683	9509	1031	1233	1422	289	309	310	121	124	115	.	.	.
0.85	•••	3086	3841	7437	773	880	1087	206	216	229	.	.	.	.	.	.
	•••	3387	4667	7438	820	962	1076	211	222	217	.	.	.	.	.	.
0.9	•••	2313	2873	5066	535	594	692	.	.	.	.	.	.	.	.	.
	•••	2539	3451	5058	562	636	681	.	.	.	.	.	.	.	.	.
0.95	•••	1352	1629	2380	.	.	.	.	.	.	.	.	.	.	.	.
	•••	1468	1878	2368	.	.	.	.	.	.	.	.	.	.	.	.

TABLE 7: ALPHA= 0.05 POWER= 0.8   EXPECTED ACCRUAL THRU MINIMUM FOLLOW-UP= 900

		DEL=.02			DEL=.05			DEL=.10			DEL=.15			DEL=.20		
FACT=		1.0 .75	.50 .25	.00 BIN	1.0 .75	.50 .25	.00 BIN	1.0 .75	.50 .25	.00 BIN	1.0 75	.50 .25	.00 BIN	1.0 .75	.50 .25	.00 BIN
PCONT=•••					REQUIRED NUMBER OF PATIENTS												
0.05	***	1772	1788	1863	384	388	395	138	140	141	81	82	82	57	57	57
	•••	1778	1810	3481	386	392	681	139	140	217	82	82	115	57	57	73
0.1	***	3690	3733	4092	712	729	762	228	231	235	123	124	125	81	81	82
	•••	3704	3806	6047	720	743	1076	230	233	310	123	125	153	81	82	93
0.15	***	5398	5475	6328	1001	1037	1117	306	313	323	158	160	163	100	101	102
	•••	5424	5615	8304	1016	1068	1422	309	317	390	159	162	186	101	102	110
0.2	***	6768	6888	8408	1237	1297	1441	371	383	399	187	191	195	117	118	120
	•••	6808	7108	10251	1262	1350	1719	377	390	458	189	193	213	117	119	124
0.25	***	7765	7937	10252	1419	1504	1724	424	441	465	210	216	222	129	131	134
	•••	7822	8253	11890	1454	1583	1967	432	452	514	213	219	235	130	132	135
0.3	***	8396	8632	11816	1548	1661	1961	464	486	519	228	235	243	138	141	144
	•••	8475	9059	13219	1594	1765	2164	474	501	557	231	239	252	140	143	143
0.35	***	8688	9001	13073	1630	1769	2147	491	518	559	240	249	259	144	147	151
	•••	8794	9554	14239	1687	1899	2313	504	536	588	244	253	263	146	149	147
0.4	***	8684	9088	14009	1670	1833	2281	507	539	586	247	257	268	147	151	155
	•••	8821	9777	14950	1737	1985	2412	522	560	606	252	262	268	149	153	149
0.45	***	8429	8940	14615	1673	1857	2362	513	548	600	249	259	271	147	151	155
	•••	8605	9768	15352	1749	2027	2461	528	570	613	253	264	268	149	153	147
0.5	***	7979	8607	14886	1644	1845	2390	508	545	600	245	255	269	144	147	152
	•••	8198	9568	15445	1728	2029	2461	525	569	606	250	262	263	145	149	143
0.55	***	7388	8137	14820	1589	1800	2363	494	531	587	237	247	260	137	141	145
	•••	7656	9212	15228	1677	1991	2412	511	556	588	242	253	252	139	143	135
0.6	***	6711	7570	14419	1511	1725	2283	470	506	560	224	233	245	128	131	135
	•••	7026	8728	14703	1601	1916	2313	487	531	557	228	239	235	129	133	124
0.65	***	5996	6938	13682	1410	1620	2149	438	471	521	205	213	224	116	119	122
	•••	6350	8136	13868	1499	1804	2164	453	493	514	209	218	213	117	120	110
0.7	***	5276	6257	12612	1289	1487	1962	396	425	467	182	189	197	100	102	105
	•••	5653	7448	12724	1373	1656	1967	410	444	458	185	193	186	101	104	93
0.75	***	4567	5534	11213	1148	1323	1722	344	368	401	154	159	165	82	83	85
	•••	4943	6662	11271	1223	1469	1719	356	383	390	156	162	153	83	84	73
0.8	***	3864	4759	9487	981	1125	1431	282	299	321	119	123	127	.	.	.
	•••	4217	5769	9509	1043	1242	1422	290	309	310	121	125	115	.	.	.
0.85	***	3144	3908	7436	783	888	1087	207	217	229	.	.	.	.	.	.
	•••	3449	4735	7438	829	968	1076	212	223	217	.	.	.	.	.	.
0.9	***	2357	2921	5067	540	599	693	.	.	.	.	.	.	.	.	.
	•••	2586	3497	5058	567	639	681	.	.	.	.	.	.	.	.	.
0.95	***	1376	1650	2381	.	.	.	.	.	.	.	.	.	.	.	.
	•••	1491	1896	2368	.	.	.	.	.	.	.	.	.	.	.	.

TABLE 7: ALPHA= 0.05 POWER= 0.8   EXPECTED ACCRUAL THRU MINIMUM FOLLOW-UP= 950

	DEL=.02			DEL=.05			DEL=.10			DEL=.15			DEL=.20		
FACT=	1.0 .75	.50 .25	.00 BIN	1.0 .75	.50 .25	.00 BIN	1.0 .75	.50 .25	.00 BIN	1.0 75	.50 .25	.00 BIN	1.0 .75	.50 .25	.00 BIN
PCONT=***			REQUIRED NUMBER OF PATIENTS												
0.05 ***	1773	1790	1863	384	388	395	139	140	141	82	82	82	57	58	58
***	1779	1812	3481	386	391	681	139	140	217	82	82	115	58	58	73
0.1 ***	3692	3738	4092	713	730	762	228	231	235	122	124	125	81	82	82
***	3707	3813	6047	721	743	1076	230	234	310	123	124	153	82	82	93
0.15 ***	5402	5484	6328	1003	1040	1117	306	314	323	158	160	163	101	102	103
***	5429	5628	8304	1019	1070	1422	310	318	390	159	162	186	101	102	110
0.2 ***	6774	6901	8409	1242	1302	1440	372	384	400	188	191	196	116	118	120
***	6817	7129	10251	1267	1354	1719	378	391	458	189	193	213	117	119	124
0.25 ***	7774	7956	10252	1426	1511	1724	425	442	466	211	216	222	129	131	134
***	7835	8284	11890	1461	1588	1967	433	452	514	213	219	235	130	133	135
0.3 ***	8409	8658	11816	1557	1670	1960	466	487	519	229	235	243	139	141	144
***	8493	9101	13219	1603	1773	2164	476	501	557	232	239	252	140	143	143
0.35 ***	8706	9035	13073	1641	1780	2147	494	520	559	241	249	258	144	148	151
***	8817	9608	14239	1698	1907	2313	506	538	588	245	254	263	146	149	147
0.4 ***	8707	9131	14009	1682	1845	2281	510	540	587	248	257	268	147	151	154
***	8851	9842	14950	1750	1996	2412	524	561	606	253	262	268	149	153	149
0.45 ***	8459	8993	14615	1687	1872	2362	516	549	600	249	259	271	147	151	155
***	8643	9844	15352	1763	2040	2461	531	572	613	254	265	268	149	153	147
0.5 ***	8016	8670	14885	1660	1860	2389	511	547	600	246	256	268	144	147	152
***	8246	9654	15445	1743	2042	2461	527	571	606	251	262	263	146	150	143
0.55 ***	7433	8210	14821	1606	1816	2363	497	533	587	238	247	260	138	141	145
***	7713	9306	15228	1694	2005	2412	514	557	588	242	253	252	139	143	135
0.6 ***	6765	7651	14419	1527	1741	2282	474	509	561	224	234	245	128	131	135
***	7093	8827	14703	1617	1929	2313	490	532	557	229	239	235	130	133	124
0.65 ***	6058	7023	13682	1427	1636	2149	441	473	520	206	214	224	116	119	122
***	6423	8237	13868	1515	1817	2164	456	495	514	210	219	213	117	120	110
0.7 ***	5343	6343	12612	1305	1501	1962	399	427	467	182	190	197	101	103	105
***	5729	7546	12724	1389	1667	1967	412	445	458	186	193	186	102	104	93
0.75 ***	4635	5617	11213	1161	1336	1722	347	369	401	154	159	165	82	84	85
***	5018	6755	11271	1236	1478	1719	357	384	390	156	162	153	83	84	73
0.8 ***	3928	4835	9487	993	1135	1431	283	300	321	120	123	127	.	.	.
***	4286	5850	9509	1055	1249	1422	291	310	310	121	125	115	.	.	.
0.85 ***	3200	3971	7436	792	895	1087	208	217	229	.	.	.	.	.	.
***	3508	4801	7438	837	973	1076	213	223	217	.	.	.	.	.	.
0.9 ***	2400	2966	5067	545	602	692	.	.	.	.	.	.	.	.	.
***	2630	3541	5058	571	641	681	.	.	.	.	.	.	.	.	.
0.95 ***	1397	1671	2381	.	.	.	.	.	.	.	.	.	.	.	.
***	1512	1913	2368	.	.	.	.	.	.	.	.	.	.	.	.

TABLE 7: ALPHA= 0.05 POWER= 0.8   EXPECTED ACCRUAL THRU MINIMUM FOLLOW-UP= 1000

		DEL=.02			DEL=.05			DEL=.10			DEL=.15			DEL=.20		
FACT=		1.0 .75	.50 .25	.00 BIN	1.0 .75	.50 .25	.00 BIN	1.0 .75	.50 .25	.00 BIN	1.0 75	.50 .25	.00 BIN	1.0 .75	.50 .25	.00 BIN
PCONT=***					REQUIRED NUMBER OF PATIENTS											
0.05	***	1774	1791	1863	385	389	395	139	140	141	81	82	82	58	58	58
	•••	1780	1813	3481	386	392	681	140	140	217	81	82	115	58	58	73
0.1	***	3695	3742	4092	715	731	762	228	231	235	123	124	125	81	81	82
	•••	3711	3819	6047	722	745	1076	230	233	310	123	125	153	81	82	93
0.15	***	5406	5492	6328	1006	1043	1117	307	314	323	158	160	163	101	101	103
	•••	5435	5641	8304	1021	1072	1422	310	318	390	160	161	186	101	102	110
0.2	***	6781	6914	8409	1246	1306	1441	373	385	400	188	191	195	117	118	120
	•••	6826	7151	10251	1271	1357	1719	379	391	458	190	193	213	118	119	124
0.25	***	7784	7975	10253	1431	1518	1724	427	443	466	211	216	222	130	131	134
	•••	7848	8315	11890	1467	1593	1967	435	453	514	214	219	235	131	133	135
0.3	***	8422	8683	11816	1565	1678	1960	468	489	519	230	236	243	139	141	144
	•••	8510	9142	13219	1612	1779	2164	477	502	557	233	240	252	140	143	143
0.35	***	8724	9069	13073	1651	1790	2146	496	521	560	242	250	259	145	148	151
	•••	8840	9660	14239	1708	1916	2313	508	538	588	246	254	263	146	150	147
0.4	***	8730	9173	14009	1694	1858	2281	513	543	586	249	258	268	148	151	155
	•••	8881	9905	14950	1762	2005	2412	526	562	606	253	263	268	150	153	149
0.45	***	8488	9045	14615	1700	1885	2362	519	551	600	251	260	271	148	151	155
	•••	8681	9919	15352	1777	2051	2461	534	573	613	255	265	268	150	153	147
0.5	***	8053	8733	14886	1675	1875	2390	515	549	601	247	256	268	145	148	152
	•••	8293	9738	15445	1759	2054	2461	530	572	606	251	262	263	146	150	143
0.55	***	7479	8281	14821	1621	1831	2363	500	536	587	239	248	260	138	141	145
	•••	7769	9397	15228	1710	2017	2412	516	558	588	243	253	252	140	143	135
0.6	***	6820	7730	14419	1543	1756	2283	477	511	561	225	234	245	129	131	135
	•••	7157	8923	14703	1633	1941	2313	493	533	557	230	239	235	130	133	124
0.65	***	6119	7106	13683	1443	1651	2149	444	475	521	207	215	224	116	119	122
	•••	6493	8334	13868	1531	1828	2164	458	496	514	211	219	213	118	120	110
0.7	***	5409	6426	12613	1321	1515	1962	401	429	467	183	190	197	101	103	105
	•••	5801	7641	12724	1403	1678	1967	414	446	458	186	193	186	102	104	93
0.75	***	4701	5697	11213	1175	1348	1722	349	371	401	155	160	165	83	84	85
	•••	5091	6842	11271	1250	1488	1719	359	385	390	157	162	153	83	85	73
0.8	***	3990	4907	9486	1004	1145	1431	285	301	321	120	123	127	.	.	.
	•••	4353	5926	9509	1065	1256	1422	293	310	310	122	125	115	.	.	.
0.85	***	3253	4031	7436	800	902	1087	209	218	229	.	.	.	.	.	.
	•••	3565	4862	7438	845	977	1076	213	223	217	.	.	.	.	.	.
0.9	***	2440	3010	5066	550	606	693	.	.	.	.	.	.	.	.	.
	•••	2672	3581	5058	575	643	681	.	.	.	.	.	.	.	.	.
0.95	***	1418	1690	2380	.	.	.	.	.	.	.	.	.	.	.	.
	•••	1533	1929	2368	.	.	.	.	.	.	.	.	.	.	.	.

TABLE 7: ALPHA= 0.05 POWER= 0.8	EXPECTED ACCRUAL THRU MINIMUM FOLLOW-UP= 1100

		DEL=.02			DEL=.05			DEL=.10			DEL=.15			DEL=.20		
FACT=		1.0 .75	.50 .25	.00 BIN	1.0 .75	.50 .25	.00 BIN	1.0 .75	.50 .25	.00 BIN	1.0 75	.50 .25	.00 BIN	1.0 .75	.50 .25	.00 BIN
PCONT=***				REQUIRED NUMBER OF PATIENTS												
0.05	•••	1776	1794	1863	385	389	395	139	140	140	81	82	82	57	58	58
	•••	1783	1816	3481	387	392	681	139	140	217	82	82	115	58	58	73
0.1	•••	3700	3751	4092	717	734	752	229	232	235	123	124	125	81	81	82
	•••	3717	3831	6047	725	746	1076	230	234	310	123	125	153	81	82	93
0.15	•••	5415	5509	6329	1011	1047	1116	308	314	323	158	160	163	101	102	103
	•••	5447	5666	8304	1027	1075	1422	312	318	390	159	162	186	101	102	110
0.2	•••	6795	6940	8409	1254	1314	1441	375	386	400	188	192	195	117	118	120
	•••	6843	7191	10251	1280	1363	1719	380	392	458	190	193	213	118	119	124
0.25	•••	7803	8012	10253	1443	1529	1724	430	445	466	213	217	222	130	131	133
	•••	7873	8374	11890	1479	1602	1967	437	455	514	214	219	235	131	133	135
0.3	•••	8449	8733	11816	1581	1693	1960	471	491	519	230	236	243	139	142	144
	•••	8545	9220	13219	1627	1791	2164	480	503	557	233	240	252	140	143	143
0.35	•••	8759	9135	13073	1669	1809	2146	500	525	559	243	250	258	145	148	151
	•••	8887	9759	14239	1728	1930	2313	511	540	588	247	254	263	147	149	147
0.4	•••	8775	9257	14008	1716	1880	2281	518	546	587	250	258	268	148	151	155
	•••	8941	10026	14950	1785	2023	2412	531	564	606	254	263	268	150	153	149
0.45	•••	8546	9147	14614	1726	1909	2362	524	555	600	252	261	271	148	151	155
	•••	8757	10059	15352	1803	2070	2461	538	576	613	257	265	268	150	153	147
0.5	•••	8126	8853	14886	1702	1902	2390	520	553	600	249	257	268	145	148	152
	•••	8386	9896	15445	1786	2075	2461	535	574	606	253	263	263	147	150	143
0.55	•••	7568	8418	14821	1651	1859	2363	506	539	587	240	249	259	138	142	145
	•••	7879	9569	15228	1739	2039	2412	521	561	588	244	254	252	140	143	135
0.6	•••	6924	7879	14419	1573	1784	2282	482	514	560	226	235	245	129	132	135
	•••	7281	9102	14703	1662	1963	2313	497	535	557	230	239	235	131	133	124
0.65	•••	6237	7262	13682	1472	1677	2149	449	478	521	208	215	224	117	119	122
	•••	6628	8515	13868	1560	1849	2164	463	498	514	212	219	213	118	120	110
0.7	•••	5534	6584	12613	1348	1539	1962	406	432	467	184	191	197	101	103	105
	•••	5941	7816	12724	1430	1697	1967	418	448	458	187	194	186	103	104	93
0.75	•••	4826	5847	11213	1200	1369	1722	352	373	401	155	159	165	82	84	85
	•••	5228	7006	11271	1273	1504	1719	362	386	390	158	162	153	83	85	73
0.8	•••	4107	5043	9486	1024	1163	1431	287	302	321	121	124	126	.	.	.
	•••	4479	6069	9509	1085	1269	1422	295	312	310	122	125	115	.	.	.
0.85	•••	3354	4144	7436	815	914	1087	210	219	229	.	.	.	.	.	.
	•••	3672	4975	7438	859	985	1076	214	224	217	.	.	.	.	.	.
0.9	•••	2516	3090	5066	559	612	692	.	.	.	.	.	.	.	.	.
	•••	2750	3656	5058	583	647	681	.	.	.	.	.	.	.	.	.
0.95	•••	1456	1726	2380	.	.	.	.	.	.	.	.	.	.	.	.
	•••	1570	1957	2368	.	.	.	.	.	.	.	.	.	.	.	.

TABLE 7: ALPHA= 0.05 POWER= 0.8   EXPECTED ACCRUAL THRU MINIMUM FOLLOW-UP= 1200

PCONT	DEL=.02 1.0/.75	.50/.25	.00/BIN	DEL=.05 1.0/.75	.50/.25	.00/BIN	DEL=.10 1.0/.75	.50/.25	.00/BIN	DEL=.15 1.0/75	.50/.25	.00/BIN	DEL=.20 1.0/.75	.50/.25	.00/BIN
0.05	1777	1797	1863	385	390	395	139	139	140	82	82	82	58	58	58
	1785	1819	3481	388	392	681	139	140	217	82	82	115	58	58	73
0.1	3704	3760	4092	719	736	762	229	232	235	123	124	124	81	82	82
	3724	3841	6047	727	747	1076	231	234	310	124	124	153	81	82	93
0.15	5424	5525	6328	1016	1051	1117	309	315	322	159	160	163	101	102	103
	5458	5689	8304	1031	1078	1422	312	319	390	160	161	186	101	102	110
0.2	6808	6965	8408	1261	1321	1441	376	387	400	189	192	195	117	118	120
	6861	7229	10251	1287	1368	1719	382	393	458	190	193	213	118	119	124
0.25	7822	8048	10252	1454	1539	1723	432	446	466	213	217	222	130	132	133
	7899	8429	11890	1490	1609	1967	439	455	514	215	220	235	131	133	135
0.3	8475	8783	11815	1594	1706	1960	474	493	519	232	237	243	139	142	144
	8580	9295	13219	1642	1801	2164	482	505	557	234	240	252	140	142	143
0.35	8794	9199	13073	1687	1825	2146	503	526	559	244	250	259	145	148	151
	8932	9853	14239	1745	1944	2313	514	541	588	247	254	263	147	150	147
0.4	8821	9338	14008	1737	1899	2281	521	548	586	251	259	268	148	151	154
	9000	10139	14950	1805	2038	2412	534	565	606	255	263	268	150	153	149
0.45	8605	9246	14614	1749	1931	2362	529	558	600	253	262	271	148	151	154
	8830	10191	15352	1826	2088	2461	542	577	613	257	266	268	150	153	147
0.5	8198	8968	14886	1728	1925	2389	525	556	601	250	258	268	145	148	151
	8476	10043	15445	1811	2094	2461	539	576	606	254	263	263	147	150	143
0.55	7655	8548	14821	1677	1884	2363	511	542	587	241	250	259	139	142	145
	7984	9727	15228	1765	2058	2412	526	562	588	245	254	252	140	143	135
0.6	7026	8020	14419	1600	1808	2283	487	517	560	228	235	244	130	132	135
	7400	9268	14703	1689	1983	2313	501	537	557	232	240	235	130	133	124
0.65	6350	7409	13682	1498	1701	2149	453	481	520	209	216	223	117	119	121
	6757	8682	13868	1585	1867	2164	466	499	514	212	220	213	118	121	110
0.7	5653	6730	12613	1373	1561	1962	409	434	467	185	190	197	101	103	105
	6073	7978	12724	1454	1713	1967	421	449	458	188	194	186	102	104	93
0.75	4943	5987	11213	1222	1388	1722	355	375	400	156	160	165	82	84	85
	5356	7155	11271	1294	1518	1719	364	387	390	158	163	153	83	85	73
0.8	4217	5169	9487	1043	1177	1430	290	304	321	121	124	127	.	.	.
	4597	6199	9509	1102	1279	1422	296	312	310	122	125	115	.	.	.
0.85	3448	4248	7436	829	924	1087	211	220	229	.	.	.	.	.	.
	3771	5077	7438	871	992	1076	215	224	217	.	.	.	.	.	.
0.9	2586	3162	5066	566	617	692	.	.	.	.	.	.	.	.	.
	2822	3724	5058	589	650	681	.	.	.	.	.	.	.	.	.
0.95	1491	1757	2380	.	.	.	.	.	.	.	.	.	.	.	.
	1605	1982	2368	.	.	.	.	.	.	.	.	.	.	.	.

PCONT=•••   REQUIRED NUMBER OF PATIENTS

TABLE 7: ALPHA= 0.05 POWER= 0.8   EXPECTED ACCRUAL THRU MINIMUM FOLLOW-UP= 1300

		DEL=.02			DEL=.05			DEL=.10			DEL=.15			DEL=.20		
FACT=		1.0 .75	.50 .25	.00 BIN	1.0 .75	.50 .25	.00 BIN	1.0 .75	.50 .25	.00 BIN	1.0 75	.50 .25	.00 BIN	1.0 .75	.50 .25	.00 BIN
PCONT=***				REQUIRED NUMBER OF PATIENTS												
0.05	***	1779	1800	1863	387	390	395	139	140	141	81	82	82	58	58	58
	•••	1787	1821	3481	388	393	681	140	141	217	81	82	115	58	58	73
0.1	***	3709	3768	4092	721	737	762	230	232	236	123	124	125	81	81	82
	•••	3730	3852	6047	728	747	1076	231	234	310	123	124	153	81	82	93
0.15	***	5432	5541	6329	1020	1055	1117	310	315	323	159	161	162	101	102	102
	•••	5470	5711	8304	1035	1081	1422	313	318	390	160	162	186	102	102	110
0.2	***	6821	6991	8409	1269	1327	1441	379	388	400	189	192	195	117	119	120
	•••	6879	7266	10251	1293	1372	1719	383	393	458	191	193	213	118	120	124
0.25	***	7842	8084	10252	1464	1548	1724	434	448	466	214	218	222	130	132	133
	•••	7924	8482	11890	1500	1617	1967	440	456	514	215	220	235	131	133	135
0.3	***	8501	8832	11816	1607	1719	1961	476	495	519	232	237	244	140	142	144
	•••	8614	9365	13219	1655	1811	2164	485	505	557	235	240	252	141	143	143
0.35	***	8829	9262	13073	1703	1840	2147	507	529	559	245	251	258	146	149	151
	•••	8979	9942	14239	1761	1956	2313	517	543	588	249	255	263	147	149	147
0.4	***	8866	9417	14009	1756	1917	2281	525	551	587	253	260	268	149	152	154
	•••	9059	10247	14950	1824	2052	2412	537	567	606	256	263	268	150	154	149
0.45	***	8662	9341	14615	1771	1952	2362	532	561	601	254	263	271	149	152	155
	•••	8903	10315	15352	1848	2104	2461	546	578	613	258	266	268	150	154	147
0.5	***	8270	9079	14886	1751	1947	2390	529	559	601	251	259	268	146	149	152
	•••	8564	10181	15445	1835	2111	2461	543	578	606	255	263	263	147	150	143
0.55	***	7741	8672	14821	1702	1906	2363	515	545	588	243	250	259	140	142	146
	•••	8087	9875	15228	1789	2076	2412	529	565	588	246	254	252	141	144	135
0.6	***	7125	8154	14419	1625	1831	2283	491	521	561	229	237	245	130	133	135
	•••	7515	9422	14703	1714	2000	2313	505	539	557	232	240	235	131	134	124
0.65	***	6458	7548	13682	1524	1722	2149	458	484	521	211	216	224	118	120	122
	•••	6879	8836	13868	1610	1884	2164	470	500	514	213	220	213	119	120	110
0.7	***	5765	6868	12613	1396	1581	1962	413	436	467	186	191	198	102	103	105
	•••	6197	8127	12724	1476	1728	1967	424	451	458	188	194	186	102	104	93
0.75	***	5055	6118	11213	1243	1405	1722	358	377	401	157	161	165	83	84	85
	•••	5476	7292	11271	1314	1530	1719	367	388	390	159	162	153	84	84	73
0.8	***	4320	5286	9487	1060	1191	1431	292	305	321	122	124	127	.	.	.
	•••	4707	6318	9509	1118	1289	1422	298	313	310	123	125	115	.	.	.
0.85	***	3537	4344	7436	841	934	1087	213	220	229	.	.	.	.	.	.
	•••	3864	5172	7438	882	999	1076	216	224	217	.	.	.	.	.	.
0.9	***	2651	3230	5067	574	622	692	.	.	.	.	.	.	.	.	.
	•••	2889	3785	5058	596	653	681	.	.	.	.	.	.	.	.	.
0.95	***	1523	1786	2381	.	.	.	.	.	.	.	.	.	.	.	.
	•••	1636	2005	2368	.	.	.	.	.	.	.	.	.	.	.	.

TABLE 7: ALPHA= 0.05 POWER= 0.8   EXPECTED ACCRUAL THRU MINIMUM FOLLOW-UP= 1400

		DEL=.02			DEL=.05			DEL=.10			DEL=.15			DEL=.20		
FACT=		1.0 .75	.50 .25	.00 BIN	1.0 .75	.50 .25	.00 BIN	1.0 .75	.50 .25	.00 BIN	1.0 75	.50 .25	.00 BIN	1.0 .75	.50 .25	.00 BIN
PCONT=***					REQUIRED NUMBER OF PATIENTS											
0.05	***	1781	1802	1863	387	390	395	139	140	141	81	81	82	58	58	58
	•••	1789	1823	3481	388	393	681	139	140	217	81	82	115	58	58	73
0.1	***	3714	3776	4092	724	738	762	230	233	235	123	124	125	81	81	82
	•••	3736	3861	6047	730	749	1076	231	234	310	123	124	153	81	82	93
0.15	***	5441	5556	6328	1025	1058	1116	311	316	323	159	161	163	101	101	102
	•••	5481	5732	8304	1039	1082	1422	313	319	390	160	162	186	101	102	110
0.2	***	6835	7015	8409	1276	1333	1441	380	388	400	190	192	195	117	119	120
	•••	6897	7300	10251	1300	1376	1719	384	394	458	191	193	213	118	119	124
0.25	***	7861	8120	10252	1473	1557	1724	436	449	465	214	218	222	130	132	134
	•••	7949	8531	11890	1509	1623	1967	442	457	514	216	220	235	131	133	135
0.3	***	8528	8879	11816	1620	1730	1960	479	496	516	233	238	243	140	142	144
	•••	8648	9432	13219	1667	1819	2164	486	507	557	235	241	252	141	143	143
0.35	***	8864	9324	13073	1718	1854	2146	509	531	559	246	252	259	147	149	151
	•••	9024	10026	14239	1776	1966	2313	520	544	588	248	255	263	148	150	147
0.4	***	8911	9494	14009	1774	1933	2281	528	553	586	254	260	268	150	152	155
	•••	9117	10347	14950	1842	2065	2412	540	569	606	256	264	268	150	153	149
0.45	***	8719	9432	14614	1790	1970	2362	536	563	600	255	262	271	150	152	155
	•••	8975	10431	15352	1867	2117	2461	549	580	613	259	267	268	150	153	147
0.5	***	8340	9185	14885	1773	1966	2390	533	562	600	252	260	269	146	149	151
	•••	8649	10310	15445	1856	2125	2461	546	579	606	255	263	263	148	150	143
0.55	***	7824	8790	14821	1725	1926	2363	519	548	587	244	251	260	140	143	145
	•••	8186	10013	15228	1811	2091	2412	533	566	588	248	255	252	141	143	135
0.6	***	7220	8280	14419	1648	1851	2283	495	523	561	230	237	245	130	133	135
	•••	7624	9565	14703	1736	2015	2313	508	540	557	234	241	235	131	134	124
0.65	***	6562	7678	13682	1546	1742	2149	460	486	521	211	217	224	118	120	122
	•••	6995	8980	13868	1631	1899	2164	472	502	514	213	220	213	119	121	110
0.7	***	5873	6997	12613	1417	1599	1962	416	438	467	186	192	197	102	103	105
	•••	6315	8265	12724	1496	1740	1967	426	451	458	189	194	186	102	104	93
0.75	***	5160	6240	11213	1262	1420	1722	360	378	401	157	161	164	83	84	86
	•••	5589	7419	11271	1332	1541	1719	369	388	390	159	163	153	84	85	73
0.8	***	4418	5395	9487	1075	1203	1431	294	306	321	122	124	127	.	.	.
	•••	4810	6428	9509	1132	1297	1422	299	313	310	123	125	115	.	.	.
0.85	***	3620	4433	7437	852	941	1087	214	221	229	.	.	.	.	.	.
	•••	3950	5257	7438	892	1004	1076	217	225	217	.	.	.	.	.	.
0.9	***	2712	3291	5066	579	626	693	.	.	.	.	.	.	.	.	.
	•••	2951	3839	5058	601	656	681	.	.	.	.	.	.	.	.	.
0.95	***	1552	1812	2380	.	.	.	.	.	.	.	.	.	.	.	.
	•••	1664	2025	2368	.	.	.	.	.	.	.	.	.	.	.	.

TABLE 7: ALPHA= 0.05 POWER= 0.8   EXPECTED ACCRUAL THRU MINIMUM FOLLOW-UP= 1500

	DEL=.02			DEL=.05			DEL=.10			DEL=.15			DEL=.20		
FACT=	1.0 .75	.50 .25	.00 BIN	1.0 .75	.50 .25	.00 BIN	1.0 .75	.50 .25	.00 BIN	1.0 75	.50 .25	.00 BIN	1.0 .75	.50 .25	.00 BIN

PCONT=***   REQUIRED NUMBER OF PATIENTS

PCONT	DEL=.02 1.0/.75	.50/.25	.00/BIN	DEL=.05 1.0/.75	.50/.25	.00/BIN	DEL=.10 1.0/.75	.50/.25	.00/BIN	DEL=.15 1.0/.75	.50/.25	.00/BIN	DEL=.20 1.0/.75	.50/.25	.00/BIN
0.05	1783	1805	1863	387	391	395	140	140	140	82	82	82	57	58	58
	1792	1825	3481	389	393	681	140	140	217	82	82	115	57	58	73
0.1	3719	3784	4092	725	740	762	230	232	235	124	124	125	82	82	82
	3742	3870	6047	731	750	1076	232	234	310	124	125	153	82	82	93
0.15	5450	5572	6328	1027	1061	1117	311	317	322	159	161	163	101	102	102
	5492	5752	8304	1042	1085	1422	314	320	390	160	162	186	101	102	110
0.2	6848	7039	8408	1282	1338	1441	380	389	399	190	192	195	118	119	120
	6914	7332	10251	1306	1380	1719	384	394	458	191	194	213	118	119	124
0.25	7880	8154	10252	1482	1564	1724	437	450	466	215	218	222	131	132	134
	7974	8579	11890	1518	1628	1967	443	457	514	217	220	235	131	133	135
0.3	8554	8926	11816	1631	1740	1960	481	498	519	233	238	244	140	142	144
	8683	9495	13219	1678	1826	2164	488	507	557	236	241	252	142	143	143
0.35	8899	9384	13073	1732	1867	2147	512	532	560	247	252	259	147	149	151
	9068	10105	14239	1790	1975	2313	521	545	588	249	255	263	148	150	147
0.4	8956	9568	14009	1790	1948	2281	532	555	587	254	260	268	150	152	155
	9173	10442	14950	1858	2075	2412	543	570	606	258	264	268	151	154	149
0.45	8775	9521	14615	1808	1985	2362	539	565	600	257	263	271	150	152	155
	9045	10540	15352	1885	2130	2461	551	581	613	260	267	268	151	154	147
0.5	8408	9287	14885	1792	1984	2390	536	563	601	253	260	268	146	149	152
	8733	10430	15445	1875	2138	2461	549	580	606	257	264	263	148	150	143
0.55	7906	8902	14821	1745	1944	2363	522	550	587	245	251	260	140	142	145
	8281	10144	15228	1832	2105	2412	535	567	588	248	255	252	142	143	135
0.6	7312	8401	14419	1670	1869	2283	499	525	560	230	237	245	130	133	135
	7730	9699	14703	1756	2029	2313	511	542	557	234	241	235	131	134	124
0.65	6661	7802	13683	1567	1760	2149	464	488	520	212	217	224	118	120	122
	7105	9113	13868	1651	1912	2164	475	503	514	215	220	213	119	121	110
0.7	5975	7119	12613	1436	1615	1962	419	440	467	187	192	197	102	104	105
	6427	8393	12724	1515	1752	1967	428	453	458	189	195	186	103	104	93
0.75	5260	6355	11213	1279	1434	1722	363	380	400	157	161	165	83	84	85
	5697	7537	11271	1347	1550	1719	371	390	390	159	163	153	83	85	73
0.8	4509	5497	9487	1089	1214	1430	295	307	322	122	125	127	.	.	.
	4907	6529	9509	1145	1304	1422	301	314	310	124	125	115	.	.	.
0.85	3697	4516	7436	862	950	1087	215	221	229	.	.	.	.	.	.
	4031	5335	7438	902	1009	1076	218	225	217	.	.	.	.	.	.
0.9	2769	3348	5066	585	630	693	.	.	.	.	.	.	.	.	.
	3009	3890	5058	605	658	681	.	.	.	.	.	.	.	.	.
0.95	1580	1835	2380	.	.	.	.	.	.	.	.	.	.	.	.
	1690	2042	2368	.	.	.	.	.	.	.	.	.	.	.	.

TABLE 7: ALPHA= 0.05 POWER= 0.8   EXPECTED ACCRUAL THRU MINIMUM FOLLOW-UP= 1600

		DEL=.02			DEL=.05			DEL=.10			DEL=.15			DEL=.20		
FACT=		1.0 .75	.50 .25	.00 BIN	1.0 .75	.50 .25	.00 BIN	1.0 .75	.50 .25	.00 BIN	1.0 75	.50 .25	.00 BIN	1.0 .75	.50 .25	.00 BIN
PCONT=•••		REQUIRED NUMBER OF PATIENTS														
0.05	•••	1785	1806	1863	388	391	395	139	140	141	82	82	82	58	58	58
	•••	1793	1827	3481	389	393	681	140	140	217	82	82	115	58	58	73
0.1	•••	3724	3792	4092	727	741	762	231	233	235	124	124	125	81	82	82
	•••	3748	3879	6047	733	750	1076	232	234	310	124	125	153	82	82	93
0.15	•••	5458	5586	6328	1031	1064	1117	312	317	323	160	161	163	101	102	103
	•••	5503	5769	8304	1046	1086	1422	314	320	390	160	162	186	102	102	110
0.2	•••	6861	7063	8409	1287	1342	1441	382	390	400	190	193	195	118	119	120
	•••	6931	7363	10251	1311	1383	1719	386	395	458	192	194	213	118	119	124
0.25	•••	7899	8188	10253	1490	1571	1724	439	451	466	215	219	222	131	132	134
	•••	8000	8623	11890	1525	1633	1967	444	458	514	217	220	235	132	133	135
0.3	•••	8580	8972	11816	1642	1749	1960	483	499	519	234	239	243	141	142	144
	•••	8717	9555	13219	1688	1833	2164	490	508	557	236	241	252	142	143	143
0.35	•••	8933	9442	13073	1745	1878	2147	514	534	559	248	253	259	147	149	151
	•••	9113	10180	14239	1803	1984	2313	524	546	588	250	256	263	148	150	147
0.4	•••	9000	9640	14009	1805	1961	2281	534	557	587	255	261	268	150	152	155
	•••	9230	10532	14950	1873	2086	2412	545	571	606	258	264	268	151	154	149
0.45	•••	8831	9607	14615	1826	2001	2362	542	567	600	257	264	271	150	152	155
	•••	9114	10643	15352	1902	2141	2461	554	583	613	260	267	268	151	154	147
0.5	•••	8476	9384	14886	1812	2000	2390	539	566	601	254	261	268	147	149	152
	•••	8813	10545	15445	1893	2151	2461	552	582	606	257	264	263	148	150	143
0.55	•••	7985	9010	14821	1765	1962	2363	526	552	587	246	252	260	140	143	145
	•••	8373	10265	15228	1850	2117	2412	538	569	588	249	256	252	142	144	135
0.6	•••	7401	8515	14419	1689	1886	2283	501	527	561	232	238	245	131	133	135
	•••	7830	9825	14703	1775	2041	2313	513	540	557	235	241	235	132	134	124
0.65	•••	6757	7919	13682	1586	1776	2149	467	490	521	213	218	224	118	120	122
	•••	7211	9239	13868	1669	1923	2164	478	504	514	215	221	213	119	121	110
0.7	•••	6073	7235	12613	1455	1630	1962	421	442	467	188	192	197	102	104	105
	•••	6533	8514	12724	1532	1763	1967	431	454	458	190	195	186	103	105	93
0.75	•••	5356	6463	11213	1295	1447	1722	365	381	401	158	161	165	84	84	85
	•••	5799	7647	11271	1362	1560	1719	373	390	390	160	163	153	84	85	73
0.8	•••	4597	5593	9487	1102	1224	1431	297	308	321	123	125	127	.	.	.
	•••	4999	6623	9509	1157	1311	1422	302	314	310	124	126	115	.	.	.
0.85	•••	3771	4594	7436	872	956	1087	216	222	229	.	.	.	.	.	.
	•••	4108	5408	7438	910	1013	1076	219	225	217	.	.	.	.	.	.
0.9	•••	2823	3401	5066	590	633	693	.	.	.	.	.	.	.	.	.
	•••	3064	3936	5058	610	660	681	.	.	.	.	.	.	.	.	.
0.95	•••	1605	1858	2380	.	.	.	.	.	.	.	.	.	.	.	.
	•••	1715	2058	2368	.	.	.	.	.	.	.	.	.	.	.	.

TABLE 7: ALPHA= 0.05 POWER= 0.8   EXPECTED ACCRUAL THRU MINIMUM FOLLOW-UP= 1700

PCONT=***	DEL=.02 1.0 .75	.50 .25	.00 BIN	DEL=.05 1.0 .75	.50 .25	.00 BIN	DEL=.10 1.0 .75	.50 .25	.00 BIN	DEL=.15 1.0 75	.50 .25	.00 BIN	DEL=.20 1.0 .75	.50 .25	.00 BIN
					REQUIRED NUMBER OF PATIENTS										
0.05 ***	1787	1808	1863	388	391	395	139	140	140	82	82	82	57	57	57
•••	1795	1829	3481	390	393	681	140	140	217	82	82	115	57	57	73
0.1 ***	3728	3798	4092	728	741	762	231	233	236	123	124	125	82	82	82
•••	3754	3886	6047	734	751	1076	231	233	310	124	124	153	82	82	93
0.15 ***	5467	5600	6328	1035	1066	1117	312	318	323	160	161	163	102	102	103
•••	5514	5786	8304	1049	1088	1422	314	320	390	160	163	186	102	102	110
0.2 ***	6874	7086	8408	1292	1346	1440	382	390	399	190	193	195	118	119	120
•••	6948	7392	10251	1316	1387	1719	386	395	458	191	194	213	119	119	124
0.25 ***	7919	8221	10253	1498	1576	1724	440	452	465	216	219	222	131	133	134
•••	8024	8665	11890	1532	1637	1967	445	458	514	216	221	235	131	133	135
0.3 ***	8606	9016	11816	1651	1757	1960	484	499	518	235	239	243	141	142	144
•••	8750	9611	13219	1698	1839	2164	492	509	557	237	241	252	141	143	143
0.35 ***	8967	9498	13073	1757	1889	2146	516	535	560	248	253	259	148	148	151
•••	9156	10252	14239	1814	1991	2313	526	546	588	250	256	263	148	150	147
0.4 ***	9044	9709	14009	1820	1974	2281	537	559	586	256	261	267	151	153	155
•••	9284	10618	14950	1886	2095	2412	547	571	606	258	265	268	152	154	149
0.45 ***	8886	9689	14615	1842	2014	2362	545	569	600	258	265	271	151	153	155
•••	9181	10741	15352	1916	2150	2461	556	583	613	261	267	268	152	154	147
0.5 ***	8542	9479	14885	1829	2016	2390	542	567	600	255	261	269	148	150	152
•••	8892	10652	15445	1910	2162	2461	554	583	606	258	265	263	148	151	143
0.55 ***	8062	9113	14821	1784	1977	2363	528	554	588	246	253	259	141	142	146
•••	8463	10380	15228	1868	2128	2412	541	569	588	250	256	252	141	144	135
0.6 ***	7486	8624	14419	1707	1902	2283	503	529	561	233	238	244	131	133	135
•••	7927	9943	14703	1792	2052	2313	515	544	557	235	241	235	131	134	124
0.65 ***	6849	8030	13683	1603	1790	2149	469	492	520	214	218	224	119	120	122
•••	7312	9356	13868	1686	1933	2164	480	505	514	216	221	213	119	121	110
0.7 ***	6166	7344	12613	1472	1644	1962	424	443	467	188	192	197	103	104	105
•••	6634	8626	12724	1547	1772	1967	432	454	458	190	195	186	103	104	93
0.75 ***	5446	6565	11213	1309	1459	1722	367	382	401	158	161	165	84	85	85
•••	5894	7749	11271	1376	1567	1719	374	391	390	160	163	153	84	85	73
0.8 ***	4680	5683	9487	1115	1234	1430	297	309	321	123	124	126	.	.	.
•••	5086	6710	9509	1168	1316	1422	303	314	310	123	125	115	.	.	.
0.85 ***	3841	4666	7436	879	962	1087	216	222	229	.	.	.	.	.	.
•••	4180	5476	7438	918	1017	1076	219	225	217	.	.	.	.	.	.
0.9 ***	2873	3451	5066	594	636	692	.	.	.	.	.	.	.	.	.
•••	3114	3978	5058	614	662	681	.	.	.	.	.	.	.	.	.
0.95 ***	1629	1877	2380	.	.	.	.	.	.	.	.	.	.	.	.
•••	1737	2073	2368	.	.	.	.	.	.	.	.	.	.	.	.

TABLE 7: ALPHA= 0.05 POWER= 0.8   EXPECTED ACCRUAL THRU MINIMUM FOLLOW-UP= 1800

		DEL=.02			DEL=.05			DEL=.10			DEL=.15			DEL=.20	
FACT=	1.0 .75	.50 .25	.00 BIN	1.0 .75	.50 .25	.00 BIN	1.0 .75	.50 .25	.00 BIN	1.0 75	.50 .25	.00 BIN	1.0 .75	.50 .25	.00 BIN
PCONT=•••						REQUIRED NUMBER OF PATIENTS									
0.05 •••	1788	1810	1863	388	391	395	139	141	141	82	82	82	57	57	57
•••	1797	1830	3481	390	393	681	139	141	217	82	82	115	57	57	73
0.1 •••	3732	3806	4092	729	742	762	231	233	235	123	125	125	81	82	82
•••	3759	3894	6047	735	751	1076	231	234	310	123	125	153	82	82	93
0.15 •••	5475	5615	6328	1037	1068	1117	312	317	323	159	162	163	101	102	102
•••	5525	5802	8304	1050	1089	1422	315	319	390	161	162	186	102	102	110
0.2 •••	6888	7107	8408	1297	1350	1441	384	390	399	191	193	195	118	119	120
•••	6966	7419	10251	1320	1388	1719	387	395	458	192	195	213	118	119	124
0.25 •••	7937	8253	10252	1504	1583	1723	441	452	465	216	219	222	132	132	134
•••	8048	8706	11890	1539	1641	1967	447	459	514	217	220	235	132	132	135
0.3 •••	8632	9060	11816	1660	1765	1961	486	501	519	235	240	243	141	143	144
•••	8783	9666	13219	1707	1845	2164	492	510	557	237	240	252	141	143	143
0.35 •••	9001	9555	13074	1770	1899	2146	519	537	559	249	253	258	147	150	150
•••	9199	10320	14239	1826	1998	2313	526	546	588	251	256	263	148	150	147
0.4 •••	9087	9777	14008	1833	1986	2281	539	560	586	256	262	267	150	153	155
•••	9339	10698	14950	1899	2103	2412	549	573	606	258	264	268	152	154	149
0.45 •••	8940	9768	14615	1857	2027	2362	548	570	600	258	264	271	150	153	155
•••	9246	10833	15352	1932	2160	2461	558	585	613	262	267	268	152	154	147
0.5 •••	8607	9568	14886	1845	2029	2389	544	569	600	255	262	269	147	150	152
•••	8968	10755	15445	1926	2171	2461	555	584	606	258	264	263	148	150	143
0.55 •••	8137	9211	14820	1800	1991	2364	531	555	587	247	253	260	141	143	145
•••	8549	10488	15228	1884	2139	2412	542	570	588	249	256	252	141	144	135
0.6 •••	7570	8727	14419	1725	1916	2283	506	531	560	233	238	245	132	132	135
•••	8020	10055	14703	1809	2062	2313	517	544	557	236	242	235	132	134	124
0.65 •••	6938	8136	13682	1620	1804	2148	471	492	521	213	218	224	119	120	121
•••	7409	9467	13868	1701	1943	2164	481	506	514	216	222	213	119	121	110
0.7 •••	6257	7447	12612	1487	1656	1962	425	444	467	189	193	197	102	103	105
•••	6731	8732	12724	1561	1781	1967	434	456	458	191	195	186	103	105	93
0.75 •••	5534	6662	11213	1323	1469	1722	368	382	400	159	162	165	83	84	85
•••	5987	7845	11271	1388	1575	1719	375	391	390	161	163	153	84	85	73
0.8 •••	4758	5769	9487	1125	1242	1431	299	309	321	123	125	127	.	.	.
•••	5169	6792	9509	1178	1322	1422	303	315	310	123	126	115	.	.	.
0.85 •••	3908	4735	7436	888	967	1086	217	222	229	.	.	.	.	.	.
•••	4248	5538	7438	924	1020	1076	219	226	217	.	.	.	.	.	.
0.9 •••	2922	3498	5067	598	639	693	.	.	.	.	.	.	.	.	.
•••	3162	4017	5058	618	663	681	.	.	.	.	.	.	.	.	.
0.95 •••	1650	1896	2380	.	.	.	.	.	.	.	.	.	.	.	.
•••	1757	2085	2368	.	.	.	.	.	.	.	.	.	.	.	.

TABLE 7: ALPHA= 0.05 POWER= 0.8   EXPECTED ACCRUAL THRU MINIMUM FOLLOW-UP= 1900

		DEL=.02			DEL=.05			DEL=.10			DEL=.15			DEL=.20		
FACT=		1.0 .75	.50 .25	.00 BIN	1.0 .75	.50 .25	.00 BIN	1.0 .75	.50 .25	.00 BIN	1.0 75	.50 .25	.00 BIN	1.0 .75	.50 .25	.00 BIN
PCONT=***		REQUIRED NUMBER OF PATIENTS														
0.05	***	1789	1812	1863	388	391	395	140	140	141	82	82	82	58	58	58
	•••	1799	1831	3481	391	393	681	140	140	217	82	82	115	58	58	73
0.1	***	3738	3813	4092	730	743	762	231	234	235	123	125	125	82	82	82
	•••	3765	3899	6047	736	752	1076	232	234	310	125	125	153	82	82	93
0.15	***	5483	5627	6328	1040	1070	1117	313	318	323	160	161	163	102	102	103
	•••	5536	5817	8304	1053	1091	1422	315	320	390	161	163	186	102	102	110
0.2	***	6901	7129	8409	1301	1353	1440	383	391	400	191	193	196	118	118	120
	•••	6982	7445	10251	1325	1390	1719	387	395	458	192	194	213	118	120	124
0.25	***	7956	8284	10252	1512	1588	1724	441	452	465	216	220	222	132	133	134
	•••	8072	8744	11890	1545	1645	1967	448	458	514	217	220	235	132	133	135
0.3	***	8658	9101	11815	1669	1773	1960	486	501	519	235	239	243	141	142	144
	•••	8816	9717	13219	1714	1850	2164	494	509	557	237	241	252	142	144	143
0.35	***	9036	9608	13073	1780	1907	2147	520	538	559	249	254	258	148	149	151
	•••	9242	10383	14239	1835	2004	2313	528	547	588	251	256	263	148	151	147
0.4	***	9131	9842	14009	1845	1996	2281	540	562	586	258	262	268	151	153	154
	•••	9390	10775	14950	1911	2110	2412	551	573	606	260	265	268	152	154	149
0.45	***	8993	9844	14614	1871	2040	2362	550	572	600	258	265	270	151	153	155
	•••	9310	10921	15352	1945	2168	2461	560	585	613	262	268	268	152	154	147
0.5	***	8670	9654	14885	1861	2042	2389	547	571	600	256	262	268	147	149	152
	•••	9043	10851	15445	1940	2180	2461	558	584	606	258	265	263	148	151	143
0.55	***	8210	9306	14821	1816	2004	2363	533	557	588	247	253	260	141	144	144
	•••	8632	10591	15228	1899	2148	2412	545	571	588	250	256	252	142	144	135
0.6	***	7651	8827	14418	1740	1930	2282	509	532	560	234	239	244	132	133	135
	•••	8110	10160	14703	1824	2072	2313	520	545	557	236	242	235	133	134	124
0.65	***	7022	8238	13682	1636	1816	2149	474	495	520	213	218	224	118	120	122
	•••	7502	9571	13868	1716	1952	2164	483	507	514	216	220	213	120	121	110
0.7	***	6343	7546	12612	1501	1667	1961	427	445	467	190	193	197	103	104	106
	•••	6823	8831	12724	1574	1788	1967	436	456	458	191	194	186	103	104	93
0.75	***	5616	6754	11213	1336	1478	1721	369	383	401	159	161	165	84	84	85
	•••	6075	7936	11271	1400	1580	1719	376	391	390	160	163	153	84	85	73
0.8	***	4835	5849	9487	1135	1249	1431	300	310	322	123	125	127	.	.	.
	•••	5247	6870	9509	1187	1327	1422	305	315	310	125	125	115	.	.	.
0.85	***	3971	4801	7436	895	972	1087	217	223	229	.	.	.	.	.	.
	•••	4313	5597	7438	931	1023	1076	220	225	217	.	.	.	.	.	.
0.9	***	2966	3541	5067	602	641	692	.	.	.	.	.	.	.	.	.
	•••	3207	4054	5058	621	665	681	.	.	.	.	.	.	.	.	.
0.95	***	1671	1913	2381	.	.	.	.	.	.	.	.	.	.	.	.
	•••	1778	2098	2368	.	.	.	.	.	.	.	.	.	.	.	.

TABLE 7: ALPHA= 0.05 POWER= 0.8   EXPECTED ACCRUAL THRU MINIMUM FOLLOW-UP= 2000

	DEL=.02			DEL=.05			DEL=.10			DEL=.15			DEL=.20		
FACT=	1.0 .75	.50 .25	.00 BIN	1.0 .75	.50 .25	.00 BIN	1.0 .75	.50 .25	.00 BIN	1.0 75	.50 .25	.00 BIN	1.0 .75	.50 .25	.00 BIN
PCONT=***				REQUIRED NUMBER OF PATIENTS											
0.05 ***	1791	1814	1862	389	392	395	140	140	141	82	82	82	57	57	57
•••	1800	1832	3481	390	394	681	140	140	217	82	82	115	57	57	73
0.1 ***	3742	3819	4092	731	745	762	231	234	235	124	125	125	81	82	82
•••	3771	3906	6047	737	752	1076	232	235	310	124	125	153	81	82	93
0.15 ***	5492	5641	6329	1042	1072	1117	314	317	322	160	161	162	101	102	102
•••	5546	5832	8304	1056	1092	1422	316	320	390	161	162	186	102	102	110
0.2 ***	6914	7151	8409	1306	1357	1441	385	391	400	191	194	195	119	119	120
•••	6999	7471	10251	1329	1394	1719	389	396	458	192	194	213	119	120	124
0.25 ***	7975	8315	10252	1517	1594	1724	444	454	466	216	219	222	131	132	134
•••	8096	8781	11890	1551	1649	1967	447	459	514	217	221	235	132	134	135
0.3 ***	8684	9142	11816	1677	1779	1960	489	502	519	236	240	244	141	142	144
•••	8847	9766	13219	1722	1855	2164	495	510	557	237	241	252	142	144	143
0.35 ***	9069	9660	13074	1790	1916	2146	521	539	560	250	254	259	147	150	151
•••	9284	10445	14239	1845	2010	2313	530	549	588	251	256	263	149	150	147
0.4 ***	9174	9905	14009	1857	2005	2281	542	562	586	257	262	267	151	152	155
•••	9442	10847	14950	1922	2117	2412	552	574	606	260	265	268	152	154	149
0.45 ***	9045	9919	14615	1885	2051	2362	551	574	600	260	265	271	151	152	155
•••	9372	11004	15352	1957	2176	2461	562	586	613	262	269	268	152	154	147
0.5 ***	8732	9739	14886	1875	2054	2390	549	572	601	256	262	269	147	150	152
•••	9115	10942	15445	1954	2189	2461	560	585	606	260	265	263	149	151	143
0.55 ***	8281	9397	14821	1831	2017	2364	536	559	587	249	254	260	141	144	145
•••	8712	10689	15228	1914	2156	2412	546	572	588	251	256	252	142	144	135
0.6 ***	7730	8922	14419	1756	1941	2282	511	534	561	234	239	245	131	134	135
•••	8196	10260	14703	1837	2080	2313	521	546	557	236	242	235	132	135	124
0.65 ***	7106	8334	13682	1651	1829	2149	475	496	521	215	219	224	119	120	122
•••	7592	9670	13868	1730	1960	2164	485	507	514	217	221	213	120	121	110
0.7 ***	6426	7641	12612	1515	1677	1962	429	446	467	190	194	197	104	104	105
•••	6912	8925	12724	1587	1796	1967	437	456	458	191	195	186	104	105	93
0.75 ***	5697	6842	11214	1347	1487	1722	371	385	401	160	162	165	84	85	85
•••	6160	8021	11271	1410	1587	1719	377	392	390	161	164	153	84	85	73
0.8 ***	4907	5926	9486	1145	1256	1431	301	310	321	124	125	127	.	.	.
•••	5324	6941	9509	1195	1331	1422	306	316	310	124	126	115	.	.	.
0.85 ***	4031	4862	7436	902	977	1087	219	224	229	.	.	.	.	.	.
•••	4375	5652	7438	936	1026	1076	221	226	217	.	.	.	.	.	.
0.9 ***	3010	3581	5066	606	644	692	.	.	.	.	.	.	.	.	.
•••	3251	4087	5058	624	666	681	.	.	.	.	.	.	.	.	.
0.95 ***	1690	1929	2380	.	.	.	.	.	.	.	.	.	.	.	.
•••	1795	2110	2368	.	.	.	.	.	.	.	.	.	.	.	.

TABLE 7: ALPHA= 0.05 POWER= 0.8   EXPECTED ACCRUAL THRU MINIMUM FOLLOW-UP= 2250

		DEL=.02			DEL=.05			DEL=.10			DEL=.15			DEL=.20		
FACT=		1.0 .75	.50 .25	.00 BIN	1.0 .75	.50 .25	.00 BIN	1.0 .75	.50 .25	.00 BIN	1.0 75	.50 .25	.00 BIN	1.0 .75	.50 .25	.00 BIN
PCONT=***					REQUIRED NUMBER OF PATIENTS											
0.05	***	1796	1816	1863	389	392	395	140	140	140	81	81	83	57	57	57
	•••	1804	1835	3481	390	393	681	140	140	217	81	81	115	57	57	73
0.1	***	3753	3833	4092	734	747	762	232	233	236	123	125	125	81	81	83
	•••	3784	3920	6047	739	753	1076	233	235	310	123	125	153	81	81	93
0.15	***	5512	5672	6328	1047	1076	1116	314	319	323	160	162	163	101	102	102
	•••	5571	5865	8304	1062	1094	1422	316	320	390	162	162	186	102	102	110
0.2	***	6946	7201	8409	1316	1364	1441	387	392	399	191	194	195	118	119	119
	•••	7039	7527	10251	1337	1397	1719	389	396	458	192	194	213	119	119	124
0.25	***	8021	8388	10253	1531	1604	1724	446	454	465	218	219	222	132	133	133
	•••	8155	8864	11890	1563	1655	1967	449	460	514	218	221	235	132	133	135
0.3	***	8746	9239	11817	1697	1794	1960	492	505	519	236	240	243	142	143	145
	•••	8926	9879	13219	1739	1864	2164	497	511	557·	239	241	252	142	143	143
0.35	***	9151	9783	13074	1812	1934	2147	525	541	559	250	254	258	149	150	151
	•••	9384	10586	14239	1867	2023	2313	533	550	588	253	257	263	149	150	147
0.4	***	9278	10054	14009	1884	2027	2280	547	565	586	258	263	268	151	153	154
	•••	9568	11016	14950	1947	2131	2412	555	575	606	261	266	268	151	154	149
0.45	***	9173	10094	14615	1915	2075	2362	556	576	600	261	266	271	151	153	154
	•••	9522	11196	15352	1985	2193	2461	565	587	613	263	268	268	151	154	147
0.5	***	8883	9933	14886	1908	2080	2389	554	575	600	257	263	268	149	150	151
	•••	9286	11153	15445	1984	2207	2461	564	587	606	260	266	263	149	150	143
0.55	***	8451	9610	14820	1866	2044	2364	539	561	587	249	254	260	142	143	145
	•••	8903	10911	15228	1945	2175	2412	550	573	588	252	257	252	143	145	135
0.6	***	7916	9145	14419	1790	1968	2282	516	536	561	235	240	244	132	133	134
	•••	8400	10487	14703	1870	2098	2313	525	548	557	237	241	235	133	134	124
0.65	***	7300	8558	13682	1683	1855	2148	479	497	520	215	219	223	119	120	122
	•••	7801	9896	13868	1760	1977	2164	489	509	514	218	222	213	119	120	110
0.7	***	6622	7858	12612	1545	1701	1962	433	448	466	191	194	196	104	104	105
	•••	7120	9139	12724	1616	1811	1967	440	457	458	192	195	186	104	105	93
0.75	***	5883	7044	11213	1374	1507	1722	374	387	401	160	163	164	84	84	86
	•••	6355	8214	11271	1434	1600	1719	379	393	390	162	164	153	84	86	73
0.8	***	5075	6103	9486	1167	1271	1430	303	312	322	123	125	126	.	.	.
	•••	5497	7104	9509	1214	1341	1422	308	316	310	125	126	115	.	.	.
0.85	***	4170	5002	7436	916	987	1087	219	223	229	.	.	.	.	.	.
	•••	4516	5775	7438	949	1032	1076	222	226	217	.	.	.	.	.	.
0.9	***	3109	3674	5067	613	648	693	.	.	.	.	.	.	.	.	.
	•••	3348	4162	5058	629	669	681	.	.	.	.	.	.	.	.	.
0.95	***	1734	1964	2381	.	.	.	.	.	.	.	.	.	.	.	.
	•••	1836	2134	2368	.	.	.	.	.	.	.	.	.	.	.	.

TABLE 7: ALPHA= 0.05 POWER= 0.8   EXPECTED ACCRUAL THRU MINIMUM FOLLOW-UP= 2500

	DEL=.02			DEL=.05			DEL=.10			DEL=.15			DEL=.20		
FACT=	1.0 .75	.50 .25	.00 BIN	1.0 .75	.50 .25	.00 BIN	1.0 .75	.50 .25	.00 BIN	1.0 75	.50 .25	.00 BIN	1.0 .75	.50 .25	.00 BIN
PCONT=***			REQUIRED NUMBER OF PATIENTS												
0.05 ***	1798	1820	1862	390	392	395	140	140	140	83	83	83	58	58	58
***	1808	1837	3481	392	393	681	140	140	217	83	83	115	58	58	73
0.1 ***	3764	3846	4092	736	748	762	233	234	236	124	124	124	81	83	83
***	3796	3933	6047	742	754	1076	233	234	310	124	124	153	83	83	93
0.15 ***	5533	5699	6327	1052	1079	1117	315	318	323	161	162	162	101	102	102
***	5596	5893	8304	1065	1096	1422	317	320	390	161	162	186	101	102	110
0.2 ***	6977	7248	8409	1323	1370	1440	387	393	399	192	193	195	118	118	120
***	7077	7577	10251	1345	1401	1719	390	396	458	193	195	213	118	120	124
0.25 ***	8067	8456	10252	1543	1614	1724	446	456	465	217	220	221	133	133	134
***	8211	8940	11890	1574	1662	1967	451	461	514	218	221	235	133	133	135
0.3 ***	8808	9331	11815	1712	1806	1961	493	506	518	237	240	243	142	143	143
***	9001	9979	13219	1754	1873	2164	499	512	557	239	242	252	142	143	143
0.35 ***	9231	9898	13073	1833	1949	2146	527	542	559	251	254	259	148	149	151
***	9481	10712	14239	1886	2034	2313	536	549	588	252	256	263	149	151	147
0.4 ***	9377	10193	14009	1909	2046	2281	549	567	587	259	264	268	151	152	154
***	9687	11167	14950	1970	2145	2412	558	576	606	261	265	268	152	154	149
0.45 ***	9293	10254	14615	1942	2096	2362	559	577	599	262	267	271	151	152	154
***	9662	11367	15352	2011	2208	2461	568	589	613	264	268	268	152	154	147
0.5 ***	9024	10114	14886	1936	2102	2390	558	577	601	259	264	268	148	149	151
***	9446	11340	15445	2011	2223	2461	567	589	606	261	265	263	149	151	143
0.55 ***	8611	9802	14820	1895	2067	2364	543	564	587	249	254	259	142	143	145
***	9079	11109	15228	1971	2190	2412	552	574	588	252	258	252	143	145	135
0.6 ***	8087	9346	14418	1820	1992	2283	520	539	561	236	240	245	133	134	136
***	8589	10690	14703	1896	2114	2313	527	548	557	239	242	235	133	134	124
0.65 ***	7479	8761	13683	1712	1876	2149	483	499	520	217	220	224	120	120	121
***	7993	10095	13868	1786	1992	2164	492	511	514	218	221	213	120	121	110
0.7 ***	6801	8054	12612	1571	1721	1962	436	449	467	192	193	196	102	104	106
***	7308	9327	12724	1639	1824	1967	442	458	458	192	195	186	104	104	93
0.75 ***	6054	7224	11214	1396	1524	1721	376	387	401	161	162	165	84	84	86
***	6533	8383	11271	1454	1611	1719	381	393	390	162	164	153	84	86	73
0.8 ***	5227	6259	9487	1184	1284	1431	304	312	321	124	124	126	.	.	.
***	5654	7245	9509	1231	1349	1422	309	317	310	124	126	115	.	.	.
0.85 ***	4296	5126	7436	929	995	1087	220	224	229	.	.	.	.	.	.
***	4643	5881	7438	961	1037	1076	221	226	217	.	.	.	.	.	.
0.9 ***	3196	3754	5067	620	652	692	.	.	.	.	.	.	.	.	.
***	3434	4226	5058	636	671	681	.	.	.	.	.	.	.	.	.
0.95 ***	1773	1993	2381	.	.	.	.	.	.	.	.	.	.	.	.
***	1871	2154	2368	.	.	.	.	.	.	.	.	.	.	.	.

TABLE 7: ALPHA= 0.05 POWER= 0.8   EXPECTED ACCRUAL THRU MINIMUM FOLLOW-UP= 2750

		DEL=.02			DEL=.05			DEL=.10			DEL=.15			DEL=.20		
FACT=		1.0 .75	.50 .25	.00 BIN	1.0 .75	.50 .25	.00 BIN	1.0 .75	.50 .25	.00 BIN	1.0 75	.50 .25	.00 BIN	1.0 .75	.50 .25	.00 BIN
PCONT=***					REQUIRED NUMBER OF PATIENTS											
0.05	***	1800	1823	1862	390	393	395	140	140	140	82	82	82	58	58	58
	•••	1811	1838	3481	391	393	681	140	140	217	82	82	115	58	58	73
0.1	***	3774	3860	4092	738	749	762	233	233	235	123	125	125	82	82	82
	•••	3808	3942	6047	744	754	1076	233	235	310	125	125	153	82	82	93
0.15	***	5553	5726	6328	1057	1082	1116	315	319	322	161	161	163	101	102	102
	•••	5620	5919	8304	1068	1098	1422	318	321	390	161	163	186	102	102	110
0.2	***	7008	7292	8409	1332	1376	1442	388	393	400	192	194	195	118	120	120
	•••	7115	7622	10251	1352	1405	1719	391	397	458	194	194	213	118	120	124
0.25	***	8110	8519	10252	1555	1622	1724	448	456	465	218	219	223	132	133	133
	•••	8263	9008	11890	1584	1666	1967	452	460	514	219	221	235	132	133	135
0.3	***	8868	9415	11816	1727	1818	1961	496	507	518	239	240	243	142	142	144
	•••	9073	10070	13219	1768	1880	2164	501	511	557	239	242	252	142	144	143
0.35	***	9308	10004	13072	1851	1964	2146	531	544	560	252	256	259	149	149	150
	•••	9573	10826	14239	1902	2041	2313	538	551	588	254	257	263	149	150	147
0.4	***	9475	10322	14009	1930	2062	2280	553	569	587	260	264	267	153	153	154
	•••	9798	11300	14950	1988	2155	2412	560	577	606	263	266	268	153	154	149
0.45	***	9410	10401	14614	1965	2113	2363	563	580	601	263	266	271	153	154	154
	•••	9795	11520	15352	2031	2219	2461	570	589	613	264	269	268	153	154	147
0.5	***	9159	10277	14886	1961	2122	2390	562	579	601	259	264	267	149	150	153
	•••	9597	11508	15445	2033	2236	2461	570	589	606	260	266	263	149	150	143
0.55	***	8761	9980	14820	1921	2088	2363	548	565	587	250	256	259	142	144	146
	•••	9245	11285	15228	1995	2205	2412	556	575	588	252	257	252	142	144	135
0.6	***	8249	9530	14418	1845	2012	2282	522	539	560	236	240	245	132	133	135
	•••	8761	10871	14703	1921	2127	2313	531	549	557	239	242	235	133	133	124
0.65	***	7646	8946	13683	1737	1895	2150	486	501	520	216	219	225	120	120	122
	•••	8170	10273	13868	1807	2003	2164	493	510	514	218	221	213	120	122	110
0.7	***	6966	8231	12614	1594	1737	1962	438	452	467	192	194	197	102	104	104
	•••	7481	9494	12724	1660	1835	1967	445	459	458	194	195	186	104	104	93
0.75	***	6209	7389	11212	1418	1538	1721	377	388	400	161	163	164	84	85	85
	•••	6694	8533	11271	1473	1618	1719	383	395	390	161	164	153	84	85	73
0.8	***	5369	6400	9487	1201	1295	1431	305	312	321	125	125	126	.	.	.
	•••	5797	7371	9509	1244	1356	1422	309	318	310	125	126	115	.	.	.
0.85	***	4412	5237	7437	940	1003	1088	221	225	228	.	.	.	.	.	.
	•••	4757	5974	7438	969	1041	1076	223	226	217	.	.	.	.	.	.
0.9	***	3275	3825	5066	625	656	692	.	.	.	.	.	.	.	.	.
	•••	3512	4282	5058	639	673	681	.	.	.	.	.	.	.	.	.
0.95	***	1806	2019	2380	.	.	.	.	.	.	.	.	.	.	.	.
	•••	1902	2172	2368	.	.	.	.	.	.	.	.	.	.	.	.

TABLE 7: ALPHA= 0.05 POWER= 0.8   EXPECTED ACCRUAL THRU MINIMUM FOLLOW-UP= 3000

		DEL=.01			DEL=.02			DEL=.05			DEL=.10			DEL=.15	
FACT=	1.0 .75	.50 .25	.00 BIN	1.0 .75	.50 .25	.00 BIN	1.0 .75	.50 .25	.00 BIN	1.0 75	.50 .25	.00 BIN	1.0 .75	.50 .25	.00 BIN
PCONT=***					REQUIRED NUMBER OF PATIENTS										
0.01 ***	947	950	955	347	347	350	115	115	115	58	58	58	40	40	40
•••	947	950	3648	347	350	1206	115	115	328	58	58	134	40	40	80
0.02 ***	2095	2105	2128	673	677	680	182	182	182	80	80	80	50	50	50
•••	2098	2113	6022	673	677	1793	182	182	419	80	80	155	50	50	89
0.05 ***	6340	6392	6647	1805	1825	1862	392	392	395	140	140	140	80	80	80
•••	6358	6463	12848	1813	1840	3481	392	392	681	140	140	217	80	80	115
0.1 ***	13990	14128	15470	3782	3868	4093	740	748	763	230	235	235	122	125	125
•••	14035	14368	23235	3820	3950	6047	745	755	1076	235	235	310	125	125	153
0.15 ***	20905	21152	24437	5570	5750	6328	1060	1085	1115	317	320	320	160	163	163
•••	20987	21617	32385	5642	5942	8304	1070	1100	1422	317	320	390	160	163	186
0.2 ***	26495	26882	32833	7037	7333	8410	1337	1378	1442	388	392	400	193	193	193
•••	26623	27617	40298	7150	7663	10251	1355	1408	1719	392	395	458	193	193	213
0.25 ***	30595	31153	40310	8155	8578	10250	1562	1630	1723	448	455	467	220	220	223
•••	30782	32215	46976	8315	9065	11890	1592	1670	1967	452	460	514	220	220	235
0.3 ***	33215	33985	46678	8927	9493	11815	1738	1825	1960	497	508	520	238	242	242
•••	33475	35435	52416	9140	10150	13219	1780	1885	2164	500	512	557	238	242	252
0.35 ***	34465	35488	51830	9385	10105	13075	1865	1975	2147	530	545	560	253	253	257
•••	34805	37385	56620	9658	10930	14239	1915	2050	2313	538	553	588	253	257	263
0.4 ***	34490	35822	55700	9568	10442	14008	1948	2075	2282	553	568	587	260	265	268
•••	34940	38222	59588	9905	11425	14950	2005	2162	2412	560	575	606	260	265	268
0.45 ***	33490	35188	58258	9520	10540	14615	1985	2128	2360	565	580	598	265	268	272
•••	34063	38120	61319	9917	11657	15352	2050	2230	2461	572	590	613	265	268	268
0.5 ***	31660	33785	59477	9287	10430	14885	1982	2140	2390	565	580	602	260	265	268
•••	32390	37250	61814	9737	11660	15445	2053	2245	2461	572	590	606	260	265	263
0.55 ***	29233	31820	59350	8900	10142	14822	1945	2105	2365	550	568	587	250	253	260
•••	30140	35770	61072	9395	11443	15228	2015	2215	2412	557	575	588	253	257	252
0.6 ***	26435	29477	57875	8402	9700	14417	1870	2027	2282	523	542	560	238	242	245
•••	27530	33805	59093	8923	11030	14703	1940	2140	2313	535	550	557	238	242	235
0.65 ***	23477	26893	55063	7802	9115	13682	1760	1910	2147	490	505	520	215	220	223
•••	24740	31445	55878	8335	10430	13868	1828	2015	2164	497	512	514	220	223	213
0.7 ***	20522	24163	50920	7120	8395	12613	1615	1753	1963	440	452	467	193	193	197
•••	21902	28750	51427	7640	9643	12724	1678	1843	1967	445	460	458	193	197	186
0.75 ***	17657	21317	45452	6355	7535	11215	1435	1550	1723	380	388	400	160	163	163
•••	19070	25727	45739	6842	8665	11271	1487	1625	1719	385	395	390	163	163	153
0.8 ***	14882	18332	38675	5495	6527	9485	1213	1303	1430	305	313	320	125	125	125
•••	16232	22337	38815	5927	7480	9509	1255	1360	1422	310	317	310	125	125	115
0.85 ***	12115	15118	30602	4517	5335	7435	950	1007	1085	220	223	227	.	.	.
•••	13303	18470	30654	4862	6055	7438	977	1045	1076	223	227	217	.	.	.
0.9 ***	9175	11465	21250	3347	3890	5065	628	658	692	.	.	.	.	.	.
•••	10097	13892	21256	3580	4330	5058	643	673	681	.	.	.	.	.	.
0.95 ***	5627	6865	10633	1835	2042	2380	.	.	.	.	.	.	.	.	.
•••	6137	8035	10622	1930	2185	2368	.	.	.	.	.	.	.	.	.
0.98 ***	2570	2945	3662	.	.	.	.	.	.	.	.	.	.	.	.
•••	2735	3230	3648	.	.	.	.	.	.	.	.	.	.	.	.

TABLE 7: ALPHA= 0.05 POWER= 0.8   EXPECTED ACCRUAL THRU MINIMUM FOLLOW-UP= 3250

		DEL=.01			DEL=.02			DEL=.05			DEL=.10			DEL=.15		
FACT=		1.0 .75	.50 .25	.00 BIN	1.0 .75	.50 .25	.00 BIN	1.0 .75	.50 .25	.00 BIN	1.0 75	.50 .25	.00 BIN	1.0 .75	.50 .25	.00 BIN
PCONT=•••					REQUIRED NUMBER OF PATIENTS											
0.01	•••	948	948	953	347	347	347	116	116	116	59	59	59	43	43	43
	•••	948	953	3648	347	347	1206	116	116	328	59	59	134	43	43	80
0.02	•••	2094	2107	2126	672	677	680	181	181	184	79	79	79	51	51	51
	•••	2099	2115	6022	677	677	1793	181	181	419	79	79	155	51	51	89
0.05	•••	6343	6397	6649	1806	1826	1863	393	393	396	141	141	141	84	84	84
	•••	6364	6470	12848	1814	1842	3481	393	393	681	141	141	217	84	84	115
0.1	•••	14002	14148	15473	3792	3881	4092	742	750	761	233	233	233	124	124	124
	•••	14051	14403	23235	3829	3959	6047	745	753	1076	233	233	310	124	124	153
0.15	•••	20924	21193	24437	5587	5774	6327	1062	1086	1116	314	319	322	160	160	160
	•••	21014	21688	32385	5661	5961	8304	1075	1099	1422	319	319	390	160	160	186
0.2	•••	26526	26945	32836	7066	7372	8407	1343	1384	1441	387	396	401	192	192	192
	•••	26669	27728	40298	7185	7700	10251	1362	1408	1719	393	396	458	192	192	213
0.25	•••	30642	31246	40311	8196	8634	10251	1571	1636	1725	449	458	466	217	222	222
	•••	30845	32381	46976	8363	9122	11890	1598	1676	1967	452	461	514	217	222	235
0.3	•••	33283	34111	46676	8984	9569	11816	1749	1834	1961	498	506	517	238	241	241
	•••	33559	35655	52416	9208	10227	13219	1790	1891	2164	501	514	557	238	241	252
0.35	•••	34550	35655	51832	9455	10199	13071	1879	1985	2148	534	547	558	254	254	257
	•••	34919	37670	56620	9744	11023	14239	1928	2058	2313	539	550	588	254	257	263
0.4	•••	34602	36040	55702	9658	10552	14010	1964	2086	2281	558	571	588	263	263	266
	•••	35086	38572	59588	10007	11535	14950	2021	2172	2412	563	579	606	263	266	268
0.45	•••	33632	35460	58259	9626	10666	14614	2004	2142	2362	566	582	599	263	266	271
	•••	34252	38539	61319	10037	11784	15352	2066	2237	2461	574	591	613	266	271	268
0.5	•••	31844	34114	59478	9406	10573	14888	2004	2151	2391	566	582	599	263	263	266
	•••	32627	37730	61814	9869	11795	15445	2069	2256	2461	574	591	606	263	266	263
0.55	•••	29464	32210	59351	9038	10292	14823	1964	2118	2362	550	566	588	249	254	257
	•••	30434	36300	61072	9541	11589	15228	2034	2224	2412	558	579	588	254	257	252
0.6	•••	26718	29914	57877	8542	9853	14419	1891	2042	2281	526	542	558	238	241	246
	•••	27874	34371	59093	9073	11177	14703	1961	2148	2313	534	550	557	238	241	235
0.65	•••	23804	27368	55066	7949	9268	13680	1777	1923	2148	490	506	523	217	222	225
	•••	25133	32026	55788	8485	10573	13868	1847	2021	2164	498	509	514	217	222	213
0.7	•••	20887	24649	50917	7261	8542	12613	1631	1766	1961	441	452	466	192	192	198
	•••	22317	29326	51427	7786	9780	12724	1693	1850	1967	449	461	458	192	198	186
0.75	•••	18032	21794	45449	6489	7673	11215	1449	1563	1720	379	387	401	160	165	165
	•••	19489	26266	45739	6977	8786	11271	1501	1631	1719	384	396	390	160	165	153
0.8	•••	15245	18771	38672	5617	6644	9487	1224	1311	1433	306	314	319	124	124	127
	•••	16629	22821	38815	6048	7578	9509	1265	1368	1422	311	319	310	124	124	115
0.85	•••	12437	15492	30604	4612	5425	7437	956	1013	1086	222	225	230	.	.	.
	•••	13653	18869	30654	4959	6129	7438	986	1046	1076	222	225	217	.	.	.
0.9	•••	9428	11743	21249	3415	3946	5067	631	661	693	.	.	.	.	.	.
	•••	10362	14167	21256	3646	4374	5058	647	677	681	.	.	.	.	.	.
0.95	•••	5766	7006	10633	1863	2061	2378	.	.	.	.	.	.	.	.	.
	•••	6283	8155	10622	1953	2199	2368	.	.	.	.	.	.	.	.	.
0.98	•••	2619	2984	3659	.	.	.	.	.	.	.	.	.	.	.	.
	•••	2781	3256	3648	.	.	.	.	.	.	.	.	.	.	.	.

TABLE 7: ALPHA= 0.05 POWER= 0.8   EXPECTED ACCRUAL THRU MINIMUM FOLLOW-UP= 3500

		DEL=.01			DEL=.02			DEL=.05			DEL=.10			DEL=.15		
FACT=		1.0 .75	.50 .25	.00 BIN	1.0 .75	.50 .25	.00 BIN	1.0 .75	.50 .25	.00 BIN	1.0 75	.50 .25	.00 BIN	1.0 .75	.50 .25	.00 BIN
PCONT=***					REQUIRED NUMBER OF PATIENTS											
0.01	***	948	951	951	347	347	347	116	116	116	58	58	58	41	41	41
	***	948	951	3648	347	347	1206	116	116	328	58	58	134	41	41	80
0.02	***	2097	2106	2129	676	676	680	181	181	181	81	81	81	50	50	50
	***	2103	2115	6022	676	676	1793	181	181	419	81	81	155	50	50	89
0.05	***	6350	6408	6647	1808	1831	1861	391	391	396	137	137	143	81	81	81
	***	6367	6478	12848	1817	1843	3481	391	391	681	137	143	217	81	81	115
0.1	***	14011	14172	15473	3803	3888	4092	741	750	764	233	233	233	125	125	125
	***	14064	14435	23235	3838	3966	6047	746	755	1076	233	233	310	125	125	153
0.15	***	20945	21233	24436	5606	5795	6329	1065	1088	1117	318	318	321	160	160	163
	***	21041	21755	32385	5681	5979	8304	1079	1100	1422	318	321	390	160	163	186
0.2	***	26559	27008	32833	7099	7405	8406	1350	1388	1441	391	396	400	190	195	195
	***	26711	27840	40298	7216	7732	10251	1368	1411	1719	391	396	458	195	195	213
0.25	***	30689	31336	40308	8236	8686	10252	1581	1639	1721	452	458	466	216	221	221
	***	30902	32535	46976	8411	9173	11890	1607	1677	1967	452	461	514	221	221	235
0.3	***	33343	34236	46678	9036	9640	11815	1761	1840	1957	501	510	519	239	242	242
	***	33641	35869	52416	9269	10293	13219	1796	1893	2164	505	513	557	239	242	252
0.35	***	34635	35820	51829	9526	10284	13075	1893	1992	2146	536	545	557	251	256	260
	***	35032	37943	56620	9823	11106	14239	1940	2062	2313	540	554	588	256	256	263
0.4	***	34714	36254	55700	9741	10660	14006	1980	2097	2281	557	571	583	260	265	268
	***	35233	38908	59588	10103	11635	14950	2033	2176	2412	566	580	606	265	265	268
0.45	***	33778	35723	58255	9727	10786	14615	2018	2155	2360	571	583	601	265	268	268
	***	34443	38935	61319	10147	11897	15352	2080	2246	2461	575	592	613	265	268	268
0.5	***	32028	34437	59476	9523	10704	14886	2024	2167	2391	566	583	601	260	265	268
	***	32868	38188	61814	9995	11923	15445	2088	2263	2461	575	592	606	265	265	263
0.55	***	29691	32588	59348	9164	10436	14820	1983	2132	2365	554	571	589	251	256	260
	***	30724	36800	61072	9675	11722	15228	2050	2234	2412	563	580	588	256	256	252
0.6	***	26991	30339	57878	8677	9998	14417	1910	2059	2281	528	545	563	239	242	242
	***	28213	34901	59093	9211	11311	14703	1975	2155	2313	536	554	557	239	242	235
0.65	***	24126	27819	55066	8082	9412	13682	1796	1936	2150	493	505	519	216	221	225
	***	25509	32573	55878	8625	10704	13868	1861	2033	2164	496	513	514	221	221	213
0.7	***	21239	25110	50919	7396	8677	12611	1648	1773	1963	443	452	466	190	195	195
	***	22718	29861	51427	7925	9902	12724	1709	1858	1967	449	461	458	195	195	186
0.75	***	18395	22245	45450	6612	7799	11211	1464	1569	1721	382	391	400	160	163	163
	***	19891	26772	45739	7108	8896	11271	1511	1639	1719	388	396	390	163	163	153
0.8	***	15590	19186	38672	5725	6752	9488	1236	1318	1429	309	312	321	125	125	125
	***	17004	23269	38815	6157	7671	9509	1275	1371	1422	312	318	310	125	125	115
0.85	***	12743	15843	30601	4701	5506	7435	965	1018	1088	221	225	230	.	.	.
	***	13976	19235	30654	5043	6192	7438	991	1053	1076	225	225	217	.	.	.
0.9	***	9663	12002	21251	3476	3996	5063	636	662	694	.	.	.	.	.	.
	***	10608	14423	21256	3704	4413	5058	650	676	681	.	.	.	.	.	.
0.95	***	5900	7134	10634	1887	2080	2377	.	.	.	.	.	.	.	.	.
	***	6416	8266	10622	1975	2211	2368	.	.	.	.	.	.	.	.	.
0.98	***	2663	3016	3660	.	.	.	.	.	.	.	.	.	.	.	.
	***	2820	3278	3648	.	.	.	.	.	.	.	.	.	.	.	.

TABLE 7: ALPHA= 0.05 POWER= 0.8   EXPECTED ACCRUAL THRU MINIMUM FOLLOW-UP= 3750

	DEL=.01			DEL=.02			DEL=.05			DEL=.10			DEL=.15		
FACT=	1.0 .75	.50 .25	.00 BIN	1.0 .75	.50 .25	.00 BIN	1.0 .75	.50 .25	.00 BIN	1.0 75	.50 .25	.00 BIN	1.0 .75	.50 .25	.00 BIN
PCONT=•••						REQUIRED NUMBER OF PATIENTS									
0.01 •••	950	950	953	350	350	350	115	115	115	59	59	59	40	40	40
•••	950	950	3648	350	350	1206	115	115	328	59	59	134	40	40	80
0.02 •••	2097	2106	2125	672	678	678	181	181	184	81	81	81	53	53	53
•••	2103	2116	6022	678	678	1793	181	184	419	81	81	155	53	53	89
0.05 •••	6353	6409	6644	1812	1831	1863	391	391	397	138	138	138	81	81	81
•••	6372	6484	12848	1822	1844	3481	391	397	681	138	138	217	81	81	115
0.1 •••	14022	14191	15472	3809	3897	4090	743	753	762	231	231	237	125	125	125
•••	14078	14468	23235	3847	3972	6047	747	756	1076	231	237	310	125	125	153
0.15 •••	20965	21275	24438	5622	5815	6325	1072	1090	1118	316	322	322	162	162	162
•••	21068	21818	32385	5697	5997	8304	1081	1103	1422	316	322	390	162	162	186
0.2 •••	26590	27072	32834	7122	7437	8406	1353	1390	1441	391	397	400	194	194	194
•••	26753	27944	40298	7250	7759	10251	1372	1413	1719	391	397	458	194	194	213
0.25 •••	30734	31428	40309	8275	8734	10253	1587	1643	1722	453	456	466	218	218	222
•••	30968	32688	46976	8453	9218	11890	1615	1681	1967	456	462	514	218	222	235
0.3 •••	33409	34362	46675	9091	9706	11815	1769	1850	1962	500	509	518	237	241	241
•••	33728	36078	52416	9331	10356	13219	1806	1897	2164	503	513	557	241	241	252
0.35 •••	34718	35988	51831	9593	10366	13072	1906	2003	2144	537	547	559	250	256	259
•••	35150	38206	56620	9897	11187	14239	1947	2065	2313	541	550	588	256	256	263
0.4 •••	34825	36466	55703	9828	10756	14009	1994	2106	2281	559	575	588	259	265	269
•••	35384	39237	59588	10193	11731	14950	2047	2187	2412	565	578	606	265	265	268
0.45 •••	33922	35988	58259	9828	10900	14613	2037	2168	2359	569	584	597	265	269	269
•••	34634	39322	61319	10253	12006	15352	2097	2253	2461	578	593	613	265	269	268
0.5 •••	32209	34756	59478	9634	10825	14884	2037	2178	2388	569	584	603	259	265	269
•••	33100	38622	61072	10113	12040	15445	2103	2272	2461	578	593	606	265	265	263
0.55 •••	29918	32950	59350	9284	10563	14819	2000	2144	2365	556	569	588	250	256	259
•••	31006	37278	61072	9803	11843	15228	2065	2243	2412	565	578	588	256	256	252
0.6 •••	27265	30743	57878	8800	10131	14416	1925	2069	2281	531	547	559	237	241	247
•••	28544	35406	59093	9344	11434	14703	1990	2163	2313	537	550	557	241	241	235
0.65 •••	24438	28250	55062	8209	9547	13681	1812	1947	2150	494	503	518	218	222	222
•••	25872	33087	55878	8759	10825	13868	1878	2037	2164	500	513	514	218	222	213
0.7 •••	21575	25553	50918	7522	8806	12612	1662	1788	1962	443	456	466	194	194	194
•••	23103	30368	51427	8056	10015	12724	1722	1863	1967	447	462	458	194	194	186
0.75 •••	18738	22672	45453	6734	7915	11215	1475	1578	1722	381	391	400	162	162	166
•••	20272	27247	45739	7225	8993	11271	1525	1643	1719	387	397	390	162	162	153
0.8 •••	15916	19572	38675	5828	6850	9484	1244	1325	1431	306	316	322	125	125	125
•••	17359	23688	38815	6259	7750	9509	1281	1375	1422	312	316	310	125	125	115
0.85 •••	13028	16175	30603	4784	5581	7437	972	1025	1084	222	228	228	.	.	.
•••	14284	19572	30654	5125	6256	7438	997	1053	1076	222	228	217	.	.	.
0.9 •••	9884	12247	21250	3528	4043	5065	640	663	691	.	.	.	.	.	.
•••	10840	14659	21256	3753	4447	5058	653	678	681	.	.	.	.	.	.
0.95 •••	6022	7253	10634	1909	2093	2378	.	.	.	.	.	.	.	.	.
•••	6537	8369	10622	1994	2219	2368	.	.	.	.	.	.	.	.	.
0.98 •••	2703	3050	3659	.	.	.	.	.	.	.	.	.	.	.	.
•••	2853	3303	3648	.	.	.	.	.	.	.	.	.	.	.	.

TABLE 7: ALPHA= 0.05 POWER= 0.8   EXPECTED ACCRUAL THRU MINIMUM FOLLOW-UP≈ 4000

		DEL=.01			DEL=.02			DEL=.05			DEL=.10			DEL=.15		
FACT=		1.0 .75	.50 .25	.00 BIN	1.0 .75	.50 .25	.00 BIN	1.0 .75	.50 .25	.00 BIN	1.0 75	.50 .25	.00 BIN	1.0 .75	.50 .25	.00 BIN
PCONT=•••				REQUIRED NUMBER OF PATIENTS												
0.01	***	947	953	953	347	347	347	113	113	113	57	57	57	43	43	43
	•••	947	953	3648	347	347	1206	113	113	328	57	57	134	43	43	80
0.02	***	2097	2107	2127	673	677	677	183	183	183	77	77	77	53	53	53
	•••	2103	2117	6022	677	677	1793	183	183	419	77	77	155	53	53	89
0.05	***	6357	6417	6647	1813	1833	1863	393	393	393	137	137	143	83	83	83
	•••	6383	6493	12848	1823	1847	3481	393	393	681	137	143	217	83	83	115
0.1	***	14033	14213	15473	3817	3907	4093	743	753	763	233	233	233	123	123	123
	•••	14097	14497	23235	3853	3977	6047	747	757	1076	233	233	310	123	123	153
0.15	***	20987	21313	24437	5643	5833	6327	1073	1093	1117	317	317	323	163	163	163
	•••	21097	21877	32385	5717	6013	8304	1083	1103	1422	317	323	390	163	163	186
0.2	***	26623	27137	32833	7153	7473	8407	1357	1393	1443	393	397	397	193	193	193
	•••	26797	28047	40298	7277	7787	10251	1373	1413	1719	393	397	458	193	193	213
0.25	***	30783	31517	40307	8313	8783	10253	1593	1647	1723	453	457	467	217	223	223
	•••	31027	32837	46976	8497	9257	11890	1617	1683	1967	457	463	514	217	223	235
0.3	***	33473	34487	46677	9143	9767	11817	1777	1853	1957	503	507	517	237	243	243
	•••	33813	36277	52416	9387	10413	13219	1813	1903	2164	507	513	557	237	243	252
0.35	***	34807	36153	51833	9657	10443	13073	1917	2007	2147	537	547	557	253	257	257
	•••	35263	38463	56620	9967	11257	14223	1957	2073	2313	543	553	588	253	257	263
0.4	***	34937	36673	55703	9903	10847	14007	2003	2117	2283	563	573	587	263	263	267
	•••	35533	39547	59588	10283	11813	14950	2057	2187	2412	567	577	606	263	267	268
0.45	***	34063	36247	58257	9917	11003	14613	2053	2177	2363	573	587	597	263	267	273
	•••	34817	39687	61319	10353	12103	15352	2107	2257	2461	577	593	613	267	267	268
0.5	***	32387	35063	59477	9737	10943	14887	2053	2187	2387	573	583	603	263	263	267
	•••	33333	39037	61814	10223	12143	15445	2117	2277	2461	577	593	606	263	267	263
0.55	***	30143	33303	59347	9397	10687	14823	2017	2157	2363	557	573	587	253	257	257
	•••	31283	37733	61072	9923	11953	15228	2083	2247	2412	563	577	588	253	257	252
0.6	***	27527	31133	57877	8923	10257	14417	1943	2077	2283	533	547	563	237	243	243
	•••	28863	35887	59093	9473	11547	14703	2003	2167	2313	537	553	557	237	243	235
0.65	***	24743	28663	55063	8333	9667	13683	1827	1957	2147	497	507	523	217	223	223
	•••	26227	33573	55878	8883	10937	13868	1887	2043	2164	503	513	514	217	223	213
0.7	***	21903	25973	50917	7643	8923	12613	1677	1797	1963	447	457	467	193	193	197
	•••	23467	30843	51427	8173	10117	12724	1733	1873	1967	453	463	458	193	197	186
0.75	***	19067	23073	45453	6843	8023	11213	1487	1587	1723	383	393	403	163	163	163
	•••	20637	27693	45739	7337	9083	11271	1533	1647	1719	387	397	390	163	163	153
0.8	***	16233	19943	38673	5927	6943	9487	1257	1333	1433	307	317	323	123	127	127
	•••	17703	24083	38815	6353	7827	9509	1293	1377	1422	313	317	310	123	127	115
0.85	***	13303	16483	30603	4863	5653	7437	977	1027	1087	223	227	227	.	.	.
	•••	14577	19893	30654	5203	6307	7438	1003	1053	1076	223	227	217	.	.	.
0.9	***	10097	12473	21253	3583	4087	5067	643	667	693	.	.	.	.	.	.
	•••	11063	14877	21256	3803	4473	5058	653	677	681	.	.	.	.	.	.
0.95	***	6137	7367	10633	1927	2107	2377	.	.	.	.	.	.	.	.	.
	•••	6653	8463	10622	2013	2227	2368	.	.	.	.	.	.	.	.	.
0.98	***	2737	3077	3663	.	.	.	.	.	.	.	.	.	.	.	.
	•••	2887	3317	3648	.	.	.	.	.	.	.	.	.	.	.	.

TABLE 7: ALPHA= 0.05 POWER= 0.8   EXPECTED ACCRUAL THRU MINIMUM FOLLOW-UP= 4250

		DEL=.01			DEL=.02			DEL=.05			DEL=.10			DEL=.15	
FACT=	1.0 .75	.50 .25	.00 BIN	1.0 .75	.50 .25	.00 BIN	1.0 .75	.50 .25	.00 BIN	1.0 75	.50 .25	.00 BIN	1.0 .75	.50 .25	.00 BIN
PCONT=•••						REQUIRED NUMBER OF PATIENTS									
0.01 •••	949	949	953	347	347	347	113	113	113	56	56	56	39	39	39
•••	949	953	3648	347	347	1206	113	113	328	56	56	134	39	39	80
0.02 •••	2100	2111	2128	677	677	677	184	184	184	82	82	82	50	50	50
•••	2107	2118	6022	677	677	1793	184	184	419	82	82	155	50	50	89
0.05 •••	6361	6425	6648	1813	1835	1863	390	390	396	141	141	141	82	82	82
•••	6382	6499	12848	1824	1845	3481	390	396	681	141	141	217	82	82	115
0.1 •••	14043	14234	15473	3828	3913	4094	747	751	762	230	237	237	124	124	124
•••	14113	14521	23235	3864	3981	6047	747	758	1076	230	237	310	124	124	153
0.15 •••	21009	21353	24441	5656	5847	6329	1076	1091	1119	315	322	322	163	163	163
•••	21126	21937	32385	5734	6028	8304	1080	1102	1422	315	322	390	163	163	186
0.2 •••	26655	27197	32834	7175	7498	8408	1363	1395	1442	390	396	400	194	194	194
•••	26835	28142	40298	7307	7813	10251	1378	1416	1719	396	396	458	194	194	213
0.25 •••	30826	31613	40308	8354	8822	10250	1597	1650	1724	453	460	464	220	220	220
•••	31092	32983	46976	8539	9300	11890	1622	1686	1967	453	460	514	220	220	235
0.3 •••	33536	34609	46679	9194	9825	11818	1788	1856	1958	503	513	517	241	241	241
•••	33901	36468	52416	9442	10462	13219	1820	1905	2164	507	513	557	241	241	252
0.35 •••	34889	36313	51832	9725	10515	13072	1926	2015	2143	538	549	560	252	258	258
•••	35374	38710	56620	10037	11323	14239	1969	2075	2313	545	556	588	252	258	263
0.4 •••	35048	36883	55703	9980	10936	14007	2015	2122	2281	560	577	588	262	262	269
•••	35675	39847	59588	10363	11893	14950	2064	2196	2412	566	581	606	262	269	268
0.45 •••	34205	36500	58260	10005	11100	14613	2064	2185	2362	577	588	598	262	269	269
•••	35002	40038	61319	10448	12194	15352	2118	2266	2461	581	592	613	269	269	268
0.5 •••	32569	35363	59475	9835	11053	14882	2068	2196	2387	570	588	598	262	262	269
•••	33561	39433	61814	10331	12243	15445	2128	2281	2461	581	592	606	262	269	263
0.55 •••	30363	33646	59348	9506	10802	14818	2033	2164	2362	560	570	588	252	258	258
•••	31553	38162	61072	10037	12056	15228	2090	2256	2412	566	581	588	252	258	252
0.6 •••	27792	31510	57878	9034	10377	14421	1958	2090	2281	534	545	560	237	241	241
•••	29173	36341	59093	9587	11652	14703	2015	2175	2313	538	556	557	241	241	235
0.65 •••	25036	29063	55062	8450	9789	13682	1841	1969	2149	496	507	517	220	220	226
•••	26566	34035	55878	9003	11036	13868	1898	2047	2164	503	513	514	220	220	213
0.7 •••	22213	26368	50918	7753	9034	12615	1693	1803	1962	449	453	464	194	194	198
•••	23824	31294	51427	8284	10207	12724	1739	1873	1967	449	460	458	194	194	186
0.75 •••	19387	23457	45450	6945	8121	11213	1495	1590	1724	386	390	400	163	163	163
•••	20981	28110	45739	7441	9166	11271	1544	1650	1719	390	396	390	163	163	153
0.8 •••	16529	20290	38672	6017	7026	9485	1261	1335	1431	311	315	322	124	124	124
•••	18023	24451	38815	6446	7891	9509	1299	1378	1422	311	315	310	124	124	115
0.85 •••	13565	16773	30603	4933	5713	7434	981	1027	1087	226	226	230	.	.	.
•••	14850	20191	30654	5267	6357	7438	1006	1055	1076	226	226	217	.	.	.
0.9 •••	10299	12683	21253	3630	4126	5065	645	666	694	.	.	.	.	.	.
•••	11270	15080	21256	3849	4502	5058	655	677	681	.	.	.	.	.	.
0.95 •••	6244	7473	10632	1948	2122	2377	.	.	.	.	.	.	.	.	.
•••	6761	8550	10622	2026	2234	2368	.	.	.	.	.	.	.	.	.
0.98 •••	2770	3099	3658	.	.	.	.	.	.	.	.	.	.	.	.
•••	2918	3333	3648	.	.	.	.	.	.	.	.	.	.	.	.

TABLE 7: ALPHA= 0.05 POWER= 0.8   EXPECTED ACCRUAL THRU MINIMUM FOLLOW-UP= 4500

		DEL=.01			DEL=.02			DEL=.05			DEL=.10			DEL=.15		
FACT=		1.0 .75	.50 .25	.00 BIN	1.0 .75	.50 .25	.00 BIN	1.0 .75	.50 .25	.00 BIN	1.0 75	.50 .25	.00 BIN	1.0 .75	.50 .25	.00 BIN
PCONT=•••						REQUIRED NUMBER OF PATIENTS										
0.01	•••	948	948	953	345	352	352	116	116	116	60	60	60	41	41	41
	•••	948	953	3648	345	352	1206	116	116	328	60	60	134	41	41	80
0.02	•••	2100	2111	2130	678	678	678	183	183	183	82	82	82	53	53	53
	•••	2107	2118	6022	678	678	1793	183	183	419	82	82	155	53	53	89
0.05	•••	6364	6431	6645	1815	1837	1864	390	390	397	138	138	138	82	82	82
	•••	6393	6506	12848	1826	1848	3481	390	397	681	138	138	217	82	82	115
0.1	•••	14059	14257	15472	3833	3918	4091	746	750	761	233	233	233	127	127	127
	•••	14126	14550	23235	3866	3990	6047	750	757	1076	233	233	310	127	127	153
0.15	•••	21030	21394	24438	5673	5865	6326	1076	1095	1117	318	318	323	161	161	161
	•••	21153	21997	32385	5752	6038	8304	1083	1106	1422	318	323	390	161	161	186
0.2	•••	26688	27262	32835	7203	7530	8407	1365	1398	1443	390	397	397	195	195	195
	•••	26880	28241	40298	7331	7838	10251	1380	1414	1719	390	397	458	195	195	213
0.25	•••	30873	31699	40312	8389	8861	10252	1605	1657	1725	453	458	465	217	221	221
	•••	31155	33123	46976	8580	9334	11890	1628	1684	1967	458	465	514	221	221	235
0.3	•••	33600	34732	46680	9240	9881	11816	1792	1864	1961	503	510	521	240	240	244
	•••	33983	36656	52416	9491	10515	13219	1826	1909	2164	510	514	557	240	244	252
0.35	•••	34980	36476	51832	9784	10583	13076	1931	2021	2145	543	548	559	255	255	255
	•••	35486	38944	56620	10106	11388	14239	1976	2078	2313	543	555	588	255	255	263
0.4	•••	35164	37083	55702	10054	11017	14010	2028	2130	2280	566	577	588	262	266	266
	•••	35823	40136	59588	10443	11966	14950	2073	2197	2412	570	581	606	262	266	268
0.45	•••	34350	36746	58256	10095	11197	14617	2073	2190	2359	577	588	600	266	266	273
	•••	35186	40373	61319	10538	12277	15352	2130	2269	2461	581	593	613	266	266	268
0.5	•••	32745	35655	59475	9930	11152	14887	2078	2208	2388	577	588	600	262	266	266
	•••	33787	39810	61814	10432	12333	15445	2141	2287	2461	581	593	606	262	266	263
0.55	•••	30581	33978	59347	9611	10909	14820	2044	2175	2366	559	570	588	255	255	262
	•••	31823	38573	61072	10144	12153	15228	2107	2258	2412	566	581	588	255	255	252
0.6	•••	28050	31875	57878	9143	10488	14419	1965	2096	2280	536	548	559	240	240	244
	•••	29478	36768	59093	9701	11748	14703	2028	2179	2313	543	555	557	240	244	235
0.65	•••	25320	29438	55065	8558	9896	13683	1853	1976	2145	498	510	521	217	221	221
	•••	26895	34473	55878	9116	11130	13868	1909	2055	2164	503	514	514	221	221	213
0.7	•••	22519	26749	50921	7856	9138	12615	1702	1808	1961	446	458	465	195	195	195
	•••	24161	31717	51427	8396	10297	12724	1751	1882	1967	453	465	458	195	195	186
0.75	•••	19691	23824	45453	7046	8216	11213	1504	1601	1725	386	390	401	161	165	165
	•••	21315	28511	45739	7534	9244	11271	1549	1657	1719	390	397	390	161	165	153
0.8	•••	16822	20625	38674	6101	7102	9487	1268	1342	1432	311	318	323	127	127	127
	•••	18334	24798	38815	6528	7957	9509	1301	1380	1422	311	318	310	127	127	115
0.85	•••	13818	17051	30603	5003	5775	7433	986	1031	1088	221	228	228	.	.	.
	•••	15116	20467	30654	5336	6398	7438	1009	1061	1076	221	228	217	.	.	.
0.9	•••	10488	12885	21248	3675	4159	5066	649	667	694	.	.	.	.	.	.
	•••	11467	15263	21256	3889	4526	5058	656	678	681	.	.	.	.	.	.
0.95	•••	6348	7568	10635	1965	2134	2381	.	.	.	.	.	.	.	.	.
	•••	6866	8625	10622	2040	2242	2368	.	.	.	.	.	.	.	.	.
0.98	•••	2798	3124	3660	.	.	.	.	.	.	.	.	.	.	.	.
	•••	2944	3349	3648	.	.	.	.	.	.	.	.	.	.	.	.

TABLE 7: ALPHA= 0.05 POWER= 0.8   EXPECTED ACCRUAL THRU MINIMUM FOLLOW-UP= 4750

		DEL=.01			DEL=.02			DEL=.05			DEL=.10			DEL=.15		
FACT=		1.0 .75	.50 .25	.00 BIN	1.0 .75	.50 .25	.00 BIN	1.0 .75	.50 .25	.00 BIN	1.0 75	.50 .25	.00 BIN	1.0 .75	.50 .25	.00 BIN

PCONT=•••   REQUIRED NUMBER OF PATIENTS

PCONT		DEL=.01 1.0/.75	.50/.25	.00/BIN	DEL=.02 1.0/.75	.50/.25	.00/BIN	DEL=.05 1.0/.75	.50/.25	.00/BIN	DEL=.10 1.0/.75	.50/.25	.00/BIN	DEL=.15 1.0/.75	.50/.25	.00/BIN
0.01	***	946	953	953	348	348	348	115	115	115	55	55	55	39	39	39
	•••	946	953	3648	348	348	1206	115	115	328	55	55	134	39	39	80
0.02	***	2098	2110	2129	673	680	680	182	182	182	79	79	79	51	51	51
	•••	2105	2117	6022	673	680	1793	182	182	419	79	79	155	51	51	89
0.05	***	6368	6440	6646	1820	1837	1860	388	395	395	139	139	139	79	79	79
	•••	6397	6511	12848	1825	1849	3481	395	395	681	139	139	217	79	79	115
0.1	***	14068	14277	15469	3839	3927	4093	744	752	763	234	234	234	122	122	122
	•••	14139	14574	23235	3875	3993	6047	752	756	1076	234	234	310	122	122	153
0.15	***	21050	21430	24435	5684	5882	6325	1077	1096	1120	317	317	324	162	162	162
	•••	21181	22048	32385	5767	6048	8304	1084	1108	1422	317	324	390	162	162	186
0.2	***	26722	27320	32830	7223	7556	8411	1369	1397	1440	395	395	400	193	193	193
	•••	26924	28330	40298	7359	7857	10251	1381	1417	1719	395	395	458	193	193	213
0.25	***	30919	31785	40307	8423	8902	10252	1607	1659	1725	455	459	467	217	222	222
	•••	31215	33253	46976	8617	9365	11890	1630	1690	1967	455	459	514	222	222	235
0.3	***	33662	34857	46677	9282	9931	11812	1801	1868	1963	502	514	519	241	241	241
	•••	34065	36832	52416	9544	10560	13219	1832	1908	2164	507	514	557	241	241	252
0.35	***	35063	36630	51830	9840	10648	13070	1944	2027	2145	542	550	562	253	257	257
	•••	35597	39179	56620	10168	11444	14239	1979	2082	2313	542	554	588	253	257	263
0.4	***	35272	37279	55702	10125	11095	14009	2034	2141	2283	566	573	585	265	265	269
	•••	35965	40414	59588	10517	12037	14950	2082	2200	2412	566	578	606	265	265	268
0.45	***	34488	36987	58255	10173	11285	14614	2086	2200	2359	578	585	602	265	269	269
	•••	35367	40699	61319	10624	12353	15352	2141	2272	2461	585	590	613	265	269	268
0.5	***	32925	35937	59473	10026	11249	14883	2093	2212	2390	573	585	602	265	265	269
	•••	34006	40177	61814	10525	12417	15445	2145	2295	2461	578	590	606	265	265	263
0.55	***	30795	34298	59347	9705	11012	14823	2058	2181	2359	562	573	585	253	257	257
	•••	32082	38970	61072	10244	12239	15228	2117	2264	2412	566	578	588	253	257	252
0.6	***	28294	32220	57875	9247	10589	14420	1979	2105	2283	538	550	562	241	241	245
	•••	29774	37184	59093	9805	11843	14703	2039	2188	2313	542	554	557	241	241	235
0.65	***	25599	29810	55060	8660	9995	13683	1868	1987	2145	495	507	519	217	222	222
	•••	27214	34885	55878	9218	11218	13868	1920	2058	2164	502	514	514	222	222	213
0.7	***	22815	27119	50916	7960	9235	12615	1713	1820	1963	447	459	467	193	193	198
	•••	24490	32118	51427	8494	10375	12724	1761	1884	1967	455	459	458	193	198	186
0.75	***	19989	24174	45449	7133	8297	11213	1516	1607	1718	388	395	400	162	162	162
	•••	21640	28883	45739	7627	9313	11271	1559	1659	1719	388	395	390	162	162	153
0.8	***	17096	20944	38673	6183	7176	9484	1279	1345	1428	312	317	324	127	127	127
	•••	18628	25124	38815	6606	8012	9509	1310	1385	1422	312	317	310	127	127	115
0.85	***	14056	17317	30605	5067	5827	7437	989	1037	1089	222	229	229	.	.	.
	•••	15370	20730	30654	5395	6440	7438	1013	1060	1076	222	229	217	.	.	.
0.9	***	10667	13078	21252	3713	4195	5067	649	668	692	.	.	.	.	.	.
	•••	11653	15441	21256	3927	4552	5058	661	680	681	.	.	.	.	.	.
0.95	***	6444	7655	10632	1979	2145	2378	.	.	.	.	.	.	.	.	.
	•••	6962	8700	10622	2058	2248	2368	.	.	.	.	.	.	.	.	.
0.98	***	2830	3143	3661	.	.	.	.	.	.	.	.	.	.	.	.
	•••	2972	3364	3648	.	.	.	.	.	.	.	.	.	.	.	.

TABLE 7: ALPHA= 0.05 POWER= 0.8   EXPECTED ACCRUAL THRU MINIMUM FOLLOW-UP= 5000

		DEL=.01			DEL=.02			DEL=.05			DEL=.10			DEL=.15	
FACT=	1.0 .75	.50 .25	.00 BIN	1.0 .75	.50 .25	.00 BIN	1.0 .75	.50 .25	.00 BIN	1.0 75	.50 .25	.00 BIN	1.0 .75	.50 .25	.00 BIN
PCONT=•••				REQUIRED NUMBER OF PATIENTS											
0.01 •••	946	954	954	346	346	346	116	116	116	58	58	58	41	41	41
•••	946	954	3648	346	346	1206	116	116	328	58	58	134	41	41	80
0.02 •••	2104	2108	2129	679	679	679	183	183	183	79	79	79	54	54	54
•••	2108	2121	6022	679	679	1793	183	183	419	79	79	155	54	54	89
0.05 •••	6371	6441	6646	1821	1833	1858	391	391	396	141	141	141	83	83	83
•••	6404	6516	12848	1829	1846	3481	391	396	681	141	141	217	83	83	115
0.1 •••	14079	14296	15471	3846	3933	4091	746	754	758	233	233	233	121	121	121
•••	14154	14596	23235	3883	3996	6047	754	758	1076	233	233	310	121	121	153
0.15 •••	21071	21471	24441	5696	5891	6329	1079	1096	1116	316	321	321	158	158	158
•••	21208	22104	32385	5779	6058	8304	1083	1104	1422	321	321	390	158	158	186
0.2 •••	26754	27383	32833	7246	7579	8408	1371	1404	1441	391	396	396	191	196	196
•••	26966	28421	40298	7383	7879	10251	1383	1421	1719	396	396	458	191	196	213
0.25 •••	30966	31879	40308	8454	8941	10254	1616	1658	1721	454	458	466	221	221	221
•••	31279	33383	46976	8654	9396	11890	1633	1691	1967	458	458	514	221	221	235
0.3 •••	33729	34971	46679	9329	9979	11816	1804	1871	1958	504	508	516	241	241	241
•••	34154	37008	52416	9591	10604	13219	1833	1908	2164	508	516	557	241	241	252
0.35 •••	35146	36783	51829	9896	10708	13071	1946	2033	2146	541	546	558	254	254	258
•••	35708	39396	56620	10229	11496	14239	1991	2083	2313	546	554	588	254	258	263
0.4 •••	35383	37479	55704	10191	11166	14008	2046	2146	2279	566	579	583	266	266	266
•••	36108	40683	59588	10591	12104	14950	2091	2204	2412	571	579	606	266	266	268
0.45 •••	34633	37221	58258	10254	11366	14616	2096	2208	2358	579	591	596	266	266	271
•••	35546	41008	61319	10708	12429	15352	2146	2279	2461	583	596	613	266	271	268
0.5 •••	33104	36216	59479	10116	11341	14883	2104	2221	2391	579	591	604	266	266	266
•••	34221	40521	61814	10616	12496	15445	2158	2296	2461	583	596	606	266	266	263
0.55 •••	31004	34608	59346	9804	11108	14821	2066	2191	2366	566	571	583	254	258	258
•••	32333	39346	61072	10341	12329	15228	2121	2266	2412	566	579	588	254	258	252
0.6 •••	28541	32558	57879	9346	10691	14416	1991	2116	2283	541	546	558	241	241	246
•••	30058	37571	59093	9904	11921	14703	2046	2191	2313	541	554	557	241	241	235
0.65 •••	25871	30158	55066	8758	10096	13683	1879	1991	2146	496	508	521	221	221	221
•••	27521	35283	55878	9316	11296	13868	1929	2066	2164	504	516	514	221	221	213
0.7 •••	23104	27471	50916	8054	9329	12608	1721	1821	1958	446	458	466	191	196	196
•••	24808	32504	51427	8591	10454	12724	1766	1883	1967	454	458	458	196	196	186
0.75 •••	20271	24508	45454	7221	8383	11216	1521	1608	1721	383	391	404	158	166	166
•••	21946	29241	45739	7716	9379	11271	1566	1658	1719	391	396	390	166	166	153
0.8 •••	17358	21246	38671	6258	7246	9483	1283	1346	1429	308	316	321	121	129	129
•••	18908	25433	38815	6683	8066	9509	1316	1383	1422	316	316	310	129	129	115
0.85 •••	14283	17571	30604	5129	5879	7433	996	1033	1083	221	229	229	.	.	.
•••	15608	20979	30654	5454	6479	7438	1016	1058	1076	229	229	217	.	.	.
0.9 •••	10841	13254	21254	3754	4229	5066	654	671	691	.	.	.	.	.	.
•••	11833	15608	21256	3966	4571	5058	658	679	681	.	.	.	.	.	.
0.95 •••	6541	7741	10633	1991	2154	2379	.	.	.	.	.	.	.	.	.
•••	7046	8766	10622	2066	2254	2368	.	.	.	.	.	.	.	.	.
0.98 •••	2854	3166	3658	.	.	.	.	.	.	.	.	.	.	.	.
•••	2996	3379	3648	.	.	.	.	.	.	.	.	.	.	.	.

TABLE 7: ALPHA= 0.05 POWER= 0.8   EXPECTED ACCRUAL THRU MINIMUM FOLLOW-UP= 5500

FACT=	DEL=.01			DEL=.02			DEL=.05			DEL=.10			DEL=.15		
	1.0 .75	.50 .25	.00 BIN	1.0 .75	.50 .25	.00 BIN	1.0 .75	.50 .25	.00 BIN	1.0 75	.50 .25	.00 BIN	1.0 .75	.50 .25	.00 BIN
PCONT=***				REQUIRED NUMBER OF PATIENTS											
0.01 ***	953	953	953	348	348	348	114	114	114	59	59	59	37	37	37
***	953	953	3648	348	348	1206	114	114	328	59	59	134	37	37	80
0.02 ***	2099	2113	2127	678	678	678	183	183	183	78	78	78	50	50	50
***	2108	2121	6022	678	678	1793	183	183	419	78	78	155	50	50	89
0.05 ***	6384	6453	6645	1824	1838	1860	394	394	394	141	141	141	78	78	78
***	6411	6521	12848	1833	1852	3481	394	394	681	141	141	217	78	78	115
0.1 ***	14103	14331	15473	3859	3942	4093	746	752	760	229	238	238	128	128	128
***	14185	14639	23235	3895	4005	6047	752	760	1076	238	238	310	128	128	153
0.15 ***	21110	21542	24438	5724	5916	6329	1082	1095	1118	320	320	320	160	160	160
***	21261	22202	32385	5806	6081	8304	1090	1104	1422	320	320	390	160	160	186
0.2 ***	26816	27504	32830	7291	7621	8410	1379	1406	1439	394	394	403	196	196	196
***	27050	28590	40298	7429	7915	10251	1393	1426	1719	394	394	458	196	196	213
0.25 ***	31057	32047	40310	8520	9010	10253	1618	1668	1723	458	458	463	215	224	224
***	31400	33628	46976	8721	9455	11890	1640	1695	1967	458	463	514	224	224	235
0.3 ***	33856	35204	46677	9414	10069	11815	1819	1879	1962	504	513	518	238	243	243
***	34324	37340	52416	9684	10679	13219	1846	1915	2164	513	513	557	243	243	252
0.35 ***	35319	37088	51828	10005	10825	13072	1962	2039	2149	545	554	559	257	257	257
***	35933	39815	56620	10344	11595	14239	1998	2085	2313	545	554	588	257	257	263
0.4 ***	35603	37858	55705	10322	11298	14007	2058	2154	2278	568	573	587	265	265	265
***	36392	41185	59588	10720	12214	14950	2108	2209	2412	573	581	606	265	265	268
0.45 ***	34907	37679	58254	10399	11518	14612	2113	2218	2360	581	587	600	265	270	270
***	35905	41598	61319	10858	12558	15352	2163	2286	2461	581	595	613	265	270	268
0.5 ***	33444	36744	59478	10275	11504	14887	2121	2237	2388	581	587	600	265	265	265
***	34645	41177	61814	10784	12640	15445	2176	2305	2461	581	595	606	265	265	263
0.55 ***	31423	35204	59349	9978	11284	14818	2085	2204	2360	568	573	587	257	257	257
***	32830	40049	61072	10523	12475	15228	2140	2278	2412	568	581	588	257	257	252
0.6 ***	29016	33202	57878	9533	10872	14419	2011	2127	2278	540	545	559	238	243	243
***	30611	38311	59093	10088	12076	14703	2066	2195	2313	545	554	557	243	243	235
0.65 ***	26395	30823	55064	8947	10275	13685	1893	2003	2149	499	513	518	215	224	224
***	28109	36020	55878	9505	11444	13868	1943	2072	2164	504	513	514	224	224	213
0.7 ***	23645	28136	50920	8232	9491	12613	1736	1833	1962	449	458	463	196	196	196
***	25405	33215	51427	8763	10591	12724	1783	1893	1967	458	463	458	196	196	186
0.75 ***	20813	25144	45453	7388	8534	11210	1535	1618	1723	389	394	403	160	160	160
***	22532	29902	45739	7874	9497	11271	1577	1668	1719	389	394	390	160	160	153
0.8 ***	17865	21811	38674	6398	7374	9483	1296	1357	1434	312	320	320	128	128	128
***	19446	26010	38815	6815	8163	9509	1324	1393	1422	312	320	310	128	128	115
0.85 ***	14716	18044	30603	5234	5971	7434	999	1040	1090	224	224	229	.	.	.
***	16064	21432	30654	5559	6549	7438	1021	1063	1076	224	229	217	.	.	.
0.9 ***	11169	13589	21248	3826	4280	5064	655	669	691	.	.	.	.	.	.
***	12164	15913	21256	4033	4610	5058	664	683	681	.	.	.	.	.	.
0.95 ***	6705	7896	10633	2017	2168	2383	.	.	.	.	.	.	.	.	.
***	7214	8886	10622	2085	2264	2368	.	.	.	.	.	.	.	.	.
0.98 ***	2905	3199	3661	.	.	.	.	.	.	.	.	.	.	.	.
***	3034	3400	3648	.	.	.	.	.	.	.	.	.	.	.	.

TABLE 7: ALPHA= 0.05 POWER= 0.8   EXPECTED ACCRUAL THRU MINIMUM FOLLOW-UP= 6000

	DEL=.01			DEL=.02			DEL=.05			DEL=.10			DEL=.15		
FACT=	1.0 .75	.50 .25	.00 BIN	1.0 .75	.50 .25	.00 BIN	1.0 .75	.50 .25	.00 BIN	1.0 75	.50 .25	.00 BIN	1.0 .75	.50 .25	.00 BIN
PCONT=•••				REQUIRED NUMBER OF PATIENTS											
0.01 •••	949 949	949 949	955 3648	349 349	349 349	349 1206	115 115	115 115	115 328	55 55	55 55	55 134	40 40	40 40	40 80
0.02 •••	2104 2110	2110 2119	2125 6022	679 679	679 679	679 1793	184 184	184 184	184 419	79 79	79 79	79 155	49 49	49 49	49 89
0.05 •••	6394 6415	6460 6529	6649 12848	1825 1834	1840 1849	1864 3481	394 394	394 394	394 681	139 139	139 139	139 217	79 79	79 79	79 115
0.1 •••	14125 14215	14365 14680	15469 23235	3865 3904	3949 4009	4090 6047	745 754	754 760	760 1076	235 235	235 235	235 310	124 124	124 124	124 153
0.15 •••	21154 21310	21619 22294	24439 32385	5749 5830	5944 6094	6325 8304	1084 1090	1099 1105	1114 1422	319 319	319 319	319 390	160 160	160 160	160 186
0.2 •••	26884 27139	27619 28744	32830 40298	7330 7474	7660 7945	8410 10251	1375 1390	1405 1420	1444 1719	394 394	394 400	400 458	190 190	190 190	190 213
0.25 •••	31150 31519	32215 33859	40309 46976	8575 8779	9064 9499	10249 11890	1630 1645	1669 1699	1720 1967	454 460	460 460	469 514	220 220	220 220	220 235
0.3 •••	33985 34489	35434 37654	46675 52416	9490 9769	10150 10744	11815 13219	1825 1855	1885 1915	1960 2164	505 505	514 514	520 557	244 244	244 244	244 252
0.35 •••	35485 36154	37384 40210	51829 56620	10105 10444	10930 11680	13075 14239	1975 2005	2050 2095	2149 2313	544 550	550 559	559 588	250 259	259 259	259 263
0.4 •••	35824 36670	38224 41659	55699 59588	10444 10849	11425 12319	14005 14950	2074 2119	2164 2215	2284 2412	565 574	574 580	589 606	265 265	265 265	265 268
0.45 •••	35185 36244	38119 42139	58255 61319	10540 11005	11659 12679	14614 15352	2125 2179	2230 2290	2359 2461	580 589	589 595	595 613	265 265	265 265	274 268
0.5 •••	33784 35059	37249 41779	59479 61814	10429 10939	11659 12769	14884 15445	2140 2185	2245 2314	2389 2461	580 580	589 595	604 606	265 265	265 265	265 263
0.55 •••	31819 33304	35770 40699	59350 61072	10144 10690	11440 12610	14824 15228	2104 2155	2215 2284	2365 2412	565 574	574 580	589 588	250 259	259 259	259 252
0.6 •••	29479 31135	33805 38980	57874 59093	9700 10255	11029 12205	14419 14703	2029 2080	2140 2200	2284 2313	544 544	550 559	559 557	244 244	244 244	244 235
0.65 •••	26890 28660	31444 36694	55060 55878	9115 9670	10429 11575	13684 13868	1909 1960	2014 2074	2149 2164	505 505	514 514	520 514	220 220	220 220	220 213
0.7 •••	24160 25969	28750 33874	50920 51427	8395 8920	9640 10705	12610 12724	1750 1795	1840 1900	1960 1967	454 454	460 460	469 458	190 190	199 199	199 186
0.75 •••	21319 23074	25729 30499	45454 45739	7534 8020	8665 9604	11215 11271	1549 1585	1624 1669	1720 1719	385 394	394 394	400 390	160 160	160 160	160 153
0.8 •••	18334 19945	22339 26530	38674 38815	6529 6940	7480 8245	9484 9509	1300 1330	1360 1390	1429 1422	310 319	319 319	319 310	124 124	124 124	124 115
0.85 •••	15115 16480	18469 21844	30604 30654	5335 5650	6055 6604	7435 7438	1009 1024	1045 1060	1084 1076	220 229	229 229	229 217	. .	. .	. .
0.9 •••	11464 12469	13894 16180	21250 21256	3889 4084	4330 4639	5065 5058	655 664	670 685	694 681	. .	. .	. .	. .	. .	. .
0.95 •••	6865 7369	8035 8995	10630 10622	2044 2110	2185 2275	2380 2368	. .	. .	. .	. .	. .	. .	. .	. .	. .
0.98 •••	2944 3079	3229 3415	3664 3648	. .	. .	. .	. .	. .	. .	. .	. .	. .	. .	. .	. .

TABLE 7: ALPHA= 0.05 POWER= 0.8   EXPECTED ACCRUAL THRU MINIMUM FOLLOW-UP= 6500

	DEL=.01			DEL=.02			DEL=.05			DEL=.10			DEL=.15		
FACT=	1.0 .75	.50 .25	.00 BIN	1.0 .75	.50 .25	.00 BIN	1.0 .75	.50 .25	.00 BIN	1.0 75	.50 .25	.00 BIN	1.0 .75	.50 .25	.00 BIN

PCONT=●●●   REQUIRED NUMBER OF PATIENTS

PCONT	DEL=.01 1.0/.75	.50/.25	.00/BIN	DEL=.02 1.0/.75	.50/.25	.00/BIN	DEL=.05 1.0/.75	.50/.25	.00/BIN	DEL=.10 1.0/.75	.50/.25	.00/BIN	DEL=.15 1.0/.75	.50/.25	.00/BIN
0.01 ●●●	947	953	953	346	346	346	118	118	118	59	59	59	43	43	43
●●●	947	953	3648	346	346	1206	118	118	328	59	59	134	43	43	80
0.02 ●●●	2107	2117	2123	677	677	677	183	183	183	76	76	76	53	53	53
●●●	2107	2123	6022	677	677	1793	183	183	419	76	76	155	53	53	89
0.05 ●●●	6397	6472	6651	1825	1841	1863	395	395	395	141	141	141	86	86	86
●●●	6429	6537	12848	1831	1847	3481	395	395	681	141	141	217	86	86	115
0.1 ●●●	14148	14402	15475	3878	3959	4089	752	752	758	232	232	232	124	124	124
●●●	14240	14717	23235	3911	4018	6047	752	758	1076	232	232	310	124	124	153
0.15 ●●●	21195	21688	24434	5773	5958	6326	1083	1099	1116	319	319	319	157	157	157
●●●	21363	22381	32385	5855	6115	8304	1093	1110	1422	319	319	390	157	157	186
0.2 ●●●	26947	27727	32836	7372	7697	8406	1386	1408	1441	395	395	401	189	189	189
●●●	27223	28887	40298	7512	7973	10251	1392	1424	1719	395	395	458	189	189	213
0.25 ●●●	31243	32381	40311	8633	9121	10248	1636	1678	1727	460	460	466	222	222	222
●●●	31643	34071	46976	8834	9543	11890	1652	1695	1967	460	460	514	222	222	235
0.3 ●●●	34113	35657	46675	9566	10226	11818	1831	1890	1961	508	514	514	238	238	238
●●●	34650	37938	52416	9842	10801	13219	1857	1922	2164	508	514	557	238	238	252
0.35 ●●●	35657	37672	51832	10199	11022	13070	1987	2058	2150	547	547	557	254	254	254
●●●	36362	40571	56620	10541	11753	14239	2020	2101	2313	547	557	588	254	254	263
0.4 ●●●	36037	38572	55699	10551	11532	14012	2085	2172	2280	573	579	590	265	265	265
●●●	36947	42098	59588	10963	12409	14950	2123	2221	2412	573	579	606	265	265	268
0.45 ●●●	35462	38539	58261	10665	11786	14613	2139	2237	2361	579	590	596	265	271	271
●●●	36583	42645	61319	11136	12777	15352	2188	2296	2461	590	596	613	265	271	268
0.5 ●●●	34113	37727	59480	10573	11792	14890	2150	2253	2393	579	590	596	265	265	265
●●●	35462	42342	61814	11087	12881	15445	2198	2318	2461	590	596	606	265	265	263
0.55 ●●●	32212	36297	59350	10291	11591	14825	2117	2221	2361	563	579	590	254	254	254
●●●	33756	41302	61072	10843	12728	15228	2166	2286	2412	573	579	588	254	254	252
0.6 ●●●	29911	34373	57877	9852	11174	14418	2042	2150	2280	541	547	557	238	238	248
●●●	31633	39606	59093	10411	12328	14703	2091	2204	2313	547	557	557	238	238	235
0.65 ●●●	27370	32023	55066	9267	10573	13677	1922	2020	2150	508	508	525	222	222	222
●●●	29190	37321	55878	9826	11688	13868	1971	2085	2164	508	514	514	222	222	213
0.7 ●●●	24646	29326	50916	8542	9777	12615	1766	1847	1961	449	460	466	189	200	200
●●●	26498	34471	51427	9072	10811	12724	1808	1906	1967	460	466	458	189	200	186
0.75 ●●●	21796	26265	45446	7675	8786	11217	1565	1630	1717	384	395	401	167	167	167
●●●	23583	31048	45739	8152	9696	11271	1597	1678	1719	395	395	390	167	167	153
0.8 ●●●	18773	22820	38669	6641	7577	9484	1311	1370	1435	313	319	319	124	124	124
●●●	20404	27012	38815	7047	8314	9509	1337	1392	1422	313	319	310	124	124	115
0.85 ●●●	15491	18871	30603	5422	6131	7437	1012	1045	1083	222	222	232	.	.	.
●●●	16872	22218	30654	5731	6657	7438	1028	1067	1076	222	232	217	.	.	.
0.9 ●●●	11743	14164	21249	3943	4376	5064	661	677	693	.	.	.	.	.	.
●●●	12751	16423	21256	4138	4668	5058	671	687	681	.	.	.	.	.	.
0.95 ●●●	7008	8152	10632	2058	2198	2377	.	.	.	.	.	.	.	.	.
●●●	7502	9088	10622	2123	2280	2368	.	.	.	.	.	.	.	.	.
0.98 ●●●	2984	3255	3661	.	.	.	.	.	.	.	.	.	.	.	.
●●●	3108	3433	3648	.	.	.	.	.	.	.	.	.	.	.	.

TABLE 7: ALPHA= 0.05 POWER= 0.8   EXPECTED ACCRUAL THRU MINIMUM FOLLOW-UP= 7000

		DEL=.01			DEL=.02			DEL=.05			DEL=.10			DEL=.15	
FACT=	1.0 .75	.50 .25	.00 BIN	1.0 .75	.50 .25	.00 BIN	1.0 .75	.50 .25	.00 BIN	1.0 75	.50 .25	.00 BIN	1.0 .75	.50 .25	.00 BIN
PCONT=***					REQUIRED NUMBER OF PATIENTS										
0.01 ***	950	950	950	344	344	344	116	116	116	57	57	57	40	40	40
•••	950	950	3648	344	344	1206	116	116	328	57	57	134	40	40	80
0.02 ***	2105	2111	2129	676	676	676	180	180	180	81	81	81	46	46	46
•••	2111	2122	6022	676	676	1793	180	180	419	81	81	155	46	46	89
0.05 ***	6410	6480	6644	1831	1842	1860	390	390	396	134	145	145	81	81	81
•••	6434	6539	12848	1831	1849	3481	390	396	681	134	145	217	81	81	115
0.1 ***	14169	14431	15475	3890	3966	4089	746	757	764	232	232	232	127	127	127
•••	14267	14757	23235	3925	4019	6047	757	757	1076	232	232	310	127	127	153
0.15 ***	21232	21757	24435	5797	5979	6329	1090	1096	1114	320	320	320	162	162	162
•••	21414	22457	32385	5874	6119	8304	1096	1107	1422	320	320	390	162	162	186
0.2 ***	27007	27836	32835	7407	7729	8405	1387	1411	1440	396	396	396	197	197	197
•••	27305	29026	40298	7547	8002	10251	1394	1422	1719	396	396	458	197	197	213
0.25 ***	31336	32537	40307	8685	9175	10249	1639	1674	1720	460	460	466	221	221	221
•••	31756	34276	46976	8895	9584	11890	1656	1702	1967	460	460	514	221	221	235
0.3 ***	34235	35869	46677	9636	10295	11817	1842	1895	1954	512	512	519	239	239	239
•••	34812	38214	52416	9910	10861	13219	1866	1930	2164	512	512	557	239	239	252
0.35 ***	35816	37945	51829	10284	11106	13077	1989	2059	2146	547	554	554	256	256	256
•••	36580	40920	56620	10627	11824	14239	2024	2105	2313	547	554	588	256	256	263
0.4 ***	36254	38907	55696	10662	11631	14005	2094	2175	2280	571	582	582	267	267	267
•••	37216	42512	59588	11065	12489	14950	2140	2227	2412	571	582	606	267	267	268
0.45 ***	35722	38931	58251	10785	11894	14617	2157	2245	2356	582	589	600	267	267	267
•••	36901	43114	61319	11257	12874	15352	2199	2297	2461	589	600	613	267	267	268
0.5 ***	34434	38190	59476	10704	11922	14886	2164	2262	2391	582	589	600	267	267	267
•••	35845	42869	61814	11211	12979	15445	2210	2321	2461	589	600	606	267	267	263
0.55 ***	32590	36796	59347	10435	11719	14816	2129	2234	2367	571	582	589	256	256	256
•••	34189	41865	61072	10977	12832	15228	2181	2297	2412	571	582	588	256	256	252
0.6 ***	30339	34900	57877	9997	11310	14414	2059	2157	2280	547	554	565	239	239	239
•••	32106	40185	59093	10557	12430	14703	2105	2216	2313	547	554	557	239	239	235
0.65 ***	27819	32572	55066	9409	10704	13679	1936	2035	2146	501	512	519	221	221	221
•••	29685	37892	55878	9962	11789	13868	1982	2087	2164	512	512	514	221	221	213
0.7 ***	25106	29860	50919	8674	9899	12611	1772	1860	1965	449	460	466	197	197	197
•••	26996	35022	51427	9199	10907	12724	1814	1901	1967	460	460	458	197	197	186
0.75 ***	22247	26769	45452	7799	8895	11211	1569	1639	1720	390	396	396	162	162	162
•••	24056	31557	45739	8271	9776	11271	1604	1674	1719	390	396	390	162	162	153
0.8 ***	19185	23269	38669	6749	7670	9490	1317	1370	1429	309	320	320	127	127	127
•••	20836	27445	38815	7151	8387	9509	1341	1394	1422	320	320	310	127	127	115
0.85 ***	15842	19237	30601	5506	6189	7431	1020	1055	1090	221	221	232	.	.	.
•••	17231	22551	30654	5815	6696	7438	1037	1072	1076	221	232	217	.	.	.
0.9 ***	11999	14425	21250	3995	4415	5062	659	676	694	.	.	.	.	.	.
•••	13014	16636	21256	4187	4695	5058	670	687	681	.	.	.	.	.	.
0.95 ***	7134	8265	10634	2076	2210	2374	.	.	.	.	.	.	.	.	.
•••	7624	9175	10622	2140	2286	2368	.	.	.	.	.	.	.	.	.
0.98 ***	3015	3277	3662	.	.	.	.	.	.	.	.	.	.	.	.
•••	3137	3452	3648	.	.	.	.	.	.	.	.	.	.	.	.

TABLE 7: ALPHA= 0.05 POWER= 0.8   EXPECTED ACCRUAL THRU MINIMUM FOLLOW-UP= 7500

		DEL=.01			DEL=.02			DEL=.05			DEL=.10			DEL=.15		
FACT=		1.0 .75	.50 .25	.00 BIN	1.0 .75	.50 .25	.00 BIN	1.0 .75	.50 .25	.00 BIN	1.0 75	.50 .25	.00 BIN	1.0 .75	.50 .25	.00 BIN
PCONT=***						REQUIRED	NUMBER	OF	PATIENTS							
0.01	***	950	950	950	350	350	350	118	118	118	61	61	61	43	43	43
	***	950	950	3648	350	350	1206	118	118	328	61	61	134	43	43	80
0.02	***	2105	2112	2124	680	680	680	181	181	181	80	80	80	50	50	50
	***	2112	2124	6022	680	680	1793	181	181	419	80	80	155	50	50	89
0.05	***	6406	6481	6643	1831	1843	1861	387	399	399	136	136	136	80	80	80
	***	6443	6549	12848	1831	1850	3481	387	399	681	136	136	217	80	80	115
0.1	***	14187	14468	15474	3893	3968	4093	755	755	762	230	237	237	125	125	125
	***	14293	14780	23235	3931	4025	6047	755	755	1076	230	237	310	125	125	153
0.15	***	21275	21818	24436	5818	5993	6324	1093	1100	1118	324	324	324	162	162	162
	***	21474	22531	32385	5893	6136	8304	1093	1111	1422	324	324	390	162	162	186
0.2	***	27068	27943	32836	7437	7756	8405	1393	1411	1437	399	399	399	193	193	193
	***	27380	29150	40298	7580	8018	10251	1400	1430	1719	399	399	458	193	193	213
0.25	***	31430	32686	40306	8731	9218	10250	1643	1681	1718	455	462	462	218	218	218
	***	31880	34468	46976	8937	9612	11890	1662	1700	1967	462	462	514	218	218	235
0.3	***	34362	36080	46674	9706	10355	11818	1850	1899	1962	511	511	518	237	237	237
	***	34974	38468	52416	9980	10906	13219	1868	1925	2164	511	518	557	237	237	252
0.35	***	35986	38206	51830	10362	11187	13074	2000	2068	2143	549	549	556	256	256	256
	***	36781	41236	56620	10711	11881	14239	2030	2105	2313	549	556	588	256	256	263
0.4	***	36462	39237	55700	10756	11731	14011	2105	2187	2281	575	575	586	268	268	268
	***	37475	42893	59588	11168	12568	14950	2143	2225	2412	575	586	606	268	268	268
0.45	***	35986	39324	58261	10899	12005	14611	2168	2255	2356	586	593	593	268	268	268
	***	37224	43561	61319	11368	12950	15352	2206	2300	2461	586	593	613	268	268	268
0.5	***	34756	38618	59480	10824	12043	14881	2180	2274	2386	586	593	605	268	268	268
	***	36211	43355	61814	11337	13074	15445	2225	2330	2461	586	593	606	268	268	263
0.55	***	32949	37280	59349	10561	11843	14818	2143	2243	2368	568	575	586	256	256	256
	***	34606	42380	61072	11105	12924	15228	2187	2300	2412	575	586	588	256	256	252
0.6	***	30743	35405	57875	10130	11431	14412	2068	2161	2281	549	549	556	237	237	249
	***	32555	40718	59093	10693	12530	14703	2112	2218	2313	549	556	557	237	237	235
0.65	***	28250	33087	55062	9549	10824	13681	1943	2037	2150	500	511	518	218	218	218
	***	30155	38431	55878	10093	11881	13868	1993	2086	2164	511	518	514	218	218	213
0.7	***	25550	30368	50918	8806	10018	12612	1786	1861	1962	455	462	462	193	193	193
	***	27474	35536	51427	9324	10993	12724	1824	1906	1967	455	462	458	193	193	186
0.75	***	22674	27249	45455	7918	8993	11218	1580	1643	1718	387	399	399	162	162	162
	***	24511	32030	45739	8386	9849	11271	1606	1681	1719	387	399	390	162	162	153
0.8	***	19568	23686	38675	6849	7749	9481	1325	1374	1430	312	312	324	125	125	125
	***	21249	27849	38815	7243	8443	9509	1343	1400	1422	312	324	310	125	125	115
0.85	***	16175	19568	30605	5581	6256	7437	1025	1055	1081	230	230	230	.	.	.
	***	17574	22868	30654	5881	6743	7438	1036	1074	1076	230	230	217	.	.	.
0.9	***	12249	14656	21249	4043	4449	5068	661	680	687	.	.	.	.	.	.
	***	13250	16843	21256	4224	4711	5058	668	687	681	.	.	.	.	.	.
0.95	***	7250	8368	10636	2093	2218	2375	.	.	.	.	.	.	.	.	.
	***	7737	9249	10622	2150	2293	2368	.	.	.	.	.	.	.	.	.
0.98	***	3050	3305	3661	.	.	.	.	.	.	.	.	.	.	.	.
	***	3162	3462	3648	.	.	.	.	.	.	.	.	.	.	.	.

TABLE 7: ALPHA= 0.05 POWER= 0.8   EXPECTED ACCRUAL THRU MINIMUM FOLLOW-UP= 8000

	DEL=.01			DEL=.02			DEL=.05			DEL=.10			DEL=.15		
FACT=	1.0 .75	.50 .25	.00 BIN	1.0 .75	.50 .25	.00 BIN	1.0 .75	.50 .25	.00 BIN	1.0 75	.50 .25	.00 BIN	1.0 .75	.50 .25	.00 BIN
PCONT=•••				REQUIRED NUMBER OF PATIENTS											
0.01 •••	953	953	953	345	345	345	113	113	113	53	53	53	45	45	45
•••	953	953	3648	345	345	1206	113	113	328	53	53	134	45	45	80
0.02 •••	2105	2113	2125	673	673	673	185	185	185	73	73	73	53	53	53
•••	2113	2125	6022	673	673	1793	185	185	419	73	73	155	53	53	89
0.05 •••	6413	6493	6645	1833	1845	1865	393	393	393	133	145	145	85	85	85
•••	6445	6553	12848	1833	1853	3481	393	393	681	133	145	217	85	85	115
0.1 •••	14213	14493	15473	3905	3973	4093	753	753	765	233	233	233	125	125	125
•••	14325	14813	23235	3933	4025	6047	753	753	1076	233	233	310	125	125	153
0.15 •••	21313	21873	24433	5833	6013	6325	1093	1105	1113	313	325	325	165	165	165
•••	21513	22593	32385	5913	6145	8304	1093	1105	1422	325	325	390	165	165	186
0.2 •••	27133	28045	32833	7473	7785	8405	1393	1413	1445	393	393	393	193	193	193
•••	27465	29273	40298	7605	8033	10251	1405	1425	1719	393	393	458	193	193	213
0.25 •••	31513	32833	40305	8785	9253	10253	1645	1685	1725	453	465	465	225	225	225
•••	31993	34645	46976	8985	9645	11890	1665	1705	1967	465	465	514	225	225	235
0.3 •••	34485	36273	46673	9765	10413	11813	1853	1905	1953	505	513	513	245	245	245
•••	35125	38705	52416	10045	10953	13219	1873	1925	2164	513	513	557	245	245	252
0.35 •••	36153	38465	51833	10445	11253	13073	2005	2073	2145	545	553	553	253	253	253
•••	36985	41545	56620	10785	11945	14239	2033	2105	2313	553	553	588	253	253	263
0.4 •••	36673	39545	55705	10845	11813	14005	2113	2185	2285	573	573	585	265	265	265
•••	37733	43253	59588	11253	12633	14950	2153	2233	2412	573	585	606	265	265	268
0.45 •••	36245	39685	58253	11005	12105	14613	2173	2253	2365	585	593	593	265	265	273
•••	37533	43973	61319	11473	13033	15352	2213	2305	2461	585	593	613	265	265	268
0.5 •••	35065	39033	59473	10945	12145	14885	2185	2273	2385	585	593	605	265	265	265
•••	36573	43813	61814	11453	13153	15445	2233	2325	2461	585	593	606	265	265	263
0.55 •••	33305	37733	59345	10685	11953	14825	2153	2245	2365	573	573	585	253	253	253
•••	35013	42865	61072	11225	13013	15228	2193	2305	2412	573	585	588	253	253	252
0.6 •••	31133	35885	57873	10253	11545	14413	2073	2165	2285	545	553	565	245	245	245
•••	32993	41225	59093	10813	12613	14703	2125	2225	2313	545	553	557	245	245	235
0.65 •••	28665	33573	55065	9665	10933	13685	1953	2045	2145	505	513	525	225	225	225
•••	30605	38925	55878	10213	11965	13868	1993	2093	2164	513	513	514	225	225	213
0.7 •••	25973	30845	50913	8925	10113	12613	1793	1873	1965	453	465	465	193	193	193
•••	27913	36013	51427	9445	11065	12724	1833	1913	1967	453	465	458	193	193	186
0.75 •••	23073	27693	45453	8025	9085	11213	1585	1645	1725	393	393	405	165	165	165
•••	24933	32453	45739	8485	9913	11271	1613	1685	1719	393	393	390	165	165	153
0.8 •••	19945	24085	38673	6945	7825	9485	1333	1373	1433	313	313	325	125	125	125
•••	21625	28213	38815	7333	8493	9509	1353	1405	1422	313	313	310	125	125	115
0.85 •••	16485	19893	30605	5053	6305	7433	1025	1053	1053	225	225	225	.	.	.
•••	17885	23153	30654	5945	6773	7438	1033	1065	1076	225	225	217	.	.	.
0.9 •••	12473	14873	21253	4085	4473	5065	665	673	693	.	.	.	.	.	.
•••	13485	17025	21256	4265	4733	5058	673	685	681	.	.	.	.	.	.
0.95 •••	7365	8465	10633	2105	2225	2373	.	.	.	.	.	.	.	.	.
•••	7845	9313	10622	2165	2293	2368	.	.	.	.	.	.	.	.	.
0.98 •••	3073	3313	3665	.	.	.	.	.	.	.	.	.	.	.	.
•••	3185	3473	3648	.	.	.	.	.	.	.	.	.	.	.	.

TABLE 7: ALPHA= 0.05 POWER= 0.8   EXPECTED ACCRUAL THRU MINIMUM FOLLOW-UP= 8500

		DEL=.01			DEL=.02			DEL=.05			DEL=.10			DEL=.15	
FACT=		1.0 .75	.50 .25	.00 BIN	1.0 .75	.50 .25	.00 BIN	1.0 .75	.50 .25	.00 BIN	1.0 75	.50 .25	.00 BIN	1.0 .75	.50 .25	.00 BIN
PCONT=***						REQUIRED NUMBER OF PATIENTS									
0.01	***	949	949	949	346	346	346	112	112	112	56	56	56	35	35	35
	•••	949	949	3648	346	346	1206	112	112	328	56	56	134	35	35	80
0.02	***	2110	2118	2131	673	673	673	184	184	184	78	78	78	48	48	48
	•••	2110	2118	6022	673	673	1793	184	184	419	78	78	155	48	48	89
0.05	***	6423	6495	6644	1833	1841	1863	388	396	396	141	141	141	78	78	78
	•••	6453	6559	12848	1841	1855	3481	396	396	681	141	141	217	78	78	115
0.1	***	14230	14520	15476	3916	3980	4094	750	758	758	240	240	240	120	120	120
	•••	14350	14838	23235	3945	4030	6047	758	758	1076	240	240	310	120	120	153
0.15	***	21349	21936	24443	5850	6028	6325	1090	1098	1119	325	325	325	163	163	163
	•••	21561	22658	32385	5921	6155	8304	1098	1111	1422	325	325	390	163	163	186
0.2	***	27193	28141	32837	7494	7813	8408	1395	1416	1438	396	396	396	197	197	197
	•••	27546	29381	40298	7635	8060	10251	1408	1430	1719	396	396	458	197	197	213
0.25	***	31613	32986	40304	8825	9300	10248	1650	1685	1727	460	460	460	218	218	218
	•••	32101	34813	46976	9024	9675	11890	1663	1706	1967	460	460	514	218	218	235
0.3	***	34609	36471	46679	9823	10461	11821	1855	1905	1961	516	516	516	240	240	240
	•••	35281	38936	52416	10100	10992	13219	1884	1926	2164	516	516	557	240	240	252
0.35	***	36309	38710	51835	10511	11319	13075	2011	2075	2139	545	558	558	261	261	261
	•••	37193	41826	56620	10865	11991	14239	2046	2110	2313	545	558	588	261	261	263
0.4	***	36883	39850	55702	10936	11893	14010	2118	2195	2280	580	580	588	261	269	269
	•••	37980	43598	59588	11340	12692	14950	2160	2237	2412	580	580	606	261	269	268
0.45	***	36500	40041	58260	11098	12190	14613	2181	2266	2365	588	588	601	269	269	269
	•••	37831	44363	61319	11566	13096	15352	2224	2309	2461	588	601	613	269	269	268
0.5	***	35366	39433	59471	11056	12246	14881	2195	2280	2386	588	588	601	261	269	269
	•••	36917	44248	61814	11553	13231	15445	2237	2330	2461	588	601	606	261	269	263
0.55	***	33645	38158	59344	10801	12055	14817	2160	2258	2365	566	580	588	261	261	261
	•••	35395	43321	61072	11340	13096	15228	2203	2301	2412	580	580	588	261	261	252
0.6	***	31506	36343	57878	10376	11651	14421	2088	2173	2280	545	558	558	240	240	240
	•••	33411	41698	59093	10928	12692	14703	2131	2224	2313	545	558	557	240	240	235
0.65	***	29063	34035	55065	9789	11035	13678	1969	2046	2152	503	516	516	218	218	226
	•••	31039	39390	55878	10320	12041	13868	2003	2096	2164	516	516	514	218	218	213
0.7	***	26364	31294	50921	9037	10206	12615	1799	1876	1961	452	460	460	197	197	197
	•••	28340	36458	51427	9547	11141	12724	1833	1918	1967	460	460	458	197	197	186
0.75	***	23453	28106	45446	8123	9165	11213	1586	1650	1727	388	396	396	163	163	163
	•••	25344	32866	45739	8578	9972	11271	1621	1685	1719	396	396	390	163	163	153
0.8	***	20286	24451	38668	7026	7890	9483	1331	1374	1430	311	311	325	120	120	120
	•••	21986	28553	38815	7409	8535	9509	1353	1408	1422	311	325	310	120	120	115
0.85	***	16772	20193	30606	5709	6360	7430	1026	1055	1090	226	226	226	.	.	.
	•••	18183	23410	30654	5998	6814	7438	1047	1068	1076	226	226	217	.	.	.
0.9	***	12679	15080	21256	4128	4498	5063	665	673	694	.	.	.	.	.	.
	•••	13691	17184	21256	4298	4753	5058	673	686	681	.	.	.	.	.	.
0.95	***	7473	8548	10631	2118	2237	2373	.	.	.	.	.	.	.	.	.
	•••	7940	9377	10622	2173	2301	2368	.	.	.	.	.	.	.	.	.
0.98	***	3095	3329	3661	.	.	.	.	.	.	.	.	.	.	.	.
	•••	3215	3478	3648	.	.	.	.	.	.	.	.	.	.	.	.

TABLE 7: ALPHA= 0.05 POWER= 0.8   EXPECTED ACCRUAL THRU MINIMUM FOLLOW-UP= 9000

	DEL=.01			DEL=.02			DEL=.05			DEL=.10			DEL=.15		
FACT=	1.0 .75	.50 .25	.00 BIN	1.0 .75	.50 .25	.00 BIN	1.0 .75	.50 .25	.00 BIN	1.0 75	.50 .25	.00 BIN	1.0 .75	.50 .25	.00 BIN
PCONT=***				REQUIRED NUMBER OF PATIENTS											
0.01 ***	951 951	951 951	951 3648	352 352	352 352	352 1206	119 119	119 119	119 328	60 60	60 60	60 134	37 37	37 37	37 80
0.02 ***	2107 2107	2121 2121	2130 6022	681 681	681 681	681 1793	186 186	186 186	186 419	82 82	82 82	82 155	51 51	51 51	51 89
0.05 ***	6427 6464	6509 6562	6644 12848	1837 1837	1851 1851	1860 3481	389 389	397 397	397 681	141 141	141 141	141 217	82 82	82 82	82 115
0.1 ***	14257 14370	14550 14865	15472 23235	3921 3952	3989 4034	4087 6047	749 757	757 757	757 1076	231 231	231 231	231 310	127 127	127 127	127 153
0.15 ***	21390 21615	21997 22717	24441 32385	5865 5946	6036 6157	6329 8304	1095 1095	1109 1109	1117 1422	321 321	321 321	321 390	164 164	164 164	164 186
0.2 ***	27262 27614	28244 29490	32834 40298	7530 7665	7836 8070	8407 10251	1401 1410	1410 1424	1446 1719	397 397	397 397	397 458	195 195	195 195	195 213
0.25 ***	31695 32212	33126 34971	40312 46976	8857 9060	9330 9704	10252 11890	1657 1671	1680 1702	1725 1967	456 456	465 465	465 514	217 217	217 217	217 235
0.3 ***	34732 35430	36659 39142	46680 52416	9884 10154	10514 11031	11819 13219	1860 1882	1905 1927	1964 2164	510 510	510 510	524 557	240 240	240 240	240 252
0.35 ***	36479 37387	38940 42090	51832 56620	10581 10927	11391 12044	13079 14239	2017 2054	2076 2107	2144 2313	546 555	555 555	555 588	254 254	254 254	254 263
0.4 ***	37086 38220	40132 43926	55702 59588	11017 11422	11962 12750	14010 14950	2130 2166	2197 2234	2279 2412	577 577	577 577	591 606	262 262	262 262	262 268
0.45 ***	36749 38121	40371 44736	58259 61319	11197 11661	12277 13169	14617 15352	2189 2234	2265 2310	2355 2461	591 591	591 600	600 613	262 262	262 262	276 268
0.5 ***	35655 37252	39809 44646	59474 61814	11152 11661	12336 13304	14887 15445	2211 2242	2287 2332	2391 2461	591 591	591 600	600 606	262 262	262 262	262 263
0.55 ***	33981 35767	38571 43755	59347 61072	10905 11445	12156 13169	14820 15228	2175 2211	2256 2310	2369 2412	569 577	577 577	591 588	254 254	254 254	262 252
0.6 ***	31875 33801	36771 42135	57876 59093	10491 11031	11751 12764	14415 14703	2099 2144	2175 2234	2279 2313	546 546	555 555	555 557	240 240	240 240	240 235
0.65 ***	29436 31447	34476 39831	55064 55878	9892 10432	11130 12111	13686 13868	1972 2017	2054 2099	2144 2164	510 510	510 510	524 514	217 217	217 217	217 213
0.7 ***	26745 28747	31717 36870	50924 51427	9141 9645	10297 11197	12615 12724	1806 1837	1882 1919	1964 1967	456 456	465 465	465 458	195 195	195 195	195 186
0.75 ***	23820 25724	28514 33239	45456 45739	8219 8669	9240 10027	11211 11271	1604 1626	1657 1680	1725 1719	389 397	397 397	397 390	164 164	164 164	164 153
0.8 ***	20625 22335	24801 28882	38670 38815	7102 7476	7957 8579	9487 9509	1342 1356	1379 1401	1432 1422	321 321	321 321	321 310	127 127	127 127	127 115
0.85 ***	17047 18465	20467 23662	30606 30654	5775 6059	6396 6832	7431 7438	1027 1041	1064 1072	1086 1076	231 231	231 231	231 217	. .	. .	. .
0.9 ***	12885 13889	15261 17340	21246 21256	4155 4326	4529 4762	5069 5058	667 667	681 681	690 681	. .	. .	. .	. .	. .	. .
0.95 ***	7566 8039	8624 9434	10635 10622	2130 2189	2242 2310	2377 2368	. .	. .	. .	. .	. .	. .	. .	. .	. .
0.98 ***	3120 3232	3345 3494	3660 3648	. .	. .	. .	. .	. .	. .	. .	. .	. .	. .	. .	. .

TABLE 7: ALPHA= 0.05 POWER= 0.8   EXPECTED ACCRUAL THRU MINIMUM FOLLOW-UP= 9500

		DEL=.01			DEL=.02			DEL=.05			DEL=.10			DEL=.15	
FACT=		1.0 .75	.50 .25	.00 BIN	1.0 .75	.50 .25	.00 BIN	1.0 .75	.50 .25	.00 BIN	1.0 75	.50 .25	.00 BIN	1.0 .75	.50 .25	.00 BIN
PCONT=***					REQUIRED NUMBER OF PATIENTS											
0.01	***	956	956	956	348	348	348	110	110	110	54	54	54	39	39	39
	***	956	956	3648	348	348	1206	110	110	328	54	54	134	39	39	80
0.02	***	2105	2120	2129	680	680	680	182	182	182	78	78	78	54	54	54
	***	2120	2120	6022	680	680	1793	182	182	419	78	78	155	54	54	89
0.05	***	6443	6514	6642	1835	1844	1859	395	395	395	134	134	134	78	78	78
	***	6466	6561	12848	1844	1859	3481	395	395	681	134	134	217	78	78	115
0.1	***	14280	14574	15468	3925	3996	4091	752	752	766	229	229	229	125	125	125
	***	14384	14883	23235	3958	4044	6047	752	752	1076	229	229	310	125	125	153
0.15	***	21429	22046	24430	5882	6048	6324	1099	1108	1123	315	324	324	158	158	158
	***	21666	22783	32385	5953	6167	8304	1099	1108	1422	315	324	390	158	158	186
0.2	***	27319	28325	32829	7559	7853	8414	1393	1417	1440	395	395	395	196	196	196
	***	27699	29584	40298	7687	8090	10251	1408	1431	1719	395	395	458	196	196	213
0.25	***	31784	33256	40310	8898	9364	10252	1654	1693	1725	458	458	467	220	220	220
	***	32330	35118	46976	9103	9729	11890	1669	1702	1967	458	467	514	220	220	235
0.3	***	34857	36828	46675	9934	10560	11810	1868	1906	1963	514	514	514	244	244	244
	***	35584	39345	52416	10204	11059	13219	1892	1930	2164	514	514	557	244	244	252
0.35	***	36629	39179	51829	10646	11439	13069	2025	2082	2144	553	553	562	253	253	253
	***	37579	42353	56620	10988	12080	14239	2049	2105	2313	553	553	588	253	253	263
0.4	***	37279	40414	55700	11098	12033	14004	2144	2200	2286	576	576	585	268	268	268
	***	38458	44229	59588	11501	12793	14950	2168	2239	2412	576	585	606	268	268	268
0.45	***	36985	40699	58250	11288	12356	14613	2200	2272	2358	585	585	600	268	268	268
	***	38395	45084	61319	11739	13220	15352	2239	2310	2461	585	600	613	268	268	268
0.5	***	35940	40177	59476	11249	12413	14883	2215	2295	2390	585	585	600	268	268	268
	***	37564	45022	61814	11748	13363	15445	2248	2334	2461	585	600	606	268	268	263
0.55	***	34301	38965	59343	11012	12238	14826	2177	2263	2358	576	576	585	253	253	253
	***	36130	44158	61072	11549	13235	15228	2224	2310	2412	576	585	588	253	253	252
0.6	***	32220	37184	57870	10584	11843	14423	2105	2191	2286	553	553	562	244	244	244
	***	34183	42552	59093	11130	12831	14703	2144	2224	2313	553	553	557	244	244	235
0.65	***	29813	34880	55059	9990	11216	13686	1987	2058	2144	505	514	514	220	220	220
	***	31840	40239	55878	10528	12175	13868	2025	2096	2164	514	514	514	220	220	213
0.7	***	27114	32116	50912	9230	10370	12618	1820	1883	1963	458	458	467	196	196	196
	***	29133	37255	51427	9729	11264	12724	1844	1915	1967	458	467	458	196	196	186
0.75	***	24169	28886	45449	8295	9316	11216	1607	1654	1716	395	395	395	158	158	158
	***	26093	33589	45739	8746	10076	11271	1630	1693	1719	395	395	390	158	158	153
0.8	***	20939	25119	38671	7179	8010	9483	1345	1384	1431	315	315	324	125	125	125
	***	22664	29180	38815	7544	8628	9509	1360	1408	1422	315	315	310	125	125	115
0.85	***	17320	20725	30605	5825	6443	7440	1037	1060	1084	229	229	229	.	.	.
	***	18730	23884	30654	6110	6870	7438	1051	1075	1076	229	229	217	.	.	.
0.9	***	13078	15444	21248	4195	4552	5065	671	680	695	.	.	.	.	.	.
	***	14075	17486	21256	4362	4780	5058	671	680	681	.	.	.	.	.	.
0.95	***	7654	8699	10632	2144	2248	2381	.	.	.	.	.	.	.	.	.
	***	8114	9483	10622	2191	2310	2368	.	.	.	.	.	.	.	.	.
0.98	***	3141	3364	3664	.	.	.	.	.	.	.	.	.	.	.	.
	***	3245	3498	3648	.	.	.	.	.	.	.	.	.	.	.	.

TABLE 7: ALPHA= 0.05 POWER= 0.8   EXPECTED ACCRUAL THRU MINIMUM FOLLOW-UP= 10000

		DEL=.01			DEL=.02			DEL=.05			DEL=.10			DEL=.15		
FACT=		1.0 .75	.50 .25	.00 BIN	1.0 .75	.50 .25	.00 BIN	1.0 .75	.50 .25	.00 BIN	1.0 75	.50 .25	.00 BIN	1.0 .75	.50 .25	.00 BIN
PCONT=•••					REQUIRED NUMBER OF PATIENTS											
0.01	•••	957	957	957	341	341	341	116	116	116	57	57	57	41	41	41
	•••	957	957	3648	341	341	1206	116	116	328	57	57	134	41	41	80
0.02	•••	2107	2116	2132	682	682	682	182	182	182	82	82	82	57	57	57
	•••	2116	2116	6022	682	682	1793	182	182	419	82	82	155	57	57	89
0.05	•••	6441	6516	6641	1832	1841	1857	391	391	391	141	141	141	82	82	82
	•••	6466	6566	12848	1841	1857	3481	391	391	681	141	141	217	82	82	115
0.1	•••	14291	14591	15466	3932	3991	4091	757	757	757	232	232	232	116	116	116
	•••	14416	14907	23235	3957	4041	6047	757	757	1076	232	232	310	116	116	153
0.15	•••	21466	22107	24441	5891	6057	6332	1091	1107	1116	316	316	316	157	157	157
	•••	21707	22832	32385	5966	6182	8304	1107	1107	1422	316	316	390	157	157	186
0.2	•••	27382	28416	32832	7582	7882	8407	1407	1416	1441	391	391	391	191	191	191
	•••	27766	29682	40298	7707	8107	10251	1407	1432	1719	391	391	458	191	191	213
0.25	•••	31882	33382	40307	8941	9391	10257	1657	1691	1716	457	457	466	216	216	216
	•••	32432	35257	46976	9141	9741	11890	1666	1707	1967	457	466	514	216	216	235
0.3	•••	34966	37007	46682	9982	10607	11816	1866	1907	1957	507	516	516	241	241	241
	•••	35732	39541	52416	10241	11091	13219	1891	1932	2164	516	516	557	241	241	252
0.35	•••	36782	39391	51832	10707	11491	13066	2032	2082	2141	541	557	557	257	257	257
	•••	37757	42591	56620	11057	12116	14239	2057	2116	2313	557	557	588	257	257	263
0.4	•••	37482	40682	55707	11166	12107	14007	2141	2207	2282	582	582	582	266	266	266
	•••	38682	44507	59588	11566	12841	14950	2166	2241	2412	582	582	606	266	266	268
0.45	•••	37216	41007	58257	11366	12432	14616	2207	2282	2357	591	591	591	266	266	266
	•••	38666	45407	61319	11816	13282	15352	2241	2316	2461	591	591	613	266	266	268
0.5	•••	36216	40516	59482	11341	12491	14882	2216	2291	2391	591	591	607	266	266	266
	•••	37882	45391	61814	11841	13416	15445	2257	2341	2461	591	591	606	266	266	263
0.55	•••	34607	39341	59341	11107	12332	14816	2191	2266	2366	566	582	582	257	257	257
	•••	36466	44541	61072	11632	13291	15228	2232	2316	2412	582	582	588	257	257	252
0.6	•••	32557	37566	57882	10691	11916	14416	2116	2191	2282	541	557	557	241	241	241
	•••	34557	42941	59093	11232	12891	14703	2157	2232	2313	557	557	557	241	241	235
0.65	•••	30157	35282	55066	10091	11291	13682	1991	2066	2141	507	516	516	216	216	216
	•••	32216	40632	55878	10616	12232	13868	2032	2107	2164	507	516	514	216	216	213
0.7	•••	27466	32507	50916	9332	10457	12607	1816	1882	1957	457	457	466	191	191	191
	•••	29507	37632	51427	9816	11316	12724	1857	1916	1967	457	466	458	191	191	186
0.75	•••	24507	29241	45457	8382	9382	11216	1607	1657	1716	391	391	407	166	166	166
	•••	26441	33932	45739	8816	10116	11271	1632	1691	1719	391	391	390	166	166	153
0.8	•••	21241	25432	38666	7241	8066	9482	1341	1382	1432	316	316	316	132	132	132
	•••	22966	29457	38815	7607	8657	9509	1366	1407	1422	316	316	310	132	132	115
0.85	•••	17566	20982	30607	5882	6482	7432	1032	1057	1082	232	232	232	.	.	.
	•••	18991	24107	30654	6157	6891	7438	1041	1066	1076	232	232	217	.	.	.
0.9	•••	13257	15607	21257	4232	4566	5066	666	682	691	.	.	.	.	.	.
	•••	14257	17616	21256	4382	4791	5058	682	682	681	.	.	.	.	.	.
0.95	•••	7741	8766	10632	2157	2257	2382	.	.	.	.	.	.	.	.	.
	•••	8191	9532	10622	2207	2316	2368	.	.	.	.	.	.	.	.	.
0.98	•••	3166	3382	3657	.	.	.	.	.	.	.	.	.	.	.	.
	•••	3266	3507	3648	.	.	.	.	.	.	.	.	.	.	.	.

TABLE 7: ALPHA= 0.05 POWER= 0.8   EXPECTED ACCRUAL THRU MINIMUM FOLLOW-UP= 11000

	DEL=.01			DEL=.02			DEL=.05			DEL=.10			DEL=.15		
FACT=	1.0 .75	.50 .25	.00 BIN	1.0 .75	.50 .25	.00 BIN	1.0 .75	.50 .25	.00 BIN	1.0 75	.50 .25	.00 BIN	1.0 .75	.50 .25	.00 BIN

PCONT=•••   REQUIRED NUMBER OF PATIENTS

PCONT	DEL=.01 1.0/.75	.50/.25	.00/BIN	DEL=.02 1.0/.75	.50/.25	.00/BIN	DEL=.05 1.0/.75	.50/.25	.00/BIN	DEL=.10 1.0/.75	.50/.25	.00/BIN	DEL=.15 1.0/.75	.50/.25	.00/BIN
0.01 ***	953	953	953	348	348	348	117	117	117	62	62	62	35	35	35
•••	953	953	3648	348	348	1206	117	117	328	62	62	134	35	35	80
0.02 ***	2108	2125	2125	678	678	678	183	183	183	73	73	73	45	45	45
•••	2108	2125	6022	678	678	1793	183	183	419	73	73	155	45	45	89
0.05 ***	6453	6525	6645	1833	1850	1860	392	392	392	145	145	145	73	73	73
•••	6480	6580	12848	1850	1850	3481	392	392	681	145	145	217	73	73	115
0.1 ***	14335	14637	15473	3940	4005	4088	750	760	760	238	238	238	128	128	128
•••	14455	14940	23235	3967	4050	6047	750	760	1076	238	238	310	128	128	153
0.15 ***	21540	22200	24438	5920	6085	6332	1090	1107	1118	320	320	320	155	155	155
•••	21798	22925	32385	5985	6195	8304	1107	1107	1422	320	320	390	155	155	186
0.2 ***	27507	28590	32825	7625	7910	8405	1410	1420	1437	392	392	403	200	200	200
•••	27903	29855	40298	7745	8120	10251	1410	1437	1719	392	392	458	200	200	213
0.25 ***	32045	33623	40305	9010	9450	10248	1668	1695	1723	458	458	458	227	227	227
•••	32633	35520	46976	9203	9780	11890	1685	1712	1967	458	458	514	227	227	235
0.3 ***	35207	37335	46675	10072	10677	11815	1877	1915	1960	513	513	513	238	238	238
•••	36005	39893	52416	10330	11145	13219	1888	1932	2164	513	513	557	238	238	252
0.35 ***	37088	39810	51828	10825	11595	13070	2042	2080	2152	557	557	557	255	255	255
•••	38122	43028	56620	11155	12190	14239	2070	2108	2313	557	557	588	255	255	263
0.4 ***	37858	41185	55705	11293	12217	14005	2152	2207	2273	568	585	585	265	265	265
•••	39123	45035	59588	11695	12932	14950	2180	2245	2412	585	585	606	265	265	268
0.45 ***	37682	41598	58252	11513	12558	14610	2218	2290	2355	585	595	595	265	265	265
•••	39195	46015	61319	11970	13372	15352	2245	2317	2461	595	595	613	265	265	268
0.5 ***	36747	41175	59473	11502	12640	14885	2235	2300	2383	585	595	595	265	265	265
•••	38480	46053	61814	11997	13520	15445	2262	2345	2461	595	595	606	265	265	263
0.55 ***	35207	40047	59352	11282	12475	14813	2207	2273	2355	568	585	585	255	255	255
•••	37115	45245	61072	11805	13400	15228	2235	2317	2412	585	585	588	255	255	252
0.6 ***	33200	38315	57878	10870	12080	14417	2125	2190	2273	540	557	557	238	238	238
•••	35245	43660	59093	11392	12998	14703	2163	2235	2313	557	557	557	238	238	235
0.65 ***	30818	36015	55062	10275	11447	13685	1998	2070	2152	513	513	513	227	227	227
•••	32925	41340	55878	10787	12338	13868	2042	2108	2164	513	513	514	227	227	213
0.7 ***	28140	33210	50920	9495	10595	12613	1833	1888	1960	458	458	458	200	200	200
•••	30202	38298	51427	9973	11403	12724	1860	1932	1967	458	458	458	200	200	186
0.75 ***	25142	29900	45448	8532	9495	11210	1613	1668	1723	392	392	403	155	155	155
•••	27095	34530	45739	8955	10210	11271	1640	1695	1719	392	392	390	155	155	153
0.8 ***	21815	26005	38672	7377	8158	9478	1355	1393	1437	320	320	320	128	128	128
•••	23547	29965	38815	7718	8725	9509	1365	1410	1422	320	320	310	128	128	115
0.85 ***	18047	21430	30598	5975	6552	7432	1035	1063	1090	227	227	227	.	.	.
•••	19460	24482	30654	6233	6937	7438	1052	1080	1076	227	227	217	.	.	.
0.9 ***	13592	15913	21248	4280	4610	5067	667	678	695	.	.	.	.	.	.
•••	14582	17855	21256	4435	4820	5058	678	678	681	.	.	.	.	.	.
0.95 ***	7900	8890	10633	2163	2262	2383	.	.	.	.	.	.	.	.	.
•••	8340	9605	10622	2218	2317	2368	.	.	.	.	.	.	.	.	.
0.98 ***	3197	3400	3665	.	.	.	.	.	.	.	.	.	.	.	.
•••	3290	3527	3648	.	.	.	.	.	.	.	.	.	.	.	.

TABLE 7: ALPHA= 0.05 POWER= 0.8   EXPECTED ACCRUAL THRU MINIMUM FOLLOW-UP= 12000

		DEL=.01			DEL=.02			DEL=.05			DEL=.10			DEL=.15		
FACT=		1.0 .75	.50 .25	.00 BIN	1.0 .75	.50 .25	.00 BIN	1.0 .75	.50 .25	.00 BIN	1.0 75	.50 .25	.00 BIN	1.0 .75	.50 .25	.00 BIN
PCONT=•••					REQUIRED NUMBER OF PATIENTS											
0.01	•••	949	949	949	349	349	349	109	109	109	49	49	49	38	38	38
	•••	949	949	3648	349	349	1206	109	109	328	49	49	134	38	38	80
0.02	•••	2108	2119	2119	679	679	679	188	188	188	79	79	79	49	49	49
	•••	2119	2119	6022	679	679	1793	188	188	419	79	79	155	49	49	89
0.05	•••	6458	6529	6649	1838	1849	1868	398	398	398	139	139	139	79	79	79
	•••	6488	6578	12848	1849	1849	3481	398	398	681	139	139	217	79	79	115
0.1	•••	14359	14678	15469	3949	4009	4088	758	758	758	229	229	229	128	128	128
	•••	14498	14978	23235	3979	4039	6047	758	758	1076	229	229	310	128	128	153
0.15	•••	21619	22298	24439	5948	6098	6319	1099	1099	1118	319	319	319	158	158	158
	•••	21878	22999	32385	6008	6199	8304	1099	1118	1422	319	319	390	158	158	186
0.2	•••	27619	28748	32828	7658	7939	8408	1399	1418	1448	398	398	398	188	188	188
	•••	28039	30008	40298	7789	8149	10251	1418	1429	1719	398	398	458	188	188	213
0.25	•••	32209	33859	40309	9068	9499	10249	1669	1699	1718	458	458	469	218	218	218
	•••	32839	35749	46976	9259	9818	11890	1688	1699	1967	458	458	514	218	218	235
0.3	•••	35438	37658	46669	10148	10748	11809	1879	1909	1958	518	518	518	248	248	248
	•••	36278	40208	52416	10418	11179	13219	1898	1939	2164	518	518	557	248	248	252
0.35	•••	37388	40208	51829	10928	11678	13069	2048	2089	2149	548	559	559	259	259	259
	•••	38468	43448	56620	11258	12248	14239	2078	2119	2313	548	559	588	259	259	263
0.4	•••	38228	41659	55699	11419	12319	13999	2168	2209	2288	578	578	589	259	259	259
	•••	39548	45529	59588	11809	12998	14950	2179	2239	2412	578	578	606	259	259	268
0.45	•••	38119	42139	58249	11659	12679	14618	2228	2288	2359	589	589	589	259	259	278
	•••	39679	46568	61319	12098	13448	15352	2258	2318	2461	589	589	613	259	278	268
0.5	•••	37249	41779	59479	11659	12769	14888	2239	2318	2389	589	589	608	259	259	259
	•••	39038	46658	61814	12139	13609	15445	2269	2348	2461	589	589	606	259	259	263
0.55	•••	35768	40699	59348	11438	12608	14828	2209	2288	2359	578	578	589	259	259	259
	•••	37729	45889	61072	11948	13489	15228	2239	2318	2412	578	578	588	259	259	252
0.6	•••	33799	38978	57878	11029	12199	14419	2138	2198	2288	548	559	559	248	248	248
	•••	35888	44318	59093	11539	13099	14703	2168	2239	2313	548	559	557	248	248	235
0.65	•••	31448	36698	55058	10429	11569	13688	2018	2078	2149	518	518	518	218	218	218
	•••	33578	41978	55878	10939	12428	13868	2048	2108	2164	518	518	514	218	218	213
0.7	•••	28748	33878	50918	9638	10699	12608	1838	1898	1958	458	458	469	199	199	199
	•••	30848	38899	51427	10118	11498	12724	1868	1928	1967	458	458	458	199	199	186
0.75	•••	25729	30499	45458	8659	9608	11209	1628	1669	1718	398	398	398	158	158	158
	•••	27698	35078	45739	9079	10279	11271	1639	1699	1719	398	398	390	158	158	153
0.8	•••	22339	26528	38678	7478	8239	9488	1358	1388	1429	319	319	319	128	128	128
	•••	24079	30428	38815	7819	8779	9509	1369	1418	1422	319	319	310	128	128	115
0.85	•••	18469	21848	30608	6049	6608	7429	1039	1058	1088	229	229	229	.	.	.
	•••	19898	24829	30654	6308	6968	7438	1058	1069	1076	229	229	217	.	.	.
0.9	•••	13898	16178	21248	4328	4639	5059	668	679	698	.	.	.	.	.	.
	•••	14869	18068	21256	4478	4838	5058	679	679	681	.	.	.	.	.	.
0.95	•••	8029	8989	10628	2179	2269	2378	.	.	.	.	.	.	.	.	.
	•••	8468	9679	10622	2228	2318	2368	.	.	.	.	.	.	.	.	.
0.98	•••	3229	3409	3668	.	.	.	.	.	.	.	.	.	.	.	.
	•••	3319	3529	3648	.	.	.	.	.	.	.	.	.	.	.	.

TABLE 7: ALPHA= 0.05 POWER= 0.8   EXPECTED ACCRUAL THRU MINIMUM FOLLOW-UP= 13000

		DEL=.01			DEL=.02			DEL=.05			DEL=.10			DEL=.15		
FACT=		1.0 .75	.50 .25	.00 BIN	1.0 .75	.50 .25	.00 BIN	1.0 .75	.50 .25	.00 BIN	1.0 75	.50 .25	.00 BIN	1.0 .75	.50 .25	.00 BIN
PCONT=***								REQUIRED NUMBER OF PATIENTS								
0.01	***	951	951	951	346	346	346	118	118	118	53	53	53	41	41	41
	***	951	951	3648	346	346	1206	118	118	328	53	53	134	41	41	80
0.02	***	2121	2121	2121	671	671	671	183	183	183	74	74	74	53	53	53
	***	2121	2121	6022	671	671	1793	183	183	419	74	74	155	53	53	89
0.05	***	6476	6541	6651	1841	1841	1861	399	399	399	139	139	139	86	86	86
	***	6488	6586	12848	1841	1861	3481	399	399	681	139	139	217	86	86	115
0.1	***	14406	14711	15479	3953	4018	4083	756	756	756	236	236	236	118	118	118
	***	14536	15003	23235	3986	4051	6047	756	756	1076	236	236	310	118	118	153
0.15	***	21686	22381	24428	5956	6119	6326	1093	1114	1114	313	313	313	151	151	151
	***	21958	23084	32385	6033	6216	8304	1093	1114	1422	313	313	390	151	151	186
0.2	***	27731	28881	32834	7691	7971	8406	1406	1418	1439	399	399	399	183	183	183
	***	28166	30148	40298	7821	8166	10251	1418	1439	1719	399	399	458	183	183	213
0.25	***	32379	34069	40309	9121	9543	10246	1678	1699	1731	464	464	464	216	216	216
	***	33029	35966	46976	9304	9836	11890	1678	1711	1967	464	464	514	216	216	235
0.3	***	35661	37936	46679	10226	10799	11818	1894	1926	1959	508	508	508	236	236	236
	***	36539	40504	52416	10486	11221	13219	1906	1938	2164	508	508	557	236	236	252
0.35	***	37676	40569	51826	11026	11753	13074	2056	2101	2154	541	561	561	248	248	248
	***	38793	43798	56620	11351	12294	14239	2068	2121	2313	561	561	588	248	248	263
0.4	***	38566	42096	55693	11526	12403	14016	2166	2219	2284	573	573	594	269	269	269
	***	39951	45964	59588	11916	13053	14950	2198	2251	2412	573	573	606	269	269	268
0.45	***	38533	42649	58261	11786	12781	14613	2231	2296	2361	594	594	594	269	269	269
	***	40146	47069	61319	12229	13529	15352	2263	2328	2461	594	594	613	269	269	268
0.5	***	37721	42336	59484	11786	12879	14894	2251	2316	2393	594	594	594	269	269	269
	***	39561	47199	61814	12273	13691	15445	2284	2349	2461	594	594	606	269	269	263
0.55	***	36291	41296	59354	11591	12728	14829	2219	2284	2361	573	573	594	248	248	248
	***	38306	46463	61072	12099	13573	15228	2251	2328	2412	573	573	588	248	248	252
0.6	***	34373	39606	57871	11168	12326	14418	2154	2198	2284	541	561	561	236	236	248
	***	36486	44903	59093	11688	13183	14703	2186	2251	2313	561	561	557	236	248	235
0.65	***	32021	37319	55064	10571	11688	13671	2024	2089	2154	508	508	529	216	216	216
	***	34178	42563	55878	11071	12501	13868	2056	2121	2164	508	508	514	216	216	213
0.7	***	29324	34471	50916	9771	10811	12619	1841	1906	1959	464	464	464	204	204	204
	***	31436	39443	51427	10246	11558	12724	1873	1926	1967	464	464	458	204	204	186
0.75	***	26269	31046	45444	8784	9694	11221	1634	1678	1711	399	399	399	171	171	171
	***	28251	35564	45739	9186	10344	11271	1646	1699	1719	399	399	390	171	171	153
0.8	***	22824	27016	38663	7581	8308	9478	1374	1386	1439	313	313	313	118	118	118
	***	24558	30831	38815	7918	8816	9509	1374	1406	1422	313	313	310	118	118	115
0.85	***	18871	22218	30603	6131	6651	7431	1049	1061	1081	216	236	236	.	.	.
	***	20289	25131	30654	6379	6996	7438	1061	1081	1076	216	236	217	.	.	.
0.9	***	14158	16421	21243	4376	4668	5058	671	691	691	.	.	.	.	.	.
	***	15133	18241	21256	4506	4851	5058	671	691	681	.	.	.	.	.	.
0.95	***	8146	9088	10636	2198	2284	2381	.	.	.	.	.	.	.	.	.
	***	8568	9738	10622	2231	2328	2368	.	.	.	.	.	.	.	.	.
0.98	***	3259	3433	3661	.	.	.	.	.	.	.	.	.	.	.	.
	***	3336	3531	3648	.	.	.	.	.	.	.	.	.	.	.	.

TABLE 7: ALPHA= 0.05 POWER= 0.8   EXPECTED ACCRUAL THRU MINIMUM FOLLOW-UP= 14000

		DEL=.01			DEL=.02			DEL=.05			DEL=.10			DEL=.15		
FACT=		1.0 .75	.50 .25	.00 BIN	1.0 .75	.50 .25	.00 BIN	1.0 .75	.50 .25	.00 BIN	1.0 75	.50 .25	.00 BIN	1.0 .75	.50 .25	.00 BIN
PCONT=•••				REQUIRED NUMBER OF PATIENTS												
0.01	•••	954	954	954	337	337	337	114	114	114	57	57	57	44	44	44
	•••	954	954	3648	337	337	1206	114	114	328	57	57	134	44	44	80
0.02	•••	2109	2122	2122	674	674	674	184	184	184	79	79	79	44	44	44
	•••	2122	2122	6022	674	674	1793	184	184	419	79	79	155	44	44	89
0.05	•••	6484	6532	6637	1842	1842	1864	394	394	394	149	149	149	79	79	79
	•••	6497	6589	12848	1842	1864	3481	394	394	681	149	149	217	79	79	115
0.1	•••	14429	14757	15479	3964	4012	4082	757	757	757	232	232	232	127	127	127
	•••	14569	15024	23235	3999	4047	6047	757	757	1076	232	232	310	127	127	153
0.15	•••	21757	22457	24439	5972	6112	6322	1094	1107	1107	324	324	324	162	162	162
	•••	22037	23157	32385	6042	6217	8304	1107	1107	1422	324	324	390	162	162	186
0.2	•••	27834	29024	32839	7722	8002	8409	1409	1422	1444	394	394	394	197	197	197
	•••	28302	30262	40298	7849	8177	10251	1422	1422	1719	394	394	458	197	197	213
0.25	•••	32537	34274	40307	9179	9577	10242	1667	1702	1724	464	464	464	219	219	219
	•••	33202	36164	46976	9354	9879	11890	1689	1702	1967	464	464	514	219	219	235
0.3	•••	35862	38207	46677	10299	10859	11817	1899	1934	1947	512	512	512	232	232	232
	•••	36772	40762	52416	10544	11257	13219	1912	1934	2164	512	512	557	232	232	252
0.35	•••	37949	40924	51822	11104	11817	13077	2052	2109	2144	547	547	547	254	254	254
	•••	39104	44144	56620	11419	12342	14239	2074	2122	2313	547	547	588	254	254	263
0.4	•••	38907	42512	55694	11629	12482	14009	2179	2227	2284	582	582	582	267	267	267
	•••	40329	46362	59588	12014	13112	14950	2192	2249	2412	582	582	606	267	267	268
0.45	•••	38929	43107	58249	11887	12867	14617	2249	2297	2354	582	604	604	267	267	267
	•••	40587	47517	61319	12329	13589	15352	2262	2332	2461	582	604	613	267	267	268
0.5	•••	38194	42862	59474	11922	12972	14884	2262	2319	2389	582	604	604	267	267	267
	•••	40049	47714	61814	12399	13764	15445	2284	2354	2461	582	604	606	267	267	263
0.55	•••	36794	41869	59347	11712	12832	14814	2227	2297	2367	582	582	582	254	254	254
	•••	38837	46992	61072	12202	13637	15228	2262	2319	2412	582	582	588	254	254	252
0.6	•••	34904	40189	57877	11314	12434	14407	2157	2214	2284	547	547	569	232	232	232
	•••	37039	45439	59093	11804	13252	14703	2179	2249	2313	547	547	557	232	232	235
0.65	•••	32572	37892	55064	10697	11782	13672	2039	2087	2144	512	512	512	219	219	219
	•••	34742	43094	55878	11187	12574	13868	2052	2109	2164	512	512	514	219	219	213
0.7	•••	29864	35022	50912	9892	10907	12609	1864	1899	1969	464	464	464	197	197	197
	•••	31977	39944	51427	10347	11629	12724	1877	1934	1967	464	464	458	197	197	186
0.75	•••	26762	31557	45452	8899	9774	11209	1632	1667	1724	394	394	394	162	162	162
	•••	28757	36002	45739	9284	10382	11271	1654	1702	1719	394	394	390	162	162	153
0.8	•••	23262	27449	38662	7674	8387	9494	1374	1387	1422	324	324	324	127	127	127
	•••	25012	31194	38815	7989	8864	9509	1387	1409	1422	324	324	310	127	127	115
0.85	•••	19237	22549	30599	6182	6694	7429	1059	1072	1094	219	232	232	.	.	.
	•••	20637	25397	30654	6427	7022	7438	1059	1072	1076	219	232	217	.	.	.
0.9	•••	14429	16634	21254	4419	4699	5062	674	687	687	.	.	.	.	.	.
	•••	15387	18397	21256	4537	4874	5058	674	687	681	.	.	.	.	.	.
0.95	•••	8269	9179	10627	2214	2284	2367	.	.	.	.	.	.	.	.	.
	•••	8667	9787	10622	2249	2332	2368	.	.	.	.	.	.	.	.	.
0.98	•••	3277	3452	3662	.	.	.	.	.	.	.	.	.	.	.	.
	•••	3369	3544	3648	.	.	.	.	.	.	.	.	.	.	.	.

TABLE 7: ALPHA= 0.05 POWER= 0.8   EXPECTED ACCRUAL THRU MINIMUM FOLLOW-UP= 15000

		DEL=.01			DEL=.02			DEL=.05			DEL=.10			DEL=.15		
FACT=		1.0 / .75	.50 / .25	.00 / BIN	1.0 / .75	.50 / .25	.00 / BIN	1.0 / .75	.50 / .25	.00 / BIN	1.0 / 75	.50 / .25	.00 / BIN	1.0 / .75	.50 / .25	.00 / BIN
PCONT=•••		REQUIRED NUMBER OF PATIENTS														
0.01	•••	947	947	947	347	347	347	122	122	122	61	61	61	47	47	47
	•••	947	947	3648	347	347	1206	122	122	328	61	61	134	47	47	80
0.02	•••	2110	2124	2124	685	685	685	174	174	174	85	85	85	47	47	47
	•••	2124	2124	6022	685	685	1793	174	174	419	85	85	155	47	47	89
0.05	•••	6474	6549	6647	1847	1847	1861	399	399	399	136	136	136	85	85	85
	•••	6511	6586	12848	1847	1861	3481	399	399	681	136	136	217	85	85	115
0.1	•••	14461	14785	15474	3961	4022	4097	760	760	760	235	235	235	122	122	122
	•••	14597	15047	23235	3999	4060	6047	760	760	1076	235	235	310	122	122	153
0.15	•••	21811	22524	24436	5986	6136	6324	1097	1111	1111	324	324	324	160	160	160
	•••	22097	23222	32385	6061	6211	8304	1097	1111	1422	324	324	390	160	160	186
0.2	•••	27947	29147	32836	7749	8011	8410	1411	1435	1435	399	399	399	197	197	197
	•••	28411	30385	40298	7885	8185	10251	1411	1435	1719	399	399	458	197	197	213
0.25	•••	32686	34472	40299	9211	9610	10247	1674	1697	1711	460	460	460	211	211	211
	•••	33385	36324	46976	9399	9886	11890	1697	1711	1967	460	460	514	211	211	235
0.3	•••	36085	38461	46674	10360	10899	11822	1899	1922	1960	511	511	511	235	235	235
	•••	36999	41011	52416	10599	11297	13219	1899	1936	2164	511	511	557	235	235	252
0.35	•••	38199	41236	51835	11185	11874	13074	2072	2110	2147	549	549	549	249	249	249
	•••	39399	44447	56620	11499	12385	14239	2086	2124	2313	549	549	588	249	249	263
0.4	•••	39235	42886	55697	11724	12572	14011	2185	2222	2274	572	586	586	272	272	272
	•••	40674	46735	59588	12099	13172	14950	2199	2260	2412	572	586	606	272	272	268
0.45	•••	39324	43561	58261	12010	12947	14611	2260	2297	2349	586	586	586	272	272	272
	•••	41011	47935	61319	12422	13636	15352	2274	2335	2461	586	586	613	272	272	268
0.5	•••	38611	43360	59485	12047	13074	14874	2274	2335	2386	586	586	610	272	272	272
	•••	40524	48160	61814	12497	13824	15445	2297	2349	2461	586	586	606	272	272	263
0.55	•••	37285	42385	59349	11836	12924	14822	2236	2297	2372	572	586	586	249	249	249
	•••	39347	47461	61072	12324	13711	15228	2260	2335	2412	572	586	588	249	249	252
0.6	•••	35410	40711	57872	11424	12535	14410	2161	2222	2274	549	549	549	235	235	249
	•••	37561	45924	59093	11911	13322	14703	2185	2236	2313	549	549	557	235	249	235
0.65	•••	33085	38424	55060	10824	11874	13674	2035	2086	2147	511	511	511	211	211	211
	•••	35274	43561	55878	11297	12647	13868	2072	2110	2164	511	511	514	211	211	213
0.7	•••	30361	35536	50911	10022	10997	12610	1861	1899	1960	460	460	460	197	197	197
	•••	32499	40397	51427	10449	11672	12724	1885	1936	1967	460	460	458	197	197	186
0.75	•••	27249	32035	45460	8986	9849	11222	1636	1674	1711	399	399	399	160	160	160
	•••	29236	36399	45739	9385	10435	11271	1660	1697	1719	399	399	390	160	160	153
0.8	•••	23686	27849	38672	7749	8447	9474	1374	1397	1435	310	324	324	122	122	122
	•••	25435	31524	38815	8072	8897	9509	1374	1411	1422	310	324	310	122	122	115
0.85	•••	19561	22861	30610	6249	6736	7435	1060	1074	1074	235	235	235	.	.	.
	•••	20972	25660	30654	6474	7060	7438	1060	1074	1076	235	235	217	.	.	.
0.9	•••	14649	16847	21249	4449	4711	5072	685	685	685	.	.	.	.	.	.
	•••	15610	18549	21256	4561	4885	5058	685	685	681	.	.	.	.	.	.
0.95	•••	8372	9249	10636	2222	2297	2372	.	.	.	.	.	.	.	.	.
	•••	8761	9849	10622	2260	2335	2368	.	.	.	.	.	.	.	.	.
0.98	•••	3310	3460	3661	.	.	.	.	.	.	.	.	.	.	.	.
	•••	3385	3549	3648	.	.	.	.	.	.	.	.	.	.	.	.

TABLE 7: ALPHA= 0.05 POWER= 0.8   EXPECTED ACCRUAL THRU MINIMUM FOLLOW-UP= 17000

		DEL=.01			DEL=.02			DEL=.05			DEL=.10			DEL=.15		
FACT=		1.0 .75	.50 .25	.00 BIN	1.0 .75	.50 .25	.00 BIN	1.0 .75	.50 .25	.00 BIN	1.0 75	.50 .25	.00 BIN	1.0 .75	.50 .25	.00 BIN
PCONT=***				REQUIRED	NUMBER	OF PATIENTS									
0.01	***	946 946	946	351 351	351	112 112	112	54 54	54	27 27	27				
	•••	946 946	3648	351 351	1206	112 112	328	54 54	134	27 27	80				
0.02	***	2110 2110	2136	665 665	665	181 181	181	70 70	70	54 54	54				
	•••	2110 2110	6022	665 665	1793	181 181	419	70 70	155	54 54	89				
0.05	***	6487 6556	6641	1839 1855	1855	394 394	394	139 139	139	70 70	70				
	•••	6530 6599	12848	1855 1855	3481	394 394	681	139 139	217	70 70	115				
0.1	***	14520 14844	15481	3980 4022	4091	750 750	750	240 240	240	112 112	112				
	•••	14647 15099	23235	4006 4065	6047	750 750	1076	240 240	310	112 112	153				
0.15	***	21941 22664	24449	6020 6147	6317	1090 1116	1116	325 325	325	155 155	155				
	•••	22239 23317	32385	6089 6232	8304	1116 1116	1422	325 325	390	155 155	186				
0.2	***	28146 29379	32837	7805 8060	8400	1414 1430	1430	394 394	394	197 197	197				
	•••	28630 30585	40298	7916 8214	10251	1414 1430	1719	394 394	458	197 197	213				
0.25	***	32991 34819	40301	9292 9675	10254	1685 1711	1727	452 452	452	224 224	224				
	•••	33714 36646	46976	9462 9930	11890	1685 1711	1967	452 452	514	224 224	235				
0.3	***	36476 38941	46676	10466 10992	11826	1897 1924	1966	521 521	521	240 240	240				
	•••	37454 41422	52416	10695 11359	13219	1924 1940	2164	521 521	557	240 240	252				
0.35	***	38702 41831	51835	11316 11996	13075	2067 2110	2136	564 564	564	266 266	266				
	•••	39961 44992	56620	11630 12464	14239	2094 2136	2313	564 564	588	266 266	263				
0.4	***	39850 43590	55702	11885 12692	14010	2195 2237	2280	580 580	580	266 266	266				
	•••	41337 47372	59588	12251 13245	14950	2221 2264	2412	580 580	606	266 266	268				
0.45	***	40046 44355	58252	12182 13101	14605	2264 2306	2365	580 606	606	266 266	266				
	•••	41789 48674	61319	12591 13739	15352	2280 2322	2461	580 606	613	266 266	268				
0.5	***	39425 44254	59469	12251 13229	14886	2280 2322	2391	580 606	606	266 266	266				
	•••	41380 48971	61814	12676 13925	15445	2306 2365	2461	580 606	606	266 266	263				
0.55	***	38150 43319	59341	12055 13101	14817	2264 2306	2365	580 580	580	266 266	266				
	•••	40275 48334	61072	12522 13824	15228	2280 2322	2412	580 580	588	266 266	252				
0.6	***	36349 41704	57870	11656 12692	14419	2179 2221	2280	564 564	564	240 240	240				
	•••	38532 46804	59093	12124 13415	14703	2195 2264	2313	564 564	557	240 240	235				
0.65	***	34027 39382	55065	11035 12039	13670	2051 2094	2152	521 521	521	224 224	224				
	•••	36237 44424	55878	11486 12735	13868	2067 2110	2164	521 521	514	224 224	213				
0.7	***	31291 36450	50926	10211 11146	12607	1881 1924	1966	452 452	452	197 197	197				
	•••	33432 41194	51427	10636 11784	12724	1897 1940	1967	452 452	458	197 197	186				
0.75	***	28104 32864	45444	9165 9972	11205	1642 1685	1727	394 394	394	155 155	155				
	•••	30101 37114	45739	9531 10509	11271	1669 1711	1719	394 394	390	155 155	153				
0.8	***	24449 28545	38660	7890 8527	9489	1371 1414	1430	309 325	325	112 112	112				
	•••	26191 32115	38815	8187 8952	9509	1387 1414	1422	325 325	310	112 112	115				
0.85	***	20199 23402	30611	6360 6811	7422	1047 1074	1090	224 224	224	. .	.				
	•••	21575 26080	30654	6572 7082	7438	1074 1074	1076	224 224	217	. .	.				
0.9	***	15072 17181	21261	4490 4745	5069	665 691	691	. .	.	. .	.				
	•••	16007 18796	21256	4617 4899	5058	691 691	681	. .	.	. .	.				
0.95	***	8554 9377	10636	2237 2306	2365	. .	.	. .	.	. .	.				
	•••	8936 9930	10622	2264 2349	2368	. .	.	. .	.	. .	.				
0.98	***	3326 3470	3666	. .	.	. .	.	. .	.	. .	.				
	•••	3411 3555	3648	. .	.	. .	.	. .	.	. .	.				

TABLE 7: ALPHA= 0.05 POWER= 0.8   EXPECTED ACCRUAL THRU MINIMUM FOLLOW-UP= 20000

		DEL=.01			DEL=.02			DEL=.05			DEL=.10			DEL=.15		
FACT=		1.0 .75	.50 .25	.00 BIN	1.0 .75	.50 .25	.00 BIN	1.0 .75	.50 .25	.00 BIN	1.0 75	.50 .25	.00 BIN	1.0 .75	.50 .25	.00 BIN
PCONT=***					REQUIRED NUMBER OF PATIENTS											
0.01	***	963	963	963	332	332	332	113	113	113	63	63	63	32	32	32
	•••	963	963	3648	332	332	1206	113	113	328	63	63	134	32	32	80
0.02	***	2113	2113	2132	682	682	682	182	182	182	82	82	82	63	63	63
	•••	2113	2132	6022	682	682	1793	182	182	419	82	82	155	63	63	89
0.05	***	6513	6563	6632	1832	1863	1863	382	382	382	132	132	132	82	82	82
	•••	6532	6613	12848	1863	1863	3481	382	382	681	132	132	217	82	82	115
0.1	***	14582	14913	15463	3982	4032	4082	763	763	763	232	232	232	113	113	113
	•••	14732	15132	23235	4013	4063	6047	763	763	1076	232	232	310	113	113	153
0.15	***	22113	22832	24432	6063	6182	6332	1113	1113	1113	313	313	313	163	163	163
	•••	22413	23463	32385	6113	6232	8304	1113	1113	1422	313	313	390	163	163	186
0.2	***	28413	29682	32832	7882	8113	8413	1413	1432	1432	382	382	382	182	182	182
	•••	28932	30832	40298	7982	8232	10251	1413	1432	1719	382	382	458	182	182	213
0.25	***	33382	35263	40313	9382	9732	10263	1682	1713	1713	463	463	463	213	213	213
	•••	34132	37032	46976	9563	9963	11890	1682	1713	1967	463	463	514	213	213	235
0.3	***	37013	39532	46682	10613	11082	11813	1913	1932	1963	513	513	513	232	232	232
	•••	38032	41963	52416	10813	11413	13219	1913	1932	2164	513	513	557	232	232	252
0.35	***	39382	42582	51832	11482	12113	13063	2082	2113	2132	563	563	563	263	263	263
	•••	40682	45663	56620	11782	12532	14239	2082	2132	2313	563	563	588	263	263	263
0.4	***	40682	44513	55713	12113	12832	14013	2213	2232	2282	582	582	582	263	263	263
	•••	42232	48182	59588	12432	13363	14950	2213	2263	2412	582	582	606	263	263	268
0.45	***	41013	45413	58263	12432	13282	14613	2282	2313	2363	582	582	582	263	263	263
	•••	42813	49613	61319	12813	13863	15352	2282	2332	2461	582	582	613	263	263	268
0.5	***	40513	45382	59482	12482	13413	14882	2282	2332	2382	582	582	613	263	263	263
	•••	42513	49982	61814	12913	14063	15445	2313	2363	2461	582	582	606	263	263	263
0.55	***	39332	44532	59332	12332	13282	14813	2263	2313	2363	582	582	582	263	263	263
	•••	41482	49382	61072	12763	13963	15228	2282	2332	2412	582	582	588	263	263	252
0.6	***	37563	42932	57882	11913	12882	14413	2182	2232	2282	563	563	563	232	232	232
	•••	39813	47882	59093	12363	13563	14703	2213	2263	2313	563	563	557	232	232	235
0.65	***	35282	40632	55063	11282	12232	13682	2063	2113	2132	513	513	513	213	213	213
	•••	37513	45463	55878	11713	12863	13868	2082	2132	2164	513	513	514	213	213	213
0.7	***	32513	37632	50913	10463	11313	12613	1882	1913	1963	463	463	463	182	182	182
	•••	34663	42182	51427	10832	11882	12724	1913	1932	1967	463	463	458	182	182	186
0.75	***	29232	33932	45463	9382	10113	11213	1663	1682	1713	382	382	413	163	163	163
	•••	31213	37982	45739	9713	10613	11271	1682	1713	1719	382	382	390	163	163	153
0.8	***	25432	29463	38663	8063	8663	9482	1382	1413	1432	313	313	313	132	132	132
	•••	27163	32813	38815	8332	9032	9509	1382	1413	1422	313	313	310	132	132	115
0.85	***	20982	24113	30613	6482	6882	7432	1063	1063	1082	232	232	232	.	.	.
	•••	22332	26582	30654	6663	7132	7438	1063	1082	1076	232	232	217	.	.	.
0.9	***	15613	17613	21263	4563	4782	5063	682	682	682	.	.	.	.	.	.
	•••	16482	19082	21256	4682	4913	5058	682	682	681	.	.	.	.	.	.
0.95	***	8763	9532	10632	2263	2313	2382	.	.	.	.	.	.	.	.	.
	•••	9113	10013	10622	2282	2332	2368	.	.	.	.	.	.	.	.	.
0.98	***	3382	3513	3663	.	.	.	.	.	.	.	.	.	.	.	.
	•••	3432	3582	3648	.	.	.	.	.	.	.	.	.	.	.	.

TABLE 7: ALPHA= 0.05 POWER= 0.8   EXPECTED ACCRUAL THRU MINIMUM FOLLOW-UP= 25000

		DEL=.01			DEL=.02			DEL=.05			DEL=.10			DEL=.15	
FACT=	1.0 .75	.50 .25	.00 BIN	1.0 .75	.50 .25	.00 BIN	1.0 .75	.50 .25	.00 BIN	1.0 75	.50 .25	.00 BIN	1.0 .75	.50 .25	.00 BIN
PCONT=•••				REQUIRED NUMBER OF PATIENTS											
0.01 ***	954	954	954	352	352	352	102	102	102	40	40	40	40	40	40
•••	954	954	3648	352	352	1206	102	102	328	40	40	134	40	40	80
0.02 ***	2102	2102	2141	665	665	665	165	165	165	79	79	79	40	40	40
•••	2102	2141	6022	665	665	1793	165	165	419	79	79	155	40	40	89
0.05 ***	6540	6579	6641	1852	1852	1852	391	391	391	141	141	141	79	79	79
•••	6540	6602	12848	1852	1852	3481	391	391	681	141	141	217	79	79	115
0.1 ***	14704	14977	15477	4016	4040	4079	766	766	766	227	227	227	102	102	102
•••	14829	15204	23235	4016	4079	6047	766	766	1076	227	227	310	102	102	153
0.15 ***	22329	23040	24454	6102	6204	6329	1102	1102	1102	329	329	329	165	165	165
•••	22641	23602	32385	6141	6266	8304	1102	1102	1422	329	329	390	165	165	186
0.2 ***	28829	30079	32829	7954	8141	8415	1415	1415	1454	391	391	391	204	204	204
•••	29352	31141	40298	8040	8266	10251	1415	1415	1719	391	391	458	204	204	213
0.25 ***	33977	35852	40290	9516	9829	10266	1704	1704	1727	454	454	454	227	227	227
•••	34766	37516	46976	9665	10016	11890	1704	1704	1967	454	454	514	227	227	235
0.3 ***	37790	40352	46665	10766	11204	11829	1915	1954	1954	516	516	516	227	227	227
•••	38852	42641	52416	10977	11477	13219	1915	1954	2164	516	516	557	227	227	252
0.35 ***	40391	43641	51829	11727	12266	13079	2102	2102	2141	540	540	540	266	266	266
•••	41727	46540	56620	11977	12641	14239	2102	2141	2313	540	540	588	266	266	263
0.4 ***	41891	45727	55704	12352	13040	14016	2227	2227	2266	579	579	579	266	266	266
•••	43477	49227	59588	12665	13454	14950	2227	2266	2412	579	579	606	266	266	268
0.45 ***	42391	46829	58266	12727	13477	14602	2290	2329	2352	602	602	602	266	266	266
•••	44227	50790	61319	13079	13977	15352	2290	2329	2461	602	602	613	266	266	268
0.5 ***	42079	46954	59477	12829	13641	14891	2329	2352	2391	602	602	602	266	266	266
•••	44102	51290	61814	13204	14204	15445	2329	2352	2461	602	602	606	266	266	263
0.55 ***	41016	46165	59352	12665	13540	14829	2290	2329	2352	579	579	579	266	266	266
•••	43165	50766	61072	13079	14102	15228	2290	2329	2412	579	579	588	266	266	252
0.6 ***	39290	44602	57891	12266	13141	14415	2204	2227	2290	540	540	540	227	227	227
•••	41540	49266	59093	12665	13704	14703	2227	2266	2313	540	540	557	227	227	235
0.65 ***	37016	42266	55079	11641	12477	13665	2079	2102	2141	516	516	516	227	227	227
•••	39227	46829	55878	12016	13016	13868	2102	2141	2164	516	516	514	227	227	213
0.7 ***	34165	39165	50915	10766	11516	12602	1891	1915	1954	454	454	454	204	204	204
•••	36290	43415	51427	11102	12016	12724	1915	1954	1967	454	454	458	204	204	186
0.75 ***	30790	35329	45454	9641	10290	11204	1665	1704	1727	391	391	391	165	165	165
•••	32727	39079	45739	9954	10704	11271	1665	1704	1719	391	391	390	165	165	153
0.8 ***	26766	30641	38665	8290	8790	9477	1391	1415	1415	329	329	329	141	141	141
•••	28454	33704	38815	8516	9102	9509	1391	1415	1422	329	329	310	141	141	115
0.85 ***	22040	24977	30602	6641	6977	7415	1079	1079	1079	227	227	227	.	.	.
•••	23329	27227	30654	6790	7204	7438	1079	1079	1076	227	227	217	.	.	.
0.9 ***	16290	18141	21266	4641	4829	5079	665	704	704	.	.	.	.	.	.
•••	17141	19454	21256	4727	4954	5058	665	704	681	.	.	.	.	.	.
0.95 ***	9040	9704	10641	2266	2329	2391	.	.	.	.	.	.	.	.	.
•••	9352	10141	10622	2290	2352	2368	.	.	.	.	.	.	.	.	.
0.98 ***	3415	3540	3665	.	.	.	.	.	.	.	.	.	.	.	.
•••	3477	3602	3648	.	.	.	.	.	.	.	.	.	.	.	.

TABLE 8: ALPHA= 0.05 POWER= 0.9   EXPECTED ACCRUAL THRU MINIMUM FOLLOW-UP= 30

	DEL=.10			DEL=.15			DEL=.20			DEL=.25			DEL=.30		
FACT=	1.0 .75	.50 .25	.00 BIN	1.0 .75	.50 .25	.00 BIN	1.0 .75	.50 .25	.00 BIN	1.0 75	.50 .25	.00 BIN	1.0 .75	.50 .25	.00 BIN
PCONT=***				REQUIRED NUMBER OF PATIENTS											
0.05 ***	180	181	195	105	106	114	74	75	80	58	59	62	48	48	51
•••	180	183	300	105	108	158	74	76	101	58	60	71	48	49	53
0.1 ***	284	286	326	150	153	173	99	101	113	73	75	83	58	60	65
•••	285	290	429	151	157	212	100	104	129	74	78	88	59	62	63
0.15 ***	366	370	447	184	188	225	117	121	142	84	87	101	65	68	77
•••	367	377	540	186	195	257	118	126	153	85	91	101	66	71	72
0.2 ***	426	432	554	208	214	270	129	134	166	91	96	115	69	73	86
•••	428	442	634	210	223	295	131	142	172	93	101	112	71	77	79
0.25 ***	465	472	645	222	230	308	136	143	185	95	101	126	72	77	93
•••	467	486	711	225	243	326	138	153	187	97	108	120	74	82	83
0.3 ***	484	493	719	228	238	337	138	147	199	96	103	134	72	78	98
•••	487	512	771	232	255	349	142	159	197	99	112	126	74	84	86
0.35 ***	485	497	775	227	240	358	137	148	209	95	104	139	71	78	100
•••	489	522	814	232	260	364	141	162	204	98	113	129	74	85	87
0.4 ***	471	487	812	220	236	371	133	146	214	92	102	141	69	77	100
•••	476	518	840	226	260	372	138	162	206	96	113	129	72	84	86
0.45 ***	445	465	831	209	228	375	127	142	214	88	99	139	66	74	98
•••	451	503	848	216	254	372	132	159	204	92	110	126	69	82	83
0.5 ***	409	435	832	194	216	372	119	135	210	83	94	135	62	70	94
•••	418	479	840	202	245	364	125	153	197	87	106	120	65	78	79
0.55 ***	367	399	813	177	202	359	110	127	201	77	88	128	57	66	88
•••	378	450	814	187	232	349	117	145	187	82	100	112	61	73	72
0.6 ***	323	361	776	160	186	339	100	117	187	70	81	117	52	60	80
•••	337	416	771	170	217	326	107	135	172	75	92	101	55	67	63
0.65 ***	279	321	721	142	168	310	90	106	169	63	73	104	46	53	70
•••	294	378	711	152	198	295	96	123	153	67	83	88	49	59	53
0.7 ***	237	281	647	124	149	273	78	93	146	54	63	88	.	.	.
•••	253	337	634	133	177	257	84	108	129	58	71	71	.	.	.
0.75 ***	197	241	555	105	127	228	66	78	118	.	.	.	.	.	.
•••	214	293	540	114	152	212	71	90	101	.	.	.	.	.	.
0.8 ***	160	199	445	85	103	175	.	.	.	.	.	.	.	.	.
•••	175	245	429	92	123	158	.	.	.	.	.	.	.	.	.
0.85 ***	123	154	317	.	.	.	.	.	.	.	.	.	.	.	.
•••	135	189	300	.	.	.	.	.	.	.	.	.	.	.	.

TABLE 8: ALPHA= 0.05 POWER= 0.9   EXPECTED ACCRUAL THRU MINIMUM FOLLOW-UP= 40

		DEL=.10			DEL=.15			DEL=.20			DEL=.25			DEL=.30		
FACT=		1.0 .75	.50 .25	.00 BIN	1.0 .75	.50 .25	.00 BIN	1.0 .75	.50 .25	.00 BIN	1.0 75	.50 .25	.00 BIN	1.0 .75	.50 .25	.00 BIN
PCONT=•••						REQUIRED NUMBER OF PATIENTS										
0.05	***	180	182	195	105	107	114	74	76	80	58	59	62	48	49	51
	•••	181	184	300	106	109	158	75	77	101	58	60	71	48	50	53
0.1	***	285	287	326	151	154	173	100	103	113	74	76	83	59	61	65
	•••	286	293	429	152	158	212	101	106	129	75	79	88	60	62	63
0.15	***	367	372	447	186	191	225	118	123	142	85	89	101	66	69	77
	•••	369	381	540	187	198	257	120	128	153	87	93	101	67	72	72
0.2	***	428	435	554	210	217	270	131	137	166	93	98	115	71	75	86
	•••	430	448	634	212	228	295	133	145	172	95	103	112	73	79	79
0.25	***	467	477	645	225	235	308	138	147	185	97	104	126	74	79	93
	•••	471	495	711	228	249	326	141	157	187	100	111	120	76	84	83
0.3	***	487	500	719	232	244	337	142	152	199	99	107	134	74	81	98
	•••	491	524	771	236	263	349	146	165	197	102	115	126	77	86	86
0.35	***	489	506	775	232	248	358	141	154	209	98	108	139	74	81	100
	•••	495	536	814	237	270	364	146	168	204	102	117	129	77	87	87
0.4	***	476	498	812	226	245	371	138	153	214	96	107	141	72	80	100
	•••	483	536	840	233	271	372	144	169	206	100	117	129	75	87	86
0.45	***	451	478	831	216	238	375	132	149	215	92	104	139	69	77	98
	•••	461	524	848	224	267	372	139	166	204	97	115	126	73	85	83
0.5	***	418	451	832	202	227	372	125	143	210	87	99	135	65	74	94
	•••	429	503	840	212	258	364	132	161	197	92	111	120	69	81	79
0.55	***	379	418	813	187	214	359	117	134	201	82	93	128	61	69	88
	•••	393	476	814	197	246	349	124	153	187	86	104	112	64	76	72
0.6	***	337	381	776	170	198	339	107	125	187	75	86	117	55	63	80
	•••	353	443	771	181	230	326	114	142	172	79	96	101	59	69	63
0.65	***	294	343	721	152	180	310	96	113	169	67	77	104	49	56	70
	•••	313	406	711	163	211	295	103	129	153	71	86	88	52	61	53
0.7	***	253	303	647	133	160	273	84	99	146	58	67	88	.	.	.
	•••	272	364	634	144	189	257	90	114	129	62	74	71	.	.	.
0.75	***	214	262	555	114	137	228	71	83	118	.	.	.	.	.	.
	•••	233	317	540	123	162	212	76	95	101	.	.	.	.	.	.
0.8	***	175	217	445	92	111	175	.	.	.	.	.	.	.	.	.
	•••	192	265	429	100	130	158	.	.	.	.	.	.	.	.	.
0.85	***	135	168	317	.	.	.	.	.	.	.	.	.	.	.	.
	•••	148	204	300	.	.	.	.	.	.	.	.	.	.	.	.

TABLE 8: ALPHA= 0.05 POWER= 0.9   EXPECTED ACCRUAL THRU MINIMUM FOLLOW-UP= 50

		DEL=.10			DEL=.15			DEL=.20			DEL=.25			DEL=.30		
FACT=		1.0 .75	.50 .25	.00 BIN	1.0 .75	.50 .25	.00 BIN	1.0 .75	.50 .25	.00 BIN	1.0 75	.50 .25	.00 BIN	1.0 .75	.50 .25	.00 BIN
PCONT=***		REQUIRED NUMBER OF PATIENTS														
0.05	***	181	182	195	106	107	114	75	76	80	58	59	62	48	49	51
	•••	181	185	300	106	109	158	75	77	101	59	60	71	49	50	53
0.1	***	285	289	326	152	155	173	101	104	113	75	77	83	59	61	65
	•••	287	295	429	153	160	212	102	107	129	76	79	88	60	63	63
0.15	***	369	374	447	187	193	225	119	124	142	86	90	101	67	70	77
	•••	371	385	540	189	201	257	121	130	153	88	94	101	68	73	72
0.2	***	430	438	554	212	220	270	133	140	166	94	100	115	72	76	86
	•••	433	454	634	215	232	295	135	147	172	96	105	112	74	80	79
0.25	***	470	482	645	228	239	308	141	150	185	99	106	126	75	80	93
	•••	474	503	711	232	255	326	144	160	187	102	113	120	77	85	83
0.3	***	490	506	719	235	250	337	145	156	199	101	110	134	76	83	98
	•••	496	534	771	241	270	349	149	169	197	105	118	126	79	88	86
0.35	***	493	514	775	236	254	358	145	158	209	101	111	139	76	83	100
	•••	500	550	814	243	278	364	150	173	204	105	120	129	79	89	87
0.4	***	482	508	812	231	253	371	142	158	214	99	110	141	75	82	100
	•••	491	551	840	239	280	372	148	174	206	104	120	129	78	89	86
0.45	***	458	491	831	222	247	375	137	154	215	96	107	139	72	80	98
	•••	470	542	848	231	277	372	144	172	204	101	118	126	75	87	83
0.5	***	427	466	832	209	237	372	130	148	210	91	103	135	68	76	94
	•••	440	523	840	220	269	364	138	167	197	96	114	120	72	83	79
0.55	***	389	435	813	195	224	359	122	140	201	85	97	128	63	71	88
	•••	406	497	814	206	257	349	130	159	187	90	108	112	67	78	72
0.6	***	349	400	776	178	208	339	112	130	187	78	90	117	58	65	80
	•••	368	465	771	190	241	326	120	148	172	83	99	101	61	71	63
0.65	***	308	362	721	161	190	310	101	118	169	70	80	104	51	58	70
	•••	329	428	711	172	221	295	108	134	153	75	89	88	54	62	53
0.7	***	268	322	647	142	169	273	89	104	146	61	69	88	.	.	.
	•••	289	385	634	153	198	257	95	118	129	64	76	71	.	.	.
0.75	***	228	279	555	121	145	228	75	87	118	.	.	.	.	.	.
	•••	248	337	540	131	169	212	80	98	101	.	.	.	.	.	.
0.8	***	188	232	445	98	118	175	.	.	.	.	.	.	.	.	.
	•••	206	281	429	106	136	158	.	.	.	.	.	.	.	.	.
0.85	***	145	179	317	.	.	.	.	.	.	.	.	.	.	.	.
	•••	159	215	300	.	.	.	.	.	.	.	.	.	.	.	.

TABLE 8: ALPHA= 0.05 POWER= 0.9   EXPECTED ACCRUAL THRU MINIMUM FOLLOW-UP= 60

		DEL=.10			DEL=.15			DEL=.20			DEL=.25			DEL=.30		
FACT=		1.0 .75	.50 .25	.00 BIN	1.0 .75	.50 .25	.00 BIN	1.0 .75	.50 .25	.00 BIN	1.0 75	.50 .25	.00 BIN	1.0 .75	.50 .25	.00 BIN
PCONT=	•••	REQUIRED NUMBER OF PATIENTS														
0.05	•••	181	183	195	106	108	114	75	76	80	59	60	62	48	49	51
	•••	182	186	300	107	110	158	76	78	101	59	61	71	49	50	53
0.1	•••	286	290	326	153	157	173	101	104	113	75	78	83	60	62	65
	•••	287	297	429	154	161	212	103	108	129	76	80	88	61	63	63
0.15	•••	370	377	447	188	195	225	121	126	142	87	91	101	68	71	77
	•••	372	388	540	191	203	257	123	131	153	89	95	101	69	73	72
0.2	•••	432	442	554	214	223	270	134	142	166	96	101	115	73	77	86
	•••	435	459	634	217	235	295	137	149	172	98	106	112	75	81	79
0.25	•••	472	486	645	230	243	308	143	153	185	101	108	126	77	82	93
	•••	477	511	711	235	259	326	147	163	187	104	114	120	79	86	83
0.3	•••	493	512	719	238	255	337	147	159	199	103	112	134	78	84	98
	•••	500	544	771	244	275	349	152	172	197	107	120	126	81	89	86
0.35	•••	497	522	775	240	260	358	148	162	209	104	113	139	78	85	100
	•••	506	561	814	248	284	364	154	177	204	108	122	129	81	91	87
0.4	•••	487	518	812	236	260	371	146	162	214	102	113	141	77	84	100
	•••	498	565	840	245	287	372	153	178	206	107	123	129	80	90	86
0.45	•••	465	503	831	228	254	375	142	159	214	99	110	139	74	82	98
	•••	478	558	848	238	285	372	149	176	204	104	121	126	77	88	83
0.5	•••	435	479	832	216	245	371	135	153	210	94	106	135	70	78	94
	•••	451	541	840	227	277	364	143	171	197	99	116	120	74	84	79
0.55	•••	399	450	813	202	232	359	127	145	201	88	100	128	66	73	88
	•••	418	516	814	214	266	349	134	163	187	93	110	112	69	79	72
0.6	•••	361	416	776	186	217	339	117	135	187	81	92	117	60	67	80
	•••	381	484	771	198	249	326	125	152	172	86	101	101	63	72	63
0.65	•••	321	378	721	168	198	310	106	123	169	73	83	104	53	59	70
	•••	343	446	711	180	229	295	113	138	153	77	91	88	56	63	53
0.7	•••	281	337	647	149	177	273	93	108	146	63	71	88	.	.	.
	•••	303	403	634	160	205	257	99	121	129	67	78	71	.	.	.
0.75	•••	241	293	555	127	152	228	78	90	118	.	.	.	.	.	.
	•••	262	352	540	137	175	212	83	100	101	.	.	.	.	.	.
0.8	•••	199	245	445	103	123	175	.	.	.	.	.	.	.	.	.
	•••	217	294	429	111	140	158	.	.	.	.	.	.	.	.	.
0.85	•••	154	189	317	.	.	.	.	.	.	.	.	.	.	.	.
	•••	168	224	300	.	.	.	.	.	.	.	.	.	.	.	.

TABLE 8: ALPHA= 0.05 POWER= 0.9   EXPECTED ACCRUAL THRU MINIMUM FOLLOW-UP= 70

	DEL=.10			DEL=.15			DEL=.20			DEL=.25			DEL=.30		
FACT=	1.0 .75	.50 .25	.00 BIN	1.0 .75	.50 .25	.00 BIN	1.0 .75	.50 .25	.00 BIN	1.0 75	.50 .25	.00 BIN	1.0 .75	.50 .25	.00 BIN
PCONT=***				REQUIRED NUMBER OF PATIENTS											
0.05 ***	181	184	195	106	108	114	75	77	80	59	60	62	49	49	51
•••	182	186	300	107	110	158	76	78	101	59	61	71	49	50	53
0.1 ***	287	291	326	153	158	173	102	105	113	76	78	83	60	62	65
•••	288	299	429	155	162	212	103	108	129	77	80	88	61	63	63
0.15 ***	371	379	447	189	196	225	122	127	142	88	92	101	68	71	77
•••	374	392	540	192	205	257	124	132	153	90	95	101	70	74	72
0.2 ***	433	445	554	215	226	270	136	143	166	97	102	115	74	78	86
•••	437	464	634	219	238	295	139	151	172	99	107	112	76	81	79
0.25 ***	475	491	645	232	246	308	145	155	185	102	109	126	78	83	93
•••	480	517	711	238	263	326	149	165	187	105	115	120	80	87	83
0.3 ***	497	518	719	242	259	337	150	162	199	105	114	134	79	85	98
•••	504	552	771	248	280	349	155	174	197	109	121	126	82	90	86
0.35 ***	502	529	775	244	265	358	151	166	209	106	116	139	80	86	100
•••	511	572	814	252	290	364	157	180	204	110	124	129	82	92	87
0.4 ***	492	527	812	241	265	371	149	166	214	104	115	141	78	86	100
•••	505	578	840	250	293	372	156	181	206	109	124	129	81	91	86
0.45 ***	472	514	831	233	261	375	145	163	214	101	113	139	76	83	98
•••	487	572	848	244	292	372	153	180	204	106	122	126	79	89	83
0.5 ***	443	492	832	222	252	372	139	157	210	97	109	135	72	80	94
•••	461	556	840	234	285	364	147	175	197	102	118	120	75	85	79
0.55 ***	409	463	813	208	240	359	131	149	201	91	102	128	67	74	88
•••	429	532	814	221	273	349	139	166	187	96	112	112	70	80	72
0.6 ***	371	430	776	192	224	339	121	139	187	84	94	117	61	68	80
•••	394	501	771	205	256	326	129	155	172	88	103	101	64	73	63
0.65 ***	332	393	721	174	205	310	110	126	169	75	85	104	54	60	70
•••	356	463	711	187	236	295	117	141	153	79	92	88	57	64	53
0.7 ***	292	351	647	155	183	273	96	111	146	65	73	88	.	.	.
•••	316	418	634	166	210	257	103	123	129	69	79	71	.	.	.
0.75 ***	252	306	555	133	157	228	81	93	118	.	.	.	.	.	.
•••	273	365	540	143	180	212	86	102	101	.	.	.	.	.	.
0.8 ***	209	256	445	108	127	175	.	.	.	.	.	.	.	.	.
•••	228	304	429	116	143	158	.	.	.	.	.	.	.	.	.
0.85 ***	161	197	317	.	.	.	.	.	.	.	.	.	.	.	.
•••	176	231	300	.	.	.	.	.	.	.	.	.	.	.	.

TABLE 8: ALPHA= 0.05 POWER= 0.9   EXPECTED ACCRUAL THRU MINIMUM FOLLOW-UP= 80

		DEL=.10			DEL=.15			DEL=.20			DEL=.25			DEL=.30		
FACT=		1.0 .75	.50 .25	.00 BIN	1.0 .75	.50 .25	.00 BIN	1.0 .75	.50 .25	.00 BIN	1.0 75	.50 .25	.00 BIN	1.0 .75	.50 .25	.00 BIN
PCONT=***					REQUIRED NUMBER OF PATIENTS											
0.05	***	182	184	195	107	109	114	76	77	80	59	60	62	49	50	51
	•••	183	187	300	107	110	158	76	78	101	59	61	71	49	50	53
0.1	***	287	293	326	154	158	173	103	106	113	76	79	83	61	62	65
	•••	289	301	429	156	163	212	104	109	129	77	81	88	61	64	63
0.15	***	372	381	447	191	198	225	123	128	142	89	93	101	69	72	77
	•••	375	395	540	193	206	257	125	133	153	90	96	101	70	74	72
0.2	***	435	448	554	217	228	270	137	145	166	98	103	115	75	79	86
	•••	439	469	634	221	240	295	140	152	172	100	108	112	77	82	79
0.25	***	477	495	645	235	249	308	147	157	185	104	111	126	79	84	93
	•••	483	523	711	240	266	326	151	166	187	107	116	120	81	87	83
0.3	***	500	524	719	244	263	337	152	165	199	107	115	134	81	86	98
	•••	508	560	771	252	284	349	157	176	197	110	122	126	83	91	86
0.35	***	506	536	775	248	270	358	154	168	209	108	117	139	81	87	100
	•••	517	582	814	256	295	364	160	182	204	112	125	129	84	92	87
0.4	***	498	536	812	245	271	371	153	169	214	107	117	141	80	87	100
	•••	511	589	840	255	299	372	159	184	206	111	126	129	83	92	86
0.45	***	478	524	831	238	267	375	149	166	215	104	115	139	77	85	98
	•••	495	584	848	249	297	372	156	183	204	108	124	126	80	90	83
0.5	***	451	503	832	227	258	372	143	161	210	99	111	135	74	81	94
	•••	470	570	840	240	291	364	150	178	197	104	120	120	77	86	79
0.55	***	418	476	813	214	246	359	134	153	201	93	104	128	69	76	88
	•••	440	546	814	227	279	349	142	169	187	98	113	112	72	81	72
0.6	***	381	443	776	198	230	339	125	142	187	86	96	117	63	69	80
	•••	405	515	771	211	262	326	132	158	172	91	105	101	66	74	63
0.65	***	343	406	721	180	211	310	113	129	169	77	86	104	56	61	70
	•••	367	477	711	193	241	295	120	143	153	81	93	88	58	65	53
0.7	***	303	364	647	160	189	273	99	114	146	67	74	88	.	.	.
	•••	327	431	634	172	215	257	105	125	129	70	80	71	.	.	.
0.75	***	262	317	555	137	162	228	83	95	118	.	.	.	.	.	.
	•••	284	377	540	148	184	212	88	104	101	.	.	.	.	.	.
0.8	***	217	265	445	111	130	175	.	.	.	.	.	.	.	.	.
	•••	237	314	429	119	146	158	.	.	.	.	.	.	.	.	.
0.85	***	168	204	317	.	.	.	.	.	.	.	.	.	.	.	.
	•••	183	238	300	.	.	.	.	.	.	.	.	.	.	.	.

TABLE 8: ALPHA= 0.05 POWER= 0.9   EXPECTED ACCRUAL THRU MINIMUM FOLLOW-UP= 90

		DEL=.10			DEL=.15			DEL=.20			DEL=.25			DEL=.30		
FACT=		1.0 .75	.50 .25	.00 BIN	1.0 .75	.50 .25	.00 BIN	1.0 .75	.50 .25	.00 BIN	1.0 75	.50 .25	.00 BIN	1.0 .75	.50 .25	.00 BIN
PCONT=***					REQUIRED NUMBER OF PATIENTS											
0.05	***	182	185	195	107	109	114	76	77	80	59	60	62	49	50	51
	•••	183	188	300	108	111	158	76	78	101	60	61	71	49	50	53
0.1	***	288	294	326	155	159	173	103	106	113	77	79	83	61	63	65
	•••	290	302	429	157	164	212	104	109	129	78	81	88	62	64	63
0.15	***	373	383	447	192	199	225	123	129	142	90	93	101	70	72	77
	•••	377	397	540	195	208	257	126	134	153	91	96	101	71	74	72
0.2	***	437	451	554	219	230	270	138	146	166	99	104	115	76	79	86
	•••	442	473	634	223	242	295	142	153	172	101	108	112	77	82	79
0.25	***	479	499	645	237	252	308	148	159	185	105	112	126	80	84	93
	•••	486	529	711	243	269	326	153	168	187	108	117	120	82	88	83
0.3	***	503	529	719	247	266	337	154	167	199	108	117	134	82	87	98
	•••	512	568	771	255	287	349	159	178	197	112	123	126	84	91	86
0.35	***	510	543	775	251	274	358	156	171	209	109	119	139	82	88	100
	•••	522	590	814	260	299	364	162	184	204	113	127	129	85	93	87
0.4	***	503	544	812	249	276	371	155	172	214	108	119	141	81	88	100
	•••	518	599	840	260	303	372	162	186	206	113	127	129	84	93	86
0.45	***	485	533	831	243	272	375	152	169	214	106	117	139	79	86	98
	•••	503	596	848	254	302	372	159	185	204	110	125	126	82	91	83
0.5	***	459	514	832	232	264	371	146	164	210	101	112	135	75	82	94
	•••	479	582	840	245	296	364	153	180	197	106	121	120	78	87	79
0.55	***	427	487	813	219	252	359	138	156	201	95	106	128	70	77	88
	•••	450	559	814	232	284	349	145	172	187	100	115	112	73	81	72
0.6	***	391	455	776	203	236	339	128	145	187	88	98	117	64	70	80
	•••	416	528	771	217	267	326	135	160	172	92	106	101	67	74	63
0.65	***	353	417	721	185	217	310	116	132	169	79	88	104	57	62	70
	•••	378	489	711	198	246	295	123	145	153	83	94	88	59	65	53
0.7	***	313	375	647	165	193	273	102	116	146	68	75	88	.	.	.
	•••	337	442	634	177	219	257	108	127	129	71	81	71	.	.	.
0.75	***	271	327	555	142	166	228	85	97	118	.	.	.	.	.	.
	•••	293	387	540	152	187	212	90	105	101	.	.	.	.	.	.
0.8	***	225	273	445	115	133	175	.	.	.	.	.	.	.	.	.
	•••	245	322	429	123	148	158	.	.	.	.	.	.	.	.	.
0.85	***	174	210	317	.	.	.	.	.	.	.	.	.	.	.	.
	•••	189	243	300	.	.	.	.	.	.	.	.	.	.	.	.

TABLE 8: ALPHA= 0.05 POWER= 0.9   EXPECTED ACCRUAL THRU MINIMUM FOLLOW-UP= 100

	DEL=.10			DEL=.15			DEL=.20			DEL=.25			DEL=.30		
FACT=	1.0 .75	.50 .25	.00 BIN	1.0 .75	.50 .25	.00 BIN	1.0 .75	.50 .25	.00 BIN	1.0 75	.50 .25	.00 BIN	1.0 .75	.50 .25	.00 BIN
PCONT=***				REQUIRED NUMBER OF PATIENTS											
0.05 ***	182	185	195	107	109	114	76	77	80	59	60	62	49	50	51
•••	183	188	300	108	111	158	77	78	101	60	61	71	49	50	53
0.1 ***	289	295	326	155	160	173	104	107	113	77	79	83	61	63	65
•••	291	303	429	157	165	212	105	109	129	78	81	88	62	64	63
0.15 ***	374	385	447	193	201	225	124	130	142	90	94	101	70	73	77
•••	378	400	540	196	209	257	127	134	153	92	97	101	71	75	72
0.2 ***	438	454	554	220	232	270	140	147	166	100	105	115	76	80	86
•••	444	476	634	225	244	295	143	154	172	102	109	112	78	83	79
0.25 ***	482	503	645	239	255	308	150	160	185	106	113	126	80	85	93
•••	489	534	711	245	271	326	154	169	187	109	118	120	82	88	83
0.3 ***	506	534	719	250	270	337	156	169	199	110	118	134	83	88	98
•••	516	574	771	258	290	349	161	180	197	113	124	126	85	92	86
0.35 ***	514	550	775	254	278	358	158	173	209	111	120	139	83	89	100
•••	527	598	814	264	302	364	165	186	204	115	127	129	86	94	87
0.4 ***	508	551	812	253	280	371	158	174	214	110	120	141	82	89	100
•••	524	608	840	264	307	372	165	188	206	114	128	129	85	93	86
0.45 ***	491	542	831	247	277	375	154	172	215	107	118	139	80	87	98
•••	510	606	848	259	307	372	162	187	204	112	126	126	83	91	83
0.5 ***	466	523	832	237	269	372	148	167	210	103	114	135	76	83	94
•••	488	593	840	250	300	364	156	182	197	108	122	120	79	88	79
0.55 ***	435	497	813	224	257	359	140	159	201	97	108	128	71	78	88
•••	459	570	814	237	289	349	148	174	187	102	116	112	74	82	72
0.6 ***	400	465	776	208	241	339	130	148	187	90	99	117	65	71	80
•••	425	539	771	222	272	326	138	162	172	94	107	101	68	75	63
0.65 ***	362	428	721	190	221	310	118	134	169	80	89	104	58	62	70
•••	388	500	711	203	250	295	125	147	153	84	95	88	60	66	53
0.7 ***	322	385	647	169	198	273	104	118	146	69	76	88	.	.	.
•••	347	452	634	181	223	257	110	129	129	72	81	71	.	.	.
0.75 ***	279	337	555	145	169	228	87	98	118	.	.	.	.	.	.
•••	302	396	540	156	190	212	92	106	101	.	.	.	.	.	.
0.8 ***	232	281	445	118	136	175	.	.	.	.	.	.	.	.	.
•••	252	329	429	125	151	158	.	.	.	.	.	.	.	.	.
0.85 ***	179	215	317	.	.	.	.	.	.	.	.	.	.	.	.
•••	194	248	300	.	.	.	.	.	.	.	.	.	.	.	.

TABLE 8: ALPHA= 0.05 POWER= 0.9   EXPECTED ACCRUAL THRU MINIMUM FOLLOW-UP= 110

	DEL=.05			DEL=.10			DEL=.15			DEL=.20			DEL=.25		
FACT=	1.0 .75	.50 .25	.00 BIN	1.0 .75	.50 .25	.00 BIN	1.0 .75	.50 .25	.00 BIN	1.0 75	.50 .25	.00 BIN	1.0 .75	.50 .25	.00 BIN
PCONT=•••			REQUIRED NUMBER OF PATIENTS												
0.05 •••	512	515	547	183	185	195	108	109	114	76	78	80	60	60	62
•••	513	521	943	184	188	300	108	111	158	77	79	101	60	61	71
0.1 •••	934	941	1055	290	296	326	156	161	173	104	107	113	77	80	83
•••	936	953	1491	292	304	429	158	165	212	105	110	129	78	81	88
0.15 •••	1292	1303	1547	375	387	447	194	202	225	125	130	142	91	94	101
•••	1296	1324	1970	379	402	540	197	210	257	127	135	153	92	97	101
0.2 •••	1569	1585	1996	440	457	554	222	234	270	141	148	166	100	105	115
•••	1574	1618	2381	446	480	634	226	246	295	144	155	172	103	109	112
0.25 •••	1761	1784	2388	484	507	645	241	257	308	151	161	185	107	113	126
•••	1768	1830	2724	492	539	711	247	273	326	156	170	187	110	119	120
0.3 •••	1871	1903	2715	509	539	719	252	273	337	158	170	199	111	119	134
•••	1882	1965	2998	520	580	771	260	293	349	163	181	197	114	125	126
0.35 •••	1907	1949	2973	518	556	775	257	281	358	160	175	209	112	121	139
•••	1921	2030	3203	532	605	814	267	305	364	167	188	204	116	128	129
0.4 •••	1878	1932	3159	513	559	812	256	284	371	160	176	214	112	121	141
•••	1896	2037	3340	530	616	840	267	311	372	167	190	206	116	129	129
0.45 •••	1795	1864	3272	497	550	831	251	281	375	157	174	214	109	119	139
•••	1818	1994	3409	517	615	848	263	311	372	164	189	204	114	127	126
0.5 •••	1669	1757	3310	473	532	832	241	273	372	151	169	210	105	115	135
•••	1699	1914	3409	496	603	840	254	304	364	159	184	197	109	123	120
0.55 •••	1512	1622	3273	442	507	813	228	261	359	143	161	201	99	109	128
•••	1550	1806	3340	468	581	814	242	293	349	151	176	187	103	116	112
0.6 •••	1338	1471	3162	408	475	776	213	245	339	133	150	187	91	100	117
•••	1385	1677	3203	434	549	771	226	276	326	140	164	172	95	107	101
0.65 •••	1160	1314	2976	370	438	721	194	225	310	121	136	169	82	90	104
•••	1215	1535	2998	397	510	711	207	254	295	127	149	153	85	96	88
0.7 •••	987	1156	2718	330	394	647	173	201	273	106	119	146	70	77	88
•••	1050	1380	2724	356	461	634	185	226	257	112	130	129	73	82	71
0.75 •••	826	998	2386	286	345	555	149	172	228	89	99	118	.	.	.
•••	892	1215	2381	310	403	540	159	192	212	94	107	101	.	.	.
0.8 •••	677	839	1981	239	288	445	120	138	175	.	.	.	.	.	.
•••	740	1035	1970	259	335	429	128	152	158	.	.	.	.	.	.
0.85 •••	534	673	1506	184	220	317	.	.	.	.	.	.	.	.	.
•••	588	833	1491	199	252	300	.	.	.	.	.	.	.	.	.
0.9 •••	383	483	959	.	.	.	.	.	.	.	.	.	.	.	.
•••	423	593	943	.	.	.	.	.	.	.	.	.	.	.	.

TABLE 8: ALPHA= 0.05 POWER= 0.9   EXPECTED ACCRUAL THRU MINIMUM FOLLOW-UP= 120

		DEL=.05			DEL=.10			DEL=.15			DEL=.20			DEL=.25	
FACT=	1.0 .75	.50 .25	.00 BIN	1.0 .75	.50 .25	.00 BIN	1.0 .75	.50 .25	.00 BIN	1.0 75	.50 .25	.00 BIN	1.0 .75	.50 .25	.00 BIN
PCONT=•••			REQUIRED NUMBER OF PATIENTS												
0.05 ***	513	516	547	183	186	195	108	110	114	76	78	80	60	61	62
•••	514	521	943	184	189	300	109	111	158	77	79	101	60	61	71
0.1 ***	935	942	1055	290	297	326	157	161	173	104	108	113	78	80	83
•••	937	955	1491	293	306	429	158	166	212	106	110	129	79	81	88
0.15 ***	1293	1305	1547	377	388	447	195	203	225	126	131	142	91	95	101
•••	1297	1328	1970	381	404	540	198	211	257	128	135	153	93	97	101
0.2 ***	1570	1588	1996	442	459	554	223	235	270	142	149	166	101	106	115
•••	1576	1623	2381	448	483	634	228	247	295	145	156	172	103	110	112
0.25 ***	1763	1788	2388	486	511	645	243	259	308	153	163	185	108	114	126
•••	1771	1838	2724	495	543	711	249	275	326	157	171	187	111	119	120
0.3 ***	1874	1908	2715	512	544	719	255	275	337	159	172	199	112	120	134
•••	1885	1976	2998	524	585	771	263	295	349	165	182	197	115	125	126
0.35 ***	1911	1956	2973	522	561	775	260	284	358	162	177	209	113	122	139
•••	1926	2044	3203	536	612	814	270	308	364	168	189	204	117	129	129
0.4 ***	1883	1942	3159	518	565	812	260	287	371	162	178	214	113	123	141
•••	1903	2054	3340	535	624	840	271	314	372	169	192	206	117	130	129
0.45 ***	1801	1877	3272	503	558	831	254	285	375	159	176	214	110	121	139
•••	1827	2016	3409	524	623	848	267	314	372	166	190	204	115	128	126
0.5 ***	1677	1772	3310	479	541	832	245	277	371	153	171	210	106	116	135
•••	1709	1939	3409	503	612	840	258	308	364	161	186	197	111	124	120
0.55 ***	1523	1641	3273	450	516	813	232	266	359	145	163	201	100	110	128
•••	1563	1834	3340	476	590	814	246	296	349	153	177	187	104	117	112
0.6 ***	1351	1493	3162	416	484	776	217	249	339	135	152	187	92	101	117
•••	1401	1708	3203	443	559	771	230	279	326	142	165	172	96	108	101
0.65 ***	1175	1338	2977	378	446	721	198	229	310	123	138	169	83	91	104
•••	1235	1566	2998	406	519	711	211	257	295	129	150	153	86	96	88
0.7 ***	1005	1181	2718	337	403	647	177	205	273	108	121	146	71	78	88
•••	1071	1412	2724	364	469	634	189	229	257	114	131	129	74	82	71
0.75 ***	845	1023	2386	293	352	555	152	175	228	90	100	118	.	.	.
•••	913	1244	2381	317	410	540	162	195	212	95	108	101	.	.	.
0.8 ***	696	862	1981	245	294	445	123	140	175	.	.	.	.	.	.
•••	760	1061	1970	265	340	429	130	154	158	.	.	.	.	.	.
0.85 ***	550	692	1506	189	224	317	.	.	.	.	.	.	.	.	.
•••	606	854	1491	204	255	300	.	.	.	.	.	.	.	.	.
0.9 ***	395	497	959	.	.	.	.	.	.	.	.	.	.	.	.
•••	436	606	943	.	.	.	.	.	.	.	.	.	.	.	.

TABLE 8: ALPHA= 0.05 POWER= 0.9   EXPECTED ACCRUAL THRU MINIMUM FOLLOW-UP= 130

		DEL=.05			DEL=.10			DEL=.15			DEL=.20			DEL=.25		
FACT=		1.0 .75	.50 .25	.00 BIN	1.0 .75	.50 .25	.00 BIN	1.0 .75	.50 .25	.00 BIN	1.0 75	.50 .25	.00 BIN	1.0 .75	.50 .25	.00 BIN
PCONT= ***					REQUIRED NUMBER OF PATIENTS											
0.05	***	513	516	547	183	186	195	108	110	114	77	78	80	60	61	62
	•••	514	522	943	184	189	300	109	111	158	77	79	101	60	61	71
0.1	***	935	943	1055	291	298	326	157	162	173	105	108	113	78	80	83
	•••	938	957	1491	294	307	429	159	166	212	106	110	129	79	81	88
0.15	***	1294	1307	1547	378	390	447	196	204	225	126	132	142	92	95	101
	•••	1298	1332	1970	382	406	540	199	212	257	129	136	153	93	97	101
0.2	***	1572	1591	1996	443	462	554	224	237	270	142	150	166	102	106	115
	•••	1578	1629	2381	450	486	634	229	248	295	146	156	172	104	110	112
0.25	***	1765	1792	2388	488	514	645	245	261	308	154	164	185	109	115	126
	•••	1774	1846	2724	498	547	711	251	277	326	158	172	187	111	120	120
0.3	***	1877	1914	2715	515	548	719	257	278	337	161	173	200	113	120	134
	•••	1889	1986	2998	527	590	771	265	297	349	166	183	197	116	126	126
0.35	***	1915	1964	2973	525	567	775	263	287	358	164	178	209	115	123	139
	•••	1931	2058	3203	541	618	814	273	311	364	170	190	204	118	130	129
0.4	***	1888	1952	3159	522	572	812	263	291	371	164	180	214	114	124	141
	•••	1909	2072	3340	541	631	840	274	317	372	171	193	206	118	131	129
0.45	***	1808	1889	3272	508	565	831	258	288	375	161	178	214	112	122	139
	•••	1835	2037	3409	530	631	848	270	317	372	168	192	204	116	129	126
0.5	***	1685	1788	3310	486	549	832	249	281	372	155	173	210	107	117	135
	•••	1720	1963	3409	510	620	840	262	311	364	163	187	197	112	125	120
0.55	***	1533	1659	3273	457	524	813	236	269	359	148	165	201	101	111	128
	•••	1577	1861	3340	483	598	814	250	299	349	155	179	187	106	118	112
0.6	***	1364	1514	3162	423	493	776	221	253	339	137	154	187	93	102	117
	•••	1417	1737	3203	451	567	771	234	282	326	144	167	172	97	109	101
0.65	***	1191	1361	2977	386	455	721	202	233	310	125	140	169	84	92	104
	•••	1253	1596	2998	413	527	711	215	260	295	131	151	153	87	97	88
0.7	***	1022	1205	2718	345	410	647	180	208	273	110	122	146	72	78	88
	•••	1091	1441	2724	371	477	634	192	231	257	115	132	129	75	83	71
0.75	***	863	1047	2386	300	359	555	155	178	228	92	101	118	.	.	.
	•••	934	1272	2381	324	417	540	165	197	212	96	108	101	.	.	.
0.8	***	713	884	1981	250	299	445	125	142	175	.	.	.	.	.	.
	•••	779	1085	1970	271	345	429	132	155	158	.	.	.	.	.	.
0.85	***	565	710	1506	193	228	317	.	.	.	.	.	.	.	.	.
	•••	622	873	1491	208	259	300	.	.	.	.	.	.	.	.	.
0.9	***	406	509	959	.	.	.	.	.	.	.	.	.	.	.	.
	•••	447	619	943	.	.	.	.	.	.	.	.	.	.	.	.

TABLE 8: ALPHA= 0.05 POWER= 0.9   EXPECTED ACCRUAL THRU MINIMUM FOLLOW-UP= 140

	DEL=.05			DEL=.10			DEL=.15			DEL=.20			DEL=.25		
FACT=	1.0 .75	.50 .25	.00 BIN	1.0 .75	.50 .25	.00 BIN	1.0 .75	.50 .25	.00 BIN	1.0 75	.50 .25	.00 BIN	1.0 .75	.50 .25	.00 BIN
PCONT=•••	REQUIRED NUMBER OF PATIENTS														
0.05 ***	513 517 547			183 186 195			108 110 113			77 78 80			60 61 62		
•••	514 523 943			185 189 300			109 111 158			77 79 101			60 61 71		
0.1 ***	936 944 1055			291 299 326			158 162 173			105 108 113			78 80 83		
•••	939 959 1491			294 307 429			160 166 212			106 110 129			79 82 88		
0.15 ***	1295 1309 1547			379 392 447			196 205 225			127 132 142			92 95 101		
•••	1300 1335 1970			384 407 540			200 212 257			129 136 153			93 98 101		
0.2 ***	1573 1594 1996			445 464 554			226 238 270			143 151 166			102 107 115		
•••	1580 1635 2381			452 488 634			231 249 295			147 157 172			104 110 112		
0.25 ***	1767 1796 2388			491 517 645			246 263 308			155 165 185			109 115 126		
•••	1777 1854 2724			501 551 711			253 278 326			159 173 187			112 120 120		
0.3 ***	1880 1920 2715			518 552 719			259 280 337			162 174 200			114 121 134		
•••	1893 1997 2998			531 595 771			267 299 349			167 184 197			117 126 126		
0.35 ***	1919 1971 2973			529 572 775			265 290 358			166 180 209			116 124 139		
•••	1936 2072 3203			545 624 814			275 313 364			172 191 204			119 130 129		
0.4 ***	1893 1962 3159			527 578 812			265 293 371			166 181 214			115 124 141		
•••	1916 2088 3340			546 637 840			277 320 372			172 194 206			119 131 129		
0.45 ***	1814 1901 3272			514 572 831			261 292 375			163 180 214			113 123 139		
•••	1843 2057 3409			536 638 848			274 320 372			170 193 204			117 130 126		
0.5 ***	1693 1803 3310			492 556 832			252 285 372			157 175 210			109 118 135		
•••	1730 1987 3409			517 628 840			266 314 364			165 188 197			113 125 120		
0.55 ***	1543 1677 3273			463 532 813			240 273 359			149 167 201			102 112 128		
•••	1590 1887 3340			491 606 814			254 302 349			157 180 187			107 118 112		
0.6 ***	1377 1535 3162			430 501 776			224 256 339			139 155 187			95 103 117		
•••	1433 1765 3203			458 575 771			238 285 326			146 168 172			98 109 101		
0.65 ***	1206 1384 2976			393 463 721			205 236 310			126 141 169			85 92 104		
•••	1271 1624 2998			421 534 711			218 262 295			133 152 153			88 97 88		
0.7 ***	1039 1228 2717			351 418 647			183 210 273			111 123 146			73 79 88		
•••	1110 1469 2724			378 484 634			195 233 257			117 133 129			76 83 71		
0.75 ***	881 1069 2386			306 365 555			157 180 228			93 102 118			. . .		
•••	953 1298 2381			331 423 540			167 198 212			97 109 101			. . .		
0.8 ***	729 904 1981			256 304 445			127 143 175			. . .			. . .		
•••	798 1108 1970			276 350 429			134 156 158			. . .			. . .		
0.85 ***	579 726 1506			197 231 317			. . .			. . .			. . .		
•••	637 892 1491			211 262 300			. . .			. . .			. . .		
0.9 ***	416 521 959			. . .			. . .			. . .			. . .		
•••	458 631 943			. . .			. . .			. . .			. . .		

TABLE 8: ALPHA= 0.05 POWER= 0.9 EXPECTED ACCRUAL THRU MINIMUM FOLLOW-UP= 150

	DEL=.05			DEL=.10			DEL=.15			DEL=.20			DEL=.25		
FACT=	1.0 .75	.50 .25	.00 BIN	1.0 .75	.50 .25	.00 BIN	1.0 .75	.50 .25	.00 BIN	1.0 75	.50 .25	.00 BIN	1.0 .75	.50 .25	.00 BIN
PCONT=***				REQUIRED NUMBER OF PATIENTS											
0.05 ***	513	517	547	184	187	195	108	110	113	77	78	80	60	61	62
•••	515	524	943	185	190	300	109	112	158	77	79	101	60	61	71
0.1 ***	937	945	1055	292	300	326	158	163	173	105	108	113	78	80	83
•••	939	961	1491	295	308	429	160	167	212	107	111	129	79	82	88
0.15 ***	1296	1311	1547	380	393	447	197	206	225	127	133	142	92	96	101
•••	1301	1339	1970	385	409	540	201	213	257	130	136	153	94	98	101
0.2 ***	1575	1597	1996	446	466	554	227	239	270	144	152	166	103	107	115
•••	1582	1640	2381	454	491	634	232	250	295	147	157	172	105	111	112
0.25 ***	1769	1801	2388	493	520	645	248	264	308	156	166	185	110	116	126
•••	1780	1861	2724	503	554	711	255	280	326	160	173	187	113	120	120
0.3 ***	1883	1925	2715	521	557	719	261	282	337	163	175	199	114	122	134
•••	1897	2007	2998	534	599	771	270	301	349	169	185	197	118	127	126
0.35 ***	1922	1979	2973	533	577	775	268	292	358	167	181	209	116	125	139
•••	1941	2085	3203	550	629	814	278	315	364	173	192	204	120	131	129
0.4 ***	1898	1971	3159	531	583	812	268	296	371	167	183	214	116	125	141
•••	1923	2105	3340	551	643	840	280	322	372	174	195	206	120	132	129
0.45 ***	1820	1914	3272	519	578	831	264	295	375	165	181	214	114	123	139
•••	1852	2076	3409	542	645	848	277	322	372	172	194	204	118	130	126
0.5 ***	1701	1817	3310	497	563	832	255	288	371	159	176	210	110	119	135
•••	1741	2009	3409	523	635	840	269	317	364	167	189	197	114	126	120
0.55 ***	1553	1695	3273	470	539	813	243	276	359	151	168	201	103	113	128
•••	1603	1912	3340	497	613	814	257	305	349	159	181	187	108	119	112
0.6 ***	1389	1555	3162	437	508	776	227	260	339	141	157	187	95	104	117
•••	1449	1791	3203	465	582	771	241	287	326	148	169	172	99	110	101
0.65 ***	1220	1405	2977	399	470	721	208	239	310	128	142	169	86	93	104
•••	1289	1651	2998	428	541	711	221	264	295	134	153	153	89	98	88
0.7 ***	1055	1250	2717	358	424	647	186	213	273	112	125	146	74	79	88
•••	1129	1495	2724	385	490	634	197	235	257	118	133	129	76	83	71
0.75 ***	897	1090	2386	312	371	555	160	182	228	94	103	118	.	.	.
•••	972	1322	2381	337	428	540	169	200	212	98	109	101	.	.	.
0.8 ***	745	923	1981	260	309	445	128	145	175	.	.	.	.	.	.
•••	815	1130	1970	281	354	429	136	157	158	.	.	.	.	.	.
0.85 ***	593	742	1506	200	235	317	.	.	.	.	.	.	.	.	.
•••	652	908	1491	215	264	300	.	.	.	.	.	.	.	.	.
0.9 ***	426	532	959	.	.	.	.	.	.	.	.	.	.	.	.
•••	469	642	943	.	.	.	.	.	.	.	.	.	.	.	.

TABLE 8: ALPHA= 0.05 POWER= 0.9   EXPECTED ACCRUAL THRU MINIMUM FOLLOW-UP= 160

		DEL=.05			DEL=.10			DEL=.15			DEL=.20			DEL=.25		
FACT=		1.0 .75	.50 .25	.00 BIN	1.0 .75	.50 .25	.00 BIN	1.0 .75	.50 .25	.00 BIN	1.0 75	.50 .25	.00 BIN	1.0 .75	.50 .25	.00 BIN
PCONT=***					REQUIRED	NUMBER	OF PATIENTS									
0.05	***	514	518	547	184	187	195	109	110	114	77	78	80	60	61	62
	•••	515	524	943	185	190	300	109	112	158	78	79	101	60	61	71
0.1	***	937	946	1055	293	301	326	158	163	173	106	109	113	79	81	83
	•••	940	963	1491	296	309	429	160	167	212	107	111	129	79	82	88
0.15	***	1297	1313	1547	381	395	447	198	206	225	128	133	142	93	96	101
	•••	1302	1343	1970	386	410	540	201	214	257	130	137	153	94	98	101
0.2	***	1576	1600	1996	448	469	554	228	240	270	145	152	166	103	108	115
	•••	1584	1645	2381	456	493	634	233	251	295	148	158	172	105	111	112
0.25	***	1771	1805	2388	495	523	645	249	266	308	157	166	185	111	116	126
	•••	1782	1869	2724	506	557	711	256	281	326	161	174	187	113	121	120
0.3	***	1885	1931	2715	524	560	719	263	284	337	165	176	199	115	122	134
	•••	1901	2017	2998	538	603	771	272	303	349	170	186	197	118	127	126
0.35	***	1926	1986	2973	536	582	775	270	295	358	168	182	209	117	125	139
	•••	1946	2098	3203	554	634	814	280	317	364	174	193	204	121	131	129
0.4	***	1903	1981	3159	536	589	812	271	299	371	169	184	214	117	126	141
	•••	1929	2121	3340	556	649	840	282	324	372	176	196	206	121	132	129
0.45	***	1827	1926	3272	524	584	831	267	297	375	166	183	215	115	124	139
	•••	1860	2095	3409	547	651	848	280	325	372	173	195	204	119	131	126
0.5	***	1709	1832	3310	503	570	832	258	291	372	161	178	210	111	120	135
	•••	1752	2030	3409	529	641	840	272	319	364	168	190	197	115	126	120
0.55	***	1563	1712	3273	476	546	813	246	279	359	153	169	201	104	113	128
	•••	1616	1935	3340	504	620	814	260	307	349	160	182	187	108	120	112
0.6	***	1401	1574	3162	443	515	776	230	262	339	142	158	187	96	105	117
	•••	1464	1816	3203	472	589	771	244	290	326	149	170	172	100	110	101
0.65	***	1235	1426	2977	406	477	721	211	241	310	129	143	169	86	93	104
	•••	1306	1677	2998	434	547	711	224	266	295	136	154	153	90	98	88
0.7	***	1071	1271	2718	364	431	647	189	215	273	114	125	146	74	80	88
	•••	1147	1520	2724	391	496	634	200	237	257	119	134	129	77	84	71
0.75	***	913	1110	2386	317	377	555	162	184	228	95	104	118	.	.	.
	•••	990	1346	2381	342	433	540	171	201	212	99	110	101	.	.	.
0.8	***	760	941	1981	265	314	445	130	146	175	.	.	.	.	.	.
	•••	831	1150	1970	285	357	429	137	158	158	.	.	.	.	.	.
0.85	***	606	757	1506	204	238	317	.	.	.	.	.	.	.	.	.
	•••	666	924	1491	218	267	300	.	.	.	.	.	.	.	.	.
0.9	***	436	542	959	.	.	.	.	.	.	.	.	.	.	.	.
	•••	479	652	943	.	.	.	.	.	.	.	.	.	.	.	.

TABLE 8: ALPHA= 0.05 POWER= 0.9   EXPECTED ACCRUAL THRU MINIMUM FOLLOW-UP= 170

		DEL=.05			DEL=.10			DEL=.15			DEL=.20			DEL=.25		
FACT=		1.0 .75	.50 .25	.00 BIN	1.0 .75	.50 .25	.00 BIN	1.0 .75	.50 .25	.00 BIN	1.0 75	.50 .25	.00 BIN	1.0 .75	.50 .25	.00 BIN
PCONT=***					REQUIRED NUMBER OF PATIENTS											
0.05	***	514	518	547	184	187	195	109	110	113	77	78	80	60	61	62
	***	515	525	943	186	190	300	109	112	158	78	79	101	60	61	71
0.1	***	938	947	1055	293	301	326	159	164	173	106	109	113	79	81	83
	***	941	965	1491	296	310	429	161	167	212	107	111	129	80	82	88
0.15	***	1298	1315	1547	382	396	447	199	207	225	128	133	142	93	96	101
	***	1304	1346	1970	387	412	540	202	214	257	131	137	153	94	98	101
0.2	***	1578	1603	1996	450	471	554	229	241	270	146	153	166	104	108	115
	***	1586	1651	2381	458	495	634	234	252	295	149	158	172	106	111	112
0.25	***	1773	1809	2388	497	526	645	251	267	308	158	167	185	111	117	126
	***	1785	1877	2724	508	560	711	258	282	326	162	174	187	114	121	120
0.3	***	1888	1937	2715	526	564	719	265	286	337	166	177	199	116	123	134
	***	1904	2027	2998	541	607	771	273	304	349	171	186	197	119	128	126
0.35	***	1930	1994	2973	540	586	775	272	297	358	170	183	209	118	126	139
	***	1951	2111	3203	558	638	814	282	319	364	176	194	204	122	132	129
0.4	***	1908	1990	3159	540	594	812	273	301	371	170	185	214	118	127	141
	***	1936	2136	3340	561	654	840	285	326	372	177	197	206	122	133	129
0.45	***	1833	1937	3272	528	590	831	269	300	375	168	184	215	116	125	139
	***	1869	2113	3409	553	657	848	282	327	372	175	196	204	120	131	126
0.5	***	1717	1846	3310	508	576	832	261	293	371	162	179	210	111	120	135
	***	1762	2051	3409	535	647	840	275	321	364	170	191	197	116	127	120
0.55	***	1573	1728	3273	481	553	813	249	282	359	154	171	201	105	114	128
	***	1628	1958	3340	510	626	814	263	309	349	162	183	187	109	120	112
0.6	***	1413	1592	3162	449	522	776	233	265	339	144	159	187	97	105	117
	***	1479	1840	3203	478	595	771	247	292	326	151	171	172	101	110	101
0.65	***	1249	1445	2976	412	483	721	214	244	310	131	144	169	87	94	104
	***	1322	1701	2998	441	553	711	227	268	295	137	154	153	90	99	88
0.7	***	1086	1291	2718	370	436	647	191	217	273	115	126	146	75	80	88
	***	1164	1544	2724	397	501	634	202	239	257	120	135	129	77	84	71
0.75	***	929	1130	2386	323	382	555	164	186	228	96	104	118	.	.	.
	***	1007	1367	2381	347	437	540	173	202	212	100	110	101	.	.	.
0.8	***	775	959	1981	269	318	445	132	147	175	.	.	.	.	.	.
	***	847	1169	1970	290	361	429	139	159	158	.	.	.	.	.	.
0.85	***	618	771	1506	207	241	317	.	.	.	.	.	.	.	.	.
	***	679	939	1491	221	269	300	.	.	.	.	.	.	.	.	.
0.9	***	445	552	959	.	.	.	.	.	.	.	.	.	.	.	.
	***	488	661	943	.	.	.	.	.	.	.	.	.	.	.	.

TABLE 8: ALPHA= 0.05 POWER= 0.9   EXPECTED ACCRUAL THRU MINIMUM FOLLOW-UP= 180

FACT=	DEL=.05 1.0 / .75	.50 / .25	.00 / BIN	DEL=.10 1.0 / .75	.50 / .25	.00 / BIN	DEL=.15 1.0 / .75	.50 / .25	.00 / BIN	DEL=.20 1.0 / 75	.50 / .25	.00 / BIN	DEL=.25 1.0 / .75	.50 / .25	.00 / BIN
PCONT=•••						REQUIRED NUMBER OF PATIENTS								
0.05 •••	514	519	547	185	188	195	109	111	114	77	78	80	60	61	62
•••	516	526	943	186	190	300	110	112	158	78	79	101	60	61	71
0.1 •••	938	949	1055	294	302	326	159	164	173	106	109	114	79	81	83
•••	942	967	1491	297	310	429	161	168	212	108	111	129	80	82	88
0.15 •••	1299	1317	1547	383	397	447	199	208	225	129	134	142	93	96	101
•••	1305	1349	1970	388	413	540	203	215	257	131	137	153	95	98	101
0.2 •••	1579	1606	1996	451	473	554	230	242	270	146	153	166	104	108	115
•••	1588	1656	2381	459	497	634	235	253	295	149	159	172	106	111	112
0.25 •••	1775	1813	2388	499	529	645	252	269	308	159	168	185	112	117	126
•••	1788	1884	2724	510	563	711	259	283	326	163	175	187	114	121	120
0.3 •••	1891	1942	2715	529	568	719	266	287	337	167	178	199	117	123	134
•••	1908	2037	2998	544	611	771	275	305	349	172	187	197	120	128	126
0.35 •••	1934	2001	2973	543	590	775	274	299	358	171	184	209	119	127	139
•••	1956	2123	3203	561	642	814	284	320	364	177	194	204	122	132	129
0.4 •••	1913	2000	3159	544	599	812	276	303	371	172	186	214	119	127	141
•••	1942	2151	3340	565	659	840	287	328	372	178	198	206	123	133	129
0.45 •••	1839	1949	3272	533	596	831	272	302	375	169	185	215	117	125	139
•••	1877	2131	3409	558	662	848	285	329	372	176	197	204	121	131	126
0.5 •••	1725	1860	3310	514	582	832	264	296	371	164	180	210	112	121	135
•••	1772	2071	3409	541	653	840	277	323	364	171	192	197	116	127	120
0.55 •••	1583	1745	3273	487	559	813	252	284	359	156	172	201	106	114	128
•••	1641	1979	3340	516	632	814	266	311	349	163	184	187	110	120	112
0.6 •••	1425	1610	3162	455	528	776	236	267	339	145	160	187	98	106	117
•••	1493	1862	3203	484	601	771	249	294	326	152	171	172	101	111	101
0.65 •••	1262	1464	2976	417	489	721	217	246	310	132	145	169	88	94	104
•••	1338	1724	2998	447	559	711	229	270	295	138	155	153	91	99	88
0.7 •••	1101	1310	2718	375	442	647	193	219	273	116	127	146	75	81	88
•••	1181	1566	2724	403	506	634	205	240	257	121	135	129	78	84	71
0.75 •••	944	1148	2386	327	387	555	166	187	228	96	105	118	.	.	.
•••	1023	1388	2381	352	441	540	175	204	212	100	111	101	.	.	.
0.8 •••	789	975	1981	273	322	445	133	148	175	.	.	.	.	.	.
•••	862	1187	1970	294	364	429	140	160	158	.	.	.	.	.	.
0.85 •••	630	785	1506	210	243	317	.	.	.	.	.	.	.	.	.
•••	692	953	1491	224	271	300	.	.	.	.	.	.	.	.	.
0.9 •••	453	561	959	.	.	.	.	.	.	.	.	.	.	.	.
•••	497	670	943	.	.	.	.	.	.	.	.	.	.	.	.

TABLE 8: ALPHA= 0.05 POWER= 0.9   EXPECTED ACCRUAL THRU MINIMUM FOLLOW-UP= 190

		DEL=.05			DEL=.10			DEL=.15			DEL=.20			DEL=.25		
FACT=		1.0 .75	.50 .25	.00 BIN	1.0 .75	.50 .25	.00 BIN	1.0 .75	.50 .25	.00 BIN	1.0 75	.50 .25	.00 BIN	1.0 .75	.50 .25	.00 BIN
PCONT=***					REQUIRED NUMBER OF PATIENTS											
0.05	***	515	519	547	185	188	195	109	111	113	77	78	80	60	61	62
	•••	516	526	943	186	191	300	110	112	158	78	79	101	61	61	71
0.1	***	939	950	1055	294	303	326	160	164	173	107	109	113	79	81	83
	•••	942	969	1491	298	311	429	162	168	212	108	111	129	80	82	88
0.15	***	1300	1319	1547	384	398	447	200	208	225	129	134	142	94	97	101
	•••	1306	1353	1970	390	414	540	203	215	257	131	137	153	95	98	101
0.2	***	1581	1609	1996	453	475	554	231	243	270	147	154	166	104	109	115
	•••	1590	1661	2381	461	499	634	236	254	295	150	159	172	106	111	112
0.25	***	1778	1817	2388	501	532	645	253	270	308	159	169	185	112	118	126
	•••	1791	1891	2724	513	565	711	260	284	326	163	175	187	115	121	120
0.3	***	1894	1948	2715	532	571	719	268	289	337	168	179	199	117	124	134
	•••	1912	2046	2998	547	614	771	277	307	349	173	187	197	120	128	126
0.35	***	1937	2009	2973	546	594	775	276	300	358	172	185	209	120	127	139
	•••	1961	2135	3203	565	646	814	286	322	364	178	195	204	123	132	129
0.4	***	1918	2009	3159	548	604	812	278	306	371	173	187	214	120	128	141
	•••	1949	2166	3340	570	664	840	289	329	372	179	198	206	123	133	129
0.45	***	1846	1961	3272	538	601	831	274	305	375	171	186	215	117	126	139
	•••	1885	2148	3409	563	667	848	287	330	372	177	198	204	121	132	126
0.5	***	1733	1874	3310	519	588	832	267	298	372	165	181	210	113	122	135
	•••	1783	2090	3409	546	658	840	280	325	364	172	193	197	117	127	120
0.55	***	1593	1760	3273	492	565	813	254	287	359	157	173	201	107	115	128
	•••	1653	2000	3340	522	638	814	268	313	349	164	184	187	111	121	112
0.6	***	1437	1628	3162	460	534	776	239	270	339	147	161	187	99	106	117
	•••	1507	1884	3203	490	606	771	252	296	326	153	172	172	102	111	101
0.65	***	1276	1483	2977	423	494	721	219	248	310	133	146	169	88	95	104
	•••	1354	1746	2998	452	564	711	231	272	295	139	156	153	91	99	88
0.7	***	1115	1329	2718	380	447	647	195	221	273	117	128	146	76	81	88
	•••	1197	1587	2724	408	510	634	207	241	257	122	135	129	78	84	71
0.75	***	958	1166	2386	332	391	555	168	189	228	97	106	118	.	.	.
	•••	1039	1408	2381	357	445	540	177	205	212	101	111	101	.	.	.
0.8	***	802	991	1981	277	325	445	135	150	175	.	.	.	.	.	.
	•••	877	1203	1970	298	367	429	141	160	158	.	.	.	.	.	.
0.85	***	641	798	1506	212	246	317	.	.	.	.	.	.	.	.	.
	•••	704	966	1491	227	272	300	.	.	.	.	.	.	.	.	.
0.9	***	461	569	959	.	.	.	.	.	.	.	.	.	.	.	.
	•••	505	678	943	.	.	.	.	.	.	.	.	.	.	.	.

TABLE 8: ALPHA= 0.05 POWER= 0.9   EXPECTED ACCRUAL THRU MINIMUM FOLLOW-UP= 200

		DEL=.05			DEL=.10			DEL=.15			DEL=.20			DEL=.25	
FACT=	1.0 .75	.50 .25	.00 BIN	1.0 .75	.50 .25	.00 BIN	1.0 .75	.50 .25	.00 BIN	1.0 75	.50 .25	.00 BIN	1.0 .75	.50 .25	.00 BIN
PCONT=•••	REQUIRED NUMBER OF PATIENTS														
0.05 •••	515	520	547	185	188	195	109	111	114	77	78	80	60	61	62
•••	516	527	943	186	191	300	110	112	158	78	79	101	61	61	71
0.1 •••	939	951	1055	295	303	326	160	165	173	107	109	113	79	81	83
•••	943	970	1491	298	311	429	162	168	212	108	111	129	80	82	88
0.15 •••	1301	1321	1547	385	400	447	201	209	225	130	134	142	94	97	101
•••	1308	1356	1970	391	415	540	204	215	257	132	138	153	95	99	101
0.2 •••	1582	1612	1996	454	476	554	232	244	270	147	154	166	105	109	115
•••	1592	1666	2381	463	501	634	237	254	295	150	159	172	107	112	112
0.25 •••	1780	1821	2388	503	534	645	255	271	308	160	169	185	113	118	126
•••	1794	1898	2724	515	568	711	262	285	326	164	176	187	115	122	120
0.3 •••	1897	1954	2715	534	574	719	270	290	337	169	180	199	118	124	134
•••	1916	2056	2998	550	617	771	278	308	349	173	188	197	121	128	126
0.35 •••	1941	2016	2973	550	598	775	278	302	358	173	186	209	120	127	139
•••	1966	2147	3203	569	650	814	288	323	364	179	196	204	123	132	129
0.4 •••	1923	2018	3159	551	608	812	280	307	371	174	188	214	120	128	141
•••	1955	2180	3340	574	668	840	292	331	372	180	199	206	124	134	129
0.45 •••	1852	1972	3272	542	606	831	277	307	375	172	187	215	118	126	139
•••	1893	2164	3409	567	672	848	289	332	372	179	198	204	122	132	126
0.5 •••	1741	1888	3310	523	593	832	269	300	372	167	182	210	114	122	135
•••	1793	2108	3409	551	663	832	282	327	364	174	194	197	118	128	120
0.55 •••	1603	1776	3273	497	570	813	257	289	359	159	174	201	108	116	128
•••	1665	2020	3340	527	643	814	271	315	349	165	185	187	111	121	112
0.6 •••	1449	1645	3162	465	539	776	241	272	339	148	162	187	99	107	117
•••	1521	1905	3203	496	611	771	254	297	326	154	173	172	103	111	101
0.65 •••	1289	1501	2977	428	500	721	221	250	310	134	147	169	89	95	104
•••	1369	1767	2998	457	568	711	234	273	295	140	156	153	92	99	88
0.7 •••	1129	1346	2718	385	452	647	198	223	273	118	129	146	76	81	88
•••	1213	1608	2724	413	515	634	209	243	257	123	136	129	79	84	71
0.75 •••	972	1183	2386	337	396	555	169	190	228	98	106	118	.	.	.
•••	1054	1426	2381	361	449	540	178	206	212	102	111	101	.	.	.
0.8 •••	815	1006	1981	281	329	445	136	151	175	.	.	.	.	.	.
•••	891	1220	1970	301	370	429	142	161	158	.	.	.	.	.	.
0.85 •••	652	810	1506	215	248	317	.	.	.	.	.	.	.	.	.
•••	715	979	1491	229	274	300	.	.	.	.	.	.	.	.	.
0.9 •••	469	577	959	.	.	.	.	.	.	.	.	.	.	.	.
•••	513	686	943	.	.	.	.	.	.	.	.	.	.	.	.

TABLE 8: ALPHA= 0.05 POWER= 0.9   EXPECTED ACCRUAL THRU MINIMUM FOLLOW-UP= 225

		DEL=.05			DEL=.10			DEL=.15			DEL=.20			DEL=.25		
FACT=		1.0 .75	.50 .25	.00 BIN	1.0 .75	.50 .25	.00 BIN	1.0 .75	.50 .25	.00 BIN	1.0 75	.50 .25	.00 BIN	1.0 .75	.50 .25	.00 BIN
PCONT=***					REQUIRED NUMBER OF PATIENTS											
0.05	***	515	521	547	185	189	195	109	111	113	78	79	80	60	61	62
	•••	517	528	943	187	191	300	110	112	158	78	79	101	61	61	71
0.1	***	941	954	1055	296	305	326	161	165	173	107	110	113	80	81	83
	•••	945	974	1491	300	313	429	163	168	212	108	111	129	80	82	88
0.15	***	1303	1325	1547	387	402	447	202	210	225	131	135	142	94	97	101
	•••	1311	1363	1970	393	417	540	206	216	257	133	138	153	96	99	101
0.2	***	1586	1619	1996	457	480	554	234	246	270	149	155	166	106	109	115
	•••	1597	1678	2381	466	505	634	239	256	295	152	160	172	107	112	112
0.25	***	1785	1832	2388	508	540	645	257	274	308	162	170	185	114	119	126
	•••	1801	1915	2724	520	573	711	264	287	326	166	177	187	116	122	120
0.3	***	1904	1967	2715	540	581	719	273	293	337	171	181	200	119	125	134
	•••	1925	2078	2998	557	624	771	282	310	349	175	189	197	122	129	126
0.35	***	1951	2034	2973	557	607	775	282	306	358	176	188	209	122	129	139
	•••	1979	2174	3203	577	658	814	292	326	364	181	197	204	125	133	129
0.4	***	1935	2041	3159	560	618	812	285	312	371	177	190	214	122	129	141
	•••	1971	2213	3340	583	678	840	296	334	372	183	200	206	125	134	129
0.45	***	1868	2000	3272	552	617	831	282	311	375	175	189	214	120	128	139
	•••	1914	2203	3409	578	682	848	295	336	372	181	200	204	123	133	126
0.5	***	1761	1920	3310	535	605	832	275	305	372	169	185	210	115	123	135
	•••	1818	2152	3409	563	674	840	288	330	364	176	195	197	119	129	120
0.55	***	1627	1813	3273	509	583	813	263	294	359	161	176	201	109	117	128
	•••	1695	2067	3340	539	654	814	276	319	349	168	187	187	113	122	112
0.6	***	1477	1685	3162	478	552	776	246	277	339	150	164	187	101	108	117
	•••	1555	1954	3203	508	622	771	260	301	326	157	174	172	104	112	101
0.65	***	1320	1543	2976	440	512	721	226	254	310	137	149	169	90	96	104
	•••	1405	1816	2998	470	579	711	239	276	295	142	158	153	93	100	88
0.7	***	1162	1388	2718	396	463	647	202	227	273	120	130	146	77	82	88
	•••	1250	1655	2724	424	524	634	213	245	257	125	137	129	79	85	71
0.75	***	1005	1222	2386	347	405	555	173	193	228	100	107	118	.	.	.
	•••	1090	1470	2381	371	457	540	182	208	212	103	112	101	.	.	.
0.8	***	845	1041	1981	289	336	445	138	153	175	.	.	.	.	.	.
	•••	923	1257	1970	309	376	429	145	162	158	.	.	.	.	.	.
0.85	***	678	838	1506	221	253	317	.	.	.	.	.	.	.	.	.
	•••	742	1007	1491	235	278	300	.	.	.	.	.	.	.	.	.
0.9	***	487	596	959	.	.	.	.	.	.	.	.	.	.	.	.
	•••	532	703	943	.	.	.	.	.	.	.	.	.	.	.	.

TABLE 8: ALPHA= 0.05 POWER= 0.9   EXPECTED ACCRUAL THRU MINIMUM FOLLOW-UP= 250

		DEL=.05			DEL=.10			DEL=.15			DEL=.20			DEL=.25		
FACT=		1.0 .75	.50 .25	.00 BIN	1.0 .75	.50 .25	.00 BIN	1.0 .75	.50 .25	.00 BIN	1.0 75	.50 .25	.00 BIN	1.0 .75	.50 .25	.00 BIN
PCONT=•••							REQUIRED NUMBER OF PATIENTS									
0.05	•••	516	522	547	186	189	195	110	111	114	78	79	80	61	61	62
	•••	518	529	943	187	191	300	110	112	158	78	79	101	61	62	71
0.1	•••	942	956	1055	298	306	326	162	166	173	108	110	114	80	81	83
	•••	947	978	1491	301	314	429	163	169	212	109	112	129	81	82	88
0.15	•••	1306	1330	1547	389	405	447	203	211	225	131	136	142	95	97	101
	•••	1314	1371	1970	396	420	540	207	217	257	133	138	153	96	99	101
0.2	•••	1590	1626	1996	461	484	554	236	248	270	150	156	166	106	110	115
	•••	1602	1689	2381	470	508	634	241	257	295	153	160	172	108	112	112
0.25	•••	1790	1842	2388	512	545	645	260	276	308	163	172	185	115	119	126
	•••	1808	1930	2724	525	578	711	267	289	326	167	177	187	117	122	120
0.3	•••	1911	1981	2715	546	588	719	276	296	337	172	183	199	120	126	134
	•••	1935	2099	2998	563	630	771	285	312	349	177	190	197	123	129	126
0.35	•••	1960	2052	2973	564	615	775	286	309	358	177	189	209	123	129	139
	•••	1991	2201	3203	584	666	814	296	328	364	183	198	204	126	134	129
0.4	•••	1947	2063	3159	569	628	812	289	316	371	179	192	214	123	130	141
	•••	1987	2244	3340	592	686	840	300	337	372	185	202	206	126	135	129
0.45	•••	1883	2026	3272	562	627	831	287	316	375	177	191	214	121	129	139
	•••	1934	2239	3409	588	692	848	299	339	372	183	201	204	124	133	126
0.5	•••	1780	1951	3310	545	616	832	279	309	372	172	186	210	117	124	135
	•••	1842	2192	3409	574	684	840	292	334	364	178	196	197	120	129	120
0.55	•••	1650	1848	3273	520	594	813	267	298	359	164	178	201	110	118	128
	•••	1723	2110	3340	551	664	814	281	322	349	170	188	187	114	122	112
0.6	•••	1504	1723	3162	489	563	776	251	281	339	153	166	187	102	108	117
	•••	1586	1998	3203	519	632	771	264	304	326	159	175	172	105	112	101
0.65	•••	1350	1582	2977	451	523	721	231	258	310	139	151	169	91	97	104
	•••	1439	1860	2998	481	588	711	243	279	295	144	158	153	94	100	88
0.7	•••	1193	1427	2718	407	473	647	206	230	273	122	131	146	78	82	88
	•••	1284	1697	2724	435	532	634	217	248	257	126	138	129	80	85	71
0.75	•••	1035	1258	2386	356	414	555	177	196	228	101	108	118	.	.	.
	•••	1123	1508	2381	380	464	540	185	209	212	104	113	101	.	.	.
0.8	•••	873	1073	1981	297	343	445	141	154	175	.	.	.	.	.	.
	•••	953	1290	1970	316	381	429	147	164	158	.	.	.	.	.	.
0.85	•••	701	864	1506	226	257	317	.	.	.	.	.	.	.	.	.
	•••	767	1033	1491	240	281	300	.	.	.	.	.	.	.	.	.
0.9	•••	503	613	959	.	.	.	.	.	.	.	.	.	.	.	.
	•••	548	718	943	.	.	.	.	.	.	.	.	.	.	.	.

TABLE 8: ALPHA= 0.05 POWER= 0.9   EXPECTED ACCRUAL THRU MINIMUM FOLLOW-UP= 275

PCONT=***	DEL=.05			DEL=.10			DEL=.15			DEL=.20			DEL=.25		
FACT=	1.0 .75	.50 .25	.00 BIN	1.0 .75	.50 .25	.00 BIN	1.0 .75	.50 .25	.00 BIN	1.0 75	.50 .25	.00 BIN	1.0 .75	.50 .25	.00 BIN
					REQUIRED NUMBER OF PATIENTS										
0.05 ***	517	523	547	186	189	195	110	111	113	78	79	80	61	61	62
***	519	530	943	188	191	300	111	112	158	78	79	101	61	62	71
0.1 ***	944	959	1055	299	307	326	162	166	173	108	110	113	80	81	83
***	949	981	1491	302	314	429	164	169	212	109	112	129	81	82	88
0.15 ***	1308	1334	1547	391	407	447	204	212	225	132	136	142	95	98	101
***	1317	1377	1970	398	421	540	208	218	257	134	139	153	96	99	101
0.2 ***	1593	1633	1996	464	488	553	238	249	270	151	157	166	107	110	115
***	1607	1699	2381	473	511	634	243	258	295	153	161	172	109	112	112
0.25 ***	1795	1852	2388	517	550	645	262	278	308	165	172	185	115	120	126
***	1815	1945	2724	530	582	711	269	290	326	168	178	187	117	123	120
0.3 ***	1918	1994	2715	551	594	719	279	299	337	174	184	199	121	126	134
***	1944	2118	2998	569	635	771	288	314	349	178	191	197	123	130	126
0.35 ***	1970	2069	2973	571	622	775	289	312	358	179	191	209	124	130	139
***	2004	2225	3203	591	672	814	299	330	364	184	199	204	127	134	129
0.4 ***	1959	2084	3159	576	636	812	293	319	371	181	194	214	124	131	141
***	2003	2273	3340	601	693	840	304	339	372	187	202	206	127	135	129
0.45 ***	1898	2052	3272	570	637	831	291	319	375	179	193	215	122	129	139
***	1953	2272	3409	597	700	848	303	341	372	185	202	204	125	134	126
0.5 ***	1799	1981	3310	554	626	832	284	313	371	174	188	210	118	125	135
***	1865	2229	3409	584	693	840	297	336	364	180	198	197	121	130	120
0.55 ***	1673	1881	3273	530	604	813	272	302	359	166	179	201	112	118	128
***	1750	2149	3340	561	673	814	285	325	349	172	189	187	115	123	112
0.6 ***	1530	1758	3162	499	573	776	256	284	339	155	167	187	103	109	117
***	1616	2038	3203	530	641	771	268	306	326	161	176	172	106	113	101
0.65 ***	1378	1618	2977	461	532	721	235	261	310	141	152	169	92	97	104
***	1471	1900	2998	491	596	711	247	281	295	146	159	153	95	100	88
0.7 ***	1222	1462	2718	416	482	647	210	233	273	123	132	146	79	83	88
***	1316	1736	2724	444	540	634	220	249	257	127	138	129	81	85	71
0.75 ***	1063	1292	2386	364	421	555	179	198	228	102	109	118	.	.	.
***	1154	1543	2381	388	470	540	188	211	212	105	113	101	.	.	.
0.8 ***	899	1103	1981	303	349	445	143	156	175	.	.	.	.	.	.
***	981	1320	1970	323	385	429	149	164	158	.	.	.	.	.	.
0.85 ***	722	887	1506	231	261	317	.	.	.	.	.	.	.	.	.
***	789	1055	1491	244	283	300	.	.	.	.	.	.	.	.	.
0.9 ***	518	628	959	.	.	.	.	.	.	.	.	.	.	.	.
***	564	731	943	.	.	.	.	.	.	.	.	.	.	.	.

TABLE 8: ALPHA= 0.05 POWER= 0.9   EXPECTED ACCRUAL THRU MINIMUM FOLLOW-UP= 300

		DEL=.05			DEL=.10			DEL=.15			DEL=.20			DEL=.25	
FACT=	1.0 .75	.50 .25	.00 BIN	1.0 .75	.50 .25	.00 BIN	1.0 .75	.50 .25	.00 BIN	1.0 75	.50 .25	.00 BIN	1.0 .75	.50 .25	.00 BIN
PCONT=•••				REQUIRED NUMBER OF PATIENTS											
0.05 •••	517	524	547	187	190	195	110	112	113	78	79	80	61	61	62
•••	520	531	943	188	192	300	111	112	158	78	79	101	61	61	71
0.1 •••	945	961	1055	300	308	326	163	167	173	108	111	113	80	82	83
•••	951	985	1491	303	315	429	165	169	212	109	112	129	81	82	88
0.15 •••	1311	1339	1547	393	409	447	205	213	225	133	136	142	96	98	101
•••	1321	1383	1970	400	423	540	209	218	257	134	139	153	97	99	101
0.2 •••	1597	1640	1996	466	491	553	239	250	270	151	157	166	107	111	115
•••	1612	1709	2381	476	513	634	244	259	295	154	161	172	109	113	112
0.25 •••	1801	1861	2388	520	554	645	264	280	308	166	173	185	116	120	126
•••	1821	1959	2724	534	585	711	271	291	326	169	178	187	118	123	120
0.3 •••	1925	2007	2715	556	599	719	282	301	337	175	185	199	122	127	134
•••	1954	2137	2998	574	640	771	290	316	349	180	191	197	124	130	126
0.35 •••	1979	2085	2973	577	629	775	292	315	358	181	192	209	125	131	139
•••	2016	2248	3203	598	678	814	302	332	364	186	199	204	127	134	129
0.4 •••	1971	2105	3159	583	643	812	296	322	371	183	195	214	125	132	141
•••	2018	2301	3340	608	700	840	307	341	372	188	203	206	128	136	129
0.45 •••	1914	2076	3272	578	645	831	295	322	375	181	194	214	123	130	139
•••	1972	2303	3409	606	707	848	307	343	372	187	203	204	126	134	126
0.5 •••	1817	2009	3310	563	635	832	288	316	371	176	189	210	119	126	135
•••	1888	2263	3409	593	700	840	300	339	364	182	198	197	122	130	120
0.55 •••	1695	1912	3273	539	613	813	276	305	359	168	181	201	112	119	128
•••	1776	2185	3340	570	681	814	289	327	349	174	190	187	115	123	112
0.6 •••	1555	1791	3162	508	582	776	259	287	339	157	169	187	104	110	117
•••	1645	2076	3203	539	648	771	272	308	326	162	177	172	107	113	101
0.65 •••	1405	1651	2977	470	541	721	239	264	310	142	153	169	93	98	104
•••	1501	1937	2998	500	603	711	250	283	295	147	160	153	95	101	88
0.7 •••	1250	1495	2717	424	490	647	213	235	273	124	133	145	79	83	88
•••	1346	1771	2724	452	546	634	223	251	257	129	139	129	81	85	71
0.75 •••	1090	1322	2386	371	428	555	182	200	228	103	109	118	.	.	.
•••	1183	1575	2381	396	475	540	190	212	212	106	113	101	.	.	.
0.8 •••	923	1129	1981	309	354	445	145	157	175	.	.	.	.	.	.
•••	1006	1346	1970	328	389	429	151	165	158	.	.	.	.	.	.
0.85 •••	742	908	1506	235	264	317	.	.	.	.	.	.	.	.	.
•••	810	1075	1491	248	286	300	.	.	.	.	.	.	.	.	.
0.9 •••	532	642	959	.	.	.	.	.	.	.	.	.	.	.	.
•••	577	743	943	.	.	.	.	.	.	.	.	.	.	.	.

TABLE 8: ALPHA= 0.05 POWER= 0.9   EXPECTED ACCRUAL THRU MINIMUM FOLLOW-UP= 325

		DEL=.05			DEL=.10			DEL=.15			DEL=.20			DEL=.25		
FACT=		1.0 .75	.50 .25	.00 BIN	1.0 .75	.50 .25	.00 BIN	1.0 .75	.50 .25	.00 BIN	1.0 75	.50 .25	.00 BIN	1.0 .75	.50 .25	.00 BIN
PCONT=***					REQUIRED NUMBER OF PATIENTS											
0.05	***	518	525	547	187	190	195	110	112	114	78	79	80	61	61	62
	***	520	532	943	188	192	300	111	112	158	79	79	101	61	62	71
0.1	***	947	964	1055	301	309	326	163	167	173	109	111	114	81	82	83
	***	953	988	1491	304	316	429	165	170	212	110	112	129	81	82	88
0.15	***	1313	1343	1547	395	411	447	207	214	225	133	137	142	96	98	101
	***	1324	1389	1970	402	424	540	210	219	257	135	139	153	97	99	101
0.2	***	1601	1647	1996	469	494	554	241	252	270	152	158	166	108	111	115
	***	1617	1719	2381	479	515	634	246	260	295	155	161	172	109	113	112
0.25	***	1806	1871	2388	524	558	645	266	281	308	167	174	185	117	121	126
	***	1828	1972	2724	538	589	711	273	292	326	170	179	187	118	123	120
0.3	***	1932	2020	2715	561	604	719	284	303	337	177	186	199	122	127	134
	***	1963	2154	2998	579	644	771	292	317	349	181	192	197	125	130	126
0.35	***	1988	2101	2973	583	635	775	295	317	358	182	193	209	125	131	139
	***	2028	2269	3203	604	683	814	305	334	364	187	200	204	128	135	129
0.4	***	1983	2125	3159	590	650	812	299	324	371	184	196	214	126	132	141
	***	2034	2326	3340	615	706	840	310	343	372	190	204	206	129	136	129
0.45	***	1928	2099	3272	586	653	831	298	325	375	183	195	214	124	131	139
	***	1991	2332	3409	614	713	848	310	346	372	188	204	204	127	135	126
0.5	***	1836	2036	3310	571	643	832	291	320	372	178	191	210	120	127	135
	***	1910	2294	3409	601	707	840	304	341	364	184	199	197	123	130	120
0.55	***	1716	1941	3273	548	622	813	279	308	359	170	182	201	114	120	128
	***	1801	2219	3340	579	688	814	292	329	349	175	190	187	116	123	112
0.6	***	1578	1822	3162	517	590	776	263	290	339	158	170	187	105	110	117
	***	1672	2110	3203	548	655	771	275	310	326	164	177	172	107	114	101
0.65	***	1431	1683	2977	478	549	721	242	267	310	144	154	169	94	98	104
	***	1529	1971	2998	508	610	711	253	285	295	148	160	153	96	101	88
0.7	***	1276	1526	2718	432	497	647	216	237	273	126	134	146	80	84	88
	***	1375	1803	2724	460	551	634	225	253	257	130	139	129	82	86	71
0.75	***	1115	1351	2385	378	434	555	184	201	228	104	110	118	.	.	.
	***	1209	1604	2381	402	479	540	192	213	212	107	114	101	.	.	.
0.8	***	946	1155	1981	315	358	445	146	158	175	.	.	.	.	.	.
	***	1030	1371	1970	334	392	429	152	166	158	.	.	.	.	.	.
0.85	***	761	928	1506	238	267	317	.	.	.	.	.	.	.	.	.
	***	829	1094	1491	251	288	300	.	.	.	.	.	.	.	.	.
0.9	***	544	654	959	.	.	.	.	.	.	.	.	.	.	.	.
	***	590	754	943	.	.	.	.	.	.	.	.	.	.	.	.

TABLE 8: ALPHA= 0.05 POWER= 0.9   EXPECTED ACCRUAL THRU MINIMUM FOLLOW-UP= 350

	DEL=.05			DEL=.10			DEL=.15			DEL=.20			DEL=.25		
FACT=	1.0 .75	.50 .25	.00 BIN	1.0 .75	.50 .25	.00 BIN	1.0 .75	.50 .25	.00 BIN	1.0 75	.50 .25	.00 BIN	1.0 .75	.50 .25	.00 BIN
PCONT=•••					REQUIRED NUMBER OF PATIENTS										
0.05 •••	519	525	547	187	190	195	111	112	113	78	79	80	61	61	62
•••	521	533	943	189	192	300	111	113	158	79	79	101	61	62	71
0.1 •••	948	966	1055	302	310	326	164	167	173	109	111	113	81	82	83
•••	954	990	1491	305	316	429	166	170	212	110	112	129	81	83	88
0.15 •••	1316	1348	1547	397	412	447	207	214	225	134	137	142	96	98	101
•••	1327	1394	1970	403	426	540	210	219	257	135	139	153	97	99	101
0.2 •••	1604	1653	1996	472	496	554	242	253	270	153	158	166	108	111	115
•••	1621	1727	2381	482	517	634	247	260	295	155	162	172	110	113	112
0.25 •••	1811	1880	2388	528	561	645	268	283	308	167	175	185	117	121	126
•••	1835	1985	2724	542	592	711	274	293	326	171	179	187	119	123	120
0.3 •••	1940	2032	2715	566	609	719	286	305	337	178	187	199	123	128	134
•••	1972	2170	2998	584	648	771	294	318	349	182	192	197	125	131	126
0.35 •••	1998	2117	2973	588	640	775	298	319	358	184	194	209	126	132	139
•••	2040	2289	3203	610	687	814	307	335	364	188	201	204	129	135	129
0.4 •••	1995	2144	3159	596	657	812	302	327	371	186	197	214	127	133	141
•••	2049	2350	3340	621	711	840	313	345	372	191	205	206	130	136	129
0.45 •••	1943	2122	3272	593	659	831	301	328	375	184	196	215	125	131	139
•••	2009	2359	3409	621	719	848	313	347	372	190	204	204	128	135	126
0.5 •••	1853	2061	3310	579	650	832	295	322	372	180	192	210	121	127	135
•••	1931	2324	3409	609	713	840	307	342	364	185	200	197	124	131	120
0.55 •••	1737	1969	3273	556	629	813	283	310	359	171	183	201	114	120	128
•••	1825	2250	3340	587	694	814	295	331	349	177	191	187	117	124	112
0.6 •••	1602	1851	3162	525	598	776	266	293	339	160	171	187	105	111	117
•••	1698	2142	3203	556	661	771	278	312	326	165	178	172	108	114	101
0.65 •••	1455	1713	2976	486	556	721	245	269	310	145	155	169	94	99	104
•••	1556	2003	2998	516	615	711	256	286	295	150	161	153	96	101	88
0.7 •••	1301	1555	2717	439	503	647	218	239	273	127	135	146	80	84	88
•••	1401	1833	2724	467	556	634	228	254	257	131	140	129	82	86	71
0.75 •••	1139	1378	2386	384	439	555	186	203	228	105	111	118	.	.	.
•••	1235	1632	2381	408	483	540	194	214	212	108	114	101	.	.	.
0.8 •••	967	1178	1981	320	362	445	148	159	175	.	.	.	.	.	.
•••	1052	1394	1970	338	395	429	153	167	158	.	.	.	.	.	.
0.85 •••	778	946	1506	242	270	317	.	.	.	.	.	.	.	.	.
•••	847	1111	1491	254	290	300	.	.	.	.	.	.	.	.	.
0.9 •••	556	665	959	.	.	.	.	.	.	.	.	.	.	.	.
•••	602	763	943	.	.	.	.	.	.	.	.	.	.	.	.

TABLE 8: ALPHA= 0.05 POWER= 0.9   EXPECTED ACCRUAL THRU MINIMUM FOLLOW-UP= 375

	DEL=.05			DEL=.10			DEL=.15			DEL=.20			DEL=.25		
FACT=	1.0 .75	.50 .25	.00 BIN	1.0 .75	.50 .25	.00 BIN	1.0 .75	.50 .25	.00 BIN	1.0 75	.50 .25	.00 BIN	1.0 .75	.50 .25	.00 BIN
PCONT=•••				REQUIRED NUMBER OF PATIENTS											
0.05 •••	519	526	547	188	190	195	111	112	114	78	79	80	61	61	62
•••	522	533	943	189	192	300	111	113	158	79	79	101	61	62	71
0.1 •••	949	968	1055	302	311	326	164	168	173	109	111	114	81	82	83
•••	956	993	1491	306	317	429	166	170	212	110	112	129	81	83	88
0.15 •••	1318	1352	1547	398	414	447	208	215	225	134	137	142	96	98	101
•••	1330	1400	1970	405	427	540	211	220	257	136	140	153	97	100	101
0.2 •••	1608	1660	1996	474	498	553	243	253	270	154	159	166	109	111	115
•••	1626	1735	2381	484	519	634	248	261	295	156	162	172	110	113	112
0.25 •••	1816	1889	2388	531	565	645	270	284	308	168	175	185	118	121	126
•••	1842	1997	2724	545	594	711	276	294	326	171	180	187	119	123	120
0.3 •••	1947	2044	2715	570	613	719	288	306	337	179	187	200	124	128	134
•••	1981	2185	2998	588	651	771	296	319	349	183	193	197	126	131	126
0.35 •••	2007	2132	2973	593	645	775	300	321	358	185	195	209	127	132	139
•••	2051	2308	3203	615	692	814	309	337	364	190	201	204	129	135	129
0.4 •••	2007	2162	3160	602	662	812	305	329	371	187	198	214	128	133	141
•••	2063	2372	3340	628	716	840	316	346	372	192	205	206	130	137	129
0.45 •••	1958	2143	3272	600	666	831	304	330	375	186	197	215	126	132	139
•••	2026	2384	3409	627	724	848	316	349	372	191	205	204	129	135	126
0.5 •••	1871	2085	3310	586	657	832	298	325	372	181	193	210	122	127	135
•••	1951	2352	3409	616	719	840	310	344	364	186	200	197	124	131	120
0.55 •••	1757	1995	3273	563	636	813	286	313	359	173	184	201	115	121	128
•••	1848	2280	3340	594	700	814	298	332	349	178	192	187	118	124	112
0.6 •••	1624	1879	3162	532	605	776	269	295	339	161	172	187	106	111	118
•••	1723	2172	3203	563	667	771	281	313	326	166	179	172	108	114	101
0.65 •••	1478	1740	2976	493	562	721	248	271	310	146	155	169	95	99	104
•••	1581	2032	2998	523	620	711	258	288	295	151	161	153	97	101	88
0.7 •••	1324	1582	2717	446	509	647	221	241	273	128	135	146	81	84	88
•••	1427	1861	2724	473	561	634	230	255	257	131	140	129	82	86	71
0.75 •••	1161	1403	2386	390	444	555	188	204	228	105	111	118	.	.	.
•••	1258	1656	2381	414	487	540	196	215	212	108	114	101	.	.	.
0.8 •••	987	1199	1981	324	366	445	149	160	175	.	.	.	.	.	.
•••	1073	1414	1970	343	398	429	154	167	158	.	.	.	.	.	.
0.85 •••	795	963	1505	245	272	317	.	.	.	.	.	.	.	.	.
•••	864	1126	1491	257	291	300	.	.	.	.	.	.	.	.	.
0.9 •••	567	676	959	.	.	.	.	.	.	.	.	.	.	.	.
•••	613	772	943	.	.	.	.	.	.	.	.	.	.	.	.

TABLE 8: ALPHA= 0.05 POWER= 0.9   EXPECTED ACCRUAL THRU MINIMUM FOLLOW-UP= 400

		DEL=.05			DEL=.10			DEL=.15			DEL=.20			DEL=.25		
FACT=		1.0 .75	.50 .25	.00 BIN	1.0 .75	.50 .25	.00 BIN	1.0 .75	.50 .25	.00 BIN	1.0 75	.50 .25	.00 BIN	1.0 .75	.50 .25	.00 BIN
PCONT=•••					REQUIRED NUMBER OF PATIENTS											
0.05	•••	520	527	547	188	191	195	111	112	114	78	79	80	61	61	62
	•••	523	534	943	189	192	300	111	113	158	79	79	101	61	62	71
0.1	•••	951	970	1055	303	311	326	165	168	173	109	111	113	81	82	83
	•••	958	995	1491	307	317	429	166	170	212	110	112	129	81	83	88
0.15	•••	1321	1356	1547	400	415	447	209	215	225	134	138	142	97	99	101
	•••	1333	1404	1970	406	428	540	212	220	257	136	140	153	98	100	101
0.2	•••	1612	1666	1996	476	501	554	244	254	270	154	159	166	109	112	115
	•••	1631	1743	2381	487	521	634	249	261	295	156	162	172	110	113	112
0.25	•••	1821	1898	2388	534	568	645	271	285	308	169	176	185	118	122	126
	•••	1848	2008	2724	548	597	711	277	295	326	172	180	187	120	124	120
0.3	•••	1954	2056	2715	574	617	719	290	308	337	180	188	199	124	128	134
	•••	1990	2200	2998	592	654	771	298	320	349	183	193	197	126	131	126
0.35	•••	2016	2147	2973	598	650	775	302	323	358	186	196	209	127	132	139
	•••	2063	2326	3203	620	695	814	311	338	364	190	202	204	130	135	129
0.4	•••	2018	2180	3159	608	668	812	307	331	371	188	199	214	128	134	141
	•••	2077	2393	3340	633	720	840	318	348	372	193	206	206	131	137	129
0.45	•••	1972	2164	3272	606	672	831	307	332	375	187	198	215	126	132	139
	•••	2043	2408	3409	634	729	848	318	350	372	192	206	204	129	135	126
0.5	•••	1888	2108	3310	593	663	832	300	327	372	182	194	210	122	128	135
	•••	1971	2378	3409	623	724	840	312	345	364	187	201	197	125	131	120
0.55	•••	1776	2020	3273	570	643	813	289	315	359	174	185	201	116	121	128
	•••	1870	2307	3340	601	705	814	300	334	349	179	192	187	118	124	112
0.6	•••	1645	1905	3162	539	611	776	272	297	339	162	173	187	107	111	117
	•••	1746	2200	3203	570	672	771	283	315	326	167	179	172	109	114	101
0.65	•••	1501	1767	2977	500	568	721	250	273	310	147	156	169	95	99	104
	•••	1606	2060	2998	529	625	711	260	289	295	151	162	153	97	102	88
0.7	•••	1346	1608	2718	452	515	647	223	243	273	129	136	146	81	84	88
	•••	1451	1887	2724	479	565	634	232	256	257	132	140	129	83	86	71
0.75	•••	1183	1426	2386	396	449	555	190	206	228	106	111	118	.	.	.
	•••	1281	1680	2381	419	490	540	197	216	212	109	115	101	.	.	.
0.8	•••	1006	1220	1981	329	370	445	151	161	175	.	.	.	.	.	.
	•••	1093	1434	1970	347	400	429	155	168	158	.	.	.	.	.	.
0.85	•••	810	979	1506	248	274	317	.	.	.	.	.	.	.	.	.
	•••	880	1140	1491	260	292	300	.	.	.	.	.	.	.	.	.
0.9	•••	577	686	959	.	.	.	.	.	.	.	.	.	.	.	.
	•••	623	780	943	.	.	.	.	.	.	.	.	.	.	.	.

TABLE 8: ALPHA= 0.05 POWER= 0.9   EXPECTED ACCRUAL THRU MINIMUM FOLLOW-UP= 425

		DEL=.05			DEL=.10			DEL=.15			DEL=.20			DEL=.25		
FACT=		1.0 .75	.50 .25	.00 BIN	1.0 .75	.50 .25	.00 BIN	1.0 .75	.50 .25	.00 BIN	1.0 75	.50 .25	.00 BIN	1.0 .75	.50 .25	.00 BIN
PCONT=***				REQUIRED NUMBER OF PATIENTS												
0.05	***	520	528	547	188	191	195	111	112	113	78	79	80	61	61	62
	•••	523	535	943	189	192	300	111	113	158	79	79	101	61	62	71
0.1	***	952	972	1055	304	312	326	165	168	173	110	111	113	81	82	83
	•••	960	997	1491	307	318	429	166	170	212	110	112	129	82	83	88
0.15	***	1323	1360	1547	401	416	447	209	216	225	135	138	142	97	99	101
	•••	1336	1409	1970	408	428	540	212	220	257	136	140	153	98	100	101
0.2	***	1615	1672	1996	478	503	554	245	255	270	155	159	166	109	112	115
	•••	1635	1751	2381	489	523	634	249	262	295	157	162	172	110	113	112
0.25	***	1827	1906	2388	537	570	645	272	286	308	170	176	185	118	122	126
	•••	1855	2018	2724	551	598	711	278	295	326	173	180	187	120	124	120
0.3	***	1960	2067	2715	578	621	719	292	309	337	180	188	200	125	129	134
	•••	1999	2213	2998	596	657	771	299	321	349	184	193	197	127	131	126
0.35	***	2025	2161	2974	603	654	775	304	324	358	187	196	209	128	133	139
	•••	2074	2343	3203	624	699	814	313	339	364	191	202	204	130	136	129
0.4	***	2030	2197	3159	613	673	812	310	333	371	189	200	214	129	134	141
	•••	2091	2413	3340	638	724	840	320	349	372	194	206	206	131	137	129
0.45	***	1986	2184	3272	612	677	832	309	334	375	188	199	214	127	132	139
	•••	2060	2431	3409	639	733	848	320	351	372	193	206	204	130	136	126
0.5	***	1904	2131	3310	599	669	832	303	329	371	183	195	210	123	128	135
	•••	1990	2402	3409	629	728	840	314	347	364	188	201	197	125	131	120
0.55	***	1795	2044	3273	577	649	813	291	317	359	175	186	201	116	121	128
	•••	1891	2333	3340	608	709	814	303	335	349	180	193	187	119	124	112
0.6	***	1665	1930	3162	546	617	776	274	299	339	163	173	187	107	112	118
	•••	1769	2227	3203	576	676	771	285	316	326	168	180	172	109	114	101
0.65	***	1522	1792	2976	506	574	721	252	275	310	148	157	169	96	99	104
	•••	1629	2086	2998	535	629	711	263	290	295	152	162	153	98	102	88
0.7	***	1368	1632	2717	458	520	647	225	244	273	129	136	146	82	85	88
	•••	1473	1911	2724	485	569	634	234	257	257	133	141	129	83	86	71
0.75	***	1203	1448	2385	401	453	555	192	206	228	107	112	118	.	.	.
	•••	1302	1701	2381	424	493	540	198	216	212	109	115	101	.	.	.
0.8	***	1024	1239	1981	332	373	445	152	162	175	.	.	.	.	.	.
	•••	1112	1451	1970	350	402	429	156	168	158	.	.	.	.	.	.
0.85	***	825	994	1506	250	276	317	.	.	.	.	.	.	.	.	.
	•••	894	1153	1491	262	294	300	.	.	.	.	.	.	.	.	.
0.9	***	587	694	959	.	.	.	.	.	.	.	.	.	.	.	.
	•••	632	787	943	.	.	.	.	.	.	.	.	.	.	.	.

TABLE 8: ALPHA= 0.05 POWER= 0.9   EXPECTED ACCRUAL THRU MINIMUM FOLLOW-UP= 450

	DEL=.05			DEL=.10			DEL=.15			DEL=.20			DEL=.25		
FACT=	1.0 .75	.50 .25	.00 BIN	1.0 .75	.50 .25	.00 BIN	1.0 .75	.50 .25	.00 BIN	1.0 75	.50 .25	.00 BIN	1.0 .75	.50 .25	.00 BIN
PCONT=***				REQUIRED NUMBER OF PATIENTS											
0.05 ***	521	528	547	189	191	195	111	112	113	78	79	80	61	61	62
***	524	535	943	190	193	300	111	113	158	79	80	101	61	62	71
0.1 ***	954	974	1055	305	312	326	165	168	173	110	111	113	81	82	83
***	961	999	1491	308	318	429	167	171	212	111	112	129	82	83	88
0.15 ***	1325	1363	1547	402	417	447	210	216	225	135	138	142	97	99	101
***	1339	1413	1970	409	429	540	213	220	257	136	140	153	98	100	101
0.2 ***	1619	1678	1995	480	505	554	246	256	270	155	160	166	109	112	115
***	1640	1757	2381	491	524	634	250	262	295	157	163	172	111	113	112
0.25 ***	1832	1914	2388	540	573	645	274	287	308	171	177	185	119	122	126
***	1861	2028	2724	554	600	711	280	296	326	173	180	187	120	124	120
0.3 ***	1967	2078	2715	581	624	719	293	310	337	181	189	199	125	129	134
***	2007	2226	2998	599	660	771	301	322	349	185	194	197	127	131	126
0.35 ***	2034	2174	2973	607	658	775	306	326	358	188	197	209	129	133	139
***	2085	2359	3203	629	702	814	315	339	364	192	203	204	131	136	129
0.4 ***	2041	2213	3159	618	678	812	312	334	371	190	200	214	129	134	141
***	2105	2432	3340	643	727	840	322	350	372	195	207	206	132	137	129
0.45 ***	2000	2203	3272	617	682	831	311	336	375	189	200	214	127	133	139
***	2076	2452	3409	645	738	848	322	352	372	194	207	204	130	136	126
0.5 ***	1920	2152	3310	605	675	832	305	330	372	185	195	210	123	129	135
***	2009	2425	3409	635	733	840	316	348	364	189	202	197	126	132	120
0.55 ***	1813	2067	3273	583	654	813	294	319	359	176	186	201	117	122	128
***	1911	2357	3340	613	713	814	305	336	349	181	193	187	119	124	112
0.6 ***	1685	1954	3162	552	622	776	277	301	339	164	174	187	108	112	117
***	1791	2251	3203	582	680	771	288	317	326	169	180	172	110	114	101
0.65 ***	1543	1816	2976	512	579	721	254	276	310	149	158	168	96	100	104
***	1651	2110	2998	541	633	711	264	291	295	153	163	153	98	102	88
0.7 ***	1388	1655	2718	463	524	647	227	245	273	130	137	145	82	85	88
***	1495	1934	2724	490	572	634	235	258	257	133	141	129	83	86	71
0.75 ***	1222	1470	2386	405	457	555	193	208	228	107	112	118	.	.	.
***	1323	1721	2381	428	496	540	200	217	212	109	115	101	.	.	.
0.8 ***	1041	1257	1981	336	375	445	153	162	175	.	.	.	.	.	.
***	1130	1468	1970	354	404	429	157	168	158	.	.	.	.	.	.
0.85 ***	838	1007	1506	253	278	317	.	.	.	.	.	.	.	.	.
***	909	1165	1491	264	295	300	.	.	.	.	.	.	.	.	.
0.9 ***	596	703	959	.	.	.	.	.	.	.	.	.	.	.	.
***	642	794	943	.	.	.	.	.	.	.	.	.	.	.	.

TABLE 8: ALPHA= 0.05 POWER= 0.9   EXPECTED ACCRUAL THRU MINIMUM FOLLOW-UP= 475

	DEL=.05			DEL=.10			DEL=.15			DEL=.20			DEL=.25		
FACT=	1.0 .75	.50 .25	.00 BIN	1.0 .75	.50 .25	.00 BIN	1.0 .75	.50 .25	.00 BIN	1.0 75	.50 .25	.00 BIN	1.0 .75	.50 .25	.00 BIN

PCONT=•••   REQUIRED NUMBER OF PATIENTS

PCONT	1.0/.75	.50/.25	.00/BIN	1.0/.75	.50/.25	.00/BIN	1.0/.75	.50/.25	.00/BIN	1.0/.75	.50/.25	.00/BIN	1.0/.75	.50/.25	.00/BIN
0.05 •••	521	529	547	189	191	194	111	112	113	79	79	80	61	61	62
•••	524	536	943	190	193	300	112	113	158	79	80	101	61	62	71
0.1 •••	955	976	1055	305	313	326	165	169	173	110	112	113	81	82	83
•••	963	1001	1491	309	319	429	167	171	212	111	112	129	82	83	88
0.15 •••	1328	1367	1547	403	419	447	210	217	225	135	138	142	97	99	101
•••	1342	1417	1970	410	430	540	213	221	257	137	140	153	98	100	101
0.2 •••	1622	1683	1996	482	506	553	247	256	270	156	160	166	110	112	115
•••	1644	1764	2381	492	525	634	251	262	295	158	163	172	111	113	112
0.25 •••	1837	1923	2388	542	575	644	275	288	308	171	177	185	119	122	126
•••	1868	2037	2724	557	602	711	281	297	326	174	181	187	121	124	120
0.3 •••	1974	2088	2715	585	627	719	295	311	337	182	189	200	125	129	134
•••	2016	2239	2998	603	662	771	302	322	349	186	194	197	127	131	126
0.35 •••	2043	2188	2973	611	662	775	308	327	358	188	197	209	129	133	139
•••	2096	2374	3203	633	704	814	317	340	364	193	203	204	131	136	129
0.4 •••	2052	2229	3159	623	682	812	314	336	371	191	201	214	130	134	141
•••	2118	2450	3340	648	731	840	324	351	372	196	207	206	132	137	129
0.45 •••	2013	2221	3272	623	687	831	314	337	375	190	200	214	128	133	139
•••	2092	2472	3409	650	741	848	324	354	372	195	207	204	130	136	126
0.5 •••	1936	2172	3310	611	680	832	308	332	371	185	196	210	124	129	135
•••	2027	2447	3409	640	737	840	319	349	364	190	202	197	126	132	120
0.55 •••	1831	2089	3273	589	659	813	296	320	359	177	187	201	117	122	128
•••	1931	2380	3340	619	717	814	307	337	349	182	193	187	119	125	112
0.6 •••	1704	1976	3162	558	628	776	279	302	339	165	175	187	108	112	117
•••	1812	2274	3203	588	684	771	289	318	326	169	180	172	110	115	101
0.65 •••	1562	1838	2977	517	584	721	256	278	310	150	158	169	96	100	104
•••	1673	2132	2998	546	637	711	266	292	295	153	163	153	98	102	88
0.7 •••	1408	1676	2718	469	528	647	228	246	273	131	137	146	82	85	88
•••	1516	1955	2724	495	575	634	237	258	257	134	141	129	83	86	71
0.75 •••	1241	1489	2385	410	460	555	194	208	228	108	112	118	.	.	.
•••	1342	1740	2381	432	498	540	201	217	212	110	115	101	.	.	.
0.8 •••	1058	1274	1981	340	378	445	153	163	175	.	.	.	.	.	.
•••	1146	1484	1970	357	406	429	158	169	158	.	.	.	.	.	.
0.85 •••	851	1020	1505	255	279	317	.	.	.	.	.	.	.	.	.
•••	922	1176	1491	266	296	300	.	.	.	.	.	.	.	.	.
0.9 •••	605	711	959	.	.	.	.	.	.	.	.	.	.	.	.
•••	650	800	943	.	.	.	.	.	.	.	.	.	.	.	.

TABLE 8: ALPHA= 0.05 POWER= 0.9   EXPECTED ACCRUAL THRU MINIMUM FOLLOW-UP= 500

PCONT=	FACT=	DEL=.05 1.0/.75	.50/.25	.00/BIN	DEL=.10 1.0/.75	.50/.25	.00/BIN	DEL=.15 1.0/.75	.50/.25	.00/BIN	DEL=.20 1.0/75	.50/.25	.00/BIN	DEL=.25 1.0/.75	.50/.25	.00/BIN
		REQUIRED NUMBER OF PATIENTS														
0.05	***	522	529	547	189	191	195	111	112	113	79	79	80	61	62	62
	***	525	536	943	190	193	300	112	113	158	79	80	101	61	62	71
0.1	***	956	978	1055	306	313	326	166	169	173	110	112	113	81	82	83
	***	964	1003	1491	309	319	429	167	171	212	111	113	129	82	83	88
0.15	***	1330	1371	1547	405	420	447	211	217	225	136	138	142	97	99	101
	***	1345	1421	1970	411	431	540	214	221	257	137	140	153	98	100	101
0.2	***	1626	1689	1996	484	508	553	248	257	270	156	160	166	110	112	115
	***	1649	1770	2381	494	526	634	252	263	295	158	163	172	111	113	112
0.25	***	1842	1930	2388	545	578	645	276	288	308	172	178	185	119	123	126
	***	1874	2046	2724	559	604	711	282	297	326	174	181	187	121	124	120
0.3	***	1981	2099	2715	588	630	719	296	312	337	183	190	199	126	129	134
	***	2024	2250	2998	606	664	771	303	323	349	186	194	197	128	132	126
0.35	***	2052	2201	2973	615	666	775	309	328	358	189	198	209	129	134	139
	***	2107	2389	3203	637	707	814	318	341	364	193	203	204	131	136	129
0.4	***	2063	2244	3159	628	686	812	316	337	371	192	202	214	130	135	141
	***	2131	2467	3340	653	734	840	325	352	372	197	207	206	133	138	129
0.45	***	2026	2239	3272	628	692	831	316	339	375	191	201	214	128	133	139
	***	2107	2491	3409	655	744	848	326	354	372	196	207	204	131	136	126
0.5	***	1951	2192	3310	616	684	832	309	333	372	186	196	210	124	129	135
	***	2044	2468	3409	645	740	840	320	350	364	191	203	197	127	132	120
0.55	***	1848	2110	3273	594	664	813	298	322	359	178	188	201	118	122	128
	***	1950	2402	3340	624	721	814	309	338	349	183	194	187	120	125	112
0.6	***	1723	1998	3162	563	632	776	281	303	339	166	175	187	108	113	118
	***	1832	2296	3203	593	688	771	291	319	326	170	181	172	110	115	101
0.65	***	1582	1860	2977	523	588	721	258	279	310	151	158	168	97	100	104
	***	1693	2154	2998	551	640	711	268	293	295	154	163	153	98	102	88
0.7	***	1427	1697	2718	473	533	647	230	248	273	131	138	146	83	85	88
	***	1536	1975	2724	499	578	634	238	259	257	134	141	129	84	87	71
0.75	***	1258	1508	2386	414	464	555	196	209	228	108	113	118	.	.	.
	***	1360	1758	2381	436	501	540	202	218	212	110	115	101	.	.	.
0.8	***	1073	1290	1981	343	381	445	154	163	175	.	.	.	.	.	.
	***	1163	1498	1970	360	408	429	159	169	158	.	.	.	.	.	.
0.85	***	864	1033	1506	257	281	317	.	.	.	.	.	.	.	.	.
	***	934	1187	1491	268	297	300	.	.	.	.	.	.	.	.	.
0.9	***	613	718	959	.	.	.	.	.	.	.	.	.	.	.	.
	***	658	806	943	.	.	.	.	.	.	.	.	.	.	.	.

TABLE 8: ALPHA= 0.05 POWER= 0.9   EXPECTED ACCRUAL THRU MINIMUM FOLLOW-UP= 550

	DEL=.02			DEL=.05			DEL=.10			DEL=.15			DEL=.20		
FACT=	1.0 / .75	.50 / .25	.00 / BIN	1.0 / .75	.50 / .25	.00 / BIN	1.0 / .75	.50 / .25	.00 / BIN	1.0 / 75	.50 / .25	.00 / BIN	1.0 / .75	.50 / .25	.00 / BIN
PCONT=***	REQUIRED NUMBER OF PATIENTS														
0.05 ***	2439	2451	2580	523	530	547	189	191	195	111	112	113	79	79	80
•••	2443	2472	4822	526	537	943	190	193	300	112	113	158	79	80	101
0.1 ***	5076	5103	5668	959	981	1055	307	314	326	166	169	173	110	112	113
•••	5085	5157	8376	968	1006	1491	310	319	429	168	171	212	111	113	129
0.15 ***	7416	7464	8766	1334	1377	1547	407	421	446	212	217	225	136	139	142
•••	7432	7559	11502	1350	1427	1970	413	432	540	215	221	257	137	140	153
0.2 ***	9281	9354	11647	1633	1699	1996	488	511	553	249	258	270	157	161	166
•••	9305	9502	14199	1658	1782	2381	498	528	634	253	264	295	158	163	172
0.25 ***	10621	10727	14201	1852	1945	2388	550	582	644	278	290	308	173	178	185
•••	10656	10938	16469	1886	2062	2724	564	607	711	283	298	326	175	181	187
0.3 ***	11445	11591	16367	1994	2118	2715	594	635	719	299	314	337	184	191	199
•••	11494	11880	18310	2040	2272	2998	612	668	771	306	324	349	187	195	197
0.35 ***	11790	11983	18108	2069	2225	2973	622	672	775	312	330	358	191	199	209
•••	11854	12368	19723	2127	2415	3203	643	712	814	321	343	364	195	203	204
0.4 ***	11708	11961	19404	2084	2274	3159	636	693	812	319	339	371	194	202	214
•••	11793	12460	20708	2156	2498	3340	661	739	840	328	353	372	198	208	206
0.45 ***	11265	11589	20243	2052	2272	3272	637	700	831	319	341	375	193	202	214
•••	11373	12223	21265	2136	2527	3409	664	750	848	329	356	372	197	208	204
0.5 ***	10527	10942	20619	1981	2228	3310	626	693	832	313	336	371	188	198	210
•••	10666	11729	21393	2077	2506	3409	655	746	840	324	352	364	192	203	197
0.55 ***	9571	10097	20529	1881	2149	3273	604	673	813	302	324	359	179	189	201
•••	9748	11047	21093	1986	2442	3340	634	727	814	312	340	349	184	195	187
0.6 ***	8475	9131	19973	1758	2038	3162	573	641	776	284	306	339	167	176	187
•••	8698	10237	20365	1870	2337	3203	602	694	771	294	321	326	171	181	172
0.65 ***	7321	8115	18952	1618	1900	2976	532	596	721	261	281	310	152	159	169
•••	7599	9344	19209	1731	2193	2998	560	646	711	270	294	295	155	164	153
0.7 ***	6189	7102	17471	1462	1735	2718	482	540	647	233	250	273	132	138	146
•••	6520	8396	17625	1573	2012	2724	507	583	634	241	260	257	135	142	129
0.75 ***	5139	6115	15532	1292	1543	2386	421	470	555	198	211	228	109	113	118
•••	5504	7403	15612	1394	1790	2381	443	505	540	204	219	212	111	115	101
0.8 ***	4190	5151	13140	1103	1320	1981	349	385	445	156	164	175	.	.	.
•••	4558	6353	13172	1192	1525	1970	365	410	429	160	169	158	.	.	.
0.85 ***	3318	4180	10300	887	1055	1506	261	283	317	.	.	.	.	.	.
•••	3653	5212	10303	958	1206	1491	271	298	300	.	.	.	.	.	.
0.9 ***	2458	3132	7018	628	731	959	.	.	.	.	.	.	.	.	.
•••	2723	3901	7006	672	816	943	.	.	.	.	.	.	.	.	.
0.95 ***	1464	1835	3297	.	.	.	.	.	.	.	.	.	.	.	.
•••	1614	2219	3280	.	.	.	.	.	.	.	.	.	.	.	.

TABLE 8: ALPHA= 0.05 POWER= 0.9   EXPECTED ACCRUAL THRU MINIMUM FOLLOW-UP= 600

		DEL=.02			DEL=.05			DEL=.10			DEL=.15			DEL=.20		
FACT=		1.0 .75	.50 .25	.00 BIN	1.0 .75	.50 .25	.00 BIN	1.0 .75	.50 .25	.00 BIN	1.0 75	.50 .25	.00 BIN	1.0 .75	.50 .25	.00 BIN
PCONT=•••					REQUIRED NUMBER OF PATIENTS											
0.05	•••	2440	2453	2580	524	531	547	190	192	194	112	112	113	79	79	80
	•••	2445	2476	4822	527	538	943	191	193	300	112	113	158	79	79	101
0.1	•••	5078	5108	5668	961	985	1055	308	315	326	167	169	173	110	112	113
	•••	5088	5166	8376	970	1009	1491	311	320	429	168	171	212	111	113	129
0.15	•••	7420	7472	8766	1339	1383	1547	409	423	446	213	218	225	136	139	142
	•••	7438	7576	11502	1356	1434	1970	415	433	540	215	221	257	137	140	153
0.2	•••	9287	9368	11647	1640	1709	1996	491	513	553	250	259	270	157	161	166
	•••	9314	9528	14199	1666	1792	2381	500	530	634	254	264	295	159	163	172
0.25	•••	10631	10746	14201	1861	1959	2388	554	585	644	280	291	308	173	178	185
	•••	10669	10976	16469	1898	2077	2724	568	610	711	285	298	326	176	181	187
0.3	•••	11458	11617	16367	2007	2137	2715	599	640	719	301	316	337	185	191	199
	•••	11511	11932	18310	2056	2292	2998	617	671	771	308	325	349	188	195	197
0.35	•••	11807	12018	18108	2085	2248	2973	629	678	775	315	332	358	192	199	209
	•••	11878	12436	19723	2147	2440	3203	650	716	814	323	344	364	196	204	204
0.4	•••	11731	12007	19404	2105	2300	3159	643	700	812	322	341	371	195	203	214
	•••	11823	12547	20708	2180	2527	3340	668	744	840	331	355	372	199	208	206
0.45	•••	11294	11648	20243	2076	2303	3272	645	707	831	322	343	375	194	203	214
	•••	11412	12332	21265	2164	2558	3409	672	756	848	332	358	372	198	208	204
0.5	•••	10565	11017	20619	2009	2263	3310	635	700	832	316	338	371	189	198	210
	•••	10716	11861	21393	2108	2541	3409	663	752	840	326	353	364	194	204	197
0.55	•••	9619	10190	20528	1912	2185	3273	613	680	813	305	327	359	181	190	201
	•••	9812	11201	21093	2020	2478	3340	643	733	814	315	341	349	185	195	187
0.6	•••	8536	9243	19972	1791	2076	3162	582	648	776	287	308	339	169	177	187
	•••	8779	10408	20365	1905	2373	3203	611	699	771	297	322	326	173	181	172
0.65	•••	7398	8245	18952	1651	1937	2977	541	603	721	264	283	310	153	160	169
	•••	7697	9527	19209	1767	2229	2998	568	650	711	273	295	295	156	164	153
0.7	•••	6282	7244	17471	1495	1771	2717	490	546	647	235	251	273	133	139	145
	•••	6634	8583	17625	1607	2045	2724	515	587	634	242	261	257	136	142	129
0.75	•••	5243	6260	15532	1322	1575	2386	428	475	555	200	212	228	109	113	118
	•••	5626	7584	15612	1426	1819	2381	449	508	540	205	219	212	111	116	101
0.8	•••	4297	5290	13140	1129	1346	1981	353	389	445	157	165	175	.	.	.
	•••	4678	6518	13172	1220	1548	1970	370	413	429	161	170	158	.	.	.
0.85	•••	3416	4301	10300	908	1075	1505	264	286	317	.	.	.	.	.	.
	•••	3761	5351	10303	979	1222	1491	274	300	300	.	.	.	.	.	.
0.9	•••	2536	3224	7018	641	743	959	.	.	.	.	.	.	.	.	.
	•••	2808	4002	7006	685	825	943	.	.	.	.	.	.	.	.	.
0.95	•••	1508	1883	3297	.	.	.	.	.	.	.	.	.	.	.	.
	•••	1661	2266	3280	.	.	.	.	.	.	.	.	.	.	.	.

TABLE 8: ALPHA= 0.05 POWER= 0.9   EXPECTED ACCRUAL THRU MINIMUM FOLLOW-UP= 650

PCONT=•••	DEL=.02			DEL=.05			DEL=.10			DEL=.15			DEL=.20		
FACT=	1.0 .75	.50 .25	.00 BIN	1.0 .75	.50 .25	.00 BIN	1.0 .75	.50 .25	.00 BIN	1.0 75	.50 .25	.00 BIN	1.0 .75	.50 .25	.00 BIN
					REQUIRED NUMBER OF PATIENTS										
0.05 •••	2442	2455	2580	525	532	547	190	192	194	112	112	114	79	80	80
•••	2446	2479	4822	528	538	943	191	193	300	112	113	158	79	80	101
0.1 •••	5081	5113	5668	964	987	1055	309	316	326	167	170	173	111	112	114
•••	5091	5176	8376	973	1011	1491	312	320	429	168	171	212	111	113	129
0.15 •••	7425	7481	8766	1343	1389	1547	411	424	447	214	219	225	137	139	142
•••	7444	7593	11502	1361	1440	1970	417	434	540	216	222	257	138	140	153
0.2 •••	9294	9381	11647	1647	1719	1996	493	515	554	252	259	270	158	162	166
•••	9323	9555	14199	1674	1801	2381	503	532	634	255	264	295	159	164	172
0.25 •••	10640	10766	14201	1871	1972	2388	558	588	645	281	292	308	174	179	185
•••	10682	11014	16469	1909	2091	2724	571	612	711	286	299	326	176	182	187
0.3 •••	11472	11643	16367	2020	2154	2715	604	644	718	303	317	337	186	192	199
•••	11529	11984	18310	2070	2310	2998	622	674	771	309	326	349	189	195	197
0.35 •••	11825	12054	18108	2101	2269	2973	635	683	775	317	334	358	193	200	209
•••	11901	12504	19723	2166	2462	3203	656	720	814	325	345	364	197	204	204
0.4 •••	11754	12053	19404	2125	2326	3159	650	705	812	324	343	371	196	204	214
•••	11854	12634	20708	2202	2553	3340	674	748	840	333	356	372	200	209	206
0.45 •••	11324	11708	20243	2099	2332	3272	653	713	831	325	346	375	195	204	214
•••	11452	12440	21265	2190	2588	3409	679	760	848	335	359	372	199	209	204
0.5 •••	10603	11091	20619	2036	2294	3310	643	707	832	320	341	372	191	199	210
•••	10767	11989	21393	2138	2572	3409	671	757	840	329	354	364	195	204	197
0.55 •••	9667	10282	20529	1941	2219	3273	622	688	813	308	329	359	182	190	201
•••	9875	11349	21093	2052	2511	3340	651	738	814	318	342	349	186	195	187
0.6 •••	8597	9354	19972	1822	2110	3162	591	655	776	290	310	339	170	177	187
•••	8859	10572	20365	1938	2407	3203	619	704	771	300	323	326	174	182	172
0.65 •••	7474	8372	18952	1683	1971	2976	549	610	721	267	285	310	154	160	168
•••	7794	9701	19209	1800	2261	2998	576	655	711	275	296	295	157	164	153
0.7 •••	6373	7380	17471	1526	1803	2718	497	552	647	237	253	273	134	139	146
•••	6743	8759	17625	1640	2075	2724	521	591	634	244	262	257	137	142	129
0.75 •••	5344	6399	15532	1351	1605	2385	434	479	555	201	213	228	110	114	118
•••	5743	7754	15612	1456	1846	2381	454	511	540	207	220	212	112	116	101
0.8 •••	4398	5420	13140	1154	1371	1981	358	392	445	158	166	175	.	.	.
•••	4792	6673	13172	1245	1570	1970	374	415	429	162	171	158	.	.	.
0.85 •••	3509	4415	10300	928	1094	1505	267	288	317	.	.	.	.	.	.
•••	3863	5481	10303	998	1237	1491	276	301	300	.	.	.	.	.	.
0.9 •••	2610	3310	7018	654	754	959	.	.	.	.	.	.	.	.	.
•••	2887	4094	7006	697	833	943	.	.	.	.	.	.	.	.	.
0.95 •••	1550	1928	3297	.	.	.	.	.	.	.	.	.	.	.	.
•••	1705	2308	3280	.	.	.	.	.	.	.	.	.	.	.	.

TABLE 8: ALPHA= 0.05 POWER= 0.9   EXPECTED ACCRUAL THRU MINIMUM FOLLOW-UP= 700

		DEL=.02			DEL=.05			DEL=.10			DEL=.15			DEL=.20		
FACT=		1.0 .75	.50 .25	.00 BIN	1.0 .75	.50 .25	.00 BIN	1.0 .75	.50 .25	.00 BIN	1.0 75	.50 .25	.00 BIN	1.0 .75	.50 .25	.00 BIN
PCONT=•••					REQUIRED NUMBER OF PATIENTS											
0.05	•••	2443	2457	2580	525	533	547	190	192	194	112	113	114	79	79	80
	•••	2448	2482	4822	528	539	943	191	193	300	112	113	158	79	79	101
0.1	•••	5083	5118	5668	966	990	1055	310	317	326	167	170	173	111	112	114
	•••	5095	5185	8376	976	1013	1491	313	320	429	169	171	212	111	113	129
0.15	•••	7429	7490	8766	1348	1395	1547	412	425	446	214	219	225	137	139	142
	•••	7450	7610	11502	1366	1444	1970	418	435	540	216	222	257	138	141	153
0.2	•••	9301	9395	11647	1653	1727	1996	496	517	554	253	260	270	158	162	166
	•••	9332	9581	14199	1682	1810	2381	506	533	634	256	265	295	160	163	172
0.25	•••	10650	10785	14201	1880	1985	2388	561	592	645	282	293	308	175	179	185
	•••	10695	11052	16469	1920	2103	2724	575	614	711	288	300	326	177	182	187
0.3	•••	11485	11670	16367	2032	2170	2715	609	648	719	305	318	337	187	192	199
	•••	11546	12035	18310	2085	2326	2998	626	677	771	311	327	349	189	196	197
0.35	•••	11842	12089	18108	2117	2289	2973	640	688	775	319	335	358	194	201	209
	•••	11925	12571	19723	2183	2482	3203	661	723	814	.327	345	364	197	205	204
0.4	•••	11777	12098	19404	2144	2350	3159	656	711	812	327	345	371	197	205	214
	•••	11884	12719	20708	2224	2576	3340	681	752	840	335	357	372	201	209	206
0.45	•••	11353	11766	20243	2122	2359	3272	660	719	831	327	347	375	196	205	215
	•••	11491	12544	21265	2215	2614	3409	686	765	848	337	360	372	200	209	204
0.5	•••	10641	11165	20619	2061	2324	3310	650	713	831	322	342	372	192	200	210
	•••	10817	12114	21393	2165	2601	3409	678	761	840	331	355	364	195	205	197
0.55	•••	9716	10372	20529	1969	2250	3273	629	694	813	310	331	359	183	191	201
	•••	9939	11491	21093	2081	2541	3340	658	742	814	320	344	349	187	196	187
0.6	•••	8658	9462	19973	1851	2142	3162	598	661	776	293	312	339	171	178	187
	•••	8937	10729	20365	1969	2437	3203	626	708	771	302	324	326	174	182	172
0.65	•••	7550	8494	18952	1713	2003	2977	556	615	721	269	286	310	155	161	169
	•••	7888	9866	19209	1831	2291	2998	582	659	711	277	297	295	158	165	153
0.7	•••	6462	7511	17470	1555	1833	2718	503	556	647	239	254	273	135	140	145
	•••	6850	8926	17625	1669	2102	2724	527	594	634	246	263	257	137	142	129
0.75	•••	5441	6530	15532	1378	1632	2385	439	483	555	203	214	228	110	114	118
	•••	5854	7914	15612	1483	1870	2381	459	514	540	208	221	212	112	116	101
0.8	•••	4496	5544	13140	1178	1394	1981	362	395	445	159	166	175	.	.	.
	•••	4901	6818	13172	1268	1589	1970	377	417	429	163	171	158	.	.	.
0.85	•••	3596	4522	10300	946	1111	1506	270	289	317	.	.	.	.	.	.
	•••	3959	5601	10303	1016	1251	1491	279	302	300	.	.	.	.	.	.
0.9	•••	2679	3391	7018	665	763	959	.	.	.	.	.	.	.	.	.
	•••	2962	4181	7006	708	840	943	.	.	.	.	.	.	.	.	.
0.95	•••	1589	1969	3297	.	.	.	.	.	.	.	.	.	.	.	.
	•••	1745	2347	3280	.	.	.	.	.	.	.	.	.	.	.	.

TABLE 8: ALPHA= 0.05 POWER= 0.9   EXPECTED ACCRUAL THRU MINIMUM FOLLOW-UP= 750

		DEL=.02			DEL=.05			DEL=.10			DEL=.15			DEL=.20		
FACT=		1.0 .75	.50 .25	.00 BIN	1.0 .75	.50 .25	.00 BIN	1.0 .75	.50 .25	.00 BIN	1.0 75	.50 .25	.00 BIN	1.0 .75	.50 .25	.00 BIN
PCONT=***		REQUIRED NUMBER OF PATIENTS														
0.05	***	2444	2459	2580	526	533	547	190	192	195	112	113	114	79	79	80
	***	2449	2485	4822	529	539	943	191	193	300	112	113	158	79	79	101
0.1	***	5086	5123	5668	968	993	1055	310	317	326	168	170	173	111	112	114
	***	5098	5194	8376	978	1015	1491	313	321	429	169	171	212	112	113	129
0.15	***	7433	7499	8766	1352	1399	1547	414	427	446	215	220	225	138	139	142
	***	7455	7627	11502	1370	1449	1970	420	435	540	217	222	257	139	141	153
0.2	***	9308	9408	11647	1660	1735	1996	499	519	553	253	261	270	159	162	166
	***	9341	9607	14199	1689	1818	2381	508	534	634	257	265	295	160	164	172
0.25	***	10660	10804	14201	1889	1997	2388	565	594	645	284	294	308	175	180	185
	***	10708	11089	16469	1930	2115	2724	578	615	711	289	300	326	177	182	187
0.3	***	11498	11696	16367	2044	2185	2715	613	651	719	306	319	337	187	193	199
	***	11564	12085	18310	2099	2341	2998	630	679	771	312	327	349	190	196	197
0.35	***	11860	12124	18108	2132	2308	2974	645	692	775	321	337	358	195	201	209
	***	11948	12637	19723	2200	2501	3203	666	726	814	328	346	364	198	205	204
0.4	***	11800	12144	19404	2162	2372	3160	663	715	812	329	346	371	198	205	214
	***	11915	12803	20708	2245	2599	3340	686	755	840	337	357	372	201	209	206
0.45	***	11383	11824	20243	2143	2384	3272	666	724	831	330	349	375	198	205	214
	***	11530	12647	21265	2239	2639	3409	692	768	848	339	361	372	201	209	204
0.5	***	10678	11239	20619	2085	2352	3310	657	719	832	325	344	371	193	200	210
	***	10867	12235	21393	2192	2628	3409	684	765	840	334	356	364	196	205	197
0.55	***	9764	10462	20529	1995	2280	3273	636	700	814	313	332	359	184	191	201
	***	10002	11629	21093	2110	2569	3340	664	746	814	322	344	349	188	196	187
0.6	***	8719	9567	19972	1879	2172	3162	605	667	776	295	313	339	172	179	187
	***	9015	10879	20365	1998	2464	3203	632	712	771	304	325	326	175	183	172
0.65	***	7624	8612	18952	1740	2032	2976	562	620	721	271	288	310	155	161	169
	***	7980	10024	19209	1860	2317	2998	589	662	711	279	298	295	158	165	153
0.7	***	6549	7636	17470	1582	1861	2718	509	561	647	241	255	273	135	140	145
	***	6953	9084	17625	1697	2126	2724	532	597	634	248	263	257	138	143	129
0.75	***	5535	6655	15532	1403	1656	2386	444	487	555	204	215	228	111	114	118
	***	5962	8065	15612	1508	1891	2381	464	516	540	209	221	212	113	116	101
0.8	***	4589	5662	13140	1199	1414	1981	366	398	445	160	167	175	.	.	.
	***	5004	6955	13172	1290	1606	1970	381	419	429	163	171	158	.	.	.
0.85	***	3680	4623	10300	963	1126	1505	272	291	317	.	.	.	.	.	.
	***	4050	5714	10303	1032	1263	1491	281	303	300	.	.	.	.	.	.
0.9	***	2745	3467	7018	676	772	959	.	.	.	.	.	.	.	.	.
	***	3033	4261	7006	718	846	943	.	.	.	.	.	.	.	.	.
0.95	***	1626	2008	3297	.	.	.	.	.	.	.	.	.	.	.	.
	***	1783	2382	3280	.	.	.	.	.	.	.	.	.	.	.	.

TABLE 8: ALPHA= 0.05 POWER= 0.9   EXPECTED ACCRUAL THRU MINIMUM FOLLOW-UP= 800

	DEL=.02			DEL=.05			DEL=.10			DEL=.15			DEL=.20		
FACT=	1.0 .75	.50 .25	.00 BIN	1.0 .75	.50 .25	.00 BIN	1.0 .75	.50 .25	.00 BIN	1.0 75	.50 .25	.00 BIN	1.0 .75	.50 .25	.00 BIN
PCONT=•••				REQUIRED NUMBER OF PATIENTS											
0.05 •••	2445	2461	2580	527	534	547	191	192	195	112	113	114	79	79	80
•••	2450	2488	4822	530	540	943	191	193	300	112	113	158	79	80	101
0.1 •••	5088	5128	5668	970	995	1055	311	317	326	168	170	173	111	112	113
•••	5101	5203	8376	980	1018	1491	314	321	429	169	172	212	112	113	129
0.15 •••	7438	7507	8766	1356	1404	1547	415	428	447	215	220	225	138	140	142
•••	7461	7643	11502	1375	1453	1970	421	436	540	217	222	257	139	141	153
0.2 •••	9314	9422	11647	1666	1743	1996	501	521	554	254	261	270	159	162	166
•••	9350	9632	14199	1696	1825	2381	510	535	634	258	265	295	161	164	172
0.25 •••	10669	10823	14201	1898	2008	2388	568	597	645	285	295	308	176	180	185
•••	10721	11126	16469	1940	2125	2724	581	617	711	290	301	326	178	182	187
0.3 •••	11511	11723	16367	2056	2200	2715	617	654	719	308	320	337	188	193	199
•••	11582	12135	18310	2112	2355	2998	634	681	771	314	328	349	190	196	197
0.35 •••	11878	12159	18108	2147	2326	2973	650	695	775	323	338	358	196	202	209
•••	11972	12703	19723	2217	2518	3203	670	728	814	330	347	364	198	205	204
0.4 •••	11823	12190	19404	2180	2393	3159	668	720	812	331	348	371	199	206	214
•••	11946	12885	20708	2264	2619	3340	691	758	840	339	358	372	202	210	206
0.45 •••	11412	11883	20243	2164	2408	3272	672	729	831	332	350	375	198	206	215
•••	11570	12746	21265	2261	2662	3409	697	771	848	340	362	372	202	210	204
0.5 •••	10716	11311	20619	2108	2378	3310	663	724	832	327	345	372	194	201	210
•••	10917	12352	21393	2217	2652	3409	690	768	840	335	357	364	197	205	197
0.55 •••	9812	10550	20529	2020	2307	3273	643	705	813	315	334	359	185	192	201
•••	10065	11761	21093	2136	2594	3340	670	750	814	324	345	349	188	196	187
0.6 •••	8779	9670	19973	1905	2200	3162	611	672	776	297	315	339	173	179	187
•••	9092	11023	20365	2025	2490	3203	638	715	771	305	326	326	176	183	172
0.65 •••	7697	8726	18952	1767	2060	2977	568	625	721	273	289	310	156	162	169
•••	8071	10173	19209	1887	2342	2998	594	665	711	280	299	295	159	165	153
0.7 •••	6634	7757	17471	1608	1887	2718	515	565	647	243	256	273	136	140	146
•••	7053	9234	17625	1723	2149	2724	537	600	634	249	264	257	138	143	129
0.75 •••	5626	6776	15532	1426	1680	2386	449	490	555	206	216	228	111	115	118
•••	6065	8208	15612	1532	1911	2381	468	518	540	210	222	212	113	116	101
0.8 •••	4678	5774	13140	1220	1434	1981	370	400	445	161	168	175	.	.	.
•••	5104	7083	13172	1310	1622	1970	384	420	429	164	171	158	.	.	.
0.85 •••	3761	4720	10300	979	1140	1506	274	292	317	.	.	.	.	.	.
•••	4138	5821	10303	1048	1274	1491	283	304	300	.	.	.	.	.	.
0.9 •••	2808	3539	7018	686	780	959	.	.	.	.	.	.	.	.	.
•••	3099	4336	7006	727	852	943	.	.	.	.	.	.	.	.	.
0.95 •••	1661	2044	3297	.	.	.	.	.	.	.	.	.	.	.	.
•••	1818	2415	3280	.	.	.	.	.	.	.	.	.	.	.	.

TABLE 8: ALPHA= 0.05 POWER= 0.9   EXPECTED ACCRUAL THRU MINIMUM FOLLOW-UP= 850

	DEL=.02			DEL=.05			DEL=.10			DEL=.15			DEL=.20		
FACT=	1.0 .75	.50 .25	.00 BIN	1.0 .75	.50 .25	.00 BIN	1.0 .75	.50 .25	.00 BIN	1.0 75	.50 .25	.00 BIN	1.0 .75	.50 .25	.00 BIN
PCONT=***	REQUIRED NUMBER OF PATIENTS														
0.05 ***	2446	2463	2580	528	535	547	191	192	195	112	113	113	79	79	80
•••	2452	2490	4822	530	540	943	191	193	300	112	113	158	79	80	101
0.1 ***	5091	5133	5668	972	997	1055	312	318	326	168	170	173	111	112	113
•••	5105	5211	8376	983	1019	1491	315	322	429	169	172	212	112	113	129
0.15 ***	7442	7516	8766	1360	1409	1547	416	428	446	216	220	225	138	140	142
•••	7467	7659	11502	1379	1457	1970	422	436	540	218	222	257	139	141	153
0.2 ***	9321	9435	11647	1672	1751	1995	503	522	554	255	261	270	159	162	166
•••	9359	9658	14199	1703	1831	2381	512	536	634	258	266	295	161	164	172
0.25 ***	10679	10843	14201	1906	2018	2388	570	598	645	286	295	308	176	180	185
•••	10733	11162	16469	1950	2135	2724	583	618	711	290	301	326	178	182	187
0.3 ***	11525	11749	16367	2067	2213	2715	621	657	718	309	321	337	188	193	199
•••	11599	12184	18310	2124	2368	2998	637	683	771	315	328	349	191	196	197
0.35 ***	11895	12194	18108	2161	2343	2974	654	699	775	324	339	358	196	202	209
•••	11995	12767	19723	2232	2534	3203	674	731	814	331	348	364	199	205	204
0.4 ***	11846	12235	19404	2197	2413	3160	673	724	813	333	349	371	200	206	214
•••	11976	12965	20708	2282	2637	3340	696	761	840	340	359	372	203	210	206
0.45 ***	11442	11940	20244	2184	2431	3272	678	733	832	334	351	375	199	206	214
•••	11609	12843	21265	2282	2683	3409	702	774	848	342	362	372	203	210	204
0.5 ***	10754	11383	20619	2130	2402	3310	669	729	832	328	346	372	195	202	210
•••	10967	12465	21393	2240	2675	3409	696	771	840	337	358	364	198	205	197
0.55 ***	9860	10637	20529	2044	2333	3273	649	709	814	317	335	359	186	192	201
•••	10127	11888	21093	2161	2618	3340	675	752	814	325	346	349	189	197	187
0.6 ***	8839	9770	19973	1930	2227	3162	617	676	776	299	316	339	173	180	187
•••	9168	11161	20365	2051	2514	3203	644	718	771	307	326	326	176	183	172
0.65 ***	7770	8837	18952	1792	2086	2976	574	629	720	275	290	310	157	162	169
•••	8159	10317	19209	1913	2365	2998	599	668	711	282	299	295	159	165	153
0.7 ***	6716	7873	17471	1632	1911	2718	520	569	647	244	257	273	136	140	146
•••	7150	9376	17625	1748	2170	2724	542	602	634	250	264	257	138	143	129
0.75 ***	5714	6890	15532	1448	1701	2385	453	493	555	206	216	228	112	115	118
•••	6164	8344	15612	1554	1929	2381	471	520	540	211	222	212	113	117	101
0.8 ***	4764	5881	13140	1239	1451	1981	373	402	445	162	168	175	.	.	.
•••	5199	7205	13172	1329	1636	1970	386	422	429	165	171	158	.	.	.
0.85 ***	3837	4811	10301	994	1153	1506	276	293	317	.	.	.	.	.	.
•••	4221	5921	10303	1062	1284	1491	284	305	300	.	.	.	.	.	.
0.9 ***	2868	3607	7018	695	787	959	.	.	.	.	.	.	.	.	.
•••	3163	4406	7006	735	856	943	.	.	.	.	.	.	.	.	.
0.95 ***	1694	2077	3297	.	.	.	.	.	.	.	.	.	.	.	.
•••	1852	2446	3280	.	.	.	.	.	.	.	.	.	.	.	.

TABLE 8: ALPHA= 0.05 POWER= 0.9   EXPECTED ACCRUAL THRU MINIMUM FOLLOW-UP= 900

		DEL=.02			DEL=.05			DEL=.10			DEL=.15			DEL=.20		
FACT=		1.0 .75	.50 .25	.00 BIN	1.0 .75	.50 .25	.00 BIN	1.0 .75	.50 .25	.00 BIN	1.0 75	.50 .25	.00 BIN	1.0 .75	.50 .25	.00 BIN
PCONT=•••					REQUIRED NUMBER OF PATIENTS											
0.05	•••	2447	2465	2580	528	535	548	191	192	195	112	113	113	79	80	80
	•••	2453	2493	4822	531	540	943	192	194	300	113	113	158	80	80	101
0.1	•••	5093	5138	5668	974	999	1055	312	318	326	168	171	173	111	112	113
	•••	5108	5220	8376	984	1020	1491	315	322	429	170	172	212	112	113	129
0.15	•••	7446	7525	8766	1363	1413	1547	417	429	447	216	220	225	138	140	142
	•••	7472	7675	11502	1383	1460	1970	423	437	540	218	222	257	139	141	153
0.2	•••	9327	9449	11647	1677	1757	1995	505	524	554	255	262	270	160	163	165
	•••	9368	9683	14199	1709	1837	2381	513	537	634	259	266	295	161	164	172
0.25	•••	10689	10862	14201	1914	2028	2388	573	600	645	287	296	308	177	180	185
	•••	10746	11198	16469	1959	2144	2724	585	620	711	291	301	326	179	182	187
0.3	•••	11538	11775	16367	2078	2226	2715	624	660	719	310	322	337	189	194	199
	•••	11617	12233	18310	2136	2380	2998	640	685	771	316	329	349	191	197	197
0.35	•••	11913	12229	18108	2174	2359	2973	658	702	775	326	339	358	197	203	209
	•••	12018	12830	19723	2248	2549	3203	678	732	814	332	348	364	199	206	204
0.4	•••	11870	12281	19404	2213	2432	3159	677	728	812	334	350	371	200	207	214
	•••	12007	13043	20708	2301	2655	3340	700	763	840	342	360	372	203	210	206
0.45	•••	11471	11998	20243	2203	2452	3272	683	738	831	335	352	375	200	207	215
	•••	11648	12938	21265	2303	2702	3409	707	777	848	343	363	372	203	210	204
0.5	•••	10792	11454	20619	2152	2425	3310	675	732	831	330	348	371	195	202	210
	•••	11017	12575	21393	2262	2696	3409	701	774	840	339	359	364	198	206	197
0.55	•••	9908	10722	20529	2067	2357	3273	654	713	813	318	336	359	186	193	201
	•••	10190	12011	21093	2186	2640	3340	681	756	814	327	347	349	190	197	187
0.6	•••	8898	9868	19972	1954	2251	3162	622	680	776	300	317	339	174	180	187
	•••	9243	11293	20365	2076	2536	3203	648	721	771	308	327	326	177	183	172
0.65	•••	7841	8945	18952	1816	2109	2976	579	633	721	276	291	310	158	163	168
	•••	8245	10454	19209	1937	2386	2998	603	671	711	283	300	295	160	165	153
0.7	•••	6797	7985	17471	1655	1934	2718	524	572	647	245	258	273	137	141	145
	•••	7244	9513	17625	1771	2190	2724	546	604	634	251	265	257	138	143	129
0.75	•••	5799	7001	15532	1469	1721	2386	457	496	555	208	217	228	112	115	118
	•••	6260	8473	15612	1575	1946	2381	475	522	540	212	222	212	113	117	101
0.8	•••	4847	5984	13140	1257	1468	1981	375	404	445	162	168	175	.	.	.
	•••	5290	7320	13172	1346	1649	1970	389	423	429	165	172	158	.	.	.
0.85	•••	3911	4898	10301	1007	1165	1505	278	295	317	.	.	.	.	.	.
	•••	4301	6016	10303	1075	1293	1491	286	305	300	.	.	.	.	.	.
0.9	•••	2925	3672	7018	703	794	959	.	.	.	.	.	.	.	.	.
	•••	3224	4473	7006	743	861	943	.	.	.	.	.	.	.	.	.
0.95	•••	1725	2109	3297	.	.	.	.	.	.	.	.	.	.	.	.
	•••	1883	2474	3280	.	.	.	.	.	.	.	.	.	.	.	.

TABLE 8: ALPHA= 0.05 POWER= 0.9   EXPECTED ACCRUAL THRU MINIMUM FOLLOW-UP= 950

PCONT		DEL=.02			DEL=.05			DEL=.10			DEL=.15			DEL=.20		
FACT=		1.0 / .75	.50 / .25	.00 / BIN	1.0 / .75	.50 / .25	.00 / BIN	1.0 / .75	.50 / .25	.00 / BIN	1.0 / 75	.50 / .25	.00 / BIN	1.0 / .75	.50 / .25	.00 / BIN

REQUIRED NUMBER OF PATIENTS

PCONT		DEL=.02 1.0/.75	.50/.25	.00/BIN	DEL=.05 1.0/.75	.50/.25	.00/BIN	DEL=.10 1.0/.75	.50/.25	.00/BIN	DEL=.15 1.0/.75	.50/.25	.00/BIN	DEL=.20 1.0/.75	.50/.25	.00/BIN
0.05	***	2448	2466	2580	528	536	547	191	192	194	112	113	114	79	80	80
	***	2454	2495	4822	532	540	943	192	194	300	112	113	158	79	80	101
0.1	***	5096	5143	5668	976	1001	1055	313	318	326	169	171	173	112	112	114
	***	5111	5227	8376	986	1022	1491	315	322	429	169	172	212	112	113	129
0.15	***	7450	7533	8766	1367	1417	1547	419	430	447	217	220	225	139	140	142
	***	7478	7690	11502	1387	1464	1970	424	438	540	219	223	257	139	141	153
0.2	***	9335	9462	11647	1683	1764	1996	506	525	553	256	262	270	160	163	166
	***	9377	9707	14199	1716	1843	2381	515	538	634	259	266	295	161	164	172
0.25	***	10698	10881	14201	1923	2037	2388	576	602	644	287	296	308	177	181	185
	***	10759	11233	16469	1968	2153	2724	587	621	711	292	302	326	179	182	187
0.3	***	11551	11801	16367	2088	2238	2715	627	662	718	311	323	337	190	194	200
	***	11635	12281	18310	2148	2392	2998	642	686	771	317	329	349	191	197	197
0.35	***	11930	12264	18108	2187	2374	2973	662	704	774	327	340	358	197	203	209
	***	12042	12891	19723	2262	2563	3203	681	735	814	333	349	364	200	205	204
0.4	***	11892	12326	19405	2229	2450	3160	682	730	812	336	351	371	201	207	214
	***	12037	13120	20708	2318	2671	3340	704	766	840	343	360	372	204	210	206
0.45	***	11501	12055	20243	2221	2472	3272	687	741	831	337	353	375	200	207	215
	***	11687	13030	21265	2322	2721	3409	711	779	848	345	363	372	203	210	204
0.5	***	10830	11524	20619	2172	2447	3310	679	736	831	332	349	371	196	202	210
	***	11066	12681	21393	2284	2716	3409	705	777	840	340	359	364	199	206	197
0.55	***	9955	10805	20528	2089	2380	3273	659	717	813	320	337	359	187	193	201
	***	10251	12130	21093	2208	2660	3340	685	758	814	328	348	349	190	197	187
0.6	***	8957	9963	19973	1976	2274	3162	628	684	776	302	318	338	175	180	187
	***	9317	11421	20365	2099	2556	3203	653	723	771	310	328	326	177	184	172
0.65	***	7911	9049	18952	1838	2132	2977	584	637	721	277	292	310	158	163	169
	***	8330	10585	19209	1960	2405	2998	608	673	711	285	300	295	160	166	153
0.7	***	6876	8093	17471	1676	1955	2718	528	575	647	247	258	273	137	141	146
	***	7335	9643	17625	1793	2208	2724	549	606	634	252	266	257	139	143	129
0.75	***	5881	7107	15532	1489	1740	2385	460	498	555	209	217	228	112	115	118
	***	6353	8596	15612	1595	1961	2381	478	523	540	213	222	212	114	116	101
0.8	***	4927	6081	13140	1274	1484	1982	378	406	445	163	169	175	.	.	.
	***	5378	7431	13172	1363	1662	1970	391	424	429	166	172	158	.	.	.
0.85	***	3982	4982	10300	1020	1176	1505	279	296	317	.	.	.	.	.	.
	***	4377	6106	10303	1088	1302	1491	287	306	300	.	.	.	.	.	.
0.9	***	2979	3733	7018	711	800	959	.	.	.	.	.	.	.	.	.
	***	3282	4535	7006	751	866	943	.	.	.	.	.	.	.	.	.
0.95	***	1755	2139	3297	.	.	.	.	.	.	.	.	.	.	.	.
	***	1914	2500	3280	.	.	.	.	.	.	.	.	.	.	.	.

TABLE 8: ALPHA= 0.05 POWER= 0.9   EXPECTED ACCRUAL THRU MINIMUM FOLLOW-UP= 1000

PCONT=***		DEL=.02			DEL=.05			DEL=.10			DEL=.15			DEL=.20		
FACT=		1.0 .75	.50 .25	.00 BIN	1.0 .75	.50 .25	.00 BIN	1.0 .75	.50 .25	.00 BIN	1.0 75	.50 .25	.00 BIN	1.0 .75	.50 .25	.00 BIN
						REQUIRED NUMBER OF PATIENTS										
0.05	***	2449	2469	2580	530	536	547	191	193	195	112	113	113	80	80	80
	•••	2456	2497	4822	532	541	943	192	194	300	113	113	158	80	80	101
0.1	***	5098	5147	5668	978	1003	1055	313	319	326	169	171	173	111	113	113
	•••	5115	5236	8376	988	1023	1491	316	322	429	170	172	212	112	113	129
0.15	***	7455	7542	8766	1371	1421	1547	420	431	446	217	221	225	138	140	142
	•••	7484	7706	11502	1391	1467	1970	425	438	540	219	223	257	139	141	153
0.2	***	9341	9475	11647	1689	1770	1996	508	526	553	257	263	270	160	163	166
	•••	9386	9731	14199	1721	1848	2381	516	538	634	260	266	295	161	165	172
0.25	***	10708	10900	14201	1930	2046	2388	578	604	645	288	297	308	178	181	185
	•••	10772	11268	16469	1976	2160	2724	590	622	711	293	302	326	179	183	187
0.3	***	11564	11828	16367	2099	2250	2715	630	665	719	312	323	337	190	195	200
	•••	11652	12328	18310	2160	2402	2998	645	688	771	318	330	349	192	197	197
0.35	***	11948	12299	18108	2201	2389	2973	666	707	775	328	341	358	198	203	209
	•••	12065	12952	19723	2276	2576	3203	685	736	814	335	349	364	200	206	204
0.4	***	11915	12371	19405	2245	2467	3160	686	734	812	337	351	371	201	207	214
	•••	12068	13196	20708	2334	2686	3340	708	768	840	344	361	372	205	211	206
0.45	***	11531	12111	20243	2239	2491	3272	691	745	831	339	355	375	201	207	215
	•••	11727	13120	21265	2341	2738	3409	715	781	848	346	364	372	204	211	204
0.5	***	10867	11593	20619	2191	2468	3310	685	740	831	333	350	371	196	203	210
	•••	11116	12785	21393	2305	2735	3409	710	779	840	341	360	364	200	206	197
0.55	***	10002	10888	20529	2110	2401	3273	665	721	813	322	338	360	188	194	201
	•••	10312	12245	21093	2230	2680	3340	690	760	814	330	348	349	191	197	187
0.6	***	9015	10057	19973	1998	2296	3162	632	688	776	303	319	339	175	181	187
	•••	9390	11543	20365	2121	2575	3203	657	726	771	311	328	326	178	184	172
0.65	***	7980	9150	18952	1860	2154	2976	588	640	721	279	293	310	158	163	168
	•••	8413	10712	19209	1982	2424	2998	611	675	711	285	301	295	161	166	153
0.7	***	6953	8198	17471	1697	1975	2718	533	578	647	248	259	273	138	141	146
	•••	7425	9768	17625	1813	2225	2724	553	608	634	253	266	257	140	143	129
0.75	***	5961	7210	15532	1508	1758	2386	464	501	555	210	218	228	113	115	118
	•••	6443	8715	15612	1614	1976	2381	481	525	540	213	223	212	114	116	101
0.8	***	5005	6176	13140	1290	1498	1981	381	408	445	163	169	175	.	.	.
	•••	5463	7536	13172	1379	1673	1970	393	425	429	166	172	158	.	.	.
0.85	***	4050	5062	10300	1033	1187	1506	281	297	317	.	.	.	.	.	.
	•••	4451	6191	10303	1100	1309	1491	288	306	300	.	.	.	.	.	.
0.9	***	3033	3791	7018	718	806	959	.	.	.	.	.	.	.	.	.
	•••	3338	4594	7006	757	870	943	.	.	.	.	.	.	.	.	.
0.95	***	1783	2167	3297	.	.	.	.	.	.	.	.	.	.	.	.
	•••	1943	2525	3280	.	.	.	.	.	.	.	.	.	.	.	.

813

TABLE 8: ALPHA= 0.05 POWER= 0.9   EXPECTED ACCRUAL THRU MINIMUM FOLLOW-UP= 1100

	DEL=.02			DEL=.05			DEL=.10			DEL=.15			DEL=.20		
FACT=	1.0 .75	.50 .25	.00 BIN	1.0 .75	.50 .25	.00 BIN	1.0 .75	.50 .25	.00 BIN	1.0 75	.50 .25	.00 BIN	1.0 .75	.50 .25	.00 BIN

PCONT=***   REQUIRED NUMBER OF PATIENTS

PCONT	DEL=.02			DEL=.05			DEL=.10			DEL=.15			DEL=.20		
0.05 ***	2451	2472	2580	530	537	547	191	192	195	112	113	114	79	80	80
•••	2458	2501	4822	533	541	943	192	194	300	113	113	158	79	80	101
0.1 ***	5103	5157	5668	981	1006	1055	314	319	326	169	170	173	111	113	114
•••	5121	5251	8376	992	1026	1491	317	323	429	170	172	212	112	113	129
0.15 ***	7463	7559	8766	1377	1427	1547	422	432	446	217	221	225	139	140	142
•••	7496	7735	11502	1398	1473	1970	426	439	540	219	223	257	140	141	153
0.2 ***	9354	9502	11647	1699	1781	1996	511	528	554	258	263	270	161	163	166
•••	9404	9779	14199	1732	1858	2381	518	540	634	261	267	295	162	164	172
0.25 ***	10727	10938	14201	1945	2062	2388	582	607	644	290	298	307	178	181	185
•••	10798	11336	16469	1992	2175	2724	593	624	711	294	302	326	180	183	187
0.3 ***	11591	11880	16367	2118	2272	2715	635	668	719	314	324	337	191	195	199
•••	11687	12420	18310	2180	2421	2998	650	690	771	319	330	349	192	197	197
0.35 ***	11983	12368	18108	2225	2415	2974	672	712	775	330	342	358	199	203	209
•••	12112	13069	19723	2302	2600	3203	690	739	814	336	350	364	201	206	204
0.4 ***	11961	12460	19404	2274	2498	3159	693	739	812	339	353	371	202	208	214
•••	12129	13341	20708	2365	2714	3340	714	771	840	346	362	372	205	211	206
0.45 ***	11589	12223	20244	2272	2527	3272	700	750	831	341	356	375	202	208	214
•••	11805	13293	21265	2376	2769	3409	723	785	848	348	365	372	205	211	204
0.5 ***	10942	11730	20619	2228	2506	3310	693	746	832	336	351	371	197	203	210
•••	11214	12983	21393	2342	2768	3409	717	783	840	343	361	364	200	206	197
0.55 ***	10097	11047	20529	2149	2441	3273	673	727	813	324	340	359	188	195	201
•••	10432	12464	21093	2270	2714	3340	698	764	814	331	349	349	191	197	187
0.6 ***	9131	10237	19973	2039	2337	3162	641	694	776	306	320	339	176	181	187
•••	9532	11776	20365	2162	2610	3203	665	730	771	313	329	326	178	184	172
0.65 ***	8115	9344	18952	1900	2193	2976	596	646	720	281	294	310	159	164	169
•••	8573	10950	19209	2023	2457	2998	619	679	711	287	301	295	161	166	153
0.7 ***	7102	8397	17471	1735	2012	2718	540	583	647	250	260	273	138	142	146
•••	7595	10003	17625	1852	2255	2724	559	611	634	254	266	257	140	144	129
0.75 ***	6115	7403	15532	1544	1790	2386	470	505	555	210	219	228	113	115	118
•••	6615	8936	15612	1648	2002	2381	486	527	540	214	224	212	114	117	101
0.8 ***	5152	6353	13140	1319	1525	1981	385	411	445	164	169	175	.	.	.
•••	5623	7732	13172	1407	1694	1970	397	426	429	167	172	158	.	.	.
0.85 ***	4180	5212	10300	1055	1206	1506	283	298	317	.	.	.	.	.	.
•••	4590	6351	10303	1121	1323	1491	290	307	300	.	.	.	.	.	.
0.9 ***	3132	3901	7017	731	816	959	.	.	.	.	.	.	.	.	.
•••	3442	4702	7006	769	876	943	.	.	.	.	.	.	.	.	.
0.95 ***	1835	2219	3297	.	.	.	.	.	.	.	.	.	.	.	.
•••	1995	2569	3280	.	.	.	.	.	.	.	.	.	.	.	.

TABLE 8: ALPHA= 0.05 POWER= 0.9   EXPECTED ACCRUAL THRU MINIMUM FOLLOW-UP= 1200

		DEL=.02			DEL=.05			DEL=.10			DEL=.15			DEL=.20		
FACT=		1.0 .75	.50 .25	.00 BIN	1.0 .75	.50 .25	.00 BIN	1.0 .75	.50 .25	.00 BIN	1.0 75	.50 .25	.00 BIN	1.0 .75	.50 .25	.00 BIN
PCONT=***		REQUIRED NUMBER OF PATIENTS														
0.05	***	2453	2476	2580	531	538	547	192	193	194	112	112	113	79	79	79
	•••	2461	2505	4822	534	541	943	192	193	300	112	113	158	79	79	101
0.1	***	5108	5166	5668	985	1009	1055	315	319	325	169	171	172	112	112	113
	•••	5128	5265	8376	995	1027	1491	317	322	429	170	172	212	112	113	129
0.15	***	7472	7576	8766	1383	1434	1546	423	433	446	218	221	225	139	140	142
	•••	7507	7764	11502	1404	1477	1970	427	439	540	220	223	257	139	141	153
0.2	***	9368	9528	11647	1709	1792	1996	513	530	553	259	264	270	161	163	166
	•••	9421	9824	14199	1743	1867	2381	521	541	634	261	267	295	162	164	172
0.25	***	10746	10976	14201	1959	2077	2388	585	610	644	291	298	307	178	181	185
	•••	10823	11401	16469	2008	2187	2724	596	625	711	295	303	326	180	183	187
0.3	***	11617	11932	16366	2137	2292	2715	640	671	718	316	325	337	191	195	199
	•••	11722	12508	18310	2200	2437	2998	654	692	771	320	331	349	193	197	197
0.35	***	12018	12436	18108	2248	2440	2973	678	716	775	332	343	358	199	204	209
	•••	12159	13183	19723	2326	2620	3203	695	742	814	337	350	364	202	206	204
0.4	***	12007	12547	19404	2300	2527	3159	700	744	812	341	355	370	203	208	214
	•••	12190	13480	20708	2393	2739	3340	720	774	840	347	362	372	205	211	206
0.45	***	11648	12332	20243	2303	2558	3271	707	756	831	343	358	375	202	208	214
	•••	11882	13456	21265	2408	2797	3409	729	789	848	350	366	372	205	211	204
0.5	***	11017	11861	20619	2263	2541	3310	700	751	832	338	353	371	198	204	210
	•••	11311	13169	21393	2377	2797	3409	724	787	840	345	361	364	201	207	197
0.55	***	10189	11200	20528	2185	2478	3273	680	733	813	327	341	359	190	195	201
	•••	10550	12668	21093	2307	2745	3340	704	768	814	334	349	349	192	198	187
0.6	***	9243	10408	19972	2076	2373	3162	648	699	776	308	322	339	177	181	187
	•••	9670	11993	20365	2200	2641	3203	671	733	771	315	330	326	179	184	172
0.65	***	8245	9527	18952	1937	2229	2977	603	650	721	283	295	310	160	164	169
	•••	8726	11172	19209	2059	2486	2998	625	682	711	289	302	295	162	166	153
0.7	***	7244	8583	17470	1771	2044	2717	546	587	646	251	261	273	139	142	145
	•••	7756	10220	17625	1887	2281	2724	565	613	634	256	267	257	140	144	129
0.75	***	6260	7584	15532	1575	1819	2386	475	508	555	212	219	228	113	115	118
	•••	6775	9139	15612	1679	2025	2381	490	529	540	215	223	212	115	117	101
0.8	***	5290	6518	13140	1346	1548	1981	388	413	445	165	170	175	.	.	.
	•••	5774	7911	13172	1433	1711	1970	400	427	429	167	172	158	.	.	.
0.85	***	4300	5351	10300	1075	1222	1505	286	300	317	.	.	.	.	.	.
	•••	4720	6496	10303	1140	1335	1491	292	307	300	.	.	.	.	.	.
0.9	***	3223	4002	7018	743	825	959	.	.	.	.	.	.	.	.	.
	•••	3538	4801	7006	780	882	943	.	.	.	.	.	.	.	.	.
0.95	***	1883	2266	3297	.	.	.	.	.	.	.	.	.	.	.	.
	•••	2044	2609	3280	.	.	.	.	.	.	.	.	.	.	.	.

TABLE 8: ALPHA= 0.05 POWER= 0.9   EXPECTED ACCRUAL THRU MINIMUM FOLLOW-UP= 1300

		DEL=.02			DEL=.05			DEL=.10			DEL=.15			DEL=.20		
FACT=		1.0 .75	.50 .25	.00 BIN	1.0 .75	.50 .25	.00 BIN	1.0 .75	.50 .25	.00 BIN	1.0 75	.50 .25	.00 BIN	1.0 .75	.50 .25	.00 BIN
PCONT=***		REQUIRED NUMBER OF PATIENTS														
0.05	***	2455	2479	2580	532	538	547	192	193	194	112	113	114	80	80	80
	•••	2463	2509	4822	535	542	943	193	194	300	113	113	158	80	80	101
0.1	***	5113	5176	5668	987	1012	1056	316	320	326	170	172	173	112	113	114
	•••	5134	5279	8376	998	1030	1491	318	323	429	171	172	212	112	113	129
0.15	***	7481	7593	8766	1389	1440	1547	424	434	447	219	222	225	139	141	142
	•••	7519	7791	11502	1410	1481	1970	429	440	540	220	224	257	140	141	153
0.2	***	9381	9555	11647	1719	1801	1995	515	531	553	259	264	271	162	163	166
	•••	9440	9869	14199	1753	1874	2381	523	541	634	262	267	295	162	164	172
0.25	***	10766	11014	14201	1972	2091	2388	588	612	644	292	299	308	179	182	185
	•••	10848	11465	16469	2021	2198	2724	600	627	711	296	303	326	180	184	187
0.3	***	11643	11984	16367	2154	2310	2715	644	674	718	317	326	337	192	195	199
	•••	11758	12593	18310	2218	2453	2998	658	694	771	321	331	349	193	198	197
0.35	***	12054	12504	18108	2269	2462	2973	682	720	774	334	344	358	200	204	209
	•••	12205	13292	19723	2348	2639	3203	700	744	814	339	351	364	202	206	204
0.4	***	12053	12634	19404	2326	2553	3159	705	748	812	343	356	370	204	209	214
	•••	12250	13613	20708	2420	2761	3340	725	777	840	349	362	372	206	211	206
0.45	***	11707	12439	20244	2332	2588	3272	713	760	831	345	359	375	204	209	214
	•••	11959	13613	21265	2438	2821	3409	734	791	848	352	367	372	206	211	204
0.5	***	11092	11989	20619	2294	2572	3309	707	757	832	341	354	371	199	204	210
	•••	11407	13346	21393	2410	2824	3409	730	790	840	347	362	364	201	207	197
0.55	***	10281	11349	20529	2219	2511	3273	687	738	813	329	342	359	190	195	201
	•••	10665	12861	21093	2341	2772	3340	710	771	814	336	350	349	193	198	187
0.6	***	9353	10572	19972	2110	2407	3162	655	704	776	310	323	339	177	182	187
	•••	9803	12197	20365	2235	2668	3203	678	736	771	316	331	326	180	185	172
0.65	***	8372	9701	18952	1971	2261	2976	609	655	721	285	297	310	160	164	168
	•••	8873	11378	19209	2094	2512	2998	630	684	711	290	303	295	162	167	153
0.7	***	7380	8760	17470	1803	2074	2718	552	591	647	253	262	273	139	142	146
	•••	7910	10422	17625	1918	2305	2724	570	616	634	257	267	257	141	144	129
0.75	***	6399	7754	15532	1605	1846	2385	479	511	555	213	220	228	114	115	118
	•••	6927	9327	15612	1708	2045	2381	494	531	540	216	224	212	115	117	101
0.8	***	5420	6673	13140	1371	1570	1981	393	415	445	166	171	175	.	.	.
	•••	5916	8077	13172	1457	1727	1970	403	429	429	168	173	158	.	.	.
0.85	***	4414	5480	10300	1094	1238	1506	288	301	317	.	.	.	.	.	.
	•••	4841	6629	10303	1157	1345	1491	294	309	300	.	.	.	.	.	.
0.9	***	3310	4094	7018	754	833	959	.	.	.	.	.	.	.	.	.
	•••	3629	4891	7006	790	887	943	.	.	.	.	.	.	.	.	.
0.95	***	1928	2308	3297	.	.	.	.	.	.	.	.	.	.	.	.
	•••	2088	2645	3280	.	.	.	.	.	.	.	.	.	.	.	.

TABLE 8: ALPHA= 0.05 POWER= 0.9   EXPECTED ACCRUAL THRU MINIMUM FOLLOW-UP= 1400

	DEL=.02			DEL=.05			DEL=.10			DEL=.15			DEL=.20		
FACT=	1.0 .75	.50 .25	.00 BIN	1.0 .75	.50 .25	.00 BIN	1.0 .75	.50 .25	.00 BIN	1.0 75	.50 .25	.00 BIN	1.0 .75	.50 .25	.00 BIN
PCONT=•••			REQUIRED NUMBER OF PATIENTS												
0.05 •••	2457 2466	2481 2512	2580 4822	533 535	539 542	547 943	192 192	193 194	194 300	113 113	113 113	114 158	80 80	80 80	80 101
0.1 •••	5118 5141	5185 5291	5668 8376	990 1000	1013 1031	1055 1491	317 318	320 323	325 429	170 171	171 172	173 212	112 112	113 113	114 129
0.15 •••	7490 7530	7610 7817	8766 11502	1395 1416	1445 1485	1547 1970	425 430	435 440	446 540	219 220	222 223	225 257	139 140	141 141	142 153
0.2 •••	9395 9457	9581 9911	11647 14199	1727 1761	1809 1880	1996 2381	517 525	533 542	554 634	260 262	265 268	270 295	162 163	164 164	165 172
0.25 •••	10784 10874	11052 11526	14201 16469	1984 2034	2103 2208	2388 2724	591 602	613 627	645 711	293 297	300 304	308 326	179 180	182 184	185 187
0.3 •••	11670 11792	12035 12675	16367 18310	2170 2235	2327 2466	2715 2998	647 661	677 696	718 771	318 322	326 332	337 349	192 194	196 198	199 197
0.35 •••	12089 12252	12571 13395	18108 19723	2289 2369	2482 2656	2973 3203	688 703	723 746	774 814	335 340	346 352	358 364	200 203	205 206	209 204
0.4 •••	12099 12310	12719 13739	19404 20708	2349 2444	2576 2781	3160 3340	710 730	752 779	812 840	345 351	357 363	371 372	205 206	209 212	214 206
0.45 •••	11766 12036	12544 13760	20243 21265	2359 2466	2614 2844	3272 3409	719 740	765 794	831 848	347 353	360 367	375 372	205 206	209 212	214 204
0.5 •••	11165 11501	12114 13513	20619 21393	2324 2440	2601 2848	3310 3409	713 735	761 793	831 840	342 348	355 363	372 364	199 202	205 207	210 197
0.55 •••	10372 10777	11491 13043	20528 21093	2250 2372	2541 2796	3273 3340	694 716	742 773	813 814	331 337	344 351	360 349	191 193	196 198	200 187
0.6 •••	9461 9932	10729 12387	19973 20365	2142 2267	2437 2692	3162 3203	661 682	708 738	776 771	311 318	324 331	339 326	178 180	182 185	187 172
0.65 •••	8494 9014	9866 11572	18952 19209	2003 2125	2291 2536	2977 2998	615 635	659 687	721 711	286 291	297 304	310 295	161 163	164 166	169 153
0.7 •••	7511 8057	8926 10610	17470 17625	1833 1948	2102 2327	2718 2724	556 574	594 618	647 634	254 258	262 268	273 257	140 141	143 144	145 129
0.75 •••	6530 7072	7913 9502	15532 15612	1632 1734	1870 2063	2385 2381	483 498	514 533	555 540	214 217	220 224	228 212	114 115	116 117	118 101
0.8 •••	5544 6049	6818 8231	13140 13172	1394 1479	1589 1741	1981 1970	395 405	417 430	444 429	166 169	171 173	175 158	. .	. .	. .
0.85 •••	4522 4954	5601 6751	10300 10303	1110 1172	1250 1354	1506 1491	290 296	302 309	317 300	. .	. .	. .	. .	. .	. .
0.9 •••	3391 3713	4181 4973	7017 7006	763 798	840 892	959 943	. .	. .	. .	. .	. .	. .	. .	. .	. .
0.95 •••	1970 2129	2347 2676	3297 3280	. .	. .	. .	. .	. .	. .	. .	. .	. .	. .	. .	. .

TABLE 8: ALPHA= 0.05 POWER= 0.9   EXPECTED ACCRUAL THRU MINIMUM FOLLOW-UP= 1500

	DEL=.02			DEL=.05			DEL=.10			DEL=.15			DEL=.20		
FACT=	1.0 .75	.50 .25	.00 BIN	1.0 .75	.50 .25	.00 BIN	1.0 .75	.50 .25	.00 BIN	1.0 75	.50 .25	.00 BIN	1.0 .75	.50 .25	.00 BIN

PCONT=•••   REQUIRED NUMBER OF PATIENTS

PCONT	DEL=.02 1.0/.75	.50/.25	.00/BIN	DEL=.05 1.0/.75	.50/.25	.00/BIN	DEL=.10 1.0/.75	.50/.25	.00/BIN	DEL=.15 1.0/.75	.50/.25	.00/BIN	DEL=.20 1.0/.75	.50/.25	.00/BIN
0.05 •••	2459	2484	2580	533	539	547	192	193	195	112	113	113	80	80	80
•••	2468	2515	4822	536	543	943	193	194	300	112	113	158	80	80	101
0.1 •••	5122	5194	5668	993	1015	1055	317	320	326	170	172	172	112	112	113
•••	5147	5303	8376	1003	1032	1491	319	323	429	170	172	212	112	113	129
0.15 •••	7499	7627	8765	1400	1449	1547	427	435	446	219	222	225	140	140	142
•••	7542	7842	11502	1421	1489	1970	431	440	540	220	224	257	140	142	153
0.2 •••	9408	9607	11647	1735	1818	1996	519	534	553	260	265	270	162	164	166
•••	9475	9952	14199	1770	1886	2381	526	543	634	262	267	295	163	165	172
0.25 •••	10804	11089	14201	1997	2115	2388	594	615	645	294	300	307	180	182	185
•••	10900	11585	16469	2045	2217	2724	604	629	711	297	304	326	181	184	187
0.3 •••	11696	12085	16367	2185	2341	2715	652	680	719	319	327	337	193	196	200
•••	11827	12754	18310	2250	2478	2998	665	697	771	323	332	349	194	198	197
0.35 •••	12124	12637	18108	2308	2501	2974	692	725	774	337	346	358	202	205	209
•••	12299	13495	19723	2389	2671	3203	707	748	814	341	352	364	203	207	204
0.4 •••	12144	12802	19404	2372	2599	3159	715	755	812	346	357	371	205	209	214
•••	12370	13860	20708	2467	2798	3340	734	781	840	352	364	372	207	212	206
0.45 •••	11825	12647	20243	2384	2639	3272	725	768	832	349	361	375	205	209	215
•••	12112	13901	21265	2491	2863	3409	744	796	848	354	367	372	207	212	204
0.5 •••	11239	12234	20619	2352	2628	3309	719	765	832	344	356	371	200	205	210
•••	11593	13672	21393	2467	2870	3409	740	795	840	350	364	364	202	207	197
0.55 •••	10462	11629	20528	2280	2569	3273	699	746	814	332	344	359	191	196	200
•••	10887	13215	21093	2402	2819	3340	721	776	814	338	352	349	194	199	187
0.6 •••	9567	10879	19972	2172	2465	3162	667	712	776	313	325	338	179	183	187
•••	10057	12567	20365	2296	2715	3203	687	740	771	319	332	326	181	185	172
0.65 •••	8612	10024	18952	2032	2317	2977	620	662	721	288	298	310	161	165	169
•••	9150	11753	19209	2153	2557	2998	640	689	711	292	304	295	163	167	153
0.7 •••	7636	9083	17470	1861	2126	2718	560	597	647	255	263	273	140	142	145
•••	8197	10787	17625	1975	2345	2724	577	620	634	259	268	257	142	144	129
0.75 •••	6655	8065	15532	1657	1891	2386	487	517	555	215	221	228	114	116	118
•••	7209	9667	15612	1758	2079	2381	500	534	540	217	225	212	115	117	101
0.8 •••	5662	6954	13140	1415	1606	1981	397	419	445	167	170	175	.	.	.
•••	6175	8374	13172	1498	1754	1970	408	431	429	169	173	158	.	.	.
0.85 •••	4623	5714	10300	1126	1263	1505	290	303	317	.	.	.	.	.	.
•••	5062	6865	10303	1187	1363	1491	297	309	300	.	.	.	.	.	.
0.9 •••	3467	4261	7018	772	847	959	.	.	.	.	.	.	.	.	.
•••	3791	5047	7006	806	895	943	.	.	.	.	.	.	.	.	.
0.95 •••	2008	2382	3297	.	.	.	.	.	.	.	.	.	.	.	.
•••	2167	2705	3280	.	.	.	.	.	.	.	.	.	.	.	.

TABLE 8: ALPHA= 0.05 POWER= 0.9   EXPECTED ACCRUAL THRU MINIMUM FOLLOW-UP= 1600

		DEL=.02			DEL=.05			DEL=.10			DEL=.15			DEL=.20	
FACT=	1.0 .75	.50 .25	.00 BIN	1.0 .75	.50 .25	.00 BIN	1.0 .75	.50 .25	.00 BIN	1.0 75	.50 .25	.00 BIN	1.0 .75	.50 .25	.00 BIN
PCONT=•••				REQUIRED NUMBER OF PATIENTS											
0.05 •••	2461	2488	2580	534	540	547	192	193	195	113	113	114	79	80	80
•••	2471	2518	4822	537	543	943	193	194	300	113	113	158	80	80	101
0.1 •••	5128	5203	5668	995	1018	1055	317	321	326	170	172	173	112	113	113
•••	5154	5315	8376	1005	1034	1491	319	324	429	171	172	212	113	113	129
0.15 •••	7507	7643	8766	1404	1453	1547	428	436	447	220	222	225	140	141	142
•••	7553	7866	11502	1426	1491	1970	432	441	540	221	224	257	140	141	153
0.2 •••	9422	9632	11647	1743	1825	1996	521	535	554	261	265	270	162	164	166
•••	9493	9991	14199	1778	1892	2381	528	544	634	263	268	295	163	165	172
0.25 •••	10823	11126	14201	2008	2125	2388	597	617	645	295	301	308	180	182	185
•••	10926	11641	16469	2057	2225	2724	606	630	711	298	304	326	181	184	187
0.3 •••	11723	12135	16367	2200	2355	2715	654	681	719	320	328	337	193	196	199
•••	11863	12829	18310	2265	2489	2998	667	698	771	324	332	349	195	198	197
0.35 •••	12159	12703	18108	2326	2518	2973	695	728	775	338	347	358	202	205	209
•••	12345	13591	19723	2407	2685	3203	711	749	814	342	352	364	203	207	204
0.4 •••	12190	12885	19404	2393	2619	3159	720	758	812	348	358	371	206	210	214
•••	12430	13975	20708	2488	2815	3340	738	783	840	353	364	372	208	212	206
0.45 •••	11883	12746	20243	2408	2662	3272	729	771	831	350	362	375	206	210	215
•••	12186	14036	21265	2515	2882	3409	749	798	848	356	368	372	208	212	204
0.5 •••	11311	12352	20619	2378	2652	3310	724	768	832	345	357	372	201	205	210
•••	11684	13822	21393	2494	2889	3409	744	797	840	351	364	364	203	208	197
0.55 •••	10550	11761	20529	2307	2594	3273	705	750	813	334	345	359	192	196	201
•••	10995	13377	21093	2429	2839	3340	725	778	814	339	352	349	194	199	187
0.6 •••	9670	11023	19973	2200	2490	3162	672	715	776	315	326	339	179	183	187
•••	10178	12737	20365	2324	2735	3203	692	743	771	320	332	326	181	185	172
0.65 •••	8726	10173	18952	2060	2342	2977	625	665	721	289	299	310	162	165	169
•••	9281	11924	19209	2181	2576	2998	644	691	711	294	304	295	163	167	153
0.7 •••	7757	9234	17471	1887	2149	2718	565	600	647	256	264	273	140	143	146
•••	8332	10952	17625	2000	2363	2724	581	621	634	260	268	257	142	144	129
0.75 •••	6776	8208	15532	1680	1911	2386	490	518	555	216	222	228	115	116	118
•••	7340	9820	15612	1780	2094	2381	504	535	540	218	225	212	115	117	101
0.8 •••	5774	7083	13140	1434	1622	1981	400	420	445	168	171	175	.	.	.
•••	6296	8508	13172	1516	1765	1970	410	432	429	169	173	158	.	.	.
0.85 •••	4720	5821	10300	1140	1274	1506	292	304	317	.	.	.	.	.	.
•••	5163	6971	10303	1200	1370	1491	298	310	300	.	.	.	.	.	.
0.9 •••	3539	4336	7018	780	852	959	.	.	.	.	.	.	.	.	.
•••	3866	5117	7006	813	899	943	.	.	.	.	.	.	.	.	.
0.95 •••	2044	2415	3297	.	.	.	.	.	.	.	.	.	.	.	.
•••	2202	2731	3280	.	.	.	.	.	.	.	.	.	.	.	.

TABLE 8: ALPHA= 0.05 POWER= 0.9   EXPECTED ACCRUAL THRU MINIMUM FOLLOW-UP= 1700

	DEL=.02			DEL=.05			DEL=.10			DEL=.15			DEL=.20		
FACT=	1.0 .75	.50 .25	.00 BIN	1.0 .75	.50 .25	.00 BIN	1.0 .75	.50 .25	.00 BIN	1.0 75	.50 .25	.00 BIN	1.0 .75	.50 .25	.00 BIN
PCONT=***							REQUIRED NUMBER OF PATIENTS								
0.05 ***	2463	2490	2579	534	539	547	192	193	194	112	114	114	80	80	80
•••	2473	2520	4822	537	543	943	193	194	300	112	114	158	80	80	101
0.1 ***	5133	5212	5668	998	1019	1055	318	322	326	170	172	173	112	112	114
•••	5161	5325	8376	1007	1035	1491	320	324	429	171	172	212	112	114	129
0.15 ***	7516	7660	8765	1409	1457	1547	428	437	446	220	222	225	140	141	142
•••	7565	7889	11502	1430	1494	1970	432	441	540	221	224	257	140	141	153
0.2 ***	9435	9658	11647	1751	1831	1995	522	537	554	261	265	270	163	163	165
•••	9510	10028	14199	1785	1897	2381	529	544	634	263	267	295	163	165	172
0.25 ***	10843	11162	14201	2018	2135	2387	598	618	645	295	301	308	180	182	185
•••	10951	11696	16469	2067	2232	2724	607	630	711	299	304	326	182	184	187
0.3 ***	11749	12184	16366	2213	2368	2715	658	683	718	321	328	337	193	197	199
•••	11898	12902	18310	2279	2499	2998	669	699	771	324	333	349	194	197	197
0.35 ***	12194	12767	18108	2343	2534	2974	699	731	775	339	347	358	202	205	209
•••	12391	13683	19723	2424	2698	3203	714	751	814	343	352	364	204	207	204
0.4 ***	12235	12964	19404	2413	2637	3160	724	760	813	348	359	371	206	210	214
•••	12488	14086	20708	2507	2829	3340	741	784	840	354	364	372	208	212	206
0.45 ***	11940	12843	20244	2431	2683	3271	733	775	832	352	362	375	206	210	214
•••	12260	14164	21265	2537	2898	3409	752	800	848	357	369	372	208	212	204
0.5 ***	11383	12465	20619	2402	2675	3310	729	771	832	346	358	372	202	205	210
•••	11774	13964	21393	2518	2907	3409	748	799	840	352	364	364	204	208	197
0.55 ***	10637	11888	20529	2333	2618	3274	709	752	814	335	346	359	192	197	201
•••	11099	13531	21093	2454	2858	3340	729	780	814	340	352	349	194	199	187
0.6 ***	9770	11160	19973	2227	2514	3162	675	718	777	316	326	339	180	182	187
•••	10295	12898	20365	2349	2753	3203	696	745	771	321	333	326	182	185	172
0.65 ***	8837	10317	18952	2086	2365	2976	629	668	720	290	299	310	163	165	169
•••	9406	12086	19209	2205	2594	2998	647	692	711	294	305	295	163	167	153
0.7 ***	7873	9377	17471	1911	2171	2718	568	602	647	257	265	273	140	143	146
•••	8459	11108	17625	2023	2379	2724	584	622	634	260	269	257	142	144	129
0.75 ***	6890	8344	15531	1701	1929	2385	493	520	554	216	222	228	114	117	118
•••	7465	9964	15612	1801	2107	2381	505	537	540	219	225	212	116	117	101
0.8 ***	5881	7205	13140	1451	1636	1982	403	422	445	168	171	175	.	.	.
•••	6410	8633	13172	1533	1775	1970	411	432	429	170	173	158	.	.	.
0.85 ***	4811	5921	10301	1153	1283	1506	293	305	318	.	.	.	.	.	.
•••	5259	7070	10303	1211	1377	1491	299	310	300	.	.	.	.	.	.
0.9 ***	3607	4406	7018	787	856	959	.	.	.	.	.	.	.	.	.
•••	3935	5182	7006	819	902	943	.	.	.	.	.	.	.	.	.
0.95 ***	2077	2446	3297	.	.	.	.	.	.	.	.	.	.	.	.
•••	2234	2755	3280	.	.	.	.	.	.	.	.	.	.	.	.

TABLE 8: ALPHA= 0.05 POWER= 0.9   EXPECTED ACCRUAL THRU MINIMUM FOLLOW-UP= 1800

PCONT		DEL=.02 1.0/.75	.50/.25	.00/BIN	DEL=.05 1.0/.75	.50/.25	.00/BIN	DEL=.10 1.0/.75	.50/.25	.00/BIN	DEL=.15 1.0/.75	.50/.25	.00/BIN	DEL=.20 1.0/.75	.50/.25	.00/BIN
0.05	***	2465	2493	2580	535	540	548	192	193	195	112	114	114	80	80	80
	•••	2476	2522	4822	537	543	943	193	195	300	112	114	158	80	80	101
0.1	***	5138	5220	5667	999	1020	1055	318	321	326	171	172	173	112	112	114
	•••	5166	5336	8376	1009	1036	1491	319	324	429	171	172	212	112	114	129
0.15	***	7525	7674	8766	1413	1460	1547	429	438	447	220	222	225	139	141	141
	•••	7577	7911	11502	1434	1496	1970	433	442	540	222	224	257	141	141	153
0.2	***	9449	9683	11647	1757	1837	1995	524	537	553	262	265	270	163	164	165
	•••	9528	10064	14199	1792	1901	2381	530	544	634	264	267	295	163	165	172
0.25	***	10862	11198	14201	2028	2144	2388	600	620	645	296	301	308	180	182	186
	•••	10977	11748	16469	2078	2240	2724	609	631	711	298	305	326	181	183	187
0.3	***	11775	12233	16366	2226	2380	2715	660	685	719	321	328	337	193	197	199
	•••	11931	12971	18310	2292	2508	2998	672	701	771	325	333	349	195	198	197
0.35	***	12228	12829	18108	2359	2549	2973	702	732	775	339	348	357	202	206	209
	•••	12437	13770	19723	2440	2709	3203	717	751	814	344	353	364	204	207	204
0.4	***	12280	13043	19404	2432	2655	3159	728	762	812	350	360	371	207	210	213
	•••	12547	14192	20708	2526	2843	3340	744	785	840	354	366	372	208	213	206
0.45	***	11998	12939	20243	2452	2702	3271	738	777	831	352	363	375	207	210	215
	•••	12332	14286	21265	2558	2913	3409	756	802	848	357	369	372	208	213	204
0.5	***	11454	12575	20619	2425	2697	3309	732	774	831	348	359	371	202	206	210
	•••	11862	14100	21393	2541	2922	3409	751	800	840	353	364	364	204	208	197
0.55	***	10722	12012	20529	2357	2640	3273	713	756	813	336	346	359	193	197	201
	•••	11200	13677	21093	2478	2874	3340	733	782	814	341	353	349	195	199	187
0.6	***	9868	11292	19972	2251	2535	3162	681	721	776	317	327	339	180	183	186
	•••	10408	13050	20365	2373	2769	3203	699	746	771	321	333	326	181	186	172
0.65	***	8945	10455	18951	2109	2386	2976	633	670	721	291	300	310	163	165	168
	•••	9528	12239	19209	2229	2610	2998	650	694	711	294	305	295	164	168	153
0.7	***	7985	9513	17471	1934	2190	2718	571	604	647	258	265	273	141	143	145
	•••	8583	11256	17625	2045	2393	2724	587	624	634	261	269	257	141	144	129
0.75	***	7001	8473	15531	1721	1946	2386	496	522	555	217	222	228	114	117	118
	•••	7584	10100	15612	1819	2119	2381	508	537	540	219	225	212	116	117	101
0.8	***	5984	7320	13140	1468	1649	1981	404	423	445	168	172	175	.	.	.
	•••	6518	8750	13172	1548	1784	1970	413	433	429	170	173	158	.	.	.
0.85	***	4898	6016	10300	1165	1293	1505	294	305	317	.	.	.	.	.	.
	•••	5352	7161	10303	1223	1383	1491	300	310	300	.	.	.	.	.	.
0.9	***	3672	4473	7017	794	861	960	.	.	.	.	.	.	.	.	.
	•••	4002	5241	7006	825	906	943	.	.	.	.	.	.	.	.	.
0.95	***	2109	2474	3297	.	.	.	.	.	.	.	.	.	.	.	.
	•••	2265	2776	3280	.	.	.	.	.	.	.	.	.	.	.	.

REQUIRED NUMBER OF PATIENTS

TABLE 8: ALPHA= 0.05 POWER= 0.9   EXPECTED ACCRUAL THRU MINIMUM FOLLOW-UP= 1900

	DEL=.02			DEL=.05			DEL=.10			DEL=.15			DEL=.20		
FACT=	1.0 .75	.50 .25	.00 BIN	1.0 .75	.50 .25	.00 BIN	1.0 .75	.50 .25	.00 BIN	1.0 75	.50 .25	.00 BIN	1.0 .75	.50 .25	.00 BIN
PCONT=•••				REQUIRED NUMBER OF PATIENTS											
0.05 •••	2466	2495	2580	535	540	547	192	193	194	113	113	114	79	79	79
•••	2478	2524	4822	538	543	943	193	194	300	113	114	158	79	79	101
0.1 •••	5143	5227	5668	1001	1022	1056	318	322	326	171	172	173	113	113	114
•••	5172	5345	8376	1010	1037	1491	320	324	429	171	172	212	113	113	129
0.15 •••	7533	7690	8766	1417	1464	1547	429	438	446	220	223	225	140	141	142
•••	7588	7931	11502	1438	1498	1970	433	441	540	222	224	257	140	141	153
0.2 •••	9462	9708	11647	1764	1843	1996	524	538	553	262	266	270	163	163	166
•••	9546	10098	14199	1797	1904	2381	531	545	634	265	268	295	163	165	172
0.25 •••	10881	11233	14201	2037	2153	2388	602	621	645	296	301	307	180	182	185
•••	11002	11798	16469	2086	2245	2724	611	631	711	299	305	326	182	184	187
0.3 •••	11801	12281	16367	2238	2391	2716	662	686	718	323	329	337	193	197	199
•••	11966	13038	18310	2303	2516	2998	673	702	771	325	332	349	196	198	197
0.35 •••	12264	12891	18108	2374	2562	2973	704	735	774	341	349	358	203	205	209
•••	12482	13854	19723	2454	2719	3203	718	752	814	344	353	364	204	208	204
0.4 •••	12326	13121	19405	2450	2671	3160	730	766	812	351	360	370	206	210	213
•••	12605	14294	20708	2545	2856	3340	747	787	840	355	365	372	209	212	206
0.45 •••	12055	13030	20243	2472	2720	3271	741	779	831	353	363	375	206	210	215
•••	12404	14403	21265	2578	2927	3409	759	802	848	358	369	372	209	212	204
0.5 •••	11525	12681	20618	2447	2716	3309	736	776	831	349	360	372	201	206	210
•••	11947	14231	21393	2562	2937	3409	755	801	840	353	365	364	204	208	197
0.55 •••	10805	12130	20528	2379	2660	3272	717	757	813	337	348	360	193	197	201
•••	11300	13817	21093	2500	2890	3340	736	783	814	342	353	349	194	199	187
0.6 •••	9963	11421	19972	2274	2557	3162	684	723	776	318	327	338	180	184	187
•••	10518	13194	20365	2396	2785	3203	703	748	771	323	332	326	182	185	172
0.65 •••	9048	10585	18952	2132	2405	2977	636	673	721	292	300	310	163	166	168
•••	9644	12383	19209	2250	2624	2998	654	695	711	296	305	295	163	167	153
0.7 •••	8093	9644	17470	1954	2207	2718	574	607	647	258	266	273	141	144	146
•••	8702	11395	17625	2065	2405	2724	590	626	634	261	269	257	142	144	129
0.75 •••	7107	8596	15532	1740	1961	2386	498	524	554	217	222	228	115	116	118
•••	7698	10228	15612	1837	2130	2381	510	538	540	220	225	212	116	117	101
0.8 •••	6081	7431	13140	1484	1662	1982	406	424	445	168	172	175	.	.	.
•••	6623	8861	13172	1562	1793	1970	414	433	429	170	173	158	.	.	.
0.85 •••	4981	6106	10300	1177	1301	1505	296	306	317	.	.	.	.	.	.
•••	5438	7248	10303	1232	1388	1491	300	311	300	.	.	.	.	.	.
0.9 •••	3733	4535	7018	800	866	959	.	.	.	.	.	.	.	.	.
•••	4065	5297	7006	830	907	943	.	.	.	.	.	.	.	.	.
0.95 •••	2139	2500	3298	.	.	.	.	.	.	.	.	.	.	.	.
•••	2294	2796	3280	.	.	.	.	.	.	.	.	.	.	.	.

TABLE 8: ALPHA= 0.05 POWER= 0.9   EXPECTED ACCRUAL THRU MINIMUM FOLLOW-UP= 2000

	DEL=.02			DEL=.05			DEL=.10			DEL=.15			DEL=.20		
FACT=	1.0 .75	.50 .25	.00 BIN	1.0 .75	.50 .25	.00 BIN	1.0 .75	.50 .25	.00 BIN	1.0 75	.50 .25	.00 BIN	1.0 .75	.50 .25	.00 BIN
PCONT=***				REQUIRED NUMBER OF PATIENTS											
0.05 ***	2469	2497	2580	536	541	547	192	194	195	112	114	114	80	80	80
•••	2480	2526	4822	539	544	943	194	194	300	114	114	158	80	80	101
0.1 ***	5147	5236	5669	1002	1024	1055	319	322	326	171	172	172	112	112	114
•••	5179	5354	8376	1012	1037	1491	320	324	429	171	172	212	112	114	129
0.15 ***	7542	7706	8766	1421	1467	1547	431	437	446	221	222	225	140	141	142
•••	7599	7951	11502	1441	1501	1970	435	442	540	222	224	257	141	141	153
0.2 ***	9475	9731	11647	1770	1849	1996	526	539	554	262	266	270	162	165	166
•••	9564	10131	14199	1804	1909	2381	532	546	634	265	269	295	164	165	172
0.25 ***	10900	11269	14201	2046	2160	2387	604	622	645	297	302	307	181	182	185
•••	11027	11846	16469	2095	2251	2724	612	632	711	300	305	326	182	184	187
0.3 ***	11827	12329	16367	2250	2402	2715	665	687	719	324	330	337	195	197	200
•••	12001	13104	18310	2315	2525	2998	675	702	771	326	334	349	196	199	197
0.35 ***	12299	12952	18109	2389	2576	2974	707	736	775	341	349	359	204	206	209
•••	12526	13936	19723	2469	2729	3203	721	754	814	345	354	364	205	207	204
0.4 ***	12371	13196	19405	2467	2686	3160	734	767	812	351	361	371	207	211	214
•••	12662	14391	20708	2561	2867	3340	750	789	840	356	366	372	209	212	206
0.45 ***	12111	13120	20244	2491	2739	3272	745	781	831	355	364	375	207	211	215
•••	12475	14515	21265	2597	2940	3409	762	805	848	359	370	372	209	212	204
0.5 ***	11594	12785	20619	2467	2735	3310	740	779	831	350	360	371	202	206	210
•••	12031	14356	21393	2582	2952	3409	759	804	840	355	365	364	205	209	197
0.55 ***	10887	12245	20529	2401	2680	3274	721	760	814	339	347	360	194	197	201
•••	11397	13951	21093	2521	2905	3340	740	785	814	342	354	349	195	199	187
0.6 ***	10057	11544	19972	2296	2575	3162	687	726	776	319	329	339	181	184	187
•••	10625	13332	20365	2417	2800	3203	705	749	771	324	334	326	182	185	172
0.65 ***	9150	10712	18952	2154	2424	2976	640	675	721	292	301	310	164	166	169
•••	9757	12521	19209	2271	2637	2998	656	696	711	296	305	295	165	167	153
0.7 ***	8197	9769	17471	1975	2225	2717	577	609	647	259	266	274	141	144	146
•••	8816	11529	17625	2084	2419	2724	592	626	634	262	270	257	142	145	129
0.75 ***	7210	8715	15532	1757	1976	2386	501	525	555	217	222	229	115	116	119
•••	7809	10350	15612	1854	2141	2381	512	539	540	220	225	212	116	117	101
0.8 ***	6176	7536	13140	1499	1674	1981	407	425	445	169	172	175	.	.	.
•••	6722	8966	13172	1576	1801	1970	416	435	429	171	174	158	.	.	.
0.85 ***	5062	6191	10300	1187	1309	1506	297	306	317	.	.	.	.	.	.
•••	5521	7330	10303	1242	1394	1491	301	311	300	.	.	.	.	.	.
0.9 ***	3791	4594	7017	806	870	959	.	.	.	.	.	.	.	.	.
•••	4124	5350	7006	835	910	943	.	.	.	.	.	.	.	.	.
0.95 ***	2167	2525	3297	.	.	.	.	.	.	.	.	.	.	.	.
•••	2321	2815	3280	.	.	.	.	.	.	.	.	.	.	.	.

TABLE 8: ALPHA= 0.05 POWER= 0.9   EXPECTED ACCRUAL THRU MINIMUM FOLLOW-UP= 2250

	DEL=.02			DEL=.05			DEL=.10			DEL=.15			DEL=.20		
FACT=	1.0 .75	.50 .25	.00 BIN	1.0 .75	.50 .25	.00 BIN	1.0 .75	.50 .25	.00 BIN	1.0 75	.50 .25	.00 BIN	1.0 .75	.50 .25	.00 BIN
PCONT=***				REQUIRED NUMBER OF PATIENTS											
0.05 ***	2473	2503	2580	537	541	547	192	194	195	112	114	114	80	80	80
***	2485	2531	4822	539	544	943	194	194	300	114	114	158	80	80	101
0.1 ***	5159	5255	5668	1006	1026	1056	319	323	326	171	171	173	112	114	114
***	5194	5374	8376	1015	1039	1491	320	325	429	171	173	212	112	114	129
0.15 ***	7564	7742	8766	1430	1473	1546	433	438	447	221	223	224	140	140	142
***	7627	7997	11502	1450	1506	1970	435	443	540	222	224	257	140	142	153
0.2 ***	9509	9790	11648	1784	1860	1995	528	539	554	264	267	269	163	164	165
***	9607	10209	14199	1818	1917	2381	534	547	634	266	268	295	164	165	172
0.25 ***	10947	11352	14202	2067	2178	2387	607	624	645	298	302	308	181	182	185
***	11089	11958	16469	2114	2264	2724	615	634	711	300	305	326	182	184	187
0.3 ***	11892	12442	16367	2278	2426	2715	669	690	718	325	330	337	195	196	199
***	12085	13254	18310	2341	2541	2998	679	704	771	327	333	349	195	198	197
0.35 ***	12385	13099	18108	2421	2606	2974	713	739	775	343	350	358	204	207	209
***	12638	14126	19723	2502	2750	3203	725	756	814	345	354	364	205	208	204
0.4 ***	12481	13376	19405	2505	2721	3160	741	772	812	354	361	371	208	210	213
***	12802	14618	20708	2598	2892	3340	755	790	840	357	367	372	209	212	206
0.45 ***	12251	13335	20243	2535	2775	3272	752	786	831	357	365	375	208	210	215
***	12646	14775	21265	2639	2968	3409	767	807	848	361	370	372	209	212	204
0.5 ***	11763	13030	20620	2516	2775	3310	748	784	832	351	361	371	204	207	209
***	12234	14643	21393	2628	2982	3409	764	806	840	357	367	364	205	208	197
0.55 ***	11086	12515	20528	2451	2722	3273	728	766	814	340	348	359	194	198	201
***	11628	14258	21093	2569	2936	3340	747	787	814	344	354	349	196	199	187
0.6 ***	10281	11832	19973	2347	2618	3162	696	731	776	320	329	339	181	184	187
***	10878	13650	20365	2465	2831	3203	711	752	761	325	334	326	182	185	172
0.65 ***	9390	11008	18952	2202	2465	2977	646	679	721	294	302	311	164	165	168
***	10023	12837	19209	2317	2667	2998	662	699	711	298	306	295	164	167	153
0.7 ***	8444	10059	17471	2021	2262	2718	583	612	646	260	267	272	142	143	146
***	9084	11832	17625	2126	2445	2724	597	628	634	263	269	257	143	145	129
0.75 ***	7450	8987	15532	1798	2008	2386	506	528	555	219	223	227	115	117	118
***	8065	10627	15612	1891	2162	2381	516	541	540	221	226	212	117	118	101
0.8 ***	6395	7778	13139	1531	1698	1981	412	427	446	170	173	176	.	.	.
***	6955	9204	13172	1605	1818	1970	419	435	429	171	174	158	.	.	.
0.85 ***	5248	6388	10301	1211	1326	1506	299	308	317	.	.	.	.	.	.
***	5714	7514	10303	1262	1405	1491	303	312	300	.	.	.	.	.	.
0.9 ***	3927	4729	7018	818	877	959	.	.	.	.	.	.	.	.	.
***	4260	5466	7006	846	915	943	.	.	.	.	.	.	.	.	.
0.95 ***	2232	2580	3297	.	.	.	.	.	.	.	.	.	.	.	.
***	2382	2856	3280	.	.	.	.	.	.	.	.	.	.	.	.

TABLE 8: ALPHA= 0.05 POWER= 0.9   EXPECTED ACCRUAL THRU MINIMUM FOLLOW-UP= 2500

	DEL=.02			DEL=.05			DEL=.10			DEL=.15			DEL=.20		
FACT=	1.0 .75	.50 .25	.00 BIN	1.0 .75	.50 .25	.00 BIN	1.0 .75	.50 .25	.00 BIN	1.0 75	.50 .25	.00 BIN	1.0 .75	.50 .25	.00 BIN

PCONT=***   REQUIRED NUMBER OF PATIENTS

PCONT	DEL=.02 1.0/.75	.50/.25	.00/BIN	DEL=.05 1.0/.75	.50/.25	.00/BIN	DEL=.10 1.0/.75	.50/.25	.00/BIN	DEL=.15 1.0/.75	.50/.25	.00/BIN	DEL=.20 1.0/.75	.50/.25	.00/BIN
0.05 ***	2477	2508	2579	537	542	546	193	193	195	114	114	114	79	79	79
•••	2489	2534	4822	540	545	943	193	195	300	114	114	158	79	79	101
0.1 ***	5171	5271	5668	1011	1029	1056	320	323	326	171	171	173	112	114	114
•••	5209	5393	8376	1018	1040	1491	321	324	429	171	173	212	112	114	129
0.15 ***	7584	7777	8765	1437	1479	1546	434	440	446	221	223	224	140	142	142
•••	7654	8039	11502	1456	1509	1970	437	443	540	223	224	257	140	142	153
0.2 ***	9542	9846	11646	1796	1870	1995	531	542	552	264	267	270	164	165	165
•••	9649	10279	14199	1829	1923	2381	536	546	634	265	268	295	164	165	172
0.25 ***	10995	11434	14201	2084	2193	2387	611	626	645	298	302	308	181	183	186
•••	11149	12062	16469	2133	2273	2724	618	634	711	301	306	326	183	184	187
0.3 ***	11958	12551	16367	2301	2445	2715	673	693	718	326	331	337	195	196	199
•••	12168	13392	18310	2364	2556	2998	683	706	771	327	334	349	196	198	197
0.35 ***	12470	13237	18108	2451	2629	2973	718	743	774	345	351	358	204	206	209
•••	12745	14298	19723	2529	2768	3203	729	758	814	348	354	364	206	208	204
0.4 ***	12590	13546	19404	2540	2749	3159	746	774	812	354	362	371	209	211	214
•••	12939	14823	20708	2631	2914	3340	761	792	840	359	367	372	211	212	206
0.45 ***	12386	13536	20243	2573	2809	3271	758	790	831	358	367	374	209	211	214
•••	12811	15012	21265	2676	2992	3409	773	809	848	362	370	372	211	212	204
0.5 ***	11926	13259	20618	2558	2811	3309	754	789	831	354	362	371	204	208	211
•••	12427	14902	21393	2668	3009	3409	770	809	840	358	367	364	206	209	197
0.55 ***	11276	12767	20529	2495	2759	3273	736	770	814	342	349	359	195	198	201
•••	11846	14536	21093	2611	2962	3340	751	790	814	346	354	349	196	199	187
0.6 ***	10492	12096	19973	2390	2654	3162	701	734	776	323	329	339	183	184	187
•••	11115	13934	20365	2506	2858	3203	717	754	771	326	334	326	183	186	172
0.65 ***	9615	11276	18952	2245	2499	2976	652	683	721	295	302	311	164	167	168
•••	10270	13120	19209	2358	2692	2998	667	701	711	299	306	295	165	167	153
0.7 ***	8673	10323	17470	2061	2293	2717	589	615	646	262	267	273	142	143	145
•••	9329	12101	17625	2164	2467	2724	601	629	634	264	270	257	143	145	129
0.75 ***	7670	9234	15533	1833	2036	2386	509	531	554	220	223	227	115	117	118
•••	8299	10873	15612	1923	2181	2381	520	542	540	221	226	212	117	117	101
0.8 ***	6596	7996	13140	1559	1720	1981	414	427	445	170	173	174	.	.	.
•••	7165	9414	13172	1631	1831	1970	421	436	429	171	174	158	.	.	.
0.85 ***	5417	6564	10299	1231	1340	1506	299	309	317	.	.	.	.	.	.
•••	5889	7676	10303	1281	1414	1491	304	312	300	.	.	.	.	.	.
0.9 ***	4049	4846	7018	829	884	959	.	.	.	.	.	.	.	.	.
•••	4384	5567	7006	854	918	943	.	.	.	.	.	.	.	.	.
0.95 ***	2287	2627	3296	.	.	.	.	.	.	.	.	.	.	.	.
•••	2436	2890	3280	.	.	.	.	.	.	.	.	.	.	.	.

TABLE 8: ALPHA= 0.05 POWER= 0.9   EXPECTED ACCRUAL THRU MINIMUM FOLLOW-UP= 2750

	DEL=.02			DEL=.05			DEL=.10			DEL=.15			DEL=.20		
FACT=	1.0	.50	.00	1.0	.50	.00	1.0	.50	.00	1.0	.50	.00	1.0	.50	.00
	.75	.25	BIN	.75	.25	BIN	.75	.25	BIN	75	.25	BIN	.75	.25	BIN
PCONT=***				REQUIRED NUMBER OF PATIENTS											
0.05 ***	2481	2511	2579	539	542	548	194	194	194	113	113	113	80	80	80
•••	2494	2538	4822	541	544	943	194	194	300	113	113	158	80	80	101
0.1 ***	5183	5288	5668	1013	1030	1055	321	322	326	171	171	173	113	113	113
•••	5223	5409	8376	1020	1041	1491	321	324	429	171	173	212	113	113	129
0.15 ***	7605	7811	8765	1443	1484	1546	434	439	446	221	223	225	140	140	142
•••	7681	8076	11502	1462	1512	1970	438	443	540	223	225	257	140	142	153
0.2 ***	9575	9899	11648	1807	1878	1995	532	542	553	264	267	269	163	164	166
•••	9692	10343	14199	1840	1930	2381	538	548	634	266	269	295	164	164	172
0.25 ***	11043	11511	14202	2100	2206	2387	613	627	644	300	304	307	181	184	185
•••	11209	12156	16469	2146	2282	2724	620	635	711	302	305	326	181	184	187
0.3 ***	12022	12655	16367	2322	2463	2715	676	696	718	326	331	336	195	197	199
•••	12249	13518	18310	2384	2567	2998	685	706	771	329	335	349	197	199	197
0.35 ***	12555	13370	18108	2478	2652	2973	721	745	775	345	352	359	204	208	209
•••	12851	14456	19723	2553	2784	3203	734	759	814	349	355	364	205	208	204
0.4 ***	12697	13708	19404	2570	2775	3158	751	778	813	357	364	370	209	211	214
•••	13069	15011	20708	2660	2931	3340	764	793	840	360	367	372	211	212	206
0.45 ***	12519	13724	20243	2608	2837	3272	762	793	831	359	367	376	209	211	214
•••	12969	15226	21265	2708	3013	3409	778	810	848	364	370	372	211	212	204
0.5 ***	12084	13473	20619	2595	2842	3310	759	792	831	355	362	370	204	208	209
•••	12611	15138	21393	2703	3030	3409	775	810	840	359	367	364	205	209	197
0.55 ***	11456	12999	20528	2533	2790	3274	740	773	813	343	350	359	195	197	201
•••	12051	14786	21093	2646	2985	3340	755	792	814	346	355	349	197	199	187
0.6 ***	10690	12340	19973	2430	2686	3162	707	738	776	324	331	338	181	185	187
•••	11336	14191	20365	2543	2879	3203	721	755	771	328	335	326	184	185	172
0.65 ***	9826	11524	18952	2284	2529	2976	658	685	721	297	304	311	164	166	168
•••	10497	13375	19209	2392	2714	2998	672	703	711	300	307	295	166	168	153
0.7 ***	8885	10565	17471	2095	2322	2717	594	618	648	263	267	273	142	144	146
•••	9558	12343	17625	2196	2485	2724	604	632	634	264	271	257	144	144	129
0.75 ***	7875	9459	15532	1864	2058	2385	514	532	555	221	225	228	116	116	118
•••	8514	11092	15612	1952	2196	2381	522	542	540	223	226	212	116	118	101
0.8 ***	6781	8193	13139	1584	1737	1981	417	429	445	170	173	175	.	.	.
•••	7358	9600	13172	1655	1844	1970	422	438	429	171	175	158	.	.	.
0.85 ***	5571	6722	10300	1247	1352	1505	302	309	318	.	.	.	.	.	.
•••	6046	7818	10303	1295	1421	1491	305	312	300	.	.	.	.	.	.
0.9 ***	4159	4953	7017	838	890	958	.	.	.	.	.	.	.	.	.
•••	4494	5654	7006	862	923	943	.	.	.	.	.	.	.	.	.
0.95 ***	2337	2669	3298	.	.	.	.	.	.	.	.	.	.	.	.
•••	2483	2920	3280	.	.	.	.	.	.	.	.	.	.	.	.

TABLE 8: ALPHA= 0.05 POWER= 0.9   EXPECTED ACCRUAL THRU MINIMUM FOLLOW-UP= 3000

		DEL=.01			DEL=.02			DEL=.05			DEL=.10			DEL=.15		
FACT=		1.0 .75	.50 .25	.00 BIN	1.0 .75	.50 .25	.00 BIN	1.0 .75	.50 .25	.00 BIN	1.0 75	.50 .25	.00 BIN	1.0 .75	.50 .25	.00 BIN
PCONT=•••							REQUIRED	NUMBER OF	PATIENTS						
0.01	•••	1310	1315	1322	482	482	485	160	160	160	80	80	80	58	58	58
	•••	1310	1318	5053	482	482	1670	160	160	455	80	80	185	58	58	110
0.02	•••	2893	2908	2950	932	935	943	253	253	253	110	110	110	73	73	73
	•••	2900	2923	8342	932	940	2484	253	253	581	110	110	215	73	73	123
0.05	•••	8758	8815	9205	2485	2515	2578	538	542	545	193	193	193	115	115	115
	•••	8777	8905	17796	2495	2540	4822	542	545	943	193	193	300	115	115	158
0.1	•••	19322	19460	21430	5192	5305	5668	1015	1033	1055	320	325	325	170	170	170
	•••	19367	19727	32183	5237	5425	8376	1022	1045	1491	320	325	429	170	170	212
0.15	•••	28858	29110	33853	7625	7843	8765	1450	1487	1547	433	440	445	223	223	223
	•••	28940	29597	44858	7705	8110	11502	1468	1513	1970	437	445	540	223	223	257
0.2	•••	36550	36935	45478	9605	9950	11645	1817	1885	1997	535	542	553	265	268	268
	•••	36680	37700	55820	9730	10400	14199	1847	1933	2381	538	550	634	265	268	295
0.25	•••	42163	42722	55835	11087	11585	14200	2113	2218	2387	613	628	643	298	302	305
	•••	42350	43825	65069	11267	12242	16469	2158	2290	2724	620	635	711	302	305	326
0.3	•••	45715	46483	64655	12085	12752	16367	2342	2477	2713	680	695	718	328	332	335
	•••	45970	47998	72605	12328	13633	18310	2402	2578	2998	688	707	771	328	335	349
0.35	•••	47342	48370	71792	12635	13495	18107	2500	2672	2972	725	748	775	347	350	358
	•••	47683	50372	78428	12950	14600	19723	2575	2795	3203	737	760	814	347	355	364
0.4	•••	47260	48602	77158	12800	13858	19405	2597	2800	3160	755	782	812	358	362	370
	•••	47705	51178	82539	13195	15182	20708	2687	2945	3340	767	797	840	362	365	372
0.45	•••	45718	47447	80695	12647	13900	20245	2638	2863	3272	767	797	830	362	365	373
	•••	46295	50668	84937	13120	15422	21265	2740	3028	3409	782	812	848	362	370	372
0.5	•••	43003	45197	82382	12235	13670	20620	2627	2870	3310	763	793	830	355	362	370
	•••	43742	49112	85622	12785	15355	21393	2735	3050	3409	778	812	840	358	365	364
0.55	•••	39415	42155	82210	11627	13213	20530	2567	2818	3272	745	775	812	343	350	358
	•••	40348	46757	84594	12245	15013	21093	2680	3005	3340	760	793	814	347	355	349
0.6	•••	35275	38612	80170	10877	12568	19970	2465	2713	3163	710	740	775	325	332	340
	•••	36440	43810	81854	11545	14425	20365	2575	2897	3203	725	755	771	328	335	326
0.65	•••	30905	34810	76273	10022	11755	18950	2315	2555	2975	662	688	722	298	302	310
	•••	32308	40427	77401	10712	13607	19209	2425	2732	2998	673	703	711	302	305	295
0.7	•••	26590	30913	70532	9085	10787	17470	2125	2345	2717	598	620	647	265	268	272
	•••	28190	36703	71235	9767	12565	17625	2225	2503	2724	610	632	634	265	272	257
0.75	•••	22510	26995	62957	8065	9665	15530	1892	2080	2387	515	535	553	220	223	227
	•••	24212	32660	63356	8713	11290	15612	1975	2210	2381	523	542	540	223	227	212
0.8	•••	18703	23030	53570	6955	8372	13138	1607	1753	1982	418	430	445	170	175	175
	•••	20375	28262	53764	7535	9767	13172	1675	1855	1970	425	437	429	170	175	158
0.85	•••	15065	18905	42392	5713	6865	10300	1262	1363	1505	302	310	317	.	.	.
	•••	16570	23365	42460	6190	7945	10303	1307	1427	1491	305	313	300	.	.	.
0.9	•••	11357	14350	29435	4262	5045	7018	845	895	958	.	.	.	.	.	.
	•••	12545	17660	29443	4592	5732	7006	868	925	943	.	.	.	.	.	.
0.95	•••	7010	8702	14728	2383	2705	3298	.	.	.	.	.	.	.	.	.
	•••	7700	10390	14713	2525	2945	3280	.	.	.	.	.	.	.	.	.
0.98	•••	3290	3850	5068	.	.	.	.	.	.	.	.	.	.	.	.
	•••	3530	4307	5053	.	.	.	.	.	.	.	.	.	.	.	.

TABLE 8: ALPHA= 0.05 POWER= 0.9   EXPECTED ACCRUAL THRU MINIMUM FOLLOW-UP= 3250

		DEL=.01			DEL=.02			DEL=.05			DEL=.10			DEL=.15		
FACT=		1.0 .75	.50 .25	.00 BIN	1.0 .75	.50 .25	.00 BIN	1.0 .75	.50 .25	.00 BIN	1.0 75	.50 .25	.00 BIN	1.0 .75	.50 .25	.00 BIN
PCONT=***							REQUIRED NUMBER OF PATIENTS								
0.01	***	1311	1314	1322	482	482	485	160	160	160	79	79	79	54	54	54
	***	1311	1319	5053	482	482	1670	160	160	455	79	79	185	54	54	110
0.02	***	2895	2911	2947	932	937	940	254	254	254	111	111	111	71	71	71
	***	2903	2922	8342	932	937	2484	254	254	581	111	111	215	71	71	123
0.05	***	8764	8821	9208	2489	2516	2578	539	542	547	192	192	192	111	111	111
	***	8786	8916	17796	2500	2541	4822	542	547	943	192	192	300	111	111	158
0.1	***	19335	19486	21428	5206	5316	5669	1018	1034	1054	319	322	328	173	173	173
	***	19384	19766	32183	5246	5438	8376	1026	1043	1491	322	322	429	173	173	212
0.15	***	28879	29150	33851	7648	7871	8764	1452	1492	1546	436	441	444	222	222	225
	***	28968	29675	44858	7729	8139	11502	1473	1517	1970	436	444	540	222	225	257
0.2	***	36581	37004	45478	9639	9999	11646	1826	1891	1996	534	542	555	266	266	271
	***	36722	37824	55820	9772	10454	14199	1855	1936	2381	539	547	634	266	271	295
0.25	***	42207	42816	55832	11134	11654	14200	2126	2229	2386	615	628	644	298	303	306
	***	42410	44002	65069	11324	12320	16469	2172	2297	2724	623	636	711	303	306	326
0.3	***	45779	46611	64656	12149	12848	16366	2359	2492	2716	680	696	718	328	331	336
	***	46055	48241	72605	12404	13737	18310	2419	2586	2998	688	709	771	331	336	349
0.35	***	47428	48541	71795	12718	13615	18108	2521	2687	2971	729	750	774	347	352	360
	***	47797	50689	78428	13051	14733	19723	2594	2809	3203	737	761	814	347	355	364
0.4	***	47371	48826	77158	12905	14002	19405	2622	2817	3158	758	783	810	360	363	371
	***	47859	51575	82539	13314	15343	20708	2708	2960	3340	769	799	840	360	368	372
0.45	***	45863	47729	80695	12770	14067	20242	2668	2887	3272	769	799	831	360	368	376
	***	46489	51158	84937	13266	15603	21265	2765	3044	3409	786	815	848	363	371	372
0.5	***	43190	45551	82385	12380	13859	20619	2659	2895	3309	769	799	831	355	363	371
	***	43986	49687	85622	12949	15554	21393	2760	3066	3409	783	815	840	360	368	364
0.55	***	39653	42586	82206	11792	13417	20529	2597	2841	3272	750	777	815	344	352	360
	***	40655	47412	84594	12429	15221	21093	2708	3020	3340	761	794	814	347	355	349
0.6	***	35569	39116	80167	11056	12778	19974	2497	2741	3163	718	742	774	328	331	339
	***	36817	44522	81854	11738	14639	20365	2606	2914	3203	729	758	771	328	336	326
0.65	***	31268	35374	76272	10211	11966	18953	2346	2581	2976	664	688	721	298	303	311
	***	32754	41172	77401	10909	13815	19209	2451	2749	2998	677	704	711	303	306	295
0.7	***	27005	31511	70531	9268	10991	17471	2156	2367	2716	599	620	647	263	266	274
	***	28687	37451	71235	9964	12762	17625	2248	2516	2724	612	631	634	266	271	257
0.75	***	22959	27595	62958	8241	9858	15532	1915	2099	2386	517	534	555	222	225	230
	***	24727	33380	63356	8899	11470	15612	1996	2221	2381	526	547	540	222	225	212
0.8	***	19153	23592	53571	7115	8537	13141	1628	1769	1980	420	433	444	173	173	176
	***	20876	28911	53764	7700	9918	13172	1693	1863	1970	425	436	429	173	173	158
0.85	***	15473	19392	42391	5844	6998	10300	1278	1371	1506	303	311	314	.	.	.
	***	17011	23906	42460	6324	8057	10303	1322	1433	1491	306	314	300	.	.	.
0.9	***	11678	14720	29434	4352	5132	7017	851	899	956	.	.	.	.	.	.
	***	12889	18051	29443	4686	5799	7006	875	929	943	.	.	.	.	.	.
0.95	***	7201	8899	14728	2424	2736	3296	.	.	.	.	.	.	.	.	.
	***	7895	10576	14713	2562	2968	3280	.	.	.	.	.	.	.	.	.
0.98	***	3358	3911	5067	.	.	.	.	.	.	.	.	.	.	.	.
	***	3597	4349	5053	.	.	.	.	.	.	.	.	.	.	.	.

TABLE 8: ALPHA= 0.05 POWER= 0.9   EXPECTED ACCRUAL THRU MINIMUM FOLLOW-UP= 3500

FACT=	DEL=.01 1.0 .75	DEL=.01 .50 .25	DEL=.01 .00 BIN	DEL=.02 1.0 .75	DEL=.02 .50 .25	DEL=.02 .00 BIN	DEL=.05 1.0 .75	DEL=.05 .50 .25	DEL=.05 .00 BIN	DEL=.10 1.0 75	DEL=.10 .50 .25	DEL=.10 .00 BIN	DEL=.15 1.0 .75	DEL=.15 .50 .25	DEL=.15 .00 BIN
PCONT=***				REQUIRED NUMBER OF PATIENTS											
0.01 ***	1310	1315	1318	484	484	484	160	160	160	81	81	81	55	55	55
•••	1315	1318	5053	484	484	1670	160	160	455	81	81	185	55	55	110
0.02 ***	2899	2911	2946	933	933	942	251	251	251	111	111	111	73	73	73
•••	2902	2925	8342	933	939	2484	251	251	581	111	111	215	73	73	123
0.05 ***	8770	8831	9208	2491	2523	2578	540	545	545	195	195	195	111	111	111
•••	8791	8922	17796	2505	2543	4822	540	545	943	195	195	300	111	111	158
0.1 ***	19343	19506	21431	5212	5331	5667	1021	1035	1056	321	321	326	172	172	172
•••	19401	19807	32183	5261	5448	8376	1026	1044	1491	321	326	429	172	172	212
0.15 ***	28898	29193	33851	7668	7898	8765	1458	1493	1546	435	440	443	221	225	225
•••	28995	29747	44858	7755	8166	11502	1476	1520	1970	440	443	540	221	225	257
0.2 ***	36613	37068	45480	9671	10048	11649	1835	1896	1998	536	545	554	265	268	268
•••	36765	37946	55820	9811	10503	14199	1861	1940	2381	540	548	634	265	268	295
0.25 ***	42256	42907	55836	11180	11722	14198	2138	2234	2386	618	633	645	300	303	309
•••	42475	44181	65069	11381	12393	16469	2181	2304	2724	624	636	711	303	303	326
0.3 ***	45838	46740	64656	12209	12935	16365	2374	2505	2715	685	697	720	326	330	335
•••	46141	48478	72605	12480	13836	18310	2430	2593	2998	688	706	771	330	335	349
0.35 ***	47515	48708	71791	12798	13726	18106	2540	2701	2972	732	750	773	347	353	356
•••	47912	51001	78428	13145	14855	19723	2613	2815	3203	741	764	814	347	356	364
0.4 ***	47483	49046	77155	13005	14137	19405	2645	2838	3161	764	785	811	361	365	370
•••	48005	51969	82539	13434	15485	20708	2733	2972	3340	773	799	840	361	365	372
0.45 ***	46010	48014	80695	12891	14225	20241	2692	2908	3270	776	799	828	361	370	373
•••	46678	51631	84937	13402	15770	21265	2788	3060	3409	785	816	848	365	370	372
0.5 ***	43371	45900	82384	12518	14032	20618	2683	2916	3310	773	799	828	356	365	370
•••	44228	50240	85622	13110	15735	21393	2788	3077	3409	785	816	840	361	370	364
0.55 ***	39885	43009	82209	11950	13603	20530	2628	2867	3275	755	781	811	347	353	361
•••	40961	48031	84594	12603	15415	21093	2736	3039	3340	767	793	814	347	356	349
0.6 ***	35863	39608	80170	11229	12973	19973	2526	2762	3161	720	746	776	326	330	338
•••	37190	45196	81854	11923	14834	20365	2631	2928	3203	732	758	771	330	335	326
0.65 ***	31620	35916	76271	10383	12165	18950	2374	2601	2978	668	694	720	300	303	309
•••	33186	41875	77401	11098	14011	19209	2479	2759	2998	680	706	711	300	309	295
0.7 ***	27411	32083	70531	9444	11185	17471	2181	2386	2718	601	624	645	265	268	274
•••	29161	38153	71235	10147	12944	17625	2272	2526	2724	615	633	634	265	268	257
0.75 ***	23395	28160	62959	8411	10033	15534	1936	2111	2386	522	536	554	221	225	230
•••	25220	34049	63356	9071	11635	15612	2018	2234	2381	528	545	540	221	225	212
0.8 ***	19576	24126	53570	7265	8691	13140	1642	1779	1980	423	431	443	172	172	172
•••	21347	29516	53764	7851	10056	13172	1703	1870	1970	426	440	429	172	172	158
0.85 ***	15858	19851	42391	5970	7116	10301	1289	1380	1508	303	309	318	.	.	.
•••	17433	24410	42460	6446	8161	10303	1333	1438	1491	309	312	300	.	.	.
0.9 ***	11981	15070	29438	4439	5212	7015	860	904	960	.	.	.	.	.	.
•••	13215	18413	29443	4771	5865	7006	881	930	943	.	.	.	.	.	.
0.95 ***	7379	9085	14729	2461	2768	3296	.	.	.	.	.	.	.	.	.
•••	8079	10748	14713	2596	2986	3280	.	.	.	.	.	.	.	.	.
0.98 ***	3418	3961	5069	.	.	.	.	.	.	.	.	.	.	.	.
•••	3655	4390	5053	.	.	.	.	.	.	.	.	.	.	.	.

829

TABLE 8: ALPHA= 0.05 POWER= 0.9   EXPECTED ACCRUAL THRU MINIMUM FOLLOW-UP= 3750

		DEL=.01			DEL=.02			DEL=.05			DEL=.10			DEL=.15		
FACT=		1.0 .75	.50 .25	.00 BIN	1.0 .75	.50 .25	.00 BIN	1.0 .75	.50 .25	.00 BIN	1.0 75	.50 .25	.00 BIN	1.0 .75	.50 .25	.00 BIN	
PCONT=***							REQUIRED NUMBER OF PATIENTS										
0.01	***	1309	1315	1319	481	484	484	156	156	156	81	81	81	59	59	59
	•••	1315	1319	5053	481	484	1670	156	156	455	81	81	185	59	59	110
0.02	***	2900	2913	2947	931	934	940	250	250	256	109	109	109	72	72	72
	•••	2903	2928	8342	934	940	2484	250	256	581	109	109	215	72	72	123
0.05	***	8772	8838	9203	2497	2525	2581	541	541	547	194	194	194	115	115	115
	•••	8797	8937	17796	2506	2547	4822	541	547	943	194	194	300	115	115	158
0.1	***	19356	19531	21428	5225	5341	5669	1019	1038	1056	322	322	325	172	172	172
	•••	19413	19844	32183	5272	5459	8376	1028	1043	1491	322	325	429	172	172	212
0.15	***	28919	29234	33850	7684	7925	8763	1465	1497	1544	438	443	447	222	222	222
	•••	29022	29825	44858	7778	8191	11502	1478	1522	1970	438	443	540	222	222	257
0.2	***	36644	37131	45481	9700	10090	11647	1840	1906	1994	537	547	550	265	269	269
	•••	36809	38065	55820	9847	10550	14199	1868	1943	2381	541	550	634	265	269	295
0.25	***	42303	43000	55834	11225	11787	14200	2150	2243	2388	622	631	644	303	303	306
	•••	42537	44350	65069	11434	12462	16469	2191	2309	2724	625	634	711	303	306	326
0.3	***	45906	46868	64656	12269	13019	16366	2388	2515	2716	687	700	719	331	331	334
	•••	46225	48709	72605	12550	13928	18310	2444	2600	2998	691	709	771	331	334	349
0.35	***	47600	48878	71791	12875	13834	18106	2556	2716	2975	734	753	775	350	353	359
	•••	48025	51303	78428	13234	14969	19723	2628	2825	3203	743	762	814	350	353	364
0.4	***	47594	49268	77159	13100	14266	19403	2665	2853	3156	766	784	813	359	363	372
	•••	48153	52347	82539	13544	15622	20708	2750	2984	3340	775	800	840	363	368	372
0.45	***	46150	48293	80697	13006	14375	20243	2716	2922	3269	781	803	831	363	368	372
	•••	46872	52090	84937	13534	15922	21265	2809	3072	3409	790	818	848	368	372	372
0.5	***	43559	46250	82384	12653	14200	20618	2712	2931	3306	775	800	831	359	363	372
	•••	44472	50772	85622	13259	15903	21393	2809	3091	3409	790	813	840	363	368	364
0.55	***	40118	43422	82206	12100	13784	20528	2656	2884	3275	756	781	813	344	353	359
	•••	41266	48631	84594	12766	15593	21093	2759	3050	3340	772	800	814	350	353	349
0.6	***	36153	40084	80168	11388	13159	19972	2553	2781	3162	725	747	775	325	331	340
	•••	37553	45841	81854	12097	15016	20365	2656	2941	3203	734	762	771	331	334	326
0.65	***	31966	36443	76272	10553	12350	18953	2403	2618	2975	672	697	719	297	303	312
	•••	33606	42547	77401	11275	14187	19209	2500	2772	2998	681	706	711	303	306	295
0.7	***	27809	32631	70531	9612	11359	17468	2200	2403	2716	606	625	644	265	269	275
	•••	29622	38819	71235	10325	13113	17625	2294	2538	2724	616	634	634	265	269	257
0.75	***	23809	28703	62956	8566	10197	15531	1956	2125	2384	522	537	556	222	222	228
	•••	25690	34681	63356	9231	11787	15612	2037	2238	2381	531	547	540	222	228	212
0.8	***	19984	24631	53572	7403	8834	13141	1656	1788	1981	425	434	443	172	172	175
	•••	21794	30087	53764	7994	10178	13172	1718	1878	1970	428	438	429	172	175	158
0.85	***	16222	20284	42391	6081	7225	10300	1300	1384	1506	306	312	316	.	.	.
	•••	17825	24884	42460	6565	8253	10303	1338	1441	1491	306	312	300	.	.	.
0.9	***	12269	15391	29434	4522	5284	7019	865	906	959	.	.	.	.	.	.
	•••	13522	18747	29443	4844	5918	7006	884	931	943	.	.	.	.	.	.
0.95	***	7544	9256	14725	2491	2791	3297	.	.	.	.	.	.	.	.	.
	•••	8247	10900	14713	2628	3003	3280	.	.	.	.	.	.	.	.	.
0.98	***	3481	4009	5069	.	.	.	.	.	.	.	.	.	.	.	.
	•••	3709	4422	5053	.	.	.	.	.	.	.	.	.	.	.	.

| TABLE 8: ALPHA= 0.05 POWER= 0.9 | | | EXPECTED ACCRUAL THRU MINIMUM FOLLOW-UP= 4000 | | | | | | | | | | | | |

		DEL=.01			DEL=.02			DEL=.05			DEL=.10			DEL=.15		
FACT=		1.0 .75	.50 .25	.00 BIN	1.0 .75	.50 .25	.00 BIN	1.0 .75	.50 .25	.00 BIN	1.0 75	.50 .25	.00 BIN	1.0 .75	.50 .25	.00 BIN
PCONT=***					REQUIRED NUMBER OF PATIENTS											
0.01	***	1313	1317	1323	483	483	483	157	157	157	83	83	83	57	57	57
	•••	1313	1317	5053	483	483	1670	157	157	455	83	83	185	57	57	110
0.02	***	2903	2913	2947	933	937	943	253	253	253	113	113	113	73	73	73
	•••	2907	2927	8342	933	937	2484	253	253	581	113	113	215	73	73	123
0.05	***	8777	8847	9207	2497	2527	2577	543	543	547	193	193	193	113	113	113
	•••	8803	8943	17796	2507	2547	4822	543	547	943	193	193	300	113	113	158
0.1	***	19367	19553	21427	5237	5353	5667	1023	1037	1053	323	323	327	173	173	173
	•••	19433	19883	32183	5283	5467	8376	1027	1047	1491	323	323	429	173	173	212
0.15	***	28943	29273	33853	7707	7953	8767	1467	1503	1547	437	443	447	223	223	223
	•••	29053	29897	44858	7797	8217	11502	1483	1523	1970	437	443	540	223	223	257
0.2	***	36677	37197	45477	9733	10133	11647	1847	1907	1997	537	547	553	267	267	267
	•••	36853	38183	55820	9883	10587	14199	1877	1947	2381	543	547	634	267	267	295
0.25	***	42347	43093	55833	11267	11847	14203	2157	2253	2387	623	633	643	303	303	307
	•••	42597	44523	65069	11487	12523	16469	2203	2313	2724	627	637	711	303	307	326
0.3	***	45967	46993	64657	12327	13103	16367	2403	2523	2713	687	703	717	327	333	337
	•••	46313	48943	72605	12623	14013	18310	2457	2607	2998	693	707	771	333	333	349
0.35	***	47683	49047	71793	12953	13937	18107	2577	2727	2973	737	753	773	347	353	357
	•••	48143	51603	78428	13327	15077	19723	2643	2833	3203	743	763	814	353	357	364
0.4	***	47707	49487	77157	13197	14393	19403	2687	2867	3157	767	787	813	363	367	373
	•••	48303	52717	82539	13653	15747	20708	2767	2993	3340	777	797	840	363	367	372
0.45	***	46297	48567	80697	13117	14513	20243	2737	2937	3273	783	803	833	363	367	373
	•••	47063	52533	84937	13663	16063	21265	2827	3083	3409	793	817	848	367	373	372
0.5	***	43743	46587	82383	12783	14357	20617	2733	2953	3307	777	803	833	357	363	373
	•••	44717	51287	85622	13403	16057	21393	2833	3103	3409	793	817	840	363	367	364
0.55	***	40347	43827	82207	12243	13953	20527	2677	2903	3273	757	783	813	347	353	357
	•••	41563	49203	84594	12923	15757	21093	2783	3063	3340	773	797	814	353	357	349
0.6	***	36437	40547	80167	11543	13333	19973	2573	2797	3163	727	747	777	327	333	337
	•••	37913	46453	81854	12263	15183	20365	2677	2953	3203	737	763	771	333	337	326
0.65	***	32307	36947	76273	10713	12523	18953	2423	2637	2977	673	697	723	303	303	307
	•••	34017	43177	77401	11443	14353	19209	2523	2783	2998	683	707	711	303	307	295
0.7	***	28193	33153	70533	9767	11527	17473	2223	2417	2717	607	627	647	267	267	273
	•••	30067	39447	71235	10487	13267	17625	2313	2547	2724	617	637	634	267	273	257
0.75	***	24213	29217	62957	8713	10347	15533	1977	2143	2387	523	537	553	223	223	227
	•••	26143	35277	63356	9387	11923	15612	2053	2247	2381	533	547	540	223	227	212
0.8	***	20377	25107	53573	7537	8967	13137	1673	1803	1983	423	433	443	173	173	173
	•••	22227	30623	53764	8127	10297	13172	1733	1883	1970	427	437	429	173	173	158
0.85	***	16573	20697	42393	6193	7327	10297	1307	1393	1507	307	313	317	.	.	.
	•••	18203	25323	42460	6673	8337	10303	1347	1443	1491	307	313	300	.	.	.
0.9	***	12547	15703	29437	4593	5347	7017	867	907	957	.	.	.	.	.	.
	•••	13813	19057	29443	4917	5967	7006	887	933	943	.	.	.	.	.	.
0.95	***	7697	9413	14727	2523	2813	3297	.	.	.	.	.	.	.	.	.
	•••	8407	11043	14713	2653	3017	3280	.	.	.	.	.	.	.	.	.
0.98	***	3533	4053	5067	.	.	.	.	.	.	.	.	.	.	.	.
	•••	3763	4457	5053	.	.	.	.	.	.	.	.	.	.	.	.

TABLE 8: ALPHA= 0.05 POWER= 0.9   EXPECTED ACCRUAL THRU MINIMUM FOLLOW-UP= 4250

		DEL=.01			DEL=.02			DEL=.05			DEL=.10			DEL=.15		
FACT=		1.0	.50	.00	1.0	.50	.00	1.0	.50	.00	1.0	.50	.00	1.0	.50	.00
		.75	.25	BIN	.75	.25	BIN	.75	.25	BIN	75	.25	BIN	.75	.25	BIN
PCONT=***							REQUIRED NUMBER OF PATIENTS									
0.01	***	1310	1314	1321	481	481	485	156	156	156	82	82	82	56	56	56
	***	1314	1321	5053	481	481	1670	156	156	455	82	82	185	56	56	110
0.02	***	2904	2914	2946	932	938	942	252	252	252	109	109	109	71	71	71
	***	2908	2929	8342	932	938	2484	252	252	581	109	109	215	71	71	123
0.05	***	8783	8854	9204	2500	2525	2578	538	545	545	194	194	194	113	113	113
	***	8811	8953	17796	2511	2553	4822	545	545	943	194	194	300	113	113	158
0.1	***	19383	19574	21427	5245	5362	5666	1023	1038	1055	322	322	326	173	173	173
	***	19447	19918	32183	5294	5475	8376	1034	1044	1491	322	326	429	173	173	212
0.15	***	28960	29311	33854	7721	7976	8762	1469	1501	1548	439	443	443	220	226	226
	***	29077	29970	44858	7823	8238	11502	1484	1523	1970	439	443	540	220	226	257
0.2	***	36713	37258	45478	9761	10171	11648	1856	1916	1994	538	545	556	269	269	269
	***	36893	38296	55820	9916	10628	14199	1884	1952	2381	545	549	634	269	269	295
0.25	***	42397	43187	55838	11312	11903	14202	2171	2256	2387	623	634	645	301	305	305
	***	42663	44685	65069	11535	12583	16469	2207	2313	2724	630	641	711	305	305	326
0.3	***	46035	47118	64656	12385	13178	16366	2415	2532	2717	687	704	719	333	333	337
	***	46396	49165	72605	12689	14092	18310	2468	2610	2998	694	708	771	333	333	349
0.35	***	47767	49218	71790	13023	14032	18108	2589	2738	2972	736	758	772	347	354	358
	***	48255	51889	78428	13412	15176	19723	2659	2840	3203	747	762	814	354	354	364
0.4	***	47820	49700	77155	13288	14506	19404	2702	2883	3159	768	789	811	358	364	368
	***	48453	53075	82539	13756	15866	20708	2780	2999	3340	779	800	840	364	368	372
0.45	***	46438	48846	80693	13231	14648	20244	2759	2957	3269	783	804	832	364	368	375
	***	47257	52962	84937	13784	16200	21265	2844	3088	3409	793	815	848	368	375	372
0.5	***	43927	46923	82383	12913	14500	20620	2755	2968	3308	783	804	832	358	364	368
	***	44958	51783	85622	13539	16206	21393	2851	3116	3409	793	815	840	364	368	364
0.55	***	40580	44224	82209	12381	14107	20531	2702	2918	3276	762	783	815	347	354	358
	***	41859	49749	84594	13072	15909	21093	2802	3074	3340	772	800	814	347	358	349
0.6	***	36723	40998	80169	11691	13497	19972	2596	2812	3163	726	751	779	326	333	337
	***	38264	47040	81854	12417	15339	20365	2695	2968	3203	740	762	771	333	337	326
0.65	***	32643	37435	76273	10862	12683	18952	2447	2653	2978	677	698	719	301	305	311
	***	34418	43782	77401	11599	14506	19209	2536	2791	2998	687	708	711	305	305	295
0.7	***	28563	33657	70532	9916	11684	17471	2245	2430	2717	609	630	645	262	269	273
	***	30497	40042	71235	10639	13412	17625	2330	2557	2724	619	634	634	269	269	257
0.75	***	24600	29711	62956	8854	10494	15530	1994	2153	2383	528	538	556	220	226	226
	***	26576	35845	63356	9527	12052	15612	2064	2256	2381	534	545	540	226	226	212
0.8	***	20754	25567	53568	7657	9088	13140	1686	1809	1979	428	432	443	173	173	173
	***	22638	31128	53764	8252	10398	13172	1746	1888	1970	428	439	429	173	173	158
0.85	***	16908	21087	42390	6293	7423	10299	1321	1399	1505	305	311	315	.	.	.
	***	18565	25737	42460	6771	8418	10303	1357	1448	1491	311	315	300	.	.	.
0.9	***	12806	15994	29434	4661	5411	7016	874	910	959	.	.	.	.	.	.
	***	14085	19351	29443	4986	6017	7006	889	932	943	.	.	.	.	.	.
0.95	***	7848	9566	14729	2553	2833	3297	.	.	.	.	.	.	.	.	.
	***	8560	11174	14713	2681	3031	3280	.	.	.	.	.	.	.	.	.
0.98	***	3584	4094	5071	.	.	.	.	.	.	.	.	.	.	.	.
	***	3807	4487	5053	.	.	.	.	.	.	.	.	.	.	.	.

TABLE 8: ALPHA= 0.05 POWER= 0.9   EXPECTED ACCRUAL THRU MINIMUM FOLLOW-UP= 4500

		DEL=.01			DEL=.02			DEL=.05			DEL=.10			DEL=.15	
FACT=	1.0 .75	.50 .25	.00 BIN	1.0 .75	.50 .25	.00 BIN	1.0 .75	.50 .25	.00 BIN	1.0 75	.50 .25	.00 BIN	1.0 .75	.50 .25	.00 BIN
PCONT=•••				REQUIRED NUMBER OF PATIENTS											
0.01 •••	1313	1313	1320	480	480	487	161	161	161	82	82	82	60	60	60
•••	1313	1320	5053	480	480	1670	161	161	455	82	82	185	60	60	110
0.02 •••	2906	2917	2951	930	937	941	251	251	255	109	109	109	71	71	71
•••	2910	2928	8342	937	937	2484	251	255	581	109	109	215	71	71	123
0.05 •••	8790	8861	9206	2501	2528	2580	543	543	548	195	195	195	116	116	116
•••	8812	8963	17796	2516	2550	4822	543	543	943	195	195	300	116	116	158
0.1 •••	19391	19601	21428	5257	5374	5666	1027	1038	1054	323	323	323	172	172	172
•••	19459	19950	32183	5302	5482	8376	1031	1050	1491	323	323	429	172	172	212
0.15 •••	28983	29355	33855	7743	7995	8767	1470	1504	1545	435	442	446	221	221	221
•••	29107	30041	44858	7845	8261	11502	1488	1522	1970	442	442	540	221	221	257
0.2 •••	36746	37324	45480	9791	10207	11647	1860	1916	1995	536	548	555	266	266	266
•••	36937	38411	55820	9953	10661	14199	1886	1954	2381	543	548	634	266	266	295
0.25 •••	42443	43282	55837	11355	11955	14201	2179	2265	2388	622	633	645	300	307	307
•••	42720	44850	65069	11584	12637	16469	2220	2321	2724	626	638	711	300	307	326
0.3 •••	46099	47246	64657	12439	13256	16365	2426	2539	2715	690	705	716	330	334	334
•••	46481	49384	72605	12754	14160	18310	2478	2618	2998	694	712	771	330	334	349
0.35 •••	47854	49384	71794	13098	14126	18104	2606	2748	2973	739	757	773	352	352	356
•••	48367	52174	78428	13496	15270	19723	2670	2850	3203	746	761	814	352	356	364
0.4 •••	47933	49920	77156	13373	14617	19403	2719	2895	3158	773	791	813	363	368	368
•••	48603	53430	82539	13856	15978	20708	2798	3007	3340	780	802	840	363	368	372
0.45 •••	46583	49114	80693	13335	14775	20242	2775	2966	3270	784	806	829	363	368	375
•••	47445	53378	84937	13901	16320	21265	2861	3101	3409	795	818	848	368	375	372
0.5 •••	44108	47253	82380	13031	14644	20618	2775	2985	3311	784	806	829	363	368	368
•••	45195	52260	85622	13672	16338	21393	2872	3124	3409	795	818	840	363	368	364
0.55 •••	40811	44614	82207	12513	14257	20528	2719	2933	3270	768	784	813	345	352	356
•••	42157	50280	84594	13215	16050	21093	2820	3079	3340	773	802	814	352	356	349
0.6 •••	37005	41437	80171	11831	13650	19972	2618	2831	3165	728	750	773	330	334	341
•••	38613	47602	81854	12570	15483	20365	2715	2973	3203	739	761	771	330	334	326
0.65 •••	32970	37905	76271	11006	12840	18953	2467	2670	2978	678	701	723	300	307	311
•••	34811	44362	77401	11753	14644	19209	2557	2805	2998	690	712	711	300	307	295
0.7 •••	28927	34140	70530	10061	11831	17468	2265	2445	2719	611	626	645	266	266	273
•••	30911	40616	71235	10785	13541	17625	2343	2561	2724	622	638	634	266	273	257
0.75 •••	24978	30187	62958	8985	10628	15533	2006	2163	2388	525	543	555	221	228	228
•••	26992	36379	63356	9667	12169	15612	2078	2265	2381	532	548	540	221	228	212
0.8 •••	21113	26002	53569	7777	9206	13136	1695	1819	1983	424	435	446	172	172	176
•••	23032	31605	53764	8373	10500	13172	1751	1893	1970	431	442	429	172	172	158
0.85 •••	17227	21457	42393	6386	7511	10301	1324	1403	1504	307	311	318	.	.	.
•••	18903	26130	42460	6866	8490	10303	1365	1448	1491	307	311	300	.	.	.
0.9 •••	13053	16264	29433	4728	5464	7016	874	915	960	.	.	.	.	.	.
•••	14351	19623	29443	5048	6056	7006	896	937	943	.	.	.	.	.	.
0.95 •••	7991	9701	14730	2580	2854	3300	.	.	.	.	.	.	.	.	.
•••	8700	11298	14713	2703	3045	3280	.	.	.	.	.	.	.	.	.
0.98 •••	3626	4132	5070	.	.	.	.	.	.	.	.	.	.	.	.
•••	3851	4508	5053	.	.	.	.	.	.	.	.	.	.	.	.

TABLE 8: ALPHA= 0.05 POWER= 0.9   EXPECTED ACCRUAL THRU MINIMUM FOLLOW-UP= 4750

	DEL=.01			DEL=.02			DEL=.05			DEL=.10			DEL=.15		
FACT=	1.0 .75	.50 .25	.00 BIN	1.0 .75	.50 .25	.00 BIN	1.0 .75	.50 .25	.00 BIN	1.0 75	.50 .25	.00 BIN	1.0 .75	.50 .25	.00 BIN
PCONT=***					REQUIRED NUMBER OF PATIENTS										
0.01 ***	1314	1314	1322	483	483	483	158	158	158	79	79	79	55	55	55
•••	1314	1314	5053	483	483	1670	158	158	455	79	79	185	55	55	110
0.02 ***	2905	2917	2948	934	934	942	253	253	253	110	110	110	75	75	75
•••	2913	2929	8342	934	942	2484	253	253	581	110	110	215	75	75	123
0.05 ***	8791	8867	9207	2502	2533	2580	542	542	550	193	193	193	110	115	115
•••	8819	8969	17796	2514	2549	4822	542	542	943	193	193	300	115	115	158
0.1 ***	19400	19621	21430	5264	5383	5668	1025	1041	1053	324	324	324	170	170	174
•••	19478	19982	32183	5312	5490	8376	1029	1048	1491	324	324	429	170	174	212
0.15 ***	29002	29394	33852	7758	8019	8767	1476	1504	1547	435	443	447	222	222	222
•••	29133	30107	44858	7865	8273	11502	1492	1523	1970	443	443	540	222	222	257
0.2 ***	36773	37386	45477	9817	10244	11645	1868	1920	1998	538	542	554	265	269	269
•••	36982	38519	55820	9983	10695	14199	1892	1955	2381	542	550	634	269	269	295
0.25 ***	42492	43371	55832	11392	12009	14199	2188	2272	2390	625	633	645	300	305	305
•••	42782	45010	65069	11634	12686	16469	2224	2324	2724	625	637	711	305	305	326
0.3 ***	46162	47373	64655	12496	13327	16367	2438	2549	2715	692	704	716	329	336	336
•••	46570	49598	72605	12817	14230	18310	2485	2620	2998	697	709	771	329	336	349
0.35 ***	47943	49550	71792	13165	14210	18105	2616	2758	2972	740	756	775	348	352	360
•••	48482	52448	78428	13577	15350	19723	2680	2853	3203	752	763	814	352	352	364
0.4 ***	48042	50132	77155	13462	14721	19407	2735	2901	3162	775	792	811	360	364	372
•••	48750	53766	82539	13957	16075	20708	2810	3012	3340	780	799	840	364	372	372
0.45 ***	46732	49384	80694	13434	14895	20243	2794	2977	3269	787	811	827	364	372	376
•••	47634	53778	84937	14016	16438	21265	2877	3107	3409	799	815	848	364	372	372
0.5 ***	44290	47579	82385	13149	14776	20618	2794	2996	3309	787	804	827	360	364	372
•••	45430	52717	85622	13795	16447	21393	2882	3131	3409	799	815	840	364	372	364
0.55 ***	41036	44990	82207	12643	14400	20528	2739	2948	3274	768	787	811	348	352	360
•••	42445	50785	84594	13351	16189	21093	2834	3091	3340	775	799	814	352	352	349
0.6 ***	37279	41855	80172	11966	13795	19970	2640	2842	3162	732	752	775	329	336	336
•••	38953	48137	81854	12710	15619	20365	2727	2984	3203	740	763	771	329	336	326
0.65 ***	33294	38360	76269	11142	12983	18949	2485	2680	2977	680	697	720	300	305	312
•••	35189	44907	77401	11895	14780	19209	2573	2810	2998	692	709	711	305	305	295
0.7 ***	29280	34607	70529	10192	11966	17472	2276	2454	2715	614	625	645	265	269	269
•••	31318	41155	71235	10921	13672	17625	2359	2573	2724	621	637	634	269	269	257
0.75 ***	25337	30641	62957	9112	10755	15529	2022	2169	2383	530	542	554	222	222	229
•••	27399	36892	63356	9793	12282	15612	2093	2264	2381	538	550	540	222	229	212
0.8 ***	21462	26413	53572	7888	9313	13142	1706	1825	1979	424	435	443	170	174	174
•••	23409	32059	53764	8487	10589	13172	1761	1896	1970	431	443	429	174	174	158
0.85 ***	17531	21810	42390	6480	7596	10299	1333	1409	1504	305	312	317	.	.	.
•••	19234	26501	42460	6955	8558	10303	1369	1452	1491	312	312	300	.	.	.
0.9 ***	13292	16526	29434	4789	5518	7014	882	918	958	.	.	.	.	.	.
•••	14598	19882	29443	5102	6095	7006	899	934	943	.	.	.	.	.	.
0.95 ***	8119	9836	14728	2604	2870	3297	.	.	.	.	.	.	.	.	.
•••	8838	11408	14713	2727	3055	3280	.	.	.	.	.	.	.	.	.
0.98 ***	3673	4164	5067	.	.	.	.	.	.	.	.	.	.	.	.
•••	3891	4532	5053	.	.	.	.	.	.	.	.	.	.	.	.

TABLE 8: ALPHA= 0.05 POWER= 0.9   EXPECTED ACCRUAL THRU MINIMUM FOLLOW-UP= 5000

		DEL=.01			DEL=.02			DEL=.05			DEL=.10			DEL=.15		
FACT=		1.0 .75	.50 .25	.00 BIN	1.0 .75	.50 .25	.00 BIN	1.0 .75	.50 .25	.00 BIN	1.0 75	.50 .25	.00 BIN	1.0 .75	.50 .25	.00 BIN
PCONT=***				REQUIRED NUMBER OF PATIENTS												
0.01	***	1316	1316	1321	483	483	483	158	158	158	79	79	79	58	58	58
	•••	1316	1316	5053	483	483	1670	158	158	455	79	79	185	58	58	110
0.02	***	2904	2921	2946	933	933	941	254	254	254	108	108	108	71	71	71
	•••	2908	2929	8342	933	941	2484	254	254	581	108	108	215	71	71	123
0.05	***	8796	8879	9204	2508	2533	2579	541	546	546	191	196	196	116	116	116
	•••	8829	8979	17796	2521	2554	4822	541	546	943	191	196	300	116	116	158
0.1	***	19416	19641	21429	5271	5391	5666	1029	1041	1054	321	321	329	171	171	171
	•••	19491	20016	32183	5321	5496	8376	1033	1046	1491	321	321	429	171	171	212
0.15	***	29021	29433	33854	7779	8041	8766	1479	1508	1546	441	441	446	221	221	221
	•••	29166	30171	44858	7879	8291	11502	1491	1529	1970	441	446	540	221	221	257
0.2	***	36808	37454	45479	9846	10279	11646	1871	1921	1996	541	546	554	266	266	271
	•••	37021	38629	55820	10016	10729	14199	1896	1954	2381	541	546	634	266	271	295
0.25	***	42533	43466	55833	11433	12058	14204	2191	2271	2383	629	633	646	304	304	308
	•••	42846	45158	65069	11679	12733	16469	2229	2321	2724	629	641	711	304	304	326
0.3	***	46229	47504	64654	12554	13391	16366	2446	2554	2716	691	704	716	329	333	333
	•••	46654	49808	72605	12879	14296	18310	2496	2629	2998	696	708	771	333	333	349
0.35	***	48029	49721	71791	13233	14296	18108	2629	2766	2971	741	758	771	354	354	358
	•••	48596	52721	78428	13654	15433	19723	2691	2858	3203	746	766	814	354	354	364
0.4	***	48154	50346	77158	13546	14821	19404	2746	2916	3158	771	791	808	358	366	371
	•••	48896	54096	82539	14046	16179	20708	2821	3021	3340	783	804	840	366	366	372
0.45	***	46871	49646	80696	13533	15008	20241	2808	2991	3271	791	808	829	366	371	371
	•••	47821	54171	84937	14121	16546	21265	2891	3116	3409	796	821	848	366	371	372
0.5	***	44471	47896	82383	13258	14904	20616	2808	3008	3308	791	808	829	358	366	371
	•••	45666	53158	85622	13916	16583	21393	2904	3141	3409	796	821	840	366	366	364
0.55	***	41266	45358	82208	12766	14533	20529	2758	2958	3271	771	791	816	346	354	358
	•••	42729	51271	84594	13479	16308	21093	2854	3096	3340	779	804	814	354	358	349
0.6	***	37554	42271	80171	12096	13933	19971	2654	2858	3158	733	754	779	329	333	341
	•••	39283	48658	81854	12846	15746	20365	2746	2991	3203	741	766	771	333	333	326
0.65	***	33608	38804	76271	11279	13121	18954	2496	2691	2979	683	704	721	304	304	308
	•••	35558	45433	77401	12033	14904	19209	2583	2816	2998	691	708	711	304	308	295
0.7	***	29621	35054	70529	10321	12104	17471	2291	2466	2716	616	629	646	266	271	271
	•••	31704	41671	71235	11058	13783	17625	2371	2579	2724	621	641	634	266	271	257
0.75	***	25691	31071	62958	9233	10871	15533	2033	2179	2383	529	541	554	221	229	229
	•••	27783	37379	63356	9916	12383	15612	2104	2271	2381	533	546	540	221	229	212
0.8	***	21796	26816	53571	7996	9416	13141	1721	1829	1979	429	433	446	171	171	171
	•••	23771	32491	53764	8591	10679	13172	1771	1896	1970	433	441	429	171	171	158
0.85	***	17829	22146	42391	6566	7679	10296	1341	1416	1504	308	308	316	.	.	.
	•••	19546	26854	42460	7033	8616	10303	1371	1454	1491	308	316	300	.	.	.
0.9	***	13521	16779	29433	4846	5566	7016	883	916	958	.	.	.	.	.	.
	•••	14841	20121	29443	5158	6129	7006	904	941	943	.	.	.	.	.	.
0.95	***	8246	9958	14729	2629	2891	3296	.	.	.	.	.	.	.	.	.
	•••	8966	11516	14713	2746	3066	3280	.	.	.	.	.	.	.	.	.
0.98	***	3708	4196	5071	.	.	.	.	.	.	.	.	.	.	.	.
	•••	3929	4554	5053	.	.	.	.	.	.	.	.	.	.	.	.

TABLE 8: ALPHA= 0.05 POWER= 0.9   EXPECTED ACCRUAL THRU MINIMUM FOLLOW-UP= 5500

	DEL=.01			DEL=.02			DEL=.05			DEL=.10			DEL=.15		
FACT=	1.0 .75	.50 .25	.00 BIN	1.0 .75	.50 .25	.00 BIN	1.0 .75	.50 .25	.00 BIN	1.0 75	.50 .25	.00 BIN	1.0 .75	.50 .25	.00 BIN

PCONT=***   REQUIRED NUMBER OF PATIENTS

PCONT	DEL=.01 1.0/.75	.50/.25	.00/BIN	DEL=.02 1.0/.75	.50/.25	.00/BIN	DEL=.05 1.0/.75	.50/.25	.00/BIN	DEL=.10 1.0/.75	.50/.25	.00/BIN	DEL=.15 1.0/.75	.50/.25	.00/BIN
0.01 ***	1315	1315	1324	485	485	485	160	160	160	78	78	78	59	59	59
•••	1315	1315	5053	485	485	1670	160	160	455	78	78	185	59	59	110
0.02 ***	2905	2919	2946	930	939	939	251	251	251	114	114	114	73	73	73
•••	2910	2933	8342	939	939	2484	251	251	581	114	114	215	73	73	123
0.05 ***	8804	8892	9208	2512	2539	2580	540	545	545	196	196	196	114	114	114
•••	8837	8988	17796	2520	2553	4822	540	545	943	196	196	300	114	114	158
0.1 ***	19438	19685	21426	5289	5408	5669	1027	1040	1054	320	325	325	169	174	174
•••	19520	20070	32183	5339	5509	8376	1035	1049	1491	325	325	429	169	174	212
0.15 ***	29063	29517	33848	7814	8075	8763	1480	1508	1544	435	444	444	224	224	224
•••	29214	30295	44858	7915	8323	11502	1494	1530	1970	444	444	540	224	224	257
0.2 ***	36873	37574	45480	9895	10344	11650	1879	1929	1998	540	545	554	265	270	270
•••	37107	38834	55820	10074	10784	14199	1901	1956	2381	545	554	634	265	270	295
0.25 ***	42629	43646	55834	11513	12159	14199	2204	2278	2388	628	636	642	306	306	306
•••	42973	45461	65069	11765	12819	16469	2237	2328	2724	628	642	711	306	306	326
0.3 ***	46355	47749	64656	12654	13520	16366	2465	2567	2713	697	705	719	334	334	334
•••	46823	50210	72605	12989	14414	18310	2512	2635	2998	697	710	771	334	334	349
0.35 ***	48198	50045	71793	13369	14455	18104	2649	2781	2974	746	760	774	353	353	361
•••	48822	53235	78428	13800	15583	19723	2713	2869	3203	752	765	814	353	353	364
0.4 ***	48376	50769	77155	13704	15010	19405	2773	2933	3158	779	793	815	361	367	367
•••	49193	54729	82539	14227	16353	20708	2850	3034	3340	788	801	840	367	367	372
0.45 ***	47158	50164	80694	13726	15225	20244	2836	3015	3268	793	807	829	367	367	375
•••	48198	54913	84937	14323	16751	21265	2919	3125	3409	801	820	848	367	375	372
0.5 ***	44834	48514	82385	13470	15134	20620	2842	3029	3309	793	807	829	361	367	367
•••	46135	54005	85622	14144	16806	21393	2924	3153	3409	801	820	840	367	367	364
0.55 ***	41713	46072	82207	12998	14785	20524	2787	2988	3276	774	793	815	348	353	361
•••	43289	52190	84594	13726	16545	21093	2878	3111	3340	779	801	814	353	353	349
0.6 ***	38091	43060	80166	12338	14194	19974	2685	2878	3158	738	752	774	334	334	339
•••	39925	49633	81854	13099	15981	20365	2773	3007	3203	746	765	771	334	334	326
0.65 ***	34219	39637	76275	11526	13374	18951	2525	2713	2974	683	705	719	306	306	312
•••	36268	46429	77401	12288	15129	19209	2616	2828	2998	691	710	711	306	306	295
0.7 ***	30287	35905	70528	10564	12343	17472	2319	2484	2718	614	628	650	265	270	270
•••	32454	42643	71235	11306	13993	17625	2396	2589	2724	623	636	634	270	270	257
0.75 ***	26363	31895	62957	9455	11092	15533	2058	2195	2383	532	540	554	224	224	229
•••	28527	38289	63356	10143	12571	15612	2121	2278	2381	540	545	540	224	224	212
0.8 ***	22435	27564	53574	8190	9601	13140	1736	1846	1984	430	435	444	174	174	174
•••	24465	33293	53764	8790	10825	13172	1783	1907	1970	430	444	429	174	174	158
0.85 ***	18388	22779	42387	6719	7819	10303	1351	1420	1503	306	312	320	.	.	.
•••	20148	27504	42460	7187	8727	10303	1384	1461	1491	312	312	300	.	.	.
0.9 ***	13952	17238	29434	4954	5655	7016	889	925	958	.	.	.	.	.	.
•••	15285	20574	29443	5262	6191	7006	903	939	943	.	.	.	.	.	.
0.95 ***	8488	10184	14730	2671	2919	3295	.	.	.	.	.	.	.	.	.
•••	9203	11710	14713	2781	3084	3280	.	.	.	.	.	.	.	.	.
0.98 ***	3785	4253	5069	.	.	.	.	.	.	.	.	.	.	.	.
•••	3991	4596	5053	.	.	.	.	.	.	.	.	.	.	.	.

TABLE 8: ALPHA= 0.05 POWER= 0.9 EXPECTED ACCRUAL THRU MINIMUM FOLLOW-UP= 6000

	DEL=.01			DEL=.02			DEL=.05			DEL=.10			DEL=.15		
FACT=	1.0 .75	.50 .25	.00 BIN	1.0 .75	.50 .25	.00 BIN	1.0 .75	.50 .25	.00 BIN	1.0 75	.50 .25	.00 BIN	1.0 .75	.50 .25	.00 BIN
PCONT=***				REQUIRED NUMBER OF PATIENTS											
0.01 ***	1315	1315	1324	484	484	484	160	160	160	79	79	79	55	55	55
***	1315	1315	5053	484	484	1670	160	160	455	79	79	185	55	55	110
0.02 ***	2905	2920	2950	934	940	940	250	250	250	109	109	109	70	70	70
***	2914	2935	8342	934	940	2484	250	250	581	109	109	215	70	70	123
0.05 ***	8815	8905	9205	2515	2539	2575	544	544	544	190	190	190	115	115	115
***	8845	9004	17796	2524	2560	4822	544	544	943	190	190	300	115	115	158
0.1 ***	19459	19729	21430	5305	5425	5665	1030	1045	1054	325	325	325	169	169	169
***	19555	20134	32183	5350	5524	8376	1039	1045	1491	325	325	429	169	169	212
0.15 ***	29110	29599	33850	7840	8110	8764	1489	1510	1549	439	445	445	220	220	220
***	29275	30415	44858	7954	8350	11502	1504	1525	1970	439	445	540	220	220	257
0.2 ***	36934	37699	45475	9949	10399	11644	1885	1930	1999	544	550	550	265	265	265
***	37195	39025	55820	10129	10834	14199	1909	1960	2381	544	550	634	265	265	295
0.25 ***	42724	43825	55834	11584	12244	14200	2215	2290	2389	625	634	640	304	304	304
***	43090	45745	65069	11845	12895	16469	2254	2335	2724	634	640	711	304	304	326
0.3 ***	46480	47995	64654	12754	13630	16369	2479	2575	2710	694	709	715	334	334	334
***	46990	50599	72605	13105	14515	18310	2524	2635	2998	700	709	771	334	334	349
0.35 ***	48370	50374	71794	13495	14599	18109	2674	2794	2974	745	760	775	349	355	355
***	49045	53725	78428	13939	15715	19723	2725	2875	3203	754	769	814	355	355	364
0.4 ***	48604	51175	77155	13855	15184	19405	2800	2944	3160	784	799	814	364	364	370
***	49489	55324	82539	14389	16510	20708	2869	3040	3340	790	805	840	364	370	372
0.45 ***	47449	50665	80695	13900	15424	20245	2860	3025	3274	799	814	829	364	370	370
***	48565	55615	84937	14515	16930	21265	2935	3139	3409	805	820	848	370	370	372
0.5 ***	45199	49114	82384	13669	15355	20620	2869	3049	3310	790	814	829	364	364	370
***	46585	54790	85622	14359	16999	21393	2950	3169	3409	805	820	840	364	370	364
0.55 ***	42154	46759	82210	13210	15010	20530	2815	3004	3274	775	790	814	349	355	355
***	43825	53050	84594	13954	16744	21093	2905	3124	3340	784	805	814	355	355	349
0.6 ***	38614	43810	80170	12565	14425	19969	2710	2899	3160	739	754	775	334	334	340
***	40549	50530	81854	13330	16189	20365	2800	3019	3203	745	769	771	334	334	326
0.65 ***	34810	40429	76270	11755	13609	18949	2554	2734	2974	685	700	724	304	304	310
***	36949	47344	77401	12520	15334	19209	2635	2839	2998	694	709	711	304	310	295
0.7 ***	30910	36700	70534	10789	12565	17470	2344	2500	2719	619	634	649	265	274	274
***	33154	43540	71235	11530	14185	17625	2419	2599	2724	625	640	634	265	274	257
0.75 ***	26995	32659	62959	9664	11290	15529	2080	2209	2389	535	544	550	220	229	229
***	29215	39124	63356	10345	12730	15612	2140	2290	2381	535	550	540	220	229	212
0.8 ***	23029	28264	53569	8374	9769	13135	1750	1855	1984	430	439	445	175	175	175
***	25105	34030	53764	8965	10960	13172	1804	1909	1970	430	439	429	175	175	158
0.85 ***	18904	23365	42394	6865	7945	10300	1360	1429	1504	310	310	319	.	.	.
***	20695	28090	42460	7330	8824	10303	1390	1465	1491	310	310	300	.	.	.
0.9 ***	14350	17659	29434	5044	5734	7015	895	925	955	.	.	.	.	.	.
***	15700	20974	29443	5350	6250	7006	910	940	943	.	.	.	.	.	.
0.95 ***	8704	10390	14725	2704	2944	3295	.	.	.	.	.	.	.	.	.
***	9415	11875	14713	2815	3100	3280	.	.	.	.	.	.	.	.	.
0.98 ***	3850	4309	5065	.	.	.	.	.	.	.	.	.	.	.	.
***	4054	4630	5053	.	.	.	.	.	.	.	.	.	.	.	.

TABLE 8: ALPHA= 0.05 POWER= 0.9   EXPECTED ACCRUAL THRU MINIMUM FOLLOW-UP= 6500

PCONT=*** — REQUIRED NUMBER OF PATIENTS

		DEL=.01			DEL=.02			DEL=.05			DEL=.10			DEL=.15		
FACT=		1.0 .75	.50 .25	.00 BIN	1.0 .75	.50 .25	.00 BIN	1.0 .75	.50 .25	.00 BIN	1.0 75	.50 .25	.00 BIN	1.0 .75	.50 .25	.00 BIN
0.01	***	1311	1321	1321	482	482	482	157	157	157	76	76	76	53	53	53
	•••	1311	1321	5053	482	482	1670	157	157	455	76	76	185	53	53	110
0.02	***	2913	2919	2946	937	937	937	254	254	254	108	108	108	70	70	70
	•••	2913	2936	8342	937	937	2484	254	254	581	108	108	215	70	70	123
0.05	***	8818	8916	9208	2513	2540	2578	541	547	547	189	189	189	108	108	108
	•••	8861	9013	17796	2529	2556	4822	541	547	943	189	189	300	108	108	158
0.1	***	19488	19765	21428	5318	5438	5666	1034	1045	1051	319	319	330	173	173	173
	•••	19586	20177	32183	5367	5530	8376	1034	1051	1491	319	319	429	173	173	212
0.15	***	29147	29677	33853	7870	8136	8763	1489	1516	1548	443	443	443	222	222	222
	•••	29326	30528	44858	7983	8373	11502	1506	1532	1970	443	443	540	222	222	257
0.2	***	37006	37824	45478	9998	10453	11646	1890	1938	1993	541	547	557	265	271	271
	•••	37282	39216	55820	10183	10882	14199	1912	1961	2381	547	547	634	265	271	295
0.25	***	42813	43999	55829	11656	12322	14197	2231	2296	2383	628	638	644	303	303	303
	•••	43219	46025	65069	11922	12962	16469	2263	2334	2724	628	638	711	303	303	326
0.3	***	46610	48241	64653	12848	13736	16368	2491	2588	2718	693	709	720	330	336	336
	•••	47162	50965	72605	13206	14613	18310	2540	2643	2998	703	709	771	330	336	349
0.35	***	48543	50688	71797	13612	14733	18107	2686	2806	2968	752	758	774	352	352	362
	•••	49275	54188	78428	14061	15832	19723	2741	2881	3203	752	768	814	352	352	364
0.4	***	48826	51572	77160	14002	15345	19407	2816	2962	3157	785	801	807	362	368	368
	•••	49778	55894	82539	14548	16651	20708	2881	3049	3340	791	801	840	362	368	372
0.45	***	47731	51160	80692	14067	15605	20242	2887	3043	3271	801	817	833	368	368	378
	•••	48933	56268	84937	14695	17089	21265	2962	3147	3409	807	823	848	368	368	372
0.5	***	45553	49687	82382	13856	15556	20616	2897	3066	3309	801	817	833	362	368	368
	•••	47032	55537	85622	14548	17171	21393	2968	3173	3409	807	823	840	362	368	364
0.55	***	42586	47412	82203	13417	15221	20528	2838	3017	3271	774	791	817	352	352	362
	•••	44357	53857	84594	14158	16927	21093	2930	3131	3340	785	801	814	352	352	349
0.6	***	39118	44519	80166	12777	14636	19976	2741	2913	3163	742	758	774	330	336	336
	•••	41150	51371	81854	13547	16374	20365	2822	3027	3203	752	768	771	336	336	326
0.65	***	35371	41172	76272	11965	13817	18952	2578	2751	2978	687	703	720	303	303	313
	•••	37591	48192	77401	12734	15513	19209	2659	2848	2998	693	709	711	303	303	295
0.7	***	31513	37451	70530	10990	12761	17473	2367	2513	2718	622	628	644	265	271	271
	•••	33821	44367	71235	11737	14353	17625	2432	2605	2724	628	638	634	271	271	257
0.75	***	27597	33382	62957	9858	11467	15529	2101	2221	2383	531	547	557	222	222	232
	•••	29872	39888	63356	10535	12881	15612	2156	2296	2381	541	547	540	222	222	212
0.8	***	23589	28913	53571	8536	9917	13141	1766	1863	1977	433	433	443	173	173	173
	•••	25712	34698	53764	9127	11077	13172	1808	1912	1970	433	443	429	173	173	158
0.85	***	19391	23908	42391	6998	8054	10297	1370	1435	1506	313	313	313	.	.	.
	•••	21211	28627	42460	7453	8910	10303	1402	1467	1491	313	313	300	.	.	.
0.9	***	14717	18048	29433	5129	5796	7014	898	931	953	.	.	.	.	.	.
	•••	16082	21331	29443	5432	6293	7006	915	937	943	.	.	.	.	.	.
0.95	***	8899	10573	14727	2735	2968	3293	.	.	.	.	.	.	.	.	.
	•••	9608	12030	14713	2838	3114	3280	.	.	.	.	.	.	.	.	.
0.98	***	3911	4349	5064	.	.	.	.	.	.	.	.	.	.	.	.
	•••	4106	4658	5053	.	.	.	.	.	.	.	.	.	.	.	.

TABLE 8: ALPHA= 0.05 POWER= 0.9   EXPECTED ACCRUAL THRU MINIMUM FOLLOW-UP= 7000

	DEL=.01			DEL=.02			DEL=.05			DEL=.10			DEL=.15		
FACT=	1.0 .75	.50 .25	.00 BIN	1.0 .75	.50 .25	.00 BIN	1.0 .75	.50 .25	.00 BIN	1.0 75	.50 .25	.00 BIN	1.0 .75	.50 .25	.00 BIN

PCONT=•••   REQUIRED NUMBER OF PATIENTS

PCONT	1.0/.75	.50/.25	.00/BIN	1.0/.75	.50/.25	.00/BIN	1.0/.75	.50/.25	.00/BIN	1.0/.75	.50/.25	.00/BIN	1.0/.75	.50/.25	.00/BIN
0.01 •••	1317	1317	1317	484	484	484	162	162	162	81	81	81	57	57	57
•••	1317	1317	5053	484	484	1670	162	162	455	81	81	185	57	57	110
0.02 •••	2910	2927	2945	932	939	939	250	250	250	110	110	110	75	75	75
•••	2916	2934	8342	939	939	2484	250	250	581	110	110	215	75	75	123
0.05 •••	8831	8919	9210	2525	2542	2577	547	547	547	197	197	197	110	110	110
•••	8866	9024	17796	2531	2560	4822	547	547	943	197	197	300	110	110	158
0.1 •••	19506	19804	21431	5331	5447	5664	1037	1044	1055	320	326	326	169	169	169
•••	19611	20224	32183	5377	5535	8376	1037	1044	1491	320	326	429	169	169	212
0.15 •••	29195	29744	33850	7897	8166	8761	1492	1516	1545	442	442	442	221	221	221
•••	29376	30630	44858	8009	8394	11502	1510	1534	1970	442	442	540	221	221	257
0.2 •••	37070	37945	45476	10050	10505	11649	1895	1936	2000	547	547	554	267	267	267
•••	37367	39386	55820	10231	10914	14199	1919	1965	2381	547	547	634	267	267	295
0.25 •••	42904	44181	55836	11719	12395	14197	2234	2304	2385	635	635	641	302	302	309
•••	43341	46281	65069	11992	13025	16469	2269	2339	2724	635	641	711	302	309	326
0.3 •••	46736	48480	64656	12937	13836	16367	2507	2595	2717	694	705	722	326	337	337
•••	47331	51315	72605	13305	14694	18310	2542	2647	2998	705	711	771	337	337	349
0.35 •••	48707	51000	71790	13725	14851	18106	2700	2811	2969	746	764	775	355	355	355
•••	49495	54640	78428	14186	15936	19723	2752	2892	3203	757	764	814	355	355	364
0.4 •••	49046	51969	77151	14134	15481	19401	2840	2969	3161	781	799	810	361	361	372
•••	50061	56425	82539	14687	16776	20708	2899	3056	3340	792	799	840	361	372	372
0.45 •••	48014	51630	80697	14221	15772	20241	2910	3056	3266	799	816	827	372	372	372
•••	49291	56897	84937	14851	17231	21265	2980	3155	3409	810	827	848	372	372	372
0.5 •••	45896	50236	82384	14029	15737	20620	2916	3074	3312	799	816	827	361	372	372
•••	47471	56232	85622	14729	17330	21393	2986	3179	3409	810	827	840	361	372	364
0.55 •••	43009	48031	82209	13602	15411	20532	2864	3039	3277	781	792	810	355	355	361
•••	44864	54605	84594	14355	17091	21093	2945	3144	3340	792	799	814	355	355	349
0.6 •••	39607	45196	80172	12972	14834	19972	2759	2927	3161	746	757	775	326	337	337
•••	41714	52161	81854	13749	16542	20365	2840	3032	3203	746	764	771	337	337	326
0.65 •••	35915	41871	76270	12167	14011	18946	2601	2759	2980	694	705	722	302	309	309
•••	38207	48987	77401	12937	15674	19209	2671	2857	2998	694	711	711	302	309	295
0.7 •••	32082	38155	70530	11187	12944	17470	2385	2525	2717	624	635	641	267	267	274
•••	34451	45137	71235	11922	14501	17625	2455	2612	2724	624	641	634	267	274	257
0.75 •••	28162	34049	62959	10032	11631	15534	2111	2234	2385	536	547	554	221	221	232
•••	30490	40605	63356	10715	13014	15612	2164	2304	2381	536	547	540	221	221	212
0.8 •••	24126	29516	53572	8691	10056	13136	1779	1866	1982	431	442	442	169	169	169
•••	26279	35326	53764	9280	11187	13172	1825	1919	1970	431	442	429	169	169	158
0.85 •••	19850	24406	42390	7116	8160	10301	1376	1440	1510	309	309	320	.	.	.
•••	21694	29125	42460	7571	8982	10303	1405	1464	1491	309	309	300	.	.	.
0.9 •••	15072	18415	29440	5209	5867	7011	904	932	956	.	.	.	.	.	.
•••	16444	21670	29443	5500	6340	7006	915	939	943	.	.	.	.	.	.
0.95 •••	9087	10750	14729	2770	2986	3295	.	.	.	.	.	.	.	.	.
•••	9787	12167	14713	2864	3126	3280	.	.	.	.	.	.	.	.	.
0.98 •••	3960	4386	5069	.	.	.	.	.	.	.	.	.	.	.	.
•••	4152	4684	5053	.	.	.	.	.	.	.	.	.	.	.	.

839

TABLE 8: ALPHA= 0.05 POWER= 0.9   EXPECTED ACCRUAL THRU MINIMUM FOLLOW-UP= 7500

	DEL=.01			DEL=.02			DEL=.05			DEL=.10			DEL=.15		
FACT=	1.0 .75	.50 .25	.00 BIN	1.0 .75	.50 .25	.00 BIN	1.0 .75	.50 .25	.00 BIN	1.0 75	.50 .25	.00 BIN	1.0 .75	.50 .25	.00 BIN
PCONT=***						REQUIRED NUMBER OF PATIENTS									
0.01 ***	1318	1318	1318	481	481	481	155	155	155	80	80	80	61	61	61
•••	1318	1318	5053	481	481	1670	155	155	455	80	80	185	61	61	110
0.02 ***	2911	2930	2949	931	943	943	249	256	256	106	106	106	68	68	68
•••	2918	2937	8342	931	943	2484	249	256	581	106	106	215	68	68	123
0.05 ***	8836	8937	9200	2525	2543	2581	537	549	549	193	193	193	118	118	118
•••	8874	9031	17796	2536	2562	4822	549	549	943	193	193	300	118	118	158
0.1 ***	19531	19843	21425	5337	5461	5668	1036	1043	1055	324	324	324	174	174	174
•••	19643	20274	32183	5393	5543	8376	1036	1043	1491	324	324	429	174	174	212
0.15 ***	29236	29825	33849	7925	8187	8761	1493	1524	1543	443	443	443	218	218	218
•••	29431	30736	44858	8037	8412	11502	1505	1531	1970	443	443	540	218	218	257
0.2 ***	37130	38068	45481	10093	10550	11649	1906	1943	1993	549	549	549	268	268	268
•••	37449	39556	55820	10280	10955	14199	1925	1962	2381	549	549	634	268	268	295
0.25 ***	42999	44349	55831	11787	12462	14199	2243	2311	2386	631	631	643	305	305	305
•••	43468	46524	65069	12061	13081	16469	2274	2349	2724	631	643	711	305	305	326
0.3 ***	46868	48706	64655	13018	13925	16362	2518	2600	2712	699	706	718	331	331	331
•••	47499	51650	72605	13393	14768	18310	2555	2656	2998	706	718	771	331	331	349
0.35 ***	48875	51305	71787	13831	14968	18106	2712	2825	2975	755	762	774	350	350	361
•••	49718	55055	78428	14300	16036	19723	2768	2893	3203	755	762	814	350	361	364
0.4 ***	49268	52343	77161	14262	15624	19400	2855	2986	3155	781	800	811	361	368	368
•••	50349	56937	82539	14825	16899	20708	2911	3061	3340	793	800	840	368	368	372
0.45 ***	48293	52093	80693	14375	15924	20243	2918	3068	3268	800	818	830	368	368	368
•••	49643	57481	84937	15012	17368	21265	2993	3162	3409	811	818	848	368	368	372
0.5 ***	46250	50768	82381	14199	15905	20618	2930	3087	3305	800	811	830	361	368	368
•••	47893	56881	85622	14900	17480	21393	3005	3193	3409	811	818	840	368	368	364
0.55 ***	43418	48631	82205	13786	15593	20525	2881	3050	3275	781	800	811	350	350	361
•••	45361	55318	84594	14536	17243	21093	2956	3155	3340	793	800	814	350	361	349
0.6 ***	40081	45837	80168	13156	15012	19974	2780	2937	3162	743	762	774	331	331	343
•••	42268	52899	81854	13936	16693	20365	2855	3043	3203	755	762	771	331	331	326
0.65 ***	36443	42549	76268	12350	14187	18950	2618	2768	2975	699	706	718	305	305	312
•••	38799	49730	77401	13118	15818	19209	2693	2862	2998	699	718	711	305	305	295
0.7 ***	32630	38818	70531	11356	13111	17468	2405	2536	2712	624	631	643	268	268	275
•••	35056	45849	71235	12099	14637	17625	2468	2618	2724	631	643	634	268	268	257
0.75 ***	28700	34681	62956	10193	11787	15530	2124	2236	2386	537	549	556	218	230	230
•••	31074	41262	63356	10868	13130	15612	2180	2311	2381	537	549	540	230	230	212
0.8 ***	24631	30087	53574	8836	10175	13137	1786	1880	1981	436	436	443	174	174	174
•••	26818	35900	53764	9418	11281	13172	1831	1925	1970	436	443	429	174	174	158
0.85 ***	20281	24886	42387	7224	8255	10299	1381	1437	1505	312	312	312	.	.	.
•••	22149	29581	42460	7674	9050	10303	1411	1468	1491	312	312	300	.	.	.
0.9 ***	15387	18743	29431	5281	5918	7018	905	931	961	.	.	.	.	.	.
•••	16775	21968	29443	5562	6380	7006	912	943	943	.	.	.	.	.	.
0.95 ***	9256	10899	14724	2787	3005	3293	.	.	.	.	.	.	.	.	.
•••	9961	12286	14713	2893	3136	3280	.	.	.	.	.	.	.	.	.
0.98 ***	4006	4418	5068	.	.	.	.	.	.	.	.	.	.	.	.
•••	4193	4700	5053	.	.	.	.	.	.	.	.	.	.	.	.

TABLE 8: ALPHA= 0.05 POWER= 0.9   EXPECTED ACCRUAL THRU MINIMUM FOLLOW-UP= 8000

		DEL=.01			DEL=.02			DEL=.05			DEL=.10			DEL=.15		
FACT=		1.0 .75	.50 .25	.00 BIN	1.0 .75	.50 .25	.00 BIN	1.0 .75	.50 .25	.00 BIN	1.0 75	.50 .25	.00 BIN	1.0 .75	.50 .25	.00 BIN
PCONT=***					REQUIRED NUMBER OF PATIENTS											
0.01	***	1313	1313	1325	485	485	485	153	153	153	85	85	85	53	53	53
	***	1313	1313	5053	485	485	1670	153	153	455	85	85	185	53	53	110
0.02	***	2913	2925	2945	933	933	945	253	253	253	113	113	113	73	73	73
	***	2925	2933	8342	933	945	2484	253	253	581	113	113	215	73	73	123
0.05	***	8845	8945	9205	2525	2545	2573	545	545	545	193	193	193	113	113	113
	***	8885	9045	17796	2533	2565	4822	545	545	943	193	193	300	113	113	158
0.1	***	19553	19885	21425	5353	5465	5665	1033	1045	1053	325	325	325	173	173	173
	***	19673	20313	32183	5405	5553	8376	1045	1045	1491	325	325	429	173	173	212
0.15	***	29273	29893	33853	7953	8213	8765	1505	1525	1545	445	445	445	225	225	225
	***	29493	30825	44858	8065	8433	11502	1513	1533	1970	445	445	540	225	225	257
0.2	***	37193	38185	45473	10133	10585	11645	1905	1945	1993	545	545	553	265	265	265
	***	37533	39713	55820	10325	10985	14199	1925	1965	2381	545	553	634	265	265	295
0.25	***	43093	44525	55833	11845	12525	14205	2253	2313	2385	633	633	645	305	305	305
	***	43585	46765	65069	12125	13133	16469	2273	2345	2724	633	645	711	305	305	326
0.3	***	46993	48945	64653	13105	14013	16365	2525	2605	2713	705	705	713	333	333	333
	***	47665	51965	72605	13473	14845	18310	2565	2653	2998	705	713	771	333	333	349
0.35	***	49045	51605	71793	13933	15073	18105	2725	2833	2973	753	765	773	353	353	353
	***	49933	55453	78428	14405	16125	19723	2773	2893	3203	753	765	814	353	353	364
0.4	***	49485	52713	77153	14393	15745	19405	2865	2993	3153	785	793	813	365	365	373
	***	50625	57413	82539	14953	17005	20708	2925	3065	3340	793	805	840	365	365	372
0.45	***	48565	52533	80693	14513	16065	20245	2933	3085	3273	805	813	833	365	373	373
	***	49993	58033	84937	15153	17485	21265	3005	3165	3409	813	825	848	373	373	372
0.5	***	46585	51285	82385	14353	16053	20613	2953	3105	3305	805	813	833	365	365	373
	***	48313	57505	85622	15065	17605	21393	3025	3193	3409	805	825	840	365	365	364
0.55	***	43825	49205	82205	13953	15753	20525	2905	3065	3273	785	793	813	353	353	353
	***	45845	55985	84594	14705	17385	21093	2973	3153	3340	793	805	814	353	353	349
0.6	***	40545	46453	80165	13333	15185	19973	2793	2953	3165	745	765	773	333	333	333
	***	42805	53585	81854	14105	16833	20365	2873	3045	3203	753	765	771	333	333	326
0.65	***	36945	43173	76273	12525	14353	18953	2633	2785	2973	693	705	725	305	305	305
	***	39365	50413	77401	13293	15953	19209	2705	2873	2998	705	713	711	305	305	295
0.7	***	33153	39445	70533	11525	13265	17473	2413	2545	2713	625	633	645	265	273	273
	***	35625	46525	71235	12265	14765	17625	2473	2625	2724	633	645	634	265	273	257
0.75	***	29213	35273	62953	10345	11925	15533	2145	2245	2385	533	545	553	225	225	225
	***	31625	41885	63356	11025	13233	15612	2193	2313	2381	545	553	540	225	225	212
0.8	***	25105	30625	53573	8965	10293	13133	1805	1885	1985	433	433	445	173	173	173
	***	27325	36433	53764	9533	11373	13172	1833	1925	1970	433	445	429	173	173	158
0.85	***	20693	25325	42393	7325	8333	10293	1393	1445	1505	313	313	313	.	.	.
	***	22573	30005	42460	7773	9113	10303	1413	1473	1491	313	313	300	.	.	.
0.9	***	15705	19053	29433	5345	5965	7013	905	933	953	.	.	.	.	.	.
	***	17085	22245	29443	5625	6405	7006	925	945	943	.	.	.	.	.	.
0.95	***	9413	11045	14725	2813	3013	3293	.	.	.	.	.	.	.	.	.
	***	10113	12393	14713	2905	3145	3280	.	.	.	.	.	.	.	.	.
0.98	***	4053	4453	5065	.	.	.	.	.	.	.	.	.	.	.	.
	***	4233	4725	5053	.	.	.	.	.	.	.	.	.	.	.	.

TABLE 8: ALPHA= 0.05 POWER= 0.9   EXPECTED ACCRUAL THRU MINIMUM FOLLOW-UP= 8500

		DEL=.01			DEL=.02			DEL=.05			DEL=.10			DEL=.15		
FACT=		1.0 .75	.50 .25	.00 BIN	1.0 .75	.50 .25	.00 BIN	1.0 .75	.50 .25	.00 BIN	1.0 75	.50 .25	.00 BIN	1.0 .75	.50 .25	.00 BIN
PCONT=***					REQUIRED NUMBER OF PATIENTS											
0.01	***	1310	1323	1323	481	481	481	155	155	155	78	78	78	56	56	56
	•••	1310	1323	5053	481	481	1670	155	155	455	78	78	185	56	56	110
0.02	***	2917	2925	2946	941	941	941	248	248	248	112	112	112	70	70	70
	•••	2917	2938	8342	941	941	2484	248	248	581	112	112	215	70	70	123
0.05	***	8854	8952	9207	2521	2556	2577	545	545	545	197	197	197	112	112	112
	•••	8896	9045	17796	2535	2564	4822	545	545	943	197	197	300	112	112	158
0.1	***	19577	19917	21426	5361	5475	5666	1034	1047	1055	325	325	325	176	176	176
	•••	19705	20350	32183	5411	5560	8376	1047	1047	1491	325	325	429	176	176	212
0.15	***	29310	29968	33857	7975	8238	8761	1501	1523	1544	439	439	439	226	226	226
	•••	29543	30911	44858	8089	8450	11502	1515	1536	1970	439	439	540	226	226	257
0.2	***	37257	38298	45481	10171	10631	11651	1918	1948	1990	545	545	558	269	269	269
	•••	37618	39871	55820	10363	11013	14199	1926	1969	2381	545	545	634	269	269	295
0.25	***	43186	44681	55838	11906	12586	14201	2258	2309	2386	630	643	643	303	303	303
	•••	43704	46990	65069	12182	13181	16469	2288	2351	2724	630	643	711	303	303	326
0.3	***	47117	49165	64656	13181	14095	16368	2535	2606	2713	707	707	715	333	333	333
	•••	47826	52268	72605	13550	14910	18310	2564	2662	2998	707	715	771	333	333	349
0.35	***	49221	51885	71788	14031	15178	18111	2734	2840	2968	758	758	771	354	354	354
	•••	50156	55838	78428	14506	16206	19723	2790	2904	3203	758	771	814	354	354	364
0.4	***	49696	53075	77151	14506	15866	19407	2883	3002	3159	792	800	813	367	367	367
	•••	50900	57878	82539	15072	17099	20708	2938	3074	3340	792	800	840	367	367	372
0.45	***	48846	52961	80692	14647	16198	20244	2960	3087	3265	800	813	835	367	375	375
	•••	50334	58571	84937	15293	17601	21265	3023	3172	3409	813	821	848	367	375	372
0.5	***	46926	51779	82379	14498	16206	20618	2968	3116	3308	800	813	835	367	367	367
	•••	48719	58090	85622	15208	17728	21393	3031	3201	3409	813	821	840	367	367	364
0.55	***	44227	49752	82209	14103	15909	20533	2917	3074	3278	779	800	813	354	354	354
	•••	46310	56603	84594	14860	17516	21093	2989	3159	3340	792	800	814	354	354	349
0.6	***	40997	47040	80169	13500	15335	19968	2811	2968	3159	750	758	779	333	333	333
	•••	43313	54236	81854	14273	16950	20365	2883	3053	3203	758	771	771	333	333	326
0.65	***	37435	43781	76272	12679	14506	18948	2649	2790	2981	694	707	715	303	303	311
	•••	39900	51070	77401	13457	16079	19209	2713	2875	2998	707	715	711	303	311	295
0.7	***	33653	40041	70535	11680	13415	17473	2428	2556	2713	630	630	643	269	269	269
	•••	36173	47146	71235	12416	14868	17625	2492	2628	2724	630	643	634	269	269	257
0.75	***	29713	35841	62956	10490	12055	15526	2152	2258	2386	537	545	558	226	226	226
	•••	32157	42463	63356	11162	13338	15612	2203	2322	2381	545	545	540	226	226	212
0.8	***	25570	31124	53564	9088	10397	13138	1812	1884	1982	431	439	439	176	176	176
	•••	27801	36938	53764	9653	11446	13172	1841	1926	1970	439	439	429	176	176	158
0.85	***	21086	25740	42386	7422	8421	10299	1395	1451	1501	311	311	311	.	.	.
	•••	22977	30401	42460	7855	9173	10303	1416	1472	1491	311	311	300	.	.	.
0.9	***	15994	19351	29437	5411	6020	7018	906	928	962	.	.	.	.	.	.
	•••	17388	22510	29443	5680	6445	7006	920	949	943	.	.	.	.	.	.
0.95	***	9568	11170	14732	2832	3031	3300	.	.	.	.	.	.	.	.	.
	•••	10256	12501	14713	2925	3151	3280	.	.	.	.	.	.	.	.	.
0.98	***	4094	4490	5071	.	.	.	.	.	.	.	.	.	.	.	.
	•••	4277	4745	5053	.	.	.	.	.	.	.	.	.	.	.	.

TABLE 8: ALPHA= 0.05 POWER= 0.9   EXPECTED ACCRUAL THRU MINIMUM FOLLOW-UP= 9000

		DEL=.01			DEL=.02			DEL=.05			DEL=.10			DEL=.15	
FACT=	1.0 .75	.50 .25	.00 BIN	1.0 .75	.50 .25	.00 BIN	1.0 .75	.50 .25	.00 BIN	1.0 75	.50 .25	.00 BIN	1.0 .75	.50 .25	.00 BIN
PCONT=•••						REQUIRED NUMBER OF PATIENTS									
0.01 ***	1311	1320	1320	479	479	487	164	164	164	82	82	82	60	60	60
•••	1320	1320	5053	479	479	1670	164	164	455	82	82	185	60	60	110
0.02 ***	2917	2931	2954	937	937	937	254	254	254	105	105	105	74	74	74
•••	2917	2940	8342	937	937	2484	254	254	581	105	105	215	74	74	123
0.05 ***	8857	8961	9209	2526	2549	2580	546	546	546	195	195	195	119	119	119
•••	8902	9051	17796	2535	2557	4822	546	546	943	195	195	300	119	119	158
0.1 ***	19604	19950	21426	5370	5482	5662	1041	1050	1050	321	321	321	172	172	172
•••	19725	20391	32183	5429	5564	8376	1041	1050	1491	321	321	429	172	172	212
0.15 ***	29355	30044	33855	7994	8264	8767	1500	1522	1545	442	442	442	217	217	217
•••	29594	30997	44858	8106	8466	11502	1514	1536	1970	442	442	540	217	217	257
0.2 ***	37320	38414	45479	10207	10657	11647	1919	1950	1995	546	546	555	262	262	262
•••	37702	40011	55820	10401	11040	14199	1927	1972	2381	546	546	634	262	262	295
0.25 ***	43282	44849	55837	11954	12637	14204	2265	2324	2391	636	636	645	307	307	307
•••	43822	47197	65069	12246	13222	16469	2287	2346	2724	636	645	711	307	307	326
0.3 ***	47242	49380	64657	13259	14159	16364	2535	2616	2715	704	712	712	330	330	330
•••	47999	52552	72605	13627	14969	18310	2580	2661	2998	704	712	771	330	330	349
0.35 ***	49380	52170	71790	14122	15270	18105	2751	2850	2976	757	757	771	352	352	352
•••	50370	56197	78428	14595	16282	19723	2796	2909	3203	757	771	814	352	352	364
0.4 ***	49920	53430	77159	14617	15981	19401	2895	3007	3156	794	802	816	366	366	366
•••	51180	58312	82539	15180	17182	20708	2940	3075	3340	794	802	840	366	366	372
0.45 ***	49110	53376	80691	14775	16319	20242	2962	3097	3269	802	816	825	366	375	375
•••	50662	59069	84937	15419	17700	21265	3030	3179	3409	816	825	848	366	375	372
0.5 ***	47256	52260	82379	14640	16341	20616	2985	3120	3314	802	816	825	366	366	366
•••	49110	58650	85622	15351	17835	21393	3052	3210	3409	816	825	840	366	366	364
0.55 ***	44610	50280	82207	14257	16049	20526	2931	3075	3269	780	802	816	352	352	352
•••	46761	57201	84594	15014	17624	21093	3007	3165	3340	794	802	814	352	352	349
0.6 ***	41437	47602	80174	13650	15486	19972	2827	2976	3165	749	757	771	330	330	344
•••	43814	54861	81854	14429	17070	20365	2895	3066	3203	757	771	771	330	330	326
0.65 ***	37905	44362	76267	12840	14640	18951	2670	2805	2976	704	712	726	307	307	307
•••	40425	51689	77401	13605	16192	19209	2729	2886	2998	704	712	711	307	307	295
0.7 ***	34139	40619	70530	11827	13537	17466	2445	2557	2715	622	636	645	262	276	276
•••	36704	47737	71235	12561	14977	17625	2504	2639	2724	636	645	634	276	276	257
0.75 ***	30187	36375	62961	10626	12165	15531	2166	2265	2391	546	546	555	231	231	231
•••	32662	43004	63356	11287	13425	15612	2211	2324	2381	546	546	540	231	231	212
0.8 ***	26002	31605	53565	9209	10500	13132	1815	1896	1986	434	442	442	172	172	172
•••	28266	37410	53764	9771	11526	13172	1851	1927	1970	434	442	429	172	172	158
0.85 ***	21457	26129	42396	7507	8489	10297	1401	1446	1500	307	307	321	.	.	.
•••	23361	30772	42460	7949	9217	10303	1424	1477	1491	307	321	300	.	.	.
0.9 ***	16260	19626	29436	5460	6059	7012	915	937	960	.	.	.	.	.	.
•••	17655	22740	29443	5730	6464	7006	929	951	943	.	.	.	.	.	.
0.95 ***	9704	11301	14730	2850	3044	3300	.	.	.	.	.	.	.	.	.
•••	10387	12592	14713	2940	3156	3280	.	.	.	.	.	.	.	.	.
0.98 ***	4132	4506	5069	.	.	.	.	.	.	.	.	.	.	.	.
•••	4304	4754	5053	.	.	.	.	.	.	.	.	.	.	.	.

TABLE 8: ALPHA= 0.05 POWER= 0.9   EXPECTED ACCRUAL THRU MINIMUM FOLLOW-UP= 9500

	DEL=.01			DEL=.02			DEL=.05			DEL=.10			DEL=.15		
FACT=	1.0 / .75	.50 / .25	.00 / BIN	1.0 / .75	.50 / .25	.00 / BIN	1.0 / .75	.50 / .25	.00 / BIN	1.0 / 75	.50 / .25	.00 / BIN	1.0 / .75	.50 / .25	.00 / BIN

PCONT=*** REQUIRED NUMBER OF PATIENTS

PCONT	DEL=.01 1.0/.75	.50/.25	.00/BIN	DEL=.02 1.0/.75	.50/.25	.00/BIN	DEL=.05 1.0/.75	.50/.25	.00/BIN	DEL=.10 1.0/75	.50/.25	.00/BIN	DEL=.15 1.0/.75	.50/.25	.00/BIN
0.01 ***	1313	1313	1322	481	481	481	158	158	158	78	78	78	54	54	54
•••	1313	1322	5053	481	481	1670	158	158	455	78	78	185	54	54	110
0.02 ***	2913	2928	2951	933	942	942	253	253	253	110	110	110	78	78	78
•••	2928	2937	8342	942	942	2484	253	253	581	110	110	215	78	78	123
0.05 ***	8865	8969	9207	2533	2548	2580	538	538	553	196	196	196	110	110	110
•••	8913	9055	17796	2548	2571	4822	538	538	943	196	196	300	110	110	158
0.1 ***	19624	19980	21429	5383	5493	5668	1037	1051	1051	324	324	324	173	173	173
•••	19752	20417	32183	5430	5564	8376	1037	1051	1491	324	324	429	173	173	212
0.15 ***	29394	30107	33850	8019	8271	8770	1503	1526	1550	443	443	443	220	220	220
•••	29646	31080	44858	8129	8470	11502	1512	1535	1970	443	443	540	220	220	257
0.2 ***	37389	38514	45473	10243	10694	11644	1915	1954	2001	538	553	553	268	268	268
•••	37778	40153	55820	10433	11074	14199	1939	1978	2381	553	553	634	268	268	295
0.25 ***	43374	45013	55828	12009	12689	14194	2272	2319	2390	633	633	648	300	300	300
•••	43944	47397	65069	12294	13259	16469	2295	2358	2724	633	648	711	300	300	326
0.3 ***	47373	49596	64654	13330	14233	16370	2548	2619	2714	704	704	719	339	339	339
•••	48157	52826	72605	13695	15016	18310	2580	2666	2998	704	719	771	339	339	349
0.35 ***	49549	52446	71788	14209	15349	18104	2761	2856	2975	752	766	775	348	348	363
•••	50579	56540	78428	14693	16355	19723	2809	2904	3203	766	766	814	348	363	364
0.4 ***	50128	53762	77155	14717	16070	19410	2904	3008	3165	790	799	814	363	372	372
•••	51449	58725	82539	15287	17273	20708	2951	3079	3340	799	799	840	363	372	372
0.45 ***	49383	53776	80694	14898	16441	20241	2975	3103	3269	814	814	823	372	372	372
•••	50998	59548	84937	15539	17795	21265	3046	3189	3409	814	823	848	372	372	372
0.5 ***	47578	52717	82380	14779	16465	20621	2999	3127	3308	799	814	823	363	372	372
•••	49501	59177	85622	15491	17938	21393	3055	3213	3409	814	823	840	363	372	364
0.55 ***	44989	50784	82205	14399	16189	20526	2951	3094	3269	790	799	814	348	348	363
•••	47198	57766	84594	15159	17733	21093	3008	3174	3340	790	799	814	348	363	349
0.6 ***	41854	48133	80172	13790	15619	19965	2842	2984	3165	752	766	775	339	339	339
•••	44285	55439	81854	14565	17178	20365	2913	3070	3203	752	766	771	339	339	326
0.65 ***	38363	44903	76268	12983	14779	18944	2675	2809	2975	695	704	719	300	300	315
•••	40928	52265	77401	13743	16299	19209	2738	2889	2998	704	719	711	300	300	295
0.7 ***	34610	41150	70529	11962	13672	17472	2453	2571	2714	624	633	648	268	268	268
•••	37208	48299	71235	12698	15073	17625	2509	2643	2724	633	648	634	268	268	257
0.75 ***	30644	36890	62953	10750	12285	15524	2168	2263	2381	538	553	553	220	229	229
•••	33147	43516	63356	11415	13505	15612	2215	2319	2381	538	553	540	229	229	212
0.8 ***	26416	32054	53572	9316	10584	13140	1820	1892	1978	434	443	443	173	173	173
•••	28696	37849	53764	9872	11582	13172	1859	1939	1970	434	443	429	173	173	158
0.85 ***	21809	26497	42385	7592	8556	10299	1408	1455	1503	315	315	315	.	.	.
•••	23733	31119	42460	8019	9269	10303	1431	1479	1491	315	315	300	.	.	.
0.9 ***	16522	19885	29433	5516	6095	7013	918	933	956	.	.	.	.	.	.
•••	17923	22973	29443	5778	6490	7006	933	942	943	.	.	.	.	.	.
0.95 ***	9839	11406	14731	2865	3055	3293	.	.	.	.	.	.	.	.	.
•••	10513	12674	14713	2960	3165	3280	.	.	.	.	.	.	.	.	.
0.98 ***	4163	4528	5065	.	.	.	.	.	.	.	.	.	.	.	.
•••	4338	4765	5053	.	.	.	.	.	.	.	.	.	.	.	.

TABLE 8: ALPHA= 0.05 POWER= 0.9   EXPECTED ACCRUAL THRU MINIMUM FOLLOW-UP= 10000

		DEL=.01			DEL=.02			DEL=.05			DEL=.10			DEL=.15		
FACT=		1.0 .75	.50 .25	.00 BIN	1.0 .75	.50 .25	.00 BIN	1.0 .75	.50 .25	.00 BIN	1.0 75	.50 .25	.00 BIN	1.0 .75	.50 .25	.00 BIN
PCONT=***					REQUIRED NUMBER OF PATIENTS											
0.01	***	1316	1316	1316	482	482	482	157	157	157	82	82	82	57	57	57
	***	1316	1316	5053	482	482	1670	157	157	455	82	82	185	57	57	110
0.02	***	2916	2932	2941	932	941	941	257	257	257	107	107	107	66	66	66
	***	2916	2941	8342	941	941	2484	257	257	581	107	107	215	66	66	123
0.05	***	8882	8982	9207	2532	2557	2582	541	541	541	191	191	191	116	116	116
	***	8916	9066	17796	2541	2566	4822	541	541	943	191	191	300	116	116	158
0.1	***	19641	20016	21432	5391	5491	5666	1041	1041	1057	316	316	332	166	166	166
	***	19782	20457	32183	5441	5566	8376	1041	1057	1491	316	332	429	166	166	212
0.15	***	29432	30166	33857	8041	8291	8766	1507	1532	1541	441	441	441	216	216	216
	***	29691	31157	44858	8141	8491	11502	1516	1532	1970	441	441	540	216	216	257
0.2	***	37457	38632	45482	10282	10732	11641	1916	1957	1991	541	541	557	266	266	266
	***	37866	40282	55820	10466	11091	14199	1941	1966	2381	541	557	634	266	266	295
0.25	***	43466	45157	55832	12057	12732	14207	2266	2316	2382	632	641	641	307	307	307
	***	44057	47591	65069	12341	13291	16469	2291	2357	2724	632	641	711	307	307	326
0.3	***	47507	49807	64657	13391	14291	16366	2557	2632	2716	707	707	716	332	332	332
	***	48316	53091	72605	13766	15066	18310	2591	2666	2998	707	716	771	332	332	349
0.35	***	49716	52716	71791	14291	15432	18107	2766	2857	2966	757	766	766	357	357	357
	***	50791	56882	78428	14782	16416	19723	2807	2907	3203	757	766	814	357	357	364
0.4	***	50341	54091	77157	14816	16182	19407	2916	3016	3157	791	807	807	366	366	366
	***	51707	59116	82539	15391	17341	20708	2966	3082	3340	791	807	840	366	366	372
0.45	***	49641	54166	80691	15007	16541	20241	2991	3116	3266	807	816	832	366	366	366
	***	51316	60007	84937	15657	17882	21265	3041	3191	3409	816	832	848	366	366	372
0.5	***	47891	53157	82382	14907	16582	20616	3007	3141	3307	807	816	832	366	366	366
	***	49866	59682	85622	15616	18041	21393	3066	3216	3409	816	832	840	366	366	364
0.55	***	45357	51266	82207	14532	16307	20532	2957	3091	3266	791	807	816	357	357	357
	***	47616	58307	84594	15291	17832	21093	3032	3182	3340	791	807	814	357	357	349
0.6	***	42266	48657	80166	13932	15741	19966	2857	2991	3157	757	766	782	332	332	341
	***	44757	55991	81854	14707	17282	20365	2916	3066	3203	757	766	771	332	332	326
0.65	***	38807	45432	76266	13116	14907	18957	2691	2816	2982	707	707	716	307	307	307
	***	41407	52816	77401	13882	16391	19209	2757	2891	2998	707	716	711	307	307	295
0.7	***	35057	41666	70532	12107	13782	17466	2466	2582	2716	632	641	641	266	266	266
	***	37691	48832	71235	12832	15166	17625	2516	2641	2724	632	641	634	266	266	257
0.75	***	31066	37382	62957	10866	12382	15532	2182	2266	2382	541	541	557	232	232	232
	***	33607	44007	63356	11532	13591	15612	2216	2332	2381	541	557	540	232	232	212
0.8	***	26816	32491	53566	9416	10682	13141	1832	1891	1982	432	441	441	166	166	166
	***	29116	38266	53764	9966	11641	13172	1866	1941	1970	441	441	429	166	166	158
0.85	***	22141	26857	42391	7682	8616	10291	1416	1457	1507	307	316	316	.	.	.
	***	24082	31432	42460	8091	9307	10303	1432	1482	1491	316	316	300	.	.	.
0.9	***	16782	20116	29432	5566	6132	7016	916	941	957	.	.	.	.	.	.
	***	18182	23166	29443	5816	6516	7006	932	941	943	.	.	.	.	.	.
0.95	***	9957	11516	14732	2891	3066	3291	.	.	.	.	.	.	.	.	.
	***	10632	12757	14713	2966	3166	3280	.	.	.	.	.	.	.	.	.
0.98	***	4191	4557	5066	.	.	.	.	.	.	.	.	.	.	.	.
	***	4366	4782	5053	.	.	.	.	.	.	.	.	.	.	.	.

TABLE 8: ALPHA= 0.05 POWER= 0.9   EXPECTED ACCRUAL THRU MINIMUM FOLLOW-UP= 11000

	DEL=.01			DEL=.02			DEL=.05			DEL=.10			DEL=.15		
FACT=	1.0 .75	.50 .25	.00 BIN	1.0 .75	.50 .25	.00 BIN	1.0 .75	.50 .25	.00 BIN	1.0 75	.50 .25	.00 BIN	1.0 .75	.50 .25	.00 BIN

PCONT=•••   REQUIRED NUMBER OF PATIENTS

PCONT	DEL=.01 1.0/.75	.50/.25	.00/BIN	DEL=.02 1.0/.75	.50/.25	.00/BIN	DEL=.05 1.0/.75	.50/.25	.00/BIN	DEL=.10 1.0/.75	.50/.25	.00/BIN	DEL=.15 1.0/.75	.50/.25	.00/BIN
0.01 ***	1310	1310	1327	485	485	485	155	155	155	73	73	73	62	62	62
•••	1310	1310	5053	485	485	1670	155	155	455	73	73	185	62	62	110
0.02 ***	2922	2933	2950	942	942	942	255	255	255	117	117	117	73	73	73
•••	2922	2933	8342	942	942	2484	255	255	581	117	117	215	73	73	123
0.05 ***	8890	8983	9203	2537	2548	2575	540	540	540	200	200	200	117	117	117
•••	8928	9082	17796	2548	2565	4822	540	540	943	200	200	300	117	117	158
0.1 ***	19680	20065	21430	5408	5507	5672	1035	1052	1052	320	320	320	172	172	172
•••	19835	20505	32183	5452	5573	8376	1035	1052	1491	320	320	429	172	172	212
0.15 ***	29515	30295	33843	8075	8323	8763	1503	1530	1547	447	447	447	227	227	227
•••	29800	31285	44858	8185	8505	11502	1520	1530	1970	447	447	540	227	227	257
0.2 ***	37572	38837	45475	10347	10787	11650	1932	1960	1998	540	557	557	265	265	265
•••	38023	40525	55820	10540	11128	14199	1943	1970	2381	540	557	634	265	265	295
0.25 ***	43650	45465	55832	12162	12822	14197	2273	2328	2383	640	640	640	310	310	310
•••	44293	47950	65069	12437	13355	16469	2300	2355	2724	640	640	711	310	310	326
0.3 ***	47747	50205	64660	13520	14417	16370	2565	2630	2713	705	705	722	337	337	337
•••	48638	53577	72605	13895	15160	18310	2592	2675	2998	705	705	771	337	337	349
0.35 ***	50040	53230	71793	14455	15583	18102	2785	2867	2977	760	760	777	348	348	365
•••	51195	57493	78428	14940	16535	19723	2823	2922	3203	760	777	814	348	348	364
0.4 ***	50772	54732	77155	15005	16353	19405	2933	3032	3153	788	805	815	365	365	365
•••	52213	59858	82539	15572	17480	20708	2977	3087	3340	805	805	840	365	365	372
0.45 ***	50167	54908	80692	15225	16755	20247	3015	3125	3263	805	815	832	365	375	375
•••	51938	60848	84937	15875	18030	21265	3070	3197	3409	815	832	848	375	375	372
0.5 ***	48517	54000	82380	15132	16810	20615	3032	3153	3307	805	815	832	365	365	365
•••	50590	60617	85622	15847	18212	21393	3087	3225	3409	815	832	840	365	365	364
0.55 ***	46070	52185	82205	14785	16545	20522	2988	3115	3280	788	805	815	348	348	365
•••	48435	59297	84594	15528	18003	21093	3043	3180	3340	788	805	814	348	365	349
0.6 ***	43055	49628	80170	14197	15985	19972	2878	3005	3153	750	760	777	337	337	337
•••	45630	57015	81854	14950	17453	20365	2933	3070	3203	760	777	771	337	337	326
0.65 ***	39635	46427	76275	13372	15132	18955	2713	2823	2977	705	705	722	310	310	310
•••	42330	53835	77401	14125	16562	19209	2768	2895	2998	705	722	711	310	310	295
0.7 ***	35905	42643	70528	12338	13988	17470	2482	2592	2713	623	640	650	265	265	265
•••	38600	49810	71235	13053	15325	17625	2537	2647	2724	640	640	634	265	265	257
0.75 ***	31890	38287	62955	11090	12575	15528	2190	2273	2383	540	540	557	227	227	227
•••	34475	44887	63356	11733	13730	15612	2235	2328	2381	540	557	540	227	227	212
0.8 ***	27562	33293	53577	9605	10825	13135	1850	1905	1987	430	447	447	172	172	172
•••	29900	39030	53764	10138	11750	13172	1877	1943	1970	430	447	429	172	172	158
0.85 ***	22777	27507	42385	7817	8725	10303	1420	1465	1503	310	310	320	.	.	.
•••	24730	32028	42460	8230	9385	10303	1437	1475	1491	310	310	300	.	.	.
0.9 ***	17233	20577	29432	5655	6195	7020	925	942	953	.	.	.	.	.	.
•••	18635	23547	29443	5903	6552	7006	925	953	943	.	.	.	.	.	.
0.95 ***	10182	11705	14730	2922	3087	3290	.	.	.	.	.	.	.	.	.
•••	10853	12888	14713	3005	3180	3280	.	.	.	.	.	.	.	.	.
0.98 ***	4253	4600	5067	.	.	.	.	.	.	.	.	.	.	.	.
•••	4407	4803	5053	.	.	.	.	.	.	.	.	.	.	.	.

TABLE 8: ALPHA= 0.05 POWER= 0.9   EXPECTED ACCRUAL THRU MINIMUM FOLLOW-UP= 12000

		DEL=.01			DEL=.02			DEL=.05			DEL=.10			DEL=.15		
FACT=		1.0 .75	.50 .25	.00 BIN	1.0 .75	.50 .25	.00 BIN	1.0 .75	.50 .25	.00 BIN	1.0 75	.50 .25	.00 BIN	1.0 .75	.50 .25	.00 BIN
PCONT=●●●					REQUIRED NUMBER OF PATIENTS											
0.01	●●●	1309	1309	1328	488	488	488	158	158	158	79	79	79	49	49	49
	●●●	1309	1309	5053	488	488	1670	158	158	455	79	79	185	49	49	110
0.02	●●●	2918	2929	2948	938	938	938	248	248	248	109	109	109	68	68	68
	●●●	2929	2929	8342	938	938	2484	248	248	581	109	109	215	68	68	123
0.05	●●●	8899	9008	9199	2539	2558	2569	548	548	548	188	188	188	109	109	109
	●●●	8948	9079	17796	2539	2569	4822	548	548	943	188	188	300	109	109	158
0.1	●●●	19729	20138	21428	5419	5528	5659	1039	1039	1058	319	319	319	169	169	169
	●●●	19879	20558	32183	5468	5588	8376	1039	1058	1491	319	319	429	169	169	212
0.15	●●●	29599	30409	33848	8108	8348	8768	1508	1519	1549	439	439	439	218	218	218
	●●●	29899	31418	44858	8209	8528	11502	1519	1538	1970	439	439	540	218	218	257
0.2	●●●	37699	39019	45469	10399	10838	11648	1928	1958	1999	548	548	548	259	259	259
	●●●	38179	40748	55820	10579	11168	14199	1939	1969	2381	548	548	634	259	259	295
0.25	●●●	43819	45739	55838	12248	12889	14198	2288	2329	2389	638	638	638	308	308	308
	●●●	44528	48289	65069	12518	13418	16469	2318	2359	2724	638	638	711	308	308	326
0.3	●●●	47989	50599	64658	13628	14509	16369	2569	2629	2708	709	709	709	338	338	338
	●●●	48938	54019	72605	14018	15229	18310	2599	2678	2998	709	709	771	338	338	349
0.35	●●●	50378	53719	71798	14599	15709	18109	2798	2869	2978	758	769	769	349	349	349
	●●●	51608	58058	78428	15079	16628	19723	2828	2918	3203	758	769	814	349	349	364
0.4	●●●	51169	55328	77149	15188	16508	19399	2948	3038	3158	799	799	818	368	368	368
	●●●	52718	60529	82539	15739	17599	20708	2989	3098	3340	799	799	840	368	368	372
0.45	●●●	50659	55609	80689	15428	16928	20239	3019	3139	3278	818	818	829	368	368	368
	●●●	52538	61628	84937	16058	18169	21265	3079	3199	3409	818	829	848	368	368	372
0.5	●●●	49118	54788	82388	15349	16999	20618	3049	3169	3308	818	818	829	368	368	368
	●●●	51289	61478	85622	16058	18349	21393	3098	3229	3409	818	829	840	368	368	364
0.55	●●●	46759	53048	82208	15008	16748	20528	3008	3128	3278	788	799	818	349	349	349
	●●●	49208	60218	84594	15758	18169	21093	3068	3188	3340	799	799	814	349	349	349
0.6	●●●	43808	50528	80168	14419	16189	19969	2899	3019	3158	758	769	769	338	338	338
	●●●	46448	57949	81854	15188	17618	20365	2948	3079	3203	758	769	771	338	338	326
0.65	●●●	40429	47348	76268	13609	15338	18949	2738	2839	2978	698	709	728	308	308	308
	●●●	43178	54769	77401	14348	16718	19209	2779	2899	2998	709	709	711	308	308	295
0.7	●●●	36698	43538	70538	12559	14179	17468	2498	2599	2719	638	638	649	278	278	278
	●●●	39439	50689	71235	13268	15458	17625	2539	2659	2724	638	638	634	278	278	257
0.75	●●●	32659	39128	62959	11288	12728	15529	2209	2288	2389	548	548	548	229	229	229
	●●●	35269	45698	63356	11918	13838	15612	2239	2329	2381	548	548	540	229	229	212
0.8	●●●	28268	34028	53569	9769	10958	13129	1849	1909	1988	439	439	439	169	169	169
	●●●	30619	39709	53764	10298	11839	13172	1879	1939	1970	439	439	429	169	169	158
0.85	●●●	23359	28088	42398	7939	8828	10298	1429	1459	1508	308	308	319	.	.	.
	●●●	25328	32539	42460	8329	9439	10303	1448	1478	1491	308	319	300	.	.	.
0.9	●●●	17659	20978	29438	5738	6248	7009	919	938	949	.	.	.	.	.	.
	●●●	19058	23888	29443	5959	6589	7006	938	949	943	.	.	.	.	.	.
0.95	●●●	10388	11869	14719	2948	3098	3289	.	.	.	.	.	.	.	.	.
	●●●	11048	13009	14713	3019	3188	3280	.	.	.	.	.	.	.	.	.
0.98	●●●	4309	4628	5059	.	.	.	.	.	.	.	.	.	.	.	.
	●●●	4459	4819	5053	.	.	.	.	.	.	.	.	.	.	.	.

TABLE 8: ALPHA= 0.05 POWER= 0.9   EXPECTED ACCRUAL THRU MINIMUM FOLLOW-UP= 13000

PCONT	DEL=.01			DEL=.02			DEL=.05			DEL=.10			DEL=.15		
FACT=	1.0 / .75	.50 / .25	.00 / BIN	1.0 / .75	.50 / .25	.00 / BIN	1.0 / .75	.50 / .25	.00 / BIN	1.0 / 75	.50 / .25	.00 / BIN	1.0 / .75	.50 / .25	.00 / BIN
PCONT=***						REQUIRED NUMBER OF PATIENTS									
0.01	1321	1321	1321	476	476	476	151	151	151	74	74	74	53	53	53
	1321	1321	5053	476	476	1670	151	151	455	74	74	185	53	53	110
0.02	2913	2934	2946	931	931	931	248	248	248	106	106	106	74	74	74
	2934	2934	8342	931	931	2484	248	248	581	106	106	215	74	74	123
0.05	8914	9011	9206	2544	2556	2576	541	541	541	183	183	183	106	106	106
	8958	9088	17796	2544	2576	4822	541	541	943	183	183	300	106	106	158
0.1	19769	20171	21426	5436	5534	5664	1049	1049	1049	313	313	334	171	171	171
	19931	20614	32183	5481	5599	8376	1049	1049	1491	313	334	429	171	171	212
0.15	29681	30526	33853	8134	8373	8763	1516	1536	1548	443	443	443	216	216	216
	29986	31534	44858	8243	8536	11502	1516	1536	1970	443	443	540	216	216	257
0.2	37818	39216	45476	10453	10876	11644	1938	1959	1991	541	541	561	269	269	269
	38338	40959	55820	10636	11201	14199	1959	1971	2381	541	541	634	269	269	295
0.25	43993	46029	55823	12326	12956	14191	2296	2328	2381	638	638	638	301	301	301
	44741	48596	65069	12598	13464	16469	2316	2361	2724	638	638	711	301	301	326
0.3	48239	50969	64651	13736	14613	16368	2588	2641	2718	703	703	724	334	334	334
	49246	54446	72605	14114	15296	18310	2609	2674	2998	703	724	771	334	334	349
0.35	50688	54186	71801	14731	15836	18111	2804	2881	2966	756	768	768	346	346	366
	51988	58586	78428	15198	16714	19723	2836	2934	3203	768	768	814	346	346	364
0.4	51566	55688	77164	15349	16649	19411	2966	3043	3161	801	801	801	366	366	366
	53191	61153	82539	15901	17701	20708	2999	3096	3340	801	801	840	366	366	372
0.45	51164	56266	80686	15609	17083	20236	3043	3141	3271	821	821	833	366	366	378
	53093	62344	84937	16238	18286	21265	3096	3206	3409	821	821	848	366	378	372
0.5	49681	55531	82376	15556	17169	20614	3064	3173	3303	821	821	833	366	366	366
	51944	62258	85622	16238	18481	21393	3108	3238	3409	821	821	840	366	366	364
0.55	47406	53861	82201	15219	16921	20528	3011	3129	3271	789	801	821	346	346	366
	49929	61044	84594	15966	18306	21093	3076	3194	3340	801	801	814	346	366	349
0.6	44513	51371	80166	14634	16368	19976	2913	3031	3161	756	768	768	334	334	334
	47231	58801	81854	15381	17754	20365	2966	3096	3203	756	768	771	334	334	326
0.65	41166	48186	76266	13821	15511	18956	2751	2848	2978	703	703	724	301	301	313
	43981	55616	77401	14548	16844	19209	2804	2913	2998	703	724	711	301	313	295
0.7	37449	44371	70534	12761	14353	17473	2511	2609	2718	626	638	638	269	269	269
	40244	51489	71235	13464	15576	17625	2556	2653	2724	638	638	634	269	269	257
0.75	33386	39886	62961	11461	12879	15523	2219	2296	2381	541	541	561	216	216	236
	36031	46419	63356	12099	13951	15612	2251	2328	2381	541	541	540	216	216	212
0.8	28913	34698	53569	9921	11071	13139	1861	1906	1971	431	443	443	171	171	171
	31286	40321	53764	10441	11936	13172	1894	1938	1970	443	443	429	171	171	158
0.85	23908	28621	42389	8048	8914	10291	1439	1471	1504	313	313	313	.	.	.
	25879	33029	42460	8438	9499	10303	1451	1483	1491	313	313	300	.	.	.
0.9	18046	21329	29433	5794	6293	7008	931	931	951	.	.	.	.	.	.
	19444	24168	29443	6033	6618	7006	931	951	943	.	.	.	.	.	.
0.95	10571	12034	14731	2966	3108	3291	.	.	.	.	.	.	.	.	.
	11221	13118	14713	3031	3194	3280	.	.	.	.	.	.	.	.	.
0.98	4343	4656	5058	.	.	.	.	.	.	.	.	.	.	.	.
	4494	4851	5053	.	.	.	.	.	.	.	.	.	.	.	.

TABLE 8: ALPHA= 0.05 POWER= 0.9   EXPECTED ACCRUAL THRU MINIMUM FOLLOW-UP= 14000

	DEL=.01			DEL=.02			DEL=.05			DEL=.10			DEL=.15		
FACT=	1.0 .75	.50 .25	.00 BIN	1.0 .75	.50 .25	.00 BIN	1.0 .75	.50 .25	.00 BIN	1.0 75	.50 .25	.00 BIN	1.0 .75	.50 .25	.00 BIN

PCONT=•••   REQUIRED NUMBER OF PATIENTS

PCONT	1.0/.75	.50/.25	.00/BIN	1.0/.75	.50/.25	.00/BIN	1.0/.75	.50/.25	.00/BIN	1.0/.75	.50/.25	.00/BIN	1.0/.75	.50/.25	.00/BIN
0.01 ***	1317	1317	1317	477	477	477	162	162	162	79	79	79	57	57	57
•••	1317	1317	5053	477	477	1670	162	162	455	79	79	185	57	57	110
0.02 ***	2927	2927	2949	932	932	932	254	254	254	114	114	114	79	79	79
•••	2927	2949	8342	932	932	2484	254	254	581	114	114	215	79	79	123
0.05 ***	8912	9017	9214	2542	2564	2577	547	547	547	197	197	197	114	114	114
•••	8969	9087	17796	2542	2564	4822	547	547	943	197	197	300	114	114	158
0.1 ***	19797	20217	21429	5447	5539	5657	1037	1037	1059	324	324	324	162	162	162
•••	19972	20659	32183	5482	5587	8376	1037	1059	1491	324	324	429	162	162	212
0.15 ***	29737	30634	33854	8164	8387	8759	1514	1527	1549	442	442	442	219	219	219
•••	30087	31627	44858	8269	8562	11502	1527	1527	1970	442	442	540	219	219	257
0.2 ***	37949	39384	45474	10509	10907	11642	1934	1969	2004	547	547	547	267	267	267
•••	38487	41134	55820	10684	11222	14199	1947	1982	2381	547	547	634	267	267	295
0.25 ***	44179	46279	55834	12399	13029	14197	2297	2332	2389	639	639	639	302	302	302
•••	44949	48882	65069	12657	13497	16469	2319	2367	2724	639	639	711	302	302	326
0.3 ***	48484	51319	64654	13834	14687	16367	2599	2647	2717	709	709	722	337	337	337
•••	49534	54819	72605	14197	15374	18310	2612	2682	2998	709	709	771	337	337	349
0.35 ***	51004	54644	71794	14849	15934	18104	2809	2892	2962	757	757	779	359	359	359
•••	52347	59067	78428	15317	16787	19723	2844	2927	3203	757	779	814	359	359	364
0.4 ***	51962	56429	77149	15479	16774	19399	2962	3054	3159	792	792	814	359	372	372
•••	53642	61727	82539	16039	17802	20708	3019	3102	3340	792	814	840	372	372	372
0.45 ***	51634	56897	80697	15772	17229	20239	3054	3159	3264	814	827	827	372	372	372
•••	53642	62987	84937	16402	18397	21265	3102	3207	3409	814	827	848	372	372	372
0.5 ***	50234	56232	82377	15737	17334	20624	3067	3172	3312	814	827	827	372	372	372
•••	52557	62974	85622	16424	18607	21393	3124	3242	3409	814	827	840	372	372	364
0.55 ***	48029	54609	82202	15409	17089	20532	3032	3137	3277	792	792	814	359	359	359
•••	50619	61819	84594	16144	18432	21093	3089	3207	3340	792	814	814	359	359	349
0.6 ***	45194	52159	80172	14827	16542	19972	2927	3032	3159	757	757	779	337	337	337
•••	47959	59592	81854	15584	17872	20365	2984	3089	3203	757	779	771	337	337	326
0.65 ***	41869	48987	76274	14009	15667	18944	2752	2857	2984	709	709	722	302	302	302
•••	44717	56394	77401	14744	16962	19209	2809	2914	2998	709	709	711	302	302	295
0.7 ***	38159	45137	70534	12937	14499	17474	2529	2612	2717	639	639	639	267	267	267
•••	40972	52229	71235	13624	15689	17625	2564	2669	2724	639	639	634	267	267	257
0.75 ***	34042	40609	62952	11629	13007	15527	2227	2297	2389	547	547	547	219	219	232
•••	36724	47084	63356	12237	14044	15612	2262	2332	2381	547	547	540	219	219	212
0.8 ***	29514	35324	53572	10054	11187	13134	1864	1912	1982	442	442	442	162	162	162
•••	31907	40889	53764	10557	11992	13172	1899	1947	1970	442	442	429	162	162	158
0.85 ***	24404	29129	42394	8164	8982	10299	1444	1457	1514	302	302	324	.	.	.
•••	26377	33447	42460	8527	9542	10303	1457	1479	1491	302	324	300	.	.	.
0.9 ***	18419	21674	29444	5867	6344	7009	932	932	954	.	.	.	.	.	.
•••	19797	24439	29443	6077	6637	7006	932	954	943	.	.	.	.	.	.
0.95 ***	10754	12167	14722	2984	3124	3299	.	.	.	.	.	.	.	.	.
•••	11362	13204	14713	3054	3207	3280	.	.	.	.	.	.	.	.	.
0.98 ***	4384	4677	5062	.	.	.	.	.	.	.	.	.	.	.	.
•••	4524	4852	5053	.	.	.	.	.	.	.	.	.	.	.	.

TABLE 8: ALPHA= 0.05 POWER= 0.9   EXPECTED ACCRUAL THRU MINIMUM FOLLOW-UP= 15000

	DEL=.01			DEL=.02			DEL=.05			DEL=.10			DEL=.15		
FACT=	1.0 .75	.50 .25	.00 BIN	1.0 .75	.50 .25	.00 BIN	1.0 .75	.50 .25	.00 BIN	1.0 75	.50 .25	.00 BIN	1.0 .75	.50 .25	.00 BIN
PCONT=***					REQUIRED NUMBER OF PATIENTS										
0.01 ***	1322	1322	1322	474	474	474	160	160	160	85	85	85	61	61	61
•••	1322	1322	5053	474	474	1670	160	160	455	85	85	185	61	61	110
0.02 ***	2935	2935	2949	947	947	947	249	249	249	99	99	99	61	61	61
•••	2935	2935	8342	947	947	2484	249	249	581	99	99	215	61	61	123
0.05 ***	8935	9024	9197	2536	2560	2574	549	549	549	197	197	197	122	122	122
•••	8972	9099	17796	2560	2574	4822	549	549	943	197	197	300	122	122	158
0.1 ***	19847	20274	21422	5461	5536	5672	1036	1036	1060	324	324	324	174	174	174
•••	20011	20686	32183	5499	5597	8376	1036	1060	1491	324	324	429	174	174	212
0.15 ***	29822	30736	33849	8185	8410	8761	1524	1524	1547	436	436	436	211	211	211
•••	30174	31735	44858	8297	8574	11502	1524	1547	1970	436	436	540	211	211	257
0.2 ***	38072	39549	45474	10547	10960	11649	1936	1960	1997	549	549	549	272	272	272
•••	38635	41311	55820	10735	11236	14199	1960	1974	2381	549	549	634	272	272	295
0.25 ***	44349	46524	55824	12460	13074	14199	2311	2349	2386	624	647	647	310	310	310
•••	45160	49135	65069	12736	13547	16469	2311	2372	2724	647	647	711	310	310	326
0.3 ***	48699	51647	64660	13922	14761	16360	2597	2649	2710	699	722	722	324	324	324
•••	49810	55172	72605	14297	15422	18310	2635	2686	2998	699	722	771	324	324	349
0.35 ***	51310	55060	71785	14972	16036	18099	2822	2897	2972	760	760	774	347	361	361
•••	52711	59522	78428	15436	16861	19723	2860	2935	3203	760	774	814	347	361	364
0.4 ***	52336	56935	77161	15624	16899	19397	2986	3061	3160	797	797	811	361	361	361
•••	54099	62274	82539	16172	17897	20708	3024	3099	3340	797	811	840	361	361	372
0.45 ***	52097	57474	80686	15924	17372	20236	3061	3160	3272	811	811	835	361	361	361
•••	54174	63610	84937	16547	18497	21265	3122	3211	3409	811	835	848	361	361	372
0.5 ***	50761	56874	82374	15910	17485	20611	3085	3197	3310	811	811	835	361	361	361
•••	53161	63647	85622	16585	18699	21393	3136	3249	3409	811	835	840	361	361	364
0.55 ***	48624	55322	82210	15586	17236	20522	3047	3160	3272	797	797	811	347	361	361
•••	51272	62522	84594	16299	18535	21093	3099	3211	3340	797	811	814	361	361	349
0.6 ***	45835	52899	80161	15010	16697	19974	2935	3047	3160	760	760	774	324	324	347
•••	48661	60310	81854	15736	17986	20365	2986	3099	3203	760	774	771	324	347	326
0.65 ***	42549	49735	76261	14185	15811	18947	2761	2860	2972	699	722	722	310	310	310
•••	45436	57099	77401	14897	17072	19209	2822	2911	2998	699	722	711	310	310	295
0.7 ***	38822	45849	70524	13111	14635	17461	2536	2611	2710	624	647	647	272	272	272
•••	41672	52899	71235	13786	15774	17625	2574	2672	2724	647	647	634	272	272	257
0.75 ***	34674	41260	62949	11785	13135	15535	2236	2311	2386	549	549	549	235	235	235
•••	37374	47686	63356	12385	14124	15612	2274	2349	2381	549	549	540	235	235	212
0.8 ***	30085	35897	53574	10172	11274	13135	1885	1922	1974	436	436	436	174	174	174
•••	32485	41386	53764	10674	12061	13172	1899	1960	1970	436	436	429	174	174	158
0.85 ***	24886	29574	42385	8260	9047	10299	1435	1472	1510	310	310	310	.	.	.
•••	26860	33835	42460	8611	9586	10303	1449	1486	1491	310	310	300	.	.	.
0.9 ***	18736	21961	29424	5911	6385	7022	924	947	961	.	.	.	.	.	.
•••	20124	24685	29443	6122	6661	7006	947	947	943	.	.	.	.	.	.
0.95 ***	10899	12286	14724	3010	3136	3286	.	.	.	.	.	.	.	.	.
•••	11522	13285	14713	3061	3211	3280	.	.	.	.	.	.	.	.	.
0.98 ***	4411	4697	5072	.	.	.	.	.	.	.	.	.	.	.	.
•••	4547	4861	5053	.	.	.	.	.	.	.	.	.	.	.	.

TABLE 8: ALPHA= 0.05 POWER= 0.9   EXPECTED ACCRUAL THRU MINIMUM FOLLOW-UP= 17000

		DEL=.01			DEL=.02			DEL=.05			DEL=.10			DEL=.15		
FACT=		1.0 .75	.50 .25	.00 BIN	1.0 .75	.50 .25	.00 BIN	1.0 .75	.50 .25	.00 BIN	1.0 75	.50 .25	.00 BIN	1.0 .75	.50 .25	.00 BIN
PCONT=•••						**REQUIRED NUMBER OF PATIENTS**											
0.01	***	1329	1329	1329	479	479	479	155	155	155	70	70	70	54	54	54
	•••	1329	1329	5053	479	479	1670	155	155	455	70	70	185	54	54	110
0.02	***	2917	2944	2944	946	946	946	240	240	240	112	112	112	70	70	70
	•••	2917	2944	8342	946	946	2484	240	240	581	112	112	215	70	70	123
0.05	***	8952	9037	9207	2561	2561	2577	537	537	537	197	197	197	112	112	112
	•••	8995	9106	17796	2561	2577	4822	537	537	943	197	197	300	112	112	158
0.1	***	19917	20342	21431	5467	5552	5664	1047	1047	1047	325	325	325	181	181	181
	•••	20087	20751	32183	5510	5595	8376	1047	1047	1491	325	325	429	181	181	212
0.15	***	29974	30909	33857	8230	8442	8766	1515	1541	1541	436	436	436	224	224	224
	•••	30330	31902	44858	8341	8596	11502	1515	1541	1970	436	436	540	224	224	257
0.2	***	38304	39876	45486	10636	11019	11656	1940	1966	1982	537	537	564	266	266	266
	•••	38899	41619	55820	10806	11290	14199	1966	1982	2381	537	537	634	266	266	295
0.25	***	44679	46990	55830	12591	13186	14206	2306	2349	2391	649	649	649	309	309	309
	•••	45545	49609	65069	12846	13611	16469	2322	2365	2724	649	649	711	309	309	326
0.3	***	49157	52260	64654	14095	14902	16374	2604	2662	2705	707	707	707	325	325	325
	•••	50347	55814	72605	14461	15524	18310	2646	2689	2998	707	707	771	325	325	349
0.35	***	51877	55830	71794	15184	16204	18116	2832	2901	2960	750	776	776	351	351	351
	•••	53407	60319	78428	15625	16985	19723	2875	2944	3203	776	776	814	351	351	364
0.4	***	53067	57870	77149	15864	17096	19407	3002	3071	3156	792	792	819	367	367	367
	•••	54937	63225	82539	16416	18031	20708	3029	3114	3340	792	819	840	367	367	372
0.45	***	52966	58576	80692	16204	17606	20241	3087	3172	3257	819	819	835	367	367	367
	•••	55150	64696	84937	16815	18669	21265	3130	3215	3409	819	819	848	367	367	372
0.5	***	51776	58082	82376	16204	17734	20624	3114	3199	3300	819	819	835	367	367	367
	•••	54284	64840	85622	16857	18881	21393	3156	3257	3409	819	819	840	367	367	364
0.55	***	49752	56595	82206	15906	17521	20539	3071	3156	3284	792	792	819	351	351	351
	•••	52472	63777	84594	16602	18727	21093	3114	3215	3340	792	819	814	351	351	349
0.6	***	47032	54241	80166	15327	16942	19960	2960	3045	3156	750	776	776	325	325	325
	•••	49949	61610	81854	16050	18175	20365	3002	3114	3203	776	776	771	325	325	326
0.65	***	43786	51070	76272	14504	16076	18940	2790	2875	2986	707	707	707	309	309	309
	•••	46735	58380	77401	15200	17240	19209	2832	2917	2998	707	707	711	309	309	295
0.7	***	40046	47144	70535	13415	14860	17479	2561	2620	2705	622	649	649	266	266	266
	•••	42952	54114	71235	14052	15949	17625	2604	2662	2724	649	649	634	266	266	257
0.75	***	35839	42469	62954	12055	13330	15524	2264	2322	2391	537	537	564	224	224	224
	•••	38575	48775	63356	12634	14249	15612	2280	2349	2381	537	537	540	224	224	212
0.8	***	31121	36944	53561	10397	11444	13144	1881	1924	1982	436	436	436	181	181	181
	•••	33544	42299	53764	10865	12166	13172	1897	1966	1970	436	436	429	181	181	158
0.85	***	25740	30399	42384	8426	9165	10296	1456	1472	1499	309	309	309	.	.	.
	•••	27695	34521	42460	8766	9659	10303	1456	1499	1491	309	309	300	.	.	.
0.9	***	19349	22510	29437	6020	6445	7024	920	946	962	.	.	.	.	.	.
	•••	20709	25086	29443	6216	6700	7006	946	946	943	.	.	.	.	.	.
0.95	***	11162	12506	14732	3029	3156	3300	.	.	.	.	.	.	.	.	.
	•••	11757	13441	14713	3087	3215	3280	.	.	.	.	.	.	.	.	.
0.98	***	4490	4745	5069	.	.	.	.	.	.	.	.	.	.	.	.
	•••	4601	4899	5053	.	.	.	.	.	.	.	.	.	.	.	.

TABLE 8: ALPHA= 0.05 POWER= 0.9   EXPECTED ACCRUAL THRU MINIMUM FOLLOW-UP= 20000

		DEL=.01			DEL=.02			DEL=.05			DEL=.10			DEL=.15		
FACT=		1.0 .75	.50 .25	.00 BIN	1.0 .75	.50 .25	.00 BIN	1.0 .75	.50 .25	.00 BIN	1.0 75	.50 .25	.00 BIN	1.0 .75	.50 .25	.00 BIN
PCONT=***							REQUIRED NUMBER OF PATIENTS								
0.01	***	1313	1313	1313	482	482	482	163	163	163	82	82	82	63	63	63
	•••	1313	1313	5053	482	482	1670	163	163	455	82	82	185	63	63	110
0.02	***	2932	2932	2932	932	932	932	263	263	263	113	113	113	63	63	63
	•••	2932	2932	8342	932	932	2484	263	263	581	113	113	215	63	63	123
0.05	***	8982	9063	9213	2563	2563	2582	532	532	532	182	182	182	113	113	113
	•••	9013	9132	17796	2563	2563	4822	532	532	943	182	182	300	113	113	158
0.1	***	20013	20463	21432	5482	5563	5663	1032	1063	1063	313	332	332	163	163	163
	•••	20182	20832	32183	5532	5613	8376	1032	1063	1491	313	332	429	163	163	212
0.15	***	30163	31163	33863	8282	8482	8763	1532	1532	1532	432	432	432	213	213	213
	•••	30563	32113	44858	8382	8613	11502	1532	1532	1970	432	432	540	213	213	257
0.2	***	38632	40282	45482	10732	11082	11632	1963	1963	1982	532	563	563	263	263	263
	•••	39263	42013	55820	10882	11332	14199	1963	1982	2381	563	563	634	263	263	295
0.25	***	45163	47582	55832	12732	13282	14213	2313	2363	2382	632	632	632	313	313	313
	•••	46113	50182	65069	12982	13682	16469	2332	2363	2724	632	632	711	313	313	326
0.3	***	49813	53082	64663	14282	15063	16363	2632	2663	2713	713	713	713	332	332	332
	•••	51082	56613	72605	14632	15613	18310	2632	2682	2998	713	713	771	332	332	349
0.35	***	52713	56882	71782	15432	16413	18113	2863	2913	2963	763	763	763	363	363	363
	•••	54332	61332	78428	15863	17132	19723	2882	2932	3203	763	763	814	363	363	364
0.4	***	54082	59113	77163	16182	17332	19413	3013	3082	3163	813	813	813	363	363	363
	•••	56063	64432	82539	16682	18213	20708	3063	3113	3340	813	813	840	363	363	372
0.45	***	54163	60013	80682	16532	17882	20232	3113	3182	3263	813	832	832	363	363	363
	•••	56482	66082	84937	17132	18863	21265	3163	3232	3409	813	832	848	363	363	372
0.5	***	53163	59682	82382	16582	18032	20613	3132	3213	3313	813	832	832	363	363	363
	•••	55763	66363	85622	17232	19113	21393	3182	3263	3409	813	832	840	363	363	364
0.55	***	51263	58313	82213	16313	17832	20532	3082	3182	3263	813	813	813	363	363	363
	•••	54113	65382	84594	16982	18932	21093	3132	3213	3340	813	813	814	363	363	349
0.6	***	48663	55982	80163	15732	17282	19963	2982	3063	3163	763	763	782	332	332	332
	•••	51632	63232	81854	16432	18382	20365	3032	3113	3203	763	763	771	332	332	326
0.65	***	45432	52813	76263	14913	16382	18963	2813	2882	2982	713	713	713	313	313	313
	•••	48463	59982	77401	15563	17463	19209	2863	2932	2998	713	713	711	313	313	295
0.7	***	41663	48832	70532	13782	15163	17463	2582	2632	2713	632	632	632	263	263	263
	•••	44632	55613	71235	14413	16132	17625	2613	2682	2724	632	632	634	263	263	257
0.75	***	37382	44013	62963	12382	13582	15532	2263	2332	2382	532	563	563	232	232	232
	•••	40132	50113	63356	12932	14413	15612	2282	2363	2381	532	563	540	232	232	212
0.8	***	32482	38263	53563	10682	11632	13132	1882	1932	1982	432	432	432	163	163	163
	•••	34913	43413	53764	11113	12282	13172	1913	1963	1970	432	432	429	163	163	158
0.85	***	26863	31432	42382	8613	9313	10282	1463	1482	1513	313	313	313	.	.	.
	•••	28813	35332	42460	8932	9763	10303	1463	1482	1491	313	313	300	.	.	.
0.9	***	20113	23163	29432	6132	6513	7013	932	932	963	.	.	.	.	.	.
	•••	21432	25582	29443	6313	6732	7006	932	963	943	.	.	.	.	.	.
0.95	***	11513	12763	14732	3063	3163	3282	.	.	.	.	.	.	.	.	.
	•••	12082	13613	14713	3113	3232	3280	.	.	.	.	.	.	.	.	.
0.98	***	4563	4782	5063	.	.	.	.	.	.	.	.	.	.	.	.
	•••	4663	4913	5053	.	.	.	.	.	.	.	.	.	.	.	.

TABLE 8: ALPHA= 0.05 POWER= 0.9   EXPECTED ACCRUAL THRU MINIMUM FOLLOW-UP= 25000

		DEL=.01			DEL=.02			DEL=.05			DEL=.10			DEL=.15	
FACT=	1.0 .75	.50 .25	.00 BIN	1.0 .75	.50 .25	.00 BIN	1.0 .75	.50 .25	.00 BIN	1.0 75	.50 .25	.00 BIN	1.0 .75	.50 .25	.00 BIN
PCONT=•••						REQUIRED NUMBER OF PATIENTS									
0.01 •••	1329	1329	1329	477	477	477	165	165	165	79	79	79	40	40	40
•••	1329	1329	5053	477	477	1670	165	165	455	79	79	185	40	40	110
0.02 •••	2915	2954	2954	954	954	954	266	266	266	102	102	102	79	79	79
•••	2915	2954	8342	954	954	2484	266	266	581	102	102	215	79	79	123
0.05 •••	9016	9079	9204	2540	2579	2579	540	540	540	204	204	204	102	102	102
•••	9040	9141	17796	2579	2579	4822	540	540	943	204	204	300	102	102	158
0.1 •••	20165	20579	21415	5516	5579	5665	1040	1040	1040	329	329	329	165	165	165
•••	20329	20915	32183	5540	5641	8376	1040	1040	1491	329	329	429	165	165	212
0.15 •••	30477	31477	33852	8352	8540	8766	1516	1540	1540	454	454	454	227	227	227
•••	30891	32352	44858	8454	8641	11502	1540	1540	1970	454	454	540	227	227	257
0.2 •••	39102	40852	45477	10852	11165	11641	1954	1977	1977	540	540	540	266	266	266
•••	39829	42477	55820	11016	11391	14199	1977	1977	2381	540	540	634	266	266	295
0.25 •••	45891	48454	55829	12915	13454	14204	2329	2352	2391	641	641	641	290	290	290
•••	46915	50954	65069	13165	13766	16469	2352	2352	2724	641	641	711	290	290	326
0.3 •••	50790	54227	64641	14579	15266	16352	2641	2665	2704	704	704	704	329	329	329
•••	52165	57665	72605	14891	15766	18310	2665	2704	2998	704	704	771	329	329	349
0.35 •••	53954	58329	71790	15766	16665	18102	2891	2915	2977	766	766	766	352	352	352
•••	55704	62665	78428	16165	17290	19723	2891	2954	3203	766	766	814	352	352	364
0.4 •••	55602	60852	77165	16579	17665	19391	3040	3102	3165	790	790	790	352	352	352
•••	57727	66040	82539	17079	18415	20708	3079	3141	3340	790	790	840	352	352	372
0.45 •••	55954	61977	80704	17016	18227	20227	3141	3204	3266	829	829	829	352	352	352
•••	58391	67915	84937	17540	19102	21265	3165	3227	3409	829	829	848	352	352	372
0.5 •••	55165	61852	82391	17079	18415	20602	3165	3227	3290	829	829	829	352	352	352
•••	57891	68352	85622	17704	19352	21393	3204	3266	3409	829	829	840	352	352	364
0.55 •••	53454	60641	82204	16829	18227	20516	3141	3204	3266	790	790	829	352	352	352
•••	56391	67477	84594	17477	19204	21093	3165	3227	3340	790	790	814	352	352	349
0.6 •••	50954	58391	80165	16290	17704	19977	3016	3079	3165	766	766	766	329	329	329
•••	54016	65391	81854	16915	18665	20365	3040	3102	3203	766	766	771	329	329	326
0.65 •••	47766	55204	76266	15415	16790	18954	2852	2915	2977	704	704	727	290	290	290
•••	50852	62079	77401	16040	17704	19209	2891	2954	2998	704	704	711	290	290	295
0.7 •••	43954	51102	70516	14266	15516	17477	2602	2665	2704	641	641	641	266	266	266
•••	46954	57579	71235	14829	16352	17625	2641	2665	2724	641	641	634	266	266	257
0.75 •••	39516	46079	62954	12790	13891	15540	2290	2329	2391	540	540	540	227	227	227
•••	42266	51852	63356	13290	14602	15612	2329	2352	2381	540	540	540	227	227	212
0.8 •••	34352	40016	53579	11016	11891	13141	1915	1954	1977	454	454	454	165	165	165
•••	36766	44852	53764	11415	12454	13172	1915	1954	1970	454	454	429	165	165	158
0.85 •••	28352	32790	42391	8852	9477	10290	1454	1477	1516	329	329	329	.	.	.
•••	30266	36391	42460	9141	9852	10303	1477	1477	1491	329	329	300	.	.	.
0.9 •••	21165	24040	29415	6266	6602	7016	954	954	954	.	.	.	.	.	.
•••	22415	26204	29443	6415	6790	7006	954	954	943	.	.	.	.	.	.
0.95 •••	11954	13079	14727	3102	3204	3290	.	.	.	.	.	.	.	.	.
•••	12454	13790	14713	3141	3227	3280	.	.	.	.	.	.	.	.	.
0.98 •••	4641	4829	5079	.	.	.	.	.	.	.	.	.	.	.	.
•••	4727	4954	5053	.	.	.	.	.	.	.	.	.	.	.	.

INDEX

Milton Keynes UK
Ingram Content Group UK Ltd.
UKHW051927141024
449569UK00027B/1384